*Analysis and Continuum Mechanics*

JAMES SERRIN

# Analysis and Continuum Mechanics

## A Collection of Papers Dedicated to J. Serrin on His Sixtieth Birthday

Invited by
S. S. Antman, H. Brezis, B. D. Coleman,
M. Feinberg, J. A. Nohel, and W. P. Ziemer

Reprinted from
Archive for Rational Mechanics and Analysis
edited by C. Truesdell

With 57 Figures

Springer-Verlag Berlin Heidelberg New York
London Paris Tokyo Hong Kong

Professor Dr. Stuart S. Antman

Department of Mathematics, University of Maryland
College Park, MD 20742, USA

Professor Dr. Haïm Brezis

Analyse Numérique, Tour 55–65, 5[e] étage, Université Pierre et Marie Curie, 4, Place Jussieu
75252 Paris Cédex 05, France

Professor Dr. Bernard D. Coleman

Department of Mechanics & Material Science, Rutgers, The State University
New Brunswick, NJ 08854, USA

Professor Dr. Martin Feinberg

Department of Chemical Engineering, University of Rochester
Rochester, NY 14627, USA

Professor Dr. John A. Nohel

Mathematics Research Center, University of Wisconsin
Madison, WI 53705, USA

Professor Dr. William P. Ziemer

Mathematics Department, Indiana University, Bloomington, IN 47405, USA

ISBN 3-540-50917-8 Springer-Verlag Berlin Heidelberg New York
ISBN 0-387-50917-8 Springer-Verlag New York Berlin Heidelberg



Printing: Weihert-Druck GmbH, 6100 Darmstadt
Binding: J. Schäffer GmbH & Co. KG, 6718 Grünstadt
2155/3150-543210 – Printed on acid-free paper

# *Preface*

This volume collects papers dedicated to *James Serrin* on his sixtieth birthday, November 1, 1986.

While in older times a great mathematician usually completed important researches in several different branches of mathematics, nowadays nearly everyone is a specialist. The mathematical community frowns at time stolen from output of publishable "results" to get to the bottom, and thence to the top, of another field. *James Serrin* is one of the few mathematicians schooled since the forties whose discoveries shine in more than one direction.

When, in the autumn of 1950, *Serrin* was writing his thesis, he often came to the office that David Gilbarg and I shared, and so I heard many of their discussions of flows with free boundaries. The boy was father to the man. Though his Midwestern youthfulness offered an irresistible butt for teasing, I chose to place his finished thesis on the very first pages of the new *Journal of Rational Mechanics and Analysis*, 1952, for his was the kind of research the *Journal* was founded to further, and he was the kind of man whose life in mathematics it was founded to foster.

*Serrin's* immediately following work led me to suggest to Siegfried Flügge that he invite him to compose the basic exposition of mathematical fluid dynamics for the *Encyclopedia of Physics*. His article, published in 1959, brought him fame at once, and it remains the basic reference for its field, especially for the Navier-Stokes theory. Grown for that article and from it came his fundamental papers on exact stability, on existence and regularity of solutions of the Navier-Stokes equations, and on uniqueness of flows of compressible fluids. Later came his analysis of the asymptotic behavior of boundary-layer solutions, later yet his determination of the dynamic structure of tornadoes (1972) and the study of dynamic phase transitions in shock layers (1984–1986).

*Serrin's* major works on elliptic partial-differential equations began to appear in 1956: a paper on isolated singularities of solutions (co-author Gilbarg) and another establishing a Harnack inequality for them. The following years saw the exterior Dirichlet problem (co-author Norman Meyers), theorems of local behavior, further studies on singularities, and Liouville theorems for non-linear Poisson equations. *Serrin* discovered (1968) that the Dirichlet problem for minimal surfaces in higher dimensions allows solutions for arbitrarily assigned, continuous boundary data only when the boundary of the domain considered has everywhere non-positive mean curvature, and he extended this remarkable theorem to other strongly nonlinear elliptic equations. He also re-introduced, refined and developed Alexandrov's technique of the moving plane for the study of partial-differential equations (1971). His studies of the calculus of variations began in 1961 with a new definition of variational integrals. They continue today: a general variational identity and extensions of the mountain-pass theorem (co-author of both, Patrizia Pucci). Recently *Serrin*, together with an important school of co-authors, has written a sequence of papers on the foundations of semilinear/quasilinear theory. Earlier he had published two basic and long monographs: one on local behavior of solutions of quasilinear equations (*Acta Mathematica*, 1964), one on the Dirichlet problem for elliptic differential equations in many variables (*Philosophical Transactions*, 1969), dedicated to the memory of Sergei Bernstein.

*Serrin's* first paper on real analysis appeared in 1961. In it he proved that a weakly differentiable function has a representative which has ordinary partial derivatives almost everywhere, and he exhibited a function (the fakir's bed of nails) that is strongly differentiable yet has a differential nowhere. In another early work (co-author Meyers) he showed that weak and strong differentiation are identical.

In 1959 the nature of constitutive relations and the restrictions they must satisfy was a subject of great moment in rational mechanics. In that year *Serrin*, influenced by the work of Stokes a century earlier, wrote an important paper on nonlinear fluids. Though isolated, it remains one of the classic sources. In 1985, partly in collaboration with Ernest Dunn, he came back to constitutive problems, proposing and developing a theory of interstitial working and a model for the dynamics of chemical mixtures. In another isolated and almost unremarked early paper (co-author K. C. Whang) he showed that while fluid entropy is increased by passage through a shock transition, the increase does not occur in the expected monotone way, but rather by first an increase to a maximum value and then a decrease to the final value.

While *Serrin's* interest in thermodynamics is evident from his article on fluids in the *Encyclopedia*, he did not take it up as an end in itself until the 1970s. Then and for many years thereafter he gave his main and intense interest to thermodynamics and the thermomechanics of deformable continua. While he has published little on that subject, he has lectured widely about it and has circulated sets of notes which have exerted influence on several major students of the field. Everyone who knows him perceives at once his close and deep familiarity with the works of Kelvin, Clausius, Gibbs, and Carathéodory, and his ability to use his scholarship in evaluating and developing thermomechanics today. His thermodynamical papers in this *Archive* (Volumes 59 and 77 (co-authors R. L. Fosdick, Bernard Coleman, and David Owen)), are well known and influential; his clarification of Mach's concept of hotness manifold and his accumulation function (Volume 70) are recognized and applied. In 1983 he organized a great conference on the thermomechanics of continua at the University of Minnesota; he edited its proceedings under the title *New Perspectives in Thermodynamics* (Springer-Verlag, Berlin Heidelberg, 1986). One of his major papers appears in it. The conference itself remains an enlightening experience for those who took part in it.

The foregoing remarks are designed to give a reader who is familiar with only a part of *Serrin's* mathematical achievement some idea of the rest.

Breadth and depth are two of the qualities in which *Serrin* recalls the great masters of the past. Another is his sure sense of definition and problem; another, his taste; yet another, his presentation. Technique he has in good measure, and he calls upon it when it is needed, but he prefers to use the simplest tools that will do the job. He writes so as to teach and to share. His works invite reading.

Finally, the Press joins me in repeating our thanks to *James Serrin* for his many years of devotion to this *Archive* and for his maintaining its position in mathematical analysis – years before, during, and after his service as co-editor.

*C. Truesdell*

# Contents

Differentiability of the Blow-up Curve for One Dimensional Nonlinear Wave Equations.
By L. A. Caffarelli & A. Friedman ............................. 1

On the Existence of Positive Entire Solutions of a Semilinear Elliptic Equation.
By W.-Y. Ding & W.-M. Ni ............................. 17

The Uniqueness of Hill's Spherical Vortex.
By C. J. Amick & L. E. Fraenkel ............................. 43

Lyapunov Functions for Thermomechanics with Spatially Varying Boundary Temperatures.
By J. M. Ball & G. Knowles ............................. 73

A Semilinear Elliptic Problem Which Is Not Selfadjoint.
By P. Pucci ............................. 85

A Multiparameter Study of a Boundary Value Problem from Chemical Reactor Theory.
By P. Fife, L. Hsiao, & T. Zhang ............................. 107

Boundary Regularity for Quasiminima.
By W. P. Ziemer ............................. 137

Positively Invariant Regions for a Problem in Phase Transitions.
By V. Roytburd & M. Slemrod ............................. 149

A Note on a Theorem of R. Duffin.
By P. H. Rabinowitz ............................. 169

The Boundary Value Problems for Non-Linear Elliptic Equations and the Maximum Principle for Euler-Lagrange Equations.
By I. J. Bakelman ............................. 181

Qualitative Properties of Large Buckled States of Spherical Shells.
By K.-G. Shih & S. S. Antman ............................. 211

Stable Equilibrium Configurations of Elastic Crystals.
By J.L. Ericksen ........ 239

Inequalities Between Dirichlet and Neumann Eigenvalues.
By H.A. Levine & H.F. Weinberger ........ 253

Embeddings of Anisotropic Sobolev Spaces.
By D.E. Edmunds & R.M. Edmunds ........ 269

Monotonic Decreasing and Asymptotic Behavior of the Kinetic Energy for Weak Solutions of the Navier-Stokes Equations in Exterior Domains.
By G.P. Galdi & P. Maremonti ........ 277

Quasilinear Hyperbolic Systems with Involutions.
By C.M. Dafermos ........ 291

A Generalized Norton-Hoff Model and the Prandtl-Reuss Law of Plasticity.
By R. Temam ........ 309

Symmetry and the Bifurcation of Capillary-Gravity Waves.
By M. Jones & J. Toland ........ 357

Smoothness of Linear Laminates.
By M. Chipot, D. Kinderlehrer, & G. Vergara Caffarelli ........ 383

On the Behavior of the Derivatives of Minimizers near Singular Points.
By E. Giusti ........ 399

Ground States and Dirichlet Problems for $-\Delta u = f(u)$ in $R^2$.
By F.V. Atkinson & L.A. Peletier ........ 409

On Phase Transitions with Bulk, Interfacial, and Boundary Energy.
By M.E. Gurtin ........ 429

Spectral Properties of the Laplacian in the Complement of a Deformed Cylinder.
By W. Littman ........ 451

A New Proof of Moser's Parabolic Harnack Inequality Using the Old Ideas of Nash.
By E.B. Fabes & D.W. Stroock ........ 459

Weakly Decaying Energy Separation and Uniqueness of Motions of an Elastic-Plastic Oscillator with Work-Hardening.
By D.R. Owen ........ 471

On Effects of Virtual Inertia During Diffusion of a Dispersed Medium in a Suspension.
By G. Capriz & P. Giovine ........ 491

Existence of Positive Solutions for Semilinear Elliptic Equations in General Domains.
By J. Smoller & A. Wasserman ........ 499

Assembling a Rearrangement.
By G. Talenti ........ 521

Asymptotic Behaviour of Solutions of Semi-Linear Elliptic Equations in $\mathbb{R}^n$.
By K. McLeod ........ 531

Existence Theorems Concerning Simple Integrals of the Calculus of Variations for Discontinuous Solutions.
By L. Cesari, P. Brandi, & A. Salvadori ........ 543

On the Uniqueness of Flow of a Navier-Stokes Fluid Due to a Stretching Boundary.
By J.B. McLeod & K.R. Rajagopal ........ 565

On the Behavior at Infinity of Solutions of Elliptic Systems with a Finite Energy Integral.
By V.A. Kondratiev & O.A. Oleinik ........ 575

A Class of Quasilinear Differential Inequalities Whose Solutions Are Ultimately Constant.
By R. Redheffer ........ 591

Singular Solutions for Some Semilinear Elliptic Equations.
By H. Brezis & L. Oswald ........ 615

Eventual $C^\infty$-Regularity and Concavity for Flows in One-Dimensional Porous Media.
By D.G. Aronson & J.L. Vazquez ........ 627

Fine Phase Mixtures as Minimizers of Energy.
By J.M. Ball & R.D. James ........ 647

Fit Regions and Functions of Bounded Variation.
By W. Noll & E.G. Virga ........ 687

The Lavrentiev Phenomenon for Invariant Variational Problems.
By A.C. Heinricher & V.J. Mizel ........ 709

Steady, Structured Shock Waves. Part 1: Thermoelastic Materials. By J.E. Dunn & R.L. Fosdick ........ 747

Published Works of James Serrin ........ 819

Places of First Publication ........ 827

# *Differentiability of the Blow-up Curve for one Dimensional Nonlinear Wave Equations*

LUIS A. CAFFARELLI & AVNER FRIEDMAN

*Dedicated to Jim Serrin*

## Introduction

Consider the Cauchy problem

$$\frac{\partial^2 u}{\partial t^2} - \sum_{i=1}^{N} \frac{\partial^2 u}{\partial x_i^2} = F(u) \qquad (x \in \mathbb{R}^N, t > 0),$$

$$u(x, 0) = f(x) \qquad (x \in \mathbb{R}^N),$$

$$u_t(x, 0) = g(x) \qquad (x \in \mathbb{R}^N)$$

with $N \leqq 3$. It is well known that if $F(u)$ is superlinear, *i.e.*,

$$F(u) \sim A\,|u|^p \quad \text{if } |u| \to \infty \qquad (A > 0, p > 1) \tag{0.1}$$

then solutions generally blow up in finite time; see [2]–[6].

In a recent paper [1] we have proved, under some assumptions on $f$ and $g$, which *a priori* formally ensure that

$$u_t > (1 + \varepsilon)\,|\nabla_x u| \quad \text{for some } \varepsilon > 0, \tag{0.2}$$

that there exists a surface $\Gamma: t = \varphi(x)$ such that $u(x, t) < \infty$ if and only if $t < \varphi(x)$ (*i.e.*, $\Gamma$ is the blow-up surface), and

$$\varphi(x) \text{ is continuously differentiable.} \tag{0.3}$$

Setting $d(x, t) = \operatorname{dist}((x, t), \Gamma)$, we also derived the asymptotic behavior of $u$ and its derivatives near $\Gamma$, such as

$$d^q(x, t)\, u(x, t) \to A_0 \left(\frac{1 - \alpha^2}{1 + \alpha^2}\right)^{1/(p-1)}, \qquad A_0^{p-1} = \frac{1}{A} \frac{2(p + 1)}{(p - 1)^2}, \tag{0.4}$$

where $(x, t) \to (x_0, t_0) \in \Gamma$, $\alpha = |\nabla\varphi(x_0)|$, $q = 2/(p - 1)$.

It should be noted that the assertion (0.2) cannot follow from local arguments only, *i.e.*, some boundedness conditions on the Cauchy data must be imposed.

Indeed, consider

$$u(x, t) = [(|x|^2 - (t - t_0)^2)^+]^{-r} \qquad (r > 0),$$

which is a solution of

$$u_{tt} - \sum_{i=1}^{N} \frac{\partial^2 u}{\partial x_i^2} = A\,|u|^p \qquad \left(p = 1 + \frac{1}{r},\ A = 2r(2r + 3 + N)\right);$$

here $\varphi(x) = t_0 - |x|$ is not differentiable at $x = 0$.

The main purpose of this paper is to remove, in case $N = 1$, all those assumptions on $f$ and $g$, which were made in [1] in order to ensure *a priori* that (0.2) holds. In fact, we shall only assume that

(0.5) the first three derivatives of $f'(x)$ and $g(x)$ are locally bounded,

and then establish the results (0.3), (0.4).

In § 1 we prove the existence of a solution $u$ in a region $\{0 < t < \varphi(x)\}$ with

$$u(x, t) \geqq c/(d(x, t))^{q/2} \qquad (c > 0) \tag{0.6}$$

near $\Gamma = \{t = \varphi(x)\}$; $\varphi(x)$ is Lipschitz continuous with constant 1. In § 2 we establish the sharp estimate

$$u(x, t) \leqq C/(d(x, t))^q. \tag{0.7}$$

Assuming further that

(0.8) the Lipschitz constant of $\varphi$ is $<1$

we prove that, near $\Gamma$,

$$u(x, t) \geqq c/(d(x, t))^q \qquad (c > 0) \tag{0.9}$$

with similar sharp estimates (analogous to (0.7), (0.9)) on $u_t$ and $u_{tt}$.

In § 3 we establish (0.8) and, thereby, the sharp estimates on $u$, $u_t$, $u_{tt}$. This enables us to extend the method of [1], thereby completing the proofs of (0.3), (0.4).

Finally, in § 4 we establish monotonicity of $\varphi(x)$ under some assumptions on the Cauchy data.

## § 1. Existence of a solution

We impose the following assumptions on $F$:

$$F(u) \geqq -0 \qquad \forall\, u \in \mathbb{R}^1;$$

$$F(u) \in C^3(\mathbb{R}^1);$$

there exists a $u_0 \geqq 0$ such that

$$F'(u) \geqq 0, \qquad F''(u) \geqq 0 \qquad \text{if} \quad u \geqq u_0; \tag{1.1}$$

$$F(u)\, u^{-p} \to A \qquad \text{if} \quad u \to \infty, \qquad A > 0, \quad p > 1;$$

$$\limsup_{u \to \infty} F'(u)\, u^{1-p} < A\left(p + \frac{p-1}{2}\right);$$

$$|F^{(j)}(u)| \leqq C u^{p-j} \qquad \text{if} \quad u \geqq u_0 + 1, \quad j = 2, 3.$$

We also assume that $f(x)$ and $g(x)$ satisfy

(1.2) $\quad f^{(j+1)}(x)$ and $g^j(x)$ are locally bounded, $\quad 0 \leqq j \leqq 3$.

Consider the Cauchy problem

$$u_{tt} - u_{xx} = F(u) \quad \text{for} \quad x \in \mathbb{R}^1, \quad t > 0, \tag{1.3}$$

$$u(x, 0) = f(x) \quad \text{for} \quad x \in \mathbb{R}^1, \tag{1.4}$$

$$u_t(x, 0) = g(x) \quad \text{for} \quad x \in \mathbb{R}^1. \tag{1.5}$$

By a smooth solution we mean a solution with four locally bounded derivatives.

In this section we prove:

**Theorem 1.1.** *There exists a unique pair of functions $u(x, t)$, $\varphi(x)$ such that*

(i) $\varphi(x) > 0$ *and $u$ is a smooth solution of* (1.3)–(1.5) *in*

$$\Omega = \{x \in \mathbb{R}^1, 0 \leqq t < \varphi(x)\};$$

(ii) *either* $\varphi(x) \equiv \infty$ *or* $\varphi(x) < \infty \quad \forall\, x \in \mathbb{R}^1$;

(iii) *if* $\varphi(x) \not\equiv \infty$ *then* $\varphi(x)$ *is Lipschitz continuous with constant* $\leqq 1$;

(iv) *if* $\varphi(x) \not\equiv \infty$ *then* $u(x, t) \to \infty$ *if* $t \uparrow \varphi(x)$; *more precisely, for any* $R > 0$ *there exist* $c > 0$, $\delta_0 > 0$ *such that*

$$u(x, t) \geqq c/(d(x, t))^{q/2} \quad \left(q = \frac{2}{p-1}\right) \tag{1.6}$$

*if* $|x| < R$, $d(x, t) \leqq \delta_0$, *where*

$$d(x, t) = \text{dist}\,((x, t), \Gamma),$$

$$\Gamma = \{(x, \varphi(x)), \quad -\infty < x < \infty\}.$$

In proving the theorem is will be convenient to work with the variables

$$\xi = t + x, \quad \eta = t - x;$$

then $u_{tt} - u_{xx} = 4u_{\xi\eta}$. We also introduce the past and future cones

$$K^-(x_0, t_0) = \{(x, t); |x - x_0| < t_0 - t, 0 < t < t_0\},$$

$$K^+(x_0, t_0) = \{(x, t); |x - x_0| < t - t_0, t_0 < t < \infty\}.$$

We begin with several lemmas for $C^2$ supersolutions:

$$\Box u \equiv u_{tt} - u_{xx} \geqq 0 \quad \text{in} \quad E \equiv \mathbb{R}^1 \times (0, \infty) \tag{1.7}$$

with

$$u(x, 0),\ u_t(x, 0) \text{ locally bounded.} \tag{1.8}$$

**Lemma 1.2.** *If $u$ is a supersolution then for any $R > 0$ there exists a constant $C \geqq 0$ such that*

$$u \geqq -C \quad \text{in} \quad E_R \equiv \{|x| \leqq R, 0 \leqq t \leqq R\}.$$

**Proof.** This follows immediately from the representation formula

$$(1.9)\qquad \begin{aligned} u(x, t) = \tfrac{1}{2}[u(x+t, 0) + u(x-t, 0)] + \tfrac{1}{2} \int_{x-t}^{x+t} u_t(\xi, 0)\, d\xi \\ + \tfrac{1}{2} \iint_{K^-(x,t)} (\Box u)(\xi, \tau)\, d\xi\, d\tau . \end{aligned}$$

**Lemma 1.3.** *If $u$ is a supersolution then for any $R > 0$ there exists a constant $c \geqq 0$ such that, in $E_R$,*

$$(1.10)\qquad u_\xi \geqq -c, \quad u_\eta \geqq -c$$

*and, consequently, also*

$$(1.11)\qquad u_t \geqq -c,$$

$$(1.12)\qquad |u_x| \leqq u_t + c.$$

**Proof.** Since $(u_\xi)_\eta \geqq 0$, $u_\xi$ is monotone increasing in $\eta$. Also $u_\xi$ is locally bounded on $\{t = 0\}$ (by (1.8)). Hence $u_\xi \geqq -c$. Similarly $u_\eta \geqq -c$.

**Lemma 1.4.** *If $u$ is a supersolution then for any $R > 0$ there exists a constant $c \geqq 0$ such that, for any $X = (x, t)$ in $E_R$,*

$$u \geqq u(X) - c \quad \text{in} \quad K^+(X) \cap E_R,$$

$$u \leqq u(X) + c \quad \text{in} \quad K^-(X).$$

**Proof.** Let $T$ be a rectangle with sides contained in the lines

$$\{\xi = \xi_0\}, \quad \{\xi = \xi_1\} \quad (\xi_0 < \xi_1)$$

$$\{\eta = \eta_0\}, \quad \{\eta = \eta_1\} \quad (\eta_0 < \eta_1)$$

and set $A_{ij} = (\xi_i, \eta_j)$. Integrating the inequality $u_{\xi\eta} = 4\Box u \geqq 0$ over $T$ we get the 4-point inequality

$$(1.13)\qquad u(A_{11}) + u(A_{00}) \geqq u(A_{10}) + u(A_{01}).$$

Taking $A_{00}$ on the $x$-axis and using also Lemma 1.2, we obtain

$$u(A_{11}) \geqq -u(A_{00}) + u(A_{01}) + u(A_{10}) \geqq u(A_{10}) - c,$$

*i.e.*. for every $X \in E_R$, $u(Y) \geqq u(X) - c$ for all $Y$ in the maximal segment $l_\xi(X)$ in the positive $\xi$-direction initiating at $X$ and contained in $E_R$.

Similarly $u(Y) \geqq u(X) - c$ for $Y$ on the maximal segment $l_\eta(X)$ in the positive $\eta$-direction initiating at $X$ and contained in $E_R$. Since further, by the previous paragraph, $u(Z) \geqq u(Y) - c$ if $Z \in l_\xi(Y)$, we get $u(Z) \geqq u(X) - 2c$ if $Z \in K^+(X) \cap E_R$, and the lemma follows.

We are now ready to construct a solution of (1.3)–(1.5) by a sequence of truncations. (The method of successive approximations used in [1] cannot be extended here since the inequality $u \geqq 0$ need not hold.) Set

$$F_n(u) = \begin{cases} F(\min(u, n)) & \text{if} \quad u > u_0 \\ F(u) & \text{if} \quad u \leqq u_0 \end{cases}$$

and let $u_n$ be the solution of

$$\begin{aligned} \Box u_n &= F_n(u_n) \quad \text{in } E, \\ u_n(x, 0) &= f(x) \quad (x \in \mathbb{R}^1), \\ \frac{\partial}{\partial t} u_n(x, 0) &= g(x) \quad (x \in \mathbb{R}^1). \end{aligned} \tag{1.14}$$

By Lemma 1.2

$$u_n \geqq -c_R \quad \text{in } E_R, \quad \forall R > 0. \tag{1.15}$$

Let $\{X_j\}$ be a dense sequence in $E$. For any sub-sequence of the $u_n$ we can choose a sub-subsequence such that

$$\lim_{n' \to \infty} u_{n'}(X_j) \text{ either exists or } = +\infty \tag{1.16}$$

for any $j$. Denote this sub-subsequence, for simplicity, by $u_n$.
If

$$\lim_{n \to \infty} u_n(X_{j_0}) = \infty$$

then, by Lemma 1.4, for any $R > 0$,

$$\lim_{n \to \infty} u_n(X) = \infty \quad \text{uniformly in } K^+(X_{j_0}) \cap E_R. \tag{1.17}$$

On the other hand, if

$$\lim_{n \to \infty} u_n(X_{j_0}) < \infty \tag{1.18}$$

then $u_n(X_{j_0}) \leqq C$ for all $n$ and, by Lemma 1.4,

$$u_n(X) \leqq C + c \quad \text{in} \quad K^-(X_{j_0}). \tag{1.19}$$

By the representation (1.9) we have, for any $X \in K^-(X_{j_0})$,

$$u_n(X) = U(X) + \tfrac{1}{2} \iint_{K^-(X)} F_n(u_n) \tag{1.20}$$

where $U$ is the solution of (1.14) with $F_n \equiv 0$. In view of (1.2), for any $R > 0$

$$|D^\alpha U(X)| \leqq C_R \quad \text{in } E_R \quad (0 \leqq |\alpha| \leqq 4).$$

Using also (1.19) we can differentiate (1.20) successively and deduce that

$$|D^\alpha u_n(X)| \leqq C \quad \text{in } K^-(X_{j_0}) \quad (0 \leqq |\alpha| \leqq 4).$$

Consequently, for a subsequence,

$$u_n \to u \text{ uniformly in } K^-(X_{j_0})$$

and $u$ is a smooth solution of (1.3)–(1.5) in $K^-(X_{j_0})$.

Recalling also the alternative (1.17) and the fact that the $X_j$ are dense in $E$, we conclude that, for a subsequence, $u_n \to u$ uniformly in compact subsets of a region $\Omega$ of the form

$$\Omega = \{x \in \mathbb{R}^1, 0 \leqq t < \varphi(x)\};$$

$u_n \to \infty$ in $\bar{E} \setminus \bar{\Omega}$, and either $\varphi(x) \equiv \infty$ or else $0 < \varphi(x) < \infty$ and $\varphi(x)$ is Lipschitz continuous with constant $\leqq 1$, for all $x \in \mathbb{R}^1$; further, $u$ is a smooth solution of (1.3)–(1.5) in $\Omega$.

We have thus established the existence of the pair $(u, \varphi)$ as asserted in (i)–(iii).

To prove (1.6) we represent $\partial u_n/\partial t$ at any point $Y \in E$:

$$\frac{\partial u_n(Y)}{\partial t} = \frac{1}{2} \iint_{K^-(Y)} \frac{\partial}{\partial t} F_n + \frac{\partial U}{\partial t}$$

where $U$ is the function appearing in (1.20), or

$$u_{n,t}(Y) = \frac{1}{2} \int_{l_1(Y) \cup l_2(Y)} F_n \cos(t, \nu)\, ds + O(1) \tag{1.21}$$

where $l_1(Y)$ ($l_2(Y)$) is the line segment from $Y$ to the $x$-axis parallel to the $\xi$- ($\eta$-) axis. By Lemma 1.3, the integrand in (1.21) is bounded by

$$\begin{aligned} C\,|u_n(Y)|^p + C &\leqq C(|u_n(Y)| + 1)^p \\ &\leqq C(u_n(Y) + C_0)^p \end{aligned}$$

where Lemma 1.2 was used in the last inequality. We conclude that

$$\frac{u_{n,t}}{(u_n + C_0)^p} \leqq C \quad \text{at } Y. \tag{1.22}$$

We now take a variable point $Y = (x_0, t)$ and integrate (1.22) with respect to $t$, $t_1 < t < t_2$, where $t_1 < \varphi(x_0) < t_2$. We obtain

$$-\frac{1}{(u_n(x_0, t_2) + C_0)^{p-1}} + \frac{1}{(u_n(x_0, t_1) + C_0)^{p-1}} \leqq C(t_2 - t_1).$$

Taking $n \to \infty$ and then letting $t_2 \uparrow \varphi(x_0)$, we get

$$(u(x_0, t_1) + C_0)^{1-p} \leqq C(\varphi(x_0) - t_1)$$

which yields the assertion (1.6).

It remains to prove uniqueness. Suppose $\bar{u}(x, t)$ is another solution with blow-up curve $y = \bar{\varphi}(x)$ satisfying (i)–(iv). By standard local uniqueness for the Cauchy problem in cones $K^-(X_0)$ it then follows that $u(x, t) = \bar{u}(x, t)$ in $\{0 < t < \min\{\varphi(x), \bar{\varphi}(x)\}\}$. In view of (1.6) and the corresponding inequality for $\bar{u}$ it then also follows that $\varphi(x) = \bar{\varphi}(x)$.

*Remark 1.1.* From the uniqueness of $(u, \varphi)$ it follows that the full sequence $u_n$ is uniformly convergent to $u$ in compact subsets of $\Omega$.

## § 2. Estimates near the blow-up curve

In the sequel we assume that $\varphi(x) \not\equiv \infty$.

**Theorem 2.1.** *For any* $R > 0$ *there exists a constant* $C > 0$ *such that*

$$u(X) \leqq C/(d(X))^q \text{ for all } X \in \Omega \cap E_R. \tag{2.1}$$

**Proof.** Let $X_2 = (x_0, t_2)$ and $X_1 = (x_0, t_1)$ $(t_2 > t_1)$ be two points in $\Omega$ within distance $\delta$ form $\Gamma$. Introduce the square $T$ in $\Omega$ with sides parallel to the axes of $\xi$ and $\eta$ having $X_1$, $X_2$ as two of its vertices. Denote the remaining two vertices by $X$ and $Y$. Then

$$u(X_2) = u(X) + u(Y) - u(X_1) + 4 \iint_T F(u).$$

Since $\delta$ is small, the number $M = u(X_1)$ is large (by (1.6)). Using Lemma 1.4 we conclude that

$$u(X_2) \geqq 2(M - c) - M + c_0 \iint_T (M - c)^p \qquad (c_0 > 0),$$

or

$$u(X_2) \geqq M - 2c + \tfrac{1}{2} c_0 M^p \frac{\sigma^2}{M^{p-1}}$$

provided $t_2 - t_1 = \sigma/M^{1/q}$. Choosing $\sigma$ such that $c_0\sigma^2 \geqq 4$ we deduce that

$$M_2 \equiv u(X_2) \geqq 2M \qquad (t_2 = t_1 + \sigma/M^{1/q}).$$

Next we repeat the process with $X_2$ and $X_3 = (x_0, t_3)$ obtaining

$$M_3 \equiv u(X_3) \geqq 2M_2 \geqq 2^2 M \qquad (t_3 = t_2 + \sigma/M_2^{1/q}).$$

More generally, if

$$M_n = u(x_0, t_n) \geqq 2M_{n-1} \geqq 2^{n-1} M,$$

$$t_{n+1} = t_n + \sigma/M_n^{1/q} \leqq t_n + \sigma/\left(2^{\frac{n-1}{q}} M^{\frac{1}{q}}\right)$$

and

$$X_{n+1} = (x_0, t_{n+1}) \in \Omega,$$

then

$$M_{n+1} \equiv u(X_{n+1}) \geqq 2M_n \geqq 2^n M.$$

Setting $t_\infty = \lim t_n$ we conclude that $(x_0, t_\infty) \notin \Omega$. Hence

$$d(X_1) \leqq t_\infty - t_1 = \sum_{n=1}^{\infty} (t_{n+1} - t_n) \leqq \sum_{n=1}^{\infty} \frac{\sigma}{2^{\frac{n-1}{q}} M^{\frac{1}{q}}}$$
$$\leqq C/M^{1/q}.$$

Since $M = u(X_1)$, (2.1) follows.

For the remainder of this section we make the assumption that, for some $R > 0$,

$$\frac{|\varphi(x) - \varphi(y)|}{|x - y|} \leqq \theta < 1 \quad \text{if } |x| < R, \quad |y| < R. \tag{2.2}$$

**Lemma 2.2.** *If* (2.2) *holds then for any* $R' < R$ *there exists a* $\delta > 0$ *and* $c > 0$ *such that*

$$u(X) \geqq \frac{c}{(d(X))^q} \qquad (X = (x, t)) \tag{2.3}$$

*if* $|x| < R'$ *and* $d(X) < \delta$.

**Proof.** From the representation (1.9) for $u_t$ we have

$$\begin{aligned} u_t(X) &= \tfrac{1}{2} \iint\limits_{K^-(X)} \frac{\partial}{\partial t} F(u) + O(1) \\ &= \tfrac{1}{2} \int\limits_{l_1(X) \cup l_2(X)} F(u) \cos(\nu, t)\, ds + O(1) \end{aligned} \tag{2.4}$$

where $l_1(X)$ $(l_2(X))$ is the line segment from $X$ to the $x$-axis parallel to the $\xi$- ($\eta$-) axis. By Lemma 1.4 and Theorem 2.1

$$u(Y) \leqq u(X) + c \quad \text{if} \quad Y \in l_i(X),$$
$$u(Y) \leqq C(d(Y))^{-q}.$$

Hence, if $d(Y)$ is small enough,

$$\begin{aligned} |F(u) \cos(\nu, t)| &\leqq C\,|u(Y)|^p \\ &\leqq C(u(X))^{p-(1+\varepsilon)/q}\,(d(Y))^{-(1+\varepsilon)}, \end{aligned} \tag{2.5}$$

for any small $\varepsilon > 0$. Condition (2.2) implies that

$$d(Y) \geqq c(d(X) + |X - Y|), \quad c > 0. \tag{2.6}$$

Substituting this into (2.5) and using the resulting estimate in (2.4), we obtain

$$\begin{aligned} u_t(X) &\leqq C(u(X))^{p-(1+\varepsilon)/q} \int_0^C \frac{d\lambda}{(d(X) + \lambda)^{1+\varepsilon}} + O(1) \\ &\leqq C u(X)^{p-(1+\varepsilon)/q}\,(d(X))^{-\varepsilon} \end{aligned}$$

i.e.,

$$\frac{u_t}{u^{p-(1+\varepsilon)/q}} \leqq \frac{C}{d^\varepsilon}. \tag{2.7}$$

Taking $X = (x_0, t)$ in (2.7) and integrating with respect to $t$, $t_0 < t < \varphi(x_0)$, we obtain

$$\frac{1}{u^{p-1-(1+\varepsilon)/q}(X)} \leqq C\, d^{1-\varepsilon}(X)$$

which is the assertion (2.3).

**Lemma 2.3.** *If* (2.2) *holds then for any* $R' < R$ *there exist a small* $\delta > 0$ *and positive constants* $C$, $c$ *such that*

$$cu^p \leqq u_{tt} \leqq Cu^p, \tag{2.8}$$

$$cu^{\frac{p+1}{2}} \leqq u_t \leqq Cu^{\frac{p+1}{2}} \tag{2.9}$$

*for all* $X = (x, t)$ *with* $|x| < R'$, $d(X) < \delta$; *consequently* (*by Theorem 2.1 and Lemma 2.2*) *also*

$$c\,d^{-q-2} \leqq u_{tt} \leqq C\,d^{-q-2}, \tag{2.10}$$

$$c\,d^{-q-1} \leqq u_t \leqq C\,d^{-q-1}. \tag{2.11}$$

**Proof.** Proceeding with (2.4) and using Lemma 2.2 we have

$$\begin{aligned} u_t(X) &\geqq c \int_{l_1(X)\cup l_2(X)} (d(Y))^{-pq} - C \\ &\geqq c \int_0^C \frac{d\lambda}{(d(X)+\lambda)^{pq}} - C \\ &\geqq \frac{c}{(d(X))^{pq-1}} = \frac{c}{(d(X))^{q+1}}. \end{aligned} \tag{2.12}$$

Using Theorem 2.1 we find that

$$u_t(X) \geqq cu^{\frac{p+1}{2}}(X).$$

Next we can represent $u_{tt}(X)$ analogously to (2.4):

$$\begin{aligned} u_{tt}(X) &= \tfrac{1}{2} \int_{l_1(X)\cup l_2(X)} F'(u)\, u_t \cos(\nu, t)\, ds + O(1) \\ &\geqq c \int_{l_1(X)\cup l_2(X)} \frac{1}{d(Y)^{(p-1)q+q+1}} \end{aligned}$$

by (2.12) and Lemma 2.1. Using (2.6), we get

$$u_{tt}(X) \geqq c\,d^{-q-2}(X)$$

or, by Theorem 2.1,

$$u_{tt} \geqq cu^p.$$

Similarly one can prove by the same method that $u_t$ and $u_{tt}$ are bounded above as asserted in (2.8)–(2.11).

*Remark 2.1.* The derivative $u_{xx}$ can be estimated from $u_{xx} = u_{tt} - F$ and Lemma 2.3. The third-order derivatives $u_{ttx}$, $u_{ttt}$ can be estimated by the same method as in Lemma 2.3, and $u_{xxx}$ from the differential equation. The fourth-order derivatives can similarly be estimated. Thus we obtain, assuming (2.2), that

$$|D^\alpha u(X)| \leqq \frac{C}{(d(X))^{q+|\alpha|}} \quad (0 \leqq |\alpha| \leqq 4). \tag{2.13}$$

## § 3. Regularity of the blow-up curve

In this section we do not make the assumption (2.2).

**Lemma 3.1.** *For any* $R > 0$ *there exists a* $\delta_0 > 0$ *and a constant* $c > 0$ *such that*

(3.1)
$$u_\xi(X) \geqq \frac{c}{(d(X))^{q/2}}$$

*if* $|x| < R,\ d(X) < \delta_0$.

**Proof.** Take any line segment $l$ initiating at $X$ and going downward in the $\eta$-direction, such that $l$ is contained in $2\delta_0$-neighborhood of $\Gamma$. Since $(u_\xi)_\eta = 4F(u) \geqq u^p$ $(c > 0)$ along $l$ (if $\delta_0$ is small enough), we conclude that

$$u_\xi(X) = \int_l u_{\xi\eta} - C \geqq \int_l cu^p - C$$

$$\geqq \int_l \frac{c}{(d(Y))^{pq/2}} - C \quad \text{(by (1.6))}$$

$$\geqq c \int_l \frac{1}{(|Y - X| + d(X))^{pq/2}} \geqq \frac{c}{d(X)^{q/2}}.$$

Set $B_\varrho(X_0) = \{|X - X_0| < \varrho\}$.

**Lemma 3.2.** *Let* $X_0 = (x_0, t_0) \in \Gamma$ *and let* $X_1 = (x_1, \tau)$, $X_2 = (x_2, \tau)$ *belong to* $\partial K^-(X_0)$ *with* $x_1 < x_2$. *Assume that* $X_1, X_2$ *belong to* $\Omega$ *and that* $|X_1 - X_2| \leqq \sigma$, $\sigma$ *sufficiently small. Then there exist* $\varepsilon_0 > 0$ *and* $\delta > 0$ *such that*

(3.2)
$$u_t + (1 \pm \varepsilon)\, u_x > 0 \text{ if } (x, t) \in B_\delta(X_0) \cap \Omega_\delta, \quad 0 < \varepsilon \leqq \varepsilon_0.$$

**Proof.** Choose points $X_3 = (x_3, \tau)$, $X_4 = (x_4, \tau)$ with $x_3 < x_1$, $x_2 < x_4$ such that $|X_4 - X_3| < 2\sigma$ and the closed interval $\overline{X_3X_4}$ lies in $\Omega$. In view of Lemma 3.1 we have, for some $c > 0$,

(3.3)
$$u_\xi > \varepsilon_0(|u_x| + |u_t|) + c \text{ on } \overline{X_3X_4}$$

provided $\varepsilon_0$ is sufficiently small.

We can represent $u_\xi$ and $\varepsilon u_x$ in the forms

$$2u_\xi(X) = u_\xi(l) + u_\xi(r) + \int_l^r u_{\xi t}\, dx + \iint_{K^-(X) \cap \{t > \tau\}} F'(u)\, u_\xi,$$

$$2\varepsilon u_x(X) = u_x(l) + u_x(r) + \varepsilon \int_l^r u_{xt}\, dx + \iint_{K^-(X) \cap \{t > \tau\}} F'(u)\, \varepsilon u_x$$

where $l$ and $r$ are the points where $\partial K^-(X)$ intersects $\{t = \tau\}$; $l$ lies to the left of $r$. Since

$$\int_l^r u_{\xi t} = \int_l^r (u_{tt} + u_{xt}) = \int_l^r (u_{xx} + F(u) + u_{xt})$$

$$= u_\xi(r) - u_\xi(l) + \int_l^r F(u),$$

$$\varepsilon \int_l^r u_{xt} = \varepsilon u_t(r) - \varepsilon u_t(l),$$

we obtain

$$(3.4)\qquad \begin{aligned} 2(u_\xi + \varepsilon u_x) = & -\varepsilon u_\eta(l) + \int_l^r F(u)\,dx + 2u_\xi(r) + \varepsilon u_\xi(r) \\ & + \iint_{K^-(X)\cap\{t>\tau\}} F'(u)\,(u_\xi + \varepsilon u_x). \end{aligned}$$

We claim that $u_\xi + \varepsilon u_x > 0$ for all $X$ in $K^-(Y)\cap\{t > \tau\}$, $Y \in B_\delta(X_0)\cap\Omega$, where $\delta$ is so small that $K^-(Y)\cap\{t = \tau\}$ is contained in $\overline{X_3X_4}$. Indeed, otherwise there is a point $X = (x, t)$ with the smallest $t$ for which $u_\xi + \varepsilon u_x = 0$; in view of (3.3), $t > \tau$. From the representation (3.4) at $(x, t)$ we then deduce, using (3.3) and the inequality $u_\xi + \varepsilon u_x > 0$ in $K^-(X)\cap\{t > \tau\}$, that the right-hand side of (3.4) is strictly positive, a contradiction.

We have thus proved that (3.2) holds for $u_t + (1 + \varepsilon)\,u_x$; the proof for $u_t + (1 - \varepsilon)\,u_x$ is the same.

The assertion (3.2) implies that the blow-up curve is a graph in any direction $(1 \pm \varepsilon, 1)$, that is, (2.2) holds for some $\theta < 1$. But then the assertions of Lemmas 2.2, 2.3 and (2.13) are valid in some neighborhood of $X_0$ and we can then apply the results of [1; §§ 4–7] in order to deduce that $\varphi(x)$ is continuously differentiable. Thus we have:

**Corollary 3.3.** *If $X_0 = (x_0, t_0) \in \Gamma$ and the two points of $\partial K^-(X_0)\cap\{t = \tau\}$ belong to $\Omega$ for some $\tau < t_0$, $t_0 - \tau$ small enough, then there exist $\delta > 0$ and $0 < \theta < 1$ such that $\varphi(x)$ is continuously differentiable if $|x - x_0| < \delta$ and $|\varphi'(x)| < \theta$.*

Using this corollary we shall now prove the main result of this paper.

**Theorem 3.4.** *The blow-up curve $y = \varphi(x)$ is continuously differentiable for all $x \in \mathbb{R}^1$, and $|\varphi'(x)| < 1$.*

**Proof.** Take $X_0 = (x_0, t_0) \in \Gamma$ and introduce points $X_1 = (x_1, \tau)$, $X_2 = (x_2, \tau)$ on $\partial K^-(X_0)\cap(t = \tau)$ with $x_1 < x_2$; $t_0 - \tau$ is sufficiently small. Suppose $X_1 \in \Gamma$. Then the segment $\overline{X_0X_1}$ is contained in $\Gamma$. Introduce points $X_{11} = (x_{11}, \tau_1)$, $X_{12} = (x_{12}, \tau_1)$ on $\partial K^-(X_1)\cap\{t = \tau_1\}$ with $x_{11} < x_{12}$ and $\tau - \tau_1$ small. If $X_{11} \in \Omega$ then, since clearly also $X_{12} \in \Omega$. Corollary 3.3 implies that $\varphi(x)$ is in $C^1(x_1 - \delta, x_1 + \delta)$ for some $\delta > 0$, with $|\varphi'(x)| < 1$; this contradicts the fact that $\varphi'(x) = 1$ if $x_1 < x < x_0$ (since $\overline{X_0X_1} \subset \Gamma$).

We conclude that if $X_1 \in \Gamma$ then also $X_{11} \in \Gamma$ and proceeding step by step we conclude that the entire line segment from $X_0$ to the $x$-axis in the $\xi$-direction belongs to $\Gamma$, a contradiction.

We have thus proved that $X_1 \notin \Gamma$, i.e., $X_1 \in \Omega$. Similarly $X_2 \in \Omega$, and Corollary 3.3 shows that $\varphi(x)$ is continuously differentiable in some interval $|x - x_0| < \delta$, with $|\varphi'(x)| < 1$. Since $X_0$ was arbitrary, the theorem follows.

*Remark 3.1.* As a by-product of the proof of the continuous differentiability of $\varphi(x)$ in [1], we deduce that if $(x, t) \to (x_0, \varphi(x_0))$ then

$$\frac{u_t^2 - u_x^2}{u^{p+1}} \to \frac{2A}{p+1}, \quad \frac{u_{tt}}{u^p} \to \frac{A}{1-\alpha^2},$$

$$\frac{u_t^2}{u^{p+1}} \to \frac{2A}{p+1} \frac{1}{1-\alpha^2}$$

where $\alpha = \varphi'(x_0)$, and (0.4) holds.

## § 4. Monotonicity of $\varphi(x)$

In this section we assume that

$$\text{(1.1) holds with } u_0 = 0, \tag{4.1}$$

and

$$f(x) \geqq \gamma_1, \quad g(x) \geqq \gamma_2 \quad \text{if} \quad |x| < R + T \tag{4.2}$$

where $\gamma_1$, $\gamma_2$ are nonnegative constants such that the solution of

$$w''(t) = F(w) \quad (t > 0),$$

$$w(0) = \gamma_1, \quad w'(0) = \gamma_2$$

satisfies: $w(t) \to \infty$ if $t \to T_1$ for some $T_1 < T$. By comparison (*cf.* [1]) it follows that $\varphi(x) < T$ if $|x| < R$. Hence, in studying $\varphi(x)$ for $|x| < R$, only the Cauchy data in $\{|x| < R + T\}$ are relevant. We shall also assume that

$$\begin{aligned} f(x) &= f(-x), \quad g(x) = g(-x), \\ f''(x) &\leqq 0, \quad g''(x) < 0 \quad \text{if} \quad |x| < R + T. \end{aligned} \tag{4.3}$$

Set

$$K_{R,T} = \bigcup_{|x|<R} K^-(x, T).$$

It will be convenient to construct $u$ in $K_{R,T}$ by successive iterations, as in [1], namely, $u_0 \equiv 0$ and

$$\begin{aligned} \Box u_{n+1} &= F(u_n), \\ u_{n+1}(x, 0) = f(x), &\quad \frac{\partial}{\partial t} u_{n+1}(x, 0) = g(x). \end{aligned} \tag{4.4}$$

Then (see [1]),

$$0 \leqq u_n \leqq u_{n+1} \uparrow u \quad \text{in } K_{R,T}.$$

From the representation formula (1.9) we get inductively that $u_m(x, t) = u_m(-x, t)$; hence also $u(x, t) = u(-x, t)$ in $K_{R,T}$ and $\varphi(x) = \varphi(-x)$ if $|x| < R$.

**Theorem 4.1.** *If* (4.1)–(4.3) *hold then*

$$\varphi'(x) > cx \quad \text{if } 0 < x < R \tag{4.5}$$

*for some positive constant c.*

We begin with several lemmas.

**Lemma 4.2.** *Suppose*

$$\Box J = aJ + h \quad \text{in } K_{R,T},$$
$$J(x, 0) = f_0(x), \quad J_t(x, 0) = g_0(x) \quad \text{if } |x| < R + T,$$

*where* $a(x, t) = a(-x, t)$, $J(x, t) = -J(-x, t)$, $h(x, t) = -h(-x, t)$, $h \leqq 0$ *in* $K_{R,T} \cap \{x > 0\}$, *and*

$$f_0(x) \leqq 0, \quad f_0'(x) \leqq 0, \quad g_0(x) \leqq 0 \quad \text{if} \quad 0 \leqq x \leqq R + T.$$

*Thus* $J(x, t) \leqq 0$ *in* $K_{R,T} \cap \{x \geqq 0\}$.

**Proof.** Let $0 \leqq x \leqq R$. By (1.9),

$$J(x, t) = \tfrac{1}{2}(f_0(x + t) + f_0(x - t)) + \tfrac{1}{2} \int_{x-t}^{x+t} g_0(\xi)\, d\xi + \tfrac{1}{2} \iint_{K^-(x,t)} (aJ + h). \tag{4.6}$$

Since $x \geqq 0$,

$$\int_{x-t}^{x+t} g_0 \leqq 0$$

and

$$f_0(x + t) + f_0(x - t) = f_0(t + x) - f_0(t - x) = 2x f_0'(\xi) \leqq 0.$$

Hence

$$J(x, t) \leqq \iint_{K^-(x,t)\setminus K_2} (aJ + h) \leqq \iint_{K^-(x,t)\setminus K_2} aJ \tag{4.7}$$

where $K_0 = K^-(x, t) \cap \{x \leqq 0\}$, $K_1$ is the reflection of $K_0$ with respect to the $t$-axis, and $K_2 = K_0 \cup K_1$.

We can now proceed by approximation. Define a sequence $J_m$ by

(4.8)

$$J_{n+1}(x, t) = \tfrac{1}{2}(f_0(x + t) + f_0(x - t)) + \tfrac{1}{2} \int_{x-t}^{x+t} g_0(\xi)\, d\xi + \tfrac{1}{2} \iint_{K^-(x,t)} (aJ_n + h)$$

with $J_0 \equiv 0$. Analogously to (4.7) we have

$$J_1(x, t) \leqq \iint_{K^-(x,t)\setminus K_2} aJ_0 = 0 \quad \text{if} \quad x \geqq 0.$$

Now assume inductively that $J_n(x, t) = -J_n(-x, t)$, and

$$J(x, t) \leqq J_n(x, t) \leqq J_{n-1}(x, t) \leqq 0 \quad \text{if } x \geqq 0.$$

Comparing the representations (1.9) for $J$, $J_{n+1}$ and $J_{n+1}$, $J_n$ we easily deduce that for $x \geqq 0$,

$$(J - J_{n+1})(x, t) \leqq \iint_{K^-(x,t)\setminus K_2} a(J - J_n) \leqq 0$$

and similarly $J_{n+1} - J_n \leqq 0$. Since $J_n \downarrow \tilde{J}$ where $\tilde{J}$ must clearly coincide with $J$, it follows that $J(x, t) \leqq 0$ if $x \geqq 0$.

Applying Lemma 4.2 to $J = u_x$ we deduce that

$$u_x \leqq 0 \quad \text{in} \quad K_{R,T} \cap \{x > 0\}. \tag{4.9}$$

This implies that $\varphi'(x) \geqq 0$ if $0 \leqq x \leqq R$.

**Lemma 4.3.** *For any* $\varepsilon > 0$ *there exists a positive constant* $N$ *such that*

$$tu_x + xu_t + Nu_x \leqq 0 \quad \text{in} \quad K_{R,T} \cap \{x \geqq 0, t \geqq \varepsilon\}. \tag{4.10}$$

**Proof.** The function

$$J = tu_x + xu_t + Nu_x$$

satisfies:

$$\Box J = F'(u)\, J.$$

Further, for any small $\varepsilon > 0$.

$$J(x, \varepsilon) = \varepsilon u_x + xu_t + Nu_x \equiv f_0(x),$$

$$J_t(x, \varepsilon) = u_x + \varepsilon u_{xt} + xu_{tt} + Nu_{xt} \equiv g_0(x).$$

Observe that

$$u_{xx}(x, \varepsilon) = u_{xx}(x, 0) + \varepsilon u_{xxt}(x, 0) + O(\varepsilon^2)$$
$$= f''(x) + \varepsilon g''(x) + O(\varepsilon^2)$$

so that, by (4.3),

$$f_0'(x) < 0 \text{ if } 0 \leqq x \leqq R + T \tag{4.11}$$

provided $N$ is sufficiently large. Since $f_0(0) = 0$, we also have

$$f_0(x) \leqq 0 \quad \text{if} \quad 0 \leqq x \leqq R + T. \tag{4.12}$$

Next,

$$u_{xt}(x, \varepsilon) = u_{xt}(x, 0) + O(\varepsilon)$$
$$= g'(x) + O(\varepsilon)$$

and $g'(x) < 0$ if $x > 0$, $g'(0) = 0$, $g''(x) < 0$. Since also

$$\frac{\partial}{\partial x} u_{xt}(0, \varepsilon) = u_{txx}(0, \varepsilon) = g''(0) + O(\varepsilon) < 0$$

by (4.3), we get

$$u_{xt}(x, \varepsilon) < -c_0 x \quad (c_0 > 0)$$

if $0 \leqq x < R + T$. It follows that

$$g_0(x) < 0 \quad \text{if} \quad 0 < x \leqq R + T$$

provided $N$ is sufficiently large. Recalling (4.11), (4.12), we can now apply Lemma 4.2 to deduce the inequality (4.10).

**Proof of Theorem 4.1.** Lemma 4.3 asserts that along any curve

$$\frac{dx}{ds} = t + N, \quad \frac{dt}{ds} = x \tag{4.13}$$

there holds: $du/ds \leqq 0$.

Let $0 < x_0 < R$, $t_0 = \varphi(x_0)$ and consider a curve (4.13) written in the form

$$\gamma\colon (t + N)^2 = x^2 + (t_0 + N - \eta)^2 - x_0^2 \quad (\eta > 0)$$

with $x > x_0$; $t = t(x) \uparrow$ if $x \uparrow$. For any constant $C > 0$, this curve intersects $\{t = t_0 + C\eta\}$ at a point

$$x = x_0 + \frac{1}{\tilde{x}_0}(t_0 + N)(C + 1)\,\eta + O(\eta^2) \quad (\eta \to 0),$$

where $\tilde{x}_0 = (x_0 + x/)2$. Hence setting

$$M = \frac{1}{\tilde{x}_0}(t_0 + N)(C + 1),$$

we conclude that

$$\varphi(x_0 + M\eta + O(\eta^2)) - \varphi(x_0) \geqq C\eta,$$

which implies, as $\eta \to 0$,

$$\varphi'(x_0) \geqq \frac{x_0}{t_0 + N}\,\frac{C}{C + 1}.$$

Taking $C \to \infty$ the assertion (4.5) follows with $c = 1/(\varphi(R) + N)$.

*Remark 4.1.* If (4.3) is replaced by

$$f'(x + t) + f'(x - t) + g(x + t) - g(x - t) > 0 \quad (|x| \leqq R, 0 \leqq t \leqq T),$$

then from the representation (1.9) for $u_{m,x} - \varepsilon u_{m,t}$ in $K_{R,T}$ for $\varepsilon$ positive and sufficiently small we deduce, inductively, that $u_{m,x} - \varepsilon u_{m,t} > 0$ in $K_{R,T}$ for all $m \geqq 1$. Hence $u_x - \varepsilon u_t \geqq 0$ in $K_{R,T}$. It follows that $\varphi'(x) > 0$ if $0 < x < R$.

This work is partially supported by National Science Foundation Grants 7406375 and MCS 7915171.

## References

1. L. A. CAFFARELLI & A. FRIEDMAN, The blow-up boundary for nonlinear wave equations, to appear.
2. R. GLASSEY, Blow-up theorems for nonlinear wave equations, *Math. Z.* **132** (1973), 183–203.
3. R. GLASSEY, Finite-time blow-up for solutions of nonlinear wave equations, *Math. Z.* **177** (1981), 323–340.
4. F. JOHN, Blow-up of solutions of nonlinear wave equations in three space dimensions, *Manuscripta Math.* **28** (1979), 235–268.
5. H. LEVINE, Instability and nonexistence of global solutions to nonlinear wave equations of the form $Pu_{tt} = -Au + F(u)$, *Trans. Amer. Math. Soc.* **192** (1974), 1–21.
6. T. KATO, Blow-up solutions of some nonlinear hyperbolic equations, *Comm. Pure Appl. Math.* **32** (1980), 501–505.

University of Chicago
Chicago, Illinois
Northwestern University
Evanston, Illinois

*(Received March 22, 1985)*

# *On the Existence of Positive Entire Solutions of a Semilinear Elliptic Equation*

WEI-YUE DING & WEI-MING NI

*Dedicated to James Serrin on the occasion of his sixtieth birthday*

## Abstract

Under suitable hypotheses we obtain various theorems concerning the existence of positive solutions of the equation

$$\Delta u - u + Q(x)\, u^p = 0$$

in $\mathbb{R}^n$, where $p > 1$ and $Q(x)$ is a given potential. If $Q$ is radially symmetric, our result is particularly simple and general. We also study symmetries of solutions of the above equation in a ball with the boundary condition $u = 0$.

## § 1. Introduction

The semilinear elliptic equation in $\mathbb{R}^n$

$$\Delta u - u + |u|^{p-1}\, u = 0, \tag{1.1}$$

$p > 1$, arises in various branches of applied mathematics (see *e.g.* [B, L] and references therein) and has been studied extensively in recent years ([N], [S], [B], [St], [B, L]). A typical existence theorem states that equation (1.1) with $1 < p < (n+2)/(n-2)$ has infinitely many solutions which tend to 0 at infinity and one of them is positive. However, all these known solutions are radial and indeed, any positive solutions which are small at infinity are necessarily radial ([G, N, Nl]). Whether equation (1.1) has a non-radial solution which is small at infinity remains an open problem. The difficulty in treating equation (1.1) arises because the domain $\mathbb{R}^n$ is unbounded and therefore we lack compact embedding theorems of Sobolev type. However, if we restrict attention to the class of radial functions, "compactness" of such a kind is regained and "standard" variational approaches work (see [N], [S], [B, L]). Naturally, this method does not seem to apply to the more general equation

$$\Delta u - u + Q(x)\,|u|^{p-1}\,u = 0 \tag{1.2}$$

where $Q(x)$ is a non-radial "potential". It should be remarked that if $Q(x) \to 0$ at infinity, once again some "compactness" exists and standard variational arguments deliver existence theorems (see [St]).

In this paper we consider equation (1.2) (and its generalization (2.1)) where the potential $Q(x)$ (or $f(x, u)$ in (2.1), respectively) is neither radial nor necessarily small at infinity. We shall consider *positive* solutions only (the so-called "ground states"). The method we use may be described roughly as follows. First, using the Mountain-Pass Lemma, we solve, for each $k$, the Dirichlet problem

$$\begin{cases} \Delta u - u + Q(x)\,u^p = 0 \\ u > 0 \quad \text{in } B_k \quad \text{and} \quad u = 0 \quad \text{on } \partial B_k \end{cases} \tag{1.2$_k$}$$

where $B_k$ is the ball of radius $k$ centered at the origin. We then establish some upper bounds, independent of $k$, of the solutions of $(1.2)_k$ we have obtained; this will ensure that a subsequence of solutions $\{u_k\}$ converges as $k \to \infty$. The major step in this approach is that we then have to show that the limit of this convergent subsequence is positive. We present three different approaches to reach this property under different assumptions on $Q$; this is done in § 3. In § 2 we derive the above-mentioned uniform upper bounds and some preliminaries. In § 4 we consider radial $Q$. A corollary of our main result in § 4 reads as follows: *if* $1 < p < (n+2)/(n-2)$ *and* $0 \leqq Q(r) \leqq C(1 + r^l)$ *where* $r = |x|$, *C is a positive constant and* $0 \leqq l < (n-1)(p-1)/2$, *for all* $r > 0$, *then equation* (1.2) *has a positive radial solution which tends to zero at* $\infty$. This conclusion seems to contain all preceding theorems in the radial case (as far as existence is concerned) and seems to be optimal. We also include in § 5 some discussions of the classical approach to (1.2) and $(1.2)_k$ through maximization, which does not seem to apply to the more general equation (2.1).

We mention that some of the ideas and methods developed in this paper also apply to the more delicate equation

$$\Delta u + K(x)\,u^{(n+2)/(n-2)} = 0, \tag{1.3}$$

$u > 0$ in $\mathbb{R}^n$, $n \geqq 3$, which arises in the problem of finding conformal metrics with prescribed scalar curvatures in Riemannian geometry (see [Ni]). Equation (1.3) with $K$ bounded between two positive constants is treated in [D, N].

While preparing this paper, we were informed that D. ZHANG [Z] had also obtained some results concerning equation (1.2). After this paper was written, we were informed kindly by P.-L. LIONS that he had some related results [L].

The major conclusions reported in this paper were obtained in 1983 while the DING was visiting Minnesota. He acknowledges the warm hospitality he received from the School of Mathematics at the University of Minnesota. He is especially grateful to JAMES SERRIN, to whom this paper is dedicated, for inviting him to Minneapolis.

## § 2. Preliminaries and uniform upper estimates

We consider the following equation in $\mathbb{R}^n$:

$$\Delta u - a(x)\, u + f(x, u) = 0, \tag{2.1}$$

which generalizes (1.2). Let $E_k$ denote $H_0^1(B_k)$, i.e. $E_k$ is the closure of smooth functions with supports contained in $B_k$ in the norm

$$\|u\|_{H^1(\mathbb{R}^n)} = \left[\int_{\mathbb{R}^n} (|\nabla u|^2 + u^2)\right]^{1/2}.$$

Let $\lambda_{1k}$ be the first eigenvalue of $-\Delta + a(x)$ on $B_k$ with Dirichlet boundary data. Assume that $a(x) \geqq 0$ is locally Hölder continuous and that there is a $\bar{\lambda} > 0$ such that

$$\lambda_{1k} \geqq \bar{\lambda} \quad \text{for all } k \geqq 1. \tag{2.2}$$

It is easy to see that (2.2) is satisfied if there is an $a_0 > 0$ such that

$$a(x) \geqq a_0 \quad \text{for all } |x| \geqq \bar{r}, \quad \text{for some } \bar{r} > 0. \tag{2.2$'$}$$

Let $E$ be the closure of smooth functions with compact support in the norm

$$\|u\| = \left[\int_{\mathbb{R}^n} (|\nabla u|^2 + au^2)\right]^{1/2}. \tag{2.3}$$

It is easy to verify that under (2.2) $E$ is a Hilbert space and $\bigcup_{k \geqq 1} E_k$ is dense in $E$. Note the obvious inclusions

$$E_1 \subseteq E_2 \subseteq \ldots \subseteq E.$$

For $f(x, u)$, we assume that $f$ is locally Hölder continuous and that

(2.4)

(a) $\lim\limits_{u \to 0+} \dfrac{f(x, u)}{u} = 0$ uniformly in $x \in \mathbb{R}^n$, and there is an open set $U \subseteq \mathbb{R}^n$ such that $\lim\limits_{u \to \infty} \dfrac{f(x, u)}{u} = \infty$ uniformly in $x \in U$.

(b) $0 \leqq f(x, u) \leqq C(1 + u^p)$ for all $x \in \mathbb{R}^n$, $u \in \mathbb{R}^+$, where $C > 0$ is a constant and $1 < p < \dfrac{n+2}{n-2}$.

(c) there is a number $0 < \theta < \frac{1}{2}$ and a function $0 \leqq A(x) \in L^1(\mathbb{R}^n) \cap C(\mathbb{R}^n)$ such that for all $x \in \mathbb{R}^n$, $u > 0$ we have $\quad A(x) + \theta u f(x, u) \geqq F(x, u) \equiv \int_0^u f(x, t)\, dt.$

*Hypotheses* (2.2) *and* (2.4) (a), (b), (c) *shall be maintained throughout* § 2 *and* § 3. It is clear that the variational functional

$$J(u) = \tfrac{1}{2} \int_{\mathbb{R}^n} (|\nabla u|^2 + au^2) - \int_{\mathbb{R}^n} F(x, u) = \tfrac{1}{2} \|u\|^2 - \int_{\mathbb{R}^n} F(x, u)$$

is well defined and is continuously differentiable on $E$. Let $J_k$ denote the restriction of $J$ to the subspace $E_k$.

Since we seek only positive solutions of (2.1), it is convenient to define $f(x, u) \equiv 0$ for $u \leqq 0$ and $x \in \mathbb{R}^n$. Then by the strong maximum principle, any nontrivial critical point $u_k$ of $J_k$ is necessarily a positive solution of

$$(2.1)_k \qquad \begin{aligned} &\Delta u - a(x)\, u + f(x, u) = 0 \quad \text{in } B_k \\ &u = 0 \quad \text{on } \partial B_k. \end{aligned}$$

Likewise, any nontrivial critical point $\bar{u} \geqq 0$ of $J$ is necessarily a positive solution of (2.1).

**Lemma 2.5.** *Assume that* (2.2) *and* (2.4) *hold. Let $u_k$ be a critical point of $J_k$ and let $u_k$ converge weakly to $\bar{u}$ in $E$. Then $\bar{u}$ is a critical point of $J$.*

The proof of this lemma is somewhat standard (assumption (2.4) (b) is used here) and is thus omitted. (See *e.g.* the arguments used in the proof of Lemma 2.19, p. 161, in [R]).

Now we shall solve $(2.1)_k$ and derive a uniform upper bound for the solutions we obtain. From (2.4) (a), (b), we see that for any $\varepsilon > 0$, there is a constant $C_\varepsilon > 0$ such that

$$0 < F(x, u) \leqq \varepsilon u^2 + C_\varepsilon u^{2n/(n-2)} \qquad \text{for all } x \in \mathbb{R}^n, \quad u > 0.$$

Choosing $\varepsilon < \bar{\lambda}/4$ ($\bar{\lambda}$ appears in (2.2)), we have, for $u \in E_k$,

$$\begin{aligned} J(u) = J_k(u) &\geqq \tfrac{1}{4} \int_{B_k} (|\nabla u|^2 + au^2) + \frac{\bar{\lambda}}{4} \int_{B_k} u^2 - \int_{B_k} (\varepsilon u^2 + C_\varepsilon |u|^{2n/(n-2)}) \\ &\geqq \tfrac{1}{4} \|u\|^2 - C'_\varepsilon \Big( \int_{B_k} |\nabla u|^2 \Big)^{n/(n-2)} \\ &\geqq \tfrac{1}{4} \|u\|^2 - C'_\varepsilon \|u\|^{2n/(n-2)} \end{aligned}$$

where the constant $C'_\varepsilon$ is independent of $k$. Since $\bigcup_{k \geqq 1} E_k$ is dense in $E$,

$$J(u) \geqq \tfrac{1}{4} \|u\|^2 - C'_\varepsilon \|u\|^{2n/(n-2)} \qquad \text{for all } u \in E.$$

Therefore, there are positive constants $\delta$ and $\bar{\alpha}$ such that

$$(2.6) \qquad J(u) \geqq \bar{\alpha} > 0 \quad \text{on } \|u\| = \delta \quad \text{in } E.$$

On the other hand, (2.4) (a) implies that (with no loss of generality, we may assume that $U \cap B_1$ is not empty) there is an $e \in E_1$ such that $e \geqq 0$, supp $e \subseteq U \cap B_1$, $\|e\| > \delta$ and $J(te) < 0$ for all $t > 1$. Now define $\Gamma(\Gamma_k)$ as the set of all continuous paths in $E(E_k$, respectively) connecting 0 and $e$, and let

$$\alpha = \inf_{\gamma \in \Gamma} \max_{u \in \gamma} J(u) \tag{2.7}$$

and

$$\alpha_k = \inf_{\gamma \in \Gamma_k} \max_{u \in \gamma} J(u). \tag{2.8}$$

Since $\Gamma_k \subseteq \Gamma_{k+1} \subseteq \Gamma$, we have

$$\alpha_k \geqq \alpha_{k+1} \geqq \alpha \geqq \bar{\alpha} > 0 \qquad \text{for all } k \geqq 1. \tag{2.9}$$

Moreover, it is not hard to prove that $\alpha_k \to \alpha$ as $k \to \infty$ since $\bigcup_{k \geqq 1} E_k$ is dense in $E$. By the well known Mountain-Pass Lemma [A, R], $\alpha_k$ is a critical value of $J_k$. Let $u_k \in E_k$ be a critical point of $J_k$ corresponding to $\alpha_k$, i.e. $J_k(u_k) = \alpha_k$ and $J_k'(u_k) = 0$. We have, in particular

$$J_k'(u_k)\, u_k = \|u_k\|^2 - \int_{B_k} u_k f(x, u_k) = 0 \tag{2.10}$$

and

$$J(u_k) = \tfrac{1}{2} \|u_k\|^2 - \int_{B_k} F(x, u_k) = \alpha_k. \tag{2.11}$$

(2.10) and (2.4) (c) together imply

$$\theta \|u_k\|^2 = \int_{B_k} \theta u_k f(x, u_k) \geqq \int_{B_k} F(x, u_k) - \int_{B_k} A(x).$$

Combining with (2.11) and (2.9), we obtain

$$(\tfrac{1}{2} - \theta) \|u_k\|^2 \leqq \alpha_k + A \leqq \alpha_1 + A$$

where $A \equiv \int_{\mathbb{R}^n} A(x)$. Thus

$$\|u_k\|^2 \leqq \frac{\alpha_1 + A}{(\frac{1}{2} - \theta)}. \tag{2.12}$$

Passing to a subsequence of $\{u_k\}$ if necessary, we may assume that $u_k$ converges to $\bar{u}$ *weakly* in $E$. Lemma 2.5 then implies that $\bar{u}$ is a critical point of $J$; thus $\bar{u} \geqq 0$ is a classical solution of (2.1) by standard theorems on elliptic regularity. Showing that $\bar{u} \not\equiv 0$ under various circumstances is the major theme of our next section. Summing up, we have the following

**Theorem 2.13.** *Assume* (2.2) *and* (2.4) *hold. For each* $k$, *the Dirichlet problem* $(2.1)_k$ *possesses a positive solution* $u_k$ *with* $J(u_k) = \alpha_k$ *and* $\alpha_k \to \alpha \geqq \bar{\alpha} > 0$, *in which* $\alpha_k$ *and* $\alpha$ *are given by* (2.7) *and* (2.8). *Moreover,* $\{u_k\}$ *is uniformly bounded in* $E$ *and so it contains a subsequence that converges weakly to* $\bar{u} \geqq 0$ *in* $E$.

A natural question arises: is $J(\bar{u}) = \alpha$? and, is $\alpha$ achieved by a path containing $\bar{u}$ in $\Gamma$? An obvious necessary condition is $\bar{u} \not\equiv 0$ (since $\alpha > 0$). The following proposition gives a sufficient condition. It will be useful in § 3 and such a result seems to be of interest in its own right.

**Proposition 2.14.** *In addition to* (2.2) *and* (2.4), *assume that*

(2.15) $$\tfrac{1}{2}uf(x,u) \geqq F(x,u) \quad \textit{for all } x \in \mathbb{R}^n, \quad u \geqq 0, \quad \textit{and}$$

(2.16) $$\frac{f(x,u)}{u} \textit{ is nondecreasing in } u > 0, \textit{ for each } x \in \mathbb{R}^n.$$

*Then, if the weak limit* $\bar{u}$ *of* $\{u_k\}$ *is nontrivial, there is a path* $\bar{\gamma} \in \Gamma$ *with* $\bar{u} \in \bar{\gamma}$ *such that*

(2.17) $$J(\bar{u}) = \max_{u \in \bar{\gamma}} J(u) = \alpha.$$

**Proof.** Since $J'(\bar{u})\,\bar{u} = 0$, i.e.

(2.18) $$\|\bar{u}\|^2 = \int_{\mathbb{R}^n} \bar{u}f(x,\bar{u})$$

we conclude that

(2.19) $$J(\bar{u}) = \int [\tfrac{1}{2}\bar{u}f(x,\bar{u}) - F(x,\bar{u})].$$

Similarly

(2.20) $$J(u_k) = \int [\tfrac{1}{2}u_k f(x,u_k) - F(x,u_k)] = \alpha_k.$$

Now, the integrands in (2.19) and (2.20) are nonnegative by (2.15). Since $u_k \to \bar{u}$ a.e. in $\mathbb{R}^n$, we conclude by Fatou's Lemma that

(2.21) $$J(\bar{u}) \leqq \liminf J(u_k) = \lim \alpha_k = \alpha.$$

Set, for $t \geqq 0$,

$$g(t) \equiv J(t\bar{u}) = \tfrac{1}{2}t^2\|\bar{u}\|^2 - \int F(x,t\bar{u}).$$

Differentiating with respect to $t$, we have

$$\begin{aligned} g'(t) &= t\|\bar{u}\|^2 - \int \bar{u}f(x,t\bar{u}) \\ &= \int [t\bar{u}f(x,\bar{u}) - \bar{u}f(x,t\bar{u})] \\ &= \int t\bar{u}^2 \left[\frac{f(x,\bar{u})}{\bar{u}} - \frac{f(x,t\bar{u})}{t\bar{u}}\right], \end{aligned}$$

where the second equality follows from (2.18). Now, (2.16) implies that $g'(t) \geqq 0$ if $t \in (0,1)$ and $g'(t) \leqq 0$ if $t \geqq 1$. Thus $g(1) = J(\bar{u})$ is the (absolute) maximum on the half-line $l = \{t\bar{u} \mid t \geqq 0\}$.

Let $V^+$ be the set $\{a\bar{u} + be \mid a \geqq 0, b \geqq 0\}$ and let $V$ be the 2-dimensional subspace of $E$ spanned by $\bar{u}$ and $e$. Let $S$ be a circle on $V$ with radius $R$ so large that $J \leqq 0$ on $S \cap V^+$ (this follows from (2.4) (a) and a standard compactness argument on $V^+$) and $\bar{u}$ and $e$ lie inside $S$. Suppose $l$ and $l_1 = \{te \mid t \geqq 0\}$ intersect $S$ at $v$ and $v_1$ respectively. Then, let $\bar{\gamma}$ be the path that consists of the segment on $l$ with endpoints 0 and $v$, the arc $S \cap V^+$ (connecting $v$ and $v_1$), and the segment on $l_1$ with endpoints $v_1$ and $e$. It is clear that $\bar{u} \in \bar{\gamma}$ and $\bar{\gamma} \in \Gamma$ and

$$J(\bar{u}) = \max_{u \in \bar{\gamma}} J(u).$$

Thus, $J(\bar{u}) \geqq \alpha$ by the definition of $\alpha$. This, together with (2.21), gives $J(\bar{u}) = \alpha$.

*Q.e.d.*

## § 3. Main existence theorems

In this section we shall prove several existence theorems concerning positive solutions (in $E$) of equation (2.1). By Theorem 2.13 we have only to show that $\bar{u} \not\equiv 0$ (where $\bar{u}$ is given by Theorem 2.13). The methods we use to achieve this are diverse, and so we divide this section into several subsections, in each of which we present one method.

§§ 3.1. In this subsection we shall prove some existence theorems by using comparison arguments and the variational approach of § 2.

Let $h(x, u) \geqq 0$, $\not\equiv 0$ (therefore with no loss of generality we assume that $h(x, u) > 0$ for some $(x, u) \in B_1 \times \mathbb{R}^+$) and that it is Hölder continuous and satisfies the following conditions:

$$(3.1) \qquad uh_u(x, u) \geqq (1 + \varepsilon)\, h(x, u) \geqq 0 \quad \text{for } x \in \mathbb{R}^n, \quad u > 0,$$

$$(3.2) \qquad h(x, u) \leqq C(1 + u^p),\ 1 < p < \frac{n+2}{n-2} \quad \text{for } x \in \mathbb{R}^n, \quad u > 0,$$

where $\varepsilon > 0$, $C > 0$ are constants independent of $x \in \mathbb{R}^n$. Again, we set $h(x, u) \equiv 0$ for $x \in \mathbb{R}^n$, $u \leqq 0$ for convenience. Set

$$I(u) = \tfrac{1}{2}\|u\|^2 - \int_{\mathbb{R}^n} H(x, u)$$

where

$$H(x, u) = \int_0^u h(x, t)\, dt.$$

We choose $e \geqq 0$ in $E_1$ (defined in § 2) so that $J(te) < 0$ and $I(te) < 0$ for all $t > 1$. Observe that this change does not alter the value $\alpha$. (See the last paragraph in the proof of Proposition 2.14.) Let

$$\beta = \inf_{\gamma \in \Gamma} \max_{u \in \gamma} I(u),$$

$$M_f = \left\{u \in E \setminus \{0\} \mid \|u\|^2 = \int_{\mathbb{R}^n} uf(x, u)\right\}$$

and

$$M_h = \left\{u \in E \setminus \{0\} \mid \|u\|^2 = \int_{\mathbb{R}^n} uh(x, u)\right\}.$$

From (3.1) we see that $h(x, u)/u$ is strictly increasing in $u > 0$ if $h(x, u) > 0$; in fact, $h(x, u)/u^{1+\varepsilon}$ is nondecreasing. Thus it is easy to verify that if $\{x \in \mathbb{R}^n \mid$

$h(x, u(x)) > 0\}$ is not empty, then the half-line $l_u = \{tu \mid t \geqq 0\}$ intersects $M_h$ at exactly one point $\bar{t}u$. Moreover, $I(\bar{t}u) > 0$ is the (absolute) maximum of $I$ on $l_u$. Using the arguments in the last paragraph of the proof of Proposition 2.14, we conclude that

$$\beta \leqq \beta^*, \tag{3.3}$$

where $\beta^* = \inf_{u \in M_h} I(u)$. Similarly, we define $\alpha^* = \inf_{u \in M_f} J(u)$.

**Theorem 3.4.** *Suppose* (2.2), (2.4) *and* (2.16) *hold. Suppose in addition that there is an* $R > 0$ *such that*

$$f(x, u) \leqq h(x, u) \quad \text{for all } x \in \mathbb{R}^n \setminus B_R,\ u > 0. \tag{3.5}$$

*Then* $\bar{u} \equiv 0$ *implies that* $\alpha \geqq \beta^*$, *where* $\bar{u}$, $\alpha$ *are given in* § 2.

**Theorem 3.6.** *Under the hypotheses of Theorem* 3.4, *if* $\alpha < \beta^*$ *then* $0 < \bar{u} \in E$ *is a solution of* (2.1).

Theorem 3.6 follows immediately from Theorem 3.4. We shall give some useful corollaries and examples to illustrate Theorem 3.6 later.

**Proof of Theorem 3.4.**

Since $\bar{u} \equiv 0$, $\{u_k\}$ converges to 0 weakly in $E$. Thus as $k \to \infty$, (a subsequence of) $u_k \to 0$ uniformly on compact sets by elliptic regularity estimates. Hence

$$0 \leqq \varepsilon_k \equiv \int_{B_R} u_k f(x, u_k) \to 0.$$

(Note that $R$ is fixed.) It was observed earlier that for each $k$, there is a unique $t_k > 0$ such that $t_k u_k \in M_h$; i.e.

$$t_k^2 \|u_k\|^2 = \int_{\mathbb{R}_n} t_k u_k h(x, t_k u_k).$$

On the other hand, by (3.5),

$$\|u_k\|^2 = \int_{\mathbb{R}_n} u_k f(x, u_k) = \varepsilon_k + \int_{|x|>R} u_k f(x, u_k)$$
$$\leqq \varepsilon_k + \int_{|x|>R} u_k h(x, u_k).$$

Therefore, if $t_k \geqq 1$, then

$$t_k^2 \varepsilon_k + t_k^2 \int_{|x|>R} u_k h(x, u_k)$$
$$\geqq \int_{|x|>R} t_k u_k h(x, t_k u_k)$$
$$\geqq \int_{|x|>R} t_k^{2+\varepsilon} u_k h(x, u_k);$$

the last inequality comes from the fact that

$$h(x, tu) \geqq t^{1+\varepsilon} h(x, u)$$

for all $x \in \mathbb{R}^n$, $t \geqq 1$ and $u \geqq 0$, which is a consequence of (3.1). Hence

$$\begin{aligned} t_k^2 \varepsilon_k &\geqq (t_k^{2+\varepsilon} - t_k^2) \int_{|x|>R} u_k h(x, u_k) \\ &\geqq (t_k^{2+\varepsilon} - t_k^2) \int_{|x|>R} u_k f(x, u_k) \\ &= (t_k^{2+\varepsilon} - t_k^2)(\|u_k\|^2 - \varepsilon_k) \\ &\geqq \alpha(t_k^{2+\varepsilon} - t_k^2), \quad \text{for large } k, \end{aligned}$$

since $\varepsilon_k \to 0$ as $k \to \infty$ and $\|u_k\|^2 \geqq 2J_k(u_k) = 2\alpha_k \geqq 2\alpha$. Thus,

$$\varepsilon_k \geqq \alpha(t_k^\varepsilon - 1),$$

and $t_k \to 1$ as $k \to \infty$. In particular, $\{t_k\}$ is a bounded sequence. *Consequently, $\{t_k u_k\}$ converges to* 0 *weakly in E.* Next, observe that $J(u_k)$ is the maximum of $J$ restricted to $l_{u_k}$ (by (2.16), see the proof of Proposition 2.14), and then

$$\begin{aligned} \alpha_k &= J(u_k) \geqq J(t_k u_k) \\ &= \tfrac{1}{2} t_k^2 \|u_k\|^2 - \int_{|x|>R} F(x, t_k u_k) - \int_{B_R} F(x, t_k u_k) \\ &\geqq \tfrac{1}{2} t_k^2 \|u_k\|^2 - \int_{|x|>R} H(x, t_k u_k) - \int_{B_R} F(x, t_k u_k) \\ &\geqq I(t_k u_k) - \int_{B_R} F(x, t_k u_k) \\ &\geqq \beta^* - \int_{B_R} F(x, t_k u_k) \end{aligned}$$

by (3.3) and the fact that $t_k u_k \in M_h$. As before, the integral

$$\int_{B_R} F(x, t_k u_k) \to 0$$

as $k \to \infty$ since $t_k u_k$ converges weakly to 0 in $E$. Letting $k \to \infty$, we obtain $\alpha \geqq \beta^*$.

*Q.e.d.*

In practical situations, the quantities $\alpha^*, \beta^*$ are often much easier to compute than $\alpha, \beta$ thus it seems desirable to have the following

**Corollary 3.7.** *Under the hypotheses of Theorem* 3.6, *suppose in addition that f satisfies* (3.1) *also. If* $\alpha^* < \beta^*$, *then* (2.1) *has a positive solution* $\bar{u}$ *in E.*

**Proof.** Since under the additional hypothesis $uf_u(x, u) \geqq (1+\varepsilon) f(x, u) \geqq 0$ for $x \in \mathbb{R}^n$, $u > 0$, we also have $\alpha \leqq \alpha^*$ (see the arguments which lead to (3.3)). Thus $\alpha^* < \beta^*$ implies $\alpha < \beta^*$ and then Theorem 3.6 applies.

*Q.e.d.*

Now we come back to the equation

$$\Delta u - u + Q(x) u^p = 0, \quad u > 0, \tag{3.8}$$

in $\mathbb{R}^n$, assuming that $Q(x) \geqq 0$ and $\not\equiv 0$, $1 < p < (n+2)/(n-2)$. For purposes of comparison we also consider

$$\Delta u - u + K(x) u^p = 0, \quad u > 0 \tag{3.9}$$

in $\mathbb{R}^n$, where $K(x) \geqq 0$. That is, we set $f(x, u) = Q(x) u^p$, and $h(x, u) = K(x) u^p$, both $Q$ and $K$ are *bounded* functions on $\mathbb{R}^n$, and $Q(x) \leqq K(x)$ for all $|x| \geqq R$, for some $R > 0$. Thus all the hypotheses of Corollary 3.7 are satisfied, and

$$I(u) = \tfrac{1}{2} \|u\|^2 - \frac{1}{p+1} \int_{\mathbb{R}^n} K u_+^{p+1}$$

$$J(u) = \tfrac{1}{2} \|u\|^2 - \frac{1}{p+1} \int_{\mathbb{R}^n} Q u_+^{p+1}$$

$$M_f = \left\{ u \in E \setminus \{0\} \mid \|u\|^2 = \int_{\mathbb{R}^n} Q u_+^{p+1} \right\}$$

$$M_h = \left\{ u \in E \setminus \{0\} \mid \|u\|^2 = \int_{\mathbb{R}^n} K u_+^{p+1} \right\}$$

where

$$\|u\|^2 = \int_{\mathbb{R}^n} (|\nabla u|^2 + u^2), \qquad u_+(x) = \max\{u(x), 0\},$$

and $E = H_0^1(\mathbb{R}^n)$. Now, for any $u \in E$, there is at most one $t > 0$ such that $tu \in M_f$. This number $t$ may be calculated explicitly as follows

$$t = \left[ \frac{\|u\|^2}{\int_{\mathbb{R}^n} Q u_+^{p+1}} \right]^{\frac{1}{p-1}}. \tag{3.10}$$

Therefore

$$\alpha^* = \inf_{u \in M_f} \left( \tfrac{1}{2} - \frac{1}{p+1} \right) \int_{\mathbb{R}^n} Q u_+^{p+1},$$

*i.e.*

$$\alpha^* = \left( \tfrac{1}{2} - \frac{1}{p+1} \right) \inf_{u \in E \setminus \{0\}} \left[ \frac{\|u\|}{\left( \int_{\mathbb{R}^n} Q u_+^{p+1} \right)^{\frac{1}{p+1}}} \right]^{\frac{2(p+1)}{p-1}} \tag{3.11}$$

Similarly,

$$\beta^* = \left(\tfrac{1}{2} - \frac{1}{p+1}\right) \inf_{u \in E \setminus \{0\}} \left[ \frac{\|u\|}{\left( \int\limits_{\mathbb{R}^n} K u_+^{p+1} \right)^{\frac{1}{p+1}}} \right]^{\frac{2(p+1)}{p-1}}. \tag{3.12}$$

Thus we have

**Corollary 3.13.** *Suppose that* $Q(x) \geqq 0$, $K(x) \geqq 0$ *in* $\mathbb{R}^n$ *with* $Q(x) \leqq K(x)$ *for* $|x| > R$, $R$ *being some positive constant. Then* (3.8) *possesses a positive solution in* $H_0^1(\mathbb{R}^n)$ *if*

$$\inf_{0 \neq u \in H_0^1(\mathbb{R}^n)} \frac{\|u\|}{\left( \int\limits_{\mathbb{R}^n} Q u_+^{p+1} \right)^{\frac{1}{p+1}}} < \inf_{0 \neq u \in H_0^1(\mathbb{R}^n)} \frac{\|u\|}{\left( \int\limits_{\mathbb{R}^n} K u_+^{p+1} \right)^{\frac{1}{p+1}}}. \tag{3.14}$$

**Corollary 3.15.** *Given* $Q(x) \geqq 0$ *in* $\mathbb{R}^n$, *equation* (3.8) *possesses a positive solution in* $H_0^1(\mathbb{R}^n)$ *if*

$$\inf_{0 \neq u \in H_0^1(\mathbb{R}^n)} \frac{\|u\|}{\left( \int\limits_{\mathbb{R}^n} Q u_+^{p+1} \right)^{\frac{1}{p+1}}} < \lim_{R \to \infty} \left\{ \inf_{0 \neq u \in H_0^1(\mathbb{R}^n)} \frac{\|u\|}{\left( \int\limits_{|x|>R} Q u_+^{p+1} \right)^{\frac{1}{p+1}}} \right\}. \tag{3.16}$$

**Proof.** Take $K(x) = \begin{cases} Q(x) & \text{if } |x| > R, \\ 0 & \text{if } |x| \leqq R; \end{cases}$

the result follows from the preceding corollary.

*Q.e.d.*

It is clear that the following corollary is another reformulation of Corollary 3.15.

**Corollary 3.17.** *Given* $Q(x) \geqq 0$ *in* $\mathbb{R}^n$, *equation* (3.8) *has a positive solution in* $H_0^1(\mathbb{R}^n)$ *if*

$$\sup_{\|u\|=1} \int\limits_{\mathbb{R}^n} Q u_+^{p+1} > \lim_{R \to \infty} \left[ \sup_{\|u\|=1} \int\limits_{|x|>R} Q u_+^{p+1} \right]. \tag{3.18}$$

We should remark that the case $Q \equiv 1$ is not included in (3.18), but $\lim\limits_{|x| \to \infty} Q(x) = 0$ is obviously included. Also, if $Q(x) \geqq a_0 > 0$ for $|x| < R_0$ and $Q(x) \leqq a_1$ for $|x| > R_1$ where $a_0 \gg a_1$, then (3.18) holds and equation (3.8) can be solved.

**Corollary 3.19.** *If*

$$\lim_{|x| \to \infty} Q(x) = \inf_{x \in \mathbb{R}^n} Q(x), \tag{3.20}$$

*then* (3.8) *has a positive solution in* $H_0^1(\mathbb{R}^n)$.

**Proof.** Set

$$m = \lim_{|x|\to\infty} Q(x) = \inf_{x\in\mathbb{R}^n} Q(x).$$

If $Q \not\equiv$ constant, then it is clear that

$$\tilde{\alpha} = \sup_{\|u\|=1} \int_{\mathbb{R}^n} Q u_+^{p+1} > \alpha_m = \sup_{\|u\|=1} \int_{\mathbb{R}^n} m u_+^{p+1}$$

since $\alpha_m$ is attained by some positive function. On the other hand, given any $\varepsilon > 0$, there is an $r$ large enough that

$$\alpha_r = \sup_{\|u\|=1} \int_{|x|>r} Q u_+^{p+1} \leqq (m+\varepsilon) \sup_{\|u\|=1} \int_{|x|>r} u_+^{p+1}$$
$$\leqq \frac{m+\varepsilon}{m}\alpha_m;$$

thus

$$\lim_{r\to\infty} \alpha_r \leqq \frac{m+\varepsilon}{m}\tilde{\alpha}.$$

Since $\varepsilon > 0$ is arbitrary and $\alpha_m < \tilde{\alpha}$, (3.18) holds.

If $Q \equiv$ constant, the result is well known (see Theorems 3.23, 4.8).

*Q.e.d.*

Let

$$\bar{Q} = \limsup_{|x|\to\infty} Q(x), \qquad Q_r = \min_{|x|\leqq r} Q,$$

$$\tau = \lim_{r\to\infty} \tau_r, \qquad \tau_r = \sup_{\|u\|_{H_0^1(B_r)}=1} \int_{B_r} u_+^{p+1}$$

**Corollary 3.21.** *If* $Q_r\tau_r > \bar{Q}\tau$ *for some* $r > 0$, *then* (3.18) *holds and* (3.8) *has a positive solution in* $H_0^1(\mathbb{R}^n)$.

The proof is straightforward since

$$\tilde{\alpha} \geqq Q_r\tau_r > \bar{Q}\tau \geqq \lim_{r\to\infty} \alpha_r.$$

*Remark. Observe that in* (3.11), (3.12), (3.14), (3.16), (3.18) *and the definitions of* $\tilde{\alpha}$, $\alpha_m$, $\tau_r$ ... etc., *one can replace* $u_+$ *by* $|u|$ *without altering any of those numbers since* $u$ *and* $|u|$ *have the same* $H^1$*-norm.*

§§ 3.2. In this subsection we use the well known technique of the "moving parallel plane" to obtain a uniform positive lower bound on $u_k$ in Theorem 2.13 and therefore we have $\bar{u} > 0$. This technique goes back to A. D. ALEXANDROV, and has been used by various authors; see, *e.g.* [Se], [G, N, N1] and [G, N, N2].

We start with a few definitions. Let $\Omega$ be a convex domain in $\mathbb{R}^n$ which is symmetric with respect to the hyperplanes $\Sigma_i \equiv \{x = (x_1, \ldots, x_n) \in \mathbb{R}^n \mid x_i = 0\}$, $i = 1, \ldots, n$. Let $e_i$ denote the unit vector pointing along the positive $x_i$-axis. Let $\varrho \geqq 0$ be a real number. Define $E(\varrho, \Omega)$ to be the set of all functions $u$ on $\Omega$ satisfying $u(y + te_i) \leqq u(y + (2\lambda - t)\, e_i)$ for all $t \geqq \lambda \geqq \varrho$ or $t \leqq -\lambda \leqq -\varrho$, $y \in \Sigma_i \cap \Omega$, $1 \leqq i \leqq n$, provided $y + te_i$ and $y + (2\lambda - t)\, e_i \in \Omega$. The following lemma is an easy consequence of Theorem 2.1′ in [G, N, N2].

**Lemma 3.22.** *Let $\Omega$ be a bounded convex domain in $\mathbb{R}^n$ which is symmetric with respect to $\Sigma_i$, for all $1 \leqq i \leqq n$, and let $g$ be a continuous function which is locally Lipschitz in $u$. Let $u \in C^2(\Omega) \cap C(\bar{\Omega})$ be a solution of $\Delta u + g(x, u) = 0$ with $u > 0$ in $\Omega$ and $u = 0$ on $\partial\Omega$. If $g(\cdot, s) \in E(\varrho, \Omega)$ for all $s > 0$ and $g(x, 0) \geqq 0$ for all $x \in \partial\Omega$, then $u \in E(\varrho, \Omega)$. In particular, $u$ attains its maximum in the cube $C_\varrho = \{x = (x_1, \ldots, x_n) \in \mathbb{R}^n \mid |x_i| \leqq \varrho,\ 1 \leqq i \leqq n\}$.*

**Theorem 3.23.** *Let (2.2) and (2.4) hold, and let $f(x, u)$ be locally Lipschitz-continuous in $u$. Assume that there is a $\varrho \geqq 0$ such that*

$$(3.24) \qquad -a(x) \in E(\varrho, \mathbb{R}^n) \quad \text{and} \quad f(x, s) \in E(\varrho, \mathbb{R}^n) \quad \text{for all } s > 0.$$

*Then equation (2.1) has a positive solution $\bar{u} \in E \cap E(\varrho, \mathbb{R}^n)$.*

**Proof.** Let $u_k$ be a solution of $(2.1)_k$ obtained in Theorem 2.13. By (3.24) and thus Lemma 3.22, we have $u_k \in E(\varrho, B_k)$ and $M_k = \max u_k$ is attained in the cube $C_\varrho$, say, at $P_k$. We claim that *there is a positive constant $\delta$ such that $M_k \geqq \delta$ for all $k$.* Indeed, from $J_k'(u_k)\, u_k = 0$ we conclude that

$$\int u_k f(x, u_k) = \|u_k\|^2 \geqq \bar{\lambda} \int u_k^2 .$$

Thus

$$\int u_k (f(x, u_k) - \bar{\lambda} u_k) \geqq 0 .$$

Now our assertion follows from (2.4) (a).

Suppose that $C_\varrho \subseteq B_{\bar{k}}$ for some $\bar{k}$. By standard arguments on regularity (and (2.4) (b)), we conclude that there is a constant $M$, depending only on $p$, $\varrho$ and $\bar{k}$, such that

$$\|u_k\|_{C^{1,\gamma}(C_\varrho)} \leqq M$$

for all $k \geqq \bar{k}$. Thus, some subsequence $\{u_{k_j}\}$ converges in $C^1(C_\varrho)$ to $u^* \in C^1(C_\varrho)$. Since $u_{k_j} \to \bar{u}$ weakly in $E$ by Theorem 2.13, $u_{k_j} \to \bar{u}$ almost everywhere, and $u^* = \bar{u}$. It is obvious that $\bar{u} \in E(\varrho, \mathbb{R}^n)$ and is a classical solution of (2.1) with $\max \bar{u} \geqq \delta$.

*Q.e.d.*

In particular, this theorem covers the well-known special case $Q \equiv 1$ in equation (3.8).

§§ 3.3. The "geometrical" theorem in §§ 3.2 is somewhat "rigid"; we shall extend it in this subsection.

**Theorem 3.25.** *Let* (2.2), (2.4) *and* (2.16) *hold. Suppose* $a_1(x)$ *and* $f_1(x, u)$ *satisfy* (2.15), (2.16) *and the hypotheses of Theorem* 3.23. *If there are functions* $b_1(x)$ *and* $b_2(x)$ *such that*

$$a_1(x) - b_1(x) \leqq a(x) \leqq a_1(x) \tag{3.26}$$

$$f_1(x, u) \leqq f(x, u) \leqq f_1(x, u) + b_2(x)\,(u + u^p) \tag{3.27}$$

*for all* $x \in \mathbb{R}^n$, $u > 0$, *where* $b_i(x) \in C(\mathbb{R}^n)$ *with*

$$b_i(x) \to 0 \text{ as } |x| \to \infty, \qquad i = 1, 2, \tag{3.28}$$

*then equation* (2.1) *has a positive solution in E.*

**Proof.** The basic idea is to compare functionals $J(u)$ (defined in § 2) and

$$J_1(u) = \tfrac{1}{2} \int_{\mathbb{R}^n} (|\nabla u|^2 + a_1 u^2) - \int_{\mathbb{R}^n} F_1(x, u)$$

where

$$F_1(x, u) = \int_0^u f_1(x, s)\, ds.$$

Note that the induced norm

$$\|u\|_1 = \left[\int_{\mathbb{R}^n} (|\nabla u|^2 + a_1 u^2)\right]^{\frac{1}{2}}$$

is equivalent to the norm $\|u\|$ defined by (2.3) in view of (3.26) and (3.28). Choose $e \geqq 0$ in $E_1$ (as in § 2) so that $J(te) < 0$ and $J_1(te) < 0$ for all $t > 1$. Observe once again that this new choice of $e$ does not change the value $\alpha$ defined in (2.7) (although $\Gamma_k$ and $\Gamma$ are changed, and we shall denote also the new ones by $\Gamma_k$ and $\Gamma$). This is clear by the last paragraph in the proof of Proposition 2.14. Let

$$\beta_k = \inf_{\gamma \in \Gamma_k} \max_{u \in \gamma} J_1(u),$$

$$\beta = \inf_{\gamma \in \Gamma} \max_{u \in \gamma} J_1(u).$$

By Theorem 3.23 and Proposition 2.14, we see that $\beta$ is a critical value of $J_1$ and there is a $\gamma_1 \in \Gamma$ such that

$$\beta = \max_{u \in \gamma_1} J_1(u). \tag{3.29}$$

It is clear that $J(u) \leqq J_1(u)$ for all $u \in E$; thus $\alpha \leqq \beta$.

*Case* 1. $\alpha = \beta$.
Then

$$\alpha \leqq \max_{u \in \gamma_1} J(u) \leqq \max_{u \in \gamma_1} J_1(u) = \beta = \alpha$$

*i.e.*

$$\alpha = \max_{u \in \gamma_1} J(u),$$

that is, also $\alpha$ is achieved in $\gamma_1 \in \Gamma$. Next we assert that *there is a* $u_1 \in \gamma_1$ *such that* $J'(u_1) = 0$ *and* $J(u_1) = \alpha$. Suppose this is not so. Then, we can find an $\varepsilon > 0$ and a $\delta > 0$ such that $\| J'(u) \| \geqq \delta$ on $\gamma_\varepsilon \equiv \gamma_1 \cap J^{-1}[\alpha - \varepsilon, \alpha + \varepsilon]$. Thus $\| J'(u] \| \geqq \delta/2$ on some neighborhood $U_\varepsilon$ of $\gamma_\varepsilon$. Consider the deformation $\eta(t, x)$ defined by the "gradient flow"

$$(3.30) \qquad \begin{aligned} &\eta_t = -\psi(\eta)\, J'(\eta)/\| J'(\eta) \| \quad \text{and} \\ &\eta(0, x) \equiv x \quad \text{for all } x \in E, \end{aligned}$$

where $\psi$ is a cut-off function so chosen that $\psi$ is positive in $U_\varepsilon$ and vanishes identically outside $U_\varepsilon$. Now, on the deformed path $\eta(1, \gamma_1)$, we have $J < \alpha$; thus

$$\max_{u \in \eta(1, \gamma_1)} J(u) < \alpha,$$

a contradiction. This proves our assertion.

A few remarks are in order. First, our argument concerning the deformation (3.30) is rather sketchy since such a deformation is somewhat standard; see, for example, the proof of Theorem 1.9 in [R], pp. 148–153. Also observe that some "compactness" conditions, such as the well-known "Palais-Smale condition", are *usually* assumed in treating this kind of deformation. In our case, the functional $J$ does not enjoy compactness of that kind in $E$. However, we get around this difficulty by constructing an explicit path $\gamma_1$ which achieves $\alpha$ and is naturally *compact*.

From our assertion above, we see that $\alpha$ is a critical value of $J$ and $u_1$ is a nontrivial critical point of $J$. Furthermore, the construction of $\gamma_1$ (as in the proof of Proposition 2.14) makes $u_1$ nonnegative in $\mathbb{R}^n$; thus, by the maximum principle, $u_1$ must be positive in $\mathbb{R}^n$.

*Case* 2. $\alpha < \beta$. Let $u_k$, $\alpha_k$ and $\bar{u}$ be as in Theorem 2.13. It suffices to show that $\bar{u} \not\equiv 0$ in $E$. Suppose the contrary, $\bar{u} \equiv 0$; *i.e.*, suppose that $u_k$ converges to 0 weakly in $E$. From (3.26) and (3.27) we deduce

$$(3.31) \qquad \begin{aligned} J(u) &\geqq \tfrac{1}{2} \int (|\nabla u|^2 + a_1 u^2) - \int F_1(x, u) - \tfrac{1}{2} \int b_1 u^2 - \int b_2 \left( \frac{u^2}{2} + \frac{|u|^{p+1}}{p+1} \right) \\ &\geqq J_1(u) - \int b(u^2 + |u|^{p+1}) \end{aligned}$$

where $b \equiv b_1 + b_2 \to 0$ at $\infty$.

Let $l_k = \{tu_k \mid t \geqq 0\}$. Since $f$ satisfies (2.16), we see from the proof of Proposition 2.14 that

$$(3.32) \qquad J(u_k) = \max_{u \in l_k} J(u).$$

Suppose $J_1|_{l_k}$ attains its maximum at $t_k u_k$. We claim that *the sequence* $t_k > 0$ *is bounded*. Set $g_k(t) = J_1(tu_k)$, $t \geqq 0$. Using (2.4) (c) for $f_1$, we have

$$
\begin{aligned}
(3.33) \qquad g_k'(t) &= t\,\|u_k\|_1^2 - \int u_k f_1(x, tu_k) \\
&\leqq t\,\|u_k\|_1^2 - \frac{1}{t\theta_1}\int F_1(x, tu_k) + \frac{1}{t\theta_1}\int A_1(x) \\
&= -\left(\frac{1}{2\theta_1} - 1\right) t\,\|u_k\|_1^2 + \frac{1}{t\theta_1} J_1(tu_k) + \frac{1}{t\theta_1}\int A_1(x) \\
&\leqq -Ct + \frac{1}{t\theta_1} g_k(t) + \frac{1}{t\theta_1}\int A_1(x)
\end{aligned}
$$

where $C = \left(\frac{1}{2\theta_1} - 1\right) C'$ and $C' > 0$ is a uniform lower bound for $\|u_k\|$, for all $k$, i.e. $\|u_k\| \geqq C' > 0$ for all $k$ since $J(u_k) = \alpha_k \geqq \alpha > 0$. On the other hand, by (2.16) applied to $f_1$, when $t \geqq 1$ we have

$$
\begin{aligned}
(3.34) \qquad g_k'(t) &\leqq t\,\|u_k\|_1^2 - t\int u_k f_1(x, u_k) \\
&\leqq t\,[\|u_k\|^2 + \int b_1 u^2 - \int u_k f(x, u_k) + \int b_2(u_k^2 + u_k^{p+1})] \\
&\leqq (p+1)\, t \int b(u_k^2 + u_k^{p+1})
\end{aligned}
$$

since $u_k \geqq 0$ and

$$\|u_k\|^2 = \int u_k f(x, u_k).$$

Next, *we claim that*

$$(3.35) \qquad \varepsilon_k = \int b(u_k^2 + u_k^{p+1}) \to 0 \quad \text{as } k \to \infty.$$

We postpone the proof of (3.35) till the end of this subsection. From (3.34) and (3.35) we observe that if $t \geqq 1$,

$$g_k(t) \leqq g_k(1) + (p+1)\,\varepsilon_k t^2.$$

By (3.31),

$$g_k(1) = J_1(u_k) \leqq J(u_k) + \varepsilon_k = \alpha_k + \varepsilon_k.$$

Now, (3.33) gives, if $t \geqq 1$

$$g_k'(t) \leqq -\left(C - \frac{(p+1)\,\varepsilon_k}{\theta_1}\right) t + \bar{C}$$

where the constant $C$ is given in (3.33) and $\bar{C}$ is some constant. By (3.35) we can find $k_0 \geqq 1$, $T > 1$ such that $g_k'(t) < 0$ for $k \geqq k_0$, $t \geqq T$, and $T$ is independent of $k \geqq k_0$. Since $g_k'(t_k) = 0$, $t_k \leqq T$ for all $k \geqq k_0$. This proves our assertion.

In view of (3.31), (3.32) and (3.35), we see that

$$J_1(t_k u_k) \leqq J(u_k) + T^{p+1} \int b(u_k^2 + u_k^{p+1})$$
$$= \alpha_k + \varepsilon_k T^{p+1}.$$

Since $\alpha_k$ converges to $\alpha$, which is smaller than $\beta$, we have

$$\max_{u \in l_k} J_1(u) \leqq \beta - \bar{\varepsilon}$$

for $k$ sufficiently large, where $\bar{\varepsilon} < \beta$ is some positive constant. Now we can construct a path $\gamma_k \in \Gamma_k$ based on $l_k$ such that

$$\max_{u \in \gamma_k} J_1(u) \leqq \beta - \bar{\varepsilon}.$$

The construction of such a path $\gamma_k$ is similar to the arguments in the last paragraph of the proof of Proposition 2.14 and is therefore omitted here. By the definition of $\beta_k$, we then have $\beta_k \leqq \beta - \bar{\varepsilon}$ for all $k$ sufficiently large. Letting $k \to \infty$, we see that $\beta \leqq \beta - \bar{\varepsilon}$, a contradiction. To finish the proof, it remains to show (3.35) (assuming $\bar{u} = 0$). Indeed, given any $\varepsilon > 0$, there is an $R > 0$ such that $b(x) \leqq \varepsilon/2C$, for $|x| \geqq R$ where $C > 0$ is a constant such that

$$\int (u_k^2 + u_k^{p+1}) \leqq C$$

for all $k \geqq 1$. (The existence of such a constant follows from (2.12) and the following well known Sobolev-type inequality

$$\|u\|_{L^{p+1}} \leqq C \|\nabla u\|_{L^2}^{1-a} \|u\|_{L^2}^{a} \qquad \text{for all } u \in C_0^\infty(\mathbb{R}^n),$$

where

$$a = 1 - \left(\tfrac{1}{2} - \frac{1}{p+1}\right) n\,.)$$

Then

$$\varepsilon_k = \left(\int_{|x| \geqq R} + \int_{|x| \leqq R}\right) (b(u_k^2 + u_k^{p+1}))$$
$$\leqq \frac{\varepsilon}{2} + \int_{|x| \leqq R} b(u_k^2 + u_k^{p+1}).$$

Since $u_k$ converges to zero weakly in $E$, the second integral over the domain $|x| \leqq R$ converges to zero as $k \to \infty$ by the Sobolev compact embedding theorem. This completes the proof.

*Q.e.d.*

## § 4. The radial case

We consider the following equation

$$\Delta u - a(r)\, u + f(r, u) = 0 \tag{4.1}$$

in $\mathbb{R}^n$, where $r = |x|$. We still need to assume that $a(r)$ and $f(r, u)$ are locally Hölder continuous and (2.2) holds. Instead of working in $E$, we shall deal with the subspace $E_r$ of $E$ which consists in all radial functions in $E$. For $f$, we assume that $f(r, u) \equiv 0$ if $u \leqq 0$ and

(4.2)
(a) $f \in C(\mathbb{R} \times \mathbb{R}^+)$ and $f(r, u)$ is locally Lipschitz-continuous in $u$, and $f \not\equiv 0$,

(b) there is a $p$ such that $1 < p < (n+2)/(n-2)$ and

$$0 \leqq f(r, u) \leqq C_1(r)(1 + u^p)$$

for all $r \geqq 0$, $u > 0$, where $C_1(r) \geqq 0$ is continuous,

(c) there are constants $q > 1$, $\delta > 1$ such that

$$f(r, u) \leqq C_2(r) u^q$$

for all $r \geqq 0$, $u \in (0, \delta)$, where $C_2(r)$ is continuous and satisfies

$$0 < C_2(r) \leqq C(1 + r^l)$$

for all $r \geqq 0$, where $C$ is a positive constant and $0 \leqq l < \frac{1}{2}(n-1)(q-1)$,

(d) there are a $\theta \in (0, \frac{1}{2})$ and a continuous $A(r) \geqq 0$ such that

$$A_1 \equiv \int_0^\infty r^{n-1} A(r)\, dr < \infty$$

and

$$A(r) + \theta u f(r, u) \geqq F(r, u)$$

for all $r \geqq 0$, $u > 0$.

Note that there is no restriction on the growth of $C_1(r)$. An example of a nonlinearity satisfying (4.2) is

$$f(r, u) = (1 + r^2)^{l/2} u^p, \qquad 0 \leqq l < \tfrac{1}{2}(n-1)(p-1).$$

Therefore it is not clear that the variational functional

$$J(u) = \tfrac{1}{2} \int_{\mathbb{R}^n} [|\nabla u|^2 + a(|x|)\, u^2] - \int_{\mathbb{R}^n} F(|x|, u)$$

where, as before

$$F(|x|, u) = \int_0^u f(|x|, s)\, ds,$$

is well defined on $E_r$. However, $J$ is well defined on each subspace $E_{kr}$, where $E_{kr}$ is the subspace of $E_k$ which consists of all radial functions in $E_k$. (The spaces $E_k$ and $E$ are defined in § 2.) Let

$$J_k = J|_{E_{kr}};$$

then the analysis (in § 2) in deriving (2.12) and in proving Theorem 2.13 carries over. That is, the sequence $\{\|u_k\|\}$, $u_k$ being a critical point of $J_k$ obtained by use of the Mountain-Pass Lemma (as in § 2), is still uniformly bounded. Thus, by passing to a subsequence if necessary, we may assume that $u_k$ converges weakly to $\bar{u}$ in $E_r$. Let $B$ be a (fixed) ball in $\mathbb{R}^n$. Using (4.2) (b) and standard *interior $L^p$-estimates* and Schauder estimates for solutions $u_k$, we conclude that (passing to a subsequence again if necessary) $u_k$ converges to $\bar{u}$ in the space $C^2(\bar{B})$. Therefore $\bar{u}$ is a nonnegative classical solution of (4.1) in $B$. Since $B$ is arbitrary, $\bar{u} \geqq 0$ is a solution of (4.1) in $\mathbb{R}^n$.

The remaining task is to show that $\bar{u} \not\equiv 0$, for then the strong maximum principle will guarantee that $\bar{u} > 0$ in $\mathbb{R}^n$ and $\bar{u}$ will be a desired solution of (4.2). Condition (4.2) (c) will be used in establishing this fact. We shall also need the following estimates due to WALTER STRAUSS (see *e.g.* [S]).

**Radial Lemma.** *For any radial function $u$ in $H^1(\mathbb{R}^n)$, $n \geqq 2$,*

$$|u(r)| \leqq C \|u\|_{H^1(\mathbb{R}^n)} r^{(1-n)/2} \quad \text{for } r \geqq 1,$$

*in which $C$ depends only on $n$.*

Note that under (2.2), $E$ is continuously embedded into $H^1(\mathbb{R}^n)$. Since $\{\|u_k\|\}$ is bounded above,

$$u_k(r) \leqq C r^{(1-n)/2} \tag{4.3}$$

for $r \geqq 1$ and for all $k$. On the other hand, we observe, just as in the proof of Theorem 3.23, that

$$\int_0^k r^{n-1} u_k(r) \left[f(r, u_k(r)) - \bar{\lambda} u_k(r)\right] dr > 0.$$

Thus, there is an $r_k \in [0, k)$ such that $f(r_k, u_k(r_k)) > \bar{\lambda} u_k(r_k)$.

Let $S = \{k \mid u_k(r_k) < \delta\}$. Then, by (4.2) (c), we have, for $k \in S$,

$$\begin{aligned} \bar{\lambda} u_k(r_k) &\leqq C_2(r_k)\, [u_k(r_k)]^q \\ &\leqq C(1 + r_k^l)\, [u_k(r_k)]^q; \end{aligned}$$

*i.e.*, for $k \in S$,

$$u_k(r_k) \geqq \left[\frac{\bar{\lambda}}{C(1 + r_k^l)}\right]^{\frac{1}{q-1}}. \tag{4.4}$$

If $r_{k_j} \to \infty$, $k_j \in S$, then by the range of $l$, we deduce from (4.4) that

$$r_{k_j}^{\frac{n-1}{2}} u_{k_j}(r_{k_j}) \geqq \text{constant} \cdot r_{k_j}^{\left(\frac{n-1}{2} - \frac{l}{q-1}\right)} \to \infty \quad \text{as } k_j \to \infty,$$

contradicting (4.3). Thus there is an $R > 0$ such that $r_k \leqq R$ for all $k \in S$. This, together with (4.4), implies that there is a positive uniform lower bound $\beta$ for $u_k(r_k)$, $k \in S$, i.e.

$$u_k(r_k) \geqq \beta > 0 \quad \text{for all } k \in S. \tag{4.5}$$

Next, we wish to obtain similar bounds for $r_k$'s and $u_k(r_k)$'s for $k \notin S$. We proceed as follows. For $k \notin S$, we have $u_k(r_k) \geqq \delta$. It follows from (4.3) immediately that there is an $R > 0$ such that $r_k \leqq R$ for all $k \notin S$. Therefore, there is a constant $R$ such that $r_k \leqq R$ for all $k$ and $u_k(r_k) \geqq \min\{\beta, \delta\} > 0$ for all $k$. By the compactness of $B_R$, there is a subsequence $r_{k_j}$ that converges to $\bar{r} \leqq R$ and

$$u_{k_j}(r_{kj}) \to \bar{u}(\bar{r})$$

as $j \to \infty$ since $u_k$ converges to $\bar{u}$ uniformly on $B_R$. Thus, $\bar{u}(\bar{r}) \geqq \min\{\beta, \delta\} > 0$ and $\bar{u} \not\equiv 0$. We have thus proved the following

**Theorem 4.6.** *Suppose* (2.2) *and* (4.2) *hold. Then equation* (4.1) *has a positive radial solution in* $E_r$.

We now return to our model equation

$$\Delta u - u + Q(r)\, u^p = 0, \qquad 1 < p < \frac{n+2}{n-2}. \tag{4.7}$$

From Theorem 4.6, we have

**Corollary 4.8.** *If* $Q \geqq 0$ *in* $[0, \infty)$ *and satisfies the growth condition*

$$Q(r) \leqq C(1 + r^l) \text{ for } r \geqq 0, \tag{4.9}$$

$C > 0$ *being some constant and* $l < (n-1)(p-1)/2$, *then equation* (4.7) *has a positive radial solution in* $H_0^1(\mathbb{R}^n)$.

The solutions obtained in Corollary 4.8 must tend to zero exponentially fast at infinity. This standard fact follows from the Radial Lemma and a well known result of T. KATO [K].

## § 5. Maximization

For equation (3.8) a natural approach exploits the classical method by maximization (or, equivalently, minimization). *That is, let*

$$M = \sup_{\|u\|_{H^1(\mathbb{R}^n)} = 1} \int_{\mathbb{R}^n} Q(x)\, |u|^{p+1} \tag{5.1}$$

*and try to show that M is attained by some element in* $H^1(\mathbb{R}^n)$, *which would then be a solution of* (3.8). For equation (4.7), the same idea applies, except we now restrict attention to the class of radial functions, letting

$$M_r = \sup_{\|u\|_{H_r^1(\mathbb{R}^n)} = 1} \int_{\mathbb{R}^n} Q(|x|)\, |u|^{p+1} \tag{5.2}$$

where $H_r^1(\mathbb{R}^n)$ is the set of all radial functions in $H^1(\mathbb{R}^n)$.

We first look at the radial case. For equation (4.7), we assume that (as in Corollary 4.8) $Q$ could actually grow at infinity. Thus it is not clear that $M_r$, given by (5.2), is finite. Even if $M_r$ is finite, it is not at all clear that $M_r$ is always assumed.

Corollary 4.8 follows from Theorem 4.6 directly; however, the approach we have been using in this paper employs the Mountain-Pass Lemma, which is quite general and therefore is good for obtaining general theorems of existence. We shall sketch a second proof for Corollary 4.8 by use of the maximization method which is simpler although the *exact homogeneity* of the term $u^p$ is used (to eliminiate a Lagrange multiplier). Furthermore, this second proof also has several by-products and one of them is the finiteness of (5.2) under the growth condition (4.9) on $Q$.

*A sketch of the proof of Corollary* 4.8 *by maximization*

Define

$$M_{k,r} = \sup_{\|u\|_{H^1_{0,r}(B_k)}=1} \int_{B_k} Q(x)\,|u|^{p+1}. \tag{5.3$_k$}$$

It is not hard to see that

(i) $M_{k,r} < \infty$ and $M_{k,r}$ is assumed by an element $u_k > 0$ in $H^1_{0,r}(B_k)$ with $\|u_k\|_{H^1_{0,r}(B_k)} = 1$, where $H^1_{0,r}(B_k)$ is the set of all radial functions in $H^1_0(B_k)$.

(ii) $u_k$ is a solution of

$$\begin{aligned} &\Delta u - u + \lambda Q u^p = 0 \quad \text{in } B_k, \\ &u > 0 \quad \text{in } B_k \quad \text{and} \quad u = 0 \quad \text{on } \partial B_k, \end{aligned} \tag{4.7$'_k$}$$

where the Lagrange multiplier $\lambda = 1/M_{k,r}$. Thus

$$w_k \equiv \left(\frac{1}{M_{k,r}}\right)^{\frac{1}{p-1}} u_k \tag{5.4}$$

solves

$$\begin{aligned} &\Delta u - u + Q u^p = 0 \quad \text{in } B_k, \\ &u > 0 \quad \text{in } B_k, \quad u = 0 \quad \text{on } \partial B_k. \end{aligned} \tag{4.7$_k$}$$

(iii) $M_{k,r}$ increases monotonically with $k$ and thus

$$\|w_k\|_{H^1_r(\mathbb{R}^n)} = \left(\frac{1}{M_{k,r}}\right)^{\frac{1}{p-1}} \leqq \left(\frac{1}{M_{1,r}}\right)^{\frac{1}{p-1}}, \tag{5.5}$$

i.e. $\{w_k\}$ is uniformly bounded (above) in $H^1_r(\mathbb{R}^n)$. Standard elliptic estimates guarantee that (by passing to a subsequence which we still denote by $\{w_k\}$) the sequence $\{w_k\}$ converges to a limit $w$ in $H^1_r(\mathbb{R}^n)$ and the convergence is uniform in $C^2(\Omega)$ for any compact subset $\Omega$ in $\mathbb{R}^n$, and $w \geqq 0$ is a solution of (4.7).

(iv) It remains to show that $w \not\equiv 0$. To this end, set

$$A_k = \{x \in B_k \mid Q(x)\, w_k^{p-1}(x) > 1\}.$$

Clearly, $A_k$ is not empty. For $x \in A_k$, we have

$$w_k(x) \geqq \left(\frac{1}{Q(x)}\right)^{\frac{1}{p-1}} \geqq \frac{C}{|x|^{\frac{l}{p-1}}} \tag{5.6}$$

for $|x| \geqq 1$, by (4.9). On the other hand, by (5.5) and the Radial Lemma, we have

$$w_k(x) \leqq \frac{C}{|x|^{\frac{n-1}{2}}} \tag{5.7}$$

for $|x| \geqq 1$. Combining (5.6), (5.7) and the condition on $l$, we see immediately that $A_k$ is contained in a *fixed* ball of radius $s$. Thus

$$|w_k|_{L^\infty(B_s)} \geqq \operatorname*{Min}_{B_s} \left(\frac{1}{Q}\right)^{\frac{1}{p-1}} > 0$$

for all $k$. Thus $w \not\equiv 0$ and the proof is complete.

As a consequence, we have the following

**Proposition 5.8.** *Let $Q \geqq 0$ in $[0, \infty)$ with $Q(r) \leqq C(1 + r^l)$ for $r \geqq 0$ where $l < (n-1)(p-1)/2$. Then the number $M_r$ defined by (5.2) is finite.*

**Proof.** Since $M_{k,r}$ (defined in $(5.3)_k$) tends monotonically to $M_r$ as $k \to \infty$, *it suffices to show that $M_{k,r}$ is bounded above.* Otherwise $M_{k,r} \to \infty$ as $k \to \infty$. Then, by (5.4)

$$\|w_k\|_{H_r^1(\mathbb{R}^n)} \to 0 \quad \text{as } k \to \infty .$$

Since $w_k$ converges to $w$ in $C^2(\Omega)$ ($\Omega$ is a compact set), we see that

$$\|w\|_{H_r^1(\Omega)} = \lim_{k\to\infty} \|w_k\|_{H_r^1(\Omega)} \leqq \lim \|w_k\|_{H_r^1(\mathbb{R}^n)} = 0,$$

i.e., $w \equiv 0$ on $\Omega$. Since $\Omega$ is arbitrary, $w \equiv 0$ on $\mathbb{R}^n$, a contradiction.

*Q.e.d.*

A natural question arises: does the conclusion of Proposition 5.8 still hold if $l > (n-1)(p-1)/2$? The answer is negative as the following example shows.

*Example* 5.9. For each integer $m > 0$, define

$$u_m(r) = \begin{cases} \dfrac{1}{m^{(n-1)/2}} \cdot e^{(r-m)/2}, & r \leqq m, \\[2ex] \dfrac{1}{r^{(n-1)/2}}, & m \leqq r \leqq m+1, \\[2ex] \dfrac{1}{(m+1)^{(n-1)/2}} \cdot e^{(m+1-r)/2}, & r \geqq m+1. \end{cases}$$

It is routine to verify that $u_m \in H^1(\mathbb{R}^n)$ and that there is a *uniform upper bound* for $\|u_m\|_{H^1(\mathbb{R}^n)}$. However, if $Q(r) = (1 + r^l)$,

$$\int_{\mathbb{R}^n} Q(x)\,|u_m|^{p+1}\,dx \geqq \int_m^{m+1} r^{l-\frac{n-2}{2}(p+1)+n-1}\,dr$$

$$= \int_m^{m+1} r^{l-(n-1)(p-1)/2}\,dr$$

$$\geqq m^{l-(n-1)(p-1)/2} \int_m^{m+1} dr \to \infty \qquad \text{as } m \to \infty;$$

thus $M_r = \infty$. Incidentally, this example also shows that the Radial Lemma is sharp.

Our second consequence of the above proof of Corollary 4.8 is the following

**Proposition 5.10.** *Suppose that* $Q \geqq 0$ *in* $[0, \infty)$, $Q \to \infty$ *as* $r \to \infty$ *with*

$$Q(r) \leqq C(1 + r^l) \qquad \textit{for } r \geqq 0,$$

*where* $l < (n-1)(p-1)/2$. *Then for every large* $k$ *the problem* $(4.7)_k$ *has a positive solution which is not radially symmetric.*

**Proof.** Define

$$M_k = \sup_{\|u\|_{H_0^1(B_k)}=1} \int_{B_k} Q(x)\,|u|^{p+1}.$$

It is clear that $M_k \to \infty$ as $k \to \infty$ since $Q(r) \to \infty$ as $r \to \infty$. Thus, for large $k$, with $M_{k,r}$ given by $(5.3)_k$, $M_{k,r} \neq M_k$. Since both $M_{k,r}$ and $M_k$ are assumed (this is a rather standard fact whose proof we omit here), say, by $u_k$ and $u_k^*$ respectively, we see that $u_k^*$ can not be radially symmetric. Rescaling, we have

$$w_k \equiv \left(\frac{1}{M_{k,r}}\right)^{\frac{1}{p-1}} u_k, \qquad w_k^* \equiv \left(\frac{1}{M_k}\right)^{\frac{1}{p-1}} u_k^*,$$

and so for large $k$ there are two positive solutions of $(4.7)_k$, one of which is radial while the other is not. (The fact that both $u_k$ and $u_k^*$ are positive in $B_k$ is also standard.)

*Q.e.d.*

*Remark* 5.11. The procedure (i)–(iii) in the above proof of Corollary 4.8 does not have to be restricted to the class of radial functions. Everything carries over without change if we simply drop the restriction to radial functions.

Finally, we take up the question whether (5.1) is assumed for bounded $Q$'s. As an easy example, we state that *if* $Q(x)$ *is bounded, radial and monotonically* *increasing, then $M$ in* (5.1) *is not realized.* This is essentially a special case of the following result, which is not restricted to radial functions.

**Proposition 5.12.** *Suppose there is an $R_0 > 0$ such that*

$$\min_{2R \leqq |x| \leqq 4R} Q(x) \geqq \max_{|x| \leqq R} Q(x) \tag{5.13}$$

*for all $R \geqq R_0$. If $Q \not\equiv$ constant, then the number*

$$M = \sup_{\|u\|_{H^1(\mathbb{R}^n)}=1} \int_{\mathbb{R}^n} Q(x) |u|^{p+1} dx$$

*is not attained.*

*Remark* 5.14. Condition (5.13) may be replaced by the following slightly more general one:

There are two sequences $R_k \uparrow \infty$, $S_k \uparrow \infty$ such that

$$\min_{R_k - S_k \leqq |x| \leqq R_k + S_k} Q(x) \geqq \max_{|x| \leqq S_k} Q(x) \quad \text{for all } k.$$

In fact, one can easily formulate various weaker assumptions from the sufficient but not necessary ones used in the proof.

**Proof of Proposition 5.12.** Suppose $M$ is attained by $u$. Define $\tilde{u}_R(x) = u(x - x_R)$, i.e., $\tilde{u}_R$ is a translation of $u$ by $x_R$, where $x_R = (3R, 0, \ldots, 0)$. Then $\tilde{u}_R$ also has $H^1$-norm one and for $R \geqq R_0$,

$$\begin{aligned} \int_{\mathbb{R}^n} Q(x) |\tilde{u}_R|^{p+1} &\geqq \int_{B_R(x_R)} Q |\tilde{u}_R|^{p+1} \\ &\geqq \left[\min_{B_R(x_R)} Q\right] \cdot \int_{B_R(x_R)} |\tilde{u}_R|^{p+1} \\ &\geqq \min_{2R \leqq |x| \leqq 4R} Q \cdot \int_{B_R(0)} |u|^{p+1} \\ &\geqq \max_{B_R(0)} Q \cdot \int_{B_R(0)} |u|^{p+1} \\ &= \int_{B_R(0)} Q(x) |u|^{p+1} + \delta(R), \end{aligned} \tag{5.15}$$

in which $\delta(R)$ is defined by the last equality. It is clear that

$$\delta(R) = \int_{B_R(0)} \left[\max_{B_R(0)} Q - Q(x)\right] |u|^{p+1} \geqq 0$$

is nondecreasing in $R$ and $\delta \not\equiv 0$ since $Q \not\equiv$ constant. Let

$$\delta(\infty) = \lim_{R\to\infty} \delta(R);$$

then there is an $R_1 \geqq R_0$ such that $\delta(R) > \frac{1}{2}\,\delta(\infty)$ and

$$\int_{B_R(0)} Q\,|u|^{p+1} > M - \tfrac{1}{2}\,\delta(\infty)$$

for $R \geqq R_1$. By (5.15), we see that

$$\int_{\mathbb{R}^n} Q\,|\tilde{u}_R|^{p+1} > M,$$

a contradiction.

*Q.e.d.*

*Acknowledgment.* The research of Ni was supported in part by NSF Grant DMS 8200033A01.

## References

[A, R] Ambrosetti, A., & P. Rabinowitz, Dual variational methods in critical point theory and applications, *J. Funct. Anal.* **14** (1973), 349–381.

[B, L] Berestycki, H., & P.-L. Lions, Nonlinear scalar field equations I, II, *Arch. Rational Mech. Anal.* **82** (1983), 313–345, 347–375.

[B] Berger, M., On the existence and structure of stationary states for a nonlinear Klein-Gordon equation, *J. Funct. Anal.* **9** (1972), 249–261.

[D, N] Ding, W.-Y., & W.-M. Ni, On the elliptic equation $\Delta u + K\,u^{\frac{n+2}{n-2}} = 0$ and related topics, *Duke Math. J.* **52** (1985), 485–506.

[G, N, N1] Gidas, B., W.-M. Ni, & L. Nirenberg, Symmetry of positive solutions of nonlinear elliptic equations in $\mathbb{R}^n$, *Advances in Math. Supplementary Studies* 7A (1981), 369–402.

[G, N, N2] Gidas, B., W.-M. Ni, & L. Nirenberg, Symmetry and related properties via the maximum principle, *Comm. Math. Phys.* **68** (1979), 209–243.

[K] Kato, T., Growth properties of solutions of the reduced wave equation with a variable coefficient, *Comm. Pure Appl. Math.* **12** (1959), 403–425.

[L] Lions, P.-L., The concentration-compactness principle in the calculus of variations. The locally compact case, II. *Annales de l'Institut Henri Poincaré—Analyse non linéaire* **1** (1984), 223–283.

[N] Nehari, Z., On a nonlinear differential equation arising in nuclear physics, *Proc. Roy. Irish Acad.* **62** (1963), 117–135.

[Ni] Ni, W.-M., On the elliptic equation $\Delta u + K(x)\,u^{\frac{n+2}{n-2}} = 0$, its generalizations and applications to geometry, *Indiana Univ. Math. J.* **31** (1982), 493–529.

[R] Rabinowitz, P., Variational methods for nonlinear eigenvalue problems, *Eigenvalues of Nonlinear Problems*, C.I.M.E. Edizioni Cremonese 1974.

[Se] SERRIN, J., A symmetry problem in potential theory, *Arch. Rational Mech. Anal.* **43** (1971), 304–318.

[S] STRAUSS, W. A., Existence of solitary waves in higher dimensions, *Comm. Math. Phys.* **55** (1977), 149–162.

[St] STUART, C. A., Bifurcation for Dirichlet problems without eigenvalues, *Proc. London Math. Soc.* (3) **45** (1982), 169–192.

[Z] ZHANG, D., Private communication.

Institute of Mathematics
Academia Sinica
Beijing, P.R.C.

and

School of Mathematics
University of Minnesota
Minneapolis

*(Received May 23, 1985)*

# *The Uniqueness of Hill's Spherical Vortex*

C. J. AMICK & L. E. FRAENKEL

*Dedicated to James Serrin on the occasion of his 60th birthday*

## 1. Introduction

### *1.1. Background*

The mathematical description of steady vortex rings, in an ideal fluid occupying the whole space $\mathbb{R}^3$, can be approached in various ways. The physical basis of the problem, its history up to 1973, and several formulations are outlined in [12], pp. 14–21. Another, quite different formulation and the plan for a corresponding existence theory are presented in [6]. Further existence theorems, variational principles and results are to be found in [3], [7], [11], [13] and [20]. Here we state only definitions and equations that seem relevant to our immediate purpose.

Consider a Stokes stream function $\Psi$, defined on the closure $\overline{\Pi}$ of the half-plane

$$\Pi = \{(r, z) \mid r > 0, -\infty < z < \infty\},$$

where $r$ and $z$ may be regarded as cylindrical co-ordinates, points of $\mathbb{R}^3$ being denoted by $X = (X_1, X_2, X_3) = (r \cos \theta, r \sin \theta, z)$. The significance of $\Psi$ is that (a) the *fluid velocity* $\boldsymbol{q}$ has cylindrical components (in the directions $r, \theta, z$ increasing, respectively) $-\Psi_z/r, 0, \Psi_r/r$; (b) streamlines in a meridional plane ($\theta =$ const.) are level curves of $\Psi$, and $2\pi(\Psi_2 - \Psi_1)$ is the volumetric flow rate, or flux, between two stream surfaces of revolution described by $\Psi(r, z) =$ const. $= \Psi_j$ $(j = 1, 2)$; (c) the *vorticity* curl $\boldsymbol{q}$ has cylindrical components 0, $-(L\Psi)/r$, 0, where

$$L = r\frac{\partial}{\partial r}\left(\frac{1}{r}\frac{\partial}{\partial r}\right) + \frac{\partial^2}{\partial z^2}.$$

Such a function $\Psi$ corresponds to a steady vortex ring if there exists a bounded open set $A \subset \Pi$, called the *cross-section* of the ring and unknown a priori, such that $\Psi \in C^1(\overline{\Pi}) \cap C^2(\Pi \setminus \partial A)$ and satisfies the equations

$$L\Psi = r\left(\frac{1}{r}\Psi_r\right)_r + \Psi_{zz} = \begin{cases} -\lambda r^2 f_0(\Psi) & \text{in } A, \\ 0 & \text{in } \Pi \setminus \overline{A}, \end{cases} \tag{1.1a}$$

$$\Psi|_{\partial A} = 0, \quad \Psi|_{r=0} = -k, \tag{1.1b, c}$$

as

$$r^2 + z^2 \to \infty \text{ in } \overline{\Pi}, \quad \Psi(r, z) \sim -\tfrac{1}{2} Wr^2 - k, \quad \Psi_z/r \to 0$$

and

$$\Psi_r/r \to -W. \tag{1.1d}$$

Here $f_0$ is a given, (strictly) positive *vorticity function*, which need be defined only on $(0, \infty)$ because of (1.2) below. We suppose for the moment that the *vortex-strength parameter* $\lambda$, the *flux constant* $k$ and the *propagation speed* $W$ are also prescribed, with $\lambda > 0$, $k \geqq 0$ and $W > 0$. (The constant $W$ is the speed of the vortex ring relative to the fluid at infinity; in (1.1) we have taken co-ordinate axes fixed in the ring and have demanded that the fluid velocity $\boldsymbol{q} \to (0, 0, -W)$ at infinity.) In most existence theorems, other sets of constants are prescribed, and 'free' elements of the set $\{\lambda, k, W\}$ are calculated a posteriori. This is illustrated by the remarks following (1.8) below.

Since $L\Psi < 0$ in $A$ and $\Psi|_{\partial A} = 0$, the maximum principle implies that $\Psi > 0$ in $A$; similarly, $\Psi < 0$ in $\Pi \setminus \bar{A}$. Therefore we define the cross-section by

$$A = \{(r, z) \in \Pi \mid \Psi(r, z) > 0\}. \tag{1.2}$$

It is often convenient to write

$$\Psi(r, z) = \psi(r, z) - \tfrac{1}{2} Wr^2 - k, \tag{1.3}$$

where $\psi$ is the stream function due to vorticity, while $-\frac{1}{2} Wr^2 - k$ represents a uniform stream. Note that the latter has zero vorticity: $L(\frac{1}{2} Wr^2 + k) = 0$. We define

$$f(t) = \begin{cases} 0, & t \leqq 0, \\ f_0(t), & t > 0. \end{cases}$$

Abbreviating the conditions at infinity, we now re-write (1.1) as

$$L\psi = -\lambda r^2 f(\Psi) \quad \text{in } \Pi, \tag{1.4a}$$

$$\psi|_{r=0} = 0, \quad \psi(r, z) \to 0 \quad \text{as} \quad r^2 + z^2 \to \infty \quad \text{in } \overline{\Pi}, \tag{1.4b, c}$$

where it is to be understood that (1.4a) need not hold pointwise on $\partial A$, and that $\psi_z/r$ and $\psi_r/r \to 0$ at infinity. Maximum principles for weak solutions show that $\psi > 0$ in $\Pi$.

### *1.2. Hill's spherical vortex*

Only one *explicit* exact solution of (1.1) or (1.4) is known: that discovered by M. J. M. HILL [17] in 1894 for the case

$$k = 0 \quad \text{and} \quad f(t) = f_H(t) = \begin{cases} 0, & t \leqq 0, \\ 1, & t > 0. \end{cases} \tag{1.5}$$

HILL observed that for this case a sphere $\{X \in \mathbb{R}^3 \mid |X| = a\}$ can serve as the boundary of a steady vortex 'ring' (Figure 1). Thus the cross-section is

$$A_H = \{(r, z) \in \Pi \mid r^2 + z^2 < a^2\}, \tag{1.6a}$$

and we let $\varrho = (r^2 + z^2)^{\frac{1}{2}}$. HILL found the solution

$$\psi_H(r, z) = \Psi_H(r, z) + \tfrac{1}{2} Wr^2 = \begin{cases} \tfrac{1}{2} Wr^2 \left(\tfrac{5}{2} - \tfrac{3}{2} \dfrac{\varrho^2}{a^2}\right), & \varrho \leqq a, \\ \tfrac{1}{2} Wr^2 \dfrac{a^3}{\varrho^3}, & \varrho \geqq a, \end{cases} \tag{1.6b}$$

where

$$\lambda a^2 / W = \tfrac{15}{2}. \tag{1.6c}$$

Anticipating the definition (1.7), we note that, for a fluid of unit density, the kinetic energy $\pi \|\psi_H\|^2$ is given by

$$\|\psi_H\|^2 = \int_\Pi \frac{1}{r^2} (\psi_{H,r}^2 + \psi_{H,z}^2)\, r\, dr\, dz = \tfrac{10}{7} W^2 a^3. \tag{1.6d}$$

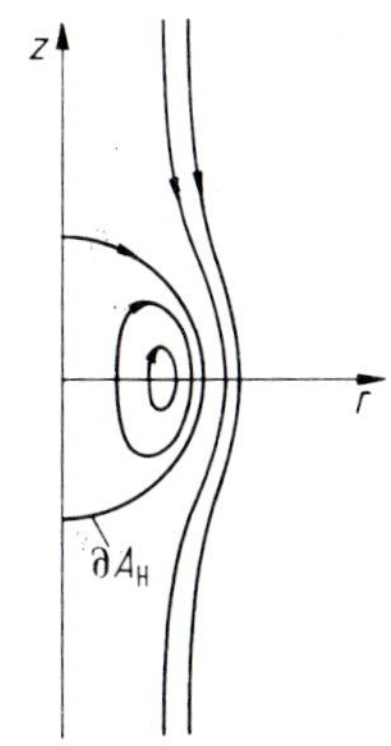


Fig. 1. Hill's spherical vortex. The streamlines in $\Pi$ are level curves of $\Psi_H$

By *Hill's problem* we mean (1.1) or (1.4) for the case $k = 0$, $f = f_H$. Solutions of Hill's problem are presented (among other solutions) in [12] and [13], but no uniqueness theorem appears in these papers, and it has been unknown whether these solutions of Hill's problem are in fact HILL's solution. Indeed, we know of no result in the literature that connects in any way the many solutions, of the basic general problem (1.1), that have been obtained by different formulations and different existence theorems.

## *1.3. Results*

In the present paper we prove that, for Hill's problem, (a) any weak solution is HILL'S solution $\psi_H$, modulo translation in the $z$-direction; (b) any solution of the isoperimetric variational problem formulated in [12] is a weak solution, indeed, any *local* maximizer is a weak solution (and is therefore HILL'S solution). The result (b) is not obvious or immediate, because $f_H$ has a simple discontinuity; consequently, the functional that we maximize is not Fréchet differentiable on the whole Hilbert space appropriate to the problem.

We now make these statements precise. The Hilbert space $H(\Pi)$ is the completion of the set $C_0^\infty(\Pi)$, of real-valued functions having derivatives of every order and compact support in $\Pi$, in the norm $\|\cdot\|$ corresponding to the inner product

$$\langle \phi, \chi \rangle = \int_\Pi \frac{1}{r^2} (\phi_r \chi_r + \phi_z \chi_z)\, r\, dr\, dz. \tag{1.7}$$

Thus $\pi \|\phi\|^2$ is the kinetic energy of the motion with stream function $\phi$; also,

$$\langle \phi, \chi \rangle = -\int_\Pi \frac{1}{r^2} \phi L \chi\, r\, dr\, dz \quad \text{if} \quad \phi, \chi \in C_0^\infty(\Pi).$$

We shall say that $\psi$ is a *weak solution of Hill's problem* if $\psi \in H(\Pi) \setminus \{0\}$ and if there exist constants $\lambda \in \mathbb{R}$ and $W > 0$ such that

$$\left.\begin{aligned} \langle \phi, \psi \rangle &= \lambda \int_{A(\psi)} \phi\, r\, dr\, dz \quad \text{for all} \quad \phi \in H(\Pi), \\ &\text{where} \\ A(\psi) &= \{(r, z) \in \Pi \mid \psi(r, z) > \tfrac{1}{2} W r^2\}. \end{aligned}\right\} \tag{1.8}$$

Setting $\phi = \psi$ in (1.8), one sees that $\lambda > 0$ and that $A(\psi)$ must have positive area. (For Theorem 1.2, we shall prescribe $\|\psi\| > 0$ and $W > 0$; then $\lambda = \|\psi\|^2 \Big/ \int_{A(\psi)} \psi\, r\, dr\, dz$.) Our first result is

**Theorem 1.1.** *If $\psi$ is a weak solution of Hill's problem, then $\psi(r, z) = \psi_H(r, z - c)$ for some $c \in \mathbb{R}$; here $\psi_H$ is as in* (1.6).

We now turn to the variational principle in [12]. To state it for Hill's problem, we define

$$t_+ = \int_0^t f_H(s)\, ds = \max\{t, 0\},$$

$$J(\phi) = \int_\Pi (\phi(r, z) - \tfrac{1}{2} W r^2)_+\, r\, dr\, dz \quad \text{for all} \quad \phi \in H(\Pi),$$

$$S(\eta) = \{\phi \in H(\Pi) \mid \|\phi\|^2 = \eta > 0\};$$

the sphere $S(\eta)$ is a surface of constant energy. The variational problem is: given $W$ and $\eta$, find $\psi \in S(\eta)$ such that $J(\psi) = \max_{\phi \in S(\eta)} J(\phi)$. For the sake of a wide uniqueness statement, we consider not only solutions of this problem, but also local maximizers of $J$ on $S(\eta)$; we shall prove

**Theorem 1.2.** *Let $\psi$ be a local maximizer of $J$ on $S(\eta)$: that is, $J(\phi) \leq J(\psi)$ for all $\phi \in S(\eta)$ in some neighbourhood of $\psi$. Then $\psi$ is a weak solution of Hill's problem* (*so that Theorem 1.1 applies*).

### *1.4. Method*

The principal steps in the paper are as follows.

(i) We make the transformation $\psi = r^2 v$ in order to prove that, when $\psi$ is a weak solution of Hill's problem, $v$ depends only on $\varrho = \{r^2 + (z - c)^2\}^{\frac{1}{2}}$ for some $c \in \mathbb{R}$ (and on the parameters). The resulting one-dimensional problem for $v$ can then be analysed without difficulty. It is a fortunate and crucial fact that

$$\psi = r^2 v \Rightarrow \frac{1}{r^2} L\psi = \frac{1}{r^3}(r^3 v_r)_r + v_{zz} = \Delta_5 v \quad \text{for } r > 0, \tag{1.9}$$

where $v(r, z)$ is now regarded as cylindrically symmetric in $\mathbb{R}^5$, that is, $r = (x_1^2 + \ldots + x_4^2)^{\frac{1}{2}}$ and $z = x_5$, while $\Delta_5 = \sum_{j=1}^{5} \partial^2/\partial x_j^2$ is the Laplace operator in $\mathbb{R}^5$. The identity (1.9) was noted by CHANDRASEKHAR ([9], p. 252), and has been used by NI [20] to prove regularity in his theory of steady vortex rings. However, NI considered only non-linearities $f$ smoother than $f_H$, so that HILL's vortex is outside the range of his theory. The transformation $\psi = r^2 v$ is exceptionally useful for HILL's problem because the equivalence, for $r > 0$, of the conditions $\psi > \frac{1}{2} W r^2$ and $v > \frac{1}{2} W$ means that (1.4) becomes

$$\Delta_5 v = -\lambda f_H(v - \tfrac{1}{2} W) \quad \text{in } \mathbb{R}^5 \setminus \{r = 0\}, \tag{1.10a}$$

$$v(x) \to 0 \quad \text{as } |x| \to \infty. \tag{1.10b}$$

In Cartesian co-ordinates $x_1, \ldots, x_5$, equation (1.10a) has constant coefficients on both sides, and this allows us to prove, in step (ii), that $v$ depends only on $\varrho$. For $k \neq 0$ or $f \neq f_H$, the right-hand member of (1.4a) does not transform to a function independent of $r$.

(ii) To prove that the weak form of (1.10) implies spherical symmetry of the function $v$, we first transform (in section 2) the definition (1.8) of weak solution, showing at the same time that the exceptional line $\{r = 0\}$ causes no difficulty. In section 3, we establish regularity properties and positivity of the weak solution $v$, and then adapt to the present problem the powerful method initiated by SERRIN in [24] and greatly enlarged by GIDAS, NI & NIRENBERG in [14]. A slight, further extension is needed here because $C^2$ solutions are considered in [24] and [14], whereas for HILL's vortex second derivatives have a finite jump across $\partial A_H$. The generalized maximum principles in GILBARG & TRUDINGER's book [15] enable us to modify the relevant parts of [14].

Different applications of the method in [24] and [14], to free-boundary problems of the same general kind as Hill's problem, have already been made in [8], [11] and [18].

(iii) To prove Theorem 1.2, we transform (in section 2) the variational principle for $\psi$ to one for $v$. In section 4, we overcome the difficulty, that the convex functional $N$ corresponding to $J$ is not Fréchet differentiable on its whole domain, by using convex analysis. Detailed examination of the subdifferential of $N$, and of its left-hand and right-hand Gateaux derivatives, leads to the result.

### 1.5. Miscellaneous remarks

(i) The *existence* of a global maximizer, of $J$ over the sphere $S(\eta)$, is not in doubt: such a function is constructed in [12], pp. 40–42 (although rather indirectly, by limiting procedures) and that particular function is easily seen to be a weak solution, for any $k \geqq 0$ and for a class of non-linearities $f$ that includes $f_H$. For Hill's problem, one can also prove more directly, by means of the transformation $\psi = r^2 v$ and symmetrisation with respect to a point in $\mathbb{R}^5$ (that is, by re-arrangement of $v$ to a spherically symmetric function $v^*$) that a global maximizer exists and is a weak solution. However, maximizers other than these are conceivable; it is for this reason, and to demonstrate the strength of the variational principle, that we present Theorem 1.2.

(ii) There is a variant of HILL's solution (not published, we believe, but well known to specialists) in which the fluid domain is a ball, say $\{X \in \mathbb{R}^3 \mid |X| < b\}$, with cross-section $D = \{(r, z) \in \Pi \mid r^2 + z^2 < b^2\}$. We set $k = 0$, $f = f_H$ as before; replace $\Pi$ by $D$ in (1.1a) and (1.4a); and replace the condition (1.1d) or (1.4c) at infinity by

$$\Psi(r, z) = -\tfrac{1}{2} W r^2 \quad \text{on } \partial D, \tag{1.11}$$

which states that the normal velocity on $\partial D$ is the normal component of $(0, 0, -W)$, and also implies the conditions (1.1c) and (1.4b) on $r = 0$ for this problem.

After extending in Appendix A the relevant theorem in [14], we show in Appendix B that the earlier results and methods carry over to this case with only minor changes. One of these is that, while transformed solutions $v$ are always spherically symmetric, existence and uniqueness depend on what constants are given. If $\lambda$, $W$ and $b$ are prescribed, there may be no solution or there may be two; if $\|\psi\|$, $W$ and $b$ are prescribed, the solution always exists and is unique.

(iii) In [22], NORBURY constructed perturbations of HILL's solution that represent genuine rings (homeomorphic to a solid torus). He solved (1.1) with $f = f_H$ and $0 < k < k_0$, where $k_0$ is small, by reducing an integral equation to a contraction mapping of a small closed ball, in a Banach space of functions capable of representing the unknown boundary $\partial A$. We shall prove in [4] that NORBURY's solutions are also unique for sufficiently small (positive) values of $k$. This is not trivial because, in NORBURY's Banach space, the closed ball forming the domain of his contraction mapping necessarily has a radius that tends to zero as $k \to 0$. Thus, for small $k > 0$, there could exist solutions close to HILL's solution that are outside the range of the local uniqueness result in [22]. It is reassuring that, in fact, a single branch of solutions emerges from HILL's solution as the parameter $k$ increases from zero. NORBURY's numerical calculations [23] suggest that this branch is defined for all $k > 0$, and represents rings of small cross-section as $k \to \infty$.

(iv) A small third contribution, in our endeavour to unify the diverse theories of steady vortex rings, will be presented in [5]. There we consider (1.1) with $k = 0$ and the power-law vorticity function $f_0(t) = t^\beta$, $\beta = \text{const.} \in (0, 5)$, and prove that for these cases the solutions in [12] coincide with those found by a wholly different variational principle in [13]. This is not a uniqueness result, but merely a proof of the equivalence of two different methods.

## 2. The transformed problem

### 2.1. *Further notation and terminology*

We define the Hilbert space $E$ to be the completion of the set $C_0^\infty(\mathbb{R}^5)$ in the norm $|||\cdot|||$ corresponding to the inner product

$$\langle u, w\rangle_E = \frac{1}{2\pi^2}\int_{\mathbb{R}^5} \nabla u(x)\cdot\nabla w(x)\,dx = \frac{1}{2\pi^2}\int_{\mathbb{R}^5}\nabla u\cdot\nabla w;$$

abbreviations as in the last expression will be used where no confusion can arise.

Any set of the form

$$F = \{x\in\mathbb{R}^5 \mid (|x'|, x_5)\in X\subset\overline{\Pi}\},\qquad \text{where}\qquad x' = (x_1,\ldots,x_4),$$

will be called a figure of revolution, or cylindrically symmetric set, with cross-section $X$. The *subscript c*, attached to the symbol for a set of functions, denotes the subset of *cylindrically symmetric* functions; that is, of functions $u$ such that $u(x) = \tilde{u}(|x'|, x_5)$ for all $x$ in the cylindrically symmetric domain of $u$. We shall sometimes write $u$ for $\tilde{u}$. The closed linear subspace of $E$, formed by completing $C_{0,c}^\infty(\mathbb{R}^5)$ in the norm $|||\cdot|||$, will be denoted by $E_c$.

By the *transformation* $\phi = r^2u$ we mean that, given $\phi\in H(\Pi)$, we define $u:\mathbb{R}^5\setminus\{x'=0\}\to\mathbb{R}$ by $u(x) = \phi(|x'|, x_5)/|x'|^2$, or that, given $u\in E_c$ with $u(x) = \tilde{u}(|x'|, x_5)$, we define $\phi:\Pi\to\mathbb{R}$ by $\phi(r,z) = r^2\tilde{u}(r,z)$. Since $\mathbb{R}^3$ does not occur in this statement, we now write $r = |x'|$, $z = x_5$ with no danger of confusion. Note that

$$\langle u, w\rangle_E = \int_\Pi (u_rw_r + u_zw_z)\,r^3\,dr\,dz \quad\text{if}\quad u, w\in E_c, \tag{2.1}$$

because the three-dimensional unit sphere $S^3 = \{y\in\mathbb{R}^4 \mid |y| = 1\}$ has area $2\pi^2$.

Given a constant $W > 0$, we define

$$P(u) = \{x\in\mathbb{R}^5 \mid u(x) > \tfrac{1}{2}W\}\quad \text{for any}\quad u\in E. \tag{2.2}$$

Of course, the elements of $E$ are really equivalence classes (of functions equal almost everywhere), and the precise form of $P(u)$ depends on the representative $u$ selected from an equivalence class $[u]\in E$, but different representatives only change $P(u)$ by sets of (five-dimensional Lebesgue) measure zero. If $u\in E_c$ and $\phi = r^2u$, then $P(u)$ is the figure of revolution with cross-section $A(\phi)$ [*cf.* (1.8)].

For sets, $|\cdot|_n$ denotes $n$-dimensional Lebesgue measure.

For any open set $\Omega\subset\mathbb{R}^n$, $W_p^m(\Omega)$ denotes the Sobolev space of functions having generalized derivatives up to order $m$ in $L_p(\Omega)$, $p\geqq 1$; its norm will be written

$$\|u\|_{m,p,\Omega} = \left\{ \sum_{|\alpha| \leq m} \int_\Omega |D^\alpha u|^p \right\}^{1/p}.$$

Note that $\|\cdot\|_{0,p,\Omega}$ denotes the norm of $L_p(\Omega)$.

## 2.2. *Transformation of weak solutions*

We begin with three simple lemmas that establish basic properties of functions in $E$.

**Lemma 2.1.** (a) *The space $E$ is embedded in $L_{10/3}(\mathbb{R}^5)$, and*

$$\|u\|_{0,10/3,\mathbb{R}^5} \leq c_1 \, |||u|||, \quad \textit{where } c_1 = (\tfrac{2}{5})^{\frac{1}{2}} \tfrac{4}{3}\pi, \quad \textit{for all } u \in E. \tag{2.3}$$

(b) *With $P(u)$ as in* (2.2),

$$|P(u)|_5 \leq c_2 W^{-10/3} \, |||u|||^{10/3}, \quad \textit{where } c_2 = (2c_1)^{10/3}, \quad \textit{for all } u \in E. \tag{2.4}$$

**Proof.** The first inequality is a standard result of Sobolev embedding ([21], p. 128), combined with the inequality between the geometric mean and the root mean square. The second inequality then follows from

$$\int_{P(u)} u^{10/3} \, dx \geq (\tfrac{1}{2} W)^{10/3} \, |P(u)|_5.$$

Q.e.d.

**Lemma 2.2.** *The spaces $H(\Pi)$ and $E_c$ are isometrically isomorphic under the transformation $\phi = r^2 u$ of any $\phi \in H(\Pi)$ or $u \in E_c$.*

**Proof.** Let $C_d^\infty(\mathbb{R}^5)$ denote the set of functions in $C_{0,c}^\infty(\mathbb{R}^5)$ that have support disjoint from the $z$-axis (here $r = |x'|$ and $z = x_5$). First we show that $C_d^\infty(\mathbb{R}^5)$ is dense in $E_c$. Let $\mu$ be a non-decreasing function in $C^\infty(\mathbb{R} \to \mathbb{R})$ such that $\mu(t) = 0$ for $t \leq \frac{1}{2}$ while $\mu(t) = 1$ for $t \geq 1$. For any $u \in C_{0,c}^\infty(\mathbb{R}^5)$ and any positive integer $n$, define $u_n \in C_d^\infty(\mathbb{R}^5)$ by $u_n(r, z) = \mu(nr) \, u(r, z)$. A calculation shows that $|||u - u_n||| \leq \text{const.}\, n^{-1}$, where the constant depends on $u$ but not on $n$; thus $C_d^\infty(\mathbb{R}^5)$ is dense in $C_{0,c}^\infty(\mathbb{R}^5)$ under the norm $|||\cdot|||$, and hence in $E_c$.

Now let $\phi = r^2 u$ and $\chi = r^2 w$, where $\phi$ and $\chi$ are in $C_0^\infty(\Pi)$ or, equivalently, $u$ and $w$ are in $C_d^\infty(\mathbb{R}^5)$. From (1.7) and (2.1) we find that

$$\langle \phi, \chi \rangle_{H(\Pi)} = \int_\Pi \{ r^3 (u_r w_r + u_z w_z) + 2(r^2 uw)_r \} \, dr \, dz = \langle u, w \rangle_E,$$

since the integral of $2(r^2 uw)_r$ vanishes.

Q.e.d.

**Lemma 2.3.** *Let $F$ be a figure of revolution having finite measure: $|F|_5 < \infty$. Let $E_c^\perp$ denote the orthogonal complement in $E$ of $E_c$. Then*

$$\int_F w \, dx = 0 \quad \textit{for all} \quad w \in E_c^\perp. \tag{2.5}$$

**Proof.** Any $u \in C_0^\infty(\mathbb{R}^5)$ has values $u(r, \xi, z)$, where $\xi \in S^3$; we define a mean-value operator $M$ by

$$(Mu)(r, z) = \frac{1}{2\pi^2} \int_{S^3} u(r, \xi, z)\, d\omega_\xi,$$

where $d\omega_\xi$ denotes the element of surface area at $\xi$, and decompose $u$ as follows:

$$u = u_0 + u_1, \quad \text{where } u_0 = Mu, \quad u_1 = u - Mu. \tag{2.6}$$

Then $u_0 \in E_c$, $Mu_1 = 0$ and $u_1(r, \xi, z) = O(r)$ as $r \to 0$. We now show that $\langle u_0, w_1 \rangle_E = 0$ for all $u, w \in C_0^\infty(\mathbb{R}^5)$; hence (2.6) corresponds, for such functions, to the unique orthogonal decomposition $E = E_c \oplus E_c^\perp$. Differentiating the equation $Mw_1 = 0$, we obtain $M(w_{1,r}) = (Mw_1)_r = 0$ and $M(w_{1,z}) = 0$. Consequently,

$$\langle u_0, w_1 \rangle_E = \frac{1}{2\pi^2} \int_\Pi r^3\, dr\, dz \int_{S^3} (u_{0,r}, w_{1,r} + u_{0,z} w_{1,z})\, d\omega_\xi$$

$$= \int_\Pi \{u_{0,r} M(w_{1,r}) + u_{0,z} M(w_{1,z})\}\, r^3\, dr\, dz = 0.$$

Since $|||Mu||| = |||u_0||| \leqq |||u|||$ for all $u \in C_0^\infty(\mathbb{R}^5)$, we extend $M$ by continuity to have domain $E$; then the result $\langle Mu, w - Mw \rangle_E = 0$ extends to all $u, w \in E$, so that $M$ and $I - M$ (where $I$ denotes the identity operator on $E$) are respectively the projection operators of $E$ onto $E_c$ and onto $E_c^\perp$. Equation (2.5) now follows from the bound

$$\left| \int_F u\, dx \right| = \left| \int_F Mu\, dx \right| \leqq c_1\, |||Mu|||\, |F|_5^{7/10} \quad \text{for all } u \in E,$$

implied by (2.3), and from the fact that $Mw = 0$ whenever $w \in E_c^\perp$.

Q.e.d.

We are now in a position to state our first objective in terms of functions in $E$.

**Theorem 2.4.** *Theorem 1.1 is equivalent to the following statement.*
(I) *If there exist* $v \in E_c \setminus \{0\}$, $\lambda \in \mathbb{R}$ *and* $W > 0$ *such that*

$$\int_{\mathbb{R}^5} \nabla u \cdot \nabla v\, dx = \lambda \int_{P(v)} u\, dx \quad \textit{for all} \quad u \in E, \tag{2.7}$$

*where* $P(v)$ *is as in* (2.2), *then* $v(x) = v_H(|x'|, x_5 - c)$ *for some* $c \in \mathbb{R}$. *Here* $v_H(r, z) = \psi_H(r, z)/r^2$ *and* $\psi_H$ *is as in* (1.6).

**Proof.** For any $u \in E$, we use the decomposition $u = u_0 + u_1$, where $u_0 \in E_c$ and $u_1 \in E_c^\perp$. Let ($\mathrm{I}_0$) denote the variant of statement (I) that results from replacing $u \in E$ by $u_0 \in E_c$ in (2.7). Lemma 2.2 shows that Theorem 1.1

and $(I_0)$ are equivalent. Moreover, $(I_0)$ and (I) are equivalent because

$$\int_{\mathbb{R}^5} \nabla u_1 \cdot \nabla v = 0 \quad \text{and} \quad \lambda \int_{P(v)} u_1 = 0 \quad \text{for all} \quad u_1 \in E_c^\perp;$$

the first integral vanishes because $v \in E_c$, and the second vanishes by (2.4) and Lemma 2.3.

Q.e.d.

### *2.3. Transformation of the variational principle*

Given $W > 0$, we define

$$N(u) = \frac{1}{2\pi^2} \int_{\mathbb{R}^5} (u - \tfrac{1}{2} W)_+ \, dx = \frac{1}{2\pi^2} \int_{P(u)} (u - \tfrac{1}{2} W) \, dx \quad \text{for all } u \in E, \tag{2.8}$$

$$S_c(\eta) = \{u \in E_c \mid |||u|||^2 = \eta > 0\}. \tag{2.9}$$

Under the isomorphism in Lemma 2.2, the variational problem stated before Theorem 1.2 becomes: given $W$ and $\eta$, find $v \in S_c(\eta)$ such that $N(v) = \max\limits_{u \in S_c(\eta)} N(u)$. Thus we have

**Theorem 2.5.** *Theorem 1.2 is equivalent to the following statement.*

(II) *Let $v$ be a local maximizer of $N$ on $S_c(\eta)$. Then $v$ is a transformed weak solution of Hill's problem; that is, there exists $\lambda \in \mathbb{R}$ such that* (2.7) *holds.*

## 3. Transformed weak solutions correspond to Hill's vortex

### *3.1. Preliminary estimates*

Here and in section 3.2, we prove the truth of statement (I) in Theorem 2.4. In fact, we prove a little more: that, if the hypothesis $v \in E_c$ in (I) is weakened to $v \in E$, then the conclusion still holds, provided that $v_H(|x'|, x_5 - c)$, for some $c \in \mathbb{R}$, is replaced by $v_H(|x' - b'|, x_5 - b_5)$, for some $b \in \mathbb{R}^5$. Thus our hypothesis is: there exist $v \in E \setminus \{0\}$, $\lambda \in \mathbb{R}$ and $W > 0$ such that

$$\int_{\mathbb{R}^5} \nabla u \cdot \nabla v \, dx = \lambda \int_{P(v)} u \, dx \quad \text{for all} \quad u \in E, \tag{3.1}$$

where $P(v)$ is as in (2.2). Setting $u = v$, we see that $\lambda > 0$ and $|P(v)|_5 > 0$. *It is to be understood throughout section 3 that $v$, $\lambda$ and $W$ have these properties.*

*Notation.* We define *balls* to be *open* and of positive, finite radius; $\mathcal{B}(c, R)$ will denote the ball with centre $c$ and radius $R$ in the space implied by the context. The non-negative and non-positive parts of a real-valued function are defined, respectively, by

$$g_+(x) = \max\{g(x), 0\}, \quad g_-(x) = \min\{g(x), 0\}; \tag{3.2}$$

note that, in contrast to the convention in integration theory, the non-positive part is non-positive. Since sections 3 and 4 concern only statements (I) and (II), we can safely ignore two previous conventions: the $\Delta_5$ in (1.9) now becomes $\Delta$, and (as is usual) $\phi$ will be used for smooth test functions rather than for elements of $H(\Pi)$.

**Lemma 3.1.** (a) $v \in W^2_{p,\mathrm{loc}}(\mathbb{R}^5) \cap C^{1+\alpha}(\mathbb{R}^5)$ *for all* $p \in (1, \infty)$ *and* $\alpha \in (0, 1)$.

(b) $v(x) \to 0$ *(pointwise) as* $|x| \to \infty$.

(c) $-\Delta v(x) = \begin{cases} \lambda & \textit{in} \quad P(v), \\ 0 & \textit{in} \quad \{x \in \mathbb{R}^5 \mid v(x) < \frac{1}{2} W\}. \end{cases}$

(d) *There exists a number* $R_v$ *such that* $P(v) \subset \mathscr{B}(0, R_v)$ *in* $\mathbb{R}^5$.

**Proof.** (a) We use AGMON's $L_p$ approach to the Dirichlet problem, applying Theorem 6.1 of [1] to the operator $\Delta$. The hypotheses of that theorem are amply satisfied, because $v \in L_{10/3}(\mathbb{R}^5)$ and because (3.1) implies that, for any ball $B$ in $\mathbb{R}^5$ and any $p \in (1, \infty)$,

$$\left| \int_B v\, \Delta\phi \right| = \lambda \left| \int_{P(v) \cap B} \phi \right| \leqq M \, \|\phi\|_{0,p',B} \quad \text{for all} \quad \phi \in C_0^\infty(B),$$

where

$$M = \lambda \, |P(v) \cap B|_5^{1/p} \quad \text{and} \quad \frac{1}{p} + \frac{1}{p'} = 1.$$

For any $x_0 \in \mathbb{R}^5$ and for $j = 0, \ldots, 3$, let $B_j = \mathscr{B}(x_0, R_j)$ with $R_0 = \frac{1}{2}$, $R_1 = 1$, $R_2 = \frac{3}{2}$ and $R_3 = 2$. (These choices of $R_j$, and the inequalities (3.3) to (3.5), are dimensionally consistent in the physicist's sense only if we use non-dimensional variables, for example, if $R_j$ is replaced by $R_j \lambda^{\frac{1}{2}} / W^{\frac{1}{2}}$ and $v$ by $v/W$.) A first application of AGMON's theorem, with $p = 10/3$, yields

$$\|v\|_{2,10/3,B_1} \leqq k_3 \{M_3(x_0) + \|v\|_{0,10/3,B_2}\}, \tag{3.3a}$$

where

$$M_3(x_0) = \lambda \, |P(v) \cap B_3|_5^{3/10}, \tag{3.3b}$$

and where $k_3$ is an absolute constant. Since $W^2_{10/3}(B_1)$ is embedded in the space $C(\bar{B}_1)$, a fortiori in $L_p(B_1)$ for all $p \geqq 1$, a second application of AGMON's theorem gives

$$\|v\|_{2,p,B_0} \leqq k_2(p)\, M_2(x_0, p) + k_1(p) \{M_3(x_0) + \|v\|_{0,10/3,B_2}\} \quad \text{for all } p \in (1, \infty), \tag{3.4a}$$

where

$$M_2(x_0, p) = \lambda \, |P(v) \cap B_2|_5^{1/p}. \tag{3.4b}$$

Finally, embedding theory gives

$$\|v\|_{C^{1+\alpha}(\bar{B}_0)} \leqq k_0(p, \alpha) \, \|v\|_{2,p,B_0} \quad \text{for } p(1-\alpha) > 5; \tag{3.5}$$

the constants $k_n$ depend only on $p$ ($n = 1, 2$) or on $p$ and $\alpha$ ($n = 0$). To obtain bounds independent of $x_0$, we merely replace $|P(v) \cap B_j|_5$ by $|B_j|_5$ and the norm of $v$ in $L_{10/3}(B_2)$ by that in $L_{10/3}(\mathbb{R}^5)$.

(b) Lemma 2.1 shows that the norm of $v$ in $L_{10/3}(\mathbb{R}^5 \setminus \mathscr{B}(0, m))$, and $|P(v) \setminus \mathscr{B}(0, m)|_5$, both tend to zero as $m \to \infty$. Hence the right-hand member of (3.4a) tends to zero as $|x_0| \to \infty$; so, therefore, does the norm of $v$ in $C^{1+\alpha}(\bar{B}_0)$, and this shows that $v(x) \to 0$ as $|x| \to \infty$.

(c) Since $v$ is continuous in $\mathbb{R}^5$, the set $P(v)$ is open. Since $v \in W^2_{p,\mathrm{loc}}(\mathbb{R}^5)$, we infer from (3.1) that

$$\int_{P(v)} \phi(\Delta v + \lambda) = 0 \quad \text{for all} \quad \phi \in C_0^\infty(P(v));$$

hence $-\Delta v = \lambda$ almost everywhere in $P(v)$. But the qualification 'almost everywhere' can be removed by means of further regularity theory, or by means of (3.6) below; in fact, $v$ is real-analytic in $P(v)$. The argument is similar for the set in which $v(x) < \frac{1}{2} W$.

(d) This follows from (b) and the definition of $P(v)$.

Q.e.d.

**Lemma 3.2.** *The function $v$ is the Newtonian potential of $P(v)$ with density $\lambda$:*

$$v(x) = \frac{\lambda}{8\pi^2} \int_{P(v)} |y - x|^{-3}\, dy \quad \text{for all} \quad x \in \mathbb{R}^5. \tag{3.6}$$

*It follows that $v > 0$ in $\mathbb{R}^5$.*

**Proof.** We choose the test function $u$ in (3.1) to be a smooth approximation to the Newtonian kernel in $\mathbb{R}^5$. Let $\mu$ be a non-decreasing function in $C^\infty(\mathbb{R} \to \mathbb{R})$ such that $\mu(t) = 0$ for $t \leqq \frac{1}{2}$ and $\mu(t) = 1$ for $t \geqq 1$, and define, for all $z \in \mathbb{R}^5$ and any positive integer $n$,

$$K(z) = \frac{1}{8\pi^2} |z|^{-3}, \qquad K_n(z) = \mu(n\,|z|)\, K(z).$$

We choose and fix any $x \in \mathbb{R}^5$, and replace the variable of integration in (3.1) by $y$. Then $K_n(\cdot - x) \in E$; we set $u(y) = K_n(y - x)$ in (3.1), and integrate the left-hand member by parts (first over a large ball $\mathscr{B}(0, R)$; the boundary integral is at most $\max_{|y|=R} |v(y)|$ and tends to zero as $R \to \infty$, by Lemma 3.1(b)). Thus

$$-\int_{\mathbb{R}^5} \Delta K_n(y - x)\, v(y)\, dy = \lambda \int_{P(v)} K_n(y - x)\, dy. \tag{3.7}$$

Now $-\Delta K_n$ is a mollifying kernel: $\Delta K_n(z) = 0$ for $|z| \geqq 1/n$, and

$$-\int_{\mathscr{B}(0,1/n)} \Delta K_n(z)\, dz = 1, \qquad \int_{\mathscr{B}(0,1/n)} |\Delta K_n(z)|\, dz \leqq \text{const.},$$

where the constant is independent of $n$. Since $v \in C(\mathbb{R}^5)$, the left-hand member of (3.7) tends (pointwise) to $v(x)$ as $n \to \infty$. In addition,

$$\int_{\mathbb{R}^5} |K(z) - K_n(z)|\, dz < \tfrac{1}{6} n^{-2},$$

so that the right-hand member of (3.7) tends to the right-hand member of (3.6) as $n \to \infty$.

Q.e.d.

**Lemma 3.3.** *Let $R_v$ be as in Lemma 3.1*(d). *There exist constants $a_0 > 0$ and $a_j$ $(j = 1, \ldots, 5)$ such that*

$$\left.\begin{aligned} v(x) &= a_0\,|x|^{-3} + \sum_{j=1}^{5} a_j x_j\,|x|^{-5} + h(x), \\ |h(x)| &\leqq \text{const.}\,|x|^{-5}, \qquad |\nabla h(x)| \leqq \text{const.}\,|x|^{-6}, \end{aligned}\right\} \quad \textit{for} \quad |x| \geqq 2R_v. \tag{3.8}$$

*In fact,*

$$a_0 = \frac{\lambda}{8\pi^2}\,|P(v)|_5, \qquad a_j = \frac{3\lambda}{8\pi^2} \int_{P(v)} y_j\,dy \qquad (j = 1, \ldots, 5),$$

*but these details will not be needed.*

**Proof.** In (3.6), with $|y| < \frac{1}{2}|x|$ for all $y \in P(v)$, we may differentiate repeatedly under the integral sign (once is sufficient here) and expand $|y - x|^{-3}$ and its derivatives, essentially in powers of $|y|/|x|$, to finitely many terms with a remainder.

Q.e.d.

*Remark.* Following the procedure in [14], we now eliminate the dipole terms in (3.8) by defining (for all $x \in \mathbb{R}^5$)

$$v_0(x) = v(x + b), \quad \text{where} \quad b_j = a_j/3a_0 \quad (j = 1, \ldots, 5). \tag{3.9}$$

Then

$$\left.\begin{aligned} v_0(x) &= a_0\,|x|^{-3} + g(x), \qquad a_0 > 0, \\ |g(x)| &\leqq \text{const.}\,|x|^{-5}, \qquad |\nabla g(x)| \leqq \text{const.}\,|x|^{-6}, \end{aligned}\right\} \quad \text{for} \quad |x| \geqq 2R_v + |b|. \tag{3.10}$$

### *3.2. The maximum principle and reflecting hyperplanes*

We note that $v$ is a Newtonian potential with an unusual property: $v(x)$ is constant on the boundary $\partial P(v)$ of the set in which the density differs from zero. This is the underlying reason that $P(v)$ will turn out to be a ball, with $v$ spherically symmetric about its centre (*cf.* [24]).

The method in [24] and [14] depends on moving hyperplanes in from infinity, reflecting the graph of a function about these hyperplanes, and then using the maximum principle. For positive solutions $v$ of certain elliptic problems set in $\mathbb{R}^n$, the arguments in [14] are of two types: (a) those which depend only on approximations to $v(x)$ for large $|x|$, and (b) those which apply the maximum principle to classical solutions. Our result (3.8) is sufficient for (a), but we shall have to use (3.1) and a generalized maximum principle in place of (b).

The following maximum principle is far more general than is needed in this section, somewhat more general than is needed in Appendix A, and considerably less general than results in [15]. We state this particular theorem because it is close to the Maximum Principle and Lemma H on p. 212 of [14]; thus it shows the feasibility of extending results in [14] to weak solutions of problems other than ours.

Let $\Omega$ be an open set in $\mathbb{R}^n$. Define, for $x \in \Omega$ and (say) $u \in C^2(\Omega)$,

$$Lu \equiv \nabla \cdot \{a(x) \cdot \nabla u\} + p(x) \cdot \nabla u + q(x)\, u,$$

where the symmetric $n \times n$ matrix $a(x)$ is uniformly positive definite:

$$\xi \cdot a(x) \cdot \xi \geqq c_0 \, |\xi|^2, \qquad c_0 = \text{const.} > 0, \qquad \text{for all } \xi \in \mathbb{R}^n \text{ and } x \in \Omega;$$

the entries $a_{ij}$ of $a$ are in $W^1_\infty(\Omega)$, the components $p_1, \ldots, p_n$ of $p$ are in $L_\infty(\Omega)$, and $q \in L_\infty(\Omega)$. Given a connected open (possibly unbounded) set $G \subset \Omega$, we say that $Lu \geqq 0$ *weakly in* $G$ if $u \in C(\overline{G}) \cap W^1_2(G)$ and

$$A(\phi, u; G) \equiv \int_G \{-\nabla\phi \cdot a(x) \cdot \nabla u + \phi p(x) \cdot \nabla u + \phi q(x)\, u\}\, dx \geqq 0$$

for all $\phi \in C_0^\infty(G)$ with $\phi \geqq 0$.

**Theorem 3.4.** (a) *Let $G \subset \Omega$ be connected and open. If $Lu \geqq 0$ weakly in $G$ and $u \leqq 0$ in $G$, then either $u \equiv 0$ or $u < 0$ in $G$.*

(b) *Let $B \subset \Omega$ be a ball, let $x_0 \in \partial B$ and let $m_0$ be a unit vector outward from $B$ at $x_0$. (That is, $m_0 \cdot (x_0 - c) > 0$, where $c$ is the centre of $B$.) If $Lu \geqq 0$ weakly in $B$, $u < 0$ in $B$, and $u(x_0) = 0$, then*

$$\liminf_{t \to 0+} \frac{u(x_0) - u(x_0 - t m_0)}{t} > 0,$$

*which implies that $m_0 \cdot (\nabla u)(x_0) > 0$ when this derivative exists.*

**Proof.** (a) follows from Theorem 8.19 of [15]; (b) follows from the proof of Lemma 3.4 and (3.11) in [15] when the classical maximum principle used there is replaced by (a). The restriction $q \leqq 0$ in $\Omega$, imposed in [15], is not necessary for the particular conclusions in (a) and (b), because we can use a perturbation of our operator $L$ that satisfies this restriction: if $A(\phi, u; G) \geqq 0$ and $u \leqq 0$ in $G$, then

$$\int_G \{-\nabla\phi \cdot a(x) \cdot \nabla u + \phi p(x) \cdot \nabla u + \phi q_-(x)\, u\}\, dx$$

$$= A(\phi, u; G) - \int_G \phi q_+(x)\, u\, dx \geqq 0$$

for all $\phi \in C_0^\infty(G)$ with $\phi \geqq 0$.

Q.e.d.

*Prospectus.* It is easy to see that the function $v_0$, defined by (3.9), satisfies (3.1); our aim is to prove that $v_0(x) = v_H(|x'|, x_5)$. Having emphasized this, we shall *omit the subscript* 0 *from* $v_0$, restoring it only in the statement of our final result.

Lemmas 3.5, 3.6 and 3.8 below are merely statements for our case of Lemmas 4.1, 4.2 and 4.4 in [14]. We include these statements for the sake of clarity, but, apart from offering in Appendix C an alternative proof of Lemma 4.1 in [14], we refer to [14] for proofs of these results. The first part of the proof of Theorem 3.9 is also to be found in [14], but we include it as an essential part of the present story.

*Reflecting hyperplanes.* Let $\gamma$ be a fixed unit vector in $\mathbb{R}^5$. For each $\mu \in \mathbb{R}$, define $T_\mu(\gamma) = \{x \in \mathbb{R}^5 \mid x \cdot \gamma = \mu\}$. We may suppose that, after a suitable rotation of axes, $\gamma = (1, 0, \ldots, 0)$; then

$$T_\mu = \{x \mid x_1 = \mu\}; \tag{3.11a}$$

also,

$$x^\mu = (2\mu - x_1, x''), \quad \text{where } x'' = (x_2 \ldots, x_5), \tag{3.11b}$$

denotes the reflection in $T_\mu$ of any point $x$. We define open half-spaces by $Y(\mu) = \{x \mid x_1 < \mu\}$ and $Z(\mu) = \{x \mid x_1 > \mu\}$.

**Lemma 3.5.** *Let $v$ be as in* (3.10), *and consider two points $y$ and $z$ such that $y_1 < z_1, y_1 + z_1 \geqq 2\mu > 0$ and $y'' = z''$. There exists a number $R(\mu)$, depending only on $v$ and* $\min\{1, \mu\}$, *such that*

$$v(y) > v(z) \quad \textit{whenever} \quad |y| \geqq R(\mu).$$

**Lemma 3.6.** *There exists a number $\mu_0 \geqq 1$ such that*

$$v(x) > v(x^\mu) \quad \textit{whenever} \quad x \in Y(\mu) \quad \textit{and } \mu \geqq \mu_0.$$

**Lemma 3.7.** *Assume that for some $\mu > 0$ we have $v(x) \geqq v(x^\mu)$ for all $x \in Y(\mu)$, and $v(x_0) \neq v(x_0^\mu)$ for some $x_0 \in Y(\mu)$. Then*

(a) $v(x) > v(x^\mu)$ *for all* $x \in Y(\mu)$, (3.12)

(b) $\dfrac{\partial v}{\partial x_1}(x) < 0$ *for all* $x \in T_\mu$. (3.13)

**Proof.** (a) We define the reflection in $T_\mu$ of any function $F$ by $F_\mu(x) = F(x^\mu)$, and set $w = v_\mu - v$. By hypothesis, $w(x) \leqq 0$ for $x \in Y(\mu)$, and we prove strict inequality by means of (3.1) and the maximum principle.

Let $Y = Y(\mu)$ and $Z = Z(\mu)$. Given $\phi \in C_0^\infty(Y)$, we note that $\phi_\mu$ has support in $Z$, and choose $u = \phi_\mu$ in (3.1) to obtain

$$\int_Z \nabla\phi_\mu(z) \cdot \nabla v(z)\, dz = \lambda \int_{Z \cap P(v)} \phi_\mu(z)\, dz.$$

Set $z = x^\mu$ in this equation; then $x \in Y$, $\partial/\partial z_1 = -\partial/\partial x_1$ and $\partial/\partial z_j = \partial/\partial x_j$ for $j = 2, \ldots, 5$. Also, $\phi_\mu(z) = \phi(z^\mu) = \phi(x)$ and $v(z) = v(x^\mu) = v_\mu(x)$. Consequently,

$$\int_Y \nabla\phi(x) \cdot \nabla v_\mu(x)\, dx = \lambda \int_{Y \cap P(v_\mu)} \phi(x)\, dx. \tag{3.14}$$

Now choose $u = \phi$ in (3.1), and subtract the resulting equation from (3.14) to obtain

$$\int_Y \nabla\phi \cdot \nabla w = \lambda \int_{Y \cap P(v_\mu)} \phi - \lambda \int_{Y \cap P(v)} \phi. \tag{3.15}$$

Since $v(x) > \frac{1}{2} W$ for $x \in P(v)$, while $v(x^\mu) > \frac{1}{2} W$ for $x \in P(v_\mu)$, our hypothesis implies that $Y \cap P(v)$ contains $Y \cap P(v_\mu)$; hence

$$\int_Y \nabla\phi \cdot \nabla w \leqq 0 \quad \text{for all} \quad \phi \in C_0^\infty(Y) \quad \text{with} \quad \phi \geqq 0.$$

As it happens, we can now apply Theorem 3.4(a) to all of $Y$, because (3.10) shows that $v \in L_2(\mathbb{R}^5)$; then $w \in C(\overline{Y}) \cap W_2^1(Y)$, by Lemma 3.1, and so $\Delta w \geqq 0$ weakly in $Y$. Since $w \not\equiv 0$ in $Y$, by hypothesis, we conclude that $w < 0$ in $Y$. (The appeal to (3.10) is not necessary. The set $X \equiv \{x \in Y \mid w(x) = 0\}$ is closed in $Y$ because $w$ is continuous, and open in $Y$ by application of Theorem 3.4(a) to a small ball about any zero of $w$ in $Y$. Therefore $X = \emptyset$ or $X = Y$, and the latter is contrary to hypothesis.)

(b) For any $x_0 \in T_\mu$ and any $R > 0$, define $B = \mathscr{B}((x_{0,1} - R, x_0''), R)$, so that $B \subset Y$ and $\partial B \cap T_\mu = \{x_0\}$. Then $w < 0$ in $B$, and $w(x_0) = 0$; by Theorem 3.4(b), $(\partial w/\partial x_1)(x_0) > 0$, since Lemma 3.1(a) ensures that this derivative exists. Finally, $(\partial w/\partial x_1)(x_0) = -2(\partial v/\partial x_1)(x_0)$.

Q.e.d.

**Lemma 3.8.** *The set* $\{\mu > 0 \mid v(x) > v(x^\mu)$ *for all* $x \in Y(\mu)\}$ *is open in* $\mathbb{R}$.

**Theorem 3.9.** *Let* $v, \lambda$ *and* $W$ *be as in* (3.1), *and* $v_0$ *as in* (3.9). *Then* $v_0(x) = v_H(|x'|, x_5)$, *where* $v_H(r, z) = \psi_H(r, z)/r^2$ *and* $\psi_H$ *is as in* (1.6).

**Proof.** Only $v_0$ is discussed in this proof; we continue to abbreviate $v_0$ to $v$. Let $(m, \infty)$, with $m \geqq 0$, be the maximal open internal such that (3.12) and (3.13) hold whenever $\mu \in (m, \infty)$. That such an interval exists follows from Lemma 3.6 and the fact that (3.12) implies (3.13). If $m > 0$, then by continuity $v(x) \geqq v(x^m)$ for all $x \in Y(m)$, and by Lemma 3.5 there exist points $x_0 \in Y(m)$ such that $v(x_0) > v(x_0^m)$. Hence Lemma 3.7 is applicable and shows that (3.12) and (3.13) hold for $\mu \geqq m$; by Lemma 3.8, $(m, \infty)$ is not maximal. We conclude that $m = 0$, whence $v(-x_1, x'') \geqq v(x_1, x'')$ whenever $x_1 \geqq 0$. Repeating the argument for the unit vector $\gamma = (-1, 0, \ldots, 0)$, we see that $v$ is an even function of $x_1$. Also, $(\partial v/\partial x_1)(x) < 0$ whenever $x_1 > 0$, because (3.13) holds for all $\mu > 0$. The same argument holds for every unit vector $\gamma$ in $\mathbb{R}^5$; therefore, $v$ depends only on $|x|$ and is strictly decreasing as $|x|$ increases.

It follows that $P(v) = \mathscr{B}(0, a)$ for some $a > 0$. Let $|x| = \varrho$ and $v(x) = \tilde{v}(\varrho)$; from Lemma 3.1 we infer that $\tilde{v} \in C^1[0, \infty)$, that $\tilde{v}$ is real-analytic in $[0, a)$ and $(a, \infty)$, and that

$$\frac{1}{\varrho^4} \frac{d}{d\varrho}\left(\varrho^4 \frac{d}{d\varrho}\right) \tilde{v} = \begin{cases} -\lambda & \text{for} \quad 0 < \varrho < a, \\ \ \ 0 & \text{for} \quad \varrho > a, \end{cases}$$

$$\tilde{v}(a) = \tfrac{1}{2} W, \quad \tilde{v}(\varrho) \to 0 \quad \text{as} \quad \varrho \to \infty.$$

This problem can be solved explicitly and easily; the solution corresponds to (1.6b) and (1.6c). (In fact, we can reach this conclusion with less a priori knowledge of $\tilde{v}$; once spherical symmetry is established, the maximum principle ensures that $P(v)$ must be a ball about the origin, otherwise $\tilde{v}$ would have a local minimum.) Q.e.d.

## 4. Local maximizers of $N$ on $S_c(\eta)$

In this section, recalling that

$$N(u) = \frac{1}{2\pi^2} \int_{\mathbb{R}^5} (u - \tfrac{1}{2} W)_+ \, dx = \frac{1}{2\pi^2} \int_{P(u)} (u - \tfrac{1}{2} W) \, dx \qquad \text{for all} \quad u \in E,$$

we prove statement (II) in Theorem 2.5: that a local maximizer of $N$ on the sphere $S_c(\eta)$ in $E_c$ is a weak solution in the sense of (2.7). If we widen the question by considering $N$ on the corresponding sphere, say $\Sigma(\eta)$, in $E$, then the same analysis shows that a local maximizer of $N$ on $\Sigma(\eta)$ is a weak solution in the sense of (3.1). However, it is not obvious that a local maximizer on $S_c(\eta)$ (arising from Hill's problem set in $\Pi$) is a local maximizer on the bigger sphere $\Sigma(\eta)$, and it does not seem worth while to pursue this point.

The functional $N$ is not Gateaux differentiable, let alone Fréchet differentiable, at all points of the space $E$ (see the remark following Lemma 4.2 below). Hence it is not obvious that any local maximizer of $N$ on $S_c(\eta)$ is a weak solution, and (as was mentioned in (iii) of section 1.4) we use convex analysis to prove this result. However, convex analysis is not needed for the global maximizers described in (i) of section 1.5; in particular, the method of spherical re-arrangement (which conserves $N(v)$ and does not increase $|||v|||$) leads directly to a weak solution of Hill's problem.

Before coming to the statement (II), we derive relevant properties of the functional $N$, and these are established without restricting $N$ to $E_c$.

**Lemma 4.1.** *The functional $N$ is convex, bounded by*

$$0 \leqq N(u) \leqq \text{const. } W^{-7/3} |||u|||^{10/3} \quad \textit{for all} \quad u \in E, \tag{4.1}$$

*and locally Lipschitz continuous:*

$$|N(u) - N(w)| \leqq \text{const. } W^{-7/3} R^{7/3} |||u - w||| \quad \textit{for all} \quad u, w \in \mathcal{B}(0, R) \textit{ in } E. \tag{4.2}$$

*Here the constants are independent of $u$, $w$, $W$ and $R$.*

**Proof.** The convexity of $N$ follows from that of the function $(\cdot)_+$; for $t \in [0, 1]$,

$$\begin{aligned} N((1-t)\,u + tw) &= \frac{1}{2\pi^2} \int_{\mathbb{R}^5} \{(1-t)\,(u - \tfrac{1}{2} W) + t(w - \tfrac{1}{2} W)\}_+ \\ &\leqq \frac{1}{2\pi^2} \int_{\mathbb{R}^5} \{(1-t)\,(u - \tfrac{1}{2} W)_+ + t(w - \tfrac{1}{2} W)_+\} \\ &= (1-t)\,N(u) + tN(w). \end{aligned}$$

The bound (4.1) results from Hölder's inequality and (2.3), (2.4). Lipschitz continuity of bounded convex functionals is a standard result ([10], pp. 12–13; [16], pp. 110, 113), but it seems worth while to give the short proof of (4.2).

Let $h = |||w - u||| > 0$ and $q = w + R(w - u)/h$. Then

$$w = \frac{h}{R+h} q + \frac{R}{R+h} u \quad \Rightarrow \quad N(w) \leqq \frac{h}{R+h} N(q) + \frac{R}{R+h} N(u)$$

by convexity, and $q \in \mathscr{B}(0, 2R)$. Accordingly,

$$N(w) - N(u) \leqq \frac{h}{R+h} N(q) - \frac{h}{R+h} N(u) \leqq \frac{h}{R} N(q) \leqq \text{const. } h W^{-\frac{7}{3}} R^{\frac{7}{3}},$$

by (4.1). We obtain a similar inequality by interchanging $w$ and $u$, and (4.2) follows.

Q.e.d.

At any point $v \in E$, the *right-hand Gateaux derivative* of $N$ in the direction $u$ is defined by

$$N'_+(v)\, u = \lim_{t \to 0+} \frac{N(v + tu) - N(v)}{t}, \qquad u \in E;$$

the left-hand derivative $N'_-(v)\, u$ is defined similarly with $t \to 0-$. It is to be expected from Lemma 4.1, and is true ([16], p. 117), that these limits exist for all $v$ and $u$ in $E$, that $N'_-(v)\, u = -N'_+(v)(-u)$ and that $N'_+(v)$ is a sublinear functional on $E$. We now calculate these one-sided derivatives.

**Lemma 4.2.** *At any* $v \in E$ *we have, respectively,*

$$2\pi^2 N'_\pm(v)\, u = \int_{P(v)} u(x)\, dx + \int_{X(v)} u_\pm(x)\, dx \quad \textit{for all} \quad u \in E,$$

*where* $P(v)$ *is as in* (2.2), $u_+$ *and* $u_-$ *are as in* (3.2), *and* $X(v) = \{x \mid v(x) = \frac{1}{2} W\}$.

**Proof.** It is sufficient to prove the result for $N'_+(v)$. We abbreviate $P(v)$ to $P$ and $X(v)$ to $X$, let $V(x) = v(x) - \frac{1}{2} W$, and define

$$Q(t) = P(v + tu) = \{x \mid V(x) + tu(x) > 0\}, \qquad t > 0.$$

Then

$$2\pi^2 \frac{N(v + tu) - N(v)}{t} = \frac{1}{t} \left\{ \int_{Q(t)} (V + tu) - \int_P V \right\}$$

$$= \int_{Q(t) \cap P} u + \frac{1}{t} \int_{Q(t) \setminus P} (V + tu) - \frac{1}{t} \int_{P \setminus Q(t)} V. \tag{4.3}$$

We consider these integrals one at a time, always taking fixed, but arbitrary, representatives $v$, $u$ of the equivalence classes $[v], [u] \in E$.

(i) First,

$$\int_{Q(t) \cap P} u = \int_P u - \int_{D(t)} u, \tag{4.4}$$

where

$$D(t) = P \setminus Q(t) = \{x \mid V(x) > 0,\ V(x) + tu(x) \leqq 0\},$$

so that $u < 0$ on $D(t)$. Consequently, $D(s) \subset D(t)$ for $0 < s \leqq t$, because $V(x) \leqq s\,|u(x)|$ implies that $V(x) \leqq t\,|u(x)|$. For any $x \in P$, define $t_0(x) = V(x)/|u(x)| > 0$; then $x \notin D(t)$ for $t < t_0(x)$. Hence $|D(t)|_5 \to 0$ as $t \to 0+$, and

$$\left| \int_{D(t)} u \right| \leqq \|u\|_{0,10/3,\mathbb{R}^5}\, |D(t)|_5^{7/10} \to 0 \quad \text{as} \quad t \to 0+. \tag{4.5}$$

(ii) Since

$$Q(t) \setminus P = \{x \mid V(x) + tu(x) > 0,\ V(x) \leqq 0\},$$

we have $u > 0$ on $Q(t) \setminus P$, and $Q(t) \setminus P = \{x \mid u(x) > 0,\ V(x) = 0\} \cup R(t)$, where

$$R(t) = \{x \mid V(x) + tu(x) > 0,\ V(x) < 0\}.$$

Then $|R(t)|_5 \to 0$ as $t \to 0+$, by the reasoning used for $D(t)$ in (i), and

$$\frac{1}{t} \int_{Q(t)\setminus P} (V + tu) = \int_X u_+ + \frac{1}{t} \int_{R(t)} (V + tu), \tag{4.6}$$

where

$$0 \leqq \frac{1}{t} \int_{R(t)} (V + tu) \leqq \int_{R(t)} u \to 0 \quad \text{as} \quad t \to 0+. \tag{4.7}$$

(iii) The third integral in (4.3) is over $D(t)$, already considered in (i), and

$$0 \leqq \frac{1}{t} \int_{P\setminus Q(t)} V \leqq \int_{D(t)} |u| \to 0 \quad \text{as} \quad t \to 0+. \tag{4.8}$$

Combining (4.4) to (4.8), we obtain the result of the lemma.

Q.e.d.

Lemma 4.2 shows that *$N$ is Gateaux differentiable at $v$ in all directions* (that is, $N'_+(v) = N'_-(v)$) *if and only if the set $\{x \mid v(x) = \frac{1}{2} W\}$ has measure zero.* The *subdifferential* of $N$ at any point $v \in E$ is the set

$$\partial N(v) = \{g^* \in E^* \mid g^*(u - v) \leqq N(u) - N(v) \quad \text{for all } u \in E\}$$

in the dual space $E^*$ of $E$. The following is another standard result ([16], p. 122).

**Lemma 4.3.** *At any point $v \in E$, a bounded linear functional $g^* \in \partial N(v)$ if and only if*

$$N'_-(v)\, u \leqq g^*(u) \leqq N'_+(v)\, u \quad \text{for all } u \in E.$$

We now consider the restriction $N|_{E_c} = N_c$, say. (The subscript $c$ is redundant in some statements, but helpful in others.) It is clear that $N_c$ enjoys the same properties relative to $E_c$ as $N$ does relative to $E$. The following lemma provides a generalization of the usual Euler-Lagrange equation, in weak form, characterizing a local maximum.

**Lemma 4.4.** *Let $v$ be a local maximizer of $N_c$ on the sphere $S_c(\eta)$ (that is, $v \in S_c(\eta)$ and $N_c(u) \leqq N_c(v)$ for all $u \in S_c(\eta)$ in some neighbourhood of $v$).*
*If $g^* \in \partial N_c(v)$, then $g^* = \alpha\langle v, \cdot\rangle_E$ for some $\alpha > 0$.*

**Proof.** There exists a unique element $g \in E_c$ such that $g^* = \langle g, \cdot\rangle_E$, and $g$ has a unique decomposition $g = \alpha v + w$, where $\alpha \in \mathbb{R}$, $w \in (\text{span}\,\{v\})^\perp$ and $(\cdot)^\perp$ denotes the orthogonal complement in $E_c$. We prove that $w = 0$. Assume the contrary, and define

$$\hat{w} = w/|||w|||, \qquad u_\beta = (\cos\beta)\, v + (\sin\beta)\,\eta^{\frac{1}{2}}\hat{w}, \tag{4.9}$$

so that $|||u_\beta|||^2 = \eta$ and $|||u_\beta - v|||^2 = 2\eta(1-\cos\beta)$. Since $v$ is a local maximizer, we may suppose that $\beta \neq 0$ and that $N_c(u_\beta) \leqq N_c(v)$ whenever $|\beta|$ is sufficiently small. By the definition of $\partial N_c(v)$,

$$\begin{aligned} 0 &\geqq N_c(u_\beta) - N_c(v) \geqq \langle g, u_\beta - v\rangle_E \\ &= \langle \alpha v + w, (\cos\beta - 1)\, v + (\sin\beta)\,\eta^{\frac{1}{2}}\hat{w}\rangle_E \\ &= \alpha\,(\cos\beta - 1)\,\eta + (\sin\beta)\,\eta^{\frac{1}{2}}\,|||w|||, \end{aligned}$$

and this is a contradiction for $\beta > 0$ and sufficiently small.

To prove that $\alpha \geqq 0$, we let $w$ be any element of $(\text{span}\,\{v\})^\perp \setminus \{0\}$, not related to $g$, and again define $u_\beta$ by (4.9). Then

$$0 \geqq N_c(u_\beta) - N_c(v) \geqq \langle \alpha v, u_\beta - v\rangle_E = \alpha\,(\cos\beta - 1)\,\eta,$$

which shows that $\alpha \geqq 0$.

Finally, suppose that $\alpha = 0$. Then $0 \in \partial N_c(v)$, so that $N_c(v) \leqq N_c(u)$ for all $u \in E_c$; choosing $u = 0$, we obtain $N_c(v) = 0$. Hence $N_c(u) = 0$ for all $u \in S_c(\eta)$ sufficiently near $v$, and, since $N_c(0) = 0$ and $N_c$ is non-negative and convex, we have $N_c(u) = 0$ in the cone

$$K_\delta = \{u \in E_c \mid 0 < |||u|||^2 \leqq \eta,\ |||\eta^{\frac{1}{2}}\hat{u} - v||| < \delta\} \qquad (\hat{u} = u/|||u|||),$$

for some $\delta > 0$. This can be shown false; we set $u = \frac{3}{4}(v + w)$, choose a point $(r_0, z_0) \in \Pi$ outside the set $\{(r, z) \mid v(r, z) \leqq -1\}$ for some representative $v$ of $[v] \in E_c$, and choose $w$ as follows: $|||w|||$ is so small that $u \in K_\delta$, but $w(r, z) \to \infty$ as $(r, z) \to (r_0, z_0)$. (For example, $w$ could be $1/r^2$ times the function in (3.4) of [12].) Then $u \in K_\delta$ but $N_c(u) > 0$.

Q.e.d.

**Theorem 4.5.** *Let $v$ be a local maximizer of $N_c$ on $S_c(\eta)$. Then $v$ is a transformed weak solution of Hill's problem; in fact,*

$$\int_{\mathbb{R}^5} \nabla u \cdot \nabla v\, dx = \frac{1}{\alpha} \int_{P(v)} u\, dx \quad \textit{for all} \quad u \in E, \tag{4.10}$$

*where $\alpha$ is as in Lemma 4.4.*

**Proof.** Combining the results of Lemmas 4.2, 4.3 and 4.4, we obtain

$$\int_{P(v)} u + \int_{X(v)} u_- \leqq \alpha \int_{\mathbb{R}^5} \nabla u \cdot \nabla v \leqq \int_{P(v)} u + \int_{X(v)} u_+ \tag{4.11}$$

for all $u \in E_c$. Define $P_0(v) = P(v) \cup X(v) = \{x \mid v(x) \geqq \frac{1}{2} W\}$, and note that our bound (2.4) for $|P(v)|_5$ applies equally well to $P_0(v)$. Also, $u_+ \in E$ when $u \in E$, with $|||u_+||| \leqq |||u|||$, and similarly for $u_-$. Therefore, we may first use Lemma 2.3 to extend (4.11) to all $u \in E$ (just as in the proof of Theorem 2.4), and then repeat the proof of Lemma 3.1; the bounds implied by (4.11) are as adequate as were those implied by (3.1). In particular, the previous arguments show that $-\alpha \Delta v(x) = 1$ in $P(v)$, that $\Delta v(x) = 0$ wherever $v(x) < \frac{1}{2} W$, and that $P_0(v)$ is bounded; hence

$$\begin{aligned} \alpha \int_{\mathbb{R}^5} \nabla\phi \cdot \nabla v &= -\alpha \int_{\mathbb{R}^5} \phi \Delta v \\ &= \int_{P(v)} \phi - \alpha \int_{X(v)} \phi \Delta v \quad \text{for all} \quad \phi \in C_0^\infty(\mathbb{R}^5). \end{aligned}$$

But $v(x)$ is constant almost everywhere on $X(v)$; then two applications, first to $\nabla v$ and then to $\Delta v$, of a known theorem ([19], p. 53) show that $\Delta v(x) = 0$ almost everywhere on $v(X)$. Thus

$$\int_{\mathbb{R}^5} \nabla\phi \cdot \nabla v = \frac{1}{\alpha} \int_{P(v)} \phi \quad \text{for all} \quad \phi \in C_0^\infty(\mathbb{R}^5),$$

and we extend this result by continuity to obtain (4.10).

Q.e.d.

*Remark.* One can also show that if $w$ is a local minimizer of $N$ on $S_c(\eta)$, then $N(w) = 0$, so that $w(x) \leqq \frac{1}{2} W$ almost everywhere.

*Acknowledgement.* The main result of this paper was obtained while both authors enjoyed the kind hospitality of the Mathematics Research Center, University of Wisconsin, Madison, during the spring of 1983.

## Appendix A. Extension to weak solutions of a theorem of Gidas, Ni & Nirenberg

The theorem in question is Theorem 2.1 of [14]. As well as giving a slight extension, we correct an oversight: that the caps $Z(\mu)$ and reflected caps $Y(\mu)$, defined in (ii) below, need not be connected, so that the maximum principle can be applied only to components of such caps. The geometrical and analytical setting is as follows.

(i) Let $\Omega$ be a bounded, connected, open set in $\mathbb{R}^n$, with smooth boundary $\partial\Omega$; of class $C^1$ is sufficient for Lemmas A.1 and A.2. For Theorem A.3, however, we assume that $\partial\Omega$ is of class $C^{2+\alpha}$ for some $\alpha \in (0, 1)$ which we may take to be the same Hölder exponent for $\partial\Omega$ and for the data mentioned after (A.2). (In fact, only $\partial\Omega \cap \{x \mid x_1 > m - \varepsilon\}$, where $m$ is defined below and $\varepsilon > 0$, need be of class $C^{2+\alpha}$.) The outward unit normal to $\partial\Omega$ is denoted by $\nu = (\nu_1, \ldots, \nu_n)$.

(ii) Let $T_\mu$ and $x^\mu$ be as in (3.11), except that $x_n$ replaces $x_5$, but now define a *cap* by $Z(\mu) = \{x \in \Omega \mid x_1 > \mu\}$ and the *reflected cap* by $Y(\mu) = \{x \in \mathbb{R}^n \mid x^\mu \in Z(\mu)\}$. Note that our earlier definitions result from replacing $\Omega$ by $\mathbb{R}^n$, but that now $Z(\mu)$ need not be connected. Indeed, $Z(\mu)$ may have infinitely many components (maximal connected subsets) even when $\partial\Omega$ is of class $C^\infty$. We define critical positions of the reflecting hyperplane $T_\mu$ by

$$M = \sup\{x_1 \mid x \in \Omega\} = \sup\{\mu \mid Z(\mu) \text{ is not empty}\},$$

$$k = \inf\{\alpha \mid \mu \in (\alpha, M) \Rightarrow Y(\mu) \subset \Omega\},$$

$$l = \inf\{\beta \mid \mu \in (\beta, M) \Rightarrow \nu_1(x) > 0 \text{ for all } x \in \partial Z(\mu) \setminus T_\mu\},$$

$$m = \max\{k, l\}.$$

Figure 2 illustrates this notation. The following lemma states facts needed for Theorem A.3, and shows that it is consistent with remarks in [14] to call $Z(m)$ the *maximal cap*, and to call $Z(k)$, when $k < l$, the *optimal cap*. The proof of the lemma is omitted. (In [24] and [14], such results are regarded as self-evident; our proof is not difficult, but it is not short.)

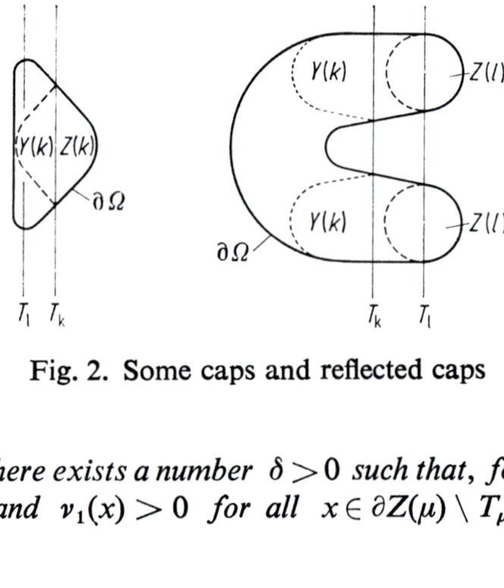


Fig. 2. Some caps and reflected caps

**Lemma A.1.** (a) *There exists a number* $\delta > 0$ *such that, for any* $\mu \in (M - \delta, M)$, *we have* $Y(\mu) \subset \Omega$ *and* $\nu_1(x) > 0$ *for all* $x \in \partial Z(\mu) \setminus T_\mu$; *hence* $k < M$ *and* $l < M$.

(b) $Y(k) \subset \Omega$.

(c) $\nu_1(x) > 0$ *for all* $x \in \partial Z(l) \setminus T_l$.

(d) *For any* $\mu > m$, *the boundaries* $\partial Y(\mu) \setminus T_\mu$ *and* $\partial\Omega$ *are disjoint.*

(e) *If* $k > l$, *then* $\partial Y(k) \setminus T_k$ *meets* $\partial\Omega$ *tangentially at some point (and* $\overline{Y(k)} \subset \overline{\Omega}$).

(f) *There exists a point* $x_0 \in T_l \cap \partial\Omega$ *such that* $\nu_1(x_0) = 0$.

Note that, if $\mu \in [m, M)$, then $Y(\mu) \subset \Omega$ and $\nu_1 > 0$ on $\partial Z(\mu) \setminus T_\mu$. The condition $k > l$ in (e) is necessary.

(iii) We consider a function $u \in C^1(\overline{\Omega})$, with $u > 0$ in $\Omega$, such that

$$\int_{\Omega} \{\nabla\phi \cdot \nabla u - \phi b_1(x)\, D_1 u\}\, dx = \int_{\Omega} \phi g(u)\, dx \quad \text{for all} \quad \phi \in C_0^\infty(\Omega), \tag{A.1}$$

$$u = 0 \quad \text{on} \quad \Gamma_m, \tag{A.2}$$

where $D_1 = \partial/\partial x_1$ and $\Gamma_\mu = \partial Z(\mu) \setminus T_\mu$. We also assume that $u \in \overline{C^{2+\alpha}(\Gamma_{m-\varepsilon})}$ for some $\varepsilon > 0$; that the coefficient $b_1 \in C^{0+\alpha}(\overline{\Omega})$ and that $b_1 \geqq 0$ on $\overline{Y(m) \cup Z(m)}$. The function $g: [0, \infty) \to \mathbb{R}$ is assumed to have a decomposition $g = g_1 + g_2$ such that $g_1 \in C^1[0, \infty)$, while $g_2$ is non-decreasing and its restriction to $[0, \beta)$ is in $C^{0+\alpha}[0, \beta)$ for some $\beta > 0$.

**Lemma A.2.** *Let $\Omega$ be as in* (i), *and $u$ as in* (iii). *Assume that, for some $\mu \in [m, M)$ and for some component $Y_0(\mu)$ of $Y(\mu)$, we have $u(x) \geqq u(x^\mu)$ for all $x \in Y_0(\mu)$, $u(x_0) \neq u(x_0^\mu)$ for some $x_0 \in Y_0(\mu)$, and $D_1 u(x) \leqq 0$ for all $x \in Z_0(\mu)$, where $Z_0(\mu)$ is the reflection in $T_\mu$ of $Y_0(\mu)$. Then*

(a) *$u(x) > u(x^\mu)$ for all $x \in Y_0(\mu)$,*

(b) *$D_1 u(x) < 0$ for all $x \in \{\partial Y_0(\mu) \cap \partial Z_0(\mu)\} \setminus \partial\Omega$.*

**Proof.** Defining $w = u_\mu - u$, where $F_\mu(x) = F(x^\mu)$ for any function $F$, we calculate as in the proof of Lemma 3.7. Let $Y_0 = Y_0(\mu)$ and $Z_0 = Z_0(\mu)$. Given $\phi \in C_0^\infty(Y_0)$, we first choose $\phi_\mu$ as the test function in (A.1), transform from $Z_0$ to $Y_0$, then choose $\phi$ itself as the test function, to obtain (in place of (3.15))

$$\int_{Y_0} \{\nabla\phi \cdot \nabla w - \phi b_1 D_1 w\} = \int_{Y_0} \phi\{g(u_\mu) - g(u) - (b_1 + b_{1,\mu})\, D_1 u_\mu\} \tag{A.3}$$

for all $\phi \in C_0^\infty(Y_0)$.

Now $g_1(u_\mu) - g_1(u) = c(x)\,(u_\mu - u)$ for some $c \in C(\overline{\Omega})$ because $g_1 \in C^1[0, \infty)$, and $g_2(u_\mu) \leqq g_2(u)$ because $g_2$ is non-decreasing and $u_\mu \leqq u$ in $Y_0$. By hypothesis, $b_1 + b_{1,\mu} \geqq 0$ in $Y_0$, and $D_1 u_\mu \geqq 0$ in $Y_0$. Hence

$$\int_{Y_0} \{\nabla\phi \cdot \nabla w - \phi b_1 D_1 w - \phi c w\} \leqq 0 \quad \text{for all} \quad \phi \in C_0^\infty(Y_0) \quad \text{with} \quad \phi \geqq 0.$$

Since $w \not\equiv 0$ in $Y_0$, by hypothesis, we conclude from Theorem 3.4(a) that $w < 0$ in $Y_0$. Given $x_0 \in \{\partial Y_0 \cap \partial Z_0\} \setminus \partial\Omega$, we define $B = \mathscr{B}((x_{0,1} - R, x_0''), R)$ and choose $R$ so small that $B \subset Y_0$. Then, since $u \in C^1(\overline{\Omega})$, Theorem 3.4(b) shows that $(D_1 w)(x_0) > 0$; finally, $(D_1 w)(x_0) = -2(D_1 u)(x_0)$.

Q.e.d.

**Theorem A.3.** (a) *Let $\Omega$ be as in* (i), *and $u$ as in* (iii). *For any $\mu \in (m, M)$, we have $u(x) > u(x^\mu)$ for all $x \in Y(\mu)$. Also, $D_1 u(x) < 0$ for all $x$ in the maximal cap $Z(m)$.*

(b) *Suppose that $(D_1 u)(x_0) = 0$ at some $x_0 \in T_m \cap \Omega$. Let $Z_0(m)$ be the component of $Z(m)$ containing $(x_{0,1} + \varepsilon, x_0'')$ for sufficiently small $\varepsilon > 0$, and let $Y_0(m)$ be the reflection in $T_m$ of $Z_0(m)$. Then $u$ is an even function of $x_1 - m$, the set $\overline{\Omega} = \overline{Y_0(m) \cup Z_0(m)}$, and $b_1 \equiv 0$ in $\Omega$.*

**Proof.** The hypotheses in (iii) ensure existence of a set $\Omega_{2\delta}=\{x\in\Omega \mid \operatorname{dist}(x,\Gamma_m)<2\delta\}$, with $\delta>0$, such that $g\circ u\in C^{0+\alpha}(\overline{\Omega}_{2\delta})$. Regularity theory ([2], pp. 667–8) now shows that $u\in C^{2+\alpha}(\overline{\Omega}_\delta)$. Then the proof of Lemma 2.1 in [14] stands, and Lemma A.2 replaces Lemma 2.2 of [14].

The remaining part of the proof is essentially as on pp. 218–219 of [14]. For the proof of (a), it is enough that Lemma A.2 refers to any component of $Y(\mu)$. For the proof of (b), the implication of (a) that $u(x)\geqq u(x^m)$ for all $x\in Y_0(m)$, the result $D_1u<0$ in $Z(m)$, the hypothesis $(D_1u)(x_0)=0$ and Lemma A.2 imply that $u(x)=u(x^m)$ for all $x\in Y_0(m)$ (otherwise, Lemma A.2(b) would be contradicted at $x_0$). Then, by continuity, $u(x)=u(x^m)=0$ for $x\in\partial Y_0(m)\setminus T_m$, and so $\partial Y_0(m)\setminus T_m\subset\partial\Omega$. Since $\Omega$ is connected, $\overline{\Omega}=\overline{Y_0(m)}\cup\overline{Z_0(m)}$. To prove that $b_1\equiv 0$ in $\Omega$, we apply (A.3) with $\mu=m$ and $Y_0=Y_0(m)$, noting that (A.3) follows from (A.1) without additional hypotheses. Since we now have $w=u_m-u\equiv 0$ in $Y_0(m)$, (A.3) reduces to

$$\int_{Y_0(m)}\phi(b_1+b_{1,m})\,D_1u_m=0 \quad \text{for all} \quad \phi\in C_0^\infty(Y_0(m)),$$

with $D_1u_m>0$ by (a), and $b_1\geqq 0$, $b_{1,m}\geqq 0$. Hence $b_1\equiv 0$ and $b_{1,m}\equiv 0$ in $Y_0(m)$.

Q.e.d.

## Appendix B. Hill's vortex in a ball

The problem has been formulated in remark (ii) of section 1.5; the cross-section of the fluid domain is now $D=\{(r,z)\in\Pi \mid r^2+z^2<b^2\}$. We denote the analogue of HILL'S solution by $\psi_h$, and let $\varrho=(r^2+z^2)^{\frac{1}{2}}$ as before; an elementary calculation yields

$$\psi_h(r,z)\equiv\Psi_h(r,z)+\tfrac{1}{2}Wr^2=\begin{cases}\frac{1}{2}\dfrac{Wr^2}{1-c}\left(\frac{5}{2}-c-\frac{3}{2}\dfrac{\varrho^2}{a^2}\right), & \varrho\leqq a,\\[2ex] \frac{1}{2}\dfrac{Wr^2}{1-c}\left(\dfrac{a^3}{\varrho^3}-c\right), & a\leqq\varrho\leqq b,\end{cases} \tag{B.1 a}$$

where

$$c=\frac{a^3}{b^3},\qquad \frac{\lambda b^2}{W}=\frac{15}{2}\frac{1}{c^{2/3}(1-c)}, \tag{B.1 b}$$

and

$$\|\psi_h\|^2=\int_D\frac{1}{r^2}(\psi_{h,r}^2+\psi_{h,z}^2)\,r\,dr\,dz=W^2b^3\frac{c}{1-c}\left(1+\frac{3}{7}\frac{1}{1-c}\right). \tag{B.1 c}$$

Here the norm is that of the Hilbert space $H(D)$, which results from replacing $\Pi$ by $D$ in (1.7) and in the sentence preceding it. Obviously, $\psi_h\to\psi_H$ as $b\to\infty$ with $a$ fixed.

Suppose that $W$ and $b$ are prescribed. Then (B.1b) shows that $\lambda(c)$, with $0 < c < 1$, has a single stationary point, a minimum at $c = 2/5$; we define $\lambda_0 = \lambda(2/5)$. Hence, if $\lambda$ is prescribed, we have no solution of HILL's type for $\lambda < \lambda_0$, one solution for $\lambda = \lambda_0$, and two solutions for $\lambda > \lambda_0$. On the other hand, the energy $\pi \|\psi_h\|^2$, as a function of $c$ on $(0, 1)$, is strictly increasing and has range $(0, \infty)$; prescription of this norm always gives exactly one solution of HILL's type.

Let $b$ be given and fixed henceforth, and let $\Omega = \mathscr{B}(0, b)$ in $\mathbb{R}^5$. The Hilbert spaces $E(\Omega)$ and $E_c(\Omega)$ are defined as $E$ and $E_c$ were, but with $\Omega$ replacing $\mathbb{R}^5$; we make the same adjustment in the definition (2.2) of $P(u)$. The transformation of weak solutions, from $H(D)$ to $E_c(\Omega)$, proceeds essentially as in section 2; perhaps a little more easily, because $E(\Omega)$ is equivalent to $\mathring{W}_2^1(\Omega)$ (functions in $E(\Omega)$ are in $L_p(\Omega)$ for $1 \leqq p \leqq 10/3$), and obviously $|P(u)|_5 \leqq |\Omega|_5$. Thus a weak solution $\psi \in H(D)$ of Hill's problem for $D$ is equivalent, under the transformation $\psi = r^2 v$, to a function $v \in E_c(\Omega)$ satisfying the hypothesis of the following theorem. We now weaken the condition $v \in E_c(\Omega)$ to $v \in E(\Omega)$.

**Theorem B.1.** *If there exist* $v \in E(\Omega) \setminus \{0\}$, $\lambda \in \mathbb{R}$ *and* $W > 0$ *such that*

$$\int_\Omega \nabla u \cdot \nabla v \, dx = \lambda \int_{P(v)} u \, dx \quad \text{for all} \quad u \in E(\Omega), \tag{B.2}$$

*then* $v(x) = v_h(|x'|, x_5)$ *and* $\lambda \geqq \lambda_0$. *Here* $v_h(r, z) = \psi_h(r, z)/r^2$ *and* $\psi_h$ *is as in* (B.1); *thus* $\psi_h$ *denotes a pair of functions for given* $\lambda > \lambda_0$, *but with distinct values of* $\|\psi_h\|$.

**Proof.** In order to apply Theorem A.3 to $v$, we must first prove that $v \in C^1(\bar{\Omega})$ and that $v > 0$ in $\Omega$. Turning to Theorem 8.1 of [1], and proceeding as in the proof of Lemma 3.1 (but now with estimates of $\|v\|_{2,p,\Omega}$), we find that $v \in C^{1+\alpha}(\bar{\Omega})$ for every $\alpha \in (0, 1)$. Then $v = 0$ on $\partial\Omega$; Theorems 8.1 and 8.19 of [15] show that $v > 0$ in $\Omega$. Alternatively, we can proceed as in the proof of Lemma 3.2 to obtain

$$v(x) = \lambda \int_{P(v)} G(x, y) \, dy \quad \text{for all} \quad x \in \Omega,$$

where $G$ is the Green function of the Dirichlet problem for $-\Delta$ in the ball $\Omega$ (it is minus the function on p. 19 of [15]). Classical estimates then show that $v \in C^{1+\alpha}(\bar{\Omega})$ for every $\alpha \in (0, 1)$, and the positivity of $v$ in $\Omega$ follows from that of $G$.

We now apply Theorem A.3 to the function $v$ and the ball $\Omega$ in $\mathbb{R}^5$, setting $b_1 = 0$, $g_1 = 0$ and $g_2(t) = \lambda f_H(t - \frac{1}{2} W)$, where $f_H$ is as in (1.5); the hypotheses are amply satisfied and the maximal cap is the half-ball in which $x_1 > 0$. Therefore $D_1 v(x) < 0$ whenever $x_1 > 0$ and $x \in \Omega$. Now choosing the unit vector $\gamma = (-1, 0, \ldots, 0)$, we see that $D_1 v(x) > 0$ whenever $x_1 < 0$ and $x \in \Omega$. By continuity, $D_1 v = 0$ on $T_0 \cap \Omega$, and the theorem now states that $v$ is an even function of $x_1$. The same argument holds for every unit vector $\gamma$ in $\mathbb{R}^5$; consequently, $v$ depends only on $|x|$, and is strictly decreasing as $|x|$ increases. The proof now concludes like that of Theorem 3.9.

Q.e.d.

The results in section 4 are not affected in any significant way when $\mathbb{R}^5$ is replaced by the ball $\Omega$ in $\mathbb{R}^5$, and $E$ by $E(\Omega)$. Let us denote by $N(u, \Omega)$ and $S_c(\eta, \Omega)$ the results of these changes in the definitions (2.8) and (2.9). Then we have

**Theorem B.2.** *Let $v$ be a local maximizer of $N(\cdot, \Omega)$ on the sphere $S_c(\eta, \Omega)$. Then $v$ is a transformed weak solution of Hill's problem for $D$; that is, there exists $\lambda \in \mathbb{R}$ such that* (B.2) *holds.*

## Appendix C. Alternative proof of a lemma of Gidas, Ni & Nirenberg

We are concerned here with Lemma 4.1 in [14], of which our Lemma 3.5 is a particular case; the following proof is somewhat different from that in [14]. We let $x'' = (x_2, \ldots, x_n)$, as elsewhere, but $r$ now denotes *spherical* radius.

**Lemma C.1.** *Assume that, outside some ball in $\mathbb{R}^n$,*

$$v(x) = a_0 r^{-m} + g(x), \qquad a_0 > 0,\, m > 0,\, r \equiv |x|,$$

*where $g(x) \to 0$ and $|\nabla g(x)| = O(r^{-m-3})$ as $r \to \infty$. Consider two points $y$ and $z$ such that $y_1 < z_1$, $y_1 + z_1 \geqq 2\mu > 0$ and $y'' = z''$. There exists a number $R(\mu)$, depending only on $v$ and* $\min\{1, \mu\}$, *such that*

$$v(y) > v(z) \text{ whenever } |y| \geqq R(\mu).$$

**Proof.** There exist positive constants $r_0$ and $K$ such that, for $r \geqq r_0$,

$$|\nabla g(x)| \leqq a_0 K r^{-m-3}, \tag{C.1}$$

$$|g(x)| = \left| \int_r^\infty \frac{\partial g}{\partial \tilde{r}}\, d\tilde{r} \right| \leqq \frac{a_0 K}{m+2} r^{-m-2}. \tag{C.2}$$

We consider only points outside the ball $\mathscr{B}(0, r_0)$, and introduce the notation (Figure 3)

$$y_1 = \alpha - h, \qquad z_1 = \alpha + h, \qquad \alpha \geqq \mu > 0,\, h > 0,$$

$$s = |y| = \{(\alpha - h)^2 + |y''|^2\}^{\frac{1}{2}}, \qquad t = |z| = \{(\alpha + h)^2 + |y''|^2\}^{\frac{1}{2}},$$

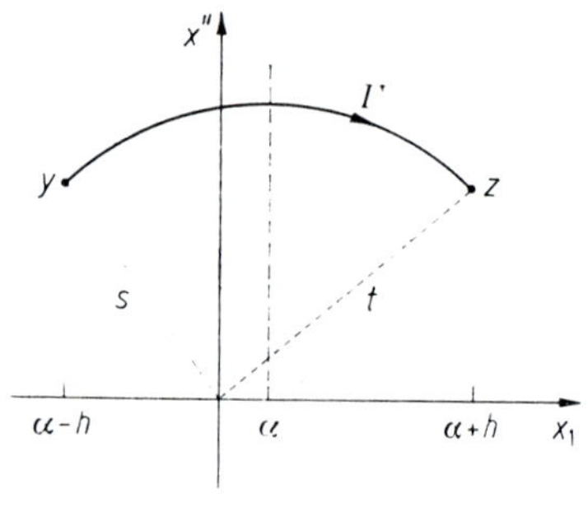


Fig. 3. Notation for the proof of Lemma C.1

so that

$$t^2 - s^2 = 4\alpha h, \quad t > s \geqq r_0. \tag{C.3}$$

The result will follow from two estimates of $|g(y) - g(z)|$ and one of $s^{-m} - t^{-m}$.

Let $P_2$ denote the two-dimensional plane containing the points 0, $y$ and $z$ (or any such plane if $y'' = 0$), and let $\Gamma$ be the circular arc in $P_2$ from $y$ to $z$, centred at $(\alpha, 0)$. Then $\Gamma$ has length $\pi h$ at most, and $r = |x| \geqq s$ for $x \in \Gamma$, so that (C.1) yields

$$|g(y) - g(z)| = \left| \int_\Gamma \nabla g(x) \cdot dx \right| \leqq \pi h a_0 K s^{-m-3}. \tag{C.4}$$

Alternatively, by (C.2),

$$|g(y) - g(z)| \leqq |g(y)| + |g(z)| \leqq \frac{2a_0 K}{m+2} s^{-m-2}. \tag{C.5}$$

For our third estimate, we first note that

$$t^m - s^m \geqq c_m t^{m-1}(t - s), \quad \text{where} \quad c_m = \min\{m, 1\}; \tag{C.6}$$

if $m \in (0, 1]$, this is true because $1 - \xi^m \geqq m(1 - \xi)$ for $0 \leqq \xi \leqq 1$ (differentiate both sides), and, if $m > 1$, because $s^m < t^{m-1} s$. Accordingly, in view of (C.3),

$$\begin{aligned} s^{-m} - t^{-m} &= \frac{t^m - s^m}{s^m t^m} \geqq \frac{c_m t^{m-1}(t - s)}{s^m t^m} = \frac{c_m}{s^m t} \frac{4\alpha h}{t + s} \\ &> \frac{2c_m \alpha h}{s^m t^2} = \frac{2c_m \alpha h}{s^m(s^2 + 4\alpha h)}. \end{aligned} \tag{C.7}$$

If $4\alpha h < s^2$, we use (C.7) and (C.4) to obtain

$$\begin{aligned} \frac{1}{a_0}\{v(y) - v(z)\} &= s^{-m} - t^{-m} + \frac{1}{a_0}\{g(y) - g(z)\} \\ &> \frac{c_m \alpha h}{s^{m+2}} - \frac{\pi K h}{s^{m+3}} \geqq 0 \quad \text{if} \quad s \geqq \frac{\pi K}{c_m \alpha}. \end{aligned}$$

If $4\alpha h \geqq s^2$, we use (C.7) and (C.5) to obtain

$$\frac{1}{a_0}\{v(y) - v(z)\} > \tfrac{1}{4} \frac{c_m}{s^m} - \frac{2K}{m+2} \frac{1}{s^{m+2}} \geqq 0 \quad \text{if} \quad s^2 \geqq \frac{8K}{c_m(m+2)}.$$

We define

$$R(\mu) = \max\left\{ r_0, \frac{\pi K}{c_m \min\{1, \mu\}}, \left(\frac{8K}{c_m(m+2)}\right)^{\frac{1}{2}} \right\},$$

and the lemma is proved.

Q.e.d.

*Remarks.* (i) If the hypothesis $|\nabla g(x)| = O(r^{-m-3})$ as $r \to \infty$ is weakened to $|\nabla g(x)| = O(r^{-m-2-\delta})$, $\delta > 0$, then an obvious variant of the foregoing proof still holds; if it is weakened to $|\nabla g(x)| = o(r^{-m-2})$, then a proof is still possible, but explicit calculation of $R(\mu)$ must be replaced by an 'epsilon argument'.

(ii) In [14], the lemma is stated for $m > 0$ (in Theorem 4, p. 211), but proved (on pp. 232–234) only for $m \geqq 1$; however, (C.6) shows that this is a small matter.

## References

1. AGMON, S., The $L_p$ approach to the Dirichlet problem. *Ann. Scuola Norm. Sup. Pisa* (3) **13** (1959), 405–448.
2. AGMON, S., DOUGLIS, A., & NIRENBERG, L., Estimates near the boundary for solutions of elliptic partial differential equations satisfying general boundary conditions. I. *Comm. Pure Appl. Math.* **12** (1959), 623–727.
3. AMBROSETTI, A., & MANCINI, G., On some free boundary problems. In *Recent contributions to nonlinear partial differential equations* (edited by H. BERESTYCKI & H. BRÉZIS). Pitman, 1981.
4. AMICK, C. J., & FRAENKEL, L. E., The uniqueness of Norbury's perturbation of Hill's spherical vortex. To appear.
5. AMICK, C. J., & FRAENKEL, L. E., Note on the equivalence of two variational principles for certain steady vortex rings. To appear.
6. BENJAMIN, T. B., The alliance of practical and analytical insights into the nonlinear problems of fluid mechanics. In *Applications of methods of functional analysis to problems of mechanics*, Lecture notes in math. **503**. Springer, 1976.
7. BERESTYCKI, H., Some free boundary problems in plasma physics and fluid mechanics. In *Applications of nonlinear analysis in the physical sciences* (edited by H. AMANN, N. BAZLEY & K. KIRCHGÄSSNER). Pitman, 1981.
8. CAFFARELLI, L. A., & FRIEDMAN, A., Asymptotic estimates for the plasma problem. *Duke Math. J.* **47** (1980), 705–742.
9. CHANDRASEKHAR, S., *Hydrodynamic and hydromagnetic stability*. Oxford, 1961.
10. EKELAND, I., & TEMAM, R., *Convex analysis and variational problems*. North-Holland, 1976.
11. ESTEBAN, M. J., Nonlinear elliptic problems in strip-like domains: symmetry of positive vortex rings. *Nonlinear Analysis, Theory, Methods and Applications* **7** (1983), 365–379.
12. FRAENKEL, L. E., & BERGER, M. S., A global theory of steady vortex rings in an ideal fluid. *Acta Math.* **132** (1974), 13–51.
13. FRIEDMAN, A., & TURKINGTON, B., Vortex rings: existence and asymptotic estimates. *Trans. Amer. Math. Soc.* **268** (1981), 1–37.
14. GIDAS, B., NI, W.-M., & NIRENBERG, L., Symmetry and related properties via the maximum principle. *Comm. Math. Phys.* **68** (1979), 209–243.
15. GILBARG, D., & TRUDINGER, N. S., *Elliptical partial differential equations of second order*. Springer, 1977.
16. GILES, J. R., *Convex analysis with application in differentiation of convex functions*. Pitman, 1982.
17. HILL, M. J. M., On a spherical vortex. *Philos. Trans. Roy. Soc. London* A **185** (1894), 213–245.

18. Keady, G., & Kloeden, P. E., Maximum principles and an application to an elliptic boundary-value problem with a discontinuous nonlinearity. Research report, Dept. of Math., University of Western Australia, 1984.
19. Kinderlehrer, D., & Stampacchia, G., *An introduction to variational inequalities and their applications*. Academic Press, 1980.
20. Ni, W.-M., On the existence of global vortex rings. *J. d'Analyse Math.* **37** (1980), 208–247.
21. Nirenberg, L., On elliptic partial differential equations. *Ann. Scuola Norm. Sup. Pisa* (3) **13** (1959), 115–162.
22. Norbury, J., A steady vortex ring close to Hill's spherical vortex. *Proc. Cambridge Philos. Soc.* **72** (1972), 253–284.
23. Norbury, J., A family of steady vortex rings. *J. Fluid Mech.* **57** (1973), 417–431.
24. Serrin, J., A symmetry problem in potential theory. *Arch. Rational Mech. Anal.* **43** (1971), 304–318.

Department of Mathematics
University of Chicago
and
Mathematics Division
University of Sussex
Brighton

*(Received May 10, 1985)*

# *Lyapunov Functions for Thermomechanics with Spatially Varying Boundary Temperatures*

J.M. BALL & G. KNOWLES

*Dedicated to James Serrin on the occasion of his 60th birthday*

## 1. Introduction

Consider a continuous body subjected to conservative body and surface forces, with a part $\partial\Omega_2$ of the boundary maintained at a temperature $\theta = \theta_0(X)$ and with the remainder of the boundary thermally insulated. A calculation of DUHEM [1911] shows that if $\theta_0$ is *constant* then the equations of motion possess a Lyapunov function, the *equilibrium free energy*, given in a standard notation (see Section 2) by

$$E = \int_{\Omega} \varrho_R(\tfrac{1}{2}|v|^2 + U + \psi - \theta_0\eta)\, dX - \int_{\partial\Omega\setminus\partial\Omega_1} t_R \cdot x\, dA. \tag{1.1}$$

The purpose of this paper is to show that for certain cases when the reference heat flux vector $q_R = \hat{q}_R(X, \theta, \operatorname{Grad}\theta)$ there is a corresponding equilibrium free energy function, namely

$$E = \int_{\Omega} \varrho_R(\tfrac{1}{2}|v|^2 + U + \psi - \phi(X)\,\eta)\, dX - \int_{\partial\Omega\setminus\partial\Omega_1} t_R \cdot x\, dA, \tag{1.2}$$

that is nonincreasing along solutions even when $\theta_0$ depends on $X$.

In (1.2) $\phi$ denotes the solution of the stationary heat equation

$$\operatorname{Div} \hat{q}_R(X, \phi, \operatorname{Grad}\phi) = 0, \qquad X \in \Omega, \tag{1.3}$$

with boundary conditions

$$\phi|_{\partial\Omega_2} = \theta_0, \qquad \hat{q}_R(X, \phi, \operatorname{Grad}\phi) \cdot N|_{\partial\Omega\setminus\partial\Omega_2} = 0. \tag{1.4}$$

In Section 2 we give a formal argument showing that if $\phi$ is any function satisfying $\phi|_{\partial\Omega_2} = \theta_0$ then, for motions satisfying the Planck inequality we have in general that

$$\dot{E} + I \leq 0, \tag{1.5}$$

where

$$I = \int_{\Omega} \operatorname{Grad}\left(\frac{\phi}{\theta}\right) \cdot q_R\, dX. \tag{1.6}$$

The argument applies in particular to thermoelasticity, when equality holds in the Planck inequality.

In Section 3 we make a detailed study of the dissipation integral (1.6) with $\phi$ given by (1.3), (1.4), showing that $I = I(\theta) \geqq 0$ for all temperature distributions $\theta(\cdot)$ satisfying the boundary conditions in the two cases

(a) $\hat{q}_R = -k(\theta) \operatorname{Grad} \theta$, where $\log k(\theta)$ is a concave function of $\log \theta$,

(b) $\hat{q}_R = -K(X) \operatorname{Grad} \theta$, where $K$ is a uniformly positive matrix.

In case (a) we show that if $\log k(\theta)$ is sufficiently convex in $\log \theta$ on some interval then $I(\theta)$ can be negative, and hence $E$ is not a Lyapunov function.

In cases when $E$ is a Lyapunov function it is natural to conjecture that successive states of the body at a sequence of times $t_j \to \infty$ will generically realize, in an appropriate sense, a minimizing sequence for the functional $E$. Consider, for example, a thermoelastic material. If the boundary conditions allow conserved quantities these should be considered as constraints, and it may then happen (*cf.* MAN [1985]) that the velocity fields of minimizing sequences do not tend to zero as $t \to \infty$. Otherwise, however, the preceding motivation leads to consideration of minimization problems for

$$\hat{E}(x) = \int_{\Omega} \varrho_R(X) \left[W(X, Dx(X)) + \psi(X, x(X))\right] dX - \int_{\partial\Omega\setminus\partial\Omega_1} t_R \cdot x \, dA, \tag{1.7}$$

where $W(X, F) \overset{\text{def}}{=} U(X, F, \phi(X)) - \phi(X)\, \eta(X, F, \phi(X))$. Under appropriate hypotheses the study of such minimization problems falls into the framework given in BALL [1977] (see BALL & MURAT [1984] for developments and additional references). For further discussion concerning the relationship between thermodynamics and minimization of $\hat{E}$ see BALL [1984], where the results in this paper were announced, and BALL & KNOWLES [1985].

It would be interesting to find Lyapunov functions for some cases when $q_R$ depends also on mechanical variables and allowing spatially varying boundary temperatures. A Lyapunov function applying to the case when the spatial heat flux vector $q$ is given by

$$q = -k(\theta) \operatorname{grad} \theta,$$

the gradient being with respect to $x$, could be relevant for the study of Bénard convection, for example.

## 2. Equilibrium Free Energy

Consider a continuous body occupying in a reference configuration the bounded open subset $\Omega \subset \mathbb{R}^n$. At time $t$ the particle occupying in the reference configuration the point $X \in \Omega$ has position $x(X, t) \in \mathbb{R}^n$ and temperature $\theta(X, t) > 0$. Assuming the external volumetric heat supply to be zero, the governing equations are

$$\varrho_R \dot{v} = \operatorname{Div} T_R + \varrho_R b, \tag{2.1}$$

$$\varrho_R \dot{U} - \operatorname{tr}(T_R \dot{F}^T) + \operatorname{Div} q_R = 0, \tag{2.2}$$

where $v = \dot{x}$ is the velocity, $\varrho_R(X)$ is the density in the reference configuration, $T_R$ is the Piola-Kirchhoff stress tensor, $b$ is the body force density, $U$ is the internal energy density, $F = Dx(X, t)$ is the deformation gradient and $q_R$ is the (reference) heat flux vector. (Here and below, Div, $D$ and Grad all refer to differentiation with respect to $X$, dots to differentiation with respect to $t$.)

We make the thermodynamic assumption that motions of the body satisfy the *Planck inequality* (see TRUESDELL [1984 p. 112])

$$\varrho_R \theta \dot{\eta} \geqq -\operatorname{Div} q_R, \tag{2.3}$$

where $\eta(X, t)$ denotes the specific entropy. We recall that the *Clausius-Duhem inequality*

$$\varrho_R \dot{\eta} \geqq -\operatorname{Div}\left(\frac{q_R}{\theta}\right) \tag{2.4}$$

follows from (2.3) and the *Fourier inequality*

$$q_R \cdot \operatorname{Grad} \theta \leqq 0. \tag{2.5}$$

For nonsmooth solutions (2.1)–(2.3) must be interpreted in an appropriate weak or distributional sense (*cf.* DAFERMOS [1983]). We suppose that the body force is conservative, so that

$$b(X, t) = -\nabla_x \psi(X, x(X, t)) \tag{2.6}$$

for some potential $\psi(X, x)$.

We impose the following boundary conditions:

*Mechanical*:

$$\begin{aligned} x &= x_0(X), \quad X \in \partial\Omega_1, \\ T_R N &= t_R(X), \quad X \in \partial\Omega \setminus \partial\Omega_1, \end{aligned} \tag{2.7}$$

*Thermal*:

$$\begin{aligned} \theta &= \theta_0(X), \quad X \in \partial\Omega_2, \\ q_R \cdot N &= 0, \quad X \in \partial\Omega \setminus \partial\Omega_2. \end{aligned} \tag{2.8}$$

Here $\partial\Omega_1$, $\partial\Omega_2$ are given subsets of the boundary $\partial\Omega$, $N = N(X)$ is the unit outward normal to $\partial\Omega$ at $X$, and $x_0$, $t_R$, $\theta_0$ are given functions.

Let $\phi = \phi(X) \geqq 0$ be a given function satisfying

$$\phi(X) = \theta_0(X), \quad X \in \partial\Omega_2. \tag{2.9}$$

It follows from (2.1)–(2.3) that

$$\frac{\partial}{\partial t}\left[\varrho_R\left(\tfrac{1}{2}|v|^2 + U + \psi - \phi\eta\right)\right] \leqq \operatorname{Div}\left[v^T T_R\right] + \left(\frac{\phi}{\theta} - 1\right) \operatorname{Div} q_R. \tag{2.10}$$

Using (2.10) and the boundary conditions (2.7), (2.8) we obtain

$$\dot{E} + I \leqq 0, \tag{2.11}$$

where

$$E = \int_\Omega \varrho_R\left(\tfrac{1}{2}|v|^2 + U + \psi - \phi\eta\right) dX - \int_{\partial\Omega \setminus \partial\Omega_1} t_R \cdot x \, dA, \tag{2.12}$$

and

$$I = \int_{\Omega} \operatorname{Grad}\left(\frac{\phi}{\theta}\right) \cdot q_R \, dX. \tag{2.13}$$

Thus $E$ will be nonincreasing along solutions provided

$$I \geqq 0. \tag{2.14}$$

An important special case is when $\theta_0 > 0$ is independent of $X$. Choosing $\phi \equiv \theta_0$ we find that

$$I = -\theta_0 \int_{\Omega} \frac{q_R \cdot \operatorname{Grad} \theta}{\theta^2} \, dX, \tag{2.15}$$

so that (2.14) holds provided (2.5) does. In fact in this case $I \geqq 0$ if we assume that (2.4) holds instead of (2.3). This result is well known (see DUHEM [1911], ERICKSEN [1966], COLEMAN & DILL [1973], for example). The corresponding function

$$E = \int_{\Omega} \varrho_R(\tfrac{1}{2}|v|^2 + U + \psi - \theta_0 \eta) \, dX - \int_{\partial\Omega \setminus \partial\Omega_1} t_R \cdot x \, dA \tag{2.16}$$

is commonly called the *equilibrium free energy*, and we carry over the same terminology to $E$ given by (2.12) whenever $\phi$ is chosen so that (2.14) holds.

As an example we consider a thermoelastic material, whose constitutive relations are given in terms of the Helmholtz free energy function $A(X, F, \theta)$ by

$$\begin{gathered} T_R = \varrho_R \frac{\partial A}{\partial F}, \qquad \eta = -\frac{\partial A}{\partial \theta}, \qquad U = A + \eta\theta, \\ q_R = \hat{q}_R(X, F, \theta, \operatorname{Grad} \theta). \end{gathered} \tag{2.17}$$

By (2.2), (2.17) we see, as is well known, that equality holds in (2.3) and that (2.4) reduces to (2.5).

## 3. The Dissipation Integral

In this section we discuss the positivity of the dissipation integral

$$I(\theta) = \int_{\Omega} \operatorname{Grad}\left(\frac{\phi}{\theta}\right) \cdot \hat{q}_R(X, \theta, \operatorname{Grad} \theta) \, dX \tag{3.1}$$

given by (2.13) when $q_R = \hat{q}_R(X, \theta, \operatorname{Grad} \theta)$. In (3.1) the admissible functions $\theta > 0$ satisfy the boundary conditions (2.8). We choose $\phi$ to be a solution of the stationary heat equation

$$\operatorname{Div} \hat{q}_R(X, \phi, \operatorname{Grad} \phi) = 0, \qquad X \in \Omega \tag{3.2}$$

subject to the same boundary conditions

$$\begin{gathered} \phi = \theta_0(X), \qquad X \in \partial\Omega_2, \\ \hat{q}_R(X, \phi, \operatorname{Grad} \phi) \cdot N = 0, \qquad x \in \partial\Omega \setminus \partial\Omega_2. \end{gathered} \tag{3.3}$$

In the examples treated below (3.2) is elliptic and $\phi$ unique. Proceeding formally for a moment, we observe that the Euler-Lagrange equation for $I$ can be written

$$\frac{\partial}{\partial X^\alpha}\left(\operatorname{Grad}\left(\frac{\phi}{\theta}\right)\cdot\frac{\partial \hat{q}_R}{\partial \theta_{,\alpha}} - \frac{\phi}{\theta^2}\hat{q}_R^\alpha\right) = \operatorname{Grad}\left(\frac{\phi}{\theta}\right)\cdot\frac{\partial \hat{q}_R}{\partial \theta} + \hat{q}_R\cdot\left(\frac{2\phi}{\theta^3}\operatorname{Grad}\theta - \frac{1}{\theta^2}\operatorname{Grad}\phi\right). \tag{3.4}$$

It is easily seen that $\theta = \phi$ is a solution of (3.4), and since $I(\phi) = 0$ we are faced with a classical question in the calculus of variations, to decide if the given solution $\phi$ is a global minimizer of $I$. The problem is not straightforward since $\phi$ is only known implicitly and the integrand may be negative.

For the remainder of this section we make the technical assumptions that $\Omega$ has a sufficiently regular boundary (it is enough that $\Omega$ is strongly Lipschitz in the sense of MORREY [1966 Section 3.4]) and that $\partial\Omega_2 \subset \partial\Omega$ is closed with positive $(n-1)$-dimensional measure. We suppose further that $\theta_0 : \partial\Omega_2 \to \mathbb{R}$ is sufficiently regular, specifically that $\theta_0$ is the boundary value on $\partial\Omega_2$ in the sense of trace of some function $\tilde{\theta} \in H^1(\Omega)$, and that there are constants $m$, $M$ such that

$$0 < m \leqq \theta_0(X) \leqq M < \infty \quad \text{for a.e. } X \in \partial\Omega_2. \tag{3.5}$$

We define a set $\mathscr{A}$ of admissible functions by

$$\mathscr{A} = \left\{\theta \in H^1(\Omega) \cap L^\infty(\Omega) : \operatorname*{ess\,inf}_{X\in\Omega} \theta(X) > 0,\ \theta|_{\partial\Omega_2} = \theta_0 \quad \text{in the sense of trace}\right\}.$$

We consider first the case

$$\hat{q}_R = -k(\theta)\operatorname{Grad}\theta, \tag{3.6}$$

where the thermal conductivity $k(\theta)$ is real-valued, continuous and strictly positive for all $\theta > 0$. By (3.1), (3.6)

$$I(\theta) = -\int_\Omega k(\theta)\operatorname{Grad}\left(\frac{\phi}{\theta}\right)\cdot\operatorname{Grad}\theta\, dX. \tag{3.7}$$

Writing $\varkappa(\theta) = \int_1^\theta k(s)\,ds$, $g(X) = \varkappa(\theta(X))$, we see that (3.2), (3.3) become

$$\Delta g = 0 \quad \text{in } \Omega, \qquad g|_{\partial\Omega_2} = \varkappa(\theta_0), \quad \left.\frac{\partial g}{\partial n}\right|_{\partial\Omega\setminus\partial\Omega_2} = 0. \tag{3.8}$$

It is easily checked that $\varkappa(\theta_0)$ is the boundary value of an $H^1(\Omega)$ function (for example of $\varkappa(\tilde{\psi})$, where $\tilde{\psi} = \max\{m, \min\{M, \tilde{\theta}\}\}$. It follows by standard theory that (3.8) has a unique weak solution $g$, i.e. $g \in H^1(\Omega)$, $g|_{\partial\Omega_2} = \varkappa(\theta_0)$, and

$$\int_\Omega \operatorname{Grad} g \cdot \operatorname{Grad} v\, dX = 0 \tag{3.9}$$

for all $v \in H^1(\Omega)$ with $v|_{\partial\Omega_2} = 0$. Defining $\phi = \varkappa^{-1}(g)$ we have that

$$\int_\Omega k(\phi) \operatorname{Grad} \phi \cdot \operatorname{Grad} v \, dX = 0 \tag{3.10}$$

for all $v \in H^1(\Omega)$ with $v|_{\partial\Omega_2} = 0$. By the maximum principle (for an appropriate version see CHICCO [1970])

$$m \leqq \phi(X) \leqq M \qquad \text{a.e. } X \in \Omega. \tag{3.11}$$

Also $\phi \in C^1(\Omega)$.

**Theorem 3.1.** *Let* $\log k(\theta)$ *be a concave function of* $\log \theta$. *Then* $I(\theta) \geqq 0$ *for all* $\theta \in \mathscr{A}$.

To prove the theorem we need some elementary lemmas.

**Lemma 3.2.** *Let $Q$ be an open interval (finite, semi-infinite, or infinite) of* $\mathbb{R}$. *Let* $h: Q \to (0, \infty)$. *Define* $f: Q \times \mathbb{R}^n \to \mathbb{R}$ *by*

$$f(w, y) = \frac{|y|^2}{h(w)}.$$

*Then $f$ is convex if and only if $h$ is concave.*

**Proof.** Let $t \in [0, 1]$, $w, \overline{w} \in Q$, $y, \overline{y} \in \mathbb{R}^n$. Then

$$\begin{aligned}
\delta f &\overset{\text{def}}{=} f(tw + (1-t)\,\overline{w}, ty + (1-t)\,\overline{y}) - tf(w, y) - (1-t)\,f(\overline{w}, \overline{y}) \\
&= \frac{1}{h(tw + (1-t)\,\overline{w})} \Bigg[ [th(w) + (1-t)\,h(\overline{w}) - h(tw + (1-t)\,\overline{w})] \\
&\quad \times \left( \frac{t\,|y|^2}{h(w)} + \frac{(1-t)\,|\overline{y}|^2}{h(\overline{w})} \right) - \frac{t(1-t)}{h(w)\,h(\overline{w})} |h(\overline{w})\,y - h(w)\,\overline{y}|^2 \Bigg].
\end{aligned}$$

If $h$ is concave then clearly $\delta f \leqq 0$, hence $f$ convex. If $f$ is convex the concavity of $h$ follows from $\delta f \leqq 0$ on choosing $\overline{y} = \dfrac{h(\overline{w})}{h(w)} y$. $\square$

We introduce the change of variable

$$w = \int_1^\theta \frac{k(s)}{s} \, ds \tag{3.12}$$

Let $\theta = \theta(w)$ denote the inverse function; thus $\theta(\cdot): Q \to \mathbb{R}$, where

$$Q = \left( -\int_0^1 \frac{k(s)}{s} \, ds, \quad \int_1^\infty \frac{k(s)}{s} \, ds \right).$$

**Lemma 3.3.** $k(\theta(\cdot))$ is concave on $Q$ if and only if $\log k(\theta)$ is a concave function of $\log \theta$.

**Proof.** Suppose $\log k(\theta)$ is concave in $\log \theta$. Then $\log k(\theta)$ is locally Lipschitz in $\log \theta$ on $\mathbb{R}$, and hence $k(\theta)$ is locally Lipschitz in $\theta$ on $(0, \infty)$. In particular $k$ is differentiable a.e. on $(0, \infty)$ with locally bounded derivative. By the chain rule the locally Lipschitz function $k(\theta(\cdot))$ has derivative

$$\frac{dk(\theta(w))}{dw} = \frac{d \log k(\theta(w))}{d \log \theta(w)} \quad \text{a.e. } w \in Q. \tag{3.13}$$

Since $d \log k(\theta)/d \log \theta$ is a.e. nonincreasing in $\log \theta$, $dk(\theta(w))/dw$ is a.e. nonincreasing in $w$. Hence $k(\theta(\cdot))$ is concave.

The converse is proved similarly. □

*Remark.* By making the identification $h(t) = k(e^t)$, $t = \log \theta$ and using a similar proof one can show that a necessary and sufficient condition for a function $h: \mathbb{R} \to (0, \infty)$ to be such that $\log h(\cdot)$ is convex (respectively concave) is that $h$ be locally integrable and for each $s \in \mathbb{R}$ there exists $\lambda(s) \in \mathbb{R}$ with

$$h(t) \geqq h(s) + \lambda(s) \int_s^t h(\tau)\, d\tau \quad \text{for all } t \in \mathbb{R},$$

(respectively $\leqq$).

**Proof of Theorem 3.1.** Let $\theta \in \mathscr{A}$. Then $w = w(X)$ defined by (3.12) belongs to $H^1(\Omega) \cap L^\infty(\Omega)$ with $\operatorname{Grad} w(X) = (k(\theta(X))/\theta(X)) \operatorname{Grad} \theta(X)$ a.e. $X \in \Omega$. Thus $I(\theta) = J(w)$, where

$$J(w) = \int_\Omega \hat{f}(X, w(X), \operatorname{Grad} w(X))\, dX, \tag{3.14}$$

and

$$\hat{f}(X, w, y) \overset{\text{def}}{=} \frac{\phi(X)}{k(\theta(w))} |y|^2 - y \cdot \operatorname{Grad} \phi(X). \tag{3.15}$$

It follows from Lemmas 3.2, 3.3 that $\hat{f}(X, \cdot, \cdot)$ is convex on $I \times \mathbb{R}^n$. Define

$$\overline{w}(X) = \int_1^{\phi(X)} \frac{k(s)}{s}\, ds.$$

Note that $\hat{f}(X, \cdot, \cdot)$ is differentiable at $(w, y)$ unless $y \neq 0$ and $\theta(w)$ belongs to the set $S$ of points where $k(\cdot)$ is not differentiable. Since $\log k(\theta)$ is concave in $\log \theta$, it follows easily that $S$ is countable. If $\phi(X) = s \in S$ on a set $M$ of positive measure then (Morrey [1966 p. 69]) $\operatorname{Grad} \phi(X) = 0$ a.e. $X \in M$. Thus $\hat{f}(X, \cdot, \cdot)$ is differentiable at $(\overline{w}(X), \operatorname{Grad} \overline{w}(X))$ for a.e. $X \in \Omega$, and by the convexity we have

$$(\hat{f}X, w(X), \operatorname{Grad} w(X)) \geqq \hat{f}(X, \overline{w}(X), \operatorname{Grad} \overline{w}(X)) + r_w(X), \quad \text{a.e. } X \in \Omega, \tag{3.16}$$

where

$$r_w(X) = \frac{\partial \hat{f}}{\partial w}(X, \overline{w}(X), \operatorname{Grad} \overline{w}(X))\,(w(X) - \overline{w}(X))$$

$$+ \frac{\partial \hat{f}}{\partial y}(X, \overline{w}(X), \operatorname{Grad} \overline{w}(X)) \cdot (\operatorname{Grad} w(X) - \operatorname{Grad} \overline{w}(X))$$

$$= -\frac{k'(\phi)}{k(\phi)}|\operatorname{Grad} \phi|^2 \int_\phi^\theta \frac{k(s)}{s}\,ds + \operatorname{Grad} \phi \cdot \operatorname{Grad}\left(\int_\phi^\theta \frac{k(s)}{s}\,ds\right),$$

since $\theta(\overline{w}(X)) = \phi(X)$. Setting $u = \int_\phi^\theta \frac{k(s)}{s}\,ds$, $cv = \frac{u}{k(\phi)}$ and noting that since $k(\cdot)$ is Lipschitz, $v \in H^1(\Omega)$ (*cf.* MARCUS & MIZEL [1972]) with $v|_{\partial\Omega_2} = 0$, we deduce from (3.10) that

$$\int_\Omega r_w(X)\,dX = 0. \tag{3.17}$$

Integrating (3.16) we thus have $I(\theta) \geqq I(\phi) = 0$ as required. □

*Remark*. The proof in fact shows that $I(\theta) \geqq 0$ for all $\theta \in \mathcal{A}_1$, where $\mathcal{A}_1 = \{\theta \in W^{1,1}(\Omega) : \theta > 0$ a.e., $w = \int_1^\theta \frac{k(s)}{s}\,ds \in H^1(\Omega)$ and $\theta|_{\partial\Omega_2} = \theta_0$ in the sense of trace$\}$.

The condition that $\log k$ be concave in $\log \theta$ is satisfied, for example, by the functions

$$k(\theta) = \mu\theta^\alpha, \quad \mu > 0, \quad \alpha \in \mathbb{R},$$

$$k(\theta) = \mu\,(\log \theta)^\alpha, \quad \mu > 0, \quad \alpha > 0,$$

the first example (for applications see KATH & COHEN [1982], LARSEN & POMRANING [1980], ZELDOVICH & RAIZER [1969]) being critical in that $\log k$ is affine in $\log \theta$. Clearly products of $k$'s satisfying the condition also satisfy it. If $k$ is $C^1$ on $(0, \infty)$ the condition takes the form that $\theta k'(\theta)/k(\theta)$ be nonincreasing in $\theta$.

To investigate how close the condition is to being necessary for $I$ to be nonnegative on $\mathcal{A}$ we compute the second variation. Suppose $k$ is $C^2$ on $(0, \infty)$. Let $u \in W^{1,\infty}(\Omega)$ with $u|_{\partial\Omega_2} = 0$. Then

$$\delta^2 I(\phi)\,(\phi u, \phi u) \overset{\text{def}}{=} \frac{d^2}{d\varepsilon^2} I(\phi(1 + \varepsilon u))|_{\varepsilon=0}$$

$$= 2 \int_\Omega \operatorname{Grad} u \cdot \operatorname{Grad}(\phi k(\phi)\,u)\,dX$$

$$= \int_\Omega [2a\,|\operatorname{Grad} u|^2 - \Delta a \cdot u^2]\,dX, \tag{3.18}$$

where $a \overset{\text{def}}{=} \phi k(\phi)$ and where we have used (3.9). Note that

$$\begin{aligned} \Delta a &= \left[\left(\frac{\phi k'(\phi)}{k(\phi)} + 1\right) k(\phi)\, \phi_{,\alpha}\right]_{,\alpha} \\ &= \left(\frac{\phi k'}{k}\right)' k(\phi)\, |\mathrm{Grad}\, \phi|^2 . \end{aligned} \tag{3.19}$$

In particular, if $\log k(\theta)$ is concave in $\log \theta$ then $\Delta a \leqq 0$ and $\delta^2 I(\phi) \geqq 0$, consistent with Theorem 3.1. The Jacobi equation, that is the Euler-Lagrange equation for (3.18), is

$$\mathrm{Div}\,(2a\, \mathrm{Grad}\, u) = -\Delta a \cdot u . \tag{3.20}$$

We now let $n = 1$, $\Omega = (0, 1)$, $\partial\Omega_2 = \partial\Omega$, so that

$$k(\phi)\, \phi_X = c, \qquad X \in [0, 1], \tag{3.21}$$

where we assume $c = \varkappa(\theta_0(1)) - \varkappa(\theta_0(0))$ is nonzero. We seek a function $u(X) = z(\tau)$, $\tau = \log \phi$, making (3.18) negative. Note that for such a function, by (3.19), (3.21),

$$\delta^2 I(\phi)\, (\phi u, \phi u) = J(z),$$

where

$$J(z) = 2c \int_{\log\theta_0(0)}^{\log\theta_0(1)} [z_\tau^2 - \tfrac{1}{2} p(\tau)\, z^2]\, d\tau \tag{3.22}$$

and

$$p(\tau) = \frac{d^2}{d\tau^2} \log k(e^\tau) . \tag{3.23}$$

Also, (3.20) becomes

$$z_{\tau\tau} + \tfrac{1}{2} p(\tau)\, z = 0 . \tag{3.24}$$

Suppose that we can find a solution $\bar{z} \neq 0$ of (3.24) on an interval $[\alpha, \beta]$ with $\bar{z}(\alpha) = \bar{z}(\beta) = 0$. Let

$$\theta_0(0) < e^\alpha, \qquad \theta_0(1) > e^\beta . \tag{3.25}$$

Employing classical reasoning (*cf.* Bolza [1904]) we set

$$z_1(\tau) = \begin{cases} \bar{z}(\tau), & \tau \in [\alpha, \beta], \\ 0 & \text{otherwise}, \end{cases}$$

and note that by (3.24)

$$\begin{aligned} J(z_1) &= 2c \int_\alpha^\beta [\bar{z}_\tau^2 - \tfrac{1}{2} p(\tau)\, \bar{z}^2]\, d\tau \\ &= 2c \bar{z} \bar{z}_\tau \big|_\alpha^\beta = 0 . \end{aligned}$$

But $z_1$ cannot be a minimizer of $J$ among $W^{1,\infty}$ functions vanishing at $\log \theta_0(0)$, $\log \theta_0(1)$ since by standard arguments $z_1$ would then be a smooth solution of (3.24) on $[\log \theta_0(0), \log \theta_0(1)]$. In particular we would have $\bar{z}_\tau(\alpha) = 0$, and hence

$\bar{z} \equiv 0$ by uniqueness of solutions to the initial-value problem for (3.24), a contradiction. Thus $J(z)$ takes negative values and so

$$\inf_{\theta \in \mathcal{A}} I(\theta) < I(\phi) = 0. \tag{3.26}$$

We give two ways of constructing an appropriate solution $\bar{z}$. First, suppose $\log k(\theta)$ convex in $\log \theta$ but not affine, equivalently $\theta k'(\theta)/k(\theta)$ nondecreasing in $\theta$ but not constant. Then $p(\tau) \geqq 0$ and $p(\tau_0) > 0$ for some $\tau_0$. Let $\bar{z}$ be the solution of (3.24) with initial data $\bar{z}(\tau_0) = 1$, $\bar{z}_\tau(\tau_0) = 0$. Since $\bar{z}_{\tau\tau} \leqq 0$ where $\bar{z} \geqq 0$ and since $\bar{z}_{\tau\tau}(\tau_0) < 0$ it follows that $\bar{z}$ has two roots $\alpha, \beta$ with $\alpha < \tau_0 < \beta$. Second, suppose that $p(\tau) \geqq 2\varepsilon^2 > 0$ on an interval of length greater than $\pi/\varepsilon$. If $\tau_0$ is the mid-point of the interval and $\bar{z}$ is the solution of (3.24) with $\bar{z}(\tau_0) = 1$, $\bar{z}_\tau(\tau_0) = 0$ then $\bar{z}$ has at least two zeros in $[\tau_0 - \pi/2\varepsilon, \tau_0 + \pi/2\varepsilon]$; this follows from Sturm's first comparison theorem (HARTMAN [1964] p. 334) using the comparison solution $w = \cos \varepsilon(\tau - \tau_0)$ of $w_{\tau\tau} + \varepsilon^2 w = 0$.

If $n > 1$ and either of the above two conditions on $k$ holds then by choosing $\Omega = (0, 1) \times \Omega'$, where $\Omega'$ is a bounded open subset of $\mathbb{R}^{n-1}$, and $\partial\Omega_2 = \{0,1\} \times \Omega'$ we can find a function $\theta = \theta(X^1)$ in $\mathcal{A}$ satisfying (3.26). We have thus proved

**Theorem 3.4.** *Let* $n \geqq 1$. *Suppose* $k$ *is* $C^2$ *on* $(0, \infty)$ *and satisfies either*

(i) $\log k(\theta)$ *is convex in* $\log \theta$ *but not affine, or*

(ii) $\dfrac{d^2 \log k(\theta)}{d(\log \theta)^2} \geqq 2\varepsilon^2 > 0$ *on an interval of length greater than* $\dfrac{\pi}{\varepsilon}$.

*Then we can find* $\Omega$, $\partial\Omega_2$, $\theta_0$ *such that*

$$\inf_{\theta \in \mathcal{A}} I(\theta) < 0.$$

As an example satisfying both (i) and (ii) one can choose $k(\theta) = e^\theta$. Note that even when (i) or (ii) hold the second variation for some boundary conditions may be positive; if so the field theory of the calculus of variations (see MORREY [1966 p. 12]) can be applied to conclude that $\phi$ is a strong local minimizer of $I$, so that $E$ is a Lyapunov function for solutions with $\sup_{t \geqq 0} \|\theta(\cdot, t) - \phi(\cdot)\|_{L^\infty(\Omega)}$ sufficiently small. This information might be useful for stability studies.

We consider next the anisotropic linear case

$$\hat{q}_R = -K(X) \operatorname{Grad} \theta, \tag{3.27}$$

where we assume that the matrix $K$ is bounded and measurable in $\Omega$ and satisfies

$$K^{\alpha\beta}(X)\,\xi_\alpha \xi_\beta \geqq k_0 |\xi|^2, \qquad \xi \in \mathbb{R}^n, \qquad \text{a.e. } X \in \Omega, \tag{3.28}$$

for some constant $k_0 > 0$. We do not need to assume $K$ is symmetric (the Onsager relations, for a critique see TRUESDELL [1984 Lecture 7]). By definition, a weak solution of (3.2), (3.3) is a function $\phi \in H^1(\Omega)$ satisfying $\phi|_{\partial\Omega_2} = \theta_0$ and

$$\int_\Omega K^{\alpha\beta} \phi_{,\beta} v_{,\alpha} \, dX = 0 \tag{3.29}$$

for all $v \in H^1(\Omega)$ with $v|_{\partial\Omega_2} = 0$. It follows from CHICCO [1970] (see also TRUDINGER [1977]) that there exists a unique such weak solution $\phi$ and that

$$m \leqq \phi(X) \leqq M \quad \text{a.e. } X \in \Omega. \tag{3.30}$$

Defining $\phi$ in this way, we have from (3.1), (3.27) that

$$I(\theta) = \int_\Omega -\left(\frac{\phi}{\theta}\right)_{,\alpha} K^{\alpha\beta}\theta_{,\beta}\, dX. \tag{3.31}$$

**Theorem 3.5.** $I(\theta) \geqq 0$ *for all* $\theta \in \mathscr{A}$.

**Proof.** Let $\theta \in \mathscr{A}$ and define $w = \log\theta - \log\phi$. Then $w \in H^1(\Omega) \cap L^\infty(\Omega)$ with $\operatorname{Grad} w = (1/\theta)\operatorname{Grad}\theta - (1/\phi)\operatorname{Grad}\phi$ a.e. in $\Omega$. Hence

$$\begin{aligned} I(\theta) &= \int_\Omega [\phi K^{\alpha\beta} w_{,\alpha} w_{,\beta} + K^{\alpha\beta}\phi_{,\beta} w_{,\alpha}]\, dX \\ &\geqq \int_\Omega K^{\alpha\beta}\phi_{,\beta} w_{,\alpha}\, dX = 0, \end{aligned}$$

where we have used (3.28) and (3.29). □

*Remark.* The proof in fact shows that $I(\theta) \geqq 0$ for all $\theta \in \mathscr{A}_2$, where $\mathscr{A}_2 = \{\theta \in H^1(\Omega) : \theta > 0 \text{ a.e., } \log\theta \in H^1(\Omega) \text{ and } \theta|_{\partial\Omega_2} = \theta_0 \text{ in the sense of trace}\}$.

Setting in particular $T_R = 0$, $U = \theta$ in (2.1), (2.2) we see that by Theorems 3.1, 3.5 and under the hypotheses of these theorems

$$\frac{d}{dt}\int_\Omega \varrho_R\,(\theta - \phi\log\theta)\, dX \leqq 0 \tag{3.32}$$

for sufficiently regular positive solutions $\theta$, satisfying $\theta|_{\partial\Omega_2} = \theta_0$, $\left.\frac{\partial\theta}{\partial n}\right|_{\partial\Omega\setminus\partial\Omega_2} = 0$, of the heat equations

$$\varrho_R \frac{\partial\theta}{\partial t} = \operatorname{Div}(k(\theta)\operatorname{Grad}\theta), \tag{3.33}$$

$$\varrho_R \frac{\partial\theta}{\partial t} = \operatorname{Div}(K(X)\operatorname{Grad}\theta), \tag{3.34}$$

respectively. Various Lyapunov functions similar to (3.32) have been used for systems of reaction-diffusion equations, (see ROTHE [1984]).

We end by noting that the hypothesis of *strict* positivity of $\theta_0$ in Theorems 3.1, 3.5 is essential. In fact, if $\hat{q}_R = -\operatorname{Grad}\theta$, $n = 1$, $\Omega = (0, 1)$, $\partial\Omega_2 = \partial\Omega$, $\theta_0(0) = 0$, $\theta_0(1) = 1$, then $\phi(X) = X$ but for $\theta = X^\alpha$, $0 < \alpha < 1$, we have

$$I(\theta) = \int_0^1 -\left(\frac{X}{X^\alpha}\right)_X \alpha X^{\alpha-1}\, dX = -\infty.$$

*Acknowledgement.* The research of J.M.B. was supported by a U.K. Science and Engineering Research Council Senior Fellowship and by visits to the Mathematics Research Center, University of Wisconsin, and the Institute for Mathematics and its Applications, University of Minnesota.

## References

J. M. BALL [1977] Convexity conditions and existence theorems in nonlinear elasticity, Arch. Rational Mech. Anal. **63**, 337–403.

J. M. BALL [1984] Material instabilities and the calculus of variations, in "Phase Transformations and Material Instabilities in Solids", ed. M. GURTIN, Academic Press.

J. M. BALL & G. KNOWLES [1985] in preparation.

J. M. BALL & F. MURAT [1984] $W^{1,p}$-quasiconvexity and variational problems for multiple integrals, J. Functional Analysis **58**, 225–253.

O. BOLZA [1904] "Lectures on the Calculus of Variations", Reprinted by Chelsea, N.Y., 1973.

M. CHICCO [1970] Principio di massimo per soluzioni di problemi al contorno misti per equazioni ellittiche di tipo variazionale, Boll. Unione Mat. Ital. (4) **3**, 384–394.

B. D. COLEMAN & E. H. DILL [1973] On thermodynamics and the stability of motion of materials with memory, Arch. Rational Mech. Anal. **51**, 1–53.

C. M. DAFERMOS [1983] Hyperbolic systems of conservation laws, in "Systems of Nonlinear Partial Differential Equations" ed. J. M. BALL, D. REIDEL, 25–70.

P. DUHEM [1911] "Traité d'Enérgetique ou de Thermodynamique Générale", Gauthier-Villars, Paris.

J. L. ERICKSEN [1966] Thermoelastic stability, Proc. 5th National Cong. Appl. Mech. 187–193.

W. L. KATH & D. S. COHEN [1982] Waiting-time behavior in a nonlinear diffusion equation, Studies in Applied Math. **67**, 79–105.

P. HARTMAN [1964] "Ordinary Differential Equations", John Wiley & Sons, New York, reprinted by Birkhauser, Boston, 1982.

E. W. LARSEN & G. C. POMRANING [1980] Asymptotic analysis of nonlinear Marshak waves, SIAM J. Appl. Math. **39**, 201–212.

C-S. MAN [1985] Dynamic admissible states, negative absolute temperature, and the entropy maximum principle, preprint.

M. MARCUS & V. J. MIZEL [1972] Absolute continuity on tracks and mappings of Sobolev spaces, Arch. Rational Mech. Anal. **45**, 294–320.

C. B. MORREY [1966] "Multiple Integrals in the Calculus of Variations", Springer.

F. ROTHE [1984] "Global Solutions of Reaction-Diffusion Systems", Springer Lecture Notes in Mathematics Vol. **1072**.

N. TRUDINGER [1977] Maximum principles for linear, non-uniformly elliptic operators with measurable coefficients, Math. Zeitschrift **156**, 291–301.

C. TRUESDELL [1984] "Rational Thermodynamics", 2nd edition, Springer-Verlag, Berlin, Heidelberg, New York, Tokyo.

Y. B. ZELDOVICH & Y. P. RAIZER [1969], "Physics of Shock Waves & High Temperature Hydrodynamic Phenomena", vol. II, Academic Press, New York.

Department of Mathematics
Heriot-Watt University
Edinburgh

&

Department of Electrical Engineering,
Imperial College of Sciene and Technology,
London

*(Received August 12, 1985)*

# *A Semilinear Elliptic Problem which is not Selfadjoint*

PATRIZIA PUCCI

*Dedicated to Professor Serrin, a great Master, on the occasion of his sixtieth birthday, in gratitude.*

## § 1. Introduction

CESARI [7] & I in a preceding paper established conditions sufficient that a semilinear elliptic boundary-value problem, possibly not selfadjoint shall have a solution.

In this paper we study a particular nonselfadjoint semilinear problem, whose not easy development could not find a place in [7]. For the convenience of the reader I make the present paper essentially selfcontained, so that no specific reference to [7] is actually needed.

The results in [7] were based on a reduction of the problem to a new alternative form, and on an application of the Schauder Fixed Point Theorem. The same process was also studied in [3] in connection with nonselfadjoint boundary-value problems for nonlinear ordinary differential equations, and some actually nonselfadjoint examples were presented there. Some of the ideas in [3] and [7] could be also traced back in [5] and [6]. In [2] CESARI presented the main argument in an abstract form and derived Landesman-Lazer type statements for nonselfadjoint problems.

I thank Professor LAMBERTO CESARI for useful discussions on the subject of this note.

## § 2. The Problem

We consider here the elliptic problem in the interval $G = [0 \leqq x \leqq a, 0 \leqq y \leqq a/2]$

$$Eu \equiv u_{xx} + u_{yy} + 2\lambda^2 u = f(x, y) + g(u), \qquad (x, y) \in G, u \in \mathbb{R}, \lambda = 2\pi/a, \tag{1}$$

$$\begin{aligned} u(0, y) &= -l[u_x(0, y) - u_x(a, y)], && 0 \leqq y \leqq a/2, \\ u(a, y) &= +l[u_x(0, y) - u_x(a, y)], && 0 \leqq y \leqq a/2, \\ u(x, 0) &= u(x, a/2) = 0, && 0 \leqq x \leqq a, \end{aligned} \tag{2}$$

where $l > 0$ is a given constant. For $l = 0$ the problem reduces to the Dirichlet problem with $u = 0$ on $\partial G$. By defining the operator $N$ by $Nu = f + g(u)$, and including the boundary conditions (2) in the operator $E$, we give the problem the abstract form $Eu = Nu$.

The underlying linear problem is

$$Eu = u_{xx} + u_{yy} + 2\lambda^2 u = 0, \quad (x, y) \in G, \tag{3}$$

with the same conditions (2).

The dual problem is

$$E^*v = v_{xx} + v_{yy} + 2\lambda^2 v = 0, \qquad (x, y) \in G, \tag{4}$$

$$\begin{aligned} &v(0, y) = v(a, y) = -l[v_x(0, y) + v_x(a, x_y)], && 0 \leqq y \leqq a/2, \\ &v(x, 0) = v(x, a/2) = 0, && 0 \leqq x \leqq a. \end{aligned} \tag{5}$$

Indeed, by integration by parts, under the boundary conditions (2) and (5), we obtain

$$\langle Eu, v \rangle = \langle u, E^*v \rangle .$$

A nonzero solution to problem (3), (2) is

$$u(x, y) = \sin \lambda x \sin \lambda y, \quad (x, y) \in G = [0, a] \times [0, a/2], \tag{6}$$

and a nonzero solution to problem (4), (5) is

$$v(x, y) = (\sin \lambda x - 2l\lambda \cos \lambda x) \sin \lambda y, \quad (x, y) \in G, \tag{7}$$

as one can easily verify. One can also readily prove that any other solution of (3), (2) is proportional to $u$, and any other solution of (4), (5) is proportional to $v$. In other words,

$$p = \dim \operatorname{Ker} E = 1, \quad q = \dim \operatorname{Ker} E^* = 1,$$

and by normalization in $G = [0, a] \times [0, a/2]$ we obtain

$$\phi(x, y) = 2^{\frac{3}{2}} a^{-1} \sin \lambda x \sin \lambda y,$$

$$\omega(x, y) = 2^{\frac{3}{2}} a^{-1} (1 + 4l^2\lambda^2)^{-\frac{1}{2}} (\sin \lambda x - 2l\lambda \cos \lambda x) \sin \lambda y,$$

$$\operatorname{Ker} E = \{c\phi(x, y)\}, \quad \operatorname{Ker} E^* = \{d\omega(x, y)\}.$$

## § 3. The Operator $H$

Let $X = Y = L_2(G)$. We shall think of the functions $u \in X = Y = L_2(G)$ as extended to $\mathbb{R}^2$ by oddness with respect to $y$ and double periodicity in $x$ and $y$ of period $a$. Let $X_0 = \operatorname{Ker} E$, $X_1 = (\operatorname{Ker} E)^\perp$, $Y_0 = \operatorname{Ker} E^*$, $Y_1 = (\operatorname{Ker} E^*)^\perp$, $X = X_0 + X_1$, $Y = Y_0 + Y_1$. Let us note that, if $D(E)$ and $R(E)$ denote the domain and the range of $E$, then $D(E) \subset X$, $R(E) = Y_1$, and $E: D(E) \cap X_1 \to Y_1$ is one-to-one and surjective. Hence, the inverse operator

$$H: Y_1 \to D(E) \cap X_1$$

is well defined and surjective. The operator $H$, or $u = Hv$, can be characterized by the following relations:

$$u_{xx} + u_{yy} + 2\lambda^2 u = v, \qquad (x, y) \in G, \tag{8}$$

$$\begin{aligned} u(0, y) &= -l[u_x(0, y) - u_x(a, y)], && 0 \leqq y \leqq a/2, \\ u(a, y) &= +l[u_x(0, y) - u_x(a, y)], && 0 \leqq y \leqq a/2, \\ u(x, 0) &= u(x, a/2) = 0, && 0 \leqq x \leqq a, \end{aligned} \tag{4}$$

$$\iint_G u(x, y)\, \phi(x, y)\, dx\, dy = 0, \qquad \text{or } \langle u, \phi \rangle = 0, \tag{9}$$

$$\iint_G v(x, y)\, \omega(x, y)\, dx\, dy = 0, \qquad \text{or } \langle v, \omega \rangle = 0. \tag{10}$$

By Fredholm's Theorem the relation (10) is a necessary and sufficient condition for the existence of a solution $u$ satisfying (8), (4), (9). Note that if a function $u: \mathbb{R}^2 \to \mathbb{R}$ of class $C^1$ which is periodic of period $a$ in $x$ and $y$, is odd in $y$, and satisfies $u(0, y) = u(a, y) = 0$, then $u_x(0, y) = u_x(a, y) = 0$, $u(0, y) = u(a, y) = 0$, and the relations (4) hold.

Note that, for $v(x, y) = \sin \lambda y$, then (10) holds since $\sin \lambda x$ and $\cos \lambda x$ are orthogonal to 1 in $[0, a]$, and

$$u(x, y) = \lambda^{-2}(1 - \cos \lambda x) \sin \lambda y, \qquad (x, y) \in G,$$

satisfies (8), (4), (9), as we can easily verify.

## § 4. The Orthonormal System $\{\phi_{mn}, \psi_{mn}: m = 0, 1, \ldots, n = 1, 2, \ldots\}$

We shall consider the orthonormal system $\{\phi_{mn}, \psi_{mn}: m = 0, 1, \ldots, n = 1, 2, \ldots\}$ of functions $v$ in $G = [0, a] \times [0, a/2]$ as follows:

$$\begin{aligned} \phi_{01}(x, y) &= 2a^{-1} \sin \lambda y, \\ \phi_{m1}(x, y) &= 2^{\frac{3}{2}} a^{-1} \cos m\lambda x \sin \lambda y, \qquad m = 2, 3, \ldots, \\ \psi_{m1}(x, y) &= 2^{\frac{3}{2}} a^{-1} \sin m\lambda x \sin \lambda y, \qquad m = 2, 3, \ldots, \\ \phi_{11}(x, y) &= 2^{\frac{3}{2}} a^{-1}(1 + 4l^2\lambda^2)^{-\frac{1}{2}} (2l\lambda \sin \lambda x + \cos \lambda x) \sin \lambda y, \\ \psi_{11}(x, y) &= 2^{\frac{3}{2}} a^{-1}(1 + 4l^2\lambda^2)^{-\frac{1}{2}} (\sin \lambda x - 2l\lambda \cos \lambda x) \sin \lambda y = \omega(x, y), \\ \phi_{0n}(x, y) &= 2a^{-1} \sin n\lambda y, \qquad n = 2, 3, \ldots, \\ \phi_{mn}(x, y) &= 2^{\frac{3}{2}} a^{-1} \cos m\lambda x \sin n\lambda y, \qquad m = 1, 2, \ldots, n = 2, 3, \ldots, \\ \psi_{mn}(x, y) &= 2^{\frac{3}{2}} a^{-1} \sin m\lambda x \sin n\lambda y, \qquad m = 1, 2, \ldots, n = 2, 3, \ldots \end{aligned} \tag{11}$$

Here $\psi_{11} = \omega$, and $\phi_{11}$ is the unique linear combination of $\sin \lambda x$ and $\cos \lambda x$ that is normal and orthogonal to $\omega$. Thus (11) is the usual trigonometrical system in $G$,

$$2a^{-1} \sin n\lambda y, \quad 2^{\frac{3}{2}} a^{-1} \cos m\lambda x \sin n\lambda y, \quad 2^{\frac{3}{2}} a^{-1} \sin m\lambda x \sin n\lambda y,$$
$$m, n = 1, 2, \ldots, \quad 0 \leqq x \leqq a, \quad 0 \leqq y \leqq a/2,$$

where the pair $2^{\frac{3}{2}} a^{-1} \cos \lambda x \sin \lambda y$, $2^{\frac{3}{2}} a^{-1} \sin \lambda x \sin \lambda y$ has been replaced by the equivalent pair $\phi_{11}, \psi_{11}$. Any function $v = v(x, y)$ in $G$, $v \in L_2(G)$, extended to $\mathbb{R}^2$ by oddness in $y$ and $a$-periodicity in $x$ and $y$, has a Fourier development in $L_2(G)$ given by

$$v(x, y) = \sum_{n=1}^{\infty} 2a^{-1} \alpha_{01} \sin n\lambda y + \sum_{m,n=1}^{\infty} 2^{\frac{3}{2}} a^{-1}(\alpha_{mn} \cos m\lambda x + \beta_{mn} \sin m\lambda x) \sin n\lambda y$$

and a corresponding development in the system (11) given by

$$v(x, y) = \sum_{m=0,n=1}^{\infty} [a_{mn}\phi_{mn}(x, y) + b_{mn}\psi_{mn}(x, y)] \tag{12}$$

with $a_{mn} = \alpha_{mn}$, $b_{mn} = \beta_{mn}$ for $m = 0, 1, \ldots, n = 1, 2, \ldots, (m, n) \neq (1, 1)$, and $\alpha_{11}^2 + \beta_{11}^2 = a_{11}^2 + b_{11}^2$. In fact $a_{11}$ and $b_{11}$ can be derived by the identity

$$a_{11}\phi_{11}(x, y) + b_{11}\psi_{11}(x, y) = 2^{\frac{3}{2}} a^{-1}(\alpha_{11} \cos \lambda x + \beta_{11} \sin \lambda x) \sin \lambda y, \qquad (x, y) \in G.$$

Indeed, by substitution and comparison we have

$$a_{11} = (1 + 4l^2\lambda^2)^{-\frac{1}{2}} (\alpha_{11} + 2l\lambda\beta_{11}),$$
$$b_{11} = (1 + 4l^2\lambda^2)^{-\frac{1}{2}} (-2l\lambda\alpha_{11} + \beta_{11}).$$

Obviously, $a_{11}^2 + b_{11}^2 = \alpha_{11}^2 + \beta_{11}^2$.

All the functions $v = \phi_{mn}$ and $v = \psi_{mn}$, $m = 0, 1, \ldots, n = 1, 2, \ldots$, but $\psi_{11} = \omega$, are orthogonal to $\omega$, that is, satisfy relation (10); hence there are corresponding functions, uniquely defined, say $u = \Phi_{mn} = H\phi_{mn}$, $m = 0, 1, \ldots$, $n = 1, 2, \ldots$, and $u = \Psi_{mn} = H\psi_{mn}$, $m = 0, 1, \ldots, n = 1, 2, \ldots, (m, n) \neq (1, 1)$, that satisfy relations (8), (4), (9). These functions are listed here, and I leave to the reader the easy verification:

$$\Phi_{01}(x, y) = 2a^{-1}\lambda^{-2} (1 - \cos \lambda x) \sin \lambda y,$$

$$\Phi_{m1}(x, y) = 2^{\frac{3}{2}} a^{-1}\lambda^{-2} (1 - m^2)^{-1} (\cos m\lambda x - \cos \lambda x) \sin \lambda y, \qquad m = 2, 3, \ldots,$$

$$\Psi_{m1}(x, y) = 2^{\frac{3}{2}} a^{-1}\lambda^{-2} (1 - m^2)^{-1} \sin m\lambda x \sin \lambda y, \qquad m = 2, 3, \ldots,$$

$$\begin{aligned} \Phi_{11}(x, y) &= 2^{\frac{3}{2}} a^{-1}\lambda^{-2} (1 + 4l^2\lambda^2)^{-\frac{1}{2}} [2^{-1}l\lambda^2 a \cos \lambda x \\ &\quad - (2^{-1}l\lambda + 2^{-2}\lambda a) \sin \lambda x - l\lambda^2 x \cos \lambda x + 2^{-1}\lambda x \sin \lambda x] \sin \lambda y \\ &= 2^{\frac{3}{2}} a^{-1}\lambda^{-1}(1 + 4l^2\lambda^2)^{-\frac{1}{2}} \left[l\lambda\left(\frac{a}{2} - x\right) \cos \lambda x \right. \\ &\quad \left. - \frac{\lambda}{2}\left(l + \frac{a}{2} - x\right) \sin \lambda x\right] \sin \lambda y, \end{aligned} \tag{13}$$

$$\Phi_{0n}(x, y) = 2a^{-1}\lambda^{-2}(1 - n^2)^{-1}\left[1 - \frac{1 - 2l\nu\lambda}{1 + \eta} e^{\nu\lambda x}\right.$$

$$\left. - \frac{\eta}{1 + \eta}(1 + 2l\nu\lambda)\, e^{-\nu\lambda x}\right] \sin n\lambda y, \qquad n = 2, 3, \ldots,$$

(13)

$$\Phi_{mn}(x, y) = 2^{\frac{3}{2}} a^{-1}\lambda^{-2}(2 - m^2 - n^2)^{-1}\left[\cos m\lambda x - \frac{1 - 2l\nu\lambda}{1 + \eta} e^{\nu\lambda x}\right.$$

$$\left. - \frac{\eta}{1 + \eta}(1 + 2l\nu\lambda)\, e^{-\nu\lambda x}\right] \sin n\lambda y,$$

$$m = 1, 2, \ldots, n = 2, 3, \ldots,$$

$$\Psi_{mn}(x, y) = 2^{\frac{3}{2}} a^{-1}\lambda^{-2}(2 - m^2 - n^2)^{-1} \sin m\lambda x \sin n\lambda y,$$

$$m = 1, 2, \ldots, n = 2, 3, \ldots,$$

where $\nu = \sqrt{n^2 - 2}$ and $\eta = e^{\nu\lambda a}$.

## § 5. The Operators $P$ and $Q$

We now introduce the orthogonal projection operators $P: X \to X$, $Q: Y \to Y$ with $PX = X_0 = \operatorname{Ker} E$, $QY = Y_0 = \operatorname{Ker} E^*$. Thus, for instance,

$$u(x, y) = Qv(x, y) = \left(\iint_G v(\xi, \eta)\, \omega(\xi, \eta)\, d\xi\, d\eta\right) \omega(x, y),$$

and analogously for $P$. Then, we have

(14) $$\|P\| = 1, \quad \|I - P\| = 1, \quad \|Q\| = 1, \quad \|I - Q\| = 1,$$

and the decompositions $X = X_0 + X_1$ and $Y = Y_0 + Y_1$.

## § 6. The Operators $\sigma$ and $S$

We first note that here we have $p = q = 1$, and

$$M = \iint_G \phi(x, y)\, \omega(x, y)\, dx\, dy$$

$$= \int_0^a \int_0^{a/2} 2^{\frac{3}{2}} a^{-1} \sin \lambda x \sin \lambda y\, 2^{\frac{3}{2}} a^{-1} (1 + 4l^2\lambda^2)^{-\frac{1}{2}}$$

$$\times (\sin \lambda x - 2l\lambda \cos \lambda x) \sin \lambda y\, dx\, dy = (1 + 4l^2\lambda^2)^{-\frac{1}{2}}.$$

We take $\sigma: X_0 \to Y$ to be the identity. Further we define $S: Y_0 \to X_0$, or $u = Sv$, for $v \in Y_0$, $v = d^+\omega$, $\|v\|_2 = |d^+|$, by taking $u = d\phi$ with $d = d^+/M$, so that

$$\|u\|_2 = \|d\phi\|_2 = |d| = |d^+/M| = (1 + 4l^2\lambda^2)^{\frac{1}{2}} |d^+| = (1 + 4l^2\lambda^2)^{\frac{1}{2}} \|v\|_2,$$

or

$$\|S\| = (1 + 4l^2\lambda^2)^{\frac{1}{2}} = c_5 \tag{15}$$

in the topology of $L_2(G)$ for both $u$ and $v$. Thus, if $u \in X_0$, or $u = c\phi$, then $\sigma u = u = c\phi$, $Q\sigma u = cM\omega$, and $SQ\sigma u = M^{-1}(cM\phi) = c\phi = u$, that is,

$$SQ\sigma: X_0 \to X_0$$

is the identity map on $X_0$.

## § 7. The Series Expression for $u = Hv$

For any $v \in Y$, that is, $v = v(x, y)$ odd in $y$, $a$-periodic in $x$ and $y$, $v \in L_2(G)$, we have the Fourier series development in $L_2(G)$

$$v(x, y) = \sum_{m=0,n=1}^{\infty} 2^{\frac{3}{2}} a^{-1}(\alpha_{mn} \cos m\lambda x + \beta_{mn} \sin m\lambda x) \sin n\lambda y, \qquad (x, y) \in G,$$

with

$$\|v\|_2^2 = \sum_{m=0,n=1}^{\infty} (\alpha_{mn}^2 + \beta_{mn}^2).$$

Let $\Sigma'$ denote any sum extended to all $m = 0, 1, \ldots, n = 1, 2, \ldots, (m, n) \neq (1, 1)$. Then, we have also

$$v = \sum{}'(a_{mn}\phi_{mn} + b_{mn}\psi_{mn}) + a_{11}\phi_{11} + b_{11}\psi_{11}, \tag{16}$$

with $a_{11}^2 + b_{11}^2 = \alpha_{11}^2 + \beta_{11}^2$, and $(I - Q)\, v$ has the same development with the term in $\psi_{11}$ missing. Finally,

$$u = H(I - Q)\, v = \sum{}'(a_{mn}\Phi_{mn} + b_{mn}\Psi_{mn}) + a_{11}\Phi_{11}. \tag{17}$$

Let us prove that $u \in L_2(G)$ and that $u_x$, $u_y$ also are of class $L_2$ in suitable subsets $G_s$ of $G$ invading $G$, say $G_s \uparrow G$. Let us take $v \in Y_1 = (I - Q)\, Y$, with development (16) without the term in $\psi_{11}$, and prove that $u = Hv$ is of class $L_2(G)$. First, for $n = 2, 3, \ldots$, we write

$$\begin{aligned} S_{1n}(x) &= -\frac{1}{1+\eta} e^{\nu\lambda x} - \frac{\eta}{1+\eta} e^{-\nu\lambda x} \\ &= -\frac{1}{1+e^{\nu\lambda a}} [e^{\nu\lambda x} + e^{\nu\lambda(a-x)}], \qquad \text{and } |S_{1n}(x)| \leq 1; \end{aligned}$$

$$\begin{aligned} S_{2n}(x) &= 2l\nu\lambda \left( \frac{e^{\nu\lambda x}}{1+\eta} - \frac{\eta}{1+\eta} e^{-\nu\lambda x} \right) \\ &= \frac{2l\nu\lambda}{1+e^{\nu\lambda a}} [e^{\nu\lambda x} - e^{\nu\lambda(a-x)}], \qquad \text{and } |S_{2n}(x)| \leq 2l\nu\lambda; \end{aligned}$$

where $\nu = \sqrt{n^2 - 2}$. Thus

$$\Phi_{0n}(x, y) = 2a^{-1}\lambda^{-2}(1 - n^2)^{-1}[1 + S_{1n}(x) + S_{2n}(x)] \sin n\lambda y, \qquad n = 2, 3, \ldots,$$

$$\Phi_{mn}(x, y) = 2^{\frac{3}{2}} a^{-1}\lambda^{-2}(2 - m^2 - n^2)^{-1}[\cos m\lambda x + S_{1n}(x) + S_{2n}(x)] \sin n\lambda y,$$
$$m = 1, 2, \ldots, n = 2, 3, \ldots$$

For the sake of computation, we decompose the functions $\Phi$, $\Psi$ as follows:

$$\Phi_{01} = \Phi_{01}^{(0)} + \Phi_{01}^{(1)},$$
$$\Phi_{m1} = \Phi_{m1}^{(0)} + \Phi_{m1}^{(1)}, \qquad m = 2, 3, \ldots,$$
$$\Phi_{0n} = \Phi_{0n}^{(0)} + \Phi_{0n}^{(1)} + \Phi_{0n}^{(2)}, \qquad n = 2, 3, \ldots,$$
$$\Phi_{mn} = \Phi_{mn}^{(0)} + \Phi_{mn}^{(1)} + \Phi_{mn}^{(2)}, \qquad m = 1, 2, \ldots, n = 2, 3, \ldots,$$

where

$$\Phi_{01}^{(0)}(x, y) = 2a^{-1}\lambda^{-2} \sin \lambda y = \lambda^{-2}\phi_{01}(x, y),$$

$$\Phi_{m1}^{(0)}(x, y) = 2^{\frac{3}{2}} a^{-1}\lambda^{-2}(1 - m^2)^{-1} \cos m\lambda x \sin \lambda y = \lambda^{-2}(1 - m^2)^{-1} \phi_{m1}(x, y),$$
$$m = 2, 3, \ldots,$$

$$\Psi_{m1}^{(0)}(x, y) = 2^{\frac{3}{2}} a^{-1}\lambda^{-2}(1 - m^2)^{-1} \sin m\lambda x \sin \lambda y = \lambda^{-2}(1 - m^2)^{-1} \psi_{m1}(x, y),$$
$$m = 2, 3, \ldots,$$

$$\Phi_{0n}^{(0)}(x, y) = 2a^{-1}\lambda^{-2}(1 - n^2)^{-1} \sin n\lambda y = \lambda^{-2}(1 - n^2)^{-1} \phi_{0n}(x, y),$$
$$n = 2, 3, \ldots,$$

$$\Phi_{mn}^{(0)}(x, y) = 2^{\frac{3}{2}} a^{-1}\lambda^{-2}(2 - m^2 - n^2)^{-1} \cos m\lambda x \sin n\lambda y$$
$$= \lambda^{-2}(2 - m^2 - n^2)^{-1} \phi_{mn}(x, y), \qquad m = 1, 2, \ldots, n = 2, 3, \ldots,$$

$$\Psi_{mn}^{(0)}(x, y) = 2^{\frac{3}{2}} a^{-1}\lambda^{-2}(2 - m^2 - n^2)^{-1} \sin m\lambda x \sin n\lambda y$$
$$= \lambda^{-2}(2 - m^2 - n^2)^{-1} \psi_{mn}(x, y), \qquad m = 1, 2, \ldots, n = 2, 3, \ldots,$$

$$\Phi_{01}^{(1)}(x, y) = 2a^{-1}\lambda^{-2}(-\cos \lambda x) \sin \lambda y,$$

$$\Phi_{m1}^{(1)}(x, y) = 2^{\frac{3}{2}} a^{-1}\lambda^{-2}(1 - m^2)^{-1}(-\cos \lambda x) \sin \lambda y, \qquad m = 2, 3, \ldots,$$

$$\Phi_{0n}^{(1)}(x, y) = 2a^{-1}\lambda^{-2}(1 - n^2)^{-1} S_{1n}(x) \sin n\lambda y, \qquad n = 2, 3, \ldots,$$

$$\Phi_{mn}^{(1)}(x, y) = 2^{\frac{3}{2}} a^{-1}\lambda^{-2}(2 - m^2 - n^2)^{-1} S_{1n}(x) \sin n\lambda y,$$
$$m = 1, 2, \ldots, n = 2, 3, \ldots$$

$$\Phi_{0n}^{(2)}(x, y) = 2a^{-1}\lambda^{-2}(1 - n^2)^{-1} S_{2n}(x) \sin n\lambda y, \qquad n = 2, 3, \ldots,$$

$$\Phi_{mn}^{(2)}(x, y) = 2^{\frac{3}{2}} a^{-1}\lambda^{-2}(2 - m^2 - n^2)^{-1} S_{2n}(x) \sin n\lambda y,$$
$$m = 1, 2, \ldots, n = 2, 3, \ldots .$$

For $u = Hv$, we have

$$u(x, y) = a_{11}\Phi_{11}(x, y) + \sum{}'[a_{mn}\Phi^{(0)}_{mn}(x, y) + b_{mn}\Psi^{(0)}_{mn}(x, y)] \\ + \sum{}' a_{mn}\Phi^{(1)}_{mn}(x, y) + \sum{}' a_{mn}\Phi^{(2)}_{mn}(x, y), \quad (x, y) \in G, \tag{18}$$

or

$$u(x, y) = \bar{u}(x, y) + u^{(0)}(x, y) + u^{(1)}(x, y) + u^{(2)}(x, y), \quad (x, y) \in G, \tag{19}$$

where $\Sigma'$ denotes a sum extended over all $(m, n) \neq (1, 1)$. Furthermore, we denote by $\Sigma''$ the same sum extended only to terms $S_{1n}$, $S_{2n}$, $n = 2, 3, \ldots$. Here

$$\bar{u}(x, y) = a_{11}\Phi_{11}(x, y) \\ = a_{11}\, 2^{\frac{3}{2}}\, a^{-1}\lambda^{-1}\, (1 + 4l^2\lambda^2)^{-\frac{1}{2}} \\ \times \left[ l\lambda \left(\frac{a}{2} - x\right) \cos \lambda x - \frac{\lambda}{2}\left(l + \frac{a}{2} - x\right) \sin \lambda x \right] \sin \lambda y,$$

$$u^{(0)}(x, y) = 2^{\frac{3}{2}}\, a^{-1}\lambda^{-2} \sum{}' (m^2 + n^2 - 2)^{-1} (a_{mn} \cos m\lambda x + b_{mn} \sin m\lambda x) \sin n\lambda y,$$

$$u^{(1)}(x, y) = 2^{\frac{3}{2}}\, a^{-1}\lambda^{-2} \sum{}'' a_{mn}(m^2 + n^2 - 2)^{-1}\, S_{1n}(x) \sin n\lambda y,$$

$$u^{(2)}(x, y) = 2^{\frac{3}{2}}\, a^{-1}\lambda^{-2} \sum{}'' a_{mn}(m^2 + n^2 - 2^{-1})\, S_{2n}(x) \sin n\lambda y.$$

Hence

$$\iint_G \bar{u}^2(x, y)\, dx\, dy \\ \leq (a_{11}^2 + b_{11}^2)\, 2^3 a^{-2}\lambda^{-2}\, (1 + 4l^2\lambda^2)^{-1} \left[ l^2\lambda^2 \frac{a^2}{4}\frac{a^2}{8} + \tfrac{1}{4}\left(l + \frac{a}{2}\right)^2 \frac{a^2}{8} \right] \\ = (a_{11}^2 + b_{11}^2)\, \lambda^{-2}\, \frac{a^2}{4}\, \frac{4l^2\lambda^2 + \left(1 + \dfrac{2l}{a}\right)^2}{1 + 4l^2\lambda^2} < +\infty,$$

and

$$\iint_G (u^{(0)}(x, y))^2\, dx\, dy \leq 2^3 a^{-2}\lambda^{-4} \sum{}' (m^2 + n^2 - 2)^{-2} (a_{mn}^2 + b_{mn}^2) \\ \leq 2^3 a^{-2}\lambda^{-4} \sum{}' (a_{mn}^2 + b_{mn}^2) < +\infty.$$

We note that

$$\int_0^a (S_{1n}(x))^2\, dx = (1 + \eta)^{-2} \int_0^a (e^{2\nu\lambda x} + 2\eta + \eta^2 e^{-2\nu\lambda x})\, dx \\ = (1 + \eta)^{-2} \left[ \frac{e^{2\nu\lambda a} - 1}{2\nu\lambda} + 2\eta a + \eta^2 \frac{1 - e^{-2\nu\lambda a}}{2\nu\lambda} \right] \\ = (1 + \eta)^{-2} \left( \frac{\eta^2 - 1}{2\nu\lambda} + 2\eta a + \frac{\eta^2 - 1}{2\nu\lambda} \right) \\ = (\nu\lambda)^{-1}\, (\eta - 1)\, (\eta + 1)^{-1} + 2\eta a(\eta + 1)^{-2} \\ = (\nu\lambda)^{-1}\, (e^{\nu\lambda a} - 1)\, (e^{\nu\lambda a} + 1)^{-1} + 2e^{\nu\lambda a}\, a(e^{\nu\lambda a} + 1)^{-2},$$

where $n = 2, 3, \ldots, \nu = \sqrt{n^2 - 2}$, and since $e^x \geqq x$ and $\sqrt{n^2 - 2} \geqq n/2$, for $n = 2, 3, \ldots,$ we have

$$2e^{\nu\lambda a}(e^{\nu\lambda a} + 1)^{-2} \leqq 2\,(e^{\sqrt{n^2-2}\,\lambda a} + 1)^{-1} \leqq 2(e^{\sqrt{n^2-2}\,\lambda a})^{-1}$$
$$\leqq 2\,(\sqrt{n^2 - 2}\,\lambda a)^{-1}$$

and hence

$$\int_0^a (S_{1n}(x))^2\,dx \leqq (\sqrt{n^2 - 2}\,\lambda)^{-1} + 2\,(\sqrt{n^2 - 2}\,\lambda)^{-1}$$
$$\leqq 6/n\lambda = (6/n)\,(a/2\pi) = (3a/\pi)\,(1/n).$$

Because the system $\{\sin n\lambda y\}$ is orthogonal on $[0, a/2]$,

$$\iint_G (u^{(1)}(x, y))^2\,dx\,dy$$
$$= 2^3 a^{-2}\lambda^{-4} \sum_{n=2}^{\infty} \int_0^{a/2} \sin^2 n\lambda y\,dy \sum_{m,m'=1}^{\infty} |a_{mn}|\,|a_{m'n}|\,(m^2 + n^2 - 2)^{-1}$$
$$\times\,(m'^2 + n^2 - 2)^{-1} \int_0^a (S_{1n}(x))^2\,dx$$
$$= 2^3 a^{-2}\lambda^{-4}\,(a/4) \sum_{n=2}^{\infty} \sum_{m,m'=0}^{\infty} |a_{mn}|\,|a_{m'n}|\,(m^2 + n^2 - 2)^{-1}\,(3a/\pi)\,(1/n)$$
$$= (6/\pi)\,\lambda^{-4} \sum_{n=2}^{\infty} \frac{1}{n}\left(\sum_{m=0}^{\infty} |a_{mn}|\,(m^2 + n^2 - 2)^{-1}\right)^2$$
$$\leqq (6/\pi)\,\lambda^{-4} \sum_{n=2}^{\infty} \frac{1}{n}\left(\sum_{m=0}^{\infty} a_{mn}^2\right)\left(\sum_{m=0}^{\infty} (m^2 + n^2 - 2)^{-2}\right).$$

For $\chi(n) = \sum_{m=0}^{\infty} (m^2 + n^2 - 2)^{-2}$ we note that $\chi(n) \leqq \int_{n^2-1}^{+\infty} d\xi/\xi^2 = (n^2 - 1)^{-1}$; in other words,

$$\iint_G (u^{(1)}(x, y))^2\,dx\,dy \leqq (6/\pi)\,\lambda^{-4} \sum_{m=0,n=2}^{\infty} a_{mn}^2/n(n^2 - 1)$$
$$\leqq \pi^{-1}\lambda^{-4} \sum_{m=0,n=2}^{\infty} a_{mn}^2 < +\infty.$$

We also note that for $n = 2, 3, \ldots$

$$\int_0^a (S_{2n}(x))^2\,dx = 4(l\nu\lambda)^2\,(1 + \eta)^{-2} \int_0^a (e^{2\nu\lambda x} - 2\eta + \eta^2 e^{-2\nu\lambda x})\,dx$$
$$= 4(l\nu\lambda)^2\,(1 + \eta)^{-2}\left[\frac{e^{2\nu\lambda a} - 1}{2\nu\lambda} - 2\eta a + \eta^2\frac{1 - e^{-2\nu\lambda a}}{2\nu\lambda}\right]$$
$$= 4(l\nu\lambda)^2\,(1 + \eta)^{-2}\left(\frac{\eta^2 - 1}{2\nu\lambda} - 2\eta a + \frac{\eta^2 - 1}{2\nu\lambda}\right)$$
$$= 4(l\nu\lambda)^2\,[(\nu\lambda)^{-1}\,(e^{\nu\lambda a} - 1)\,(e^{\nu\lambda a} + 1)^{-1} - 2e^{\nu\lambda a}\,a(e^{\nu\lambda a} + 1)^{-2}]$$
$$\leqq 4(l\nu\lambda)^2\,[(\sqrt{n^2 - 2}\,\lambda)^{-1} + 2\,(\sqrt{n^2 - 2}\,\lambda)^{-1}] = 12l^2\lambda\,(n^2 - 2)^{\frac{1}{2}}$$
$$\leqq 12l^2\lambda n.$$

Therefore

$$\iint_G (u^{(2)}(x,y))^2\,dx\,dy \leqq 2^3 a^{-2}\lambda^{-4}(a/4)\sum_{n=2}^{\infty}\sum_{m,m'=0}^{\infty}|a_{mn}|\,|a_{m'n}|\,(m^2+n^2-2)^{-1}$$

$$\times(m'^2+n^2-2)^{-1}(12l^2\lambda n)$$

$$=24a^{-1}\lambda^{-3}l^2\sum_{n=2}^{\infty}n\left(\sum_{m=0}^{\infty}|a_{mn}|\,(m^2+n^2-2)^{-1}\right)^2$$

$$\leqq 24a^{-1}\lambda^{-3}l^2\sum_{n=2}^{\infty}n\left(\sum_{m=0}^{\infty}a_{mn}^2\right)\left(\sum_{m=0}^{\infty}(m^2+n^2-2)^{-2}\right)$$

$$=24a^{-1}\lambda^{-3}l^2\sum_{m=0,n=2}^{\infty}a_{mn}^2 n\chi(n),$$

for $\chi(n)=\sum_{m=0}^{\infty}(m^2+n^2-2)^{-2}$. Then, since $\chi(n)\leqq(n^2-1)^{-1}$, as shown above, we obtain

$$\iint_G (u^{(2)}(x,y))^2\,dx\,dy \leqq 24a^{-1}\lambda^{-3}l^2\sum_{m=0,n=2}^{\infty}na_{mn}^2/(n^2-1)$$

$$\leqq 2^4a^{-1}\lambda^{-3}l^2\sum_{m=0,n=2}^{\infty}a_{mn}^2<+\infty.$$

We have proved that for any function $v\in L_2(G)$ satisfying (10), that is, $\iint_G v(x,y)\,\omega(x,y)\,dx\,dy=0$, $\omega=\psi_{11}$, as given by the series (12) with $b_{11}=0$, then $u=Hv$, as given by the series (18), is of class $L_2(G)$.

Recalling that $\lambda=2\pi/a$, from the relations obtained above we have

$$\|\bar u\|_2 \leqq \left(\frac{a^2}{4\pi^2}\,\frac{a^2}{4}\right)^{\frac12}\left[\frac{4l^2\lambda^2+\left(1+\frac{2l}{a}\right)^2}{1+4l^2\lambda^2}\right]^{\frac12}\|v\|_2=\frac{a^2}{4\pi}\left[\frac{4l^2\lambda^2+\left(1+\frac{2l}{a}\right)^2}{1+4l^2\lambda^2}\right]^{\frac12}\|v\|_2;$$

$$\|u^{(0)}\|_2\leqq 2^{\frac32}a^{-1}\lambda^{-2}\left[\textstyle\sum'(a_{mn}^2+b_{mn}^2)\right]^{\frac12}\leqq 2^{\frac32}\frac{1}{a}\,\frac{a^2}{4\pi^2}\|v\|_2=2^{-\frac12}\frac{a}{\pi^2}\|v\|_2;$$

$$\|u^{(1)}\|_2\leqq \pi^{-\frac12}\lambda^{-2}\left(\sum_{m=0,n=2}^{\infty}a_{mn}^2\right)^{\frac12}\leqq\pi^{-\frac12}\frac{a^2}{4\pi^2}\|v\|_2=\frac{a^2}{4\pi^{\frac52}}\|v\|_2;$$

$$\|u^{(2)}\|_2\leqq 2^2a^{-\frac12}\lambda^{-\frac32}l\left(\sum_{m=0,n=2}^{\infty}a_{mn}^2\right)^{\frac12}\leqq 2^2a^{-\frac12}\frac{a^{\frac32}}{2^{\frac32}\pi^{\frac32}}l\|v\|_2=\frac{2^{\frac12}}{\pi^{\frac32}}al\|v\|_2.$$

Thus, by the identity (19) we finally obtain

$$\|u\|_2\leqq\left\{\frac{a^2}{4\pi}\left[\frac{4l^2\lambda^2+\left(1+\frac{2l}{a}\right)^2}{1+4l^2\lambda^2}\right]^{\frac12}+\frac{a}{2^{\frac12}\pi^2}+\frac{a^2}{4\pi^{\frac52}}+\frac{2^{\frac12}al}{\pi^{\frac32}}\right\}\|v\|_2;$$

in other words

$$\|Hv\| \leqq L\,\|v\|, \qquad v \in L_2(G),$$

where $L$ can be chosen as the number

$$L = \frac{a^2}{4}\left[\frac{4l^2\lambda^2 + \left(1 + \frac{2l}{a}\right)^2}{1 + 4l^2\lambda^2}\right]^{\frac{1}{2}} + \frac{a}{2^{\frac{1}{2}}\pi^2} + \frac{a^2}{4\pi^{\frac{5}{2}}} + \frac{2^{\frac{1}{2}}al}{\pi^{\frac{3}{2}}}.$$

## § 8. Study of the Derivatives of $u = Hv$

We recall that by (18) or (19) we have $u = \bar{u} + u^{(0)} + u^{(1)} + u^{(2)}$ in $G$, where it is immediately seen that $\bar{u}$ and its first-order partial derivatives $\bar{u}_x$ and $\bar{u}_y$ are of class $L_2(G)$. Furthermore

$$u_x^{(0)}(x, y) = 2^{\frac{3}{2}}\, a^{-1}\lambda^{-2} \sum{}' \lambda m(m^2 + n^2 - 2)^{-1} \times(-a_{mn}\sin m\lambda x + b_{mn}\cos m\lambda x)\sin n\lambda y,$$

$$\iint_G (u_x^{(0)}(x, y))^2\, dx\, dy = 2^3 a^{-2}\lambda^{-4}\lambda^2\,(a/2)\,(a/4) \sum{}' (a_{mn}^2 + b_{mn}^2)\, m^2(m^2 + n^2 - 2)^{-2}$$
$$= \lambda^{-2} \sum{}' (a_{mn}^2 + b_{mn}^2)\, m^2(m^2 + n^2 - 2)^{-2} < +\infty,$$

$$u_y^{(0)}(x, y) = 2^{\frac{3}{2}}\, a^{-1}\lambda^{-2} \sum{}' \lambda n(m^2 + n^2 - 2)^{-1}\,(a_{mn}\cos m\lambda x + b_{mn}\sin m\lambda x)\cos n\lambda y,$$

$$\iint_G (u_y^{(0)}(x, y))^2\, dx\, dy = 2^3 a^{-2}\lambda^{-4}\lambda^2(a^2/8) \sum{}' (a_{mn}^2 + b_{mn}^2)\, n^2(m^2 + n^2 - 2)^{-2}$$
$$= \lambda^{-2} \sum{}' (a_{mn}^2 + b_{mn}^2)\, n^2(m^2 + n^2 - 2)^{-2} < +\infty,$$

$$u_x^{(1)}(x, y) = 2^{\frac{3}{2}}\, a^{-1}\lambda^{-2} \sum{}'' a_{mn}(m^2 + n^2 - 2)^{-1}\, S_{1n}'(x)\sin n\lambda y,$$

$$S_{1n}'(x) = \nu\lambda\,[-e^{\nu\lambda x}\,(1 + \eta)^{-1} + \eta e^{-\nu\lambda x}\,(1 + \eta)^{-1}] = -S_{2n}(x)/2l,$$

$$\int_0^a (S_{1n}'(x))^2\, dx = (1/4l^2) \int_0^a (S_{2n}(x))^2\, dx \leqq 3\lambda n,$$

$$\iint_G (u_x^{(1)}(x, y))^2\, dx\, dy \leqq 2^3 a^{-2}\lambda^{-4}\, 3\lambda(a/4) \sum_{n=2}^{\infty} n\left(\sum_{m=0}^{\infty} a_{mn}^2\right)\left(\sum_{m=0}^{\infty}(m^2 + n^2 - 2)^{-2}\right)$$
$$\leqq 6a\lambda^{-3} \sum_{m=0,n=2}^{\infty} n a_{mn}^2/(n^2 - 1) < +\infty,$$

$$u_y^{(1)}(x, y) = 2^{\frac{3}{2}}\, a^{-1}\lambda^{-2} \sum{}'' a_{mn}(m^2 + n^2 - 2)^{-1}\, S_{1n}(x)\,\lambda n\cos n\lambda y,$$

$$\iint_G (u_y^{(1)}(x, y))^2\, dx\, dy \leqq 2^3 a^{-2}\lambda^{-4}\lambda^2\,(3a/\pi)\,(a/4) \sum_{n=2}^{\infty} n\left(\sum_{m=0}^{\infty} |a_{mn}|\,(m^2 + n^2 - 2)^{-1}\right)^2$$
$$\leqq (6/\pi)\,\lambda^{-2} \sum_{m=0,n=2}^{\infty} n a_{mn}^2/(n^2 - 1) < +\infty.$$

Because

$$S_{2n}(x) = 2l\nu\lambda(1+\eta)^{-1}(e^{\nu\lambda x} - \eta e^{-\nu\lambda x}), \qquad S'_{2n}(x) = 2l(\nu\lambda)^2(1+\eta)^{-1}(e^{\nu\lambda x} + \eta e^{-\nu\lambda x}),$$

for every $\delta$, $0 < \delta < a/2$, we have

$$\begin{aligned}
\int_\delta^{a-\delta} (S'_{2n}(x))^2\,dx
&= 4l^2(\nu\lambda)^4(1+\eta)^{-2}\int_\delta^{a-\delta}(e^{2\nu\lambda x} + 2\eta + \eta^2 e^{-2\nu\lambda x})\,dx \\
&= 4l^2(\nu\lambda)^4(1+\eta)^{-2}\left[\frac{e^{2\nu\lambda(a-\delta)} - e^{2\nu\lambda\delta}}{2\nu\lambda} + 2\eta(a-2\delta) + \eta^2\frac{e^{-2\nu\lambda\delta} - e^{-2\nu\lambda(a-\delta)}}{2\nu\lambda}\right] \\
&= 4l^2(\nu\lambda)^4(1+e^{\nu\lambda a})^{-2}\left[\frac{e^{2\nu\lambda(a-\delta)} - e^{2\nu\lambda\delta}}{2\nu\lambda} + 2\eta(a-2\delta) + \frac{e^{2\nu\lambda(a-\delta)} - e^{2\nu\lambda\delta}}{2\nu\lambda}\right] \\
&= 4l^2(\nu\lambda)^4\left[\frac{1}{\nu\lambda}\frac{e^{2\nu\lambda(a-\delta)} - e^{2\nu\lambda\delta}}{(1+e^{\nu\lambda a})^2} + 2(a-2\delta)\frac{e^{\nu\lambda a}}{(1+e^{\nu\lambda a})^2}\right],
\end{aligned}$$

where the parenthesis approaches zero exponentially as $\nu = \sqrt{n^2-2} \to +\infty$.

Thus, if $\delta_1 > \delta_2 > \ldots > \delta_s > \ldots$ is any sequence converging to zero, then for every $s$ there is an integer $N_s$ such that

$$4l^2(\nu\lambda)^4\left[\frac{1}{\nu\lambda}\frac{e^{2\nu\lambda(a-\delta_s)} - e^{2\nu\lambda\delta_s}}{(1+e^{\nu\lambda a})^2} + 2(a-2\delta_s)\frac{e^{\nu\lambda a}}{(1+e^{\nu\lambda a})^2}\right] \leqq 1;$$

for $\nu = \sqrt{n^2-2}$ and $n \geqq N_s$. Let $M_s$ denote the maximum of the above numbers for all $\nu = \sqrt{n^2-2}$, $n = 2, 3, \ldots, N_s$. We also have for $n \geqq N_s$

$$\begin{aligned}
\int_{\delta_s}^{a-\delta_s}(S_{2n}(x))^2\,dx &= 4(l\nu\lambda)^2(1+\eta)^{-2}\int_{\delta_s}^{a-\delta_s}(e^{2\nu\lambda x} - 2\eta + \eta^2 e^{-2\nu\lambda x})\,dx \\
&\leqq 4(l\nu\lambda)^2\left[\frac{1}{\nu\lambda}\frac{e^{2\nu\lambda(a-\delta_s)} - e^{2\nu\lambda\delta_s}}{(1+e^{\nu\lambda a})^2} + 2(a-2\delta_s)\frac{e^{\nu\lambda a}}{(1+e^{\nu\lambda a})^2}\right] \\
&\leqq 1/(\nu\lambda)^2.
\end{aligned}$$

Also we denote by $M'_s$ the maximum of the above numbers for all $\nu = \sqrt{n^2-2}$ $n = 2, 3, \ldots, N_s$.

Now, for every $s$, we have

$$u_x^{(2)}(x,y) = 2^{\frac{3}{2}}a^{-1}\lambda^{-2}\left(\sum_{\substack{m\geqq 0\\ 2\leqq n\leqq N_s}} + \sum_{\substack{m\geqq 0\\ n>N_s}}\right)a_{mn}(m^2+n^2-2)^{-1}S'_{2n}(x)\sin n\lambda y,$$

or

$$u_x^{(2)}(x,y) = U^{(2,s)}(x,y) + V^{(2,s)}(x,y),$$

where

$$\iint_{G_s} (U^{(2,s)}(x, y))^2\, dx\, dy$$

$$= 2^3 a^{-2}\lambda^{-4} \sum_{n=2}^{N_s} \int_0^{a/2} \sin^2 n\lambda y\, dy \sum_{m,m'=0}^{\infty} |a_{mn}|\, |a_{m'n}|$$

$$\times (m^2 + n^2 - 2)^{-1} (m'^2 + n^2 - 2)^{-1} \int_{\delta_s}^{a-\delta_s} (S'_{2n}(x))^2\, dx$$

$$\leqq 2^3 a^{-2}\lambda^{-4}(a/4) \sum_{n=2}^{N_s} \sum_{m,m'=0}^{\infty} |a_{mn}|\, |a_{m'n}|\, (m^2 + n^2 - 2)^{-1} (m'^2 + n^2 - 2)^{-1} M_s$$

$$\leqq 2a^{-1}\lambda^{-4} \sum_{n=2}^{N_s} \left(\sum_{m=0}^{\infty} |a_{mn}|\, (m^2 + n^2 - 2)^{-1}\right)^2 M_s$$

$$\leqq 2a^{-1}\lambda^{-4} M_s \sum_{n=2}^{N_s} \left(\sum_{m=0}^{\infty} a_{mn}^2\right)\left(\sum_{m=0}^{\infty} (m^2 + n^2 - 2)^{-2}\right)$$

$$\leqq 2a^{-1}\lambda^{-4} M_s \sum_{m=0,n=2}^{\infty} a_{mn}^2/(n^2 - 1) < +\infty,$$

and

$$\iint_{G_s} (V^{(2,s)}(x, y))^2\, dx\, dy$$

$$\leqq 2^3 a^{-2}\lambda^{-4}(a/4) \sum_{n>N_s} \sum_{m,m'=0}^{\infty} |a_{mn}|\, |a_{m'n}|\, (m^2 + n^2 - 2)^{-1} (m'^2 + n^2 - 2)^{-1} \cdot 1$$

$$\leqq 2a^{-1}\lambda^{-4} \sum_{m=0,n=N_s}^{\infty} a_{mn}^2/(n^2 - 1) < +\infty.$$

Thus we see that $U^{(2,s)}$ and $V^{(2,s)}$ are $L_2$ functions in $G_s = [\delta_s, a - \delta_s] \times [0, a/2]$, and this holds for all $s = 1, 2, \ldots$, with $\delta_s \to 0$ as $s \to +\infty$.

Analogously,

$$u_y^{(2)}(x, y) = 2^{\frac{3}{2}} a^{-1}\lambda^{-2} \left(\sum_{\substack{m\geqq 0\\ 2\leqq n\leqq N_s}} + \sum_{\substack{m\geqq 0\\ n>N_s}}\right) a_{mn}(m^2 + n^2 - 2)^{-1} S_{2n}(x)\, \lambda n \cos n\lambda y,$$

or

$$u_y^{(2)}(x, y) = \bar{U}^{(2,s)}(x, y) + \bar{V}^{(2,s)}(x, y),$$

where

$$\iint_{G_s} (\bar{U}^{(2,s)}(x, y))^2\, dx\, dy = 2^3 (a\lambda)^{-2} \sum_{n=2}^{N_s} n^2 \int_0^{a/2} \cos^2 n\lambda y\, dy \sum_{m,m'=0}^{\infty} |a_{mn}|\, |a_{m'n}|$$

$$\times (m^2 + n^2 - 2)^{-1} (m'^2 + n^2 - 2)^{-1} \int_{\delta_s}^{a-\delta_s} (S_{2n}(x))^2\, dx$$

$$\leqq 2a^{-1}\lambda^{-2} M'_s \sum_{m=0,n=2}^{\infty} n^2 a_{mn}^2/(n^2 - 1) < +\infty,$$

and

$$\iint_{G_s} (\bar{V}^{(2,s)}(x, y))^2 \, dx \, dy$$

$$\leq 2^3 a^{-2} \lambda^{-4} \lambda^2 (a/4) \sum_{n > N_s} n^2 \sum_{m,m'=0}^{\infty} |a_{mn}| \, |a_{m'n}| \, (m^2 + n^2 - 2)^{-1}$$

$$\times (m'^2 + n^2 - 2)^{-1} \, \lambda^{-2} (n^2 - 1)^{-1}$$

$$\leq 2a^{-1} \lambda^{-4} \sum_{m=0, n=N_s}^{\infty} n^2 a_{mn}^2 / (n^2 - 2)\,(n^2 - 1) < +\infty.$$

Thus $\bar{U}^{(2,s)}$ and $\bar{V}^{(2,s)}$ are $L_2$ functions in $G_s = [\delta_s, a - \delta_s] \times [0, a/2]$ and this holds for all $s = 1, 2, \ldots$, with $\delta_s \to 0$ as $s \to +\infty$.

We conclude that $u = Hv$ has partial derivatives $u_x$ and $u_y$ of class $L_2$ in each domain $G_s$ for every $s = 1, 2, \ldots$ with $G_s \uparrow G$.

## § 9. Reduction of the Problem to an Alternative Form

We initiate the study of problem (1), (2), where we assume $f$ of class $L_2(G)$, odd in $y$ and $a$-periodic in $x$ and $y$; and $g$ odd in $\mathbb{R}$. Throughout this section we suppose that there are positive constants $c$, $C$, $k$, $\varrho$, $R_0$, $D$ such that

(20) $$\|f\|_2 \leq c, \quad |g(u)| \leq C, \quad \text{for every } u \in \mathbb{R},$$

(21) $$|u - kg(u)| \leq \varrho R_0 \quad \text{whenever } |u| \leq R_0,$$

(22) $$|g(u) - g(\tilde{u})| \leq D\,|u - \tilde{u}| \quad \text{for all } u, \tilde{u} \in \mathbb{R}.$$

Let $X$, $Y$, $X_0$, $X_1$, $Y_0$, $Y_1$ be chosen as in Section 3. Let $R$ and $r$ be two positive constants such that

(23) $$R = 2^{-\frac{3}{2}} a R_0, \quad L\left(c + 2^{-\frac{1}{2}} aC\right) \leq r,$$

where $L$ denotes the $L_2$-norm of $H$ and hence of $H(I - Q)$, by relation (14).

Let us consider the sets $\Omega = S_0 \times S_1$, $S_0 = \{u_0 \in X_0, u_0 = \gamma\phi : \|u_0\|_2 = |\gamma| \leq R\}$, $S_1 = \{u_1 \in X_1 : \|u_1\|_2 \leq r\}$. Note that the operators $P$, $Q$, $E$ and $H$ satisfy the relations:

k$_1$) $$H(I - Q)\, Eu = (I - Q)\, u \quad \text{for all } u \in D(E),$$

k$_2$) $$QEu = EPu \quad \text{for all } u \in D(E),$$

k$_3$) $$EH(I - Q)\, Nu = (I - Q)\, Nu \quad \text{for all } u \in D(E) \cap D(N),$$

and the original problem $Eu = Nu$ is equivalent to the following system of the two equations:

$$u_1 = H(I - Q)\,(f + g(u_0 + u_1)), \quad Q(f + g(u_0 + u_1)) = 0.$$

As usual we say that the problem (1), (2) has been reduced to an alternative form, and we shall regard this problem as the fixed-point problem of the transformation

$T: (u_0, u_1) \mapsto (\bar{u}_0, \bar{u}_1)$ defined by

$$T: \begin{cases} \bar{u}_1 = T_1 u = H(I-Q)\,(f + g(u_0 + u_1)) \\ \bar{u}_0 = T_0 u = u_0 - kSQ(f + g(u_0 + u_1)), \end{cases} \tag{24}$$

where $k > 0$ is the constant in (21). The map $T_0$ can be also written in the form

$$\begin{aligned} \bar{u}_0 = T_0 u = (u_0 - SQ\sigma u_0) + SQ(\sigma(u_0 - kg(u_0)) \\ - kSQf + kSQ(g(u_0) - g(u_0 + u_1)), \end{aligned}$$

where the first term in the last expression is zero (see Section 6).

In (24) we understand that $u_1$ describes the set $S_1$ of all functions $u_1$ of class $L_2(G)$, odd in $y$, doubly $a$-periodic in $x$, $y$, orthogonal to $\phi$, $\|u_1\|_2 \leqq r$, that $u_0$ describes all element $u_0 = \gamma\phi$, $\|u_0\|_2 \leqq R$, that $g(u_0 + u_1)$ is a measurable bounded function, odd in $y$, doubly $a$-periodic in $x$, $y$, that $(I-Q)\,(f \times g(u_0 + u_1))$ is a function of class $L_2(G)$, odd in $y$, doubly $a$-periodic in $x$, $y$, orthogonal to $\omega$, that $\bar{u}_1 = H(I-Q)\,(f + g(u_0 + u_1))$ is a function of class $L_2(G)$, odd in $y$, doubly $a$-periodic in $x$, $y$, orthogonal to $\phi$, expressed in $G$ by a series (17), and that $\bar{u}_0$ is a function of the form $\bar{\gamma}\phi$.

Let us note that, by assumptions (20) and (23),

$$\|\bar{u}_1\|_2 \leqq L(c + 2^{\frac{1}{2}} aC) \leqq r;$$

in other words $T_1$ maps $\Omega = S_0 \times S_1$ into $S_1$. Furthermore, for $u_0 \in S_0$,

$$u_0 = \gamma\phi = \gamma\, 2^{\frac{3}{2}} a^{-1} \sin \lambda x \sin \lambda y, \qquad \|u_0\|_2 = |\gamma| \leqq R,$$

$$\|u_0\|_\infty = 2^{\frac{3}{2}} a^{-1} |\gamma| \leqq 2^{\frac{3}{2}} a^{-1} R = R_0.$$

Thus, from (21) we know that $|u_0(x, y) - kg(u_0(x, y))| \leqq \varrho R_0$ for $u_0 \in S_0$ and hence $\|u_0 - kg(u_0)\|_2 \leqq \varrho R_0 2^{-\frac{1}{2}} a$. From (15) we obtain that

$$\|SQ(\sigma(u_0 - kg(u_0))\|_2 \leqq c_5 \varrho R_0\, 2^{-\frac{1}{2}} a = 2c_5\varrho R, \qquad \|kSQf\|_2 \leqq kc_5 \|f\|_2 = kc_5 c.$$

From (22) we have $|g(u_0(x, y)) - g(u_0(x, y) + u_1(x, y))| \leqq D\,|u_1(x, y)|$, and hence

$$\|kSQ(g(u_0) - g(u_0 + u_1))\|_2 \leqq kc_5\, D\, \|u_1\|_2 \leqq kc_5\, Dr.$$

Thus, we finally obtain that

$$\|\bar{u}_0\|_2 = \|T_0 u\|_2 \leqq 0 + 2c_5\varrho R + kc_5 c + kc_5 Dr \leqq R,$$

provided

$$kc_5(c + Dr) \leqq (1 - 2c_5\varrho)\, R, \tag{25}$$

or

$$kc_5(c + Dr) \leqq (1 - 2c_5\varrho)\, 2^{\frac{3}{2}} a^{-1} R_0.$$

Then, $T_0$ maps $\Omega = S_0 \times S_1$ into $S_0$.

Thus we have shown that the transformation $T$ maps $S$ into $S$. Moreover, under conditions (20), (21), (22), (23) and (25) we have seen that $T_1$ maps $\Omega$ into $S_1$. In addition, for every $v \in L_2(G)$ with $Qv = 0$, we have shown that $u = Hv$ not only is of class $L_2(G)$ but also $u_x$, $u_y$ are in $L_2(G_s)$, $G_s \uparrow G$, that is, $u$ is of class $W^{1,2}(G_s)$ in every $G_s = [\delta_s, a - \delta_s] \times [0, a/2]$, $\delta_s \downarrow 0$. If $[u_n]$ is any sequence in $S_1$ and $u_n = Hv_n$ then $\|u_n\|_2 \leqq r$, hence there is a subsequence, say still $[u_n]$, such that $[u_n]$ converges weakly to some function $u$ of class $L_2(G)$, or $u_n \xrightarrow{w} u$ in $L_2(G)$. Now, $[u_n]$ restricted to $G_s$ is a relatively weakly compact subset of $W^{1,2}(G_s)$, because the partial derivatives $(u_n)_x$, $(u_n)_y$, $n = 1, 2, \ldots$, not only are in $L_2(G_s)$ but their $L_2$-norms in every $G_s$ are equibounded with respect to $n$, as shown in Section 8. Hence, $[u_n]$ restricted to each $G_s$ possesses a subsequence weakly convergent to some function $u$ in $W^{1,2}(G_s)$. From the Sobolev Imbedding Theorem (ADAMS [1], pp. 97–99. Case B for $W^{1,2}(G_s)$, $G_s \subset \mathbb{R}^2$) we also know that each $u_n$ is of class $L_q(G_s)$ for every $s = 1, 2, \ldots$, and for every $q$, $1 \leqq q \leqq \infty$. From the Rellich-Kondrachov Theorem (ADAMS [1], pp. 144–145, Part I for $W^{1,2}(G_s)$) we know that, for every $s$ and $q$ as above, the sequence $[u_n]$ possesses a subsequence which is strongly convergent in $L_q(G_s)$. For any fixed $q$ as above, therefore, by the usual diagonal process, there are a function $u$ and a subsequence $[u_{n_k}]$ which converges to $u$ weakly in $L_2(G)$ and converges to $u$ strongly in $L_q(G_s)$ for every $s$. We could even extract the subsequence in such a way that $[u_{n_k}]$ converges to $u$ pointwise a.e. in $G$. However, we cannot state that $[u_{n_k}]$ converges strongly in $L_2(G)$. To avoid this difficulty, the following argument is needed (*cf.* [4]).

Let ¶ denote the square $[0 \leqq x \leqq T;\ 0 \leqq y \leqq T]$ in $\mathbb{R}^2$. For every $F \in L_2(¶)$ expressed as a double series $\sum_{m,n=0}^{\infty} A_{mn}(x, y)$, $(x, y) \in ¶$, and for every pair of integers $\mu, \nu \geqq 1$, let us consider the partial sums and the Cesaro means defined by

$$(R_{\mu\nu}F)(x, y) = \sum_{m=0}^{\mu-1} \sum_{n=0}^{\nu-1} A_{mn}(x, y),$$

$$(R^*_{\mu\nu}F)(x, y) = (\mu\nu)^{-1} \sum_{m=0}^{\mu-1} \sum_{n=0}^{\nu-1} (R_{mn}F)(x, y).$$

For the sake of simplicity we shall write $R_M F$ and $R^*_M F$ instead of $R_{MM}F$ and $R^*_{MM}F$. If $\sum A_{mn}$ is the Fourier series of $F$ in ¶, then the following statements are well known (*cf.*, for instance, [8] and [9]):

(26) $\|R^*_M F\|_{L_2(¶)} \leqq \|F\|_{L_2(¶)}$;

(27) if $|F| \leqq C$, then $|R^*_M F| \leqq C$;

(28) $|(R^*_M F)(x, y)| \leqq (R^*_M |F|)(x, y)$;

if $0 \leqq F \leqq \tilde{F}$, then $0 \leqq (R^*_M F)(x, y) \leqq (R^*_M \tilde{F})(x, y)$;

(29) if $F \in L_2(¶)$, then $\|R^*_M F - F\|_{L_2(¶)} \to 0$ as $M \to \infty$;

(30) let $d(P, A)$ denote the distance of a point $P$ from a set $A$, let $F$ be doubly periodic of period $T$, for any $(x, y)$ and $\delta$, $0 < \delta < T/2$, let $U = [(\xi, \eta): |\xi - x| \leqq \delta]$, $V = [(\xi, \eta): |\eta - y| \leqq \delta]$, and $U$ $[V]$ denotes also the set of all points $(\xi', \eta)$ with $\xi' \equiv \xi \pmod T$ and $(\xi, \eta) \in U$ $[(\xi, \eta')$ with $\eta' \equiv \eta \pmod T$ and $(\xi, \eta) \in V]$. If $\operatorname{supp} F \subset U \cap V$ and $\|F\|_\infty \leqq C$, then $|(R_M^* F)(\xi, \eta)| \leqq \theta M^{-2} C(d(\xi, U))^{-2} (d(\eta, V))^{-2}$ for all $(\xi, \eta) \in \mathbb{R}^2 - (U \cap V)$, and where $\theta$ is a constant which depends only on $T$. If $\operatorname{supp} F \subset U$, then $|(R_M^* F)(\xi, \eta)| \leqq \theta M^{-1} C(d(\xi, U))^{-2}$ for all $(\xi, \eta) \in \mathbb{R}^2 - U$.

For any integer $M \geqq 1$ we shall now consider the transformation $T_M$ defined as in (24) with $H$ replaced by $HR_M^*$:

$$T_M: \begin{cases} \bar{u}_1 = T_1 u = HR_M^*(I - Q)(f + g(u_0 + u_1)) \\ \bar{u}_0 = T_0 u = u_0 - kSQ(f + g(u_0 + u_1)). \end{cases}$$

Here we understand that $u_1$ still describes the set $S_1$ of all functions $u_1$ of class $L_2(G)$, odd in $y$, doubly $a$-periodic in $x, y$, orthogonal to $\phi$, $\|u_1\|_2 \leqq r$, that $u_0$ describes all elements $u_0 = \gamma\phi$, $\|u_0\|_2 \leqq R$, that $g(u_0 + u_1)$ is a measurable bounded function, odd in $y$, doubly $a$-periodic in $x, y$, that $R_M^*(I - Q)(f + g(u_0 + u_1))$ is a trigonometric polynomial, odd in $y$, orthogonal to $\omega$, that the function

$$\bar{u}_1 = HR_M^*(I - Q)(f + g(u_0 + u_1))$$

is of class $L_2(G)$, odd in $y$, doubly $a$-periodic in $x, y$, orthogonal to $\phi$, expressed in $G$ by a *finite* sum (17) with $m \leqq M$, $n \leqq M$, and that, as before, $\bar{u}_0$ is a function of the form $\bar{\gamma}\phi$.

Now both $T_1$ and $T_0$ have finite-dimensional ranges; hence they are compact operators, and so is $T_M$. All previous estimates for $T$ hold also for $T_M$ and uniformly with respect to $M$. In particular $T_M$ maps $\Omega = S_0 \times S_1$ into itself. Thus, by Schauder's Fixed Point Theorem, $T_M$ has at least one fixed point $u_M = u_{M0} + u_{M1}$ in $\Omega$, that is, $u_{M0} \in S_0$, $u_{M1} \in S_1$. As before, from the sequence $[u_M]$ we can extract a subsequence $[u_{M_k}]$ such that

$$\begin{aligned} &u_{M_k} \xrightarrow{w} u \quad &&\text{weakly in } L_2(G), \\ &u_{M_k} \rightarrow u \quad &&\text{strongly in } L_2(G_s) \text{ for every } s, \\ &u_{M_k} \longrightarrow u \quad &&\text{pointwise a.e. in } G. \end{aligned}$$

Let us prove that $u$ is a weak solution of the original problem (1), (2). First, we note that, for every $M$, since $T_M u_M = u_M$, $u_M = u_{M0} + u_{M1}$, it is in particular $Q(f + g(u_M)) = 0$ and hence also

$$u_{M1} = HR_M^*(I - Q)(f + g(u_M)),$$

$$0 = R_M^* Q(f + g(u_M)),$$

and we know that $u_{M0} \in X_0 = \operatorname{Ker} E$ or $Eu_{M0} = u_{M0xx} + u_{M0yy} + 2\lambda^2 u_{M0} = 0$. Then,

$$\begin{aligned} Eu_M &= u_{Mxx} + u_{Myy} + 2\lambda^2 u_M = u_{M1xx} + u_{M1yy} + 2\lambda^2 u_{M1} \\ &= R_M^*(I - Q)(f + g(u_M)) = R_M^*(f + g(u_M)). \end{aligned}$$

If $v$ is any function of class $C^2$ satisfying (4) and (5), then we have by multiplication

$$\langle Eu_M, v\rangle - \langle R_M^*(f + g(u_M)), v\rangle = 0.$$

Since $u_M$ satisfies the boundary conditions (2) and $v$ the boundary conditions (5), by duality we have $\langle Eu_M, v\rangle = \langle u_M, E^*v\rangle$, and finally

$$\iint_G [u_M(v_{xx} + v_{yy} + 2\lambda^2 v) - vR_M^*(f + g(u_M))]\, dx\, dy = 0. \tag{31}$$

We have to prove that, as $M \to \infty$, this relation yields

$$\iint_G [u(v_{xx} + v_{yy} + 2\lambda^2 v) - v(f + g(u))]\, dx\, dy = 0. \tag{32}$$

First, $u_M \rightharpoonup u$ weakly in $L_2(G)$; hence

$$\langle u_M, v_{xx} + v_{yy} + 2\lambda^2 v\rangle \to \langle u, v_{xx} + v_{yy} + 2\lambda^2 v\rangle.$$

Now we have

$$|\langle v, R_M^* f\rangle - \langle v, f\rangle| = |\langle v, R_M^* f - f\rangle| \leqq \|v\|_2\, \|R_M^* f - f\|_2,$$

and by (29),

$$\langle v, R_M^* f\rangle \to \langle v, f\rangle \quad \text{as } M \to \infty.$$

For any given function $h = h(x, y)$, $(x, y) \in G$, and any $s$, we consider the decomposition $h = h' + h''$, where

$$h' = h, \quad h'' = 0 \quad \text{for } 0 \leqq x \leqq \delta_{s+1} \quad \text{and } a - \delta_{s+1} \leqq x \leqq a, \quad 0 \leqq y \leqq a/2,$$

$$h' = 0, \quad h'' = h \quad \text{for } \delta_{s+1} < x < a - \delta_{s+1}, \quad 0 \leqq y \leqq a/2,$$

and we know already that $0 < \delta_{s+1} < a/2$. For simplicity we denote by $g(u) = g' + g''$, $g(u_M) = g_M' + g_M''$ the analogous decompositions for $g(u)$ and $g(u_M)$. Then

$$\begin{aligned}
\Delta &= \iint_G [vR_M^* g(u_M) - vg(u)]\, dx\, dy \\
&= \iint_G [vR_M^* g(u_M) - vR_M^* g(u)]\, dx\, dy + \iint_G v[R_M^* g(u) - g(u)]\, dx\, dy \\
&= \iint_G [vR_M^* g_M' - vR_M^* g']\, dx\, dy + \iint_G [vR_M^* g_M'' - vR_M^* g'']\, dx\, dy \\
&\quad + \iint_G v[R_M^* g(u) - g(u)]\, dx\, dy \\
&= i_1 + i_2 + i_3.
\end{aligned}$$

Hence $|g'_M| \leqq C$, $|g'| \leqq C$ in $G - G_{s+1}$, $g'_M = g' = 0$ in $G_{s+1}$, and $|g'_M - g'| \leqq 2C$, $|R^*_M(g'_M - g')| \leqq 2C$. On the other hand, the points $(x, y) \in G_s$ are distant at least to $\delta_s - \delta_{s+1}$ from $G - G_{s+1}$. Thus, by (27) and (30),

$$\begin{aligned} |i_1| &\leqq \|v\|_{L_2(G)} \|R^*_M(g'_M - g')\|_{L_2(G)} \\ &\leqq \|v\|_{L_2(G)} [\|R^*_M(g'_M - g')\|_{L_2(G-G_s)} + \|R^*_M(g'_M - g')\|_{L_2(G_s)}] \\ &\leqq \|v\|_{L_2(G)} [2C\,|G - G_s|^{\frac{1}{2}} + \theta a\, 2C(\delta_s - \delta_{s+1})^{-2} M^{-1}]. \end{aligned}$$

Analogously, from (27) and (26) we obtain

$$\begin{aligned} |i_2| &\leqq \|v\|_{L_2(G)} \|R^*_M(g''_M - g'')\|_{L_2(G)} \\ &\leqq \|v\|_{L_2(G)} [\|R^*_M(g''_M - g'')\|_{L_2(G-G_s)} + \|R^*_M(g''_M - g'')\|_{L_2(G_s)}] \\ &\leqq \|v\|_{L_2(G)} [2C\,|G - G_s|^{\frac{1}{2}} + \|g''_M - g''\|_{L_2(G)}], \end{aligned}$$

and now, since $g''_M - g'' = g_M - g$ in $G_{s+1}$ and $g''_M - g'' = 0$ in $G - G_{s+1}$, by (22) we finally have

$$\begin{aligned} |i_2| &\leqq \|v\|_{L_2(G)} [2C\,|G - G_s|^{\frac{1}{2}} + \|g_M - g\|_{L_2(G_{s+1})}] \\ &\leqq \|v\|_{L_2(G)} [2C\,|G - G_s|^{\frac{1}{2}} + D\,\|u_M - u\|_{L_2G(_{s+1})}], \end{aligned}$$

where, for any fixed $s$, $\|u_M - u\|_{L_2(G_{s+1})} \to 0$ as $M \to \infty$.

Since $g = g(u)$ is a measurable bounded function, $|g(u)| \leqq C$, we find

$$|i_3| \leqq \|v\|_{L_2(G)} \|R^*_M g - g\|_{L_2(G)}.$$

Thus

$$\begin{aligned} |\Delta| \leqq \|v\|_{L_2(G)} [4C\,|G - G_s|^{\frac{1}{2}} + \theta\, a\, 2C(\delta_s - \delta_{s+1})^{-2} M^{-1} \\ + D\,\|u_M - u\|_{L_2(G_{s+1})} + \|R^*_M g - g\|_{L_2(G)}]. \end{aligned} \tag{33}$$

Given $\varepsilon > 0$, we can take $s$ sufficiently large that

$$4C\,\|v\|_{L_2(G)}\,|G - G_s|^{\frac{1}{2}} \leqq \varepsilon/4.$$

Having so fixed $s$, we can now take $M_0$ sufficiently large that, for $M \geqq M_0$, the remaining three terms in (33) are each less than or equal to $\varepsilon/4$. Now, for $s$ fixed as stated, and $M \geqq M_0$, we have $|\Delta| \leqq \varepsilon$. Thus, $\Delta \to 0$ as $M \to \infty$, and we conclude that (31) yields (32). We have proved that $u$ is a weak solution of the original problem (1), (2).

In summary, we have shown that for $f$ odd in $y$, $a$-periodic in $x$ and $y$, of class $L_2(G)$, $g$ odd and Lipschitzian in $\mathbb{R}$ and satisfying (20), (21), (22), (23), and (25), problem (1), (2) has at least weak a solution $u = u_0 + u_1$.

## § 10. Example

As usual, we consider the elliptic nonselfadjoint problem (1), (2) with $f = f(x, y)$ of class $L_2(G)$, odd in $y$ and doubly $a$-periodic, and $g = g(u)$ odd and Lipschitzian in $\mathbb{R}$. If we take, for instance, $a = 0.1$ and $l = 0.01$, then we have $\lambda = 2\pi/a = 62.8318$ and

$$L = \frac{a^2}{4\pi}\left[\frac{4l^2\lambda^2 + \left(1 + \frac{2l}{a}\right)^2}{1 + 4l^2\lambda^2}\right]^{\frac{1}{2}} + \frac{a}{2^{\frac{1}{2}}\pi^2} + \frac{a^2}{4\pi^{\frac{5}{2}}} + \frac{2^{\frac{1}{2}} al}{\pi^{\frac{3}{2}}}$$

$$= \frac{0.01}{12.56637} 1.0819431 + \frac{0.1}{(1.414213)\,(9.8696028)} + \frac{0.01}{4(17.493407)}$$

$$+ \frac{(1.414213)\,(0.001)}{5.5683249}$$

$$= 0.00086 + 0.00716 + 0.00014 + 0.00025 = 0.00841;$$

$$c_5 = (1 + 4l^2\lambda^2)^{\frac{1}{2}} = [1 + 4(0.0001)\,(3947.8315)]^{\frac{1}{2}} = (2.57913)^{\frac{1}{2}} = 1.60596.$$

If we choose $c = 0.014$, $C = 1$, $k = 1$, $D = 2$ and $\varrho = 0.1$, then we first evaluate

$$r = L(c + 2^{\frac{1}{2}} aC) = (0.00841)\left(0.014 + \frac{0.1}{1.414213}\right)$$

$$= (0.00841)\,(0.08471) = 0.0007124$$

and hence

$$kc_5(c + Dr) = 1(1.60596)\,[0.014 + 2(0.0007124)]$$

$$= (1.60596)\,(0.0154248) = 0.0247716;$$

$$1 - 2c_5\varrho = 1 - 2(1.60596)\,(0.1) = 1 - 0.321192 = 0.678808,$$

and

$$0.0247716 \leqq (0.678808)\,R;$$

in other words, if $R = 0.0364928$, then relations (23) and (25) are satisfied if

$$R_0 = 2^{\frac{3}{2}} a^{-1} R = (2.828426)\,(10)\,(0.0364928) = 1.0321718,$$

and hence $\varrho R_0 = 0.10321718$.

Now, if we take the functions $f$ and $g$ above in such a way that $\|f\|_2 \leqq 0.014$; $|g(u)| \leqq 1$, $|g(u) - g(\tilde{u})| \leqq 2\,|u - \tilde{u}|$ for every $u, \tilde{u} \in \mathbb{R}$, $|u - g(u)| \leqq 0.103$ whenever $|u| \leqq 1.033$; so that $g(1) \leqq 1.136$ and $g(1) \geqq 0.930$, and hence $0.93 \leqq g(1) \leqq 1$; then assumptions (20), (21) and (22) are certainly satisfied. In conclusion, as shown in the last section, problem (1), (2) has at least a weak solution $u = u(x, y)$ which is odd in $y$ and doubly $a$-periodic. A geometric picture of a possible function $g$ is shown in the illustration below.

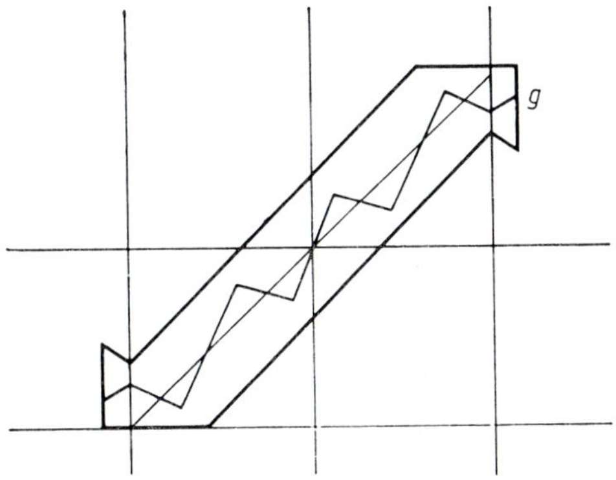


*Acknowledgment.* The research reported here was carried out in the "Gruppo Nazionale per l'Analisi Funzionale e le sue Applicazioni del Consiglio Nazionale delle Ricerche".

**References**

1. R. A. Adams, *Sobolev Spaces*, Academic Press, New York, 1975.
2. L. Cesari, "Nonlinear Analysis. New Arguments and Results", Rend. Accad. Naz. Lincei, I and II. **76** (1984), 339–345; **77** (1984), 13–20.
3. L. Cesari & T. T. Bowman, "Existence of Solutions to Nonselfadjoint Boundary Value Problems for Ordinary Differential Equations", *Nonlinear Analysis*, **9** (1985), 1211–1225.
4. L. Cesari & R. Kannan, "Solutions of Nonlinear Hyperbolic Equations at Resonance, *Nonlinear Analysis* **6** (1982), 751–805.
5. L. Cesari & R. Kannan, "Periodic Solutions of Nonlinear Wave Equations", *Archive Rational Mech. Anal.* **82** (1983), 295–312.
6. L. Cesari & P. Pucci, Global Periodic Solutions of the Nonlinear Wave Equation, *Archive Rational Mech. Anal.*, **89** (1985), 187–209.
7. L. Cesari & P. Pucci, "Existence Theorems for Nonselfadjoint Semilinear Elliptic Boundary Value Problems", *Nonlinear Analysis*, **9** (1985), 1227–1241.
8. L. Tonelli, *Serie Trigonometriche*, Zanichelli, Bologna, 1928, viii + 526.
9. A. Zygmund, *Trigonometric Series*, Vols. 1 and II, 2nd ed., Cambridge University Press, New York, 1959.

Dipartimento di Matematica
Università degli Studi
Perugia

*(Received April 30, 1985)*

# *A Multiparameter Study of a Boundary Value Problem from Chemical Reactor Theory*

PAUL FIFE, HSIAO LING, & ZHANG TONG

*Communicated by* J. B. McLEOD

*Dedicated to Tames Serrin on the occasion of his 60th birthday*

## 1. Introduction

We investigate the steady-state solutions of the following boundary value problem, which arises in the theory of catalyst particles and chemical reactors:

$$\left.\begin{aligned} D_1\,\Delta T + Yf(T) &= 0 \\ D_2\,\Delta Y - Yf(T) &= 0 \end{aligned}\right\} \quad x \in \Omega \subset \mathbb{R}^3, \tag{1.1a}$$

$$\left.\begin{aligned} -\partial_n T + a_1(T - T_e) &= 0 \\ -\partial_n Y + a_2(Y - Y_e) &= 0 \end{aligned}\right\} \quad x \in \Gamma \equiv \text{bdry of } \Omega. \tag{1.1b}$$

Here the function $f$ is positive and bounded, and $\partial_n$ represents the inward normal derivative at the boundary.

Our results are of three types:

(i) The order of magnitude of solutions is obtained as a function of the parameters $D_i$, $a_i$, $T_e$, $Y_e$, which themselves are allowed to be of any order of magnitude independently of each other. The results are stated succinctly in the form of estimates (2.1–3), which are to be interpreted in conjunction with the definitions of $u$ and $v$ in (1.4). These estimates are shown in sec. 8 to be the best possible.

(ii) Conditions are obtained under which $T(x)$ and/or $Y(x)$ are relatively constant; *i.e.* equal to a constant times a factor of the form $(1 + w(x))$, where $|w| \ll 1$ and is estimated, again, in terms of the parameters. In some cases the constant can be determined explicitly.

(iii) Conditions are obtained under which $T$ and/or $Y$ have "boundary layers", *i.e.* undergo relatively sharp variation in a neighborhood of the boundary $\Gamma$.

The method of proof is to obtain *a priori* estimates for solutions in the uniform and Hölder norms, and to apply the Leray-Schauder principle. This principle requires that the problem be reformulated as a fixed point problem for a certain nonlinear mapping. To obtain the desired parameter dependence, the particular reformulation must be chosen carefully. The results generally assert the existence of a solution with the stated properties. In many cases, however, it is easily shown that all solutions have these properties. It is well known that solutions of (1.1) are not in general unique.

In applications, $\Omega$ represents a chemical reactor in which a reactant $A$ undergoes a simple exothermic reaction $A \to P$, is transported through the boundary into the reactor from an outside pool by a mechanism with transport coefficient $a_2$, and can diffuse within $\Omega$. Its concentration is denoted by $Y$, and the temperature by $T$. Heat also diffuses and leaves through $\Gamma$ with transport coefficient $a_1$. The temperature and reactant concentrations external to the reactor are denoted by $T_e$ and $Y_e$ respectively. Typically, if the reaction is first order, the reaction rate will be given by an Arrhenius expression $BYe^{-\theta/T}$ for some positive constants $B$ and $\theta$. In this case, we set $f(T) = \exp(-\theta/T)$ and write the reaction-diffusion equations as

$$D_1' \Delta T + QBYf(T) = 0,$$

$$D_2' \Delta Y - BYf(T) = 0,$$

where $Q$ is an appropriate measure of the heat released by the reaction and $D_i'$ are diffusion coefficients. Then (1.1) is obtained by setting $D_1 = D_1'/QB$, $D_2 = D_2'/B$.

Since $T \geqq T_e$, we shall have

$$\begin{aligned} e^{-\theta/T_e} &\leqq f(T) < 1, \\ f'(T) &< \theta/T^2. \end{aligned} \tag{1.2}$$

For most of our results, we simply take $f$ to be any function satisfying

$$0 < g_1 \leqq f(T) < 1. \tag{1.3}$$

Then with $g_1$ (and of course the domain $\Omega$) fixed, the properties of the solutions are given as functions of all the remaining parameters in the problem. However, one of our results requires that $f$ also satisfy an inequality similar to (1.2), and that the ratio $\theta/T_e$ be related to the parameters by a certain inequality.

When only (1.3) is used, it is possible, and therefore appropriate, to reduce the number of free parameters in the problem to three. We do so by setting

$$\begin{aligned} T &= T_e + \frac{D_2 Y_e}{D_1} u, \\ Y &= Y_e v, \\ g(u) &= f\left(T_e + \frac{D_2 Y_e}{D_1} u\right), \end{aligned} \tag{1.4}$$

and

$$b = D_2.$$

Then (1.1,3) become

$$\left.\begin{aligned} b\,\Delta u + v g(u) &= 0 \\ b\,\Delta v - v g(u) &= 0 \end{aligned}\right\} \quad \text{in } \Omega, \qquad \begin{aligned}&(1.5\text{a})\\&(1.5\text{b})\end{aligned}$$

$$\left.\begin{aligned} -\partial_n u + a_1 u &= 0 \\ -\partial_n v + a_2(v-1) &= 0 \end{aligned}\right\} \quad \text{on } \Gamma, \qquad \begin{aligned}&(1.5\text{c})\\&(1.5\text{d})\end{aligned}$$

$$g_1 \leqq g(u) < 1. \tag{1.6}$$

The significance of these boundary value problems in chemical engineering is amply brought out in ARIS [1], which also contains an excellent survey of past mathematical results bearing upon them. Matters of existence, uniqueness, and stability are, in fact, the main concern of Volume II of that work. It is evident from examining the book of ARIS that the theory is best developed by far in the case $a_1 = a_2$ (which could be $\infty$, hence the Dirichlet problem), when the system can be reduced to a single equation. Our goal, however, is to exhibit the properties of the solutions when all restrictions regarding the magnitude of, or relations among, the transport parameters are eliminated.

The results are stated in Section 2 and proved in Sections 4–8. The core of the argument is a series of *a priori* estimates based on the lemmas in Section 3 and the maximum principle. We draw attention to Lemma 3.1; although it is basic, we have not found a direct proof in the literature (even for $\varepsilon = 0$)*. Also in Section 8, a simple example is worked out to show that our results provide the best possible order of magnitude estimates.

## 2. Statement of results

Our basic magnitude parameters will be $\beta$ and $\gamma$, defined in terms of $b$, $a_i$ by

$$\gamma \equiv \operatorname{Min}\,(1, a_2 \operatorname{Max}\,(b^{\frac{1}{2}}, b)) \tag{2.1a}$$

$$\beta \equiv \begin{cases} \dfrac{\gamma}{b \operatorname{Min}\,(1, a_1)}, & b \geqq 1 \\[2ex] \dfrac{\gamma}{\operatorname{Min}\,(1, a_1 b^{\frac{1}{2}})}, & b \leqq 1. \end{cases} \tag{2.1b}$$

In all of the following, we assume that $\Omega$ is sufficiently smooth and that $g$ is continuously differentiable and satisfies (1.6). We use the notation $|u|_0 = \sup\limits_{\Omega} |u(x)|$.

Our results concern the existence of solutions of (1.5) with certain properties depending on $b$, $a_i$, which are allowed to vary within the interval $(0, \infty]$. In all such cases, at least one solution of (1.5) exists, as our results show and as an easy

---

* *Added in proof:* Prof. O. OLEINIK has informed us that this result is known and will appear, for example, in her forthcoming book on partial differential equations.

separate argument using the Schauder fixed point theorem would also show. The cases when one or more of these three parameters is zero are not considered, but are easily handled separately. For example if $a_1 = 0$, $b < \infty$, and $a_2 \neq 0$, no solution exists; in fact (1.5a, e) and the fact that $\int_\Omega \Delta u = -\int_\Gamma \partial_n u$ would imply that $v \equiv 0$, which would contradict (1.5d). And if $a_1 = a_2 = 0$, $b < \infty$, the solution set is $(u, v)$ with $v \equiv 0$, $u \equiv$ const.

**Theorem 2.1.** *There exists a (classical) solution $(u, v)$ of (1.5) with the representation*

$$\begin{aligned} u &= \beta u^*, \\ v &= \gamma v^*, \end{aligned} \tag{2.2}$$

*with $u^*$ and $v^*$ satisfying*

$$|u^*|_0,\ |v^*|_0 < K \tag{2.3}$$

*for some constant $K$ depending only on $\Omega$ and $g_1$. If $b \leq 1$, then every solution satisfies* (2.2,3).

**Theorem 2.2.** *Assume $b > 1$. There exist constants $u_0$, $v_0$ and functions $\tilde{u}$, $\tilde{v}$, satisfying*

$$\begin{aligned} &\frac{1}{K} < u_0 < K, \\ &\frac{1}{K} < v_0 < K, \\ &|\tilde{u}|_0,\ |\tilde{v}|_0 < K \end{aligned} \tag{2.4}$$

*for some $K$ depending only on $\Omega$ and $g_1$, such that there is a solution of the form* (2.2) *with*

$$v^* = v_0 + b^{-1}\tilde{v}. \tag{2.5}$$

*Furthermore if $a_1 < 1$, then*

$$u^* = u_0 + a_1\tilde{u} \tag{2.6}$$

*and*

$$u_0 = \operatorname{Max}(1, a_2, b) - a_2 b v_0. \tag{2.7}$$

*Thus if $a_2 b > 1$,*

$$|v_0 - 1| < K/a_2 b \tag{2.8a}$$

*and if $a_2 b < 1$,*

$$|u_0 - 1| < a_2 b K. \tag{2.8b}$$

**Remark.** Representation (2.5), with (2.4), holds also in the case $b < 1$; this trivially follows from (2.3). However, (2.5) is of no interest in that case. A similar remark holds in connection with (2.6) (which also holds for $a_1 > 1$) and (2.9) below.

**Theorem 2.3.** *Assume* $b < 1$ *and* $a_1 b^{\frac{1}{2}} < 1$, *and let* $(u^*, v^*)$ *be as in Thm. 2.1.*

*If* $a_1 = a_2$, *then*

$$u^* = u_0 + a_1 b^{\frac{1}{2}} \tilde{u}, \tag{2.9}$$

*where* $u_0$ $(= \text{const})$ *and* $\tilde{u}$ *satisfy* (2.4).

*If* $\Omega$ *is a ball, then there exists a radially symmetric solution* $(u^*, v^*)$ *satisfying* (2.2, 3, 9, 4).

*Suppose that* $g$ *satisfies*

$$|g'(u)| \leqq \frac{\omega\mu}{(1 + \omega u)^2} \tag{2.10}$$

*for some positive* $\mu$ *and* $\omega$, *and all* $u > 0$. *If*

$$\mu < \omega, \tag{2.11}$$

*then* (2.9), (2.4) *hold. Moreover, there is a constant* $\nu$, *depending only on* $\Omega$ *and* $g_1$, *such that the same conclusion holds if* (2.11) *is replaced by*

$$\mu < \nu \operatorname{Max}\left[1, \frac{1}{a_2 b^{\frac{1}{2}}}, \frac{\omega}{a_1 b^{\frac{1}{2}}}, \frac{a_1 b^{\frac{1}{2}}}{\omega}, \frac{a_1}{\omega a_2^2 b^{\frac{1}{2}}}\right]. \tag{2.12}$$

**Remark.** In terms of the original problem (1.1) with $f$ given by the Arrhenius function $\exp(-\theta/T)$, it follows from (1.2) that (2.10) holds with

$$\omega = \frac{V_e D_2}{T_e D_1}, \qquad \mu = \frac{\theta}{T_e}.$$

In this case, (2.12) becomes an inequality relating all the original parameters. It will hold, for example, if $\mu$ is sufficiently small or if $\mu$ is of moderate size and $\dfrac{\omega}{a_1 b^{\frac{1}{2}}}$ is either small enough or large enough.

**Theorem 2.4.** *Assume* $b < 1$. *There are constants* $K_i$, $\tau_i$ $(i = 1, 2)$ *depending only on* $\Omega$ *and* $g_1$, *such that for every solution of* (1.5), *the function* $v^*$ *in* (2.2) *satisfies*

$$0 < v^* \leqq K_2 \exp[-\tau_2 n/\varepsilon], \tag{2.13}$$

*where* $n(x) =$ *distance from* $\Gamma$, *and* $\varepsilon = b^{\frac{1}{2}}$. *Also the lower bound*

$$v^* > K_1 \exp[-\tau_1 n/\varepsilon] \tag{2.14}$$

*holds in the boundary strip* $u \leqq n(x) \leqq \varepsilon^{\frac{1}{2}}$.

*Now assume, in addition, that* $\Omega$ *is a ball and we are considering symmetric solutions, or that* $a_1 = a_2$. *If* $a_1 b^{\frac{1}{2}} \equiv a' \leqq 1$. *then there exists a constant* $\tilde{u}_0$ *such that* (2.13), (2.14) *hold with* $v^*$ *in those inequalities replaced by* $\tilde{u}_0 - \tilde{u}$. *If* $a' > 1$, *then there exists a constant* $u_0^*$ *such that they hold with* $v^*$ *replaced by* $u_0^* - u^*$.

**Remark.** It can also be proved that $u^*$ has a boundary layer, with appropriate estimates analogous to (2.13, 14), in the case $a_1 = \infty$ and $b$ is small. This can be done by using the first equation in (6.1), estimates (2.13, 14) for $v^*$, and the maximum principle.

## 3. Lemmas

**Lemmas 3.1.** *Let* $u \in C^1(\bar{\Omega}) \cap C^2(\Omega)$, $\phi$, *and* $f \in C(\bar{\Omega})$ *satisfy*

$$\begin{aligned} \Delta u &= f, \qquad x \in \Omega \\ -\partial_n u + \varepsilon u &= \phi, \qquad x \in \Gamma \end{aligned} \tag{3.1}$$

*where* $\int_\Omega f = \int_\Gamma \phi$, $\int_\Gamma u = 0$, *and* $\varepsilon \geqq 0$.

*There is a constant* $M_1$, *depending only on* $\Omega$, *such that*

$$|u|_0 \leqq M_1 \, (|f|_0 + |\phi|_0). \tag{3.2}$$

**Proof.** We show first that it suffices to prove the lemma for the case $f = 0$. In fact, let $w$ be the solution of

$$\begin{aligned} \Delta w &= f \qquad \text{in } \Omega \\ w &= 0 \qquad \text{in } \Gamma. \end{aligned}$$

(If $f$ is Hölder continuous in $\bar{\Omega}$, $w \in C^2(\bar{\Omega})$; if $f$ is not Hölder continuous, we express it as the uniform limit of a sequence $f_n$ of smooth functions, then define $w = \lim_{n \to \infty} w_n$, where $\Delta w_n = f_n$, $w_n|_\Gamma = 0$. The maximum principle shows this limit to exist uniformly; and the estimates for $w$ below hold in any case.)

By comparing $w$ to the solutions $w_\pm$ of $\Delta w_\pm = \pm |f|_0$, $w_\pm|_\Gamma = 0$ by use of the maximum principle, it is easy to see that

$$|w|_0 \leqq C\,|f|_0, \qquad |\partial_n w|_0^\Gamma \leqq C\,|f|_0, \tag{3.3}$$

$C$ depending only on $\Omega$.

Let $v = u - w$, so $\Delta v = 0$, $-\partial_n v + \varepsilon v = \phi + \partial_n w \equiv \hat{\phi}$, $\int_\Gamma \hat{\phi} = 0 = \int_\Gamma v$.

If the lemma holds for $f = 0$, we know

$$|v|_0 \leqq M_1\,|\hat{\phi}|_0 \leqq M_1(|\phi|_0 + C\,|f|_0),$$

so that $|u|_0 \leqq |v|_0 + |w|_0$ satisfies (3.2) (with possibly a different $M_1$).

Thus assume $f = 0$. By normalizing the pair $(u, \phi)$ so that $|u|_0 = 1$, we see that it suffices to prove that for any such normalized solution pair, $|\phi|_0 \geqq M_1^{-1} > 0$. Let $\{(u_n, \phi_n, \varepsilon_n)\}$ be a sequence of triples satisfying (3.1) with $f = 0$, $|u_n|_0 = 1$, $\int_\Gamma u_n = 0$, and $0 \leqq \varepsilon_n$. We shall show that it cannot happen that $|\phi_n|_0 \to 0$. Suppose, on the contrary, that the latter limit relation were true. Let $\Omega'$ be a proper subdomain: $\bar{\Omega}' \subset \Omega$. By interior Schauder estimates [2], the functions $u_n$

and their second derivatives are equicontinuous on $\Omega'$, and a subsequence converges to a limit $u_\infty$ in $\Omega'$ satisfying $\Delta u_\infty = 0$ in $\Omega'$. By further refinement, we may suppose that each $u_n$ attains the value 1 somewhere, and hence somewhere on $\Gamma$. (If this is not possible a subsequence exists attaining the value $-1$ on $\Gamma$; we then replace $u_n$ and $\phi_n$ by their negatives.) By Green's formula,

$$\int_{\Omega'} |\nabla u_n|^2 \leqq \int_{\Omega} |\nabla u_n|^2 = \int_{\Gamma} u_n \partial_n u_n = -\int_{\Gamma} u_n(\phi_n - \varepsilon_n u_n)$$

$$\leqq - \int_{\Gamma} u_n \phi_n \leqq C\,|u_n|_0\,|\phi_n|_0 \to 0\,.$$

Passing to the limit, we find that $\int_{\Omega'} |\nabla u_\infty|^2 = 0$, so $u_\infty = \text{const}$ in $\Omega'$, and of course $|u_\infty| \leqq 1$. In particular, either $u_\infty = 1$ or $u_\infty < 1$. If the latter is true, let $n$ be so large that $u_n < \frac{1}{2}(1 + u_\infty) \equiv v < 1$ in $\Omega'$, and let $\bar{u}$ be the solution of

$$\Delta \bar{u} = 0 \quad \text{in } \Omega \setminus \Omega',$$

$$\bar{u} = 1 \quad \text{on } \Gamma, \quad \bar{u} = v \quad \text{on } \partial\Omega'.$$

By the maximum principle, $\bar{u} \geqq u_n$ in $\Omega \setminus \Omega'$ and $|\partial_n \bar{u}| \leqq |\partial_n u_n|$ at the point $\bar{x}$ on $\Gamma$ where $u_n = 1 = \bar{u}$. But the boundary condition in (3.1) implies $|\partial_n u_n| = -\partial_n u_n = \phi_n - \varepsilon u_n = \phi_n - \varepsilon \leqq |\phi_n|_0$ at $\bar{x}$, so

$$|\phi_n|_0 \geqq |\partial_n \bar{u}(x)|\,.$$

But the Hopf maximum principle implies that the right hand side is bounded away from zero, which contradicts the assumption that $|\phi_n|_0 \to 0$. A similar argument, but with a lower solution in place of the upper solution $\bar{u}$, holds in the other case that $u_\infty = 1$. For that, we use $\int_{\Gamma} u_n = 0$, which implies that $u_n$ attains negative values on $\Gamma$.

We therefore see that $|\phi|_0$ is indeed bounded away from zero, which completes the proof of the lemma. □

**Lemma 3.2.** *Let* $u \in C^2(\Omega) \cap C^1(\bar{\Omega})$ *satisfy*

$$\Delta u = f, \quad x \in \Omega,$$

$$-\varepsilon\, \partial_n u + u = 0, \quad x \in \Gamma,$$

*where* $0 \leqq \varepsilon \leqq 1$. *There is a constant* $M_2$, *depending only on* $\Omega$, *such that*

$$|u|_0 \leqq M_2\,|f|_0\,.$$

**Proof.** Define the functions $u_\pm$ as solutions of

$$\Delta u_\pm = \pm |f|_0 \quad \text{in } \Omega,$$

$$u_\pm(x) = \mp \alpha\varepsilon \quad \text{on } \Gamma,$$

where $\alpha > 0$ is to be determined. It is easy to see by the maximum principle that $|u_\pm|_0 \leqq C(|f|_0 + \alpha\varepsilon)$ for some constant $C$ depending only on $\Omega$.

Clearly $\Delta u_- \leqq f$ and also $-\varepsilon\,\partial_n u_- + u_- \geqq -C\varepsilon\,|f|_0 + \alpha\varepsilon = \varepsilon(-C\,|f|_0 + \alpha)$ on $\Gamma$. This latter quantity is zero if we choose $\alpha = C\,|f|_0$. Therefore with this choice, $u_-$ is an upper solution, and $u \leqq u_- \leqq C(1 + \varepsilon C)\,|f|_0$. A similar argument gives a lower bound and establishes the lemma.

**Lemma 3.3.** *Let $(u, v)$ be a solution of (1.5) in $C^{2+\alpha}(\overline{\Omega})$, $0 < \alpha < 1$. Assume that g is differentiable with bounded derivative. There is a constant C, depending on $\Omega$, $a_i$, b, and g, such that*

$$|u|_{2+\alpha} + |v|_{2+\alpha} \leqq C(1 + |u|_0 + |v|_0).$$

**Proof.** The Schauder estimates imply that

$$|u|_{2+\alpha} + |v|_{2+\alpha} \leqq C_1(1 + |vg(u)|_\alpha) \leqq C_2(1 + |v|_\alpha + |v|_0\,|g(u)|_\alpha).$$

Also, $|g(u)|_\alpha \leqq |g'(u)|_0\,|u|_\alpha \leqq C_3\,|u|_\alpha$ and by the maximum principle, $0 \leqq v \leqq 1$. Therefore $|u|_{2+\alpha} + |v|_{2+\alpha} \leqq C_4(1 + |u|_\alpha + |v|_\alpha)$.

On the other hand, an interpolation inequality [2] shows that

$$|u|_\alpha + |v|_\alpha \leqq \delta(|u|_{2+\alpha} + |v|_{2+\alpha}) + C_5(\delta)\,(|u|_0 + |v|_0).$$

Choosing $\delta$ sufficiently small and combining this with the previous inequality establishes the conclusion of the lemma.

## 4. The case when $b$ is large

With $\lambda < 1$ to be specified later, we assume

$$b > \lambda^{-1} \tag{4.1}$$

and seek a solution of (1.5) in the form (2.2) satisfying the conclusions of Thms. 2.1 and 2.2. For this purpose, we define coefficients

$$\beta_0 = \begin{cases} \dfrac{\gamma}{a_1 b}; & a_1 \leqq \lambda \\ 0, & a_1 > \lambda, \end{cases}$$

$$\beta_1 = \gamma/b.$$

The representations in those theorems prove to be equivalent to

$$u = \beta_0 u_0' + \beta_1 \tilde{u}'(x), \qquad v = \gamma v_0' + \beta_1 \tilde{v}'(x), \tag{4.2}$$

which will be more convenient for purposes of the proof. Here $u_0'$ and $v_0'$ are constants, normalized so that $\int_\Gamma \tilde{v}' = 0$, and in the case $\beta_0 \neq 0$, $\int_\Gamma \tilde{u}' = 0$. If $\lambda$ is chosen to depend only on $\Omega$ and $g_1$, and $u_0'$, $v_0'$, $\tilde{u}'$, $\tilde{v}'$ satisfy estimates like (2.4), then it follows easily that the conclusions of those theorems hold in this case. In fact, once the quantities in (4.2) are determined, we may set $v_0 = v_0'$, $\tilde{v} = \tilde{v}'$, and for $a_1 \leqq \lambda$ we set $u_0 = u_0'$, $\tilde{u} = \tilde{u}'$. For $a_1 \in (\lambda, 1)$ we set $u_0 = 1$,

$\tilde{u} = \tilde{u} - a_1^{-1}$, and for $a \geqq 1$, set $u^* = \tilde{u}'$, with $u_0$ and $u$ remaining undefined. Then (4.2) is the same as (2.2, 5, 6). In all cases, the estimates (2.3, 4) will follow from the corresponding estimates for the primed quantities.

From this point on, for convenience we omit the prime superscripts from the symbols in (4.2). And for symmetry purposes, we use the symbol $\gamma_0$ in place of $\gamma$.

Inserting (4.2) into (1.5) results in the problem

P1:

$$\Delta\tilde{u} + (v_0 + b^{-1}\tilde{v})\, g(\beta_0 u_0 + \beta_1 \tilde{u}) = 0, \tag{4.3a}$$

$$\Delta\tilde{v} - (v_0 + b^{-1}\tilde{v})\, g(\ldots) = 0, \tag{4.3b}$$

$$\left.\begin{aligned} -\partial_n\tilde{u} + a_1\tilde{u} + (a_1\beta_0/\beta_1)\, u_0 &= 0 \quad (4.3c)\\ -\partial_n\tilde{v} + a_2\tilde{v} + (a_2/\beta_1)(\gamma_0 v_0 - 1) &= 0 \quad (4.3d)\end{aligned}\right\} \text{ on } \Gamma,$$

$$\int_\Gamma \tilde{v} = \beta_0 \int_\Gamma \tilde{u} = 0. \tag{4.3e}$$

There are two solvability conditions for (4.3). First of all, we see that

$$0 = \int_\Omega (\Delta\tilde{u} + \Delta\tilde{v}) = -\int_\Gamma (\partial_n\tilde{u} + \partial_n\tilde{v}) = -\int_\Gamma \left[\frac{a_1\beta_0 u_0}{\beta_1} + a_1\tilde{u} + \frac{a_2}{\beta_1}(\gamma_0 v_0 - 1)\right].$$

In the case $a_1 \leqq \lambda$, we have $\int_\Gamma \tilde{u} = 0$, $\dfrac{a_1\beta_0}{\beta_1} = 1$, $\dfrac{a_2}{\beta_1} = \dfrac{a_2 b}{\gamma_0} = \operatorname{Max}[1, a_2 b]$, so this condition is

$$u_0 = \operatorname{Max}[1, a_2 b] - a_2 b v_0. \tag{4.4}$$

In the other case $a_1 > \lambda$, we have $\beta_0 = 0$ and $u_0$ is irrelevant. For simplicity we define $u_0$ in terms of $v_0$ by (4.4) in *any* case.

The other solvability condition comes from (4.3b), (4.3d), and (4.4):

$$\int_\Omega (v_0 + b^{-1}\tilde{v})\, g(\ldots) = \int_\Omega \Delta\tilde{v} = -\int_\Omega \partial_n\tilde{v} = \frac{\Gamma a_2}{\beta_1}(1 - \gamma_0 v_0) = \Gamma u_0, \tag{4.5}$$

where $\Gamma \equiv \int_\Gamma 1$.

Boundary conditions (4.3c, d), in view of (4.4), now become

$$-\partial_n\tilde{u} + a_1\tilde{u} + \nu u_0 = 0, \tag{4.6a}$$

$$-\partial_n\tilde{v} + a_2\tilde{v} - u_0 = 0, \tag{4.6b}$$

where

$$\nu = \begin{cases} 1, & a_1 \leqq \lambda, \\ 0, & a_1 > \lambda. \end{cases}$$

We define $w = \tilde{u} + \tilde{v}$ (so $\Delta w = 0$) and

$$\phi \equiv \begin{cases} -\partial_n w|_\Gamma & \text{if } a_1 \leqq \lambda, \\ \left.\left(w - \frac{1}{a_1}\partial_n w\right)\right|_\Gamma & \text{if } a_1 > \lambda. \end{cases} \tag{4.7}$$

From (4.6) and the definitions of $w$ and $\phi$, we also have the representation

$$\phi = \begin{cases} -a_1(w - \tilde{v})|_\Gamma - u_0 - \partial_n \tilde{v}|_\Gamma, & a_1 \leqq \lambda, \\ \left.\left(\tilde{v} - \frac{1}{a_1}\partial_n \tilde{v}\right)\right|_\Gamma, & a_1 > \lambda. \end{cases} \tag{4.8}$$

Note that in the case $a_1 \leqq \lambda$, it follows from (4.6) and the fact that $\int_\Gamma \tilde{u} = \int_\Gamma \tilde{v} = 0$ that $\int_\Gamma \phi = 0$, which is the solvability condition for $w$.

By introducing $w$ and $\phi$ this way, we have obtained an alternative formulation: P1 is equivalent to

P2: Find constants $u_0$, $v_0$, and functions $\phi$, $\tilde{v}$, and $w$, satisfying

$$\Delta w = 0 \quad \text{in } \Omega \tag{4.9a}$$

$$-\partial_n w|_\Gamma = \phi; \quad \int_\Gamma \phi = \int_\Gamma w = 0 \quad \text{if } a_1 \leqq \lambda, \tag{4.9b}$$

$$-\partial_n w + a_1(w - \phi) = 0 \quad \text{on } \Gamma \text{ if } a_1 > \lambda, \tag{4.9c}$$

$$u_0 = \text{Max}\,[1, a_2 b] - a_2 b v_0, \tag{4.9d}$$

$$\int_\Omega (v_0 + b^{-1}\tilde{v})\, g(\beta_0 u_0 + \beta_1 \tilde{u}) = \Gamma u_0, \tag{4.9e}$$

$$\Delta \tilde{v} - (v_0 + b^{-1}\tilde{v})\, g(\ldots) = 0 \quad \text{on } \Omega, \tag{4.9f}$$

$$-\partial_n \tilde{v} + a_2 \tilde{v} - u_0 = 0 \quad \text{on } \Gamma, \tag{4.9g}$$

$$\phi = \begin{cases} -a_1(w - \tilde{v}) - u_0 - \partial_n \tilde{v} \quad (a_1 \leqq \lambda) & (4.9\text{h}) \\ \tilde{v} - \frac{1}{a_1}\partial_n \tilde{v} \quad (a_1 > \lambda) & (4.9\text{i}) \end{cases} \quad \text{on } \Gamma.$$

Next, we identify (classical) solutions of P2 as fixed points of a certain mapping $T$. We do this first for the case $a_1 \leqq \lambda$.

We use the notation $C^{l+\alpha}$ to denote the usual Hölder spaces [2], with associated norms $|u|_{l+\alpha}$. When the domain of the functions needs to be made explicit, it appears as a superscript in the norm symbol.

Let $X$ be the following Banach space of triples, where $\alpha \in (0, 1)$:

$$X = \left\{(\phi, v_0, \tilde{v}) \in C^{1+\alpha}(\Gamma) \times R \times C^\alpha(\bar{\Omega}) : \int_\Gamma \phi = \int_\Gamma \tilde{v} = 0\right\}.$$

For constants $K_1$, $Q_1$ to be determined later, let $S$ be the subset defined by

$$S = \{(\phi, v_0, \tilde{v}) \in X:$$

$$|\phi|_0 \leqq K_1, \tag{4.10}$$

$$0 \leqq v_0 \leqq K_1,\ 0 \leqq u_0 \leqq K_1, \tag{4.11}$$

$$|\tilde{v}|_0 \leqq K_1, \tag{4.12}$$

$$|Q|^{\Gamma}_{1+\alpha} + |\tilde{v}|^{\Omega}_{\alpha} \leqq Q_1\}, \tag{4.13}$$

where the $u_0$ in (4.11) is defined by (4.4). For $(\phi, v_0, \tilde{v}) \in S$, we define $T(\phi, v_0, \tilde{v}) = (\phi', v_0', \tilde{v}')$ as follows:

1. Define $w$ by (4.9a, b).
2. Define $u_0$ by (4.9d), then find $v_0'$ from the analog of (4.9e):

$$\int_\Omega (v_0' + b^{-1}\tilde{v})\, g(\beta_0 u_0 + \beta_1 (w - \tilde{v})) = \Gamma u_0' \tag{4.14}$$

where

$$u_0' = \text{Max}\,[1, a_2 b] - a_2 b v_0'. \tag{4.15}$$

3. Solve, for $\tilde{v}'$, the following linear problem corresponding to (4.9f, g):

$$\Delta\tilde{v}' = f(x) \quad \text{in } \Omega, \tag{4.16}$$

$$-\partial_n\tilde{v}' + a_2\tilde{v}' = u_0' \quad \text{on } \Gamma, \tag{4.17}$$

where

$$f(x) \equiv (v_0' + b^{-1}\tilde{v})\, g(\beta_0 u_0 + \beta_1(w - \tilde{v})). \tag{4.18}$$

Note that (4.14), (4.16–18) automatically ensure that the solution shall satisfy

$$\int_\Gamma \tilde{v}' = 0. \tag{4.19}$$

4. Define $\phi'$ by the analog of (4.9h):

$$\phi' = -a_1(w - \tilde{v}') - u_0' - \partial_n\tilde{v}'. \tag{4.20}$$

We shall estimate $\phi_1' v_0'$, $u_0'$, and $\tilde{v}'$ in $C^0$ norms. The single symbol $C$ denotes many different constants depending only on $g_1$ and $\Omega$. We also use the notation $\Gamma = \int_\Gamma 1$, $\Omega = \int_\Omega 1$.

**Lemma 4.1.** *Assume*

$$\lambda \geqq \min\left[\frac{\Gamma}{K_1\Omega}, \frac{1}{K_1}\right]. \tag{4.21}$$

*Then*

$$|w|_0 \leqq CK_1, \tag{4.22}$$

$$0 < v_0' \leqq C, \tag{4.23}$$

$$0 < u_0' \leqq C, \tag{4.24}$$

$$|\tilde{v}'|_0 \leqq C, \tag{4.25}$$

$$|\phi'|_0 < C(1 + \lambda K_1). \tag{4.26}$$

**Proof.** First, (4.22) follows from (4.9a, b) and Lemma 3.1. From (4.14), (4.15), we obtain

$$v_0'\left(\int_\Omega g(\ldots) + \Gamma a_2 b\right) = \Gamma \max(1, a_2 b) - b^{-1} \int_\Omega \tilde{v} g. \tag{4.27}$$

Because of (4.21), the last term on the right may be estimated by either

$$b^{-1}K_1 \int g \leqq \int g \quad \text{or} \quad b^{-1}K_1 \Omega \leqq \Gamma.$$

Accordingly, we have from (4.27)

$$0 \leqq \frac{\Gamma(\operatorname{Max}(1, a_2 b) - 1)}{\Gamma a_2 b + \int g} < v_0' \leqq \frac{\Gamma \max(1, a_2 b) + \int g}{\Gamma a_2 b + \int g} < C,$$

which provides (4.23). We also obtain from this that

$$\frac{1}{\Gamma}\int g(v_0' - 1) < \operatorname{Max}(1, a_2 b) - a_2 b v_0' < 1 + \frac{1}{\Gamma} v_0' \int g < C.$$

In view of (4.15), the right inequality yields the right one of (4.24), and the left inequality can be written

$$u_0' > \frac{1}{\Gamma}\int g(v_0' - 1) > \frac{1}{\Gamma a_2 b}\int g(a_2 b v_0' - \operatorname{Max}(1, a_2 b)) = -\frac{\int g}{a_2 b} u_0',$$

which implies the left part of (4.24).

We now estimate $\tilde{v}'$. First, consider the case $a_2 \leqq 1$. Then (4.16, 17) and Lemma 3.1 imply

$$|\tilde{v}'|_0 \leqq C(|f|_0 + |u_0'|). \tag{4.28}$$

From (4.18), (4.23), and (4.21) we obtain $|f|_0 \leqq C$. Hence from this and (4.24), we get (4.25) in this case.

Next, consider the case $a_2 \geqq 1$. Let $v^* = \tilde{v}' - \frac{1}{a_2} u$, and rewrite (4.16, 17) as:

$$\Delta v^* = f,$$

$$-\frac{1}{a_2}\partial_n v^* + v^* = 0 \quad \text{on } \Gamma.$$

Lemma 3.2 now yields (4.28), hence (4.25) in this case too.

Finally, we must obtain (4.26). From (4.20) and (4.17), we get

$$\phi' = -a_1(w - \tilde{v}') - a_2 \tilde{v}'. \tag{4.29}$$

Let $\hat{v}$ be defined by $\Delta\hat{v} = -1$ in $\Omega$, $\hat{v} = 0$ on $\Gamma$. Then $\bar{v} = |f|_0 \hat{v} + \frac{u_0' + q|f|_0}{a_2}$ is an upper solution for (4.16, 17), where $q = \sup_\Gamma \partial_n \hat{v}$. Therefore on $\Gamma$, $\tilde{v}' \leqq \bar{v} = \frac{1}{a_2}(u_0' + q|f|_0) \leqq C/a_2$. Thus $0 < |a_2 \tilde{v}'| < C$ on $\Gamma$, and this, together with (4.29), (4.25), (4.22), and $a_1 \leqq \lambda < 1$ yields (4.26). □

In the following, the various constants $C$ given above will be designated by subscripts associated with the equation in which they appear. For example, the estimate constant in (4.22) will be written $C_{22}$.

**Lemma 4.2.** *If $\lambda$ and $K_1$ satisfy*

$$K_1 > \text{Max}\,[C_{23}, C_{24}, C_{25}, 2C_{26}] \tag{4.30}$$

*and*

$$\lambda < \text{Min}\left[\frac{1}{2C_{16}}, \frac{\Gamma}{\Omega K_1}, \frac{1}{K_1}\right] \tag{4.31}$$

*and if $Q_1$ is large enough, then $T$ has no fixed point on the boundary of $S$.*

**Proof.** Let $(\phi, v_0, \tilde{v})$ be a fixed point. Then Lemma 4.1 holds with primes removed. We must verify the strict inequalities corresponding to (4.10–13). But (4.11, 12) follow immediately from (4.23–25) and (4.30). As regards (4.10), we have from (4.26), (4.30), (4.31) that

$$|\phi|_0 \leqq C_{26} + (C_{26}\lambda)\,K_1 < C_{26} + \tfrac{1}{2}K_1 < K_1\,.$$

Finally, the strict version of (4.13) follows from Lemma 3.3 in the following way. Estimates (4.22–25), together with (4.2), give *a priori* bounds for $|u_0|$ and $|v|_0$, this time in terms of $a_i$ and $b$ as well as on $\Omega$ and $g_1$. Lemma 3.3 now provides *a priori* estimates for $|u|_{2+\alpha}$ and $|v|_{2+\alpha}$ as well, hence for $|\tilde{v}|_\alpha$ and (by (4.8)) for $|\phi|_{1+\alpha}$. We simply choose $Q_1$ in (4.13) to be larger than the sum of these last two estimates.

For every $\tau \in [0, 1]$, we now define the functions $g^\tau(u) \equiv \tau g(u) + (1-\tau)\,g_1$. With $g^\tau$ in place of $g$, the transformation analogous to $T$ will be denoted by $T^\tau$. Clearly $g_1 \leqq g^\tau \leqq 1$, which is the only property of $g$ used, and so all the previous lemmas hold for $T^\tau$ in place of $T$. In particular, $T^\tau$ has no fixed point on the boundary of $S$.

**Lemma 4.3.** *$T^\tau$ is compact in the space $X$, continuous, and depends continuously on $\tau$.*

**Proof.** Given any $(\phi, v_0, \tilde{v})$ in $S$, let $(\phi', v_0', \tilde{v}') = T^\tau(\phi, v_0, \tilde{v})$. We have seen that $|\phi'|_0$, $v_0'$, and $|\tilde{v}'|_0$ are subject to uniform bounds. Furthermore (4.16, 17) and Schauder estimates show that $\tilde{v}' \in C^{2+\alpha}(\bar{\Omega})$, and (4.20) shows that $\phi' \in C^{1+\alpha}(\Gamma)$. In fact, if we express the last term $\partial_n \tilde{v}'$ in (4.20) in terms of $\tilde{v}'$ through (4.17), we see that $\phi \in C^{2+\alpha}(\Gamma)$. The corresponding norms can be estimated uniformly. This establishes the compactness of $T^\tau$.

To get continuity, we again use Schauder estimates and the linearity of problem (4.9a, b) to show that the mapping $\phi \to w$ is continuous from $C^{1+\alpha}(\Gamma)$ to $C^{2+\alpha}(\bar{\Omega})$. Hence the mapping $(\phi, \tilde{v}) \to \tilde{u} = w - \tilde{v}$ is continuous from $C^{1+\alpha}(\Gamma) \times C^\alpha(\bar{\Omega})$ to $C^\alpha(\bar{\Omega})$, and since $g^\tau$ is $C^1$, we see that the mapping $(\phi, v_0, \tilde{v}) \to g^\tau = g^\tau(u(x))$ is continuous from $S$ to $C^\alpha(\Omega)$. Now (4.4, 14, 15) show the mapping $S \to v_0'$ is continuous, and Schauder estimates applied to

(4.16–18) show $(v_0', g^\tau) \to \tilde{v}'$, as a mapping from $\mathbb{R} \times C^\alpha$ to $C^\alpha$, is continuous. Finally (4.20) and the above results imply $(\phi, v_0, \tilde{v}) \to \phi'$, as a mapping from $X$ into $C^{1+\alpha}(\Gamma)$, is also continuous. Putting all these results together shows the continuity of $T^\tau$.

Its continuity in $\tau$ follows from the continuity of $\tau \to g^\tau$ from $\mathbb{R}$ to $C^\alpha(\bar{\Omega})$, together with the above arguments.

**Lemma 4.4.** *Let $K_1$ and $\lambda$ satisfy the hypotheses of Lemma 4.2. If $b \geqq \lambda^{-1}$ and $a_1 \leqq \lambda$, then there exists a solution of* (1.5) *satisfying* (2.2–8) *for some constant $K$ depending only on $\Omega$, $g_1$, $K_1$, and $\lambda$.*

**Proof.** We use the Leray-Schauder principle with respect to the family $T^\tau$ of compact mappings on the closed convex set $S$. By Lemma 4.2, there is no fixed point, for any $\tau \in [0, 1]$, on the boundary of $S$. We wish to prove the existence of a fixed point when $\tau = 1$; all that remains is to show that when $\tau = 0$, there is a unique fixed point in $S$ with index $\neq 0$. But when $\tau = 0$, $g^0(u) = g_1$ is independent of $u$, so that $T(\phi, v_0, \tilde{v})$ is independent of $(\phi, v_0, \tilde{v})$. Therefore $T^0$ is a constant mapping from $S$ to a point in $S$, so $T^0 - I$ has degree $-1$ on $S$. Therefore $T = T^1$ has a fixed point in $S$. It is the solution sought. The bounds in (4.11, 12, 22) imply (2.3, 4) for an obvious choice of constant $K$. Finally, (2.7) comes from (4.4), and (2.8) from (2.4, 7).

This settles the case when $b \geqq \lambda^{-1}$ and $a_1 \leqq \lambda$. There is an analogous argument for the case $b \geqq \lambda^{-1}$, $a_1 > \lambda$, which we now sketch.

A Banach space $X$ is defined as before, except that the requirement that $\int_\Gamma \phi = 0$ is now removed. For constants $K_2$, $K_3$, and $Q_2$ to be determined later, the set $S \subset X$ is now defined as

$$S = \{(\phi, v_0, \tilde{v}) \in X : |\phi|_0 \leqq K_2, \tag{4.10'}$$

$$0 \leqq v_0 \leqq K_3, \quad 0 \leqq u_0 \leqq K_3, \tag{4.11'}$$

$$|\tilde{v}_0| \leqq K_3, \tag{4.12'}$$

$$|\phi|_{1+\alpha}^\Gamma + |\tilde{v}|_\alpha^\Omega \leqq Q_2\}. \tag{4.13'}$$

The mapping $T$ is defined as before, except that $w$ satisfies the boundary condition (4.9c) instead of (4.9b), and in place of (4.20), we have

$$\phi' = \tilde{v}' - \frac{1}{a_1}\partial_n \tilde{v}' = \tilde{v}' - \frac{1}{a_1}(u_0' - a_2 \tilde{v}'). \tag{4.20'}$$

**Lemma 4.1'.** *Assume*

$$\lambda \leqq \operatorname{Min}\left[\frac{\Gamma}{K_3 \Omega}, \frac{1}{K_3}\right]. \tag{4.21'}$$

*Then*

$$|w|_0 \leqq K_2, \tag{4.22$'$}$$

$$0 < v_0' < C, \tag{4.23$'$}$$

$$0 < u_0' < C, \tag{4.24$'$}$$

$$|\tilde{v}'|_0 < C, \tag{4.25$'$}$$

$$|\phi'|_0 < C\left(1 + \frac{1}{\lambda}\right]. \tag{4.26$'$}$$

**Proof.** Estimate (4.22)′ easily follows from the maximum principle; the proofs of (4.23, 24)′ are the same as in the original lemma. The same is true of (4.25)′. Finally, (4.26)′ follows from (4.20)′, the fact that $|a_2 v| < C$ on $\Gamma$ (as before), and the fact that $a_1 > \lambda$. □

**Lemma 4.2′.** *If $\lambda$, $K_2$, and $K_3$ satisfy*

$$K_3 > \text{Max}\,[C_{23}', C_{24}', C_{25}'],$$

$$\lambda < \text{Min}\left[\frac{\Gamma}{K_3 \Omega}, \frac{1}{K_3}\right],$$

*and*

$$K_2 > C_{26}'\left(1 + \frac{1}{\lambda}\right),$$

*and $Q_2$ is large enough, then no fixed point of $T$ exists on the boundary of $S$.*

**Proof.** Any fixed point satisfies (4.22–26)′ with primes removed. It is then routine to show that the strict versions of (4.10–12)′ are satisfied. As before, the strict version of (4.13)′ will then hold by virtue of Lemma 3.3 for large enough $Q_2$. □

In the present context, the statement of Lemma 4.3, and its proof as well, still hold.

**Lemma 4.5.** *Let $K_3$ and $\lambda$ satisfy the hypotheses of Lemma 4.2′. If $b \geqq \lambda^{-1}$ and $a_1 > \lambda$, then there exists a solution of* (1.5) *satisfying* (2.2–5) *for some constant $K$ depending only on $\Omega$, $g_1$, $K_3$, $\lambda$. If $a_1 < 1$,* (2.6–8) *also hold.*

**Proof.** As before, the Leray-Schauder technique can be used to show the existence of a solution of P2 satisfying (4.10–12)′ and (4.22)′. It follows that

$$|\tilde{u}|_0 = |w - \tilde{v}|_0 \leqq |w|_0 + |\tilde{v}|_0 \leqq K_2 + K_3.$$

Therefore the bounds (2.4) hold with $K = K_2 + K_3$, and this can be selected to depend only on the quantities indicated. And (2.7, 8) follow as before.

## 5. The case when $b$ is small; Theorem 2.3

Our most difficult task lies in the proof of Theorem 2.3, where it is assumed that $b$ and $a_1 b^{\frac{1}{2}}$ are both small, and it is shown that under certain conditions $u^*$ is relatively constant. It is an open question as to whether $u^*$ is relatively constant when these other conditions are not met. The present section is devoted to the analysis relevant to Theorem 2.3; it will then be an easy matter, building on this analysis, to establish Theorem 2.1. That will essentially be done in Section 6.

Defining

$$\varepsilon = b^{\frac{1}{2}}, \qquad a = \varepsilon a_2, \qquad a' = \varepsilon a_1,$$

we may rewrite the basic problem (1.5) as

P1.
$$\varepsilon^2 \Delta u + vg(u) = 0,$$
$$\varepsilon^2 \Delta u - vg(u) = 0,$$
$$\left.\begin{aligned} -\varepsilon\, \partial_n u + a'u &= 0 \\ -\varepsilon\, \partial_n v + a(v-1) &= 0 \end{aligned}\right\} \text{ on } \Gamma. \tag{5.2}$$

With $\lambda$ denoting a constant as yet undetermined, we assume

$$\varepsilon \leqq \lambda, \qquad a' \leqq \lambda \tag{5.3}$$

and seek a solution in the form (2.2, 9), which we rewrite as

$$u = \beta u_0 + \gamma u, \qquad v = \gamma v^*, \tag{5.4}$$

where

$$\gamma = \text{Min}\,[1, a],$$
$$\beta = \gamma / a'.$$

We normalize $u_0$ so that $\int_\Gamma \tilde{u} = 0$.

Problem P1 will be reformulated in two different ways. First, insert (5.4) into P1 to obtain

P2
$$\varepsilon^2 \Delta \tilde{u} + v^* g(u) = 0, \tag{5.5a}$$
$$\varepsilon^2 \Delta v^* - v^* g(u) = 0, \tag{5.5b}$$
$$\left.\begin{aligned} -\varepsilon\, \partial_n \tilde{u} + a'\tilde{u} + u_0 &= 0 \\ -\varepsilon\, \partial_n v^* + a(v^* - e) &= 0 \end{aligned}\right\} \text{ on } \Gamma, \tag{5.5d}$$
$$\int_\Gamma \tilde{u} = 0, \tag{5.5e}$$

where

$$e \equiv \gamma^{-1} \equiv \text{Max}\,[1, 1/a]. \tag{5.6}$$

The argument in this section implicitly assumes that $a_2 < \infty$, but may be adapted to the case $a_2 = \infty$ as well. We leave this for the reader. Also for notational convenience we drop the asterisk from $v^*$, cautioning the reader not to confuse this function with the original $v$ in Sections 1 and 2.

We shall associate, with any solution of P2, auxiliary functions $w$ (defined on $\bar{\Omega}$), $p$ (defined on $\Gamma$) and a constant $v_1$, defined by

$$w = \tilde{u} + v, \tag{5.7}$$

$$v_\Gamma = v_1 + p(x), \qquad \int_\Gamma p = 0. \tag{5.8}$$

As we shall show, P2 is equivalent to P3:

P3. Find functions $\phi \in C^1(\Gamma)$, $w \in C^2(\bar{\Omega})$, and $v \in C^2(\bar{\Omega})$, and a positive constant $u_0$, satisfying the following ($v_1$, $p$, and $\tilde{u}$ being defined by (5.7), (5.8)):

$$\left.\begin{aligned} \Delta w = 0, \quad & x \in \Omega \\ -\partial_n w + a_1(w - \phi) = 0 \quad & \text{on } \Gamma, \end{aligned}\right\} \tag{5.9}$$

$$\left.\begin{aligned} -\varepsilon^2 \Delta v - v g(\beta u_0 + \gamma \tilde{u}) = 0 \quad & \text{in } \Omega, \\ -\varepsilon\, \partial_n v + a(v - e) = 0 \quad & \text{on } \Gamma, \end{aligned}\right\} \tag{5.10}$$

$$\phi = \frac{-a_2}{a_1} p + v_\Gamma, \tag{5.11}$$

$$u_0 = a(e - v_1). \tag{5.12}$$

**Lemma 5.1.** P2 *is equivalent to* P3, *together with* (5.7, 8).

**Proof.** Given a solution ($\phi$, $w$, $v$, $u_0$) of P3, we let $\tilde{u} = w - v$ and $u = \beta u_0 + \gamma \tilde{u}$. Then we must show that (5.5) hold.

Equations (5.5b) and (5.5d) are unchanged in the new formulation. To verify (5.5a), we note from (5.9), (5.10) that

$$\varepsilon^2 \Delta \tilde{u} + v g(u) = \varepsilon^2 \Delta w - \varepsilon^2 \Delta v + v g(u) = 0.$$

For (5.5c), we observe that on $\Gamma$,

$$\begin{aligned} -\varepsilon\, \partial_n \tilde{u} + u_0 + a' \tilde{u} &= -\varepsilon\, \partial_n w + \varepsilon\, \partial_n v + u_0 + a' w - a' v \\ &= \varepsilon(-\partial_n w + a_1 w) + \varepsilon\, \partial_n v - a' v + u_0 \\ &= \varepsilon\, a_1 \phi + a(v - e) - a' v + u_0 \\ &= \varepsilon a_1 \left( -\frac{a_2}{a_1} p + v \right) + av - ae - a'v + ae - av_1 \\ &= -ap + av - av_1 \\ &= av - a(v_1 + p) = 0. \end{aligned}$$

Finally, to verify (5.5e), we have $\left(\text{setting } \Gamma = \int_\Gamma 1\right)$

$$\int_\Gamma \tilde{u} = \int_\Gamma w - \int_\Gamma (p + v_1) = \int_\Gamma w - v_1 \Gamma = \int_\Gamma \phi + \frac{1}{a} \int_\Gamma \phi + \frac{1}{a} \int_\Gamma \partial_n w - v_1 \Gamma .$$

But from (5.11), $\int_\Gamma \phi = \int_\Gamma v = v_1 \Gamma$; and from (5.9), $\int_\Gamma \partial_n w = - \int_\Omega \Delta w = 0$, so (5.5e) holds.

Similarly, a solution of P3 can be constructed from any given solution of P2. □

We now reformulate P3 as a fixed point problem. For some positive $K_4$, $Q_4$ to be determined later and for fixed $\alpha \in (0, 1)$, we define a set $S = S(K_4, Q_4)$ in the Banach space $X = C^{1+\alpha}(\Gamma) \times \mathbb{R} \times C^\alpha(\overline{\Omega})$ as follows:

$$S = \{\phi, u_0, v) \in X \colon \operatorname{Min}\,[1, a_1]\, |\phi|_0 \leqq K_4, \tag{5.13}$$

$$\frac{1}{K_4} \leqq u_0 \leqq K_4, \tag{5.14}$$

$$|v|_0 \leqq K_4, \tag{5.15}$$

$$\int_\Gamma \phi = \int_\Gamma v, \tag{5.16}$$

$$|\phi|^\Gamma_{1+\alpha} + |v|^\Omega_\alpha \leqq Q_4\}. \tag{5.17}$$

Given such a triple, we define a new triple $(\phi', u_0', v') \equiv T(\phi, u_0, v)$ as follows.

1. A function $w$ is determined by

$$\Delta w = 0, \tag{5.18a}$$

$$-\partial_n w + a_1 (w - \phi) = 0 \quad \text{on } \Gamma. \tag{5.18b}$$

With $v_1$, $p$ defined by (5.8), note from (5.16) that

$$\int_\Gamma w = \int_\Gamma \phi = v_1 \Gamma. \tag{5.18c}$$

2. Define $\tilde{u} = \tilde{w} - v$.
3. Define $v'$ by

$$\varepsilon^2 \Delta v' - v' g(\beta u_0 + \gamma \tilde{u}) = 0, \tag{5.19a}$$

$$-\varepsilon\, \partial_n v' + a(v' - e) = 0 \quad \text{on } \Gamma, \tag{5.19b}$$

and $v_1'$, $p'$ by

$$v'_\Gamma = v_1' + p', \quad v_1' = \text{const}, \qquad \int_\Gamma p' = 0. \tag{5.20}$$

4. Define

$$\phi' = -\frac{a_2}{a_1} p' + v'_\Gamma \tag{5.21}$$

5. Define

$$u_0' = a(e - v_1'). \tag{5.22}$$

Clearly, fixed points of $T$ will be solutions of P3. We now derive some estimates for $(\phi', u_0', v')$ defined above. The symbol $C$ will represent various constants depending only on $\Omega$ and $g_1$.

**Lemma 5.2.**

$$|w|_0 \leqq CK_4 \tag{5.23}$$

*and*

$$|\tilde{u}|_0 \leqq CK_4. \tag{5.24}$$

**Proof.** Let $\hat{w} = w - v_1$, $\hat{\phi} = \phi - v_1$. Then $\Delta\hat{w} = 0$, $-\partial_n\hat{w} + a_1(\hat{w} - \hat{\phi}) = 0$, and $\int_\Gamma \hat{\phi} = 0$, by (5.16). If $a_1 \leqq 1$, we have by Lemma 3.1 that

$$|\hat{w}|_0 \leqq Ca_1 |\hat{\phi}|_0,$$

so by (5.13),

$$|w|_0 \leqq Ca_1(|v_1| + |\phi|_0) + |v_1| \leqq CK_4.$$

But if $a_1 \geqq 1$, we conclude by using the maximum principle that $|w|_0 \leqq |\phi|_0$. Hence by (5.13), we obtain (5.23) in any case. Finally, (5.24) follows from (5.23), (5.15), and the definition of $\tilde{u}$. □

For $x \in \Omega$, we let $n(x) = \text{dist}(x, \Gamma)$. In an internal neighborhood $\mathcal{N}$ of $\Gamma$, say $\mathcal{N} = \{x: n(x) \leqq \varrho\}$, we may represent points $x$ by $x = (n, s)$, $s$ representing the (unique) point on $\Gamma$ closest to $x$. In terms of the coordinates $(n, s)$, the operator $\Delta$ may be expressed as

$$\Delta = \frac{\partial^2}{\partial n^2} + \varkappa(x)\frac{\partial}{\partial n} + L_s, \tag{5.25}$$

where $L_s$ is a second order differential operator which is "tangential" in the sense that it annihilates all differentiable functions of $n$ alone. We define

$$\varkappa \equiv \sup_x |\varkappa(x)|.$$

For definiteness, we choose $\varrho$ to be half the supremum of all widths of strips $\mathcal{N}$ in which (5.25) is valid. Then $\varrho$ and $\varkappa$ will depend only on $\Omega$.

**Lemma 5.3.** *If*

$$\lambda < \frac{\sqrt{g_1}}{2\varkappa}, \tag{5.26}$$

*then*

$$\frac{1}{Ca} \leqq e - v' \leqq \frac{C}{a} \quad \text{on } \Gamma, \tag{5.27}$$

$$0 < v' \leqq C, \tag{5.28}$$

*and*

$$\frac{1}{C} < u_0' \leqq C. \tag{5.29}$$

**Proof.** We define lower and upper solutions

$$\underline{v}(x) = \begin{cases} \underline{\gamma}(e^{-\bar{\tau}n/\varepsilon} - e^{-\tau\varrho/\varepsilon}), & x \in \mathcal{N}, \\ 0 \qquad , & x \in \Omega \setminus \mathcal{N} \end{cases}$$

$$\bar{v}(x) = \begin{cases} \bar{\gamma}(e^{-\underline{\tau}n/\varepsilon} + e^{-\underline{\tau}(\varrho-n)/\varepsilon}), & n \leqq \varrho/2, \\ 2\bar{\gamma}e^{-\underline{\tau}\varrho/2\varepsilon} \qquad , & n \geqq \varrho/2 \text{ and in } \Omega \setminus \mathcal{N}, \end{cases} \tag{5.30}$$

$$\underline{\gamma} = \frac{e}{\frac{\bar{\tau}}{a} + (1 - e^{-\bar{\tau}\varrho/\varepsilon})}, \qquad \bar{\gamma} = \frac{e}{\frac{\underline{\tau}}{a}(1 - e^{-\underline{\tau}\varrho/\varepsilon}) + (1 + e^{-\underline{\tau}\varrho/\varepsilon})}, \tag{5.31}$$

$$\underline{\tau} = \sqrt{\underline{g}} - \frac{\varepsilon\varkappa}{2}, \qquad \bar{\tau} = \sqrt{\bar{g}} + \varepsilon\varkappa, \tag{5.32}$$

and where $\underline{g}$, $\bar{g}$ are the minimum and maximum values attained by the function $g(u) = g(\beta u_0 + \gamma\tilde{u})$.

It is routine to verify, using (5.25, 26), that $\underline{v}$ and $\bar{v}$ are indeed lower and upper solutions of the linear problem (5.19), and that $\underline{v} < \bar{v}$. Therefore (5.19) has a unique solution $v'$ satisfying

$$\underline{v} \leqq v' \leqq \bar{v}. \tag{5.33}$$

We note that on $\Gamma$,

$$0 < \frac{e}{1 + \bar{\sigma}} < v' < \frac{e}{1 + \bar{\sigma}}, \tag{5.34}$$

where

$$\bar{\sigma} = \frac{\bar{\tau}}{a(1 - \exp(-\bar{\tau}\varrho/\varepsilon))}, \qquad \underline{\sigma} = \frac{\underline{\tau}}{a} \frac{(1 - \exp(-\tau\underline{\varrho}/\varepsilon))}{(1 + \exp(-\tau\underline{\varrho}/\varepsilon))} \tag{5.35}$$

From (5.26, 3, 32, 35), we know that

$$\frac{1}{Ca} < \underline{\sigma} < \bar{\sigma} < \frac{C}{a}. \tag{5.26}$$

From (5.34, 36) we see that $v'$ satisfies

$$0 < v' < \frac{e}{1 + \frac{1}{Ca}} < C;$$

hence (5.28) is proved.

To prove (5.27), we note from (5.34, 36) that

$$\frac{ea}{a+C} \leqq v' \leqq \frac{ea}{a+C^{-1}} \quad \text{on } \Gamma,$$

so that for $a \geqq 1$ (so $e = 1$),

$$\frac{1}{1+c} \leqq a - \frac{a^2}{a+C^{-1}} = 1 - \frac{a}{a+C^{-1}} \leqq a\left(\frac{1}{a} - v'\right) \leqq 1 - \frac{a}{a+C} < 1 .$$

Thus (5.27) holds in both cases. Finally, (5.29) follows from (5.27) and (5.22). □

**Lemma 5.4.** *If*

$$\lambda < \frac{\sqrt{g_1}}{4\varkappa}, \qquad a' \leqq \lambda, \qquad \varepsilon \leqq \lambda,$$

*and* $a_1 = a_2$ *or* $\Omega$ *is a ball, then there exists a solution of* (1.5) *in the form* (2.2, 9), *satisfying* (2.3, 4) *for some* $K$ *depending only on* $\Omega$ *and* $g_1$. *If* $a_1 = a_2$, *then every solution has these properties.*

**Proof.** First, consider the case that $a_1 = a_2$. Then from (5.21),

$$\phi' = v'_\Gamma - p' \equiv v'_1,$$

so from (5.28) we have

$$\operatorname{Min}(1, a_1)\,|\phi'|_0 \leqq |\phi'|_0 \leqq C_{28} .$$

This means that if $K_4$ is larger than the constants in (5.28, 29) and also larger than 1, then $\phi'$, $u_0'$, $v'$ all satisfy the strict inequalities corresponding to (5.13–15). As in the proof of Lemma 4.2, it follows from Lemma 3.3 that if $Q_4$ is large enough, $T$ has on no fixed point the boundary of $S$, and by the same Leray-Schauder technique as in Lemma 4.4, the existence of a solution follows, subject to the bounds (5.14–15) and (5.24). Clearly $K_4$ can be chosen to depend only on $\Omega$ and $g_1$, and we may take $K = C_{24}K_4$ so that (2.3, 4) hold.

Any solution (1.5), in fact, may be identified with a fixed point of $T$, and the above *a priori* estimates show that every fixed point lies in $S$ when $K_4$, $Q_4$ are chosen as indicated. This establishes the final assertion of the lemma.

Next, suppose $\Omega$ is a ball. Then we modify the definition of $X$ by admitting only radially symmetric functions. In particular, $\phi$ will always be a constant, and $p'$ is always zero. Thus the *a priori* estimates described above hold, and the existence of a radially symmetric solution with the right properties follows. Furthermore every radially symmetric solution has these properties, but we cannot claim that others do as well.

We now explore the consequences of assumption (2.10).

**Lemma 5.5.** *Assume* (2.10) *holds. Let* $k_0$, $k$ *be constants such that*

$$0 < k_0 \leqq u_0, \qquad |\tilde{u}|_0 \leqq k . \tag{5.37}$$

*If*

$$a' < \frac{k_0}{2k}, \tag{5.38}$$

*then*

$$|\phi|_0 \leqq C\left(1 + \frac{1}{a_1} + \frac{\mu k}{e}\,\mathrm{Min}\left[\frac{1}{2k_0}, y, \frac{4}{yk_0^2}\right]\right), \tag{5.39}$$

*where* $y = \dfrac{w}{ea'}$.

**Proof.** Let $g_0 = g(\beta u_0)$. Then from (2.10),

$$|g(\beta u_0 + \gamma\tilde{u}) - g_0| \leqq \frac{\omega\mu k\gamma}{(1 + \omega(\beta k_0 - \gamma k))^2}.$$

Now (5.38) implies $\gamma k \leqq \frac{1}{2}\beta k_0$, so it follows that $|g(u) - g_0| \leqq h$, where

$$h = \frac{\omega\mu e^{-1}k}{(1 + \frac{1}{2}\omega e^{-1}(a')^{-1}k_0)^2}. \tag{5.40}$$

Thus

$$|\bar{g} - \underline{g}| \leqq 2h. \tag{5.41}$$

It follows from (5.34, 36) that

$$|p'|_0 \leqq \frac{e}{1 + \underline{\sigma}} - \frac{e}{1 + \bar{\sigma}} = \frac{e(\bar{\sigma} - \underline{\sigma})}{(1 + \underline{\sigma})(1 + \bar{\sigma})} \leqq \frac{e(\bar{\sigma} - \underline{\sigma})}{\left(1 + \frac{1}{Ca}\right)^2}, \tag{5.42}$$

and from (5.35, 32, 40) that

$$\begin{aligned} \bar{\sigma} - \underline{\sigma} &\leqq \frac{C}{a}\left(\bar{\tau} - \underline{\tau} + \exp\left(-\frac{1}{C\varepsilon}\right)\right) \\ &\leqq \frac{C}{a}(h + \varepsilon). \end{aligned} \tag{5.43}$$

Combining these, we have

$$|p'|_0 \leqq \frac{C}{ae}(h + \varepsilon). \tag{5.44}$$

Now from (5.21, 41, 28) we obtain

$$\begin{aligned} |\phi'|_0 &\leqq C\left(1 + \frac{a_2}{a_1}|p'|_0\right) \leqq C\left(1 + \frac{a_2\varepsilon}{a_1 ea} + \frac{a_2}{aa_1 e}h\right) \\ &= C\left(1 + \frac{\varepsilon}{ea'} + \frac{1}{ea'}h\right). \end{aligned}$$

From this and the facts that $e \geqq 1$, $a' = \varepsilon a_1$, and (5.40) we obtain

$$|\phi'|_0 \leqq C\left(1 + \frac{1}{a_1} + y\mu e^{-1}k\,(1 + \tfrac{1}{2}k_0 y)^{-2}\right).$$

Finally (5.39) follows, since

$$\frac{y}{(1+\alpha y)^2} \leqq \mathrm{Min}\left(\frac{1}{4\alpha}, y, \frac{1}{\alpha^2 y}\right).$$

**Lemma 5.6.** *There are constants* $\lambda_0$ *and* $v$ *depending only on* $\Omega$ *and* $g_1$ *such that if*

$$\lambda < \lambda_0, \quad a' \leqq \lambda, \quad \varepsilon \leqq \lambda, \tag{5.45}$$

*and* (2.10) *hold, and if* (2.11) *or* (2.12) *holds, then a solution of* (1.5) *satisfying* (2.2, 2.9, 2.4) *exists for some K depending only on* $\Omega$ *and* $g_1$.

**Proof.** Distinguish the preceding constants $C$ by appropriate subscripts, and set

$$K_0 = C_{29}^{-1}, \qquad k = C_{24}K_4. \tag{5.46}$$

We require

$$\lambda_0 \leqq \mathrm{Min}\left[\frac{\sqrt{g_1}}{4\varkappa}, \frac{1}{2C_{29}C_{24}K_4}\right], \tag{5.47}$$

and we shall impose further restrictions on it later. As in the proof of Lemma 5.4, $K$ may be chosen so that any fixed point satisfies strict inequalities corresponding to (5.13–15). First, we consider (5.13). By (5.46, 47) we see that (5.38) holds; hence (5.39) does as well. If (2.11) holds, then from (5.39) we have

$$\mathrm{Min}\,(1, a_1)\,|\phi|_0 \leqq C\left(1 + \frac{\omega k}{e}\frac{4}{yk_0^2}\right) \leqq C(1 + a'K_4). \tag{5.48}$$

If we require

$$C_{48}\lambda_0 \leqq \tfrac{1}{2}, \tag{5.49}$$

then the right side is bounded by $C_{48} + \frac{1}{2}K_4$, which is less than $K_4$ provided that $K_4 > 2C_{48}$. Therefore we select $K_4 > 2C_{48}$, and also to be larger than $C_{28}$ and $C_{29}$. Then for sufficiently large $Q_4$, there can be no fixed point on the boundary of $S$, and the existence of a solution follows, with bounds for $u_0$, $\tilde{u}$, and $v$ dependent only on $\Omega$, $g_1$, and $K_4$. But the latter can be chosen to depend on the first two.

Next, we suppose that (2.12) holds for some $v$. Then $\mu < vq$, where $q$ is one of the five quantities in brackets. From (5.39), we have (recall $y = \omega/ea'$)

$$\mathrm{Min}\,(1, a_1)\,|\phi|_0 \leqq C\left(1 + v\frac{qK_4}{e}M\right),$$

where $M = \mathrm{Min}\,[1, y, y^{-1}]$.

Consider the various cases in turn:

(a) $q = 1$. Then since $a \geqq 1$ and $M \leqq 1$, we have $qM/e \leqq 1$.
(b) $q = a^{-1}$. Same conclusion, since $1/ae \leqq 1$.
(c) $q = \omega/a'$. Then using $M \leqq ea'/\omega$, we have $qM/e \leqq 1$.
(d) $q = a'/\omega$. Same, since $M \leqq \omega/ea'$.
(e) $q = a_1/\omega_2^2 b^{\frac{1}{2}}$. Since $M \leqq \omega/ea'$, we have $qM/e \leqq (ae)^{-2} \leqq 1$.

Thus, in all cases, we conclude that

$$\text{Min}\,(1, a_1)\,|\phi|_0 \leqq C(1 + \nu K_4). \tag{5.50}$$

If we now specify $\nu \leqq \frac{1}{2C_{50}}$ and $K_4 > 2C_{50}$ the argument proceeds as in the preceding case. □

## 6. The case when $b$ is small: Theorem 2.1

For a constant $\lambda < 1$ still to be determined, we continue to assume

$$\varepsilon < \lambda.$$

Substituting the form (2.2) into (5.2), we obtain the following, where we define $\delta \equiv \text{Min}\,(1, a')$:

$$\left.\begin{aligned} \varepsilon^2 \Delta u^* + \delta v^* g(\beta u^*) &= 0 \\ e^2 \Delta v^* - v^* g(\beta u^*) &= 0 \end{aligned}\right\} \quad \text{in } \Omega$$
$$\left.\begin{aligned} -\partial_n u^* + a_1 u^* &= 0 \\ -\varepsilon\, \partial_n v^* + a(v^* - e) &= 0 \end{aligned}\right\} \quad \text{on } \Gamma. \tag{6.1}$$

Defining

$$w = u^* + \delta v^*, \tag{6.2}$$

we can show that this problem is equivalent to

$$\begin{aligned} &\Delta w = 0, \\ &-\partial_n w + a_1(w - \phi) = 0 \quad \text{on } \Gamma, \\ &\varepsilon^2 \Delta v^* - v^* g(\beta(w - \delta v^*)) = 0, \\ &-\varepsilon\, \partial_n v^* + a(v^* - e) = 0 \quad \text{on } \Gamma, \\ &\phi = -\frac{\delta a_2}{a_1}(v^*_\Gamma - e) + \delta v^*_\Gamma. \end{aligned} \tag{6.3}$$

**Lemma 6.1.** *If* (5.26) *holds, then every solution of* (6.3) *satisfies* (5.27, 28) *with $v'$ replaced by $v^*$, and also satisfies*

$$|\phi|_0 \leqq C. \tag{6.4}$$

**Proof.** The proof of (5.27, 28) is the same as that in Lemma 5.3. The proof of (6.4) follows from (5.27, 28), the last equation in (6.3), and the observations that $\frac{\delta a_2}{a_1} \leqq a$ and $\delta \leqq 1$.

**Lemma 6.2.** *There is a $\lambda_0$ depending only on $\Omega$ and $g_1$, such that if $\varepsilon \leqq \lambda_0$, then there exists a solution of* (1.5) *in the form* (2.2), *satisfying* (2.3) *for some K depending only on $\Omega$ and $g_1$. Furthermore, every solution has these properties.*

**Proof.** We again interpret (6.1) as a fixed point problem analogous to that in Section 5, but in a different space. This time, $X = C^{1+\alpha}(\Gamma) \times C^{\alpha}(\bar{\Omega})$ and $S = \{(\phi, v^*)\ X: |\phi|_0 \leqq K_5, |v^*|_0 \leqq K_5, |\phi|_{1+\alpha} + |v^*|_\alpha \leqq Q_5\}$, with $K_5$, $Q_5$ to be determined. Here no mention is made of $u_0$, and no integral condition on $\phi$ is required. The mapping $T: (\phi, v^*) \rightarrow (\phi', v^{*\prime})$ is constructed on the basis of (6.3) as in Section 5. Existence then follows from the *a priori* estimates in Lemma 6.1 in the manner by now familiar to us. The final statement also follows from those estimates. □

## 7. The case when $b$ is intermediate in magnitude

The main object in this section will be to derive the relevant representations (2.2, 9) when $b$ is in the intermediate range $\lambda^2 \leqq b \leqq \lambda^{-1}$. (Note, however, that Theorem 2.3 requires $b < 1$.)

Our point of departure will be P2 (5.5) of Section 5, which is based on (5.4) and the expressions for $\gamma$, $\beta$ given there. This agrees with (2.1, 2) only for $b \geqq 1$, $a_1 b^{\frac{1}{2}} \leqq 1$. Nevertheless, we allow $b$ to lie in this extended range. We rewrite P2 as follows:

$$b\, \Delta \tilde{u} + v^* g(u) = 0, \tag{7.1}$$

$$b\, \Delta v^* - v^* g(u) = 0, \tag{7.2}$$

$$-\partial_n \tilde{u} + a_1 \tilde{u} + b^{-\frac{1}{2}} u_0 = 0 \tag{7.3}$$

$$-\partial_n v^* + a_2(v^* - e) = 0, \tag{7.4}$$

$$u_0 = b^{-\frac{1}{2}} \int_\Omega v^* g(u), \tag{7.5}$$

$u = \beta u_0 + \gamma \tilde{u}$. (Using the fact that $\int_\Omega \Delta \tilde{u} = -\int_\Gamma \partial_n \tilde{u}$, it is easy to see that (7.5) is equivalent to (5.5e), which it replaces.).

**Lemma 7.1.** *Assume, for some $\gamma$, that*

$$\lambda^2 < b < \lambda^{-1} \tag{7.6}$$

*and*

$$a \equiv a_1 b^{\frac{1}{2}} \leqq \lambda^{-1}. \tag{7.7}$$

*Then any solution of* (7.1–5) *must satisfy*

$$|\tilde{u}|_0,\ |v^*|_0,\ u_0 \leqq C'.$$

*where $C'$ depends only on $\Omega$, $g_1$, and $\lambda$.*

**Proof.** All constants $C'$ below depend only on these three quantities. Let $\hat{v}$ be the solution of

$$\lambda^{-1} \Delta\hat{v} - g_1\hat{v} = 0,$$

$$\hat{v} = 1 \quad \text{on } \Gamma.$$

Then $\hat{v} \leqq 1$ in the interior and by the (Hopf) maximum principle, there is an $h(g_1\lambda) > 0$ such that

$$-\partial_n\hat{v} > h$$

on $\Gamma$. Let $\bar{v} = k\hat{v}$, where

$$k = \frac{ea_2}{h + a_2} = \frac{ea}{b^{\frac{1}{2}}h + a} < \lambda^{-1}(b^{\frac{1}{2}}h)^{-1} < (\lambda^2 h)^{-1} = C'.$$

Thus $\bar{v}$ is an upper solution for (7.2, 4), and we conclude that

$$0 < v^* \leqq \bar{v} < C'. \tag{7.8}$$

Now let $f(x) = v^*g(u(x)) \leqq C'$, and write (7.1) as

$$\Delta\tilde{u} = -b^{-1}f(x).$$

When $a_1 \leqq 1$, we use Lemma 3.1 to conclude that

$$|\tilde{u}|_0 \leqq C(b^{-1}|f|_0 + b^{-\frac{1}{2}}u_0) \leqq C'(1 + u_0). \tag{7.9}$$

And when $a_1 > 1$, we let $\hat{u} = \tilde{u} + b^{-\frac{1}{2}}a_1^{-1}u_0$, so that

$$\Delta\hat{u} = -b^{-1}f,$$

$$-a_1^{-1}\partial_n\hat{u} + \hat{u} = 0,$$

and Lemma 3.2 implies

$$|\hat{u}|_0 \leqq Cb^{-1}|f|_0,$$

so

$$|\tilde{u}|_0 \leqq C(b^{-1}|f|_0 + b^{-\frac{1}{2}}a_1^{-1}u_0) < C'(1 + u_0)$$

in this case as well.

From (7.5, 6) and the above bound for $|v^*|_0$, we obtain

$$0 < u_0 \leqq C'.$$

This, together with (7.8, 9), completes the proof.

**Lemma 7.2.** *Let* (7.6, 7) *hold. Then there exists a solution of* (1.5) *satisfying* (2.2, 3, 9, 4) *for some K depending only on* $\Omega$, $g_1$, *and* $\lambda$. *Furthermore, every solution of* (1.5) *has these properties.*

**Proof.** First, we restrict $b$ to the interval $(\lambda^2, 1)$, so that (7.1–5) is equivalent to (1.5), (2.2, 9). The problem can be set up as a fixed point problem on the space of triples $(\tilde{u}, v^*, u_0) \in C^\alpha(\bar{\Omega}) \times C^\alpha(\bar{\Omega}) \times \mathbb{R}$. The *a priori* bounds in Lemma 7.1 make for an easy application of the Leray-Schauder technique along the lines of the proofs in the preceding sections.

Next, we suppose $b \in (1, \lambda^{-1})$. We still obtain a solution $(\tilde{u}, v^*, u_0)$ of (7.1–5) with the required bounds. We set $u^* = u_0 + a'\tilde{u}$. However in this case, the values of $\gamma$ and $\beta$ (given following (5.4)) are not those which are defined in (2.1) and appear in (2.2). Nevertheless, they differ from the correct values only by factors depending on $\lambda$. Hence our solution $(u^*, v^*)$, when multiplied by such factors, yields a new pair $(u^*, v^*)$ satisfying (2.2) with the correct values of $\gamma$, $\beta$. Since at this point we allow $K$ to depend on $\lambda$, the existence part of the lemma holds in this case too.

The final assertion of the lemma follows from Lemma 7.1. □

**Lemma 7.3.** *Let* $\lambda^2 \leqq b \leqq \lambda^{-1}$ and $a_1 b^{\frac{1}{2}} \geqq 1$. *Then there exists a solution of* (1.5) *satisfying* (2.2, 3) *for some K depending only on* $\Omega$, $g_1$, *and* $\lambda$. *Furthermore, every solution of* (1.5) *has these properties.*

**Proof.** Substitute (2.2) into (1.5) to obtain a problem simpler than (7.1–5) ($u_0$ does not enter it). The argument proceeds as above. □

## 8. Completion of the proofs; an example

We now suppose that the constants $\lambda$, $K_1$, and $K_3$ have been selected in accordance with the requirements stated in Lemma 4.4, 4.5, 5.4, 5.6, and 6.2. From the nature of these requirements, it is clear that this can be done so that the constants depend only on $\Omega$ and $g_1$, and we assume this has been done.

**Proof of Theorem 2.1.** In the case $b \leqq \lambda^2$, the conclusions follow directly from Lemma 6.2.

In the case $b \in [\lambda^2, \lambda^{-1}]$, they follow from Lemmas 7.2 and 7.3, and the understanding that $\lambda$ depends only on $\Omega$ and $g_1$.

In the case $b > \lambda^{-1}$, they follow in a similar manner from Lemma 4.4 and 4.5. □

**Proof of Theorem 2.2.** For $b \in [\lambda^{-1}, \infty)$, the conclusions of the theorem follow from Lemmas 4.4 and 4.5 and the fact that $K_1$ and $\lambda$ have been chosen as described above.

It remains to prove (2.6–8) in the case $b \in [1, \lambda^{-1}]$, $a_1 < 1$. These inequalities imply $a_1 b^{\frac{1}{2}} \leqq \lambda^{-\frac{1}{2}} < \lambda^{-1}$, so Lemma 7.1 applies, and (2.9, 4) follow. Defining $\tilde{u}' = b^{\frac{1}{2}}\tilde{u}$, we rewrite (2.9) in the form (2.6) with $\tilde{u}$ replaced by $\tilde{u}'$. But $\tilde{u}'$ satisfies the same estimates (2.4) as does $\tilde{u}$, so (2.6) is established.

In (2.5, 6) we may always normalize $u_0$ and $v_0$ so that $\int_\Gamma \tilde{u} = \int_\Gamma \tilde{v} = 0$. But in the case, we have seen in Section 4 that (4.4) necessarily holds. This is (2.7), and (2.8) follows from (2.7) and (2.4). □

**Proof of Theorem 2.3.** In the case $b < \lambda^2$, the conclusions follow from Lemmas 5.4 and 5.6. In the remaining case $b \in [\lambda^2, 1]$, they follow (even without the extra assumptions such as (2.11, 12)) from Lemma 7.2.

**Proof of Theorem 2.4.** The upper and lower bounds $\bar{v}$, $\underline{v}$ in (5.30) were derived in the context $b \leqq \lambda^2$, $a' \leqq \lambda$, but in fact their derivation did not require the latter inequality, and so (5.30) holds whenever $b \leqq \lambda^2$. The constants $\underline{\gamma}$ and $\bar{\gamma}$, which we may identify with $K_1$ and $K_2$ in Theorem 2.4, are essentially the second and fourth terms in (5.34), and (5.36) shows them to be subject to positive lower and upper bounds depending only on $\Omega$ and $g_1$. It is clear that if now choose $\tau_1 < \underline{\tau}$, $\tau_2 > \bar{\tau}$ appropriately, then (2.13, 14) will follow from (5.30).

The remaining case is when $b \in [\lambda^2, 1]$. Then $\varepsilon > \lambda$ and (2.13) follows directly from (2.3). Proving (2.14) in this case simply amounts to showing there is a lower bound for $v^*$ which depends only on $\Omega$ and $g_1$. This may be done in a manner analogous to the proof of the upper bound (7.8). In fact, we define $\hat{v}_1$ as the solution of $\lambda^2 \Delta \hat{v}_1 - \hat{v}_1 = 0$ in $\Omega$, $\hat{v}_1 = 1$ on $\Gamma$, define $h_1(\lambda) > 0$ as a lower bound for $-\partial_n \hat{v}$ on $\Gamma$, and $\underline{v} = k\hat{v}$, $k = ea_2/(h_1 + a_2)$. Then $\underline{v}$ is a lower solution which provides a positive lower bound for $v^*$, as required.

It remains to prove the estimates for $\tilde{u}$ or $u^*$ embodied in the final assertion of Theorem 2.4. If $a' < \lambda$, then $\tilde{u}$ is characterized as satisfying (5.5). It was noted in the proof of Lemma 5.4 that if $a_1 = a_2$ of if $\Omega$ is a ball, then $\phi$ is constant. It then follows from (5.9) that $w \equiv \phi \equiv$ const. Relabel this constant $\tilde{u}_0$. Then $v^* = \tilde{u}_0 - \tilde{u}$ from (5.7), and the boundary layer estimate for $\tilde{u}$ follows immediately from (2.13, 14). If $a' \geqq \lambda$, then we work in the context of problem (6.3). As before, $\phi \equiv \text{const} \equiv u_0^*$ if $a_1 = a_2$ or $\Omega$ is a ball, and (6.2) with $\delta = \text{Min}(1, a')$ implies that $\delta^{-1}(u_0^* - u) = v^*$ satisfies (2.13, 14). Since $\lambda \leqq \delta \leqq 1$, the same is true (with different $K_i$) for $u_0^* - u^*$.

We now provide a simple example of (1.5) in one dimension which retains all three free parameters but is completely solvable. Take $\Omega = (-1, 1)$ and $g \equiv 1$. Then the solutions are necessarily even in $x$, and the problem reduces to

$$\left.\begin{aligned} bu'' + v &= 0 \\ bv'' - v &= 0 \end{aligned}\right\} \quad \text{in } \Omega,$$

$$u'(0) = v'(0) = 0,$$

$$u'(1) + a_1 u(1) = 0,$$

$$v'(1) + a_2(v(1) - 1) = 0.$$

Solving first for $v$, we obtain the following, where $\sigma \equiv b^{-\frac{1}{2}}$:

$$v = \frac{a_2 \cosh \sigma x}{a_2 \cosh \sigma + \sigma \sinh \sigma}.$$

Thus $v_{\max} = v(1) = \left(1 + \dfrac{\tanh \sigma}{a_2 b^{\frac{1}{2}}}\right)^{-1}$.

We use the notation $f \sim g$ if $C^{-1} f \leqq g \leqq Cf$ for some $C > 0$ independent of $a_1$ and $b$. Then for $b \leqq 1$, $\tanh \sigma \sim 1$ and $v_{\max} \sim \text{Min}(1, a_2 b)$. And for $b \geqq 1$, $\tanh \sigma \sim b^{-\frac{1}{2}}$, so $v_{\min} \sim \text{Min}(1, a_2 b)$. Thus in all cases,

$$v_{\max} \sim \gamma, \tag{8.1}$$

where $\gamma$ is defined in (2.1). Also, $v_{\min} = v_{\max}/\cosh \sigma$, so

$$v_{\max} - v_{\min} = v_{\max} \frac{\cosh \sigma - 1}{\cosh \sigma} \sim v_{\max}\sigma^2 = v_{\max} b^{-1} \sim \gamma b^{-1}. \tag{8.2}$$

We see from (8.1, 2) that for our function $v$, the representations (2.2) and (2.5), with $v^*$, $v_0$, $\tilde{v}$ bounded independently of $a_i$, $b$, cannot be improved.

In a similar manner, we may obtain

$$u(x) = \frac{a_2}{a_2 \cosh \sigma + \sigma \sinh \sigma} \left( \cosh \sigma + \frac{\sinh \sigma}{a_1 b^{\frac{1}{2}}} - \cosh \sigma x \right). \tag{8.3}$$

From this, we may verify that (2.1b) is the best possible magnitude parameter for (2.2). Also, we obtain from (8.3)

$$\frac{u_{\max}}{u_{\min}} = 1 + \frac{(\cosh \sigma - 1)}{\sinh \sigma} a_1 b^{\frac{1}{2}}.$$

In the case $b > 1$, $\dfrac{\cosh \sigma - 1}{\sinh \sigma} \sim b^{-\frac{1}{2}}$, which shows that (2.6) cannot be improved in this case. And for $b < 1$, this ratio is bounded and near 1 for small $b$; this shown (2.9) also to be strict. Finally, Theorem 2.4 can be verified in this case.

*Acknowledgment*. This research was supported by a grant from the Committee on Scholarly Communication with the People's Republic of China and by NSF Grant MCS-8202056. Part of this research was done while FIFE held a visiting appointment at Kyoto University.

## References

1. R. ARIS, *The Mathematical Theory of Diffusion and Reaction in Permeable Catalysts*, I and II. Clarendon Press: Oxford, 1975.
2. D. GILBARG & N. TRUDINGER, *Elliptic Partial Differential Equations of Second Order*. Springer, Berlin Heidelberg New York, 1977.

University of Arizona,
Tucson
and
Academia Sinica
People's Republic of China

*(Received March 31, 1984)*

# *Boundary Regularity for Quasiminima*

WILLIAM P. ZIEMER

*Dedicated to James Serrin, with affection and admiration, on the occasion of his sixtieth birthday*

## 1. Introduction

In [GG], GIAQUINTA & GIUSTI introduced the notion of quasiminima, a concept which is very useful in unifying the $C^{0,\alpha}$-regularity theory of elliptic equations in divergence form. In particular, their treatment provides regularity results for minima of nondifferentiable variational integrals. They also show that weak solutions of nonlinear elliptic equations in divergence form, under natural and general structural conditions, are quasiminima as are solutions to a wide class of obstacle problems. Because of the great interplay between quasiminima and the Hölder theory of partial differential equations, it is important to develop the subject fully. Indeed, significant progress has already been made. Recently, for example, Harnack inequalities have been established in [DBT] for sub and super-quasiminima which run parallel to the results obtained by TRUDINGER [T] for weak sub and supersolutions of quasilinear equations in divergence form. In particular, the results in [DBT] imply that quasiminima are Hölder continuous, a result first established by GIAQUINTA & GIUSTI, [GG].

The purpose of this paper is to investigate the behavior of quasiminima at the boundary of an arbitrary bounded domain $\Omega \subset R^n$. The results of this paper are similar in spirit to those obtained in [GZ2], in which the boundary behavior of weak solutions to a wide class of quasilinear equations was investigated. The $p$-capacity, $\gamma_p$, is employed to describe a sufficient condition for continuity of a quasiminimum at a boundary point $x_0$ of an arbitrary domain $\Omega$. For convenience, it is assumed that $\Omega$ is bounded. The $p$-capactity $(1 < p < n)$ of an arbitrary set $E \subset R^n$ is defined by

$$(1) \qquad \gamma_p(E) = \inf\{\textstyle\int |\nabla v|^p : E \subset \operatorname{int}\{v \geqq 1\},\ v \in W^{1,p}(R^n)\},$$

where $W^{1,p}(R^n)$ denotes the Sobolev space of $L^p(R^n)$ functions whose first derivatives are also elements of $L^p(R^n)$.

The main results are as follows. Let $u \in W^{1,p}_{\text{loc}}(\Omega)$ be a quasiminimum, or $Q$-minimum for brevity (see § 2.1), such that $u - \beta \in W^{1,p}_0(\Omega)$ where $\beta \in W^{1,p}(R^n)$

is assumed to be continuous on $R^n - \Omega$. Let $x_0 \in \partial\Omega$ and define

$$\delta(r) = \frac{\gamma_p[B(x_0, r) - \Omega]}{\gamma_p[B(x_0, r)]} \tag{2}$$

where $B(x_0, r)$ denotes the ball centered at $x_0$ with radius $r$. It is shown that there is a $C > 0$ (which depends on the structure in (7) and $\|u\|_{p;\Omega}$) such that if

$$\int_0^1 \exp - [C\delta(r)^{-p'}]\frac{dr}{r} = \infty, \quad p' = \frac{p}{p-1}, \tag{3}$$

then

$$\lim_{\substack{x \to x_0 \\ x \in \Omega}} u(x) = \beta(x_0). \tag{4}$$

In particular, if $\delta(r) \geqq a > 0$ for all small $r$, then (3) holds, thus establishing regularity at $x_0$, a result that was obtained in [I]. Condition (3) can be regarded as a Wiener-type condition, although it is a weaker condition for regularity (it assumes more) than the condition found in [GZ2] which applies to a smaller class of functions, namely, those that are weak solutions of quasilinear equations in divergence form with measurable coefficients. Indeed, the condition obtained in [GZ2] was

$$\int_0^1 \delta(r)^{1/(p-1)}\frac{dr}{r} = \infty. \tag{5}$$

In an attempt to obtain a condition similar to (5) for quasiminima, the $q$-capacity, $\gamma_q$, $1 < q < p$, is considered instead of the $p$-capacity. In this case I am able to show that there is a positive number $\lambda$ such that if

$$\int_0^1 [\delta(r)]^{\lambda}\frac{dr}{r} = \infty,$$

then

$$\lim_{\substack{x \to x_0 \\ x \in \Omega}} u(x) = \beta(x_0).$$

Here, $\delta_q(r)$ is as in (2) with $p$ replaced by $q$.

The proof that (3) implies boundary regularity for quasiminima relies on the weak Harnack inequalities of [DBT] and they, in turn, depend upon DeGiorgi-type estimates and a technique of KRYLOV & SAFONOV [KS] which was used in their treatment of equations in non-divergence form.

## 2. Definitions and Preliminaries

In this section functionals of the form

$$I(u, \Omega) = \int_\Omega F(x, u, \nabla u) \tag{6}$$

are considered where $F(x, z, \xi)$ is a Carathéodory function on $\Omega \times R \times R^n$, that is measurable in $x$ for every $(z, \xi)$ and continuous in $(z, \xi)$ for almost all $x \in \Omega$. The function $F$ is also required to satisfy the following structural constraints:

$$|\xi|^p - a|z|^p - g(x) \leqq F(x, z, \xi) \leqq a[|\xi|^p + |z|^p] + g(x) \tag{7}$$

where $1 < p < n$, $1 \leqq a < np/(n-p)$, and $g \in L^q$, $q > \dfrac{n}{p}$. The case of $n \geqq p$ needs a slight adjustment and will not be considered here.

The closure in the Sobolev norm of smooth functions compactly supported in $\Omega$ is denoted by $W_0^{1,p}(\Omega)$. In the sequel I will implicitly use the following well-known facts concerning Sobolev functions; *cf.* [FZ]. If $u \in W^{1,p}(R^n)$, then $u$ has a Lebesgue point at all points except possibly for a set of $\gamma_p$-capacity 0. Indeed, for $\gamma_p$-a.e. $x_0 \in R^n$,

$$\lim_{r \to 0} \alpha(n)^{-1} r^{-n} \int_{B(x_0,r)} |u(x) - u(x_0)|\, dx = 0,$$

where $\alpha(n)$ denotes the volume of the unit $n$-ball. Therefore, if $u \in W_0^{1,p}(\Omega)$, then $u = 0$ $\gamma_p$-a.e. on $R^n - \Omega$. Conversely, if $u \in W^{1,p}(R^n)$ and $u = 0$ $\gamma_p$-a.e. on $R^n - \Omega$, then $u \in W_0^{1,p}(\Omega)$, [B].

**2.1. Definition.** *A function* $u \in W_{\mathrm{loc}}^{1,p}(\Omega)$ *is called a sub Q-minimum (super Q-minimum) for* (6) *with structure* (7) *if* $Q \geqq 1$ *and*

$$I(u, \{\varphi \neq 0\}) \leqq QI(u + \varphi, \{\varphi \neq 0\}) \tag{8}$$

*for every* $\varphi \leqq 0$, $(\varphi \geqq 0)$, $\varphi \in W_0^{1,p}(\Omega, R^n)$. *A Q-minimum is a function u such that* (8) *is satisfied for each* $\varphi \in W_0^{1,p}(\Omega, R^n)$ *and is thus both a sub and super Q-minimum.*

For a measurable function $u$ defined in $R^n$, I adopt the notation

$$\begin{aligned} A^+(k, r) &= B(x_0, r) \cap \{x : (u(x) - k)^+ > 0\}, \\ A^-(k, r) &= B(x_0, r) \cap \{x : (u(x) - k)^- > 0 \end{aligned} \tag{9}$$

where $x_0$ is a point that is fixed and thus will need not be indicated in the notation.

For brevity, let $\fint_E f$ denote the integral average

$$|E|^{-1} \int_E f$$

where $|E|$ denotes the Lebesgue measure of $E$.

## 3. A Weak Harnack Inequality At The Boundary

The first result follows directly from the methods of [DBT] and is analogous to a result first established by TRUDINGER for solutions of equations in divergence form [T].

Consider an open bounded set $\Omega \subset R^n$ and let $\beta \in W^{1,p}(R^n)$ have the property that $\beta|(R^n - \Omega)$ is continuous. Let $u \in W^{1,p}_{\text{loc}}(\Omega)$ be a sub $Q$-minimum of (6) that weakly assumes the boundary values $\beta$; that is, assume

(10) $$u - \beta \quad W^{1,p}_0(\Omega).$$

The function $u$ is extended to all of $R^n$ by setting $u(x) = \beta(x)$ for $x \in R^n - \Omega$. Let $x_0 \in \partial\Omega$ and define

$$\bar{\beta}(r) = \sup\{\beta(x) : x \in B(x_0, r) - \Omega\}.$$

Choose $k > \beta(3r)$ and define

$$u_k = (u - k)^+$$
$$\bar{u}_k(r) = \sup\{u_k(x) : x \in B(x_0, r)\}.$$

Note that if $t < r$ and $\eta$ is a smooth function such that $\eta \equiv 1$ on $B(x_0, 3t)$ and $\operatorname{spt}\eta \subset B(x_0, 3r)$, then $\eta u_k \in W^{1,p}_0(\Omega)$.

Recall from [DBT] the fact that sub $Q$-minima satisfy the following weak Harnack inequality:

$$\sup_{B(r/2)} u \leq C\left[\left(\fint_{B(r)} u^p\right)^{1/p} + r^\alpha\right]$$

whenever $B[r) \subset \Omega$. In particular, $u$ is bounded on compact subsets of $\Omega$, the bound depending only o n $\|u\|_{p;\Omega}$ and the structure (7). This remains true at the boundary of $\Omega$ provided (10) holds. The upper bound, of course, will depend also on the maximum of $\beta$ in $\partial\Omega$. Indeed, it is easy to see that the methods of [DBT] can be used without any essential change to establish a weak Harnack inequality for $u_k$.

I am now in a position to state and prove the weak Harnack inequality for super $Q$-minima at the boundary.

**3.1. Theorem.** *There exist positive constants $C$ and $p_0$ depending only on the structure* (6) *and* (7) *with $C = C(\|u\|_{p;\Omega})$ in addition, such that*

$$r^\alpha + \bar{u}_k(2r) - \bar{u}_k(r) \geq C\left(\fint_{B(x_0,r)} [\bar{u}_k(2r) - u_k]^{p_0}\right)^{\frac{1}{p_0}}$$

**Proof.** Let $v(x) = \bar{u}_k(2r) - u_k(x)$ and for $0 < h < \bar{u}_k(2r)$ define a test function $\varphi$ by

(11) $$\varphi = \eta(v - h)^-$$

where $0 \leq \eta \leq 1$, $\operatorname{spt}\eta \subset B(x_0, 2sr)$, and $\eta \equiv 1$ on $B(x_0, 2tr)$, $0 < t < s < 1$. Note that $v = \bar{u}_k(2r)$ on $B(x_0, 3r) - \Omega$ and therefore $\varphi \in W^{1,p}_0[\Omega \cap B(x_0, 3r)]$. Also note that

(12) $$\bar{v}(2r) \leq \bar{u}_k(2r)$$

where $\bar{v}(2r) = \sup\{v(x) : x \in B(x_0, 2r)\}$. If (11) is employed in (6) it follows that

(13) $$\int_{\{\varphi \neq 0\}} |\nabla u|^p \leq aQ \cdot \int_{\{\varphi \neq 0\}} |\nabla(u - \varphi)|^p + aQ \int_{\{\varphi \neq 0\}} |u - \varphi|^p + \int_{\{\varphi \neq 0\}} |u|^p + (1 + Q) \int_{\{\varphi \neq 0\}} g.$$

But $\nabla u = -\nabla v$, and therefore

(14) $$\int_{\{\varphi \neq 0\}} |\nabla v|^p \leqq aQ \int_{\{\varphi \neq 0\}} |\nu v - \eta \nabla v + \nabla \eta (v - )h^-|^p + C|A^-(h, 2sr)|^{1-1/q}.$$

where $C$ depends on the structure (7) and $\sup |u|$ which in turn depends of $\int_\Omega |u|^p$. (14) can be written in the form

(15) $$\int_{A^-(h,t2r)} |\nabla v|^p \leqq aQ \int_{A^-(h,s2r)} (1-\eta)^p |\nabla v^p| + |\nabla \eta|^p (v-h)^{-p} + C|A^-(h, s2r)|^{1-1/q}$$

where $A^-(h, \varrho) = B(x_0, \varrho) \cap \{x : (v-h)^-(x) > 0\}$. By use of the well-known "hole-filling" device and Lemma 3.2 of [GG] it is easy to see that inequality (15) implies that $v$ is an element of the DeGiorgi class $DG_p^-(\Omega)$ as defined in [DBT]. Therefore, it follows that there exists $p_0 > 0$ such that

$$r^\alpha + \inf_{B(x_0,r)} v \geqq C\left\{ \int_{B(x_0,r} [\bar{u}_k(2r) - u_k(x)]^{p_0} \, dx \right\}^{1/p_0}$$

or equivalently,

$$r^\alpha + \bar{u}_k(2r) - \bar{u}_k(r) \geqq C\left\{ \int_{B(x_0,r)} [\bar{u}_k(2r) - u_k(x)]^{p_0} \, dx \right\}^{1/p_0}$$

which proves the theorem.

We now establish two lemmas that were proved by IVERT [I]. For convenience of the reader we sketch their proofs. We also recall another formulation of $p$-capacity which is equivalent to $\gamma_p$ on all Suslin sets; *cf.* [ME]:

(16) $$C_p(E) = \sup \{\| R_1 * \mu \|_{p'}^{-1} : \operatorname{spt} \mu \subset E, \, \mu(E) = 1\}.$$

Here, $\mu$ is a positive Borel measure, $p' = \dfrac{p}{p-1}$ and $R_1(x) = |x|^{1-n}$ is the Riesz kernel of order 1. The proof of the first lemma is very similar to that of the Poincaré inequality, *cf.* [P], [GZ1].

**3.2. Lemma.** *Let* $u \in W^{1,p}(B)$ *where* $B$ *is an open ball of radius* $r$ *and let* $\mu$ *be a probability measure supported in* $B \cap \{x : u(x) = 0\}$. *Then, if* $k > 0$,

(17) $$k\,|\{x : |u(x)| \geqq k\}| \leqq Cr \int_B |\nabla u| + Cr^n \int_B (R_1 * \mu) \cdot |\nabla u|.$$

**Proof.** Without loss of generality, it may be assumed that $u$ is smooth. If $x, y \in B$ with $u(y) = 0$, then

$$|u(x)| \leqq |u(x) - u(z)| + |u(z) - u(y)|$$

whenever $z \in B$. Therefore, by employing polar coordinates, it easily follows that

(18) $$u(x) \leqq C[R_1 * (\chi_B \cdot |\nabla u|)(x) + R_1 * (\chi_B \cdot |\nabla u|)(y)]$$

where $\chi_B$ denotes the characteristic function of $B$. If (18) is integrated with respect to $d\mu(y)$, then by Fubini's Theorem,

$$|u(x)| \leqq C\left[R_1*(\chi_B\cdot|\nabla u|)(x) + \int_B (R_1*\mu)\cdot|\nabla u|\right]. \tag{19}$$

Now integrating (19) over the set $\{|u| \geqq k\}$ yields the desired result.

In the following lemma, the notation and assumptions introduced at the beginning of this section will be in force. In addition, let

$$k_j = \bar{u}(2r) - 2^{-j}\bar{u}_k(2r), \quad j = 1, 1, \ldots, \tag{20}$$

$$A^+(k_j, r) = B(x_0, r) \cap \{x : (u - k_j)^+(x) > 0\}$$

where

$$\bar{u}(2r) = \sup\{u(x) : x \in B(x_0, 2r)\}.$$

Also, recall the definition of $\delta(r)$ from (2).

**3.3. Lemma.** *There is a constant $C > 0$ depending only on the structure* (7) *and $\|u\|_{p;\Omega}$ such that for any positive integer $s$,*

$$\frac{|A^+(k_s, r)|}{|B(x_0, r)|} \leqq C\delta(r)^{-1}\, s^{\frac{1-p}{p}}\left(1 + \frac{2^s f(r)}{\bar{u}_k(2r)}\right) \tag{21}$$

*where $f(r) = (|k| + \bar{u}(2r))\, r + Cr^{\varepsilon}$.*

**Proof.** Let $\mu$ be a positive measure of unit mass supported in $B(x_0, r) - A^+(k, r)$ with the property that

$$\int_{R^n} (R_1*\mu)^{p'} \leqq 2\delta(r)^{-p'} r^{\frac{n-p}{1-p}}. \tag{22}$$

Now apply Lemma 3.2 to $u_{k_j} = (u - k_j)^+$ and the set $A^+(k_{j+1}, r)$ to obtain

$$\begin{aligned} &|k_{j+1} - k_j| \cdot |A^+(k_{j+1}, r)| \\ &\leqq Cr \int_{T(k_j, k_{1+j}; r)} |\nabla u| + Cr^n \int_{T(k_j, k_{j+1}; r)} |\nabla u|\,(R_1*\mu) \\ &\leqq C\left(\int_{A^+(k_j, r)} |\nabla u|^p\right)^{\frac{1}{p}} \left(r \cdot |T(k_j, k_{j+1}; r)|^{1-\frac{1}{p}} + r^n \left\{\int_{T(k_j, k_{j+1}; r)} (R_1*\mu)^{p'}\right\}^{1-\frac{1}{p}}\right) \end{aligned} \tag{23}$$

where

$$T(k_j, k_{j+1}; r) = A^+(k_j, r) - A^+(k_{j+1}, r).$$

Consider the test function

$$\varphi = -\eta(u - k_j)^+$$

where $\eta$ is a cut-off function supported in $B(x_0, 2r)$ and identically 1 on $B(x_0, r)$.

If $\varphi$ is employed in (8), then the argument of [GG, § 4] or [DBT] yields

$$(24)\qquad \int\limits_{A^+(k_j,r)} |\nabla u|^p \leqq C\,[r^{n-p}(\bar{u}(2r)-k_j)^p + (k_j^p + \|g\|_q)\,|A^+(k_j,r)|^{1-\frac{1}{q}}]$$

$$\leqq C r^{n-p}(k_{j+1}-k_j)^p \left[1+\frac{r^\varepsilon}{\bar{u}(2r)-k_s}\right]^p$$

where $s \geqq j+1$ and $\varepsilon = \left(p-\dfrac{n}{q}\right)p^{-1}$. Here I have used the fact $k_{j+1}-k_j = (1/2)\,[\bar{u}(2r)-k_j]$. Combining (23) and (24) yields

$$(25)\qquad |A^+(k_{j+1},r)| \leqq C r^{\frac{n}{p}}\left(1+\frac{r^\varepsilon}{\bar{u}(2r)-k_s}\right) \times \left(|T(k_j,k_{j+1};r)|^{1-1/p} + r^{n-1}\left[\int\limits_{T(k_j,k_{j+1};r)} (R_1 * \mu)^{p'}\right]^{1-1/p}\right).$$

$$(26)\qquad |A^+(k_s,r)|^{p/(p-1)} \leqq |A^+(k_{j+1},r)|^{p/(p-1)}$$

$$\leqq C r^{n/(p-1)}\left(1+\frac{r^\varepsilon}{\bar{u}(2r)-k_s}\right)^{p/(p-1)} \left(|T(k_j,k_{j+1};r)| + r^{p(n-1)/(p-1)} \int\limits_{T(k_j,k_{j+1};r)} (R_1 * \mu)^{p'}\right).$$

Summing (26) over $j$ from 0 to $s-1$ implies

$$s\,|A^+(k_s,r)|^{p/p-1} \leqq C r^{n/p-1}\left(1+\frac{r^\varepsilon}{\bar{u}(2r)-k_s}\right)^{p/(p-1)} \cdot (r^n + r^n\delta(r)^{-p'})$$

$$\leqq C\delta(r)^{-p'}\, r^{np/(p-1)}\left(1-\frac{2^s r^\varepsilon}{\bar{u}_k(2r)}\right)^{p/(p-1)}$$

and hence, the conclusion.

**3.4. Theorem.** *Let $u \in W^{1,p}(\Omega)$ be a sub Q-minimum such that $u-\beta \in W_0^{1,p}(\Omega)$ where $\beta \in W^{1,p}(R^n)$ is continuous on $R^n-\Omega$. There is a constant $C>0$ depending only on the structure* (7) *and $\|u\|_{p;\Omega}$ such that if*

$$\int\limits_0^1 \exp\left(-[C\delta(r)^{-p'}]\right)\frac{dr}{r} = \infty,$$

*then*

$$(27)\qquad \limsup_{\substack{x\to x_0\\ x\in\Omega}} u(x) \leqq \beta(x_0)$$

**Proof.** Referring to the notation at the beginning of this section, assume, contrary to (27), that there are numbers $k>\beta(x_0)$ and $\theta>0$ such that

$$(28)\qquad \lim_{r\to 0} \bar{u}_k(2r) \geqq \theta > 0.$$

Because $\beta|(R^n - \Omega)$ is assumed to be continuous, it is clear that

(29) $$k > \bar{\beta}(3r).$$

for all small $r > 0$. In what follows, only such $r$ will be considered. From Theorem 3.1 it follows that

(30) $$\left( \fint_{B(x_0,r)} [\bar{u}_k(2r) - u_k]^{p_0} \right)^{1/p_0} \leqq C[\bar{u}_k(2r) - \bar{u}_k(r) - r^{\alpha}\}$$

for some $p_0 > 0$. Let

(31) $$\omega(r) = \bar{u}_k(2r) - \bar{u}_k(r) + r^{\alpha}$$

and note that

(32) $$\int_0^1 \omega(r) \frac{dr}{r} < \infty.$$

In the notation of (20), it follows from (30) that

$$\omega(r)^{p_0} \geqq \alpha(n)^{-1} r^{-n} \int_{A^+(k_j,r)} [\bar{u}_k(2r) - u_k]^{p_0} + \alpha(n)^{-1} r^{-n} \int_{A^-(k_j,r)} [\bar{u}_k(2r) - u_k]^{p_0}$$

$$\geqq \alpha(n)^{-1} r^{-n} \int_{A^-(k_j,r)} [\bar{u}_k(2r) - (k_j - k)]^{p_0} = \alpha(n)^{-1} r^{-n} \int_{A^-(k_j,r)} [\bar{u}(2r) - k_j]^{p_0}$$

$$\geqq [2^{-j} \bar{u}_k(2r)]^{p_0} \left[ \frac{|A^-(k_j, r)|}{\alpha(n)\, r^n} \right], \quad \text{for any integer } j \geqq 1.$$

Here $\alpha(n)$ denotes the volume of the unit $n$-ball. Thus, by (28)

$$\frac{|A^-(k_j, r)|}{|B(x_0, r)|} \leqq (\theta^{-1} 2^j \omega(r))^{p_0}$$

and therefore,

$$\frac{|A^+ k_j, r)|}{|B(x_0, r)|} \geqq 1 - (\theta^{-1} 2^j \omega(r))^{p_0}$$

Now refer to (21) and obtain

(33) $$\delta(r)\, [1 - (\theta^{-1} 2^s \omega(r))^{p_0}] \leqq C s^{\frac{1-p}{p}} [1 + \theta^{-1} 2^s r^{\varepsilon}].$$

for any positive integer $s$. For each small $r > 0$, let $s = s(r)$ be the largest integer such that

(34) $$(\theta^{-1} 2^{s(r)} \omega(r))^{p_0} \leqq 1/2.$$

Without loss of generality, it may be assumed that $\omega(r) \geqq r^{\varepsilon}$ and therefore (33) and (34) imply

(35) $$\delta(r) \leqq C s(r)^{\frac{1-p}{p}}.$$

Because $s(r)$ is the largest integer to satisfy (34) it follows that

$$(\theta^{-1}2^{s(r)}\omega(r))^{p_0} \geqq (1/2)^{p_0+1}$$

and therefore that

$$s(r) + \ln[\omega(r)] \geqq \ln M, \qquad M = M(\theta, p_0).$$

Thus, from (35) it follows that

$$\omega(r) \geqq M \exp(-[C\delta(r)^{-p'}]).$$

Because of (32), this yields a contradiction, and the theorem is proved.

As mentioned in the introduction, a sharper condition for boundary regularity can be obtained in the case of weak solution of quasilinear equations in divergence form. Indeed, condition (5) is sufficient for regularity at the boundary. In an attempt to obtain a similar form for the condition in the case of $Q$-minima, I consider the implications of replacing the $p$-capacity, $\gamma_p$, in the above analysis by $\gamma_q$ where $1 < q < p$. Recall that

$$\gamma_p(E) \leqq C[\gamma_q(E)]^{\frac{n-p}{n-q}}. \tag{36}$$

It will be shown below (Theorem 3.6) that if $\gamma_q$ is employed, then the condition in Theorem 3.4 can be replaced by one that allows faster convergence of $\delta_q(r)$ to 0, where

$$\delta_q(r) = \frac{\gamma_q[B(x_0, r) - \Omega]}{\gamma_q[B(x_0, r)]}.$$

The analysis is begun by invoking the notation introduced at the beginning of this section and by again assuming (28) and (29). The number $k$ thus remains fixed for the remainder of the discussion.

First, recall the following well-known version of the Poincaré inequality; *cf.* [GZ1].

**3.5. Lemma.** *With $u$ and $k$ chosen as above, there is an absolute constant $K$ such that*

$$\gamma_q[A^-(k,r)] \int_{A^+(k,r)} |u_k|^q \leqq Kr^n \int_{A^+(k,r)} |\nabla u_k|^q. \tag{37}$$

Now recall from Theorem 3.1 that

$$\omega(r) \geqq C\left( \fint_{B(x_0,r)} [\bar{u}_k(2r) - u_k]^{p_0} \right)^{1/p_0} \tag{38}$$

where $\omega(r) = \bar{u}_k(2r) - \bar{u}_k(r) + r^\alpha = \bar{u}(2r) - \bar{u}(r) + r^\alpha$. For each $r > 0$ choose $l = l(r)$ so that $k < 1 < \bar{u}(2r)$ and note that

$$\omega(r) \geqq C\left( \alpha(n)^{-1} r^{-n} \int_{T(k,l;r)} [\bar{u}_k(2r) - u_k]^{p_0} \right)^{1/p_0}. \tag{39}$$

However, $\bar{u}_k(2r) - u_k \geqq \bar{u}(2r) - 1$ on $T(k, l; r)$ and therefore (39) implies

$$\omega(r)^{p_0} \geqq C[\bar{u}(2r) - l]^{p_0} \frac{|T(k, l; r)|}{r^n}. \tag{40}$$

Choose $0 < q_0 < p_0$. If $l = 1(r)$ is chosen so that

$$\bar{u}(2r) - l \geqq \omega(r)^{q_0/p_0}, \tag{41}$$

then

$$\omega(r)^{p_0 - q_0} \geqq C \frac{|T(k, l; r)|}{r^n}. \tag{42}$$

Let $0 < \sigma < 1$, and in addition to (41), assume that $l$ is chosen to satisfy

$$\bar{u}(2r) - l \leqq (\omega r)^{\sigma \left(\frac{q_0}{p_0}\right)}. \tag{43}$$

It follows from (37) that

$$\delta_q(r)\, r^{-n} \int_{A^+(k,r)} |u_k|^q \leqq K r^{q-n} \left[ \int_{T(k,l;r)} |\nabla u_k|^q + \int_{A^+(l,r)} |\nabla u_k|^q \right]. \tag{44}$$

Now

$$\int_{T(k,l;r)} |\nabla u_k|^q \leqq \left( \int_{T(k,l;r)} |\nabla u_k|^p \right)^{\frac{q}{p}} |T(k, l; r)|^{\frac{p-q}{p}} \tag{45}$$

and Hölder's inequality applied again along with (24) and (43) yields

$$r^{q-n} \int_{A^+(l,r)} |\nabla u|^q \leqq C[(\bar{u}(2r) - l)^q + r^\beta] \leqq C\,[\omega(r)^{q\sigma \left(\frac{q_0}{p_0}\right)} + r^\beta] \tag{46}$$

where $\beta > 0$ depends only on the structure (6) and $q$. It follows from (45) and (42) that

$$\begin{aligned} r^{q-n} \int_{T(k,l;r)} |\nabla u|^q &\leqq C r^{q-n} \cdot (r^{n-p})^{\frac{q}{p}} |T(k, l; r)|^{\frac{p-q}{p}} \\ &\leqq C \left( \frac{|T(k, l; r)|}{r^n} \right)^{\frac{p-q}{p}} \leqq C\omega(r)^{(p_0 - q_0)\frac{(p-q)}{p}}. \end{aligned} \tag{47}$$

Combining (46) and (47) we see that (44) now implies

$$\left( \delta_q(r) \fint_{B(x_0,r)} |u_k|^q \right)^\lambda \leqq C[\omega(r) + r^{\beta\lambda}] \tag{48}$$

where $\lambda = \lambda(p, q, p_0, q_0, \sigma)$. Because $u_k$ is bounded, (38) and (28) imply that

$$\lim_{r \to 0} \fint_{B(x_0,r)} |u_k|^q \geqq \theta^q > 0.$$

Thus, (48) now yields

$$[\delta_q(r)\, \theta^q]^\lambda \leqq C[\omega(r) + r^{\beta\lambda}].$$

In view of the fact that

$$\int_0^1 [\omega(r) + r^{\beta\lambda}] \frac{dr}{r} < \infty,$$

the following theorem is thus established.

**3.6. Theorem.** *Let* $u \in W^{1,p}(\Omega)$ *be a sub Q-minimum such that* $u - \beta \in W_0^{1,p}(\Omega)$ *where* $\beta \in W^{1,p}(R^n)$ *is continuous on* $R^n - \Omega$. *Let* $1 < q < p$. *Then there is a constant* $\lambda > 0$ *such that if*

$$\int_0^1 [\delta_q(r)]^\lambda \frac{dr}{r} = \infty$$

*then*

$$\limsup_{\substack{x \to x_0 \\ x \in \Omega}} u(x) \leqq \beta(x_0).$$

Of course, both Theorems 3.4 and 3.6 have analogous statements for super $Q$-minima and therefore, one can conclude the desired result for continuity at the boundary for $Q$-minima as stated in the introduction.

Observe that Theorems 3.4 and 3.6 are independent in the sense that neither one implies the other. It would be interesting to know whether Theorem 3.4 admits a significantly sharper result.

*Acknowledgment.* This research was supported in part by a grant from the National Science Foundation.

## References


[B] BAGBY, T., *Quasitopologies and rational approximation*, J. Functional Analysis, **10** (1972) 259–268.

[DBT] DIBENEDETTO, E., & N. S. TRUDINGER, *Harnack inequalities for quasi-minima of variational integrals*, Australian National University Research report, 1984.

[FZ] FEDERER, H., & W. P. ZIEMER, *The Lebesgue set of a function whose partial derivatives are p-th power summable*, Indiana Univ. Math. J., **22** (1972), 139–158.

[GZ1] GARIEPY, R., & W. P. ZIEMER, *A regularity condition at the boundary for solutions of quasilinear elliptic equations*, Arch. Rational Mech. Anal., **67** (1977), 25–39.

[GZ2] GARIEPY, R., & W. ZIEMER, *Behavior at the boundary of solutions of quasilinear elliptic equation*, Arch. Rational Mech. Analysis, **56** (1974) 372–384.

[GG] GIAQUINTA, M., & E. GIUSTI, *Quasi-Minima*, Ann. L'Inst. Henri Poincaré Analyse Non Linéaire, **1** (1984).

[I] IVERT, PER-ANDERS, *Continuity of quasiminima under the presence of irregular obstacles*, preprint, (1984),

[KS] KRYLOV, N. V., & M. V. SAFONOV, *Certain properties of solutions of parabolic equations with measurable coefficients*, Math. USSR Izv., **16** (1981).

[ME] MEYERS, N. G., *A theory of capacities for potentials of functions in Lebesgue spaces*, Math. Scand. **26** (1970), 255–292.

[P] POLKING, J., *Approximation in $L^p$ by solutions of elliptic partial differential equation*, Amer. J. Math., **94** (1972), 1231–1244.

[T] TRUDINGER, N. S., *On Harnack type inequalities and their application to quasi-linear elliptic equations*, Comm. Pure Appl. Math., **20** (1967), 721–747.

Mathematics Department
Indiana University
Bloomington

*(Received July 15, 1985)*

# *Positively Invariant Regions for a Problem in Phase Transitions*

V. ROYTBURD & M. SLEMROD

*Dedicated to Professor James Serrin on the occasion of his sixtieth birthday*

**Abstract**

Positively invariant regions for the system $v_t + p(w)_x = \varepsilon v_{xx}$, $w_t - v_x = \varepsilon w_{xx}$ are constructed where $p' < 0$, $w < \alpha$, $w > \beta$, $p'(w) = 0$, $\alpha \leqq w \leqq \beta$, $\varepsilon > 0$. Such a choice of $p$ is motivated by the Maxwell construction for a van der Waals fluid. The method of an analysis is a modification of earlier ideas of CHUEH, CONLEY, & SMOLLER [1]. The results given here provide $\varepsilon$ independent $L_\infty$ bounds on the solution $(w, v)$.

## 1. Introduction

The purpose of this paper is to construct positively invariant regions in the $wv$-plane for the system

$$\begin{aligned} v_t + p(w)_x &= \varepsilon v_{xx}, \\ w_t - v_x &= \varepsilon w_{xx}, \end{aligned} \qquad -\infty < x < \infty, \qquad t > 0, \tag{1.1}$$

with initial conditions

$$w(x, 0) = w_0(x), \qquad v(x, 0) = v_0(x), \qquad -\infty < x < \infty \tag{1.2}$$

where $\varepsilon > 0$ and $p$ has the graph shown in Figure 1a. By a *positively invariant region* we mean a set $D$ (independent of $\varepsilon$) in $wv$–plane such that if $(w_0(x), v_0(x))$ lies in $D$ for all $x \in R$ then the solution of (1.1), (1.2) $(w(x, t), v(x, t)) \in D$ for all $x \in R$, $t > 0$.

The motivation for the problem with this choice of $p$ lies in the theory of phase transitions. First of all (1.1) with $\varepsilon = 0$ denotes the Lagrangian form of the balance of mass and linear momentum for an elastic fluid (solid) with stress taken as $-p(w)$. However, unlike the more typical case say for an ideal fluid (or an elastic solid) where $p' < 0$ we consider the non-standard Figure 1a. One may *observe* such a $p$ (= pressure) versus $w$ (= specific volume (or the deformation gradient)) diagram in the *equilibrium* configurations of materials undergoing first order phase transitions.

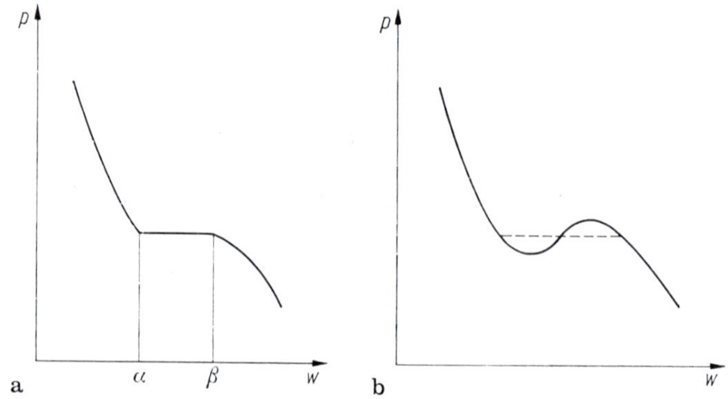


Fig. 1. The model (a) and van der Waals (b) $p$–$w$ diagrams.

Classical equilibrium theories of phase transitions have reconciled such observed $p$–$w$ graphs with non-monotone constitutive relations for $p$ such as the van der Waals fluid. (A typical isotherm is shown in Figure 1b). Some reconciliations for equilibrium theory are based on the following ideas:

(1) The Maxwell line—equal area rule—construction based on the principle that the total work done in a reversible cycle is zero (see *e.g.* FERMI [2].) The Maxwell line is sketched in Figure 1b. This construction claims that the only thermodynamically admissible states will be those seen in Figure 1a.

(2) The equilibrium distribution is a minimizer of the non-convex total free energy for a van der Waals like fluid. The Weierstrass-Erdmann corner condition of the calculus of variations yields the Maxwell condition. (See ERICKSEN [3], DACOROGNA [4], and BALL [5].)

(3) The solutions of the equilibrium problem should be taken from a regularized higher order problem which includes the effect of interfacial surface energy. Such an approach again will recover the Maxwell construction in the standard regularization. (See VAN DER WAALS [6] and references given there. For more recent ideas on such regularizations, see AIFANTIS & SERRIN [7], [8], CARR, GURTIN, & SLEMROD [9], [10], and GURTIN [11].)

(4) The equilibrium $p$–$w$ diagram as determined by statistical mechanics for a fluid whose interparticle potential is $\phi(r) = \psi_{hc}(r) + \phi_{att}(r)$, where $r$ is the distance from a particle, $\phi_{hc}(r) = \infty$ if $r < r_0$, $\phi_{hc}(r) = 0$ if $r > r_0$ and $\phi_{att}(r) = \gamma e^{-\gamma r}$. In this case the equation of state is given by the van der Waals equation $p(w) = R\theta/(w - b) - a/w^2$, $\theta$ the temperature, $a > 0, b > 0$, *with* the Maxwell construction. (See THOMPSON [12], KAC, UHLENBECK, & HEMMER [13].)

Our problem, however, is to study *non-equilibrium* dynamics. Consistently with idea (3) above, we introduce the singular right hand side of (1.1) with $\varepsilon > 0$. This has the effect incorporating both viscosity and interfacial capillarity into the dynamics consistent with VAN DER WAALS' original ideas in [6]. (See SLEMROD [14] and the references cited there.) Also and most importantly while our goal is to understand the dynamics with a non-monotone equations of state such as the van der Waals fluid, we choose for the time being the known *equilibrium* $p$–$w$ graph in Figure 1a as our equation of state. Although this choice lacks physical consistency vis-a-vis non-monotone constitutive assumptions like Figure 1b, from a mathematical viewpoint it provides a starting point to attempt to understand non-strictly monotone constitutive relations for which *both* Figures 1a and 1b are examples.

The role of positively invariant regions in this program follows from recent developments in the theory of partial differential equations. The existence of a positively invariant region for (1.1), (1.2) will provide us with an $\varepsilon$ independent $L_\infty$ $(R \times R^+)$ bound for solutions $(w_\varepsilon, v_\varepsilon)$ of (1.1), *i.e.*

$$\sup_{t>0} \sup_{x\in R} (|w_\varepsilon(x,t)| + |v_\varepsilon(x,t)|) \leqq \text{const.}$$

where the constant depends only on $(w_0, v_0)$ and is independent of $\varepsilon$ for initial data $(w_0, v_0) \in D$. Hence $\{w_\varepsilon, v_\varepsilon\}$ possesses a subsequence which converges weak-star $L^\infty$ as $\varepsilon \to 0$. We then hope to show that TARTAR's method of compensated compactness [15] coupled with recent ideas of DIPERNA [16] will serve to show that this limit is a weak solution of (1.1) with $\varepsilon = 0$. We hope to carry out this last step in future research.

## 2. A monotone mollification of the "flat" $p$

In this section we construct a $C^\infty$ monotone approximation $p_\delta$ to the "flat" $p$ shown in Figure 1 where $p_\delta$ contains precisely one inflection point at $w_i = \dfrac{\alpha+\beta}{2}$.

To begin with we assume the "flat" $p$ has the following form (see Figure 1a):

(1) $p \in C(-\infty, \infty)$;

(2) $p(w) = \text{const.}, \ \alpha \leqq w \leqq \beta$;

(3) the non-flat pieces of $p$ are smooth:

$$p_1 \equiv p|_{(-\infty,\alpha]} \in C^k(-\infty, \alpha];$$
$$p_2 \equiv p|_{[\beta,\infty)} \in C^k[\beta, \infty),$$

where $k \geqq 2$;

(4) $-M_1 < p_1'(w) < 0, \quad w < \alpha,$
$\quad -M_1 < p_2'(w) < 0, \quad w > \beta,$

(5) $\quad M_2 > p_1''(w) > 0, \quad w < \alpha,$
$\quad -M_2 < p_2''(w) < 0, \quad w > \beta,$

(6) on $(-\infty, \alpha] \cup [\beta, \infty)$, either $p''$ is uniformly continuous or $|p'''| \leqq \text{const.}$

Some remarks are due concerning our choice of $p$. We require $p$ to be concave to the right of the domain where $p$ is flat (see (5)). This requirement is consistent with van der Waals isotherms for absolute temperatures in some interval ($\theta^*$, $\theta_{\text{critical}}$). This convexity-concavity switch also seems to agree with the constitutive relation suggested by ERICKSEN in his paper [3].

The assumptions (1)–(6) are formulated on $-\infty < w < \infty$. In our motivating physical examples $p$ does not have this global form. However, if the form of $p$ locally near the flat piece satisfies (1)–(6), one could then extend the graphs of $p$ in the examples to obtain a $p$ with the above features. Since we shall show that regions in the neighborhood of the flat piece are invariant, *a posteriori* the extensions away from the flat piece will prove irrelevant.

Obviously $p \notin C^k(-\infty, \infty)$ for $k \geqq 1$. The first step in our analysis will be a mollification on $p$.

**Proposition 2.1.** Let $\phi \in C_0^\infty(-1, 1)$, $\phi \geqq 0$, $\int \phi \, dx = 1$. Denote by $J_\delta$ the mollification operator

$$(J_\delta p)(w) = \frac{1}{\delta} \int_{-\infty}^{\infty} \phi\left(\frac{w - x}{\delta}\right) p(x)\, dx.$$

Then $J_\delta p$ is in $C^\infty(R)$ and

(i) $J_\delta p$ approximates $p$ uniformly:

$$\sup_w |J_\delta p - p| \leqq M_1 \delta$$

where

$$M_1 = \sup |p'|;$$

(ii) away from the points of discontinuity of $p'$ derivatives of $J_\delta p$ approximate derivatives of $p$:

$$\sup_{w \in W_l} |(J_\delta p)^{(k)}(w) - p^{(k)}(w)| \leqq \delta \sup_{w \in W_r} |p^{(k+1)}(w)|$$

where

$$W_l = \{w; |w - \alpha| > \nu\} \cap \{w; |w - \beta| > \nu\}$$

and

$$W_r = \{w; |w - \alpha| > \nu - \delta\} \cap \{w; |w - \beta| > \nu - \delta\},$$

for $\nu > 0$, $0 < \delta < \nu$;

(iii) $-M_1 < (J_\delta p)'(w) \leqq 0$ with strict inequality for $w < \alpha$ or $w > \beta$;

(iv) $(J_\delta p)''(w) \geqq 0$ for $w < \alpha + \delta$ with strict inequality for $w < \alpha$,

$(J_\delta p)''(w) \leqq 0$ for $w > \beta - \delta$ with strict inequality for $w > \beta$,

$(J_\delta p)''(w) = 0$ for $\alpha + \delta \leqq w \leqq \beta - \delta$.

Here $\nu$, $\delta$, $0 < \delta < \eta$ are fixed and small.

**Proof.** From the definition of $J_\delta$ we have

$$(J_\delta p)(w) - p(w) = \int_{-\infty}^{\infty} \frac{1}{\delta} \phi\left(\frac{w-x}{\delta}\right) p(x)\, dx - p(w)$$
$$= \int_{w-\delta}^{w+\delta} \frac{1}{\delta} \phi\left(\frac{w-x}{\delta}\right) [p(x) - p(w)]\, dx. \tag{2.1}$$

If the whole interval $(w-\delta, w+\delta)$ lies in the domain where $p'$ exists, then $|p(x) - p(w)| = |p'(\theta)(x-w)| \leqq M_1 \delta$. It is easy to see that this same estimate is true even if $x$ and $w$ are on different branches of $p$. If, say, $x < \alpha < w$, then

$$|p(x) - p(w)| \leqq |p(x) - p(\alpha)| + |p(\alpha) - p(w)| = |p'(\theta)|\,|x-\alpha| \leqq M_1 \delta.$$

Substitution of this estimate into (2.1) yields assertion (i) of the proposition.
Assertion (ii) is obtained by $k$ integrations by parts, *i.e.*,

$$|(J_\delta p)^{(k)}(w) - p^{(k)}(w)|$$
$$= \left| \frac{1}{\delta} \int_{w-\delta}^{w+\delta} (\partial_w)^k \phi\left(\frac{w-x}{\delta}\right) p(x)\, dx - \frac{1}{\delta} \int_{w-\delta}^{w+\delta} \phi\left(\frac{w-x}{\delta}\right) p^{(k)}(w)\, dx \right|$$
$$= \left| \frac{1}{\delta} \int_{w-\delta}^{w+\delta} \phi\left(\frac{w-x}{\delta}\right) (p^{(k)}(x) - p^{(k)}(w))\, dx \right|$$
$$\leqq \sup_{w-\delta \leqq x \leqq w+\delta} |x-w| \cdot \sup_{\xi \in W_r} |p^{(k+1)}(\xi)| \leqq \delta \sup_{w \in W_r} |p^{(k+1)}(w)|.$$

In the proofs of (iii) and (iv) we distinguish between the two cases:

(A) $(w-\delta, w+\delta)$ does not contain either $\alpha$ or $\beta$.

(B) $w-\delta < \alpha < w+\delta$ or $w-\delta < \beta < w+\delta$.

In Case A:

$$(J_\delta p)'(w) = \frac{1}{\delta} \int_{w-\delta}^{w+\delta} \phi_w \left(\frac{w-x}{\delta}\right) p(x)\, dx$$
$$= -\frac{1}{\delta} \int_{w-\delta}^{w+\delta} \phi_x \left(\frac{w-x}{\delta}\right) p(x)\, dx$$
$$= \frac{1}{\delta} \int_{w-\delta}^{w+\delta} \phi\left(\frac{w-x}{\delta}\right) p'(x)\, dx,$$

and (iii) follows since $-M_1 < p'(x) < 0$. Assertion (iv) is proved with two integrations by parts.

In Case B we consider for definiteness $w - \delta < \beta < w + \delta$. Then

$$(J_\delta p)'(w) = -\frac{1}{\delta}\int_{w-\delta}^{w+\delta} \phi_x\left(\frac{w-x}{\delta}\right) p(x)\,dx$$

$$= -\frac{1}{\delta}\left(\int_{w-\delta}^{\beta} + \int_{\beta}^{w+\delta}\right) \phi_x\left(\frac{w-x}{\delta}\right) p(x)\,dx$$

$$= \frac{1}{\delta}\left(\int_{w-\delta}^{\beta} + \int_{\beta}^{w+\delta}\right) \phi\left(\frac{w-x}{\delta}\right) p'(x)\,dx$$

$$- \left(\phi p|_{w-\delta}^{\beta} + \phi p|_{\beta}^{w+\delta}\right)$$

$$= \frac{1}{\delta}\left(\int_{w-\delta}^{\beta} + \int_{\beta}^{w+\delta}\right) \phi\left(\frac{w-x}{\delta}\right) p'(x)\,dx$$

(where the boundary terms vanish since $\phi(1) = \phi(-1) = 0$). Hence again (iii) follows. Integrating by parts twice, we obtain

$$(J_\delta p)''(w) = \frac{1}{\delta}\left(\int_{w-\delta}^{\beta} + \int_{\beta}^{w+\delta}\right) \phi\left(\frac{w-x}{\delta}\right) p''(x)\,dx$$

$$+ \frac{1}{\delta}\left[\phi_x p|_{w-\delta}^{\beta} + \phi_x p|_{\beta}^{w+\delta} - \phi p'|_{w-\delta}^{\beta} - \phi p'|_{\beta}^{w+\delta}\right]$$

$$= \frac{1}{\delta}\left(\int_{w-\delta}^{\beta} + \int_{\beta}^{w+\delta}\right) \phi\left(\frac{w-x}{\delta}\right) p''(x)\,dx$$

$$+ \frac{1}{\delta}\phi\left(\frac{w-\beta}{\delta}\right)[p'(\beta+0) - p'(\beta-0)] \leqq 0.$$

The proposition is proved.

Now let us consider a bell-shaped function

$$Q(x) = \begin{cases} \exp\left(\dfrac{1}{x^2-1}\right), & |x| < 1 \\ 0, & |x| \geqq 1. \end{cases}$$

It is easily seen that $Q''(x) > 0$ for $x_0 < |x| < 1$, where $x_0 = 3^{-1/4}$, and $Q'(x) \leqq 0$ for $x \geqq 0$. The desired approximation to $p(x)$ is described in the following proposition.

**Proposition 2.2.** Consider

$$p_\delta(w) = (J_\delta p)(w) + \delta\left[Q\left(\frac{w-\alpha+a}{\frac{1}{2}(\alpha+\beta)-\alpha+a}\right) - Q\left(\frac{w-\beta-a}{\beta+a-\frac{1}{2}(\alpha+\beta)}\right)\right]$$

$$\equiv (J_\delta p)(w) + \delta q(w)$$

where $a$ is so chosen that the inflection points $x_1$ and $x_2$ of the bell-shaped functions centered at $\alpha - a$ and $\beta + a$ lie outside of $[\alpha, \beta]$. (See Fig. 2.)

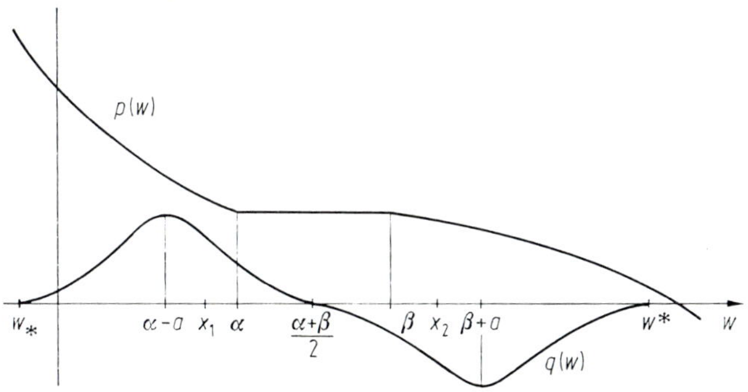


Fig. 2

Then there is a $\delta_0 > 0$ such that for $\delta < \delta_0$

(i) $p_\delta'(w) < 0, \quad w \neq (\alpha + \beta)/2$

(ii) $|p_\delta'(w)| \leqq L$

(iii) $p_\delta''(w) > 0, \quad w < (\alpha + \beta)/2,$

$p_\delta''(w) < 0, \quad w > (\alpha + \beta)/2,$

(iv) $\sup_w |p_\delta(w) - p(w)| < \text{const.}\ \delta,$

$\sup_{w \in W_l} |p_\delta'(w) - p'(w)| < \text{const.}\ \delta$

where $W_l$ is defined in Proposition 2.1.

**Proof.** Assertion (ii) is obvious since

$$|p_\delta'(w)| \leqq |(J_\delta p)'(w)| + \delta \max |q'(w)| \leqq M_1 + \delta_0 \max_w |q'(w)|.$$

Both (i) and (iii) are trivial on the interval $(x_1, x_2)$. Outside of $(x_1, x_2)$ we choose a $\delta_0$ such that the following inequalities hold:

$$\min_{w_* \leqq w \leqq x_1} |(J_\delta p)'(w)| > \delta_0 \max_{w_* \leqq w \leqq x_1} q'(w),$$

$$\min_{w_* \leqq w \leqq x_1} |(J_\delta p)''(w)| \geqq \delta_0 \max_{w_* \leqq w \leqq x_1} q''(w),$$

as well as similar inequalities in $[x_2, w^*]$. Here $w_*$ and $w^*$ are equidistant from the point $\frac{1}{2}(\beta + \alpha)$. Finally (iv) follows from Proposition 2.1 (i).

## 3. Solutions of the parabolic system

In this section we discuss existence and regularity of solutions of the system

$$U_t + F(U)_x = \varepsilon U_{xx} \tag{3.1}$$

for $t > 0$, $-\infty < x < \infty$ with initial data

$$U(x, 0) = U_0(x). \tag{3.2}$$

Here $U(x, t) \in \mathbb{R}^2$, $F: \mathbb{R}^2 \to \mathbb{R}^2$, and $F$ is globally Lipschitz continuous in $R^2$, i.e., $|F(U) - F(V)| \leqq L|U - V|$ for some constant $L > 0$.

We use a semigroup approach based on the variation of constants formula to solve (3.1), (3.2) in a certain weak sense. We then investigate regularity of this solution.

We call a function $U \in C([0, T]; C(-\infty, \infty))$ a *mild solution* of (3.1), (3.2) if $U$ satisfies the associated integral equation

$$U(\cdot, t) = G(t) * U_0(\cdot) + \int_0^t G(t - \tau) * F(U(\cdot, \tau))_x \, d\tau, \quad 0 \leqq t \leqq T. \tag{3.3}$$

In (3.3) the dependence on the space variable $x$ is suppressed, $*$ denotes the space convolution on $(-\infty, \infty)$, and $G$ is the kernel of the Green's function for the diffusion equation $U_t = \varepsilon U_{xx}$,

$$G(x, t) = (4\pi\varepsilon t)^{-1/2} \exp\left[-\frac{x^2}{4\varepsilon t}\right] \cdot I$$

where $I$ is the $2 \times 2$ identity matrix.

**Theorem 3.1.** *For any time interval* $[0, T_0)$ *there exists a unique mild solution of* (3.1), (3.2).

**Proof.** Consider the integral operator

$$(KU)(\cdot, t) = G(t) * U_0(\cdot) + \int_0^t G(t - \tau) * F(U(\cdot, \tau))_x \, d\tau.$$

In a standard fashion we will show that $K$ is a contraction for $T_0$ sufficiently small.

Let $\|U(\cdot, t)\| = \sup_x |U(x, t)|$. Then

$$\|(KU)(\cdot, t) - (KV)(\cdot, t)\| = \left\| \int_0^t G(t - \tau) * (F(U(\cdot, \tau))_x - F(V(\cdot, \tau))_x) \, d\tau \right\|.$$

Use of the explicit form of $G$ shows

$$\begin{aligned}
&|G(t - \tau) * (F(U(\cdot, \tau))_x - F(V(\cdot, \tau))_x)| \\
&= \left| \int_{-\infty}^{\infty} (4\pi\varepsilon(t - \tau))^{-\frac{1}{2}} \exp\left[-\frac{(x - y)^2}{4\varepsilon(t - \tau)}\right] (F(U(y, \tau))_y - F(V(y, \tau))_y) \, dy \right| \\
&= \left| \int_{-\infty}^{\infty} (4\pi\varepsilon(t - \tau))^{-\frac{1}{2}} \exp\left[-\frac{(x - y)^2}{4\varepsilon(t - \tau)}\right] \frac{2(x - y)}{4\varepsilon(t - \tau)} [F(U(y, \tau)) - F(V(y, \tau))] \, dy \right| \\
&\leqq (4\pi\varepsilon(t - \tau))^{-\frac{1}{2}} L \|U - V\| \int_{-\infty}^{\infty} \exp\left[-\frac{(x - y)^2}{4\varepsilon(t - \tau)}\right] \left| \frac{2(x - y)}{4\varepsilon(t - \tau)} \right| dy \\
&\leqq L \cdot C(\varepsilon(t - \tau))^{-\frac{1}{2}} \|U - V\|.
\end{aligned}$$

Here $C$ is a constant independent of $L$ and $\varepsilon$. Thus

$$\|(KU)(\cdot, t) - (KV)(\cdot, t)\| \leqq LC\, t^{\frac{1}{2}} \varepsilon^{-\frac{1}{2}} \|U - V\|$$

and for $LCT_0^{\frac{1}{2}} \varepsilon^{-\frac{1}{2}} < 1$,

$$\|KU(\cdot, t) - KV(\cdot, t)\| \leqq c \|U(\cdot, t) - V(\cdot, t)\|$$

where $0 < c < 1$. Finally endow the closed subspace $S = \{U \in C([0, T_0]; C(-\infty, \infty)); U(x, 0) = U_0(x)\}$ with the norm $|||U||| = \sup_{0 \leqq t \leqq T_0} \|U(\cdot, t)\|$. We then see $K: S \to S$ and is a contraction on $S$ and hence possesses a unique fixed point in $S$. Since $F$ is globally Lipschitz-continuous, we can repeat the construction with $U(T_0)$ as initial data, *etc.* to show existence on any finite time interval $[0, T]$. This proves the theorem.

The mild solution turns out to be a classical solution if the nonlinear function $F$ is sufficiently regular. This is described in Theorem 3.2 and the subsequent corollary.

**Theorem 3.2.** *Let* $F \in BC^{k+1}(\mathbb{R}^2)$, $U_0 \in BC^{k+1}$. *Then the mild solution of* (3.1), (3.2) *is $k$ times differentiable with respect to $x$.*

Here $BC^j(\mathbb{R}^n)$ denotes the class of functions possessing $j$ continuous bounded derivatives on $\mathbb{R}^n$.

*Remark.* The theorem is true under the slightly weaker hypothesis: $F \in BC^k(\mathbb{R}^2)$, $U_0 \in BC^k(\mathbb{R})$, with the $k^{\text{th}}$ derivatives uniformly continuous.

**Proof.** We set $k = 1$. The result for larger $k$'s can be obtained by a standard induction argument.

The proof of differentiability in $x$ proceeds in two steps. In the first step we follow an idea of DiPerna [17] to obtain bounds on the difference quotient

$$U_\Delta(x, t) = \frac{1}{\Delta}[U(x + \Delta, t) - U(x, t)], \quad 0 < t < T_0.$$

These bounds do not depend on $\Delta$ for $0 < \Delta < \Delta_0$. In the second step we show that the functions $U_\Delta$, $\Delta \to 0$, form a Cauchy sequence.

We rewrite (3.3) integrating by parts in the second integral, *i.e.*

$$U(x, t) = \int_{-\infty}^{\infty} G(x - y, t)\, U_0(y)\, dy + \int_0^t \int_{-\infty}^{\infty} G_x(x - y, t - \tau)\, F(U(y, \tau))\, dy\, d\tau. \tag{3.3'}$$

From the integral equation (3.3′) we obtain the following identities:

$$U_\Delta(x,t) = \frac{1}{\Delta}\int_{-\infty}^{\infty} [G(x+\Delta-y,t) - G(x-y,t)]\, U_0(y)\, dy$$

$$+ \frac{1}{\Delta}\int_0^t \int_{-\infty}^{\infty} [G_x(x+\Delta-y,t-\tau) - G_x(x-y,t-\tau)]\, F(U(y,\tau))\, dy\, d\tau$$

$$= -\int_{-\infty}^{\infty} G(x-y,t)\left[\frac{U_0(y+\Delta)-U_0(y)}{\Delta}\right] dy$$

$$-\frac{1}{\Delta}\int_0^t \int_{-\infty}^{\infty} G_x(x-y,t-\tau)\,[F(U(y+\Delta,\tau)) - F(U(y,\tau))]\, dy\, d\tau.$$

If $\phi(t)$ denotes $\sup_x |U_\Delta(x,t)|$, then

$$\phi(t) \leqq \int_{-\infty}^{\infty} |G(x-y,t)|\, \phi(0)\, dy + \int_0^t \int_{-\infty}^{\infty} |G_x(x-y,t-\tau)| \sup |F'|\, \phi(\tau)\, dy\, d\tau.$$

Use of the explicit form of $G$ shows

$$\phi(t) \leqq \phi(0) + \text{const.} \sup |F'| \int_0^t (t-\tau)^{-\frac{1}{2}}\, \phi(\tau)\, d\tau$$

and hence from Gronwall's inequality we see that

$$\phi(t) \leqq \phi(0) \exp[\text{const.}\ t^{\frac{1}{2}}] \leqq \phi(0) \exp[\text{const.}\ T_0^{\frac{1}{2}}] \equiv C_0,$$

where const. denotes constants independent of $\Delta$. Note that $\phi(0)$ is finite:

$$\phi(0) = \sup_x \left|\frac{1}{\Delta}[U_0(x+\Delta) - U_0(x)]\right| \leqq \sup_x |U_0'(x)| \leqq \text{const.}$$

In a similar fashion we estimate the difference

$$|U_{\Delta_1}(x,t) - U_{\Delta_2}(x,t)| \leqq \int_{-\infty}^{\infty} |G(x-y,t)|\, |U_{0_{\Delta_1}}(y) - U_{0_{\Delta_2}}(y)|\, dy$$

$$+ \int_0^t \int_{-\infty}^{\infty} |G_x(x-y,t)| \left\{\left|\frac{1}{\Delta_1}[F(U(y+\Delta_1,\tau)) - F(U(y;\tau))]\right.\right.$$

$$\left.\left.-\frac{1}{\Delta_2}[F(U(y+\Delta_2,\tau)) - F(U(y,\tau))]\right|\right\} dy\, d\tau. \tag{3.4}$$

Now if we write $F = (F_1, F_2)$ we have by use of the mean value theorem

$$\frac{1}{\Delta_1}[F_1(U(y+\Delta_1,\tau)) - F_1(U(y,\tau))] = F_1'(z_{11})\, U_{\Delta_1}(y,\tau),$$

where $z_{11} = \theta_{11} U(y + \Delta_1, \tau) + (1 - \theta_{11}) U(y, \tau)$ for some $\theta_{11}(y, \tau, \Delta_1)$, $0 \leqq \theta_{11} \leqq 1$. Similarly we see

$$F_1(U(y + \Delta_2, \tau)) - F_1(U(y, \tau)) = F_1'(z_{12}) U_{\Delta_2}(y, \tau) \Delta_2,$$

$$F_2(U(y + \Delta_1, \tau)) - F_2(U(y, \tau)) = F_2'(z_{21}) U_{\Delta_1}(y, \tau) \Delta_1,$$

$$F_2(U(y + \Delta_2, \tau)) - F_2(U(y, \tau)) = F_2'(z_{22}) U_{\Delta_2}(y, \tau) \Delta_2,$$

where $z_{ij} = \theta_{ij} U(y + \Delta_j, \tau) + (1 - \theta_{ij}) U(y, \tau)$ for some $\theta_{ij}(y, \tau, \Delta_j)$, $0 \leqq \theta_{ij} \leqq 1$, $1 \leqq i, j \leqq 2$. Hence

$$\begin{aligned}
&\left| \frac{1}{\Delta_1} [F(U(y + \Delta_1, \tau)) - F(U(y, \tau))] - \frac{1}{\Delta_2} [F(U(y + \Delta_2, \tau)) - F(U(y, \tau))] \right|^2 \\
&= |F_1'(z_{11}) U_{\Delta_1}(y, \tau) - F_1'(z_{12}) U_{\Delta_2}(y, \tau)|^2 + |F_2'(z_{21}) U_{\Delta_1}(y, \tau) - F_2'(z_{22}) U_{\Delta_2}(y, \tau)|^2 \\
&= |(F_1'(z_{11}) - F_1'(z_{12})) U_{\Delta_1}(y, \tau) + F_1'(z_{12}) (U_{\Delta_2}(y, \tau) - U_{\Delta_1}(y, \tau))|^2 \\
&\quad + |(F_2'(z_{21}) - F_2'(z_{22})) U_{\Delta_1}(y, \tau) + F_2'(z_{22}) (U_{\Delta_2}(y, \tau) - U_{\Delta_1}(y, \tau))|^2 \\
&\leqq 2(|F_1'(z_{11}) - F_1'(z_{12})|^2 + |F_2'(z_{21}) - F_2'(z_{22})|^2) |U_{\Delta_1}(y, \tau)|^2 \\
&\quad + 2(|F_1'(z_{12})|^2 + |F_2'(z_{22})|^2) |U_{\Delta_2}(y, \tau) - U_{\Delta_1}(y, \tau)|^2 .
\end{aligned} \tag{3.5}$$

Since

$$\begin{aligned}
|z_{11} - z_{12}| \leqq |U(y + \Delta_1, \tau) - U(y + \Delta_2, \tau)| &\leqq |U(y + \Delta_2, \tau) - U(y, \tau)| \\
&\quad + |U(y + \Delta_2, \tau) - U(y, \tau)|
\end{aligned}$$

(see Figure 3), we have

$$|z_{11} - z_{12}| \leqq \Delta_1 |U_{\Delta_1}(y, \tau)| + \Delta_2 \| U_{\Delta_2}(y, \tau)| .$$

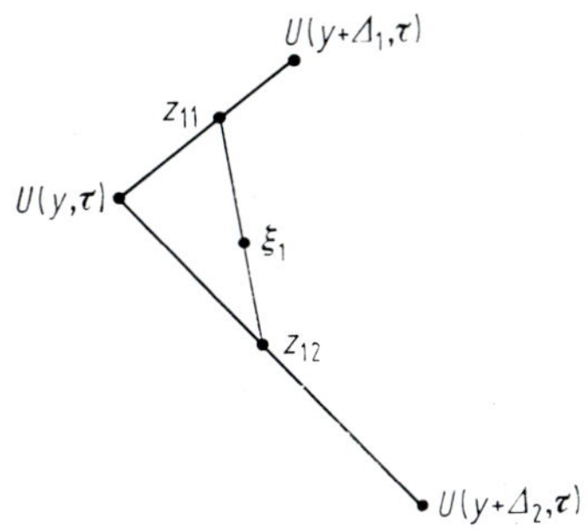


Fig. 3

Hence we obtain the inequality

$$\begin{aligned}
|F_1'(z_{11}) - F_1'(z_{12})| &\leqq |F_1''(\xi_1)| \{\Delta_1 |U_{\Delta_1}(y, \tau)| + \Delta_2 |U_{\Delta_2}(y, \tau)|\} \\
&\leqq 2 \sup |F''| \, C_0 \max(\Delta_1, \Delta_2) .
\end{aligned}$$

A similar estimate holds for $|F_2'(z_{21}) - F_2'(z_{22})|$. Thus we see that

$$\left|\frac{1}{\Delta_1}[F(U(y+\Delta_1,\tau)) - F(U(y,\tau))] - \frac{1}{\Delta_2}[F(U(y+\Delta_2,\tau)) - F(U(y,\tau))]\right|^2$$
$$\leqq 2(4 \sup|F''|\, C_0 \max(\Delta_1,\Delta_2))^2\, |U_{\Delta_1}(y,\tau)|^2$$
$$+ 2(2\sup|F'|^2)\, |U_{\Delta_2}(y,\tau) - U_{\Delta_1}(y,\tau)|^2$$
$$\leqq 32(\sup|F''| \max(\Delta_1,\Delta_2)\, C_0^2)^2 + 4 \sup|F'|^2\, |||U_{\Delta_2}(y,\tau) - U_{\Delta_1}(y,\tau)|^2 .$$

Substitution of this estimate into (3.4) shows

$$\sup_x |U_{\Delta_1}(x,t) - U_{\Delta_1}(x,t)| \leqq \sup_x |U_{0_{\Delta_1}}(x) - U_{0_{\Delta_2}}(x)|$$
$$+ \text{const.} \sqrt{t} \max(\Delta_1,\Delta_2) + \text{const.} \int_0^t (t-\tau)^{-\frac{1}{2}} \sup_y |U_{\Delta_1}(y,\tau) - U_{\Delta_2}(y,\tau)|\, d\tau$$

where the constants are independent of $\Delta_1$, $\Delta_2$. It is only left to notice that

$$|U_{0_{\Delta_1}}(x) - U_{0_{\Delta_2}}(x)| \leqq 2 \sup_x |U''(x)| \max(\Delta_1,\Delta_2).$$

(A similar estimate holds if $U_0'$ is just uniformly continuous.) Thus if

$$\phi(t) = \sup_y |U_{\Delta_1}(y,t) - U_{\Delta_2}(y,t)|,$$

then

$$\phi(t) \leqq \text{const.} \max(\Delta_1,\Delta_2) + \text{const.} \int_0^t (t-\tau)^{-\frac{1}{2}} \phi(\tau)\, d\tau$$

and Gronwall's inequality yields

$$\phi(t) \leqq \text{const.} \max(\Delta_1,\Delta_2) \exp(\text{const.} \sqrt{t})$$

which shows that $\{U_\Delta(x,t)\}$ forms a Cauchy sequence and hence the limit as $\Delta \to 0$ exists.

In order to get higher derivatives (say the second derivative) one should differentiate the integral equation (3.3') with respect to $x$, putting the differentiation on $F$ in the second integral

$$U_x(x,t) = \int_{-\infty}^{\infty} G(x-y,t)\, U_0'(y)\, dy$$
$$+ \int_0^t \int_{-\infty}^{\infty} G_x(x-y,t-\tau)\, F'(U(y,\tau))\, U_y(y,\tau)\, dy\, d\tau .$$

If we consider this as an integral equation for $U_x$, we can then apply the technique developed for the proof of differentiability of $U$. The theorem is proved.

**Corollary.** *If* $F \in BC^{k+1}(\mathbb{R}^2)$, $U_0 \in BC^{k+1}(\mathbb{R})$ *with* $k \geqq 2$, *then the mild solution of the Cauchy problem* (3.1), (3.2) *is classical.*

**Proof.** Differentiate both sides of (3.3′) with respect to $t$ in the distributional sense. Then one sees that

$$U_t = -F(U)_x + \varepsilon U_{xx}$$

and the right hand side is continuous in $(x, t)$. Hence so is the left hand side $U_t$. Thus $U$ is continuously differentiable in $t$ and $U$ is a classical solution of (3.1), (3.2).

Next we will show that our mild solutions are continuous with respect to the flux.

**Theorem 3.3.** *Let $U$ and $U_\delta$ be mild solutions of the systems*

$$U_t + F(U)_x = \varepsilon U_{xx},$$

$$U_t + F_\delta(U)_x = \varepsilon U_{xx},$$

*respectively with the same initial data* $U(x, 0) = U_0(x)$, $\delta_0 \geqq \delta > 0$. *Let*

(i) $\sup_{V \in \mathbb{R}^2} |F_\delta(V) - F(V)| = \varrho(\delta) \to 0$ *as* $\delta \to 0$,

(ii) *$F$ and $F_\delta$ be globally Lipschitz continuous with the same Lipschitz constant $L$. Then*

$$|||U - U_\delta||| = \sup_{0 \leqq t \leqq T_0} \left( \sup_x |U(x, t) - U_\delta(x, t)| \right) \to 0$$

*as* $\delta \to 0$. *Here $T_0$ is any finite time.*

**Proof.** The proof goes along the same lines as the proof of Theorem 3.1.

$$\begin{aligned}
\| U(\cdot, t) - U_\delta(\cdot, t)\| &= \sup_x \|U(x, t) - U_\delta(x, t)| \\
&= \|(KU)(\cdot, t) - (K_\delta U_\delta)(\cdot, t)\| \\
&= \left\| \int_0^t G_x(t-\tau) * [F(U(\cdot, \tau)) - F_\delta(U_\delta(\cdot, \tau))]\, d\tau \right\| \\
&\leqq \int_0^t |G_x(t-\tau) * [\|F(U(\cdot, \tau)) - F_\delta(U[(\cdot, \tau))\| + \|F_\delta(U(\cdot, \tau)) - F_\delta(U_\delta(\cdot, \tau))\|]\, d\tau \\
&\leqq \int_0^t |G_x(t-\tau)| * [\varrho(\delta) + L\,\| U(\cdot, \tau) - U_\delta(\cdot, \tau)\|]\, d\tau \\
&= \text{const.}\, \sqrt{t}\, \varrho(\delta) + L \cdot \text{const.} \int_0^t (t-\tau)^{-\frac{1}{2}} \| U(\cdot, \tau) - U_\delta(\cdot, \tau)\|\, d\tau .
\end{aligned}$$

By Gronwall's inequality we have

$$\| U(\cdot, t) - U_\delta(\cdot, t)\| \leqq \text{const.}\, \sqrt{t} \exp(L \cdot \text{const.}\, \sqrt{t})\, \varrho(\delta).$$

The proof is completed.

As a consequence of Theorem 3.3 we can give some information on the mild solutions of the Cauchy problem with the "flat" $p$,

$$\begin{aligned} v_t + p(w)_x &= \varepsilon v_{xx} \\ w_t - v_x &= \varepsilon w_{xx} \end{aligned} \tag{1.1}$$

$$w(x, 0) = w_0(x), \qquad v(x, 0) = v_0(x), \qquad -\infty < x < \infty. \tag{1.2}$$

**Corollary.** *Consider the Cauchy problem for* (1.1), (1.2) *with* $(v_0, w_0) \in L^\infty(\mathbb{R}) \times L^\infty(\mathbb{R})$. *The mild solution of this problem (as given by Theorem 3.1) is a weak solution in the sense of distributions.*

**Proof.** We know that $p$ can be approximated by functions $p_\delta \in BC^3$ (although the bounds for $p_\delta''$ and $p_\delta'''$ tend to $\infty$ as $\delta \to 0$). By Theorem 3.2 solutions of the problem with $p_\delta$ are classical. Thus by Theorem 3.3 the solution of the "flat" problem with $p$ is a uniform limit of classical solutions. Therefore it is at least a solution in the sense of distributions.

*Remark.* From the proof of Theorem 3.2 we know in addition that the solution of the "flat" problem is Lipschitz continuous in $x$.

## 4. Invariant domains

In this section we assign domains in the $wv$–plane to the "flat" $p$ system and its approximation with $p_\delta$. From the results of CHUEH, CONLEY, & SMOLLER [1] we know the domains for the approximating $p_\delta$ systems are positively invariant. We then use a continuity argument to show that the domain corresponding to the "flat" $p$ system is positively invariant as well.

First we describe the construction of CHUEH, CONLEY, & SMOLLER. Consider the system

$$\begin{aligned} v_t + p_\delta(w)_x &= \varepsilon v_{xx}, \\ w_t - v_x &= \varepsilon w_{xx}. \end{aligned} \tag{4.1}$$

There are two Riemann invariants

$$s = v + \int^w (-p_\delta'(\xi))^{\frac{1}{2}}\, d\xi, \qquad r = v - \int^w (-p_\delta'(\xi))^{\frac{1}{2}}\, d\xi$$

which are concave and convex, respectively, to the left of the inflection point $w_i = (\alpha + \beta)/2$ of $p_\delta$. To the right of $w_i$ the concavity and convexity of $r$ and $s$ switch.

We fix some point $(w_a, v_a)$ in the $wv$ plane, $w_a < \alpha$. The positively invariant domain $D_\delta$ is determined by the inequalities

$$\begin{aligned} r \leqq r_a,\ s \geqq s_a \quad &\text{for } w_a \leqq w \leqq w_i \\ r \geqq r_b,\ s \leqq s_b \quad &\text{for } w_i \leqq w \leqq w_b. \end{aligned}$$

(This is the CHEUH, CONLEY, & SMOLLER construction, which follows from applying the maximum principle to each Riemann invariant.) A graphical representation is given in Figure 4.

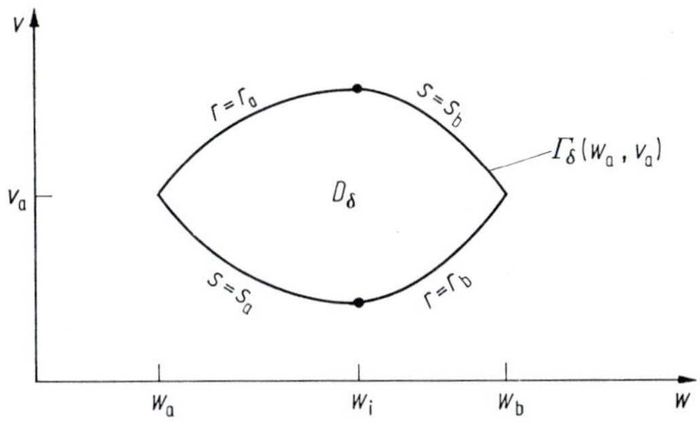


Fig. 4

Of course positively invariant here means that if we are given initial data $v(x,0) = v_0(x)$, $w(x,0) = w_0(x)$, $x \in \mathbb{R}$, where $(v_0(x), w_0(x)) \in D_\delta$, then $(v(x,t), w(x,t)) \in D_\delta$ for $-\infty < x < \infty$, $t > 0$.

Alternatively we see that $D_\delta$ is described by the inequalities

$$v - \int_{w_a}^{w} (-p_\delta'(\xi))^{\frac{1}{2}}\, d\xi \leqq v_a \leqq v + \int_{w_a}^{w} (-p_\delta'(\xi))\, d\xi, \qquad w_a \leqq w \leqq w_i,$$

$$v - \int_{w}^{w_b} (-p_\delta'(\xi))^{\frac{1}{2}}\, d\xi \leqq v_b \leqq v + \int_{w}^{w_b} (-p_\delta'(\xi))\, d\xi, \qquad w_i \leqq w \leqq w_b.$$

For given $(w_a, v_a)$ it is easy to see that the $r = r_a$ curve for $w_a \leqq w \leqq w_i$ is described by

$$v = v_a + \int_{w_a}^{w} (-p_\delta'(\xi))^{\frac{1}{2}}\, d\xi$$

and the $s = s_b$ curve for $w_i \leqq w \leqq w_b$ is given by

$$v = v_b - \int_{w_b}^{w} (-p_\delta'(\xi))^{\frac{1}{2}}\, d\xi.$$

Equating the above two expressions at $w = w_i$ we see

$$v_a - v_b = -\int_{w_b}^{w_i} (-p_\delta'(\xi))^{\frac{1}{2}}\, d\xi - \int_{w_a}^{w_i} (-p_\delta'(\xi))^{\frac{1}{2}}\, d\xi.$$

A similar computation made on the $r = r_b$ and $s = s_a$ curves shows

$$v_a - v_b = \int_{w_a}^{w_i} (-p_\delta'(\xi))^{\frac{1}{2}}\, d\xi + \int_{w_b}^{w_i} (-p_\delta'(\xi))^{\frac{1}{2}}\, d\xi$$

so that $v_a = v_b$. Furthermore we find $w_b$ from the relation

$$\int_{w_a}^{w_i} (-p'_\delta(\xi))^{\frac{1}{2}}\, d\xi = \int_{w_i}^{w_b} (-p'_\delta(\xi))^{\frac{1}{2}}\, d\xi .$$

The two middle vertices of $D_\delta$ are given by

$$\left(w_i, v_a \pm \int_{w_a}^{w_i} (-p'(\xi))^{\frac{1}{2}}\, d\xi\right).$$

We can repeat the preceding construction for the flat $p$ case since all the integrals and inequalities make sense. One obtains the hexagonal domain $D$ with vertices

$$(w_a, v_a), \qquad \left(\alpha, v_a \pm \int_{w_a}^{\alpha} (-p'(\xi))^{\frac{1}{2}}\, d\xi\right),$$

$$\left(\beta, v_a \pm \int_{w_b}^{\beta} (-p'(\xi))^{\frac{1}{2}}\, d\xi\right), \qquad (w_b, v_a),$$

where $w_b$ is defined by the same equation as before. (See Fig. 5.) The difficulty is that $p$ is no longer $C^1$ and maximum principle type arguments of CHUEH, CONLEY, & SMOLLER no longer apply.

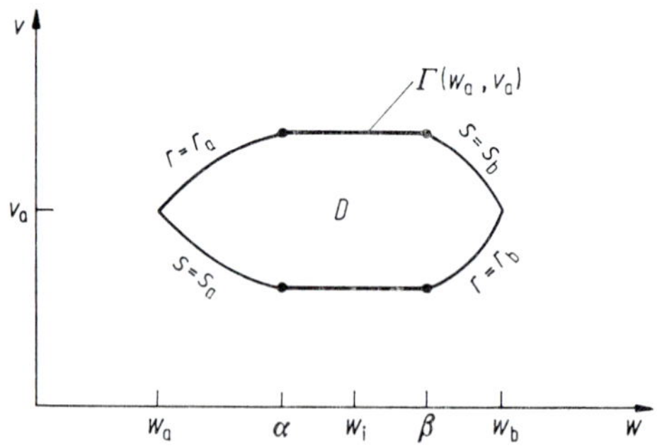


Fig. 5

Obviously $D$ depends continuously on $(w_a, v_a)$. Thus if we let $\Gamma(w, v)$ denote the boundary of the domain $D$ with left-most vertex at $(w, v)$, then for any $\varrho > 0$ there exists a neighborhood $N$ of $(w_a, v_a)$ such that $\operatorname{dist}(\Gamma(w, v), \Gamma(w_a, v_a)) < \varrho$ if $(w, v) \in N$.

Another simple observation is that the domains $D_\delta$ converge to $D$ as $\delta \to 0$. Indeed if $w < \alpha$ then

$$|\Delta| = \left| v_a + \int_{w_a}^{w} (-p'_\delta(\xi))^{\frac{1}{2}}\, d\xi - \left[ v_a + \int_{w_a}^{w} (-p'(\xi)^{\frac{1}{2}}\, d\xi \right] \right|$$

$$\leq \sup_{\xi} |(-p'_\delta(\xi))^{\frac{1}{2}} - (-p'(\xi))^{\frac{1}{2}}|\, |w - w_a| \to 0$$

since $p'_\delta \to p'$ away from $w = \alpha, \beta$ by Proposition 2.1 (ii).

If $\alpha < w < w_i$ then

$$|\Delta| = \left| \int_{w_a}^{w} (-p'_\delta(\xi))^{\frac{1}{2}} \, d\xi - \int_{w_a}^{\alpha} (-p'(\xi))^{\frac{1}{2}} \, d\xi \right|$$

$$\leqq \int_{w_a}^{\alpha-\nu} |(-p'_\delta(\xi))^{\frac{1}{2}} - (-p'(\xi))^{\frac{1}{2}}| \, d\xi + \int_{\alpha+\nu}^{w} (-p'_\delta(\xi))^{\frac{1}{2}} \, d\xi$$

$$+ \int_{\alpha-\nu}^{\alpha} |(-p'_\delta(\xi))^{\frac{1}{2}} - (-p'(\xi))^{\frac{1}{2}}| \, d\xi + \int_{\alpha}^{\alpha+\nu} (-p'_\delta(\xi))^{\frac{1}{2}} \, d\xi .$$

The first two integrals are bounded by const. $\delta$ as follows from Proposition 2.1 (ii). The last two integrals are bounded by const. $\nu$. Choosing $\nu$ small and then selecting $\delta$ correspondingly small, we can make $|\Delta|$ arbitrarily small.

We collect all the results in the following proposition. (See Fig. 6.)

**Proposition 4.1.** Let $D(w_a, v_a)$ be the hexagonal domain with left most vertex $(w_a, v_a)$. Let $D_\delta(w_a, v_a)$ be the positively invariant domain corresponding to $p_\delta$. Then for any $\varrho > 0$

(i) there exist $(w_1, v_1)$ and $\varrho_1$, $0 < \varrho_1 < \varrho$ such that $D(w_1, v_1) \supset D(w_a, v_a)$ and

$$\varrho_1 < \operatorname{dist}\{\Gamma(w_1, v_1), \Gamma(w_a, v_a)\} < \varrho;$$

(ii) there exist $\delta_0$ such that for any $0 < \delta < \delta_0$

$$D_\delta(w_a, v_a) \subset D(w_1, v_1),$$

$$\operatorname{dist}\{\Gamma_\delta(w_a, v_a), \Gamma(w_1, v_1)\} > \varrho/2,$$

$$\operatorname{dist}\{\Gamma_\delta(w_a, v_a), \Gamma(w_a, v_a)\} < \varrho/2.$$

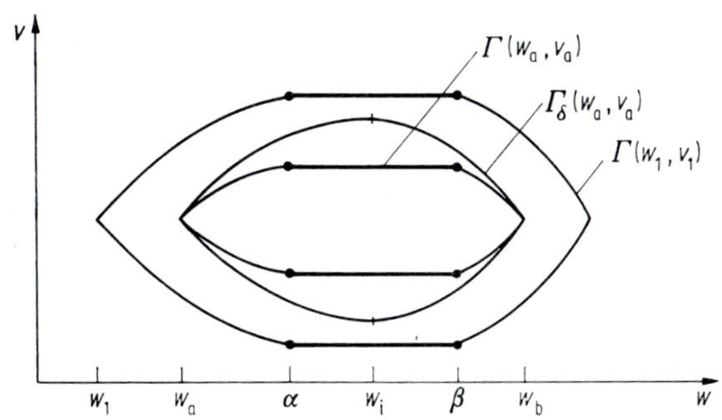


Fig. 6

After these preparations we now state and prove our main result.

**Theorem 4.1.** *The hexagonal domain $D(w_a, v_a)$ is positively invariant. That is, if initial data $w(x, 0) = w_0(x)$, $v(x, 0) = v_0(x)$, $x \in R$, for (1.1) is in $D(w_a, v_a)$ then $(w(x, t), v(x, t))$ is in $D(w_a, v_a)$ for all $x \in R$, $t > 0$.*

**Proof.** Consider a nested set of domains

$$D_0 = D(w_a, v_a) \subset D_1 = D(w_1, v_1) \subset D_2 = D(w_2, v_2) \subset \dots$$

with boundaries $\Gamma_0, \Gamma_1, \Gamma_2, \dots$ so that, for $0 < q < 1$, dist $\{\Gamma_{k-1}, \Gamma_k\} < q^k$. Such a set can be constructed by Proposition 4.1.

Now we will show that if $w(x, t), v(x, t)$ is a mild solution of the "flat" problem with initial data $(w(x, 0), v(x, 0)) \in D_0$ then $(w(x, 1), v(x, 1)) \in D_1$. To see this we select a $\delta_0 > 0$ based on the following two considerations:

(1) We desire $(w(x, 0), v(x, 0)) \in D_{\delta_0}(w_a, v_a) \subset D_1$ so that dist $\{\Gamma_\delta(w_a, v_a), \Gamma_1) > q/2$.

(2) We desire

$$\sup_{0 \leq t \leq 1} \sup_{x \in R} \{|w_{\delta_0}(x, t) - w(x, t)| + |v_{\delta_0}(x, t) - v(x, t)|\} < q/2$$

where $w_{\delta_0}(x, t), v_{\delta_0}(x, t)$ is the solution of the approximating $p_{\delta_0}$ problem (4.1) with initial data $w(x, 0), v(x, 0)$.

We note that (1) follows from Proposition 4.1 and (2) follows by Theorem 3.3 for sufficiently small $\delta_0$.

Now since $D_{\delta_0}(w_a, v_a)$ is positively invariant $w_{\delta_0}, v_{\delta_0}$ never leaves that domain. Therefore by (2) dist $\{(w(x, t), v(x, t)), \Gamma_{\delta_0}(w_a, v_a)\} < q/2$ for $0 \leq t \leq 1$. It follows by (1) that $(w(x, 1), v(x, 1)) \in D_1$ for all $x \in R$.

Next we repeat our process with $(w(x, 1), v(x, 1)) \in D_1$ viewed as new initial data. That is, we select a new $\delta_1 < \delta_0$ based on the considerations:

(1') We desire $(w(x, 1), v(x, 1)) \in D_{\delta_1}(w_1, v_1) \subset D_2$ such that

$$\text{dist}\,\{\Gamma_{\delta_1}(w_1, v_1), \Gamma_2\} > q^2/2.$$

(2') We desire

$$\sup_{1 \leq t \leq 2} \sup_{x \in R} \{|w_{\delta_1}(x, t) - w(x, t)| + |v_{\delta_1}(x, t) - v(x, t)|\} < q^2/2$$

where $w_{\delta_1}(x, t), v_{\delta_1}(x, t)$ is the solution of the approximating $p_{\delta_1}$ problem (4.1) with data $w(x, 1), v(x, 1)$ at $t = 1$.

Again (1') follows from Proposition 4.1 and (2') follows by Theorem 3.3 for sufficiently small $\delta_1$. Since $D_{\delta_1}(w_1, v_1)$ is positively invariant $w_{\delta_1}, v_{\delta_1}$ never leaves that domain. Therefore by (2') dist $\{(w(x, t), v(x, t)), \Gamma_{\delta_1}(w_1, v_1)\} < q^2/2$ for $1 \leq t \leq 2$. Thus by (1') we have $(w(x, 2), v(x, 2)) \in D_2$ for all $x \in R$.

By induction we can show that the solution starting in $D_k$ at time $t = k$ cannot leave $D_{k+1}$ for $k \leq t \leq k + 1$. Since $q + q^2 + \dots = q/(1 - q)$ we see that the mild solution of the "flat" $p$ problem (1.1) cannot leave a $q/(1 - q)$ tubular neighborhood of $D_0 = D(w_a, v_a)$. Finally since $q \in (0, 1)$ is arbitrary, we may pass to the limit $q \to 0$ to see that $D_0 = D(w_a, v_a)$ is positively invariant.

*Acknowledgment.* ROYTBURD's work was supported in part by NSF Grant No. DMS-8408260. SLEMROD's research was sponsored in part by the Air Force Office of Scientific Research, Air Force Systems Command, USAF, Contract/Grant No. AFOSR-81-0172. The U.S. Government's right to retain a nonexclusive royalty free license and and to copyright this paper for governmental purposes is acknowledged.

## References


1. CHUEH, K. N., C. C. CONLEY, & J. A. SMOLLER, Positively invariant regions for systems of nonlinear diffusion equations, Indiana University Math. J. **26**, 372–411 (1977).
2. FERMI, E., Thermodynamics, New York: Dover 1956.
3. ERICKSEN, J. L., Equilibrium of bars, J. Elasticity **5**, 191–201 (1975).
4. DACOROGNA, B., Quasiconvexity and relaxation of non-convex problems in the calculus of variations, J. Functional Analysis **46**, 102–118 (1982).
5. BALL, J., Material instabilities and the calculus of variations in "Phase transformations and material instabilities in solids", M. E. GURTIN, editor, New York, Academic Press, 1–21 (1984).
6. VAN DER WAALS, J. D., Verhandel. Konik. Akad. Weten. Amsterdam 1, No. 8 (1883); translation of J. D. VAN DER WAALS' "The thermodynamic theory of capillarity under the hypothesis of a continuous variation of density" by S. ROWLINSON, J. Statistical Physics **20**, 197–244 (1979).
7. AIFANTIS, E., & J. SERRIN, The mechanical theory of fluid interfaces and Maxwell's rule, J. Colloidal Interface Sci. **96**, 517–529 (1983).
8. AIFANTIS, E., & J. SERRIN, Equilibrium solutions in the mechanical theory of fluid microstructures, J. Colloidal Interface Sci. **96**, 530–547 (1983).
9. CARR, J., M. GURTIN, & M. SLEMROD, Structural phase transitions on a finite interval, Arch. Rational Mech. Anal. **86**, 317–351 (1984).
10. CARR, J., M. GURTIN, & M. SLEMROD, One dimensional structured phase transformations under prescribed loads, J. Elasticity **15**, 133–142 (1985).
11. GURTIN, M. E., On a theory of phase transitions with interfacial energy, Arch. Rational Mech. Anal. **87**, 187–212 (1985).
12. THOMPSON, C. J., Mathematical Statistical Mechanics, Princeton University Press, 1972.
13. KAC, M., G. E. UHLENBECK, & P. C. HEMMER, On the van der Waals theory of vapor-liquid equilibrium I. Discussion of a one-dimensional model, J. Math. Physics **4**, 216–228 (1963).
14. SLEMROD, M., Dynamics of first order phase transitions in "Phase Transformations and Material Instabilities in Solids", M. E. GURTIN, ed. New York: Academic Press 163–204 (1984).
15. TARTAR, L., Compensated compactness and applications to partial differential equations, in Research Notes in Mathematics, Nonlinear Analysis and Mechanics: Heriot-Watt Symposium, vol. 4, ed. R. J. KNOPS, Pitman 1979.
16. DIPERNA, R. J., Convergence of approximate solutions to conservation laws, Arch. Rational Mech. Anal. **82**, 27–70 (1983).
17. DIPERNA, R. J., Convergence of the viscosity method for isentropic gas dynamics, Comm. Math. Phys. **91**, 1–30 (1983).



Department of Mathematical Sciences
Rensselaer Polytechnic Institute
Troy, New York


*(Received July 18, 1985)*

# *A Note on a Theorem of R. Duffin*

PAUL H. RABINOWITZ

*Dedicated to James Serrin on the occasion of his 60th birthday*

## § 1. Introduction

In [1], R. DUFFIN studied the transverse vibrations of a string of $n$ beads stretched between two fixed end points. On the assumption that longitudinal motion was constrained by channels and that tension only depends on the extension of the string segment, the motion was modelled by the system of equations

$$m_i x_i'' = f_i(x_{i-1} + x_i) + f_{i+1}(x_i + x_{i+1}), \quad 1 \leqq i \leqq n \tag{1.1}$$

where the positive coordinate directions are as indicated in the figure below, $x_0 \equiv 0 \equiv x_{n+1}$, $m_i > 0$ is the mass of the $i^{\text{th}}$ bead, and $f_i$ is the force due to the tension in the segment joining $x_{i-1}$ and $x_i$. Let $P$ denote the cone of positive vectors in $R^n$, *i.e.* $P = \{x = (x_1, \ldots, x_n) \mid x_i \geqq 0, 1 \leqq i \leqq n\}$. Thus in the above coordinate system, if at time $t$, the beads lie in $P$, $x_1(t)$ lies above the line, $l$, joining the endpoints of the string, $x_2$ lies below $l$, *etc.* and in particular adjacent points have displacements of opposite sign relative to $l$. DUFFIN used an elegant argument based on the Brouwer Fixed Point Theorem to prove

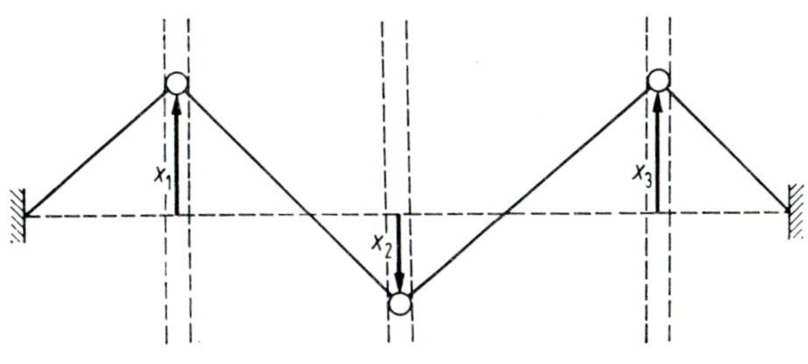


**Theorem 1.2.** *Suppose the functions $f_i$ are Lipschitz continuous and there is a constant $h > 0$ such that*

$$-f_i(s) \geqq hs, \quad 1 \leqq i \leqq n + 1 \tag{1.3}$$

*for all* $s \geqq 0$. *Let* $S \subset R^n$ *be a surface bounding a starshaped neighborhood of* 0. *Then there is an* $x^* \in S \cap P$, *a* $T^* > 0$ *and a solution* $x(t)$ *of* (1.1) *such that* $x(0) = x^*$, $x'(0) = 0$, $x(t)$ belongs to the interior of $P$ *for* $t \in [0, T)$, *and* $x(T^*) = 0$.

Stated more informally, on any such set of initial data, $S \cap P$, there is a point $x^*$ and corresponding $T^* > 0$ such that if the string starts from rest at $x^*$, then it remains in $P$ until time $T^*$ when all of the beads lie on $l$. DUFFIN also proved:

**Corollary 1.4.** *If also* $f_i(s) = -f_i(-s)$, $1 \leqq i \leqq n+1$, *then* $x(t)$ *is a* $4T$ *periodic solution of* (1.1).

DUFFIN called the type of solution obtained in Theorem 1.2 an "oppositional mode of vibration". For linear $f_i$, this mode has the highest frequency.

During a recent visit to Carnegie-Mellon University, DUFFIN posed the question of whether there exist solutions of the type obtained above which start on $S$ outside of $P$ and after some time interval all lie on $l$. In § 3 we will give some partial answers to this question. In § 2, some sharper versions of Theorem 1.2 will be obtained for a more general class of equations than (1.1). In particular we will prove that (1.1) possesses an unbounded connected set of initial data in $P$ of the type given by Theorem 1.2.

## § 2. Continua of positive solutions

Let $x = (x_1, \ldots, x_n)$ and let $g(x) = (g_1(x), \ldots, g_n(x))$ satisfy

$$(g_1) \qquad g \in C(R^n, R^n)$$

and

$$(g_2) \qquad g: P \setminus \{0\} \to P \setminus \{0\}.$$

Note that the $g$ defined by (1.1) and (1.3) satisfies $(g_2)$. Consider the system of ordinary differential equations:

$$m_i x_i'' + g_i(x) = 0, \qquad 1 \leqq i \leqq n \tag{2.1}$$

where $m_i > 0$. In this section more general versions of Theorem 1.2 will be established for (2.1). By replacing $g_i(x)$ by $g_i(x)\, m_i^{-1}$, we can assume $m_i = 1$ and study

$$x'' + g(x) = 0. \tag{2.2}$$

We will seek solutions of (2.2) which pass through the origin at say $t = 0$ and at a second value of $t$, say $t = T$, have zero velocity, *i.e.* $x'(T) = 0$. If $x(t)$ is such a solution and $y(t) = x(T-t)$, then $y(0) = x(T)$, $y'(0) = x'(T) = 0$, $y(T) = x(0) = 0$, and $y''(t) = x''(T-t) = -g(x(T-t)) = -g(y(t))$. Therefore $y$ is a solution of the type DUFFIN found. Our main result will be the existence of an unbounded connected set of data in $P$ for which there exist solutions of

Duffin type for (2.2). To state this more precisely, note that the time $T$ at which $x' = 0$ is not known *a priori*. By rescaling the time variable, $t \to T^{-1}t$, we can make the new time interval [0, 1]. Still calling $t$ the independent variable, we put (2.2) into the form

$$x'' + T^2 g(x) = 0 \tag{2.3}$$

and

$$x(0) = 0 = x'(1). \tag{2.4}$$

Thus we seek $T \in (0, \infty)$ and $x$ in the class of functions satisfying (2.4) such that $(T, x)$ satisfies (2.3). Using (2.4) to integrate (2.3) shows that

$$x'(t) = T^2 \int_t^1 g(x(\tau))\, d\tau \tag{2.5}$$

and

$$x(t) = T^2 \int_0^t \left( \int_s^1 g(x(\tau))\, d\tau \right) ds \equiv T^2 Kx. \tag{2.6}$$

Solutions of (2.2) of the desired type will be obtained from the equivalent formulation (2.6).

Let $E$ denote the set of $x \in C([0, 1], R^n)$ under the usual maximum norm. We will prove the following generalization of Theorem 1.2:

**Theorem 2.7.** *Let $g$ satisfy $(g_1)$ and $(g_2)$. Then there exists an unbounded connected set $\mathscr{D} \subset P$ such that $0 \in \mathscr{D}$ and for any $\eta \in \mathscr{D} \setminus \{0\}$, there exists a solution $(T, x)$ of (2.6) such that $x(1) = \eta$, and $x(t) \in P$ for $t \in [0, 1]$.*

To prove Theorem 2.7, note first that the form of $K$ in (2.6) and $(g_1)$ imply $K\colon E \to E$ and $K$ is compact. Set

$$\mathscr{P} = \{y \in E \mid y_i(t) \geqq 0, \quad t \in [0, 1],\ 1 \leqq i \leqq n\}.$$

Then $\mathscr{P}$ is a closed cone in $E$ and (2.6) and $(g_2)$ show $K\colon \mathscr{P} \to \mathscr{P}$. We are interested in solutions of (2.6) or equivalently

$$x = \mu Kx \tag{2.8}$$

where $\mu = T^2$. By a solution of (2.8) we mean a pair $(\mu, x) \in R^+ \times E$ satisfying (2.8) where $R^+ = \{r \in R \mid r \geqq 0\}$. There is a useful theorem about operator equations of the form (2.8) that can be exploited for this purpose.

**Proposition 2.9.** *Let $\mathscr{E}$ be a real Banach space and $\mathscr{K}\colon \mathscr{E} \to \mathscr{E}$ be compact. Suppose there exists a closed cone $\mathscr{P}$ in $\mathscr{E}$ such that $\mathscr{K}\colon \mathscr{P} \to \mathscr{P}$. Then the operator equation $x = \mu\mathscr{K}x$ has a component*—i.e. *a maximal closed connected set—of solutions, $\mathscr{C}$, such that $(0, 0) \in \mathscr{C}$ and $\mathscr{C}$ is unbounded in $R^+ \times \mathscr{P}$.*

**Proof.** Proposition 2.9 probably exists in the literature but we do not know a reference. Here is a brief proof. Consider $\mathscr{K}$ restricted to $\mathscr{P}$. By the Dugundji

Extension Theorem [2], there exists an extension $\hat{\mathscr{K}}$ of $\mathscr{K}$ to $\mathscr{E}$ which is compact and whose range lies in $\mathscr{P}$. Consider the equation

$$x = \mu \hat{\mathscr{K}} x. \tag{2.10}$$

By Theorem 3.2 of [3], (2.10) has a component $\mathscr{C}$ of solutions meeting (0, 0) and unbounded in $R^+ \times \mathscr{E}$. Since the range of $\hat{\mathscr{K}}$ lies in $\mathscr{P}, \mathscr{C} \subset R^+ \times \mathscr{P}$. But in $\mathscr{P}$, $\hat{\mathscr{K}} = \mathscr{K}$ so the proposition obtains.

Applying Proposition 2.9 to (2.8) and our $\mathscr{P}$ shows that (2.8) possesses a component of solutions, $\mathscr{C}$, meeting (0, 0) and unbounded in $R^+ \times \mathscr{P}$. Unfortunately this does not suffice to get Theorem 2.7. Indeed if $\mathscr{D} = \{x(1) \mid (\mu, x) \in \mathscr{C}\}$ then $\mathscr{D}$ is connected and $0 \in \mathscr{D}$ but *e.g.* if $g(0) = 0$, possibly $C = \{(\mu, 0) \mid \mu \in R^+\}$ and $\mathscr{D} = \{0\}$. To get around this difficulty we use an approximation argument. In the process we will see that even if $g(0) = 0$, $\mathscr{C}$ contains lots of nontrivial solutions. Let $\varepsilon > 0$ and define $\underline{\varepsilon} \in R^n$ by $\underline{\varepsilon} = (\varepsilon, \ldots, \varepsilon)$. Now set

$$(K_\varepsilon x)(t) = \int_0^t \left( \int_s^1 (g(x(\tau)) + \underline{\varepsilon}) \, d\tau \right) ds. \tag{2.11}$$

Clearly $K_\varepsilon$ is compact and $K_\varepsilon : \mathscr{P} \to \mathscr{P}$. Consider

$$x = \mu K_\varepsilon x. \tag{2.12}$$

Let $\mathscr{C}_\varepsilon$ denote the connected set obtained by applying Proposition 2.9 to (2.12). This set and a limit argument will be used to obtain Theorem 2.7. First we need two preliminary results.

**Lemma 2.13.** *For each* $\varepsilon > 0$, *set* $X_\varepsilon = \{x \in E \mid (\mu, x) \in \mathscr{C}_\varepsilon\}$. *Then* $X_\varepsilon$ *contains* 0 *and is unbounded and connected.*

**Proof.** By Proposition 2.9, and above remarks, $\mathscr{C}_\varepsilon$ is unbounded and connected in $R^+ \times E$ and $(0, 0) \in \mathscr{C}_\varepsilon$. Therefore $X_\varepsilon$, its projection on $E$ is connected and $0 \in X_\varepsilon$. If $X_\varepsilon$ is bounded, there is a sequence of points $(\mu_m, x_m) \in \mathscr{C}_\varepsilon$ such that $\mu_m \to \infty$ and $(x_m)$ is bounded in $E$. The equation satisfied by $(\mu_m, x_m)$ and $(g_2)$ yield

$$\frac{x_{mi}(t)}{\mu_m} = \int_0^t \left( \int_s^1 (g_i(x_m(\tau)) + \varepsilon) \, d\tau \right) ds \geq \int_0^t \varepsilon(1 - s) \, ds = -\tfrac{1}{2} \varepsilon (1 - s)^2 \Big|_0^t \tag{2.14}$$

for each component $x_{mi}$ of $X_m$. Setting $t = 1$ and letting $m \to \infty$ shows

$$0 \geq \tfrac{1}{2} \varepsilon, \tag{2.15}$$

a contradiction. Therefore $X_\varepsilon$ is unbounded in $E$.

**Lemma 2.16.** *For each* $\varepsilon > 0$, *set* $Y_\varepsilon = \{x(1) \mid x \in X_\varepsilon\}$. *Then* $Y_\varepsilon$ *contains* 0 *and is unbounded and connected in* $P$.

**Proof.** Clearly $0 \in Y_\varepsilon \subset P$ and $Y_\varepsilon$ is connected. Equation (2.14) shows that if $(\mu, x) \in \mathscr{C}_\varepsilon$, then $x_i''(t) < 0$ in (0, 1), $x_i'(t) > 0$ in (0, 1), and $x_i(t) > 0$ in (0, 1) for $1 \leqq i \leqq n$. Therefore each component achieves its maximum in $[0, 1]$ at $t = 1$. Consequently

$$|x(1)| = \max_{t \in [0,1]} |x(t)| \equiv \|x\|_{L^\infty}. \tag{2.17}$$

Hence, by Lemma 2.13, $Y_\varepsilon$ is unbounded.

**Proof of Theorem 2.7.** A standard theorem from point set topology [4, Chapter 1] guarantees the existence of $\mathscr{D}$ provided that for any neighborhood $\mathcal{O}$ of 0 in $R^n \cap P$, there is a $\xi \in \partial\mathcal{O}$, $\mu(\xi) > 0$, and a solution $x$ of (2.3)–(2.4) with $T^2 = \mu$ and $x(1) = \xi$. Thus choose any such set $\mathcal{O}$ and a decreasing sequence $\varepsilon_m \to 0$. By Lemma 2.16 and previous remarks, for each $m$, $Y_{\varepsilon_m}$ is an unbounded connected subset of $P$ which intersects $\mathcal{O}$. Hence there is a $(\mu_m, x_m) \in \mathscr{C}_{\varepsilon_m}$ with $x_m(1) \in \partial\mathcal{O}$. Since $\partial\mathcal{O}$ is bounded in $P$ and the maximum of $|x_m(t)|$ occurs at $t = 1$, $(x_m)$ is a bounded sequence in $L^\infty$. Suppose for the moment that the $\mu_m$'s are also bounded. Then differentiating (2.14) shows that $(\mu_m, x_m)$ is bounded in $R^+ \times C^2$. Consequently using (2.14) again we can assume $(\mu_m, x_m)$ converges in $R^+ \times C^2$ to some $(\mu, x)$, a solution of (2.6) with $x \in P$ and $x(1) \in \partial\mathcal{O}$. Clearly $\mu > 0$ for otherwise $x = 0$ contrary to $x(1) \in \partial\mathcal{O}$. Thus the theorem is proved once we establish that $(\mu_m)$ is a bounded sequence.

If $(\mu_m)$ is not bounded, we can assume $\mu_m \to \infty$ as $m \to \infty$. Consider $(x_m(1)) \subset \partial\mathcal{O}$. Therefore there is an $\alpha > 0$ and a subsequence of $x_m$ such that for at least one component of $x_m$, say $x_{mj}$, we have $x_{mj}(1) \geqq \alpha$ along the subsequence. Since $x_{mj}(t)$ is a concave function, by Lemma 2.16, $x_{mj}(t) \geqq \alpha/2$ for all $t \in [\frac{1}{2}, 1]$. By the boundedness of $\mathcal{O}$ again, there is a constant $M > 0$ such that $\|x_m\|_{L^\infty} \leqq M$ independently of $m$. Hypothesis $(g_2)$ implies there is a $\beta > 0$ such that

$$\sum_1^n |g_i(\xi)| \geqq \beta$$

for all $\xi$ such that $\alpha/2 \leqq \xi_j \leqq M$ and $0 \leqq \xi_i \leqq M$, $i \neq j$. Therefore by (2.14),

$$\begin{aligned} \mu_m^{-1} \sum_{i=1}^n x_{mi}(1) &\geqq \int_0^1 \left( \int_s^1 \left( \sum_{i=1}^n g_i(x_m(\tau)) \right) d\tau \right) ds \\ &\geqq \int_0^{\frac{1}{2}} \left( \int_{\frac{1}{2}}^1 \left( \sum_{i=1}^n g_i(x_m(2)) \right) d\tau \right) ds \geqq \beta/4. \end{aligned} \tag{2.18}$$

Letting $m \to \infty$ in (2.17) leads to

$$0 \geqq \beta/4, \tag{2.19}$$

a contradiction. Therefore $\mu_m$ is a bounded sequence and Theorem 2.7 is proved.

For each $\xi \in \mathscr{D} \setminus \{0\}$, there is a $\mu > 0$ and $x \in \mathscr{P}$ such that $(\mu, x)$ is a solution of (2.6) with $x(1) = \xi$. It is natural to ask whether there is a component, $\mathscr{C}$, of solutions of (2.6) in $R^+ \times \mathscr{P}$ such that $Y \equiv \{x(1) \mid (\mu, x) \in \mathscr{C}\} = \mathscr{D}$. I do not believe such a simple statement holds without more assumptions. However in this direction we have

**Corollary 2.20.** *For each* $\xi \in \mathscr{D} \setminus \{0\}$, *there exists an unbounded connected set* $\mathscr{A}_\xi$ *of solutions of* (2.6) *in* $R^+ \times \mathscr{P}$ *and a point* $(\mu, x) \in \mathscr{A}_\xi$ *such that* $\xi = x(1)$.

**Proof.** Let $\xi \in \mathscr{D}$. By construction, $\xi = \lim \xi_m$ where $\xi_m = x_m(1)$ and $(\mu_m, x_m) \in \mathscr{C}_{\varepsilon_m}$. Moreover $(\mu_m, x_m)$ converges in $R^+ \times C^2$ to $(\mu, x) \in R^+ \times \mathscr{P}$ and $x(1) = \xi$. Consider any bounded open set $\Omega$ in $R^+ \times \mathscr{P}$ which contains $(\mu, x)$. Then $(\mu_m, x_m) \in \Omega$ for large $m$. Since $\mathscr{C}_{\varepsilon_m}$ is unbounded, $\mathscr{C}_{\varepsilon_m} \cap \partial\Omega \neq \emptyset$ for all large $m$. It follows as in the proof of Theorem 2.7 that $\partial\Omega$ contains a solution of (2.6) and there exists $\mathscr{A}_\varepsilon$ as stated above.

If $g(0) = 0$, (2.6) possesses the family of trivial solutions $\{(\mu, 0) \mid \mu \in R^+\}$ in $R^+ \times \mathscr{P}$. If $g(0) \neq 0$, we have

**Corollary 2.21.** *If* $g(0) \neq 0$, (2.6) *possesses an unbounded component of nontrivial solutions,* $\mathscr{C}$, *in* $R^+ \times \mathscr{P}$ *which meets* $(0, 0)$ *and makes* $Y$ *unbounded in* $P$.

**Proof.** Let $\Omega$ be any bounded open neighborhood of $(0, 0)$ in $R^+ \times \mathscr{P}$. The argument of Corollary 2.20 or Theorem 2.7 shows that $\partial\Omega \cap \mathscr{C}_{\varepsilon_m} \neq \emptyset$ for all $m$. Hence as earlier there is an unbounded component $\mathscr{C}$ of solutions of (2.6) in $R^+ \times \mathscr{P}$ which meets $(0, 0)$. Equation (2.6) shows if $(\mu, x) \in \mathscr{C}$ and $\mu \neq 0$, then $x(t) > 0$ for $t > 0$ so $C \setminus \{(0, 0)\}$ consists of nontrivial solutions. Lastly (2.18) shows that $Y$ cannot be bounded.

*Remark* 2.22. If $\mathscr{S}$ denotes the set of nontrivial solutions of (2.6) in $R^+ \times \mathscr{P}$, as we approach infinity along $\mathscr{S}$, $x(1) \to 0$ or $x(1) \to \infty$, for otherwise the estimate (2.18) shows the corresponding $\mu$'s must be bounded. Then (2.6) implies the $x$'s are also bounded, a contradiction.

To study the behavior of $\mathscr{S}$ more closely, in particular near $x = 0$ or $\infty$ requires more information about $g$ near 0 and infinity. We will give a simple example of such an analysis. Let $|x|_{l_1} = \sum_{i=1}^{n} |x_n|$. By (2.6),

$$\frac{x(1)}{|x(1)|_{l_1}} = \mu \int_0^t \left( \int_s^1 \frac{g(x(\tau))}{|x(1)|_{l_1}} d\tau \right) ds. \tag{2.23}$$

Suppose that

$$|g(\xi)|_{l_1} \geqq a |\xi|^\alpha. \tag{2.24}$$

(Note that this holds for $\alpha = 1$ for the $g$ obtained from (1.1) via (1.3).) Then (2.23)–(2.24) show there is a constant $a_1$ such that

$$1 \geqq \mu a_1 |x(1)|_{l_1}^{\alpha - 1}. \tag{2.25}$$

If $\alpha > 1$, $\mu \to 0$ as $x(1) \to \infty$ while if $\alpha = 1$, $\mu \leqq a_1^{-1}$. Since (2.25) also holds for (2.12) uniformly in $\varepsilon > 0$, we see if $\alpha = 1$, $\mathscr{C}_\varepsilon$ and therefore any set obtained from it by taking limits lies in $[0, a_1^{-1}] \times \mathscr{P}$ and contains $(0, 0)$.

In [1], DUFFIN conjectured that if $S$ is as in Theorem 1.2, then there exists $\eta \in S$, $T(\eta) > 0$, and a solution $x(t)$ of (1.1) such that $x(0) = 0$, $x'(0) = \eta$,

$x'(T) = 0$, and $x(t) \in P$ for all $t \in [0, T]$. One cannot expect this to occur in the generality of (2.6). For example consider

$$x'' + V'(x) = 0 \tag{2.26}$$

where $V \in C^2(P, R)$ and $g = V'$ satisfies $(g_2)$. Suppose further that $V(x) \to 0$ uniformly as $|x| \to \infty$. (*E.g.* take $n = 1$ and $V(x) = -(x+1)^{-1}$). Since (2.26) is a Hamiltonian system

$$\tfrac{1}{2}|x'(t)|^2 + V(x(t)) \equiv \text{constant} \tag{2.27}$$

independent of $t$ for any solution $x$ of (2.26). Let $\mathcal{T} \equiv \{\eta \in P|$ there is $T > 0$ and $x$, a solution of (2.1) with $x(0) = 0$, $x'(0) = \eta$, $x'(t) = 0$, and $x(t) \in P$ for all $t \in [0, T]\}$. We claim $\mathcal{T}$ is bounded in $P$. Indeed let $\eta \in \mathcal{T}$ with corresponding $x(t)$. By (2.26),

$$\tfrac{1}{2}|x'(0)|^2 + V(x(0)) = \tfrac{1}{2}\eta^2 + V(0) = V(x(T)). \tag{2.28}$$

If $\mathcal{T}$ were unbounded, (2.28) shows $V(P)$ must also be unbounded, contrary to our assumption on $V$ at $\infty$.

Next we will show that DUFFIN's conjecture holds in his setting.

**Theorem 2.29.** *Suppose g satisfies* $(g_1)$, $(g_2)$ *and* (2.24) *with* $\alpha = 1$. *Then* $\mathcal{T}$ *contains an unbounded component containing* 0.

**Proof.** These hypotheses, Theorem 2.7, and the remarks following (2.22) show (2.6) possesses a component of solutions, $\mathscr{C}$, unbounded in $[0, a_1^{-1}] \times \mathscr{P}$ with $(0, 0) \in \mathscr{C}$. For any $(\mu, x) \in \mathscr{C}$, the change of time scale $t \to \tau = \mu^{\frac{1}{2}} t$, $x(t) = y(\tau) = x(\mu^{-\frac{1}{2}}\tau)$ transforms solutions of (2.3)–(2.4) to solutions of (2.2) with $y(0) = 0 = y'(\mu^{\frac{1}{2}})$. Consider

$$\hat{\mathcal{T}} = \{y'(0) = \mu^{-\frac{1}{2}} x'(0) \mid (\mu, x) \in \mathscr{C}\}.$$

*A priori*, $y'(0)$ is not defined for $(0, 0) \in \mathscr{C}$. But if we approach $(0, 0)$ along $\mathscr{C}$, by (2.5)

$$\mu^{-\frac{1}{2}} x'(0) = \mu^{\frac{1}{2}} \int_0^1 g(x(\tau))\, ds \to 0.$$

so $0 \in \hat{\mathcal{T}}$. It follows that the map $(\mu, x) \to \mu^{-\frac{1}{2}} x'(0)$, $\mathscr{C} \to \hat{\mathcal{T}}$ is continuous. Therefore $\hat{\mathcal{T}}$ is connected. Since $(\mu, x) \in \mathscr{C}$ implies $\mu \in [0, a_1^{-1}]$ while $\{x'(0)\}$ is unbounded via (2.5) and the unboundedness of $\{x(0)\}$ in $E$, $\hat{\mathcal{T}}$ must be unbounded and the Theorem is proved.

## § 3. Some remarks on other solutions of (2.2)

Another question DUFFIN posed in [1] is whether (1.1) possesses solutions of the type found earlier other than "positive" ones. Some partial answers will be given in this section.

To begin suppose $g$ satisfies

$$(g_4) \qquad g(\xi) = M\xi + o(|\xi|) \quad \text{as } \xi \to 0$$

where $M$ is an $n \times n$ matrix and $M: P \to P$. Associated with (2.3)–(2.4) is linear eigenvalue problem

$$(3.1) \qquad y'' + \varrho^2 My = 0, \quad 0 < t < 1$$

$$(3.2) \qquad y(0) = 0 = y'(1).$$

An eigenfunction of (3.1)–(3.2) must have the form $y(t) = (\sin(k - \frac{1}{2})\pi t)\, c$ where $k \in N$, $c \in R^n$, and satisfies

$$(3.3) \qquad (k - \tfrac{1}{2})^2 \pi^2 c = \varrho^2 Mc.$$

If $\varrho^2$ in a simple (or odd multiplicity) eigenvalue of (3.1)–(3.2), a global bifurcation theorem from [3] provides the existence of nontrivial solutions of (2.2) bifurcating from $(\varrho^2, 0)$.

Rather than work in this generality, we specialize to DUFFIN's case where $f_j(s) = -a_j s + o(s)$ as $s \to 0$ in $R$, $1 \leqq j \leqq n+1$ where $a_j \geqq h > 0$. Then (3.1) becomes

$$(3.4) \qquad y_i'' = -\frac{a_i}{m_i}(y_{i-1} + y_i) - \frac{a_{i+1}}{m_i}(y_i + y_{i+1}),$$

$1 \leqq i \leqq n$ and $y_0 = 0 = y_{n+1}$. Since $m_i$, $a_j > 0$, $M$ is a tridiagonal matrix with positive entries along its main and two neighboring diagonals. A computation shows that the determinants of the principle minors of $M$ are all positive. Consequently $M$ is an oscillation matrix [5] and its eigenvalues are all positive and simple: $\nu_1 > \ldots > \nu_n > 0$ and any eigenvector $c_j$ corresponding to $\nu_j$ has $j-1$ sign changes. (This means if $c_j = (c_{j1}, \ldots, c_{jn})$, the sequence of components changes sign exactly $j-1$ times). In particular $c_1$ has no sign changes and we can assume $c_1 \in P$.

The eigenvalues of (3.1)–(3.2) are then

$$(3.5) \qquad \left\{\frac{(k-\frac{1}{2})^2\pi^2}{\nu_i} \,\middle|\, 1 \leqq i \leqq s_i,\ k \in N\right\}.$$

These eigenvalues need not be simple; possibly $(k - \frac{1}{2})^2 \nu_i^{-1} = (j - \frac{1}{2})^2 \nu_p^{-1}$ for some choice of indices. However for most choices of $a_i$, $m_j$ they will be simple. In any event we have

**Theorem 3.6.** *Let $f_i \in C^1$ and satisfy* (1.3), $1 \leqq i \leqq n+1$. *Suppose $\varrho^2$ is an eigenvalue of* (3.1)–(3.2) *of odd multiplicity. Then* (1.1) *possesses a component $\hat{\mathscr{C}}$ of nontrivial solutions in $R^+ \times E$ which contains $(\varrho^2, 0)$ and either is unbounded or meets $(r^2, 0)$ for some other eigenvalue $r^2$ of* (3.1)–(3.2).

**Proof.** The system (1.1) is equivalent to the compact operator equation (2.6) where $Kx = Lx + o(\|x\|)$ as $x \to 0$ and

$$(Lx)(t) = \int_0^t \left( \int_s^1 Mx(\tau)\, d\tau \right) ds .$$

Since $\varrho^2$ is of odd multiplicity the conclusion follows immediately from the global bifurcation theorem of [3].

*Remark 3.7.* Since $\nu_j > \nu_1$ for $j > 1$, (3.5) shows $\mu_1 \equiv (\mu\nu_1)^{-1}\pi^2$ is the smallest eigenvalue of (3.1)–(3.2) and is simple. A standard bifurcation theorem then says that near $(\mu_1, 0)$, $\hat{\mathscr{C}}$ is a curve of the form $\mu(s) = \mu_1 + o(1)$, $x(s) = sy_1 + o(|s|)$ for $s$ near 0. Since $c_1 \in P$, $y_1 = \sin(\frac{1}{2}\pi t)\, c_1 \in \mathscr{P}$. In fact $y_1$ belongs to an open subset $\mathscr{Q}$ of $\mathscr{P}$ in the $C^1$ topology namely

$$\mathscr{Q} \equiv \{y \in C^1([0, 1], R^n) \mid y \in \mathscr{P},\ y_i(0) > 0 \quad \text{and} \quad y_i(t) > 0 \quad \text{for}$$
$$t \in (0, 1],\ 1 \leq i \leq n\}.$$

Then for $s > 0$ and small, $(\mu(s), x(s))$ lies in $R^+ \times \mathscr{Q}$. Under a stronger assumption on $g$ as defined by (1.1), this curve extends to an unbounded connected set of solutions of (1.1) in $R^+ \times \mathscr{Q}$.

**Corollary 3.8.** *Under the hypotheses of Theorem 3.6, suppose* $g\colon P \setminus \{0\}$ *into the interior of P. Then* (1.1) *contains a component of solutions* $\mathscr{C}^+$ *unbounded in* $R^+ \times \mathscr{Q}$ *and which meets* $(\mu_1, 0)$.

**Proof.** By Remark 3.7 and Theorem 3.6, we need only show that $\hat{\mathscr{C}} \cap (R^+ \times \mathscr{Q})$ has the desired property. Results from [3] or [7] show that the curve mentioned in Remark 3.7 either is part of an unbounded connected set of solutions of (1.1) in $R^+ \times \mathscr{Q}$ or it meets a point $(\mu, x) \in R^+ \times \partial\mathscr{Q}$ other than $(\mu_1, 0)$. If the latter case occurs, there is a sequence $(\mu_m, x_m) \in \hat{\mathscr{C}} \cap (R^+ \times \mathscr{Q})$ such that $(\mu_m, x_m) \to (\mu, x)$. This implies some component $z$ of $x$ satisfies $z'(0) = 0$ or $z(t) = 0$ for some $t \in [0, 1]$. But (2.5)–(2.6) and our additional assumption on $g$ show this is only possible if $z = 0$ and $x = 0$. Then $(\mu_m, x_m) \to (\mu, 0)$ where $\mu$ lies in the set defined by (3.5). Writing (2.6) as

$$\frac{x_m}{\|x_m\|} = \mu_m \frac{Kx_m}{\|x_m\|} \tag{3.9}$$

and letting $m \to \infty$, we see $x_m \|x_m\|^{-1}$ converges to a solution of

$$y = \mu L y \tag{3.10}$$

where $y \in \mathscr{P}$, $\|y\| = 1$, $\mu \neq \mu_1$. We claim this is impossible. Indeed $y$ has the form

$$y = \Sigma\, \alpha_p (\sin (k_p - \tfrac{1}{2})\, \pi t)\, c_{i_p}$$

where the $\alpha_p's$ are scalars, each $c_j$ appears at most once, and if $c_{i_p} = c_1, k_p > 1$. Let $c_1^*$ denote an eigenvalue of $M^*$, the adjoint of $M$ corresponding to $\nu_1$. We can assume $c_1^* \in P$. Let $y_1^* = \left(\sin \frac{\pi}{2} t\right) c_1^*$. Since $c_1^*$ is orthogonal to $c_k$ if $k \neq 1$, taking the inner product of (3.9) with $y_1^*$ yields

$$\int_0^1 y \cdot y_1^*, \, dt = 0. \tag{3.11}$$

But $y_1^* \in \mathcal{Q}$ and $\|y\| = 1$ implies the left hand side of (3.11) must be positive. Thus $\hat{\mathscr{C}} \cap (R^+ \times \mathcal{Q})$ is unbounded and the proof is complete.

*Remark 3.12.* As was noted earlier, the eigenvalues (3.5) of (3.1)–(3.2) are simple in general and then the argument of Remark 3.7 and [3] or [7] give curves and continua of solutions of (1.1) emanating from the corresponding bifurcation points. Globally the argument of Corollary 3.8 cannot be used to keep these solution branches from meeting each other. However we see that in general there are many distinct small amplitude solutions of (1.1) which do not lie in $R^+ \times \mathscr{P}$. In particular the small amplitude solutions associated with the eigenvalues $(4\nu_j)^{-1} \pi^2$ will have the same "nodal" shapes as the eigenvectors $c_j$, $i \leqq j \leqq n$.

Variational methods can also be used to obtain information about small solutions of (1.1) irrespective of multiplicity considerations for the eigenvalues of (3.1)–(3.2). A brief sketch of how this can be done will be given. Let

$$F_i(s) = \int_0^s f_i(z) \, dz, \qquad 1 \leqq i \leqq n+1$$

and let

$$g_i(x) = -f_i(x_{i-1} + x_i) - f_i(x_i + x_{i+1})$$

for $1 \leqq i \leqq n$ where $x_0 = x_{n+1} = 0$. Note that if

$$V(x) = -\sum_{i=1}^{n+1} F_i(x_{i-1} + x_i),$$

then $\dfrac{\partial V}{\partial x_i}(x) = g_i(x)$. Consider the variational problem: Find critical points of

$$I(x) = \int_0^2 V(x(t)) \, dt \tag{3.13}$$

subject to the constraint

$$\tfrac{1}{2} \int_0^2 \sum_{i=1}^n m_i (x_i'(t))^2 \, dt = r^2 \tag{3.14}$$

where $x$ lies in the class of functions

$$\{x(t) \in W^{1,2}([0, 2], R^n) \mid x(0) = 0 \quad \text{and} \quad x(1+s) = x(1-s), \ s \in [0, 1]\}.$$

The left hand side of (3.14) can be taken as the square of the norm in this space. The Euler equations satisfied by any smooth critical point of this problem are

$$m_i x_i'' + \mu g_i(x) = 0, \qquad 1 \leqq i \geqq n \tag{3.15}$$

*i.e.* (2.3) where $\mu^{-1}$ is the corresponding Lagrange multiplier. The choice of function space shows $x$ also satisfies (2.4).

A bifurcation theorem due to BÖHME [8] or MARINO [9] can be applied to the above situation and shows that each eigenvalue of (3.1)–(3.2) is a bifurcation point for (1.1). However it does not give us information, for a multiple eigenvalue, about the shape of the corresponding solutions of (1.1).

*Remark 3.17.* Results of VAN GROESEN [10] (based on work of EKELAND & LASRY [11]) show that if

$$H(x, x') = \tfrac{1}{2} \Sigma\, m_i\, |x_i'|^2 + V(x)$$

and $V$ is even, convex, and satisfies other technical geometrical conditions, then for each $c > 0$, the energy surface $H = c$ contains at least $n$ distinct solutions of (1.1) of the type we seek. No information is obtained however about their shape.

*Acknowledgment.* This research was sponsored by the United States Army under Contract No. DAAG29-80-C-0041 and by the National Science Foundation under Grant No. MCS-8110556. Reproduction in whole or in part is permitted for any purpose of the United States Government.

## References

1. DUFFIN, R. J., Vibration of a beaded string analyzed topologically, Arch. Rational Mech. Anal. **56** (1974), 287–293.
2. SCHWARTZ, J. T., Nonlinear Functional Analysis, lecture notes, Courant Inst. of Math. Sc. New York Univ., 1965.
3. RABINOWITZ, P. H., Some global results for nonlinear eigenvalue problems, J. Funct. Anal., **7**, (1971), 487–513.
4. WHYBURN, G. T., Topological Analysis, Princeton Math. Series, no. 23, Princeton Univ. Press, Princeton, N.J., 1958.
5. GANTMACHER, F. R., & M. G. KREIN, Oszillationmatrizen, Oscillationskerne, und Kleine Schwingungen Mechanischer Systeme, Akademie-Verlag, Berlin, 1960.
6. CRANDALL, M. G., & P. H. RABINOWITZ, Bifurcation from simple eigenvalues, J. Funct. Anal. **8** (1971), 321–340.
7. DANCER, E. N., On the structure of solutions of nonlinear eigenvalue problems, Indiana Univ. Math. J., **23**, (1974), 1069–1076.
8. BÖHME, R., Die Lösung der Verzweigungsgleichungen für nichtlineare Eigenwertprobleme. Math. Z., **127**, (1972), 105–126.
9. MARINO, A., La biforcazione nel caso variazionale, Conf. Sem. Mat. dell Univ. Bari, **132** (1977).

10. VAN GROESEN, E. W. C., Existence of multiple normal mode trajectories on convex energy surfaces of even, classical Hamiltonian systems, to appear J. Diff. Eq.
11. EKELAND, I., & J. M. LASRY, On the number of closed trajectories for a Hamiltonian flow on a convex energy surface, Ann. Math. **112**, (1980), 283–319.

Department of Mathematics and
Mathematics Research Center
University of Wisconsin
Madison

*(Received August 6, 1985)*

# *The Boundary Value Problems for Non-Linear Elliptic Equations and the Maximum Principle for Euler-Lagrange Equations*

ILYA J. BAKELMAN

*Dedicated to James Serrin on his sixtieth birthday*

In this paper we investigate the boundary value problems for non-linear elliptic equations connected with infinite convex hypersurfaces. The asymptotic cone of such hypersurfaces is very important in our considerations. We study the problem of existence of solutions for Monge-Ampère equations on the entire space with prescribed asymptotic cone and the maximum principle and estimates for Euler-Lagrange equations. These problems have natural mutual connections.

## § 1. Infinite convex hypersurfaces and their asymptotic cones

### *1.1. The main definitions*

Let $E^{n+1}$ be an $(n+1)$-dimensional Euclidean space with Cartesian coordinates $x_1, x_2, \ldots, x_n, z \equiv x_{n+1}$ and let $E^n$ be the hyperplane $z = 0$ in $E^{n+1}$.

The graphs $S_u$ of convex functions $z = u(x)$, $x \in R^n$ are called *complete infinite convex hypersurfaces* in $E^{n+1}$. Thus from this definition it follows that we consider only convex hypersurfaces in $E^{n+1}$ projected one-to-one on the entire hyperplane $E^n$.

Let $H_u$ be the convex body defined by equation

$$z \geqq u(x) \tag{1.1}$$

for all $x \in E^n$, where $u(x)$ is a convex function. Clearly

$$\partial H_u = S_u. \tag{1.2}$$

Let $A$ be any point of $H_u$. The union of all closed rays starting from $A$ and belonging to $H_u$ is called the *asymptotic cone* of $H_u$ at the point $A$. Let $\phi_\lambda : E^{n+1} \to E^{n+1}$ be the homothety of $E^{n+1}$ with the center $A$ and coefficient

$\lambda > 0$. Then

$$K_u(A) = \lim_{\lambda \to 0} \phi_\lambda(H_u). \tag{1.3}$$

For every two points $A$, $A' \in H_u$ the convex cones $K_u(A)$ and $K_u(A')$ are congruent and can coincide by parallel translation in the direction of the $z$-axis. The well known proof is based on the formula (1.3). Therefore we will denote the asymptotic cone $K_u(A)$ by $K_u$, if we do not need to use a special choice of the point $A$.

The asymptotic cone $K_u$ is called *non-degenerate* if $\partial K_u$ is projected one-to-one on the entire hyperplane $E^n$. For non-degenerate asymptotic cones it is convenient to replace them by their boundary. In this case the notation $K_u$ will be used for the boundaries of a solid asymptotic cone and we will say that $K_u$ is the asymptotic cone of the function $u(x)$.

### *1.2. Normal mapping and R-curvature of convex functions*

The normal mapping and $R$-curvature are the natural generalizations of the tangential mapping and the integral Gaussian curvature for arbitrary continuous convex functions (hypersurfaces). I introduced and studied these concepts in [1] (for more detailed exposition see [2], [3]).

The normal mapping of a convex function $u(x)$ is constructed by supporting hyperplanes of its graph $S_u$. Let $P^n$ be an $n$-dimensional Euclidean space with Cartesian coordinates $p_1, p_2, \ldots, p_n$ and let $p = (p_1, p_2, \ldots, p_n)$ be a point of $P^n$. If

$$\alpha : z = p_1^0 x_1 + p_2^0 x_2 + \ldots + p_n^0 x_n \tag{1.4}$$

is a supporting hyperplane of $S_u$, then the point

$$\chi(\alpha) = (p_1^0, p_2^0, \ldots, p_n^0) \tag{1.5}$$

is called the *normal image* of the hyperplane $\alpha$. For any set $e \subset E^n$ the *normal image* $\chi_u(e)$ is defined as

$$\chi_u(e) = \bigcup_{\alpha} x(\alpha) \tag{1.6}$$

where $\alpha$ is any supporting hyperplane of $S_u$ whose intersection with $S_u$ has points with projections in the set $e$.

We shall use the following well known properties of the normal mapping.

*Property 1.*

$$\chi_u(E^n) = \chi_{K_u}(E^n) \tag{1.7}$$

for every convex function $u(x)$ defined on the entire hyperplane $E^n$.

*Property 2.* For every Borel subset $e$ of $E^n$ the set $\chi_u(e)$ is Lebesgue measurable.

*Property 3.* Let $R(p) > 0$ be locally summable in $E^n$; then

$$\omega(R, u, e) = \int_{x_u(e)} R(p)\, dp \tag{1.8}$$

is a completely additive non-negative set function on the ring of Borel subsets of $E^n$.

If $u(x) \in C^2(E^n)$, then

$$\omega(R, u, e) = \int_e R(Du) \det(u_{ij})\, dx.$$

The set function $\omega(R, u, e)$ is called the *R-curvature* of a convex function $u(x)$. Thus the $R$-curvature is the generalization of the Monge-Ampère operator $R(Du) \det(u_{ij})$ for the set of all convex functions.

If

$$R(p) \equiv 1,$$

then

$$\omega(1, u, e) = \operatorname{meas} \chi_u(e), \tag{1.9}$$

where the right part of (1.9) is the $n$-dimensional Lebesgue measure in the space $P^n$.

If $R(p) = (1 + |p|^2)^{-(n+1)/2}$, then $\omega(R, u, e)$ is the area (Lebesgue measure) of the image of $e$ concerning the Gaussian mapping of $S_u$ in the $n$-dimensional unit sphere $S^n$.

### 1.3. Remarks

In this subsection we consider a few properties of the normal mapping of convex functions.

*Property 4*. Let $u(x)$ be any convex function; then the function

$$v(x) = u(x) + \sum_{i=1}^{n} a_i x_i + b,$$

where $a_1, a_2, \ldots, a_n, b$ are constants, is convex and the set $\chi_v(e)$ can be obtained from $\chi_u(e)$ by parallel translation of $p_n$ along the vector $a = (a_1, a_2, \ldots, a_n)$.

*Property 5*. The normal image of any convex cone $K$ is a closed convex set in $P^n$ whose dimension can take the values $0, 1, 2, \ldots, n$.

*Property 6*. If $K$ is a non-degenerate convex cone, then $\chi_k(E^n)$ is a bounded closed $n$-dimensional convex domain with interior points.

The following Remarks follow from Properties 4, 5, 6.

It is sufficient to consider only convex functions $u: E^n \to R$ whose normal images contain the origin of $P^n$. If the asymptotic cone of such function is not degenerate then we can additionally assume that the origin $0'$ of $P^n$ is an interior point of $\chi_{K_u}(E^n)$. If we also assume that the vertex of this non-degenerate convex cone $K_u$ is at the origin $0$ of $E^n$, then the equation of $K_u$ is as follows

$$z = k(x),$$

where $k(x)$ is non-negative convex function in $E^n$ and $k(x) = 0$ only at the point $0$. Since any point of $E^n$ can be taken as origin, it is sufficient to consider asymptotic non-degenerate convex cones whose vertices are projected onto the point $\theta \in E$. The equations of such cones are as follows:

$$z = k(x) + b,$$

where $b$ is any constant.

## § 2. The second boundary value problem for elliptic solutions of Monge-Ampère equations

### *2.1. The statement of the second boundary value problem*

We consider elliptic solutions $u(x)$ of the Monge-Ampère equation

$$\det(u_{ij}) = f(x, u, Du) \tag{2.1}$$

in the entire space $E^n$. We assume that $f(x, u, p)$ is continuous in $E^n \times R \times P^n$. If $u(x)$ is a $C^2$-elliptic solution of equation (2.1), then $u(x)$ is necessarily either a convex or a concave function. It is sufficient to consider only convex solutions of equation (2.1). We should impose the *assumption of positiveness* of $f(x, u, p)$ in $E^n \times R \times P^n$ if we investigate convex solutions of equation (2.1).

The convex function $u(x)$ is called a *generalized solution* of equation (2.1) if $u(x)$ satisfies (2.1) almost everywhere and the set function $\omega(1, u, e)$ is absolutely continuous. Note that

$$\omega(1, u, e) = \text{meas}\, \chi_u(e)$$

for any Borel subset $e$ of $E^n$.

The statement of the second boundary value problem is as follows. Let $K$ be a non-degenerate convex cone. Find broad sufficient conditions for equation (2.1) to have at least one generalized solution for which $K$ is the asymptotic cone.

This problem was formulated by ALEXANDROV & POGORELOV (Conference on Global Differential Geometry, Leningrad, 1958); I solved it in 1984 (see [4]). Up to 1984 only two special cases of (2.1) had been investigated. The first one was studied by ALEXANDROV [5] in 1942; it is related to existence and uniqueness of a general complete infinite hypersurface with prescribed area of Gaussian mapping. I established [7] the second Theorem in 1956; it is related to generalized solutions of the Monge-Ampère equation

$$\det(u_{ij}) = \frac{g(x)}{R(Du)}. \tag{2.2}$$

The particular case $R(Du) = [1 + (Du)^2]^{-(n+1)/2}$ corresponds to ALEXANDROV's Theorem.

The necessary and sufficient condition for the second boundary value problem for equation (2.2) to have a solution is as follows:

$$\int_{E^n} g(x)\, dx = \int_{\chi_k(R^n)} R(p)\, dp \tag{2.3}$$

where $g(x) \geqq 0$, $g(x) \in L(E^n)$, $R(p) > 0$, $R(p) \in L_{\text{loc}}(P^n)$ and $\chi_k(P^n)$ is the normal image of prescribed non-degenerate convex asymptotic cone $K$. Note that the desired solution is defined to within an additive constant.

The solution of the second boundary value problem for the general Monge-Ampère equation

$$\det(u_{ij}) = f(x, u, Du)$$

is essentially more difficult because the simple necessary and sufficient condition (2.3) for solubility of the same problem for equation (2.2) must be replaced by the complicated implicit necessary condition

$$\text{meas}\, \chi_k(P^n) = \int_{E^n} f(x, u(x), Du(x))\, dx. \tag{2.4}$$

Moreover, the application of fixed-point theorems offers difficulties because the set $E^n$ is not compact. The proof of my recent existence theorem for equation (2.1), presented in 2.3, is based on the construction of a new Monge-Ampère equation in some specially introduced Banach space. Significantly, this construction is based upon asymptotic properties of the function $f(x, u, p)$ and its first derivatives as $|x| \to +\infty$ and $|u| \to +\infty$.

### *2.2. The Second boundary problem for Monge-Ampère equations $R(Du) \det(u_{ij}) = \phi(x)$*

Let $g(x) \geqq 0$ be a summable function in $E^n$ and let $R(p) > 0$ be a locally summable function in $P^n$.

**Theorem 1.** *Let $K$ be a non-degenerate convex cone in $E^{n+1} = E^n \times R$ and let*

$$z = k(x), \qquad x \in R^n$$

*be the equation of $K$. Let*

$$\int_{E^n} g(x)\, dx = \int_{\chi_k(E^n)} R(p)\, dp. \tag{2.5}$$

*Then the second boundary value problem for the Monge-Ampère equation*

$$\det(u_{ij}) = \frac{g(x)}{R(Du)} \tag{2.6}$$

*has a generalized convex solution $u(x)$ and this solution is unique to within an additive constant.*

As we noted in Subsection 2.1 I proved this Theorem in [7].

The proof of the statement of uniqueness to within an additive constant is based on the same ideas and facts as the proof of the uniqueness of convex generalized solutions of the Dirichlet problem for equation (2.6) (see [8], Chapter IV, § 23).

The assertion of existence of generalized solutions made in Theorem 1 can be proved by the approximation of convex polyhedra. First we replace equation (2.6) by the equation

$$\omega(R, u, e) = \mu(e) \tag{2.7}$$

which can now be considered for all convex functions defined in $E^n$. In (2.7) $\mu(e)$ is any non-negative, completely additive set function subordinated to the condition

$$\mu(E^n) = \int_{\chi_k(E^n)} R(p)\, dp\,. \tag{2.8}$$

If we consider the polyhedron case, then set functions $\mu(e)$ and prescribed convex cone $K$ must satisfy the following necessary assumptions.

*Assumption A.* The convex cone $K$ is a non-degenerate convex polyhedral angle. We can assume in addition that the origin $\theta'$ of $P^n$ lies inside $\chi_k(P^n)$ and therefore that the equation of $K$ has the form

$$z = k(x) + b \tag{2.9}$$

where $k(x)$ is a convex continuous piecewise linear homogeneous function (of degree 1) in $E^n$, satisfying the conditions $k(\theta) = 0$, $k(x) > 0$ if $x \neq \theta$, and $b$ is any constant.

*Assumption B.* The set function $\mu(e)$ satisfies the following condition: there are a finite system of points $A_1, A_2, \ldots, A_m$ and a finite system of non-negative numbers $\mu_1, \mu_2, \ldots, \mu_m$ such that

$$\mu(A_i) = \mu_i \tag{2.10}$$

and

$$\mu(e) = \sum_{A_i \in e} \mu(A_i) \tag{2.11}$$

for every subset $e$ of $E^n$.

Theorem 1 can be reduced to the following theorem for convex polyhedra.

**Theorem 2.** *If the polyhedral angle $K$ satisfies Assumption A, and if the set function $\mu(e)$ satisfies Assumption B, and if*

$$\mu(E^n) = \int_{P^n} R(p)\, dp\,, \tag{2.12}$$

*then there are convex polyhedra*

$$S_u : z = u(x), \qquad x \in E^n$$

*such that the vertices of $S_u$ are projected only upon the points $A_1, A_2, \ldots, A_m$; $K$ is the asymptotic cone of $S_u$; and*

$$\omega(R, u, A_i) = \mu_i, \qquad i = 1, 2, \ldots, m\,. \tag{2.13}$$

*Moreover if $u_0(x)$ is a convex function defining one of those polyhedra, then all the others can be obtained by the formula*

$$u(x) = u_0(x) + C, \tag{2.14}$$

*where $C$ is an arbitrary constant.*

*Remarks.* 1) Theorem 1 can be obtained by approximation of convex polyhedra which are constructed with the help of Theorem 2. This is the traditional way now. In the theory of convex bodies and hypersurfaces the approximation method by polyhedra was developed by MINKOWSKI, ALEXANDROV, POGORELOV and others (see [9], [6], [10], [11], [2], [8]). Therefore we note only that the conditions of Theorem 1 suffice to obtain the sequence of convex polyhedra

$$S_{u_j}: z = u_j(x), \quad x \in E^n$$

such that: a) $u_j(\theta) = 0$, $j = 1, 2, \ldots, m$; b) all functions $u_j(x)$ satisfy the Lipschitz conditions of degree 1 with a common constant; c) $\omega(R, u_j, e)$ converge weakly to the function $\mu(e)$ and

$$\lim_{j\to\infty} \omega(R, u_j, E^n) = \int_{\chi_k(E^n)} R(p)\, dp\,.$$

From convex polyhedra $u_j(x)$ it is possible to extract a convergent subsequence whose limit is a solution of the second boundary value problem for equation (2.7) (or equation (2.6)) with a prescribed non-degenerate convex asymptotic cone. This completes the proof of Theorem 1.

Finally in this subsection we present a new variant of the proof of Theorem 2 based on a variational method which up to now has not been used in non-compact problems related to infinite convex hypersurfaces.

**Proof of Theorem 2.** We denote by $W^+ (A_1, A_2, \ldots, A_m, K)$ the set of all convex polyhedra, whose vertices are projected only onto the points $A_1, A_2, \ldots, A_m$ and whose asymptotic cones coincide with prescribed non-degenerate convex polyhedral angle $K$. Let $T$ be the set of convex polyhedra $P: z = u(x)$ such that

a) $T$ is a subset of $W^+ (A_1, A_2, \ldots, A_m, K)$;
b) If $u(x) \in T$, then the inequalities

$$0 \leqq \omega(R, u, A_i) \leqq \mu_i \tag{2.15}$$

hold for all $i = 2, 3, \ldots, m$ and

$$\omega(R, u, A_1) = \int_{\chi_k(E^n)} R(p)\, dp - \sum_{i=2}^{m} \omega(R, u, a_i) \tag{2.16}$$

c) $u(A_1) = a_1$, (2.17)

where $a_1$ is an arbitrary fixed constant.

The set $T$ is not empty, because the convex cone $K$ with the vertex $(A_1, a_1)$ belongs to $T$.

The system of real numbers

$$\xi_1 = u(A_1), \quad \xi_2 = u(A_2), \ldots, \quad \xi_m = u(A_m), \tag{2.18}$$

taken from every convex polyhedron $u(x) \in W^+(A_1, A_2, \ldots, A_m, K)$, defines this polyhedron point by point. The metric

$$d(u(x), v(x)) = \left\{\sum_{i=1}^{m} [u(A_i) - v(A_i)]^2\right\}^{\frac{1}{2}} \tag{2.19}$$

introduced in $W^+(A_1, A_2, \ldots, A_m, K)$ shows that the (2.18) maps $W^+(A_1, A_2, \ldots, A_m, K)$ isometrically on to some closed convex subset of the Euclidean space $R^m = \{\xi = (\xi_1, \xi_2, \ldots, \xi_m)\}$. Since $w^+(A_1, A_2, \ldots, A_m, K)$ can be identified with a corresponding closed subset of $R^m$, $T$ is also a closed subset of $R^m$. From (2.15–17) it follows that $T$ is a bounded subset of $R^m$. *Thus $T$ is a compact set in $R^m$.*

The function $f: T \to R$ defined by the formula

$$f(u) = \sum_{i=1}^{m} u(A_i) \tag{2.20}$$

is continuous in $T$. Hence

$$\inf_T f(u) = f_0 > -\infty$$

and there is a polyhedron $u_0(x) \in T$ such that

$$f(u_0(x)) = f_0. \tag{2.21}$$

Now we prove that $u_0(x)$ is the desired solution of the second boundary value problem satisfying the condition (2.17).

If our assertion is incorrect, then

$$\omega(R, u_0, A_s) < \mu_s \tag{2.22}$$

at least at one point $A_s$ $(s = 2, 3, \ldots, m)$, because

$$\omega(R, u_0, A_s) \leqq \mu_s$$

from the definition of the set $T$. Now consider the convex polyhedron

$$\tilde{P}: z = \tilde{u}(x) \in W^+(A_1, A_2, \ldots, A_m, K)$$

such that

$$\tilde{u}(A_1) = a_1, \quad \tilde{u}(A_2) = u_0(A_2), \ldots$$

$$\tilde{u}(A_{s-1}) = u_0(A_{s-1}), \quad \tilde{u}(A_s) = u_0(A_s) - \varepsilon,$$

$$\tilde{u}(A_{s+1}) = u_0(A_{s+1}), \ldots, \tilde{u}(A_m) = u_0(A_m),$$

where $\varepsilon > 0$ is a number small enough that the inequality

$$\omega(R, \tilde{u}, A_s) < \mu_s \tag{2.23}$$

holds. Since

$$\sum_{i=1}^{m} \omega(R, \tilde{u}, a_i) = \int_{\chi_k(E^n)} R(p)\, dp \tag{2.24}$$

and

$$\omega(R, u, A_l) \leqq \omega(R, u_0, A_l)$$

for $l = 1, 2, \ldots, s-1, s+1, \ldots, m$, then from (2.23–24) it follows that $\tilde{u}(x) \in T$. But

$$f(u(x)) = \sum_{i=1}^{m} \tilde{u}(A_i) = \sum_{i=1}^{m} u_0(A_i) - \varepsilon = f_0 - \varepsilon < f_0 .$$

This last inequality is incompatible with (2.21). Hence $u_0(x)$ is the desired solution of the second boundary value problem for the equation (2.13).

Now let convex polyhedra $P_1$ and $P_2$ be the graphs of convex solutions $u_1(x)$, $u_2(x) \in W^+(A_1, A_2, \ldots, A_m, K)$ of (2.13) that satisfy the additional condition

$$u_1(A_1) = u_2(A_1) = a_1 .$$

If $u_1(x)$ and $u_2(x)$ are different functions, then according to the lemma proved in [6], Chapter IX, there is at least one point $A_j$, where $j$ is one of the integers $1, 2, \ldots, m$, such that the polyhedral angles $V_1 \subset P_1$ and $V_2 \subset P_2$ are projected onto the point $A_j$ and

$$V_1 \subset V_2 \quad \text{and} \quad V_1 \setminus V_2 \neq \emptyset .$$

Hence

$$\operatorname{meas} \chi_{u_1}(A_j) > \operatorname{meas} \chi_{u_2}(A_j) .$$

Thus

$$\mu_j = \int_{\chi_{u_1}(A_j)} R(p)\, dp > \int_{\chi_{u_2}(A_j)} R(p)\, dp = \mu_j .$$

Therefore our assumption is incorrect. The proof of Theorem 2 is completed.

### *2.3. The second boundary value problem for general Monge-Ampère equations*

In this subsection we investigate the second boundary value problem for the Monge-Ampère equation

$$\det(u_{ij}) = f(x, u, Du) \tag{2.25}$$

in the class of convex generalized solutions. We divide the presentation into a few separate parts.

#### *A) The main Assumptions*

*A.1 Admissible convex cones.* A nondegenerate convex cone in $E^{n+1}$ is called *admissible* if the equation of $K$ is

$$z = k(x) \tag{2.26}$$

where $k(x)$ is a continuous, convex function in $E^n$, homogeneous of order 1, satisfying the conditions

a) $k(\theta) = 0$,
b) $k(x) > 0$ for all $x \neq \theta$ in $E^n$.

It is sufficient to consider only admissible convex cones (see the final part of Section 1). We denote by $K^*$ the normal image of a convex cone $K$, *i.e.*

$$K^* = \chi_k(E^n). \tag{2.27}$$

For every admissible convex cone $K$ the set $K^*$ is closed, bounded and convex in the space $P^n$ and the point $\theta'$ is an interior point of $K^*$, where $\theta'$ is the origin of $P^n$.

*A.2 The properties of the function $f(x, u, p)$ and its derivatives.* The function $f(x, u, p)$ is continuous in $E^n \times R \times P^n$ together with its derivatives $\frac{\partial f}{\partial u}$ and $\frac{\partial f}{\partial p_i}$, $(i = 1, 2, \ldots, n)$ and

$$f(x, u, p) > 0, \tag{2.28}$$

$$\frac{\partial f(x, u, p)}{\partial u} > 0 \tag{2.29}$$

in $E^n \times R \times P^n$.

We also assume that the inequalities

$$\left|\frac{\partial f}{\partial u}\right| \leq \frac{C_0}{|x|^{n+2+\alpha}}, \tag{2.30}$$

$$\left|\frac{\partial f}{\partial p_i}\right| \leq \frac{C_1}{|x|^{\eta+\alpha}}, \quad (i = 1, 2, \ldots, n) \tag{2.31}$$

hold for all $(x, u, p) \in E^n \times R \times K^*$ with $|x| \geq m_0$, where $\alpha = \text{const.} > 0$, $C_0 = \text{const.} \geq 0$, $C_1 = \text{const.} \geq 0$, $m_0 = \text{const.} \geq 1$.

*A.3 Estimators and their properties.* For every admissible convex cone $K$ there are two functions $\lambda_K(x, u)$, and $\phi_K(x, u)$, depending only on the prescribed cone $K$, such that:

a) $\lambda_K(x, u)$ and $\phi_K(x, u)$ are positive and continuous in $E^n \times R$ and increase with respect to $u$ for every fixed $x \in E^n$;

b) the inequalities

$$\lambda_K(x, u) \leq f(x, u, p) \leq \phi_K(x, u) \tag{2.32}$$

hold for all $x \in E^n$, $u \in R$, $p \in K^*$.

The functions $\lambda_K(x, u)$ and $\phi_K(x, u)$ are called *estimators*.

*B) The statement of the main Theorem and the scheme of its proof*

**Theorem 3.** *Let K be an admissible convex cone described in Assumption 1 and let $\lambda_k(x, u)$, $\phi_k(x, u)$ be estimators satisfying Assumption 2 (see part A of Subsection 2.3). If there are two numbers*

$$-\infty < a_k < b_k < +\infty$$

*such that*

$$a) \int_{E^n} \phi_k(x, k(x) + q)\, dx < +\infty \tag{2.33}$$

*for all* $q \in [a_k, b_k]$;

$$b) \int_{E^n} \phi_K(x, k(x) + a_k)\, dx < \text{meas } K^*; \tag{2.34}$$

$$\text{c)} \inf_{\gamma \in K^*} \int_{E^n} \lambda_K(x, (x, \gamma) + b_k)\, dx > \text{meas } K^* \tag{2.35}$$

*where* $z = k(x)$ *is the equation of the cone K and* $(x, \gamma) = \sum_{i=1}^{n} \gamma_i x_i$, $x = (x_1, \ldots, x_n) \in E^n$, $\gamma = (\gamma_1, \ldots, \gamma_n) \in K^*$, *then equation* (2.25) *has at least one generalized solution* $u(x)$ *with asymptotic cone K, and*

$$a_k < u(0) < b_k. \tag{2.36}$$

**Scheme of the proof.** Since we consider unbounded convex functions on the entire space $E^n$, then application of fixed point principles requires us to construct a special function space in which the second boundary value problem can be investigated. In this special function space we study some modification of the Monge-Ampère operator det $(u_{ij})$ and its inverse, which is induced by the original equation (2.25) and by the prescribed admissible convex cone $K$. The final part of the proof for Theorem 3 is based on application of the Schauder principle to the inverse of the modified Monge-Ampère equation.

*C) The function space of the second boundary value problem.* We denote by $C^0(E^n)$ the set of all continuous functions $u: E^n \to R$ and by $C^{0,1}(E^n)$ the subset of $C^0(E^n)$ consisting in all Lipschitz functions $u: E^n \to R$, i.e. $u(x) \in C^{0,1}(E^n)$ if and only if

$$L(u) = \sup_{x, y \in E^n} \frac{|u(y) - u(x)|}{|y - x|} < +\infty, \tag{2.37}$$

where $|y - x| = \text{dist}\,(x, y)$ in $E^n$.

Let $U_1 \subset U_2 \subset \ldots \subset U_m$ ... be the sequence of $n$-balls:

$$U_m: |x| \leqq m, \tag{2.38}$$

$m = 1, 2, 3, \ldots$ We denote by $\|u\|_m$ the number $\sup_{U_m} |u(x)|$ and by $\|u\|_A$ the number

$$\|u\|_A = |u(\theta)| + \sum_{m=1}^{\infty} \frac{\|u\|_m}{m^{2+\varkappa}}, \tag{2.39}$$

where $\varkappa = \text{const.} > 0$ (see Assumption 2).

Let $A$ be the subset of $C^0(E^n)$ consisting in all functions $u(x)$ such that

$$\|u\|_A < +\infty.$$

Clearly

$$\|u\|_A \leq |u(0)| + L(u) \sum_{m=1}^{\infty} \frac{1}{m^{2+\varkappa}} < +\infty$$

for every $u(x) \in C^{0,1}(E^n)$. On the other hand $\|e^{x_1 + \ldots + x_n}\|_A = +\infty$. Thus $A$ is a proper non-empty subset of $C^0(E^n)$.

The functional $\|u\|_A$ is a norm of the set $A$. Clearly $A$ is a Banach space with respect to this norm. We denote by $A_1$ the subspace of $A$ consisting in all functions $u(x) \in C^{0,1}(E^n)$.

Now we introduce the equivalence relation $\Gamma$ in $A$ setting $\tilde{u}(X)\, \Gamma v(x)$ if and only if $u(x) - v(x) = \text{const}$ in $E^n$. Clearly the factor space $B = A/\Gamma$ is again a Banach space with respect to the induced norm

$$\|\xi\|_B = \|\tilde{u}\|_A$$

where $\xi$ is any element of $B$ generated by the class of functions $\{u(x) + q\}$, where $u(x) \in A$ and $q$ is any real number. We use the notation $u(x) = u(x) - u(\theta)$. We call the function $u(x) \in A$ a *basic representative* of the element $\xi \in B$, if $u(\theta) = 0$. We shall use the notation $\xi_u$ for elements of the space $B$ generated by basic representatives $u(x) \in A$. Clearly $\tilde{u}(x)$ is a basic representative for every $u(x) \in A$.

Every convex function $u(x)$ defined in $E^n$ whose graph has an admissible asymptotic cone is an element of $A_1$. Therefore $u(x) \in A$. This statement follows directly from the compactness of $\chi_u(E^n)$ in the space $P^n$.

Let $T_k$ be a subset of $B$ consisting in all elements $\xi \in B$ such that $\xi = \xi_u$, where $u(x)$ is any convex function defined in $E^n$ and satisfying two conditions:

a) $u(\theta) = 0$;

b) the asymptotic cone of $u(x)$ is a fixed admissible cone $K$, considered in the second boundary value problem for equation (2.25).

The set $T_k$ is not empty, since $\xi_k \in T_k$, where $z = k(x)$ is the equation of the admissible convex cone $K$ mentioned above. The following lemmas are correct for any admissible convex cone $K$.

**Lemma 1.** $T_k$ is a convex subset of $B$.

**Lemma 2.** $T_k$ is a closed subset of $B$.

**Lemma 3.** $T_k$ is a compact subset of $B$.

**Proof of Lemma 1.** Let $\lambda$ and $\mu$ be any positive numbers such that $\lambda+\mu=1$. If $\xi_f$ and $\xi_g$ are any elements of $T_k$ generated by the basic representatives $f(x)$ and $g(x)$, then the functions $f(x)$ and $g(x)$ are convex in $E^n$, have one and the same asymptotic cone $K$, which is admissible, and $f(\theta)=g(\theta)=0$.

Clearly the element $\lambda\xi_f+\mu\xi_g$ is generated by the function $\lambda f(x)+\mu g(x)$. Since $\lambda f(x)+\mu g(x)$ is convex and

$$\lambda f(\theta)+\mu g(\theta)=0,$$

then

$$\lambda\xi_f+\mu\xi_g=\xi_{\lambda f+\mu g}. \tag{2.40}$$

If we prove that $K$ is the asymptotic cone of $\lambda f(x)+\mu g(x)$, then from this fact and from equality (2.40) it follows that $T_k$ is a convex set in $B$. Clearly the statement concerning the asymptotic cone of $\lambda f(x)+\mu g(x)$ need be established only for $C^2$ convex functions $f(x)$ and $g(x)$, because the case of general convex functions can be obtained by the simple approximation of corresponding $C^2$ convex functions.

Thus our convex functions $f(x)$ and $g(x)$, introduced in the beginning of the proof of Lemma 1, are twice differentiable in $E^n$.

Clearly

$$\operatorname{grad}(\lambda f+\mu g)=\lambda \operatorname{grad} f+\mu \operatorname{grad} g \tag{2.41}$$

at any point $x_0\in E^n$. From (2.41) it follows that

$$\chi(\gamma)=\lambda\chi(\gamma_1)+\mu\chi(\gamma_2),$$

where $\gamma$, $\gamma_1$, $\gamma_2$ are correspondingly the normal images of tangent hyperplanes of the graphs of the functions $\lambda f(x)+\mu g(x)$, $f(x)$ and $g(x)$ at the points

$$(x,\lambda f(x)+\mu g(x)),\ (x,f(x)),\ (x,g(x)).$$

Since $\chi(\gamma_1)$, $\chi(\gamma_2)$ are points of the convex set $\chi_k(E^n)$, then

$$\chi(\gamma)\in\chi_k(E^n). \tag{2.42}$$

If $K'$ is the asymptotic cone of the convex function $\lambda f(x)+\mu g(x)$, then from (2.42) it follows that

$$\chi_{k'}(E^n)\subset\chi_k(E^n), \tag{2.43}$$

where $z=k'(x)$ is the equation of the asymptotic cone $K_1$, whose vertex is at the point $\theta\in E^n$. From (2.43) we obtain

$$0\leqq k'(x)\leqq k(x) \tag{2.44}$$

for all $x\in E^n$.

Now we prove that

$$k'(x)=k(x)$$

for all $x\in E^n$. Let $l$ be any axis in $E^n$, passing through the point $\theta$, and $s$ be the Cartesian coordinate in $l$ such that

$$|s|=[x_1^2+x_2^2+\ldots+x_n^2]^{\frac{1}{2}}$$

for any point $x=(x_1,x_2,\ldots,x_n)\in l$.

Let $\tilde{k}'(s)$, $\tilde{k}(s)$, $\tilde{f}(s)$, $\tilde{g}(s)$ be functions which are generated by $k'(x)$, $k(x)$, $f(x)$, $g(x)$ on the axis $l$.

Then

$$k'(s) = \begin{cases} -k_1's & \text{if } s \leqq 0; \\ k_2's & \text{if } s > 0 \end{cases}$$

and

$$k(s) = \begin{cases} -k_1 s & \text{if } s \leqq 0; \\ k_2 s & \text{if } s > 0, \end{cases}$$

where $0 \leqq k_1' \leqq k_1$ and $0 \leqq k_2' \leqq k_2$. Thus we obtain the following chain of equalities

$$k_2' + k_1' = \int_{-\infty}^{+\infty} \left(\lambda \frac{d^2 f}{ds^2} + \mu \frac{d^2 g}{ds^2}\right) ds = \lambda \int_{-\infty}^{+\infty} \frac{d^2 f}{ds^2} ds + \mu \int_{-\infty}^{+\infty} \frac{d^2 g}{ds^2} ds$$

$$= \lambda(k_2 + k_1) + \mu(k_2 + k_1) = k_2 + k_1,$$

because $\lambda \geqq 0$; $\mu \geqq 0$ and $\lambda + \mu = 1$. Thus $k_2' + k_1' = k_2 + k_1$. Since $0 \leqq k_2' \leqq k_2$ and $0 \leqq k_1' \leqq k_1$, then

$$k_2' = k_2 \quad \text{and} \quad k_1' = k_1. \tag{2.45}$$

Hence

$$k'(x) = k(x) \tag{2.46}$$

for all $x \in E^n$, because $l$ is an arbitrary axis in $E^n$ passing through $\theta$.

From (2.46) it follows that

$$K' = K$$

and the proof of Lemma 1 is completed.

**Proof of Lemma 2.** Let $\bar{T}_k$ be the closure of $T_k$ in the space $B$. If $\xi \in \bar{T}_k$, then there exist the elements $\xi_1, \xi_2, \ldots, \xi_q, \ldots$ of the set $T_k$ such that

$$\lim_{q \to \infty} \|\xi_q - \xi\|_B = 0.$$

The basic representatives $u_q(x)$ of $\xi_q$ are convex functions on $E^n$ with one and the same asymptotic cone $K$, which is admissible, and $u_q(\theta) = 0$, $q = 1, 2, 3, \ldots$ Clearly

$$\lim_{q \to \infty} \|u_q(x) - u(x)\|_A = 0,$$

where $u(x)$ is the basic representative of $\xi \in \bar{T}_k$. Hence $u_q(x)$ converges uniformly to $u(x)$ in every closed ball $U_m : |x| \leqq M$. Thus $u(x)$ is a convex function with asymptotic cone $K$ and $\xi \in T_k$. The proof of Lemma 2 is completed.

**Proof of Lemma 3.** Let $\xi$ be any element of $T_k$ and $u(x)$ be the basic representative of $\xi$. Then $u(\theta) = 0$ and the convex function $u(x)$ satisfies a Lipschitz condition in $E^n$ with the constant $d_0 = \operatorname{diam} K^*$. Thus

$$\|\xi\|_B = \|u(x)\|_A = \sum_{m=1}^{\infty} \frac{\|u(x)\|_m}{m^{2+\alpha}}$$

$$\leqq d_0 \sum_{m=1}^{\infty} \frac{1}{m^{1+\alpha}} = \text{const.} < +\infty .$$

Let $\{u_\gamma(x)\}$ be the collection of basic representatives of $\{u_\gamma(x)\}$. From the last estimate follows the existence of a sequence of convex functions $u_{\gamma_q}(x)$ which is subsequence of $\{u_\gamma(x)\}$ and which is uniformly convergent to some convex function $u_0(x)$ in every ball $U_m: |x| \leqq m$. Clearly $u_0(\theta) = 0$ and $K$ is the asymptotic cone of $u_0(x)$. Hence

$$\xi_{u_0} \in T_k .$$

If we establish that

$$\lim_{q\to\infty} \|u_{\gamma_q}(x) - u_0(x)\|_A = 0,$$

then the proof of Lemma 3 will be completed.

Let $\varepsilon > 0$ be any number. We fix a positive integer $m_0$ such that

$$\sum_{m=m_0}^{\infty} \frac{1}{m^{1+\alpha}} < \frac{\varepsilon}{4 \operatorname{diam} K^*} .$$

Then

$$\|u_{\gamma_q} - u_0\|_A = \sum_{m=1}^{m_0-1} \frac{\|u_{\gamma_\theta} - u_0\|_m}{m^{2+\alpha}} + \sum_{m=m_0}^{\infty} \frac{\|u_{\gamma_q} - u_0\|_m}{m^{2+\alpha}}$$

$$\leqq (m_0 - 1)\|u_{\gamma_q} - u_0\|_m + 2 \operatorname{diam} K^* \sum_{m=m_0}^{\infty} \frac{1}{m^{1+\alpha}}$$

$$< (m_0 - 1)\|u_{\gamma_q} - u_0\|_m + \frac{\varepsilon}{2} .$$

Since $\lim \|u_{\gamma_q} - u_0\|_m = 0$, there is a positive integer $Q_0$ such that

$$(m_0 - 1)\|u_{\gamma_q} - u_0\|_m < \frac{\varepsilon}{2}$$

for $q > q_0$. Thus $\|u_{\gamma_q} - u_0\|_m < \varepsilon$ for $q > q_0$, where $\varepsilon > 0$ is any given number. Lemma 3 is proved.

**D) The proof of Theorem 3.** Now we return to the proof of Theorem 3, which is the main existence Theorem of § 2.

Let $\xi$ be any element of the set $T_k$ and let $u(x)$ be its basic representative. We consider the collection of convex functions

$$u_a(x) = u(x) + a \tag{2.47}$$

where $a \in (-\infty, +\infty)$. The function

$$F_{u_a}(x) = f(x, u_a(x), \nabla u_a(x))$$
$$= f(x, u(x) + a, \nabla u(x))$$

is non-negative for all $x \in E^n$. Let $z = k(x)$ be the equation of the cone $K$. Then

$$u_a(x) = u(x) + a \leqq k(x) + a \tag{2.49}$$

for all $x \in E^n$. From Assumption A.3 it follows that

$$F_{u_a}(x) \leqq \phi_k(x, u(x) + a) \leqq \phi_k(x, k(x) + a) \tag{2.50}$$

for all $x \in E^n$ and all real values of $a$. From the conditions of Theorem 3 it follows that

$$\int_{E^n} F_{u_a}(x)\, dx \leqq \int_{E^n} \phi_k(x, k(x) + a)\, dx < +\infty$$

for all $a \in [a_k, b_k]$. From Assumption A.3 we obtain the inequality

$$F_{u_a}(x) \geqq \inf_{\gamma \in K^*} \lambda_k(x, (x, \gamma) + a) \tag{2.51}$$

for all $x \in E^n$ and all $x \in R$.

Now we introduce the function

$$\psi(a) = \int_{E^n} F_{u_a}(x)\, dx \tag{2.52}$$

in $[a_k, b_k]$. Since

$$\psi(a) = \int_{E^n} f(x, u(x) + a, \nabla u(x))\, dx \tag{2.53}$$

for $a \in [a_k, b_k]$, then from Assumptions A.2, A.3 and the conditions of Theorem 3 it follows that $\psi(a)$ is continuous, $\psi'(a)$ exists on $[a_k, b_k]$, and the inequalities

$$\psi(a_k) \leqq \int_{E^n} \phi_k(x, k(x) + a_k)\, dx < \text{meas } K^*$$

and (2.54)

$$\psi(b_k) \geqq \inf_{k^*} \int_{E^n} \lambda_k(x, (x, \gamma) + b_k)\, dx > \text{meas } K^*$$

hold. Since

$$\psi'(a) = \int_{E^n} \frac{\partial f(x, u(x) + a, \nabla u(x))}{\partial u}\, dx > 0,$$

then from (2.54) it follows that only one number $a^* \in [a_k, b_k]$ exists such that

$$\psi(a^*) = \int_{E^n} F_{u_{a^*}}(x)\, dx = \text{meas } K^*. \tag{2.55}$$

Now we consider the second boundary value problem for the equation

$$\det(z_{ij}) = F_{u_{a*}}(x) \tag{2.56}$$

with prescribed asymptotic cone $K$.

Since all conditions of Theorem 1 are fulfilled, this boundary value problem has only one convex generalized solution $z(x)$ satisfying the condition $z(0) = a^*$ and having $K$ as the asymptotic cone of its graph.

Let $\tilde{u}(x) = u_{a*}(x) - a^*$ and $\tilde{z}(x) = z(x) - z(\theta)$ be the basic representatives of the elements $\xi_u$ and let $\eta_z$ belong to the convex set $T_k$. Clearly the second boundary value problem for equation (2.56) and the prescribed admissible convex cone $K$ generate some operator

$$G : T_k \to T_k$$

such that $\eta_z = G(\xi_u)$.

The remainder of this section establishes the continuity and compactness of the operator $G$ on the convex compact set $T_k$. They will permit us to apply the Schauder fixed point theorem to the operator $G : T_k \to T_k$. The existence theorem 3 is the final result of these investigations.

**Lemma 4.** *The functional* $a^* : T_k \to R$ *is continuous.*

**Proof.** Let $\xi$ be any element of $T_k$ and $u(x) \in A_1 \in A$ be a basic representative of $\xi$. Then $u(x)$ is a convex function whose graph has $K$ as its asymptotic cone, and $u(\theta) = 0$.

The real number $a^* = a^*(u)$ is the root of the equation

$$\int_{E^n} f(x, u(x) + a, \nabla u(x))\, dx = dx = \text{meas}\, K^* . \tag{2.57}$$

We proved above that this equation has only one root $a^* = a^*(u) \in [a_k, b_k]$. Now we must prove that $a^*(u_q)$ converges to $a^*(u)$ if

$$\lim_{q \to \infty} \| u_q - u \|_A = 0$$

and $\xi_{u_q}$, $\xi_u \in T_k$. From (2.57) it follows that

$$\int_{E^n} f(x, u(x) + a^*(u), \nabla u(x))\, dx = \int_{E^n} f(x, u_q(x) + a^*(u_q), \nabla u_q(x))\, dx = \text{meas}\, K^* .$$

Hence

$$\int_{E^n} \{[u(x) - u_q(x)] + [a^*(u) - a^*(u_q)]\} \left( \int_0^1 \frac{\partial f}{\partial u}\bigg|_{v_t} dt \right) dx$$

$$= - \sum_{i=1}^{n} \int_{E^n} \left[ \frac{\partial u(x)}{\partial x_i} - \frac{\partial u_q(x)}{\partial x_i} \right] \left( \int_0^1 \frac{\partial f}{\partial u_i}\bigg|_{v_t} dt \right) dx ,$$

where

$$v_t(x) = (1 - t)\, u_q(x) + t u(x) + (1 - t)\, a^*(u_q) + t a^*(u)$$

and

$$\frac{\partial v_t(x)}{\partial x_i} = (1-t)\frac{\partial u_q(x)}{\partial x_i} + t\frac{\partial u(x)}{\partial x_i},$$

$0 \leqq t \leqq 1, \ i = 1, 2, \ldots, n.$

Since $a^*(u)$ and $a^*(u_q)$ do not depend on $x$, from (2.58) we obtain

$$a^*(u) - a^*(u_q) = -\frac{\int\limits_{E^n} \{u(x) - u_q(x)\}\left(\int\limits_0^1 \frac{\partial f}{\partial u}\bigg|_{v_t} dt\right) dx}{\int\limits_{E^n}\left(\int\limits_0^1 \frac{\partial f}{\partial u}\bigg|_{v_t} dt\right) dx}$$

$$- \frac{\int\limits_{E^n} \sum\limits_{i=1}^n \left(\frac{\partial u(x)}{\partial x_i} - \frac{\partial u_q(x)}{\partial x_i}\right)\left[\int\limits_0^1 \frac{\partial f}{\partial u}\bigg|_{v_t} dt\right] dx}{\int\limits_{E^n}\left(\int\limits_0^1 \frac{\partial f}{\partial u}\bigg|_{v_t} dt\right) dx}.$$

Since $\partial f/\partial u$ is positive and continuous in $E^n \times R \times K^*$ (see Assumption A.2), for any compact set $Q$ in $E^n \times R \times K^*$ there is a constant $h(Q) > 0$ such that

$$\frac{\partial f(x, u, p)}{\partial u} \geqq h(Q) > 0.$$

For our purpose it is sufficient to consider the compact set

$$Q_0 = U_1 \times [\delta_1, \delta_2] \times K^*,$$

where $U_1$ is the unit ball $|x| \leqq 1$ in $E^n$ and the numbers $\delta_1, \delta_2$ are determined by conditions

$$\delta_1 \leqq v_t(x) \leqq \delta_2.$$

The finite values of $\delta_1, \delta_2$ depend only on $\|u(x)\|$, the numbers $a_k$, $b_k$ and the integer $N_1 > 0$ such that

$$\|u_q(x) - u(x)\|_A < 1$$

if $q \geqq N_1$. If $x \in U_1$, then

$$(v_t(x), \nabla v_t(x)) \in [\delta_1, \delta_2] \times K^*$$

for all $t \in [0, 1]$, because* $\nabla v_t(x) = (1-t)\nabla u(x) + t\nabla u_q(x) \in K^*$. Here we take into account that $\nabla u(x)$ and $\nabla u_q(x)$ are points of the convex set $K^*$. Thus we

---

* Since $u(x)$, $u_q(x)$ and $v_t(x)$ are convex functions, the notation $\nabla u(x)$ *etc.* is used also for supporting hyperplanes if the graph of $u(x)$ has no tangent hyperplane **at** the point $(x, u(x))$.

obtain the inequality

$$\int_{E^n}\left(\int_0^1 \frac{\partial f}{\partial u}\bigg|_{v_t} dt\right) dx \geqq h(Q_0) \text{ meas } Q_0 > 0 \tag{2.59}$$

for all $q \geqq N_1$.

According to Assumption A.2 the inequalities

$$0 < \frac{\partial f}{\partial u} \leqq \frac{C_0}{|x|^{n+2+\alpha}}$$

hold for all $x \in E^n$ with $|x| \geqq m_0$, $u \in R$ and $p \in K^*$, where $\alpha = \text{const.} > 0$ and $m_0 = \text{const.} \geqq 1$; without losing generality we can assume that $m_0$ is any positive integer greater than $(2^{\frac{1}{2+\alpha}} - 1)^{-1}$. Let

$$I_1 = \left|\int_{E^n}\left\{[u(x) - u_q(x)] \int_0^1 \frac{\partial f}{\partial u}\bigg|_{v_t} dt\right\} dx\right|.$$

Then

$$I_1 \leqq \int_{|x| \leqq m_0}\left\{|u(x) - u_q(x)| \int_0^1 \frac{\partial f}{\partial u}\bigg|_{v_t} dt\right\} dx + \int_{|x| > m_0}\left\{|u(x) - u_q(x)| \cdot \int_0^1 \frac{\partial f}{\partial u}\bigg|_{v_t} dt\right\} dx.$$

Let

$$C_2 = \sup_{|x| \leqq m_0} \frac{\partial f}{\partial u}\bigg|_{v_t} < +\infty. \tag{2.61}$$

Clearly $C_2$ depends only on $\|u(x)\|_{m_0}$, $K^*$ and the numbers $a_k$, $b_k$. Since $u(x)$ and $u_q(x)$ are basic representatives of elements of the set $T_k$,

$$u(\theta) = u_q(\theta) = 0 \tag{2.62}$$

and

$$\begin{aligned} |u(x)| &\leqq |x| \cdot \text{diam } K^*, \\ |u_q(x)| &\leqq |x| \cdot \text{diam } K^*. \end{aligned} \tag{2.63}$$

From (2.60–2.63) it follows that

$$I_1 \leqq C_2 \|u(x) - u_q(x)\|_{m_0} + \frac{C_0 \sigma_{n-1}}{2+\alpha} \sum_{m=m_0}^{\infty} \|u(x) - u_q(x)\|_{m+1} \left(\frac{1}{m^{2+\alpha}} - \frac{1}{(m+1)^{2+\alpha}}\right),$$

where $\sigma_{n-1}$ is the area of the unit sphere $S^{n-1}$. Since

$$\|u(x) - u_q(x)\|_{m_0} \leqq m_0^{2+\alpha} \|u(x) - u_q(x)\|_A$$

and

$$\frac{1}{(m+1)^{2+\alpha}} > \frac{1}{m^{2+\alpha}} - \frac{1}{(m+1)^{2+\alpha}}$$

for $m \geqq m_0 > \left(2^{\frac{1}{2+\alpha}} - 1\right)^{-1}$,

$$I_1 \leqq \left[C_2 m_0^{2+\alpha} + \frac{C_0 \sigma_{n-1}}{2+\alpha}\right] \|u(x) - U_q(x)\|_A, \tag{2.64}$$

where the integer $m_0$ satisfies the inequality

$$m_0 > \left(2^{\frac{1}{2+\alpha}} - 1\right)^{-1}$$

and constants $C_0$ and $C_2$ are independent of $q$.

Let

$$I_2 = \left| \int_{E^n} \left\{ \sum_{i=1}^{n} \left[\frac{\partial u(x)}{\partial x_i} - \frac{\partial u_q(x)}{\partial x_i}\right] \left( \int_0^1 \frac{\partial f}{\partial u_i}\bigg|_{v_t} dt \right) \right\} dx \right|.$$

Clearly

$$I_2 \leqq \sum_{i=1}^{u} \int_{R^n} \left\{ \left|\frac{\partial u(x)}{\partial x_i} - \frac{\partial u_q(x)}{\partial x_i}\right| \left( \int_0^1 \frac{\partial f}{\partial u}\bigg|_{v_t} dt \right) \right\} dx.$$

According to Assumption A.2 the inequalities

$$\left|\frac{\partial f}{\partial u_i}\right| \leqq \frac{C_1}{|x|^{n+\alpha}}, \qquad i = 1, 2, \ldots, n \tag{2.66}$$

hold for all $x \in E^n$ with $|x| \geqq m_0$, $u \in R$, and $p \in K^*$. The functions $\frac{\partial u_q(x)}{\partial x_i}$ converge to $\frac{\partial u(x)}{\partial x_i}$ almost everywhere in $E^n$ and

$$\nabla u_q(x) \in K^*, \qquad \nabla u(x) \in K^* \tag{2.67}$$

for all $x \in K^*$.

From (2.65–67) it follows that

$$I_2 \leqq \sum_{i=1}^{n} \left( \sup_{|x| \leqq m} \int_0^1 \left|\frac{\partial f}{\partial u_i}\bigg|_{v_t}\right| dt \right) \cdot \int_{|x| \leqq m} \left|\frac{\partial u(x)}{\partial x_i} - \frac{\partial u_q(x)}{\partial x_i}\right| dx$$

$$+ 2C_1\, n \operatorname{diam} K^* \int_{|x|>m} \frac{dx}{|x|^{n+\alpha}}, \tag{2.68}$$

where $m \geqq m_0$ is an arbitrary positive integer. Since

$$\int_{|x|>m} \frac{dx}{|x|^{n+\alpha}} = \frac{\sigma_{n-1}}{\alpha m^{\alpha}},$$

we can find $m^* > 0$ such that for every integer $m \geqq \max\{m_0, m^*\}$ the inequality

$$2C_1 n \operatorname{diam} K^* \frac{\sigma_{n-1}}{\alpha m^{\alpha}} < \frac{\varepsilon}{2} \tag{2.69}$$

holds, where $\varepsilon > 0$ is a given positive arbitrary number. We fix some integer $m \geqq \max\{m_0, m^*\}$. Then

$$\sup_{|x| \leqq m} \int_0^1 \left|\frac{\partial f}{\partial u_i}\right|_{v_t} dt \leqq Cm < +\infty$$

($i = 1, 2, \ldots, n$), where the constant $C_m$ depends only on $m$, $a_k$, $b_k$ and diam $K^*$. Really $\dfrac{\partial f}{\partial u_i}$ are continuous functions according to Assumption $A_2$ and we consider the supremum of $\left|\dfrac{\partial f}{\partial u_i}\right|$ in the compact set:

$$\underset{p \in K^*}{|x|} \leqq m,\ a_k - |x| \cdot \operatorname{diam} K^* \leqq u \leqq b_k + |x| \operatorname{diam} K^*.$$

Since $u(x)$ and $u_q(x)$ are convex functions,

$$\lim_{q \to \infty} \int_{|x| \leqq m} \left|\frac{\partial u(x)}{\partial x_i} - \frac{\partial u_q(x)}{\partial x_i}\right| dx = 0.$$

Therefore we can find an $N_2$ such that

$$\sum_{i=1}^{n} \left(\sup_{|x| \leqq m} \int_0^1 \left|\frac{\partial f}{\partial u_i}\right|_{v_t} dt\right) \int_{|x| \leqq m} \left|\frac{\partial u(x)}{\partial x_i} - \frac{\partial u_q(x)}{\partial x_i}\right| dx < \frac{\varepsilon}{2}.$$

if $q \geqq N_2$. Thus

$$I_2 < \varepsilon, \tag{2.70}$$

if $q \geqq N_2$.

Now from (2.59), (2.64), (2.70) it follows that

$$\lim a^*(u_q) = a^*(u)$$

if $\|u_q - u\|_A \to 0$. Lemma 4 is proved.

**Lemma 5.** *The operator* $G: T_k \to T_k$ is continuous.

**Proof.** Let the sequence $\xi_q \in T_k$ converge to the element $\xi_0 \in T_k$ in the space $B$. We should prove that

$$\lim_{q \to \infty} \|\eta_q - \eta_0\|_B = 0, \tag{2.71}$$

where $\eta_q = G(\xi_q)$ and $\xi_0 = G(\xi_0)$.

Since $T_k$ is a compact subset of $B$ (see Lemma 3), then there is a subsequence $\eta_{q_i}$ convergent to some element $\bar{\eta} \in B$, i.e.

$$\lim_{j \to \infty} \|\eta_{q_j} - \bar{\eta}\|_B = 0.$$

Since $T_k$ is closed in $B$, $\eta \in T_k$. It is well known that the set functions $\omega(1, \bar{v}_{q_j}, e)$ converge weakly* to the set function $\omega(1, v, e)$ in $E^n$, where $v_{q_j}(x)$ and $v(x)$ are representatives of $\eta_{q_j}$ and $\eta$ satisfies the conditions

$$v_{q_i}(\theta) = a^*(u_{q_i}), \qquad v(\theta) = a^*(u_0).$$

On the other hand all the functions**

$$F_{u_{q_j}+a^*(x)} = f(x, u_{q_j} + a^*(u_{q_j}), \nabla u_{q_j}) \tag{2.72}$$

are non-negative and satisfy the inequality

$$F_{u_{q_j}+a^*(x)} \leqq \phi_k(x, k(x) + b_k) \tag{2.73}$$

for all $x \in E^n$, where $z = k(x)$ is the equation of admissible convex cone $K$ (see Assumption A.1), prescribed for all functions $u_{q_j}(x)$.

Note that the following facts hold:

a) $\lim\limits_{j \to \infty} \| u_{q_j} - u_0 \|_A = 0$;

b) $\nabla u_{q_j}(x) \in K^*$ for all $x \in E^n$ and all integers $q_j$;

c) $\dfrac{\partial u_{q_j}}{\partial x_i}$ converges to $\dfrac{\partial u_0}{\partial x_i}$ almost everywhere in $E^n$, $(i = 1, 2, \ldots, n)$;

d) $\phi_M(x, k(x) + b_k)$ is a non-negative summable function in $E^n$;

e) The estimates (2.59) are correct for the functions $u_{q_j}(x)$ and $u_0(x)$.

Now we use the Lebesgue Theorem and obtain

$$\begin{aligned}\omega(1, v, e) &= \lim_{q_j \to \infty} \omega(1, u_{q_j}, e) \\ &= \lim_{q_j \to \infty} \int_e F_{u_{q_j}+a^*(u_{q_j})}(x)\, dx = \int_e F_{u_0+a^*(u_0)}(x)\, dx,\end{aligned}$$

where $e$ is any Borel subset of $E^n$. Note that we used continuity of the functional $a^* : T_k \to R$ in deriving these equalities.

Thus $v(x)$ is a convex generalized solution of the equation

$$\det\left(\frac{\partial^2 \bar{v}}{\partial x_i\, \partial x_j}\right) = F_{u_0+a^*(u_0)}(x)$$

and the function $v_0(x)$ is also a convex generalized solution of the equation

$$\det\left(\frac{\partial^2 v_0}{\partial x_i\, \partial x_j}\right) = F_{u_0+a^*(u_0)}(x).$$

Since

$$v(0) = v_0(\theta) = a^*(u_0)$$

---

* $\omega(1, u, e)$ is the $R$-curvature of a convex function $u(x)$ with $R(p) \equiv 1$ for all $p \in P^n$.

** The functions $u_{q_j}(x)$ and $u_0(x)$ are basic representatives of the elements $\xi_{q_j}$ and $\zeta_0$.

and the admissible convex cone $K$ is the asymptotic cone for both functions $\bar{v}(x)$ and $v_0(x)$, then $\bar{v}(x) = v_0(x)$ for all $x \in E^n$.

Thus (20.71) is correct. Lemma 5 is proved.

Now we can finish the proof of Theorem 3. Since $T_k$ is a compact set in $B$, then $G(T_k)$ is also compact in $B$. Moreover $G(T_k) = T_k$. Hence the operator $G$ has at least one fixed point $\xi \in T_k$. But the function $u(x) + a^*(\tilde{u}(x))$ is the representative for both $\xi$ and $G(\xi)$. Therefore $u(x)$ is the desired solution of the second boundary value problem for the equation

$$\det(u_{ij}) = f(x, u, Du).$$

Theorem 3 is proved.

## § 3. The maximum principle and estimates of solutions for elliptic Euler-Lagrange equations

The concept and techniques of the global theory of convex hypersurfaces and functions can successfully be applied to the Dirichlet problem for elliptic Euler-Lagrange equations. In this section we establish the sharp, interlocked necessary and sufficient conditions of two-sided estimates for solutions of such equations. These conditions are not subjected to a number of essential traditional limitations (see for example [12], [13], [14]). The interesting special cases of such Euler-Lagrange equations arise in problems of elasticity, plasticity, and global differential geometry (see [15], [16], [17]).

### *3.1. The main Assumptions*

Let $F(x, u; p)$ be a $C^2$-function in $\tilde{\Omega} \times R \times P^n$, where $\Omega$ is a bounded open domain in $E^n$, and $\partial\Omega$ is a closed $C^2$-hypersurface. It is well known that any $C^2$-function $u(x)$ minimizing (maximizing) the multiple integral

$$I(u) = \int_{\Omega} F(x, u, Du)\, dx \tag{3.1}$$

is necessarily a solution of the Euler-Lagrange equation

$$\sum_{i=1}^{n} \frac{d}{dx_i}\left(\frac{dF}{\partial p_i}\right) - \frac{\partial F}{\partial z} = 0. \tag{3.2}$$

This equation can also be written in the following form

$$\sum_{i,k=1}^{n} \frac{\partial^2 F}{\partial p_i\, \partial p_k} u_{ik} + \sum_{i=1}^{n} \frac{\partial^2 F}{\partial p_i\, \partial u} u_i + \sum_{i=1}^{n} \frac{\partial^2 F}{\partial p_i\, \partial x_i} - \frac{\partial F}{\partial z} = 0. \tag{3.3}$$

Equation (3.3) is elliptic if and only if the quadratic form

$$\sum_{i,k=1}^{n} \frac{\partial^2 F(x, u, p)}{\partial p_i\, \partial p_k} \xi_i \xi_k \tag{3.4}$$

is positive (negative) definite in $\bar{\Omega}\times R\times P^n$. The Dirichlet problem for wide classes of Euler-Lagrange equations is investigated by many authors (see [12], [18], [13], [16], [15], and other papers.

*Assumption A.4*. Let $F(x, u, p)$ be a $C^2$-function in $\bar{\Omega}\times R\times P^n$ and let the quadratic form (3.4) is positive definite in $\Omega\times R\times P^n$. Then there are functions $\phi(p)$, $\phi(x, u)$, $R_1(p)$, $R_2(p)$, $g_1(x, u)$, $g_2(x, u)$ satisfying the following conditions:

1) $\phi(p)$ is a strictly convex $C^2$-function in $P^{n\star}$ and $\lambda(x, u)$ is positive and the inequality

$$\lambda(x, u)\det\left(\frac{\partial^2\phi(p)}{\partial p_i\,\partial p_j}\right)\leq\det\left(\frac{\partial^2 F(x, u, p)}{\partial p_i\,\partial p_j}\right) \tag{3.5}$$

holds for every point $(x, u, p)\in\Omega\times R\times P^n$;

2) The functions $R_1(p)$, $R_2(p)$ are positive and locally summable with degree $n$ in $P^n$, the functions $g_1(x, u)$, $g_2(x, u)$ are non-negative in $\Omega\times R$ and the inequalities

$$-\frac{g_1(x, u)}{R_1(p)}\leq D(x, u, p)\leq\frac{g_2(x, u)}{R_2(p)} \tag{3.6}$$

hold for every point $(x, u, p)\in\Omega\times R\times P^n$, where

$$D(x, u, p)=-\sum_{i=1}^{n}\frac{\partial^2 F(x, u, p)}{\partial p_i\,\partial u}p_i-\sum_{i=1}^{n}\frac{\partial^2 F(x, u, p)}{\partial p_i\,\partial x_i}+\frac{\partial F(x, u, p)}{\partial u}. \tag{3.7}$$

3) The functions

$$\psi_1(x, u)=\frac{g_1^n(x, u)}{n^n\lambda(x, u)} \tag{3.8}$$

and

$$\psi_2(x, u)=\frac{g_2^n(x, u)}{n^n\lambda(x, u)} \tag{3.9}$$

are non-decreasing with respect to $u$ for every fixed $x\in\Omega$.

### *3.2. Estimates of solutions of the Dirichlet problems for the Euler-Lagrange equations*

We consider the Dirichlet problem

$$\sum_{i=1}^{n}\frac{d}{dx_i}\left(\frac{\partial F}{\partial p_i}\right)-\frac{\partial F}{\partial u}=0 \tag{3.10}$$

$$u|_{\partial\Omega}=h(X)\in C(\partial\Omega). \tag{3.11}$$

---

⋆ A convex $C^2$-function $\phi(p)$ is called *strictly convex* if $\det\left(\frac{\partial^2\phi(p)}{\partial p_i\,\partial p_j}\right)>0$ in $P^n$.

In Subsections 3.2 and 3.3 it is sufficient to consider only open bounded domains with continuous $\partial\Omega$.

Let $M = \sup_{\partial\Omega} h(x)$, $m = \inf_{\partial\Omega} h(x)$. Clearly $-\infty < m \leqq M < +\infty$.

**Theorem 4.** *Let* $u(x) \in V_2^n(\Omega) \cap C(\bar{\Omega})$ *be a solution of the Dirichlet problem* (3.10–11) *and let all conditions of Assumption* 4 *be fulfilled.** *Then the inequalities*

$$\int_{\Omega} \psi_1(x, m)\, dx < \int_{K^*} \tilde{R}_1^n(q)\, dq, \tag{3.12}$$

$$\int_{\Omega} \psi_2(x, M)\, dx < \int_{K^*} \tilde{R}_2^n(q)\, dq \tag{3.13}$$

*provide the estimates*

$$m - C(R_2, \omega_2) \operatorname{diam} \Omega \leqq u(x) \leqq M + C(R_2, \omega_1) \operatorname{diam} \Omega \tag{3.14}$$

*for all* $x \in \bar{\Omega}$, *where*

$$\omega_1 = \int_{\Omega} \psi_1(x, m)\, dx, \qquad \omega_2 = \int_{\Omega} \psi_2(x, M)\, dx \tag{3.15}$$

*and the values of the constants* $C(R_i, \omega_i)$ $(i = 1, 2)$ *will be described in the end of the proof of Theorem 4.*

**Proof.** First of all we explain the notations on the right-hand sides of inequalities (3.12), (3.13). $K^*$ is the normal image of the asymptotic cone $K$ of the strictly convex $C^2$-function $\phi(p)$; $P^n \to Q^n$ is the tangential mapping, generated by the convex function $\Phi(p)$;

$$\tilde{R}_i(q) = R_i(\chi_\Phi^{-1}(q)), \quad (i = 1, 2) \tag{3.16}$$

and $dq$ is the element of the Lebesgue measure in the Euclidean space $Q^n$.

Let $G = Co\Omega$ and $\bar{G}$ be the closure of $G$. Clearly $\bar{G} = \overline{Co\Omega}$, where $\overline{Co\Omega}$ is the closed convex hull of $\Omega$.

Let $H_M = \bar{G} \times \{M\}$ and $H_M = \bar{G} \times \{m\}$. We construct the convex body $Z$ which is the closed convex hull of the set $H_M \cup H_m \cup S_u$, where $S_u$ is the graph of $u(x)$, $x \in \bar{\Omega}$. Clearly the part of $\partial Z$ which is projected onto the open convex domain $G$ consists in the graphs $S_{v_1}$ and $S_{v_2}$, where $v_1(x)$ is a concave function in $G$. Since

$$m \leqq u|_{\partial\Omega} \leqq M \tag{3.17}$$

and

$$v_2(x) \leqq u(x) \leqq v_1(x) \tag{3.18}$$

for all $x \in G$, then it is sufficient to establish the inequalities

$$v_2(x) \geqq m - C(R_2, \omega_2) \operatorname{diam} \Omega \tag{3.19}$$

and

$$v_1(x) \leqq M + C(R_1, \omega_1) \operatorname{diam} \Omega. \tag{3.20}$$

---

* $V_2^n(\Omega)$ is the Sobolev space with functions having all second generalized derivatives summable with degree $n$ in $\Omega$.

Moreover, it is sufficient to establish only (3.19), because (3.20) can be proved in the same way as (3.19). Clearly the only interesting case is that in which there are points $x \in \Omega$ such that

$$u(x) < m. \tag{3.21}$$

We denote by $Q_2 \subset \Omega$ the set where

$$v_2(x) = u(x).$$

Clearly $Q_2$ is a Lebesgue measurable set and

$$\operatorname{meas} \bar{Q}_2 = \operatorname{meas} Q_2,$$

because $\bar{Q}_2 \subset \bar{\Omega}$, $\operatorname{meas} \partial\Omega = 0$ and

$$Q_2 = \bar{Q}_2 \setminus \partial\Omega.$$

The inequality

$$n[\det(u_{ij})]^{\frac{1}{n}} \left[\det\left(\frac{\partial^2 F}{\partial p_i\, \partial p_j}\right)\right]^{\frac{1}{n}} \leqq \sum_{i=1}^{n} \frac{\partial^2 F}{\partial p_i\, \partial p_j} u_{ij} = D(x, u(x), \nabla u(x))$$

holds at any point $x \in Q_2$. Applying Assumption A.4 to inequality (3.22), we obtain

$$R_2^n(\nabla u(x)) \det \frac{\partial^2 \phi(\nabla u(x))}{\partial p_i\, \partial p_j} \det(u_{ij}(x)) \leqq \frac{g_2^n(x, u)}{n^n \lambda(x, u)} = \psi_2(x, u(x)) \leqq \psi_2(x, M) \tag{3.23}$$

at every point $x \in Q_2$.

Since $u(x) \in V_2^n(\Omega) \cap C(\bar{\Omega})$, then the following formula

$$\omega\left(R_2^n \det\left(\frac{\partial^2 \phi}{\partial p_i\, \partial p_j}\right), v_2, G\right) = \int_{Q_2} R_2^n(\nabla u(x)) \det\left(\frac{\partial^2 \phi(\nabla u(x))}{\partial p_i\, \partial p_j}\right) \det(u_{ij}(x))\, dx$$

is established in [3] for the total $R_2^n \det\left(\frac{\partial^2 p}{\partial p_i\, \partial p_j}\right)$-curvature of the convex function $v_2(x)$. Thus from (3.24) and (3.23) it follows that

$$\omega\left(R_2^n \det\left(\frac{\partial^2 \phi}{\partial p_i\, \partial p_j}\right), v_2, G\right) \leqq \int_{\Omega} \psi_2(x, M)\, dx = \omega_2. \tag{3.25}$$

Note that

$$\int_{P^n} R_2^n(p) \det\left(\frac{\partial^2 \phi}{\partial p_i\, \partial p_j}\right) dp = \int_{\chi_\phi(P^n)} \tilde{R}_2^n(q)\, dq = \int_{K^*} \tilde{R}_2^n(q)\, dq; \tag{3.26}$$

all notations were explained at the beginning of the proof of Theorem 4.

BAKELMAN's Theorem (see [3], [15]), inequality (3.25) and the conditions of Theorem 4 (see inequality (3.13)) imply the validity of the estimate

$$m - C(R_2, \omega_2) \operatorname{diam} \Omega \leqq v_2(x) \leqq u(x) \tag{3.27}$$

for all $x \in \bar{\Omega}$.

Now we describe the way of computing of the constant $C(R_2, \omega_2)$ (see [3]). First we consider the function

$$\tau_{R_2}(\varrho) = \int_{|p| \leqq \varrho} R_2^n(p) \det \left( \frac{\partial^2 \phi(p)}{\partial p_i \, \partial p_j} \right) dp \tag{3.28}$$

for all $\varrho \in [0, +\infty)$. Since $\tau_{R_2}(\varrho)$ is continuous and strictly increasing, then it has an inverse $\varrho = T_{R_2}(\tau)$ for all $\tau \in [0, A(R_2))$, where

$$A(R_2) = \int_{K^*} \tilde{R}_2^n(q) \, dq .$$

Now $C(R_2, \omega_2)$ is defined by the formula

$$C(R_2, \omega_2) = T_{R_2}(\omega_2),$$

where $\omega_2 = \int_{\Omega} \psi_2(x, M) \, dx < A(R_2)$.

The estimate $u(x) \leqq v_1(x) \leqq M + C(R_1, \omega_1) \operatorname{diam} \Omega$ can be obtained in a similar way.

The proof of Theorem 4 is complete.

### *3.3. The important special cases and examples*

*A) The geometric maximum principle.* If the function $F$ does not depend on $x$ and $u$, then $\omega_1 = \omega_2 = 0$. Clearly $C(R_1, 0) = C(R_2, 0) = 0$. Then from (3.14) it follows that

$$m \leqq u(x) \leqq M$$

for all $x \in \bar{\Omega}$.

Therefore Theorem 3 is called the geometric maximum principle.

*B) The Euler-Lagrange equations for the functionals* $I(u) = \int_{\Omega} [F(\nabla u) + f(x, u)] \, dx$.

The Euler-Lagrange equation for the functionals

$$I(u) = \int_{\Omega} [F(\nabla u) + f(x, u)] \, dx$$

has the following form:

$$\sum_{i,k=1}^{n} \frac{\partial^2 F(\nabla u)}{\partial p_i \, \partial p_k} u_{ik} = \frac{\partial f(x, u)}{\partial u} . \tag{3.29}$$

According to Assumption A.4 $F(p) = \phi(p)$ is a strictly convex function in $P^n$; $\phi(x, u) \equiv 1$ and $\frac{\partial^2}{\partial u^2} f(x, u) \geqq 0$ in $\bar{\Omega} \times R$ (i.e. $f(x, u)$ is a convex $C^2$-function of $u$ for every fixed $x \in \bar{\Omega}$. Let

$$\frac{\partial}{\partial u} f(x, u) = f_u^+(x, u) - f_u^-(x, u)$$

where $f_u^+(x, u) \geqq 0$ and $f_u^-(x, u) \geqq 0$ are positive and negative parts of $\frac{\partial}{\partial u} f(x, u)$.

Thus inequalities (3.12) and (3.13) (see the statement of Theorem 3) can be written in the following forms:

$$\frac{1}{n^n} \int_\Omega [f_u^-(x, m)]^n \, dx < \text{meas } K^* \tag{3.12'}$$

and

$$\frac{1}{n^n} \int_\Omega [f_u^+(x, m)]^n \, dx < \text{meas } K^* . \tag{3.13'}$$

Since

$$\text{meas } K^* = \begin{cases} \text{a finite positive number if the asymptotic} \\ \text{cone of } F(p) \text{ is non-degenerate;} \\ +\infty \text{ if this cone is degenerate,} \end{cases}$$

the inequalities (3.12′) and (3.13′) are non-trivial restrictions only for the case of a non-degenerate asymptotic cone of $F(p)$.

The special case

$$f(x, u) = f(x) \cdot u,$$

which relates to many applications (see [3], [14], [13], [15]), leads to the inequalities

$$\frac{1}{n^n} \int_\Omega f_\pm^n(x) \, dx < \text{meas } K^* .$$

These inequalities provide estimates of solutions of the Dirichlet problem for elliptic equations

$$\sum_{i,j=1}^{n} \frac{\partial^2 F(\nabla u)}{\partial p_i \, \partial p_j} u_{ik} = f(x) .$$

### *Final Remarks*

1. The uniqueness of generalized convex solutions with prescribed asymptotic cones (see Section 2) will be studied in my forthcoming papers. The regularity of generalized solutions follows from results for the Dirichlet problem, obtained by CAFFARELLI, NIRENBERG, & SPRUCK [20], POGORELOV [22], CHENG & YAU [21].

2. The more detailed investigations for elliptic Euler-Lagrange equations will be developed in my next papers. I am planning to obtain various *a priori* estimates for solutions, establish new existence theorems and consider applications to elasticity-plasticity and differential geometry problems. These investigations will be connected with SERRIN's well known paper [16] and BAKELMAN's papers [3], [4], [15], [17].

## References

1. I. BAKELMAN, Generalized solutions of the Monge-Ampère equations, Dokl. Akad. Nauk USSR, **111** (1957), 1143–1145.
2. I. BAKELMAN, Geometric methods of solution of elliptic equations, Monograph, Nauka, Moscow, (1965), 1–340.
3. I. BAKELMAN, *R*-curvature, estimates and stability of solutions for elliptic equations, Journal of Diff. Equations, **43**, 1 (1982), 106–133.
4. I. BAKELMAN, The boundary value problems for *N*-dimensional Monge-Ampère equation; *N*-dimensional plasticity equation, IHES, Bures-sur-Yvette, (1984), preprint, 1–107.
5. A. D. ALEXANDROV. Existence and uniqueness of convex surface with prescribed integral curvature, Dokl. Akad. Nauk USSR, **35**, 8. (1942), 143–147.
6. A. D. ALEXANDROV. Convex polyhedra, Monograph, Moscow-Leningrad (1950), 1–428.
7. I. BAKELMAN, On the theory of Monge-Ampère equations, Vestnik LGU, N1, (1958), 25–38.
8. I. BAKELMAN, A. WERNER, & B. KANTOR, Introduction to global differential geometry, Monograph, Nauka, Moscow, (1973), 1–440.
9. H. BUSEMANN, Convex surfaces, Monograph, Interscience Publishers, Inc., New York, (1958), 1–238.
10. A. ALEXANDROV, Intrinsic geometry of convex surfaces, Monograph, Moscow-Leningrad (1948), 1–427.
11. A. POGORELOV, Exterior geometry of convex surfaces, Monograph, Moscow, Nauka, (1969), 1–759; Engl. transl. Transl. Math. Monographs, v. 35, Amer. Math. Soc. 1973.
12. S. N. BERNSTEIN, Collection of papers, vol. 3, Akad. Nauk of USSR, (1960), 1–439.
13. O. A. LADYZHENSKYA & N. N. URALTZEVA, Linear and quasilinear elliptic equations, Nauka, (1973), 2nd ed., 1–576.
14. D. GILBARG & N. S. TRUDINGER, Elliptic Partial Differential Equations of Second Order, (monograph), Second Edition, Springer Verlag, (1983), 1–511.
15. I. BAKELMAN, Geometric problems in quasilinear elliptic equations, Uspekhi Mathem. Nauk, 25 : 3, (1970), 42–112; Russian Mathem. Surveys (1970), **25**, N° 3, 45–109.
16. J. SERRIN, The problem of Dirichlet for quasilinear elliptic differential equations with many independent variables. Philos. Trans. Roy. Soc., (1969), 413–496.
17. I. BAKELMAN, Notes concerning the torsion problem of hardening rods, IMA Preprint Series, (1986), 1–38.
18. CH. MORREY, Multiple integrals, Monograph. Springer Verlag (1966).
19. M. PROTTER & H. F. WEINBERGER, Maximum principles in Differential equations. Monograph, Prentice-Hall, Inc, (1967), 1–261.

20. L. CAFFARELLI, L. NIRENBERG, & J. SPRUCK, The Dirichlet problem for Non-linear Second Order Elliptic Equations I. Monge-Ampère equations, Comm. pure and Appl. Mathematics, (1984).
21. S. Y. CHENG & S. T. YAU, On the regularity of the solution of the Monge-Ampère equations $\det(u_{ij}) = F(x, u)$. Comm. pure and Appl. Mathem., **29**, (1976), 495–516.
22. A. V. POGORELOV, The Minkowski multidimensional problem, New York, J. Wiley, (1978).

Department of Mathematics
Texas A. & M. University
College Station

*(Received July 23, 1985)*

# *Qualitative Properties of Large Buckled States of Spherical Shells*

KE-GANG SHIH & STUART S. ANTMAN

*This paper is dedicated to James B. Serrin on the occasion of his sixtieth birthday*

**Contents**

1. Introduction . . . . . . . . . . . . . . . . . . . . . . . . . . . . . 357
2. Formulation of the Governing Equations . . . . . . . . . . . . . . . . . . 358
3. Trivial Solutions . . . . . . . . . . . . . . . . . . . . . . . . . . 362
4. The Linearized Equations . . . . . . . . . . . . . . . . . . . . . . 362
5. Global Bifurcation . . . . . . . . . . . . . . . . . . . . . . . . . 365
6. Nodal Properties of Solutions . . . . . . . . . . . . . . . . . . . . 371
7. Properties of $\mathscr{G}$ . . . . . . . . . . . . . . . . . . . . . . . . . 378
8. Conclusion . . . . . . . . . . . . . . . . . . . . . . . . . . . . 382
Acknowledgment . . . . . . . . . . . . . . . . . . . . . . . . . . . 383
9. References . . . . . . . . . . . . . . . . . . . . . . . . . . . . 383

## 1. Introduction

In this paper we study the global qualitative behavior of axisymmetric buckled states of homogeneous isotropic nonlinearly elastic shells that can suffer flexure, compression, and shear. Our model is *geometrically exact* in the sense that a geometric quantity, such as $\sin\theta$, is not replaced by an approximation, such as $\theta$ or $\theta - \theta^3/6$. (The usual justification for such a replacement is that $\theta$ is known to be small for the physical situation under study. But it is quite possible that a *mathematical* model with exact geometry permits only small solutions, while those with approximate geometry have large solutions.) We allow the material properties to be described by a very general class of nonlinear constitutive relations. Consequently our governing equations form a quasilinear sixth-order system of ordinary differential equations.

Our main result is that nontrivial branches of solutions pairs can be globally characterized by a novel nodal pattern relating shear and bending effects. The detection of this pattern is greatly aided by the generality of our theory, which does not obscure the simple mathematical structure of the governing equations with approximate geometrical relations.

Indeed, the beautiful numerical results of BAUER, REISS, & KELLER (1970) for their technical theory of shells indicate that their solution branches do not enjoy the nodal properties we discover for our exact theory. We surmise that no such technical theory would have these properties. Consequently, the nature of the bifurcation diagram away from the trivial solution for these technical theories could differ markedly from ours.

There is a large literature on the buckling of spherical shells. The articles by BAUER, REISS, & KELLER (1970), ANTMAN (1971), LANGE & KRIEGSMANN (1981), and SHIH (1985) give extensive lists of references. Our results differ markedly from those developed in these references (except for the work of ANTMAN) by virtue of the generality of our theory and the global character of our theorems.

The heart of our paper is in Sections 5 and 6. In Section 5 we show that our equations meet the hypotheses of RABINOWITZ's Global Bifurcation Theorem. In Section 6 we prove a basic uniqueness theorem that enables us to show that solution branches preserve the nodal pattern they inherit from the trivial branch. The use of the isotropy of the shell material is crucial in enabling us to overcome the difficulties posed by the polar singularities.

*Notation.* Partial derivatives are denoted by subscripts. Ordinary derivatives with respect to the variable $s$ are denoted by primes. If $f$ and $g$ are functions of $x$ and $y$, then $\dfrac{\partial(f, g)}{\partial(x, y)}$ denotes the *matrix* of partial derivatives of $f$ and $g$ with respect to $x$ and $y$.

## 2. Formulation of the Governing Equations

Let $\{\boldsymbol{i}, \boldsymbol{j}, \boldsymbol{k}\}$ be a fixed right-handed orthonormal basis for the Euclidean 3-space $E^3$. For each real $\phi$ we define

$$\boldsymbol{e}_1(\phi) = \cos\phi\,\boldsymbol{i} + \sin\phi\,\boldsymbol{j}, \quad \boldsymbol{e}_2(\phi) = -\sin\phi\,\boldsymbol{i} + \cos\phi\,\boldsymbol{j}, \quad \boldsymbol{e}_3 = \boldsymbol{k}. \tag{2.1}$$

To each $(s, \phi) \in [0, \pi] \times [0, 2\pi]$ corresponds exactly one point on the sphere of radius 1 centered at the origin with position vector of the form

$$\boldsymbol{r}_*(s, \phi) = \sin s\, \boldsymbol{e}_1(\phi) - \cos s\, \boldsymbol{e}_3. \tag{2.2}$$

Note that $s$ measures the arc length to $\boldsymbol{r}_*(s, \phi)$ from the *south* pole of the sphere. This convention will simplify some of the formulas. We interpret the sphere defined by (2.2) to be the natural reference state of the midsurface of a spherical shell. The coordinates $(s, \phi)$ identify material points on this surface.

An *axisymmetric configuration* of a spherical shell that can suffer flexure, extension, and shear is determined by a pair of vector functions

$$(s, \phi) \mapsto \boldsymbol{r}(s, \phi), \quad \boldsymbol{b}(s, \phi) \tag{2.3}$$

with $\boldsymbol{r}(s, \cdot)$ and $\boldsymbol{b}(s, \cdot)$ having period $2\pi$ and with

$$(2.4)\qquad \boldsymbol{r}(s, \phi) \cdot \boldsymbol{e}_2(\phi) = 0, \qquad \boldsymbol{b}(s, \phi) \cdot \boldsymbol{e}_2(\phi) = 0, \qquad |\boldsymbol{b}(s, \phi)| = 1.$$

$$\boldsymbol{r}_\phi = \boldsymbol{k} \times \boldsymbol{r}, \qquad \boldsymbol{b}_\phi = \boldsymbol{k} \times \boldsymbol{b}.$$

The reference configuration of the shell is given by $\boldsymbol{r} = \boldsymbol{r}_*$, $\boldsymbol{b} = -\boldsymbol{r}_*$. The vector $\boldsymbol{b}(s, \phi)$ is interpreted as characterizing the deformed configuration of the material points that in the natural state of the shell lie along the ray determined by $\boldsymbol{r}_*(s, \phi)$.

We set

$$(2.5)\qquad \boldsymbol{r}(s, \phi) \equiv r(s)\, \boldsymbol{e}_1(\phi) + z(s)\, \boldsymbol{k},$$

$$(2.6)\qquad \boldsymbol{a}(s, \phi) \equiv \cos \psi(s)\, \boldsymbol{e}_1(\phi) + \sin \psi(s)\, \boldsymbol{k},$$

$$\boldsymbol{b}(s, \phi) \equiv -\sin \psi(s)\, \boldsymbol{e}_1(\phi) + \cos \psi(s)\, \boldsymbol{k},$$

$$(2.7)\qquad \boldsymbol{r}_s(s, \phi) \equiv \nu(s)\, \boldsymbol{a}(s, \phi) + \eta(s)\, \boldsymbol{b}(s, \phi),$$

$$(2.8)\qquad \tau(s) \equiv \frac{r(s)}{\sin s},$$

$$\mu(s) \equiv \psi'(s) - 1,$$

$$\sigma(s) \equiv \frac{\sin \psi(s)}{\sin s} - 1.$$

The strain variables for our problem are

$$(2.9)\qquad \boldsymbol{w} \equiv (\nu, \tau, \eta, \mu, \sigma).$$

Let $\boldsymbol{n}_1(s, \phi)$ and $\boldsymbol{m}_1(s, \phi)$ denote the resultant contact force and contact couple per unit reference length of the circle $\phi \mapsto \boldsymbol{r}_*(s, \phi)$ of radius $\sin s$ that are exerted across this circular section at $\boldsymbol{r}_*(s, \phi)$. Let $\boldsymbol{n}_2(s, \phi)$ and $\boldsymbol{m}_2(s, \phi)$ denote the resultant contact force and contact couple per unit reference length of the circle $s \mapsto \boldsymbol{r}_*(s, \phi)$ of radius 1 that are exerted across this section at $\boldsymbol{r}_*(s, \phi)$. Since we seek axisymmetric solutions, we require these resultants to have the form

$$(2.10)\qquad \boldsymbol{n}_1(s, \phi) = \hat{N}(s)\, \boldsymbol{a}(s, \phi) + \hat{H}(s)\, \boldsymbol{b}(s, \phi),$$

$$(2.11)\qquad \boldsymbol{n}_2(s, \phi) = \hat{T}(s)\, \boldsymbol{e}_2(\phi),$$

$$(2.12)\qquad \boldsymbol{m}_1(s, \phi) = -\hat{M}(s)\, \boldsymbol{e}_2(\phi),$$

$$(2.13)\qquad \boldsymbol{m}_2(s, \phi) = \hat{\Sigma}(s)\, \boldsymbol{a}(s, \phi).$$

(The reason for taking $\boldsymbol{m}_2(s, \phi) \cdot \boldsymbol{b}(s, \phi) = 0$ is explained by ANTMAN (1978, Sec. 10).) Then the equilibrium equations for the shell under an external hydrostatic pressure of intensity $\lambda$ per unit actual area are

$$(2.14)\qquad [\sin s \hat{N}(s)]' - \hat{T}(s) \cos \psi(s) - \sin s \hat{H}(s)\, \psi'(s) - \lambda r(s)\, \eta(s) = 0,$$

$$(2.15)\qquad [\sin s \hat{H}(s)]' + \hat{T}(s) \sin \psi(s) + \sin s \hat{N}(s)\, \psi'(s) + \lambda r(s)\, \nu(s) = 0,$$

$$(2.16)\qquad [\sin s \hat{M}(s)]' - \hat{\Sigma}(s) \cos \psi(s) + \sin s [\nu(s)\, \hat{H}(s) - \eta(s)\, \hat{N}(s)] = 0.$$

We can combine (2.14) and (2.15) to obtain

$$\{\sin s[\sin \psi(s)\, \hat{N}(s) + \cos \psi(s)\, \hat{H}(s)] + \lambda r(s)^2/2\}' = 0, \tag{2.17}$$

$$\{\sin s[-\cos \psi(s)\, \hat{N}(s) + \sin \psi(s)\, \hat{H}(s)]\}' + \hat{T}(s) + \lambda r(s)\, z'(s) = 0. \tag{2.18}$$

The material of the shell is homogeneous and nonlinearly elastic if there are constitutive functions $N$, $T$, $H$, $M$, $\Sigma$ such that

$$\hat{N}(s) = N(\boldsymbol{w}(s)), \quad \text{etc.}, \tag{2.19}$$

when (2.3)–(2.8) hold.

We assume that these constitutive functions are thrice continuously differentiable on their common domain of definition

$$\mathscr{D} = \{\boldsymbol{w} \in R^5 : \nu > h\,|\mu + 1|,\ \tau > h\,|\sigma + 1|\} \tag{2.20}$$

where $h$ is a given number in (0, 1), which may be interpreted as half the thickness of the shell. The inequalities in (2.20) ensure that in a standard three-dimensional interpretation of our deformation variables the local ratio of deformed to reference volume never be zero.

We require the constitutive functions to satisfy the *monotonicity conditions*:

$$\frac{\partial(N, H, M)}{\partial(\nu, \eta, \mu)}, \quad \frac{\partial(T, \Sigma)}{\partial(\tau, \sigma)} \quad \text{are positive-definite}. \tag{2.21}$$

(This is the strict form of the strong-ellipticity condition for axisymmetric deformations of axisymmetric shells. *Cf.* ANTMAN (1978).) These conditions ensure, *e.g.*, that an increase in the bending strain $\mu$ is accompanied by a corresponding increase in the bending couple $M$. We also use special forms of the following very natural inequalities

$$N_\nu + N_\tau > 0, \quad T_\tau + T_\nu > 0. \tag{2.22}$$

We impose compatible growth conditions ensuring that extreme strains are accompanied by extreme resultants:

(2.23) $N(\boldsymbol{w}) \to +\infty$ as $\nu \to \infty$ if $\tau - h\,|1 + \sigma|$ has a positive lower bound and if $\eta, \mu, \sigma$ are bounded,

$N(\boldsymbol{w}) \to -\infty$ as $\nu \to h\,|1 + \mu|$ if $\tau$ is bounded above and if $\eta, \mu, \sigma$, are bounded,

$T(\boldsymbol{w}) \to +\infty$ as $\tau \to \infty$ if $\nu - h\,|1 + \mu|$ has a positive lower bound and if $\eta, \mu, \sigma$ are bounded,

$T(\boldsymbol{w}) \to -\infty$ as $\nu \to h\,|1 + \sigma|$ if $\nu$ is bounded above and if $\eta, \mu, \sigma$ are bounded,

$H(\boldsymbol{w}) \to \pm\infty$ as $\eta \to \pm\infty$ if $(\nu, \tau, \mu, \sigma)$ lies in a compact subset of $\{(\nu, \tau, \mu, \sigma) : \nu > h\,|1 + \mu|,\ \tau > h\,|1 + \sigma|\}$,

$M(\boldsymbol{w}) \to \pm\infty$ as $\mu \to \pm\infty$ if $\nu > h\,|1+\mu|$ and if $(\tau, \eta, \sigma)$ lies in a compact subset of $\{(\tau, \eta, \sigma)\colon \tau > h\,|1+\sigma|\}$,

$\Sigma(\boldsymbol{w}) \to \pm\infty$ as $\sigma \to \pm\infty$ if $\tau > h\,|1+\sigma|$ and if $(\nu, \eta, \mu)$ lies in a compact subset of $\{(\nu, \eta, \mu)\colon \nu > h\,|1+\mu|\}$.

We finally require that the shell be isotropic and have a stress-free natural state:

(2.24) $\eta \mapsto H(\boldsymbol{w})$ is odd, $\eta \mapsto N(\boldsymbol{w}), T(\boldsymbol{w}), M(\boldsymbol{w}), \Sigma(\boldsymbol{w})$ are even,

$$M(\boldsymbol{w}) = 0 = \Sigma(\boldsymbol{w}) \quad \text{if } \mu = 0 = \sigma,$$

$$N_\sigma(\boldsymbol{w}) = N_\mu(\boldsymbol{w}) = T_\sigma(\boldsymbol{w}) = T_\mu(\boldsymbol{w}) = 0 \quad \text{if } \eta = \mu = \sigma = 0,$$

$$N(\nu, \tau, 0, \mu, \sigma) = T(\tau, \nu, 0, \sigma, \mu),$$

$$M(\nu, \tau, 0, \mu, \sigma) = \Sigma(\tau, \nu, 0, \sigma, \mu),$$

$$N(1, 1, 0, 0, 0) = T(1, 1, 0, 0, 0) = 0..$$

Our boundary conditions essentially require that the deformation be regular at the poles and that there be no net rigid displacement in the $z$-direction:

(2.25a, b) $$r(0) = 0 = r(\pi),$$

(2.26a, b) $$\eta(0) = 0 = \eta(\pi),$$

(2.27a, b) $$\psi(0) = 0, \quad \psi(\pi) = \pi.$$

(2.28) $$\int_0^\pi z(s)\, ds = 0.$$

Substituting (2.25a) into the integral of (2.17) we obtain

(2.29) $$\sin s[\hat{N}(s) \sin \psi(s) + \hat{H}(s) \cos \psi(s)] + \lambda r(s)^2/2 = 0.$$

If $\hat{N}$ and $\hat{H}$ are bounded and if $\lambda > 0$, then (2.27) implies that (2.25b) holds. If we further assume that $\tau$ is bounded, then (2.29) and (2.27) imply that $\hat{H}(0) = 0 = \hat{H}(\pi)$. The properties of $H$ then ensure that (2.26) holds. By comparing the integrals of (2.18) over $[0, s]$ and $[s, \pi]$ we readily obtain

(2.30) $$\int_0^\pi \hat{T}(s)\, ds = -\lambda \int_0^\pi r(s)\, z'(s)\, ds.$$

Note that the integral on the right side is the signed area between the curve $(r, z)$ and the $z$-axis.

Our boundary value problem consists of the strain-configuration equations (2.5)–(2.8), the equilibrium equations (2.14)–(2.16), the constitutive equations (2.19), and the boundary conditions (2.25)–(2.28). A regular solution of the remaining equations would automatically satisfy (2.25b) and (2.26).

The shell theory we employ was originated by the COSSERATS (1909). (*Cf.* NAGHDI (1972) and LIBAI & SIMMONDS (1983).) Our formulation was foreshadowed by several papers of REISSNER, typical of which is his paper of 1963. The variables we employ, and the very simple form of constitutive equations they induce, were introduced by ANTMAN (1971).

## 3. Trivial Solutions

We now seek solutions of our boundary value problem in which the shell remains spherical, unsheared, and uniformly stretched, so that $\nu = \tau = k$ (const), $\eta = 0$, $\psi(s) = s$. Thus $\mu = 0 = \sigma$. Under these conditions, (2.22) reduces (2.14)–(2.17) to

$$N(k, k, 0, 0, 0) = T(k, k, 0, 0, 0) = -\lambda k^2/2. \tag{3.1}$$

Then (2.22) and (2.23) imply that $k \mapsto N(k, k, 0, 0, 0) + \lambda k^2/2$ strictly increases from $-\infty$ to $\infty$ as $k$ increases from 0 to $\infty$ provided that $\lambda \geqq 0$. Thus (3.1) has a unique solution for $k$, denoted by $k(\lambda)$, with

$$[0, \infty) \ni \lambda \mapsto k(\lambda) \in (0, 1] \tag{3.2}$$

thrice continuously differentiable and strictly decreasing.

## 4. The Linearized Equations

We adopt the convention that if $R$ is any constitutive function (such as $N$, $N_\nu$, ...), then

$$R^0(\lambda) \equiv R(k(\lambda), k(\lambda), 0, 0, 0). \tag{4.1}$$

The linearization of our boundary value problem about the trivial solution is equivalent to the following boundary value problem for the linearized variables $\nu_1, \eta_1, \psi_1$:

$$r_1(s) = \int_0^s \{\nu_1(t) \cos t - [\eta_1(t) + k\psi_1(t)] \sin t\} \, dt = \tau_1(s) \sin s, \tag{4.2}$$

(4.3)
$$N_\nu^0[(\nu_1 \sin s)' - \tau_1 \cos s] = [(N_\tau^0 + H_\eta^0 + \lambda k)\, \eta_1 + (N_\tau^0 k + \lambda k^2/2)\, \psi_1] \sin s,$$

$$H_\eta^0(\eta_1 \sin s)' = -(N_\nu^0 + N_\tau^0 + \lambda k)\, (\nu_1 + \tau_1) \sin s + (\lambda k^2/2)\, (\psi_1 \sin s)', \tag{4.4}$$

(4.5)
$$M_\mu^0[(\psi_1{}' \sin s)' - \psi_1/\sin s] + (M_\mu^0 - M_\sigma^0)\, \psi_1 \sin s = -[H_\eta^0 k + \lambda k^2/2]\, \eta_1 \sin s,$$

$$\eta_1(0) = 0 = \eta_1(\pi), \qquad \psi_1(0) = 0 = \psi_1(\pi), \tag{4.6}$$

$$\int_0^\pi z_1(s)\, ds = \int_0^\pi \{\nu_1(t) \sin t + [\eta_1(t) + k\psi_i(t)] \cos t\} \, dt = 0. \tag{4.7}$$

In deriving this system, we relied crucially on (2.24).

From this system, we wish to extract a system for $\eta_1$ and $\psi_1$ alone. To accomplish this efficiently we write (4.3), (4.4), and the derivative of (4.4) in the form

(4.8)
$$(\nu_1 \sin s)' - \tau_1 \cos s = f, \quad (\nu_1 + \tau_1) \sin s = g, \quad (\nu_1 \sin s)' + \nu_1 \cos s = h,$$

from which we immediately deduce that

$$g = (h - f) \tan s. \tag{4.9}$$

We write out this equation, replacing $(\psi_1' \sin s)' - \psi_1/\sin s$ by its expression from (4.5) and supplement the resulting equation with (4.5), (4.6) to get the system

$$(L\eta_1)(s) + A\eta_1(s) \sin s = -a\psi_1(s) \sin s, \tag{4.10a}$$

$$(L\psi_1)(s) + B\psi_1(s) \sin s = -b\eta_1(s) \sin s, \tag{4.10b}$$

$$\eta_1(0) = 0 = \eta_1(\pi), \quad \psi_1(0) = 0 = \psi_1(\pi), \tag{4.11}$$

where

$$(Lu)(s) \equiv [u'(s) \sin s]' - u(s)/\sin s, \tag{4.12}$$

(4.13)
$$A = (H_\eta^0)^{-1} [\tfrac{1}{2} \lambda k^2 b + (N_\nu^0 + N_\tau^0 + \lambda k)(N_\nu^0)^{-1} (N_\tau^0 - N_\nu^0 + \lambda k + H_\eta^0)],$$
$$a = (H_\eta^0)^{-1} [\tfrac{1}{2} \lambda k^2 B + (N_\nu^0 + N_\tau^0 + \lambda k)(N_\nu^0)^{-1} (kN_\tau^0 - kN_\nu^0 + \tfrac{1}{2} \lambda k^2)],$$
$$B \equiv 1 - M_\sigma^0/M_\mu^0, \quad b \equiv [H_\eta^0 k + \lambda k^2/2]/M_\mu^0.$$

We now convert (4.10), (4.11), to a problem for a single complex-valued function

$$\phi = \alpha\eta_1 + \beta\psi_1 \tag{4.14}$$

where $\alpha$ and $\beta$ are complex constants to be adjusted so that (4.10a, b) can be combined into a single complex equation. Thus we require that

$$(\alpha A + \beta b)\eta_1 + (\alpha a + \beta B)\psi_1 = C\phi = C(\alpha\eta_1 + \beta\psi_1) \tag{4.15}$$

where $C$ is a complex constant to be determined. It follows from (4.15) that

$$\alpha A + \beta b = \alpha C, \quad \alpha a + \beta B = \beta C \tag{4.16}$$

so that

$$a\alpha^2 + (B - A)\alpha\beta - b\beta^2 = 0. \tag{4.17}$$

Using (4.15), we reduce (4.10), (4.11) to

$$(L\phi)(s) + C\phi \sin s = 0, \quad \phi(0) = 0 = \phi(\pi), \tag{4.18}$$

which has a nontrivial solution if and only if

$$C = n(n + 1), \quad n = 1, 2, \ldots, \tag{4.19}$$

in which case

$$\phi(s) = \phi_n(s) = \text{const}\, P_n^1(\cos s) \tag{4.20}$$

where $P_n^1$ is the associated Legendre function of the first kind of degree $n$ and order 1. When (4.19) holds, system (4.16) has a nontrivial solution if and only if

$$g(\lambda; n) \equiv [A - n(n+1)]\,[B - n(n+1)] - ab = 0. \tag{4.21}$$

Note that $A$, $B$, $a$, $b$ each depend on the eigenvalue parameter $\lambda$. For a given integer $n$ there can be none, one, or many real solutions $\lambda$ of (4.21). Noting that $N^0(\lambda) = -\lambda k(\lambda)^2/2$, we could readily fashion conditions on the constitutive assumptions to ensure any of these possibilities. (*Cf.* ANTMAN & ROSENFELD (1978) for a discussion of this issue.) We limit our attention to positive solutions $\lambda$ of (4.21). Negative solutions correspond to shear instabilities. (*Cf.* ANTMAN & CARBONE (1977).)

For $\lambda > 0$, (2.21) implies that $b > 0$. Thus (4.16) has no nontrivial solutions $(\alpha, \beta)$ with $\alpha = 0$. We can accordingly adopt the special normalization that $\alpha = 1$. In this case, the roots $\beta$ of (4.17) are

$$\beta_\pm = \frac{B - A \pm \sqrt{(A-B)^2 + 4ab}}{2b} = \frac{B - A \pm [A + B - 2n(n+1)]}{2b}, \tag{4.22}$$

the second equality coming from (4.21). Thus $\beta_\pm$ are real when (4.21) holds. In this case, $\beta_\pm$ are distinct if and only if

$$A + B \neq 2n(n+1). \tag{4.23}$$

When (4.21) and (4.23) hold, (4.14) and (4.20) reduce to a system of the form

$$\eta_1 + \beta_\pm \psi_1 = D_\pm P_n^1(\cos s). \tag{4.24}$$

If we take $D_\pm$ real and not both 0, then the distinctness of $\beta_\pm$ ensures that (4.24) can be solved uniquely for $\eta_1$ and $\psi_1$. These solutions are proportional to $P_n^1(\cos s)$. To find $D_\pm$, *i.e.*, to find the constant of proportionality, we normalize $\psi_1$ by setting

$$\psi_1(s) = P_n^1(\cos s), \qquad \eta_1(s) = E P_n^1(\cos s). \tag{4.25}$$

Then a comparison of (4.10) and (4.11) with (4.18) implies that

$$A + a/E = n(n+1) = B + bE. \tag{4.26}$$

whence

$$E = \frac{n(n+1) - B}{b} = \frac{a}{n(n+1) - A}. \tag{4.27}$$

The compatibility of these two expressions for $E$ is ensured by (4.21).

Now we study the special case in which

$$A + B = 2n(n+1) \qquad n = 1, 2, \ldots. \tag{4.28}$$

Then (4.22) implies that $\beta$ is unique with

$$2b\beta = B - A, \qquad b\beta^2 = a. \tag{4.29a, b}$$

Consequently (4.20) yields

(4.30) $$\eta_1 + \beta\psi_1 = DP_n^1(\cos s).$$

Note that (4.29b) implies that there are no such solutions if $a < 0$.

First we treat the even more special subcase that

(4.31) $$A = B = n(n+1).$$

Then $\beta = 0$, $\eta_1 = DP_n^1(\cos s)$. Moreover, (4.16) implies that $a = 0$. Since $b > 0$, the alternative theorem implies that (4.10b) has no solution unless $D = 0$ and hence $\eta_1 = 0$. In this case, $\psi_1 = \text{const}\, P_n^1(\cos s)$. Note that if we make the eminently reasonable assumption that $B < 2$, then (4.31) cannot hold. Likewise, (4.31) cannot hold if $a \neq 0$.

Now we assume that (4.28) holds, but that

(4.32) $$A \neq B \neq n(n+1).$$

Hence $\beta \neq 0$. Using (4.30), we convert (4.10b) in to

(4.33) $$L\psi_1 + n(n+1)\,\psi_1 = -bDP_n^1(\cos s)\sin s,$$

since (4.28) and (4.29) imply that $B - b\beta = n(n+1)$. Since $b \neq 0$, the alternative theorem implies that (4.33) has no solution unless $D = 0$, in which case (4.30) yields

(4.34) $$\eta_1 = -\beta\psi_1.$$

In summary, the eigenvalues of (4.10), (4.11) are solutions $\lambda$ of (4.21). We limit our attention to positive eigenvalues. If $B < 2$, then the corresponding eigenfunctions are

(4.35) $$(\eta_1, \psi_1) = ([n(n+1) - B]/b, 1)\,P_n^1(\cos s).$$

Now the function $P_n^1(\cos s)$ has exactly $n+1$ zeros on $[0, \pi]$, including those at 0 and $\pi$, each of which is simple. *Thus every nontrivial solution* $(\eta_1, \psi_1)$ *of* (4.10), (4.11) *is characterized by the fact that* $\eta_1$ *and* $\psi_1$ *have exactly the same zeros.*

## 5. Global Bifurcation

To show that our equations meet the hypotheses of the bifurcation theorem and uniqueness theorem, which form the basis of our analysis, we must transform them into alternative forms that enable us to control the polar singularity manifested by the presence of $\sin s$ as a coefficient of the highest derivatives of our differential equations. We substitute (2.19) into (2.14)–(2.17), carry out the differentiation of the leading terms, use Cramer's rule (justified by (2.21)) to solve the resulting equations for $\nu' \sin s$, $\eta' \sin s$, $\mu' \sin s$, and finally force this latest version of the system into a form suggested by the linear equations (4.3)–(4.5). Using (4.12) we obtain

$$(5.1)\qquad (\nu \sin s)' = \tau \cos s + (N_\nu^0)^{-1}(N_\tau^0 + H_\tau^0 + \lambda k)\,\eta \sin s + (N_\nu^0)^{-1}(N_\tau^0 k + \lambda k^2/2)(\psi - s)\sin s + n,$$

$$(5.2)\qquad (\eta \sin s)' = -(H_\eta^0)^{-1}(N_\nu^0 + N_\tau^0 + \lambda k)(\nu + \tau)\sin s + (H_\eta^0)^{-1}(\lambda k^2/2)\,[(\psi - s)\sin s]' + h,$$

$$(5.3)\qquad L(\psi - s) = -B(\psi - s)\sin s - b \sin s + m,$$

where

$$(5.4)\qquad n \equiv \Delta^{-1}D_1 + (\nu - \tau)\cos s - (N_\nu^0)^{-1}(N_\tau^0 + H_\eta^0 + \lambda k)\,\eta \sin s - (N_\nu^0)^{-1}(N_\tau^0 k + \lambda k^2/2)(\psi - s)\sin s,$$

$$(5.5)\qquad h \equiv \Delta^{-1}D_2 + \eta \cos s + (H_\eta^0)^{-1}(N_\nu^0 + N_\tau^0 + \lambda k)(\nu + \tau)\sin s - (H_\eta^0)^{-1}(\lambda k^2/2)\,[(\psi - s)\sin s]',$$

$$(5.6)\qquad m \equiv \Delta^{-1}D_3 + (\psi - s)'\cos s - \frac{\psi - s}{\sin s} + B(\psi - s)\sin s + b\eta \sin s,$$

$$(5.7\text{a})\qquad \Delta \equiv \det \frac{\partial(N, H, M)}{\partial(\nu, \eta, \mu),},$$

(5.7b) $D_j$ is the determinant obtained from $\Delta$ by replacing its $j^{\text{th}}$ column with

$$\begin{pmatrix} \alpha \\ \beta \\ \gamma \end{pmatrix},$$

$$(5.8\text{a})\qquad \alpha \equiv -(N_\tau \tau' + N_\sigma \sigma')\sin s + T\cos\psi - N\cos s + H\psi' \sin s + \lambda r\eta,$$

$$(5.8\text{b})\qquad \beta \equiv -(H_\tau \tau' + H_\sigma \sigma')\sin s - T\sin\psi - H\cos s - N\psi' \sin s - \lambda r\nu,$$

$$(5.8\text{c})\qquad \gamma \equiv -(M_\tau \tau' + M_\sigma \sigma')\sin s + \Sigma\cos\psi - M\cos s - (\nu H - \eta N)\sin s.$$

From (2.5)–(2.8) we find that

(5.9)

$$Lr = (\nu \sin s)'\cos\psi + (\nu \sin s)(\cos\psi)' - (\eta \sin s)'\sin\psi - (\eta \sin s)(\sin\psi)' - \tau,$$

$$(5.10)\qquad (z' \sin s)' = (\nu \sin s)'\sin\psi + (\nu \sin s)(\sin\psi)' + (\eta \sin s)'\cos\psi + (\eta \sin s)(\cos\psi)'.$$

We now introduce new variables $(u_1, u_2, u_3) \equiv \boldsymbol{u}$ by

$$(5.11\text{a})\qquad u_1(s)(\sin s)^{\frac{1}{2}} = (Lr)(s),$$

$$(5.11\text{b})\qquad u_2(s)(\sin s)^{\frac{1}{2}} = (z' \sin s)',$$

$$(5.11\text{c})\qquad u_3(s)(\sin s)^{\frac{1}{2}} = L(\psi - s)(s).$$

Let us assume that $\boldsymbol{u}$ is continuous. Then we can solve (5.11a, c), (2.25), (2.27) in terms of $u_1$ and $u_3$. Now from (5.11b) we see that $u_2$ must satisfy

$$\int_0^\pi u_2(s)(\sin s)^{\frac{1}{2}}\,ds = 0. \tag{5.12}$$

We thus find that

$$\begin{aligned} z'(s) &= (\sin s)^{-1} \int_0^s u_2(t)(\sin t)^{\frac{1}{2}}\,dt \\ &= -(\sin s)^{-1} \int_s^\pi u_2(t)(\sin t)^{\frac{1}{2}}\,dt. \end{aligned} \tag{5.13}$$

This representation shows that $z$ satisfies the boundary conditions $z'(0) = 0 = z'(\pi)$, which are consequences of (2.5)–(2.8), (2.26), (2.27). We integrate one of the equations of (5.13) and then use (2.28) to evaluate the constant of integration. We obtain

$$r(s) = \int_0^\pi K(s,t)\,u_1(t)(\sin t)^{\frac{1}{2}}\,dt \equiv (Gu_1)(s), \tag{5.14a}$$

$$\begin{aligned} z(\mathrm{s}) &= \int_0^\pi \frac{[\pi^{-1}y - 1 + \chi_{[0,s]}(y)]}{\sin y} \int_0^y u^2(t)(\sin t)^{\frac{1}{2}}\,dt\,dy \\ &\equiv (Zu_2)(s), \end{aligned} \tag{5.14b}$$

$$\psi(s) - s = \int_0^\pi K(s,t)\,u_3(t)(\sin t)^{\frac{1}{2}}\,dt \equiv (Gu_3)(s) \tag{5.14c}$$

where $\chi_{[0,s]}$ is the characteristic function for $[0, s]$ and $K$ is Green's function for $L$ subject to Dirichlet data:

$$K(s,t) \equiv \begin{cases} -\frac{1}{2}\dfrac{1+\cos s}{1+\cos t}\dfrac{\sin t}{\sin s} & \text{if } t < s, \\[2ex] -\frac{1}{2}\dfrac{1+\cos t}{1+\cos s}\dfrac{\sin s}{\sin t} & \text{if } t > s. \end{cases} \tag{5.15}$$

(We could express (5.14b) in terms of a Green's function, but doing so would interfere with our analysis.) We also record that

$$r'(s) = \int_0^\pi K_s(s,t)\,u_1(t)(\sin t)^{\frac{1}{2}}\,dt \equiv (Gu_1)'(s), \tag{5.16a}$$

$$\psi'(s) - 1 = \int_0^\pi K_s(s,t)\,u_3(t)(\sin t)^{\frac{1}{2}}\,dt \equiv (Gu_3)'(s) \tag{5.16b}$$

where

$$K_s(s,t) = \begin{cases} \frac{1}{2}\dfrac{1+\cos s}{1+\cos t}\dfrac{\sin t}{\sin s} & \text{for } t < s, \\ -\frac{1}{2}\dfrac{1+\cos t}{1+\cos s}\dfrac{1}{\sin t} & \text{for } t > s. \end{cases} \tag{5.17}$$

$z'$ is given by (5.13). From (2.7), (2.8) we then obtain

$$\nu = (Gu_1)' \cos(s + Gu_3) + (Zu_2)' \sin(s + Gu_3), \tag{5.18a}$$

$$\eta = -(Gu_1)' \sin(s + Gu_3) + (Zu_2)' \cos(s + Gu_3), \tag{5.18b}$$

$$\tau = (Gu_1)/\sin s, \qquad \sigma = [\sin(s + Gu_3)/\sin s] - 1. \tag{5.18c, d}$$

We now convert our differential equations (5.1)–(5.3) into integral equations for $\boldsymbol{u}$: We replace the left sides of (5.9), (5.10), and (5.3) with (5.14). We next substitute (5.1), (5.2) into the right sides of (5.9), (5.10). Then we replace the variables appearing on the right sides of (5.9), (5.10), (5.3) with their representations from (5.14)–(5.18). Finally, as a precautionary measure, we subtract from the right side of the resulting form of (5.11) its mean value, a move suggested by (5.12). We denote the resulting system of integral equations by

$$\boldsymbol{u} = \boldsymbol{f}(\lambda, \boldsymbol{u}). \tag{5.19}$$

Let

$$\mathscr{U} = \{\boldsymbol{u} \in C^0([0,\pi])^3 : (5.12) \text{ holds}\}. \tag{5.20a}$$

We wish to show that $\boldsymbol{f}(\lambda, \cdot)$ is a compact mapping from

$$\mathscr{U} \cap \{\boldsymbol{u} : \boldsymbol{w} \in \mathscr{D}\} \equiv \mathscr{E} \tag{5.20b}$$

into itself ($\mathscr{D}$ is defined in (2.20)).

Let $\boldsymbol{w} \mapsto R(\boldsymbol{w})$ be any constitutive function. If $R \in C^1(\mathscr{D})$, then

$$R(\boldsymbol{y}) = R(\boldsymbol{x}) + R_i(\boldsymbol{x}, \boldsymbol{y})(y_i - x_i) \tag{5.21a}$$

with

$$R_i(\boldsymbol{x}, \boldsymbol{y}) \equiv \int_0^1 R_{w_i}(t\boldsymbol{y} + (1-t)\boldsymbol{x})\, dt. \tag{5.21b}$$

Here and below, twice repeated indices are summed from 1 to 5. If $R \in C^2(\mathscr{D})$, then

$$R(\boldsymbol{y}) = R(\boldsymbol{x}) + R_{w_i}(\boldsymbol{x})(y_i - x_i) + \tfrac{1}{2} R_{ij}(\boldsymbol{x}, \boldsymbol{y})(y_i - x_i)(y_j - x_j) \tag{5.21c}$$

with

$$R_{ij}(\boldsymbol{x}, \boldsymbol{y}) = \int_0^1 R_{w_i w_j}(t\boldsymbol{y} + (1-t)\boldsymbol{x})\, dt. \tag{5.21d}$$

We let $\bar{R}_{w_i}(\boldsymbol{w})$ and $\bar{R}_{w_iw_j}(\boldsymbol{w})$ be *any* values of $R_i$ and $R_{ij}$ in which the arguments consist solely of entries from $\boldsymbol{w}$. E.g., $R_{w_i}(\boldsymbol{w})$ could be taken to be $R_i((\tau, \nu, 0, \sigma, \mu), (\nu, \tau, \eta, \mu, \sigma))$.

Since we shall use only the continuity of $\{\bar{R}_{w_i}, \bar{R}_{w_iw_j}\}$ and not the specific functional forms, we allow their meaning to vary even in the same expression.

In the spirit of (5.21) we define the continuously differentiable functions $C$ and $S$ by

$$1 - \cos\theta = \theta^2 C(\theta), \tag{5.22a}$$

$$\theta - \sin\theta = \theta^3 S(\theta). \tag{5.22b}$$

Then

$$\cos\psi - \cos s = -(\psi - s)\sin s + U, \tag{5.23a}$$

$$U \equiv -(\psi - s)^2\, C(\psi - s)\cos s + (\psi - s)^3\, S(\psi - s)\sin s, \tag{5.23b}$$

$$\sin\psi - \sin s = (\psi - s)\cos s + V, \tag{5.23c}$$

$$V \equiv -(\psi - s)^2\, C(\psi - s)\sin s - (\psi - s)^3\, S(\psi - s)\cos s. \tag{5.23d}$$

Thus

(5.23e)

$$\tau'\sin s = (\nu - \tau)\cos s - (\psi - s)(\nu\sin s + \eta\cos s) - \eta\sin s + \nu U - \eta V,$$

$$\sigma'\sin s = (\mu - \sigma)\cos s + \psi'[U - (\psi - s)\sin s]. \tag{5.23f}$$

Let

$$\boldsymbol{w}^* = (\tau, \nu, 0, \sigma, \mu). \tag{5.24}$$

By using the formulas (5.21)–(5.23) we can convert $\alpha$, given by (5.8a), to

$$\begin{aligned}\alpha = {} & T(\boldsymbol{w})\,[U - (\psi - s)\sin s] + \tfrac{1}{2}\bar{T}_{\eta\eta}\eta^2\cos s \\ & - [N_\nu(\boldsymbol{w}^*)(\nu - \tau) + N_\mu(\boldsymbol{w}^*)(\mu - \sigma) + \tfrac{1}{2}N_{ij}(\boldsymbol{w}^*, \boldsymbol{w})(w_i - w_i^*) \\ & (w_j - w_j^*) + \bar{N}_{\tau w_i}(\nu - \tau)(w_i - w_i^*) + \bar{N}_{\sigma w_i}(\mu - \sigma)(w_i - w_i^*)]\cos s \\ & + N_\tau(\boldsymbol{w})\,[(\psi - s)(\nu\sin s + \eta\cos s) + \eta\sin s - \nu U + \eta V] \\ & - N_\sigma(\boldsymbol{w})\,\psi'[U - (\psi - s)\sin s] + \bar{H}_\eta\eta\psi'\sin s + \lambda r\eta.\end{aligned} \tag{5.25}$$

We obtain similar representations for $\beta$ and $\gamma$. Note that $\alpha + [N_\nu(\boldsymbol{w}^*)(\nu - \tau) + N_\mu(\boldsymbol{w}^*)(\mu - \sigma)] - \eta(\bar{H}_\eta\psi'\sin s + \lambda r)$ consists of sums of products of continuous functions of $(\boldsymbol{w}, \psi - s)$ with quadratic functions of $\nu - \tau$, $\eta$, $\mu - \sigma$, $\psi - s$. By a lengthy computation we find that the terms in $\boldsymbol{f}$ consist of products of continuous functions of $(\boldsymbol{w}, \psi - s)$ with quadratic functions of $(\sin s)^{-\frac{1}{4}}(\nu - \tau)$, $(\sin s)^{-\frac{1}{4}}\eta$, $(\sin s)^{-\frac{1}{4}}(\mu - \sigma)$, and with a number of other such functions of $(\boldsymbol{w}, \psi - s)$. Let us examine $(\sin s)^{-\frac{1}{4}}(\mu - \sigma)$, which (5.23c), (5.14c), (5.16b) reduce to

$$\text{(5.26a)} \qquad (\sin s)^{-\frac{1}{4}}[\psi'(s) - 1 - (\psi(s) - s)\cot s - V(s)\csc s] = \int_0^\pi J(s,t)\, u_3(t)\, dt - (\sin s)^{-\frac{5}{4}} V(s)$$

where

$$\text{(5.26b)} \qquad J(s,t) = \begin{cases} \frac{1}{2}\dfrac{(1+\cos s)^2}{1+\cos t}\left(\dfrac{\sin s}{\sin t}\right)^{\frac{9}{4}}(\sin t)^{-\frac{3}{4}} & \text{if } t < s, \\ -\frac{1}{2}\dfrac{(1-\cos s)^2}{1-\cos t}\left(\dfrac{\sin t}{\sin s}\right)^{\frac{9}{4}}(\sin t)^{-\frac{3}{4}} & \text{if } t > s. \end{cases}$$

The Arzelà-Ascoli Theorem supports a straightforward proof that the kernel $J$ generates a compact and continuous operator from $C^0([0, \pi])$ to itself. Note that the kernel $(\sin s)^{-\frac{1}{4}} K_s$ (see (5.17)) does not. It is the cancellation of terms in $J$ that is responsible for this compactness.

Now (5.23d) implies that

$$\text{(5.26c)} \qquad -(\sin s)^{-\frac{5}{4}} V = [(\sin s)^{-\frac{1}{8}}(\psi - s)]^2 C(\psi - s) + [(\sin s)^{-\frac{5}{12}}(\psi - s)]^3 S(\psi - s)\cos s\,.$$

Since the Arzelà-Ascoli Theorem implies that the integral representation for $(\sin s)^{-\frac{5}{12}}(\psi - s)$ obtained from (5.14c) is a compact mapping of $\boldsymbol{u}$, it follows that (5.26c) defines a compact mapping. Since compositions of continuous and compact mappings are compact, we find that (5.26c), (5.26a), and $(\sin s)^{-\frac{1}{2}}(\mu - \sigma)^2$ correspond to compact operators taking $C^0([0, \pi])$ into itself. We emphasize that the appearance of terms such as $(\mu - \sigma)$ in $\boldsymbol{f}$ is a consequence of the isotropy conditions.

By extensive use of arguments such as these, we find that *$\boldsymbol{f}$ is a compact and continuous mapping of $\mathscr{E}$ into itself. Moreover, the form of $\boldsymbol{f}$ shows that it is Fréchet differentiable at the trivial branch of solutions*. Let $\boldsymbol{u}_0(\lambda)$ be the trivial solution of (5.19) for fixed $\lambda$; $\boldsymbol{u}_0$ corresponds to the trivial solutions discussed in Section 3. The linearization of (5.19) about the trivial state $\boldsymbol{u}_0$, namely

$$\text{(5.27)} \qquad \boldsymbol{v} = \boldsymbol{f}_v(\lambda, \boldsymbol{u}_0(\lambda))\,\boldsymbol{v}\,,$$

is equivalent to the equations of Section 4. We accordingly can invoke the global bifurcation theory of RABINOWITZ (1971) and the local theory of CRANDALL & RABINOWITZ (1971) to obtain

**5.28. Theorem.** *For a given positive integer $n$ let $\bar{\lambda}_n$ be a solution of* (4.21) *of odd algebraic multiplicity*, i.e., *let there be an odd positive integer $k$ such that*

$$\text{(5.29)} \qquad \frac{\partial^j}{\partial\lambda^j} g(\bar{\lambda}_n, n) = 0, \quad j = 0, \ldots, k-1, \quad \frac{\partial^k}{\partial\lambda^k} g(\bar{\lambda}_n, \mathrm{n}) \neq 0.$$

*Then bifurcating from the point* $(\bar{\lambda}_n, \boldsymbol{u}_0(\bar{\lambda}_n))$ *on the trivial branch of solutions are two connected sets* $\mathscr{C}^{\pm}(\bar{\lambda}_n)$ *of nontrivial solution pairs* $(\lambda, \boldsymbol{u})$ *of* (5.19) *that enjoy at least one of the following properties*:

(i) $\mathscr{C}^{\pm}(\bar{\lambda}_n)$ *cannot be confined to any compact subset of* $\mathscr{E}$.

(ii) *The closure of* $\mathscr{C}^{\pm}(\bar{\lambda}_n)$ *contains another point* $(\zeta, \boldsymbol{u}_0(\zeta))$ *where* $\zeta$ *satisfies* (4.21) *with a possibly different positive integer n.*

*Suppose, moreover, that* $\bar{\lambda}_n$ *is simple*, i.e., *that* $k = 1$ *in* (5.29). *Let* $\boldsymbol{v}_n$ *be a normalized eigenfunction of* (5.27). *Let* $P(\bar{\lambda}_n)$ *be the projection of* $\mathscr{U}$ *onto span* $(\boldsymbol{v}_n)$. *Then there is a number* $\bar{\varepsilon} > 0$ *and there are continuous functions* $\varkappa_n\colon [-\bar{\varepsilon}, \bar{\varepsilon}] \mapsto R$, $\boldsymbol{y}_n\colon [-\bar{\varepsilon}, \bar{\varepsilon}] \mapsto [I - P(\bar{\lambda}_n)]\,\mathscr{U}$ *with* $\varkappa_n(0) = 0$, $\boldsymbol{y}_n(0) = \boldsymbol{0}$ *such that* $(\bar{\lambda}_n + \varkappa_n(\varepsilon), \boldsymbol{u}_0(\bar{\lambda}_n + \varkappa_n(\varepsilon)) + \varepsilon[\boldsymbol{v}_n + \boldsymbol{y}_n(\varepsilon)])$ *is a solution pair of* (5.19) *for* $|\varepsilon| < \bar{\varepsilon}$. *Moreover, there is a neighborhood* $\mathscr{N}_n$ *of* $(\bar{\lambda}_n, \boldsymbol{u}_0(\bar{\lambda}_n))$ *such that if* $(\lambda, \boldsymbol{u})$ *is a solution pair of* (5.19) *lying in* $\mathscr{N}_n$, *then either* $(\lambda, \boldsymbol{u}) = (\lambda, \boldsymbol{u}_0(\lambda))$ *or else there is an* $\varepsilon \in [-\bar{\varepsilon}, \bar{\varepsilon}]$ *such that* $(\lambda, \boldsymbol{u}) = (\bar{\lambda}_n + \varkappa_n(\varepsilon), \boldsymbol{u}_0(\bar{\lambda}_n + \varkappa_n(\varepsilon)) + \varepsilon[\boldsymbol{v}_n + \boldsymbol{y}_n(\varepsilon)])$.

Note also that (5.14)–(5.18) imply that $\boldsymbol{w}$ corresponding to any solution of (5.19) is continuous and, in particular, $\psi \in C^1([0, \pi])$. Moreover,

$$\lim_{s \to 0, \pi} [\nu(s) - \tau(s)] = 0, \qquad \lim_{s \to 0, \pi} [\mu(s) - \sigma(s)] = 0. \tag{5.30}$$

It follows from (5.1)–(5.3) that $\boldsymbol{u}$ generates a classical solution to our boundary value problem. In fact, the requirement that the constitutive functions $N$, $T$, $H$, $M$, $\Sigma$ be thrice continuously differentiable on $\mathscr{D}$ implies that the strains $\boldsymbol{w}$ generated by $\boldsymbol{u}$ are twice continuously differentiable on the *open* interval $(0, \pi)$. To determine the boundary behavior of $\eta'$, which is needed in the sequel, we divide (2.29) by $\sin^2 s$ and use (2.19) and (5.21) to obtain

$$N(\boldsymbol{w}(s))\,\frac{\sin \psi(s)}{\sin s} + \bar{H}_\eta(\boldsymbol{w})\,\frac{\eta(s)}{\sin s}\cos \psi(s) + \lambda \tau(s)^2/2 = 0 \tag{5.31}$$

on $(0, \pi)$. Since $\eta$ satisfies (2.26), we can let $s \to 0, \pi$ in (5.31) to conclude that $\eta$ is differentiable on $[0, \pi]$. To show that $\eta \in C^1([0, \pi])$ we use the expression for $\eta'$ obtained from (5.2) and examine its limits as $s \to 0, \pi$. This process is simplified by the observation that the limiting forms for $\eta'$ can equivalently be obtained by substituting (2.19) into (2.15), carrying out the differentiations, dividing the resulting equation by $\sin s$, and exploiting (2.24) and (5.30) to show that the limits $\eta'(0)$ and $\eta'(\pi)$ agree with those found from (5.31).

An essentially equivalent version of Theorem 5.28 can be obtained by using the theory of Fitzpatrick & Pejsachowicz (1986), which generalizes that of Crandall & Rabinowitz. The former theory does not require the use of Green functions. The role of isotropy still remains paramount.

## 6. Nodal Properties of Solutions

The determination of nodal properties of solutions of our boundary value problem proves to be a more subtle question than that for the corresponding problem for plates. The main reason for this difference is that in plate theory (*cf.* ANTMAN (1978)) $\eta$ is the product of a certain function with $\sin \psi$ and so the nodal properties of $\eta$ are determined directly from those of $\psi$. On the other hand, in our shell theory $\nu, \tau, \eta, \psi$ are all coupled by nonlinear differential equations, which lack an obvious nodal pattern. To approach the question for shells we first need a candidate for a nodal pattern that is preserved on solution branches.

The solutions of the linearized problem, treated in Section 4, have the property that $\eta_1$ and $\psi_1$ have the same zeros, which are simple. Perturbation solutions, determined by SHIH (1985), also have this feature. Thus we are motivated to study whether the number of simultaneous zeros of $\eta$ and $\psi - s$ is fixed on each bifurcating branch.

We immediately confront a serious difficulty: Let $\mathscr{S}_n$, $n = 1, 2, \ldots$ be the collection of all real-valued functions $\varrho$ in $C^1([0, \pi])$ having zeros at $0, \pi$ and having exactly $n - 1$ zeros on $(0, \pi)$, each zero being simple. Then $\mathscr{S}_n$ is open in $C^1([0, \pi])$. Given any function in $\mathscr{S}_n$, it follows that all sufficiently nearby functions enjoy the same nodal properties, *i.e.*, have the same number $n + 1$ of zeros, each of which is simple. Moreover, the boundary $\partial \mathscr{S}_n$ consists of functions having at least one double zero on $[0, \pi]$. These observations support the nodal theory introduced by CRANDALL & RABINOWITZ (1970) for bifurcation problems governed by nonlinear Sturm-Liouville equations. To study pairs of functions with simultaneous zeros, we let $\mathscr{T}_n$ be the collection of all complex-valued functions in $[C^1([0, \pi])]^2$ having zeros at $0, \pi$ and having exactly $n - 1$ zeros on $(0, \pi)$, each zero being simple. But $\mathscr{T}_n$ is not open in $[C^1([0, \pi])]^2$ and there is no useful characterization of $\partial \mathscr{T}_n$. To circumvent this difficulty, we must exploit the fact that $(\eta, \psi - s)$ is not merely an arbitrary element of $\mathscr{T}_n$, but is also a solution of a boundary value problem.

Our strategy is as follows. We first describe an auxiliary condition ensuring that if there is a solution of our boundary value problem for which $\eta, \eta', \psi - s$, $(\psi - s)'$ all vanish at a point $s_0$ in $[0, \pi]$, then $\nu - \tau$ also vanishes at $s_0$. Next we prove a basic uniqueness theorem stating that if there is a solution of our boundary value problem for which $\nu - \tau, \eta, \psi - s, (\psi - s)'$ all vanish at a point $s_0$ in $[0, \pi]$, then the solution is trivial. This result does not enable us to assert that on a nontrivial solution branch $(\eta, \psi - s)$ remains in $\mathscr{T}_n$ if $(\eta, \psi - s)$ is in $\mathscr{T}_n$ at any solution pair on the branch. To remedy this difficulty, we set

$$\eta(s) = \varrho(s) \cos \theta(s), \tag{6.1}$$

$$\psi(s) - s = \varrho(s) \sin \theta(s).$$

If $\eta(s_0) = 0 = \psi(s_0) - s_0$, then $\varrho(s_0) = 0$. Thus the zeros of the real-valued function $\varrho$ are the simultaneous zeros of $\eta$ and $\psi - s$. If $\eta$ and $\psi$ are part of a nontrivial solution of our boundary value problem for which the auxiliary conditions hold, we use the uniqueness theorem to prove that $\varrho$ can be defined so that it belongs to $C^1([0, \pi])$. We can accordingly determine where $\varrho \in \mathscr{S}_n$. In particular, if at a point on a nontrivial solution branch $\varrho \in \mathscr{S}_n$, then $\varrho$ remains in $\mathscr{S}_n$ until it has a double zero (by virtue of our discussion of $\partial \mathscr{S}_n$).

By the remarks at the beginning of the last paragraph, we conclude that the solution is trivial if $\varrho$ has a double zero (and hence $\eta$ and $\psi - s$ have simultaneous double zeros) provided the auxiliary condition is met. Let $\bar{\lambda}_n$ be a simple solution of (4.21) and let $\mathscr{C}^{\pm}(\bar{\lambda}_n)$ be the two nontrivial branches bifurcating from $(\bar{\lambda}_n, \boldsymbol{u}_0(\bar{\lambda}_n))$. We wish to show that on $\mathscr{C}^{\pm}(\bar{\lambda}_n)$ the function $\varrho$ remains in $\mathscr{S}_n$ as long as the auxiliary condition is met. To do this, we must show that $\varrho$ admits a linearization and enjoys properties like those stated in the last part of Theorem 5.28. We thereby demonstrate that on $\mathscr{C}^{\pm}(\bar{\lambda}_n)$, the function $\varrho$ inherits its nodal properties from those of its linearization and preserves them as long as the auxiliary condition is met, *i.e.*, on $\mathscr{C}^{\pm}(\bar{\lambda}_n)$, $\varrho \in \mathscr{S}_n$ as long as the auxiliary condition is met. Thus on $\mathscr{C}^{\pm}(\bar{\lambda}_n)$ the function $\eta$ and $\psi - s$ have $n-1$ simultaneous zeros on $(0, \pi)$ as long as the auxiliary condition is met.

We now proceed to carry out this program. We assume throughout the ensuing discussion that $(\lambda, \psi, \boldsymbol{w})$ generates a solution pair for our boundary value problem.

Suppose that there is an $s_0 \in [0, \pi]$ such that $\eta(s_0) = \eta'(s_0) = \psi(s_0) - s_0 = \psi'(s_0) - 1 = \mu(s_0) = 0$. Then (2.24) ensures that (2.15), (2.19), (2.29) reduce to

$$N + T + \lambda\tau(s_0)\,\nu(s_0) = 0, \tag{6.2}$$

$$N + \tfrac{1}{2}\lambda\tau(s_0)^2 = 0 \tag{6.3}$$

where the arguments of $N$ and $T$ are $(\nu(s_0), \tau(s_0), 0, 0, 0)$. We know from Section 3 that (6.2), (6.3) admit a solution of the form $\nu(s_0) = \tau(s_0) = k(\lambda)$ and that this is the only solution of these equations for which $\nu(s_0) = \tau(s_0)$. By virtue of (2.20) the variables $\nu$, $\tau$, $\lambda$ are confined to $(h, \infty) \times (h, \infty) \times (0, \infty)$. Let $\mathscr{G}$ be the subset of this region for which the only solution of (6.2), (6.3) is $\nu(s_0) = \tau(s_0) = k(\lambda)$. We then have the following result, which is virtually a tautology.

**6.4. Lemma.** *If* $\eta(s_0) = \eta'(s_0) = \psi(s_0) - s_0 = \psi'(s_0) - 1 = 0$ *and if* $(\nu(s_0), \tau(s_0), \lambda) \in \mathscr{G}$, *then* $\nu(s_0) = \tau(s_0) = k(\lambda)$.

In the next section we shall study the nature of $\mathscr{G}$. The "auxiliary condition" alluded to above is just the requirement that $(\nu(s_0), \tau(s_0), \lambda) \in \mathscr{G}$. Our basic uniqueness theorem is

**6.5. Theorem.** *If there is an* $s_0 \in [0, \pi]$ *such that*

$$\nu(s_0) - k(\lambda) = \tau(s_0) - k(\lambda) = \eta(s_0) = \psi(s_0) - s_0 = \psi'(s_0) - 1 = 0, \tag{6.6}$$

*then the solution is trivial.*

**Proof.** Let

$$\begin{aligned} \boldsymbol{v} &= (v_1, v_2, v_3, v_4, v_5) \\ &= ((\nu - \tau)\sin s,\ (\nu + \tau - 2k)\sin s,\ \eta\sin s,\ \mu\sin s,\ \psi - s), \end{aligned} \tag{6.7a}$$

$$v = |v_1| + |v_2| + |v_3| + |v_4| + |v_5|. \tag{6.7b}$$

Then we can combine (5.1)–(5.3), (2.8), and (5.23) to obtain the system

(6.8a) $$v_1' = \nu(v_5 \sin s - U) + v_3 + \eta(v_5 \cos s + V) + \Delta^{-1} D_1,$$

(6.8b)
$$v_2' = (v_1 + v_2) \cot s - \nu(v_5 \sin s - U) - v_3 - \eta(v_5 \cos s + V) + \Delta^{-1} D_1,$$

(6.8c) $$v_3' = v_3 \cot s + \Delta^{-1} D_2,$$

(6.8d) $$v_4' = v_4 \cot s + \Delta^{-1} D_3,$$

(6.8e) $$v_5' = v_4 (\sin s)^{-1}.$$

If $s_0 \in (0, \pi)$ and if $\boldsymbol{v}$ satisfies (6.6) and (6.8), then $\boldsymbol{v} = \mathbf{0}$, because the right-hand side of (6.8) is a well-behaved function of $s$ and $\boldsymbol{v}$. Consequently the initial-value problem has a unique solution, which is the trivial solution $\boldsymbol{v} = \mathbf{0}$.

Thus we need only concern ourselves with the cases $s_0 = 0, \pi$, in which the right-hand side of (6.8) is singular in $s$. We just treat the former case. Throughout our discussion we let $\boldsymbol{v}$ represent an arbitrary, given solution of our boundary value problem (known to exist by the results of Section 5) which we subject to the additional requirement that it satisfy (6.6). In the analysis to follow, we let $c$ denote a typical positive constant depending on the fixed but arbitrary solution $\boldsymbol{v}$. The meaning of $c$ can change from place to place. We tacitly restrict $s$ to lie in an interval of the form $[0, \varepsilon]$ where $\varepsilon$ is small and positive.

Since $\sin s \geqq s \cos s$ on $[0, \varepsilon]$, we find that

(6.9) $$\int_0^s (\sin t)^{-\frac{1}{2}}\, dt \leqq 2 (\sin s)^{\frac{1}{2}} / \cos s < c (\sin s)^{\frac{1}{2}}.$$

Using this result in (5.13)–(5.18) and using the fact that $|\boldsymbol{u}| \leqq c$, we obtain after careful analysis the inequalities

(6.10) $$|\nu - \tau| + |\eta| + |\mu - \sigma| \leqq c (\sin s)^{\frac{1}{2}},$$
$$|\psi - s| \leqq c \sin s, \quad |\mu| + |\sigma| \leqq c.$$

Since (5.23) implies that

(6.11) $$|\sigma| \leqq c\, |v_5|\, (\sin s)^{-1},$$

we find from (5.25) and related expressions that

(6.12) $$|\alpha| + |\beta| + |\gamma| \leqq c (\sin s)^{\frac{1}{2}}.$$

Let us now examine (6.8a). From (6.10) we immediately obtain

(6.13) $$|v_1' - \Delta^{-1} D_1| \leqq c(|v_3| + |v_5|) \leqq cv \leqq cv (\sin s)^{-\frac{1}{2}}.$$

Note that (2.21) and (2.23) imply that

(6.14) $$\Delta^{-1} \leqq c.$$

Accounting for (2.24) by using the notation introduced after (5.21), we obtain

$$D_1 = \alpha H_\eta M_\mu - \alpha \overline{H}_{\mu\eta} \overline{M}_{\eta\eta} \eta^2 - (\beta M_\mu - \gamma H_\mu) \overline{N}_{\eta\eta} \eta + \beta N_\mu \overline{M}_{\eta\eta} \eta - \gamma H_\eta (\overline{N}_{\mu\eta} \eta + \overline{N}_{\mu\mu} \mu + \overline{N}_{\mu\sigma} \sigma), \tag{6.15}$$

$$\Delta^{-1} - (N_\nu H_\eta M_\mu)^{-1} \equiv Q = \overline{Q}_\eta \eta + \overline{Q}_\mu \mu + \overline{Q}_\sigma \sigma . \tag{6.16}$$

Inequalities (6.10)–(6.12), (6.14)–(6.16) imply that

$$|\Delta^{-1} D_1 - \alpha (N_\nu)^{-1}| \leqq cv(\sin s)^{-\frac{1}{2}} . \tag{6.17}$$

If in (5.25) we replace $N_\nu(\boldsymbol{w}^*)$ with $N_\nu(\boldsymbol{w}) + \overline{N}_{\nu w_i}(w_i^* - w_i)$ and replace $N_\mu(\boldsymbol{w}^*)$ with $\overline{N}_{\mu\eta} \eta + \overline{N}_{\mu\mu} \mu + \overline{N}_{\mu\sigma} \sigma$, then we obtain from (6.10) and (6.11) that

$$|\alpha + N_\nu (\nu - \tau) \cos s| \leqq cv(\sin s)^{-\frac{1}{2}} . \tag{6.18}$$

Inequalities (6.13), (6.17), (6.18) imply that

$$|v_1' + v_1 \cot s| \leqq cv(\sin s)^{-\frac{1}{2}} . \tag{6.19}$$

A virtually identical analysis yields

$$|v_2' - v_2 \cot s| \leqq cv(\sin s)^{-\frac{1}{2}} . \tag{6.20}$$

The treatment of (6.8c) is just slightly different. Using ideas the same as those leading to (6.17), we obtain

$$|\Delta^{-1} D_2| \leqq c |\beta| + cv(\sin s)^{-\frac{1}{2}} . \tag{6.21}$$

Since $H_\tau = \overline{H}_{\tau\eta} \eta$, $H_\sigma = \overline{H}_{\sigma\eta} \eta$ by (2.24), equation (5.8) yields

$$|\beta| \leqq cv(\sin s)^{-\frac{1}{2}} + |T + N + \lambda v \tau| \sin s . \tag{6.22}$$

Equation (3.1) implies that

$$T + N + \lambda \tau v \equiv X = \overline{X}_\nu (\nu - k) + \overline{X}_\tau (\tau - k) = \tfrac{1}{2} [\overline{X}_\nu (v_1 + v_2) + \overline{X}_\nu (-v_1 + v_2)] (\sin s)^{-1} . \tag{6.23}$$

Combining (6.8c), (6.21)–(6.23), we obtain

$$|v_3' - v_3 \cot s| \leqq cv(\sin s)^{-\frac{1}{2}} . \tag{6.24}$$

We remark that the only place that the variable $v_2$ appears in (6.8a, c, d, e) is in the right side of (6.8c). Similar techniques produce

$$|v_4' - v_4 \cot s| \leqq cv(\sin s)^{-\frac{1}{2}} . \tag{6.25}$$

Combining (6.19), (6.20), (6.24), (6.25), (6.8e) we obtain

$$v' \leqq |v_1'| + |v_2'| + |v_3'| + |v_4'| + |v_5'| \leqq v \cot s + cv(\sin s)^{-\frac{1}{2}} , \tag{6.26}$$

the first inequality being a standard consequence of the absolute continuity of $\boldsymbol{v}$. Since (6.6) implies that

$$v(s)/s \to 0 \quad \text{as } s \to 0, \tag{6.27}$$

we can invoke the standard uniqueness theorems associated with differential inequalities (*cf.* HARTMAN (1964, Sec. III. 6)) to conclude that $v = 0$. □

It is important to note that if we were to insert a constant exceeding 1 in front of $v \cot s$ in (6.26), then we could not conclude that $v = 0$. That the coefficient in front of $v \cot s$ is 1 is a consequence of the isotropy condition.

We are now ready to study the polar coordinates $\varrho$ and $\theta$ introduced in (6.1). We state their properties in a series of lemmas.

**6.28. Lemma.** *Let the boundary value problem have a nontrivial solution pair with* $(\nu, \tau, \lambda)$ *taking values only in* $\mathscr{G}$ *and with* $\varrho$ *and* $\theta$ *defined in any manner consistent with* (6.1). *Then the zeros of* $\varrho$ *on* $[0, \pi]$ *are isolated.*

**Proof.** If not, they would accumulate at a point $s_0 \in [0, \pi]$, at which $\eta$ and $\psi - s$ would consequently have simultaneous double zeros. By Lemma 6.4 and Theorem 6.5, the solution would then be trivial. □

Now (6.1) implies that

$$\tan \theta = (\psi - s)/\eta. \tag{6.29}$$

Suppose that the solution is not trivial and that $(\nu, \tau, \lambda)$ take values only in $\mathscr{G}$. Let $s_1$ and $s_2$ be a pair of adjacent zeros of $\varrho$. If $\eta$ vanishes nowhere on $(s_1, s_2)$, then (6.29) implies that $\tan \theta$ is continuously differentiable on $(s_1, s_2)$ and we can define $\theta$ itself (mod $\pi$) to be continuously differentiable on $(s_1, s_2)$. If $\eta$ vanishes at $s_3 \in (s_1, s_2)$, then we define

$$\theta(s) = \pi/2 + \int_{s_3}^{s} \frac{\{\eta(t)\,[\psi'(t) - 1] - [\psi(t) - t]\,\eta'(t)\}}{\eta(t)^2 + [\psi(t) - t]^2}\,dt \qquad (\text{mod } \pi) \tag{6.30}$$

for $s \in (s_1, s_2)$. If $s_4$ is a zero of $\varrho$ and if $\eta'(s_4)$ and $\psi'(s_4) - 1$ do not both vanish, then l'Hôpital's rule enables us to deduce from (6.29) that

$$\tan \theta(s) \to \frac{\psi'(s_4) - 1}{\eta'(s_4)} \quad \text{as } s \to s_4, \tag{6.31}$$

the limit existing as an extended real number. (If $\eta'(s_4) = 0 = \psi'(s_4) - 1$, then we could use Lemma 6.4 and Theorem 6.5 to show that the solution is trivial when the condition on $\mathscr{G}$ holds.) Thus $\theta(s_4)$ is defined mod $\pi$. By selecting the value of $\theta$ at such zeros of $\varrho$ appropriately, we can define

$$\varrho = \eta \cos \theta + (\psi - s) \sin \theta. \tag{6.32}$$

Hence Lemma 6.4 and Theorem 6.5 imply

**6.33. Lemma.** *Let the boundary value problem have a nontrivial solution pair with* $(\nu, \eta, \lambda)$ *taking values only in* $\mathscr{G}$. *Then* $\varrho$ *and* $\theta$ *satisfying* (6.1) *can be chosen to be continuous on* $[0, \pi]$.

Since $\eta, \psi \in C^1([0, \pi])$, the constructions we have made above ensure that $\varrho$ and $\theta$ are continuously differentiable where $\varrho$ does not vanish provided the hypotheses of Lemma 6.33 hold. Differentiating (6.1), we obtain

$$\varrho' = \eta' \cos\theta + (\psi' - 1)\sin\theta, \tag{6.34a, b}$$

$$\varrho\theta' = -\eta' \sin\theta + (\psi' - 1)\cos\theta$$

where $\varrho$ does not vanish. The continuity of the right sides of (6.34) then yields

**6.35. Lemma.** *If the hypotheses of Lemma 6.33 hold, then*

$$\varrho', \varrho\theta' \in C^0([0, \pi]).$$

We require the following refinement of this lemma:

**6.36. Lemma.** *If $s_0$ is a zero of $\varrho$ and if $\eta'(s_0)$ and $\psi'(s_0) - 1$ are not both zero, then $\varrho(s_0)\,\theta'(s_0) = 0$.*

**Proof.** The Mean Value Theorem yields

$$\eta(s) = \left(\int_0^1 \eta'(\alpha s + (1-\alpha)s_0)\,d\alpha\right)(s - s_0), \tag{6.37}$$

$$\psi(s) - s = \left(\int_0^1 \mu(\alpha s + (1-\alpha)s_0)\,d\alpha\right)(s - s_0).$$

On any small one-sided neighborhood of $s_0$, (6.1) and (6.34b) thus yield

$$\pm\varrho(s)\,\theta'(s) = \frac{\mu(s)\int_0^1 \eta'\,d\alpha - \eta'(s)\int_0^1 \mu\,d\alpha}{\left(\left[\int_0^1 \eta'\,d\alpha\right]^2 + \left[\int_0^1 \mu\,d\alpha\right]^2\right)^{\frac{1}{2}}} \tag{6.38}$$

where the arguments of $\eta'$ and $\mu$ in the integrands are $\alpha s + (1-\alpha)s_0$. Since $\eta'$ and $\mu$ are uniformly continuous on $[0, \pi]$, we find that $\int_0^1 \eta'(\alpha s + (1-\alpha)s_0)\,d\alpha \to \eta'(s_0)$ as $s \to s_0$, etc. Thus in the limit as $s \to s_0$, the numerator on the right side of (6.38) approaches 0, while the denominator approaches a positive number. □

In view of Lemma 6.35 we can examine whether $\varrho \in \mathscr{S}_n$. By the introductory remarks of this section, functions $\varrho$ in $\partial\mathscr{S}_n$ have a double zero. But this fact alone does not guarantee that $\eta$ and $\psi - s$ have a simultaneous double zero, as an examination of (6.1) and (6.34a) shows. But Lemma 6.36 coupled with (6.34) implies that at a double zero of $\varrho$ the functions $\eta$ and $\psi - s$ have a simultaneous double zero, a result that enables us to invoke Lemma 6.4 and Theorem 6.5. Marshalling these facts in the manner proposed in the introductory remarks in this section, we obtain

**6.39. Theorem.** *Let $\mathscr{K}$ be a connected set of nontrivial solution pairs with the property that $(\nu, \tau, \lambda)$ takes values only in $\mathscr{G}$ for each solution pair in $\mathscr{K}$. If for one solution pair in $\mathscr{K}$ the function $\varrho \in \mathscr{S}_n$, then $\varrho$ remains in $\mathscr{S}_n$ for every solution pair in $\mathscr{K}$. Thus $\eta$ and $\psi - s$ have exactly $n + 1$ simultaneous zeros everywhere on $\mathscr{K}$.*

Let $\bar{\lambda}_n$ be a simple root of (4.21) (so that (5.29) holds with $k = 1$). Then the last part of Theorem 5.28 together with the representations (5.14c) and (5.18b) imply that on the branch $\mathscr{C}(\bar{\lambda}_n)$ near the bifurcation point $(\bar{\lambda}_n, \mathbf{0})$, $\lambda, \eta, \psi$ have representations of the form

$$\lambda = \bar{\lambda}_n + \varkappa(\varepsilon), \qquad \eta = \varepsilon\eta_1 + o(\varepsilon), \tag{6.40}$$
$$\psi - s = \varepsilon\psi_1 + o(\varepsilon)$$

as $\varepsilon \to 0$, where $\varkappa$ is continuous and $\varkappa(0) = 0$. If we substitute (6.40) into (6.29) and use (4.25), we obtain

$$\tan\theta = \psi_1/\eta_1 + o(1) = E^{-1} + o(1). \tag{6.41}$$

Thus we may take

$$\theta = \arctan(E^{-1}) + o(1). \tag{6.42}$$

Then (6.32) implies that

$$\varrho = \varepsilon\varrho_1 + o(\varepsilon) = \varepsilon[\varrho_1 + o(\varepsilon)\,\varepsilon^{-1}] \tag{6.43}$$

with

$$\varrho_1 = (1 + E)^{-\frac{1}{2}}\,(E\eta_1 + \psi_1) = (1 + E)^{\frac{1}{2}}\,P_n^1(\cos s). \tag{6.44}$$

It follow that $\varrho_1 \in \mathscr{S}_n$ and that $\varrho \in \mathscr{S}_n$ for $\varepsilon$ sufficiently small. (Incidentally, we show in the next section that $(\nu, \tau, \lambda)$ generically belongs to $\mathscr{G}$ for $\varepsilon$ sufficiently small essentially because $\nu - \tau = 0$ on the trivial branch.) We accordingly have

**6.45. Theorem.** *Let $\bar{\lambda}_n$ be a simple root of (4.21). Let $\mathscr{K}_n^\pm$ be the maximal connected subset of $\mathscr{C}^\pm(\bar{\lambda}_n)$ with the properties that $(\nu, \tau, \lambda)$ takes values only in $\mathscr{G}$ for for each solution pair in $\mathscr{K}_n^\pm$ and that $(\bar{\lambda}_n, \mathbf{u}_0(\bar{\lambda}_n))$ belongs to the closure of $\mathscr{K}_n^\pm$. Then $\eta$ and $\psi - s$ have exactly $n + 1$ simultaneous zeros everywhere on ${}_n^\pm$.*

## 7. Properties of $\mathscr{G}$

We now study (6.2), (6.3), which we rewrite as

$$N(\nu, \tau) = -\tfrac{1}{2}\lambda\tau^2, \tag{7.1}$$

$$N(\nu, \tau) + T(\nu, \tau) = -\lambda\nu\tau, \tag{7.2}$$

from which we obtain

$$T(\nu, \tau) = \lambda\tau(\tfrac{1}{2}\tau - \nu). \tag{7.3}$$

Here we have suppressed the last three arguments (0, 0, 0) of the functions $N$ and $T$. The set $\mathscr{G}$ consists of all $(\nu, \tau, \lambda)$ in $(h, \infty)\times(h, \infty)\times(0, \infty)$ for which the only solution of (7.1), (7.2) is $\nu = \tau = k(\lambda)$. We make a few remarks about (7.1), (7.2) in general before studying a specific example.

Since $N_\nu > 0$ by (2.21) and since $N$ satisfies the growth conditions of (2.23), equation (7.1) can be solved uniquely for $\nu$ in terms of $\tau$ and $\lambda$:

$$\nu = f(\tau; \lambda), \quad \text{with } 1 = f(1; 0). \tag{7.4}$$

Since $f_\tau = -(N_\tau + \lambda\tau)/N_\nu$, we find that $f_\tau < 0$ if $N_\tau > 0$ and that $f_\tau(1; 0) > -1$ if $N_\tau(1, 1) < N_\nu(1, 1)$.

The isotropy condition (2.24) ensures that the curve in the $(\nu, \tau)$-plane defined by (7.2) is symmetric about the line $\nu = \tau$. If $T_\nu + N_\nu > 0$, if $T + N \to -\infty$ as $\nu \to 0$ for fixed $\tau$, and if $\liminf (T + N) \geqq 0$ as $\nu \to \infty$ for fixed $\tau$, then (7.2) can be uniquely solved for $\nu$ in terms of $\tau$ and $\lambda$:

$$\nu = g(\tau; \lambda) \quad \text{with } 1 = g(1; 0). \tag{7.5}$$

The results of Section 3 ensure that the graphs of $f(\cdot; \lambda)$ and $g(\cdot; \lambda)$ intersect on the line $\nu = \tau$. Since

$$g_\tau = -\frac{(\lambda g + N_\tau + T_\tau)}{(\lambda\tau + N_\nu + T_\nu)}, \tag{7.6}$$

it follows that $g_\tau < 0$ under our assumptions and, not surprisingly, that $f_\tau(\tau; \lambda) = -1$ when $\tau = g(\tau; \lambda)$. Thus the graphs of $f(\cdot; \lambda)$ and $g(\cdot; \lambda)$ cross transversally on the line $\nu = \tau$ when $f_\tau \neq -1$ here, *i.e.*, when

$$N_\nu(\tau, \tau) - N_\tau(\tau, \tau) - \lambda\tau \neq 0. \tag{7.7a}$$

Using (7.1), we can rewrite (7.7a) as a condition involving only the constitutive function $N$:

$$N_\nu(\tau, \tau) - N_\tau(\tau, \tau) + 2N(\tau, \tau)/\tau \neq 0. \tag{7.7b}$$

Wherever (7.7) holds, there is a neighborhood about $(\tau, \tau, \lambda)$ lying in $\mathscr{G}$. (See the parenthetical remark preceding Theorem 6.45.)

There are several ways to analyze (7.1), (7.2) further by making general restrictions on $N$ and $T$ leading to suitable estimates. It is also illuminating to view (7.1) and (7.2) as each describing the intersection of two graphs over the $(\nu, \tau)$-plane. Alternatively we could assume that $N$ and $T$ are derivable from a stored energy function (as in (7.9) below), characterize the equations as Euler-Lagrange equations for a constrained problem, and study the extremization of a real-valued function on a constraint curve. There are several ways to choose the constraint. It is more illuminating, however, to analyze a specific class of materials.

Let $A$, $B$, $C$, $\alpha$ be positive numbers and let

$$W(\nu, \tau) = A\alpha^{-1}\nu^{-\alpha}\tau^{-\alpha} + \tfrac{1}{2} B(\nu + \tau)^2 + \tfrac{1}{2} C(\nu^2 + \tau^2), \tag{7.8}$$

$$N(\nu, \tau) = W_\nu(\nu, \tau) - W_\nu(1, 1), \tag{7.9}$$

$$T(\nu, \tau) = W_\tau(\nu, \tau) - W_\tau(1, 1).$$

These equations give a reasonable model for material behavior in compression for $h = 0$. When (7.8), (7.9) hold, equations (7.1), (7.2) reduce to

$$(7.10) \qquad -Av^{-(\alpha+1)}\tau^{-(\alpha+1)}\tau + B(v+\tau) + Cv + A - 2B - C = -\tfrac{1}{2}\lambda\tau^2,$$

$$(7.11) \qquad -Av^{-(\alpha+1)}\tau^{-(\alpha+1)}(v+\tau) + (2B+C)(v+\tau) + 2A - 2(2B+C) = -\lambda v\tau.$$

We now solve (7.11) for $-A(v\tau)^{-(\alpha+1)}$ and substitute the resulting expression into (7.10), obtaining an equation with a factor $v - \tau$. Since we wish to study the locations of possible solutions of (7.1), (7.2) with $v \neq \tau$, we cancel this factor obtaining

$$(7.12) \qquad (B+C)(v+\tau) + A - 2B - C = \tfrac{1}{2}\lambda\tau^2.$$

Let us denote the solution of (7.12) for $v$ in terms of $\tau$ by $v^\#(\tau, \lambda)$. If we replace $v$ in (7.10) with $v^\#(\tau)$, we obtain

$$(7.13) \qquad A\tau^{-(\alpha+1)} = (\lambda\tau - C)\,[v^\#(\tau, \lambda)]^{\alpha+1}.$$

For $\lambda > 0$, trivial solutions of our boundary value problem are characterized by $v = \tau = k(\lambda) \in (0, 1)$, $N = T = -\tfrac{1}{2}\lambda k(\lambda)^2$. We accordingly limit our attention to a neighborhood (possibly quite large) of the trivial solutions by seeking solutions of (7.13) for which

$$(7.14\text{a, b, c, d}) \qquad \tau \in (0, 1), \quad v^\#(\tau, \lambda) \in (0, 1),$$
$$N(v^\#(\tau, \lambda), \tau) < 0, \quad T(v^\#(\tau, \lambda), \tau) < 0$$

for $\lambda > 0$. We wish to characterize where (7.13) has solutions satisfying (7.14). Such solutions lie in the complement of $\mathscr{G}$. (If (7.14) is not imposed, then (7.13) always has a solution.)

Note that (7.14c) is automatically satisfied by virtue of (7.1). We now use (7.12)–(7.14) to show that the set of points $(v, \tau) \in (0, 1) \times (0, 1)$ at which (7.1), (7.2) has solutions for (7.8), (7.9) can be quite small. Call this set $\mathscr{H}(\lambda)$. Clearly

$$(7.15) \qquad \mathscr{H}(\lambda) = \{(v, \tau) \in (0, 1) \times (0, 1) : v = v^\#(\tau, \lambda),$$
$$\tau \text{ satisfies (7.13)}, \quad T(v^\#(\tau, \lambda), \tau) < 0\}.$$

We now obtain a number of inequalities that elements of $\mathscr{H}(\lambda)$ must satisfy.

Equation (7.3) implies that (7.14d) is satisfied if and only if

$$(7.16) \qquad v^\#(\tau, \lambda) > \tau/2.$$

Equation (7.12) implies that

$$(7.17) \qquad v^\#(\tau, \lambda) > -\tau + 1 + \frac{B - A}{B + C},$$

which can represent a very severe or an insignificant restriction depending on the relative size of $A$, $B$, $C$. Equation (7.12) and conditions (7.14a, b) imply that

$$(7.18) \qquad \tau \leq [2(C + A)]^{\frac{1}{2}} \lambda^{-\frac{1}{2}}.$$

A simple graphical analysis of (7.13) based on (7.14a, b) shows that

$$\tau \geqq \min\left(\frac{A+C}{\lambda}, 1\right). \tag{7.19}$$

Note that if $B - A$ and $\lambda$ are sufficiently large, then the set of $\tau$'s satisfying (7.14), (7.17), (7.18) is empty. If $\lambda$ is small enough, then (7.19) implies that $\tau \geqq 1$.

By substituting (7.16) into (7.12) we obtain

$$\tau \notin [\tau_-, \tau_+], \quad \lambda\tau_\pm = \tfrac{3}{2}(B + C) \pm [\tfrac{9}{4}(B + C)^2 + 2\lambda(A - 2B - C)]^{\frac{1}{2}}, \tag{7.20}$$

the interval $[\tau_-, \tau_+]$ being regarded as empty if $\tau_\pm$ are not real. If $A - 2B - C > 0$ and if $\lambda$ is large enough, then (7.20) delivers a lower bound for $\tau$ that is larger than that of (7.19).

We seek similar estimates independent of $\lambda$. Let us use (7.18) to get a lower bound for $A\tau^{-(\alpha+1)}$. A sketch shows that a lower bound for $\tau$ is the solution of

$$A\left[\frac{2(C + A)}{\lambda}\right]^{-(\alpha+1)/2} = \lambda\tau - C. \tag{7.21}$$

Thus

$$\tau \geqq \frac{C}{\lambda} + \frac{A\lambda^{(\alpha-1)/2}}{[2(C + A)]^{(\alpha+1)/2}}. \tag{7.22}$$

If $\alpha \in (0, 1)$, then (7.22) delivers a lower bound dependent on $\lambda$, but nevertheless better than those of (7.19) and (7.20) when $\lambda$ is large enough. If $\alpha = 1$, then (7.22) gives bound that is clearly better than (7.19) and that itself has a further lower bound independent of $\lambda$. If $\alpha > 1$, then

$$\begin{aligned} 2\tau &\geqq \xi x^{-2} + (1 - \xi)\, x^{\alpha-1} \\ &= \frac{(\alpha + 1)\, C^{(\alpha-1)/(\alpha+1)} A^{2/(\alpha+1)}}{(\alpha - 1)^{(\alpha-1)/(\alpha+1)}\, 2^{-2(\alpha+2)/(\alpha+1)}\, (C + A)} \end{aligned} \tag{7.23}$$

with $x \equiv \left[\dfrac{2C}{(\alpha - 1)A}\right]^{1/(\alpha+1)}$, $\xi = \dfrac{C}{C + A}$, the expressions on the right side of (7.23) being the minimum of the right side of (7.22) for $\lambda \in (0, \infty)$. An upper bound for the rightmost term of (7.23) can be obtained by the Young inequality. The right side of (7.23) can be made arbitrarily large by taking $\alpha - 1$ sufficiently small and positive.

It is not difficult to obtain further estimates, *e.g.*, estimates for the neighborhood discussed in the comments following (7.7b). We refrain from doing so.

The basic message of this section is that the set $\mathcal{G}$ can be very large and that precise information about it can be gained directly from degenerate constitutive equations. Moreover, we should be able to strengthen the global results of Section 7 even more by obtaining estimates for the location of solutions of our boundary value problem and restricting $\mathcal{G}$ to reflect such estimates. We do not attempt such estimates here.

## 8. Conclusion

In an unshearable shell $\eta$ is constrained to be zero. The shear resultant $H$, not defined constitutively, is the Lagrange mutiplier maintaining the constraint of unshearability. It is a fundamental unknown of the constrained problem. The same analysis as that used above can treat unshearable shells if $H$ is treated as $\eta$ is.

Some of our techniques are reminiscent of those of ANTMAN (1978) for plates. The difference in boundary conditions forced us to give the coordinate functions $r$ and $z$ a central role in the analysis of Section 5, which they lacked in the plate theory. The issues treated in Section 7 never arose for plates. Of course the nodal properties of branches for shells differ markedly from those for plates and for circular arches (*cf.* ANTMAN & DUNN, (1980).) Incidentally, the uniqueness theorem of ANTMAN (1978, Section 7) is faulty; it can be readily corrected by following the development of Section 6 above.

It is illuminating to compare our results with the numerical results of BAUER, REISS, & KELLER (1970). Treating an engineering theory of shells, they discovered the remarkable fact that all the bifurcating branches they analyzed are connected. Our results point in the diametrically opposite direction. We showed that nodal properties of branches bifurcating from simple eigenvalues could only lose their nodal structure at certain special places in solution-parameter space. In particular, if we could obtain delicate pointwise estimates showing that $\nu \in (0, 1)$, $\tau \in (0, 1)$, $N < 0$, $T < 0$, everywhere on a family of bifurcating branches, then the results of Sections 6 and 7 show that nodal properties cannot switch along such branches. Consequently such branches could not connect bifurcation points with simple eigenvalues having different eigenfunctions.

Now Section 7 shows that for fixed $\lambda$ the complement of $\mathscr{G}$ is generically a finite number of points. On the basis of such genericity arguments we could expect that on a given bifurcating branch there would be no $s_0 \in [0, \pi]$ and $\lambda > 0$ such that $(\nu(s_0), \tau(s_0), \lambda) \notin \mathscr{G}$. (Such genericity arguments are dangerous: The nodal structure we found for $\eta$ and $\psi - s$ is not generic for pairs of functions.)

The chief source of the suspected variance between our results and those of BAUER, REISS, & KELLER is likely the fact that various geometric terms regarded as small were discarded in their theory (as is typical of all technical shell theories). Consequently these terms could not come into play in the large solutions they computed. A conceivable alternative explanation of the connectedness of all the bifurcating branches found by BAUER, REISS, & KELLER is that they correspond to spurious solution branches of the sort often found in numerical studies of bifurcation problems. We tend to dismiss this explanation because it does not seem compatible with the consistent pattern they found. BAUER, REISS, & KELLER did not give enough data for us to determine what sort of nodal properties their solutions possess.

Our approach may also be compared with that of LANGE & KRIEGSMANN (1981), who studied continuation properties for very thin shells by using asymptotic expansions. Our global bifurcation theory gives an approach to the continuation of branches that does not require special hypotheses on thickness, or more generally, on constitutive functions. Presumably their detailed approach could be applied to our geometrically exact theory for constitutive functions corresponding to very thin shells. We could also introduce the thickness $h$ as a second bifurcation parameter and study the role of thickness by using global multiparameter bifurcation theory (*cf.* ALEXANDER & ANTMAN (1981)).

We mention that the perturbation results of SHIH (1985) show that the local disposition of bifurcating branches depends critically on the nature of our nonlinear constitutive equations. In particular, the local structure can be quite different from that found in technical shell theories, which essentially use linear constitutive relations.

As we have indicated above, the process of numerically constructing solution branches can be quite tricky. It would seem that nodal properties on solution branches could give an extra control on numerical computation. But, to our knowledge, no such use of nodal properties has been attempted.

*Acknowledgment.* Part of the work presented here was developed by SHIH (1985) for his doctoral dissertation in the Department of Mathematics of the University of Maryland. The work of ANTMAN was partially supported by a grant from the National Science Foundation.

## References

J. C. ALEXANDER & S. S. ANTMAN (1981): Global Behavior of Bifurcating Multidimensional Continua of Solutions for Multiparameter Nonlinear Eigenvalue Problems. *Arch. Rational Mech. Anal.* **76**, 339–354.

S. S. ANTMAN (1981): Existence and Nonuniqueness of Axisymmetric Equilibrium States of Nonlinearly Elastic Shells. *Arch. Rational Mech. Anal.* **40**, 329–372.

S. S. ANTMAN (1978): Buckled States of Nonlinearly Elastic Plates, *Arch. Rational Mech. Anal.* **67**, 111–149.

S. S. ANTMAN & E. R. CARBONE (1977): Shear and Necking Instabilities in Nonlinear Elasticity, *J. Elasticity* **7**, 125–151.

S. S. ANTMAN & J. E. DUNN (1980): Qualitative Behavior of Buckled Nonlinearly Elastic Arches, *J. Elasticity* **10**, 225–239.

S. S. ANTMAN & M. REEKEN (1986): The Drawing and Whirling of Strings: Singular Global Multiparameter Eigenvalue Problems, *SIAM J. Math. Anal.*, to appear.

S. S. ANTMAN & G. ROSENFELD (1978): Global Behavior of Buckled States of Nonlinearly Elastic Rods. *SIAM Rev.* **20**, 513–566.

L. BAUER, E. L. REISS & H. B. KELLER (1970): Axisymmetric Buckling of Hollow Spheres and Hemi-Spheres. *Comm. Pure & Appl. Math.* **23**, 529–568.

E. & F. COSSERAT (1909): *Théorie des Corps Déformables.* Hermann.

M. G. CRANDALL & P. H. RABINOWITZ (1970): Nonlinear Sturm-Liouville Problems and Topological Degree, J. Math. Mech. **19**, 1083–1102.

M. G. CRANDALL & P. H. RABINOWITZ (1971): Bifurcation from Simple Eigenvalues. *J. Funct. Anal.* **8**, 321–340.

P. M. FITZPATRICK & J. PEJSACHOWICZ (1986): An Extension of the Leray-Schauder Degree for fully Nonlinear Elliptic Problems, to appear.

P. HARTMAN (1964): *Ordinary Differential Equations.* Wiley, New York.

C. G. LANGE & G. A. KRIEGSMANN (1981): The Axisymmetric Branching Behavior of Complete Spherical Shells. *Quart. Appl. Math.* **39**, 145–178.

A. LIBAI & J. G. SIMMONDS (1983): Nonlinear Elastic Shell Theory, *Advances in Applied Mechanics* **23**, Academic Press, 271–371.

P. M. NAGHDI (1972): The Theory of Plates and Shells. In *Handbuch der Physik* VIA/2 (Ed. C. TRUESDELL) Springer-Verlag, 425–640.

P. H. RABINOWITZ (1971): Some Global Results for Nonlinear Eigenvalue Problems. *J. Funct. Anal.* **7**, 487–513.

E. REISSNER (1963): On the Equations for Finite Symmetric Deflection of Thin Shells of Revolution. In *Progress in Applied Mechanics* (The Prager Anniversary Volume) D. C. DRUCKER, ed. Macmillian, 171–178.

K.-G. SHIH (1985): *Global Bifurcation Problems of Axisymmetric Buckling of Complete Spherical Shells*, Doctoral Dissertation, University of Maryland.

ICASE
Langley Research Center
Hampton, Virginia

and

Department of Mathematics
University of Maryland
College Park

*(Received August 5, 1985)*

# *Stable Equilibrium Configurations of Elastic Crystals*

J. L. ERICKSEN

*For J. B. Serrin on his 60th birthday*

## 1. Introduction

Nonlinear thermoelasticity theory for crystals is generating some knotty questions in the calculus of variations and in the theory of related partial differential equations. This stems from invariance assumptions which are suggested by molecular theory, and seem to be needed to analyze some commonly observed phenomena, such as twinning. What might be regarded as the simplest problem is to characterize the most stable configurations of unloaded crystals. Even this is difficult, because these are not all unique, but form infinite sets. A number of these do seem to match configurations observed in real crystals. After elaborating this a bit, I will present methods for constructing solutions which are special, but resemble configurations which are observed.

## 2. Crystal Elasticity

Kinematical descriptions of crystal configurations and their symmetries are discussed in detail by PITTERI [1, 2]. Some of the simpler kinds are adequately described by a set of three linearly independent vectors $e_a$ $(a = 1, 2, 3)$, called lattice vectors. Then, two such sets $e_a$ and $\bar{e}_a$ describe the same configuration provided

$$\bar{e}_a = m_a^b e_b, \tag{2.1}$$

where the $m_a^b$ are integers such that

$$\det m = \pm 1, \quad m \stackrel{\text{def}}{=} \|m_a^b\|. \tag{2.2}$$

Under the obvious rule for composing these linear transformations, they form a group $G$,

$$m \in G, \tag{2.3}$$

which is infinite, discrete and not compact. Essentially, molecular theories of elasticity are schemes for calculating $\phi$, the Helmholtz free energy per unit mass, as a function of lattice vectors and temperature $T$,

$$\phi = \hat{\phi}(e_a, T). \tag{2.4}$$

In some situations, for example for the quartz problems discussed by JAMES [3], one finds that additional variables are needed, for an adequate description. Other difficulties are encountered in dealing with long range interactions. However, (2.4) seems to apply to a variety of crystals. When it does, one should get a single value of $\phi$ for a single configuration, so $\phi$ should be invariant under $G$,

$$\hat{\phi}(m_a^b e_b, T) = \phi(e_a, T), \qquad m \in G. \tag{2.5}$$

The notion that rotating a configuration should not alter this energy gives rise to the additional invariance that, for any rotation $R$,

$$R^{-1} = R^T, \qquad \det R = 1, \tag{2.6}$$

$$\hat{\phi}(Re_a, T) = \phi(e_a, T). \tag{2.7}$$

To get to elasticity or thermoelasticity theory, the usual practice is to introduce the Born hypothesis, which is discussed in detail by ERICKSEN [4]. Briefly, one picks some reference configuration, including a definite choice of (constant) reference lattice vectors $E_a$. Then, the assumption is that a macroscopic deformation $x = \chi(X)$, with gradient

$$F = \nabla\chi, \qquad \det F > 0, \tag{2.8}$$

takes $E_a$ to $e_a$, a possible set of lattice vectors in the deformed configuration, given by the formula

$$e_a = FE_a. \tag{2.9}$$

Thus

$$\hat{\phi}(e_a, T) = \hat{\phi}(FE_a, T) = \tilde{\phi}(F, T), \tag{2.10}$$

the $E_a$ being regarded as fixed. To describe the invariance of $\tilde{\phi}$, we define, for each $m \in G$, a linear transformation $H(m)$ by the rule

$$H(m)\, E_a = m_a^b E_b, \tag{2.11}$$

such $H$ satisfying

$$\det H = \pm 1, \tag{2.12}$$

and generating a group $\bar{G}$ conjugate to $G$. Then, by easy calculation,

$$\tilde{\phi}(RFH, T) = \tilde{\phi}(F, T), \qquad H \in \bar{G}. \tag{2.13}$$

Commonly, such invariances are obscured by approximations made in calculating thermodynamic potentials from molecular theory. However, examples of molecular theories consistent with the above description can be found, for example in the works of EFTIS, MACDONALD & ARKILIC [5] and ERICKSEN [6].

The indicated invariance causes $\tilde{\phi}$ to have some unusual properties. Since $G$ and $\bar{G}$ are not compact, one can pick any $F$ in the domain of $\tilde{\phi}$, and find a sequence $H_n \in \bar{G}$ with $FH_n$ unbounded as $n \to \infty$. Thus, the assumption that $\tilde{\phi} \to \infty$

when $F$ grows large is untenable. Similarly, the invariance prevents $\tilde{\phi}$ from satisfying the Legendre-Hadamard condition for all kinematically possible values of $F$. These and other idiosyncracies are collected and discussed by FONSECA [7], who includes relevant references. At least to those familiar to the calculus of variations, this might be enough to indicate that we face unusual difficulties, in trying to apply it.

To sketch one of the simplest problems, consider that of determining the stablest equilibrium configurations of an unloaded crystal, with $T$ considered as a control parameter. Much the same problems arise, for crystals subject to an hydrostatic pressure. Mathematically, we should find the absolute minima of the functional

$$\int_{\Omega} \tilde{\phi}(F, T)\, \varrho\, dv, \tag{2.14}$$

where $\varrho$ is a positive constant, the mass density, and $\Omega$ is the region occupied by a body in the reference configuration. Make $\Omega$ as simple and smooth as you like, assume $\tilde{\phi}$ very smooth, *etc.* to simplify matters as much as is possible. Here, there are no constraints on deformation, so we can begin by trying to minimize the integrand. This might well be possible. Suppose that for some constant $F = F_0$, with $\det F_0 > 0$, we have

$$\tilde{\phi}(F, T) \geqq \tilde{\phi}(F_0, T), \tag{2.15}$$

for all $F$ in the domain of $\tilde{\phi}$. This gives us, as a very simple minimizer, a homogeneous configuration with constant lattice vectors. Of course, with (2.13), we have an infinite collection of similar minimizers, with deformation gradients of the form

$$RF_0H. \tag{2.16}$$

Were $\bar{G}$ a continuous group, we might find a smooth inhomogeneous deformation, with $F$ taking values in this set, but one can show that here, this is impossible. One could have additional minimizers not so related by invariance and, physically, this can happen at isolated temperatures, but we ignore this.

A complication is suggested by observations of twinning. Physically, it seems quite reasonable to regard twinned configurations as minimizers. Generalizing this idea, we should allow configurations for which $F$ is piecewise continuous, as long as the displacement remains continuous. If $R_1F_0H_1$ and $R_2F_0H_2$ are two values in neighboring regions, meeting at a smooth surface with normal $N$ in the reference configurations, the usual kinematical conditions of compatibility require that there exist a vector $A$ satisfying

$$R_2F_0H_2 = R_1F_0H_1(1 + A \otimes N). \tag{2.17}$$

One can normalize this by requiring that $N$ be a unit vector, but I find it more convenient to use different normalizations in different cases. I will just assume that $A \otimes N = 0$ only if $A = 0$. With $F_0$, the $R$'s and $H$'s constant, it is clear that $A$ and $N$ can be taken as constant so, where smooth, such surfaces are planes. As is discussed by JAMES [8, 9], one can deduce additional restrictions, if more than two such regions should meet. Briefly, there are many possibility for so tacking together these homogeneous deformations and, to some extent, they depend on the function $\tilde{\phi}$, or at least on the values of its minimizers. What we would like to do is to get a good intuitive picture of all of the possibilities. Presently, I don't know how best to accomplish this, so we'll try to better understand some of the subsets.

In terms of analyzing (2.17), there is no real loss of generality in setting $R_1 = H_1 = 1$, since $F_0$ is simply one of the set of minimizers. Looked at differently, if we set

$$R = R_1^T R_2, \qquad H = H_2 H_1^{-1}$$
$$\bar{A} = H_1 A, \qquad \bar{N} = H_1^{-T} N, \tag{2.18}$$

(2.17) reduces to much the same thing, *viz.*

$$R F_0 H = F_0 (1 + \bar{A} \otimes \bar{N}). \tag{2.19}$$

I will assume that $H$ belongs to the subgroup $\bar{G}^+$ of $\bar{G}$ defined by

$$\det H = 1 \quad \Leftrightarrow \quad \det m = 1, \tag{2.20}$$

there being no real loss of generality in doing this. Then, by taking determinants of both sides of (2.19), we find that

$$\det (1 + \bar{A} \otimes \bar{N}) = 1 \quad \Leftrightarrow \quad \bar{A} \cdot \bar{N} = 0 \tag{2.21}$$

By setting

$$a = F_0 \bar{A}, \qquad n = F_0^{-T} \bar{N} \quad \Rightarrow \quad a \cdot n = 0, \tag{2.22}$$

we can rewrite (2.19) as

$$R F_0 H = (1 + a \otimes n) F_0. \tag{2.23}$$

One can think of picking values of $F_0$ and $H$, and solving the latter for $R$, a and $n$, if possible. As is established by ERICKSEN [10], it is impossible unless

$$\det (J J^T - 1) = 0, \tag{2.24}$$

where

$$J \stackrel{\text{def}}{=} F_0 H^{-1} F_0^{-1}. \tag{2.25}$$

If (2.24) holds, then either

$$J J^T = 1 \quad \Rightarrow \quad a = 0 \tag{2.26}$$

or there are two non-trivial solutions.* Given $(R, a, n)$ for one, one can get corresponding values $(\bar{R}, \bar{a}, \bar{n})$ for the other by using the following prescriptions:

$$2\bar{a} = -\|n\|^2 a + n, \tag{2.27}$$
$$(\|a^2\| \, \|n^2\| + 4)\, \bar{n} = 4a + 2\, \|n\|^2 n. \tag{2.28}$$

* For this purpose, we count solutions obtained by multiplying a by a scalar, and dividing $n$ by it, as the same.

Then, set

$$\bar{R} = \hat{R}R, \tag{2.29}$$

where

$$\hat{R} = (1 + \bar{a} \otimes \bar{n})(1 - a \otimes n). \tag{2.30}$$

It is straightforward to verify that $\hat{R}$ is a rotation, and that the prescriptions deliver the second solution. This does give one characterization of solutions, and a way of constructing numerous examples, but we would like some better way of picturing the possibilities.

At least for intuitive purposes and for making contact with works on crystallography, it can be useful to revert to the description in terms of lattice vectors, eliminating the rather arbitrary reference configuration. Set

$$e_a = F_0 E_a \tag{2.31}$$

and use (2.11) and (2.23) to get the equivalent

$$\begin{aligned} RF_0HE_a &= RF_0m_a^bE_b = Rm_a^be_b \\ &= Se_a, \end{aligned} \tag{2.32}$$

where

$$S = 1 + a \otimes n. \tag{2.33}$$

That is, we can get lattice vectors on the other side of the plane by applying this simple shear to $e_a$, or we can get them by applying $R$ and $m$ as indicated. Some special cases are now well understood. Cases summarized in (2.26) translate to the condition that

$$Rm_a^be_b = e_a. \tag{2.34}$$

This means that $R$ is an element of the crystallographic point group for these lattice vectors. In the terminology used by ERICKSEN [11], $m$ belongs to the corresponding lattice group. From the theory of these groups, we know that they are finite, conjugate and different for different choices of $e_a$. Also, we must have

$$R^N = m^N = 1, \qquad N = 1, 2, 3, 4 \text{ or } 6. \tag{2.35}$$

This covers the trivial solutions, involving no shear. There is another infinite set of solutions for which $R = 1$, so that

$$Se_a = m_a^be_b. \tag{2.36}$$

There are the so-called lattice invariant shears, discussed in detail by ERICKSEN [4], things considered by metallurgists seeking to determine what deformations have taken place in certain phase transformations. For these, the $x$-ray crystallographer sees the lattice vectors as continuous, but one can need to consider them to analyze shapes of crystallites. Here, such $m$'s do not depend on the choice of $e_a$.

The more common forms of twinning involve choices of $m$ such that $m^2 = 1$. Then, as is discussed in detail by PITTERI [12], (2.31) can be solved for any choice of $e_a$ and solutions have the property that

$$m^2 = 1 \quad \Rightarrow \quad R^2 = 1. \tag{2.37}$$

Solutions can be of the form (2.34) or be non-trivial, depending on the choice of $m$ and $e_a$. PITTERI also includes an interesting example showing that

$$R^2 = 1 \quad \not\Rightarrow \quad m^2 = 1, \tag{2.38}$$

pointing out that one will either have $m^2 = 1$ or $m^N$ unbounded as $N \to \infty$, if $R^2 = 1$. Further, PITTERI has shown me a proof that, given any $m \in G$, with $\det m = 1$, we can find choices of $e_a$ such that (2.13) can be satisfied or, equivalently, such that (2.24) holds, for the corresponding arguments.

A characterization of solutions possible with $H$ similar to a rotation is given by ERICKSEN [10], who notes that this is equivalent to the assumption that

$$H^N = 1 \quad \Leftrightarrow \quad m^N = 1, \qquad N = 1, 2, 3, 4 \text{ or } 6. \tag{2.39}$$

The case $N = 1$ is trivial and $N = 2$, covered by (2.36) is well understood, so cases with $N > 2$ are of more interest. At first, I thought that this might be a way of characterizing certain kinds of observed patterns, the "rotation twins" which are described by BARRETT & MASSALSKI [13, p. 406]. It is not, but there is some connection. Later, I will elaborate this. In the words of BARRETT & MASSALSKI, "Crystals are *rotation twins* if a two-, three-, four-, or six-fold rotation of one crystal about a *twinning axis* produces the orientation of the other. The rotation axis lies either in the twinning plane or normal to it and is not a symmetry element of the lattice of the individual crystals."

## 3. Examples

Here, I will present examples with $m^4 = 1$, most of which do not fit the aforementioned description of rotation twins. Pick any set of (linearly) independent) lattice vectors $e_a$ such that

$$\|e_1\| = \|e_2\| = 1, \qquad e_3 \cdot (e_1 + e_2) = 0, \tag{3.1}$$

and let $e^a$ denote the corresponding reciprocal lattice vectors (the dual basis). Let $e$ denote the unit vector given by

$$e = (e_1 + e_2)/\|e_1 + e_2\| = (e_1 + e_2)/\sqrt{2(1 + e_1 \cdot e_2)}. \tag{3.2}$$

One 180° rotation is defined by

$$R_1 = -1 + 2e \otimes e. \tag{3.3}$$

By an elementary analysis, one can verify that, with (3.1) we can also write

$$e = (e^1 + e^2)/\|e^1 + e^2\| = (e^1 + e^2)\,\|e_1 + e_2\|/2, \tag{3.4}$$

thus, (3.3) is equivalent to

$$\begin{aligned} R_1 &= -1 + (e_1 + e_2) \otimes (e^1 + e^2) \\ &= -1 + (e^1 + e^2) \otimes (e_1 + e_2). \end{aligned} \tag{3.5}$$

This means that $R_1$ belongs to the point group for such $e_a$. Another 180° rotation is given by

$$R_2 = -1 + 2e_2 \otimes e_2 . \tag{3.6}$$

Composing these, we obtain the rotation

$$R = R_2 R_1 \tag{3.7}$$

with axis parallel to the vector

$$(e_1 + e_2) \wedge e_2 \parallel e^3 .$$

The angle of rotation $\theta$ can be calculated, using the relation

$$\operatorname{tr} R = 1 + 2\cos\theta , \tag{3.9}$$

and, by using the definition of $R$, we find that

$$\cos\theta = e_1 \cdot e_2 . \tag{3.10}$$

We can choose $e_1$ and $e_2$, consistent with (3.1), to get any desired angle, excepting $\cos\theta = \pm 1$, which would contradict the requirement that the $e_a$ be linearly independent.

Now, define $m \in G$ as follows:

$$\begin{aligned} m_1^a e_a &= -e_2 , \\ m_2^a e_a &= e_1 , \\ m_3^a e_a &= e_3 , \end{aligned} \tag{3.11}$$

it being straightforward to show that

$$m^4 = 1 . \tag{3.12}$$

By elementary calculations, we then find that

$$\begin{aligned} Rm_1^a e_a &= e_1 - 2e_1 \cdot e_2 e_2 , \\ Rm_2^a e_a &= e_2 , \\ Rm_3^a e_a &= e_3 - 2e_1 \cdot e_3 e_2 . \end{aligned} \tag{3.13}$$

We then have

$$Rm_b^a e_a = (1 + a \otimes n)\, e_b , \tag{3.14}$$

with

$$\begin{aligned} a &= 2e_2 , \\ n &= e^2 - e_2 . \end{aligned} \tag{3.15}$$

To have $n \neq 0$, we need

$$e^2 \neq e_2 ,$$

which will hold unless $e_1 \cdot e_2 = e_2 \cdot e_3 = e_1 \cdot e_3 = 0$, that is, unless the vectors are all orthogonal. If they are all orthogonal, (3.13) reduces to the form (2.34), $R$ then being in the associated point group, the solutions becoming trivial. For one thing, the examples show that the analog of (2.37) does not hold, *i.e.*

$$m^4 = 1 \quad \not\Rightarrow \quad R^4 = 1\,. \tag{3.16}$$

From (3.10), we can have $R^4 = 1$, a 90° rotation, by assuming that $e_1 \cdot e_2 = 0 \neq e_1 \cdot e_3$. Then $n$ is orthogonal to $e_1$ and $e_2$, hence parallel to $e^3$ which, from (3.8), is the axis of rotation. This particular case conforms to the description of rotation twins given by BARRETT & MASSALSKI [13, p. 406]. Another special case satisfies their description in part. By choosing $e_1$ and $e_2$ so that $2e_1 \cdot e_2 = -1$, we see from (3.10) that $R$ will satisfy $R^3 = 1$. Their description also dictates that the axis of rotation, here $e^3$, should be either parallel or perpendicular to $n$. Checking this, one finds that it is not parallel, and is perpendicular only if $e_1 \cdot e_3 = e_2 \cdot e_3 = 0$. Of course, one can pick $e_a$ not satisfying these conditions. A somewhat similar example is given by PITTERI [12], with $R^2 = 1$, with the axis of $R$ neither parallel nor perpendicular to $n$. As far as I know, nothing excludes the possibility that any lattice vectors might minimize some potential function $\tilde{\phi}$, and occur in nature.

## 4. Some Constructions

Here, we deduce some fairly general procedures for constructing examples of solutions fitting the aforementioned description of rotation twins. First, consider solutions of (2.32) with the special property that

$$Ra = a \neq 0, \qquad R \neq 1, \tag{4.1}$$

so that $R$ has a unique axis parallel to $a$, hence perpendicular to $n$. Then, (2.32) is equivalent to

$$\begin{aligned} M &\stackrel{\text{def}}{=} m_a^b e_b \otimes e^a \\ &= R^T S \\ &= R^T + a \otimes n\,. \end{aligned} \tag{4.2}$$

Further, since $n$ is perpendicular to $a$, the axis of $R$, there is a vector $b$ such that

$$n = (R - 1)\, b \neq 0, \qquad b \cdot a = 0\,. \tag{4.3}$$

Let

$$\bar{S} = 1 + a \otimes b, \tag{4.4}$$

and we find that

$$\begin{aligned} \bar{S} R^T \bar{S}^{-1} &= R^T + a \otimes Rb - a \otimes b \\ &= R^T + a \otimes n \\ &= M, \end{aligned} \tag{4.5}$$

so $M$ is similar to $R^T$. This implies that

$$M^N = R^N = 1, \qquad N = 2, 3, 4 \text{ or } 6. \tag{4.6}$$

Briefly, taking traces gives

$$\begin{aligned} \operatorname{tr} M &= m_a^a = \text{integer} \\ &= \operatorname{tr} R^T = 1 + 2\cos\theta, \end{aligned} \tag{4.7}$$

where $\theta$ is the angle of rotation. Insert the possible integers, bearing in mind that $M$ is similar to $R^T$, and you get (4.6), which is well known to crystallographers. It can be helpful to note that any rotation $R$ is similar to its transpose: it is easy to verify that

$$R^T = \tilde{R}R\tilde{R}^T, \tag{4.8}$$

where $\tilde{R}$ is a 180° rotation, with axis perpendicular to that of $R$. Thus, $M$ is also similar to $R$. With (4.5) and (4.2), we have

$$Me_a = m_a^b e_b = \bar{S}R^T\bar{S}^{-1}e_a. \tag{4.9}$$

Let

$$\bar{e}_a = \bar{S}^{-1}e_a, \qquad e_a = \bar{S}e_a, \tag{4.10}$$

and we have, as the equivalent of (4.9),

$$Rm_a^b\bar{e}_b = \bar{e}_a. \tag{4.11}$$

Referring to (2.34), this means that $m$ is in the lattice group for the lattice vectors $\bar{e}_a$, $R$ being the corresponding element of the associated point group, also similar to $m$. To construct examples, we can pick any $m \in G$ such that

$$m^N = 1, \qquad N = 2, 3, 4 \text{ or } 6. \tag{4.12}$$

As is discussed by Ericksen [11], this guarantees that there will exist vectors $\bar{e}_a$ and a rotation $R$ satisfying (4.11), with $R^N = 1$. It is in fact fairly easy to determine them, given $m$. Then pick $a \neq 0$ parallel to the axis of $R$, $b \neq 0$ any vector perpendicular to it. With $\bar{S}$ then defined be (4.4), we can use (4.10) to determine $e_a$, satisfying (4.9). Use (4.3) to define $n$ which, by the construction, will be perpendicular to $a$. Routine calculations then show that we have a solution of (2.32), satisfying (4.1).

With solutions of this kind, we can construct configurations consisting of symmetry-related wedges, fit together like the sections of an orange, special cases of the topological possibilities discussed by James [8, 9]. Later, I will discuss this a bit more.

Much the same kind of analysis applies to solutions of (2.32) such that

$$Rn = n, \qquad R \neq 1, \tag{4.13}$$

which also fit the description of rotation twins. Again define $M$ as above. Instead of (4.2), we use

$$M^{-T} = R^T(1 - n \otimes a) = R^T - n \otimes a. \tag{4.14}$$

Here, we set

$$a = (1 - R)\,c, \qquad c \cdot n = 0, \tag{4.15}$$

and similarly conclude that

$$M^{-T} = \bar{\bar{S}} R^T \bar{\bar{S}}^{-1}, \tag{4.16}$$

with

$$\bar{\bar{S}} = 1 + n \otimes c, \tag{4.17}$$

or, equivalently,

$$M = \bar{\bar{S}}^{-T} R^T \bar{\bar{S}}^T. \tag{4.18}$$

Then, setting

$$\bar{\bar{e}}_a = \bar{\bar{S}}^T e_a, \tag{4.19}$$

we obtain, as the analog of (4.11),

$$R m_a^b \bar{\bar{e}}_b = \bar{\bar{e}}_a. \tag{4.20}$$

In a similar way, this can be read backwards to construct examples, and $\bar{\bar{S}}^T$ is also interpretable as a simple shear. These solutions cannot be used to generate the "orange section" configurations referred to above, but do fit the description of rotation twins.

Bearing in mind (4.8), we note that, for all of these examples, $M$ is similar to $R$. Suppose, conversely, that $M$ is similar to $R$ and

$$M = R^T(1 + a \otimes n). \tag{4.21}$$

In particular, we must have

$$\operatorname{tr} M = \operatorname{tr} R^T = \operatorname{tr} R, \tag{4.22}$$

implying that

$$R^T a \cdot n = 0. \tag{4.23}$$

Also $M^{-T}$ must be similar to $R$, and

$$M^{-T} = R^T(1 - n \otimes a), \tag{4.24}$$

similarly implying that

$$R^T n \cdot a = Ra \cdot n = 0. \tag{4.25}$$

Of course, we must also have $a \cdot n = 0$.

By elementary reasoning, left to the reader, one finds that the three equations can hold only if at least one of the following four conditions holds:

$$Ra = \pm a, \qquad Rn = \pm n. \tag{4.26}$$

If $Ra = -a$, $R$ is a 180° rotation, so $R^2 = 1$. With $M$ similar, we also have $M^2 = 1$. These are then standard twinning solutions, referred to in (2.37). Similar remarks apply to the possibility that $Rn = -n$. For all such twinning solutions, it is known that either $Ra = a$ and $Rn = -n$ or $Rn = n$ and $Ra = -a$. Thus, when $M$ is similar to $R$, (4.26) reduces to the statement that either $Ra = a$ or $Rn = n$, the possibilities analyzed above.

Thus, we have a rather nice characterization of a subset of the rotation twins, at least. There might or might not be other solutions fitting their description. I have not made a careful study of this question, but do not known of any examples.

## 5. Composition

Suppose that we have two solutions of (2.32) for the same choice of $e_a$, say

$$Se_a = (1 + a \otimes n)\, e_a = Rm_a^b e_b \tag{5.1}$$

and

$$\bar{S}e_a = (1 + \bar{a} \otimes \bar{n})\, e_a = \bar{R}\bar{m}_a^b e_b. \tag{5.2}$$

One way of composing these is obtained by noting that

$$\begin{aligned} Rm_a^b(\bar{S}e_b) &= R\bar{S}R^T(Rm_a^b e_b) \\ &= R\bar{S}R^T(Se_a) \\ &= R\bar{R}m_a^b\bar{m}_b^c e_c \\ &= R\bar{R}m_a^b\bar{m}_b^c(R^T S(m^{-1})_c^d\, e_d). \end{aligned} \tag{5.3}$$

In particular, we then have

$$\tilde{S}(Se_a) = \tilde{R}\tilde{m}_a^b(Se_b). \tag{5.4}$$

with

$$\begin{aligned} \tilde{S} &= R\bar{S}R^T = 1 + (R\bar{a}) \otimes (R\bar{n}), \\ \tilde{R} &= R\bar{R}R^T, \\ \tilde{m} &= m\bar{m}m^{-1} \in G. \end{aligned} \tag{5.5}$$

So, (5.1) links $e_a$ by an admissible discontinuity to lattice vectors $Se_a$. With (5.4) we link the latter by a similar discontinuity to lattice vectors $\tilde{S}(Se_a)$. One can similarly compose to link the latter to a fourth set, *etc.* This is a first step toward building configurations involving several regions tacked together. It needs to be supplemented by considerations such as are discussed by JAMES [8, 9], relevant when the various planes of discontinuity intersect.

In particular, nothing prevents us from taking the two solutions indicated by (5.1) and (5.2) to be the same, iterating this as many times as we desire. This gives a sequence of the form

$$\begin{aligned} Se_a &= Rm_a^b e_b \\ RSR^T(Se_a) &= Rm_a^b(Se_b) \\ R^2S(R^T)^2\,(RSR^TS)\, e_a &= Rm_a^b(RSR^TS)\, e_a, \end{aligned}$$

the N[th] step giving

$$R^N S(R^T)^N e_a^{N-1} = R m_a^b e_b^{N-1}$$
$$= R^{N+1}(m^{N+1})_a^b e_b^{N-1}, \tag{5.6}$$

with

$$e_a^{N-1} = \prod_{M=0}^{N-1} R^M S(R^T)^M e_a. \tag{5.7}$$

Here, $(5.6)_2$ follows immediately from $(5.3)_3$ for $N=1$ and is easily established by induction for $N>1$.

Sometimes, after a finite number of steps, this iteration will return the lattice vectors to their original values, the condition for this being that, for some $N = V < \infty$,

$$R^V (M^V)_a^b e_b = e_a. \tag{5.8}$$

One can continue the iteration, reproducing the collection $e_a^{N-1}$ in a periodic fashion, so I call these *periodic solutions*.

Recalling (2.33), we see that (5.8) implies that $m^V$ is in the lattice group for $e_a$, with $R^V$ in the corresponding point group. These groups are finite and conjugate, so we have

$$R^V \text{ similar to } m^V \text{ similar to } M^V, \tag{5.9}$$

with $M$ defined as in (4.2). That these groups are finite also implies that there are smallest integers $P$ and $Q$ such that

$$R^P = 1, \quad m^Q = M^Q = 1. \tag{5.10}$$

Also, it implies that there exists a rotation $\hat{R}$ such that

$$M \text{ is similar to } \hat{R} \quad \Rightarrow \quad \hat{R}^Q = 1. \tag{5.11}$$

First, suppose that $P=Q$, in which case we can take $R=\hat{R}$. giving us the solutions discussed in § 4. For example, we might pick one with $Ra=a$, $P=Q=3$. The iteration generates the three directions, $n$, $Rn$ and $R^2 n$, making 120° angles with each other. Clearly, we can construct three half planes whose normals have these directions, their edges meeting on a line with direction $a$. In the wedges thus formed, we can assign lattice vectors $e_a$, $Rm_a^b e_b$ and $R^2(m^2)_a^b e_b$, and have all the desired jump conditions satisfied on the planes of discontinuity. Clearly, (5.8) holds, with $M=3$. Such triplets are called by some "trills". Clearly, one can construct similar solutions with four or six wedges, resembling the structure seen in oranges. If one uses the solutions with $Rn=n$, one gets parallel planes, so one can't form such wedges, but one can put together any number of strips, with the lattice vectors changing in a periodic fashion, as one traverses them. The width of the strips can be assigned arbitrarily. Such patterns of plane defects often occur in parts of real crystals. Frequently, these are the standard twins, described by (2.37) but, clearly, they need not be. From casual observation, one might be misled into thinking that they are, when they are not.

The aforementioned description of rotation twins does serve to warn us against making this mistake, to look closely at the crystal structure.

They are the remaining possibilities, with $P \neq Q$. In § 3, we encountered examples with $P = 3, Q = 4$, enough to indicate that we are not discussing the null set. From (5.8)–(5.11) we must have

$$\operatorname{tr} R = 1 + 2\cos(2\pi/P),$$
$$\operatorname{tr} M = \operatorname{tr} \hat{R} = 1 + 2\cos(2\pi/Q) = \text{integer}. \tag{5.12}$$

and

$$\cos(2V\pi/P) = \cos(2V\pi/Q). \tag{5.13}$$

If $P = 1$, the only solutions of (5.1) are the lattice invariant shears noted in (2.36), and none of these satisfy (5.10). Similarly, $Q = 1$ yields no nontrivial possibilities. As is discussed by PITTERI [12], possibilities consistent with (5.10) are such that $P = 2 \Rightarrow Q = 2$ and vice versa. The remaining possible values of $P$ and $Q$ are 3, 4 and 6. From (5.13), we must have

$$V(1/P - 1/Q) = \text{integer}, \tag{5.14}$$

so it is straightforward to find the smallest possible values of $V$. With (2.18), we easily find that

$$\begin{aligned} (P, Q) &= (3, 4) \text{ or } (4, 3) &\Rightarrow\quad V &= 12, \\ (P, Q) &= (3, 6) \text{ or } (6, 3) &\Rightarrow\quad V &= 6, \\ (P, Q) &= (4, 6) \text{ or } (6, 4) &\Rightarrow\quad V &= 12. \end{aligned} \tag{5.15}$$

In all these cases, (5.8) is satisfied in the trivial way, by having $R^V = m^V = 1$. Since all solutions with $m^Q = 1$ are fairly well characterized by ERICKSEN [10], one might better characterize the (proper) subset satisfying (5.10), with $P \neq Q$, and determine the kinds of patterns which they can generate. Here, I'll not pursue this.

These rather naive and elementary analyses give us snapshots of what some of our minimizers can look like, a means of constructing certain kinds of examples, but not enough to give us a good picture of the whole set. To some degree, metallurgy is the art of controlling these and more complex morphologies, albeit imperfectly. Clearly, we have a long way to go, to develop mathematical theory to apply to such problems.

*Acknowledgment.* This material is based on work supported by the National Science Foundation under Grant No. MEA-8304750. Also, I thank Dr. MARIO PITTERI for helpful comments.

## References

1. PITTERI, M., On $\nu + 1$ lattices, *J. Elasticity* **15**, 3–25 (1985).
2. PITTERI, M., On crystallographic space groups and generalized lattice groups, pending publication.

3. James, R. D., The stability and metastability of quartz, to appear in *Proc. Workshop on Metastability and Partial Differential Equations*, Minneapolis, 1985, to be published by Springer-Verlag.
4. Ericksen, J. L., "The Cauchy and Born Hypotheses for Crystals", in *Phase Transformations and Material Instabilities in Solids* (ed. M. E. Gurtin), New York: Academic Press, 1984.
5. Eftis, J., MacDonald, D. E. & Arkilic, G. M., Theoretical calculations on the pressure variation of second-order elastic coefficients for alkali metals, *Mater. Sci. Eng.* **7**, 141–150 (1971).
6. Ericksen, J. L., "Special Topics in Nonlinear Elastostatics", in Advances in Applied Mechanics (ed. C.-S. Yih) vol. 17, New York: Academic Press, 1977.
7. Fonseca, I. M. Q. C., *Variational Methods for Elastic Crystals*, Ph.D. Thesis, Univ. Minnesota, 1985.
8. James, R. D., *Mechanics of coherent phase transformations in solids*, MRL Report, Brown University, Division of Engineering, October, 1982.
9. James, R. D., Stress-free joints and polycrystals, *Arch. Rational Mech. Anal.* **86**, 13–37 (1984).
10. Ericksen, J. L., Some surface defects in unstressed thermoelastic solids, *Arch. Rational Mech. Anal.* **88**, 337–345 (1985).
11. Ericksen, J. L., On the symmetry of deformable crystals, *Arch. Rational Mech. Anal.* **72**, 1–13 (1979).
12. Pitteri, M., On type II twins, to appear in *Int. J. Plasticity*.
13. Barrett, C., & Massalski, T. B., *Structure of Metals*, 3rd ed. McGraw Hill, Inc. 1966.

Department of Aerospace
Engineering and Mechanics
and School of Mathematics
University of Minnesota
Minneapolis

*(Received August 19, 1985)*

# *Inequalities between Dirichlet and Neumann Eigenvalues*

HOWARD A. LEVINE & HANS F. WEINBERGER

*To James Serrin on his sixtieth birthday*

## 1. Introduction

The purpose of this paper is to derive some inequalities of the form

$$\mu_{k+R} < \lambda_k \quad \text{for } k = 1, 2, \ldots \tag{1.1}$$

between the eigenvalues $\lambda_1 < \lambda_2 \leqq \ldots$ of the Dirichlet problem

$$\begin{aligned} \Delta u + \lambda u &= 0 \quad \text{in } D \subset R^N, \\ u &= 0 \quad \text{on } \partial D \end{aligned} \tag{1.2}$$

and the eigenvalues $0 = \mu_1 < \mu_2 < \ldots$ of the Neumann problem

$$\begin{aligned} \Delta v + \mu v &= 0 \quad \text{in } D, \\ \frac{\partial v}{\partial \nu} &= 0 \quad \text{on } \partial D \end{aligned} \tag{1.3}$$

for some classes of $N$-dimensional domains $D$. Here $\partial/\partial\nu$ denotes the outward normal derivative.

It is an immediate consequence of the variational formulation of the eigenvalue problems that

$$\mu_k \leqq \lambda_k \text{ for } k = 1, 2, \ldots$$

PÓLYA [6] proved that

$$\mu_2 < \lambda_1 .$$

PAYNE [5] showed that when $N = 2$ and $D$ is a convex domain whose boundary $\partial D$ is twice continuously differentiable,

$$\mu_{k+2} < \lambda_k \quad \text{for } k = 1, 2, \ldots$$

Convexity in two dimensions is characterized by the fact that the curvature, which is a scalar defined on the boundary $\partial D$, is nonnegative. In higher dimensions there are $N - 1$ principal curvatures at each point of $\partial D$, and hence there are a number of possible generalizations of PAYNE's result.

In Section 2 we shall derive a generalization of the form (1.1) for each integer $R$ on the interval $[1, N]$ under conditions which depend upon the principal curvatures of $\partial D$. The most important of these results is the fact that the inequality

$$\mu_{k+N} < \lambda_k \quad \text{for } k = 1, 2, \ldots$$

is valid for all smooth bounded convex domains.

We also obtain the inequality

$$\mu_{k+1} < \lambda_k \quad \text{for } k = 1, 2, \ldots$$

when $\partial D$ has nonnegative mean curvature, which is a recent result of AVILES [1].

In Section 3 we show that the conditions on the principal curvatures can be stated in terms of their symmetric functions, which are the coefficients of the characteristic polynomial of the curvature matrix. A simpler but more restrictive set of such conditions is also presented.

The above results are derived under the hypothesis that the boundary $\partial D$ has Hölder continuous second derivatives. In Section 4 we establish a semicontinuity result for the Neumann eigenvalues, which allows us to obtain the nonstrict version of the inequality (1.1) for some classes of domains whose boundary is not smooth.

In particular, we shall establish the inequality

$$\mu_{k+N} \leqq \lambda_k \quad \text{for } k = 1, 2, \ldots$$

for all bounded convex domains.

## 2. Inequalities for smooth domains

If $D$ is a bounded domain in $R^N$ with smooth boundary, the curvature matrix of a point $P$ of the boundary $\partial D$ is defined as follows: Let $\nu(x)$ be a continuously differentiable vector field defined in a neighborhood of $P$ with the property that when $x$ lies on $\partial D$, $\nu$ is an outward unit normal vector to $\partial D$ at $x$. The matrix

$$K_{ij} = \nu_{i,j} - \sum_{k=1}^{N} \nu_{i,k}\nu_k\nu_j \tag{2.1}$$

evaluated at $x$ in called the *curvature matrix* of of $\partial D$ at $x$. It has the eigenvalue 0 with the eigenvector $\nu$. The $N-1$ other eigenvalues of $K_{ij}$ are called the *principal curvatures*

$$\varkappa_1 \leqq \varkappa_2 \leqq \ldots \leqq \varkappa_{N-1}$$

of $\partial D$ at $x$. (The symbol $_{,j}$ denotes differentiation with respect to $x_j$.)

It is easy to see that the matrix $K_{ij}$ is symmetric and that the $ij$ entry is a tangential derivative of $\nu_i$.

Our results are based on the following proposition:

**Proposition 2.1.** *Let $D$ be a bounded domain whose boundary $\partial D$ is of class $C^{2,\alpha}$ for some $\alpha \in (0, 1)$. If for some integer $R \in [1, N]$ the principal curvatures* *of* $\partial D$ *at each of its points have the property that the sum of any* $N-R+1$ *of the set of* $N$ *numbers*

$$\left\{\varkappa_1, \varkappa_2, \dots, \varkappa_{N-1}, \sum_{j=1}^{N-1} \varkappa_j\right\} \tag{2.2}$$

*is nonnegative, then*

$$\mu_{k+R} < \lambda_k \quad \text{for } k = 1, 2, \dots \tag{2.3}$$

**Proof.** We recall that [3, p. 399] if $\phi$ is continuously differentiable and

$$\int_D \phi v_i \, dx = 0 \quad \text{for } i = 1, 2, \dots, k+R-1, \tag{2.4}$$

where $v_1, v_2, \dots,$ is an orthonormal set of eigenfunctions of (1.3) corresponding to the Neumann eigenvalues $0 = \mu_1 < \mu_2 \leqq \dots,$ then

$$\int_D |\nabla\phi|^2 \, dx \geqq \mu_{k+R} \int_D \phi^2 \, dx. \tag{2.5}$$

We consider the set of functions

$$\phi = \sum_{i=1}^{k} a_i u_i + \sum_{p=1}^{N} b_p u_{k,p} \tag{2.6}$$

where $\{u_1, u_2, \dots\}$ is an orthonormal set of eigenfunctions of (1.2) corresponding to the Dirichlet eigenvalues $0 < \lambda_1 < \lambda_2 \leqq \dots$ Because the $u_i$ are orthonormal and satisfy (1.2), integration by parts shows that

$$\int_D |\nabla\phi|^2 \, dx - \lambda_k \int_D \phi^2 \, dx = \sum_{i=1}^{k} a_i^2(\lambda_i - \lambda_k) + \oint_{\partial D} w \frac{\partial w}{\partial \nu} \, dS, \tag{2.7}$$

where we have defined

$$w = \sum_{p=1}^{N} b_p u_{k,p}.$$

The sum on the right of (2.7) is clearly nonpositive. In order to see whether the integral is also nonpositive, we note that

$$\begin{aligned} \frac{\partial w}{\partial \nu} &= \sum_{q,r=1}^{N} b_q u_{k,qr} \nu_r \\ &= \Sigma\, b_q \left(\nu_r \frac{\partial}{\partial x_q} - \nu_q \frac{\partial}{\partial x_r}\right) u_{k,r} - \lambda_k u_k \, \Sigma\, b_q \nu_q. \end{aligned} \tag{2.8}$$

Since $u_k$ vanishes on the boundary, the last term is zero and

$$u_{k,r} = \nu_r \frac{\partial u_k}{\partial \nu} \quad \text{for } l = 1, \dots, N. \tag{2.9}$$

Because the directional derivatives which appear in (2.8) are in directions tangent to $\partial D$, we can apply them to both sides of (2.9). Then (2.8) yields

$$w\frac{\partial w}{\partial \nu} = \sum_{p,q,r=1}^{N} b_p \nu_p \frac{\partial u_k}{\partial \nu} b_q \left(\nu_r \frac{\partial}{\partial x_q} - \nu_q \frac{\partial}{\partial x_r}\right)\left(\nu_r \frac{\partial u_k}{\partial \nu}\right)$$

or

$$\begin{aligned} w\frac{\partial w}{\partial \nu} &= \tfrac{1}{2}\Sigma\, b_p b_q \left(\nu_r \frac{\partial}{\partial x_q} - \nu_q \frac{\partial}{\partial x_r}\right)(\nu_p \nu_r\, |\nabla u_k|^2) \\ &\quad - \tfrac{1}{2}\Sigma\, b_p b_q\, |\nabla u_k|^2 \left[\nu_r \left(\nu_r \frac{\partial}{\partial x_q} - \nu_q \frac{\partial}{\partial x_r}\right)\nu_p \right. \\ &\quad \left. - \nu_p \left(\nu_r \frac{\partial}{\partial x_q} - \nu_q \frac{\partial}{\partial x_r}\right)\nu_r\right]. \end{aligned} \tag{2.10}$$

We see from the definition (2.1) of the curvature matrix $K_{pq}$ that since $|\nu| = 1$ on $\partial D$,

$$\sum_{r=1}^{N}\left[\nu_r \left(\nu_r \frac{\partial}{\partial x_q} - \nu_q \frac{\partial}{\partial x_r}\right)\nu_p - \nu_p \left(\nu_r \frac{\partial}{\partial x_q} - \nu_q \frac{\partial}{\partial x_r}\right)\nu_r\right] = K_{pq} + \Sigma\, K_{rr}\nu_p \nu_q .$$

The divergence theorem shows that the integral over $\partial D$ of the first sum in (2.10) is zero. Thus

$$\int_{\partial D} w \frac{\partial w}{\partial \nu}\, dS = -\tfrac{1}{2}\sum_{p,q=1}^{N} b_p b_q \int_{\partial D} J_{pq}(x)\, dS \tag{2.11}$$

where

$$J_{pq} = \left(\frac{\partial u_k}{\partial \nu}\right)^2 [K_{pq} + (\Sigma\, K_{rr})\,\nu_p \nu_q]. \tag{2.12}$$

We recall that the vector $\nu$ lies in the null space of the curvature matrix. It follows that $J_{pq}$ has the eigenvalues

$$\left\{\left(\frac{\partial u_k}{\partial \nu}\right)^2 \varkappa_1, \ldots, \left(\frac{\partial u_k}{\partial \nu}\right)^2 \varkappa_{N-1}, \left(\frac{\partial u_k}{\partial \nu}\right)^2 \sum_{r=1}^{N-1} \varkappa_r\right\}. \tag{2.13}$$

We rotate the $x$-coordinates so that the matrix

$$\int_{\partial D} J_{pq}\, dS \tag{2.14}$$

on the right of (2.8) is diagonal, with its diagonal elements arranged in non-decreasing order. By the first montonicity principle [8, p. 58] the ordered eigenvalues of the $(N - R + 1)\times(N - R + 1)$ matrix $\bar{J}(x)$ defined by

$$\bar{J}_{\alpha\beta}(x) = J_{\alpha\beta}(x) \quad \text{for } \alpha, \beta = 1, \ldots, N - R + 1$$

are upper bounds for the $N - R + 1$ lowest eigenvalues of the matrix $J$. We add these inequalities and use the hypothesis that the sum of any $N - R + 1$ of the numbers (2.2) is nonnegative to see that

$$\operatorname{tr}(\bar{J}) = \sum_{p=1}^{N-R+1} J_{pp}(x) \geqq 0 . \tag{2.15}$$

Since this is true at each boundary point, we have

$$\sum_{p=1}^{N-R+1} \oint_{\partial D} J_{pp}\, dS \geqq 0. \tag{2.16}$$

To show that this inequality is strict, we apply the Gauss-Bonnet formula

$$\oint_{\partial D^*} \varkappa\, dS = \omega_{N-1}$$

to the convex hull $D^*$ of $D$. Here

$$\varkappa = \varkappa_1 \dots \varkappa_{N-1}$$

is the Gaussian curvature of $\partial D^*$.

It is easily seen that $\varkappa = 0$ on $\partial D^* \setminus \partial D$. Since the area $\omega_{N-1}$ of the unit sphere is positive, there must be an open subset $S$ of $\partial D^* \cap \partial D$ on which $\varkappa > 0$. Since $D^*$ is convex, all the $\varkappa_j$ must be positive on $S$. We therefore see from (2.12) and (2.16) that the inequality (2.16) is strict unless $\partial u_k/\partial \nu \equiv 0$ on $S$. In this case, the Green's function representation shows that the solution of $\Delta u + \lambda_k u = 0$ in a sufficiently small ball $B$ centered at a point of $S$ with boundary values $u_k$ in $D \cap \partial B$ and zero outside $D$ coincides with $u_k$ in $B \cap D$ and vanishes outside $D$. Because solutions of $\Delta u + \lambda_k u = 0$ are analytic, this would imply that $u_k \equiv 0$ in $D$. Since this would contradict the definition of an eigenfunction, we conclude that equality cannot hold in (2.16). That is,

$$\sum_{p=1}^{N-R+1} \oint_{\partial D} J_{pp}\, dS > 0.$$

This clearly implies that the largest of the terms in this sum if positive. Because the elements of the diagonal matrix (2.14) appear in nondecreasing order, we conclude that

$$\int_{\partial D} J_{pp}\, dS > 0 \quad \text{for } p = N-R+1, N-R+2, \dots, N. \tag{2.17}$$

We now choose the $k+N$ numbers $a_i$ and $b_p$ so that they are not all zero and that they satisfy the $k+R-1$ conditions (2.4) and the $N-R$ conditions $b_1 = \dots = b_{N-R} = 0$. We then see from (2.11) and (2.17) that the integral on the right of (2.7) is nonpositive. The sum on the right is also nonpositive. Thus we have

$$\int_D |\nabla\phi|^2\, dx \leqq \lambda_k \int_D \phi^2\, dx. \tag{2.18}$$

If $\phi \equiv 0$, equality holds, so that the integral in (2.7) must vanish, which, in turn, implies that $b = 0$. Since the $u_i$ are orthonormal, $\phi = \Sigma\, a_i u_i \equiv 0$ would imply that $a$ as well as $b$ is zero, contrary to our construction. Therefore $\phi \not\equiv 0$, and (2.5) and (2.18) imply that

$$\mu_{k+R} \leqq \lambda_k. \tag{2.19}$$

Equality holds if and only if both (2.5) and (2.16) are equalities. Equality in (2.5) is valid only if $\phi$ is an eigenfunction of the Neumann problem, so that $\Delta\phi + \mu_{k+R}\phi = 0$ in $D$ and $\partial\phi/\partial\nu = 0$ on $\partial D$. As we have already seen, equality in (2.18) implies that $b = 0$ so that $\phi = 0$ on $\partial D$. As we showed in the derivation of (2.17), these conditions imply that $\phi \equiv 0$, which contradicts what we proved above. Therefore equality cannot hold in (2.18), and the statement (2.3) of the Proposition is proved.

There are two important cases in which the conditions of Proposition 2.1 are easily verified.

**Theorem 2.1.** *Let $D$ be a convex domain in $R^N$ whose boundary has Hölder-continuous second derivatives. Then the Neumann and Dirichlet eigenvalues satisfy the inequalities*

$$\mu_{k+N} < \lambda_k \quad \text{for } k = 1, 2, \ldots \tag{2.20}$$

**Proof.** Since $D$ is convex, all the principal curvatures $\varkappa_j$ are nonnegative at each boundary point. Hence each of the numbers in the set (2.2) is nonnegative, so that the hypotheses of Proposition 2.1 hold with $R = N$. Thus Proposition 2.1 implies (2.20).

The sum of the $N$ numbers in (2.2) is

$$2 \sum_{j=1}^{N-1} \varkappa_j = 2(N-1)\, H,$$

where $H$ is called the mean curvature. Hence for $R = 1$ Proposition 2.1 yields a result which is contained in a recent paper of AVILES [1].

**Theorem 2.2** (AVILES). *Let the boundary $\partial D$ of the domain $D$ be of class $C^{2,\alpha}$ and let its mean curvature be nonnegative at all its points. Then*

$$\mu_{k+1} < \lambda_k \quad \text{for } k = 1, 2, \ldots \tag{2.21}$$

The conditions of Proposition 2.1 can also be simplified for other values of $T$.

**Theorem 2.3.** *If the inequalities*

$$2 \sum_{i=1}^{N-1} \varkappa_i - \varkappa_j \geqq 0 \quad \text{for } j = 1, \ldots, N-1 \tag{2.22}$$

*are satisfied and $\partial D \in C^{2,\alpha}$, then*

$$\mu_{k+2} < \lambda_k \quad \text{for } k = 1, 2, \ldots \tag{2.23}$$

**Proof.** By adding the inequalities (2.22) we find that

$$\sum_{i=1}^{N-1} \varkappa_i \geqq 0. \tag{2.24}$$

The left-hand side of the latter inequality is the sum of the first $N-1$ elements of the set (2.2). The left-hand side of (2.22) is the sum of all the elements of (2.2) except for $\varkappa_j$. Thus (2.22) implies the hypotheses of Proposition 2.1 with $R = 2$, and the Theorem is proved.

**Theorem 2.4.** *If* $R > 2$, *if every sum of* $N - R + 1$ *members of the set*

$$\{\varkappa_1, \varkappa_2, \ldots, \varkappa_{N-1}\} \tag{2.25}$$

*is nonnegative, and if* $\partial D \in C^{2,\alpha}$, *then*

$$\mu_{k+R} < \lambda_k \quad \text{for } k = 1, 2, \ldots \tag{2.26}$$

**Proof.** Since $\varkappa_1 \leqq \varkappa_2 \leqq \ldots \leqq \varkappa_{N-1}$, the above condition is equivalent to

$$\sum_{i=1}^{N-R+1} \varkappa_i \geqq 0. \tag{2.27}$$

At least one of the terms in the sum must be nonnegative, so that

$$0 \leqq \varkappa_{N-R+1} \leqq \varkappa_{N-R+2} \leqq \cdots \leqq \varkappa_{N-1}.$$

Therefore

$$\sum_{i=1}^{N-1} \varkappa_i \geqq \varkappa_{N-1}$$

and the set (2.2) is ordered. Thus the hypotheses of Proposition 2.1 follow from (2.27). This proves the Theorem.

*Remark.* It is easily seen that each of the Theorems 2.1 through 2.4 implies Proposition 2.1 for the corresponding values of $R$.

## 3. Conditions involving curvature invariants

The eigenvalues of the curvature matrix $K_{ij}$ are the zeros of the characteristic polynomial

$$\det(\lambda\, \delta_{ij} - K_{ij}) = \sum_{k=0}^{N-1} (-1)^k\, S_k\,(\varkappa_1, \ldots, \varkappa_{N-1})\, \lambda^{N-k}. \tag{3.1}$$

$S_k(\varkappa_1, \ldots, \varkappa_{N-1})$ is the $k^{\text{th}}$ elementary symmetric function, which is defined as the sum of all products of $k$ distinct elements of the set $\{\varkappa_1, \ldots, \varkappa_{N-1}\}$. By definition $S_0(\varkappa_1, \ldots, \varkappa_{n-1}) = 1$.

At least in principle, these symmetric functions can be found from the equation 3.1 without computing the eigenvalues $\varkappa_j$. The hypotheses of Theorem 2.1 are easily seen to be equivalent to

$$S_l(\varkappa_1, \ldots, \varkappa_{N-1}) \geqq 0 \quad \text{for } l = 1, \ldots, N-1. \tag{3.2}$$

The condition of Theorem 2.2 is, of course, just

$$S_1(\varkappa_1, \ldots, \varkappa_{N-1}) \geqq 0. \tag{3.3}$$

We may restate Theorem 2.3 in a similar fashion.

**Theorem 3.1.** *Let* $\partial D \in C^{2,\alpha}$. *If* (3.2) *is valid then*

$$\mu_{k+N} < \lambda_k \quad \text{for } k = 1, 2, \ldots \tag{3.4}$$

*If* (3.3) *is valid, then*

$$\mu_{k+1} < \lambda_k \quad \text{for } k = 1, 2, \ldots \tag{3.5}$$

*If*

$$\sum_{l=0}^{N-1-j} (-1)^l \binom{N-1-l}{j} S_l(\varkappa_1, \ldots, \varkappa_{N-1}) \, [2S_1(\varkappa_1, \ldots, \varkappa_{N-1})]^{N-1-j-l} \geqq 0, \tag{3.6}$$
$$\text{for } j = 0, \ldots, N-2,$$

*then*

$$\mu_{k+2} < \lambda_k \quad \text{for } k = 1, 2, \ldots$$

**Proof.** We have already shown that (3.2) and (3.3) imply (3.4) and (3.5), respectively.

The conditions of Theorem 2.3 are that the $(N-1)\times(N-1)$ diagonal matrix whose eigenvalues are $2S_1(\varkappa, \ldots, \varkappa_{N-1}) - \varkappa_j$ has nonnegative eigenvalues. This is again equivalent to the condition that the coefficients of its characteristic polynomial have alternating signs. It is easily seen that this characteristic polynomial can be obtained from the characteristic polynomial of the $(N-1)\times(N-1)$ matrix with eigenvalues $\varkappa_j$ by replacing the variable $\lambda$ by $2S_1(\varkappa_1, \ldots, \varkappa_{N-1}) - \lambda$. The characteristic polynomial of the latter matrix is just (3.1) divided by $\lambda$. In this way we obtain the conditions (3.6) and the Theorem is proved.

For other values of $R$ we can use similar reasoning. The conditions of Theorem 2.4 are equivalent to the nonnegativity of the symmetric functions of the $\binom{N-1}{R-2}$ sums of $N-R+1$ distinct elements of $\{\varkappa_1, \ldots, \varkappa_{N-1}\}$.

**Theorem 3.2.** *If* $R \geqq 3$, $\partial D \in C^{2,\alpha}$, *and*

$$S_j\left(\left\{\sum_{i=1}^{N-R+1} \varkappa_{l_i} : l_1 < l_2 < \ldots < l_{N-R+1}\right\}\right) \geqq 0 \quad \text{for } j = 1, \ldots, \binom{N-1}{R-2}, \tag{3.7}$$

*then*

$$\mu_{k+R} < \lambda_k \quad \text{for } k = 1, 2, \ldots$$

We remark that because the left-hand sides of (3.7) are symmetric in the $\varkappa_i$, they can be written as polynomials in the elementary symmetric functions $S_i(\varkappa_1, \ldots, \varkappa_{N-1})$. When $R = 3$, the sums are all of the form $S_1(\varkappa_1, \ldots, \varkappa_{N-1}) - \varkappa_j$ and there are $N-1$ of them. By a derivation like that of (3.6) we find the following result:

**Theorem 3.3.** *If*

(3.8)
$$\sum_{l=0}^{N-1-j} (-1)^l \binom{N-1-l}{j} S_l(\varkappa_1, \ldots, \varkappa_{N-1}) [S_1(\varkappa_1, \ldots, \varkappa_{N-1})]^{N-1-l-j} \geqq 0$$
$$\text{for } j = 1, 2, \ldots, N-2,$$

*then*

$$\mu_{k+3} < \lambda_k \quad \text{for } k = 1, 2, \ldots$$

We can now write down the conditions for the important case of 3 dimensions in terms of the mean curvature $H = (\varkappa_1 + \varkappa_2)/2$ and the Gaussian curvature $\varkappa = \varkappa_1 \varkappa_2$. We see that

(3.9)
$$H \geqq 0$$

implies

(3.10)
$$\mu_{k+1} < \lambda_k .$$

The conditions

(3.11)
$$H \geqq 0, \quad \varkappa \geqq 0$$

imply that

(3.12)
$$\mu_{k+3} < \lambda_k .$$

Finally, the intermediate conditions

(3.13)
$$H \geqq 0, \quad 8H^2 + \varkappa \geqq 0$$

imply that

(3.14)
$$\mu_{k+2} < \lambda_k .$$

It is easily seen from the proofs that for a fixed $R$, Theorem 3.1 or 3.2 is equivalent to Proposition 2.1. While it is therefore clear that the conditions of Theorem 3.2 for a larger $R$ imply those for a smaller $R$, the number

$$\binom{N-1}{R-2}$$

of these conditions is not monotone in $R$.

We shall provide a partial remedy for this situation by establishing a slightly weaker result in which the number of conditions increases with $R$. We begin with the following algebraic lemma.

**Lemma 3.1.** *Suppose that for some (not necessarily ordered) set of numbers* $\mu_1, \ldots, \mu_\nu$ *and some positive integer* $l \leqq \nu$

(3.15)
$$S_j(\mu_1, \ldots, \mu_\nu) \geqq 0 \quad \text{for } j = 1, \ldots, l.$$

*Then*

(3.16)
$$S_j(\mu_1, \ldots, \mu_{\nu-1}) \geqq 0 \quad \text{for } j = 1, \ldots, l-1.$$

**Proof.** It is easily seen that

$$S_j(\mu_1, \ldots, \mu_\nu) = S_j(\mu_1, \ldots, \mu_{\nu-1}) + \mu_\nu S_{j-1}(\mu_1, \ldots, \mu_{\nu-1}) \tag{3.17}$$

where, as always, we define $S_0(\mu_1, \ldots, \mu_{\nu-1}) = 1$. By using this identity, we find that

$$\begin{aligned} S_j(\mu_1, \ldots, \mu_\nu)\, S_j(\mu_1, \ldots, \mu_{\nu-1}) &- S_{j+1}(\mu_1, \ldots, \mu_\nu)\, S_{j-1}(\mu_1, \ldots, \mu_{\nu-1}) \\ &= S_j(\mu_1, \ldots, \mu_{\nu-1})^2 - S_{j+1}(\mu_1, \ldots, \mu_{\nu-1})\, S_{j-1}(\mu_1, \ldots, \mu_{\nu-1}) \\ &= \frac{\nu}{(j+1)(\nu-j)} S_j(\mu_1, \ldots, \mu_{\nu-1})^2 \\ &\quad + \left\{\frac{j(\nu-j-1)}{(j+1)(\nu-j)} S_j(\mu_1, \ldots, \mu_{\nu-1})^2 \right. \\ &\quad \left. - S_{j+1}(\mu_1, \ldots, \mu_{\nu-1})\, S_{j-1}(\mu_1, \ldots, \mu_\nu)\right\}. \end{aligned}$$

Newton's inequality [4, p. 104, Theo. 144] states that the term in braces is non-negative. Therefore

$$\begin{aligned} S_j(\mu_1, \ldots, \mu_\nu)\, S_j(\mu_1, \ldots, \mu_{\nu-1}) \geqq S_{j+1}(\mu_1, \ldots, \mu_\nu)\, S_{j-1}(\mu_1, \ldots, \mu_{\nu-1}) \\ + \frac{\nu}{(j+1)(\nu-j)} S_j(\mu_1, \ldots, \mu_{\nu-1})^2. \end{aligned}$$

We now see from the hypothesis (3.15) that if $j \leqq l-1$ and if $S_{j-1}(\mu_1 \ldots, \mu_{\nu-1}) \geqq 0$, then $S_j(\mu_1, \ldots, \mu_{\nu-1}) \geqq 0$. (Note that $S_j(\mu_1, \ldots, \mu_\nu) = 0$ implies that $S_j(\mu_1, \ldots, \mu_{\nu-1}) = 0$.) Since $S_0(\mu_1, \ldots, \mu_{\nu-1}) = 1 > 0$, the Lemma is proved by induction.

**Theorem 3.4.** *If* $\partial D \in C^{2,\alpha}$, $R \geqq 3$, *and at each point of* $\partial D$

$$S_j(\varkappa_1, \ldots, \varkappa_{N-1}) \geqq 0 \quad \text{for } j = 1, \ldots, R-1, \tag{3.18}$$

*then*

$$\mu_{k+R} < \lambda_k \quad \text{for } k = 1, 2, \ldots \tag{3.19}$$

**Proof.** Apply Lemma 3.1 $R-2$ times to find that

$$S_1(\varkappa_1, \ldots, \mu_{N-R+1}) \geqq 0.$$

Since $\varkappa_1 \leqq \varkappa_2 \leqq \ldots \leqq \varkappa_{N-1}$, this give the conditions of Theorem 2.4, and the Theorem is proved.

When $R = 2$ we find the following simpler result.

**Theorem 3.5.** *If* $\partial D \in C^{2,\alpha}$ *and if*

$$\begin{aligned} S_1(\varkappa_1, \ldots, \varkappa_{N-1}) &\geqq 0, \\ S_2(\varkappa_1, \ldots, \varkappa_{N-1}) + S_1(\varkappa_1, \ldots, \varkappa_{N-1})^2 &\geqq 0, \end{aligned} \tag{3.20}$$

*then*

$$\mu_{k+2} < \lambda_k \quad \text{for } k = 1, 2, \ldots$$

**Proof.** Clearly

$$S_1\left((\varkappa_1, \ldots, \varkappa_{N-2}, \sum_{i=1}^{N-1} \varkappa_i, \varkappa_{N-1}\right) 2S_1(\varkappa_1, \ldots, \varkappa_{N-1}) \geqq 0$$

while by (3.17) and (3.20)

$$S_2(\varkappa_1, \ldots, \varkappa_{N-2}, \Sigma\, \varkappa_i, \varkappa_{N-1}) = S_2(\varkappa_1, \ldots, \varkappa_{N-1}) + S_1(\varkappa_1, \ldots, \varkappa_{N-1})^2 \geqq 0.$$

Thus Lemma 3.1 shows that

$$S_1(\varkappa_1, \ldots, \varkappa_{N-2}, \Sigma\, \varkappa_i) \geqq 0,$$

and the result follows from Theorem 2.3.

We note that when $\varkappa_1 = -2, \varkappa_2 = \ldots = \varkappa_{n-2} = 0, \varkappa_{N-1} = 5$, the conditions (3.6) are satisfied while the conditions (3.20) are not, which shows that Theorem 3.5 is strictly weaker than Theorem 3.1. Similarly, the example $\varkappa_1 = -2, \varkappa_2 = \ldots \ldots = \varkappa_{N-R} = 0, \varkappa_{N-R+1} = \ldots = \varkappa_{N-1} = 3$ shows that when $3 \leqq R \leqq N-1$, Theorem 3.4 is strictly weaker than Theorem 3.2.

## 4. Extension to nonsmooth domains

We shall extend the inequalities (2.3) to more general domains by means of a limiting process. We will have to give up the strictness of such inequalities in the process.

We begin with an elementary lemma.

**Lemma 4.1.** *Let $a, b > 0$. If $v$ is a $C^1$ function on $[0, a+b]$, then*

$$\int_a^{a+b} v^2\, dt \leqq 2b \coth a \int_0^{a+b} (v^2 + v'^2)\, dt. \tag{4.1}$$

**Proof.** A standard variational argument shows that the maximum $\mu$ of the ratio of the integral on the left to the integral on the right is the largest root of the equation

$$(\mu^{-1} - 1)^{1/2} \tan (\mu^{-1} - 1)^{1/2}\, b = \tanh a. \tag{4.2}$$

The well known estimates

$$\sin \varrho \leqq \varrho, \qquad \cos \varrho \geqq 1 - \tfrac{1}{2}\varrho^2$$

show that

$$\mu \leqq \frac{b\,(1 + \frac{1}{2}\, b \tanh a)}{\tanh a + b\,(1 + \frac{1}{2}\, b \tanh a)} \leqq b \coth a$$

which yields (4.1) of the Lemma.

As usual, we define the Minkowski distance $d(A, B)$ between the two point sets $A$ and $B$ to be the infimmum of numbers $\delta$ such that every point of $A$ is within distance $\delta$ of $B$ and every point of $B$ is within distance $\delta$ of $A$.

We shall impose the following conditions on the domain $D$:

**Condition $\mathcal{N}$.** The boundary $\partial D$ can be covered by a finite collection of open sets $Z_\nu$ such that

a) For some $\theta > 0$

$$\bigcup_\nu (Z_\nu) \supset \{x: d(x, \partial D) \leqq \theta\}.$$

b) Each $Z_\nu$ is the image of the cylinder $B_{N-1} \times (0, 1)$, where $B_{N-1}$ is the open unit ball in $R^{N-1}$, under an invertible differentiable mapping $x = \gamma^\nu(\eta, t)$ whose Jacobian is bounded above and below by positive numbers. (Here $\eta \in B_{N-1}$, $t \in (0, 1)$).

c) There is a continuous function $g^\nu$ on $B_{N-1}$ such that

$$\begin{aligned} D \cap Z_\nu &= \{x = \gamma^\nu(\eta, t): \eta \in B_{N-1}, 0 < t < g^\nu(\eta)\}, \\ \partial D \cap Z_\nu &= \{x = \gamma^\nu(\eta, t): \eta \in B_{N-1}, t = g^\nu(\eta)\}, \\ &\alpha \leqq g^\nu(\eta) \leqq \beta, \\ &0 < \alpha \leqq \beta < 1 \end{aligned} \tag{4.3}$$

and

d) There is a positive constant $s$ such that if $x \in Z_\nu \setminus \bar{D}$, $d(x, \bar{D})$ is the Euclidean distance from $x$ to $\bar{D}$, and $x = \gamma^\nu(\eta, t)$, then

$$t - g^\nu(\eta) \leqq s\, d(x, \bar{D}).$$

Our conclusions on convergence will be based on the following semicontinuity lemma.

**Lemma 4.2.** *Let the domain $D$ satisfy the conditions $\mathcal{N}$. Let $D_n$ $n = 1, 2, \ldots$, be a sequence of domains, each of which satisfies a cone condition, and such that*

$$D_n \supset D \tag{4.4}$$

*and*

$$\lim_{n\to\infty} d(D_n, D) = 0. \tag{4.5}$$

*If $\mu_l(D_n)$ is the $l^{\text{th}}$ Neumann eigenvalue of $D_n$ and $\mu_l(D)$ is the corresponding eigenvalue of $D$, then*

$$\liminf_{n\to\infty} \mu_l(D_n) \geqq \mu_l(D). \tag{4.6}$$

**Proof.** We see from (4.3a) and (4.5) that when $n$ is sufficiently large, $D_n$ lies in the union of $D$ and the $Z_\nu$. For such an $n$ we shall derive a bound for the difference in the integrals of $\psi^2$ over $D$ and over $D_n$ for any smooth function defined on $D_n$ in terms of the integral of $\psi^2 + |\nabla\psi|^2$ over $D_n$.

In order to do this we note that

$$(4.7)\qquad \begin{aligned}\int_{(D_n-D)\cap Z_\nu} \psi^2\,dx &= \int_{\gamma^{-1}[(D_n-D)\cap Z_\nu]} [\psi(\gamma^\nu(t,\eta))]^2\,J\,dt\,d\eta\\ &\leqq \int_{B_{N-1}} \int_{g^\nu(\eta)}^{g^\nu(\eta)+r(\eta)} [\psi(\gamma^\nu(t,\eta))]^2\,J\,dt\,d\eta\,.\end{aligned}$$

Since $J$ is bounded above and $r(\eta) \leqq s\,d(D_n, D)$ by (4.3d), Lemma 4.1 shows that there is a constant $c_1$ such that

$$\int_{g^\nu(\eta)}^{g^\nu(\eta)+r(\eta)} \psi^2 J\,dt \leqq c_1\,d(D_n, D) \int_0^{g^\nu+r} (\psi^2 + \psi_t^2)\,dt\,.$$

We substitute this in (4.7), transform back to the $x$-coordinates, and use the fact that $J$ has a positive lower bound to see that there is a constant $c_2$ such that

$$\int_{(D_n-D)\cap Z_\nu} \psi^2\,dx \leqq c_2\,d(D_n, D) \int_{D_n\cap Z_\nu} (\psi^2 + |\nabla\psi|^2)\,dx\,.$$

We add these inequalities to find that

$$(4.8)\qquad \int_{D_n} \psi^2\,dx - \int_D \psi^2\,dx \leqq c\,d(D_n, D) \int_{D_n} (\psi^2 + |\nabla\psi|^2)\,dx\,,$$

where $c = Lc_2$ with $L$ the number of sets $Z_\nu$.

Let $w_1, \ldots, w_l$ be a set of infinitely differentiable functions which are orthogonal on $D_n$ and which approximate the first $l$ Neumann eigenfunctions of $D_n$ so well that any linear combination

$$(4.9)\qquad \psi(x) = \sum_{i=1}^{l} c_i w_i(x)$$

satisfies

$$(4.10)\qquad \int_{D_n} |\nabla\psi|^2\,dx < (\mu_l(D_n) + \varepsilon) \int_{D_n} \psi^2\,dx\,.$$

Select the constants in (4.9) so that they are not all zero and

$$\int_D \psi v_i\,dx = 0, \qquad i = 1, \ldots, l-1\,.$$

Then

$$(4.11)\qquad \mu_l(D) \int_D \psi^2\,dx \leqq \int_D |\nabla\psi|^2\,dx\,.$$

We observe that

$$(4.12)\qquad \int_D |\nabla\psi|^2\,dx \leqq \int_{D_n} |\nabla\psi|^2\,dx \leqq (\mu_l(D_n) + \varepsilon) \int_{D_n} \psi^2\,dx\,,$$

while by (4.8) and (4.10)

$$(4.13)\qquad \int_D \psi^2\,dx \geqq (1 - c\,d(D_n, D)\,[1 + \mu_l(D_n) + \varepsilon]) \int_{D_n} \psi^2\,dx\,.$$

Since the Dirichlet eigenvalue $\lambda_l(D)$ decreases as $D$ grows [3, p. 409, Theorem 3], we see that

$$\mu_l(D_n) \leqq \lambda_l(D_n) \leqq \lambda_l(D).$$

Thus, if $n$ is so large that $c\, d(D_n, D)\, [1 + \lambda_l(D) + \varepsilon] < 1$, we see from (4.11), (4.12), and (4.13) that

$$\mu_l(D) \leqq \frac{\mu_l(D_n) + \varepsilon}{1 - c\, d(D_n, SD)\, [1 + \lambda_l(D) + \varepsilon]}.$$

Since this is true for any positive $\varepsilon$, we have

$$\mu_l(D) \leqq \frac{\mu_l(D_n)}{1 - c\, d(D_n, D)\, [1 + \lambda_l(D)]}.$$

We now let $n \to \infty$ through a sequence to obtain the statement (4.6) of the Lemma.

*Remarks 1.* If $D$ is bounded and convex and 0 is any of its points, then a closed ball of some radius $R_1$ centered at 0 lies in $D$ and an open ball of radius $R_2$ centered at 0 contains $\bar{D}$. If $(r, \omega)$ are polar coordinates centered at 0, we can divide the annular region between the two balls into finitely many "cylinders" with the coordinates $t = (r - R_1)/(R_2 - R_1)$ and $\eta = \eta(\omega)$. Thus *any bounded convex domain satisfies condition $\mathcal{N}$.*

2. If $D$ and $D_n$ are any convex domains with $d(D_n, D)$ small, there are dilations of $D$ with constants near 1, one of which takes $\bar{D}$ inside $D_n$ while the other makes $\bar{D}$ contain $D_n$. The above proof then shows that *for bounded convex domains $d(D_n, D) \to 0$ implies that*

$$\mu_l(D_n) \to \mu_l(D).$$

*3.* We recall [3, p. 423, Theorem 11] that the Dirichlet eigenvalues are continuous in the sense that $d(D_n, D) \to 0$ implies that $\lambda_l(D_n) \to \lambda_l(D)$.

We can now obtain limiting forms of Proposition 2.1.

**Theorem 4.1.** *Let the domain $D$ satisfy the condition $\mathcal{N}$ and suppose that there is a sequence of domains $D_n \supset D$ such that each $D_n$ satisfies the hypotheses of Proposition 2.1, and let*

$$\lim_{n \to \infty} d(D_n, D) = 0.$$

*Then the eigenvalues of $D$ satisfy the inequalities*

$$\mu_{k+R} \leqq \lambda_k \qquad \text{for } k = 1, 2, \ldots$$

**Proof.** Apply lemma 4.2 to the inequalities

$$\mu_{k+R}(D_n) < \lambda_k(D_n) \leqq \lambda_k(D).$$

A theorem of MINKOWSKI [2, p. 35ff.] states that a convex domain $D$ can be approximated in the sense of set distance by smooth convex domains $D_n \supset D$. This and Remark 1 after Lemma 4.2 yield the following result:

**Theorem 4.2.** *If $D$ is any nonempty bounded convex domain, then*

$$\mu_{k+N} \leqq \lambda_k \text{ for } k = 1, 2, \ldots \tag{4.14}$$

It is not known whether there is a convex domain in $N > 1$ dimensions for which equality holds in (4.14).

It is easy to find nonconvex domains for which the condition $\mathcal{N}$ can be verified, but it is difficult to determine whether a nonconvex domain is the limit of a sequence of larger domains which satisfy the conditions of Proposition 2.1 for some $R < N$. We present one class of domains where this can be done.

Let $N = 3$ and let $D$ be obtained by rotating a two-dimensional domain $D_0$ about a coplanar line which does not intersect the closure of $D_0$. Suppose that the boundary of $D_0$ is of class $C^{2,\alpha}$ with the exception of a subset of the interior of $\partial D \cap \partial D^*$ where $D^*$ is the convex hull of $D$. By the Minkowski construction we construct a smooth convex domain $D_0' \supset D_0$ which is arbitrarily close to the convex hull $D_0^*$ of $D_0$. We then construct a convex domain $D_0''$ with smooth boundary such that $\partial D_0''$ coincides with $\partial D_0$ on the part of $\partial D_0 \cap \partial D^*$ near the closure of $\partial D_0 \setminus \partial D^*$, and with $\partial D_0'$ near the set where $\partial D_0$ is not smooth. Let $D_0'''$ be the subdomain of $D_0''$ whose boundary consists of $\partial D_0 \setminus \partial D_0^*$ and a part of $\partial D_0''$. Let $D'''$ be obtained by rotating $D_0'''$. Then $D'''$ contains and is arbitrarily close to $D$, and it has nonnegative mean curvature if this is true of $\partial D \setminus \partial D^*$. It is easily verified that $D$ satisfies Condition $\mathcal{N}$.

The same reasoning also works when $D$ is obtained by rotating $D_0$ about a line of symmetry.

We conclude from Theorem 4.1 that *if $D$ is a domain of revolution and if $\partial D \setminus \partial D^*$ is smooth and has nonnegative mean curvature, then*

$$\mu_{k+1} \leqq \lambda_k \text{ for } k = 1, 2, \ldots$$

The same result follows for an $N$-dimensional domain which is obtained by rotating a two-dimensional domain about an $(N-2)$-dimensional hyperplane.

It would be interesting to find a larger class of domains to which Theorem 4.1 can be applied.

It was observed by JOSEPH HERSCH that if a domain $D'$ is obtained from a domain $D$ by removing a set of measure zero, then $\lambda_k(D') \geqq \lambda_k(D)$ while $\mu_k(D') \leqq \mu_k(D)$. Therefore if $D$ satisfies the conditions of Proposition 2.1, then the eigenvalues of $D'$ still satisfy the inequalities (2.3). Since the boundary of $D'$ may not even have nonnegative mean curvature, this observation makes one wonder whether the inequalities (2.3) are not, in fact true for all domains.

However, numerical computation shows that $\mu_3 > \lambda_1$ for the two-dimensional annular sector

$$D = \{(r, \theta) \colon 1 < r < 2, 0 < \theta < 3\pi/2\}.$$

By Lemma 4.2 one can find smooth domains containing this $D$ for which the same inequality is valid. Thus the inequality (2.3) is not true for all smooth two-dimensional domains.

Computation shows that $\mu_3 < \lambda_1 < \mu_4$ for a disc and that $\mu_4 < \lambda_1 < \mu_5$ for a three-dimensional ball. Thus $\mu_{N+k}$ cannot be replaced by $\mu_{N+k+1}$ in the inequality (2.20). On the other hand, one can show that there are constants $a > 0$ and $c > 1$ such that for the $N$-ball

$$\mu_{ac^N} < \lambda_1 .$$

This suggests that perhaps (2.20) can be replaced by a better inequality of the form

$$\mu_{\phi(N,k)} < \lambda_k$$

for convex $N$-dimensional domains.

*Acknowledgement.* LEVINE was supported in part by the two NSF Institutes, by the C.N.R. (Italy), and by AFOSR Grant 84-0252. WEINBERGER was supported by grants from the NSF and the AFOSR.

## References

1. P. AVILES, Symmetry theorems related to Pompeiu's problem. *American J. of Math.* (in press).
2. T. BONESSEN & W. FENCHEL, *Theorie der Konvexen Körper*. Springer, Berlin, 1934, Chelsea, New York, 1948.
3. R. COURANT & D. HILBERT, *Methods of Mathematical Physics*, vol. I, Interscience, New York (1962).
4. G. H. HARDY, J. E. LITTLEWOOD, & G. POLYA, *Inequalities*, 2nd ed., Cambridge U. Press, London, 1952.
5. L. E. PAYNE, Inequalities for eigenvalues of plates and membranes, *J. Rational Mech. Anal.* **4** (1955), 517–529.
6. G. POLYA, Remarks on the foregoing paper, *J. of Math. and Phys.* **31** (1952), 55–57.
7. M. H. PROTTER & H. F. WEINBERGER, *Maximum Principles in Differential Equations*. Prentice Hall, Englewood Cliffs, 1967, Springer, New York, 1984.
8. H. F. WEINBERGER, *Variational Methods for Eigenvalue Approximation*, C.B.M.S. Regional Conf. Ser. # 15, S.I.A.M., Philadelphia, 1974.

Iowa State University
Ames

and

Institute for Mathematics
and its Applications
University of Minnesota
Minneapolis

*(Received November 4, 1985)*

# *Embeddings of Anisotropic Sobolev Spaces*

D. E. EDMUNDS & R. M. EDMUNDS

*Dedicated to Professor James Serrin on the occasion of his 60th birthday*

## 1. Introduction

During the last twenty years a great deal of effort has been put into the analysis of embeddings of Sobolev spaces from the standpoint of approximation and entropy numbers. These numbers enable classifications of compact embedding maps to be made, and their intimate connection with eigenvalues of elliptic operators helps to explain the emphasis placed upon them by the strong group in the Soviet Union centred around BIRMAN and SOLOMJAK. Estimates have been obtained for the embeddings of the Sobolev space $W^{k,p}(\Omega)$ in suitable Lebesgue spaces $L^q(\Omega)$ when $\Omega$ is a bounded open subset of $\mathbb{R}^n$, and even for certain types of unbounded sets $\Omega$. When $kp = n$, estimates for these numbers are known for the compact embedding of $W^{k,p}(\Omega)$ in particular Orlicz spaces.

The position as regards anisotropic Sobolev spaces is less satisfactory in that while bounds for the numbers have been obtained for embeddings of such spaces in Lebesgue spaces, nothing along these lines has been done, so far as we are aware, about the embeddings in Orlicz spaces which occur in critical cases. The results of this paper go some way towards making good this deficiency.

## 2. Preliminaries

Let $\Omega$ be an open subset of $\mathbb{R}^n$, denote points of $\mathbb{R}^n$ by $x = (x_i) = (x_1, \ldots, x_n)$, set $D_i = \partial/\partial x_i$, let $\boldsymbol{r} = (r_1, \ldots, r_n) \in \mathbb{N}^n$, let $p \in [1, \infty)$ and denote by $W^{\boldsymbol{r},p}(\Omega)$ the anisotropic Sobolev space of all (real-valued or complex-valued) functions $u$ in $L^p(\Omega)$ which, for all $i \in \{1, 2, \ldots, n\}$, have generalised derivatives $D_i^{r_i}u$ in $L^p(\Omega)$. When furnished with the norm $\|\cdot\|_{\boldsymbol{r},p,\Omega}$ defined by

$$\|u\|_{\boldsymbol{r},p,\Omega} = \|u\|_{p,\Omega} + \sum_{i=1}^{n} \|D_i^{r_i}u\|_{p,\Omega},$$

where $\|v\|_{p,\Omega} = \left(\int_\Omega |v(x)|^p \, dx\right)^{1/p}$, $W^{\boldsymbol{r},p}(\Omega)$ becomes a Banach space (*cf.* [2],

§ 9.1). The closure in $W^{r,p}(\Omega)$ of $C_0^\infty(\Omega)$, the space of all infinitely differentiable functions with compact support in $\Omega$, will be denoted by $W_0^{r,p}(\Omega)$. When $r_1 = r_2 = \ldots = r_n = k$, the spaces $W^{r,p}(\Omega)$ and $W_0^{r,p}(\Omega)$ coincide with the familiar Sobolev spaces $W^{k,p}(\Omega)$ and $W_0^{k,p}(\Omega)$ respectively. The books [2] give a comprehensive account of these anisotropic spaces.

We shall need certain facts about Orlicz spaces. An Orlicz function is a map $\phi : [0, \infty) \to \mathbb{R}$ which is continuous, convex and such that $\lim_{t\to 0^+} \phi(t)/t = 0$ and $\lim_{t\to\infty} \phi(t)/t = \infty$. Given an Orlicz function $\phi$, the Orlicz space $L^\phi(\Omega)$ is the linear hull of the set of all functions $u : \Omega \to \mathbb{R}$ (or $\mathbb{C}$; functions equal almost everywhere being identified) such that $\int_\Omega \phi(|u(x)|)\, dx < \infty$. With the norm $\|\cdot\|_{\phi,\Omega}$ given by

$$\|u\|_{\phi,\Omega} = \inf\left\{\lambda > 0 : \int_\Omega \phi(|u(x)|/\lambda)\, dx \leq 1\right\},$$

$L^\phi(\Omega)$ is a Banach space (*cf.* [11], §§ 3.8, 3.9).

Now let $X$ and $Y$ be Banach spaces over the same field, let $B_X$ be the closed unit ball in $X$ and let $T \in \mathscr{B}(X, Y)$, the family of all bounded linear maps from $X$ to $Y$. For each $m \in \mathbb{N}$ define

$$a_m(T) = \inf\{\|T - F\| : F \in \mathscr{B}(X, Y), \dim F(X) < m\},$$

$$e_m(T) = \inf\{\varepsilon > 0 : T(B_X) \text{ can be covered by } 2^{m-1} \text{ closed balls with radius } \varepsilon\};$$

$a_m(T)$ is called the $m^{\text{th}}$ approximation number of $T$, while $e_m(T)$ is the $m^{\text{th}}$ entropy number. Details of the main properties of these numbers are given in [14], but it may be helpful to recall here some of their interesting connections with analytical entities related to operators. Thus if $X = Y$ and $T$ is compact, with $(\lambda_m(T))$ standing for the sequence of all eigenvalues of $T$, each repeated according to algebraic multiplicity and ordered so that $|\lambda_n(T)| \geq |\lambda_{n+1}(T)|$ for all $n \in \mathbb{N}$, then CARL & TRIEBEL [7] have shown that for all $m, n \in \mathbb{N}$,

$$|\lambda_n(T)| \leq \left(\prod_{j=1}^n |\lambda_j(T)|\right)^{1/n} \leq (\sqrt{2})^{(m-1)/n}\, e_m(T).$$

(If $T$ has only $n$ non-zero eigenvalues put $\lambda_m(T) = 0$ for $m > n$.) The choice of $m = n + 1$ gives the rather surprising inequality

$$|\lambda_n(T)| \leq \sqrt{2}\, e_{n+1}(T),$$

first discovered by CARL [6]. Moreover, it is known [7] that for all $n \in \mathbb{N}$,

$$|\lambda_n(T)| \leq 16(a_n(T))^{1/n}$$

and that Weyl's inequality holds in the generalised form

$$\sum_{j=1}^n |\lambda_j(T)|^p \leq C_p \sum_{j=1}^n (a_j(T))^p$$

for all $n \in \mathbb{N}$ and all $p \in (0, \infty)$, with

$$C_p = \begin{cases} 2e/\sqrt{p} & \text{if } 0 < p < 1, \\ \sqrt{2e}\, 2^{1/p} & \text{if } p \geqq 1 \end{cases}$$

(*cf.* KÖNIG [13]). KÖNIG [12] has also shown that for all $n \in \mathbb{N}$,

$$|\lambda_n(T)| = \lim_{k \to \infty} (a_n(T^k))^{1/k};$$

while if, in addition, $X$ is a Hilbert space, it is classical that $a_n(T) = \lambda_n(|T|)$, where $|T|$ is the positive square root of $T^*T$.

Next we come to embedding maps. These repay study for many reasons; note in particular that the process of reduction of an elliptic boundary-value problem to an operator equation gives rise to an operator which is the composition of an embedding map and a continuous map, and that the entropy and approximation numbers are sub-multiplicative in the sense that

$$a_{m+n-1}(T_1 \circ T_2) \leqq a_m(T_1)\, a_n(T_2),$$

a similar inequality holding for the entropy numbers. Let $\Omega$ be bounded, let $p, q \in (1, \infty)$ and let $k \in \mathbb{N}$ be such that $k/n > p^{-1} - q^{-1}$; let $I_0 : W^{k,p}(\Omega) \to L^q(\Omega)$ be the natural embedding. Then it is known that (*cf.* [3], [8]) $a_s(I_0) = O(s^{-h})$ as $s \to \infty$, where $h = kn^{-1} - \max(0, p^{-1} - q^{-1})$; the same holds for the embedding $I$ of $W^{k,p}(\Omega)$ in $L^q(\Omega)$ if the boundary $\partial\Omega$ of $\Omega$ is minimally smooth in the sense of STEIN [15]. The work of BIRMAN & SOLOMJAK [3] shows that $e_s(I_0) = O(s^{-k/n})$ as $s \to \infty$, the same estimate holding for $e_s(I)$ if $\partial\Omega$ is minimally smooth. When $kp = n$, $W_0^{k,p}(\Omega)$ may be compactly embedded in any Orlicz space $L^{\phi_\nu}(\Omega)$, where $\phi_\nu(t) = \exp(t^\nu) - 1$ and $1 < \nu < n/(n-k)$; it turns out for this embedding map $J_0$ we have $a_s(J_0) = O((\log s)^{1-(k/n)-(1/\nu)})$ as $s \to \infty$, the same estimate holding for the embedding $J$ of $W^{k,p}(\Omega)$ in $L^{\phi_\nu}(\Omega)$ if $\partial\Omega$ is minimally smooth (*cf.* [8], [10]). Identical estimates hold for $e_s(J_0)$ and $e_s(J)$ (*cf.* [9]). Analogous results for the embedding of $W^{\mathbf{r},p}(\Omega)$ in $L^q(\Omega)$ have been obtained by BORZOV [4], [5]: for example, he has shown that if $\Omega$ is a box, the $s^{\text{th}}$ entropy number of this embedding is $O(s^{-r/n})$ as $s \to \infty$, provided that $p \in [1, \infty)$, $q \in [1, \infty]$, $r/n > p^{-1} - q^{-1}$, where $nr^{-1} = \sum_{i=1}^{n} r_i^{-1}$. This result can be extended to more general open sets $\Omega$ by use of an anisotropic analogue of CALDERÓN's extension theorem (*cf.* [2]) if $\Omega$ satisfies what is called a strong $\mathbf{r}$-horn condition. To explain this condition some notation is desirable.

Given any $h \in (0, \infty]$, any $\varepsilon \in (0, \infty)$ and any $\mathbf{a} = (a_1, \ldots, a_n) \in \mathbb{R}^n$ with each $a_i \neq 0$, the set

$$V(\mathbf{r}, h, \varepsilon, \mathbf{a}) = \bigcup_{0<v<h} \{x \in \mathbb{R}^n : x_i/a_i > 0,\ v < (x_i/a_i)^{r_i} < (1+\varepsilon)\, v \quad \text{for } i = 1, 2, \ldots, n\}$$

is called an $\mathbf{r}$-*horn* of radius $h$ and opening $\varepsilon$. An open set $\Omega \subset \mathbb{R}^n$ is said to satisfy a *weak* $\mathbf{r}$-*horn* condition if there is a positive integer $N$ such that for each

$j \in \{1, 2, \dots, N\}$, there are an open set $\Omega_j$ and an $\boldsymbol{r}$-horn $V_j(\boldsymbol{r}, h, \varepsilon, \boldsymbol{a}^{(j)})$ such that

$$\Omega = \bigcup_{j=1}^{N} \Omega_j = \bigcup_{j=1}^{N} (\Omega_j + V_j(\boldsymbol{r}, h, \varepsilon, \boldsymbol{a}^{(j)})).$$

If, in addition, there exists $\delta > 0$ such that $\Omega = \bigcup_{j=1}^{N} \Omega_j^{(\delta)}$, where $\Omega_j^{(\delta)} = \{x \in \Omega_j:$ $\operatorname{dist}(x, \Omega \setminus \Omega_j) > \delta\}$, then $\Omega$ is said to satisfy a *strong* $\boldsymbol{r}$*-horn* condition. Note that if $r_1 = r_2 = \dots = r_n$, every $\boldsymbol{r}$-horn is a cone. More details about these conditions on $\Omega$ will be found in [2], § 8; every box satisfies a strong $\boldsymbol{r}$-horn condition, for all $\boldsymbol{r}$.

Finally, we shall need the following two theorems:

**Theorem A** ([2], Theorem 9.6). *Suppose that $\Omega$ satisfies a strong $\boldsymbol{r}$-horn condition and that $p \in (1, \infty)$. Then $W^{\boldsymbol{r},p}(\Omega)$ is the set of all functions which are the restrictions to $\Omega$ of elements of $W^{\boldsymbol{r},p}(\mathbb{R}^n)$; there is a bounded, linear extension map $E: W^{\boldsymbol{r},p}(\Omega) \to W^{\boldsymbol{r},p}(\mathbb{R}^n)$.*

**Theorem B** ([2], Remark 10.6). *Suppose that $\Omega$ satisfies a weak $\boldsymbol{r}$-horn condition, let $p \in (1, \infty)$, $p' = p/(p-1)$, suppose that $rp = n$, where $n/r = \sum_{i=1}^{n} 1/r_i$, and define $\Phi$ by*

$$\Phi(t) = \exp(t^{p'}) - \sum_{0 \leqq j < p-1} t^{p'j}/j! \quad (t \geqq 0).$$

*Then $W^{\boldsymbol{r},p}(\Omega)$ is continuously embedded in $L^{\Phi}(\Omega)$.*

If $\Psi$ is any Orlicz function which is dominated by $\Phi$ globally in the sense that there exists $k > 0$ such that $\Psi(t) \leqq \Phi(kt)$ for all $t \geqq 0$, then by [1], Theorem 8.12, $L^{\Phi}(\Omega)$ is continuously embedded in $L^{\Psi}(\Omega)$ and so, by Theorem B, $W^{\boldsymbol{r},p}(\Omega)$ is continuously embedded in $L^{\Psi}(\Omega)$. If $\Omega$ is bounded, the same holds under the weaker condition that there exist positive constants $k$ and $t_0$ such that $\Psi(t) \leqq \Phi(kt)$ for all $t \geqq t_0$ (see Theorem 8.12 of [1] again).

## 3. The results

**Theorem 1.** *Let $\Omega$ be bounded and satisfy a strong $\boldsymbol{r}$-horn condition, let $p \in (1, \infty)$ be such that $rp = n$, where $n/r = \sum_{j=1}^{n} 1/r_j$, let $\phi$ be defined by $\phi(t) = t^{\sigma} \exp(t^{\nu})$ $(t \geqq 0)$, where $\sigma \in (p, \infty)$ and $\nu \in (1, p')$, with $p' = p/(p-1)$; and let $I$ be the embedding of $W^{\boldsymbol{r},p}(\Omega)$ in $L^{\phi}(\Omega)$. Then as $k \to \infty$,*

$$e_k(I) = O((\log k)^{-1/(\gamma\nu)}),$$

*where $\gamma = \max\{\sigma/\nu, p'/(p'-\nu)\}$. The same estimate holds for the embedding of $W_0^{\boldsymbol{r},p}(\Omega)$ in $L^{\phi}(\Omega)$ without the hypothesis relating to a strong $\boldsymbol{r}$-horn condition.*

**Proof.** First suppose that $\Omega$ is an open box $Q$ with sides parallel to the coordinate axes and of lengths $\mu_1, \mu_2, \ldots, \mu_n$; put $a_{ij}(Q) = \mu_i^{r_i/r_j}/\mu_j$, $\boldsymbol{\mu} = (\mu_i)$. Let $q \in (1, \infty)$. A close examination of the proof of Lemma 5 of BORZOV [5] (see also the proof of Lemma 2.12 of [3] and lemma 10.1 of [2]) shows that given any $u \in W^{r,p}(Q)$, there is a polynomial $P_Q u$ in $x_1, x_2, \ldots, x_n$, with degree in $x_i$ less than $r_i$ $(i = 1, 2, \ldots, n)$, such that

$$\|u - P_Q u\|_{q,Q} \leqq C\,|Q|^{1/q} \left[\prod_{i=1}^{n} \sum_{j=1}^{n} a_{ji}(Q)\right]^{1/p} (q/p)^{1-(1/p)-(1/q)}\,(1/q)^{1/q}\,\|u\|_{r,p,Q},$$

where $C$ is a constant independent of $u$, $Q$ and $q$. Let $m$ be the least common multiple of $r_1, r_2, \ldots, r_n$ and let $d_1, d_2, \ldots, d_n \in \mathbb{N}$ be such that $m = d_1 r_1 = \ldots \ldots = d_n r_n$. Subdivide the $i^{\text{th}}$ edge of $Q$ into $N^{d_i}$ parts of equal length $\mu_i/N^{d_i}$ $(i = 1, \ldots, n;\ N \in \mathbb{N})$. This gives $N^{mp}$ boxes $Q_s$, each with measure $|Q|\,N^{-mp}$. For each $s$, $a_{ij}(Q_s) = \mu_i^{r_i/r_j}/\mu_j$ and so

$$\prod_{i=1}^{n} \sum_{j=1}^{n} a_{ji}(Q_s) = c(\boldsymbol{\mu}, \boldsymbol{r}),$$

a constant independent of $s$.

Now put

$$(P_{N,Q} u)(x) = \sum_{s=1}^{N^{mp}} \chi_{Q_s}(x)\,(P_{Q_s} u)(x) \qquad (x \in \mathbb{R}^n)$$

and write $U = u - P_{N,Q}\, u$. Then

$$\|U\|_{q,Q} \leqq C N^{-mp/q}\,|Q|^{1/q}\,[c(\boldsymbol{\mu}, \boldsymbol{r})]^{1/p}\,(q/p)^{(1/p')-(1/q)}\,(1/q)^{1/q}\,\|u\|_{r,p,Q}.$$

It follows that for any $\lambda > 0$,

$$\begin{aligned}\int_Q \Phi(|U(x)|/\lambda)\,dx &= \int_Q \sum_{j=1}^{\infty} (1/j!)\,(|U(x)|/\lambda)^{\sigma+\nu j}\,dx \\ &\leqq \sum_{j=1}^{\infty} (\|U\|_{\sigma+\nu j,Q}\,\lambda^{-1})^{\sigma+\nu j}/j! \\ &\leqq N^{-mp}\,|Q|\,p^{-1}\,z^{\sigma/\nu} \sum_{j=1}^{\infty} z^j\{(\sigma+\nu j)/p\}^{(\sigma+\nu j)/p'}/j!,\end{aligned}$$

where $z = (C\lambda^{-1}\,c^{1/p}\,\|u\|_{r,p,Q})^{\nu}$. From this it is elementary to deduce that

$$\int_Q \Phi(|U(x)|/\lambda)\,dx \leqq N^{-mp}\,|Q|\,p^{-1}\,z^{\sigma/\nu}\{(\nu+\sigma)\,e/p\}^{\sigma/p'} \sum_{j=1}^{\infty} \{z(\nu/p)^{\nu/p'}\,e^{\sigma/p'}\}^j\,j^{\nu j/p'}/j!.$$

We now use the fact (*cf.* [10], Appendix) that there is a constant $c_0 > 0$ such that for all $\omega \in (0, 1)$ and all $a > 0$,

$$\sum_{j=1}^{\infty} j^{\omega j} a^j/j! \leqq c_0 \exp\{(2e)^{\omega/(1-\omega)}\,a^{1/(1-\omega)}\}.$$

Thus since $\nu/p' < 1$,

$$\int_Q \Phi(|U(x)|/\lambda)\,dx \leqq K_1 N^{-mp}\,z^{\sigma/\nu}\,\exp(K_2 z^{p'/(p'-\nu)}),$$

where

$$K_1 = |Q|\, p^{-1}\{(\nu+\sigma)\, e/p\}^{\sigma/p'}\, c_0, \qquad K_2 = (2\nu p^{-1}\, e^{1+\sigma/\nu})^{\nu/(p'-\nu)}.$$

Hence

$$\|U\|_{\phi,Q} \leqq \delta\, \|u\|_{r,p,Q},$$

where

$$\delta = K\{\log(N^{mp}K_1^{-1})\}^{-1/(\gamma\nu)}$$

and $K$ is a positive constant, independent of $u$. This shows that $\|P_{N,Q}u\|_{\phi,Q} \leqq \|I\| + \delta$ if $\|u\|_{r,p,Q} \leqq 1$, where $\|I\|$ denotes the operator norm of the embedding $I$. Observe that $\delta \to 0$ as $N \to \infty$; we shall henceforth assume that $N$ is so large that $\delta < 1$.

Since the map $P_{N,Q}$ is linear, with rank at most $MN^{mp}$, where $M = \prod_{i=1}^{n} r_i$, there is a linear subspace $F$ of $L^{\phi}(Q)$ such that $\dim F = MN^{mp}$ and $P_{N,Q}(B) \subset F \cap B_{L^{\phi}(Q)}(0, \|I\| + \delta)$, where $B$ is the open unit ball in $W^{r,p}(Q)$ and $B_X(a, R)$ denotes the open ball in $X$ with centre $a$ and radius $R$. Now suppose that all the spaces involved are real: the argument for complex spaces is exactly the same. Since $F$ is topologically isomorphic to $\mathbb{R}^{MN^{mp}}$, given any $\varepsilon \in (0, 1)$, $P_{N,Q}(B)$ can be covered by $K_3\{3[(\|I\| + \delta)/\varepsilon]\}^{MN^{mp}}$ balls $B_j$ of radius $\varepsilon$, lying in $F$; here $[A]$ stands for the integer part of $A$. Given any $u \in B$, $P_{N,Q}\, u \in F$, $\|u - P_{N,Q}\, u\|_{\phi,Q} \leqq \delta$ and there exists $j$ such that $P_{N,Q}\, u \in B_j = B_F(x_j, \varepsilon)$. Thus $u \in B_{L^{\phi}}(x_j, \delta + \varepsilon)$, and hence $I(B)$ can be covered by $K_3\{[(\|I\| + \delta)/\varepsilon]\}^{MN^{mp}}$ balls of radius $\delta + \varepsilon$. It follows that $I(B)$ can be covered by $K_3(K_4[1/\delta])^{MN^{mp}}$ balls of radius $2\delta$. Since $\delta = K\{\log(N^{mp}\, K_1^{-1})\}^{-1/(\gamma\nu)}$, $I(B)$ can be covered by

$$K_3(K_4[1/\delta])^{MK_1 \exp\{(\delta/K)^{-\gamma\nu}\}} := A(\delta)$$

balls of radius $2\delta$. Consideration of the inequality $A(\delta) \leqq 2^{k-1}$ now shows that

$$e_k(I) = O((\log k)^{-1/(\gamma\nu)}),$$

as required.

To deal with the more general case in which $\Omega$ is merely required to be bounded and to satisfy a strong $r$-horn condition, we observe that by Theorem A, there is an extension map $E\colon W^{r,p}(\Omega) \to W^{r,p}(\mathbb{R}^n)$ and a constant $c$ such that for all $u \in W^{r,p}(\Omega)$,

$$\|u\|_{r,p,\Omega} \leqq \|Eu\|_{r,p,\mathbb{R}^n} \leqq c\, \|u\|_{r,p,\Omega}.$$

Let $Q$ be an open box such that $\bar{\Omega} \subset Q$. Since $W^{r,p}(Q)$ coincides with the set of restrictions to $Q$ of functions in $W^{r,p}(\mathbb{R}^n)$, given any $u \in W^{r,p}(\Omega)$ we may define $\tilde{u} = Eu|_Q \in W^{r,p}(Q)\colon \tilde{u}(x) = u(x)$ for all $x \in \Omega$ and

$$\|\tilde{u}\|_{r,p,Q} \leqq \|Eu\|_{r,p,\mathbb{R}^n} \leqq c\, \|u\|_{r,p,\Omega}.$$

Then

$$\left\| u - \sum_{j=1}^{N^{mp}} \chi_{Q_j \cap \Omega} P_{Q_j} \tilde{u} \right\|_{\phi,Q} \leqq \left\| \tilde{u} - \sum_{j=1}^{N^{mp}} \chi_{Q_j} P_{Q_j} \tilde{u} \right\|_{\phi,Q}$$

and hence

$$\left\| u - \sum_{j=1}^{N^{mp}} \chi_{Q_j \cap \Omega} P_{Q_j} \tilde{u} \right\|_{\Phi,Q} \leqq K \{\log (N^{mp} K_1^{-1})\}^{-1/(\gamma\nu)} \|u\|_{r,p,\Omega},$$

from which the estimate for $e_k(I)$ follows as before. The rest is clear.

**Theorem 2.** *Under the same assumptions as in Theorem 1,*

$$a_k(I) = O((\log k)^{-1/(\gamma\nu)})$$

*as* $k \to \infty$.

**Proof.** First suppose that $\Omega = Q$. Then as shown in the proof of Theorem 1,

$$\|u - P_{N,Q} u\|_{\Phi,Q} \leqq K\{\log (N^{mp} K_1^{-1})\}^{-1/(\gamma\nu)} \|u\|_{r,p,Q}.$$

Hence for all $k \in \mathbb{N}$ of the form $MN^{mp}$,

$$a_k(I) \leqq \text{const.} (\log k)^{-1/(\gamma\nu)}.$$

But given any $k \in \mathbb{N}$, there exists $N \in \mathbb{N}$ such that $MN^{mp} \leqq k \leqq M(N+1)^{mp}$; and as $a_k(I)$ decreases monotonically as $k$ increases, the result follows, when $\Omega = Q$. The general case is handled just as in the proof of Theorem 1.

At the expense of a little technical complication in the proof, the hypothesis that $\sigma \in (p, \infty)$ in Theorems 1 and 2 may be weakened to $\sigma \in (1, \infty)$.

## References


1. Adams, R. A.: *Sobolev spaces*. New York, London: Academic Press 1975.
2. Besov, O. V., V. P. Il'in, & S. N. Nikol'skii: *Integral representations of functions and imbedding theorems*, I and II. Washington, New York, Toronto, London, Sydney: Winston/Wiley 1978 and 1979.
3. Birman, M. S., & M. Z. Solomjak: Quantitative analysis in Sobolev embedding theorems and applications to spectral theory, *Amer. Math. Soc. Translations* **114** (1980).
4. Borzov, V. V.: On some applications of piecewise polynomial approximations of functions of anisotropic classes $W_p^r$, *Soviet Math. Dokl.* **12** (1971), 804–807.
5. Borzov, V. V.: Some applications of theorems on piecewise polynomial approximations of functions in anisotropic classes $W_p^r$, *Problemy Mat. Fiz.*, vyp. **6**, Izdat. Leningrad Univ., Leningrad (1973), 53–67 (Russian).
6. Carl, B.: Entropy numbers, *s*-numbers and eigenvalues problem, *J. Functional Analysis* **41** (1981), 290–306.
7. Carl, B., & H. Triebel: Inequalities between eigenvalues, entropy numbers and related quantities in Banach spaces, *Math. Ann.* **251** (1980), 129–133.
8. Edmunds, D. E.: Embeddings of Sobolev spaces, Proc. Spring School 'Nonlinear analysis, function spaces and applications', Teubner-Texte Math. **19** (Teubner, Leipzig 1979), 38–58.
9. Edmunds, D. E., & R. M. Edmunds: Entropy and approximation numbers of embeddings in Orlicz spaces, *J. Lond. Math. Soc.* **32** (1985), 528–538.

10. EDMUNDS, D. E., & V. B. MOSCATELLI: Fourier approximation and embeddings of Sobolev spaces, *Dissertationes Mathematicae* **145** (1977), 1–50.
11. FUČIK, S., O. JOHN, & A. KUFNER: *Function spaces.* Prague: Academia 1977.
12. KÖNIG, H.: A formula for the eigenvalues of a compact operator, *Studia Math.* **65** (1979), 141–146.
13. KÖNIG, H.: Some inequalities for the eigenvalues of a compact operator, to appear.
14. PIETSCH, A.: *Operator ideals.* Berlin: VEB Deutscher Verlag der Wissenschaften 1978.
15. STEIN, E. M.: *Singular integrals and differentiability properties of functions.* Princeton: Princeton University Press 1970.

Mathematics Division
University of Sussex
and
Department of Pure Mathematics
University College, Cardiff

*(Received November 13, 1985)*

# *Monotonic Decreasing and Asymptotic Behavior of the Kinetic Energy for Weak Solutions of the Navier-Stokes Equations in Exterior Domains*

GIOVANNI P. GALDI & PAOLO MAREMONTI

*To James Serrin on his 60th birthday*

## Introduction

Existence of weak solutions of the three-dimensional Navier-Stokes problem was first proved by J. LERAY in the case of the Cauchy problem, *cf.* LERAY (1934). As is well known, these solutions are important in that they are the only solutions which, so far, are known to exist for all times, without restriction on the data. Unfortunately, however, the question of whether they are classical, in an ordinary sense, is still open, even though partial conclusions regarding regularity are available: LERAY (1934), SCHEFFER (1976, 1980), CAFFARELLI, KOHN, & NIRENBERG (1982). Subsequently, E. HOPF (1951), using a different technique, constructed weak solutions for a general initial-boundary value problem. However, HOPF's solution (even for the Cauchy problem) has weaker properties than LERAY's solution. Among others, we refer to the "energy inequality" for the velocity field $\boldsymbol{u}(x, t)$, *i.e.*

$$\text{(I)} \qquad \int_\Omega u^2(x, t)\, dx - \int_\Omega u^2(x, s)\, dx \leqq -2 \int_s^t \int_\Omega \nabla \boldsymbol{u} : \nabla \boldsymbol{u}\, dx\, d\tau$$

*for* $s = 0$, *for almost all* $s > 0$ *and for all* $t \geqq s$.
Actually, unlike those of LERAY, the solutions constructed by HOPF satisfy (I) if $\Omega$ is a bounded domain, while if $\Omega$ is unbounded, they obey the following weaker relation[1]

$$\text{(II)} \qquad \int_\Omega u^2(x, t)\, dx - \int_\Omega u^2(x, 0)\, dx \leqq -2 \int_0^t \int_\Omega \nabla \boldsymbol{u} : \nabla \boldsymbol{u}\, dx\, d\tau$$

*for all* $t \geqq 0$.

---

[1] Notice that if $\boldsymbol{u}(x, t)$ is a regular solution in the ordinary sense, it certainly verfies (I) with the *equality* sign. So far, regular global solutions are known to exist only for "small" data: LADYZHENSKAYA (1969).

On the other hand the validity of the inequality (I) has several remarkable implications. In particular, it ensures that the kinetic energy $(1/2) \int_\Omega u^2(x, t)\, dx$ is a monotonically decreasing function of time, which is what has to be expected on physical grounds. This should be contrasted with (II), which does not rule out the possibility of an (unacceptable) increasing energy. Moreover, the existence of weak solutions with decreasing kinetic energy allows one to formulate asymptotic stability theorems in a very large class of perturbations: GALDI (1975), MASUDA (1975), HEYWOOD (1980), MAREMONTI (1984). In this respect, it is worth noticing that the authors of the above papers suppose that HOPF'S solutions satisfy (I) instead of (II), in the case of an exterior domain[2]. Because of this oversight, the results of GALDI (1975), MASUDA (1975), HEYWOOD (1980), MAREMONTI (1984a, b) remain formal for such domains. Finally, we observe that the validity of (I) is intimately related to the regularity of weak solutions through the fundamental uniqueness teorem of LERAY-SATHER-SERRIN (see SERRIN (1962)).

This paper concerns some properties of solutions to Navier-Stokes equations in exterior (three-dimensional) domains. Our motivation is twofold. On the one hand, the aim is to prove the existence of global (weak) solutions verifying the energy relation (I). On the other, we wish to study the asymptotic behavior in time of their kinetic energy. Precisely, employing a method introduced by LERAY (1934) and with the aid of some estimates for solutions to the *linear* Navier-Stokes equations given in SOLONNIKOV (1977) (see Section 1), in Section 2 we prove the existence of global solutions corresponding to arbitrary initial data from a Sobolev space of suitable fractional order. These solutions, which are more regular than HOPF's solutions (having time derivatives and second-order spatial derivatives) satisfy the Navier-Stokes equations almost everywhere and possess a kinetic energy which decreases monotonically in time. We note that for the Cauchy problem their existence was established by LADYZENSKAYA (1969). Thereafter we study the asymptotic behavior in time of the kinetic energy of a weak solution satisfying (I). As is known, this problem was set by LERAY (1934, p. 248) a long time ago, and only recently has it begun to receive satisfactory answers: KATO (1984), MAREMONTI (1984b), MASUDA (1984), GALDI & RIONERO (1985), SCHONBEK (1985). In Section 3 we prove that the kinetic energy tends to zero and, what is more, we conclude that the order of decay is related to the summability of the initial data. These results, which rely on those of MAREMONTI (1986) and on a new estimate for weak solutions in exterior domains (*cf.* Lemma 3.1), show that if the initial data belong to the Lebesgue space $L^2(\Omega) \cap L^q(\Omega)$, $1 \leqq q \leqq 3/2$, the kinetic energy decays like $t^{-2\beta}$ where $\beta = (2 - q)/4q$. This behavior is further improved if $\Omega$ is the whole space. We notice that our results either improve or contain as particular cases those of KATO (1984), MAREMONTI (1984), MASUDA (1984), SCHONBEK (1985), MAREMONTI (1986).

---

[2] The validity of (I) in unbounded domains has been recently explicitly questioned in MASUDA (1984).

## 1. Preliminaries and notations

Let $\Omega$ be a domain of the three-dimensional Euclidean space $\mathbb{R}^3$, exterior to $\nu$ ($\geqq 0$) $C^2$-smooth, compact subregions, We designate by $R_0$ the diameter of $\bar{\Omega}^c$. For $\Omega' \subseteq \Omega$ and $s > 0$ we set $\Omega'_s \equiv \Omega' \times (0, s)$. By $L^p(\Omega')$ $p \in [1, \infty]$ we denote the Lebesgue space of functions on $\Omega'$. The symbol $L^p(\Omega'_s)$ has an analogous meaning. The norm in $L^p(\Omega)$ [respectively, in $L^p(\Omega_s)$] will be indicated by $|\cdot|_p$ [respectively, by $|\cdot|_{p,\Omega_s}$]. $W^{m,p} = W^{m,p}(\Omega)$ $m \geqq 0$ denotes the Sobolev space of order $(m, p)$ of functions on $\Omega$, and $|\cdot|_{m,p}$ is its associated norm. Furthermore, $\mathscr{W}_p(\Omega_s)$ is the space of functions $u$ on $\Omega_s$ having $p^{\text{th}}$ power summable (generalized) derivatives of the first and second order with respect to $x \in \Omega$ and of the first order with respect to $t \in (0, s)$. An equivalent norm [SOLONNIKOV (1977)], in $\mathscr{W}_p(\Omega_s)$ is

$$|u|_{\mathscr{W}_p(\Omega_s)} \equiv |u|_{p,\Omega_s} + \sum_{i,j} \left|\frac{\partial^2 u}{\partial x_i \, \partial x_j}\right|_{p,\Omega_s} + \left|\frac{\partial u}{\partial t}\right|_{p,\Omega_s}.$$

For an open set $A$ in $\mathbb{R}^n$ we let $C_0^\infty(A)$ be the set of indefinitely differentiable functions of compact support in $A$. Moreover, denoting by $\mathscr{C}_0^\infty(\Omega)$ the class of solenoidal functions from $C_0^\infty(\Omega)$, we indicate by $\mathring{J}(\Omega)$ the completion of $\mathscr{C}_0^\infty(\Omega)$ in $L^2(\Omega)$ and by $\mathring{J}^{2-2/p,p}, (\Omega)$ the completion of $\mathscr{C}_0^\infty(\Omega)$ in the norm

$$\|u\|_{2-2/p,p} \equiv \begin{cases} |u|_p + \langle u \rangle^{(2-2/p)} & \text{for } p \neq 3/2 \\ |u|_p + \langle u \rangle^{(2-2/p)} + |u \varrho^{\frac{2}{p}-2}|_p & \text{for } p = 3/2 \end{cases}$$

where $\varrho = \varrho(x)$ is the distance from $x$ to $\partial\Omega$ and

$$\langle u \rangle^{2-2/p} \equiv \begin{cases} \left( \int\limits_\Omega \int\limits_\Omega \dfrac{|u(x) - u(y)|^p}{|x - y|^{1+2p}} \, dx \, dy \right)^{\frac{1}{p}} & \text{for } p < 2 \\ \left( \int\limits_\Omega \int\limits_\Omega \dfrac{|\nabla u(x) - \nabla u(y)|^p}{|x - y|^{1+p}} \, dx \, dy \right)^{\frac{1}{p}} & \text{for } p > 2 \\ |\nabla u|_2 & \text{for } p = 2. \end{cases}$$

For the elementary properties of the space introduced above, we refer the reader to MIRANDA (1978) and SOLONNIKOV (1977). Finally, if $X$ is a Banach space, by $L^p(0, s; X)$ we denote the class of functions $u$ from $(0, s)$ in $X$ such that $\int\limits_0^s |u|_X^p \, d\xi < \infty$, where $|\cdot|_x$ is the $X$-norm. We remark explicitly that, depending on the context, we shall use the same symbol for a space of scalar, vector or tensor-valued functions. Also, by the symbol $C$ we denote a generic constant whose numerical value is inessential to our aims; it may have several values in a single computation. Typically, we may have $2C < C$ in the same line.

Now, for $T > 0$, consider in $\Omega$ and during the time interval $[0, T]$ the motion of a viscous, incompressible fluid, governed by the Navier-Stokes equations.

If $u(x, t), \pi(x, t)$ represent velocity and pressure fields of that motion, they must satisfy the system

$$(1.1)\qquad \begin{aligned} &\frac{\partial u}{\partial t} + u \cdot \nabla u = -\nabla \pi + \Delta u \\ &\nabla \cdot u = 0 \quad \text{in } \Omega_T, \end{aligned}$$

where the kinematic viscosity has been set equal to one, without loss of generality. Along with (1.1) we shall consider the following initial conditions and boundary conditions:

$$(1.2)\qquad \begin{aligned} u(x, 0) &= u_0(x) \quad x \in \Omega, \\ u(y, t) &= 0 \quad (y, t) \in \partial\Omega \times (0, T). \end{aligned}$$

In this section, however, we will be interested in giving some results concerning the *linearized* version of problem (1.1)–(1.2), which we shall write in the following form:

$$(1.3)\qquad \begin{aligned} &\frac{\partial w}{\partial t} = \Delta w - \nabla p + f, \\ &\nabla \cdot w = 0 \quad \text{in } \Omega_T, \\ &w(x, 0) = w_0(x) \quad x \in \Omega, \\ &w(y, t) = 0 \quad (y, t) \in \partial\Omega \times (0, T), \end{aligned}$$

where $f = f(x, t)$ is a given "body force". We have the following results.

**Lemma 1.1.** *For any* $f \in L^q(\Omega_T)$, $w_0 \in \mathring{J}^{2-2/q,q}(\Omega)$, $q > 1$, *problem* (1.3) *admits for all* $T > 0$ *a unique solution in* $\mathscr{W}_q(\Omega_T)$ *satisfying the following estimate:*

$$(1.4)\qquad |w|_{\mathscr{W}_q(\Omega_T)} + |\nabla p|_{q,\Omega_T} \leqq C(T)\{|f|_{q,\Omega_T} + \|w_0\|_{2-2/q,q}\},$$

*where* $C$ *depends only on* $T$.

**Proof.** It is a consequence of Theorem 4.2 and Remark 2 on p. 495 of the paper of SOLONNIKOV (1977).

**Lemma 1.2.** *Assume* $f \in C_0^\infty(\Omega_T)$ *and* $w_0 \in \mathscr{C}_0^\infty(\Omega)$. *Then there exists a unique (smooth) solution* $w(x, t)$ *of* (1.3) *which is in* $\mathscr{W}_q(\Omega_T)$ *and satisfies the estimate:* (1.4) *for all* $q > 1$.

**Proof.** In the light of Lemma 1.1 the proof is reduced to show the following uniqueness result. Let $w_1, p_1$ and $w_2, p_2$ be two solutions of (1.3) corresponding to the same data $f$ and $w_0$ and satisfying (1.4) with exponents $q_1$ and $q_2$ respectively; then, $w_1 = w_2$ and $\nabla p_1 = \nabla p_2$. Assume $q_1 < q_2$ and set $w_1 - w_2 = w$, $p_1 - p_2 = p$. We thus conclude that $w, p$ is a solution of the problem (1.3) with $f = w_0 = 0$. Denote by $\omega$ a smooth neighborhood of $\partial\Omega$; obviously $\nabla p$,

$\dfrac{\partial^2 \boldsymbol{w}}{\partial x_i\, \partial x_j} \in L^{q_1}(0, T; L^{q_1}(\Omega \cap \omega))$ for all $i, j = 1, 2, 3$. Moreover, since $p$ satisfies

$$\Delta p = 0 \quad \text{in} \quad (\Omega - \omega) \times \{t\},$$

$$\frac{dp}{dn} = \Delta \boldsymbol{w} \cdot \boldsymbol{n} \quad \text{on} \quad \partial(\Omega - \omega) \times \{t\},$$

from SOLONNIKOV (1977) [lemma 2.1], we conclude that $|\nabla p|_{q_1, \Omega_T} < C$. Uniquenes, therefore, follows as a paricular case of the theorems proved by GALDI & MAREMONTI (1984), GALDI & MAREMONTI (1986).

We end this section by proving an inequality of Sobolev type.

**Lemma 1.3.** *Let $f\colon \Omega \to \mathbb{R}$ with $\nabla f \in L^p(\Omega)$, $p < 3$. Then there is a constant $f_0$ such that $\psi = f - f_0$ satisfies the following inequalities:*

$$R^{3-p} \int_{S(1)} |\psi(R, \gamma)|^p \, d\gamma \leqq C \int_{|x| \geqq R} |\nabla f|^p \, dx, \tag{1.5}$$

$$|\psi|_q \leqq C |\nabla f|_p, \tag{1.6}$$

*where $1 < p < 3$, $q = 3p/(3 - p)$ and $S(1)$ is the unit sphere.*

**Proof.** The first part of the lemma and the estimate (1.5) are a generalization of an analogous result proved by PAYNE & WEINBERGER (1957) for $p = 2$. However, the proof given by these authors remains unchanged in our case $1 < p < 3$ and therefore, it will be omitted. As far as (1.6) is concerned, we proceed as follows. Denote by $\varphi$ a smooth "cut-off" function such that $\varphi(x) = 0$ for $|x| \leqq R$ and $\varphi(x) = 1$ for $|x| \geqq 2R$ $(R > R_0)$, and set $\tilde{\psi} = \varphi\psi$. For $\tilde{\psi}$ we have the representation

$$\tilde{\psi}(x) = \frac{1}{4\pi} \sum_{j=1}^{3} \int_{|x-y| \leqq \varrho} \frac{\partial \tilde{\psi}}{\partial y_j} \frac{\partial}{\partial y_j} \left( \frac{1}{|x - y|} \right) dy - \frac{1}{4\pi} \int_{|x-y| = \varrho} \tilde{\psi} \frac{\partial}{\partial n} \left( \frac{1}{|x - y|} \right) d\sigma, \tag{1.7}$$

for $\varrho > R$. In virtue of (1.5), (1.7) furnishes in the limit as $\varrho \to \infty$

$$\tilde{\psi}(x) = \frac{1}{4\pi} \sum_{j=1}^{3} \int_{\mathbb{R}^3} \frac{\partial \tilde{\psi}}{\partial y_j} \frac{\partial}{\partial y_j} \left( \frac{1}{|x - y|} \right) dy.$$

By well known estimates on Riesz potentials, *cf.* MIRANDA (1978), we thus obtain the Sobolev inequality for $\tilde{\psi}$,

$$|\tilde{\psi}|_q \leqq C |\nabla \tilde{\psi}|_p, \quad q = \frac{3p}{3 - p}. \tag{1.8}$$

From (1.8) we easily have

$$\left( \int_{|x| \geqq 2R} |\psi|^q \, dx \right)^{\frac{1}{q}} \leqq C \left\{ |\nabla f|_p + \left( \int_{R \leqq |x| \leqq 2R} |\psi|^p \, dx \right)^{\frac{1}{p}} \right\}. \tag{1.9}$$

Applying Poincaré's inequality, *cf.* MIRANDA (1978), to the second integral on the right-hand side of (1.9), we deduce

$$\left(\int_{|x|\geq 2R}|\psi|^q\,dx\right)^{\frac{1}{q}}\leq C\left\{|\nabla f|_p+\left(\int_{\{|x|=R\}\cup\{|x|=2R\}}|\psi|^p\,d\sigma\right)^{\frac{1}{p}}\right\}. \tag{1.10}$$

Therefore, from (1.10) and (1.5) it follows that

$$\left(\int_{|x|\geq 2R}|\psi|^q\,dx\right)^{\frac{1}{q}}\leq C\,|\nabla f|_p. \tag{1.11}$$

To complete the proof of the theorem we now choose another smooth "cut-off" function $\xi(x)$ such that $\xi(x)=1$ for $|x|\leq 2R$ and $\xi(x)=0$ for $|x|\geq 3R$. Since $\xi\psi$ is of compact support in $\bar{\Omega}$ we can apply the results of SOLONNIKOV & SCADILOV (1973) [Lemma 1 and the remark following it], to obtain

$$|\xi\psi|_q\leq C\,|\nabla(\xi\psi)|_p$$

from which it follows that

$$\left(\int_{|x|\leq 2R}|\psi|^q\,dx\right)^{\frac{1}{q}}\leq C\left\{|\nabla f|_p+\left(\int_{2R\leq|x|\leq 3R}|\psi|^p\,dx\right)^{\frac{1}{p}}\right\}\leq C\,|\nabla f|_p, \tag{1.12}$$

where in the last step we have used the Hölder inequality (recall that $q>p$) and (1.11). Therefore, estimate (1.6) is a consequence of (1.11) and (1.12).

*Remark.* We notice that lemma 1.2 remains valid whatever be the number $n$ of space dimensions, provided $p<n$ and $q=np/(n-p)$.

## 2. Existence theorem

The aim of this section is to prove a theorem of existence of global (weak) solutions to the problem (1.1). Precisely, denoting by $H^{9/10,5/4}(\Omega)$ the completion of $\mathscr{C}_0^\infty(\Omega)$ in the norm of $W^{9/10,5/4}(\Omega)$, we have

**Theorem 2.1.** *Assume* $\boldsymbol{u}_0\in H^{9/10,5/4}(\Omega)$. *Then there are functions* $\boldsymbol{u}(x,t)$, $p(x,t)$ *enjoying the following properties for all* $T>0$

(i) $\boldsymbol{u}\in L^\infty(0,T;\mathring{J}(\Omega))\cap L^2(0,T;\mathring{J}^{1,2}(\Omega))$;
$\boldsymbol{u}\in \mathscr{W}_{5/4}(\Omega_T)$; $\nabla p\in L^{5/4}(\Omega_T)$;

(ii) $\boldsymbol{u},p$ *satisfy* (1.1) *a.e. in* $\Omega_T$;

(iii) $\boldsymbol{u},p$ *satisfy the "energy inequality" in the following form*:

$$|\boldsymbol{u}(t)|_2^2\leq|\boldsymbol{u}(s)|_2^2-2\int_s^t|\nabla\boldsymbol{u}(\tau)|_2^2\,d\tau$$

*for* $s=0$, *for almost all* $s>0$ *and for all* $t\geq s$;

(iv) $\boldsymbol{u}(x, t)$ *can be redefined on a set of zero t-measure in such a way that* $\lim_{t \to s^+} |\boldsymbol{u}(t) - \boldsymbol{u}(s)|_2 = 0$ *for all* $s \in [0, T]$.

Before proving the theorem, we wish to make the following remarks.

(a) The solutions provided by theorem 2.1 are more regular than HOPF's solutions [HOPF (1951)], and, in particular, their kinetic energy decreases monotonically with time. To our knowledge, this is the first example of global solutions which enjoy this property in an exterior domain $\Omega \neq \mathbb{R}^3$ *without restrictions on the "size" of the data*. For the Cauchy problem, solutions satisfying (iii) were constructed by LERAY (1934).

(b) In the case $\Omega = \mathbb{R}^3$ and $\Omega$ bounded, theorem 2.1 has already been proved by O. A. LADYZHENSKAYA (1969).

(c) The regularity of the initial data can be weakened to $\boldsymbol{u}_0 \in Y(\Omega) = \mathring{J}^{2/5,5/4}(\Omega) \cap \mathring{J}(\Omega)$, provided one can prove that $\mathscr{C}_0^\infty(\Omega)$ is dense in $Y(\Omega)$ (endowed with its natural norm).

(d) As far as the smoothness of our solutions is concerned, we can give only results of partial regularity. In fact, on the one hand, because of the validity of (iii) and of the results of HEYWOOD (1980) concerning existence of classical solutions (local in time or global for small data), it is possible to prove a "théorème de structure" in the sense of LERAY (*cf.* HEYWOOD (1980)). On the other hand, by suitably modifying the construction, one can show that our solutions verify a "generalized energy inequality" in the sense specified in CAFFARELLI, KOHN & NIRENBERG (1982). Therefore they would have further (partial) regularities along the lines of CAFFARELLI, KOHN & NIRENBERG (1982) (Theorem B).

The method we shall employ to prove theorem 2.1 is that introduced by J. LERAY (1934, chapter V), to construct his "solutions turbolentes". We thus begin to consider for all $n = 1, 2, \ldots,$ the following initial-boundary value problem

$$\begin{aligned} &\frac{\partial \boldsymbol{U}}{\partial t} + \boldsymbol{U}_{(n)} \nabla \boldsymbol{U} = \Delta \boldsymbol{U} - \nabla \tau, \\ &\nabla \cdot \boldsymbol{U} = 0 \quad \text{in } \Omega_T, \\ &\boldsymbol{U}(x, 0) = \boldsymbol{U}_{0n}(x) \quad x \in \Omega, \\ &\boldsymbol{U}(y, t) = 0 \quad (y, t) \in \partial\Omega \times (0, T), \end{aligned} \tag{2.1}$$

where

$$\boldsymbol{U}_{(n)}(x, t) \equiv \int_{\mathbb{R}^3} J_{1/n}(x - y)\, \boldsymbol{U}(y, t)\, dy$$

is a (spatial) "mollification" of $\boldsymbol{U}$, and $\{\boldsymbol{U}_{0n}(x)\}$ is a sequence of functions from $\mathscr{C}_0^\infty(\Omega)$ converging to $\boldsymbol{u}_0(x)$ in the space $H^{9/10,5/4}$. For the system (2.1) it is not hard to prove the existence of a global regular solution. In fact, recalling that

$$\sup_{\Omega} |\boldsymbol{U}_{(n)}(x, t)| \leqq c(n)\, |\boldsymbol{U}(t)|_2 \tag{2.2}$$

where $c(n)$ depends only on ($c(n) \to \infty$ as $n \to \infty$), we may show that $U(x, t)$ obeys the following (formal) *a priori* estimates

$$\tfrac{1}{2} \frac{d}{dt} |U(t)|_2^2 = -|\nabla U(t)|_2^2,$$

$$\text{(2.3)} \qquad \frac{d}{dt} |\nabla U(t)|_2^2 \leq - \left| \frac{\partial U}{\partial t} \right|_2^2 + c^2(n) \, |U(t)|_2^2 \, |\nabla U(t)|_2^2,$$

$$\frac{d}{dt} |\nabla U(t)|_2^2 \leq -|P \Delta U|_2^2 + c^2(n) \, |U(t)|_2^2 \, |\nabla U(t)|_2^2$$

$P$ being the orthogonal projection of $L^2(\Omega)$ onto $\mathring{J}(\Omega)$ (LADYZHENSKAYA (1969)). Relations (2.3) are easily obtained after multiplication of (2.1) by $U$ (estimate (2.3)), by $\frac{\partial U}{\partial t}$ (estimate (2.3)) and by $-P \Delta U$ (estimate (2.3)) and employing inequality (2.2). Using (2.3) and the method of HEYWOOD (1980), one thus proves the existence of classical solutions $U(x; t, n)$, $\tau(n; x, t)$ of (2.1) satisfying (in particular) the following properties for all $n$:

$$U \in L^\infty(0, T; \mathring{J}) \cap L^\infty([0, T]; \mathring{J}^{1,2}),$$

$$\frac{\partial U}{\partial t}, \; \frac{\partial^2 U}{\partial x_i \, \partial x_j}, \; \nabla \tau \in L^2(\Omega_T)$$

and the estimate (2.3).

We now wish to obtain estimates of the solution $U, \tau$ which hold *uniformly in n*. To this end, let us consider the linearized problem (1.3) with $f(n; x, t) = U_{(n)} \cdot \nabla U$ and $w_0 = U_{0n}$. It is well known that (*cf.* LADYZHENSKAYA (1969)),

$$\text{(2.4)} \qquad |f|_{5/4, \Omega_T} \leq \sqrt{4} \, |U_{0n}|_2^{\frac{5}{2}};$$

therefore, the system (1.3) will admit a (unique) solution $w = w(n; x, t)$ satisfying the estimate (1.4) with $q = 5/4$. Let us prove that $w = U$. To this end, we shall follow an argument of LADYZHENSKAYA (1969). Setting $v = w - U$, one readily proves that $v$ satisfies the following identity:

$$\text{(2.5)} \qquad \int_{\Omega_T} v \cdot \left( \frac{\partial \phi}{\partial t} + \Delta \phi \right) dx \, dt = 0$$

for all solenoidal $\phi$ such that $\phi$, $\frac{\partial \phi}{\partial t}$, $\frac{\partial^2 \phi}{\partial x_i \, \partial x_j} \in L^q(\Omega_T)$ for all $q > 1$, $\phi$ vanishes on $\partial\Omega$ and $\phi(x, T) = 0$. In virtue of lemma 1.2 we may choose $\phi(x, t)$ as a solution of the adjoint problem of (1.3) corresponding to a "body force" $F(x, t) \in C_0^\infty(\Omega \times (0, T))$ and zero data at time $T$. Evidently, from (2.5) follows

$$\int_{\Omega_T} v(x, t) \cdot F(x, t) \, dx \, dt = 0$$

which, by the arbitrariness of $F$, implies $v \equiv 0$. From what we have just proved, we deduce that for all $n$, $U(n; x, t)$ satisfies the estimate (1.4), which by (2.4) can be written as

$$(2.6)\qquad \begin{aligned} |U|_{\mathscr{W}_{5/4}(\Omega_T)} + |\nabla\tau|_{5/4,\Omega_T} &\leqq C(T)\{|U_{0n}|_2^2 + \|U_{0n}\|_{2/5,5/4}\} \\ &\leqq C(T)\{|U_{0n}|_{9/10,5/4}^2 + |U_{0n}|_{9/10,5/4}\}, \end{aligned}$$

where $C$ depends only on $T$. In the last step of the inequality (2.6) we have used the embedding $H^{9/10,5/4} \hookrightarrow \overset{\circ}{J}$ (*cf.* MIRANDA (1978)).

We wish now to derive an estimate which, along with (2.3), will allow us to prove the energy inequality (iii) of the theorem. Following LERAY (1934), we introduce the "cut-off" function $g \in C^\infty(\mathbb{R}^3)$, $g(\xi) = 0$ for $|\xi| < 1$ and $g(\xi) = 1$ for $|\xi| \geqq 2$, and set $g_R(x) = g(x/R)\,(R > R_0)$. Multiplying (2.1) by $g_R U$ and integrating by parts, we obtain

$$(2.7)\qquad \begin{aligned} \tfrac{1}{2}\frac{d}{dt}|g_R^{1/2}U(t)|_2^2 \leqq{}& -|g_R^{1/2}\nabla U(t)|_2^2 + \tfrac{1}{2}\||\Delta g_R|^{1/2}U(t)|_2^2 \\ &+ \int_\Omega \nabla g_R\cdot U_{(n)}|U|^2\,dx + \int_\Omega \nabla g_R\cdot U\tau\,dx. \end{aligned}$$

The following inequalities hold:

$$\text{(a)}\qquad \begin{aligned} \int_{\Omega_T} \nabla g_R\cdot U_{(n)}|U|^2\,dx\,dt &\leqq \frac{C}{R}\int_0^T |U_{(n)}(t)|_3\,|U(t)|_3^2\,dt \leqq \frac{C}{R}\int_0^T |U(t)|_3^3\,dt \\ &\leqq \frac{C}{R}\int_0^T |\nabla U(t)|_2^{3/2}\,|U(t)|_2^{3/2}\,dt < \frac{C}{R}T^{1/4}|U_{0n}|_2^3 \end{aligned}$$

$$\text{(b)}\qquad \int_0^T \||\Delta g_R|^{1/2}U(t)|_2^2\,dt \leqq \frac{C}{R^2}T|U_{0n}|_2^2.$$

In the preceding inequalities we employed the following facts: $|\nabla g_R| \leqq C/R$, $|\Delta g_R| \leqq C/R^2$, the embedding $W^{1,2}(\Omega) \hookrightarrow L^3(\Omega)$ (*cf.* MIRANDA (1978)) and the relation $(2.3)_1$. We wish now to estimate the last integral on the right-hand side of (2.7). Since $\nabla\tau \in L^{5/4}(\Omega_T)$, because of lemma 1.3 we can redefine $\tau$ (without loss of generality) by adding a suitable function of $t$, in such a way that $\tau \in L^{4/5}(0, T; L^{15/7}(\Omega))$. We thus obtain

$$\text{(c)}\qquad \begin{aligned} \int_{\Omega_T} \nabla g_R\,U\tau\,dx\,dt &\leqq \frac{C}{R}\left(\int_0^T |\tau(t)|_{5/7}^{5/4}\,dt\right)^{4/5}\left(\int_0^T |U(t)|_{15/8}^5\,dt\right)^{1/5} \\ &\leqq \frac{C}{R}T^{1/5}\{|U_{0n}|_{9/10,5/4}^2 + |U_{0n}|_{9/10,5/4}\}\{U_{0n}|_{9/10,5/4}^{11/9} + |U_{0n}|_{9/10,5/4}\} \end{aligned}$$

where in the last step we have used (2.6), $L^p$ interpolation and the embedding $H^{9/10,5/4}(\Omega) \hookrightarrow L^2(\Omega)$ (*cf.* MIRANDA (1978)). Integrating (2.7) with respect to time, recalling that $\int_{|x|\geqq R} |U_{0n}(x)|^2\,dx$ and $|U_{0n}|_{9/10,5/4}$ can be increased by $\int_{|x|\geqq R} |u_0(x)|^2 dx$ and $|u_0|_{9/10,5/4}$, respectively, and using (a)–(c), we deduce

$$(2.8)\qquad \int_{|x|\geqq R+1} |U(n; x, t)|^2\,dx \leqq \int_{|x|\geqq R} |u_0(x)|^2\,dx + D/R,$$

where $D$ depends only on $|U_0|_{9/10,5/4}$.

Thanks to the estimate (2.6) it is now routine to show that from the sequence $\{U(n; x, t)\}$ it is possible to select a subsequence $\{U(n_k; x, t)\}$ converging to a function $\boldsymbol{u}(x, t)$ verifying the statement (i) of theorem 2.1. In particular one shows that $\boldsymbol{U} \to \boldsymbol{u}$ weakly in $L^2(0, T; \mathring{J}^{1,2})$ and weakly in $\mathring{J}(\Omega)$, uniformly with respect to $t$. This last fact in turn implies that $\boldsymbol{u}$ is weakly continuous in time in $\mathring{J}(\Omega)$. Moreover, since $\boldsymbol{U}$ satisfies (2.1) it also follows that $\boldsymbol{u}$ and $\pi$ satisfy statement (ii). For the details, we refer the reader to LERAY (1934), HOPF (1951), LADYZHENSKAYA (1969). We shall, however, briefly sketch the deduction of the energy inequality (iii). To this end, we shall again follow an argument of LERAY (1934). First of all, employing, for example, FRIEDRICHS' inequality (*cf.* MIRANDA (1978)), one proves that $\boldsymbol{U}(n_k; x, t)$ converges strongly in $L^2(0, T; L^2(\Omega'))$ for any compact subdomain $\Omega' \subset \Omega$. This implies that we can select another subsequence, again labelled $n_k$, which converges in $L^2(\Omega')$ for almost all $s \in [0, T]$. Now, multiplying (2.3) by $\boldsymbol{U} = \boldsymbol{U}(n_k; x, t)$ and integrating over $\Omega \times (s, t)$, we obtain

$$(2.9) \qquad 2 \int_s^t |\nabla \boldsymbol{U}(\xi)|_2^2 \, d\xi + |\boldsymbol{U}(t)|_2^2 = \int_{\Omega_R} |\boldsymbol{U}(x, s)|^2 \, dx + \int_{\Omega - \Omega_R} |\boldsymbol{U}(x, s)|^2 \, dx$$

where $\Omega_R = \Omega \cap \{|x| \leqq R\}$ $(R > R_0)$. Furthermore letting $n_k \to \infty$ in (2.9) by the weak convergence of $\boldsymbol{U}$ to $\boldsymbol{u}$ in $L^2(0, T; \mathring{J}^{1,2}(\Omega))$ and in $L^2(\Omega)$, we obtain

$$(2.10) \; 2 \int_s^t |\nabla \boldsymbol{u}(\xi)|_2^2 \, d\xi + |\boldsymbol{u}(t)|_2^2 \leqq \overline{\lim_{n_k \to \infty}} \left[ \int_{\Omega_R} |\boldsymbol{U}(x, s)|^2 \, dx + \int_{\Omega - \Omega_R} |\boldsymbol{U}(x, s)|^2 \, dx \right].$$

But, by (2.8) and the strong convergence of $\boldsymbol{U}$ to $\boldsymbol{u}$ in $L^2$ on compact sets, the right-hand side of (2.10) tends for almost all $s$ to $|u(s)|_2^2$ which, implies the validity of (iii). Finally, the validity of (iv) is a consequence of (iii) and of the weak continuity with respect to time of $\boldsymbol{u}$ in $\mathring{J}(\Omega)$. The theorem is therefore completely proved.

## 3. Decay of the kinetic energy

In this section we will prove the following theorem.

**Theorem 3.1.** *Assume that* $\boldsymbol{u}(x, t)$ *is a suitable weak solution of* (1.1)–(1.2)*: namely, it enjoys the following conditions for all* $T > 0$*:*

(a) $$\boldsymbol{u} \in L^2(0, T; \mathring{J}^{1,2}(\Omega)) \cap L^\infty(0, T; \mathring{J}(\Omega));$$

(b)

$$\int_{\Omega_T} \left[ -\boldsymbol{u} \left( \frac{\partial \boldsymbol{\phi}}{\partial t} + \Delta \boldsymbol{\phi} \right) + \boldsymbol{u} \cdot \nabla \boldsymbol{u} \cdot \boldsymbol{\phi} \right] dx \, dt = \int_\Omega \boldsymbol{u} \cdot \boldsymbol{\phi}(x, 0) \, dx - \int_\Omega \boldsymbol{u}(x, T) \, \boldsymbol{\phi}(x, T) \, dx$$

*for all* $\boldsymbol{\phi} \in L^2(0, T; \mathring{J}^{1,2}(\Omega) \cap W^{2,2}(\Omega))$ *with* $\dfrac{\partial \boldsymbol{\phi}}{\partial t} \in L^2(\Omega_T)$.

(c) *The energy inequality* (iii) *of theorem 2.1 holds.*

*Then, if* $u_0 \in \mathring{J}(\Omega) \cap L^q(\Omega)$ *for some* $1 \leqq q \leqq 3/2$, *the kinetic energy* $E(t) = \frac{1}{2}|u(t)|_2^2$ *(is non-increasing and) decays in time according to the following law:*

$$E(t) = O(t^{-2\beta}) \tag{3.1}$$

*where* $\beta = (2 - q)/4q$. *In particular, if* $\Omega \equiv \mathbb{R}^3$, $\beta = 3(2 - q)/4q$ *for* $q \in [4/3, 3/2]$ *and* $\beta = 3/8$ *if* $q \in [1, 4/3]$. *Finally if* $q = 2$ *we have* $E(t) = o(1)$.

Before giving a proof of the theorem we make some remarks. To this end we denote by $\mathcal{I}$ the class of weak solutions satisfying the assumptions of the theorem.

(a) The hypothesis $\mathcal{I} \neq \emptyset$ is basic to several results on behavior asymptotic in time in exterior domains: GALDI (1975), MASUDA (1975), HEYWOOD (1980), MAREMONTI (1984a). However, as already noticed in section 2, remark (a), $\mathcal{I} \neq \emptyset$ was known to happen, so far, either when $\Omega = \mathbb{R}^3$ [LERAY (1934)] or for (smooth) arbitrary $\Omega$ but for initial data whose amplitude is suitably small: LADYZHENSKAYA (1969), MASUDA (1975), HEYWOOD (1980). On the other hand, theorem 2.1 proves $\mathcal{I} \neq 0$ for a $C^2$-exterior domain without further restrictions; and so it allows us to extend the results of GALDI (1975), MASUDA (1975), HEYWOOD (1980), MAREMONTI (1984) to such a case.

(b) As observed in the previous remark, LERAY proves $\mathcal{I} \neq \emptyset$ with data $u \in \mathring{J}(\mathbb{R}^3)$ only. Thus, if $u \in \mathring{J}(\mathbb{R}^3) \cap L^1(\mathbb{R}^3)$ theorem 3.1 furnishes, as a particular case, for LERAY's solutions $E(t) = O(t^{-3/4})$ which improves on the estimate given in SCHOENBEK (1985), which shows that $E(t) = O(t^{-1/2})$.

To prove theorem 3.1 we need some preliminary lemmas.

**Lemma 3.1.** *Assume* $a_0 \in \mathring{J}^{1,2}(\Omega) \cap L^q(\Omega)$, $q \in [1, 2)$ *with* $|a_0|_2 \, |\nabla a_0(a_0)|_2 \leqq M$, *for a suitable numerical constant* $M$. *Then there is a unique regular solution* $a(x, t)$, $\varphi(x, t)$ *of* (1.1)–[1.2) *with* $a(x, 0) = a_0$ *such that* $|a(t)|_2^2 = O(t^{-\beta})$ *where* $\beta = (2 - q)/4q$. *In particular if* $\Omega \equiv \mathbb{R}^3$ *we can choose* $\beta = 3(2 - q)/4q$ *for* $q \in [4/3, 2)$ *and* $\beta = 3/8$ *if* $a \in [1, 4/3]$. *Finally, if* $q = 2$, $|a(t)|_2^2 = o(1)$.

**Proof.** See MAREMONTI (1986), Teorema 1.2 and n.5.

The next result is interesting in itself and concerns an estimate of the $L^q$-norm of *any* weak solution satisfying (a) and (b) of theorem *3.1*. This estimate becomes trivial when $\Omega$ is a bounded domain.

**Lemma 3.2.** *Let* $u, p$ *be two functions verifying* (a) *and* (b) *of theorem* 3.1. *Then if* $u_0 \in L^q(\Omega)$, *we have* $u \in L^\infty(0, T; L^q(\Omega))$ *for all* $T > 0$ *and for all* $q \in (1, 3/2]$.

**Proof.** Let $\phi(x, t) \equiv \psi(x, \tau - t)$, where $\psi$ is a solution of problem (1.3) with $f(x, t) \equiv 0$ and $\psi_0(x) = \psi(x, 0) \in \mathscr{C}_0^\infty(\Omega)$. On the strength of lemma 1.2, such a $\phi$ can be substituted into (b) of Theorem 3.1 to obtain

$$\int_\Omega u(x, \tau), \psi_0(x) \, dx = \int_{\Omega_\tau} u \cdot \nabla u \cdot \phi \, dx \, dt + \int_\Omega u_0(x) \cdot \psi(x, \tau) \, dx. \tag{3.2}$$

From a result of SOLONNIKOV (1977, Theorem 5.1, relation (5.6)), it follows that the solution $\psi$ obeys the estimate

$$\sup_{s\in[0,\tau]} |\psi(s)|_r \leqq C(\tau)\,|\psi_0|_r, \quad \text{for all } r > 1, \tag{3.3}$$

where $C(\tau)$ is an increasing function of $\tau$ only, which is bounded on every compact set. Now, by the Hölder inequality we have

$$\int_{\Omega_\tau} \boldsymbol{u}\cdot\nabla\boldsymbol{u}\cdot\boldsymbol{\phi}\,dx\,dt \leqq \int_0^T |\boldsymbol{u}(t)|_{\frac{2q'}{q'-2}}\,|\nabla\boldsymbol{u}(t)|_2\,|\boldsymbol{\phi}(t)|_{q'}\,dt \tag{3.4}$$

where $q' = q/(q-1)$. On the other hand, since $q \in (1, 3/2]$, we know that $2q'/(q'-2) \in (2, 6]$ and thus from the Sobolev inequality we have (*cf.* MIRANDA (1978)),

$$|\boldsymbol{u}|_{\frac{2q'}{q'-2}} \leqq C\,|\nabla\boldsymbol{u}|_2^{\frac{a}{2}}\,|\boldsymbol{u}|_2^{\frac{1}{2}-a}\ a = 3/q'. \tag{3.5}$$

Substituting (3.5) into (3.4), recalling that $\boldsymbol{u} \in (L^\infty(0, T; \mathring{J}(\Omega))$ and employing (3.3), we obtain

$$\int_{\Omega_\tau} |\boldsymbol{u}\cdot\nabla\boldsymbol{u}\cdot\boldsymbol{\phi}|\,dx\,dt \leqq C(T)\sup_{t\in[0,T]} |\boldsymbol{u}(t)|_2^{\frac{1}{2}-a}\,|\psi_0|_{q'}\int_0^T |\nabla\boldsymbol{u}|_2^{\frac{1}{2}+a}\,dt, \tag{3.6}$$

where $C(T)$ depends only on $T$. Moreover, using (3.3) again we deduce

$$\int_\Omega \boldsymbol{u}_0(x)\cdot\psi(x,\tau)\,dx \leqq C(T)\,|\boldsymbol{u}_0|_q\,|\psi_0|_{q'}. \tag{3.7}$$

Therefore, collecting (3.6) and (3.7) and recalling that $a < 1$, from (3.2), we obtain

$$\int_\Omega \boldsymbol{u}(x,\tau)\cdot\psi_0(x)\,dx \leqq C\,|\psi_0|_{q'} \tag{3.8}$$

where $C$ depends on $T$ and on the norms of $\boldsymbol{u}$ in $L^\infty(0,T;\mathring{J}(\Omega))$ and $L^2(0,T;\mathring{J}^{1,2}(\Omega))$. The arbitrariness of $\psi_0$ in $\mathscr{C}_0^\infty(\Omega)$ and the Helmholtz decomposition theorem of $L^q(\Omega)$ [see SOLONNIKOV (1977)] allow us to conclude from (3.8) that $\boldsymbol{u} \in L^\infty(0, T; L^q(\Omega))$, which completes the proof of the lemma.

We are now in a *position to prove theorem* 3.1. By assumption (c) it is easy to prove the existence of a time $\bar{t} > 0$ such that $|\boldsymbol{u}(\bar{t})|_2\,|\nabla\boldsymbol{u}(\bar{t})|_2 \leqq M$ and

$$|\boldsymbol{u}(t)|_2^2 \leqq |\boldsymbol{u}(\bar{t})|_2^2 - 2\int_{\bar{t}}^t |\nabla\boldsymbol{u}(s)|_2^2\,ds \quad \text{for all } t \geqq \bar{t}. \tag{3.9}$$

Moreover, since $\boldsymbol{u}_0 \in L^q(\Omega)$, by lemma 3.2 we also know that $|\boldsymbol{u}(\bar{t})|_q < \infty$. Therefore, if we take $\bar{t}$ as the initial time, the hypotheses of lemma 3.1 are satisfied and, consequently, there is a regular solution $\boldsymbol{u}^*(x, t)$, $\boldsymbol{p}^*(x, t)$ having $\boldsymbol{u}(x, \bar{t})$ as initial data and satisfying the estimate (3.1). On the other hand, since the weak solution $u(x, t)$ satisfies (3.9), by a well known uniqueness theorem of SATHER and SERRIN [SERRIN (1962)], we conclude that $u^*(x, t) = u(x, t)$ for all $t \geqq \bar{t}$. The theorem is therefore completely proved.

*Acknowledgments.* It is a pleasure to thank Professor V. A. SOLONNIKOV for discussions on linearized Navier-Stokes equations. This work was done under the auspices of GNFM of Italian C.N.R. and an MPI 40% Contract at the University of Napoli.

## References

L. CAFFARELLI, R. KOHN & L. NIRENBERG, *Partial regularity of suitable weak solutions of the Navier-Stokes equations*, Comm. on Pure and Applied Math., **35** (1982) pp. 771–831.

G. P. GALDI, *Sulla stabilità asintotica in media dei moti stazionari idrodinamici e magneto-idrodinamici in domini limitati o meno*, Ricerche di Matem., **24** (1975) pp. 137–151.

G. P. GALDI & P. MAREMONTI, *A uniqueness theorem for viscous flows in exterior domains with summability assumptions on the gradient of pressure*, Rend. Acc. Naz. dei Lincei, **76** (1985) pp. 28–32.

G. P. GALDI & P. MAREMONTI, *A uniqueness theorem for viscous fluid motions in exterior domains*, Arch. Rational Mech. and Anal., **91** (1985) pp. 375–384.

G. P. GALDI & S. RIONERO, *Weighted energy methods in fluid dynamics and elasticity*, Lecture Notes in Mathematics, No. 1134, Springer-Verlag, 1985.

J. G. HEYWOOD, *The Navier-Stokes equations: on the existence, regularity and decay of solutions*, Indiana Univ. Math. J., **29** (1980) pp. 639–681.

E. HOPF, *Über die Anfangswertaufgabe für die hydrodynamischen Grundgleichungen*, Math. Nachr., **4** (1951) pp. 213–231.

T. KATO, *Strong $L^p$-solutions of the Navier-Stokes equations in $R^m$, with applications to weak solutions*, Math. Z., **187** (1984) pp. 471–480.

O. A. LADYZHENSKAYA, *The mathematical theory of viscous incompressible flow*, 2nd ed., Gordon and Breach, New York, 1969.

J. LERAY, *Sur le mouvement d'un liquide visqueux emplissant l'espace*, Acta Math., **63** (1934) pp. 193–248.

P. MAREMONTI, *Asymptotic stability theorems for viscous fluid motions in exterior domains*, Rend. Sem. Matem. dell'Univ. di Padova, **71** (1984a) pp. 35–72.

P. MAREMONTI, *On the energy asymptotic decay of viscous incompressible fluids in exterior domains*, Proceedings of the workshop on mathematical aspects of fluid and plasma dynamics, Trieste, May 30–June 2, (1984b).

P. MAREMONTI, *Stabilità asintotica in media per moti fluidi viscosi in domini esterni*, Ann. Mat. Pura ed Appl., (1986 to appear).

K. MASUDA, *On the stability of incompressible viscous fluid motions past objects*, J. Math. Soc. Japan, **27** (1975) pp. 294–327.

K. MASUDA, *Weak solutions of Navier-Stokes equations*, Tohoku Math. J., **36** (1984) pp. 623–646.

C. MIRANDA, *Istituzioni di analisi funzionale lineare*, vol. I–II, Unione Matem. Italiana and C. N. R., Tip. Oderisi Ed., Gubbio (Italy), 1978.

L. E. PAYNE & H. S. WEINBERGER, *Note on a lemma of Finn and Gilbarg*, Acta Math., **98** (1957) pp. 297–299.

V. SCHEFFER, "*Turbulence and Hausdorff dimension*", in *Turbulence and the Navier-Stokes equations*, Lecture Notes in Math., No. 565, Springer-Verlag, 1976, pp. 94–112.

V. SCHEFFER, *The Navier-Stokes equation son a bounded domain* Comm. Math. Physics, **73** (1980) pp. 1–42.

M. E. SCHONBEK, *$L^2$-Decay for weak solutions of the Navier-Stokes equations*, Arch. Rational Mech. and Anal., **89** (1985) pp. 209–222.

J. SERRIN, *The initial value problem for the Navier-Stokes equations*, R. E. Langer Ed., Univ. of Wisconsin Press, 1963, pp. 69–98.

V. A. SOLONNIKOV, *Estimates for solutions of nonstationary Navier-Stokes equations*, J. of Soviet Math., **8** (1977) pp. 467–528.

V. A. SOLONNIKOV & V. E. SCADILOV, *On a boundary value problem for a stationary system of Navier-Stokes equations*, Trudy Math. Inst. Steklov, **125** (1973) pp. 186–199.

Istituto Matematico
Università di Ferrara
and
Dipartimento di Matematica
"R. Caccioppoli"
Università di Napoli

*(Received October 29, 1985)*

# *Quasilinear Hyperbolic Systems with Involutions*

CONSTANTINE M. DAFERMOS

*Dedicated to James Serrin on his sixtieth birthday*

## 1. Introduction

The evolution of $n$-dimensional continuous media with elastic response is generally governed by quasilinear hyperbolic systems of partial differential equations

$$\partial_t U + \sum_{\alpha=1}^{m} \partial_\alpha G_\alpha(U) = 0, \tag{1.1}$$

which may express, as applicable, the conservation laws of mass, momentum, energy, electric charge, *etc.* Here $x$ takes values in $R^m$ and $\partial_\alpha$ stands for the operator $\partial/\partial x_\alpha$. The state vector $U$ takes values in an open subset $\mathcal{O}$ of $R^n$ and

$$G_\alpha : \mathcal{O} \to R^n, \qquad \alpha = 1, \ldots, m, \tag{1.2}$$

are given, smooth, constitutive functions.

As a model, consider the referential formulation of the equations of hyperelasticity in $m$ space dimensions:

$$\begin{cases} \partial_t F_{i\alpha} = \partial_\alpha v_i, & i = 1, \ldots, m, \quad \alpha = 1, \ldots, m \\ \partial_t v_i = \sum_{\alpha=1}^{m} \partial_\alpha T_{i\alpha}(F), & i = 1, \ldots, m, \end{cases} \tag{1.3}$$

where $v$ denotes velocity, $F$ is the deformation gradient, and $T$ stands for the Piola-Kirchhoff stress, which is the gradient of the stored energy function $\sigma$, *i.e.*,

$$T_{i\alpha}(F) = \frac{\partial\sigma(F)}{\partial F_{i\alpha}}, \qquad i = 1, \ldots, m, \quad \alpha = 1, \ldots, m. \tag{1.4}$$

A second example is provided by Maxwell's equations:

$$\begin{cases} \partial_t B = -\operatorname{curl} E \\ \partial_t D = \operatorname{curl} H \end{cases} \tag{1.5}$$

where $B$ is the magnetic induction, $D$ is the electric displacement, $E$ is the electric field and $H$ is the magnetic field. Here we postulate general, nonlinear, constitutive relations

$$E = E(B, D), \qquad H = H(B, D), \tag{1.6}$$

which satisfy the *lossless condition*

$$\frac{\partial H}{\partial D} = \frac{\partial E}{\partial B}. \tag{1.7}$$

In continuum physics, the system (1.1) of conservation laws is supplemented by an inequality of the form

$$\partial_t \eta(U) + \sum_{\alpha=1}^{m} \partial_\alpha q_\alpha(U) \leqq 0 \tag{1.8}$$

which expresses, explicitly or implicitly, the Second Law of thermodynamics. In the context of thermodynamics proper, (1.8) is called the Clausius-Duhem inequality, $-\eta$ is entropy and $-q$ is entropy flux. We adopt here the standard terminology in the literature on quasilinear hyperbolic systems (*cf.* [11]) and call $\eta$ the *entropy* and $q$ the *entropy flux*, even though these quantities do not generally correspond to the physical quantities with the same name. For example, in the context of hyperelasticity the proper choice of "entropy" and "entropy flux" is (*cf.* [6])

$$\eta(F, v) = \tfrac{1}{2}|v|^2 + \sigma(F), \tag{1.9}$$

$$q_\alpha(F, v) = -\sum_{i=1}^{m} v_i T_{i\alpha}(F), \tag{1.10}$$

*i.e.*, "entropy" is mechanical energy. Similarly, in electromagnetism the natural choice of "entropy" and "entropy flux" is

$$\eta(B, D) = \frac{1}{4\pi} \int \{H \cdot dB + E \cdot dD\}, \tag{1.11}$$

$$q(B, D) = \frac{1}{4\pi} E \times H, \tag{1.12}$$

that is, "entropy" is electromagnetic field energy and "entropy flux" is the Poynting vector.

A fundamental premiss of continuum physics is that the Clausius-Duhem inequality should be satisfied automatically by any smooth physical process that complies with the conservation laws. Accordingly, we postulate*

$$\nabla q_\alpha(U) = \nabla \eta(U) \nabla G_\alpha(U), \qquad \alpha = 1, \ldots, m, \tag{1.13}$$

so that every smooth solution of (1.1) satisfies identically (1.8), as an equality. It is clear that (1.13) does indeed hold in hyperelasticity, when $\eta$ and $q$ are defined by (1.9), (1.10), and in electromagnetism, with $\eta$ and $q$ given by (1.11), (1.12).

* Throughout the paper $\nabla$ denotes gradient with respect to $U$.

By virtue of (1.13), the entropy inequality (1.8) does not impose any restriction on smooth solutions of (1.1). The situation changes when we consider weak solutions of (1.1). This is often necessary, because, when (1.1) is hyperbolic, even solutions starting out from smooth initial data generally develop discontinuities (shock waves) in a finite time. Within the context of weak solutions, (1.8) assumes an important role as an admissibility criterion employed in order to rule out solutions that are physically meaningless (*cf.* [11, 6]).

It is clear that when $\eta$ is an affine function of $U$ then (1.8) does not impose any constraints on (even weak) solutions of (1.1). The more "nonlinear" $\eta$ is, the further it restricts weak solutions of (1.1). The maximal effectiveness of (1.8) is attained when $\eta$ is a convex function of $U$. In fact the assumption of convexity of $\eta$ is often embodied into the definition itself of entropy (*cf.* [11]).

An entropy inequality (1.8) with $\eta$ convex exerts a smoothing influence on weak solutions of (1.1). This is manifested in a number of known properties, including the following: (a) The entropy inequality suppresses the oscillation in the weak limit of a family of admissible solutions of (1.1) (*cf.* [8, 9]). (b) Relatively weak shocks compatible with (1.8) satisfy the Lax shock stability criterion (*cf.* [11, 6]). (c) When they exist, smooth solutions of the Cauchy problem for (1.1) depend continuously upon the initial data, within the class of weak solutions with bounded variation that comply with (1.8) (*cf.* [7, 6]).

Nevertheless, quite often the laws of physics rule out convexity of the entropy. For example, the "entropy" (1.9) in our model system (1.3) would be convex only if the stored energy function $\sigma$ were convex. However, global convexity of $\sigma$ is generally incompatible with the principle of material frame-indifference (*cf.* [15]), according to which $\sigma$ is invariant under rigid rotations, *i.e.*,

$$\sigma(OF) = \sigma(F) \quad \text{for all orthogonal matrices } O. \tag{1.14}$$

At the same time, the conservation laws of physical systems are often endowed with special geometric structure which manifests itself in that a system of $k$ linear differential equations

$$\sum_{\alpha=1}^{m} A_\alpha \partial_\alpha U = 0, \tag{1.15}$$

with $A_\alpha$, $\alpha = 1, \ldots, m$, fixed $k \times n$ matrices, holds automatically for solutions of (1.1), so long as it is satisfied by the initial data. We call (1.15) *involutions* of (1.1). For instance,

$$\partial_\beta F_{i\alpha} - \partial_\alpha F_{i\beta} = 0, \quad \alpha, \beta = 1, \ldots, m, \quad i = 1, \ldots, m, \tag{1.16}$$

are involutions of the equations of motion (1.3) of hyperelastic media while

$$\operatorname{div} B = 0, \quad \operatorname{div} D = 0 \tag{1.17}$$

are involutions of Maxwell's equations (1.5). These involutions are inherited by the extensions of the theory of elasticity into the realm of thermodynamics and electrodynamics (thermoelasticity, magnetoelasticity, *etc.*).

In the presence of involutions, oscillations in certain directions are suppressed and wave amplitudes are confined on a proper subcone of $R^n$. It is thus plausible that the entropy inequality (1.8) may play an effective role as a regulator of waves, providing $\eta$ is convex just in the direction of the cone of amplitudes. In the context of hyperelasticity, the entropy (1.9) is convex in the direction of the cone of amplitudes if the stored energy function $\sigma$ is rank-one convex (*cf.* Section 2). This condition is not incompatible with the principle of material frame-indifference; it guarantees that the system (1.3) is hyperbolic.

The aim of this paper is to provide evidence in support of the conjecture that effective entropies need only be convex in the direction of the cone of amplitudes. As a first step towards that goal, we show, in Section 3, that under strict convexity of entropy in the direction of the cone of amplitudes, the Lax shock admissibility criterion (*cf.* [10, 6]) for weak shocks is equivalent to the entropy inequality (1.8). Under the same hypothesis, we demonstrate, in Section 4, that whenever the Cauchy problem for (1.1) has a classical solution then this solution is unique and stable within the broader class of weak solutions with bounded variation, which satisfy the entropy inequality (1.8), in the sense of measures, and have moderately strong shocks.

Readers familiar with the subject will not fail to notice similarities, in formalism as well as in spirit, between the ideas developed here and the theory of *compensated compactness* (*cf.* [14]).

## 2. The Cone of Amplitudes

We consider the system (1.1), assuming it is *hyperbolic* that is, for any unit vector $\nu$ in $R^m$ and every $U$ in $\mathcal{O}$ the matrix

$$\Lambda(U;\nu) := \sum_{\beta=1}^{m} \nu_\beta \nabla G_\beta(U) \tag{2.1}$$

has real eigenvalues (characteristic speeds) $\lambda_1(U;\nu), \ldots, \lambda_n(U;\nu)$, not necessarily distinct, and $n$ linearly independent eigenvectors (characteristic directions) $r_1(U;\nu), \ldots, r_n(U;\nu)$.

We assume that, for some positive integer $k$, there is a collection $\{A_\alpha : \alpha = 1, \ldots, m\}$ of $k \times n$ matrices with the property

$$A_\alpha G_\beta(U) + A_\beta G_\alpha(U) = 0, \qquad \alpha, \beta = 1, \ldots, m, \tag{2.2}$$

for all $U$ in $\mathcal{O}$.

Under the above hypothesis, if $U(x, t)$ is any bounded measurable (weak) solution of (1.1) then

$$\partial_t \sum_{\beta=1}^{m} A_\beta \partial_\beta U = \sum_{\alpha,\beta=1}^{m} \partial_\alpha \partial_\beta \{A_\beta G_\alpha(U)\} = 0. \tag{2.3}$$

In particular, solutions of the Cauchy problem for (1.1) will satisfy the involuti on (1.15), provided the initial data do so.

From (2.2) it follows easily that

$$\sum_{\alpha,\beta=1}^{m} \nu_\alpha \nu_\beta A_\alpha \nabla G_\beta(U) = 0, \tag{2.4}$$

for any unit vector $\nu$ in $R^m$. We set

$$N(\nu) := \sum_{\alpha=1}^{m} \nu_\alpha A_\alpha; \tag{2.5}$$

(2.4) implies that any nonzero row of $N(\nu)$ is a left eigenvector of $\Lambda(U;\nu)$, with zero eigenvalue. Consequently, systems endowed with involutions typically have zero characteristic speeds with high multiplicity and so they are not strictly hyperbolic.

Next we turn to shock waves. The Rankine-Hugoniot condition for a shock that propagates in the direction of the unit vector $\nu$ with speed $s$ reads

$$\sum_{\beta=1}^{m} \nu_\beta [G_\beta(U)] = s[U], \tag{2.6}$$

where, as usual, a bracket denotes the jump (left minus right limit) of the enclosed quantity across the shock.

By combining (2.6) with (2.2) we get

$$s\sum_{\alpha=1}^{m} \nu_\alpha A_\alpha [U] = \sum_{\alpha,\beta=1}^{m} \nu_\alpha \nu_\beta A_\alpha\, [G_\beta(U)] = 0. \tag{2.7}$$

Therefore, if $s \neq 0$, (2.7) and (2.5) give

$$N(\nu)\,[U] = 0. \tag{2.8}$$

Note that (2.8) can be derived, even when $s = 0$, as the Rankine-Hugoniot jump condition of the involution, assuming, of course, that (1.15) holds at the initial time.

It follows from (2.8) that the amplitude $[U]$ of the shock lies on the null space $\ker N(\nu)$ of $N(\nu)$. In particular, $\ker N(\nu)$ contains every characteristic direction with nonzero characteristic speed, because all nonzero rows of $N(\nu)$ are left eigenvectors of $\Lambda(U;\nu)$ associated with the zero eigenvalue.

Motivated by the above discussion, we now introduce the *cone of amplitudes* $\mathscr{K}$ as the union of all $\ker N(\nu)$, *i.e.*, as the set of $V$ in $R^n$ with

$$N(\nu)\,V = 0, \tag{2.9}$$

for some unit vector $\nu$ in $R^m$. In particular, $\mathscr{K}$ contains every characteristic direction with nonzero characteristic speed.

At this point it may be instructive to illustrate the above in the context of the model system (1.3), which is hyperbolic if the *acoustic tensor* (*cf.* [15])

$$Q_{ij}(F;\nu) = \sum_{\alpha,\beta=1}^{m} \frac{\partial^2 \sigma(F)}{\partial F_{i\alpha}\,\partial F_{j\beta}} \nu_\alpha \nu_\beta \tag{2.10}$$

is positive definite, for all unit vectors $\nu$ in $R^m$. The $m^2 + m$ characteristic speeds are the $2m$ square roots of the $m$ eigenvalues of $Q$ and zero, with multiplicity

$m^2 - m$. All zero characteristic speeds are induced by the involution (1.16), *i.e.*, the left eigenspace of the zero eigenvalue is spanned by the rows of $N(\nu)$, and (2.9) takes the form

$$\nu_\beta S_{i\alpha} - \nu_\alpha S_{i\beta} = 0, \quad i = 1, \dots, m, \quad \alpha, \beta = 1, \dots, m. \tag{2.11}$$

It is clear that (2.11) holds if and only if $S_{i\alpha} = \zeta_i \nu_\alpha$, $i = 1, \dots, m$, $\alpha = 1, \dots, m$, for some $\zeta$ in $R^m$. In particular, the $m \times m$ matrix $S$ has rank one. Consequently, the cone of amplitudes $\mathscr{K}$ consists of vectors $(\zeta \otimes \nu, w)$ in $R^{m^2+m}$ with arbitrary $\zeta, \nu$ and $w$ in $R^m$.

For the case of Maxwell's equations (1.5), with involutions (1.17), condition (2.9) reads

$$\sum_{\alpha=1}^{m} \nu_\alpha B_\alpha = 0, \quad \sum_{\alpha=1}^{m} \nu_\alpha D_\alpha = 0. \tag{2.12}$$

When $m = 3$, (2.12) is satisfied for any two vectors $B$, $D$ in $R^3$ and so the cone of amplitudes $\mathscr{K}$ covers the entire phase space $R^6$. By contrast, if $m = 2$, *i.e.*, $B$ and $D$ depend only upon two space variables $(x_1, x_2)$ and time $t$, then (2.12) holds only when the projections of $B$, $D$ on the plane $(x_1, x_2)$ are collinear and so the cone of amplitudes is a proper subset of the phase space $R^6$.

We now return to the general system (1.1), with involutions (1.15), endowed with an entropy $\eta$. We are pursuing the idea that entropy need only be convex in the direction of the cone of amplitudes $\mathscr{K}$.

**Definition 2.1.** We say $\eta$ is *strictly $\mathscr{K}$-convex* if for any $U$ in $\mathscr{O}$

$$Z^T \Theta(U) Z > 0 \quad \text{for all } Z \in \mathscr{K} \setminus \{0\}, \tag{2.13}$$

where $\Theta$ is the Hessian $\nabla(\nabla\eta)^T$.

Referring to system (1.3), for which the structure of the cone of amplitudes was discussed above, we easily see that the entropy (1.9) is strictly $\mathscr{K}$-convex if the stored energy function is *rank-one convex*, *i.e.*,

$$\sum_{i,j=1}^{m} \sum_{\alpha,\beta=1}^{m} \frac{\partial^2 \sigma(F)}{\partial F_{i\alpha}\, \partial F_{j\beta}} \nu_\alpha \nu_\beta \zeta_i \zeta_j > 0 \tag{2.14}$$

holds for all unit vectors $\zeta$ and $\nu$ in $R^m$. By virtue of (2.10), $\sigma$ is rank-one convex if and only if the acoustic tensor $Q$ is positive definite, *i.e.*, if and only if the system (1.3) is hyperbolic. The implications of (2.14) in elastostatics and elastodynamics have been studied extensively (*cf.* [1, 15]).

## 3. The Lax Stability Criterion

We fix $U_-$ in $\mathscr{O}$ and consider shocks with left state $U_-$ which propagate in the direction of a fixed unit vector $\nu$ in $R^m$. The Rankine-Hugoniot condition (2.6) implies that when the amplitude of the shock is small then the speed $s$ is near a characteristic speed $\lambda(U_-; \nu)$ and the amplitude $[U]$ points close to a corresponding characteristic direction $r(U_-; \nu)$.

As we saw in Section 2, $[U]$ and $r(U_-;\nu)$ lie on the null space $\ker N(\nu)$ of the matrix $N(\nu)$, defined by (2.5), which is invariant under $\Lambda(U;\nu)$. We assume $\lambda(U_-;\nu)$ is a simple eigenvalue of the restriction of $\Lambda(U_-;\nu)$ on $\ker N(\nu)$. Then the bifurcation theorem of CRANDALL & RABINOWITZ [3] yields the conclusion that every state $U_+$ in a small neighborhood of $U_-$ which can be joined to $U_-$ with a shock of speed close to $\lambda(U_-;\nu)$ lies on a smooth curve $U(\cdot)$: $(-1,1)\to\mathcal{O}$ through $U_-$, *i.e.*,

$$s(\tau)\{U(\tau)-U_-\}=\sum_{\alpha=1}^{m}\nu_\alpha\{G_\alpha(U(\tau))-G_\alpha(U_-)\},\qquad -1<\tau<1, \tag{3.1}$$

$$U(0)=U_-. \tag{3.2}$$

Differentiating (3.1) with respect to $\tau$ yields

$$\dot{s}(\tau)\{U(\tau)-U_-\}=\left\{\sum_{\alpha=1}^{m}\nu_\alpha\nabla G_\alpha(U(\tau))-s(\tau)\right\}\dot{U}(\tau). \tag{3.3}$$

In particular, setting $\tau=0$ in (3.3) we deduce

$$s(0)=\lambda(U_-;\nu),\ \dot{U}(0)=r(U_-;\nu). \tag{3.4}$$

As is well known, the shock that joins $U_-$ to $U_+=U(\tau_+)$, for some $\tau_+\in(-1,1)$, will be unstable unless it satisfies the Lax condition [10]:

$$\lambda(U_-;\nu)>s>\lambda(U_+;\nu). \tag{3.5}$$

The entropy inequality (1.8) induces another restriction on the same shock, namely,

$$-s\{\eta(U_+)-\eta(U_-)\}+\sum_{\alpha=1}^{m}\nu_\alpha\{q_\alpha(U_+)-q_\alpha(U_-)\}\leqq 0. \tag{3.6}$$

When the shock has small amplitude, it was shown by LAX [11] that (3.5) and (3.6) are equivalent, providing $\eta$ is strictly convex and the characteristic field is *genuinely nonlinear* at $U_-$, that is,

$$\nabla\lambda(U_-;\nu)\,r(U_-;\nu)\neq 0. \tag{3.7}$$

We proceed to show how these assumptions may be relaxed, in the presence of involutions.

**Proposition 3.1.** *Assume $\eta$ is strictly $\mathcal{K}$-convex and $s(\tau)$ is nondegenerate, in the sense that $\dot{s}(\tau)$ may have an at most finite set of zeros on* $(-1,1)$. *Then a shock with small amplitude satisfies the Lax stability conditions* (3.5) *if and only if it conforms with the entropy inequality* (3.6).

**Proof.** For $\tau\in(-1,1)$ we set

$$\Phi(\tau):=-s(\tau)\{\eta(U(\tau))-\eta(U_-)\}+\sum_{\alpha=1}^{m}\nu_\alpha\{q_\alpha(U(\tau))-q_\alpha(U_-)\}, \tag{3.8}$$

in which case (3.6) reduces to $\Phi(\tau_+) \leqq 0$. We differentiate $\Phi$ with respect to $\tau$ and use (1.13) and (3.3) to get

$$\begin{aligned}(3.9)\qquad \dot{\Phi}(\tau) &= -\dot{s}(\tau)\{\eta(U(\tau)) - \eta(U_-)\} + \left\{\sum_{\alpha=1}^{m} \nu_\alpha \nabla q_\alpha(U(\tau)) - s(\tau)\nabla\eta(U(\tau))\right\}\dot{U}(\tau) \\ &= -\dot{s}(\tau)\{\eta(U(\tau)) - \eta(U_-)\} + \nabla\eta(U(\tau))\left\{\sum_{\alpha=1}^{m} \nu_\alpha \nabla G_\alpha(U(\tau)) - s(\tau)\right\}\dot{U}(\tau) \\ &= -\dot{s}(\tau)\{\eta(U(\tau)) - \eta(U_-) - \nabla\eta(U(\tau))\,[U(\tau) - U_-]\}.\end{aligned}$$

Since $U(\tau) - U_-$ lies on $\mathscr{K}$ and $\eta$ is strictly $\mathscr{K}$-convex, the expression in braces on the right-hand side of (3.9) is negative, for any $\tau \neq 0$.

Assume, for definiteness, $\tau_+ > 0$. Since $s(\tau)$ is nondegenerate, if $\tau_+$ is sufficiently small, $\dot{s}(\tau)$ does not change sign on the open interval $(0, \tau_+)$. By virtue of (3.4) and (3.9), $s(\tau_+) < \lambda(U_-;\nu)$ is equivalent to $\dot{s}(\tau) < 0$, $0 < \tau < \tau_+$, and this, in turn, is equivalent to $\Phi(\tau_+) < 0$.

A symmetrical argument shows that (3.6) is also equivalent to the right half of (3.5). This completes the proof.

It is known (*cf.* [10]) that, under the proper parametrization, it is

$$(3.10)\qquad \ddot{s}(0) = \nabla\lambda(U_-;\nu)\, r(U_-;\nu)$$

and so the assumption of genuine nonlinearity (3.7) implies our hypothesis that $s(\tau)$ is nondegenerate.

In the context of hyperelasticity, Proposition 3.1 is related to earlier work by MALEK-MADANI [13] (*cf.* also [5]).

There are more discriminating shock stability criteria (*cf.* [12]) and entropy admissibility criteria (*cf.* [6]), which were designed in order to test shocks with large amplitude. It can be shown that these criteria are also equivalent, at least for shocks of moderate strength, provided the entropy is strictly $\mathscr{K}$-convex.

## 4. Uniqueness of Solutions

By a *classical solution* of (1.1) on a strip $R^m \times [0, \tau)$ we mean a uniformly Lipschitz continuous function $\overline{U}(x, t)$, defined on $R^m \times [0, \tau)$ and taking values in $\mathcal{O}$, which satisfies (1.1) almost everywhere. Classical solutions start out from smooth initial data and may survive on a short time interval; however, when the $G_\alpha$ are nonlinear, singularities generally develop in a finite time.

No existence theorem for globally defined weak solutions of the Cauchy problem for (1.1) is presently known, unless $n = 1$ or $m = 1$. It is natural to seek such solutions in the set $BV$ of functions with bounded variation (*cf.* [16]). A *weak solution* of class $BV$ is a bounded measurable function $U(x, t)$, defined on a strip $R^m \times [0, \tau)$ and taking values in $\mathcal{O}$, with distributional derivatives $\partial_t U$, $\partial_\alpha U$, $\alpha = 1, \ldots, m$, that are locally finite Borel measures, which satisfies (1.1) in the sense of distributions.

It can be shown, by means of elementary examples, that the Cauchy problem for (1.1), with fixed initial data, may have several solutions of class $BV$. In fact, uniqueness of solutions is not generally attained even after imposing the additional requirement that they satisfy an entropy inequality (1.8), with $\eta$ convex. However, as we shall see in this section, whenever a classical solution happens to exist then the entropy inequality, with $\eta$ strictly $\mathscr{K}$-convex, manages to rule out all other weak solutions of class $BV$ with shocks of moderate strength.

In the sequel we will be assuming that there is a collection $\Gamma_\alpha$, $\alpha = 1, \ldots, m$, of $n \times n$ matrices with the property

$$A_\alpha \Gamma_\beta + A_\beta \Gamma_\alpha = 0, \qquad \alpha, \beta = 1, \ldots, m \tag{4.1}$$

and such that, for any unit vector $\nu$ in $R^m$, the rows of the matrix $N(\nu)$, defined by (2.5), span the null space of the adjoint of the matrix

$$\Xi(\nu) := \sum_{\beta=1}^{m} \nu_\beta \Gamma_\beta . \tag{4.2}$$

For instance, in the typical case where the rows of $N(\nu)$ span the left eigenspace of the zero eigenvalue of $\Lambda(U; \nu)$* we may choose $\Gamma_\alpha = \nabla G_\alpha(\hat{U})$ for some fixed $\hat{U}$ in $\mathcal{O}$. However, $\Gamma_\alpha$ with the above properties may exist even in the exceptional case where the zero eigenvalue has higher multiplicity**.

For $y \in R^m$ and $p > 0$, we let

$$K_p(y) := \{x \in R^m : |x_\alpha - y_\alpha| < p, \alpha = 1, \ldots, m\} \tag{4.3}$$

denote the cube of side $2p$ centered at $y$. A distribution on $R^m$ will be called *$2p$-periodic* if it is periodic, with period $2p$, in $x_\alpha$, $\alpha = 1, \ldots, m$.

The following proposition shows that vector fields satisfying involutions are induced by a potential.

**Lemma 4.1.** *Assume that* $V \in [L^2_{\text{loc}}(R^m)]^n$ *is $2p$-periodic and that satisfies*

$$\int_{K_p(0)} V \, dx = 0, \tag{4.4}$$

$$\sum_{\alpha=1}^{m} A_\alpha \, \partial_\alpha V = 0. \tag{4.5}$$

*Then there exists a $2p$-periodic distribution* $\Psi \in [H^1_{\text{loc}}(R^m)]^n$ *such that*

$$V = \sum_{\beta=1}^{m} \Gamma_\beta \, \partial_\beta \Psi, \tag{4.6}$$

$$\|\Psi\|_{L^2(K_p(0))} \leqq cp \, \|V\|_{H^{-1}(K_p(0))}, \tag{4.7}$$

*where the constant $c$ depends solely on the $\Gamma_\alpha$. Conversely, any $V$ determined through* (4.6), *by a potential $\Psi$ with the above properties satisfies* (4.4), (4.5).

---

* As noted in Section 2, the equations of hyperelasticity have this property.

** The equations of adiabatic thermoelasticity provide such an example.

**Proof.** It is clear that (4.6) implies (4.4) and (4.5), the latter by virtue of (4.1).
Conversely, assume (4.4) and (4.5) hold. We expand $V$ in Fourier series:

$$V(x) = \sum_{\xi} \exp\left\{\frac{i\pi}{p}(\xi \cdot x)\right\} X(\xi), \tag{4.8}$$

where the summation runs over all $n$-tuples of integers. On account of (4.4), $X(0) = 0$. Substituting $V$ from (4.8) into (4.5) we deduce

$$\left(\sum_{\alpha=1}^{m} \xi_\alpha A_\alpha\right) X(\xi) = 0, \tag{4.9}$$

which shows that $X(\xi)$ lies on the complexification of the null space of $N(\nu)$, where

$$\nu = \frac{1}{|\xi|}\xi. \tag{4.10}$$

We attempt to determine the potential $\Psi$ by means of its expansion in Fourier series:

$$\Psi(x) = \sum_{\xi} \exp\left\{\frac{i\pi}{p}(\xi \cdot x)\right\} Y(\xi). \tag{4.11}$$

Substituting $\Psi$ from (4.11) into (4.6) and using (4.10), (4.2) and (4.8), we conclude

$$\Xi(\nu)\, Y(\xi) = -\frac{ip}{\pi}\frac{1}{|\xi|} X(\xi), \qquad \xi \neq 0. \tag{4.12}$$

Since $X(\xi)$ lies on the (complexification of the) null space of $N(\nu)$ and the rows of $N(\nu)$ span the null space of the adjoint of $\Xi(\nu)$, it follows that (4.12) has a solution $Y(\xi)$ for any $\xi \neq 0$ and

$$|Y(\xi)| \leq cp\frac{1}{|\xi|}|X(\xi)|, \qquad \xi \neq 0, \tag{4.13}$$

where $c$ is a constant that depends solely on the $\Gamma_\alpha$. From (4.13) and the basic theory of Sobolev spaces of periodic distributions (*cf.*, *e.g.*, [2]) it follows that $\Psi \in [H^1(K_p(0))]^n$ and (4.7) holds. This completes the proof.

The next two propositions may be viewed as generalizations of the standard Gårding inequality for strongly elliptic operators.

**Lemma 4.2.** *Assume $P$ is a symmetric $n \times n$ matrix which is positive in the direction of the cone of amplitudes $\mathcal{K}$, that is,*

$$Z^T P Z \geq \mu |Z|^2, \qquad Z \in \mathcal{K}, \tag{4.14}$$

*with* $\mu > 0$. *Then*

(4.15)
$$\int_{K_p(0)} V(x)^T PV(x)\,dx \geqq \mu \,\|V\|^2_{L^2(K_p(0))}$$

*for all* $2p$-*periodic* $V \in [L^2_{\text{loc}}(R^m)]^n$ *which satisfy* (4.4) *and* (4.5).

**Proof.** We again write down the Fourier expansion (4.8) of $V$. By Parseval's relation,

(4.16)
$$\|V\|^2_{L^2(K_p(0))} = (2p)^m \sum_{\xi} |X(\xi)|^2,$$

(4.17)
$$\int_{K_p(0)} V(x)^T PV(x)\,dx = (2p)^m \sum_{\xi} X(\xi)^* PX(\xi).$$

As noted in the proof of Lemma 4.1, since $V$ satisfies (4.4), (4.5), $X(\xi)$ lies on the complexification of the null space of $N(\nu)$, with $\nu$ given by (4.10), and hence, *a fortiori*, on the complexification of $\mathscr{K}$. Then (4.14) yields

(4.18)
$$X(\xi)^* PX(\xi) \geqq \mu\, |X(\xi)|^2.$$

Combining (4.16), (4.17) and (4.18), we arrive at (4.15). This completes the proof.

**Lemma 4.3.** *Assume $P$ is a bounded measurable symmetric $n \times n$ matrix-valued function, defined on $R^m$, which is uniformly positive in the direction of the cone of amplitudes $\mathscr{K}$, that is,*

(4.19)
$$Z^T P(x)\, Z \geqq \mu\, |Z|^2, \qquad Z \in \mathscr{K}, \qquad x \in R^m,$$

*with* $\mu > 0$, *and whose local oscillation is lower than* $\mu$, *in the sense*

(4.20)
$$\lim_{p \to 0} \sup_{x \in K_p(y)} |P(x) - P(y)|_{\mathscr{L}(R^n)} < \mu - 2\delta, \qquad y \in R^m,$$

*where* $\delta > 0$. *Suppose $K$ is any open cube in* $R^m$. *Then*

(4.21)
$$\int_K V(x)^T P(x)\, V(x)\,dx \geqq \delta\, \|V\|^2_{L^2(K)} - C\, \|V\|^2_{H^{-1}(K)}$$

*holds for any* $V \in [L^2(K)]^n$ *with compact support in $K$ which satisfies*

(4.22)
$$\int_K V(x)\,dx = 0,$$

(4.23)
$$\sum_{\alpha=1}^{m} A_\alpha\, \partial_\alpha V = 0,$$

*in the sense of distributions, where $C$ is a constant independent of $V$.*

**Proof.** By virtue of (4.20), there is a covering of $\bar{K}$ by the union of a finite set $K_1, \ldots, K_S$ of cubes, say $K_I = K_{p_I}(y_I)$, $I = 1, \ldots, S$, with the property

(4.24)
$$\sup_{x \in K_I} |P(x) - P(y_I)|_{\mathscr{L}(R^n)} \leqq \mu - 2\delta, \qquad I = 1, \ldots, S.$$

With the above covering we associate a partition of unity induced by $C^\infty$ functions $\phi_1, \dots, \phi_S$ on $R^m$ such that $\operatorname{supp} \phi_I \subset K_I \cap K$, $I = 1, \dots, S$, and

$$\sum_{I=1}^{S} \phi_I^2(x) = 1, \qquad x \in \operatorname{supp} V. \tag{4.25}$$

Then

$$\begin{aligned} \int_K V(x)^T P(x) V(x)\, dx &= \sum_{I=1}^{S} \int_{K_I} \phi_I^2(x) V(x)^T P(x) V(x)\, dx \\ &= \sum_{I=1}^{S} \int_{K_I} \phi_I^2(x) V(x)^T P(y_I) V(x)\, dx \\ &+ \sum_{I=1}^{S} \int_{K_I} \phi_I^2(x) V(x)^T [P(x) - P(y_I)] V(x)\, dx. \end{aligned} \tag{4.26}$$

On account of (4.24) and (4.25),

$$\sum_{I=1}^{S} \int_{K_I} \phi_I^2(x) V(x)^T [P(x) - P(y_I)] V(x)\, dx \geqq -(\mu - 2\delta) \int_K |V(x)|^2\, dx. \tag{4.27}$$

Next, recalling Lemma 4.1, we consider the potential $\Psi \in [H^1(K)]^n$ which induces $V$ through (4.6) and set

$$\phi_I(x) V(x) = U_I(x) + W_I(x), \qquad x \in K, \tag{4.28}$$

$$U_I(x) := \sum_{\beta=1}^{m} \Gamma_\beta\, \partial_\beta\{\phi_I(x) \Psi(x)\}, \tag{4.29}$$

$$W_I(x) := -\left\{\sum_{\beta=1}^{m} \partial_\beta \phi_I(x)\, \Gamma_\beta\right\} \Psi(x). \tag{4.30}$$

By Lemma 4.1,

$$\int_{K_I} U_I(x)\, dx = 0, \qquad I = 1, \dots, S, \tag{4.31}$$

$$\sum_{\alpha=1}^{m} A_\alpha\, \partial_\alpha U_I = 0, \qquad I = 1, \dots, S, \tag{4.32}$$

in the sense of distributions. Therefore, Lemma 4.2 yields

$$\int_{K_I} U_I(x)^T P(y_I) U_I(x)\, dx \geqq \mu \int_{K_I} |U_I(x)|^2\, dx, \qquad I = 1, \dots, S. \tag{4.33}$$

Furthermore, from (4.30) and (4.7) we get

$$\int_{K_I} |W_I(x)|^2\, dx \leqq c_1 \|V\|^2_{H^{-1}(K_I)}, \qquad I = 1, \dots, S, \tag{4.34}$$

where $c_1$ is a constant independent of $V$.

We now return to (4.26). By using (4.28), (4.33), and (4.34) we conclude that

$$(4.35)\quad \int_{K_I} \phi_I^2(x)\, V(x)^T P(y_I)\, V(x)\, dx$$
$$\geqq \left(1 - \frac{\delta}{2\mu}\right) \int_{K_I} U_I(x)^T P(y_I)\, U_I(x)\, dx - \frac{2\mu}{\delta} \int_{K_I} W_I(x)^T P(y_I)\, W_I(x)\, dx$$
$$\geqq \left(\mu - \frac{\delta}{2}\right) \int_{K_I} |U_I(x)|^2\, dx - c_2 \|V\|^2_{H^{-1}(K_I)}, \qquad I = 1, \ldots, S.$$

Again by (4.28) and (4.34),

$$(4.36)\quad \int_{K_I} |U_I(x)|^2\, dx \geqq \left(1 - \frac{\delta}{2\mu}\right) \int_{K_I} \phi_I^2(x)\, |V(x)|^2\, dx - \frac{2\mu}{\delta} \int_{K_I} |W_I(x)|^2\, dx$$
$$\geqq \left(1 - \frac{\delta}{2\mu}\right) \int_{K_I} \phi_I^2(x)\, |V(x)|^2\, dx - c_3 \|V\|^2_{H^{-1}(K_I)}.$$

Combining (4.26), (4.27), (4.35), (4.36) and using (4.25) we arrive at (4.21). This completes the proof.

We have now laid the preparation for proving the main result of this section establishing uniqueness and stability of classical solutions of (1.1) within the class of weak solutions with bounded variation which satisfy the entropy inequality and have moderately strong shocks. The proof combines ideas developed in [7] and [4].

**Theorem 4.1.** *Assume $\bar{U}(x, t)$ is a classical solution of* (1.1) *defined on a strip $R^m \times [0, \tau)$ and taking values in a convex compact subset $\mathscr{C}$ of $\mathcal{O}$. Let $U(x, t)$ be any BV weak solution of* (1.1) *which is defined on $R^m \times [0, \tau)$, takes values in $\mathscr{C}$, coincides with $\bar{U}(x, t)$, for all $x$ outside a fixed cube $K$ in $R^m$ and $t \in [0, \tau)$*, and such that*

$$(4.37)\quad \sum_{\alpha=1}^{m} A_\alpha\, \partial_\alpha [U(x, 0) - \bar{U}(x, 0)] = 0,$$

$$(4.38)\quad \int_{R^m} \{U(x, 0) - \bar{U}(x, 0)\}\, dx = 0.$$

*Suppose, further, $U(x, t)$ satisfies an entropy inequality* (1.8) *with $\eta$ strictly $\mathscr{K}$-convex. Under these hypotheses, there is a $\varkappa > 0$ with the property that if the strength of the shocks of $U(x, t)$ does not exceed $\varkappa$, in the sense*

$$(4.39)\quad \lim_{\rho \to 0} \sup_{x \in K_\rho(y)} |U(x, t) - U(y, t)| < \varkappa, \qquad y \in R^m, \qquad 0 \leqq t < \tau,$$

*then, for $t \in [0, \tau)$,*

(4.40)

$$\int_{R^m} |U(x, t) - \bar{U}(x, t)|^2\, dx \leqq C(1 + t) \exp(\gamma t + c t^2) \int_{R^m} |U(x, 0) - \bar{U}(x, 0)|^2\, dx,$$

---

* For instance the support of both $U(x, t)$ and $\bar{U}(x, t)$ may be contained in $K \times [0, \tau]$.

*where $C$ and $c$ depend solely on $\mathscr{C}$ while $\gamma$ depends on the Lipschitz constant of $\bar{U}(x, t)$ as well as on $\mathscr{C}$. In particular, if $U(x,0) \equiv \bar{U}(x, 0)$ on $R^m$ then $U(x, t) \equiv \bar{U}(x, t)$ on $R^m \times [0, \tau)$.*

**Proof.** For any vectors $U, \bar{U}$ in $\mathscr{C}$ let us define

$$\omega(U, \bar{U}) := \eta(U) - \eta(\bar{U}) - \nabla\eta(\bar{U})[U - \bar{U}], \tag{4.41}$$

$$\chi_\alpha(U, \bar{U}) := q_\alpha(U) - q_\alpha(\bar{U}) - \nabla\eta(\bar{U})[G_\alpha(U) - G_\alpha(\bar{U})], \tag{4.42}$$

$$\psi_\alpha(U, \bar{U}) := G_\alpha(U) - G_\alpha(\bar{U}) - \nabla G_\alpha(\bar{U})[U - \bar{U}]. \tag{4.43}$$

These functions are of quadratic order in $|U - \bar{U}|$ and so

$$|\omega(U, \bar{U})| \leqq C_1 |U - \bar{U}|^2, \tag{4.44}$$

$$|\psi_\alpha(U, \bar{U})| \leqq C_2 |U - \bar{U}|^2, \quad \alpha = 1, \dots, m, \tag{4.45}$$

where $C_1$ and $C_2$ depend solely on $\mathscr{C}$. Furthermore,

$$\omega(U, \bar{U}) = [U - \bar{U}]^T \hat{P}(U, \bar{U})[U - \bar{U}] \tag{4.46}$$

where

$$\hat{P}(U, \bar{U}) := \int_0^1 \int_0^w \Theta(\bar{U} + z(U - \bar{U}))\, dz\, dw, \tag{4.47}$$

with $\Theta$ denoting the Hessian of $\eta$. In particular, since $\eta$ is strictly $\mathscr{K}$-convex,

$$Z^T \hat{P}(U, \bar{U}) Z \geqq \mu |Z|^2, \quad Z \in \mathscr{K}, \tag{4.48}$$

where $\mu$ is a positive constant depending solely upon $\mathscr{C}$.

We now consider the two given functions $U(x, t)$ and $\bar{U}(x, t)$. They both satisfy Equation (1.1); $U(x, t)$ also satisfies (1.8), generally as an inequality, while $\bar{U}(x, t)$, being Lipschitz continuous, satisfies (1.8) as an equality. Combining the above properties and noting that the usual product rule for differentiation applies to the product of a Lipschitz function with a $BV$ function, we derive easily the inequality

$$\begin{aligned} &\partial_t \omega(U(x, t), \bar{U}(x, t)) + \sum_{\alpha=1}^m \partial_\alpha \chi_\alpha(U(x, t), \bar{U}(x, t)) \\ &\quad \leqq \partial_t \nabla\eta(\bar{U}(x, t))[U(x, t) - \bar{U}(x, t)] \\ &\qquad + \sum_{\alpha=1}^m \partial_\alpha \nabla\eta(\bar{U}(x, t))[G_\alpha(U(x, t)) - G_\alpha(\bar{U}(x, t))]. \end{aligned} \tag{4.49}$$

We now note that (1.13) implies the symmetry condition

$$\Theta(\bar{U}) \nabla G_\alpha(\bar{U}) = \nabla G_\alpha(\bar{U})^T \Theta(\bar{U}), \quad \alpha = 1, \dots, m. \tag{4.50}$$

Consequently, since $\bar{U}(x, t)$ is a Lipschitz continuous solution of (1.1) we get

$$\begin{aligned} \partial_t \nabla\eta(\bar{U}(x,t)) &= -\partial_t \bar{U}(x,t)^T \Theta(\bar{U}(x,t)) \\ &= -\sum_{\alpha=1}^{m} \partial_\alpha \bar{U}(x,t)^T \nabla G_\alpha(\bar{U}(x,t))^T \Theta(\bar{U}(x,t)) \\ &= -\sum_{\alpha=1}^{m} \partial_\alpha \nabla\eta(\bar{U}(x,t)) \nabla G_\alpha(\bar{U}(x,t)) \end{aligned} \tag{4.51}$$

and so (4.49), (4.43) yield

$$\begin{aligned} &\partial_t \omega(U(x,t), \bar{U}(x,t)) + \sum_{\alpha=1}^{m} \partial_\alpha \chi_\alpha(U(x,t), \bar{U}(x,t)) \\ &\leqq \sum_{\alpha=1}^{m} \partial_\alpha \nabla\eta(\bar{U}(x,t))\, \psi_\alpha(U(x,t), \bar{U}(x,t)). \end{aligned} \tag{4.52}$$

Integrating (4.52) over $R^m \times [0, t]$, $0 < t < \tau$, we obtain

$$\begin{aligned} \int_{R^m} \omega(U(x,t), \bar{U}(x,t))\, dx &\leqq \int_{R^m} \omega(U(x,0), \bar{U}(x,0))\, dx \\ &+ \int_0^t \int_{R^m} \sum_{\alpha=1}^{m} \partial_\alpha \nabla\eta(\bar{U}(x,s))\, \psi_\alpha(U(x,s), \bar{U}(x,s))\, dx\, ds. \end{aligned} \tag{4.53}$$

By virtue of (4.37), (4.38),

$$\sum_{\alpha=1}^{m} A_\alpha \partial_\alpha[U(x,t) - \bar{U}(x,t)] = 0, \qquad x \in R^m, \quad 0 \leqq t < \tau, \tag{4.54}$$

$$\int_{R^m} \{U(x,t) - \bar{U}(x,t)\}\, dx = 0, \qquad 0 \leqq t < \tau. \tag{4.55}$$

Moreover, taking account of (4.47), we infer that when (4.39) holds with $\varkappa$ sufficiently small then, for all $y \in R^m$, $t \in [0, \tau)$,

$$\lim_{\rho \to 0} \sup_{x \in K_\rho(y)} |\hat{P}(U(x,t), \bar{U}(x,t)) - \hat{P}(U(y,t), \bar{U}(y,t))| < \mu - 2\delta \tag{4.56}$$

where $\mu$ is the positive constant in (4.48) and $\delta > 0$. Thus, combining (4.46), (4.48), (4.56), (4.54), (4.55), recalling that $U(\cdot, t) - \bar{U}(\cdot, t)$ is supported in the cube $K$ and making appeal to Lemma 4.3, we deduce

$$\begin{aligned} &\int_{R^m} \omega(U(x,t), \bar{U}(x,t))\, dx \\ &\geqq \delta \| U(\cdot, t) - \bar{U}(\cdot, t)\|^2_{L^2(R^m)} - C_3 \| U(\cdot, t) - \bar{U}(\cdot, t)\|^2_{H^{-1}(R^m)}. \end{aligned} \tag{4.57}$$

We estimate the second term on the right-hand side of (4.57) as follows:

$$\text{(4.58)} \qquad \|U(\cdot,t)-\bar{U}(\cdot,t)\|_{H^{-1}(R^m)} \leqq \|U(\cdot,0)-\bar{U}(\cdot,0)\|_{L^2(R^m)} + \int_0^t \|\partial_t\{U(\cdot,s)-\bar{U}(\cdot,s)\}\|_{H^{-1}(R^m)}\,ds,$$

$$\text{(4.59)} \qquad \|\partial_t\{U(\cdot,s)-\bar{U}(\cdot,s)\}\|_{H^{-1}(R^m)} = \left\|\sum_{\alpha=1}^m \partial_\alpha\{G_\alpha(U(\cdot,s))-G_\alpha(\bar{U}(\cdot,s))\}\right\|_{H^{-1}(R^m)} \leqq \sum_{\alpha=1}^m \|G_\alpha(U(\cdot,s))-G_\alpha(\bar{U}(\cdot,s))\|_{L^2(R^m)} \leqq C_4\|U(\cdot,s)-\bar{U}(\cdot,s)\|_{L^2(R^m)}$$

where $C_4$ depends only on $\mathscr{C}$.

Combining (4.53) with (4.57), (4.58), (4.59), (4.44) and (4.45), we arrive at an estimate of the form

$$\text{(4.60)} \qquad \|U(\cdot,t)-\bar{U}(\cdot,t)\|^2_{L^2(R^m)} \leqq C_5\|U(\cdot,0)-\bar{U}(\cdot,0)\|^2_{L^2(R^m)} + \gamma\int_0^t \|U(\cdot,s)-\bar{U}(\cdot,s)\|^2_{L^2(R^m)}\,ds + 2c\left\{\int_0^t \|U(\cdot,s)-\bar{U}(\cdot,s)\|_{L^2(R^m)}\,ds\right\}^2$$

where $C_5$ and $c$ depend solely on $\mathscr{C}$ while $\gamma$ depends on the Lipschitz constant of $\bar{U}(x,t)$ as well as on $\mathscr{C}$. We majorize the last term on the right-hand side of (4.60) by Schwarz's inequality and then we apply the generalized Gronwall inequality, thus verifying (4.40). The proof of the theorem is now complete.

We have thus demonstrated that the entropy inequality manages to preserve uniqueness of smooth motions. On the other hand, it is known that the entropy inequality alone cannot guarantee uniqueness of motions with strong shocks. The results of Section 3 seem to indicate that the entropy inequality may be able to secure uniqueness of motions with weak shocks. A result in that direction has been established, by DIPERNA [7], for piecewise smooth solutions of a genuinely nonlinear system of two conservation laws in one space dimension.

*Acknowledgment.* This work was supported by the National Science Foundation through Grant No. DMS-8025355, by the U.S. Army under Contract No. DAAG-29-83-K-0029 and by the Office of Naval Research under Contract No. N00014-83-K-0542.

## References

1. BALL, J. M., Strict convexity, strong ellipticity and regularity in the calculus of variations. *Math. Proc. Cambridge Phil. Soc.* **87** (1980), 501–513.
2. BERS, L., F. JOHN, & M. SCHECHTER, *Partial Differential Equations*. New York: Interscience 1964.
3. CRANDALL, M. G., & P. H. RABINOWITZ, Bifurcation from simple eigenvalues. *J. Functional Anal.* **8** (1971), 321–340.

4. DAFERMOS, C. M., The second law of thermodynamics and stability. *Arch. Rational Mech. Anal.* **70** (1979), 167–179.
5. DAFERMOS, C. M., The equations of elasticity are special. *Trends in Applications of Pure Mathematics to Mechanics.* Vol. III, ed. R. J. KNOPS, London: Pitman 1981, pp. 96–103.
6. DAFERMOS, C. M., Hyperbolic systems of conservation laws. *Systems of Nonlinear Partial Differential Equations*, ed. J. M. BALL. NATO ASI Series C, No. 111. Dordrecht: D. Reidel 1983, pp. 25–70.
7. DIPERNA, R. J., Uniqueness of solutions to hyperbolic conservation laws. *Indiana U. Math. J.* **28** (1979), 137–188.
8. DIPERNA, R. J., Convergence of approximate solutions to conservation laws. *Arch. Rational Mech. Anal.* **82** (1983), 27–70.
9. DIPERNA, R. J., Compensated compactness and general systems of conservation laws. Trans. Am. Math. Soc. **242** (1985), 383–420.
10. LAX, P. D., Hyperbolic systems of conservation laws. *Comm. Pure Appl. Math.* **10** (1957), 537–566.
11. LAX, P. D., Shock waves and entropy. *Contributions to Functional Analysis*, ed. E. A. ZARANTONELLO. New York: Academic Press 1971, pp. 603–634.
12. LIU, T.-P., The entropy condition and the admissibility of shocks. *J. Math. Anal. Appl.* **53** (1976), 78–88.
13. MALEK-MADANI, R., Energy criteria for finite hyperelasticity. *Arch. Rational Mech. Anal.* **77** (1981), 177–188.
14. TARTAR, L., Compensated compactness and applications to partial differential equations. *Nonlinear Analysis and Mechanics* Vol. IV, ed. R. J. KNOPS, London: Pitman 1979, pp. 136–212.
15. TRUESDELL, C., & W. NOLL, *The Nonlinear Field Theories of Mechanics. Handbuch der Physik* III/3, ed. S. FLÜGGE. Berlin Heidelberg New York: Springer 1965.
16. VOLPERT, A. I., The spaces *BV* and quasilinear equations. *Mat. Sbornik* **73** (1967), 255–302. English transl. *Math. USSR Sbornik* **2** (1967), 225–267.



Division of Applied Mathematics
Brown University
Providence, Rhode Island


*(Received November 25, 1985)*

# *A generalized Norton-Hoff Model and the Prandtl-Reuss Law of Plasticity*

ROGER TEMAM

*Dedicated to James Serrin on the occasion of his 60th Birthday*

## Abstract

The aim of this article is to study the quasistatic evolution of a three-dimensional elastic-perfectly plastic solid which satisfies the Prandtl-Reuss law. The evolution of the field of stresses $\sigma$—which solves a time dependent variational inequality—and that of the field of displacements $u$, have been described in previous works [15], [26], [35], [36], [37] but it was not shown there that $\sigma$ and $u$ satisfy indeed the Prandtl-Reuss constitutive law. In this article we find $\sigma$ and $u$ in a class of functions which are sufficiently regular for the Prandtl-Reuss law to make sense and we prove that $\sigma$ and $u$ satisfy the constitutive law. This result is attained by considering the elastic-perfectly plastic model as the limit of a family of elastic-visco-plastic models like those of NORTON and HOFF. The Norton-Hoff type models which we introduce depend on a viscosity parameter $\alpha > 0$; we study the perturbed models (*i.e.* $\alpha > 0$ fixed) and then we pass to the limit $\alpha \to 0$.

## Table of Contents

Introduction . . . . . 138
1. The equations of plasticity . . . . . 139
   1.1. The general equations . . . . . 139
   1.2. The constitutive law . . . . . 140
   1.3. Function spaces . . . . . 142
   1.4. Hypotheses and the safe load assumption . . . . . 143
   1.5. The evolution problem for $\sigma$ . . . . . 144
2. The modified Norton-Hoff models . . . . . 145
   2.1. Preliminaries . . . . . 145
   2.2. The Norton-Hoff model . . . . . 147
3. Regularized equations . . . . . 147
   3.1. The perturbed problems . . . . . 147
   3.2. Existence of solutions . . . . . 148
   3.3. More regularity . . . . . 150

4. Proof of theorem 2.1 . . . . . . . . . . . . . . . . . . . . . . . . . 152
4.1. *A priori* estimates . . . . . . . . . . . . . . . . . . . . . . . . . 153
4.2. Passage to the limit $\lambda \to 0$ . . . . . . . . . . . . . . . . . . . 154
4.3. Proof of (4.29). . . . . . . . . . . . . . . . . . . . . . . . . . 156
4.4. Proof of uniqueness . . . . . . . . . . . . . . . . . . . . . . . 159
4.5. More regularity . . . . . . . . . . . . . . . . . . . . . . . . . 160
4.6. Estimates independent of $\alpha$ . . . . . . . . . . . . . . . . . . . 164
5. Convergence of the generalized Norton-Hoff model to the Prandtl Reuss model: the main results . . . . . . . . . . . . . . . . . . . . . . . . . 165
5.1. The main results for the stresses . . . . . . . . . . . . . . . . . . 165
5.2. Proof of theorem 5.1 . . . . . . . . . . . . . . . . . . . . . . . 166
5.3. Proof of (5.13). . . . . . . . . . . . . . . . . . . . . . . . . . 169
5.4. Proof of theorem 5.2 . . . . . . . . . . . . . . . . . . . . . . . 171
5.5. The strains and the constitutive law . . . . . . . . . . . . . . . . 172

Appendix: Function spaces and trace theorems . . . . . . . . . . . . . . 177
References . . . . . . . . . . . . . . . . . . . . . . . . . . . . . 181

## Introduction

The problems of plasticity have been recently brought to mathematical treatment and were in the last years the object of important developments. This was made possible on the one hand thanks to a new formulation of the laws of plasticity [21], [22] and on the other hand thanks to the recent progress in convex analysis [8], [23], [31] and in nonlinear analysis (variational inequalities and monotone operators [18], [7]).

Two aspects of the problems of plasticity are now fairly well understood. One of them is the static model of HENCKY which is in fact a nonlinear elasticity model (*cf.* W. NOLL [25] and C. TRUESDELL & W. NOLL [42], P. GERMAIN [12]) but however has several common features with plasticity. A full account on these questions and the corresponding references can be found in R. TEMAM [38]: the aim of [38] and of the work related to [38] was to create a function setting which allows for discontinuous fields of displacements appearing in phenomena of plasticity and more generally when cracks appear. The appropriate spaces of functions whose derivatives are measures have been studied and as well the duality between these generalized fields of displacements and the natural field of stresses so that the virtual work theorem [10] and the constitutive law can be proved (see [38]).

The second aspect of plasticity which has been studied is the quasistatic evolution of the field of stresses $\sigma$ and the field of displacements $u$, for a solid governed by the Prandtl-Reuss law; *cf.* G. DUVAUT & J. L. LIONS [7], C. JOHNSON [15], [16], P. SUQUET [35], [36], [37], R. TEMAM [38] and the references therein. The evolution of $\sigma$ is totally (and uniquely) described by a time dependent variational inequality, and some useful information on the field of displacements $u$ were obtained; in particular $u$ belongs for each $t$ to the function space $BD(\Omega)$ which was introduced and studied in relation with the static problems mentioned before (for all that *cf.* [38]). However, as far as I know, it was not proved in the evolutionary case that the functions $\sigma$ and $u$ which were obtained do indeed satisfy the Prandtl-

Reuss constitutive law, one of the reasons beeing that $\sigma$ and $u$ were not found in a class of sufficiently regular functions and the terms appearing in the constitutive law did not make sense. My aim here is to fill this gap: I derive enough regularity properties for $\sigma$ and $u$ to give mathematical meaning to the expressions appearing in the Prandtl-Reuss law and actually to prove it. Therefore this paper is totally devoted to the time-dependent case (quasistatic evolution) although it will use extensively the tools and ideas developed in [38] and in relation with [38].

In order to achieve the task a totally new proof of existence for $\sigma$ and $u$ is given here. We consider the elastic perfectly plastic model of PRANDTL and REUSS as the limit of an appropriate elastic viscoplastic model when a viscosity parameter $\alpha$ converges to 0. The elasto-viscoplastic model that we consider generalizes that of NORTON and HOFF [26] and is similar to that studied by A. FRIAA [9]; nevertheless a major difference with the model in [9] is that the model proposed here leads to more regular solutions which have a better behavior as $\alpha \to 0$. We study for fixed $\alpha$ the existence and uniqueness of solutions $\sigma_\alpha$, $u_\alpha$ to the Norton-Hoff model and then we pass to the limit $\alpha \to 0$: it seems that this passage to the limit $\alpha \to 0$ is interesting by itself but of course the primary motivation is to obtain the informations that we are looking for in the limit case (*cf.* (5.13) and its consequences).

The paper is organized as follows: Section 1 gives the notations, the assumptions and the formulation of the elastic-perfectly plastic model. Section 2 contains the description of the Norton-Hoff type models and the results for these models; these results are proved in Sections 3 and 4. Finally in Section 5 we pass to the limit $\alpha \to 0$ and derive the results concerning the Prandtl Reuss law, which are the *main results*. They are given in Theorems 5.2 and 5.3: Theorem 5.2 improves the previous existence and uniqueness results [7], [15], [37] while Theorem 5.3 proves the principle of maximal dissipation of the Prandtl-Reuss law. Finally the Appendix contains technical results.

## 1. The equations of plasticity

In this section we recall the Prandtl-Reuss law [28] which describes the plastic behavior of solids.

### *1.1. The general equations*

A solid body occupies at rest a region $\Omega$ of $\mathbb{R}^3$ with boundary $\Gamma$. This solid undergoes time-dependent deformations under the action of volume forces of density $f$ inside $\Omega$ and surface forces of surface density $F$ on some part $\Gamma_1$ of $\Gamma$; $f$ and $F$ depend on $x$ and $t$, $x = (x_1, x_2, x_3) \in \Omega$ (for $f$) or $\Gamma_1$ (for $F$). The state of the deformed material is described by the tensor field $\sigma = \sigma(x, t)$ and the vector field $u = u(x, t)$; $\sigma(x, t)$ is the Cauchy stress tensor at point $x$ at time $t$ and $u(x, t)$ is the displacement at time $t$ of the material particle which is at $x$ when the body is at rest. Under the assumption of small displacements, the motion of the body is governed by the classical equation

$$\varrho \ddot{u} = f + \operatorname{div} \sigma, \quad x \in \Omega, \; t > 0, \tag{1.1}$$

where $\varrho$ is the density of the solid and div $\sigma$ is the vector with components

$$\sigma_{ij,j} = \sum_{j=1}^{3} \frac{\partial \sigma_{ij}}{\partial x_j}, \quad i = 1, 2, 3, \tag{1.2}$$

$\sigma_{ij}$ denoting the components of $\sigma$. The Einstein index summation convention is used and $\varphi_{,j} = \dfrac{\partial \varphi}{\partial x_j}$. Partial differentiation with respect to $t$ is denoted by a dot so that $\ddot{u} = \dfrac{\partial^2 u}{\partial t^2}, \ldots$

In the quasistatic case, the evolution is slow, the acceleration term $\varrho \ddot{u}$ is negligible and (1.1) is replaced by

$$\operatorname{div} \sigma + f = 0. \tag{1.3}$$

Several other equations are necessary to describe the evolution of the body. First the continuity of forces on $\Gamma_1$ which implies

$$\sigma \cdot \nu = F, \quad x \in \Gamma_1, \quad t > 0, \tag{1.4}$$

where $\nu = (\nu_1, \nu_2, \nu_3)$ is the unit outward normal on $\Gamma$ and $\sigma \cdot \nu$ is the vector with components

$$\sigma_{ij} \nu_j = \sum_{j=1}^{3} \sigma_{ij} \nu_j, \quad i = 1, 2, 3. \tag{1.5}$$

Then the displacement $u$ is given on the complement $\Gamma_0$ of $\Gamma_1$ on $\Gamma$

$$u = U, \quad x \in \Gamma_0. \quad t > 0, \tag{1.6}$$

and for an initial-value problem the initial state of displacements, velocities and stresses is given

$$u(x, 0) = u_0(x), \quad x \in \Omega \tag{1.7}$$

$$\dot{u}(x, 0) = u_1(x), \quad x \in \Omega \tag{1.8}$$

$$\sigma(x, 0) = \sigma_0(x), \quad x \in \Omega. \tag{1.9}$$

### *1.2. The constitutive law*

The general equations above are completed by the constitutive relation connecting the stresses $\sigma$ to the field of strains $\varepsilon = \varepsilon(u)$,

$$\varepsilon_{ij}(u) = \tfrac{1}{2} \left( \frac{\partial u_i}{\partial x_j} + \frac{\partial u_j}{\partial x_i} \right).$$

Let E denote the space of symmetric tensors of order 2 and $\mathrm{E}^{\mathrm{D}}$ the subspace of tensors with a vanishing trace; if $\xi = (\xi_{ij})$ belongs to E then its deviatoric

$\xi^D \in E^D$ is $\xi - \frac{1}{3}(\operatorname{tr}\xi) I$, or

$$\xi^D_{ij} = \xi_{ij} - \tfrac{1}{3}(\xi_{ll})\,\delta_{ij}. \tag{1.10}$$

If $\xi$, $\eta \in E$, their scalar product in E is denoted $\xi \cdot \eta = \xi_{ij}\eta_{ij}$ and $|\xi| = \{\xi \cdot \xi\}^{\frac{1}{2}}$. Whence

$$\xi \cdot \eta = \xi^D \cdot \eta^D + \tfrac{1}{3}(\operatorname{tr}\xi)(\operatorname{tr}\eta). \tag{1.11}$$

In linear elasticity the relations between stresses and strains are linear at every point and every time:

$$\sigma_{ij} = 2\mu\varepsilon_{ij}(u) + \lambda\varepsilon_{kk}(u)\,\delta_{ij} \tag{1.12}$$

or

$$\sigma^D_{ij} = 2\mu\varepsilon^D_{ij}(u), \qquad \sigma_{kk} = (2\mu + 3\lambda)\,\varepsilon_{kk}(u) \tag{1.13}$$

where $\lambda$, $\mu$ are the Lamé coefficients. Inverting these relations we can write

$$\varepsilon(u) = A\sigma \tag{1.14}$$

where $A$ is the invertible positive definite operator in E defined by

$$(A\xi)_{ij} = \frac{1}{2\mu}\xi_{ij} - \frac{\lambda}{2\mu(2\mu + 3\lambda)}\xi_{kk}\delta_{ij}, \qquad \forall\, \xi \in \mathrm{E}. \tag{1.15}$$

For the Prandtl-Reuss law we are given a convex set $\mathrm{K} \subset \mathrm{E}$ in which the stress tensors $\sigma(x, t)$ must remain; K is of the form $\mathrm{K}^D \otimes \mathbb{R}I$, with

(1.16) $\mathrm{K}^D \subset \mathrm{E}^D$ is a closed convex bounded set containing 0 in its interior.

For plastic behavior it is first assumed that

$$\sigma(x, t) \in \mathrm{K}, \qquad \forall\, x, \qquad \forall\, t, \tag{1.17}$$

while (1.14) is no longer valid, and instead we write

$$\varepsilon(\dot{u}) = A\dot{\sigma} + \boldsymbol{\lambda}. \tag{1.18}$$

The Prandtl-Reuss law gives a set of conditions satisfied by $\boldsymbol{\lambda}$ when $\boldsymbol{\lambda} \neq 0$ and according to J. J. Moreau [21] those conditions are equivalent to

$$\boldsymbol{\lambda} \cdot (\tau - \sigma) \leqq 0, \qquad \forall\, \tau \in K \tag{1.19}$$

where (1.19) holds at every point $x \in \Omega$ and every time $t$. Also it is of course assumed that (1.17) holds at time $t = 0$.

The relations (1.18), (1.19) can be also written in alternate forms which are sometime convenient. For instance if $\chi_K$ is the indicator function of K in $\mathrm{E}(\chi_K(\xi) = 0$ if $\xi \in \mathrm{K}$, $= +\infty$ if $\xi \in \mathrm{E} \setminus \mathrm{K})$, and $\partial\chi_K$ is its subdifferential, *cf.* [8], then (1.19) is equivalent to

$$\varepsilon(\dot{u}) - A\dot{\sigma} \in \partial\chi_K(\sigma). \tag{1.20}$$

Also if $\tau \in \mathrm{K}$ we denote by $C_K(\tau)$ the cone

$$C_K(\tau) = \{\xi, \xi = r(\eta - \tau), \qquad \eta \in \mathrm{K}, \qquad r > 0\}.$$

If $\tau \in \overset{\circ}{K}$ (= the interior of K) then $C_K(\tau) = E$, while if $\tau \in \partial K$, $C_K(\tau)$ is the cone of tangents at K at point $\tau$. It is easy to see that $\partial\chi_K(\tau)$ is nothing else than the polar cone $C_K(\tau)^0$ of $C_K(\tau)$ and thus (1.20) is equivalent to the normality law:

$$\varepsilon(\dot{u}) - A\dot{\sigma} \in C_K(\sigma)^0. \tag{1.21}$$

## 1.3. Function spaces

We assume that $\Omega$ is an open connected bounded set of class $\mathscr{C}^2$ of $\mathbb{R}^3$ and that $\Gamma_0$ and $\Gamma_1$ are two open disjoint parts of the boundary $\Gamma$ of $\Omega$ such that

$$\Gamma_0 \neq \emptyset, \ \bar{\Gamma}_0 \cup \bar{\Gamma}_1 = \Gamma \quad \text{and} \quad \Gamma^* = \Gamma \setminus (\Gamma_1 \cup \Gamma_0) \text{ is a } \mathscr{C}^2 \text{ curve on } \Gamma. \tag{1.22}$$

We denote by $L^2(\Omega)$ (respectively $L^2(\Omega; E)$) the space of $L^2$ functions from $\Omega$ into $\mathbb{R}$ (respectively into E) and we denote by $H^1(\Omega)$ the Sobolev space of order 1

$$H^1(\Omega) = \left\{u \in L^2(\Omega), \ \frac{\partial u}{\partial x_i} \in L^2(\Omega), \quad i = 1, 2, 3\right\}.$$

We set $\mathbb{L}^2(\Omega) = L^2(\Omega)^3$, $\mathbb{H}^1(\Omega) = H^1(\Omega)^3$. The scalar product and the norm on either $L^2(\Omega)$, $\mathbb{L}^2(\Omega)$ or $L^2(\Omega; E)$ are written $(.,.)$ and $|.|$. The scalar product and the norm on $H^1(\Omega)$ or $\mathbb{H}^1(\Omega)$ are written $((.,.))$, $\|.\|$. Finally we denote by $L^p(\Omega)$ (respectively $\mathbb{L}^p(\Omega)$, respectively $L^p(\Omega; E)$) $1 \leq p \leq \infty$, the spaces of $L^p$ functions from $\Omega$ into $\mathbb{R}$ (respectively $\mathbb{R}^3$, E) which are Banach spaces when endowed with their natural norm.

The space of $L^2$ real functions on $\Gamma$ for the superficial measure $d\Gamma$ is denoted by $L^2(\Gamma)$ and $\mathbb{L}^2(\Gamma) = L^2(\Gamma)^3$; $H^{\frac{1}{2}}(\Gamma) \subset L^2(\Gamma)$ is the space of traces on $\Gamma$ of the functions $u$ in $H^1(\Omega)$, and $H^{-\frac{1}{2}}(\Gamma)$ is the dual of $H^{\frac{1}{2}}(\Gamma)$ (*cf.* J. L. LIONS & MAGENES [19]).

We recall (*cf.* R. TEMAM [38], [39]) that if $\tau \in L^2(\Omega; E)$ and $\operatorname{div}\tau \in \mathbb{L}^2(\Omega)$ then we can define the trace of $\tau \cdot \nu$ on $\Gamma$ as an element of $\mathbb{H}^{-\frac{1}{2}}(\Gamma)$, and there is a constant $c_1$ depending only on $\Omega$ such that

$$\|\tau \cdot \nu\|_{\mathbb{H}^{-\frac{1}{2}}(\Gamma)} \leq c_1\{|\tau|_{L^2(\Omega;E)} + |\operatorname{div}\tau|_{\mathbb{L}^2(\Omega)}\}, \quad \forall\, \tau \in L^2(\Omega; E),\ \operatorname{div}\tau \in \mathbb{L}^2(\Omega). \tag{1.23}$$

Furthermore we have the generalized Green's formula

$$\begin{aligned} \int_\Omega \tau \cdot \varepsilon(v)\, dx &= \int_\Omega \tau_{ij}\varepsilon_{ij}(v)\, dx = \int_\Omega \tau_{ij} v_{i,j}\, dx \\ &= \int_\Gamma \tau_{ij} v_i \nu_j\, d\Gamma - \int_\Omega \tau_{ij,j} v_i\, dx \\ &= \int_\Gamma \tau \cdot v \otimes \nu\, d\Gamma - \int_\Omega v \operatorname{div}\tau\, dx^1 \end{aligned} \tag{1.24}$$

valid for every $v \in \mathbb{H}^1(\Omega)$ and $\tau \in L^2(\Omega; E)$ with $\operatorname{div}\tau \in \mathbb{L}^2(\Omega)$.

---

[1] $\int_\Gamma \tau \cdot v \otimes \nu\, d\Gamma = \int_\Gamma (\tau \cdot \nu) \cdot v\, d\Gamma$ is an improper notation for the scalar product $\langle \tau \cdot \nu, v\rangle$ in the duality between $\mathbb{H}^{-\frac{1}{2}}(\Gamma)$ and $\mathbb{H}^{\frac{1}{2}}(\Gamma)$.

### *1.4. Hypotheses and the safe load assumption*

The initial-value problem which we will solve in a weak sense is the problem $\mathscr{P}$:

(1.25) $\mathscr{P}$ consists in (1.3), (1.4), (1.6) to (1.9), (11.7) to (1.19).

The assumptions on the data $f$, $F$, $U$ are the following

(1.26) $$f \text{ and } \dot{f} \text{ belong to } L^\infty(0, T; \mathbb{L}^2(\Omega));$$

(1.27) $$F \text{ and } \dot{F} \text{ belong to } L^\infty(0, T; \mathbb{L}^2(\Gamma_1));$$

(1.28) $$U, \dot{U} \text{ and } \ddot{U} \text{ belong to } L^\infty(0, T; \mathbb{H}^1(\Omega)) \text{ and } U = 0 \text{ on } \Gamma_1;$$ [1]

(1.29) $$u_0, u_1 \in \mathbb{H}^1(\Omega), u_0 = U(0), \text{ and } u_1 = \dot{U}(0) \text{ on } \Gamma_1$$
$$\sigma_0 \in L^2(\Omega; E), \quad \sigma_0^D(x) \in K^D \text{ a.e. } x \in \Omega.$$

Given $U_0 \in \mathbb{H}^1(\Omega)$, $f_0 \in \mathbb{L}^2(\Omega)$, $F_0 \in \mathbb{L}^2(\Gamma_1)$ the set of the corresponding kinematically admissible displacement fields (*cf.* P. GERMAIN [10]) is

(1.30) $$\mathscr{C}_{\mathrm{ad}}(U_0) = \{u \in \mathbb{H}^1(\Omega), u = U_0 \text{ on } \Gamma_0\}$$

and the corresponding set of statically admissible tensor fields is

(1.31)
$$\mathscr{S}_{\mathrm{ad}}(f_0, F_0) = \{\sigma \in L^2(\Omega; E), \operatorname{div} \sigma + f_0 = 0 \text{ in } \Omega, \sigma \cdot \nu = F_0 \text{ on } \Gamma_1\}.$$

Thus under appropriate regularity assumptions (1.3), (1.4), (1.6) are equivalent to

(1.32) $$u(t) \in \mathscr{C}_{\mathrm{ad}}(U(t))$$

(1.33) $$\sigma(t) \in \mathscr{S}_{\mathrm{ad}}(f(t), F(t)) \quad \forall\, t,$$

where $u(t), f(t), \ldots,$ is the function $x \to u(x, t)$, $x \to f(x, t), \ldots$

Finally we set

(1.34) $$\mathscr{C}_{\mathrm{ad}}^0 = \mathscr{C}_{\mathrm{ad}}(0) \quad \text{and} \quad \mathscr{S}_{\mathrm{ad}}^0 = \mathscr{S}_{\mathrm{ad}}(0, 0). \quad \square$$

If it exists, the solution $u$, $\sigma$, of problem $\mathscr{P}$ is such that $\sigma(t)$ belongs to

(1.35) $$\mathscr{K}_{\mathrm{ad}} \cap \mathscr{S}_{\mathrm{ad}}(f(t), F(t)), \quad \forall\, t$$

where

(1.36) $$\mathscr{K}_{\mathrm{ad}} = \{\tau \in L^2(\Omega; E), \tau(x) \in K \text{ a.e. } x\}.$$

However, the sets (1.35) may well be empty (for more details see R. TEMAM [38], Chapter I, Section 5). The safe load condition is an assumption which guarantees that these sets are not empty and we formulate this assumption in the following

[1] The boundary value $U$ is extended inside $\Omega$; (1.28) is equivalent to assuming that $U_{|\Gamma}$, $\dot{U}_{|\Gamma}$, $\ddot{U}_{|\Gamma}$ belong to $L^\infty(0, T; \mathbb{H}^{\frac{1}{2}}(\Gamma))$.

form:

$$(1.37)\quad \begin{cases} \text{There is a } \chi \text{ such that } \chi, \dot{\chi}, \ddot{\chi} \text{ belong to} \\ L^\infty((0,T)\times\Omega; E) = L^\infty(0,T; L^\infty(\Omega; E)),\ \chi(.,t) \in \mathscr{S}_{ad}(f(t), F(t))\ \forall t \\ \text{and, for some } \delta > 0,\ \chi(x,t) \in \dfrac{K}{1+\delta},\ \forall x \in \Omega,\ \forall t \in (0,T). \end{cases}$$

*1.5. The evolution problem for σ*

Assuming that the solution of problem $\mathscr{P}$ is sufficiently regular, we can derive from $\mathscr{P}$ an evolution equation involving only $\sigma$ and which, we will see, determines completely $\sigma$.

Let us assume for instance that

$$(1.38)\qquad \sigma \text{ and } \dot{\sigma} \in L^2(0,T; L^2(\Omega; E)).$$

Then as mentioned before

$$(1.39)\qquad \sigma(t) \in \mathscr{S}_{ad}(f(t), F(t)) \cap \mathscr{K}_{ad},\ \forall t$$

$$(1.40)\qquad \sigma(0) = \sigma_0.$$

Now let us consider $\tau$ such that

$$(1.41)\qquad \tau, \dot{\tau} \in L^2(0,T; L^2(\Omega; E))$$

$$(1.42)\qquad \tau(t) \in \mathscr{S}_{ad}(f(t), F(t)) \cap \mathscr{K}_{ad},\ \forall t.$$

We multiply (1.18) by $(\tau - \sigma)(x,t)$, integrate over $\Omega$ and then, due to (1.19), we find

$$(A\dot{\sigma}, \tau - \sigma) \geqq (\varepsilon(\dot{u}), \tau - \sigma).$$

Thanks to (1.6), (1.24), (1.39), (1.42) the right-hand side of this relation is equal to $\int_{\Gamma_0} \dot{U} \otimes \nu \cdot (\tau - \sigma)\, d\Gamma$. Thus for every $\tau$ satisfying (1.41), (1.42), we have

$$(1.43)\qquad (A\dot{\sigma}, \tau - \sigma) \geqq \int_{\Gamma_0} \dot{U} \otimes \nu \cdot (\tau - \sigma)\, d\Gamma,\ \forall t;$$

(1.38)–(1.40) with (1.43) constitute an evolution variational inequation which allows us to determine $\sigma$ for all time (see Sec. 5.1–5.4). This evolution problem for $\sigma$ has been studied by G. DUVAUT & J. L. LIONS [7], C. JOHNSON [15], [16].

*Remark 1.1.* Once $\sigma$ is determined from this evolution problem, we still have to determine $u$ satisfying (1.18), (1.19), (1.7), (1.8) and (it is to be hoped) (1.6). This is not obvious and the main contribution here is the verification of (1.19).

Concerning (1.6), by comparison with the static case (deformation theory *i.e.* the Hencky law see [38], [13]) we do not expect in general $u$ to satisfy this boundary condition, since slip lines due to plastification may appear on $\Gamma_0$. The generalization of the Prandtl-Reuss law allowing for discontinuities on $\Gamma_0$ as in [38] remains to be an open problem.

## 2. The modified Norton Hoff models

The existence of solution to the problem $\mathscr{P}$ is obtained by considering the elasto-plastic model corresponding to $\mathscr{P}$ as the limit of a family of visco-elastic models when the viscosity tends to 0. The viscoelastic model used here, which departs from that used by JOHNSON [15] for a similar purpose, is a generalized Norton-Hoff law as in A. FRIAA [9] but the version proposed here is more appropriate for the mathematical treatment.

### *2.1. Preliminaries*

For $\xi$ in E let $d(\xi)$ denote the distance of $\xi$ to $K$, $\text{dist}\,(\xi, K) = \text{dist}\,(\xi^D, K^D)$ and let

$$\theta(\xi) = \{1 + d^2(\xi)\}^{\frac{1}{2}}. \tag{2.1}$$

Clearly $\theta(\xi) = \theta(\xi^D)$ and

$$\theta(\xi) \geqq 1, \quad \theta(\xi) = 1 \Leftrightarrow \xi \in K. \tag{2.2}$$

The function $\theta$ is continuous on E. By a result of E. H. ZARANTONELLO [43], $d^2(\xi)$ is differentiable on E with differential $2(\xi - \Pi_K\xi)$, $\Pi_K$ the projector in E onto K; thus $\theta$ is differentiable with differential

$$\theta'(\xi) = \{1 + d^2(\xi)\}^{-\frac{1}{2}}\,(\xi - \Pi_K\xi). \tag{2.3}$$

Let $\alpha > 0$ be a fixed parameter, $0 < \alpha \leqq 1$.[1] We set

$$\gamma_\alpha(\xi) = \frac{\alpha}{\alpha + 1}\,\theta(\xi)^{\frac{\alpha+1}{\alpha}}. \tag{2.4}$$

The function $\gamma_\alpha$ is continuously differentiable with differential

$$\gamma'_\alpha(\xi) = \theta(\xi)^{\frac{1}{\alpha}}\,\theta'(\xi) = \{1 + d^2(\xi)\}^{\frac{1}{2\alpha} - \frac{1}{2}}\,(\xi - \Pi_K\xi). \tag{2.5}$$

Since $d$ is convex and $\gamma_\alpha$ is a convex increasing function of $d$, $\gamma_\alpha$ is convex and its differential $\gamma'_\alpha$ is a monotone operator, *i.e.*

$$(\gamma'_\alpha(\xi) - \gamma'_\alpha(\eta)) \cdot (\xi - \eta) \geqq 0, \quad \forall\, \xi, \eta \in \text{E}. \tag{2.6}$$

It follows immediately from (2.5) that $\gamma'_\alpha(\xi) = 0$ if and only if $\xi = \Pi_K\xi$, *i.e.* $\xi \in \text{K}$:

$$\gamma'_\alpha(\xi) = 0 \Leftrightarrow \xi \in \text{K}\ (\Leftrightarrow \xi^D \in K^D). \tag{2.7}$$

---

[1] $\alpha$ is intended to go to 0.

Since $K^D$ is bounded, there exists $k_1 > 0$ such that $K^D$ is included in the ball of $E^D$ centered at 0 of radius $k_1$. It is then elementary that

$$|\xi^D| - k_1 \leqq d(\xi) \leqq |\xi^D|, \ \forall \xi, \tag{2.8}$$

and consequently

$$|\gamma'(\xi)| = \{1 + d^2(\xi)\}^{\frac{1}{2\alpha} - \frac{1}{2}} d(\xi) \leqq (1 + |\xi^D|^2)^{\frac{1}{2\alpha}}, \ \forall \xi \in E. \tag{2.9}$$

Also since $\Pi_K\xi \cdot (\xi - \Pi_K\xi) \geqq 0$, we have

$$\gamma_\alpha'(\xi) \cdot \xi = \{1 + d^2(\xi)\}^{\frac{1}{2\alpha} - \frac{1}{2}} \xi(\xi - \Pi_K\xi)$$

$$\gamma_\alpha'(\xi) \cdot \xi \geqq \{1 + d^2(\xi)\}^{\frac{1}{2\alpha} - \frac{1}{2}} d^2(\xi). \tag{2.10}$$

We consider now the function

$$\tau \in L^{\frac{1+\alpha}{\alpha}}(\Omega; E) \to \int_\Omega \gamma_\alpha(\tau(x))\, dx.$$

This function is finite everywhere and convex and therefore it is a continuous function. It is Gateaux differentiable on $L^{(1+\alpha)/\alpha}(\Omega; E)$ and its differential is denoted $\gamma_\alpha'(\tau)$ (although this is a misnomer). If $\tau$ belongs to $L^{(1+\alpha)}(\Omega; E)$, $\gamma_\alpha'(\tau)$ is in the dual space $L^{1/\alpha}(\Omega, E)$ (thanks to (2.9)) and $\gamma_\alpha'$ is a monotone operator from the first space into the second one.

The vector $\xi - \Pi_K\xi$ has the direction of a normal vector $n_K(\Pi_K\xi)$ of K at point $\Pi_K\xi$. The unit outward normal vector $n_K(\zeta)$ to K at $\zeta \in \partial K$ depends continuously on $\zeta$ but in general the function $\zeta \to n_K(\zeta)$ is not differentiable but this function is continuously differentiable when $\partial K$ is of class $\mathscr{C}^2$ (*cf.* J. SERRIN [32]). We will therefore make one of the two following assumptions:

(2.11) The boundary $\partial K^D$ of $K^D$ is of class $\mathscr{C}^2$

or

(2.12) There exists a decreasing family of convex sets $K_j = K_j^D \otimes \mathbb{R}I$, such that $\partial K_j^D$ is of class $\mathscr{C}^2$ for every $j \in \mathbb{N}$ and $K^D = \bigcap_{j=1}^{\infty} K_j^D$.

In the case (2.12) we set $d_j(\xi) = d(\xi, K_j) = d(\xi^D, K_j^D)$ and consider the functions $\theta_j$ and $\gamma_{\alpha_j}$ defined by the analog of formulas (2.1), (2.4). The function $\gamma_{\alpha_j}$ is twice continuously differentiable and, clearly as $j \to \infty$, $d_j$, $\theta_j$, $\gamma_{\alpha_j}$, $\gamma_{\alpha_j}'$ converge to $d$, $\theta$, $\gamma_\alpha$, $\gamma_\alpha'$. Also since $K_j \supset K$, (2.7) implies that

$$\gamma_{\alpha_j}'(\xi) = 0, \quad \forall \xi \in K_j \text{ (and thus } \forall \xi \in K). \tag{2.13}$$

### 2.2. *The Norton-Hoff model*

The Norton-Hoff model we shall consider is a modification of problem $\mathscr{P}$ where the constraint $\sigma(x, t) \in \mathrm{K}$ is relaxed and, in (1.18), $\boldsymbol{\lambda}$ is replaced by $\gamma'_\alpha(\sigma)$. Thus we are looking for $\sigma_\alpha$ and $u_\alpha$ satisfying (1.3), (1.4), (1.6)–(1.9) and

$$A\dot{\sigma}_\alpha + \gamma'_\alpha(\sigma_\alpha) = \varepsilon(\dot{u}_\alpha). \tag{2.14}$$

Our aim is to show that this problem, called $\mathscr{P}_\alpha$, is well set for any $\alpha > 0$. In Sec. 4 we shall prove the following

**Theorem 2.1.** *Under the above assumptions, for every $\alpha > 0$, there exists a unique solution $\sigma_\alpha$, $u_\alpha$ of Problem $\mathscr{P}_\alpha$, such that*

$$\sigma_\alpha \in \mathrm{L}^\infty(0, \mathrm{T}; \mathrm{L}^2(\Omega; \mathrm{E})), \qquad \sigma_\alpha^{\mathrm{D}} \in L^{\frac{\alpha+1}{\alpha}}(0, \mathrm{T}; \mathrm{L}^{\frac{\alpha+1}{\alpha}}(\Omega; \mathrm{E}^{\mathrm{D}})) \tag{2.15}$$

$$u_\alpha \in \mathrm{L}^\infty(0, \mathrm{T}; \mathbb{W}^{1,\alpha+1}(\Omega)), \qquad \operatorname{div} u_\alpha \in \mathrm{L}^\infty(0, \mathrm{T}; \mathrm{L}^2(\Omega)). \tag{2.16}$$

*Furthermore if one of the assumptions* (2.11), (2.12) *is satisfied then*

$$\dot{\sigma}_\alpha \in \mathrm{L}^\infty(0, \mathrm{T}; \mathrm{L}^2(\Omega; \mathrm{E})) \tag{2.17}$$

$$\dot{u}_\alpha \in \mathrm{L}^{\alpha+1}(0, \mathrm{T}; \mathbb{W}^{1,\alpha+1}(\Omega)), \qquad \operatorname{div} u_\alpha \in \mathrm{L}^2(0, \mathrm{T}; \mathrm{L}^2(\Omega)). \tag{2.18}$$

## 3. Regularized equations

Although the problems $\mathscr{P}_\alpha$ can be studied directly, we shall obtain better informations by introducing a regularization $\mathscr{P}_{\alpha\lambda}$ of these problems ($\lambda > 0$) and obtaining the solutions $\sigma_\alpha$, $u_\alpha$ of $\mathscr{P}_\alpha$ as the limits as $\lambda \to 0$ of the solutions $\sigma_{\alpha\lambda}$, $u_{\alpha\lambda}$ of the regularized problems $\mathscr{P}_{\alpha\lambda}$.

### 3.1. *The perturbed problems*

In order to describe the perturbed problems $\mathscr{P}_{\alpha\lambda}$ we make the following remark: if $u_\alpha$, $\sigma_\alpha$ are solutions of $\mathscr{P}_\alpha$ and $w \in \mathbb{H}^1(\Omega)$ then, taking the scalar product in $\mathrm{L}^2(\Omega; \mathrm{E})$ of each side of (1.3) with $\varepsilon(w)$ and using the generalized Green's formula (1.24) we find

$$(\sigma_{\alpha\lambda}, \varepsilon(w)) = (f(t), w) + \int_\Gamma \sigma_{\alpha\lambda} \cdot w \otimes \nu \, d\Gamma.$$

If furthermore $w \in \mathscr{C}^0_{\mathrm{ad}}$ (*i.e.*, $w = 0$ on $\Gamma_0$) and we utilize (1.4), we arrive at

$$(\sigma_{\alpha\lambda}, \varepsilon(w)) = L(t, w) \tag{3.1}$$

where

$$L(t, w) = \int_\Omega f(t)\, w \, dx + \int_\Gamma F(t)\, w \, d\Gamma. \tag{3.2}$$

Given $\alpha > 0$, $\lambda > 0$, the perturbed problem $\mathscr{P}_{\alpha\lambda}$ is: *to find* $\sigma_{\alpha\lambda}$, $u_{\alpha\lambda}$ *satisfying*

$$(3.3) \qquad (A\dot{\sigma}_{\alpha\lambda}, \tau) + (\gamma'_\alpha(\sigma_{\alpha\lambda}), \tau) = (\varepsilon(\dot{u}_{\alpha\lambda}), \tau), \quad \forall\, \tau \in L^2(\Omega) \times L^{\frac{\alpha+1}{\alpha}}(\Omega, E^D).^1$$

$$(3.4) \qquad \lambda(\varepsilon(\dot{u}_{\alpha\lambda}), \varepsilon(w)) + (\sigma_{\alpha\lambda}, \varepsilon(w)) = L(t\,; w), \quad \forall\, w \in \mathscr{C}^0_{\mathrm{ad}},$$

and the initial conditions (1.8), (1.9) and the boundary condition

$$(3.5) \qquad \dot{u}_{\alpha\lambda} = \dot{U} \quad \text{on } \Gamma_0.$$

Note that (3.3) is nothing else than the equation obtained by taking the scalar product of each side of (2.14) with $\tau$ (and replacing $\sigma_\alpha$, $u_\alpha$ by $\sigma_{\alpha\lambda}$, $u_{\alpha\lambda}$);[2] (3.4) is an appropriate perturbation of (3.1) and (3.5) is simply derived from (1.6) by time differentiation. Also we can set $v_{\alpha\lambda} = \dot{u}_{\alpha\lambda} - \dot{U}$ and replace (3.4), (3.5) by

$$(3.4)' \qquad \lambda(\varepsilon(v_{\alpha\lambda}), \varepsilon(w)) + (\sigma_{\alpha\lambda}, \varepsilon(w)) = L(t, w) - \lambda(\varepsilon(\dot{U}), \varepsilon(w)) \quad \forall\, w \in \mathscr{C}^0_{\mathrm{ad}}$$

$$(3.5)' \qquad v = 0 \quad \text{on } \Gamma_0$$

so that $v(t) \in \mathscr{C}^0_{\mathrm{ad}}$, $\forall\, t$.

### 3.2. Existence of solutions

The evolution equations (3.3), (3.4)′, (3.5)′ completed with the initial conditions (1.8), (1.9) can be solved by the general methods applicable to monotone nonlinear evolution equations (*cf.* J. L. LIONS [18], Chapter II, Section 1). We briefly recall the proof of existence of solutions for such equations using the Faedo-Galerkin method. Let $(\tau_j)_{j\in\mathbb{N}}$ and $(w_j)_{j\in\mathbb{N}}$ be a total and free family of elements of $L^2(\Omega) \times L^{\frac{\alpha+1}{\alpha}}(\Omega; E^D)$ and of $\mathscr{C}^0_{\mathrm{ad}}$ which are separable spaces. We omit the indices $\alpha$, $\lambda$, for simplicity and for every integer $m$ we look for $\sigma_m \,(= \sigma_{\alpha\lambda m})$ and $v_m \,(= v_{\alpha\lambda m})$, taking their values in Span $(\tau_1, \ldots, \tau_m)$ and Span $(w_1, \ldots, w_m)$ and such that ($\dot{u}_m = v_m + \dot{U}$):

$$(3.6) \qquad (A\dot{\sigma}_m, \tau_j) + (\gamma'_\alpha(\sigma_m), \tau_j) = (\varepsilon(\dot{u}_m), \tau_j), \qquad 1 \leqq j \leqq m.$$

$$(3.7) \qquad \lambda(\varepsilon(\dot{u}_m), \varepsilon(w_j)) + (\sigma_m, \varepsilon(w_j)) = L(t, w_j), \qquad 1 \leqq j \leqq m.$$

$$(3.8) \qquad \sigma_m(0) = \sigma_{0m}, \qquad v_m(0) = v_{1m},$$

where $\sigma_{0m} \in$ Span $(\tau_1, \ldots, \tau_m)$, $v_{1m} \in$ Span $(w_1, \ldots, w_m)$ and $\sigma_{0m} \to \sigma_0$ in $L^2(\Omega) \times L^{\frac{\alpha+1}{\alpha}}(\Omega; E^D)$, $v_{1m} \to v_1 = u_1 - \dot{U}(0)$ in $H^1(\Omega)$ as $m \to \infty$.

---

[1] This corresponds to the decomposition of $E = \mathbb{R}I \otimes E^D$; $\gamma'_\alpha(\xi)$ takes its values in $E^D$.

[2] The scalar product is that of $L^2(\Omega) \times L^{\frac{\alpha+1}{\alpha}}(\Omega; E^D)$ with $L^2(\Omega) \times L^{\alpha+1}(\Omega; E^D)$.

*A priori* estimates for $\sigma_m$ and $v_m$ are obtained by replacing (this is licit) $\tau_j$ and $w_j$ by $\sigma_m$ and $v_m$ in (3.6), (3.7). After adding the relations that we obtain, we find

$$\frac{1}{2}\frac{d}{dt}(A\sigma_m, \sigma_m) + \lambda\,|\varepsilon(v_m)|^2 + (\gamma'_\alpha(\sigma_m), \sigma_m) \tag{3.9}$$

$$= L(t, v_m) - (\varepsilon(\dot{U}), \sigma_m) - \lambda(\varepsilon(\dot{U}), \varepsilon(v_m)).$$

The terms In the right-hand side of (3.8) are bounded as follows:

$$\begin{aligned}
|L(t, v_m)| &\leqq \left| \int_\Omega f v_m\,dx + \int_{\Gamma_1} F v_m\,d\Gamma \right| \\
&\leqq |f(t)|\,|v_m(t)| + |F(t)|_{L^2(\Gamma_1)}\,|v_m(t)|_{L^2(\Gamma_1)} \\
&\leqq \text{(by Lemma 3.1 below) } c'_1(|f(t)| + |F(t)|_{L^2(\Gamma_1)})\,|\varepsilon(v_m)(t)| \\
&\leqq \text{(by the assumptions (1.24), (2.25)) } c'_2\,|\varepsilon(v_m)| \\
&\leqq \frac{\lambda}{4}|\varepsilon(v_m)|^2 + \frac{c'_3}{\lambda}
\end{aligned}$$

where the $c_i$, $c'_j$ are various positive constants;

$$\begin{aligned}
|(\varepsilon(\dot{U}), \sigma_m)| &\leqq |\varepsilon(\dot{U})|\,|\sigma_m| \\
&\leqq \text{(by (1.26) and since A is positive-definite) } c'_4\,|\sigma_m| \leqq c'_5(A\sigma_m, \sigma_m)^{\frac{1}{2}} \\
&\leqq (A\sigma_m, \sigma_m) + c'_6,
\end{aligned}$$

$$\begin{aligned}
\lambda\,|(\varepsilon(\dot{U}), \varepsilon(v_m))| &\leqq \lambda\,|\varepsilon(\dot{U})|\,|\varepsilon(v_m)| \\
&\leqq \frac{\lambda}{4}|\varepsilon(v_m)|^2 + \lambda\,|\varepsilon(\dot{U})|^2 \\
&\leqq \frac{\lambda}{4}|\varepsilon(v_m)|^2 + c'_7\lambda.
\end{aligned}$$

Finally (3.9) implies

$$\frac{d}{dt}(A\sigma_m, \sigma_m) + \lambda\,|\varepsilon(v_m)|^2 + 2(\gamma'_\alpha(\sigma_m), \sigma_m) \tag{3.10}$$

$$\leqq 2(A\sigma_m, \sigma_m) + c'_8\left(1 + \lambda + \frac{1}{\lambda}\right)$$

It follows from (3.10), (2.10) and Gronwall's Lemma that as $m \to \infty$, $\sigma_m = \sigma_{\alpha\lambda m}$ remains in a bounded set of $L^\infty(0, T; L^2(\Omega; E))$, $\sigma_m^D$ in a bounded set of $L^{\frac{\alpha+1}{\alpha}}(0, T; L^{\frac{\alpha+1}{\alpha}}(\Omega; E))$ and $v_m$ remains in a bounded set of $L^2(0, T; H^1(\Omega))$. This is sufficient to pass to the limit in (3.6)–(3.8) and to show using the classical methods for monotone equations [18] that a subsequence $\sigma_{m'}$, $v_{m'}$ extracted from

$\sigma_m$, $v_m$, converges to functions $\sigma_{\alpha\lambda}$, $v_{\alpha\lambda}$ which satisfy

$$\sigma_{\alpha\lambda} \in L^\infty(0, T; L^2(\Omega; E)) \cap L^{\frac{\alpha+1}{\alpha}}(0, T; L^{\frac{\alpha+1}{\alpha}}(\Omega; E)), \tag{3.11}$$

$$\dot{u}_{\alpha\lambda} \in L^2(0, T; \mathbb{H}^1(\Omega)) \tag{3.12}$$

together with (3.3), (3.4), (1.8), (1.9) ($v_{\alpha\lambda} = \dot{u}_{\alpha\lambda} - \dot{U}$). This proves the existence of solution for $\mathscr{P}_{\alpha\lambda}$.

This step of the proof will be completed with

**Lemma 3.1.** *$|\varepsilon(v)|$ is a norm on $\mathscr{C}^0_{\text{ad}}$, equivalent to the norm induced by $\mathbb{H}^1(\Omega)$.*

**Proof.** Since meas $\Gamma_0 > 0$, it follows from Korn and Poincaré inequalities that $|\varepsilon(v)| + |v|_{L^2(\Gamma_0)}$ is a norm on $\mathbb{H}^1(\Omega)$ equivalent to the natural norm (see for instance [38]); the statement for $\mathscr{C}^0_{\text{ad}}$ follows.

### 3.3. More regularity

If $\gamma'_\alpha$ is differentiable (assumption (2.11)) we can establish further regularity properties of the solutions of $\mathscr{P}_{\alpha\lambda}$.

The (ordinary) differential system (3.6), (3.7) can be differentiated with respect to time and we find that:

$$(A\ddot{\sigma}_m, \tau_j) + (\gamma''_\alpha(\sigma_m)\, \dot{\sigma}_m, \tau_j) = (\varepsilon(\ddot{u}_m), \tau_j) \qquad 1 \leqq j \leqq m \tag{3.13}$$

$$\lambda(\varepsilon(\ddot{u}_m), \varepsilon(w_j)) + (\dot{\sigma}_m, \varepsilon(w_j)) = \dot{L}(t, w_j) \qquad 1 \leqq j \leqq m \tag{3.14}$$

where

$$\dot{L}(t, w) = \int_\Omega \dot{f}(t)\, w\, dx + \int_\Gamma \dot{F}(t)\, w\, d\Gamma. \tag{3.15}$$

In (3.13) $\tau_j$ can be replaced by $\dot{\sigma}_m$ and in (3.14) $w_j$ can be replaced by $\dot{v}_m (\ddot{u}_m = \ddot{U} + \dot{v}_m)$; after adding the relations that we obtain we find

$$\tfrac{1}{2}\frac{d}{dt}(A\dot{\sigma}_m, \dot{\sigma}_m) + \lambda\, |\varepsilon(\dot{v}_m)|^2 + (\gamma''_\alpha(\sigma_m)\, \dot{\sigma}_m, \dot{\sigma}_m) = \dot{L}(t, \dot{v}_m) - (\varepsilon(\ddot{U}), \dot{\sigma}_m) - \lambda(\varepsilon(\ddot{U}), \varepsilon(\dot{v}_m)). \tag{3.16}$$

The term $(\gamma''_\alpha(\sigma_m)\, \dot{\sigma}_m, \dot{\sigma}_m)$ is non-negative (since $\gamma_\alpha$ is convex) and the terms in the right-hand side of (3.16) can be estimated exactly as we estimated the terms in the right-hand side of (3.9)

$$\begin{aligned}
|\dot{L}(t, \dot{v}_m)| &\leqq |\dot{f}(t)|\, |\dot{v}_m(t)| + |\dot{F}(t)|_{L^2(\Gamma_1)}\, |\dot{v}_m(t)|_{L^2(\Gamma_1)} \\
&\leqq c'_9(|\dot{f}(t)| + |\dot{F}(t)|_{L^2(\Gamma_1)})\, |\varepsilon(\dot{v}_m(t))| \\
&\leqq \text{(by the assumptions (1.26), (1.27))}\ c'_{10}\, |\varepsilon(\dot{v}_m)| \\
&\leqq \frac{\lambda}{4}\, |\varepsilon(\dot{v}_m)|^2 + \frac{c'_{11}}{\lambda},
\end{aligned}$$

$$|(\varepsilon(\dot{U}), \dot{\sigma}_m)| \leqq |\varepsilon(\dot{U})| \, |\dot{\sigma}_m| \leqq c'_{12}(A\dot{\sigma}_m, \dot{\sigma}_m)^{\frac{1}{2}}$$
$$\leqq (A\dot{\sigma}_m, \dot{\sigma}_m) + c'_{13},$$
$$\lambda \, |(\varepsilon(\ddot{U}), \varepsilon(\dot{v}_m))| \leqq \lambda \, |\varepsilon(\ddot{U})| \, |\varepsilon(\dot{v}_m)|$$
$$\leqq \frac{\lambda}{4} |\varepsilon(\dot{v}_m)|^2 + c'_{14}.$$

Finally (3.16) implies

$$\frac{d}{dt}(A\dot{\sigma}_m, \dot{\sigma}_m) + \lambda \, |\varepsilon(\dot{v}_m)|^2 \leqq 2(A\dot{\sigma}_m, \dot{\sigma}_m) + c'_{15}\left(1 + \lambda + \frac{1}{\lambda}\right). \tag{3.17}$$

It follows from (3.17) and Gronwall's Lemma that, as $m \to \infty$, $\dot{\sigma}_m = \dot{\sigma}_{\alpha\lambda m}$ remains in a bounded set of $L^\infty(0, T; L^2(\Omega; E))$, $\varepsilon(\dot{v}_m)$ remains in a bounded set of $L^2(0, T; L^2(\Omega; E))$ or, using Lemma 3.1 and since $\dot{v}_m(t) \in \mathscr{C}^0_{\text{ad}}$ $\forall\, t$, $\dot{v}_m$ remains in a bounded set of $L^2(0, T; \mathbb{H}^1(\Omega))$. Because of these estimates and by properly choosing the subsequence $m' \to \infty$, we can obtain a solution $\sigma_{\alpha\lambda}$, $u_{\alpha\lambda}$ of $\mathscr{P}_{\alpha\lambda}$ which satisfies not only (3.11), (3.12), but also

$$\dot{\sigma}_{\alpha\lambda} \in L^\infty(0, T; L^2(\Omega; E)) \tag{3.19}$$

$$\ddot{u}_{\alpha\lambda} \, (= \dot{v}_{\alpha\lambda} + \ddot{U}) \in L^2(0, T; \mathbb{H}^1(\Omega)). \tag{3.20}$$

In conclusion we have

**Proposition 3.1.** *Under the general assumptions above, for every* $\alpha > 0$, $\lambda > 0$, *the problem* $\mathscr{P}_{\alpha\lambda}$ *possesses a unique solution* $\sigma_{\alpha\lambda}$, $u_{\alpha\lambda}$, *which satisfies* (3.11), (3.12). *If we further assume* (2.11), *then* $\sigma_{\alpha\lambda}$, $u_{\alpha\lambda}$ *satisfy* (3.19), (3.20).

**Proof.** The only point which is not yet proved is the uniqueness.[1] Let then $\sigma^1_{\alpha\lambda}$, $u^1_{\alpha\lambda}$, $\sigma^2_{\alpha\lambda}$, $u^2_{\alpha\lambda}$, be two pairs of solutions of $\mathscr{P}_{\alpha\lambda}$ and let $\sigma_{\alpha\lambda} = \sigma^1_{\alpha\lambda} - \sigma^2_{\alpha\lambda}$, $u_{\alpha\lambda} = u^1_{\alpha\lambda} - u^2_{\alpha\lambda}$. We subtract respectively the equations (3.3) and (3.4) for $\sigma^1_{\alpha\lambda}$ and $\sigma^2_{\alpha\lambda}$, $u^1_{\alpha\lambda}$ and $u^2_{\alpha\lambda}$, which leads to:

$$(A\dot{\sigma}_{\alpha\lambda}, \tau) + (\gamma'_\alpha(\sigma^1_{\alpha\lambda}) - \gamma'_\alpha(\sigma^2_{\alpha\lambda}), \tau) = (\varepsilon(\dot{u}_{\alpha\lambda}), \tau), \tag{3.21}$$

$$\lambda(\varepsilon(\dot{u}_{\alpha\lambda}), \varepsilon(w)) + (\sigma_{\alpha\lambda}, \varepsilon(w)) = 0, \quad \forall\, w \in \mathscr{C}^0_{\text{ad}}. \tag{3.22}$$

We can take $\tau = \sigma_{\alpha\lambda}$ $(= \sigma_{\alpha\lambda}(t))$ in (3.21) and (due to (3.5)), $w = \dot{u}_{\alpha\lambda}$ $(= \dot{u}_{\alpha\lambda}(t))$ in (3.22). Then adding these relations we obtain

$$\frac{1}{2}\frac{d}{dt}(A\sigma_{\alpha\lambda}, \sigma_{\alpha\lambda}) + (\gamma'_\alpha(\sigma^1_{\alpha\lambda}) - \gamma'_\alpha(\sigma^2_{\alpha\lambda}), \sigma^1_{\alpha\lambda} - \sigma^2_{\alpha\lambda})$$
$$+ \lambda \, |\varepsilon(\dot{u}_{\alpha\lambda})|^2 = 0$$

---

[1] We have proved the existence of $\dot{u}_{\alpha\lambda}$. Of course, in order to obtain $u_{\alpha\lambda}$ we just integrate $\dot{u}_{\alpha\lambda}$ with respect to $t$, taking into account the relations (1.6) and (1.7) for $u_{\alpha\lambda}(u_{\alpha\lambda}(0) = u_0$, $u_{\alpha\lambda} = U$ on $\Gamma_0)$.

and using the monotonicity of $\gamma'_\alpha$,

$$(\gamma'_\alpha(\sigma^1_{\alpha\lambda}) - \gamma'_\alpha(\sigma^2_{\alpha\lambda}), \sigma^1_{\alpha\lambda} - \sigma^2_{\alpha\lambda}) \geqq 0.$$

We obtain easily that $\sigma_{\alpha\lambda}(t) = 0$, $u_{\alpha\lambda}(t) = 0$, $\forall\, t$.

*Remark 3.1.* The analogue of Remark 2.1 can be made for the problems $\mathscr{P}_{\alpha\lambda}$.

## 4. Proof of theorem 2.1

The proof of Theorem 2.1 is based on a passage to the limit $\lambda \to 0$ and we need for that purpose *a priori* estimates independent of $\lambda$. However all the *a priori* estimates used to prove Proposition 3.1 actually depend on $\lambda$ and vanish as $\lambda \to 0$. We are then going to establish new estimates; some of them will be in fact independent of $\lambda$ *and* $\alpha$.

### 4.1. *A priori estimates*

If $\tau \in L^2(\Omega; E)$ and $\operatorname{div} \tau \in L^2(\Omega)$ we can use the generalized Green's formula (1.24) and write the right-hand side of (3.3) as:

$$(4.1)\qquad (\varepsilon(\dot u_{\alpha\lambda}), \tau) = \int_\Gamma \dot u_{\alpha\lambda} \otimes \nu \cdot \tau \, d\Gamma - \int_\Omega \dot u_{\alpha\lambda} \operatorname{div} \tau \, dx$$

$$= \text{(by (3.5))} \int_{\Gamma_0} \dot U \otimes \nu \cdot \tau \, d\Gamma + \int_{\Gamma_1} \dot u_{\alpha\lambda} \otimes \nu \cdot \tau \, d\Gamma - \int_\Omega \dot u_{\alpha\lambda} \operatorname{div} \tau \, dx.$$

Now if $\tau \in \mathscr{S}^0_{\text{ad}}$ there remains

$$(4.2)\qquad (\varepsilon(\dot u_{\alpha\lambda}), \tau) = \int_{\Gamma_0} \dot U \otimes \nu \cdot \tau \, d\Gamma$$

and (3.3) implies

(4.3)

$$(A\dot\sigma_{\alpha\lambda}, \sigma, \tau) + (\gamma'_\alpha(\sigma_{\alpha\lambda}), \tau) = \int_{\Gamma_0} \dot U \otimes \nu \cdot \tau \, d\Gamma, \quad \forall\, \tau \in \mathscr{S}^0_{\text{ad}}, \tau^D \in L^{\frac{\alpha+1}{\alpha}}(\Omega; E^D).$$

We replace in (4.3) $\tau$ by $\tau_{\alpha\lambda}(t) = \sigma_{\alpha\lambda}(t) - \chi(t) \in \mathscr{S}^0_{\text{ad}} \cap L^2(\Omega) \times L^{\frac{\alpha+1}{\alpha}}(\Omega; E^D)$ and whence

$$(4.4)\qquad \tfrac{1}{2} \frac{d}{dt} (A\tau_{\alpha\lambda}, \tau_{\alpha\lambda}) + (\gamma'_\alpha(\sigma_{\alpha\lambda}), \tau_{\alpha\lambda}) = \int_{\Gamma_0} \dot U \otimes \nu \cdot \tau_{\alpha\lambda} \, d\Gamma - (A\dot\chi, \tau_{\alpha\lambda}).$$

By (2.6), (2.7) and (1.37)

$$(4.5)\qquad (\gamma'_\alpha(\sigma_{\alpha\lambda}), \tau_{\alpha\lambda}) \geqq (\gamma'_\alpha(\chi), \tau_{\alpha\lambda}) = 0;$$

by (1.23) and since $\operatorname{div} \tau_{\alpha\lambda} = 0$:

$$\left| \int_{\Gamma_0} \dot{U} \otimes \nu \cdot \tau_{\alpha\lambda} \, d\Gamma \right| \leqq c_1' \, |\dot{U}|_{H^1(\Omega)} \, |\tau_{\alpha\lambda}| \tag{4.6}$$

$$\leqq c_1' \, |\dot{U}|_{H^1(\Omega)} \, (A\tau_{\alpha\lambda}, \tau_{\alpha\lambda})^{1/2}$$

$$\leqq (\text{with } (1.28)) \; c_2'(A\tau_{\alpha\lambda}, \tau_{\alpha\lambda})^{1/2}$$

$$\leqq \tfrac{1}{2} (A\tau_{\alpha\lambda}, \tau_{\alpha\lambda}) + c_3' .$$

Clearly

$$|(A\dot{\chi}, \tau_{\alpha\lambda})| \leqq \tfrac{1}{2} (A\tau_{\alpha\lambda}, \tau_{\alpha\lambda}) + \tfrac{1}{2} (A\dot{\chi}, \dot{\chi})$$

$$\leqq (\text{with } (1.37)) \; \tfrac{1}{2} (A\tau_{\alpha\lambda}, \tau_{\alpha\lambda}) + c_4',$$

and we infer from (4.4) that

$$\frac{d}{dt} (A\tau_{\alpha\lambda}, \tau_{\alpha\lambda}) + 2(\gamma_\alpha'(\sigma_{\alpha\lambda}), \tau_{\alpha\lambda}) \leqq 2(A\tau_{\alpha\lambda}, \tau_{\alpha\lambda}) + c_5' \tag{4.7}$$

where $c_1', \ldots, c_5'$ are independent of $\alpha$ and $\lambda$. We omit the nonnegative term $(\gamma_\alpha'(\sigma_{\alpha\lambda}), \sigma_{\alpha\lambda})$, apply Gronwall's Lemma and we deduce from (4.7) that

(4.8) $\tau_{\alpha\lambda}$ and $\sigma_{\alpha\lambda} = \tau_{\alpha\lambda} + \chi$ are bounded in $L^\infty(0, T; L^2(\Omega; E))$ independently of $\alpha$ and $\lambda$.

We then apply Lemma 4.1 below with $\xi = \sigma_{\alpha\lambda}(x, t)$, $\eta = \chi(x, t)$, $\forall x \in \Omega$, $t > 0$ and we can improve (4.5):

$$(\gamma_\alpha'(\sigma_{\alpha\lambda}), \tau_{\alpha\lambda}) \geqq \delta_1 \int_\Omega \{1 + d^2(\sigma_{\alpha\lambda})\}^{\frac{1}{2\alpha} - \frac{1}{2}} d(\sigma_{\alpha\lambda}) \, dx \tag{4.9}$$

where $\delta_1 > 0$ depends only on $K$ and $\chi$ and $d(\xi) = |\xi^D - \Pi_{K^D} \xi^D|$ $(\xi \in E)$ has the same meaning as in Section 2.1. Thus after integration with respect to $t$, (4.7), (4.9) give

$$\delta_1 \int_0^T \int_\Omega \{1 + d^2(\sigma_{\lambda\alpha})\}^{\frac{1}{2\alpha} - \frac{1}{2}} d(\sigma_{\alpha\lambda}) \, dx \, dt \tag{4.10}$$

$$\leqq \tfrac{1}{2} (A(\sigma_0 - \chi_0)), \sigma_0 - \chi(0)) + \int_0^T (A\tau_{\alpha\lambda}, \tau_{\alpha\lambda}) \, dt + c_5' T \leqq c_6',$$

and

(4.11) the left-hand side of (4.10) is bounded independently of $\alpha$ and $\lambda$.

We return to (4.7) which we write in the form

$$\tfrac{1}{2} \frac{d}{dt} (A\tau_{\alpha\lambda}, \tau_{\alpha\lambda}) + (\gamma_\alpha'(\sigma_{\alpha\lambda}), \sigma_{\alpha\lambda}) \leqq c_7' + (\gamma_\alpha'(\sigma_{\alpha\lambda}), \chi) .$$

With (2.9), (2.10) this gives after integrating with respect to $t$ and using (1.37)

$$\text{(4.12)} \quad \int_0^T \int_\Omega \{1 + d^2(\sigma_{\alpha\lambda})\}^{\frac{1}{2\alpha}-\frac{1}{2}} d^2(\sigma_{\alpha\lambda})\, dx\, dt$$

$$\leqq c_7'T + (A(\sigma_0 - \chi(0)), \sigma_0 - \chi(0)) + c_8' \int_0^T \int_\Omega \{1 + d^2(\sigma_{\alpha\lambda})\}^{\frac{1}{2\alpha}-\frac{1}{2}} d(\sigma_{\alpha\lambda})\, dx\, dt.$$

We use statement (4.11) and improve it:

(4.13) the left-hand side of (4.12) is bounded independently of $\alpha$ and $\lambda$. □

We conclude this paragraph with

**Lemma 4.1.** *Given* $\delta > 0$, *there is a* $\delta_1 > 0$ *such that for every* $\xi \in E$, $\eta \in \dfrac{K}{1+\delta}$,

$$\text{(4.14)} \quad \begin{aligned} &(\xi - \Pi_K\xi)(\xi - \eta) \geqq \delta_1 |\xi - \Pi_K\xi|, \\ &\gamma'(\xi)\cdot(\xi - \eta) \geqq \delta_1 |\gamma_\alpha'(\xi)|. \end{aligned}$$

**Proof.** Since $\dfrac{K}{1+\delta}$ is contained in the interior of K, the distance $\delta_1$ of this set to $\partial K$ (or $\complement K$) is positive, $\delta_1 > 0$.

The second statement in (4.14) follows immediately from the first one and (2.5), and we just have to prove the first inequality. If $\xi = \Pi_K\xi$ $(\xi \in K)$, the inequality is obvious; if not, we observe that $\eta + \delta_1 \dfrac{\xi - \Pi_K\xi}{|\xi - \Pi_K\xi|}$ belongs to K since the ball centered at $\eta$ of radius $\delta_1$ is included in K. Then by the elementary properties of the projector $\Pi_K$:

$$(\xi - \Pi_K\xi)\cdot\left(\xi - \eta - \delta_1 \frac{\xi - \Pi_K\xi}{|\xi - \Pi_K\xi|}\right) \geqq 0$$

and (4.14) follows. □

### 4.2. *Passage to the limit* $\lambda \to 0$

We now pass to the limit $\lambda \to 0$. We have (4.8) and, for fixed $\alpha$, the conjunction of (4.13) and (2.8) implies that

(4.15) $\sigma_{\alpha\lambda}^D$ remains in a bounded set of $L^{\frac{\alpha+1}{\alpha}}(0, T; L^{\frac{\alpha+1}{\alpha}}(\Omega; E^D))$ as $\lambda \to 0$.

Then, with (2.8), (2.9), we see that

(4.16) $\gamma_\alpha'(\sigma_{\alpha\lambda})$ remains in a bounded set of $L^{\alpha+1}(0, T; L^{\alpha+1}(\Omega; E^D)$ as $\lambda \to 0$.

Concerning $u_{\alpha\lambda}$, we observe that equation (3.3) is equivalent to

$$A\dot{\sigma}_{\alpha\lambda} + \gamma'_{\alpha}(\sigma_{\alpha\lambda}) = \varepsilon(\dot{u}_{\alpha\lambda}), \tag{4.17}$$

and by integration with respect to $t$, taking into account the initial conditions (1.7), (1.9) (for $u_{\alpha\lambda}$, $\sigma_{\alpha\lambda}$):

$$A(\sigma_{\alpha\lambda}(t) - \sigma_0) + \int_0^t \gamma'_{\alpha}(\sigma_{\alpha\lambda})(s)\,ds = \varepsilon(u_{\alpha\lambda})(t) - \varepsilon(u_0) \quad 0 < t < \mathrm{T}. \tag{4.18}$$

Thus $\varepsilon(u_{\alpha\lambda})$ remains in a bounded set of $\mathrm{L}^{\infty}(0, \mathrm{T}; \mathrm{L}^{\alpha+1}(\Omega; \mathrm{E}))$ as $\lambda \to 0$ and, due to the boundary condition (1.6) for $u_{\alpha\lambda}$:

(4.19) $u_{\alpha\lambda}$ remains in a bounded set of $\mathrm{L}^{\infty}(0, \mathrm{T}; \mathbb{W}^{1+\alpha}(\Omega))$ as $\lambda \to 0$.

With these estimates we see that there exist $\sigma_{\alpha} \in \mathrm{L}^{\infty}(0, \mathrm{T}; \mathrm{L}^2(\Omega; \mathrm{E}))$ with $\sigma_{\alpha}^{\mathrm{D}} \in \mathrm{L}^{\frac{\alpha+1}{\alpha}}(0, \mathrm{T}; \mathrm{L}^{\frac{\alpha+1}{\alpha}}(\Omega; \mathrm{E}))$, $\varrho_{\alpha} \in \mathrm{L}^{\alpha+1}(0, \mathrm{T}; \mathrm{L}^{\alpha+1}(\Omega; \mathrm{E}))$, $u_{\alpha} \in \mathrm{L}^{\infty}(0, \mathrm{T}; \mathbb{W}^{1+\alpha}(\Omega))$, and a subsequence (also denoted by $\lambda$) $\lambda \to 0$ such that

$$\begin{aligned}
&\sigma_{\alpha\lambda} \to \sigma_{\alpha} \text{ in } \mathrm{L}^{\infty}(0, \mathrm{T}; \mathrm{L}^2(\Omega; \mathrm{E})) \text{ weak star},\\
&\sigma_{\alpha\lambda}^{\mathrm{D}} \to \sigma_{\alpha}^{\mathrm{D}} \text{ in } \mathrm{L}^{\frac{\alpha+1}{\alpha}}(0, \mathrm{T}; \mathrm{L}^{\frac{\alpha+1}{\alpha}}(\Omega; \mathrm{E}^{\mathrm{D}})) \text{ weakly},\\
&\gamma'_{\alpha}(\sigma_{\alpha\lambda}) \to \varrho_{\alpha} \text{ in } \mathrm{L}^{\alpha+1}(0, \mathrm{T}; \mathrm{L}^{\alpha+1}(\Omega; \mathrm{E}^{\mathrm{D}}) \text{ weakly},\\
&u_{\alpha\lambda} \to u_{\alpha} \text{ in } \mathrm{L}^{\infty}(0, \mathrm{T}; \mathbb{W}^{1+\alpha}(\Omega)) \text{ weak star}.
\end{aligned} \tag{4.20}$$

The relation $u_{\alpha\lambda} = U$ on $\Gamma_0 \times (0, T)$ is conserved and, at the limit

$$u_{\alpha} = U \text{ on } \Gamma_0 \times (0, \mathrm{T}). \tag{4.21}$$

Passing to the limit in (4.18), we get

$$A(\sigma_{\alpha} - \sigma_0) + \int_0^t \varrho_{\alpha}(s)\,ds = \varepsilon(u_{\alpha})(t) - \varepsilon(u_0), \qquad 0 < t < T. \tag{4.22}$$

Similarly we integrate (3.4) with respect to $t$ and pass to the limit $\lambda \to 0$ in the relation that we find. Whence

$$\begin{aligned}
&\lambda(\varepsilon(u_{\alpha\lambda})(t), \varepsilon(w)) - \lambda(\varepsilon(u_0), \varepsilon(w)) + \int_0^t (\sigma_{\alpha\lambda}(s), \varepsilon(w))\,ds\\
&\quad = \int_0^t L(s; w)\,ds, \qquad \forall\, w \in \mathscr{C}^0_{\mathrm{ad}}, \quad 0 < t < \mathrm{T},
\end{aligned}$$

$$\int_0^t (\sigma_{\alpha}(s), \varepsilon(w))\,ds = \int_0^t L(s; w)\,ds, \qquad \forall\, w \in \mathscr{C}^0_{\mathrm{ad}}, \quad 0 < t < \mathrm{T}$$

and differentiating again with respect to $t$:

$$(\sigma_{\alpha}(t), \varepsilon(w)) = L(t; w), \qquad \forall\, w \in \mathscr{C}^0_{\mathrm{ad}}, \quad 0 < t < T. \tag{4.23}$$

Using the Green's formula (1.24), it is elementary to deduce from (4.23) that

$$\operatorname{div} \sigma_\alpha(t) + f(t) = 0 \quad \text{in } \Omega, \tag{4.24}$$

$$\sigma_\alpha(t) \cdot \nu = F(t) \quad \text{on } \Gamma_1 \tag{4.25}$$

or

$$\sigma_\alpha(t) \in \mathscr{S}_{\text{ad}}(f(t), F(t)), \quad \tau_\alpha(t) = \sigma_\alpha(t) - \chi(t) \in \mathscr{S}^0_{\text{ad}}, \quad 0 < t < T. \tag{4.26}$$

We conclude that $\sigma_\alpha$, $u_\alpha$ is a solution of $\mathscr{P}_\alpha$, if we show that $\varrho_\alpha = \gamma'_\alpha(\sigma_\alpha)$ and $\sigma_\alpha(0) = \sigma_0$ (and $u_\alpha(0) = u_0$). We integrate (4.3) with respect to $t$, and pass to the limit $\lambda \to 0$:

$$(A(\sigma_{\alpha\lambda}(t) - \sigma_0), \tau) + \int_0^t (\gamma'_\alpha(\sigma_{\alpha\lambda})(s), \tau)\, ds = \int_{\Gamma_0} (U(t) - U(0)) \otimes \nu \cdot \tau \, d\Gamma,$$

$$(A(\sigma_\alpha(t) - \sigma_0), \tau) + \int_0^t (\varrho_\alpha(s), \tau)\, ds = \int_{\Gamma_0} (U(t) - U(0)) \otimes \nu \cdot \tau \, d\Gamma, \tag{4.27}$$

$$\forall \tau \in \mathscr{S}^0_{\text{ad}}, \quad 0 < t < \mathrm{T}.$$

This shows that $\sigma_\alpha(0) = \sigma_0$[1] (and then $u_\alpha(0) = u_0$ by (4.22)), and differentiating (4.27) again we find

$$(A\dot{\sigma}_\alpha(t), \tau) + (\varrho_\alpha(t), \tau) = \int_{\Gamma_0} \dot{U}(t) \otimes \nu \cdot \tau \, d\Gamma, \tag{4.28}$$

$$\forall \tau \in \mathrm{L}^2(\Omega) \times \mathrm{L}^{\frac{\alpha+1}{\alpha}}(\Omega; \mathrm{E}^\mathrm{D}) \cap \mathscr{S}^0_{\text{ad}}.$$

The proof of existence of Theorem 2.1 will be complete if we show that

$$\varrho_\alpha = \gamma'_\alpha(\sigma_\alpha). \tag{4.29}$$

### 4.3. *Proof of* (4.29)

Let

$$\mathrm{W}_\alpha = \mathrm{L}^2(\Omega) \times \mathrm{L}^{\frac{\alpha+1}{\alpha}}(\Omega; \mathrm{E}^\mathrm{D}) \cap \mathscr{S}^0_{\text{ad}},$$

which is a Banach space for its natural norm and let $\mathrm{W}'_\alpha$ be its dual; $\mathscr{S}^0_{\text{ad}}$ being endowed with the norm of $\mathrm{L}^2(\Omega; \mathrm{E})$ which makes it a Hilbert space, we have

$$\mathrm{W}_\alpha \subset \mathscr{S}^0_{\text{ad}} \subset \mathrm{W}'_\alpha \tag{4.30}$$

where the injections are continuous and each space is dense in the following one. We can rewrite (4.28) in the form

$$(A\dot{\tau}_\alpha(t), \tau) + (A\dot{\chi}(t), \tau) + (\varrho_\alpha(t), \tau) = \int_{\Gamma_0} \dot{U}(t) \otimes \nu \cdot \tau \, d\Gamma, \quad \forall \tau \in \mathrm{W}_\alpha \tag{4.31}$$

---

[1] At $t = 0$, (4.27) gives $(A(\sigma_\alpha(0) - \sigma_0), \tau) = 0 \ \forall \tau \in \mathscr{S}^0_{\text{ad}}$, and this implies $\sigma_\alpha(0) = \sigma_0$ since $\sigma_\alpha(0) - \sigma_0$ belongs to $\mathscr{S}^0_{\text{ad}}$.

from which it follows easily that

$$\dot{\tau}_\alpha(t) \in \mathrm{L}^{\alpha+1}(0, \mathrm{T}; \mathrm{W}'_\alpha). \tag{4.32}$$

In the other hand, the preceding results show that

$$\tau_\alpha \in \mathrm{L}^{\frac{\alpha+1}{\alpha}}(0, \mathrm{T}; \mathrm{W}_\alpha). \tag{4.33}$$

Since the spaces $\mathrm{L}^{\frac{\alpha+1}{\alpha}}(0, \mathrm{T}; \mathrm{W}_\alpha)$ and $\mathrm{L}^{\alpha+1}(0, \mathrm{T}; \mathrm{W}'_\alpha)$ are dual, a result in J. L. LIONS [18] (*cf.* also Chapter III, Section 1 in R. TEMAM [39]) shows that $\tau_\alpha$ is continuous with values in $\mathscr{S}^0_{\mathrm{ad}}$ and

$$(A\dot{\tau}_\alpha, \tau_\alpha) = \tfrac{1}{2} \frac{d}{dt} (A\tau_\alpha, \tau_\alpha). \tag{4.34}$$

It then follows from (4.31) that

$$\begin{aligned} &\tfrac{1}{2}(A\tau_\alpha(t), \tau_\alpha(t)) + \int_0^t (A\dot{\chi}(s), \tau_\alpha(s))\, ds + \int_0^t (\varrho_\alpha(s), \tau_\alpha(s))\, ds \\ &= \tfrac{1}{2}(A\tau_0, \tau_0) + \int_0^t \int_{\Gamma_0} \dot{U}(s) \otimes \nu \cdot \tau_\alpha(s)\, d\Gamma\, ds, \qquad 0 \leqq t \leqq T. \end{aligned} \tag{4.35}$$

The proof of (4.29) follows now a classical argument in the theory of monotone operators [4], [20]. Let $\tau$ be any element of $\mathrm{L}^{\frac{\alpha+1}{\alpha}}(0, \mathrm{T}; \mathrm{W}_\alpha)$ and let

$$X_\lambda = \int_0^{\mathrm{T}} (\gamma'_\alpha(\tau_{\alpha\lambda}(t) + \chi(t)) - \gamma'_\alpha(\tau(t) + \chi(t)), \tau_{\alpha\lambda}(t) - \tau(t))\, dt.$$

Due to the monotonicity of $\gamma'_\alpha$, $X_\lambda \geqq 0$. But, setting $\tau = \tau_{\alpha\lambda}(t) = \sigma_{\alpha\lambda}(t) - \chi(t)$ in (3.3) and $w = v_{\alpha\lambda}(t) = \dot{u}_{\alpha\lambda}(t) - \dot{U}(t)$ in (3.4), we deduce from (3.3), (3.4) that

$$\begin{aligned} &\tfrac{1}{2}\frac{d}{dt}(A\tau_{\alpha\lambda}, \tau_{\alpha\lambda}) + (A\dot{\chi}, \tau_{\alpha\lambda}) + (\gamma'_\alpha(\tau_{\alpha\lambda} + \chi), \tau_{\alpha\lambda}) \\ &= (\varepsilon(\dot{u}_{\alpha\lambda}), \tau_{\alpha\lambda}) \\ &= (\tau_{\alpha\lambda}, \varepsilon(\dot{U})) - (\chi, \varepsilon(\dot{u}_{\alpha\lambda}) - \varepsilon(\dot{U})) \\ &\quad + L(t, \dot{u}_{\alpha\lambda} - \dot{U}) - \lambda(\varepsilon(\dot{u}_{\alpha\lambda})), \varepsilon(\dot{u}_{\alpha\lambda}) - \varepsilon(\dot{U})). \end{aligned} \tag{4.36}$$

We drop the non-negative term $\lambda\, |\varepsilon(\dot{u}_{\alpha\lambda})|^2$ and we note that, due to (1.24) and (3.5):

$$(\chi, \varepsilon(\dot{u}_{\alpha\lambda}) - \varepsilon(\dot{U})) = L(t; \dot{u}_{\alpha\lambda} - \dot{U}).$$

Whence the right-hand side of (4.36) is bounded by

$$(\tau_{\alpha\lambda}, \varepsilon(\dot{U})) + \lambda(\varepsilon(\dot{u}_{\alpha\lambda}), \varepsilon(\dot{U}))$$

and by integration from 0 to $t$:

$$\int_0^T (\gamma'_\alpha(\tau_{\alpha\lambda}+\chi), \tau_{\alpha\lambda})\, dt \leqq -\tfrac{1}{2}(A\tau_{\alpha\lambda}(T), \tau_{\alpha\lambda}(T)) + \tfrac{1}{2}(A\tau_0, \tau_0) \tag{4.37}$$

$$+ \int_0^T \{(\tau_{\alpha\lambda}, \varepsilon(\dot{U}) - A\dot{\chi}) + \lambda(\varepsilon(\dot{u}_{\alpha\lambda}), \varepsilon(\dot{U}))\}\, dt\,.$$

We pass to the upper limit in this inequality. It can be shown with (4.28) that $\tau_{\alpha\lambda}(\mathrm{T})$ converges weakly to $\tau_\alpha(\mathrm{T})$ in $\mathrm{L}^2(\Omega; \mathrm{E})$ and therefore

$$(A\tau_\alpha(\mathrm{T}), \tau_\alpha(\mathrm{T})) \leqq \liminf_{\lambda\to 0} (A\tau_{\alpha\lambda}(\mathrm{T}), \tau_{\alpha\lambda}(\mathrm{T}))\,.$$

By integration by part and using (1.28) and (4.19), one can check that

$$\int_0^T \lambda(\varepsilon(\dot{u}_{\alpha\lambda}), \varepsilon(\dot{U}))\, dt \to 0\,.$$

Thus

$$\limsup_{\lambda\to 0} \int_0^T (\gamma'_\alpha(\tau_{\alpha\lambda}+\chi), \tau_{\alpha\lambda})\, dt \tag{4.38}$$

$$\leqq \tfrac{1}{2}(A\tau_0, \tau_0) - \tfrac{1}{2}(A\tau(\mathrm{T}), \tau(\mathrm{T})) + \int_0^T (\tau_\alpha, \varepsilon(\dot{U}) - A\dot{\chi})\, dt\,.$$

Since $\tau_\alpha(t) \in \mathscr{S}^0_{\mathrm{ad}}$, $\forall\, t$ (by (4.23), (4.24)), we infer from (1.24) that

$$\int_0^T (\tau_\alpha, \varepsilon(\dot{U}))\, dt = \int_0^T \int_{\Gamma_0} \tau_\alpha \cdot \dot{U} \otimes \nu\, d\Gamma\, dt\,.$$

Use of (4.35) shows that the right-hand side of (4.38) is equal to

$$\int_0^T (\varrho_\alpha(t), \tau_\alpha(t))\, dt$$

and

$$\limsup_{\lambda\to 0} \int_0^T (\gamma'_\alpha(\tau_{\alpha\lambda}+\chi), \tau_{\alpha\lambda})\, dt \leqq \int_0^T (\varrho_\alpha(t), \tau_\alpha(t))\, dt\,. \tag{4.39}$$

Using (4.39) and the weak convergences (4.20), we now easily show that

$$\limsup_{\lambda\to 0} X_\lambda \leqq \int_0^T (\varrho_\alpha - \gamma'_\alpha(\tau+\chi), \tau_\alpha - \tau)\, dt. \tag{4.40}$$

Since

$$\int_0^T (\varrho_\alpha - \gamma'_\alpha(\tau+\chi), \tau_\alpha - \tau)\, dt \geqq 0$$

for every $\tau$ in $\mathrm{L}^{\frac{\alpha+1}{\alpha}}(0, \mathrm{T}; \mathrm{W}_\alpha)$, we write this inequality with $\tau = \tau_\alpha \pm \eta\tilde{\tau}$, $\tilde{\tau} \in \mathrm{L}^{\frac{\alpha+1}{\alpha}}(0, \mathrm{T}; \mathrm{W}_\alpha)$, $\eta > 0$, and passing to the limit $\eta \to 0$, we find

$$\int_0^T (\varrho_\alpha - \gamma'_\alpha(\tau_\alpha+\chi), \tilde{\tau})\, dt = 0, \qquad \forall\, \tilde{\tau},$$

which is nothing else than (4.29).

*4.4. Proof of uniqueness*

Let $\sigma_\alpha^1$, $u_\alpha^1$, $\sigma_\alpha^2$, $u_\alpha^2$ be two pairs of solutions of $\mathscr{P}_\alpha$ satisfying (2.15), (2.16) and let $\sigma_\alpha = \sigma_\alpha^1 - \sigma_\alpha^2$, $u_\alpha = u_\alpha^1 - u_\alpha^2$.

We first show that $\sigma_\alpha = 0$. We subtract the relations (4.28), (4.29) for $\sigma_\alpha^1$ and $\sigma_\alpha^2$. We find

$$(A\dot{\sigma}_\alpha, \tau) + (\gamma'_\alpha(\sigma_\alpha^1) - \gamma'_\alpha(\sigma_\alpha^2), \tau) = 0, \quad \forall \tau \in W_\alpha. \tag{4.41}$$

Due to (4.25) $\sigma_\alpha(t) \in \mathscr{S}_{\text{ad}}^0$ and thus, $\sigma_\alpha(t) \in W_\alpha$, $\forall t$, and we can take $\tau = \sigma_\alpha(t)$ in (4.41); since $\sigma_\alpha \in \mathrm{L}^{\frac{\alpha+1}{\alpha}}(0, \mathrm{T}; \mathrm{W}_\alpha)$ and $\dot{\sigma}_\alpha \in \mathrm{L}^{\alpha+1}(0, \mathrm{T}; \mathrm{W}'_\alpha)$ by (4.41), we have, exactly, as for (4.34),

$$(A\dot{\sigma}_\alpha, \sigma_\alpha) = \tfrac{1}{2} \frac{d}{dt} (A\sigma_\alpha, \sigma_\alpha).$$

Whence

$$\tfrac{1}{2} \frac{d}{dt} (A\sigma_\alpha, \sigma_\alpha) + (\gamma'_\alpha(\sigma_\alpha^1) - \gamma'_\alpha(\sigma_\alpha^2), \sigma_\alpha) = 0.$$

Due to the monotonicity of $\gamma'_\alpha$, we can write

$$\frac{d}{dt} (A\sigma_\alpha, \sigma_\alpha) \leqq 0$$

and since $\sigma_\alpha(0) = 0$, $\sigma_\alpha(t) = 0$, $\forall t$.

Having proved that $\sigma_\alpha^1 = \sigma_\alpha^2$, it follows from the relations (4.22), (4.29) for $\sigma_\alpha^1$, $u_\alpha^1$ and $\sigma_\alpha^2$, $u_\alpha^2$ that

$$\varepsilon(u_\alpha^1) = \varepsilon(u_\alpha^2).$$

This, combined with (4.21), implies $u_\alpha^1 = u_\alpha^2$.

*Remark 4.1.* The equation (4.31) (with $\varrho_\alpha = \gamma'_\alpha(\sigma_\alpha)$) combined with (4.24), (4.25) constitutes an evolution equation for $\tau_\alpha = \sigma_\alpha - \chi$, which can be studied by the methods used in Section 3 and allows the determination of $\sigma_\alpha$ for all time, independently of $u_\alpha$. The reason for introducing the perturbed problems of Section 3 is that this allows us to obtain also the function $u_\alpha$ satisfying (2.16), (2.18) and (2.21), (4.22), (4.29).

*Remark 4.2.* By use of (4.29), (4.28) becomes

(4.42)

$$(A\dot{\sigma}_\alpha(t), \tau) + (\gamma'_\alpha(\sigma_\alpha(t)), \tau) = \int_{\Gamma_0} \dot{U}(t) \otimes \eta \cdot \tau \, d\Gamma, \quad \forall \tau \in \mathrm{W}_\alpha, \quad 0 < t < \mathrm{T}.$$

Given any function $\tau$ satisfying (compare to (4.32), (4.33))

$$\tau \in \mathrm{L}^{\frac{\alpha+1}{\alpha}}(0, \mathrm{T}; \mathrm{W}_\alpha), \dot{\tau} \in \mathrm{L}^{\alpha+1}(0, \mathrm{T}; \mathrm{W}'_\alpha), \tag{4.43}$$

we can replace $\tau$ by $\sigma_\alpha(t)-\tau(t),\ \forall t$ in (4.42):

$$(A\dot{\sigma}_\alpha(t),\sigma_\alpha(t)-\tau(t))+(\gamma'_\alpha(\sigma_\alpha(t)),\sigma_\alpha(t)-\tau(t)) = \int_{\Gamma_0}\dot{U}(t)\otimes\nu\cdot(\sigma_\alpha(t)-\tau(t))\,d\Gamma,$$

whence we obtain after integration using (4.34)

$$\tfrac{1}{2}(A\sigma_\alpha(t),\sigma_\alpha(t))+\int_0^t(\gamma'_\alpha(\sigma_\alpha(s)),\sigma_\alpha(s)-\tau(s))\,ds \tag{4.44}$$
$$=\tfrac{1}{2}(A\sigma_0,\sigma_0)+(A\sigma_\alpha(t),\tau(t))-(A\sigma_0,\tau_0))$$
$$-\int_0^t(A\sigma_\alpha(s),\dot{\tau}(s))\,ds+\int_0^t\int_{\Gamma_0}\dot{U}(s)\otimes\nu\cdot(\sigma_\alpha(s)-\tau(s))\,d\Gamma\,ds$$

which is valid for $0<t<T$ for every $\tau$ satisfying (4.43). Alternatively, using (2.6) we have

$$(\gamma'_\alpha(\sigma_\alpha(s)),\sigma_\alpha(s)-\tau(s)\geqq(\gamma'_\alpha(\tau(s)),\sigma_\alpha(s)-\tau(s))$$

and we infer from (4.44) that

$$\tfrac{1}{2}(A\sigma_\alpha(t),\sigma_\alpha(t))+\int_0^t(\gamma'_\alpha(\tau(s)),\sigma_\alpha(s)-\tau(s))\,ds \tag{4.45}$$
$$\leqq\tfrac{1}{2}(A\sigma_0.\sigma_0)+(A\sigma_\alpha(t),\tau(t))-(A\sigma_0,\tau(0))$$
$$-\int_0^t(A\sigma_\alpha(s),\dot{\tau}(s))\,ds+\int_0^t\int_{\Gamma_0}\dot{U}(s)\otimes\nu\cdot(\sigma_\alpha(s)-\tau(s))\,d\Gamma\,ds$$

for $0<t<T$ and for the same $\tau$'s as in (4.44).

These consequences of (4.42) will be used in Section 5. □

### *4.5. More regularity*

We have indicated in Section 4.4 that $\sigma_\alpha$ could be determined directly and independently of $u_\alpha$ by solving the initial value problem (4.31), (4.29), (4.24), (4.25) for $\sigma_\alpha$ (or $\tau_\alpha=\sigma_\alpha-\chi$). We will describe briefly how this could be done by the implementation of a Faedo-Galerkin method and then this will allow us to obtain more regularity on $\sigma_\alpha$, $u_\alpha$ when one of the assumptions (2.11), (2.12) is satisfied.

*Assumption* (2.11). In terms of $\tau_\alpha$ the initial-value problem is expressed by (4.31)–(4.33) (with $\varrho_\alpha=\gamma'_\alpha(\sigma_\alpha)$), *i.e.*

$$\tau_\alpha\in L^{\frac{\alpha+1}{\alpha}}(0,T;W_\alpha),\qquad \dot{\tau}_\alpha\in L^{\alpha+1}(0,T;W'_\alpha), \tag{4.46}$$

$$(A\dot{\tau}_\alpha(t),\tau)+(A\dot{\chi}(t),\tau)+(\gamma'_\alpha(\tau_\alpha(t)+\chi(t)),\tau) = \int_{\Gamma_0}\dot{U}(t)\otimes\nu\cdot\tau\,d\Gamma,\qquad\forall\tau\in W_\alpha. \tag{4.47}$$

$$\tau_\alpha(0)=\tau_0=\sigma_0-\chi(0). \tag{4.48}$$

General results concerning monotone operators [4], [20], [18] apply to this problem. We consider a total and free family $(\tau_k)_{k\in N}$ of elements of W which is a separable space. For each integer $m$, we look for $\tau_{\alpha m}$ $(= \sigma_{\alpha m} - \chi)$ taking its values in Span $(\tau_1, \ldots, \tau_m)$ and such that

$$(4.49)\qquad (A\dot{\tau}_{\alpha m}, \tau_k) + (A\dot{\chi}, \tau_k) + (\gamma'_\alpha(\tau_{\alpha m} + \chi), \tau_k) = \int_{\Gamma_0} \dot{U} \otimes \nu \cdot \tau_k \, d\Gamma, \quad k = 1, \ldots, m,$$

$$(4.50)\qquad \tau_{\alpha m}(0) = \tau_{0m},$$

$\tau_{0m} \in$ Span $(\tau_1, \ldots, \tau_m)$ and $\tau_{0m} \to \tau_0$ in $W_\alpha$ as $m \to \infty$.

General results [4], [24], [18] show that $\tau_{\alpha m}$ exists $\forall\, m$, and $\tau_{\alpha m}$ is bounded independently of $m$ in $L^{\frac{\alpha+1}{\alpha}}(0, T; W_\alpha)$. As $m \to \infty$, $\tau_\alpha$ converges weakly in this space to the solution $\tau_\alpha$ of (4.46)–(4.48), which is unique by Section 4.4.

Now if the assumption (2.11) is verified, we can differentiate (4.49) with respect to $t$:

$$(4.51)\qquad (A\ddot{\tau}_{\alpha m}, \tau_k) + (A\ddot{\chi}, \tau_k) + (\gamma''_\alpha(\tau_{\alpha m} + \chi)(\dot{\tau}_{\alpha m} + \dot{\chi}), \tau_k) = \int_{\Gamma_0} \ddot{U} \otimes \nu \cdot \tau_k \, d\Gamma, \quad 1 \leqq k \leqq m.$$

We can replace $\tau_k$ by $\dot{\tau}_{\alpha m}$ in this relation. We find

$$\tfrac{1}{2}\frac{d}{dt}(A\dot{\tau}_{\alpha m}, \dot{\tau}_{\alpha m}) + (A\ddot{\chi}, \dot{\tau}_{\alpha m}) + (\gamma''_\alpha(\tau_{\alpha m} + \chi) \cdot (\dot{\tau}_{\alpha m} + \dot{\chi}), \dot{\tau}_{\alpha m}) = \int_{\Gamma_0} \ddot{U} \otimes \nu \cdot \dot{\tau}_{\alpha m} \, d\Gamma.$$

Since $\gamma_\alpha$ is convex, $(\gamma''_\alpha(\xi) \cdot \eta) \cdot \eta \geqq 0$, $\forall\, \xi, \eta \in E$ and therefore

$$(\gamma''_\alpha(\tau_{\alpha m} + \chi) \cdot (\dot{\tau}_{\alpha m} + \dot{\chi}), \dot{\tau}_{\alpha m}) \geqq -(\gamma''_\alpha(\tau_{\alpha m} + \chi) \cdot (\dot{\tau}_{\alpha m} + \dot{\chi}), \dot{\chi}) = -\left(\frac{d}{dt}\gamma'_\alpha(\tau_{\alpha m} + \chi), \dot{\chi}\right).$$

Also with (1.37)

$$|(A\ddot{\chi}, \dot{\tau}_{\alpha m})| \leqq \tfrac{1}{2}(A\dot{\tau}_{\alpha m}, \dot{\tau}_{\alpha m}) + \tfrac{1}{2}(A\ddot{\chi}, \ddot{\chi}) \leqq c'_1 + \tfrac{1}{2}(A\dot{\tau}_{\alpha m}, \dot{\tau}_{\alpha m})$$

and by (1.23) and (1.28)

$$\begin{aligned}\left|\int_{\Gamma_0} \ddot{U} \otimes \nu \cdot \dot{\tau}_{\alpha m} \, d\Gamma\right| &\leqq c'_2 \,\|\ddot{U}\|_{H^1(\Omega)} \,\|\dot{\tau}_{\alpha m} \cdot \nu\|_{H^{-\frac{1}{2}}(\Gamma)} \\ &\leqq c'_3 \,\|\ddot{U}\|_{H^1(\Omega)} \,|\dot{\tau}_{\alpha m}|_{L^2(\Omega;E)} \\ &\leqq c'_4 (A\dot{\tau}_{\alpha m}, \dot{\tau}_{\alpha m})^{\frac{1}{2}} \\ &\leqq \tfrac{1}{2}(A\dot{\tau}_{\alpha m}, \dot{\tau}_{\alpha m}) + c'_5.\end{aligned}$$

Whence

$$\frac{d}{dt}(A\dot{\tau}_{\alpha m}, \dot{\tau}_{\alpha m}) \leqq 2(A\dot{\tau}_{\alpha m}, \dot{\tau}_{\alpha m}) + c_6' + \left(\frac{d}{dt}\gamma_\alpha'(\sigma_{\alpha m}), \dot{\chi}\right).$$

and, by integration in $t$,

$$(A\dot{\tau}_{\alpha m}(t), \dot{\tau}_{\alpha m})) \leqq 2\int_0^t (A\dot{\tau}_{\alpha m}, \dot{\tau}_{\alpha m})\, ds + \varphi_{\alpha m}(t) \tag{4.52}$$

with

$$\varphi_{\alpha m}(t) = c_6' T + (A\dot{\tau}_{0m}, \dot{\tau}_{0m}) + (\gamma_\alpha'(\sigma_{\alpha m}(t)), \dot{\chi}(t)) - (\gamma_\alpha'(\sigma_{\alpha m}(0)), \dot{\chi}(0)) - \int_0^t (\gamma_\alpha'(\sigma_{\alpha m}), \ddot{\chi})\, ds. \tag{4.53}$$

Gronwall's lemma will imply that

(4.54) $\dot{\tau}_{\alpha m}$ and $\dot{\sigma}_{\alpha m} = \dot{\tau}_{\alpha m} + \dot{\chi}$ are bounded in $L^\infty(0, T; L^2(\Omega; E))$ independently of $m$

provided we show that

(4.55) $\int_0^T |\varphi_{\alpha m}(t)|\, dt$ is bounded independently of $m$.

For proving

$$(A\dot{\tau}_{\alpha m}(0), \dot{\tau}_{\alpha m}(0)) \leqq \text{const.} \tag{4.56}$$

we notice that $\tau_0 = \sigma_0 - \chi(0) \in K - \chi(0)$ and therefore we can choose $\tau_{0m}$ in $K - \chi(0)$, so that $(\tau_{0m} + \chi(0)) \in K$ and $\gamma_\alpha'(\tau_{0m} + \chi(0)) = 0$. In such case replacing $\tau_k$ by $\tau_{\alpha m}$ in (4.49), and writing this relation at time $t = 0$ we find

$$(A\dot{\tau}_{\alpha m}(0), \dot{\tau}_{\alpha m}(0) = -(A\dot{\chi}(0), \dot{\tau}_{\alpha m}(0)) + \int_{\Gamma_0} \dot{U}(0) \otimes \nu \cdot \dot{\tau}_{\alpha m}(0)\, d\Gamma;$$

because of (1.37), (1.28), (1.23) and since $\dot{\tau}_{\alpha m}(0) \in \text{Span}(\tau_1, \ldots, \tau_m) \subset \mathscr{S}_{\text{ad}}^0$),

$$(A\dot{\tau}_{\alpha m}(0), \dot{\tau}_{\alpha m}(0)) \leqq \text{const.}\, |\dot{\tau}_{\alpha m}(0)| \leqq \text{const.}\, (A\dot{\tau}_{\alpha m}(0), \dot{\tau}_{\alpha m}(0))^{1/2};$$

(4.56) follows.

For (4.55) we notice that $\gamma_\alpha'(\sigma_{\alpha m}(0)) = 0$ with the previous choice of $\tau_{0m}$ and we are left with the estimates of

$$\int_0^T |(\gamma_\alpha'(\sigma_{\alpha m}), \dot{\chi})|\, dt \quad \text{and} \quad \int_0^T \left| \int_0^t (\gamma_\alpha'(\sigma_{\alpha m}), \ddot{\chi})\, ds \right| dt. \tag{4.57}$$

Due to (4.11) and (1.37)

$$\int_0^T \int_\Omega |\gamma_\alpha'(\sigma_{\alpha m}) \cdot \dot{\chi}|\, dx\, dt \leqq |\dot{\chi}|_{L^\infty(0,T;L^\infty(\Omega;E))} \int_0^T \int_\Omega |\gamma_\alpha'(\sigma_{\alpha m})|\, dx\, dt \leqq \text{const.}$$

The estimate of the second integral is similar and finally we have proved (4.55) and (4.54).

Letting $m \to \infty$ makes it clear with (4.54) and (4.20) that $\dot{\tau}_\alpha$ and $\dot{\sigma}_\alpha = \dot{\tau}_\alpha + \dot{\chi}$ belong to $L^\infty(0, T; L^2(\Omega; E))$. Then by differentiation with respect to $t$, (4.22) (with (4.29)) gives

$$A\dot{\sigma}_\alpha + \gamma'_\alpha(\sigma_\alpha) = \varepsilon(\dot{u}_\alpha),$$

whence $\varepsilon(\dot{u}_\alpha) \in L^{\alpha+1}(0, T; L^{\alpha+1}(\Omega; E))$, $\operatorname{div} \dot{u}_\alpha \in L^2(0, T; L^2(\Omega))$ and with (4.21) $\dot{u}_\alpha \in L^{\alpha+1}(0, T; W^{\alpha+1}(\Omega))$.

*Assumption* (2.12). The proof is similar when (2.11) is replaced by (2.12): we cannot differentiate (4.49) since $\gamma'_\alpha$ is not differentiable but we can replace (4.49), (4.50) by the similar system with $\gamma'_\alpha$ replaced by $\gamma'_{\alpha j}$. Whence we have $\tau_{\alpha mj}$ taking its values in Span $(\tau_1, \ldots, \tau_m)$ such that

$$(4.58) \qquad (A\dot{\tau}_{\alpha mj}, \tau_k) + (A\dot{\chi}, \tau_k) + (\gamma'_{\alpha j}(\tau_{\alpha mj} + \chi), \tau_k) = \int_{\Gamma_0} \dot{U} \otimes \nu \cdot \tau_k \, d\Gamma, \qquad k = 1, \ldots, m.$$

$$(4.59) \qquad \tau_{\alpha mj}(0) = \tau_{0m}.$$

It is elementary to verify that the solutions of (4.58), (4.59) converge as $j \to \infty$ to the solution of (4.49), (4.50). The idea is then to show that

(4.60) $\dot{\tau}_{\alpha mj}$ is bounded in $L^\infty(0, T; L^2(\Omega; E))$ independently of $j$ and $m$.

Whence, at the limit $j \to \infty$ we obtain (4.54) which allows us to conclude as before that $\dot{\tau}_\alpha$ and $\dot{\sigma}_\alpha$ are in $L^\infty(0, T; L^2(\Omega; E))$.

Computations similar to those leading to (4.8) show that the solution $\tau_{\alpha mj}$ of (4.58), (4.59) is bounded in $L^\infty(0, T; L^2(\Omega; E))$ independently of $j$ (and $m$ and $\alpha$). In finite dimensions this is sufficient to pass to the limit $j \to \infty$ ($m$, $\alpha$ fixed) and show that $\tau_{\alpha mj} \to \tau_{\alpha m}$.

Before proving (4.60) we observe that a reasoning similar to that leading to (4.11), (4.13) gives:

$$(4.61) \qquad \int_0^T \int_\Omega \{1 + d_j^2(\tau_{\alpha mj} + \chi)\}^{\frac{1}{2\alpha} - \frac{1}{2}} d_j^l(\tau_{\alpha mj} + \chi) \, dx \, dt \leq \text{const.}, \qquad l = 1 \text{ or } 2,$$

where the constant $\alpha$ is independent of $j$ (and $m$ and $\alpha$).

Then the proof of (4.60) is parallel to that of (4.54): we have $(\gamma''_{\alpha j}(\xi) \cdot \eta) \cdot \eta \geqq 0$, $\forall \xi, \eta \in E$, $\forall j$; the analogue of (4.56) for $\tau_{\alpha mj}$ is proved with the same choice of $\tau_{0m}$, observing that $\tau_{0m} + \chi(0) \in K \subset K_j$, $\forall j$ (assumption (2.12)). Thus $\gamma'_{\alpha j}(\tau_{\alpha mj}(0) + \chi(0)) = 0$ and for the proof of the analog of (4.55) there remains only to estimate the integrals similar to (4.57). They are estimated in the same manner, (4.61) replacing (4.11).

The proof of Theorem 2.1 is complete.

### 4.6. *Estimates independent of* $\alpha$

We complete this section by proving some estimates independent of $\alpha$ that will be useful in the sequel.

First we deduce immediately from (4.8), (4.11), (4.13), (4.20) that

(4.61) $\tau_\alpha$ and $\sigma_\alpha = \tau_\alpha + \chi$ are bounded in $L^\infty(0, T; L^2(\Omega; E))$ independently of $\alpha$.

(4.62) $\int_0^T \int_\Omega \{1 + d^2(\sigma_\alpha)^{\frac{1}{2\alpha} - \frac{1}{2}} d^l(\sigma_\alpha)\, dx\, dt \leqq$ const. independent of $\alpha$, $l = 1, 2$.

Due to (2.9) we have the pointwise inequality

$$|\gamma'_\alpha(\sigma_\alpha(x, t))| \leqq \{1 + d^2(\sigma_\alpha(x, t))\}^{\frac{1}{2\alpha} - \frac{1}{2}} d(\sigma_\alpha(x, t)), \qquad \forall\, x, t$$

and we conclude with (4.62) that

(4.63) $\gamma'_\alpha(\sigma_\alpha)$ is bounded in $L^1(0, T; L^1(\Omega; E))$, independently of $\alpha$.

Then if we assume (2.11) or (2.12), one can check that the proof of (4.54) in the case (2.11) (respectively (4.60) in the case (3.12)) shows that this estimate is independent of $m$ and $\alpha$ (respectively independent of $j$, $m$ and $\alpha$). Thus as $m \to \infty$ (respectively $j \to \infty$ then $m \to \infty$) we conclude that

(4.64) $\dot{\tau}_\alpha$ and $\dot{\sigma}_\alpha = \dot{\tau}_\alpha + \dot{\chi}$ are bounded in $L^\infty(0, T; L^2(\Omega,; E))$ independently of $\alpha$.

Due to the second inequality (4.14) and (1.37) we have the pointwise inequality, $x \in \Omega$, $t \in (0, T)$:

$$\delta_1 |\gamma'_\alpha(\sigma_\alpha(x, t))| \leqq \gamma'_\alpha(\sigma_\alpha(x, t)) \cdot (\sigma_\alpha(x, t) - \chi(x, t)).$$

Integrating over $\Omega$ we see that

$$\delta_1 \int_\Omega |\gamma'_\alpha(\sigma_\alpha)|\, dx \leqq \int_\Omega \gamma'_\alpha(\sigma_\alpha) \cdot (\sigma_\alpha - \chi)\, dx. \tag{4.65}$$

But according to (4.28) written with $\tau = \sigma_\alpha(t) - \chi(t)$ and (4.29):

$$\int_\Omega \gamma'_\alpha(\sigma_\alpha(t))\, (\sigma_\alpha(t) - \chi(t))\, dx = \int_{\Gamma_0} \dot{U}(t) \otimes \nu \cdot (\sigma_\alpha(t) - \chi(t))\, d\Gamma - (A\dot{\sigma}_\alpha(t), \sigma_\alpha(t) - \chi(t)).$$

By an argument used several times and by (4.61), (4.64) we prove easily that the right-hand side of this equation is bounded uniformly in $t$ and $a$. Hence with (4.62) we can now improve (4.63):

(4.66) $\gamma'_\alpha(\sigma_\alpha)$ is bounded in $L^\infty(0, T; L^1(\Omega; E))$ independently of $\alpha$.

Finally, after differentiation with respect to $t$, (4.22), (4.29) give (2.14), and with (4.64), (4.66) we obtain

(4.67) $\varepsilon(\dot{u}_\alpha)$ is bounded in $L^\infty(0, T; L^1(\Omega; E))$ independently of $\alpha$.

With (4.21) and an appropriate Poincaré inequality (*cf.* [38], Chapter II, Prop. 1.4), this implies

(4.68) $\dot{u}_\alpha$ is bounded in $L^\infty(0, T; W^1(\Omega))$ independently of $\alpha$.

Also taking the trace (in the matrix sense) of (2.14) and since $\operatorname{tr} \gamma'_\alpha(\xi) = 0 \ \forall \xi$, we obtain

(4.69) $$\operatorname{div} \dot{u}_\alpha = \operatorname{tr} (A\dot{\sigma}_\alpha)$$

and thus

(4.70) $\operatorname{div} \dot{u}_\alpha$ and $\operatorname{div} u_\alpha$ are bounded in $L^2(0, T; L^2(\Omega))$ independently of $\alpha$.

## 5. Convergence of the generalized Norton-Hoff model to the Prandtl-Reuss model: The main results

Our object is now twofold: to describe the passage to the limit $\alpha \to 0$ in the Norton-Hoff model and to prove the existence of solution for the Prandtl-Reuss model.

### *5.1. The main results for the stresses*

We recall that for every $\alpha > 0$, we have proved the existence and uniqueness of $\sigma_\alpha$, $u_\alpha$ which satisfy (2.15), (2.16),

(5.1) $$\tau_\alpha(t) = \sigma_\alpha(t) - \chi(t) \in \mathscr{S}^0_{\text{ad}}, \ \forall t,$$

(5.2) $$(A\dot{\sigma}_\alpha(t), \tau) + (\gamma'_\alpha(\sigma_\alpha(t)), \tau) = \int_{\Gamma_0} \dot{U}(t) \otimes \nu \cdot \tau \, d\Gamma, \quad \forall \tau \in W_\alpha, \quad 0 < t < T,$$

(5.3) $$A\dot{\sigma}_\alpha + \gamma'_\alpha(\sigma_\alpha) = \varepsilon(\dot{u}_\alpha)$$

(5.4) $$\operatorname{tr} A\dot{\sigma}_\alpha = \operatorname{div} \dot{u}_\alpha$$

(5.5) $$u_\alpha = U \text{ on } \Gamma_0 \times (0, T)$$

(5.6) $$u_\alpha(0) = u_0, \ \dot{u}_\alpha(0) = u_1, \ \sigma_\alpha(0) = \sigma_0$$

and (4.61)–(4.63). If furthermore (2.11) is satisfied, then (2.17), (2.18), (4.64), (4.66)–(4.68) hold true.

The Prandtl-Reuss model that we desire to solve has been described in Section 1 and in particular Section 1.5. The problem splits into two problems:
(i) The determination of $\sigma$ through (1.38)–(1.43);
(ii) The determination of $u$ satisfying (1.18), (1.19), (1.6)–(1.9).
We are going to solve the first problem and then in Section 5.5, using further technical results given in the Appendix, we will solve (ii). □

When (2.11) or (2.12) is not assumed, we shall consider a weaker form of $\mathscr{P}$ for $\sigma$ which is easily deduced from (1.38)–(1.43). If $\sigma$ satisfies (1.38)–(1.43), then integrating (1.43) and integrating by parts, we find that

$$(5.7)\qquad \tfrac{1}{2}(A\sigma(t),\sigma(t)) \leqq \tfrac{1}{2}(A\sigma_0,\sigma_0) + (A\sigma(t),\tau(t)) - (A\sigma_0,\tau(0)) - \int_0^t (A\sigma(s),\dot{\tau}(s))\,ds + \int_0^t \int_{\Gamma_0} \dot{U}(s)\otimes\nu\cdot(\sigma(s)-\tau(s))\,d\Gamma\,ds,$$

for $0<t<T$, for every $\tau$ satisfying (1.41), (1.42). Now if

$$(5.8)\qquad \sigma \in \mathrm{L}^\infty(0,\mathrm{T};\mathrm{L}^2(\Omega;\mathrm{E}))$$

$$(5.9)\qquad \sigma(t)\in \mathscr{S}_{\mathrm{ad}}(f(t),F(t)),\quad \forall\, t$$

$$(5.10)\qquad \sigma(t)\in \mathscr{K}_{\mathrm{ad}},\ \forall\, t$$

then (5.7) makes sense for almost every $t$ and for the $\tau$'s considered. A function $\sigma$ satisfying these conditions will be called a weak solution of $\mathscr{P}$.

**Theorem 5.1.** *Under the general assumptions of Section 1, when* $\alpha\to 0$, *the solutions* $\sigma_\alpha$ *of* $\mathscr{P}_\alpha$ *converges to a limit* $\sigma$ *which is a solution of* $\mathscr{P}$ (i.e. *satisfies* (5.7)–(5.9)), *and the convergence holds in the following topology*:

$$(5.11)\qquad \sigma_\alpha \to \sigma \ \textit{in}\ \mathrm{L}^\infty(0,\mathrm{T};\mathrm{L}^2(\Omega;\mathrm{E}))\ \textit{weak star}.$$

*If furthermore* (2.11) *or* (2,12) *is satisfied then* $\sigma$ *is solution of* $\mathscr{P}$ *in the sense* (1.38)–(1.43) *and*:

$$(5.12)\qquad \sigma_\alpha\to\sigma \ \text{in}\ \mathscr{C}([0,\mathrm{T}];\mathrm{L}^2(\Omega;\mathrm{E}))$$

$$(5.13)\qquad \sigma_\alpha\to\sigma \ \text{in}\ \mathrm{L}^p(\Omega\times(0,\mathrm{T});\mathrm{E})\ \ \forall\, p<\infty$$

$$(5.14)\qquad \dot{\sigma}_\alpha\to\dot{\sigma} \ \text{in}\ \mathrm{L}^\infty(0,\mathrm{T};\mathrm{L}^2(\Omega;\mathrm{E}))\ \text{weak-star}.$$

**Theorem 5.2.** *Under the general assumptions of Section 1, problem* $\mathscr{P}$ *possesses a unique weak solution* $\sigma$ (i.e. $\sigma$ *satisfies* (5.7)–(5.9)).

*If furthermore* (2.11) *or* (2.12) *are satisfied, then* $\sigma$ *is solution of* $\mathscr{P}$ *in the sense of* (1.38), (1.43). *Besides* (5.7)–(5.9), $\sigma$ *satisfies*:

$$\sigma\in \mathrm{L}^p(\Omega\times(0,\mathrm{T});\mathrm{E}),\ \forall\, p<\infty,\ \dot{\sigma}\in\mathrm{L}^\infty(0,\mathrm{T};\mathrm{L}^2(\Omega;\mathrm{E})).$$

### *5.2. Proof of Theorem 5.1*

Thanks to (4.61) we can extract a subsequence (still denoted $\alpha$) $\alpha\to 0$, such that

$$(5.15)\qquad \sigma_\alpha\to\sigma\ \text{in}\ \mathrm{L}^\infty(0,\mathrm{T};\mathrm{L}^2(\Omega;\mathrm{E}))\ \text{weak-star}.$$

Given $\tau$ satisfying (1.41), (1.42), then (4.43) is satisfied $\forall\,\alpha>0$ and we are allowed to write (4.45). We observe that $\gamma'_\alpha(\tau(t))=0\ \ \forall\, t$ because of (1.42), (2.7) and thus

$$(5.16)\qquad \tfrac{1}{2}(A\sigma_\alpha(t),\sigma_\alpha(t)) \leqq \tfrac{1}{2}(A\sigma_0,\sigma_0) + (A\sigma(t),\tau(t)) - (A\sigma_0,\tau(0)) - \int_0^t (A\sigma_\alpha,\dot{\tau})\,ds + \int_0^t\int_{\Gamma_0} \dot{U}(s)\otimes\nu\cdot(\sigma_\alpha(s)-\tau(s))\,d\Gamma\,ds.$$

It is easy to pass to the lower limit in (5.16) using (5.15).[1] We obtain precisely (5.7.)

In order to prove that $\sigma$ is a weak solution of $\mathscr{P}$ we still have to prove (5.9), (5.10). For (5.9) we simply observe that (5.15) implies

$$\sigma_\alpha - \chi \to \sigma - \chi \text{ in } L^\infty(0, T; L^2(\Omega; E))$$

and since by (4.26) $\sigma_\alpha - \chi$ takes its values in $\mathscr{S}^0_{\text{ad}}$ which is a closed subspace of $L^2(\Omega; E)$, we conclude that the same is true for $\sigma - \chi$ and (5.9) is proved.

For (5.10) we start from (4.62) written with $l = 1$ and observe that for every fixed $p$, $1 < p < \infty$, as soon as $1/2\alpha > p$, we have

$$(5.17) \quad \int_0^T \int_\Omega \{1 + d^2(\sigma_\alpha)\}^p \, d(\sigma_\alpha) \, dx \, dt \leqq \int_0^T \int_\Omega \{1 + d^2(\sigma_\alpha)\}^{\frac{1}{2\alpha} - \frac{1}{2}} d(\sigma_\alpha) \, dx \, dt \leqq c_1',$$

where the constant $c_1'$ (the right-hand side of (4.62)) is independent of $\alpha$ and $p$. As $\alpha \to 0$, we find by convexity and lower semicontinuity

$$(5.18) \quad \int_0^T \int_\Omega \{1 + d^2(\sigma)\}^p \, d(\sigma) \, dx \, dt \leqq c_1'.$$

Letting $p$ go to infinity in (5.18), we conclude that $d(\sigma(x, t)) = 0$ for a.e. $x$ and a.e. $t$, which is equivalent to (5.10).

This proves the first part of Theorem 5.1.[2]

We now assume that (2.11) or (2.12) is satisfied and prove the complementary statements.

Because of (4.64) we have besides (5.15)

$$(5.19) \quad \dot\sigma_\alpha \to \dot\sigma \text{ in } L^\infty(0, T; L^2(\Omega; E)) \text{ weak star}.$$

The relation

$$\sigma_\alpha(t) = \sigma_0 + \int_0^t \dot\sigma_\alpha(s) \, ds$$

gives at the limit

$$\sigma(t) = \sigma_0 + \int_0^t \dot\sigma(s) \, ds$$

---

[1] We proceed as follows: given a real continuous function $\varphi$ on $[0, T]$, $\varphi(t) \geqq 0$, $\forall\, t$, we multiply (5.15) by $\varphi(t)$ and integrate over $(0, T)$. We then pass to the lower limit $\alpha \to 0$, observing that by convexity

$$\int_0^T (A\sigma(t), \sigma(t)) \, \varphi(t) \, dt \leqq \liminf_{\alpha \to 0} \int_0^T (A\sigma_\alpha(t), \sigma_\alpha(t)) \, \varphi(t) \, dt.$$

We obtain an inequality of the form $\int_0^T g(t) \varphi(t) \, dt \geqq 0 \;\; \forall\, \varphi \geqq 0$ which implies that $g(t) \leqq 0$ a.e.

[2] It follows from Theorem 5.2 that a weak solution of $\mathscr{P}$ is unique, whence with a standard argument by contradiction we see that (5.15) is valid for the whole family $\alpha \to 0$ and not only for a subsequence. The rest of the proof of Theorem 5.1 is related to the whole family $\alpha$.

Of course the conclusion of uniqueness in Theorem 5.2 is totally independent of Theorem 5.1 (while existence relies heavily on Theorem 5.1).

showing that $\sigma(0) = \sigma_0$ (*i.e.* (1.40)) and

$$\sigma_\alpha(t) \to \sigma(t) \text{ weakly in } L^2(\Omega; E), \ \forall t. \tag{5.20}$$

Now we wish to recover (1.43) from (5.7). Because the function $\tau$ satisfies (1.41), (1.42), we notice that

$$t \to \sigma(t)\,(1 - \lambda(t)) + \tau(t)\,\lambda(t) \tag{5.21}$$

satisfies the same conditions for every $\mathscr{C}^1$ real function $\lambda$ from [0, T] into [0, 1].

We then write (5.7) with $t = T$ and $\tau$ replaced by (5.21) and we find:

$$\int_0^T (A\sigma, \dot\tau - \dot\sigma)\,\lambda\,dt \leqq (A\sigma(T), \tau(T) - \sigma(T))\,\lambda(T) - (A\sigma_0, \tau(0) - \sigma_0)\,\lambda(0)$$

$$- \int_0^T (A\sigma, \tau - \sigma)\,\dot\lambda\,dt + \int_0^T \int_{\Gamma_0} \lambda(\sigma - \tau)\cdot \dot U \otimes \nu\,d\Gamma\,dt.$$

For fixed $t_0 \in (0, T)$ and $m \in \mathbb{N}$, we take $\lambda(t) = \lambda_m(t) = 0$ for $t \leqq t_0 - 1/m$ or $t \geqq t_0 + 1/m$, $\lambda(t_0) = 1$, $\lambda$ linear and continuous on $[t_0 - 1/m, t_0]$ and $[t_0, t_0 + 1/m]$. We multiply by $m$ the inequality that we obtain and thus

$$m \int_0^T (A\sigma, \dot\tau - \dot\sigma)\,\lambda_m\,dt \leqq -m \int_0^T (A\sigma, \tau - \sigma)\,\dot\lambda_m\,dt \tag{5.22}$$

$$+ m \int_0^T \int_{\Gamma_0} \lambda_m(\sigma - \tau)\cdot \dot U \otimes \nu\,d\Gamma\,dt.$$

When $m \to \infty$, $m \int_0^T \varphi(t)\,\lambda_m(t)\,dt \to 2\varphi(t_0)$ if $\varphi \in L^1(0, T)$ and $t_0$ is a Lebesgue point of $\varphi$. Now if $\varphi \in L^1(0, T)$, $\dot\varphi \in L^1(0, T)$ then

$$m \int_0^T \varphi(t)\,\dot\lambda_m(t)\,dt = -m \int_0^T \dot\varphi(t)\,\lambda_m(t)\,dt \quad \to \quad -2\dot\varphi(t_0),$$

provided $t_0$ is a Lebesgue point of $\dot\varphi$.

Passing to the limit $m \to \infty$ in (5.19) we find that

$$-(A\dot\sigma(t_0), \tau(t_0) - \sigma(t_0)) \leqq \int_{\Gamma_0} (\sigma(t_0) - \tau(t_0))\cdot \dot U(t_0) \otimes \nu\,d\Gamma,$$

for every $t_0$ which is a Lebesgue point of $\sigma$, $\dot\sigma$ and $\tau$, *i.e.* for almost every $t_0$. This is exactly (1.43) which holds true for a.e. $t \in (0, T)$, for every $\tau$ satisfying (1.41), (1.42).

The last step in the proof of Theorem 5.1 is then to improve the convergence of $\sigma_\alpha$ to $\sigma$ and establish (5.12). We now prove the uniform convergence in (5.12) while the $L^p$-convergence of (5.13) relying on different ideas will be proved in Sec. 5.3.

First we prove that

$$\sigma_\alpha(t) \to \sigma(t) \text{ strongly in } L^2(\Omega; E), \ \forall t. \tag{5.23}$$

It is legitimate to write (4.44), (4.45) with $\tau = \sigma$. With (4.45), observing that $\gamma'_\alpha(\sigma(s)) = 0$, $\forall\, s$, we obtain an inequality in which we can pass to the upper limit as $\alpha \to 0$; we find

$$\limsup_{\alpha\to 0} \tfrac{1}{2}\,(A\sigma_\alpha(t), \sigma_\alpha(t)) \leqq -\tfrac{1}{2}\,(A\sigma_0, \sigma_0) + (A\sigma(t), \sigma(t)) - \int_0^t (A(\sigma(s), \dot\sigma(s))\, ds$$
$$= \tfrac{1}{2}\,(A\sigma(t), \sigma(t));$$

hence,

$$\limsup_{\alpha\to 0}\, (A\sigma_\alpha(t), \sigma_\alpha(t)) \leqq (A\sigma(t), \sigma(t)) \tag{5.24}$$

and since by (5.19) and lower semicontinuity

$$(A\sigma(t), \sigma(t)) \leqq \liminf_{\alpha\to 0}\, (A\sigma_\alpha(t), \sigma_\alpha(t)) \tag{5.25}$$

we conclude that

$$(A\sigma_\alpha(t), \sigma_\alpha(t)) \to (A\sigma(t), \sigma(t)), \ \forall\, t, \tag{5.26}$$

and (5.23) follows.

We now obtain the uniform convergence announced in (5.12) by application of the Ascoli theorem. The family $\sigma_\alpha(.)$ is clearly equicontinuous in $L^2(\Omega; E)$:

$$\begin{aligned} |\sigma_\alpha(t) - \sigma_\alpha(s)|_{L^2(\Omega;E)} &\leqq |t-s|\, |\dot\sigma_\alpha|_{L^\infty(0,T;L^2(\Omega;E))} \\ &\leqq c_1\, |t-s|, \ \forall\, s, t, \ \forall\, \alpha. \end{aligned} \tag{5.27}$$

Also for every $t$ and for every sequence $\alpha_j \to 0$, the family $\sigma_{\alpha_j}(t)$ is compact in $L^2(\Omega; E)$ since it converges to $\sigma(t)$ in this space. Thus $\sigma_{\alpha_j}$ converges to $\sigma$ in $\mathscr{C}([0,T]; L^2(\Omega; E))$ and since this convergence holds for every sequence $\alpha_j \to 0$, we have proved (5.12).

*Remark 5.1.* As indicated above, we are allowed to write (4.44) with $\tau = \delta$. We pass to the limit $\alpha \to 0$ in the relation which we thus obtain, using (5.23). After an easy computation we find (for $t = T$) that

$$\limsup_{\alpha\to 0} \int_0^T (\gamma'_\alpha(\sigma_\alpha(s)), \sigma_\alpha(s) - \sigma(s))\, dxs = 0. \tag{5.28}$$

### 5.3. *Proof of* (5.13)

We infer from (5.17) and (2.8) that for every $p < \infty$, when $\alpha \to 0$:

$$\sigma_\alpha^D \text{ remains in a bounded set of } L^p(\Omega\times(0,T); E^D). \tag{5.29}$$

It follows from (5.12) that $\sigma_\alpha$ converges to $\sigma$ in $L^2(\Omega\times(0,T); E)$ as $\alpha \to 0$ and thus for a subsequence $\alpha_j \to 0$

$$\sigma_\alpha \to \sigma \quad \text{a.e. in } \Omega\times(0,T). \tag{5.30}$$

Now for every $N > 0$, we write

$$\int_0^T \int_\Omega |\sigma^D_{\alpha_j} - \sigma^D|^p \, dx \, dt \leqq \iint_{|\sigma_{\alpha_j} - \sigma^D| \leqq N} |\sigma^D_{\alpha_j} - \sigma^D|^p \, dx \, dt + \iint_{|\sigma_{\alpha_j} - \sigma^D| > N} |\sigma^D_{\alpha_j} - \sigma^D|^p \, dx \, dt. \tag{5.31}$$

As $\alpha_j \to 0$, the first integral in the right-hand side of (5.31) converges to 0 by the Lebesgue dominated convergence theorem and (5.30). The second integral is majorized, thanks to Hölder's inequality, by

$$(\text{meas}\,\{|\sigma^D_{\alpha_j} - \sigma^D| \geqq N\})^{\frac{1}{p+1}} \left( \iint_{|\sigma^D_{\alpha_j} - \sigma^D| \leqq N} |\sigma^D_{\alpha_j} - \sigma^D|^{p+1} \, dx \, dt \right)^{\frac{p}{p+1}}$$

$$\leqq (\text{meas}\,\{|\sigma^D_{\alpha_j} - \sigma^D| \geqq N\})^{\frac{1}{p+1}} \left( \int_0^T \int_\Omega |\sigma^D_{\alpha_j} - \sigma^D|^{p+1} \, dx \, dt \right)^{\frac{p}{p+1}}$$

$$\leqq (\text{by (5.29)})\ c_2' \,(\text{meas}\,\{|\sigma^D_{\alpha_j} - \sigma^D| \geqq N\})^{\frac{1}{p+1}}.$$

But (5.29) gives also

$$\int_0^T \int_\Omega |\sigma^D_{\alpha_j} - \sigma^D|^p \, dx \, dt \leqq \text{const} = c_3'$$

$$(\text{meas}\,\{|\sigma^D_{\alpha_j} - \sigma^D| \geqq N\}) \cdot N^p \leqq c_3',$$

whence (5.31) gives

$$\int_0^T \int_\Omega |\sigma^D_\alpha - \sigma^D|^p \, dx \, dt \leqq o(1) + c_2' \left( \frac{c_3'}{N^p} \right)^{\frac{1}{p+1}} \tag{5.32}$$

and as $\alpha_j \to 0$,

$$\limsup_{\alpha_j \to 0} \int_0^T \int_\Omega |\sigma^D_{\alpha_j} - \alpha^D|^p \, dx \, dt \leqq c_2' \left( \frac{c_3'}{N^p} \right)^{\frac{1}{p+1}}.$$

Since $N > 0$ can be chosen arbitrarily large, we conclude that $\sigma^D_{\alpha_j}$ converges to $\sigma^D$ in $L^p(\Omega \times (0, T); E^D)$ as $\alpha_j \to 0$. Then with a standard argument by contradiction, we see that the whole family $\sigma^D_\alpha$ converges to $\sigma^D$ in this space

$$\sigma^D_\alpha \to \sigma^D \text{ in } L^p(\Omega \times (0, T); E^D) \text{ as } \alpha \to 0, \quad \forall\, p < \infty. \tag{5.33}$$

Finally, the proof of (5.13) follows now from Proposition A.2 of the Appendix. We apply that Proposition to $(\sigma_\alpha(t) - \sigma(t))$ with $p = 2$, $r < n\,(= 3$ here) but arbitrarily close to $n$ so that $r^*$ is arbitrarily large and $s = r^*$. Thus (A.11) gives

$$|\text{tr}\,(\sigma_\alpha(t) - \sigma(t))|^{r^*}_{L^{r^*}(\Omega)}$$
$$\leqq c_4' \{ |\sigma_\alpha(t) - \sigma(t)|^{r^*}_{L^2(\Omega; E)} + |\sigma^D_\alpha(t) - \sigma^D(t)|^{r^*}_{L^{r^*}(\Omega; E^D)} \}$$

(we recall that $\operatorname{div}(\sigma_\alpha(t) - \sigma(t)) = 0$). After integration with respect to $t$, using (5.12) and (5.25), we see that, for $\alpha \to 0$,

$$\operatorname{tr}\sigma_\alpha \to \operatorname{tr}\sigma \text{ in } L^p(\Omega \times (0, T)) \tag{5.34}$$

for every $p < \infty$. This statement together with (5.33) gives (5.13).

The proof of Theorem 5.1 is complete. □

*5.4. Proof of Theorem 5.2*

The existence part of Theorem 5.2 in the general case and in the case where (2.11) or (2.12) are assumed follows completely from Theorem 5.1. Therefore the only result which remains to be proved is the uniqueness of a weak solution of $\mathscr{P}$; this improves the theorems of uniqueness in [7], [15].

Let $\sigma_1$ and $\sigma_2$ be two weak solutions of $\mathscr{P}$ and let $\sigma = \sigma_1 - \sigma_2$. We consider a mollification $\tau_\eta$ of $(\sigma_1 + \sigma_2)/2$ which is obtained as follows $(\eta > 0)$:

$$\eta\dot{\tau}_\eta + \tau_\eta = \tfrac{1}{2}(\sigma_1 + \sigma_2) \qquad 0 < t < T \tag{5.35}$$

$$\tau_\eta(0) = \sigma_0 . \tag{5.36}$$

It is clear that $\tau_\eta$ and $\dot{\tau}_\eta$ are in $L^\infty(0, T; L^2(\Omega; E))$, that $\tau_\eta$ satisfies (1.42) and (1.41) and it is elementary to check that when $\eta \to 0$,

$$\tau_\eta \to \tfrac{1}{2}(\sigma_1 + \sigma_2) \text{ strongly in } L^2(0, T; L^2(\Omega; E)) . \tag{5.37}$$

We replace $\tau$ by $\tau_\eta$ in the relation (5.7) satisfies by $\sigma^1$ and $\sigma^2$. Then we add these relations and find

$$\tfrac{1}{2}(A\sigma_1(t), \sigma_1(t)) + \tfrac{1}{2}(A\sigma_2(t), \sigma_2(t)) \leqq -(A\sigma_0, \sigma_0) - (A(\sigma_1(t) + \sigma_2(t)), \tau_\eta(t))$$
$$- \int_0^t (A(\sigma_1 + \sigma_2), \dot{\tau}_\eta)\, ds + \int_0^t \int_{\Gamma_0} \dot{U} \otimes \nu \cdot \left(\frac{\sigma_1 + \sigma_2}{2} - \tau_\eta\right) d\Gamma\, ds;$$

by use of (5.35) and after integration we show that the left-hand side of the foregoing

$$\leqq (A(\sigma_1(t) + \sigma_2(t)), \tau_\eta(t)) - 2\eta \int_0^t (A\dot{\tau}_\eta, \dot{\tau}_\eta)\, ds$$
$$- (A\tau_\eta(t), \tau_\eta(t)) + \int_0^t \int_{\Gamma_0} \dot{U} \otimes \nu \cdot \left(\frac{\sigma_1 + \sigma_2}{2} - \tau_\eta\right) d\Gamma\, ds .$$

For almost every $t$ we can pass to the limit $\eta \to 0$ in this inequality.[1] There remains

$$(A(\sigma_1(t) - \sigma_2(t)), \sigma_1(t) - \sigma_2(t)) \leqq 0 \tag{5.38}$$

[1] We omit the non negative term $\int_0^t (A\dot{\tau}_\eta, \dot{\tau}_\eta)\, ds$ and for the integral on $\Gamma_0$ we use (5.37), (1.23) and the fact that $\operatorname{div}(\tfrac{1}{2}(\sigma_1 + \sigma_2)(t) - \tau_\eta(t)) = 0 \ \forall\, t$.

from which we conclude that $\sigma_1(t) = \sigma_2(t)$ for almost every $t$ and the uniqueness is proved. □

*Remark* 5.2. The proof of uniqueness is of course much simpler for the more regular solutions, *i.e.* those satisfying (1.38) (1.43). In this case we are allowed to take directly (*i.e.* without mollification) $\tau = \frac{1}{2}(\sigma_1 + \sigma_2)$ in the relations (1.43) for $\sigma_1$ and $\sigma_2$. By adding these relations we then immediately obtain (5.38) (*cf.* [7], [15]).

### 5.5. *The strains and the constitutive law*

Our final task is to pass to the limit $\alpha \to 0$ for the field of displacements $u_\alpha$. As before the motivations are twofold:
- To study the passage to the limit $\alpha \to 0$, *i.e.* to describe the passage from the Norton-Hoff model to perfect plasticity;
- To obtain the existence and the properties of the field of displacements $u$ for the problem of perfect plasticity. Among the properties, the most important we have in view is (1.19), which corresponds to the **constitutive law** of the material.

From now on one of the hypotheses (2.11)–(2.12) is assumed.

*The displacements.* The *a priori* estimate (4.68), as formulated, is useless in the passage to the limit $\alpha \to 0$ since the balls of $W^{1,1}(\Omega)$ (and thus of $L^\infty(0, T; W^{1,1}(\Omega))$ are not weakly compact.

At this point we recall the definition and some properties of the space $BD(\Omega)$

$$BD(\Omega) = \{u \in L^1(\Omega)^3, \varepsilon_{ij}(u) \in M_1(\Omega), \quad i, j = 1, 2, 3\}$$

where $M_1(\Omega)$ (respectively $M_1(\Omega; E)$) is the space of bounded measures on $\Omega$ with values in $\mathbb{R}$ (respectively in $E$). We equip this space with the natural norm

$$\|u\|_{BD(\Omega)} = \|u\|_{L^1(\Omega)^3} + \sum_{i,j=1}^{3} \int_\Omega |\varepsilon_{ij}(u)|$$

which makes it a Banach space and we refer the reader to G. STRANG & R. TEMAM [33] and R. TEMAM [38] for the properties of this space. At the moment we will use the fact that $W^{1,1}(\Omega)$ is included in $BD(\Omega)$ with a continuous injection (obvious) and that the space $BD(\Omega)$ is the dual of some normed space X (see [38], [33]), so that the balls of $BD(\Omega)$ are weak-star compact (*i.e.* compact for the weak topology associated to the duality between $BD(\Omega)$ and X). This implies that $L^\infty(0, T; BD(\Omega))$ is the dual of $L^1(0, T; X)$ and that the balls of $L^\infty(0, T; BD(\Omega))$ are weak-star compact (*i.e.* compact for the weak topology associated to the duality between $L^\infty(0, T; BD(\Omega))$ and $L^1(0, T; X)$).

We now deduce from (4.68) that

(5.39) $\quad \dot{u}_\alpha$ is bounded in $L^\infty(0, T; BD(\Omega))$ independently of $\alpha$

and by integration in $t$ using (1.7), (1.29)

(5.40) $\quad u_\alpha$ is bounded in $L^\infty(0, T; BD(\Omega))$ independently of $\alpha$.

Due to (5.39), (5.40), (4.70), there is a $u$ satisfying

(5.41) $$u, \dot{u} \in \mathrm{L}^\infty(0, \mathrm{T}; \mathrm{BD}(\Omega))$$

(6.42) $$\operatorname{div} u, \operatorname{div} \dot{u} \in \mathrm{L}^2(0, \mathrm{T}; \mathrm{L}^2(\Omega))$$

and a subsequence (still denoted $\alpha$) $\alpha \to 0$, such that

(5.43) $$u_\alpha \to u \quad \text{and} \quad \dot{u}_\alpha \to \dot{u} \text{ in } \mathrm{L}^\infty(0, \mathrm{T}; \mathrm{BD}(\Omega)) \text{ weak-star}$$

(5.44) $$\operatorname{div} u_\alpha \to \operatorname{div} u \quad \text{and} \quad \operatorname{div} \dot{u}_\alpha \to \operatorname{div} \dot{u} \quad \text{weakly in } \mathrm{L}^2(0, \mathrm{T}; \mathrm{L}^2(\Omega)).$$

Since $\mathrm{BD}(\Omega) \subset \mathbb{L}^{3/2}(\Omega)$ with a continuous imbedding [38] we infer from (5.43) that

(5.45) $$u_\alpha \to u \text{ in } \mathrm{L}^\infty(0, \mathrm{T}; \mathbb{L}^{3/2}(\Omega)) \text{ weak-star}$$

(5.46) $$\dot{u}_\alpha \to \dot{u} \text{ in } \mathrm{L}^\infty(0, \mathrm{T}; \mathbb{L}^{3/2}(\Omega)) \text{ weak-star}.$$

Similarly (4.66) implies that

(5.47) $\gamma'_\alpha(\sigma_\alpha)$ is bounded in $\mathrm{L}^\infty(0, \mathrm{T}; \mathrm{M}_1(\Omega; \mathrm{E}))$ independently of $\alpha$

and the subsequence $\alpha \to 0$ above can be chosen so that

(5.48) $\gamma'_\alpha(\sigma_\alpha)$ converges weakly in $\mathrm{L}^\infty(0, \mathrm{T}; \mathrm{M}_1(\Omega; \mathrm{E}))$ to $\boldsymbol{\lambda}$

where $\boldsymbol{\lambda}$ is some measure in $\mathrm{L}^\infty(0, \mathrm{T}; \mathrm{M}_1(\Omega; \mathrm{E}))$.

The passage to the limit in (5.3) is now easy (using (5.14), (5.44), (5.48)). We find precisely (1.18) at the limit. It is easy to see that (5.6) implies (1.7) at the limit but (1.8) and (1.7) are not guaranteed.

*Remark 5.3.* i) The uniqueness of $u$ (and $\boldsymbol{\lambda}$) for the problem $\mathscr{P}$ is not certain (and is not conjectured). Therefore the convergence above holds only for subsequences $\alpha_j$ converging to 0.

ii) The problem is open to derive as in [38] the properties of $u$ on the parts of $\Gamma_0$ where (1.6) is not satisfied.

*The measure* $\varepsilon(v) \cdot \tau$. Let there be given $v \in \mathrm{BD}(\Omega)$ satisfying $\operatorname{div} v \in \mathrm{L}^2(\Omega)$ and $\tau \in \mathrm{L}^2(\Omega; \mathrm{E})$, such that $\tau^\mathrm{D} \in \mathrm{L}^\infty(\Omega; \mathrm{E}^\mathrm{D})$ and $\operatorname{div} \tau \in \mathbb{L}^n(\Omega)$. The products

(5.49) $$\varepsilon(v) \cdot \tau \text{ and } \varepsilon^\mathrm{D}(v) \cdot \tau^\mathrm{D} = \varepsilon(v) \cdot \tau - \tfrac{1}{3} (\operatorname{div} u)(\operatorname{tr} \sigma)$$

do not make sense in the ordinary sense but they are defined in [17], [38] as measures, through the following formula (which define them in a first step as distributions on $\Omega$):

For every test function $\phi \in \mathscr{C}_0^\infty(\Omega)$ we set

(5.50) $$\langle \varepsilon(v) \cdot \tau, \phi \rangle = -\int_\Omega (\operatorname{div} \tau)\, v \phi \, dx - \int_\Omega \tau \cdot (v \otimes \operatorname{grad} \phi)\, dx$$

(5.51) $$\langle \varepsilon^\mathrm{D}(v) \cdot \tau^\mathrm{D}, \phi \rangle = -\tfrac{1}{3} \int_\Omega (\operatorname{div} v)(\operatorname{tr} \sigma) \phi \, dx - \int_\Omega (\operatorname{div} \tau)\, v \phi \, dx - \int_\Omega \tau \cdot (v \otimes \operatorname{grad} \phi)\, dx.$$

Furthermore (see [17], [38]), $\varepsilon^D(v) \cdot \tau^D$ is absolutely continuous with respect to $|\varepsilon^D(v)|$ and

$$\left| \int_\Omega \phi \varepsilon^D(v) \cdot \tau^D \right| \leqq k_2 \int_\Omega |\phi| \, |\varepsilon^D(v)|, \quad \forall \phi \in \mathscr{C}_0^\infty(\Omega) \tag{5.52}$$

$$\int_\Omega |\varepsilon^D(v) \cdot \tau^D| \leqq k_2 \int_\Omega |\varepsilon^D(v)| \tag{5.53}$$

where $k_2$ is larger than or equal to the diameter of $K^D$ in $E^D$.

Now let us consider a function $\tau$ which enjoys the properties

$$\tau \in L^\infty(0, T; L^2(\Omega; E)), \operatorname{div} \tau \in L^2(0, T; \mathbb{L}^3(\Omega)) \tag{5.54}$$

$$\tau^D \in L^\infty(0, T; L^\infty(\Omega; E^D)), \tag{5.55}$$

and let $u$ satisfy (5.41), (5.42). First, by a result of [17]

(5.56) $\{\varkappa \in L^2(\Omega; E), \varkappa^D \in L^2(\Omega; E^D), \operatorname{div} \varkappa \in \mathbb{L}^3(\Omega)\}$

is included in $L^p(\Omega; E)$, $\forall p < \infty$, with a continuous embedding,

so that

$$\tau \in L^\infty(0, T; L^p(\Omega; E)), \ \forall p < \infty. \tag{5.57}$$

Then we notice that for almost every $t$, $u(t)$ and $\tau(t)$ enjoy the properties requested for (5.50), (5.51) and we can define the measures $\varepsilon^D(u(t)) \cdot \tau^D(t)$ and $\varepsilon(u(t)) \cdot \tau(t)$ with

$$\int_\Omega |\varepsilon^D(u(t)) \cdot \tau^D(t)| \leqq k_2 \int_\Omega |\varepsilon^D(u(t))|$$

$$\int_\Omega |\varepsilon(u(t)) \cdot \tau(t)| \leqq k_2 \int_\Omega |\varepsilon^D(u(t))| + \tfrac{1}{3} |\operatorname{div} u(t)| \, |\operatorname{tr} \tau(t)| \, dx.$$

Thus

$$\varepsilon^D(u(.)) \cdot \tau^D(.) \text{ and } \varepsilon(u(.)) \cdot \tau(.) \text{ are in } L^\infty(0, T; M_1(\Omega)). \tag{5.58}$$

The same properties hold if we replace $u$ by $\dot{u}$.

*The measure* $\varepsilon(\dot{u}) \cdot \tau$, $\varepsilon(\dot{u}) \cdot \sigma$. Let $\tau$ satisfy (5.54) and (5.55). Then the measure $\varepsilon(\dot{u}) \cdot \tau$ makes sense and a natural question is whether the product $\varepsilon(\dot{u}_\alpha) \cdot \tau$ which is defined in the ordinary sense, converges to $\varepsilon(\dot{u}) \cdot \tau$ as $\alpha \to 0$. Similarly we can replace $\tau$ by $\sigma$ which satisfies (5.54) and (5.55) and another natural question is whether the product $\varepsilon(\dot{u}_\alpha) \cdot \sigma_\alpha$ which is defined in the ordinary sense converges to $\varepsilon(\dot{u}) \cdot \sigma$ (defined as a measure as indicated above) when $\alpha \to 0$. *We are going to give a positive answer to these questions and establish a weak form of* (1.19). □

Let $\Phi \in {}^\infty(\Omega \times (0, T))$. Then we write (5.50) with $v$, $\tau$, $\phi$ replaced by $\dot{u}(., t)$, $\tau(., t)$, $\Phi(., t)$ and integrate the relation that we obtain with respect to $t$:

$$\int_0^T \langle \varepsilon(\dot{u}(t)) \cdot \tau(t), \Phi(t) \rangle \, dt = - \int_0^T \int_\Omega \operatorname{div} \tau \dot{u} \Phi \, dx \, dt - \int_0^T \int_\Omega \tau \cdot \dot{U} \otimes \operatorname{grad} \Phi \, dx \, dt. \tag{5.59}$$

Using Green's formula, we see on the other hand that for every $\alpha$

$$(5.60)\qquad \int_0^T \int_\Omega \varepsilon(\dot{u}_\alpha) \cdot \tau\Phi \, dx \, dt = -\int_0^T \int_\Omega \operatorname{div} \tau \dot{u}_\alpha \Phi \, dx \, dt - \int_0^T \int_\Omega \tau \cdot \dot{U}_\alpha \otimes \operatorname{grad} \Phi \, dx \, dt \,.$$

As $\alpha \to 0$ the right-hand side of (5.60) converges to the right-hand side of (5.59), thanks to (5.44), (5.46) and (5.57). At the limit we precisely obtain the right-hand side of (5.59) and thus we have shown for every $\tau$ satisfying (5.54), (5.55):

$$(5.61)\qquad \varepsilon(\dot{u}_\alpha) \cdot \tau \text{ converges weakly in } L^\infty(0, T; M_1(\Omega)) \text{ to } \varepsilon(\dot{u}) \cdot \tau \text{ as } \alpha \to 0.$$

Also since

$$\varepsilon^D(\dot{u}_\alpha) \cdot \tau_D = \varepsilon(\dot{u}_\alpha) \cdot \tau - \tfrac{1}{3} \operatorname{div} \dot{u}_\alpha \cdot \operatorname{tr} \tau$$

(5.44) implies

$$(5.62)\qquad \varepsilon^D(\dot{u}_\alpha) \cdot \tau^D \text{ converges weakly in } L^\infty(0, T; M_1(\Omega)) \text{ to } \varepsilon^D(\dot{u}) \cdot \dot{\tau}^D \text{ as } \alpha \to 0.$$

Now replacing $\tau$ by $\sigma_\alpha$, we convert (5.59) and (5.60) into

$$(5.63)\qquad \int_0^T \langle \varepsilon(\dot{u}(t)) \cdot \sigma(t), \Phi(t) \rangle \, dt = -\int_0^T \int_\Omega \operatorname{div} \sigma \dot{u} \Phi \, dx \, dt - \int_0^T \int_\Omega \sigma \cdot \dot{u} \otimes \operatorname{grad} \Phi \, dx \, dt$$

$$(5.64)\qquad \int_0^T \int_\Omega \varepsilon(\dot{u}_\alpha) \cdot \sigma_\alpha \Phi \, dx \, dt = -\int_0^T \int_\Omega \operatorname{div} \sigma_\alpha \dot{u}_\alpha \Phi \, dx \, dt - \int_0^T \int_\Omega \sigma_\alpha \cdot \dot{u}_\alpha \otimes \operatorname{grad} \Phi \, dx \, dt \,.$$

Though this is more delicate than before[1], it can be proved that, for $\alpha \to 0$, the right-hand side of (5.64) converges to the right-hand side of (5.63). Indeed the convergence

$$\int_0^T \int_\Omega \operatorname{div} \sigma_\alpha \dot{u}_\alpha \Phi \, dx \, dt \to \int_0^T \int_\Omega \operatorname{div} \sigma \dot{u} \Phi \, dx \, dt$$

follows from (5.46) and the fact that $\operatorname{div} \sigma_\alpha = \operatorname{div} \sigma = -f$. The convergence

$$\int_0^T \int_\Omega \sigma_\alpha \cdot \dot{u}_\alpha \otimes \operatorname{grad} \Phi \, dx \, dt \to \int_0^T \int_\Omega \sigma \dot{u} \otimes \operatorname{grad} \Phi \, dx \, dt$$

follows from (5.13) and (5.46).

Since $\Phi$ is arbitrary this convergence implies that, as $\alpha \to 0$,

$$(5.65)\qquad \varepsilon(\dot{u}_\alpha) \cdot \sigma_\alpha \to \varepsilon(\dot{u}) \cdot \sigma \text{ weakly in } L^\infty(0, T; M_1(\Omega)) \,.$$

---

[1] This point is precisely the main motivation of the results proved in the Appendix.

Also

$$\varepsilon^{\mathrm{D}}(\dot{u}_\alpha) \cdot \sigma_\alpha^{\mathrm{D}} = \varepsilon(\dot{u}_\alpha) \cdot \sigma_\alpha - \tfrac{1}{2} \operatorname{div} \dot{u}_\alpha \cdot \operatorname{tr} \sigma_\alpha$$

and $\operatorname{div} \dot{u}_\alpha \cdot \operatorname{tr} \sigma_\alpha$ converges to $\operatorname{div} \dot{u} \cdot \operatorname{tr} \sigma$ in $L^\infty(0, T; L^1(\Omega))$ as $\alpha \to 0$ and

$$\varepsilon^{\mathrm{D}}(\dot{u}_\alpha) \cdot \sigma_\alpha^{\mathrm{D}} \to \varepsilon^{\mathrm{D}}(\dot{u}) \cdot \sigma^{\mathrm{D}} \text{ weakly in } L^\infty(0, T; M_1(\Omega)) \text{ as } \alpha \to 0. \tag{5.66}$$

Assume now that instead of (5.55), $\tau$ satisfies

$$\tau^{\mathrm{D}}(t) \in \mathscr{K}_{\mathrm{ad}} \text{ a.e. } t. \tag{5.67}$$

Then (5.55) is obviously satisfied and we deduce from (5.3), (2.6), (5.67) the inequality valid almost everywhere $x \in \Omega$, $t \in (0, T)$:

$$(\varepsilon(\dot{u}_\alpha) - A\dot{\sigma}_\alpha) \cdot (\sigma_\alpha - \tau) = \gamma'_\alpha(\sigma_\alpha) \cdot (\sigma_\alpha - \tau) \geqq \gamma'_\alpha(\tau)(\sigma_\alpha - \tau) = 0.$$

Hence for every $\Phi \in \mathscr{C}_0^\infty(\Omega \times (0, T))$

$$\int_0^T \int_\Omega (\varepsilon(\dot{u}_\alpha) - A\dot{\sigma}_\alpha)(\sigma_\alpha - \tau) \Phi \, dx \, dt \geqq 0.$$

We have shown that

$$\int_0^T \int_\Omega \varepsilon(\dot{u}_\alpha) \cdot (\sigma_\alpha - \tau) \Phi \, dx \, dt \to \int_0^T \langle \varepsilon(\dot{u}(t)) \cdot (\sigma(t) - \tau(t), \Phi(t) \rangle \, dt.$$

Because of (5.12) and (5.14)

$$\int_0^T \int_\Omega A\dot{\sigma}_\alpha \cdot (\sigma_\alpha - \tau) \Phi \, dx \, dt \to \int_0^T \int_\Omega A\dot{\sigma} \cdot (\sigma - \tau) \Phi \, dx \, dt.$$

We conclude that

$$(\varepsilon(\dot{u}) - A\dot{\sigma}) \cdot (\tau - \sigma) \geqq 0 \tag{5.68}$$

in the sense of measures on $\Omega \times (0, T)$ for every $\tau$ satisfying (5.54), (5.67). This is the weak form announced of (1.18).

We recapitulate the results we proved here in the

**Theorem 5.3.** *The hypotheses are the general assumptions of Section 1 and* (2.11) *or* (2.12).

*The problem* $\mathscr{P}$ *possesses a solution* $u$ *which satisfies* (5.41), (5.42) *and* (1.7). *Furthermore for every* $\tau$ *satisfying* (5.54), (5.55) (*and in particular* $\tau = 0$ *given by Theorem* 5.2), $\varepsilon(\dot{u}) \cdot \tau$ *and* $\varepsilon^{\mathrm{D}}(\dot{u}) \cdot \tau^{\mathrm{D}}$ *as defined through* (5.60), (5.63) *are bounded measures belonging to* $L^\infty(0, T; M_1(\Omega))$. *Moreover if* $\tau$ *satisfies* (5.67) *then we have the constitutive law*

$$(\varepsilon(\dot{u}) - A\dot{\sigma})(\sigma - \tau) \leqq 0$$

*satisfied in the sense of measures on* $\Omega \times (0, T)$.

**Theorem 5.4.** *The hypotheses are the general assumptions of Section 1 and* (2.11) *or* (2.12).

*As* $\alpha \to 0$, $u_\alpha$ *converges to* $u$ *for the topologies of* (5.43)–(5.46), *where* $u$ *is a weak solution of* $\mathscr{P}$ *in the sense given in Theorem 5.3. Furthermore*

$$\varepsilon(\dot{u}_\alpha)\cdot\tau \to \varepsilon(\dot{u})\cdot\tau, \qquad \varepsilon^{\mathrm{D}}(\dot{u}_\alpha)\cdot\tau^{\mathrm{D}} \to \varepsilon^{\mathrm{D}}(\dot{u})\cdot\tau^{\mathrm{D}}$$

$$\varepsilon(\dot{u}_\alpha)\cdot\sigma_\alpha \to \varepsilon(\dot{u}_\alpha)\cdot\sigma_\alpha, \qquad \varepsilon^{\mathrm{D}}(\dot{u}_\alpha)\cdot\sigma_\alpha^{\mathrm{D}} \to \varepsilon^{\mathrm{D}}(\dot{u})\cdot\sigma^{\mathrm{D}}$$

*weakly in* $\mathrm{L}^\infty(0, \mathrm{T}; \mathrm{M}_1(\Omega))$.

## Appendix: Function spaces and trace theorems

We assume that $\Omega$ is an open connected bounded set of class $\mathscr{C}^2$ of boundary $\Gamma$ in $\mathbb{R}^n$.

**Lemma A.1** (*Regularization*). *For every* $\sigma \in \mathrm{L}^p(\Omega; \mathrm{E})$, *with* $\operatorname{div}\sigma \in \mathbb{L}^r(\Omega)$ *and* $\sigma^{\mathrm{D}} \in \mathrm{L}^s(\Omega; \mathrm{E}^{\mathrm{D}})$, $1 \leqq r \leqq p \leqq s < \infty$, *there is a sequence* $\sigma_m$ *of functions in* $\mathscr{C}^2(\bar{\Omega}; \mathrm{E}) \cap \mathscr{C}^\infty(\Omega; \mathrm{E})$, *such that, as* $m \to \infty$,

$$\sigma_m \to \sigma \text{ in } \mathrm{L}^p(\Omega; \mathrm{E}) \tag{A.1}$$

$$\sigma_m^{\mathrm{D}} \to \sigma^{\mathrm{D}} \text{ in } \mathrm{L}^s(\Omega; \mathrm{E}) \tag{A.2}$$

$$\operatorname{div}\sigma_m \to \operatorname{div}\sigma \text{ in } \mathbb{L}^r(\Omega). \tag{A.3}$$

**Proof.** Exactly the same proof as for Proposition 1.3, Chapter I and Lemma 7.1, Chapter II of [38]. We consider a partition of unity subordinated to an appropriate covering of $\bar{\Omega}$ by $\Omega$ and balls $\mathcal{O}_i$ centered on $\Gamma$, $i \in I$ (a finite set)

$$1 = \phi_0 + \sum_{i\in I} \phi_i,$$

$\phi_0 \in \mathscr{C}_0^\infty(\Omega)$, $\phi_i \in \mathscr{C}_0^\infty(\mathcal{O}_i)$. By the assumptions on $r, p, s$, $\phi_0\sigma$ and $\phi_i\sigma$ satisfy the same properties as $\sigma$, $\forall\, i$. We approximate these functions by mollification as indicated in [38], the only difference from [38] is related to the convergence of the deviator which follows from the simple remark that $(\varrho_m * (\phi_j\sigma))^{\mathrm{D}} = \varrho_m * (\phi_j\sigma)^{\mathrm{D}}$, $\forall\, j$. The function $\sigma_m$ is finally obtained by summation of the functions approximating $\phi_0\sigma$ and the $\phi_i\sigma$, $i \in \mathrm{I}$. □

For $1 \leqq p, r, s \leqq \infty$, let

$$\mathrm{X}(\Omega) = \mathrm{X}_{p,r,s}(\Omega) = \{\sigma \in \mathrm{L}^p(\Omega; \mathrm{E}), \operatorname{div}\sigma \in \mathbb{L}^r(\Omega), \quad \sigma^{\mathrm{D}} \in \mathrm{L}^s(\Omega; \mathrm{E}^{\mathrm{D}})\}.$$

This is a Banach space for the natural norm

$$\|\sigma\|_{\mathrm{L}^p(\Omega;\mathrm{E})} + \|\operatorname{div}\sigma\|_{\mathbb{L}^r(\Omega)} + \|\sigma^{\mathrm{D}}\|_{\mathrm{L}^s(\Omega;\mathrm{ED})}$$

and Lemma A.1 asserts that $\mathscr{C}^2(\bar{\Omega}; \mathrm{E}) \cap \mathscr{C}^\infty(\Omega; \mathrm{E})$ is dense in $\mathrm{X}(\Omega)$ if $1 \leqq r \leqq p \leqq s$.

The following proposition is a generalization of the trace theorem recalled in (1.23) and a trace result in [17] (Lemma 2.4). As before we denote by $\nu$ the unit outward normal on $\Gamma$ and by $u_\tau = u - (u\cdot\nu)\nu$ the tangential component of a vector field defined on $\Gamma$.

**Proposition A.1** (*A trace property*). *We assume that* $1 < p = r < n$ *and* $s \geqq r = \dfrac{nr}{n-r}$. *Then the mapping*

$$\sigma \to (\sigma \cdot \nu)_\tau$$

*from* $\mathscr{C}^2(\bar{\Omega}; \mathrm{E})$ *into* $\mathscr{C}^2(\Gamma)^n$ *can be uniquely extended as a linear continuous mapping* $\gamma$ *from* $X(\Omega)$ *into* $\mathrm{W}^{-1/r^*, r^*}(\Gamma) = \mathrm{W}^{1/r^*, q}(\Gamma))'$ *the dual of* $\mathrm{W}^{1/r^*, q}(\Gamma)$, $r_* = q/(q-1)$ *and*

$$\langle \gamma(\sigma), u_\tau \rangle = \int_\Omega \operatorname{div} \sigma\, u\, dx + \int_\Omega \sigma^{\mathrm{D}} \varepsilon^{\mathrm{D}}(u)\, dx, \tag{A.4}$$

*for every* $u \in \mathrm{Z}(\Omega)$

$$\mathrm{Z}(\Omega) = \{u \in \mathrm{W}^{1,q}(\Omega),\ \operatorname{div} u = 0 \text{ in } \Omega,\ u \cdot \nu = 0 \text{ on } \Gamma\}. \tag{A.5}$$

**Proof.** i) We first characterize the traces on $\Gamma$ of the functions in $\mathrm{Z}(\Omega)$, *i.e.* the space $\gamma_0 \mathrm{Z}(\Omega)$.

It is clear that $\gamma_0 \mathrm{Z}(\Omega)$ is included in the space

$$\gamma_0\{u \in \mathrm{W}^{1,q}(\Omega),\ u \cdot \nu = 0 \text{ on } \Gamma\}$$

which is nothing else (*cf.* R. S. ADAMS [1], J. L. LIONS & E. MAGENES [19]) than $(1/r^* = 1 - 1/q)$:

$$\{\varphi \in \mathrm{W}^{1/r_*, q}(\Gamma),\ \varphi \cdot \nu = 0 \text{ on } \Gamma\}. \tag{A.6}$$

On the other hand if $\varphi$ belongs to the space (A.6) then, using the results of L. CATTABRIGA [5], we can solve the linear Stokes problem

$$\begin{aligned} -\Delta \Phi + \operatorname{grad} \Pi &= 0 \quad \text{in } \Omega, \\ \operatorname{div} \Phi &= 0 \quad \text{in } \Omega, \\ \Phi &= \varphi \quad \text{on } \Gamma \end{aligned}$$

and $\Phi \in \mathrm{W}^{1,q}(\Omega)$, $\Pi \in \mathrm{L}^q(\Omega)$. Clearly $\gamma_0 \Phi = \varphi \in \gamma_0 \mathrm{Z}(\Omega)$. The space $\gamma_0 \mathrm{Z}(\Omega)$ is the space (A.6).

Using again the Stokes operator and [5], we see that there is a continuous lifting operator $l_1 : \varphi \to \Phi$ from the space (A.6) into $\mathrm{Z}(\Omega)$.

ii) We recall the classical Green's formula (similar to (1.24)) which is valid for $\sigma$ and $u$ sufficiently regular:

$$\begin{aligned} \int_\Gamma (\sigma \cdot \nu)\, u\, d\Gamma &= \int_\Gamma \sigma_{ij} u_i \nu_j\, d\Gamma = \int_\Omega (\sigma_{ij} u_i)_{,j}\, dx \\ &= \int_\Omega \sigma_{ij} u_{i,j}\, dx + \int_\Omega \sigma_{ij,j} u_i\, dx \\ &= \int_\Omega \operatorname{div} \sigma u\, dx + \int_\Omega \sigma \cdot \varepsilon(u)\, dx \\ &= \int_\Omega \operatorname{div} \sigma u\, dx + \int_\Omega \sigma^{\mathrm{D}} \varepsilon^{\mathrm{D}}(u)\, dx + \frac{1}{n} \int_\Omega \operatorname{tr} \sigma \operatorname{div} u\, dx. \end{aligned}$$

Let $\sigma$ be an element of $X(\Omega)$. Then for every $\varphi$ in the space (A.6) we choose $u \in Z(\Omega)$ such that $\gamma_0 u = \varphi$ and we set

(A.7) $$M_\sigma(\varphi) = \int_\Omega \operatorname{div} \sigma u \, dx + \int_\Omega \sigma^D \varepsilon^D(u) \, dx .$$

The integrals in the right-hand side of (A.4) make sense since $\sigma^D$ and $\varepsilon^D(u)$ are in dual spaces ($L^s$ and $L^q$, $s \geqq r_* = q'$) and similarly $\operatorname{div} \sigma$ and $u$: $\operatorname{div} \sigma \in L^r$ and $u \in L^{r'}$ since, by the Sobolev imbedding theorem

(A.8) $$W^{1,q}(\Omega) \subset L^{r'}(\Omega) \left( \frac{1}{r'} = 1 - \frac{1}{r} = \frac{1}{q} - \frac{1}{n} \right).$$

Furthermore

(A.9) $$\begin{aligned} |M_\sigma(\varphi)| &\leqq |\operatorname{div} \sigma|_{L^r} |u|_{L^{r'}} + |\sigma^D|_{L^{r_*}} |\varepsilon^D(u)|_{L^q} \\ &\leqq c_1'(|\operatorname{div} \sigma|_{L^r} + |\sigma^D|_{L^{r_*}}) \|u\|_{W^{1,q}(\Omega)} \\ &\leqq c_2' \|\sigma\|_{X(\Omega)} \|u\|_{Z(\Omega)} . \end{aligned}$$

The definition in (A.7) makes sense if we show that the right-hand side is independent of the choice of $u$ in $Z(\Omega)$ such that $\gamma_0 u = \varphi$. Let $v$ be another element of $Z(\Omega)$ with $\gamma_0 v = \varphi$ and let $w = u - v$. Then $w \in W^{1,q}(\Omega)$ with $\gamma_0 w = 0$ on $\Gamma$, $\operatorname{div} w = 0$. It can be proved (*cf.* [39]) that $w$ is the limit in $W^{1,q}(\Omega)$ of functions $w_j \in \mathscr{C}_0^\infty(\Omega)^n$ with $\operatorname{div} w_j = 0$. Then, it is easy to see that

$$0 = \int_\Omega \operatorname{div} \sigma w_j \, dx + \int_\Omega \sigma^D \varepsilon^D(w_j) \, dx \underset{j \to \infty}{\to} \int_\Omega \operatorname{div} \sigma w \, dx + \int_\Omega \sigma^D \varepsilon^D(w) \, dx .$$

Now if we choose $u = l_1 \varphi$, then $\varphi \to M_\sigma(\varphi)$ is linear and continuous because of (A.9) (and the continuity of $l_1$). Thus $M_\sigma$ is an element in the dual of (A.3), *i.e.* [1], [19]:

(A.10) $$\{\varphi \in W^{-1/r_*, r_*}(\Gamma), \quad \varphi \cdot \nu = 0 \text{ on } \Gamma\},$$

and its norm in this space is majorized by $c_2' \|l_1\| \|\sigma\|_{X(\Omega)}$.

Finally we call $\gamma$ the mapping $\sigma \in X(\Omega) \to M_\sigma \in$ space (A.7), which is linear and continuous with a norm $\leqq c_2' \|l_1\|$. The generalized Green's formula (A.1) is just a rewriting of (A.4). The uniqueness of $\gamma$ follows from the classical Green's formula and the density of $\mathscr{C}^2(\bar{\Omega}; E)$ into $X(\Omega)$ (Lemma A.1). The proof is complete. □

The last result is an embedding property which generalizes Proposition 2.5 of [17].

**Proposition A.2** (*Embedding*). *We assume that* $\sigma \in L^p(\Omega; E)$, *with* $\operatorname{div} \sigma \in \mathbb{L}^r(\Omega)$ *and* $\sigma^D \in L^s(\Omega; E)$, $1 < p$, $1 < r < n$, $s \geqq r^* = \dfrac{nr}{n-r}$. *Then* $\operatorname{tr} \sigma \in L^{r^*}(\Omega)$ (*i.e.* $\sigma \in L^{r^*}(\Omega; E)$) *and there is a constant* $c = c(\Omega, r, s)$ *such that*

(A.11) $$|\operatorname{tr} \sigma|_{L^{r^*}(\Omega)} \leqq c\{|\sigma|_{L^p(\Omega; E)} + |\operatorname{div} \sigma|_{L^r(\Omega)} + |\sigma^D|_{L^s(\Omega; E^D)}\}.$$

**Proof.** We first consider the case where $r = p$; the case $r \neq p$ will be treated subsequently.

We need to show that

$$\left|\int_\Omega \psi \operatorname{tr} \sigma \, dx\right| \leq C(\sigma) \, |\psi|_{L^q(\Omega)} \tag{A.12}$$

for every $\psi \in \mathscr{C}^2(\bar{\Omega})$, where $C(\sigma)$ is like the right-hand side of (A.11) ($q' = r^*$).

Consider first the case where $\int_\Omega \psi \, dx = 0$, and let $\Psi$ be the solution of the Neumann problem

$$\Delta \Psi = \psi \text{ in } \Omega, \quad \frac{\partial \Psi}{\partial \nu} = 0 \quad \text{on } \Gamma$$

and let $u = \operatorname{grad} \Psi$; notice that $u$ is (at least) in $W^{1,l}(\Omega)$, $\forall \, l < \infty$.

Let $\sigma_m$ be the approximating sequence of $\sigma$ given by Lemma A.1 (with $r$ replaced by $p$). Applying Green's formula to $\sigma_m$ and $u$ we find

$$\begin{aligned} \frac{1}{n} \int_\Omega (\operatorname{tr} \sigma_m)(\operatorname{div} u) \, dx &= \frac{1}{n} \int_\Omega (\operatorname{tr} \sigma_m) \, \psi \, dx \\ &= \int_\Gamma (\sigma_m \cdot \nu) \, u \, d\Gamma - \int_\Omega \operatorname{div} \sigma_m u \, dx - \int_\Omega \sigma_m^{\mathrm{D}} \varepsilon^{\mathrm{D}}(u) \, dx . \end{aligned} \tag{A.13}$$

Since $u \cdot \nu = 0$ on $\Gamma$ and $u \in W^{1,l}(\Omega)$, $\forall \, l < \infty$, the trace $\gamma_0 u$ of $u$ on $\Gamma$ is in the space (A.6) ($= \gamma_0 Z(\Omega)$) and

$$\int_\Gamma (\sigma_m \cdot \nu) \, zu \, d\Gamma = \int_\Gamma (\sigma_m \cdot \nu)_\tau \, u_\tau \, d\Gamma = \langle \gamma(\sigma_m), u_\tau \rangle .$$

By Proposition A.1, as $m \to \infty$, this integral converges to $\langle \gamma(\sigma), u_\tau \rangle$.

It is easy to pass to the limit $m \to \infty$ in the other integrals in (A.13) and we find

$$\frac{1}{n} \int_\Omega \operatorname{tr} \sigma \cdot \psi \, dx = \langle \gamma(\sigma), u_\tau \rangle - \int_\Omega \operatorname{div} \sigma u \, dx - \int_\Omega \sigma^{\mathrm{D}} \varepsilon^{\mathrm{D}}(u) \, dx . \tag{A.14}$$

We have

$$\begin{aligned} |\langle \gamma(\sigma), u_\tau \rangle| &\leq |\gamma(\sigma)|_{W^{-1/r^*, r^*}(\Gamma)} \, |\gamma_0 u_\tau|_{W^{1/r^*, q}(\Gamma)} \\ &\leq (\text{by Proposition A.1}) \; c_1' (|\operatorname{div} \sigma|_{L^r} + |\sigma^{\mathrm{D}}|_{L^{r^*}}) \, \|u\|_{W^{1,q}(\Omega)} \\ &\leq c_2' (|\operatorname{div} \sigma|_{L^p} + |\sigma^{\mathrm{D}}|_{L^s}) \, |\Psi|_{W^{2,q}(\Omega)} \\ &\leq (\text{by the regularity results for the Neumann problem}) \\ &\quad\; c_3' (|\operatorname{div} \sigma|_{L^p} + |\sigma^{\mathrm{D}}|_{L^s}) \, |\psi|_{L^q(\Omega} \end{aligned}$$

$$\begin{aligned} \left|\int_\Omega \operatorname{div} \sigma u \, dx\right| &\leq |\operatorname{div} \sigma|_{L^p(\Omega)} \, |u|_{L^{p'}(\Omega)} \\ &\leq (\text{thanks to (A.8) with } p = r) \; \leq c_4' \, |\operatorname{div} \sigma|_{L^p(\Omega)} \, \|u\|_{W^{1,q}(\Omega)} \\ &\leq c_5' \, |\operatorname{div} \sigma|_{L^p(\Omega)} \, |\psi|_{L^q(\Omega)} ; \end{aligned}$$

$$\left| \int_\Omega \sigma^D \varepsilon^D(u)\, dx \right| \leq |\sigma^D|_{L^{r_*}} \, |\varepsilon^D(u)|_{L^q}$$
$$\leq c_6' \, |\sigma^D|_{L^s} \, \|u\|_{W^{1,q}(\Omega)}$$
$$\leq c_7' \, |\sigma^D|_{L^s} \, |\psi|_{L^q(\Omega)}.$$

Combining these inequalities with (A.14) we find (A.12) with $C(\sigma)$ given by

(A.15) $$C_1(\sigma) = c_8'(|\operatorname{div} \sigma|_{L^p(\Omega)} + |\sigma^D|_{L^s(\Omega;E)}).$$

Now if $\int_\Omega \psi\, dx \neq 0$, we write (A.12), (A.15) with $\psi$ replaced by

$$\phi = \psi - \frac{1}{|\Omega|} \int_\Omega \psi\, dx$$

and we obtain:

$$\left| \int_\Omega \operatorname{tr} \sigma\psi\, dx \right| \leq \frac{1}{|\Omega|} \int_\Omega |\operatorname{tr} \sigma|\, dx \cdot \int_\Omega |\psi|\, dx + C_1(\sigma)\, |\phi|_{L^q(\Omega)},$$

whence (A.12) with $C(\sigma) = C_2(\sigma)$

$$C_2(\sigma) = c_9'\{|\sigma|_{L^1(\Omega;E)} + |\operatorname{div} \sigma|_{L^p(\Omega)} + |\sigma^D|_{L^1(\Omega;E)}\},$$

and we have proved (A.11) if $r = p$.

If $r > p$, then $\operatorname{div} \sigma \in \mathbb{L}^p(\Omega)$ and the reasoning above applies and shows that $\operatorname{tr} \sigma$ and $\sigma \in L^{p^*}(\Omega; E)$, $\frac{1}{p^*} = \frac{1}{p} - \frac{1}{n} < \frac{1}{n}$. We repeat the same reasoning with this time $p$, $r$ replaced by $p^*$, and we find that $\sigma \in L^{p^{**}}(\Omega; E)$, $\frac{1}{p^{**}} = \frac{1}{p} - \frac{2}{n}$ and we continue until we find an exponent $p^{k^*} > r$ in which case we can replace $p$ by $r$ and so conclude.

Finally the case in which $p > r$ is easier: we replace $p$ by $r$ and we obtain (A.11) which may or may not improve the initial information depending on whether $p < r_*$ or $p \geqq r_*$.

The proof is complete.

*Remark A.1.* Proposition A.2 applies however $p$ close is to one ($p > 1$). The result is very likely true for $p = 1$ but this improvement is not needed in this article since we have $p = 2$.

## References


1. R. S. Adams, *Sobolev Spaces*, Academic Press, New-York, 1975.
2. G. Anzelotti, On the existence of the rates of stress and displacement for Prandtl-Reuss plasticity, *Quart. Appl. Math.*, (1983), p. 181–208.
3. G. Anzelotti & M. Giaquinta, On the existence of the fields of stresses and displacements for an elasto-perfectly plastic body in static equilibrium, *J. Math. Pures Appl.*, **61** (1982), p. 219–244.

4. F. BROWDER, Nonlinear maximal monotone operators in Banach spaces, *Math. Ann.*, **175** (1968), 89–113.
5. L. CATTABRIGA, Su un problema al contorno relativo al sistema di equazioni di Stokes, *Rend. Sem. Math. Univ. Padova*, **31** (1961), 308–340.
6. F. DEMENGEL, Fonctions à hessien borné, *Ann. Institut Fourier Grenoble*, **XXXIV** (2) (1984), 155–190.
7. G. DUVAUT et J. L. LIONS, *Les inéquations en mécanique et en physique*, Dunod, Paris, 1972.
8. I. EKELAND & R. TEMAM, *Convex analysis and variational problems*, North-Holland, Amsterdam, New-York, 1976.
9. A. FRIAÂ, La loi de Norton-Hoff généralisée en pasticité et viscoplasticité, Thèse, Université de Paris VI, 1979.
10. P. GERMAIN, *Cours de mécanique des milieux continus*, Tome I, Masson, Paris, 1973.
11. P. GERMAIN, Cours de mécanique, Ecole Polytechnique, 1979.
12. P. GERMAIN, Commentaires sur l'article "Duality and relaxation in the variational problems of plasticity de R. Temam et G. Strang", *J. Mécanique*, **19** (1980), 529–538.
13. H. HENCKY, *Z. Angew. Math. Phys.*, **4** (1924), 323.
14. H. J. HOFF, Approximate analysis of structures in presence of moderately large creep deformations, *Quart. Appl. Math.*, **12** (1) (1954), p. 49.
15. C. JOHNSON, Existence theorems for plasticity problems, *J. Math. Pures Appl.*, **55** (1976), 431–444.
16. C. JOHNSON, On plasticity with hardening, *J. Math. Anal. Appl.*, **62** (1978), 325–336.
17. R. KOHN & R. TEMAM, Dual spaces of stresses and strains with applications to Hencky plasticity, *App. Math. Optim.*, **10** (1983), p. 1–35.
18. J. L. LIONS, *Quelques méthodes de résolution des problèmes aux limites non linéaires*, Dunod et Gauthier-Villars, Paris, 1969.
19. J. L. LIONS & E. MAGENES, *Non homogeneous boundary value problems and applications*, Spinger-Verlag, Berlin, Heidelberg, New-York, 1972.
20. G. J. MINTY, On the generalization of the direct method of the calculus of variations, *Bull. Amer. Math. Soc.*, **73** (1967), 315–321.
21. J. J. MOREAU, La notion de sur-potentiel et les liaisons unilatérales en élastostatique, *C.R. Acad. Sci. Paris*, série A, **267** (1968), 954–957.
22. J. J. MOREAU, Application of convex analysis to the treatment of elastoplastic systems, dans *Applications of Methods of Functional Analysis to Problems of Mechanics*, P. GERMAIN, B. NAYROLES Ed., Lecture Notes in Math., vol. 503, Springer-Verlag, 1976, 56–89.
23. J. J. MOREAU, *Fonctionnelles convexes*, Séminaire Equations aux Dérivées Partielles, Collège de France, 1966.
24. W. NOLL, On the continuity of the solid and fluid states, *J. Rational Mech. Anal.*, **4** (1955), p. 3–81.
25. W. NOLL, *The Foundations of Mechanics and Thermodynamics*, Springer-Verlag, 1974.
26. F. H. NORTON, *The creep of steel at high temperature*, McGraw Hill, New-York, 1929.
27. W. PRAGER, *Introduction to mechanics of continua*, Ginn and Company, 1961.
28. W. PRAGER, *Problèmes de plasticité théoriques*, Dunod, Paris, 1961.
29. W. PRAGER & P. HODGE, *Theory of perfectly plastic solids*, J. Wiley, New-York, 1951.
30. A. REUSS, *Zeitschrift für Angewandte Mathematik and Physik*, **10** (1939), p. 266.
31. R. T. ROCKAFELLAR, *Convex analysis*, Princeton University Press, 1970.
32. J. SERRIN, The problem of Dirichlet for quasilinear elliptic equations with many independent variables, *Phil. Trans. Royal Soc. London*, **A 264** (1969), 413–496.
33. G. STRANG et R. TEMAM, Existence de solutions relaxées pour les équations de la placticité: étude d'un espace fonctionnel, *C. R. Acad. Sci. Paris*, **287**, série A (1978),

515–518, and Functions of bounded deformations, *Arch. Rational Mech. Anal.*, **75** (1980), 7–21.

34. G. Strang & R. Temam, Duality and relaxation in the variational problems of plasticity, *J. Mécanique*, **19** (1980), 493–527.
35. P. Suquet, Existence and regularity of solutions for plasticity problems, dans *Variational Methods in Solid Mechanics*, S. Nemat Nasser Ed., Pergamon Press, 1980.
36. P. Suquet, Sur les équations de la plasticité: existence et régularité des solutions, *J. Mécanique*, **20** (1981), 3–39.
37. P. Suquet, *Plasticité et homogénéïsation*, Thèse, Université de Paris VI, 1982.
38. R. Temam, *Mathematical Problems in Plasticity*, Dunod, Paris, 1983 (in French), and Gauthier-Villars, Paris–New York, 1984 (in English).
39. R. Temam, *Navier-Stokes equations, Theory and numerical analysis*, 3rd ed., North-Holland, Amsterdam, New York, 1984.
40. R. Temam, Principe de Dissipation maximale pour la loi de Prandtl Reuss en plasticité, *C. R. Acad. Sci. Paris*, **302**, Série I, (1986) p. 79–82.
41. C. Truesdell, *A first course in rational continuum mechanics*, vol. I., Academic Press, New York, 1977.
42. C. Truesdell & W. Noll, The non-linear field theories of mechanics, *Handbuch der Physik*, vol. III.3, Springer-Verlag, 1965.
43. E. H. Zarantonello, Projections on convex sets in Hilbert spaces and spectral theory, in *Contributions to Nonlinear Functional Analysis*, E. H. Zarantonello Ed., Academic Press, New York, 1971.

Laboratoire d'Analyse Numérique
Université Paris-Sud
Orsay, France

*(Received September 24, 1985; revised October 28, 1985)*

# *Symmetry and the Bifurcation of Capillary-Gravity Waves*

MARK JONES & JOHN TOLAND

*This Paper is Dedicated to Professor James Serrin on the Occasion of his Sixtieth Birthday*

## 1. Introduction

### *1.1 Introductory remarks*

CHEN & SAFFMAN [4] have given a weakly nonlinear theory of steady periodic two-dimensional irrotational waves on the surface of a perfect fluid of infinite depth which takes account of the effects of gravity and surface tension. They remark that the question of existence of such waves is "a non-trivial and still incompletely solved problem". It is our purpose in the present paper to give a rigorous mathematical account of the existence theory for such waves of small amplitude. In so doing we are able to give a firm basis for the more formal aspects of CHEN & SAFFMAN's work, and to vindicate many of their main conclusions. Our work will be couched throughout in the language of modern bifurcation theory, and is a consequence of the Lyapunov-Schmidt reduction procedure in the presence of certain symmetry considerations. Other studies have contributed to the rigorous theory, but it is only possible to comment further once the problem has been described in detail.

### *1.2 Capillary-gravity waves*

Consider the possible steady, two-dimensional irrotational motion of an inviscid, incompressible fluid of constant density in a channel of infinite depth, in the presence of gravity with surface tension acting at the free surface. Clearly one such motion is a uniform, horizontal, laminar flow of constant speed $c$, in which case the free surface is flat. Here we consider the possibility of the free surface being composed of a permanent $2\pi$-periodic wave train while the flow below it is asymptotic to a uniform horizontal flow of speed $c$ at infinite depth. We take the point of view that this is a bifurcation problem by seeking small amplitude solutions where the flow is close to being uniform and the free surface is almost horizontal. Certain symmetries intrinsic in the physical set-up lead to results about bifurcation, secondary bifurcation and $n$-fold changes in the minimal period of the free surface which we will see presently. The speed $c$ of the flow at infinite depth plays the role of the bifurcation parameter for each fixed value of the surface tension. Then by changing the surface tension different bifurcation diagrams are obtained. We will see that there are critical values of the surface tension where bifurcation diagrams undergo radical changes: where primary bifurcations and secondary bifurcations exchange roles for example, or where secondary bifurcations appear or disappear.

Our study is based on a recent integral equation formulation of the free boundary problem (TOLAND & JONES [18]): to find a domain $S$ in the plane of the form $\{(x, y) : y \leqq H(x) : x \in \mathbb{R}\}$ where $H$ is a smooth, $2\pi$-periodic even function and to find a function $\psi : S \to \mathbb{R}$ with

$$\Delta\psi = 0 \quad \text{on } S,$$

$$\psi_x(\pm\pi, y) = 0, \quad y < H(\pm\pi),$$

$$(\psi_y, -\psi_x) \to (-c, 0) \quad \text{as } y \to -\infty, (x, y) \in S,$$

$$\psi(x, H(x)) = 0, \quad x \in \mathbb{R} \quad \text{and} \quad \psi(x, y) \to \infty \quad \text{as} \quad y \to -\infty, \quad (x, y) \in S,$$

$$\tfrac{1}{2}\,|\nabla\psi(x, H(x))|^2 + gH(x) - T\left(\frac{H'(x)}{(1 + (H'(x))^2)^{\frac{1}{2}}}\right)' = \text{constant}, \quad x \in \mathbb{R}.$$

Here $T$, $c$ and $g$ denote the surface tension, the speed at infinite depth and the force of gravity. The existence of solutions of this free boundary value problem is then equivalent to an integral equation for a $2\pi$-periodic, odd, function $\theta$ of a real variable

$$\gamma\theta(s) = \left(\int_0^\pi \exp(\tau(w))\, dw\right)^{-1} \left\{\nu \int_0^s \int_0^\pi \sinh(\tau(x) - \tau(w))\, dw\, dx \right.$$
$$\left. + \int_0^s \int_0^\pi \int_y^x \exp\{\tau(w) + \tau(x) + \tau(y)\} \sin\theta(w)\, dw\, dy\, dx\right\}, \qquad (\mathcal{N})$$

(where $-\tau$ is the conjugate of $\theta$ in the $L_2$ sense (see § 1.3 and ZYGMUND [19])) which may be abbreviated as

$$\gamma\theta(s) = \nu A\theta(s) + B\theta(s) + Q\theta(s) + \nu C_1\theta(s) + C_2\theta(s) + H(\nu, \theta) = G(\nu, \theta) \qquad (\mathcal{N})$$

where $A$ and $B$ are linear, $Q$ is quadratic, $C_1$ and $C_2$ are cubic and $\|H(\nu, \theta)\| = O(\|\theta\|^4)$ as $\theta \to 0$ in $C^1$ (see TOLAND & JONES [18; p. 395]). If $\theta$ satisfies $(\mathcal{N})$ then the free surface of a wave is given by

$$\{(x, H(x)) : x \in \mathbb{R}\} = \left\{\left(-\int_0^t e^\tau \cos\theta, -\int_0^t e^\tau \sin\theta\right); t \in \mathbb{R}\right\},$$

where $\theta$ denotes the angle between the free surface and the horizontal. The magnitude of the surface tension is given by $\gamma = T/g$ and the speed of the flow at infinite depth is given by $\nu = c^2/g$. Linearised about the zero function (corresponding to uniform horizontal flow) $(\mathcal{N})$ becomes

$$\gamma\theta = \nu A\theta + B\theta. \qquad (\mathcal{L})$$

In § 2 we observe that for fixed $\gamma$, the eigenvalues $\nu$ of this problem are given by $\nu = \gamma k + 1/k$ in which case $\sin ks$ is an eigenfunction. If $\gamma \neq (MN)^{-1}$ for distinct natural numbers $M$ and $N$ the theory is classical bifurcation from a simple eigenvalue (Crandall & Rabinowitz [6]). If, however, $\gamma = (MN)^{-1}$ then $\nu = N^{-1} + M^{-1}$ is a double eigenvalue and when $\gamma$ is close to $(MN)^{-1}$ there are two simple eigenvalues very close to each other. These are the circumstances that are investigated here.

Let us define $\gamma_{MN} = (MN)^{-1}$ and $\nu_{MN} = M^{-1} + N^{-1}$. Our aim is to describe the solution set $(\gamma, \nu, \theta)$ of equation $(\mathcal{N})$ in an open neighbourhood of the point $(\gamma_{MN}, \nu_{MN}, 0) \in \mathbb{R} \times \mathbb{R} \times C^1$. But since the surface tension is a naturally preassigned parameter in the capillary-gravity wave problem we will study in particular the bifurcations which may occur as the parameter $\nu$ (the speed at infinite depth) changes while $\gamma$ (the surface tension) is held fixed. Then we study the effect of changing $\gamma$ on the shape of the bifurcation diagrams.

We will find a large variety of diagrams depending on: the values of $M$ and $N$; whether $\gamma > \gamma_{MN}$, or not; $\nu > \nu_{MN}$, or not. For a summary of the qualitative bifurcation types which may occur, see § 6. The diagrams there are intended to represent distinct solutions of the capillary-gravity wave problem as the various parameters are adjusted. In § 1.4 we observe that there are more solutions of $(\mathcal{N})$ than there are waves because symmetry considerations lead to the observation that distinct solutions of $(\mathcal{N})$ may correspond to the same wave.

The results here depend for their proofs on the careful analysis of symmetries in the mathematics which reflect symmetries inherent in the physical set-up. We return to a detailed account of these in § 1.4. In § 2 we discuss the reduction of $(\mathcal{N})$ to a finite dimensional problem $(\mathcal{B})$, which is along standard lines. Then § 4 and 5 are devoted to an exposition of the bifurcation theory for problem $(\mathcal{B})$ from which solutions of $(\mathcal{N})$ may be obtained. (Recall that the Lyapunov-Schmidt reduction is effective because it shows that an infinite dimensional problem such as $(\mathcal{N})$ is *equivalent* to a finite dimensional problem ($(\mathcal{B})$ in this case), and nothing is lost from a qualitative point of view by studying $(\mathcal{B})$ rather than $(\mathcal{N})$.)

There are two basic tools in the analysis: the Implicit Function Theorem and the Blowing-Up Lemma. The latter is given in § 3.

### *1.3 Related work*

The theory of bifurcation problems such as this, in which a double eigenvalue splits, under perturbation, into two simple eigenvalues has been extensively studied by Shearer [16, 17] and in a systematic way by Golubitsky & Schaeffer [7, 12]. Secondary bifurcations are observed to be a consequence of group invariance properties of the equations. The capillary-gravity wave problem is of particular interest in this context since it is a single physical problem in which many different bifurcation phenomena may be observed.

In the case $M = 2$, $N = 1$ REEDER & SHINBROT [11] have given an analysis of phenomena observed formally by CHEN & SAFFMAN. Theirs is the first rigorous theory of "Wilton ripples". Following them, TOLAND & JONES [18] introduced equation ($\mathcal{N}$) as a formulation of the problem, and gave a somewhat more systematic description of the Wilton ripple phenomenon as an aspect of bifurcation theory in the presence of symmetries. The equation ($\mathcal{N}$) is derived by a method analogous to that for NEKRASOV's integral equation for periodic water-waves when surface tension effects are neglected though, so far as we are aware, it is new in the present context.

SEKERZH-ZEN'KOVICH also reduces the capillary-gravity wave problem to an integral equation for $\theta$ which is different from ours. In [13] or [15] he considers the existence of capillary-gravity waves over a wavy bottom, or induced by a periodic distribution of pressure on the free surface, respectively. In [14] he reduces the pure capillary gravity wave problem to an integral equation for which certain small amplitude solutions are obtained by a power series method.

BEALE [2] considered the analogous problem and obtained the existence of long waves of small amplitude when the surface tension differs from $T_{MN}$, $M, N \in \mathbb{N}$. It is now possible to interpret his results in terms of bifurcation from a simple eigenvalue.

The most recent numerical work on capillary-gravity waves is due to HOGAN [8]. HUNTER & VANDEN-BROECK [9] have investigated numerically the possible convergence of periodic waves to solitary waves in the long-wave limit when the depth is finite.

There is also recent work of AMICK & KIRCHGÄSSNER [1] who give a rigorous proof of the existence of solitary waves when the Bond number lies in a certain range of values using the centre manifold theorem.

### *1.4 Symmetries*

We begin with some notation. The set of real numbers is denoted by $\mathbb{R}$, and the Banach space of *real-valued continuously differentiable $2\pi$-periodic odd functions on* $\mathbb{R}$ will be denoted by $C^1$ with norm $\|\cdot\|$. Let $\mathbb{N}$ denote the natural numbers. Then for each $N \in \mathbb{N}$

$$X_N = \{u \in C^1 : u(t) = u(t + (2\pi/N)), t \in \mathbb{R}\}.$$

It is clear that $X_{mN} \subset X_N$ for all $m, N \in \mathbb{N}$ since $X_N$ is the subspace of $C^1$ consisting of $(2\pi/N)$-periodic functions. Now let $S_N : X_N \to X_N$ be defined by

$$(S_N u)(t) = u(t + (\pi/N)), \qquad t \in \mathbb{R}.$$

Then $S_N u = u$, $u \in X_N$, implies that $u \in X_{2N}$.

A function $u \in C^1$ may be written as the sum of its Fourier series

$$u(t) = \sum_{k=1}^{\infty} A_k \sin kt, \qquad t \in \mathbb{R},$$

where $\{A_k\}$ denotes its sequence of Fourier coefficients. For such $u$ let $\mathsf{C}u$ denote the function whose Fourier expansion is

$$(\mathsf{C}u)(t) = -\sum_{k=1}^{\infty} A_k \cos kt, \qquad t \in \mathbb{R}.$$

It is elementary that $\mathsf{C}u \in L_2(-\pi, \pi)$ and a theorem of PRIVALOV (see ZYGMUND [19]) gives that $\mathsf{C}u$ is Hölder continuous with exponent $\alpha$ for any $\alpha \in (0, 1)$ when $u \in C^1$ ($\mathsf{C}u$ need not be differentiable, however).

In order to have a convenient notation later, we introduce the following abbreviations for sine and cosine,

$$s_k(t) = \sin kt, \qquad c_k(t) = \cos kt, \qquad t \in \mathbb{R}.$$

We will then write

$$u \sim \sum_{k=1}^{\infty} A_k s_k, \qquad u \in C^1.$$

(We avoid "=" here since the convergence of the right-hand side is not necessarily with respect to the $C^1$ norm.) Thus

$$\mathsf{C}u \sim -\sum_{1}^{\infty} A_k c_k \quad \text{when} \quad u \sim \sum_{1}^{\infty} A_k s_k, \qquad u \in C^1.$$

Note that

$$S_N \circ \mathsf{C}(u) = \mathsf{C} \circ S_N(u), \qquad u \in X_N.$$

We recall the equation $(\mathcal{N})$ in the abbreviated form given in § 1.2 is

$$\gamma\theta = G(\nu, \theta), \qquad \theta \in C^1,$$

where $G(\nu, \theta)$ is defined in terms of $\tau$, and $\tau = -\mathsf{C}\theta$.

It has been proved (TOLAND & JONES [18]) that $G$ commutes with $S_N$ in the following sense:

$$S_N(G(\nu, \theta)) = G(\nu, S_N\theta), \qquad \theta \in X_N.$$

As a consequence $(\gamma, \nu, S_N\theta)$ satisfies $(\mathcal{N})$ when $(\gamma, \nu, \theta)$ satisfies $(\mathcal{N})$. As we observed earlier $S_N\theta$ and $\theta$ are then distinct solutions of $(\mathcal{N})$ unless $\theta \in X_{2N}$. *However*, it is immediate from the earlier formula giving the wave profile for given solution $(\gamma, \nu, \theta)$ that both $\theta$ and $S_N\theta$ correspond to the same wave differently parameterised, the difference being only a shift of origin.

This symmetry of equation $(\mathcal{N})$ is beautifully reflected in the finite dimensional bifurcation equations; and the symmetry which they enjoy is expressed explicitly in § 2. Because $S_N\theta$ and $\theta$, $\theta \in X_N$, correspond to the same wave, it will suffice to identify waves as equivalence classes of solutions of $(\mathcal{N})$ where the equivalence relation is defined by the symmetry.

*Therefore*, in illustrating the results in the bifurcation diagrams of § 6 we illustrate only the equivalence classes corresponding to *different* waves. For example, a pitchfork curve of solutions of $(\mathcal{N})$ is illustrated as a one-pronged fork of solutions of the water-wave problem because of these symmetry considerations;

both branches of a fork of solutions of ($\mathcal{N}$) correspond to the same set of capillary-gravity waves.

To finish § 1 we emphasize again that our purpose here is to vindicate the formal methods of CHEN & SAFFMAN [4], and we pay special attention to the rigorous identification of *bifurcation points* where the solution set of the water-wave problem branches, for fixed values of the surface tension when the speed at infinite depth is regarded as the bifurcation parameter. There is no guarantee that all bifurcation points have indeed been found, but it is certain that all those guaranteed by the symmetries inherent in the problem have been identified. There is no obvious reason why there should be any others.

## 2. The Lyapunov-Schmidt Reduction

### 2.1 Introductory remarks

For any choice of $\gamma$ and $\nu$ the zero function is a solution of ($\mathcal{N}$) corresponding to uniform horizontal laminar flow with a flat free surface. To obtain the existence of other, non-trivial flows we investigate the solution set of ($\mathcal{N}$) using standard methods of bifurcation theory. When only small solutions $\theta$ are sought, and $\gamma$ and $\nu$ ar close to $\gamma_{MN}$ and $\nu_{MN}$ respectively, the problem is reduced to one or two scalar equations for four unknowns by the Lyapunov-Schmidt procedure. The symmetries observed in § 1.4 are reflected in the form of these two equations. What follows might be regarded as the mathematical theory which underlies the weakly nonlinear theory of CHEN & SAFFMAN [4], and is more-or-less typical of perturbed bifurcation theory at a double eigenvalue.

### 2.2 The reduction

The linearisation of ($\mathcal{N}$) about $\theta = 0$ in $C^1$ is

$$\gamma\theta = \nu A\theta + B\theta. \tag{2.1}$$

If $\gamma \in \mathbb{R}$ and $\nu = \gamma k + k^{-1}$ then $s_k$ is a solution of (2.1) for any $k \in \mathbb{N}$, and otherwise (2.1) has no non-trivial solutions. To each $\gamma > 0$ corresponds a set of eigenspeeds $\{\gamma k + k^{-1} : k \in \mathbb{N}\}$, the set of corresponding eigenfunctions being $\{s_k : k \in \mathbb{N}\}$. If $\gamma = \gamma_{MN} = (MN)^{-1}$ then $\nu_{MN} = (M^{-1} + N^{-1})$ is an eigenspeed with two distinct eigenfunctions $s_M$ and $s_N$ if $M \neq N$.

From now on we suppose that $M$ and $N$ are *fixed* natural numbers with $M > N$, and we introduce the variables

$$\gamma = \gamma_{MN} + \beta, \qquad \nu = \nu_{MN} + \alpha.$$

Then equation ($\mathcal{N}$) takes the form

$$(\gamma_{MN} + \beta)\,\theta = G(\nu_{MN} + \alpha, \theta), \tag{2.2}$$

and we seek solutions $(\alpha, \beta, \theta)$ in a neighbourhood of the origin in $\mathbb{R}\times\mathbb{R}\times C^1$. The Lyapunov-Schmidt procedure now goes as follows.

Let $E = \text{span}\{s_N, s_M\} \subset C^1$ and let $F = \{u \in C^1 : \langle u, s_n \rangle = 0, n \in \{N, M\}\}$ where $\langle , \rangle$ denotes the usual $L_2$ inner product. Then $C^1 = E \oplus F$, and when $\alpha = \beta = 0$ the linear equation (2.1) has no non-zero solution in $F$. Let $P$ denote the usual orthogonal projection of $C^1$ onto $E$, and for simplicity let $(a, b)$ denote the element $as_N + bs_M \in E$ (to avoid ambiguity we will always write $(a, b) \in E$ when using this convention). The equation $(\mathcal{N})$ is equivalent to

$$\begin{aligned} (\gamma_{MN} + \beta)\, e &= PG(\nu_{MN} + \alpha, e + f), \\ (\gamma_{MN} + \beta)\, f &= (I - P)\, G(\nu_{MN} + \alpha, e + f), \end{aligned} \qquad \theta = e + f \in E \oplus F.$$

If $(\alpha, \beta, e)$ is in a neighbourhood $U$ of the origin $\mathbb{R} \times \mathbb{R} \times E$, then the second equation can be solved by the Implicit Function Theorem giving $f$ as a smooth function from $U$ into $F$. Moreover the only solutions $(\gamma, \nu, \theta)$ of $(\mathcal{N})$ in a neighbourhood $V$ of the origin in $\mathbb{R} \times \mathbb{R} \times C^1$ are of the form $(\gamma_{MN} + \beta, \nu_{MN} + \alpha, e + f(\alpha, \beta, e))$ where $(\alpha, \beta, e)$ is a solution of the finite dimensional problem

$$\beta e = PG(\nu_{MN} + \alpha, e + f(\alpha, \beta, e)) - \gamma_{MN}\, e.$$

This is called the *bifurcation equation*. It may be written in abbreviated form as

$$\beta(a, b) = (g_N(\alpha, \beta, (a, b)), g_M(\alpha, \beta, (a, b))), \tag{2.3}$$

to be satisfied by $(\alpha, \beta, (a, b)) \in U$, a neighbourhood of the origin in $\mathbb{R} \times \mathbb{R} \times E$. Here $g_N$ and $g_M$ are smooth, real-valued functions on $U$ whose Taylor polynomials can be obtained to any required order from the operator $G$ (the calculations are, however, formidable). We will present the third-order Taylor polynomial in due course, but first we observe the properties of the bifurcation equations which may be inferred from the symmetry of $G$ noted earlier.

Since the analysis and Lyapunov-Schmidt reduction of $(\mathcal{N})$ may be done in $X_N$ or $X_M$ it follows that

$$f(\alpha, \beta, (0, b)) \in X_M, \quad \langle f(\alpha, \beta, (0, b)), s_M \rangle = 0$$

and

$$f(\alpha, \beta, (a, 0)) \in X_N, \quad \langle f(\alpha, \beta, (a, 0)), s_N \rangle = 0,$$

and from the definitions of $g_N$ and $g_M$ we obtain

$$g_N(\alpha, \beta, (0, b)) = 0 \quad \text{for all } N \in \mathbb{N}$$

and

$$g_M(\alpha, \beta, (a, 0)) = 0 \quad \text{if } M \neq kN, k \in \mathbb{N}.$$

*Remark 1.* Note that $g_N$ and $g_M$ depend on both $M$ and $N$, even though the $M$ dependence of the former and the $N$ dependence of the latter have been suppressed for the sake of having a convenient notation.

*Remark 2.* If $M = kN$, $k = 2, 3, 4, 5$ $g_M(\alpha, \beta, (a, 0)) \neq 0$ for $(\alpha, \beta, (a, 0)) \in U$ (see § 2 for $k = 2, 3$ and JONES [10] for $k = 4, 5$). Numerical evidence suggests that this is a general observation, and though we have no proof, the problem can be formulated as one about the existence of zeros of a sequence generated by a nonlinear recursion.

*Remark 3.* It clearly follows that $bs_M + f(\alpha, \beta, (0, b)) \in X_M$ and $as_N + f(\alpha, \beta, (a, 0)) \in X_N$ for all $(\alpha, \beta, (a, b)) \in U$. Conversely if $as_N + bs_M + f(\alpha, \beta, (a, b)) \in X_M$ then $a = 0$, and if $as_N + bs_M + f(\alpha, \beta, (a, b)) \in X_N$ and $M \neq kN$ for all $k \in \mathbb{N}$, then $b = 0$. In particular if $M \neq kN$ for all $k \in \mathbb{N}$ then solutions of $(\mathcal{N})$ in $X_N$ or $X_M$ arise from solutions of (2.3) with $ab = 0$.

To see the symmetry of (2.3) corresponding to the fact that $G$ commutes with $S_K$ on $X_K$ let $K$ denote any common divisor of $M$ and $N$ ($K$ may be 1). Then $E \subset X_K$ and so $f(\alpha, \beta, e) \in X_K$ when $e \in E$. Since $P$ commutes with $S_K$ on $X_K$ it follows that

$$(\gamma_{MN} + \beta)\, S_K f = (I - P)\, G((\nu_{MN} + \alpha), S_K(e + f)), \qquad e \in E$$

whence, by uniqueness,

$$f(\alpha, \beta, S_K e) = S_K f(\alpha, \beta, e).$$

Now by writing $e = as_N + bs_M$ as before we obtain the relations

$$f(\alpha, \beta, ((-1)^{N/K} a, (-1)^{M/K} b)) = S_K f(\alpha, \beta, (a, b)), \qquad (a, b) \in E$$

where $K$ is any common divisor of $M$ and $N$. An analogous argument gives that

$$(\gamma_{MN} + \beta)\, S_N f = (I - P)\, G((\nu_{MN} + \alpha), S_N(e + f)) \qquad \text{when } e = (a, 0) \in E,$$

$$(\gamma_{MN} + \beta)\, S_M f = (I - P)\, G((\nu_{MN} + \alpha), S_M(e + f)) \qquad \text{when } e = (0, b) \in E,$$

whence

$$f(\alpha, \beta, (-a, 0)) = S_N f(\alpha, \beta, (a, 0)), \qquad (a, 0) \in E,$$

$$f(\alpha, \beta, (0, -b)) = S_M f(\alpha, \beta, (0, b)), \qquad (0, b) \in E.$$

When these observations are substituted into the bifurcation equations we obtain for any common divisor $K$ of $M$ and $N$

$$g_N(\alpha, \beta, ((-1)^{N/K} a, (-1)^{M/K} b)) = (-1)^{N/K} g_N(\alpha, \beta, (a, b)), \tag{2.4}$$

$$g_M(\alpha, \beta, ((-1)^{N/K} a, (-1)^{M/K} b)) = (-1)^{M/K} g_M(\alpha, \beta, (a, b)), \tag{2.5}$$

$$g_N(\alpha, \beta, (-a, 0)) = -g_N(\alpha, \beta, (a, 0)), \tag{2.6}$$

$$g_M(\alpha, \beta, (0, -b)) = -g_M(\alpha, \beta, (0, b)). \tag{2.7}$$

It is easy to see that no information is lost by taking $K$ to be the greatest common factor of $M$ and $N$.

*Remark 4.* These observations about symmetry simplify the bifurcation diagrams conceptually. It is important to note at this stage that the extra solutions which arise as a result do not yield different solutions of the water-wave problem since $\theta$ and $S_K\theta$, though possibly distinct, give different parameterisations of the same wave when $\theta \in X_K$.

It is now time to write down the cubic Taylor expansion of $(g_N, g_M)$ in $U$. More precisely the bifurcation equations (2.3) can be written as follows:

$$\beta a = A\alpha a + Bab + Ca^3 + Dab^2 + Ja^2b + u(\alpha, \beta, a, b) \qquad (\mathscr{B})\ (a)$$

$$\beta b = Y\alpha b + Za^2 + Sa^2b + Rb^3 + Ta^3 + v(\alpha, \beta, a, b) \qquad (\mathscr{B})\ (b)$$

where

$$|(u, v)| = O(|(\alpha, \beta, a, b)|^4),$$

for $(\alpha, \beta, a, b) \in U \subset \mathbb{R}^4$ (which henceforth we identify with $\mathbb{R}\times\mathbb{R}\times E$). The coefficients have numerical values as follows:

$$A = N^{-1}; \qquad Y = M^{-1};$$

$$B = -\tfrac{3}{4}\left(\frac{1-\delta}{N^2}\right); \qquad Z = -\tfrac{3}{8}\left(\frac{1-\delta}{N^2}\right);$$

$$C = \frac{1}{8N^2}\left(\frac{9\delta N}{2N-M} + \frac{N+M}{M}\right); \qquad R = \frac{8N^2 + MN + 2M^2}{8NM^2(2M-N)};$$

$$D = \left(\frac{2M^2 + 6MN + N^2}{4M^2N^2} - \frac{\delta(2M-N)(M+N)}{4MN^2(M-2N)}\right)$$

$$S = \left(\frac{M^2 + 6MN + 2N^2}{4M^2N^2} - \frac{\delta(M^2 + 2MN - 2N^2)(M+N)}{4M^2N^2(M-2N);}\right)$$

$$J = -\frac{41\varepsilon}{8N^2}; \qquad T = -\frac{41\varepsilon}{24N^2};$$

where

$$\varepsilon = 1 \quad \text{if } M = 3N, \qquad \varepsilon = 0 \quad \text{if } M \neq 3N;$$

$$\delta = 0 \quad \text{if } M = 2N, \qquad \delta = 1 \quad \text{if } M \neq 2N.$$

The calculation of these coefficients is given in JONES [10] and confirms similar calculations in CHEN & SAFFMAN [4]. It is reassuring to check that the general symmetries observed earlier are observed in these cubic Taylor polynomials, and as a consequence for all $M, N$,

$$u(\alpha, \beta, -a, 0) = -u(\alpha, \beta, a, 0),$$

$$v(\alpha, \beta, 0, -b) = -v(\alpha, \beta, 0, b),$$

$$u(\alpha, \beta, 0, b) = 0;$$

moreover

$$v(\alpha, \beta, a, 0) = 0 \quad \text{if } M \neq kN \text{ for all } k \in \mathbb{N}.$$

### *2.3 Some preliminary calculations*

Now is a good time to place on record the results of certain arithmetic observations about the signs of certain combinations. The purpose of these may not be immediately clear but should become so in § 4. Suppose that $M \neq kN$ for

$k = 2$ or 3. Then

$$A - Y = N^{-1} - M^{-1} > 0 \quad \text{for all } M \text{ and } N, \tag{2.8}$$

$$AS - CY = \frac{10M + N}{8M^2N^2} > 0 \quad \text{for all } M \text{ and } N, \tag{2.9}$$

$$RC - SD$$
$$= \frac{16(N^6 - M^6) + 18M^5N + 33M^4N^2 - 720M^3N^3 + 68M^2N^4 + 160MN^5}{64M^4N^3(M-2N)^2(2M-N)}$$
$$< 0 \text{ for all } M \text{ and } N, \tag{2.10}$$

$$AR - DY = \frac{2M^4 - 7M^3N + 48M^2N^2 - 28MN^3 - 4N^4}{8M^3N^2(M-2N)(2M-N)}$$
$$\begin{cases} > 0 & \text{if } M > 2N, \\ < 0 & \text{if } N < M < 2N. \end{cases} \tag{2.11}$$

In particular,

$$\frac{R-D}{AR-DY} - \frac{S-C}{AS-CY} = \frac{(A-Y)(RC-DS)}{(AR-DY)(AS-CY)}$$
$$\begin{cases} > 0 & \text{if } N < M < 2N \\ < 0 & \text{if } M > 2N. \end{cases} \tag{2.12}$$

(To check the signs of (2.10) and (2.11) write $M = N + k$.)

### *2.4 Bifurcation points*

Having thus reduced problem ($\mathcal{N}$) to an equivalent finite dimensional system in which certain symmetries are apparent, it suffices now to analyse the solution set of ($\mathcal{B}$). We need some definitions. Let $\mathcal{T}$ denote the line of trivial solutions of ($\mathcal{B}$) $\{(\alpha, (0, 0)) : \alpha \in \mathbb{R}\} \subset \mathbb{R} \times E$, for any fixed $\beta$.

**Definition.** *If $\beta$ is fixed, and if there exists a sequence $\{(\alpha_n, (a_n, b_n))\}$ of solutions of ($\mathcal{B}$) with $(a_n, b_n) \neq (0, 0)$ and $(\alpha_n, (a_n, b_n)) \to (\alpha_0, (0, 0))$ as $n \to \infty$, then $\alpha_0$ is called a primary bifurcation point. If there exists a continuous curve $\mathcal{C} = \{(\alpha(t), (a(t), b(t))) : t \in (-1, 1)\}$ of solutions of ($\mathcal{B}$) and an open set $W \subset \mathbb{R} \times E$ such that $W \cap \mathcal{C} \cap \mathcal{T} = \{(\alpha_0, (0, 0))\}$ then $\mathcal{C}$ is called a primary curve bifurcating at $\alpha_0$.*

**Definition.** *If $\mathcal{C}$ is a primary curve bifurcating at $\alpha_0$ (for fixed $\beta$) and if $\mathcal{D} = \{(\alpha(t), (a(t), b(t))) : t \in (-1, 1)\}$ is a curve of solutions of ($\mathcal{B}$) such that for some open set $W' \subset \mathbb{R} \times E$, $W' \cap \mathcal{D} \cap \mathcal{C}$ is a singleton different from $(\alpha_0, (0, 0))$ then $\mathcal{D}$ is called a secondary branch bifurcating from $\mathcal{C}$.*

The terms *sub-critical* and *super-critical* bifurcation refer to the value of $\alpha$ on the bifurcating branch relative to its value at the bifurcation point.

In the bifurcation analysis which follows there are two principal cases $M = kN$, and $M \neq kN$ for all $k \in \mathbb{N}$. Each of these in turn has a number of sub-cases.

In everything which follows we are considering the bifurcation of solutions of $(\mathscr{B})$ relying on the observations made in this section to infer analogous results about the integral equation $(\mathscr{N})$.

## 3. The Blowing-Up Lemma

There are two important basic tools in the bifurcation analysis. One is the familiar Implicit Function Theorem and we will dwell on it no further now. The other is a Blowing-Up Lemma of BUCHNER, MARSDEN & SCHECHTER [3] in the general form given in SHEARER [16]. First we need some notation. Suppose $m \geqq n + 1$ and $\mathsf{m}$ and $\mathsf{q}$ denote $m$-tuples of positive and non-negative integers, respectively, while $\mathsf{n}$ denotes an $n$-tuple of positive integers. Let $|\mathsf{q}| = \sum_{i=1}^{m} q_i$ if $\mathsf{q} = \{q_i\}$ and if $f: \mathbb{R}^m \to \mathbb{R}$ is smooth let $f_{\mathsf{q}} = \dfrac{\partial^{|\mathsf{q}|} f}{\partial x_1^{q_1} \ldots \partial x_m^{q_m}}$.

**The Blowing-Up Lemma.** *Suppose* $\mathsf{g}: \mathbb{R}^m \to \mathbb{R}^n$ *is* $C^\infty$ *and is given by* $\mathsf{g} = (g_1, \ldots, g_n)$, *and that there exists* $\mathsf{m}$ *and* $\mathsf{n}$ *as above with the property that for each* $i$, $1 \leqq i \leqq n$,

$$(g_i)_{\mathsf{q}}(0) = 0 \quad \text{if } \mathsf{m} \cdot \mathsf{q} = \sum_{j=1}^{m} m_j q_j \leqq n_i - 1.$$

*Define the Taylor polynomial* $\mathsf{f}$ *by*

$$f_i(x) = \sum_{\mathsf{m} \cdot \mathsf{q} = n_i} \{(q_1!\, q_2! \ldots q_m!)^{-1} (g_i)_{\mathsf{q}}(0)\}\, x_1^{q_1} \ldots x_m^{q_m}.$$

*If* $\mathsf{f}(x) = 0$ *and* $x \neq 0$ *implies that* $D\mathsf{f}(x): \mathbb{R}^m \to \mathbb{R}^n$ *has rank* $n$, *then there are neighbourhoods* $U$ *and* $V$ *of the origin in* $\mathbb{R}^m$ *and a homeomorphism* $\phi$ *from* $\mathsf{g}^{-1}(0) \cap U$ *onto* $\mathsf{f}^{-1}(0) \cap V$ *such that* $\phi(0) = 0$. *Moreover,* $(\mathsf{g}^{-1}(0) \cap U) \setminus \{0\}$ *is a* $C^\infty$ *submanifold of* $\mathbb{R}^m$ *and* $\phi$ *is a diffeomorphism of it onto* $(\mathsf{f}^{-1}(0) \cap V) \setminus \{0\}$.

*Remark 5.* We will use this to study the solution set of $(\mathscr{B})$ in $(\alpha, \beta, a, b)$ space. It is worth pointing out that the bifurcation question *for fixed* $\beta$ is not automatically answered by the lemma. The homeomorphism $\phi$ can change bifurcation diagrams dramatically for each fixed $\beta$, while leaving the structure of the solution set in $(\alpha, \beta, a, b)$ space qualitatively the same.

When the Blowing-Up Lemma is used for fixed $\beta$ in $(\alpha, a, b)$ space it can detect the presence or absence of bifurcation points in the solution set.

## 4. Bifurcation When $\mathbf{M} \neq \mathbf{kN}$

In this $\varepsilon = 0$, $\delta = 1$ and since $u$ and $v$ are real-analytic the observations of the previous section ensure that $(u, v) = (a\hat{u}, b\hat{v})$ where $|(\hat{u}, \hat{v})| = O(|(\alpha, \beta, a, b)|^3)$ for $(\alpha, \beta, (a, b)) \in U$. The first result is an elementary consequence of looking at the situation from the point of view of bifurcation from a simple eigenvalue (see SHEARER [17]). Let $R_d = (-d, d) \times (-d, d)$.

**Theorem 1.** *Suppose $M \neq kN$ for all $k \in \mathbb{N}$. Then there exists $d > 0$ and real-valued functions $\alpha_M, \alpha_N, \sigma_M$ and $\sigma_N$ continuous on $R_d$ with $\alpha_N(\beta, 0) = N\beta$, $\alpha_M(\beta, 0) = M\beta$, $\sigma_M(\beta, 0) = \sigma_N(\beta, 0) = 0$ such that the curves*

$$C_\beta^N = \{(\alpha_N(\beta, t), \beta, (t + t\sigma_N(\beta, t), 0)) : |t| < d\}$$

*and*

$$C_\beta^M = \{(\alpha_M(\beta, t), \beta, (0, t + t\sigma_M(\beta, t))) : |t| < d\}$$

*are primary curves of solutions of ($\mathcal{B}$) bifurcating at $\alpha = N\beta$ $\alpha = M\beta$ respectively.*

*Moreover, there is a neighbourhood U of the origin in $\mathbb{R}^4$ such that if $(\alpha, \beta, (a, 0)) \in U$ is a solution of ($\mathcal{B}$) with $a \neq 0$ then $(\alpha, \beta, (a, 0)) \in C_\beta^N$, and if $(\alpha, \beta, (0, b)) \in U$ is a solution with $b \neq 0$, then $(\alpha, \beta, (0, b)) \in C_\beta^M$.*

*Remark 6.* Curves $\mathscr{C}_\beta^N$ and $\mathscr{C}_\beta^M$ of solutions of ($\mathcal{N}$) are easily recovered from $C_\beta^N$ and $C_\beta^M$ by putting

$$\mathscr{C}_\beta^N = \{(\gamma, \nu, \theta) = (\gamma_{MN} + \beta, \nu_{MN} + \alpha, e + f(\alpha, \beta, e)) : (\alpha, \beta, e) \in C_\beta^N\}$$

and similarly for $\mathscr{C}_\beta^M$. Note $\mathscr{C}_\beta^N \subset \mathbb{R} \times \mathbb{R} \times X_N$ and $\mathscr{C}_\beta^M \subset \mathbb{R} \times \mathbb{R} \times X_M$, and these curves contain the totality of all non-trivial solutions of ($\mathcal{N}$) in $(\mathbb{R} \times \mathbb{R} \times (X_N \cup X_M)) \cap U$.

*Remark 7.* A straightforward calculation yields the asymptotic form of $\alpha_N$ and $\alpha_M$:

$$\alpha_N = N\beta - \frac{(8M^2 + MN + 2N^2)\, t^2}{8MN(2N - M)} + O(t^2 |(\beta, t)|),$$

$$\alpha_M = M\beta - \frac{(2M^2 + MN + 8N^2)\, t^2}{8MN(2M - N)} + O(t^2 |(\beta, t)|).$$

Hence on account of the symmetry noted earlier, $C_\beta^M$ is always a sub-critical pitchfork bifurcation, while $C_\beta^N$ is sub-critical pitchfork if $M < 2N$ and a super-critical pitchfork if $M > 2N$.

This theorem accounts for all solutions of ($\mathcal{N}$) in a neighbourhood of $(\gamma_{MN}, \nu_{MN}, 0)$ in $\mathbb{R} \times \mathbb{R} \times (X_N \cup X_M)$. Consequently no solutions are lost if ($\mathcal{B}$) (a) is divided by $a$ and ($\mathcal{B}$) (b) is divided by $b$. This yields the two equations

$$\begin{aligned} \beta &= A\alpha + Ca^2 + Db^2 + \hat{u}(\alpha, \beta, a, b), \\ \beta &= Y\alpha + Sa^2 + Rb^2 + \hat{v}(\alpha, \beta, a, b). \end{aligned} \tag{4.1}$$

We begin by noting that a qualitative description of the solution set of (4.1) can be obtained from the Blowing-Up Lemma. If we take $m = 4$, $n = 2$, $\mathsf{m} = (2, 2, 1, 1)$ and $\mathsf{n} = (2, 2)$, the hypothesis of the Blowing-Up Lemma is seen to hold by a calculation on the coefficients. Hence in a precise qualitative sense the solution set of (4.1) in $(\alpha, \beta, a, b)$ space is 'the same' as that of the truncated equation

$$\begin{aligned} \beta &= A\alpha + Ca^2 + Db^2, \\ \beta &= Y\alpha + Sa^2 + Rb^2. \end{aligned} \tag{4.2}$$

To obtain the existence of bifurcations in the solution set *for fixed* $\beta$, however, we must look further. We will consider three cases.

### 4.1 The case $\beta = 0$

In addition to the curves $C_0^M$ and $C_0^N$ which bifurcate at $\alpha = 0$ when $\beta = 0$ there may be further primary curves, or there may be secondary bifurcations on $C_0^M$ or $C_0^N$. To investigate these possibilities we apply the Blowing-Up Lemma to equation $(\mathscr{B})$ with $\beta = 0$. Let $m = 3$, $n = 2$, $\mathsf{m} = (2, 1, 1)$, $\mathsf{n} = (3, 3)$. The Taylor polynomial equation which results is

$$\begin{pmatrix} A\alpha a + Ca^3 + Dab^2 \\ Y\alpha b + Sa^2 b + Rb^3 \end{pmatrix} = \mathsf{f}(\alpha, a, b) = 0, \tag{4.3}$$

and $D\mathsf{f}$ is given by

$$\begin{pmatrix} Aa, A\alpha + 3Ca^2 + Db^2, 2Dab \\ Yb, 2Sab, Y\alpha + Sa^2 + 3Rb^2 \end{pmatrix}.$$

If $a = 0$ and $b \neq 0$ or *vice-versa*, the rank of $D\mathbf{f}$ is clearly 2 when $\mathbf{f}(\alpha, a, b) = 0$ since $AR \neq DY$ and $AS \neq CY$. So suppose $ab \neq 0$ and $\mathsf{f}(\alpha, a, b) = 0$. Then $D\mathsf{f}$ may be re-written

$$\begin{pmatrix} Aa, 2Ca^2, 2Dab \\ Yb, 2Sab, 2Rb^2 \end{pmatrix}$$

which has rank 2 since $AS \neq CY$. Hence by the Blowing-Up Lemma the solutions of $(\mathscr{B})$ when $\beta = 0$ and $M \neq kN$ are qualitatively the same as the solution set of $\mathsf{f} = 0$ in a neighbourhood of the origin. The solution set of (4.3) contains the two curves

$$\begin{aligned} a &= 0, \quad \alpha = (-R/Y)\, b^2, \\ b &= 0, \quad \alpha = (-C/A)\, a^2, \end{aligned}$$

which correspond to $C_0^M$ and $C_0^N$ respectively. Any further solutions satisfy

$$a^2 = \left(\frac{DY - AR}{CR - DS}\right)\alpha, \quad b^2 = \left(\frac{SA - CY}{CR - DS}\right)\alpha.$$

If $M > 2N$ it follows from the formulae (2.9)–(2.11) that there are no solutions other than $(\alpha, a, b) = (0, 0, 0)$. If $N < M < 2N$, then there are two parabolae of solutions of (4.3) with $ab \neq 0$.

By the Blowing-Up Lemma the only primary bifurcating curves are $C_0^M$ and $C_0^N$, and there are no secondary bifurcations when $M > 2N$. If $N < M < 2N$, there are an additional two primary bifurcations, which by the symmetry of (4.1) and an elementary calculation are sub-critical pitchforks; there are no secondary bifurcations on any of these curves.

### *4.2 The case $\beta \neq 0$*

All small solutions with $ab = 0$ are accounted for by the trivial solution set and $C_\beta^M \cup C_\beta^N$, and since $M\beta$ and $N\beta$ are the only eigenvalues of the linearised problem these are the only primary bifurcations near $\alpha = 0$. Now the object is to enumerate all the secondary bifurcations which occur on $C_\beta^M \cup C_\beta^N$ in a neighbourhood of the origin. Solutions other than those on $C_\beta^M \cup C_\beta^N$ have $ab \neq 0$ and so satisfy the equation

$$\begin{aligned} \beta &= A\alpha + Ca^2 + Db^2 + \hat{u}(\alpha, \beta, a, b), \\ \beta &= Y\alpha + Sa^2 + Rb^2 + \hat{v}(\alpha, \beta, a, b), \end{aligned} \tag{4.4}$$

where $|(\hat{u}, \hat{v})| = O(|(\alpha, \beta, a, b)|^3)$. With simple arithmetic it is easy to see that

$$C_1 \, |(\alpha, \beta)| \leqq |(a, b)|^2 \leqq C_2 \, |(\alpha, \beta)|$$

for positive constants $C_1$ and $C_2$. Moreover if $ab = 0$ then

$$d_1 \, |\beta| \leqq |\alpha| \leqq d_2 \, |\beta|$$

for some positive constants $d_1$ and $d_2$. Therefore all small solutions of (4.4) with $ab = 0$ may be written in the form

$$(\alpha, \beta, a, b) = (\sigma^2 \varrho, \pm\sigma^2, \sigma x, \sigma y) \tag{4.5}$$

where $|\varrho| \in [d_1, d_2]$ and $|x| \in [c_1', c_2']$, $y = 0$, or $|y| \in [c_1', c_2']$, $x = 0$; $c_i' > 0$, $i = 1, 2$ are fixed constants, and $\sigma > 0$ is small. These are the only possible secondary bifurcation points on $C_\beta^M \cup C_\beta^N$.

A substitution of (4.5) into (4.4) yields

$$\begin{aligned} \pm 1 &= A\varrho + Cx^2 + Dy^2 + u_\pm(\sigma, \varrho, x, y) \\ \pm 1 &= Y\varrho + Sx^2 + Ry^2 + v_\pm(\sigma, \varrho, x, y) \end{aligned} \tag{4.6$^\pm$}$$

where

$$u_\pm(\sigma, \varrho, x, y) = \sigma^{-2} \, \hat{u}(\sigma^2, \pm\sigma^2, \sigma x, \sigma y), \; \sigma \neq 0$$

and by analytic continuation as 0 when $\sigma = 0$, and $v_\pm$ is defined similarly. Note that at 0 all the first derivatives of $u_\pm$, $v_\pm$ with respect to $\sigma, \varrho, x, y$ are zero, and both $u_\pm$ and $v_\pm$ are real-analytic. We will find all secondary bifurcation points by finding all solutions $(\varrho, x, y)$ of (4.6) with $\sigma = 0$, and then using the Implicit Function Theorem to find all curves of solutions of $(4.6)^{\pm}$ when $\sigma$ is small which intersect $xy = 0$. This we can do because of the *a priori* bounds described earlier.

With $\sigma = 0$, the system (4.6) is equivalent to

$$\pm\left(\frac{S-C}{AS-YC}\right) = \varrho - \left(\frac{RC-DS}{AS-YC}\right) y^2,$$
$$\pm\left(\frac{R-D}{AR-DY}\right) = \varrho + \left(\frac{RC-DS}{AR-DY}\right) x^2. \tag{4.7$^{\pm}$}$$

The formulae (2.8)–(2.12) now dictate the various cases which occur.

(a) $M > 2N$. If $M > 2N$ then $(4.7)^{-}$ has no solutions whereas $(4.7)^{+}$ has a closed loop of solutions passing through the four points

$$\varrho = \frac{R-D}{AR-DY}, \quad x = 0, \quad y^2 = \frac{A-Y}{AR-DY}$$

$$\varrho = \frac{S-C}{AS-CY}, \quad x^2 = \frac{A-Y}{AS-CY}, \quad y = 0.$$

Without loss of generality suppose that this simple closed loop is parameterised by $\{(\hat{\varrho}(t), \hat{x}(t), \hat{y}(t)) : t \in [0, 2\pi]\}$ which passes through the four distinguished points at $t = k\dfrac{\pi}{2}$, $k = 0, 1, 2, 3$. Let $\mathscr{C}$ denote the Banach space of all $2\pi$-periodic functions $(\varrho, x, y) : [0, 2\pi] \to \mathbb{R}^3$, and define $\mathscr{F} : \mathbb{R} \times \mathscr{C} \to \mathscr{C}$ by

$$\mathscr{F}(\sigma, (\varrho, x, y)) = \begin{pmatrix} A\varrho + Cx^2 + Dy^2 + u_+(\sigma, \varrho, x, y) \\ Y\varrho + Sx^2 + Ry^2 + v_+(\sigma, \varrho, x, y) \\ 2(CR-DS)\,\hat{x}\hat{y}\varrho + (DY-AR)\,\hat{y}x + (AS-CY)\,\hat{x}y \end{pmatrix}.$$

Since the derivative at $(0, (\hat{\varrho}, \hat{x}, \hat{y}))$ of $\mathscr{F}$ with respect to $(\varrho, x, y) \in \mathscr{C}$ is a nowhere singular matrix it is immediate, by the Implicit Function Theorem that the equation

$$\mathscr{F}(\sigma, (\varrho, x, y)) = \begin{pmatrix} 1 \\ 1 \\ \hat{x}\hat{y}(2(CR-DS)\hat{\varrho} + DY - AR + AS - CY) \end{pmatrix}$$

has a unique solution close to $(\hat{\varrho}, \hat{x}, \hat{y})$ for $\sigma > 0$ sufficiently small. Let us denote it by $(\varrho_\sigma, x_\sigma, y_\sigma) \in \mathscr{C}$. Clearly $x_\sigma$ and $y_\sigma$ have each exactly two zeros close to the zeros of $\hat{x}$ and $\hat{y}$. Therefore when $\beta = +\sigma^2$, there is a closed curve $(\alpha_\sigma, a_\sigma, b_\sigma) = (\sigma^2\varrho_\sigma, \sigma x_\sigma, \sigma y_\sigma)$ of solutions of (4.1) which has $a_\sigma = 0$ and $b_\sigma = 0$ twice each. Thus, by Theorem 1 this curve of solutions intersects $C_\beta^M$ and $C_\beta^N$ exactly twice. These two secondary bifurcation points on $C_\beta^N$ and $C_\beta^M$, $\beta > 0$, are related to one another by the symmetry considerations of the previous section. There are no other secondary bifurcation points on $C_\beta^M \cup C_\beta^N$, $\beta > 0$. From (2.12) a calculation yields that the secondary bifurcation is super-critical from $C_\beta^M$ and sub-critical from $C_\beta^N$ if $\beta > 0$. If $\beta < 0$ there are no secondary bifurcation points on $C_\beta^M \cup C_\beta^N$.

This completes the treatment of the case when $M > 2N$.

(b) $N < M < 2N$. In this case both equations $(4.7)^+$ and $(4.7)^-$ have non-trivial solutions; indeed the formulae (2.8)–(2.12) yield that each has two unbounded curves of solutions. In the case of $(4.7)^+$ there is a curve passing through each of the two points

$$\varrho = \frac{S - C}{AS - CY}, \qquad x^2 = \frac{A - Y}{AS - CY}, \qquad y = 0,$$

and in the case of $(4.7)^-$ there is a curve of solutions passing through each of the two points

$$\varrho = \frac{D - R}{AR - DY}, \qquad x = 0, \quad y^2 = \frac{A - Y}{DY - AR}.$$

Let the two curves of solutions of $(4.7)^+$ be denoted by $\Gamma_i^+ = \{(\varrho_i^+(t), x_i^+(t), y_i^+(t)) : t \in (-1, 1)\}$, where $y_i^+(0) = 0$, $i = 1, 2$, and let $\Gamma_i^- = \{(\varrho_i^-(t), x_i^-(t), y_i^-(t)) : t \in (-1, 1)\}$ where $x_i^-(0) = 0$, $i = 1, 2$. Now we can apply the Implicit Function Theorem in the space of curves in $\mathbb{R}^3$ rather than $\mathscr{C}$. The result is that for $\sigma > 0$ sufficiently small there are two curves of solutions of $(4.7)^+$ and two of $(4.7)^-$ close to $\Gamma_i^+$, $i = 1, 2$, and $\Gamma_i^-$, $i = 1, 2$, respectively. Clearly the former two curves each passes through a point where $y = 0$ and the latter two each passes through a point where $x = 0$. Upon rescaling they yield two secondary bifurcations on $C_\beta^N$ when $\beta > 0$ and two secondary bifurcations on $C_\beta^M$ when $\beta < 0$. There are no secondary bifurcations on $C_\beta^N$ when $\beta < 0$ and none on $C_\beta^M$ when $\beta > 0$. A calculation now yields that all secondary bifurcations are sub-critical pitchforks. As before, the two bifurcation points on each branch are related to one another by symmetry.

## 5. Bifurcation When $M = kN$

A few remarks are needed to set the scene. Let us suppose throughout this section that $k$ denotes the (fixed) natural number $M/N$. The highest common divisor of $M$ and $N$ is $N$ in this case and the symmetry results of § 2 reduce to the following

$$-g_N(\alpha, \beta, (a, b)) = g_N(\alpha, \beta, (-a, (-1)^k b))$$

$$(-1)^k g_M(\alpha, \beta, (a, b)) = g_M(\alpha, \beta, (-a, (-1)^k b))$$

and

$$g_N(\alpha, \beta, (0, b)) = 0$$

for all $(\alpha, \beta, (a, b)) \in U$. However, $g_M(\alpha, \beta, (a, 0))$ is not in general zero, and certain of the symmetry arguments which led to the secondary bifurcation results of the previous section fail in this case.

This difficulty means that the bifurcation equations ($\mathscr{B}$) may be regarded as a problem in the subspace $\{(0, b) : b \in \mathbb{R}\} \subset \mathbb{R}^2$, not in the subspace $\{(a, 0) : a \in \mathbb{R}\}$. In the former subspace $\alpha = M\beta$ is a simple eigenvalue, and the first result is a simple consequence of this observation.

**Theorem 2.** *Suppose* $M = kN$ *for some* $k \in \mathbb{N}$. *Then there exist* $d > 0$ *and real-valued functions* $\alpha_M$ *and* $\sigma_M$, *continuous on* $R_d$ *with* $\alpha_M(\beta, 0) = M\beta$, $\sigma_M(\beta, 0) = 0$, *such that the curve*

$$C_\beta^M = \{(\alpha_M(\beta, t), \beta, (0, t + t\sigma_M(\beta, t))) : |t| < d\}$$

*is a primary curve of solutions of* ($\mathscr{B}$) *bifurcating at* $\alpha = M\beta$. *Moreover there is a neighbourhood of the origin in* $\mathbb{R} \times \mathbb{R} \times E$ *such that if* $(\alpha, \beta, (0, b)) \in U$ *is a solution of* ($\mathscr{B}$) *with* $b \neq 0$, *then* $(\alpha, \beta, (0, b)) \in C_\beta^M$.

*Remark 8.* As with Theorem 1 a branch $\mathscr{C}_\beta^M$ of solutions of ($\mathscr{N}$) can be recovered from $C_\beta^M$ by the usual transformation. In all cases the bifurcation is a sub-critical pitchfork, of the form given in Remark 7.

*Remark 9.* When $\beta \neq 0$ there are only two eigenvalues in a neighbourhood of $\alpha = 0$, namely $\alpha = N\beta$ and $\alpha = M\beta$, both are simple and the classical theory of bifurcation from a simple eigenvalue obtains. It gives the existence, for each $\beta \neq 0$ of two primary bifurcations which we denote by $C_\beta^M$ and $C_\beta^N$, which bifurcate respectively from $\alpha = M\beta$ and $\alpha = N\beta$. When $\beta = 0$, $\alpha = 0$ is a double eigenvalue and no such simple observation is valid in this case.

*Remark 10.* Since $M = kN$, $X_M \subset X_N \subset C^1$. In this case the Lyapunov-Schmidt reduction and the analysis of this section are valid in every detail in $X_N$ or in $C^1$. So, by our uniqueness theorems all the solutions obtained here for ($\mathscr{B}$) correspond to solutions of ($\mathscr{N}$) in $X_N$, which correspond in turn to solutions of the water-wave problem of minimal period at most $2\pi/N$. This is in contrast to the case where $M \neq kN$ for all $k \in \mathbb{N}$ where we have seen secondary bifurcations on $C_\beta^M$ or $C_\beta^N$ leading to waves of minimal period $2\pi/K$, where $K$ is the highest common factor of $M$ and $N$.

Now all solutions in $U$ with $a = 0$ have been accounted for. It remains to study all other possible solutions, *i.e.* those with $a \neq 0$. It is necessary to consider various cases separately.

### *5.1 Bifurcations when $M/N \geqq 4$*

In this case $B = Z = T = J = 0$ and so the bifurcation equations look deceptively like those of § 4 when $M > 2N$; they are

$$\beta a = A\alpha a + Ca^3 + Dab^2 + u(\alpha, \beta, a, b), \tag{5.1 (a)}$$

$$\beta b = Y\alpha b + Sa^2 b + Rb^3 + v(\alpha, \beta, a, b). \tag{5.1 (b)}$$

The significant difference now is that $v(\alpha, \beta, a, 0)$ need not be zero (and indeed we believe it is never zero). As usual we need to consider various cases separately.

(a) $\beta = 0$. When $\beta = 0$ an application of the Blowing-Up Lemma shows that there are exactly two bifurcations from $\alpha = 0$ when $\beta = 0$. The usual symmetry considerations and a now familiar application of scaling techniques shows that one of these is $C_0^M$ which is a sub-critical pitchfork bifurcation; recall that if $(\alpha, 0, (a, b)) \in C_0^M$, then $a = 0$. The other is a super-critical pitchfork which we denote by $C_0^N$; there is no reason for $b = 0$ when $(\alpha, 0, (a, b)) \in C_0^N$.

With this important proviso the qualitative bifurcation picture is similar to that of the case $M \neq kN$, $M > 2N$, $\beta = 0$ in § 4.

(b) $\beta \neq 0$. When $\beta \neq 0$ we seek bifurcations and secondary bifurcations by the scaling methods described in detail in § 4. All solutions on $C_\beta^M$ correspond to solutions of (5.1) (a) with $a = 0$. Upon dividing (5.1) (a) by $a$ and then using the same scaling as in § 4 (which is justified in this case also, for the same reason) we obtain the equations

$$\pm 1 = A\varrho + Cx^2 + Dy^2 + u_\pm(\sigma, \varrho, x, y), \qquad (5.2)\ (\mathrm{a})^\pm$$

$$\pm y = Y\varrho y + Sx^2 y + Ry^3 + \tilde{v}_\pm(\sigma, \varrho, x, y) \qquad (5.2)\ (\mathrm{b})^\pm$$

where $u_\pm$ is defined as before and $\tilde{v}_\pm = \sigma^{-3} v(\sigma^2\varrho, \pm\sigma^2, \sigma x, \sigma y)$ for $\sigma \neq 0$, and $\tilde{v}_\pm = 0$ when $\sigma = 0$.

When $\sigma = 0$ a curve of solutions of $(5.2)^\pm$ is given by $\{(\varrho, x, y) = (\pm A^{-1} - CA^{-1} x^2, x, 0) : x \in \mathbb{R}\}$. An application of the Implicit Function Theorem proves that when $\sigma \neq 0$ a primary bifurcation occurs at $\alpha = N\beta$, which fact we already knew from the theory of bifurcation from a simple eigenvalue (see Remark 9). Both bifurcations are symmetric pitchforks, $C_\beta^M$ being sub-critical and $C_\beta^N$ being super-critical.

Now $(5.2)^-$ has no other solutions when $\sigma = 0$. As a consequence there are no secondary bifurcation points on $C_\beta^M$ or $C_\beta^N$ when $\beta < 0$.

However there is a closed loop of solutions of $(5.2)^+$ (the equations are the same as those treated in § 4.1 when $\sigma = 0$ once the second has been divided through by $y$). Let the curve be denoted by

$$\Gamma = \{(\hat{\varrho}(t), \hat{x}(t), \hat{y}(t)) : t \in [0, 2\pi]\} \text{ with } \hat{y}(0) = \hat{y}(\pi) = \hat{x}(\pi/2) = \hat{x}(3\pi/2) = 0.$$

Now we use the Implicit Function Theorem to study the problem

$$\mathscr{F}(\sigma, (\varrho, x, y)) = \begin{pmatrix} 1 \\ 0 \\ \hat{x}\hat{y}(2(CR - DS)\hat{\varrho} + DY - AR + AS - CY) \end{pmatrix},$$

where

$$\mathscr{F}(\sigma, (\varrho, x, y)) = \begin{pmatrix} A\varrho + Cx^2 + Dy^2 + u_+(\sigma, \varrho, x, y) \\ Y\varrho y - y + Sx^2 y + Ry^3 + \tilde{v}_+(\sigma, \varrho, x, y) \\ 2(CR - DS)\,\hat{x}\hat{y}\varrho + (DY - AR)\,\hat{y}x + (AS - CY)\,\hat{x}y \end{pmatrix}.$$

A calculation shows that the Jacobian determinant of $\mathscr{F}$ evaluated at $(\sigma, \varrho, x, y) = (0, \hat{\varrho}, \hat{x}, \hat{y})$ is $2\hat{y}(\hat{x}^2(AS - CY)^2 + \hat{y}^2(AR - DY)^2 + 4\hat{x}^2\hat{y}^2(CR - DS)^2)$. This is non-zero if $\hat{y} \neq 0$. Hence the Implicit Function Theorem gives, for $\sigma \neq 0$

sufficiently small, the existence of two curves of solutions, parameterised by $t \in [\pi/4, 3\pi/4]$ and $t \in [5\pi/4, 7\pi/4]$ say, close to $\Gamma$. Clearly the arguments of § 4 give that these two curves correspond to two secondary bifurcations on $C_\beta^M$ when $\beta > 0$, and both are super-critical.

*Remark 11.* Since we believe that in general the bifurcation equations have no solutions with $b = 0$, $a \neq 0$, the failure of the Implicit Function Theorem to apply when $\hat{y} = 0$ in the above analysis was to be expected. There is no evidence for secondary bifurcation on $C_\beta^N$.

### 5.2 Bifurcations when $M/N = 3$

In this case $B = Z = 0$ and the bifurcation equations are

$$\beta a = \left(\frac{1}{N}\right) \alpha a - \left(\frac{77}{24N^2}\right) a^3 - \left(\frac{41}{8N^2}\right) a^2 b - \left(\frac{23}{36N^2}\right) ab^2 + \hat{u}, \quad (5.3)\text{ (a)}$$

$$\beta b = \left(\frac{1}{3N}\right) \alpha b - \left(\frac{41}{24N^2}\right) a^3 - \left(\frac{23}{36N^2}\right) a^2 b + \left(\frac{29}{360N^2}\right) b^3 + \hat{v}. \quad (5.3)\text{ (b)}$$

As usual we consider various cases separately.

(a) $\beta = 0$. When $\beta = 0$ we use the Blowing-Up Lemma to obtain all solutions bifurcating at $\alpha = 0$, which in this case is the only possible bifurcation point. Let $m = 3$, $n = 2$, $\mathsf{m} = (2, 1, 1)$ and $\mathsf{n} = (3, 3)$. The resulting Taylor polynomial is that obtained from (5.3) by putting $\hat{u} = \hat{v} = \beta = 0$. Now $Df$ is

$$\begin{pmatrix} Aa, A\alpha + 3Ca^2 + Db^2 + 2Jab, 2Dab + Ja^2 \\ Yb, 2Sab + 3Ta^2, Y\alpha + 3Rb^2 + Sa^2 \end{pmatrix}.$$

If $a = 0$ and $b \neq 0$ and $\mathsf{f}(\alpha, a, b) = 0$, then $Df$ has rank 2 since $AR \neq DY$. There are no solutions of $\mathsf{f} = 0$ with $a \neq 0$, $b = 0$. If $\mathsf{f}(\alpha, a, b) = 0$ and $ab \neq 0$, then $Df$ reduces to

$$\begin{pmatrix} Aa, 2Ca^2 + Jab, 2Dab + Ja^2 \\ Yb, 2Sab + 3Ta^2, 2Rb^2 - Ta^3b^{-1} \end{pmatrix}.$$

A necessary condition for the rank to be less than two is that the sub-determinants of first and second, and first and third columns, be zero. This amounts to there being a common root $w = (b/a)$ of the equations

$$JYw^2 + 2(YC - SA)\, w - 3TA = 0,$$
$$2(RA - DY)\, w^3 - JYw^2 - TA = 0.$$

However, substituting the formulae of § 2 for the coefficients reduces these equations to

$$123w^2 + 62w - 369 = 0,$$
$$\left(\frac{634}{1080}\right) w^3 + \frac{41}{24} w^2 + \frac{41}{24} = 0.$$

They do not have a common root so $Df$ has rank 2. By the Blowing-Up Lemma it is sufficient to study the structure of the solution set of $\mathsf{f} = 0$ in order to obtain all the bifurcations and secondary bifurcations of ($\mathscr{B}$) when $\beta = 0$ and $M = 3N$.

We immediately observe a branch $\{a = 0,\ Y\alpha + Rb^2 = 0\}$ corresponding to the solution set $C_0^M$ already noted. All other solutions $(\alpha, a, b)$ are solutions of

$$\begin{aligned} A\alpha + Ca^2 + Db^2 + Jab &= 0, \\ Y\alpha b + Rb^3 + Sa^2 b + Ta^3 &= 0, \end{aligned} \tag{5.4}$$

and it suffices to find all solutions $w = (b/a)$ of the equation

$$(RA - DY)\,w^3 - JYw^2 + (SA - CY)\,w + TA = 0 \tag{5.5}$$

and to each solution $w_i$ corresponds a branch

$$(\alpha, a, b) = \{(-A^{-1}\,a^2(C + Dw_i^2 + Jw_i),\ a,\ aw_i),\ a \in \mathbb{R}\}$$

of solutions of (6.4). Now $M = 3N$, so by the formulae of § 2 equation (5.5) is equivalent to

$$\left(\frac{317}{1080}\right) w^3 + \left(\frac{41}{24}\right) w^2 + \left(\frac{31}{72}\right) w - \left(\frac{41}{24}\right) = 0.$$

There are three solutions

$$w_1 \sim 0.83, \qquad w_2 \sim -1.31, \qquad w_3 \sim -5.34.$$

Now we use the Implicit Function Theorem and standard arguments to obtain three primary bifurcations from $\alpha = 0$ when $\beta = 0$ and $M = 3N$, two of which bifurcate sub-critically and one super-critically. In addition $C_0^M$ bifurcates sub-critically.

(b) $\beta \neq 0$. In addition to the primary bifurcation $C_\beta^M$ observed earlier there is another primary bifurcation point at $\alpha = N\beta$ from which a curve $C_\beta^N$ bifurcates according to the theory of bifurcation from a simple eigenvalue.

To obtain the existence of secondary bifurcation points on $C_\beta^M$ we use the same scaling argument as before after dividing the first equation by $a$. The resulting system is

$$\begin{aligned} \pm 1 &= \frac{\varrho}{N} - \left(\frac{77}{24N^2}\right) x^2 - \left(\frac{41}{8N^2}\right) xy - \left(\frac{23}{36N^2}\right) y^2 + u_\pm(\sigma, \varrho, x, y) \\ \pm y &= \left(\frac{\varrho}{3N}\right) y - \left(\frac{41}{24N^2}\right) x^3 - \left(\frac{23}{36N^2}\right) x^2 y + \left(\frac{29}{360N^2}\right) y^3 + \hat{v}_\pm(\sigma, \varrho, x, y) \end{aligned} \tag{5.6$^\pm$}$$

where $u_\pm$ and $\tilde{v}_\pm$ are defined earlier.

When $\sigma = 0$ equation $(5.6)^-$ has no solutions with $(x, y) \neq (0, 0)$. Hence by arguments used in the preceding cases there are no secondary bifurcations on $C_\beta^M$ when $\beta < 0$. When $\sigma = 0$ equation $(5.6)^+$ has solutions $\varrho = (777/317)\,N^2$, $y^2 = (720/317)\,N^2$, $x = 0$ and $\varrho = N$, $x = 0$, $y = 0$.

Upon application of the Implicit Function Theorem and a rescaling the latter corresponds to a primary bifurcation, namely the curve $C_\beta^N$. However, the former corresponds to two secondary bifurcation points on $C_\beta^M$, $\beta > 0$, by arguments which are by this stage familiar. The two secondary bifurcations are related to each other by symmetry and a calculation reveals that they are transcritical. This completes our account of the known bifurcations when $M = 3N$.

### 5.3 *Bifurcation in the case* $M = 2N$

In this case $T = J = 0$ and the bifurcation equations are

$$\begin{aligned} \beta a &= \frac{\alpha a}{N} - \left(\frac{3}{4N^2}\right) ab + \left(\frac{3}{16N^2}\right) a^3 + \left(\frac{21}{16N^2}\right) ab^2 + u(\alpha, \beta, a, b), \\ \beta b &= \frac{\alpha b}{2N} - \left(\frac{3}{8N^2}\right) a^2 + \left(\frac{3}{16N^2}\right) b^3 + \left(\frac{9}{8N^2}\right) a^2 b + v(\alpha, \beta, a, b). \end{aligned} \tag{5.7}$$

Once again there is the primary bifurcation $C_\beta^M$ for all $\beta$. Now we find the other bifurcations and secondary bifurcations.

*The case* $\beta = 0$. To obtain the qualitative structure of the solution set when $\beta = 0$ we use the Blowing-Up Lemma with $m = 3$, $n = 2$, $\mathsf{m} = (1, 1, 1)$ and $\mathsf{n} = (2, 2)$. The resulting polynomial equation is

$$\begin{aligned} \frac{\alpha a}{N} - \frac{3ab}{4N^2} &= 0, \\ \frac{\alpha b}{2N} - \frac{3a^2}{8N^2} &= 0. \end{aligned} \tag{5.8}$$

There are three bifurcating curves of solutions $\{a = \alpha = 0,\ b \in \mathbb{R}\}$ and $\{b = (4/3)\, N\alpha,\ a^2 = (16/9)\, N^2\alpha^2\}$. The former corresponds to $C_0^M$ and the latter to two other primary bifurcations at $\alpha = 0$ when $M = 2N$. We denote them by $C_0'$ and $C_0''$. A further calculation reveals that $C_0^M$ is a sub-critical pitchfork while $C_0' \cup C_0''$ consists in two transcritical bifurcations at $\alpha = 0$. (We will see from the symmetry that these are better regarded as two degenerate parabolae, one super-critical and one sub-critical, bifurcating at $\alpha = 0$.)

*The case* $\beta \neq 0$. Since $\alpha = N\beta$ is a simple eigenvalue there is a bifurcation point at $\alpha = N\beta$ when $\beta \neq 0$. We call the curve bifurcating from it $C_\beta^N$.

Now we proceed to find all secondary bifurcation points by a scaling argument as before. However, a simple investigation reveals that due to the quadratic terms the way to find all secondary bifurcation points on $C_\beta^M$ in this case is to use the scaling $(\alpha, \beta, a, b) = \sigma(\varrho, \pm 1, x, y)$ after dividing the first equation in (5.7) by $a$. The resulting equation is

$$\begin{aligned} \pm 1 &= \frac{\varrho}{N} - \frac{3y}{4N^2} + \underline{u}_\pm(\sigma, \varrho, x, y), \\ \pm y &= \frac{\varrho y}{2N} - \frac{3x^2}{8N^2} + \underline{v}_\pm(\sigma, \varrho, x, y), \end{aligned} \tag{5.9}$$

where $\underline{u}_\pm$, $\underline{v}_\pm$ are defined in an obvious way.

When $\sigma = 0$ each of $(5.9)^{\pm}$ has two solution curves. $(5.9)^{+}$ has

$$\left\{x^2 = \frac{16N^2}{9}(\varrho - 2N)(\varrho - N),\ y = \frac{4N}{3}(\varrho - N),\ \varrho \geqq 2N \text{ and } \varrho \leqq N\right\} \quad (5.10)^{+}$$

which represents two curves, one of which passes through the point

$$\varrho = N, \quad x = y = 0$$

and the other through the point

$$\varrho = 2N, \quad x = 0, \quad y = \frac{4N^2}{3}.$$

$(5.9)^{-}$ has the solution curves

$$\left\{x^2 = \frac{16N^2}{9}(\varrho + 2N)(\varrho + N),\ y = \frac{4N}{3}(\varrho + N),\ \varrho \geqq -N \text{ and } \varrho \leqq -2N\right\} \quad (5.10)^{-}$$

which pass through

$$\varrho = -N, \quad x = y = 0$$

and

$$\varrho = -2N, \quad x = 0, \quad y = -\frac{4N^2}{3},$$

respectively.

Then the Implicit Function Theorem applied in a now familiar way shows that in each case the curve passing through $x = y = 0$ corresponds to the primary curve $\mathscr{C}_{\beta}^{N}$. Furthermore, it is clear from $(5.10)^{\pm}$ that when $\beta > 0$ this is a sub-critical pitchfork while when $\beta < 0$ it is a super-critical pitchfork. In both cases the other solution curve corresponds to a single secondary bifurcation point on $\mathscr{C}_{\beta}^{M}$ and the formulae $(5.10)^{\pm}$ show that when $\beta > 0$ this is a super-critical pitchfork while when $\beta < 0$ it is a sub-critical pitchfork.

## 6. Diagrams of Solution Sets of the Water-Wave Problem

The thrust of this paper so far has been to give a detailed analysis of solutions of problem $(\mathscr{B})$, rigorously justifying each step. The outcome is a vindication of much of CHEN & SAFFMAN'S work. In this section we content ourselves with clarifying our results by drawing classical bifurcation diagrams for solutions of the capillary-gravity wave problem. To do this we identify two solutions of $(\mathscr{N})$ (or $(\mathscr{B})$) related by symmetry which correspond to the same wave. The diagrams

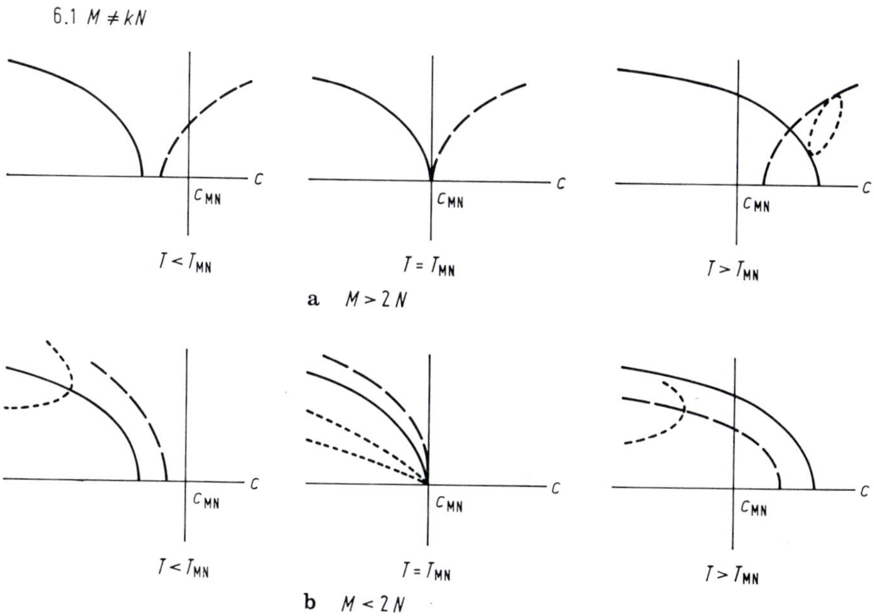


Fig. 1

are of a qualitative nature only, and serve to illustrate the observation that many of the beautiful possibilities in bifurcation theory are present in one problem: the capillary-gravity wave problem.

Throughout we return to the original notation $\gamma = T/g$, $\nu = c^2/g$, and let $\gamma_{MN} = T_{MN}/g$, $\nu_{MN} = c^2_{MN}/g$. The following key may be useful in interpreting the diagrams.

*Key*

——— curve of solutions with corresponding $\theta$ of period $2\pi/M$

– – – curve of solutions with corresponding $\theta$ of period $2\pi/N$

....... curve of solutions with corresponding $\theta$ of period $2\pi$.

The ordinate denotes a measure of the wave amplitude and the abscissa the speed of the flow at infinite depth. There is no suggestion that the diagrams are to scale; a qualitative description of the solution set is all that is being attempted. Once again we emphasize that the main goal of the work has been to detect the presence of bifurcation points when the surface tension is fixed and the speed at infinite depth is regarded as the bifurcation parameter.

*Acknowledgement*. MARK JONES acknowledges the support of an SERC Studentship during the preparation of this paper.

6.2 $M = kN$

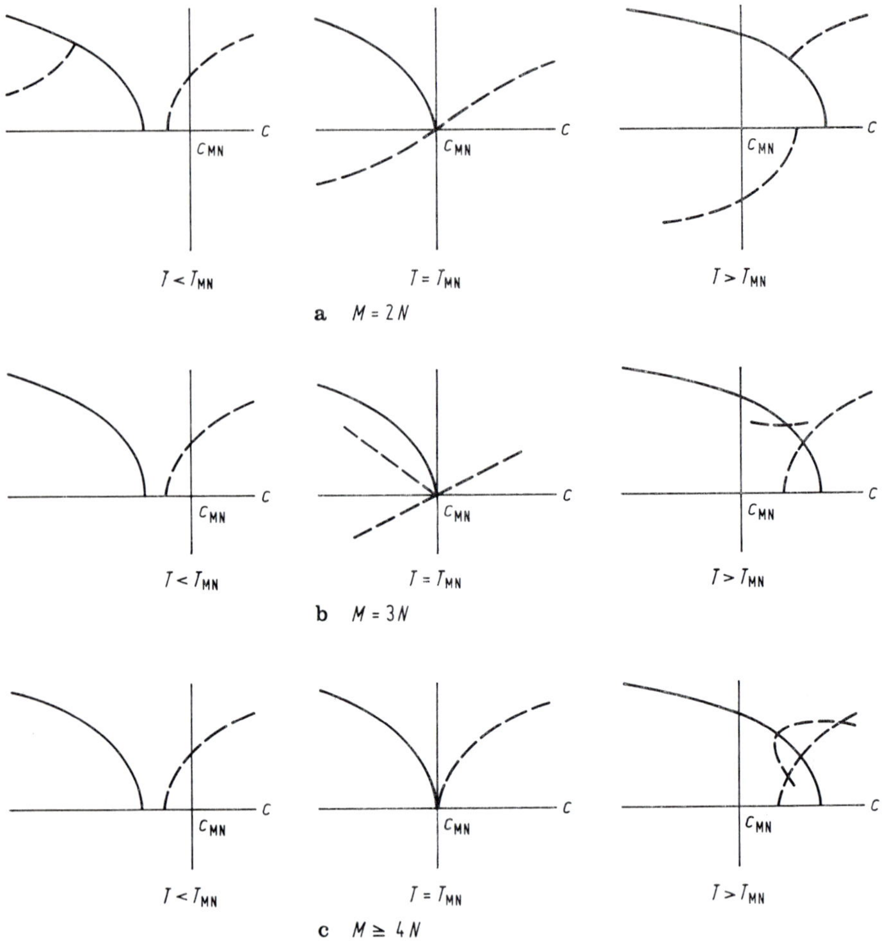


Fig. 2

## References

1. Amick, C. J., & Kirchgässner, K., On solitary water-waves in the presence of surface tension. (To appear.)
2. Beale, J. T., The existence of cnoidal waves with surface tension. *J. Diff. Eqns.*, **31** (1979), 230–263.
3. Buchner, M., Marsden, J., & Schecter, S., Applications of the blowing-up construction and algebraic geometry to bifurcation problems. *J. Diff. Eqns.*, **3** (1983) 404–433.
4. Chen, B., & Saffman, P. G., Steady gravity-capillary waves on deep water – I. weakly nonlinear waves. *Studies Appl. Math.*, **60** (1979), 183–210.
5. Chen, B., & Saffman, P. G., Steady gravity-capillary waves on deep water – II. numerical results for finite amplitude. *Studies Appl. Math.*, **62** (1980), 95–111.

6. CRANDALL, M. G., & RABINOWITZ, P. H., Bifurcation from a simple eigenvalue. *J. Funct. Anal.*, **8** (1971), 321–340.
7. GOLUBITSKY, M., & SCHAEFFER, D., Imperfect bifurcation in the presence of symmetry. *Comm. Math. Phys.*, **67** (1979), 205–232.
8. HOGAN, S. J., Some effects of surface tension on steep water waves. Parts 1, 2 and 3. *J. Fluid Mech.*, **91** (1979), 167–180; **96** (1980), 417–445; **110** (1981), 381–410.
9. HUNTER, J. K., & VANDEN-BROECK, J.-M., Solitary and periodic gravity-capillary waves of finite amplitude. *J. Fluid Mech.*, **134** (1983), 205–219.
10. JONES, M. C. W., The bifurcation and secondary bifurcation of capillary-gravity waves in the presence of symmetry. Ph. D. Thesis, University of Bath, UK, 1986.
11. REEDER, J., & SHINBROT, M., On Wilton ripples, II: rigorous results. *Arch. Rational Mech. Anal.*, **77** (1981), 321–347.
12. SCHAEFFER, D., & GOLUBITSKY, M., Boundary conditions and mode jumping in the buckling of a rectangular plate. *Comm. Math. Phys.*, **69** (1979), 209–236.
13. SEKERZH-ZEN'KOVICH, IA. I., Steady capillary-gravitational waves of finite amplitude at the surface of an infinitely deep fluid. *Tr. Mosk. Gidrofiz. Inst. Akad. Nauk USSR*, **27** (1963).
14. SEKERZH-ZEN'KOVICH, IA. I., On the theory of stationary capillary-gravitational waves of finite amplitude. *Dokl. Akad. Nauk. SSSR*, **180** (2) (1968).
15. SEKERZH-ZEN'KOVICH, IA. I., On a form of steady capillary-gravitational waves of finite amplitude. *PMM*, **34** (6) (1970), 1085–1096.
16. SHEARER, M., Secondary bifurcation for one-parameter families of bifurcation problems. Report No. 97, *Fluid Mechanics Research Institute, University of Essex, UK*, (1978).
17. SHEARER, M., Secondary bifurcation near a double eigenvalue. *SIAM J. Math. Anal.*, **11** (2) (1980), 365–389.
18. TOLAND, J. F., & JONES, M. C. W., The bifurcation and secondary bifurcation of capillary-gravity waves. *Proc. Roy. Soc. Lond., A*, **399** (1985), 391–417.
19. ZYGMUND, A., *Trigonometric Series* I and II., Cambridge University Press (1959).



Department of Pure Mathematics
Queen's University
Belfast BT 7 INN
U. K.

and

The School of Mathematics
University of Bath
BATH
BA 2 7 AY
U. K.


*(Received December 9, 1985)*

# Smoothness of Linear Laminates

MICHEL CHIPOT, DAVID KINDERLEHRER, &
GIORGIO VERGARA CAFFARELLI

*Dedicated to James Serrin on his sixtieth birthday*

## 1. Introduction

Linearly elastic laminates are examples of materials whose equilibrium equations may have only bounded measurable coefficients, yet whose solutions may be fairly smooth. This is quite different from general experience, where regularity of solutions is determined by the closeness of the system to a diagonal one. A particular situation where a laminate may appear is a highly twinned elastic or ferroelectric crystal, and there are questions related to these materials which make it useful to know some properties of these special systems of equations.

Let $\Omega \subset \mathbb{R}^n$ be a bounded domain with smooth boundary $\partial\Omega$ and with $\omega_l$, $l = 1, \ldots, N+1$, satisfying

$$-\infty = \omega_1 < \omega_2 < \ldots < \omega_{N+1} = +\infty$$

let

$$\Omega_l = \{x = (x_1, \ldots, x_n) \in \Omega : \omega_l < x_n < \omega_{l+1}\}. \tag{1.1}$$

Suppose that we are given real constants $a^l_{ijhk}$ $(i, h = 1, \ldots, m; j, k = 1, \ldots, n; l = 1, \ldots, N)$ and set

$$a_{ijhk}(x) = a^l_{ijhk} \quad \text{in } \Omega_l, \ l = 1, \ldots, N. \tag{1.2}$$

Clearly the functions $a_{ijhk}$ are bounded and depend on $x_n$ alone. We assume there exist positive $M$ and $\nu$ independent of $N$ such that

$$|a_{ijhk}(x)| \leqq M \tag{1.3}$$

and

$$a_{ijhk}(x)\, \xi_i \xi_h \eta_j \eta_k \geqq \nu\, |\xi|^2\, |\eta|^2,$$

where $|\xi|$ denotes the Euclidean length of $\xi \in \mathbb{R}^n$[1]. We further suppose that

$$\int_\Omega |\nabla \zeta|^2\, dx \leqq C_0 \int_\Omega a_{ijhk} \zeta^i_{x_j} \zeta^h_{x_k}\, dx, \qquad \zeta \in H^1_0(\Omega), \tag{1.4}$$

---

[1] Summation convention is employed throughout.

where $C_0$ is independent of $N^{(2)}$. Actually, the condition (1.3) is an elementary consequence of (1.4), but this does not concern us here.

For example, (1.4) commonly holds when $n = m$ and the $a^l_{ijl.k}$ are the coefficients of a stiffness matrix of an elastic material, for one imposes the algebraic condition, for each $l$,

$$a^l_{ijhk}\varepsilon_{ij}\varepsilon_{hk} \geqq \text{const.}\, |\varepsilon_{ij}|^2 \text{ for all matrices } \varepsilon = \varepsilon^T. \tag{1.5}$$

It might be worthwhile recalling that (1.4) holds as well when $m = n + 1$ and the $a^l_{ijhk}$ constitute the stiffness matrix of a piezoelectric material.

Let us define the tensor (matrix)

$$\sigma_{ij}(v) = a_{ijhk}v^h_{x_k}, \qquad 1 \leqq i \leqq m, l \leqq j \leqq n, \tag{1.6}$$

and suppose that

$$u \in H^1(\Omega)\colon \int_\Omega \sigma(u) \cdot \nabla\zeta\, dx = \int_\Omega f \cdot \zeta\, dx, \qquad \zeta \in H^1_0(\Omega). \tag{1.7}$$

That is, $u$ is a local solution of the system

$$-\operatorname{div} \sigma(u) = f \quad \text{ in } \Omega. \tag{1.8}$$

Our object is to show that if $f$ is a smooth vector, then $u \in H^{1,\infty}(\Omega)$ locally, namely, $u$ has a locally bounded gradient in $\Omega$, and that the size of $\nabla u$ may be estimated independently of the number $N$ of laminates. Moreover, the energy density will converge to the limit energy density weak * in $L^\infty$, a property not known for a general homogenization process pertaining to systems of equations. This, of course, will permit us to pass to the limit as $N$ tends to infinity, showing that the homogenized or $G$-convergent problem has a solution with a locally bounded gradient. Vice-versa, given $a_{ijhk}$ which are functions of $x_n$ alone, we may find a sequence of problems, each with a finite number of lamina, whose solutions converge to the solution of the original problem. There is an extensive literature about homogenization and its applications; *cf.* [1] or [13].

We shall give some examples to show how the linear response of a highly twinned crystal may be calculated, even in the absence of a good interpretation of its equilibrium displacement. Perhaps a measure-valued solution, or Young measure (*cf.* [15]), may be employed to explain most phenomena of interest. Such a crystal is, in essence, a nonlinear laminate. Here the finite energy density is explicitly calculated and it relationship to the homogenized one which arises from the linear theory is determined.

Finally, let us note that if the laminate is construed as $N$ elastic bodies in contact, our results guarantee that they are in strict contact in the sense of [14]. This means, roughly, that the tangential displacement and the traction do not experience discontinuities across the interfaces.

To describe the idea of the proof, recall that a solution of (1.7), say for $f = 0$, is in $H^{1,p}(\Omega')$, $\Omega'$ compact in $\Omega$, for some $p > 2$. This is a consequence of either Riesz-Thorin interpolation or the reverse Hölder inequality, which hold because

[2] $H^k(\Omega)$ denotes distributions whose derivatives through order $k$ are square summable in $\Omega$. $H^k_0(\Omega)$ is the subspace of $H^k(\Omega)$ whose elements vanish on $\partial\Omega$. We do not distinguish between a space $H$ and its $n$-fold product $(H)^n$.

of (1.4) ([1], [5], [11], [12]). Due to the particular structure of the system we are considering here, the derivatives $\partial u/\partial x_\mu$, $\mu = 1, \dots, n-1$, are also in $H^{1,p}(\Omega')$. It will follow from the variational principle that the functions $\sigma_{in} \in H^{1,p}(\Omega')$, $i = 1, \dots, m$. This and some accounting lead to the conclusion that $p$ above may be taken equal to the Sobolev exponent $2^*$,

$$1/2^* = 1/2 - 1/n, \quad n \geqq 3, \tag{1.9}$$

or any positive number when $n = 2$. In the jargon of homogenization theory, the quantities

$$\partial u^i/\partial x_\mu \quad \text{and} \quad \sigma_{in} \text{ are nonoscillatory.}$$

We introduce the abbreviations

$$\nabla' u = (\partial u^i/\partial x_\mu)_{1 \leqq i \leqq m, 1 \leqq \mu \leqq n-1} \quad \text{and} \quad \sigma_n = (\sigma_{in})_{1 \leqq i \leqq m} \tag{1.10}$$

and state our principal analytical results.

**Theorem 1.** *Assume that* (1.3) *and* (1.4) *hold. Let $u$ satisfy* (1.7) *with $f \in H^k(\Omega)$, $k + 1 \leqq n/2$. Then*

$$u \in H^{1,p}(\Omega'), \quad \Omega' \subset \Omega, \tag{1.11}$$

*for all $p$, $1/p > \frac{1}{2} - (k+1)/n$, and for each $\Omega' \subset\subset \Omega$ there is a constant $C = C(M, \nu, \Omega, \Omega')$ such that*

$$\|u\|_{H^{1,p}(\Omega')} \leqq C(\|u\|_{H^1(\Omega)} + \|f\|_{H^k(\Omega)}). \tag{1.12}$$

Next, we state

**Theorem 2.** *Assume that* (1.3) *and* (1.4) *hold and let $u$ be a solution of* (1.7). *If $f \in H^k(\Omega)$ where the integer $k$ satisfies $k \geqq [\frac{1}{2} n]$, then there is a $p > n$ such that*

$$u \in H^{1,\infty}(\Omega') \quad \textit{and} \quad \nabla' u, \sigma_n \in H^{1,p}(\Omega'), \quad \Omega' \subset\subset \Omega. \tag{1.13}$$

*For each $\Omega' \subset\subset \Omega$, there exists a $C = C(M, \nu, \Omega, \Omega')$ such that*

$$\begin{aligned} \|\nabla' u\|_{H^{1,p}(\Omega')} + \|\sigma_n\|_{H^{1,p}(\Omega')} &\leqq C(\|u\|_{H^1(\Omega)} + \|f\|_{H^k(\Omega)}) \\ \|u\|_{H^{1,\infty}(\Omega')} &\leqq C(\|u\|_{H^1(\Omega)} + \|f\|_{H^k(\Omega)}). \end{aligned} \tag{1.14}$$

The notation $[a]$ stands for the integral part of $a$.

## 2. Proofs of Theorems 1 and 2

Several lemmas will be helpful.

**Lemma 2.1.** *Assume that* (1.3) *and* (1.4) *hold. Let $u$ be a solution of* (1.7) *with $f \in L^2(\Omega)$. Then*

$$\nabla' u \in H^1(\Omega'), \quad \Omega' \subset\subset \Omega,$$

*and for each* $\Omega' \subset\subset \Omega$ *there exists a constant* $C = C(M, \nu, \Omega, \Omega')$ *such that*

$$\|\nabla' u\|_{H^1(\Omega')} \leqq C(\|u\|_{H^1(\Omega)} + \|f\|_{L^2(\Omega)}). \tag{2.1}$$

*Moreover,* $\partial u/\partial x_\mu$ *is a solution of the system* (1.7) *with right-hand side* $\partial f/\partial x_\mu$, *namely,*

$$-\operatorname{div} \sigma(\partial u/\partial x_\mu) = \partial f/\partial x_\mu \quad \text{in } \Omega, \quad \mu = 1, \ldots, n-1. \tag{2.2}$$

**Proof.** The proof is quite standard. Let $\alpha_\delta(x)$, $\delta > 0$, be a family of mollifiers which depend only on $x' = (x_1, \ldots, x_{n-1})$ and set

$$u_\delta(x) = (\alpha_\delta * u)(x) = \int_{\mathbb{R}^{n-1}} \alpha_\delta(x' - y')\, u(y', x_n)\, dy'.$$

Thus $u_\delta$ is a smooth function of $x'$ and converges to $u$ in $H^1(\Omega')$, $\Omega' \subset\subset \Omega$. Since the coefficients $a_{ijhk}$ are independent of $x'$,

$$-\operatorname{div} \sigma(u_\delta) = f_\delta \quad \text{in } \Omega$$

and (2.3)

$$-\operatorname{div} \sigma(\partial u_\delta/\partial x_\mu) = \partial f_\delta/\partial x_\mu \quad \text{in } \Omega.$$

Given $\Omega' \subset\subset \Omega$, choose $\eta \in C_0^\infty(\Omega)$ such that $\eta = 1$ on $\Omega'$ and $0 \leqq \eta \leqq 1$. Temporarily, we suppress the dependence of the various quantities on $\delta$. Multiplying (2.3) by $\zeta = \eta^2\, \partial u/\partial x_\mu$, we obtain

$$-\int_\Omega \operatorname{div} \sigma(u_{x_\mu}) \cdot \eta^2 u_{x_\mu}\, dx = \int_\Omega f_{x_\mu} \cdot \eta^2 u_{x_\mu}\, dx. \tag{2.4}$$

Let LHS denote the left-hand side of (2.4). After integrating by parts[3],

$$\begin{aligned} \text{LHS} &= \int_\Omega \{\eta\sigma(u_{x_\mu}) \cdot \nabla(\eta u_{x_\mu}) + \sigma(u_{x_\mu}) \cdot \eta u_{x_\mu} \otimes \nabla\eta\}\, dx \\ &= \int_\Omega \eta\sigma(u_{x_\mu}) \cdot \{\nabla(\eta u_{x_\mu}) + u_{x_\mu} \otimes \nabla\eta\}\, dx. \end{aligned}$$

Now

$$\sigma(\eta v) = \eta\sigma(v) + a_{ijhk} v^h \eta_{x_k} = \eta\sigma(v) + a(v \otimes \nabla\eta),$$

so

$$\begin{aligned} \text{LHS} = &\int_\Omega \sigma(\eta u_{x_\mu}) \cdot \{\nabla(\eta u_{x_\mu}) + u_{x_\mu} \otimes \nabla\eta\}\, dx \\ &- \int_\Omega a(u_{x_\mu} \otimes \nabla\eta) \cdot \{\nabla(\eta u_{x_\mu}) + u_{x_\mu} \otimes \nabla\eta\}\, dx. \end{aligned}$$

Hence

$$\begin{aligned} \int_\Omega \sigma(\eta u_{x_\mu}) \cdot \nabla(\eta u_{x_\mu})\, dx &= -\int_\Omega \sigma(\eta u_{x_\mu}) \cdot u_{x_\mu} \otimes \nabla\eta\, dx \\ &+ \int_\Omega a(u_{x_\mu} \otimes \nabla\eta) \cdot \{\nabla(\eta u_{x_\mu}) + u_{x_\mu} \otimes \nabla\eta\}\, dx + \int_\Omega f_{x_\mu} \cdot \eta^2 u_{x_\mu}\, dx \\ &\leqq \tau \|\nabla(\eta u_{x_\mu})\|^2_{L^2(\Omega)} + C\tau^{-1}(\|\nabla u\|^2_{L^2(\Omega)} + \|f\|^2_{L^2(\Omega)}), \end{aligned}$$

for some $C > 0$ and each $\tau > 0$. On the other hand, by (1.4)

$$\int_\Omega |\nabla(\eta u_{x_\mu})|^2\, dx \leqq C_0 \int_\Omega \sigma(\eta u_{x_\mu}) \cdot (\nabla\eta u_{x_\mu})\, dx.$$

---

[3] For vectors $a$ and $b$, the matrix $a \otimes b = (a^i b^j)$.

By choosing $\tau$ sufficiently small, we obtain (2.1) for $u_\delta$. Now it is well known that

$$\|\nabla' u_\delta\|_{H^1(\Omega')} \leqq C(\|u_\delta\|_{H^1(\Omega)} + \|f_\delta\|_{L^2(\Omega)}) \leqq C(\|u\|_{H^1(\Omega)} + \|f\|_{L^2(\Omega)}).$$

Thus, passing to the limit, we obtain (2.1) for $u$. We may now pass to the limit as $\delta \to 0$ in (2.3), which gives (2.2). □

Next we have

**Lemma 2.2.** *Assume that* (1.3) *and* (1.4) *hold. Let $u$ be a solution of* (1.7) *with $f \in L^p(\Omega)$ for some $p \geqq 2$ and that*

$$\nabla' u \in H^{1,p}(\Omega'), \qquad \Omega' \subset\subset \Omega, \tag{2.5}$$

*with*

$$\|\nabla' u\|_{H^{1,p}(\Omega)} \leqq C_1(\|u\|_{H^1(\Omega)} + K), \tag{2.6}$$

*where $C_1 = C_1(M, \nu, \Omega', \Omega)$, $K \geqq 0$. Then*

$$u \in H^{2,p}(\Omega' \cap \Omega_1), \qquad \Omega' \subset\subset \Omega, \qquad l = 1, \ldots, N, \tag{2.7}$$

*and there is a $C_2 = C_2(M, \nu, \Omega', \Omega)$ such that*

$$\|\partial^2 u/\partial x_i\, \partial x_j\|_{L^p(\Omega' \cap \Omega_l)} \leqq C_2(\|u\|_{H^1(\Omega)} + \|f\|_{L^p(\Omega)} + K). \tag{2.8}$$

**Proof.** The estimate (2.8) holds for $i = 1, \ldots, n-1$ and $j = 1, \ldots, n$ by (2.6), so it remains only to prove that the vector $\partial^2 u/\partial x_n^2 \in L^p(\Omega' \cap \Omega')$ and that (2.8) holds for $i = j = n$. This is quite easy owing to the invertibility of $(a_{inhn})$.

From (1.7) or (1.8) we know that

$$-\partial/\partial x_j(a_{ijhk} u^h_{x_h}) = f_i, \qquad i = 1, \ldots, m, \qquad \text{in } \Omega$$

in the sense of distributions. Thus, since the $a_{ijhk}$ are constant in $\Omega_l$,

$$a_{inhn} u^h_{x_n x_n} = -f_i - \sum_{(j,k) \neq (n,n)} a_{ijhk} u^h_{x_k x_j} \qquad \text{in } \Omega_l, \tag{2.9}$$

$l = 1, \ldots, N$. Now (2.9) constitutes $m$ equations in the $m$ unknowns $\partial^2 u^h/\partial^2 x_n$. This system may be solved because by (1.3),

$$a_{inhn} \xi_i \xi_h = a_{ijhk} \xi_i \xi_h\, \delta_{jn}\, \delta_{kn} \geqq \nu\, |\xi|^2. \tag{2.10}$$

Moreover, the inverse matrix $(a_{inhn})^{-1}$ is bounded independently of $l = 1, \ldots, N$ The result now follows from (2.6) and (2.9). □

As a final bit of accounting, let us note

**Lemma 2.3.** *Under the same hypotheses as Lemma* 2.2,

$$\sigma_{in}(u) \in H^{1,p}(\Omega'), \qquad \Omega' \subset\subset \Omega, \qquad i = 1, \ldots, m,$$

*and for a constant $C_2 = C_2(M, \nu, \Omega', \Omega)$,*

$$\|\sigma_{in}(u)\|_{H^{1,p}(\Omega')} \leqq C_2(\|u\|_{H^1(\Omega)} + \|f\|_{L^p(\Omega)} + K), \tag{2.11}$$

$i = 1, \ldots, m$.

**Proof.** A consequence of the preceding lemma is that for any $\Omega', \Omega' \subset\subset \Omega$,

$$\sigma_{in}(u) \in H^{1,p}(\Omega' \cap \Omega_l), \quad l = 1, \dots, N, \quad i = 1, \dots, m,$$

so that the traces of $\sigma_{in}(u)$ on the planes $x_n = \omega_l$ and $x_n = \omega_{l+1}$ make sense [9]. We remind the reader of an elementary fact, *e.g.* [8]. Let $D \subset \mathbb{R}^n$ be a domain and

$$D^+ = \{x \in D: x_n < \omega\} \quad \text{and} \quad D^- = \{x \in D: x_n < \omega\}.$$

Suppose that $v^\pm \in H^{1,p}(D^\pm)$ and

$$v^+ = v^- \quad \text{on } x_n = \omega.$$

Then the function

$$v(x) = \begin{cases} v^+(x) & x \in D^+ \\ v^-(x) & x \in D^- \end{cases}$$

is in $H^{1,p}(D)$ and

$$\|v\|_{H^{1,p}(D)} = \|v\|_{H^{1,p}(D^+)} + \|v\|_{H^{1,p}(D^-)}.$$

Thus to prove our lemma, we need only show that the jump $[\![\sigma_n]\!]$ across $x_n = \omega_l$ of $\sigma_n$ vanishes. But this is the natural boundary condition across this hyperplane, since in each $\Omega_l$ the equation holds pointwise. Let $B$ be an open ball in $\Omega'$ with center on $x_n = \omega_l$ and $B \subset \Omega_{l-1} \cup \Omega_l$. For $\zeta \in C_0^\infty(B)$,

$$\int_B \sigma(u) \cdot \nabla\zeta \, dx = \int_B f \cdot \zeta \, dx. \tag{2.12}$$

With $B^+ = B \cap \Omega_l$, $B^- = B \cap \Omega_{l-1}$, and $\Gamma = B \cap \{x_n = \omega_l\}$, we have

$$\int_{B^+} \sigma(u) \cdot \nabla\zeta \, dx - \int_{B^+} f \cdot \zeta \, dx = \int_B (-\operatorname{div} \sigma(u) - f) \cdot \zeta \, dx + \int_\Gamma \sigma_n(u) \cdot \zeta \, dx'$$
$$= \int_\Gamma \sigma_n(u) \cdot \zeta \cdot dx',$$

$dx' = dx_1 \dots dx_{n-1}$, according to the equilibrium equation. Analogously for $B^-$ with the sign reversed in the boundary integral. The conclusion now follows by adding and applying (2.12). □

As an intermediate result, let us state the

**Proposition 2.4.** *Assume that* (1.3) *and* (1.4) *hold. Let u satisfy* (1.7) *with* $f \in L^2(\Omega)$. *Then*

$$u \in H^{1,2^*}(\Omega'), \quad \Omega' \subset\subset \Omega,$$

*and, for* $\Omega' \subset\subset \Omega$, *there is a constant* $C = C(M, \nu, \Omega, \Omega')$ *such that*

$$\|u\|_{H^{1,2^*}(\Omega')} \leq C(\|u\|_{H^1(\Omega)} + \|f\|_{L^2(\Omega)}). \tag{2.13}$$

**Proof.** From Lemma 2.1, we have

$$\partial u/\partial x_\mu \in H^1(\Omega'), \mu = 1, \dots, n-1,$$

with

$$\|\nabla' u\|_{H^1(\Omega')} \leq C(\|u\|_{H^1(\Omega)} + \|f\|_{L^2(\Omega)})$$

for each $\Omega' \subset\subset \Omega$. This allows us to apply Lemma 2.3 with $p = 2$ so

$$\sigma_n(u) \in H^1(\Omega')$$

and

$$\|\sigma_n(u)\|_{H^1(\Omega')} \leqq C(\|u\|_{H^1(\Omega)} + \|f\|_{L^2(\Omega)}).$$

Thus $\nabla' u \in L^{2*}(\Omega')$ and $\sigma_n(u) \in L^{2*}(\Omega')$ and

$$\|\nabla' u\|_{L^{2*}(\Omega')} + \|\sigma_n(u)\|_{L^{2*}(\Omega')} \leqq C(\|u\|_{H^1(\Omega)} + \|f\|_{L^2(\Omega)}). \tag{2.14}$$

Solving the system

$$a_{inhn}\, \partial u^h/\partial x_n = \sigma_{in}(u) - \sum_{k<n} a_{inhk}\, \partial u^h/\partial x_k,$$

*cf.* (2.10), we obtain $\partial u/\partial x_n \in L^{2*}(\Omega')$ with the estimate (2.13). □

**Proof of Theorem 1.** We argue by induction on $k$. For $k = 0$, the result holds owing to the preceding Proposition. Assume it is valid up to order $k - 1 \geqq 0$. For fixed $\mu$, $1 \leqq \mu \leqq n - 1$, $\partial u/\partial x_\mu$ satisfies

$$\int_\Omega \sigma(u_{x_\mu}) \cdot \nabla \zeta \, dx = \int_\Omega f_{x_\mu} \cdot \zeta \, dx, \qquad \zeta \in C_0^\infty(\Omega), \tag{2.15}$$

thanks to (2.2). Thus we may apply the induction assumption to $\partial u/\partial x_\mu$ and $\partial f/\partial x_\mu \in H^{k-l}(\Omega)$, which tells us that $\partial u/\partial x_\mu \in H^{1,p}(\Omega')$, $\Omega' \subset\subset \Omega$, for all $p$ such that $1/p > \frac{1}{2} - k/n$ and provides us with the estimate

$$\|\nabla' u\|_{H^{1,p}(\Omega')} \leqq C(\|u\|_{H^1(\Omega)} + \|f\|_{H^k(\Omega)}).$$

According to the Sobolev imbedding theorem, $H^k(\Omega) \subset L^p(\Omega)$, with

$$\|f\|_{L^p(\Omega)} \leqq \text{const.}\ \|f\|_{H^k(\Omega)}.$$

We now apply Lemma 2.3 to deduce that $\sigma_{in}(u) \in H^{1,p}(\Omega')$ and

$$\begin{aligned}\|\sigma_{in}(u)\|_{H^{1,p}(\Omega')} &\leqq C_3(\|u\|_{H^1(\Omega)} + \|f\|_{H^k(\Omega)} + \|f\|_{L^p(\Omega)}) \\ &\leqq C_3(\|u\|_{H^1(\Omega)} + \|f\|_{H^k(\Omega)}),\end{aligned}$$

$i = 1, \ldots, m$. Now again by the Sobolev imbedding theorem $H^{1,p}(\Omega') \subset L^{p^*}(\Omega')$, $1/p^* = 1/p - 1/n$, so we deduce the analogue of (2.14):

$$\|\nabla' u\|_{L^{p^*}(\Omega')} + \|\sigma_n(u)\|_{L^{p^*}(\Omega')} \leqq C(\|u\|_{H^1(\Omega)} + \|f\|_{H^k(\Omega)}),$$

and thus, solving once again for $\partial u/\partial x_n$,

$$\|\nabla u\|_{L^{p^*}(\Omega')} \leqq C(\|u\|_{H^1(\Omega)} + \|f\|_{H^k(\Omega)}).$$

The result now follows since this is true for all $p$, $1/p > \frac{1}{2} - k/n$, that is, for all $p^*$ such that $1/p^* > \frac{1}{2} - (k + 1)/n$. □

**Proof of Theorem 2.** We shall show that $\partial u/\partial x_\mu \in H^{1,p}(\Omega')$, $\Omega' \subset\subset \Omega$, $1 \leqq \mu \leqq n - 1$, with $p > n$. To start, note that (2.15) holds so Theorem 1 may be applied with

$$f \in H^{k-1}(\Omega), \quad \text{where } k \text{ satisfies } (k-1)+1 \leqq n/2,$$

whence

$$\partial u/\partial x_\mu \in H^{1,p}(\Omega') \text{ for all } p, \quad 1/p > \tfrac{1}{2} - k/n$$

and

$$\|\partial u/\partial x_\mu\|_{H^{1,p}(\Omega')} \leqq C(\|u\|_{H^1(\Omega)} + \|f\|_{H^k(\Omega)}),$$

$\mu = 1, \dots, n-1$. Also $f \in L^p(\Omega)$. By Lemma 2.3, $\sigma_n(u) \in H^{1,p}(\Omega')$ and

$$\|\sigma_n(u)\|_{H^{1,p}(\Omega')} \leqq C(\|u\|_{H^1(\Omega)} + \|f\|_{H^k(\Omega)}).$$

However, $k$ is now so large that

$$1/n > \tfrac{1}{2} - k/n;$$

in fact, we may as well suppose that $k = [n/2]$, and so we may choose $p > n$. Thus $H^{1,p}(\Omega') \subset C(\Omega') \subset L^\infty(\Omega')$. Consequently, solving for $\partial u/\partial x_n$ by (2.10) as usual, we obtain that $\nabla u \in L^\infty(\Omega')$ and

$$\|\nabla u\|_{L^\infty(\Omega')} \leqq C(\|u\|_{H^1(\Omega)} + \|f\|_{H^k(\Omega)}).$$

QED

Note that since the system of equations used to retrieve $\partial u/\partial x_n$ is different in each $\Omega_l$ we do not prove that $\partial u/\partial x_n$ is continuous.

## 3. G-Convergence and related applications

Let us first consider a situation where we are given

$$\sigma_{ij}(v) = a_{ijhk} v^h_{x_k}, \quad 1 \leqq i \leqq m, \quad 1 \leqq j \leqq n \tag{3.1}$$

where the $(a_{ijhk})$ are bounded measurable functions of $x_n$ alone which satisfy

$$\int_\Omega |\nabla \zeta|^2 \, dx \leqq C_0 \int_\Omega a_{ijhk} \zeta^i_{x_j} \zeta^h_{x_k} \, dx, \qquad \zeta \in H^1_0(\Omega), \tag{3.2}$$

for some $C_0 > 0$, much as in § 1. For simplicity, we consider the Dirichlet problem. Let $\phi \in H^1(\Omega)$ and set

$$V = \{v \in H^1(\Omega) : v = \phi \text{ on } \partial\Omega\}.$$

Let $f \in L^2(\Omega)$ and suppose

$$u \in V: \int_\Omega \sigma(u) \cdot \nabla \zeta \, dx = \int_\Omega f \cdot \zeta \, dx, \zeta \in H^1_0(\Omega). \tag{3.3}$$

It follows from (3.2) that such a $u$ exists and is unique. From Theorems 1 and 2 and Proposition 2.4 we have

**Theorem 3.** *Let* $(a_{ijhk})$ *be bounded measurable functions of* $x_n$ *alone and suppose that* (3.1) *and* (3.2) *hold. Let* $u$ *denote the solution of* (3.3).

i. *If* $f \in L^2(\Omega)$, *then* $u \in H^{1,2^*}(\Omega')$, $\Omega' \subset\subset \Omega$, $1/2^* = \tfrac{1}{2} - 1/n$.

ii. *If* $f \in H^k(\Omega)$, $k+1 \leqq n/2$, *then* $u \in H^{1,p}(\Omega')$, $\Omega' \subset\subset \Omega$, *for all p satisfying* $1/p > \frac{1}{2} - (k+1)/n$.

iii. *If* $f \in H^k(\Omega)$ *where* $k \geqq [n/2]$, *then u is has a locally bounded gradient in* $\Omega$.

**Proof.** We proceed in the obvious manner. Set

$$a = \inf\{x_n : x \in \Omega\} \quad \text{and} \quad b = \sup\{x_n : x \in \Omega\}$$

and for any positive integer $N$, subdivide $(a, b)$ into $2^N$ intervals $I_{l,N}(l = 1, \ldots, 2^N)$ of length $(b-a)\cdot 2^{-N}$. Define $(a_{ijhk}^{lN})$ by

$$a_{ijhk}^{lN} = (b-a)^{-1} \cdot 2^N \int_{I_{l,N}} a_{ijhk}(x_n)\, dx_n \quad \text{for} \quad x \in \Omega \text{ such that } x_n \in I_{l,N}.$$

Finally, define $(a_{ijhk}^N)$ according to (1.2). Let $u^N$ be the solution of (3.3) where $(a_{ijhk})$ is replaced by $(a_{ijhk}^N)$. Now it is routine to show that

$$\|u^N\|_{H^1(\Omega)} \leqq C\{\|f\|_{L^2(\Omega)} + \|\phi\|_{H^1(\Omega)}\}$$

for a constant $C$ which does not depend on $N$, and so there is a subsequence of the $u^N$ which converges weakly to some $u^*$ in $H^1(\Omega)$. What is important in the present situation is that the bounded functions

$$(a_{ijhk}^N) \to (a_{ijhk}) \quad \text{in } L^1(\Omega) \quad \text{as } N \to \infty,$$

from which it follows that $u^* = u$, the unique solution of (3.3).

The desired conclusion holds because it holds for each $u^N$ with bounds independent of $N$. □

To describe a typical homogenization or $G$-convergence process, suppose that for each $N$ we are given

$$-\infty = \omega_l^N < \omega_2^N < \ldots < \omega_{N+1}^N = +\infty$$

and we determine domains $\Omega_l^N$ by (1.1). Suppose we are given real constants $(a_{ijhk}^{lN})$, $1 \leqq i, h \leqq m$, $1 \leqq j, k \leqq n$, $1 \leqq l \leqq N$, and that $(a_{ijhk}^N)$ are defined by (1.2). We assume that (1.3) and (1.4) hold for each $N$. For simplicity we confine ourselves to a Dirichlet problem as before, so give $f \in L^2(\Omega)$ and $\phi \in H^1(\Omega)$, let

$$u^N \in V\colon \int_\Omega \sigma^N(u^N)\cdot \nabla\zeta\, dx = \int_\Omega f\cdot\zeta\, dx, \quad \zeta \in H_0^1(\Omega). \tag{3.4}$$

Once again, it is routine to check that

$$\|u^N\|_{H^1(\Omega)} + \|\sigma^N(u^N)\|_{L^2(\Omega)} \leqq C\{\|f\|_{L^2(\Omega)} + \|\phi\|_{H^1(\Omega)}\}, \tag{3.5}$$

with $C$ independent of $N$. In particular, the weak closure of the $\{u^N\}$ in $H^1(\Omega)$ is not empty. The idea of $G$-convergence is to identify and characterize the limit equation. Assuming that $f \in H^k(\Omega)$, $k = n/2$, we may apply Theorem 2 to conclude that any limit point $u$ of the $\{u^N\}$ has the property $u \in H^1(\Omega) \cap H^{1,\infty}(\Omega')$, $\Omega' \subset\subset \Omega$.

It might be helpful to review just how the limit equation is obtained. In the "periodic case", when

$$(a^N_{ijhk}(x_n)) = (\alpha_{ijhk}(Nx_n)),$$

for a given collection of periodic functions $(\alpha_{ijhk}(t))$, this is among the situations that have been worked out by MCCONNELL. The limit equation is unique and has constant coefficients. To organize the more general case, we write

$$A^N = (a^N_{inhn}) \quad \text{an } m\times m \text{ matrix}$$

$$B^N = (a^N_{inh\mu}) \quad \text{a second order tensor}.$$

For any $v = (v^1, \ldots, v^m)$, we may express

$$\sigma^N_n(v) = B^N \nabla' v + A^N v_{x_n}, \tag{3.6}$$

so that

$$v_{x_n} = (A^N)^{-1} \sigma^N_n(v) - (A^N)^{-1} B^N \nabla' v. \tag{3.7}$$

In particular,

$$u^N_{x_n} = (A^N)^{-1} \sigma^N_n(u^N) - (A^N)^{-1} B^N \nabla' u^N, \tag{3.8}$$

and also for some tensors $\Gamma^N_\mu$, $\Delta^N_\mu$, $C^N_\mu$, and $D^N_\mu$,

$$\begin{aligned}\sigma^N_\mu(v) &= (\sigma^N_{1\mu}(v), \ldots, \sigma^N_{m\mu}(v))\\ &= \Gamma^N_\mu \nabla' v + \Delta^N_\mu v_{x_n}\\ &= \Gamma^N_\mu \nabla' v + \Delta^N_\mu((A^N)^{-1} \sigma^N_n(v) - (A^N)^{-1} B^N \nabla' v)\\ &= C^N_\mu \nabla' v + D^N_\mu \sigma^N_n(v).\end{aligned} \tag{3.9}$$

All of the matrices and tensors

$$(A^N)^{-1}, (A^N)^{-1} B^N, C^N_\mu, D^N_\mu \tag{3.10}$$

are uniformly bounded. Consequently, we now have the option of choosing a $G$-limit equation to be defined by the weak * limits of items in (3.10) and associating to it a $u$ in the weak closure of the $\{u^N\}$, or vice versa. Since it does not matter, let us assume that we have selected tensors $A$, $B$, $C_\mu$, and $D_\mu$ such that

$$(A^N)^{-1} \to A^{-1}, \quad (A^N)^{-1} B^N \to A^{-1}B, \quad C^N_\mu \to C_\mu, \quad \text{and } D^N_\mu \to D_\mu$$

in $L^\infty(\Omega)$ weak * as well as $u \in H^{1,\infty}(\Omega') \cap H^1(\Omega)$, $\sigma_n \in H^{1,p}(\Omega') \cap L^2(\Omega)$ with

$$\nabla' u^N \to \nabla' u \text{ and } \sigma^N_n(u^N) \to \sigma_n \quad \text{uniformly in } \Omega'$$

and

$$\partial u^N/\partial x_n \to \partial u/\partial x_n \quad \text{in } L^\infty(\Omega') \text{ weak } *.$$

This identifies the limit equation inasmuch as we may take

$$\begin{aligned}\sigma_n(v) &= Av_{x_n} + B\nabla' v\\ \sigma_\mu(v) &= C_\mu \nabla' v + D_\mu \sigma_n(v) = (C_\mu + D_\mu B)\nabla' v + D_\mu A v_{x_n},\end{aligned} \tag{3.11}$$

$\mu = 1, \ldots, n-1$. One easily verifies that with this definition,

$$\sigma_n^N(u^N) \to \sigma_n(u) \quad \text{uniformly in } \Omega'$$

and

$$\sigma_\mu^N(u^N) \to \sigma_\mu(u) \quad \text{in } L^\infty(\Omega') \text{ weak } *, \quad \mu = 1, \ldots, n-1.$$

It follows that $u$ is a solution of (3.3). As a consequence of a theorem of TARTAR, or the discussion below, the limit equation is elliptic. We arrive at

**Theorem 4.** *Assume that* (1.3) *and* (1.4) *hold for each N. Let* $f \in H^k(\Omega)$ *for some* $k \geqq [\frac{1}{2} n]$. *Let* $u^N$ *denote the solution of* (3.4) *and let u be any weak limit of the* $\{u^N\}$. *Then there is a* $p > n$ *such that*

$$u \in H^{1,\infty}(\Omega') \quad \text{and} \quad \nabla' u, \sigma_n \in H^{1,p}(\Omega'), \quad \Omega' \subset\subset \Omega. \tag{3.12}$$

*For each* $\Omega' \subset\subset \Omega$, *there exists a* $C = C(M, \nu, \Omega, \Omega')$ *such that*

$$\|\nabla' u\|_{H^{1,p}(\Omega)} + \|\sigma_n\|_{H^{1,p}(\Omega)} \leqq C(\|u\|_{H^l(\Omega)} + \|f\|_{H^k(\Omega)}) \tag{3.13}$$
$$\|u\|_{H^{1,\infty}(\Omega)} \leqq C(\|u\|_{H^1(\Omega)} + \|f\|_{H^k(\Omega)}).$$

*Also, there is a subsequence of the* $u^N$ *and a tensor* $\sigma$ *such that*

$$-\operatorname{div} \sigma(u) = f \quad \text{in } \Omega$$
$$u = \phi \quad \text{on } \partial\Omega, \tag{3.14}$$

*and for any (scalar valued)* $\zeta \in L^1(\Omega)$,

$$\int_{\Omega'} \sigma(u) \cdot \nabla u\, \zeta\, dx = \lim_{N \to \infty} \int_{\Omega'} \sigma^N(u^N) \cdot \nabla u^N \zeta\, dx. \tag{3.15}$$

To check the assertion (3.15) given $\zeta \in L^1(\Omega)$, write

$$\int_{\Omega'} \sigma^N(u^N) \cdot \nabla u^N \zeta\, dx = \int_{\Omega'} \{\sigma_n^N(u^N) \cdot u_{x_n}^N + \sigma_\mu^N(u^N) \cdot u_{x_\mu}^N\} \zeta\, dx,$$

so that each term is the product of a weakly convergent factor with a uniformly convergent one.

## 4. Elastic crystals

The small displacements from a natural state of a highly twinned elastic crystal may be discussed within the present framework. Let $\mathfrak{M}$ denote the group of unimodular matrices with integer coefficients. The invariance of the energy density $W(F)$ of an elastic crystal under change of lattice basis leads to the properties

$$W(QF) = W(F), \quad Q^TQ = \mathbf{1}, \quad \det Q = 1,$$

and

$$W(FH) = W(F), \quad H \in \mathfrak{M}_L, \quad \det F > 0, \tag{4.1}$$

where $\mathfrak{M}_L = L\mathfrak{M}L^{-1}$, for a fixed choice of lattice basis $l_1, l_2, l_3$ appearing as the columns of $L$ (*cf.* ERICKSEN [2], [3]). Thus, for example, if $F = \mathbf{1}$ is a local minimum for $W$ (a natural state), so is $F = H$ for any $H \in \mathfrak{M}_L$. Now if $F = \mathbf{1}$ in part of $\Omega$ and $F = H$ in its complement, different phases may coexist in the same body. Suppose that $y(x)$ is a deformation of $\Omega$ with $F = \nabla y$ and

$$F = H_l = \mathbf{1} + a_l \otimes n_l \quad \text{in } \Omega_l, \quad l = 1, \ldots, N,$$
$$a_l \cdot n_l = 0, \quad H_l \in \mathfrak{M}_L. \tag{4.2}$$

In such a situation, the body is said to be "twinned". The energy density

$$W(F) = W(\mathbf{1}) \quad \text{in } \Omega$$

and the Piola-Kirchhoff stress $S(F) = \partial W / \partial F$ satisfies

$$S(F) = 0 \quad \text{in } \Omega.$$

On the other hand, for general $F$, by (4.1),

$$S(FH)\, H^T = S(F), \quad H \in \mathfrak{M}_L, \tag{4.3}$$

so the stiffness tensors at $FH$ and $F$ are related by

$$S'(FH)\,[\xi] = S'(F)\,[\xi H^{-1}]\, H^{-T} \tag{4.4}$$

for any matrix $\xi$. Thus the linearized equation is different in each $\Omega_l$. Viewing this assembly as a linear material, we are confronted with a laminate. Can this be of any assistence in studying the nonlinear properties?

As a brief example, when the $a_l$ in (4.2) are all parallel to a unit vector $a$, say, and $n_l = n$ for all $l$. We write

$$F^N(x) = \mathbf{1} + f_N(x_n)\, a \otimes n, \quad x \in \Omega,$$

where $f_N$ is constant on each $\Omega_l$ and assumed bounded independently of $N$. Now there is an $f \in L^\infty(\Omega)$ such that

$$f_N \to f \quad \text{in } L^\infty(\Omega) \text{ weak * as } N \to \infty,$$

but although $W(F^N) = W(\mathbf{1})$ for each $N$, in general, for the limit $F$ of the $F^N$, the function $f$ does not attain integer values, so

$$W(F) = W(\mathbf{1} + f(x)\, a \otimes n) > W(\mathbf{1}).$$

Nonetheless, it is reasonable to take the limit energy density to be

$$W(\mathbf{1}) = \lim_{N \to \infty} W(F^N).$$

We shall use the results of the peceding section to illustrate that the assertion above has a certain stability. For this let us start by stating the problem

$$-\operatorname{div} S(F^N + t\, \nabla u^N) = tf \quad \text{in } \Omega$$
$$u^N = \phi \quad \text{on } \partial\Omega,$$

where $f$ and $\phi$ are smooth functions. As this is too difficult, we replace it by the linearized system

$$-\operatorname{div} S'(F^N)\,[\nabla u^N] = f \quad \text{in } \Omega,$$
$$u^N = \phi \quad \text{on } \partial\Omega. \tag{4.5}$$

According to our Theorem 3, under appropriate hypotheses about the tensor $S'(\mathbf{1})$, $u^N$ exists and

$$\|\nabla u^N\|_{L^\infty(\Omega')} \leqq C(\Omega', \Omega) \tag{4.6}$$

independently of $N$. Thus the sequence of functions

$$w^N(x, t) = W(F^N(x) + t\,\nabla u^N(x)), \qquad x \in \Omega', \qquad |t| \text{ small},$$

is uniformly bounded and (a subsequence) has a weak * limit $w(x, t)$. Now

$$\begin{aligned} w^N(x, t) &= W(F^N) + tS(F^N) \cdot \nabla u^N + \tfrac{1}{2} t^2 S'(F^N)\,[\nabla u^N] \cdot \nabla u^N + O(t^3), \\ &= \tfrac{1}{2} t^2 S'(F^N)\,[\nabla u^N] \cdot \nabla u^N + O(t^3), \qquad \text{in } \Omega' \subset\subset \Omega. \end{aligned} \tag{4.7}$$

That the remainder stays bounded owes, of course, to (4.6). Consequently for every $\Omega' \subset\subset \Omega$,

$$\int_{\Omega'} w^N(x, t)\, dx \to \tfrac{1}{2} t^2 \int_{\Omega'} \sigma(\nabla u) \cdot \nabla u\, dx + O(t^3),$$

where $\sigma$ is the homogenized stiffness matrix (appropriate to the subsequence), or

$$\int_{\Omega'} w(x, t)\, dx = \tfrac{1}{2} t^2 \int_{\Omega'} \sigma(\nabla u) \cdot \nabla u\, dx + O(t^3), \qquad \text{for } \Omega' \subset\subset \Omega. \tag{4.8}$$

We have partially extended to the finite theory a fact well known for the infinitesimal one: if each $y^N + tu^N$ has small energy density, then the limit configuration also has small energy density. Actually more is true. The sequence $F^N$ defines a Young measure which permits the evaluation of any continuous state function of the body in terms of some distribution applied to $f$ and $\phi$. One might consider this Young measure as defining the equilibrium configuration, much as is the practice in the study of hyperbolic conservation laws. In general, the boundary value problems associated with crystals are not easily confronted by means of established methods of the calculus of variatons; *cf.* ERICKSEN [2], [3], FONSECA [4], or [7].

We should like to carry our thoughts one step further in a way which does not require specific knowledge of Young measures on the part of the reader by considering an example. Given $\Omega$ as before and $\theta$, $0 < B < 1$, fix $\omega_2$,

$$-\infty = \omega_1 < \omega_2 < \omega_3 = +\infty,$$

such that

$$|\Omega_1| = (1 - \theta)\,|\Omega| \quad \text{and} \quad |\Omega_2| = \theta\,|\Omega|.$$

Let $f$ denote the characteristic function of $\Omega_2$ and extend it periodically in the $x_3$ direction. Set $f^N(x) = f(x', Nx_3)$ and let

$$F^N(x) = \mathbf{1} + f^N(x)\, a \otimes n \tag{4.9}$$

where $n = e_3$ and $a \cdot n = a \cdot e_3 = 0$ with $\mathbf{1} + a \otimes n \in \mathfrak{M}_L$. It is elementary to check that for each $N$ there is a piecewise homogeneous deformation $y^N(x)$ with $F^N = \nabla y^N$.

Now $W(F^N) = 0$ for each $N$ and since $f^N$ is periodic in $x_3$, [1],

$$f^N \to \theta \quad \text{in } L^\infty(\Omega) \text{ weak } *, \tag{4.10}$$

$$F^N \to F^\infty = \mathbf{1} + \theta a \otimes n \quad \text{in } L^\infty(\Omega) \text{ weak } *. \tag{4.11}$$

As we have remarked, it is unlikely that $W(F^\infty) = \lim W(F^N)$. Indeed, for any continuous function $g(F)$,

$$g(F^N(x)) \to \bar{g}(x) = (1 - \theta)\, g(\mathbf{1}) + \theta g(\mathbf{1} + a \otimes n) \text{ in } L^\infty(\Omega) \text{ weak } * \tag{4.12}$$

by the reasoning that leads to (4.10). This means that the Young measure ([15]) corresponding to the sequence $(F^N)$ is given by

$$\nu = (1 - \theta)\, \delta_{\mathbf{1}} + \theta\, \delta_{\mathbf{1}+a\otimes n}, \tag{4.13}$$

where $\delta_F$ is the Dirac measure concentrated at $F$. In particular,

$$\overline{W}(x) = (1 - \theta)\, W(\mathbf{1}) + \theta W(\mathbf{1} + a \otimes n) = 0,$$

$$\overline{S}(x) = (1 - \theta)\, S(\mathbf{1}) + \theta S(\mathbf{1} + a \otimes n) = 0,$$

$$\overline{S'}(x)\, [\xi] = (1 - \theta)\, S'(\mathbf{1})\, [\xi] + \theta S'(\mathbf{1} + a \otimes n)\, [\xi],$$

for any matrix $\xi$, which may be calculated using (4.4). Or, for example, if $v(x)$ is a fixed displacement with the property that $y^N(x) + v(x)$ is admissible, *i.e.*, $\det F^N + \nabla v) > 0$ for all $N$, then

$$\int_{\Omega'} W(F^N + \nabla v)\, dx \to \int_{\Omega'} \{(1 - \theta)\, W(\mathbf{1} + \nabla v) + \theta W(\mathbf{1} + a \otimes n + \nabla v)\}\, dx$$

for any $\Omega' \subset\subset \Omega$.

In addition, for the solution $u^N$ of (4.5), we may write

$$\partial u^N / \partial x_n = A(F^N)^{-1}\, \sigma_n^N - A(F^N)^{-1}\, B(F^N)\, \nabla' u^N. \tag{4.14}$$

adopting a notation based on (3.8), where $A(F)$ and $B(F)$ are smooth. Rewriting this we easily find that there is a smooth tensor-valued function $Z(F)$ such that

$$\nabla u^N = Z(F^N)\, [\xi^N] \quad \text{with } \xi^N = (\nabla' u^N, \sigma_n^N),$$

so that

$$F^N + t\, \nabla u^N = F^N + Z(F^N)\, [\xi^N].$$

According to our Theorem 4, $\xi^N \to \xi^\infty$ uniformly on compact subsets of $\Omega$, so for a continuous $g$,

$$\begin{aligned} g(F^N + t\, \nabla u^N) &= g(F^N + Z(F^N)\, [\xi^N]) \\ &= g(F^N + Z(F^N)\, [\xi^\infty]) + \varepsilon_N, \end{aligned}$$

where $\varepsilon_N \to 0$ as $N \to \infty$ on compact subsets of $\Omega$. From this it follows easily that the weak * limit $\hat{g}(x, t)$ of the sequence $(g(F^N + t\,\nabla u^N))$ is given by the formula

$$\hat{g}(x, t) = (1 - \theta)\, g(\mathbf{1} + tZ(\mathbf{1})\,[\xi^\infty]) + \theta g(\mathbf{1} + a \otimes n + tZ(\mathbf{1})\,[\xi^\infty]). \quad (4.15)$$

A particular case of this is the function $w(x, t)$ defined in (4.8); it is given explicitly by (4.15) with $g$ replaced by the energy density $W$. Thus,

$$w(x, t) = (1 - \theta)\, W(\mathbf{1} + tZ(\mathbf{1})\,[\xi^\infty]) + \theta W(\mathbf{1} + a \otimes n + tZ(\mathbf{1})\,[\xi^\infty]), \quad (4.16)$$

and it may be developed to second order in $t$ according to (4.8).

*Acknowledgements.* Our interest in this topic was stimulated by the workshop Homogenization and Effective Moduli of Materials and Media, October 22–26, 1984 at the I.M.A. We refer to the forthcoming proceedings of this conference for additional information. The authors thank J. Ericksen, F. Murat, & L. Tartar for their interest in this work. This research was done while the authors were in residence at the I.M.A. in Minneapolis. It was partially funded by the NSF and the CNR.

## References

1. A. Bensoussan, J.-L. Lions, & G. Papanicolaou, *Asymptotic analysis for periodic structures*, North Holland, (1978).
2. J. L. Ericksen, Ill posed problems in thermoelasticity theory, *Systems of nonlinear partial differential equations*, J. Ball (ed), Reidel, (1983), 71–93.
3. J. L. Ericksen, Twinning in crystals, I.M.A. preprint **95**, (1984).
4. I. Fonseca, *Variational methods for elastic crystals*, Thesis, Univ. of Minnesota, (1985).
5. M. Giaquinta, *Multiple integrals in the calculus of variations and nonlinear elliptic systems*, Princeton, (1983).
6. D. Gilbarg & N. Trudinger, *Elliptic partial differential equations of second order*, Springer-Verlag, (1984).
7. D. Kinderlehrer, *Twinning in crystals* II, I.M.A. preprint **106**, (1984).
8. D. Kinderlehrer & G. Stampacchia, *An introduction to variational inequalities and their applications*, Academic Press, (1980).
9. J.-L. Lions, *Problèmes aux limites dans les équations aux dérivées partielles*, Presse Université de Montréal, (1962).
10. W. H. McConnell, On the approximation of elliptic operators with discontinuous coefficients, *Ann. Scuola Norm. Sup. Pisa*, (1976), 121–137.
11. N. G. Meyers, An $L^p$ estimate for the gradient of solutions of second order elliptic divergence equations, *Ann. Scuola Norm. Sup. Pisa*, **17** (1963), 189–206.
12. N. G. Meyers & A. Elcrat, Some results on regularity for solutions of nonlinear elliptic systems and quasi-regular functions, *Duke Math. J.*, **42** (1975), 121–136.
13. F. Murat, Compacité par compensation, *Ann. Scuola Norm. Sup. Pisa*, (1978), 489–507.
14. P. Podio-Guidugli, G. Vergara Caffarelli, & E. Virga, The role of ellipticity and normality assumptions in formulating live boundary conditions in elasticity, to appear *Quart. Appl. Math.*

15. L. TARTAR, Compensated compactness and applications to partial differential equations, Nonlinear analysis and mechanics: Heriot-Watt symposium, R. J. KNOPS (ed), Research notes in math. **39**, Pitman, (1979), 136–212.

Département de Mathématiques
Ile de Saulcy
Metz

&

School of Mathematics
University of Minnesota
Minneapolis

&

Dipartimento di Matematica
Università di Pisa

*(Received December 15, 1985)*

# *On the Behavior of the Derivatives of Minimizers near Singular Points*

ENRICO GIUSTI

*Dedicated to J. Serrin for his sixtieth birthday*

## 0. Introduction

In spite of its relevance to the calculus of variations, the regularity of minimizers of variational integrals as such (*i.e.* as opposed to stationary points) has been studied only recently in [2]. Since then, several papers dedicated to the study of that problem have appeared.

In the scalar case, *i.e.* when $u$ is a real-valued function on some open set $\Omega \subset \mathbb{R}^n$, that minimizes an integral of the type

$$\mathscr{F}(u; \Omega) = \int_\Omega f(x, u, Du)\, dx,$$

one can prove Hölder regularity under quite general assumptions. In contrast, when $u$ is a vector-valued minimizer ($u : \Omega \to \mathbb{R}^N$, $N > 1$) then singularities may occur, and efforts to estimate the size of the singular set have been made.

In this paper I wish to consider a slightly different problem, always within the context of regularity: what happens to the derivatives $D^k u(x)$ of a minimizer when $x$ approaches a singular point?

To provide a solution of the problem in a very special case, we shall consider quadratic functionals with splitting coefficients, namely functionals of the form

$$\mathscr{F}(u; \Omega) = \int_\Omega \gamma^{\alpha\beta}(x)\, g_{ij}(x, u)\, D_\alpha u^i\, D_\beta u^j\, dx$$

with the usual ellipticity conditions on $\gamma$ and $g$. Moreover, we shall limit attention to the case of three independent variables ($n = 3$), in which it is known that the singularities of bounded minimizers are isolated. For this special situation we prove that if $x_0$ is a singular point for $u$ (necessarily isolated), then

$$\limsup_{x \to x_0} |x - x_0|^k\, |D^k u(x)| < +\infty$$

for every integer $k$. Examples show that this estimate is the best possible.

Functionals of the type considered here are of interest in the theory of harmonic maps of Riemannian manifolds.

A similar result is proved for minimal hypersurfaces in $\mathbb{R}^8$.

## 1. Preliminaries

The aim of this section is to derive estimates for minimizers of quadratic functionals. Most of the methods and the results are not new; for instance we may refer to the book of GIAQUINTA [1]. On the other hand we need to examine how the various estimates depend on the coefficients, an investigation which was not required in [2] or elsewhere. Since to indicate the necessary changes in the statements and proofs of the relevant theorems in [2] would be long and tedious, I prefer to give the complete proof of the main result in a form suitable for the present purpose.

We consider a quadratic functional of the form:

$$\mathscr{F}(u;\Omega)=\int_{\Omega} A_{ij}^{\alpha\beta}(x,u)\,D_\alpha u^i\,D_\beta u^j\,dx, \tag{1.1}$$

where the coefficients $A_{ij}^{\alpha\beta}$ satisfy the inequalities

$$|A(x,u)|\leqq L, \tag{1.2}$$

$$A_{ij}^{\alpha\beta}(x,u)\,\xi_\alpha^i\xi_\beta^j\geqq|\xi|^2\qquad\forall\,\xi\in\mathbb{R}^{nN}. \tag{1.3}$$

Moreover we shall suppose that the coefficients are continuous in $\bar\Omega\times\mathbb{R}^N$. This implies that for every $M$ there exists a continuous, increasing, non-negative, bounded, concave function $\omega_M(t)$ on $t\geqq 0$ such that $\omega_M(0)=0$ and for every $x, x_0\in\bar\Omega$ and $u, u_0\in\mathbb{R}^N$, $|u_0|\leqq M$, we have

$$|A(x,u)-A(x_0,u_0)|\leqq\omega_M(|x-x_0|^2+|u-u_0|^2). \tag{1.4}$$

It is easy to see that we may suppose that $\omega_M$ increases with $M$.

The following lemma is well known. A proof can be found for instance in [2].

**Lemma 1.1** (CACCIOPPOLI inequality). *Let* $u\in H^{1,2}_{\text{loc}}(\Omega;\mathbb{R}^N)$ *be a minimizer of the functional* (1.1). *Then for each pair of concentric balls* $B_\varrho\subset B_R\subset\Omega$

$$\int_{B_\varrho}|Du|^2\,dx\leqq\frac{c_1(L)}{(R-\varrho)^2}\int_{B_R}|u-u_R|^2\,dx. \tag{1.5}$$

**Lemma 1.2** ($L^p$-estimates). *Under the same assumptions, the minimizer u belongs to* $H^{1,p}_{\text{loc}}(\Omega;\mathbb{R}^N)$ *for some* $p>2$, *and for the concentric balls* $B_R\subset B_{2R}\subset\Omega$ *we have*

$$\left(\fint_{B_R}|Du|^p\,dx\right)^{\frac1p}\leqq c_2(L)\left(\fint_{B_{2R}}|Du|^2\,dx\right)^{\frac12}. \tag{1.6}$$

Here we have indicated by $\fint_A$ the average on $A$:

$$\fint_A=\frac{1}{\operatorname{meas}A}\int_A.$$

Moreover we have set $u_R=\fint_{B_R}u\,dx$, and we have denoted by $c_1(L), c_2(L)\ldots$ constants depending only on $L$ in (1.2) and possibly on $n$, $N$, but not otherwise on the coefficients or on the minimizer $u$. A proof of Lemma 1.2 may be found in [1], Chapter V.

We introduce now the quantity

$$U(x_0, R) = \fint_{B(x_0,R)} |u - u_{x_0,R}|^2 \, dx. \tag{1.7}$$

**Lemma 1.3.** *Let $u$ be a minimizer of* (1.1). *Then for concentric balls $B_\varrho \subset B_R \subset \Omega$, centered at $x_0$, we have*

$$\begin{aligned} &\int_{B(x_0,\varrho)} |u - u_{x_0,\varrho}|^2 \, dx \\ &\leqq c_3(L) \left\{ \left( \frac{\varrho}{R} \right)^{n+2} + \omega_M^{1-2/p}[q(L)\,(R^2 + U(x_0, R))] \right\} \int_{B(x_0,R)} |u - u_{x_0,R}|^2 \, dx \end{aligned} \tag{1.8}$$

*where $M \geqq |u_{x_0,R}|$.*

**Proof.** Let $v \in H^{1,2}(B_R, \mathbb{R}^N)$ be the minimizer of the frozen functional

$$\mathscr{F}^0(v; B_R) = \int_{B_R} A_{ij}^{\alpha\beta}(x_0, u_R)\, D_\alpha v^i \, D_\beta v^j \, dx \qquad (u_R = u_{x_0,R})$$

with $v = u$ on $\partial B_R$. The function $v$ is a solution of an elliptic system with constant coefficients, and hence we have the estimates

$$\int_{B_R} |Dv|^m \, dx \leqq c_4(L) \int_{B_R} |Du|^m \, dx \tag{1.9}$$

provided $u \in H^{1,m}(B_R, \mathbb{R}^N)$. In particular (1.9) holds for $m = 2$ and $m = p$. Moreover, for every $\varrho < R$

$$\int_{B_\varrho} |v - v_\varrho|^2 \, dx \leqq c_5(L) \left( \frac{\varrho}{R} \right)^{n+2} \int_{B_R} |v - v_R|^2 \, dx \tag{1.10}$$

(see [1], page 78). Coming back to $u$, we note that

$$\begin{aligned} \int_{B_\varrho} |u - u_\varrho|^2 \, dx &\leqq c_6(L) \left\{ \left( \frac{\varrho}{R} \right)^{n+2} \int_{B_R} |u - u_R|^2 \, dx + \int_{B_R} |u - v|^2 \, dx \right\} \\ &\leqq c_7(L) \left\{ \left( \frac{\varrho}{R} \right)^{n+2} \int_{B_R} |u - u_R|^2 \, dx + R^2 \int_{B_R} |D(u - v)|^2 \, dx \right\}. \end{aligned} \tag{1.11}$$

The last integral can be estimated as in [1] page 208; for simplicity we write $w = u - v$:

$$\begin{aligned} &\int_{B_R} |Dw|^2 \, dx \\ &\leqq \mathscr{F}^0(u, B_R) - \mathscr{F}^0(v, B_R) \\ &\leqq \int_{B_R} \{ |A(x, u) - A(x_0, u_R)| \, |Du|^2 + |A(x, v) - A(x_0, u_R)| \, |Dv|^2 \} \, dx \\ &\leqq \int_{B_R} \omega_M(R^2 + |u - u_R|^2) \, |Du|^2 \, dx + \int_{B_R} \omega_M(R^2 + |v - u_R|^2) \, |Dv|^2 \, dx. \end{aligned}$$

We now estimate the last integral; the oher can be treated similarly. Using (1.6) and the concavity of $\omega_M$, we conclude that

$$\fint_{B_R} \omega_M |Dv|^2\, dx \leqq \left(\fint_{B_R} \omega_M^{p/(p-2)}\, dx\right)^{1-2/p} \left(\fint_{B_R} |Dv|^p\, dx\right)^{2/p}$$

$$\leqq (2L)^{2/p} \left(\fint_{B_R} \omega_M\, dx\right)^{1-2/p} \left(\fint_{B_R} |Du|^p\, dx\right)^{2/p}$$

$$\leqq c_8(L) \left\{\omega_M\left(\fint_{B_R} (R^2 + |v - u_R|^2)\, dx\right)\right\}^{1-2/p} \fint_{B_{2R}} |Du|^2\, dx.$$

On the other hand

$$\int_{B_R} |v - u_R|^2\, dx \leqq 2 \int_{B_R} (|w|^2 + |u - u_R|^2)\, dx \leqq cR^2 \int_{B_R} (|Dw|^2 + |Du|^2)\, dx$$

$$\leqq cR^2 \int_{B_R} |Du|^2\, dx \leqq c \int_{B_R} |u - u_{2R}|^2\, dx$$

and therefore

$$\int_{B_R} \omega_M |Dv|^2\, dx \leqq cR^{-2} \omega_M^{1-2/p}(q\{R^2 + U(x_0, 2R)\}) \int_{B_{4R}} |u - u_{4R}|^2\, dx.$$

A similar estimate holds for the first integral. Inserting these estimates in (1.11) and writing $R$ instead of $4R$, we obtain the required inequality (1.8).

At this point we choose $0 < \tau < 1$ and $\varrho = \tau R$, where $\tau$ is such that

$$c_3(L)\, \tau^{2-2\delta} \leqq \tfrac{1}{2}.$$

With this choice of $\tau$, let $\varepsilon_0 > 0$ be such that

$$2c_3(L)\, \omega^{1-2/p}(q(L)\, \varepsilon) \leqq \tau^{n+2\delta}$$

for any $\varepsilon \leqq \varepsilon_0$. Suppose now that $|u_{x_0,R}| \leqq M/2$ and

$$\gamma(R) = R^2 + U(x_0, R) \leqq \varepsilon_1 = \min\left(\varepsilon_0, \tau^n (1 - \tau^\delta)^2 \frac{M^2}{4}\right). \tag{1.12}$$

We shall prove by induction that

$$|u_{x_0,\tau^k R}| \leqq |u_{x_0,R}| + \tau^{-n/2} U^{1/2}(x_0, R) \sum_{j=0}^{k-1} \tau^{i\delta}, \tag{1.13}$$

$$U(x_0, \tau^k R) \leqq \tau^{2k\delta}\, U(x_0, R). \tag{1.14}$$

Certainly (1.13) and (1.14) hold when $k = 0$. Suppose that they hold for some $k$. Then

$$|u_{x_0,\tau^k R}| \leqq \frac{M}{2} + \frac{\tau^{-n/2} \varepsilon_1^{1/2}}{1 - \tau^\delta} \leqq M$$

and

$$U(x_0, \tau^k R) \leqq \varepsilon_1 \leqq \varepsilon_0,$$

so that we deduce from (1.8)

$$U(x_0, \tau^{k+1} R) \leqq \tau^{2\delta} U(x_0, \tau^k R),$$

namely (1.14) for $k+1$. On the other hand

$$|u_{x_0,\tau R} - u_{x_0,R}| = \left| \fint_{B_\tau R} (u - u_{x_0,R})\, dx \right| \leqq U^{1/2}(x_0, R)\, \tau^{-n/2}$$

and hence (1.13) holds for $k+1$.

In conclusion, if for some $x \in \Omega$ and some $R < \operatorname{dist}(x, \partial\Omega)$ we have

$$|u_{x,R}| \leqq M/2, \qquad R^2 + U(x, R) \leqq \varepsilon_1, \tag{1.15}$$

then for every $k$

$$U(x, \tau^k R) \leqq \tau^{2k\delta} U(x, R)$$

and hence for every $\varrho < R$

$$U(x, \varrho) \leqq c \left(\frac{\varrho}{R}\right)^{2\delta} U(x, R). \tag{1.16}$$

Suppose now that

$$|u_{x_0,R}| \leqq M/4$$

$$R^2 + U(x_0, R) \leqq \varepsilon_1/2 \quad .$$

Then there is a ball $B_0$ centered at $x_0$ such that (1.15) is satisfied for every $x \in B_0$. We wish to estimate the radius of $B_0$. We have

$$u_{x,R} = \fint_B u(x + Ry)\, dy$$

and therefore

$$|Du_{x,R}| = \left| \fint_B Du(x + Ry)\, dy \right| \leqq \fint_{B(x,R)} |Du(z)|\, dz,$$

and so by use of (1.5) we conclude that

$$|Du_{x,R}| \leqq cR^{-1} U^{\frac{1}{2}}(x, 2R) \leqq c_1 R^{-1} U^{\frac{1}{2}}(x_0, 3R)$$

if $|x - x_0| \leqq R$. In a similar way

$$DU^{\frac{1}{2}}(x, R) \leqq \left( \fint_{B(x,R)} |Du|^2\, dz \right)^{\frac{1}{2}} \leqq cR^{-1} U^{\frac{1}{2}}(x_0, 3R),$$

and so the radius $r_0$ of $B_0$ can be estimated in terms of $\varepsilon_1$, $M$, $R$ and $U(x_0, 3R)$. From (2.16) we deduce

$$U(x, \varrho) \leqq c \left(\frac{\varrho}{R}\right)^{2\delta} U(x, R)$$

for every $x \in B_0$ and $\varrho < R$. This implies (see [1], page 72) that $u \in C^{0,\delta}(B_0, \mathbb{R}^N)$ and that the norm of $u$ in $C^{0,\delta}(B_0, \mathbb{R}^N)$ can be estimated in terms of $R$ and $U(x_0, 3R)$

and therefore in terms of $R$ and the $L^2$-norm of $u$

$$\|u\|^2_{C^{0,\delta}(B_0;\mathbb{R}^N)} \leqq cR^{-n-2\delta}\,\|u\|^2_{L^2(\Omega;\mathbb{R}^N)}.$$

We have thus proved the following result:

**Theorem 1.1.** *Let* $u\in H^{1,2}_{\mathrm{loc}}(\Omega;\mathbb{R}^N)$ *be a minimizer of the functional* $\mathcal{F}(u,\Omega)$, *whose coefficients* $A^{\alpha\beta}_{ij}$ *satisfy* (1.2), (1.3) *and* (1.4). *For every* $M>0$ *there is a constant* $\varepsilon_1$ *depending only on* $M, L$, *and on the modulus of continuity* $\omega$ *of the coefficients such that if for some* $x_0\in\Omega$ *and some* $R<\frac{1}{4}\operatorname{dist}(x_0,\partial\Omega)$ *we have*

$$|u_{x_0,R}|<M/4,\qquad R^2+U(x_0,R)<\varepsilon_1/2, \tag{1.17}$$

*then* $u\in C^{0,\delta}(B(x_0,r_0),\mathbb{R}^N)$ *and*

$$\|u\|_{C^{0,\delta}(B_0)} \leqq cR^{-n-2\delta}\,\|u\|_{L^2(\Omega)}. \tag{1.18}$$

*The radius* $r_0$ *of* $B_0$ *depends only on* $\varepsilon_1$, $M, L, R$ *and on the* $L^2$*-norm of* $u$, *but not otherwise on the coefficients and on* $u$.

In particular, $u$ will be Hölder-continuous in the vicinity of every point $x_0$ such that

$$\sup_{R>0}|u_{x_0,R}|<+\infty \qquad\text{and}\qquad \limsup_{R\to 0} U(x_0,R)=0.$$

It is possible to show that the above relations hold except possibly for $x_0$ in a set $\Sigma$ of Hausdorff dimension less than $n-2$ (see [1], Chapter IV, 2). This implies that $u$ is Hölder-continuous (with any exponent $\delta<1$) in an open set $\Omega_0\subset\Omega$, and

$$\dim(\Omega-\Omega_0)<n-2.$$

If we suppose that the coefficients are regular, we can derive estimates for the derivatives of $u$ in $\Omega_0$.

**Theorem 1.2.** *Let the coefficients* $A^{\alpha\beta}_{ij}(x,u)$ *are of class* $C^{k,\alpha}$ *in* $\Omega\times\mathbb{R}^N$ $(R\geqq 1)$ *and let* $u$ *be a minimizer of* $\mathcal{F}$. *Assume that* $u$ *is Hölder-continuous in* $B(x_0,r_0)$. *Then* $u$ *is in* $C^{k+1,\alpha}(B_0,\mathbb{R}^N)$, *and the norm of* $u$ *in* $B(x_0,\varrho)$ $(\varrho<r_0)$ *can be estimated in terms of* $L$, $r_0-\varrho$, $\|u\|_{C^{0,\alpha}}$ *and the corresponding norms of the coefficients.*

**Proof.** Suppose first $k=1$. Then theorem 1.5 of [1] (page 186) proves that $u$ is in $C^{1,\alpha}$, with the corresponding estimate. Now we write the Euler equation

$$D_\alpha(A^{\alpha\beta}_{sj}(x,u)\,D_\beta u^j)=\frac{\partial A^{\alpha\beta}_{ij}(x,u)}{\partial u^s}\,D_\alpha u^i\,D_\beta u^j. \tag{1.19}$$

This is an elliptic system with $C^{1,\alpha}$ coefficients and Hölder-continuous right-hand side, and therefore $u\in C^{2,\alpha}$ with the appropriate estimates.

If $k>1$, the result follows from the standard elliptic estimates applied to the system (1.19).

We remark that all the estimates depend on the coefficients and $u$ only through the stated quantities.

In particular, if a sequence of coefficients $A_{ij}^{\alpha\beta(\nu)}$ converges uniformly on compact subsets of $\bar{\Omega}\times\mathbb{R}^N$, while a sequence of minimizers $u^{(\nu)}$ of the corresponding functionals converges weakly in $L^2$, then the constant $\varepsilon_1$ and the radius $r_0$ appearing in Theorem 1.1 can be chosen independently of $\nu$; as a consequence the estimate (1.18) is uniform.

For such sequences we have the following result.

**Theorem 1.3.** *Let $A_{ij}^{\alpha\beta(\nu)}(x, u)$ be a sequence of coefficients satisfying* (1.2), (1.3) *and* (1.4) *uniformly with respect to $\nu$, and converging uniformly on compact subsets of $\bar{\Omega}\times\mathbb{R}^N$ to $A_{ij}^{\alpha\beta}(x, u)$. For each $\nu$, let $u^{(\nu)}$ be a minimizer of the functional*

$$\mathscr{F}^\nu(w, \Omega) = \int_\Omega A^\nu(x, w)\, Dw\, Dw\, dx,$$

*and suppose that $u^{(\nu)} \rightharpoonup u$ weakly in $L^2_{\mathrm{loc}}(\Omega, R^N)$. Then $u$ is a minimizer of the limit functional*

$$\mathscr{F}(w, \Omega) = \int_\Omega A(x, w)\, Dw\, Dw\, dx.$$

*Moreover, if $x_\nu$ is a singular point for $u^{(\nu)}$, and $x_\nu \to x_0$ $\Omega$, then $x_0$ is a singular point for $u$. In addition, if $u$ is Hölder-continuous in an open set $\Omega_0 \subset \Omega$, and $K$ is a compact set in $\Omega_0$, then for $\nu$ large enough the functions $u^{(\nu)}$ are Hölder-continuous in $K$, and $u^{(\nu)} \to u$ uniformly in $K$. Finally, if the coefficients $A_{ij}^{\alpha\beta(\nu)}$ are equibounded in $C^{k,\alpha}(\bar{\Omega}\times\mathbb{R}^N)$, then $u^{(\nu)} \to u$ in $C^{k+1}(K, \mathbb{R}^N)$.*

**Proof.** For the first part of the theorem see [1], page 259. The last statement depends on the estimate (1.18). Actually if $x_0 \in K$, $u$ is regular in a neighborhood of $x_0$, and therefore satisfies (1.17). By Lemma 1.1 and Rellich's theorem, $u^{(\nu)} \to u$ strongly in $L^2_{\mathrm{loc}}(\Omega; \mathbb{R}^N)$, so that for $\nu$ large enough $u^{(\nu)}$ satisfies (1.17). This means that $u^{(\nu)}$ is regular in $B(x_0, r_0)$, with $r_0$ independent of $\nu$, and satisfies (1.18). The conclusion follows by covering $K$ by a finite number of such balls $B(x_0, r_0)$ and then using Theorem 1.1 and 1.2.

Although it is not relevant to our present purposes, I remark that the above theorem shows the dimension of the singularities to be upper semi-continuous with respect to uniform convergence of the coefficients.

## 2. The behavior of derivatives near a singular point

We shall now derive some consequences from the regularity theorems of Section 1. In particular, we shall be interested in functionals $\mathscr{F}$ whose coefficients are of the special form

$$A_{ij}^{\alpha\beta}(x, u) = \gamma^{\alpha\beta}(x)\, g_{ij}(x, u) \tag{2.1}$$

with $\gamma^{\alpha\beta} = \gamma^{\beta\alpha}$. As is known, functionals of this type occour in the theory of harmonic maps of riemannian manifolds.

A special feature of such functionals is that in the dimension 3 bounded minimizers of the functional $\mathscr{F}$ can have only isolated singularities; more generally, the dimension of the singular set is not greater than $n-3$ (see [3]). Here we prove

**Theorem 2.1.** *Let* $n=3$ *and let* $u$ *be a bounded minimizer for the functional*

$$\mathscr{F}(u;\Omega)=\int_{\Omega}\gamma^{\alpha\beta}(x)\,g_{ij}(x,u)\,D_\alpha u^i\,D_\beta u^j\,dx.$$

*Suppose that the coefficients* $A=\gamma\otimes g$ *are of class* $C^{k,\alpha}(\Omega\times\mathbb{R}^N)$ *and satisfy* (1.2) *and* (1.3). *Let* $x_0$ *be a (necessarily isolated) singular point for* $u$. *Then for any* $j\leqq k+1$

$$\limsup_{x\to x_0}|x-x_0|^j\,|D^j u(x)|<+\infty. \tag{2.2}$$

**Proof.** We can suppose $x_0=0$. If (2.2) does not hold, we can find a sequence $x_\nu\to 0$ such for some $j\leqq k+1$

$$|x_\nu|^j\,|D^j u(x_\nu)|\to+\infty. \tag{2.3}$$

We set $R_\nu=2\,|x_\nu|$ and

$$v_\nu(y)=u(R_\nu y).$$

The function $v_\nu$ is a minimizer of the functional

$$\mathscr{F}_\nu(x,B)=\int_B A^\nu(y,w)\,Dw\,Dw\,dy$$

where

$$A^\nu(x,w)=A(R_\nu y,w).$$

Moreover, setting $y_\nu=x_\nu/R_\nu$, we have $\lim_{\nu\to\infty}|D^j v_\nu(y_\nu)|=+\infty$. Passing to a subsequence and using the boundedness of $u$, we can assume that $v_\nu\to v$ weakly in $L^2(B;\mathbb{R}^N)$, and $y_\nu\to y_0$ with $|y_0|=\frac{1}{2}$. From Theorem 1.2 we deduce that $v$ is a minimizer of the limit functional, with coefficients $A(0,w)=\gamma(0)\otimes g(0,w)$. Moreover, the function $v$ is homogeneous of degree zero.

Since we are in dimension 3, the function $v$ cannot have singular points other than the origin, since otherwise it would exhibit a line of singularities, contradicting the regularity theorems of [3]. In particular, $v$ must be Hölder-continuous near $y_0$ and therefore by Theorem 1.2 $v$ is of class $C^{k+1,\alpha}$ in a neighborhood $V$ of $y_0$.

By Theorem 1.3 we conclude that $v_\nu\to v$ in $C^{k+1}(V;\mathbb{R}^N)$, a result in contrast with the assumption that $|D^j v_\nu(y_\nu)|$ diverges.

I remark that the assumption $n=3$ was used only to deduce that minimizers of the functional can only have isolated singularities. For special functionals, this property may hold in higher dimension. This is the case, for instance, of the energy of maps from the unit disc $D^n$ into the unit sphere $S^N$, $N\geqq n$. In stereographic coordinates, the energy of such maps is given by

$$\mathscr{E}(u,D)=\int_D\frac{|Du|^2}{(1+|u|^2)^2}\,dx. \tag{2.4}$$

It is known ([4], [7]) that a minimizer $u$ of $\mathscr{E}$, satisfying $|u| \leqq 1$ (that means that $D$ is mapped in a closed hemisphere) is regular in dimension $n \leqq 6$ and has at most isolated singularities in dimension 7.

To such minimizers we may apply the argument of Theorem 2.1, concluding that if $u$ is singular at 0, then for every integer $k$ we have

$$\limsup_{x \to 0} |x|^k \, |D^k u(x)| < +\infty .$$

This is exactly the behavior of the equatorial map: $u = x/|x|$, which is a minimizer for $n \geqq 7$ ([6]).

I conclude this note by the remark that a similar result holds for area minimizing hypersurfaces (*i.e.* surfaces of codimension 1) in $\mathbb{R}^8$. These surfaces can have only isolated singularities (see [5], Theorem 11.7). Moreover, if a sequence $E_j$ of minimal sets converges to a minimal set $E$, and the points $x_j \in \partial E_j$ converge to a regular point $x \in \partial E$, then (for $j$ large) $x_j$ are regular and moreover the normal vectors $\nu^{E_j}(x_j)$ converge to $\nu^E(x)$ ([5], Theorem 9.4).

This means that in a neighborhood of $x$, $\partial E$ and $\partial E_j$ can be represented as graphs of functions $u$, $u_j$, defined in some open set $U \subset \mathbb{R}^7$, and satisfying the equation of minimal surfaces there.

By the standard elliptic theory (theorem 1.3 above) we deduce that $D^m u_j \to D^m u$ uniformly in $U$, for every $m$; or equivalently that $\delta^m_{E_j} \nu^{E_j}(x_j) \to \delta^m_E \nu^E(x)$. (For the definition of the operator $\delta$ see [5], 10.3).

Now suppose that $F$ is a minimal set in $\mathbb{R}^8$ with an isolated singularity at the origin, and such that for some $m$

$$\lim_{j \to \infty} |x_j|^m \, |\delta^m_F \nu^F(x_j)| = +\infty$$

for some sequence $x_j \to 0$. Setting $R_j = 2\,|x_j|$ and

$$F_j = \{x \in \mathbb{R}^8 : R_j x \in F\},$$

we obtain a sequence of sets $F_j$ and of points $y_j$ with $|y_j| = \frac{1}{2}$ such that

$$\lim_{j \to \infty} |\delta^m_{F_j} \nu^{F_j}(y_j)| = +\infty . \tag{2.5}$$

Passing to a subsequence, we can suppose that $y_j \to y_0$ and that $F_j$ converges to a minimal cone $C$ ([5], Theorem 9.3) which is regular in $\mathbb{R}^8 - \{0\}$. In particular $C$ is regular near $y_0$, and hence $\delta^m_{F_j} \nu^{F_j}(y_j) \to \delta^m_C \nu^C(y_0)$, contradicting (2.5). We have thus proved

**Theorem 2.2.** *Let $E$ be a minimal set in $\mathbb{R}^8$ and let $x_0$ be a singular point (necessarily isolated) of $\partial E$. Then for every integer $m$:*

$$\limsup_{x \to x_0} |x - x_0|^m \, |\delta^m_E \nu^E(x)| < +\infty .$$

## References


1. GIAQUINTA, M. Multiple integrals in the calculus of variations and nonlinear elliptic systems. Annals of Math. Studies **105**. Princeton Univ. Press 1983.
2. GIAQUINTA, M., & E. GIUSTI, On the regularity of the minima of variational integrals. Acta Math. **148** (1982) 31–46.

3. GIAQUINTA, M., & E. GIUSTI, The singular set of the minima of certain quadratic functionals. Ann. Sc. Norm. Sup. Pisa **11** (1984) 45–55.
4. GIAQUINTA, M., & J. SOUCEK, Harmonic maps into a hemisphere. To appear in Ann. Sc. Norm. Sup. Pisa.
5. GIUSTI, E., Minimal surfaces and functions of bounded variation. Birkhauser Boston. 1984.
6. JAGER, W., & J. KAUL, Rotationally symmetric harmonic maps from a ball into a sphere and the regularity problem for weak solutions of elliptic systems. J. Reine Angew. Math. **343** (1983) 146–161.
7. SCHOEN, R., & K. UHLENBECK, Regularity of minimizing harmonic maps into the sphere. Invent. Math. **78** (1984) 89–100.



Istituto Matematico U. Dini
Università di Firenze


*(Received December 2, 1985)*

# Ground States and Dirichlet Problems for $-\Delta u = f(u)$ in $R^2$

F. V. ATKINSON & L. A. PELETIER

*Dedicated to James Serrin on his sixtieth birthday*

## 1. Introduction

In an earlier paper [1] we considered the existence of solutions of the problem

$$\text{(I)} \qquad \begin{aligned} -\Delta u = f(u), \quad u > 0, \quad \text{in } R^N \\ u(x) \to 0 \quad \text{as } |x| \to \infty \end{aligned}$$

in which $f(u)$ is to be positive for large $u$, but not for all $u > 0$. Such solutions are sometimes called "ground states", a term borrowed from the physical context (nonlinear field equations) in which Problem I arises. Following in part the approach of [5], we used a "shooting method" (in place of variational arguments) to prove under suitable conditions the existence of such a ground state. The principal difficulty lay in showing that if $u(0)$ were chosen sufficiently large, then the associated radially symmetric solution had a zero, *i.e.* that the Dirichlet problem on some finite ball had a solution.

The solubility of Dirichlet problems for

$$-\Delta u = f(u),$$

where $f(u)$ is positive for large $u$, or perhaps for all $u > 0$ depends very critically on the rate of growth of $f(u)$ as $u \to \infty$. Here the cases $N \geqq 3$ and $N = 2$ are strikingly different. In the case $N \geqq 3$ one finds the condition (see for instance [1, 4, 5, 9, 10, 11])

$$f(u) = O(u^p) \quad \text{as } u \to \infty, \quad 1 < p < \frac{N+2}{N-2}, \tag{1.1}$$

and for $N = 2$ conditions such as (see [10, 12])

$$\log f(u) = o(u^2) \quad \text{as } u \to \infty. \tag{1.2}$$

There has been much interest recently in borderline cases for $N \geqq 3$, involving the "critical exponent" $(N+2)/(N-2)$. In particular the case

$$f(u) = u^q + u^{(N+2)/(N-2)}, \quad 1 \leqq q < (N+2)/(N-2) \tag{1.3}$$

has been studied in [6] and in [2], where the approach through ordinary differential equations used in [1] was shown to be effective for this case.

In the present paper we use the same approach to fill in a gap in our earlier treatment [1] of the case $N = 2$, and to deal with some borderline cases corresponding to (1.2). In [1] we did not cover the whole range (1.2); writing

$$F(u) = \int_0^u f(s)\, ds, \tag{1.4}$$

we asked that

$$\liminf_{u \to \infty} \frac{1}{u} \log F(u) = 0. \tag{1.5}$$

This condition allows less rapid growth than (1.2) when the growth is regular, though the "lim inf" and integral features do permit quite erratic growth. Here we confine attention to the case of regular growth, and we shall ask in particular that $\log f(u)$ be *convex* for large $u$.

Coming to specifics, we shall make the following general hypotheses about $f$.

*H*1. $f(u)$ is locally Lipschitz continuous on $[0, \infty)$;

*H*2. $f(0) = 0$ and there exists a number $\zeta > 0$ such that

$$F(u) < 0 \quad \text{for } 0 < u < \zeta \text{ and } F(\zeta) = 0 \tag{1.6}$$

$$f(u) > 0 \quad \text{for } u \geqq \zeta. \tag{1.7}$$

*H*3. There exists a number $y_0 \geqq 0$ such that

(i)
$$f(u) > 0 \quad \text{for } u \geqq y_0, \tag{1.8}$$

and the function

$$g(u) \overset{\text{def}}{=} \log f(u) \qquad u \geqq y_0 \tag{1.9}$$

satisfies

(ii)
$$g \in C^2([y_0, \infty));$$

(iii)
$$g'(u) > 0, \quad g''(u) \geqq 0 \quad \text{for } u \geqq y_0. \tag{1.10}$$

We shall always assume that $f$ satisfies H1 and H3. When considering the existence of ground states we shall assume H2 as well. In that case $y_0$ will have to be a positive number.

We formulate our main result for *ground states* first in its most general form in which the growth condition on $f$ is rèplaced by an inequality involving the function

$$h(u) = \{g(u) - \tfrac{1}{2} u g'(u)\} - \tfrac{1}{2} y_0 g'(u)\, [e^{\{g(u) - g(y_0)\}/2} - 1]^{-1} \tag{1.11}$$

and the lower bound of $f$

$$\inf\{f(u) \colon u > 0\} = -M. \tag{1.12}$$

**Theorem 1.** *Suppose $f$ satisfies H1, H2 and H3. If there exists a $\gamma > \max\{y_0, \zeta\}$ such that*

$$h(\gamma) > \log M + 1, \tag{1.13}$$

*then Problem I, with $N = 2$, has a solution $u$ in $R^2$ with $u \leqq \gamma$.*

The behaviour of $f(u)$ or $g(u)$ for $u > \gamma$ is, of course, irrelevant. All that is needed is that there should be one $\gamma$ satisfying (1.13).

Less precise, but simpler criteria can be obtained by ensuring that (1.13) is asymptotically satisfied, and in particular, by making $h(u)$ take suitably large values for large $u$. Thus we have the following corollary to Theorem 1.

**Theorem 2.** *Suppose H1, H2 and H3 are satisfied. If*

$$\limsup_{u\to\infty} \{g(u) - \tfrac{1}{2} u g'(u)\} > \log M + 1, \tag{1.14}$$

*then Problem I has a solution in $R^2$.*

To indicate the scope of these results we conclude with some examples If

$$g(u) = u^q + u^r, \qquad 0 < r < q$$

then (1.14) is satisfied if $q < 2$ but not if $q > 2$, in conformity with (1.2).

The borderline case, in which $g(u)$ behaves like $u^2$ as $u \to \infty$ can also be investigated. Thus

$$g(u) = u^2 + \log u$$

satisfies (1.14), and so does

$$g(u) = u^2 + b$$

provided $b > \log M + 1$.

An example of a function $f$, which satisfies all the requirements is

$$f(u) = (u^2 - u)\, e^{u^2}. \tag{1.15}$$

Another example, in a similar vein, is given by

$$f(u) = (u^2 - au)\, e^{u^q},$$

where $q > 1$ and $a > 0$. Here $\log M$ will become large and negative as $a \to 0$ so that for any given $\gamma > 0$, (1.13) will be satisfied if $a$ is chosen small enough.

The proofs of these results are based on a study of the asymptotically constant solutions of a class of highly nonlinear, generalized Emden-Fowler equations

$$y'' + e^{g(y)-t} = 0 \qquad t \in R$$

$$y(t) \to \infty \qquad \text{as } t \to \infty.$$

For $g$ increasing and convex and $\gamma$ large, the graphs of such solutions turn out to have a very characteristic shape (see Fig. 1).

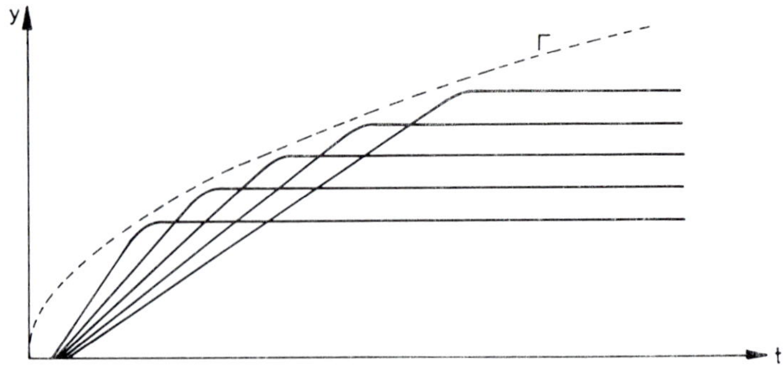


Fig. 1. Solution graphs

This property is of some intrinsic interest. The final section is therefore devoted to obtaining some first quantitative results about it.

## 2. Preliminary Transformation and Shooting Argument

We are concerned with proving the existence of spherically symmetric solution of Problem I when $N = 2$, that is with solutions $u(r)$ of the equation

$$\frac{d^2u}{dr^2} + \frac{1}{r}\frac{du}{dr} + f(u) = 0 \quad \text{on } (0, \infty). \tag{2.1}$$

Here we transform the singularity at $r = 0$ to one at $t = +\infty$ by taking

$$t = -2 \log\left(\tfrac{1}{2} r\right), \quad y(t) = u(r). \tag{2.2}$$

We thus replace Problem I by Problem II, namely

$$\frac{d^2y}{dt^2} + e^{-t} f(y) = 0 \quad t \in R \tag{2.3}$$

(II)
$$y(t) > 0 \quad t \in R \tag{2.4}$$

$$\sup\{y(t): t \in R\} < \infty \tag{2.5}$$

$$y(t) \to 0 \quad \text{as } t \to -\infty. \tag{2.6}$$

This does not, of course, change the nature of the problem, but brings it into line with the clasiscal theory of the generalized Emden-Fowler equation [7, 8, 14] and the arguments of [1, 2].

In particular, it follows from (2.3)–(2.5) that as $t \to \infty$, $y'(t)$ tends to a limit, which can only be zero and, using (2.3) again, that $y(t)$ also tends to a limit which must be finite. We shall generally write

$$\lim_{t\to\infty} y(t) = \gamma. \tag{2.7}$$

On the other hand, for any $\gamma \in (0, \infty)$ and for sufficiently large $t$ there is a solution $y(t) = y(t, \gamma)$ of (2.3) with the property (2.7); this solution will be unique, in view of the Lipschitz condition on $f$.

The subsequent argument centres on the behavior of $y(t, \gamma)$ as $t$ decreases from $+\infty$. If $y(t, \gamma)$ reaches the value 0 for some $t$ we denote this value of $t$ by $T = T(\gamma)$:

$$T(\gamma) = \inf \{t \in R: y(\cdot, \gamma) > 0 \quad \text{on } (t, \infty)\}. \tag{2.8}$$

We then have a solution of a certain Dirichlet problem which we term Problem III, namely

$$\text{(III)} \qquad \begin{aligned} &\frac{d^2 y}{dt^2} + e^{-t} f(y) = 0 \qquad t \geqq T(\gamma) \\ &y(t, \gamma) > 0, \qquad t > T(\gamma) \\ &y(T(\gamma), \gamma) = 0. \end{aligned}$$

It may of course happen that $T(\gamma)$ does not exist, *i.e.*, that $y(t, \gamma) > 0$ for all $t \in R$ and $T(\gamma) = -\infty$. We denote by $S$ the set $\gamma \in R^+$ for which $T(\gamma)$ does exist.

From [1] (see also [5]) we need

**Lemma 0.** *Suppose H1 and H2 are satisfied. If S is nonempty, Problem II has a solution $y(t)$ with*

$$\lim_{t \to \infty} y(t) \leqq \inf S.$$

We thus have to show that Problem III has a solution for some $\gamma > 0$ and $T(\gamma) > -\infty$, to ensure the existence of a ground state.

## 3. Estimates for the Dirichlet Problem

Our approach to Problem III is based on viewing $\gamma = y(\infty)$ as a parameter. We first ascertain the existence of $T(\gamma)$, the "first zero" of $y$ as $t$ decreases from $+\infty$, and then estimate $T(\gamma)$ in terms of $\gamma$.

For the standard Dirichlet problem

$$-\Delta u = f(u) \quad \text{in } B_R \tag{3.1}$$

$$\text{(IV)} \qquad u > 0 \quad \text{in } B_R \tag{3.2}$$

$$u = 0 \quad \text{on } \partial B_R, \tag{3.3}$$

where $B_R = \{x \in R^2 : |x| < R\}$, the existence of $T(\gamma)$ means that there is a solution $u(r)$ with

$$u(0) = \gamma \quad \text{and} \quad R = R(\gamma) \overset{\text{def}}{=} 2e^{-T(\gamma)/2}, \tag{3.4}$$

in view of (2.2).

Our main result concerning the Dirichlet problem is the following.

**Theorem 3.** *Suppose $f$ satisfies H1 and H3 and $\gamma$ satisfies* (1.13). *Then $T(\gamma)$ exists and*

$$T(\gamma) > h(\gamma) + \log \{\tfrac{1}{2} g'(\gamma)\} - 1. \tag{3.5}$$

*If $M = 0$ in* (1.12), *the term $-1$ may be omitted.*

Theorem 3 can be used to determine when $T(\gamma) \to \infty$ (and therefore $R(\gamma) \to 0$) as $\gamma \to \infty$, We given a few examples.

*Example* 1. $$f(u) = \pi(u)\, e^{u^q}, \quad u \geqq y_0, \quad q > 1. \tag{3.6}$$

Here $\pi(u)$ is a polynomial which is positive for $u \geqq y_0$. The requirement $q > 1$ ensures the convexity (1.10) for suitable $y_0$. Theorem 3 yields

$$T(\gamma) > (1 - \tfrac{1}{2} q)\, \gamma^q + O(\log \gamma) \quad \text{as } \gamma \to \infty \tag{3.7}$$

provided either $q < 2$, in which case (1.13) is certainly satisfied for large $\gamma$, or else $M = 0$ that is to say $\pi(u) \geqq 0$ for all $u > 0$.

It follows from (3.7) that

$$\lim_{\gamma \to \infty} T(\gamma) = \infty \qquad \text{if } 1 < q < 2.$$

The possibility that $T(\gamma) \to -\infty$ as $\gamma \to \infty$ if $q > 2$ is left open by (3.7) but does not seem supported by numerical evidence.

In the following examples we examine the borderline cases $q = 2$ and $q = 1$.

*Example* 2. $$f(u) = u^m e^{u^2}, \quad m \in R. \tag{3.8}$$

Here $M = 0$ and the existence of $T(\gamma)$ for $\gamma > 0$ is automatic. We now conclude from Theorem 3 that

$$T(\gamma) > (m + 1) \log \gamma + O(1) \quad \text{as } \gamma \to \infty \tag{3.9}$$

so that $T(\gamma) \to \infty$ as $\gamma \to \infty$ logarithmically if $m > -1$. (The departure from Lipschitz continuity at $u = 0$ does not affect this.)

The case $q = 1$ is noteworthy for the presence of an exact solution.

*Example* 3. $$f(u) = e^{a + bu}, \quad a \in R, \quad b \in R^+. \tag{3.10}$$

Here the solution $y(t)$ which converges to $\gamma$ as $t \to \infty$ is given by

$$y(t, \gamma) = \gamma - \frac{2}{b} \log\,(1 + \tfrac{1}{2}\, b e^{a + b\gamma - t}). \tag{3.11}$$

Hence in this case

$$T(\gamma) = a + \tfrac{1}{2}\, b\gamma + \log\,(\tfrac{1}{2}\, b) - \log\,(1 - e^{-b\gamma/2}). \tag{3.12}$$

Theorem 3 gives, if we take $y_0 = 0$:

$$T(\gamma) > a + \tfrac{1}{2}\, b\gamma + \log\,(\tfrac{1}{2}\, b). \tag{3.13}$$

Thus (3.13) is correct within a bounded error as $\gamma \to \infty$, though it fails to show that $T(\gamma) \to \infty$ as $\gamma \to \infty$; the latter will be a general feature of the situation when $f(0) > 0$.

These remarks can be interpreted in terms of the nonlinear eigenvalue problem

$$\text{(V)} \qquad \begin{aligned} -\Delta u &= \lambda f(u) && \text{if } |x| < 1 && (3.14)\\ u &> 0 && \text{if } |x| < 1 && (3.15)\\ u &= 0 && \text{if } |x| = 1. && (3.16) \end{aligned}$$

If Problem III has a solution $y(t, \gamma)$ with first zero $T(\gamma)$, then Problem V has a solution $u$ with $u(0) = \gamma$ if $\lambda = R^2(\gamma)$ is related to $T(\gamma)$ by (3.4). Thus, in the above situations in which $T(\gamma) \to \infty$ as $\gamma \to \infty$, we shall have

$$R(\gamma) \to 0 \quad \text{and hence } \lambda \to 0$$

when $\gamma = \max_{B_1} u \to \infty$.

If $f(0) > 0$, we shall have the same behaviour as $\gamma \to 0$, and so we shall have $\lambda \to 0$ both as $\gamma \to \infty$ and as $\gamma \to 0$. This leads to the situation that there exists a $\lambda_0 > 0$ such that Problem V has no solution if $\lambda > \lambda_0$ and at least two if $0 < \lambda < \lambda_0$.

The case $f(u) = e^u$ in general domains is discussed by BANDLE [3] and WESTON [13].

We pass now to the proof of our theorems, which will be accomplished by proving Theorem 3.

## 4. Asymptotically Constant Solutions of a Highly Nonlinear Equation

For the proof of Theorem 3 we need detailed estimates concerning solutions of the problem

(VI)
$$y'' + e^{g(y)-t} = 0 \qquad y \geqq y_0 \tag{4.1}$$
$$\lim_{t\to\infty} y(t) = \gamma \tag{4.2}$$

in which $y_0 \geqq 0$, $\gamma \in (y_0, \infty)$ and $g$ satisfies

*A*1. $g \in C^2([y_0, \gamma])$

*A*2. $g' > 0$ and $g'' \geqq 0$ on $[y_0, \gamma]$.

Here $g(u) = \log f(u)$, as in (1.9); the behavior of $g(u)$ for $u > \gamma$ will be irrelevant.

Thanks to the rapid decay of the coefficient $e^{-t}$ in (4.1) as $t \to \infty$, a solution $y = y(t, \gamma)$ of Problem VI will certainly exist for large $t$; if it is continued backwards for decreasing $t$, it will necessarily reach the value $y_0$ for some $t = T_0 = T_0(\gamma)$, with slope $y'(T_0)$. We need mainly to estimate $T_0$ and $y'(T_0)$ in order to ascertain whether $y(t)$ will reach the value zero at some $T(\gamma) \leqq T_0(\gamma)$.

We recall that if $g''(u) \equiv 0$, so that $g(u) = a + bu$ with constants $a$ and $b$, an exact solution can be given (*cf.* (3.11)). In the sequel we obtain upper and lower bounds for $y$ which coincide with the exact solution when the latter is valid.

In what follows it will be convenient to write

$$g = g(\gamma) \quad \text{and} \quad g' = g(\gamma).$$

We then have

**Lemma 1.** *For* $T_0 \leqq t < \infty$

(i)
$$y(t, \gamma) \leqq \gamma - \frac{2}{g'} \log\left(1 + \tfrac{1}{2} g' e^{g-t}\right), \tag{4.4}$$

(ii)
$$g(y(t, \gamma)) \geqq g - 2 \log\left(1 + \tfrac{1}{2} g' e^{g-t}\right). \tag{4.5}$$

As we just observed, these bounds are exact if $g$ is linear.

**Proof.** Writing $y(t)$ (or $y$) for $y(t, \gamma)$ with fixed $\gamma$, we note that $y' > 0$ and $y'' < 0$ for $t \geqq T_0$, and that $y' \to 0$ as $t \to \infty$. Thus, from (4.1) we obtain on integration

$$y'(t) = \int_t^\infty e^{g(y(s))-s}\, ds \tag{4.6}$$

and so, using the monotonicity of $y(t)$ and $g(y)$, that

$$e^{g(y(t))-t} < y'(t) < e^{g-t}. \tag{4.7}$$

This implies in particular that

$$\log y'(t) + t - g \to 0 \quad \text{as } t \to \infty. \tag{4.8}$$

We now multiply (4.1) by $\{t - g(y(t))\}'$. This yields

$$E'(t) = -\tfrac{1}{2} y'^3 g''(y), \tag{4.9}$$

where

$$E(t) = y' - \tfrac{1}{2} y'^2 g'(y) - e^{g(y)-t}. \tag{4.10}$$

Thus, since $g''(u) \geqq 0$, $E(t)$ is a non-negative, non-increasing function which tends to zero as $t \to \infty$.

We next divide (4.10) by $y'$ and get, using (4.1):

$$\begin{aligned} 0 \leqq 1 - \tfrac{1}{2} y' g'(y) + \frac{y''}{y'} &\leqq \frac{1}{2y'(t)} \int_t^\infty y'^3 g''(y)\, ds \\ &\leqq \tfrac{1}{2} y'(t) \int_t^\infty y' g''(y)\, ds \\ &= \tfrac{1}{2} y' \{g' - g'(y)\}. \end{aligned}$$

These inequalities yield upon integration

$$0 \leqq [t - \tfrac{1}{2} g(y) - \log y']_t^\infty \leqq \tfrac{1}{2} (\gamma - y)\, g' - \tfrac{1}{2} \{g - g(y)\}.$$

Computing the limit as $t \to \infty$ in the middle term by means of (4.8), we arrive at the two inequalities:

$$\tfrac{1}{2} \{g + g(y(t))\} - t - \log y'(t) \geqq 0 \tag{4.11}$$

and

$$g - t - \log y'(t) \leqq \tfrac{1}{2} \{\gamma - y(t)\}\, g'. \tag{4.12}$$

The first of these, namely (4.11), implies that

$$y' e^{-g(y)/2} \leqq e^{(g/2)-t}. \tag{4.13}$$

Because $g'(y) > 0$ and $g''(y) \geqq 0$ we have $0 < g'(y) \leqq g'$ so that (4.13) yields

$$y' g'(y)\, e^{-g(y)/2} \leqq g' e^{(g/2)-t}.$$

Integration over $(t, \infty)$ now gives

$$2\{e^{-g(y)/2} - e^{-g/2}\} \leqq g' e^{(g/2)-t},$$

which leads after some manipulation to the desired lower bound (4.5).

The second inequality, namely (4.12), may be rewritten as

$$y' e^{(\gamma - y)(g'/2)} \geqq e^{g-t} \tag{4.14}$$

from which we may derive (4.4) by an integration over $(t, \infty)$.

We note a number of further results, all valid for $T_0 \leqq t < \infty$. In the first one we rewrite (4.4) as a lower bound for $t$ in terms of $y(t)$.

**Lemma 2.** *We have*

$$t \geqq g - \tfrac{1}{2}(\gamma - y) g' + \log(\tfrac{1}{2} g') - \log\{1 - e^{(y-\gamma)(g'/2)}\} \tag{4.15}$$

$$> g - \tfrac{1}{2}(\gamma - y) g' + \log(\tfrac{1}{2} g'). \tag{4.16}$$

As a simple but significant illustration we have the following result.

**Corollary 1.** *If* $g(u) = u^m$, $m \leqq 2$ *and* $y_0 = 0$, *then*

$$T_0(\gamma) \to \infty \quad \text{as} \quad \gamma \to \infty. \tag{4.17}$$

We discuss such cases in more detail in Section 7.

We can of course use (4.5) similarly to obtain an upper bound for $t$ in terms of $y$, namely

$$t \leqq \tfrac{1}{2}\{g + g(y)\} + \log(\tfrac{1}{2} g') - \log[1 - e^{\{g(y) - g\}/2}] \tag{4.18}$$

but this does not seem adequate in the situation of Corollary 1.

A number of additional inequalities, of interest in their own right, should be noted at this point.

**Lemma 3.** *For* $T_0 \leqq t < \infty$

(i) $$y' g'(y) < 2 \tag{4.19}$$

(ii) $$2 g'(y)\, e^{g(y) - t} < 1 \tag{4.20}$$

(iii) $$y'(t) \leqq e^{(g/2) + (g(y)/2) - t} \tag{4.21}$$

(iv) $$y'(t) \geqq e^{g - (\gamma - y)(g'/2) - t}. \tag{4.22}$$

**Proof.** (i) Since $E(t) \geqq 0$, we have

$$y' - \tfrac{1}{2} y'^2 g'(y) \geqq e^{g(y) - t} > 0$$

and the result follows at once since $y' > 0$.

(ii) We write $E$ as

$$E = -\tfrac{1}{2} g'(y) \left\{ y' - \frac{1}{g'(y)} \right\}^2 + \frac{1}{2 g'(y)} - e^{g(y) - t}.$$

Since $E \geqq 0$, the last two terms must together be positive; this gives (ii). (iii) and (iv) follow respectively from (4.13) and (4.14).

We note that the upper bound for $y'$ given in (4.21) is intermediate between the bounds given in (4.7). The lower bound given by (4.22) is intermediate when $y$ is close to $\gamma$.

In later discussions (see Section 7) the curve

$$\Gamma = \{(t, u) \colon t = g(u)\}, \tag{4.23}$$

will play a critical role. Choosing $u$ to be the ordinate and $t$ the abscissa, the solution graph $\{(t, y(t)) \colon t \geqq T_0\}$ will lie to the right of $\Gamma$ for large $t$ (see Fig. 1). Considering this graph as $t$ decreases from $\infty$, we see from (4.20) that it can only cross $\Gamma$, if it does so at all, at points $(t, y)$ where $g'(y) < \frac{1}{2}$ and so, in cases of main interest when $g'(u) \to \infty$ as $u \to \infty$, in a bounded range of $y$-values.

## 5. Further Bounds for $y'(t)$

We continue assembling inequalities for the proof of Theorem 3 and need mainly a good lower bound for $y'(t), t \geqq T_0$. The lower bound (4.22) has the disadvantage of involving both $y$ and $t$ on the right, compounded by the lack of a good upper bound for $t$ in terms of $y$, or lower bound for $y$ in terms of $t$.

**Lemma 4.** *For* $t \geqq T_0$,

$$y'(t) \geqq \frac{2}{g'}\left(1 - \frac{1}{1 + \frac{1}{2} g' e^{g-t}}\right) = (e^{t-g} + \tfrac{1}{2} g')^{-1}. \tag{5.1}$$

We remark first that this lower bound for $y'$ exhibits what appears to be true behaviour in a certain range of cases, in that the lower bound is exponentially small for large $t$, and tends to $2/g'$ for small $t$, with a transition zone centered on

$$T_c = g + \log(\tfrac{1}{2} g'). \tag{5.2}$$

For the proof of Lemma 4 we insert in (4.6) the bound (4.5) for $g(y)$. This gives

$$y'(t) \geqq \int_t^\infty \exp\{g - 2\log(1 + \tfrac{1}{2} g' e^{g-s}) - s\}\, ds$$

or, with the substitution $s - g = r$,

$$y'(t) \geqq \int_{t-g}^\infty e^{-r}(1 + \tfrac{1}{2} g' e^{-r})^{-2}\, dr$$

which yields (5.1).

For the more detailed discussion of the graph of $y(t)$ in Section 7 we need a corresponding upper bound for $y'$, valid in some suitable asymptotic sense as $t$ decreases through the transition zone around $T_c$, that is to say for $T_c - t$ possibly large. For this we need an upper bound for $g(y)$ and note that, by (4.4),

$$g(y) = g - 2\log(1 + \tfrac{1}{2} g' e^{g-t}) + 2g''(\xi)(g')^{-2} \log^2(1 + \tfrac{1}{2} g' e^{g-t}) \tag{5.3}$$

for some $\xi \in (y, \gamma)$. Inserting this in (4.6) again we now obtain the desired upper bound for $y'$.

**Lemma 5.** *For* $t \geqq T_0$

$$y'(t) \leqq \frac{2L}{g'}\left(1 - \frac{1}{1 + \frac{1}{2} g' e^{g-t}}\right) \tag{5.4}$$

*where*

$$L = \exp\left\{\frac{2}{|g'|^2} \log^2 (1 + \tfrac{1}{2} g' e^{g-t}) \sup_I g''(\xi)\right\} \tag{5.5}$$

*and $I$ is the set of values $\xi$ such that*

$$g - 2 \log (1 + \tfrac{1}{2} g' e^{g-t}) \leqq g(\xi) \leqq g. \tag{5.6}$$

Note that in (5.6) we have extended the range for $\xi$ on the basis of the lower bound of $g(y)$ given by (4.5).

## 6. Proof of Theorem 3

We first take the case $y_0 = 0$, for which the existence of $T(\gamma) = T_0(\gamma)$ is ensured, and (1.13) is not needed; it is formally satisfied with $M = 0$.

By Lemma 2 we have for $y = 0$

$$T(\gamma) \geqq g - \tfrac{1}{2} \gamma g' + \log (\tfrac{1}{2} g') - \log (1 - e^{-\gamma g'/2}). \tag{6.1}$$

Setting $y_0 = 0$ in the definition (1.11) of $h$ we can write (6.1) as

$$T(\gamma) \geqq h(\gamma) + \log (\tfrac{1}{2} g'(\gamma)). \tag{6.2}$$

This completes the proof for $y_0 = 0$.

We proceed on the basis that $y_0 > 0$ and that for some $M \geqq 0$

$$f(u) \geqq -M \quad \text{for} \quad 0 \leqq u \leqq y_0. \tag{6.3}$$

We continue to denote $g(\gamma)$ and $g'(\gamma)$ by respectively $g$ and $g'$ and by $T_0$ the value of $t$ such that $y(t) = y_0$. We also write $g_0$ for $g(y_0)$.

It is intuitively evident that if $T_0$ and $y'(T_0)$ are suitably large, then, as $t$ decreases from $T_0$ there will come a $T$ such that $y(T) = 0$. This thought, slightly extended, is embodied in the next lemma.

**Lemma 6.** *Suppose that for some $s \geqq T_0$ and $M > 0$, (6.3) holds and*

$$s - \frac{y(s)}{y'(s)} + \log y'(s) > \log M + 1. \tag{6.4}$$

*Then $T(\gamma) > -\infty$.*

**Proof.** It follows from the differential equation that

$$y''(t) \leqq M e^{-t} \qquad t \leqq s$$

and so, by two integrations,

$$y'(t) \geqq y'(s) - M e^{-t}$$
$$y(s) - y(t) \geqq (s - t)\, y'(s) - M e^{-t},$$

these being valid in any interval $[T', s]$ in which $y \geqq 0$. It follows that if there exists a $t^* < s$ such that

$$(s - t^*) y'(s) - Me^{-t^*} > y(s) \tag{6.5}$$

then $T(\gamma) > -\infty$ and, moreover,

$$t^* < T(\gamma) < T_0 . \tag{6.6}$$

We take

$$t^* = \log M - \log y'(s) . \tag{6.7}$$

Then, by (6.4), $t^* < s$ as required and substitution in (6.5) yields

$$s + \log y'(s) - \log M - 1 > \frac{y(s)}{y'(s)}$$

which is equivalent to (6.4). This proves Lemma 6.

In the following Lemma we derive a lower bound for $T(\gamma)$.

**Lemma 7.** *Let* $M > 0$ *and let* $s$ *be as in Lemma* 6. *Then*

$$T(\gamma) > s - \frac{y(s)}{y'(s)} - 1 . \tag{6.8}$$

*If* $M = 0$ *in* (6.3), $s \geqq T_0$ *may be arbitrary the term* $-1$ *may be omitted.*

**Proof.** The last remark in Lemma 7 is obvious because if $M = 0$, $y'' \leqq 0$ for $t \leqq s$. Thus we proceed on the basis that $M > 0$.

Our criterion (6.5) for the existence of $T(\gamma)$ may be written as

$$s - \frac{y(s)}{y'(s)} - t - M \frac{e^{-t}}{y'(s)} > 0 , \tag{6.9}$$

to be satisfied for some $t < s$. The left hand side of (6.9) attains its maximum value at $t = t^*$; it is positive by virtue of assumption (6.4). For $t > t^*$ the left of (6.9) is decreasing, but we assert that for

$$t^{**} = s - \frac{y(s)}{y'(s)} - 1$$

it is still positive. Note that by (6.4) and (6.7), $t^{**} > t^*$. To prove the assertion, observe that

$$s - \frac{y(s)}{y'(s)} - t^{**} - M \frac{e^{-t^{**}}}{y'(s)} = 1 - M \frac{e^{-t^{**}}}{y'(s)} > 1 - M \frac{e^{-t^*}}{y'(s)} = 0 .$$

Thus $T(\gamma) > t^{**}$, as was to be proved.

Having established these two preliminary lemmas, we now turn to the proof of Theorem 3. Since the case $y_0 = 0$ has been dealt with we shall take $y_0 > 0$.

We begin by using Lemma 6 to show that $T(\gamma) > -\infty$. Setting $s = T_0$ we require an estimate from below for the expression

$$J = T_0 + \log y'(T_0) - \frac{y_0}{y'(T_0)}\,. \tag{6.10}$$

We have from (5.1) that

$$e^t y'(t) \geqq e^g(1 + \tfrac{1}{2} g' e^{g-t})^{-1}$$

and hence, that

$$T_0 + \log y'(T_0) \geqq g - \log(1 + \tfrac{1}{2} g' e^{g-T_0})\,. \tag{6.11}$$

In the remaining term on the left of (6.10) we use once more the lower bound (5.1) for $y'$ and so get

$$\begin{aligned} -\frac{y_0}{y'(T_0)} &\geqq -y_0 e^{T_0-g} - \tfrac{1}{2} g' y_0 \\ &\geqq -y_0 e^{T_0-g} - \tfrac{1}{2} g'\gamma + \log(1 + \tfrac{1}{2} g' e^{g-T_0}) \end{aligned} \tag{6.12}$$

by (4.4). Combining (6.11) and (6.12) we obtain

$$J \geqq g - \tfrac{1}{2}\gamma g' - y_0 e^{T_0-g}\,.$$

Finally, we observe that by (4.18)

$$e^{T_0-g} \leqq \tfrac{1}{2} g' e^{(g_0-g)/2}\{1 - e^{(g_0-g)/2}\}^{-1}\,. \tag{6.13}$$

This shows, in the notation of (1.11), that $J \geqq h(\gamma)$.

This completes the proof of Theorem 3 so far as the existence of $T(\gamma)$ is concerned; if $M = 0$ the existence is automatic; and if $M > 0$ the condition $h(\gamma) \geqq \log M + 1$ ensures that $J > \log M + 1$, and hence that (6.4) is satisfied with $s = T_0$. This also completes the proof of Theorem 1 on the existence of a ground state when hypothesis H2 holds as well.

To complete the proof of Theorem 3 we need to discuss the lower bound (3.5) for $T(\gamma)$ in the case $y_0 > 0$. By Lemma 7

$$T(\gamma) > T_0 - \frac{y_0}{y'(T_0)} - 1\,, \tag{6.14}$$

where the term $-1$ may be omitted if $M = 0$. From (6.12) we obtain

$$\begin{aligned} T_0 - \frac{y_0}{y'(T_0)} &\geqq T_0 + \log(1 + \tfrac{1}{2} g' e^{g-T_0}) - y_0 e^{T_0-g} - \tfrac{1}{2}\gamma g' \\ &= g - \tfrac{1}{2}\gamma g' - y_0 e^{T_0-g} + \log(\tfrac{1}{2} g') + \log\left(1 + \frac{2}{g'} e^{T_0-g}\right). \end{aligned} \tag{6.15}$$

Thus, using (6.13), and dropping the last term of (6.15) we arrive at

$$T_0 - \frac{y_0}{y'(T_0)} \geqq h(\gamma) + \log(\tfrac{1}{2} g')\,.$$

Substitution in (6.14) yields the desired lower bound, whence the proof is complete.

## 7. The Graph of y(t) when γ is Large

If the limit value $\gamma = \lim_{t\to\infty} y(t)$ becomes large the solution of Problem VI:

$$\text{(VI)} \qquad \begin{gathered} y'' + e^{g(y)-t} = 0, \\ y(t) \to \gamma \quad \text{as } t \to \infty \end{gathered}$$

acquires some striking features, which first became apparent on the perusal of computer-generated graphics; we are indebted to S. Angenent in this connection. We describe these features briefly in this section and show how they can be used to obtain improved bounds for $T_0$. A more detailed discussion is left to a subsequent paper.

For simplicity we begin with the explicitly soluble case

$$g(y) = ay, \qquad a \in R^+ \tag{7.1}$$

with solution (see (3.10)):

$$y(t, \gamma) = \gamma - \frac{2}{a} \log\left(1 + \frac{a}{2} e^{a\gamma - t}\right). \tag{7.2}$$

Setting $T_c = a\gamma + \log(a/2)$ (see (5.2)) we readily see that

$$y(t,\gamma) \approx \begin{cases} \gamma & \text{if} \quad t > T_c & (7.3) \\ \gamma + \dfrac{2}{a}(t - T_c) & \text{if} \quad t < T_c & (7.4) \end{cases}$$

provided $|t - T_c|$ is not too small. The graph of $y(t, \gamma)$ is given in Fig. 2.

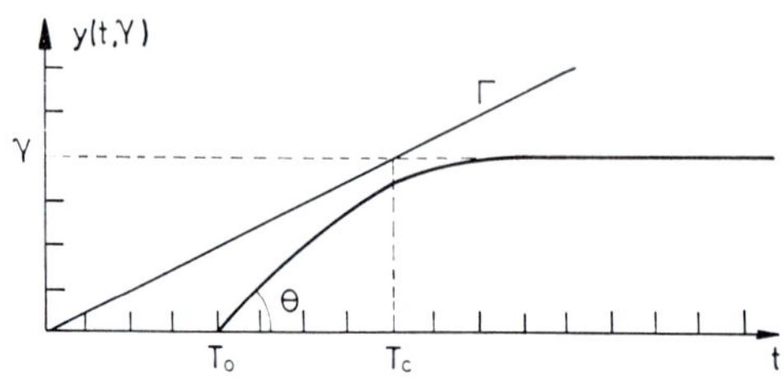


Fig. 2. The graph of $y$ when $a = 2$, $\gamma = 4$.

One has

$$y'(t) = \left(e^{t - a\gamma} + \frac{a}{2}\right)^{-1}$$

so that $y'(1) = o(1)$ or $2a^{-1} + o(1)$ if $|t - a\gamma|$ is large, according to whether $t > a\gamma$ or $t < a\gamma$.

Thus, shooting from $t = \infty$, the solution comes in nearly horizontally and is "reflected" by the curve $\Gamma$ (see (4.23)), in this case the line $y = t/a$. After reflection, and passing through a transition zone, it proceeds again in nearly a straight line, but now at an angle $\theta$ to the $t$-axis, where

$$\tan\theta = 2/a, \tag{7.5}$$

so that the slope of the solution graph after reflection is twice that of the curve $\Gamma$. This illustrates the general pattern according to which the slopes of incident and reflected rays have as their mean the slope of the tangent to $\Gamma$ at the point of incidence, roughly speaking.

The special case $g(u) = u^2$ deserves comment. If the reflection pattern sketched above were followed exactly, the solutions which come in horizontally from $t = \infty$ would all be reflected through the vertex (0, 0) of the parabola $y^2 = t$ (rather than through the focus, as in optical reflection). Actually, as may be seen from (4.16), or from Theorem 3, $T(\gamma)$ in this case grows logarithmically as $\gamma \to \infty$. Cases such as $g(u) = u^m$, $m > 2$ appear to lead to multiple reflections of the solution before it reaches the $t$-axis.

We now sketch a strategy for justifying these observations. For simplicity, we confine attention to the case $y_0 = 0$ and to cases similar to $g(u) = u^m$ $1 < m < 2$. Specifically we shall assume about $g$, in addition to A1 and A2 (see section 4):

*A*3. $g(u) - \frac{1}{2} ug'(u) > 0$ for all $u > 0$.

*A*4. $g \in C^3([0, \infty))$ and there exist nonzero constants $L_p$ such that

$$\lim_{u \to \infty} \frac{ug^{(p+1)}(u)}{g^{(p)}(u)} = L_p \quad \text{for } p = 0, 1, 2.$$

Here $g^{(p)}$ denotes the $p^{\text{th}}$ derivative of $g$.

The strategy consists in choosing a point $T_1$ below the transition point (see (5.2))

$$T_c = g + \log\left(\tfrac{1}{2} g'\right) \tag{7.6}$$

at which (i) $y'$ can be shown—by means of Lemma 4 and 5—to be close to $2/g'$ and (ii) $t - g(y)$ can be shown—by means of Lemma 1—to be large. The second requirement will ensure that $y''$ is small and $y'$ does not change much for values of $t$ below $T_1$. Together with an estimate for $y$ at $t = T_1$, this strategy enables us to obtain an estimate for $T_0$.

Thus set

$$T_1(\gamma) = g + \log\left(\tfrac{1}{2} g'\right) - \delta(\gamma), \tag{7.7}$$

where $\delta > 0$. We shall assume throughout that

(i) $$\delta(\gamma) \to \infty \quad \text{as } \gamma \to \infty$$

(ii) $$\delta^2(\gamma) = o(g(\gamma)) \quad \text{as } \gamma \to \infty.$$

We begin by estimating $y$ and $g(y)$ at $t = T_1$, then we estimate $y'$ at $t = T_1$ and finally we estimate $y'$ on $[T_0, T_1]$.

**Lemma 8.** *Suppose g satisfies A*1–4. *Then*

(a) $$y(T_1) = \gamma - \frac{2}{g'}\delta + O\left(\frac{\delta^2}{g}\right) \quad \text{as } \gamma \to \infty, \tag{7.8}$$

(b) $$g(y(T_1)) = g - 2\delta + O\left(\frac{\delta^2}{g}\right) \quad \text{as } \gamma \to \infty. \tag{7.9}$$

**Proof.** By Lemma 4.1, as $\gamma \to \infty$

$$y(T_1) \leqq \gamma - \frac{2}{g'} \log(1 + e^{\delta})$$

$$\leqq \gamma - \frac{2}{g'}\delta + O\left(\frac{2}{g'} e^{-\delta}\right) \tag{7.10}$$

and

$$g(y(T_1)) \geqq g - 2\delta + O(e^{-\delta}). \tag{7.11}$$

Since $g$ is increasing we obtain

$$g(y(T_1)) \leqq g\left(\gamma - \frac{2}{g'}\delta + O\left(\frac{2}{g'} e^{-\delta}\right)\right)$$

$$= g - 2\delta + O\left(\frac{\delta^2}{g}\right)$$

and similarly

$$y(T_1) \geqq g^{-1}(g - 2\delta + O(e^{-\delta}))$$

$$= \gamma - \frac{2}{g'}\delta + O\left(\frac{\delta^2}{g}\right),$$

in view of the regularity assumption A4.

**Lemma 9.** *If* $g$ *satisfies* A1–4, *then*

$$y'(T_1) = \frac{2}{g'} \frac{1}{1 + e^{-\delta}}\left[1 + O\left(\frac{\delta^2}{g}\right)\right] \quad \text{as } \gamma \to \infty. \tag{7.12}$$

**Proof.** By Lemma 4

$$y'(T_1) \geqq \frac{2}{g'} \frac{1}{1 + e^{-\delta}}$$

and by Lemma 5

$$y'(T_1) \leqq \frac{2}{g'} \frac{L}{1 + e^{-\delta}},$$

where

$$L = \exp\left[\tfrac{1}{2}\left\{\frac{2}{g'} \log(1 + e^{\delta})\right\}^2 \sup_{\xi} g''(\xi)\right]$$

and $\xi$ satisfies, in view of Lemma 8,

$$\gamma - \frac{4}{g'}\delta < \xi < \gamma.$$

Hence, by assumption A4,

$$L = \exp[O(\delta^2/g)] = 1 + O\left(\frac{\delta^2}{g}\right),$$

from which the assertion follows.

**Lemma 10.** *For* $T_0 \leqq t \leqq T_1$:

$$y'(t) = \frac{2}{g'}\left[1 + O\left(\frac{\delta^2}{g}\right) + O(e^{-\delta})\right] + O(\gamma e^{\gamma(g'/2)-g}) + O(\gamma e^{-\delta}) \quad \text{as } \gamma \to \infty \tag{7.13}$$

*uniformly with respect to t.*

**Proof.** Integrating equation (4.1) over $(t, T_1)$, we obtain

$$y'(t) = y'(T_1) + \int_t^{T_1} e^{g(y(s))-s}\, ds. \tag{7.14}$$

By (4.10)

$$g(y) - t < \{g(y) - \tfrac{1}{2} yg'\} - (g - \tfrac{1}{2}\gamma g') - \log(\tfrac{1}{2} g').$$
$$:= \psi(y).$$

Hence, because $\psi'' \geqq 0$,

$$g(y) - t \leqq \max\{\psi(y(T_0)), \psi(y(T_1))\}.$$

Now

$$\psi(t(T_0)) = \psi(0) = g(0) - (g - \tfrac{1}{2}\gamma g') - \log(\tfrac{1}{2} g')$$

and, by Lemma 8,

$$\psi(y(T_1)) = -2\delta + \{\gamma - y(T_1)\}\tfrac{1}{2} g' - \log(\tfrac{1}{2} g') + O\left(\frac{\delta^2}{g}\right)$$
$$= -\delta - \log(\tfrac{1}{2} g') + O\left(\frac{\delta^2}{g}\right) + O\left(\frac{\delta^2}{\gamma}\right).$$

Thus, for $g$ sufficiently large,

$$g(y) - t \leqq -\log(\tfrac{1}{2} g') + \max\{-(g - \tfrac{1}{2}\gamma g'), -\delta\}.$$

It follows that for $t \in [T_0, T_1]$

$$\int_t^{T} e^{g(y(s))-s}\, ds \leqq \frac{2}{g'} \max\{e^{-(g-\gamma(g'/2))}, e^{-\delta}\}(T_1 - T_0).$$

By Theorem 3, since $g - \frac{1}{2}\gamma g' > 0$, $T_0 \geqq \log(\frac{1}{2} g')$ and hence $T_1 - T_0 \leqq g$. Therefore, when $T_0 \leqq t \leqq T_1$

$$y'(t) - y'(T_1) = O(\gamma e^{-(g-\gamma(g'/2))}) + O(\gamma e^{-\delta}).$$

The desired estimate now follows from Lemma 9.

It remains to make a choice for $\delta$. If we set

$$\delta(\gamma) = k \log \gamma, \quad k > 0$$

and choose $k$ sufficiently large the error terms involving $e^{-\delta}$ become small compared to $O(\delta^2/g)$, whence (7.7), (7.8) and (7.13) become

$$T_1 = g + \log(\tfrac{1}{2} g') - k \log \gamma, \tag{7.15}$$

$$y(T_1) = \gamma - \frac{2k \log \gamma}{g'} + O\left(\frac{\log^2 \gamma}{g}\right), \tag{7.16}$$

$$y'(t) = \frac{2}{g'}\left[1 + O\left(\frac{\log^2 \gamma}{g}\right)\right] + O(\gamma e^{\gamma(g'/2)-g}), \tag{7.17}$$

for $T_0 \leqq t \leqq T_1$. By an elementary computation they yield an estimate for $T_0$. We make this the content of the next theorem.

**Theorem 4.** *Let $y(t, \gamma)$ be the solution of Problem VI in which $g$ satisfies A1–4 with $y_0 = 0$. Then the first zero $T(\gamma)$ of $y$ satisfies*

$$T(\gamma) = g(\gamma) - \tfrac{1}{2}\gamma g'(\gamma) + \log\left(\tfrac{1}{2} g'(\gamma)\right) + O\left(\frac{\log^2 \gamma}{\gamma}\right) \quad \text{as } \gamma \to \infty.$$

*Remark.* When

$$g(u) = u^m, \qquad 1 < m < 2,$$

Theorem 4 yields

$$T(\gamma) = (1 - \tfrac{1}{2} m)\gamma^m + (m-1)\log\gamma - \log(2/m) + O(\gamma^{-1}\log^2\gamma) \text{ as } \gamma \to \infty.$$

This work was partially supported by a grant from the Netherlands Organization for the Advancement of Pure Research (Z.W.O.) and by the Institute for Mathematics and its Applications at the University of Minnesota, and the National Sciences and Engineering Research Counal of Canada (Grant A 3979).

## References

1. ATKINSON, F. V., & L. A. PELETIER, Ground states of $-\Delta u = f(u)$ and the Emden-Fowler equation. Arch. Rational Mech. Anal. **93** (1986), 103–127.
2. ATKINSON, F. V., & L. A. PELETIER, Emden-Fowler equations involving critical exponents. To appear in Nonlinear Analysis, TMA.
3. BANDLE, C., Existence theorems, qualitative results and a priori bounds for a class of nonlinear Dirichlet problems. Arch. Rational Mech. Anal. **58** (1975), 219–238.
4. BERESTYCKI, H., & P.-L. LIONS, Existence of solutions for nonlinear scalar field equations, Part I, The ground state, Arch. Rational Mech. Anal. **82** (1983), 313–345.
5. BERESTYCKI, H., P.-L. LIONS & L. A. PELETIER, An ODE approach to the existence of positive solutions for semilinear problems in $R^n$. Indiana Univ. Math. J. **30** (1981), 141–157.
6. BREZIS, H., & L. NIRENBERG, Positive solutions of nonlinear elliptic equations involving critical Sobolev exponents. Comm. Pure and Appl. Math. **36** (1983), 437–477.
7. FOWLER, R. H., The form near infinity of real continuous solutions of a certain differential equation of the second order. Quart. J. Math. (Cambridge Series) **45** (1914), 289–350.
8. FOWLER, R. H., Further studies of Emden's and similar differential equations. Quart. J. Math. (Oxford series), **2** (1931), 259–288.
9. NI, W.-M., & J. SERRIN, Existence and non-existence theorems for ground states of quasilinear partial differential equations. The anomalous case. To appear in Accad. Naz. Lincei, Rendiconti.
10. POHOZAEV, S. I., Eigenfunctions of the equation $\Delta u + \lambda f(u) = 0$. Dokad. Akad. Nauk SSSR **165** (1965), 36–39 (in Russian) and Sov. Math. **6** (1965) 1408–1411 (in English).
11. STRAUSS, W. A., Existence of solitary waves in higher dimensions, Comm. Math. Phys. **55** (1977), 149–162.
12. TRUDINGER, N. S., On imbeddings into Orlicz spaces and some applications. Indiana Univ. Math. J. **17** (1967), 473–483.

13. WESTON, V. H., On the asymptotic solution of a partial different equation with an exponential nonlinearity. SIAM J. Math. Anal. **9** (1978), 1030–1053.
14. WONG, J. S. W., On the generalized Emden-Fowler equation, SIAM Review **17** (1975), 339–360.

Department of Mathematics
University of Toronto
Toronto, Canada
and
Mathematical Institute
University of Leiden
Leiden, The Netherlands

*(Received December 9, 1985)*

# *On Phase Transitions with Bulk, Interfacial, and Boundary Energy*

MORTON E. GURTIN

*This paper is dedicated to James Serrin*

## 1. Introduction

Consider a fluid which has free energy[1] $\psi(u)$ a prescribed function of density $u$, and which occupies a fixed container $\Omega$, with $\Omega$ a bounded, open region in $\mathbb{R}^N$.

Our purpose here is to develop further the theory of equilibrium phase transitions introduced in [1985], a theory whose underlying idea is the requirement that the density not lie in the spinodal region between phases. Specifically, we assume that $\psi$ has the form of a van der Waals free energy with its *spinodal removed* (Figure 1): the domain of $\psi$ is the union of *disjoint*, possibly unbounded, open intervals $A$ and $B$, with $\psi$ convex on $A$ and on $B$.

To ensure that $\psi(u(x))$ have meaning, the admissible density fields $u: \Omega \to \mathbb{R}$ are required to have $u(x) \in A \cup B$ for all $x \in \Omega$. This condition allows us to define complementary subsets

$$\begin{aligned} \Omega_A(u) &:= \{x \in \Omega : u(x) \in A\}, \\ \Omega_B(u) &:= \{x \in \Omega : u(x) \in B\}, \end{aligned} \tag{1.1}$$

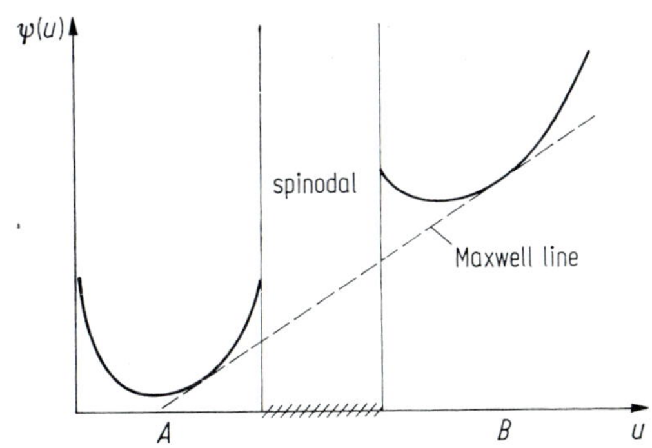


Fig. 1. Free energy $\psi(u)$ with spinodal removed

[1] We assume isothermal conditions. Here $\psi(u)$ is the free energy per unit volume; $\psi(u)/u$ is the free energy per unit mass.

which represent the regions occupied by phase $A$ and phase $B$. Further,

$$I(u) := \text{area of } \partial\Omega_A(u) \cap \partial\Omega_B(u) \tag{1.2}$$

is the **interfacial area** between phases, while

$$S_i(u) := \text{area of } \partial\Omega_i(u) \cap \partial\Omega, \quad i = A, B \tag{1.3}$$

is the **contact area** between phase $i$ and the container. (The term "area" will be made precise in subsequent sections.)

The **bulk energy** associated with an admissible field $u$ is

$$\Psi(u) := \int_\Omega \psi(u(x))\, dx. \tag{1.4}$$

We will also allow for superficial energies

$$\sigma I(u) \quad \text{and} \quad \sigma_i S_i(u),$$

respectively, between phases and between phase $i$ and the walls of the container, and this leads to a **superficial energy** of the form

$$\Phi(u) := \sigma I(u) + \sigma_A S_A(u) + \sigma_B S_B(u). \tag{1.5}$$

Here, of course, $\sigma$ and $\sigma_i$ are nonnegative constants.

We assume that there are no other forms of energy. Thus the **energy** of an admissible field $u$ is given by

$$E(u) := \Psi(u) + \Phi(u). \tag{1.6}$$

If the fluid has mass $m$, and fluid is neither added to $\Omega$ nor removed from it, then the allowable density fields must be consistent with the constraint

$$M(u) := \int_\Omega u(x)\, dx = m.$$

We are therefore led to the two-phase problem:

$$\mathcal{P}_m \begin{cases} \text{minimize } E(u) \text{ over all "admissible"} \\ u \text{ that satisfy } M(u) = m. \end{cases}$$

In discussing this and related problems we shall reserve the terms **solution** and **minimizer** for global minimizer.

The Maxwell line—which convexifies the free energy—is defined by points $\alpha_0 \in A$, $\beta_0 \in B$ with

$$\psi'(\alpha_0) = \psi'(\beta_0),$$
$$\psi(\beta_0) - \psi(\alpha_0) = \psi'(\alpha_0)\,(\beta_0 - \alpha_0).$$

In view of the constraint $M(u) = m$, Problem $\mathcal{P}_m$ is invariant[1] when $\psi(u)$ is replaced by

$$\psi(u) - \psi(\alpha_0) - \psi'(\alpha_0)\,(u - \alpha_0).$$

[1] The infimum, of course, changes by an additive constant.

Thus we may assume, without loss in generality, that (Figure 2)

$$\psi(\alpha_0) = \psi(\beta_0) = 0, \quad \psi(u) > 0 \text{ otherwise}.$$

The free energy, as sketched in Figure 1, has finite limits as the spinodal is approached from $A$ and from $B$. Let $\psi^*$ denote such a limit, or more generally, the limit of $\psi(u)$ as $u$ approaches one of the boundary points[1] of $A \cup B$. If $\ell$ is a typical dimension of $\Omega$, then an appropriate dimensionless form for this limit is $(\ell/\sigma)\,\psi^*$. One generally expects $(\ell/\sigma)\,\psi^*$ to be large, since bulk energies are usually large compared to surface energies, at least for containers of reasonable size. This and a similar argument lead to the assumption[2] that

$$\psi(u) \text{ and } \psi'(u) \text{ become unbounded as } u \to \partial(A \cup B).$$

We will show, under the above assumptions, and for surface energies consistent with

$$|\sigma_A - \sigma_B| \leqq \sigma, \tag{1.7}$$

that $\mathscr{P}_m$ *has a solution*, and each solution $u$ has the following properties:

(i) *$u$ is constant in each phase;*

(ii) *$\psi'(u(x))$ is identically constant;*

(iii) *the interface for $u$ minimizes superficial energy over all phase configurations with phase volumes equal to those of $u$.*

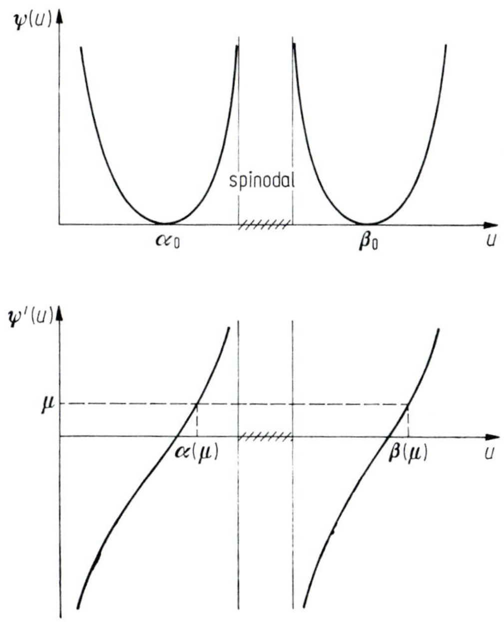


Fig. 2. Normalized free energy $\psi(u)$ and its derivative $\psi'(u)$

[1] For most models of compressible fluids, $u = 0$ is such a boundary point and $\psi(0^+) < \infty$.

[2] This assumption can be replaced by the assumption that $\sigma$ be sufficiently small, but the proof of existence is more complicated (*cf.* [1985]).

From a physical point of view, the assumption (1.7) is not restrictive. If, say, $\sigma_A > \sigma + \sigma_B$, then on the portion of the boundary in contact with phase $A$ we could add an "infinitesimal layer" of phase $B$ and thereby lower the contact energy from

$$\sigma_A S_A(u) \quad \text{to} \quad (\sigma + \sigma_B)\, S_A(u)$$

(*cf.* Figure 3). Therefore, if such layers are allowed, then for $\sigma_A > \sigma + \sigma_B$ we simply replace $\sigma_A$ by $\sigma + \sigma_B$. We will make this observation precise by showing that:

(iv) *for* $\sigma_A < \sigma + \sigma_B$, *the function obtained by replacing* $\sigma_A$ *in* $E(u)$ *by* $\sigma + \sigma_B$ *is the lower semicontinuous envelope of* $E(u)$ *over all* $u$ *with* $M(u) = m$.

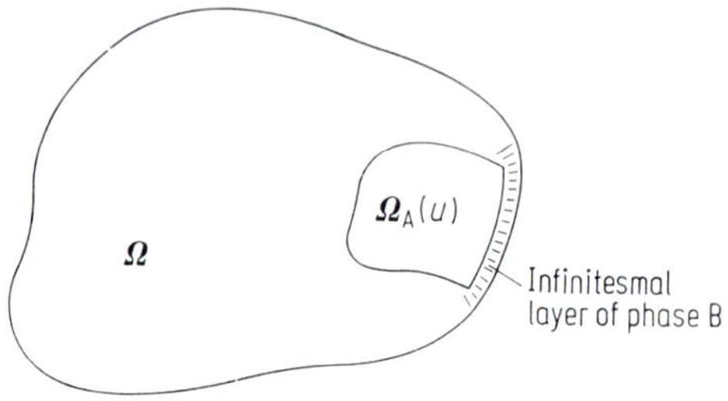


Fig. 3. Adding an infinitesimal layer of Phase B

Because of (iii), the interface is the solution of a certain isoparametric problem,[1] namely that of determining the free surface of an incompressible liquid-drop in a container $\Omega$. This problem is simply $\mathscr{P}_m$ with both materials incompressible, and hence with bulk-energy constant, and therefore reduces to minimizing the superficial energy $\Phi(u)$. By (1.2), (1.3), and (1.5), $\Phi(u)$ can be written as a function of the set $\Gamma = \Omega_A(u)$; namely,

$$\sigma I(\Gamma) + (\sigma_A - \sigma_B)\, S(\Gamma) + C \tag{1.8}$$

with

$$I(\Gamma) := \text{area of } \partial\Gamma \cap \Omega, \quad S(\Gamma) := \text{area of } \partial\Gamma \cap \partial\Omega, \tag{1.9}$$

and $C$ the *constant*

$$C = \sigma_B \text{ (area of } \partial\Omega).$$

Therefore, modulo a constant, the superficial energy is equal to $\sigma F(\Gamma)$, where $F(\Gamma)$ is the set function

$$F(\Gamma) := I(\Gamma) + \lambda S(\Gamma) \tag{1.10}$$

with

$$\lambda := (\sigma_A - \sigma_B)/\sigma. \tag{1.11}$$

Moreover, because of the constraint of incompressibility, specifying the total mass is equivalent to specifying the volume, vol $(\Gamma)$, of $\Gamma$. The liquid-drop problem therefore has the statement:

$$\mathscr{F}_v \begin{cases} \text{minimize } F(\Gamma) \text{ over "all"} \\ \text{subsets } \Gamma \text{ of } \Omega \text{ with vol}(\Gamma) = v. \end{cases}$$

[1] *Cf.*, *e.g.*, MASSARI & PEPE [1974b], GIUSTI [1981].

Because the liquid-drop problem is so crucial to our analysis of the two-phase problem, we will present a fairly complete theory. Here for (say) $\sigma_A > \sigma + \sigma_B\,(\lambda > 1)$ we have the same phenomenon as was discussed previously for the two phase problem, but in this simpler setting its resolution is more straightforward. Here $\sigma_A$ replaced by $\sigma + \sigma_B$ is equivalent to replacing $\lambda$ by one and yields $F(\Gamma)$ replaced by

$$F^*(\Gamma) = \text{area of } \partial\Gamma;$$

we will show that for $\lambda > 1$, *$F^*(\Gamma)$ is the lower semicontinuous envelope of $F(\Gamma)$ over the family of $\Gamma \subset \Omega$ with* $\mathrm{vol}\,(\Gamma) = v$.

## 2. Preliminaries

In this section we introduce some general ideas which are useful to our analysis. While most of what we present is well known, it does not seem to be readily available in the literature.

### *2a. The lower semicontinuous envelope and the natural regularization of a minimization problem*

Throughout this section $g: T \to \mathbb{R}$ with $T$ a topological space.

We say that $g$ is **lsc** (on $T$) if, given any $t \in T$ and any sequence $\{t_n\}$ in $T$ converging to $t$,

$$g(t) \leqq \liminf g(t_n).$$

Here and in what follows we use the shorthand "**lsc**" for the phrase "**lower semicontinuous**".

A function $g^*: T \to \mathbb{R}$ is an **lsc envelope of** $g$ (over $T$) if, given any $t \in T$,

$$g^*(t) \leqq \liminf g(t_n) \tag{2.1}$$

for any sequence $\{t_n\}$ in $T$ converging to $t$, and for at least one such sequence

$$g^*(t) = \lim g(t_n). \tag{2.2}$$

Clearly, $g$ can have at most one lsc envelope $g^*$.

A sequence $\{t_n\}$ in $T$ is a **minimizing sequence for** $g$ (over $T$) if

$$\inf g = \liminf g(t_n)$$

(which has meaning even when $\inf g = -\infty$).

**Proposition 2.1.** *Let $g^*$ be the lsc envelope of $g$. Then:*

(i) $g^* \leqq g$;

(ii) $\inf g^* = \inf g$;

(iii) $g^*$ *is lsc.*

**Proof.** Assertion (i) is immediate. Consider (ii). Let $\{t_n\}$ be a minimizing sequence for $g^*$ and, for each fixed $n$, choose $\{t_{nk}\}$ in $T$ with $t_{nk} \to t_n$ and $g(t_{nk}) \to g^*(t_n)$ as $k \to \infty$ (*cf.* (2.2)). Then

$$\inf g^* = \liminf_{n\to\infty} g^*(t_n) = \liminf_{n\to\infty} \lim_{k\to\infty} g(t_{nk}) \geqq \inf g,$$

which, with (i), implies (ii).

Choose $t \in T$ and $\{t_n\}$ in $T$ converging to $t$. To prove (iii) we must show that

$$g^*(t) \leqq \liminf g^*(t_n).$$

Fix $\varepsilon > 0$. Choose $\{t_{nk}\}$ with $t_{nk} \to t_n$ and $g(t_{nk}) \to g^*(t_n)$. Then for each $n$ there is an integer $k_n$ such that

$$g^*(t_n) \geqq g(t_{nk_n}) - \varepsilon,$$

and such that $t_{nk_n} \to t$ as $n \to \infty$. Thus

$$\liminf g^*(t_n) \geqq g^*(t) - \varepsilon.$$

Since $\varepsilon > 0$ is arbitrary, this is the desired result. □

Consider the problem:

$$\mathscr{P}: \text{minimize } g \text{ over } T.$$

Suppose that $g$ has an lsc envelope $g^*$. Then we will refer to the problem

$$\mathscr{P}^*: \text{minimize } g^* \text{ over } T$$

as the **natural regularization** of $\mathscr{P}$. Motivation for this terminology is furnished by

**Proposition 2.2.** *The natural regularization $\mathscr{P}^*$ of $\mathscr{P}$ has the following properties:*

(i) *every solution of $\mathscr{P}$ solves $\mathscr{P}^*$;*

(ii) *every solution of $\mathscr{P}^*$ is the limit of a minimizing sequence for $\mathscr{P}$* (i.e., *the limit of a minimizing sequence for $g$*);

(iii) *$\mathscr{P}$ and $\mathscr{P}^*$ have the same infimum.*

**Proof.** Let $t$ solve $\mathscr{P}$. Then, by Proposition 2.1(i, ii),

$$g^*(t) \leqq g(t) = \inf g = \inf g^*,$$

and (i) follows. Consider (ii). Let $t$ solve $\mathscr{P}^*$. Then

$$g^*(t) = \inf g^* = \inf g. \tag{2.3}$$

Moreover, there exists a sequence $\{t_n\}$ in $T$ such that (2.2) holds, and (2.2), (2.3) render $\{t_n\}$ a minimizing sequence for $g$. Finally, (iii) follows from Proposition 2.1(iii). □

*2b. Sets of finite perimeter. The family of parts of $\Omega$*

The variational problems under consideration, namely $\mathscr{P}_m$ and $\mathscr{F}_v$, are amenable to analysis when the subsets of $\Omega$ occupied by the individual phases are taken to be sets of finite perimeter. We now give the definition and some of the basic properties of this class of sets.[1]

Let $\Sigma$ and $\Gamma$ be subsets of $\mathbb{R}^N$ with $\Sigma$ *open*. The set $\partial\Gamma \cap \Sigma$ is that portion of the boundary of $\Gamma$ that lies within $\Sigma$. We wish to define the "area" of this set, even for sets $\Gamma$ which are quite wild. To motivate our definition, assume for the moment that $\partial\Gamma$ is smooth. Let

$$g \in C^\infty(\mathbb{R}^N, \mathbb{R}^N), \quad |g| \leqq 1, \quad g \text{ compactly supported in } \Sigma. \tag{2.4}$$

Then the inner product of $g$ with the outward unit normal on $\partial\Gamma$ is $\leqq 1$ and, by the divergence theorem,

$$\int_\Gamma \operatorname{div} g \leqq \text{area of } \partial\Gamma \cap \Sigma,$$

with area here equal to $(N-1)$-dimensional Hausdorff measure. In fact, one can show that the supremum of the left side over all such $g$ is actually equal to the right side, an observation which allows us to extend this notion of area to measurable $\Gamma$.

Assume now that $\Sigma$ is open and $\Gamma$ measurable. We call

$$\operatorname{per}(\Gamma, \Sigma) := \sup \left\{ \int_\Gamma \operatorname{div} g \colon g \text{ obeys } (2.4) \right\}$$

the **perimeter of $\Gamma$ in $\Sigma$**, and we say that $\Gamma$ has **finite perimeter** if $\operatorname{per}(\Gamma, \Sigma) < \infty$ for every bounded, open $\Sigma$. Roughly speaking, $\operatorname{per}(\Gamma, \Sigma)$ gives the "area" of that portion of the "boundary" of $\Gamma$ that lies within $\Sigma$ (*cf.* Figure 4), so that

$$\operatorname{per}(\Gamma) := \operatorname{per}(\Gamma, \mathbb{R}^N)$$

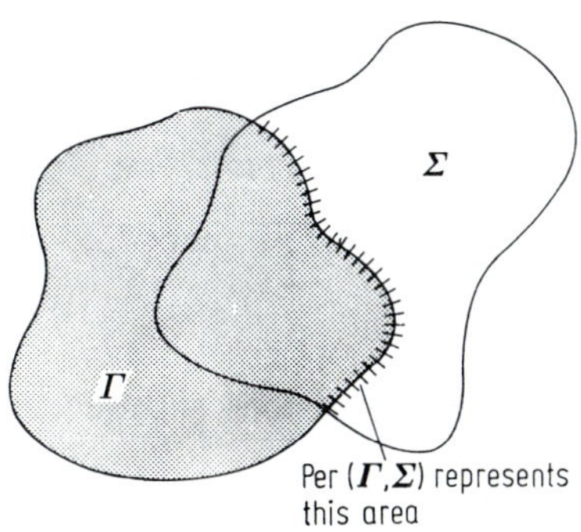


Fig. 4. The perimeter of $\Gamma$ in $\Sigma$

[1] *Cf., e.g.*, the books of FEDERER [1969], GIUSTI [1984], and MASSARI & MIRANDA [1984], and the basic papers of DE GIORGI [1954, 1955].

gives the "area" of the "boundary" of $\Gamma$; in fact, for $\partial\Gamma \cap \Sigma$ and $\partial\Gamma$ Lipschitz continuous,

$$\operatorname{per}(\Gamma) = \text{area of } \partial\Gamma, \qquad \operatorname{per}(\Gamma, \Sigma) = \text{area of } \partial\Gamma \cap \Sigma.$$

Let vol denote $N$-dimensional Lebesgue measure. We assume throughout that:

(A1) $\Omega$ is a bounded, open region in $\mathbb{R}^N$ with $\partial\Omega$ Lipschitz continuous and $\operatorname{vol}(\Omega) = 1$.

The constraint $\operatorname{vol}(\Omega) = 1$ amounts to choosing the length scale.

For convenience, we use the term **part** (of $\Omega$) for any subset of $\Omega$ with finite perimeter; and we endow the family of parts with a metric structure[1] through the distance function

$$\operatorname{dist}(\Gamma, \Lambda) := \operatorname{vol}(\Gamma \Delta \Lambda), \qquad \Gamma \Delta \Lambda := (\Gamma \setminus \Lambda) \cup (\Lambda \setminus \Gamma). \tag{2.5}$$

Then

$$\text{the set functions } \Gamma \mapsto \operatorname{per}(\Gamma, \Omega) \text{ and } \Gamma \mapsto \operatorname{per}(\Gamma) \text{ are lsc on the family of parts.} \tag{2.6}$$

Also, for $\Gamma$ a part,

$$\begin{aligned}
&\operatorname{per}(\Gamma, \Omega) = \operatorname{per}(\Omega \setminus \Gamma, \Omega),\\
&\operatorname{per}(\Gamma) + \operatorname{per}(\Omega \setminus \Gamma) = \operatorname{per}(\Omega) + 2\operatorname{per}(\Gamma, \Omega),\\
&\operatorname{per}(\Gamma, \Omega) \leqq \operatorname{per}(\Gamma),\\
&\operatorname{per}(\Gamma, \Omega) = \operatorname{per}(\Gamma) \quad \text{if } \Gamma \subset\subset \Omega.
\end{aligned} \tag{2.7}$$

Here $\Gamma \subset\subset \Omega$ signifies that the closure of $\Gamma$ is a compact subset of $\Omega$.

Given an arbitrary part $\Gamma$, MASSARI & PEPE [1974a, b] have established the following sequences $\{\Gamma_n\}$ of parts:

$$\begin{gathered}
\text{A sequence } \{\Gamma_n\} \text{ such that } \operatorname{vol}(\Gamma_n) = \operatorname{vol}(\Gamma),\\
\operatorname{per}(\Gamma_n) = \operatorname{per}(\Gamma_n, \Omega), \quad \Gamma_n \to \Gamma, \quad \operatorname{per}(\Gamma_n) \to \operatorname{per}(\Gamma).
\end{gathered} \tag{2.8}$$

$$\begin{gathered}
\text{A sequence } \{\Gamma_n\} \text{ such that } \Gamma_n \subset \Gamma,\\
\Gamma_n \subset\subset \Omega, \quad \Gamma_n \to \Gamma, \quad \operatorname{per}(\Gamma_n) \to \operatorname{per}(\Gamma).
\end{gathered} \tag{2.9}$$

## 3. The liquid-drop problem $\mathscr{F}_v$

In this section we shall discuss the liquid-drop problem under the assumptions (A1) and:

(A2) $\qquad \sigma > 0; \quad \sigma_A, \sigma_B \geqq 0.$

[1] Parts $\Gamma$ and $\Lambda$ with $\operatorname{vol}(\Gamma \Delta \Lambda) = 0$ are considered equivalent.

For this problem to have meaning (*cf.* the Introduction), we must give a precise sense to the set functions $I$ and $S$ which appear in (1.10). In view of the discussion in the second and third paragraphs of subsection 2b, and bearing in mind (1.9), for every part $\Gamma$ we define

$$\begin{gathered} F(\Gamma) := I(\Gamma) + \lambda S(\Gamma), \\ I(\Gamma) := \operatorname{per}(\Gamma, \Omega), \qquad S(\Gamma) := \operatorname{per}(\Gamma) - \operatorname{per}(\Gamma, \Omega). \end{gathered} \tag{3.1}$$

With these definitions, the **liquid-drop problem** becomes

$$\mathscr{F}_v \begin{cases} \text{minimize } F(\Gamma) \text{ over all} \\ \text{parts } \Gamma \text{ with } \operatorname{vol}(\Gamma) = v. \end{cases}$$

We shall consider $\mathscr{F}_v$ with $v \in [0, 1]$. (The solutions for $v = 0$ and $v = 1$ are, of course, $\Gamma = \emptyset$ and $\Gamma = \Omega$, respectively.)

The next theorem, for the most part well known, shows that $\mathscr{F}_v$ has a solution, at least for $|\lambda| \leqq 1$ (*i.e.*, for $|\sigma_A - \sigma_B| \leqq \sigma$). The theorem shows also that the function

$$f(v) := \inf\{F(\Gamma) \colon \Gamma \text{ is a part with } \operatorname{vol}(\Gamma) = v\} \tag{3.2}$$

is well behaved on [0, 1]; this fact will be useful when we turn to the more difficult problem $\mathscr{P}_m$.

**Theorem 3.1.** *Assume that* $|\lambda| \leqq 1$. *Then for each* $v \in [0, 1]$, *Problem* $\mathscr{F}_v$ *has a solution, and each solution* $\Gamma$ *has* $\partial\Gamma \cap \Omega$ *a* $C^\infty$ *surface. Moreover,*

(i) *For* $n = 1$, $f(v)$ *is constant on* (0, 1) *and lower semi-continuous on* [0, 1];

(ii) *for* $n \geqq 2$, $f(v)$ *is continuous on* [0, 1].

The first sentence of the theorem is known,[1] while assertion (i) is obvious. Assertion (ii) is proved[2] in the Appendix.

The case $|\lambda| > 1$ is more interesting. Suppose, without loss in generality, that

$$\lambda > 1 \tag{3.3}$$

($\sigma_A > \sigma + \sigma_B$). Then by adding a "thin layer" of phase $B$ to the surface of contact between phase $A$ and the container wall we can lower the contact energy from

$$\sigma_A S(\Gamma) \quad \text{to} \quad (\sigma + \sigma_B)\, S(\Gamma).$$

We should therefore expect that $\mathscr{F}_v$ has no solution with

$$S(\Gamma) \neq 0.$$

---

[1] *Cf.* Massari & Pepe [1974b] for a proof of existence. Regularity is discussed by Miranda [1973]; Massari [1974]; Gonzalez, Massari, & Tamanini [1981, 1983]. Existence and regularity follow also from the general results of Almgren [1976].

[2] Almgren (private communication) shows that $f(v)$ has right and left derivatives at each $v$ in (0, 1).

These observations tell us that the correct problem to study is not $\mathscr{F}_v$ but rather the problem $\mathscr{F}_v^*$ in which $\sigma_A$ is replaced by[1]

$$\sigma_A^* := \sigma + \sigma_B;$$

or equivalently, in which $\lambda$ is replaced by one and $F(\Gamma)$ by

$$F^*(\Gamma) := \operatorname{per}(\Gamma) \tag{3.4}$$

(*cf.* (3.11)):

$$\mathscr{F}_v^* \begin{cases} \text{minimize per}\,(\Gamma) \text{ over all} \\ \text{parts } \Gamma \text{ with vol}\,(\Gamma) = v. \end{cases}$$

The next theorem[2] justifies these assertions.

**Theorem 3.2.** *Assume that* $\lambda > 1$. *Then for each* $v \in [0, 1]$, *Problem* $\mathscr{F}_v^*$ *is the natural regularization of Problem* $\mathscr{F}_v$. *Moreover, the solution set of* $\mathscr{F}_v$ *consists exactly in those solutions* $\Gamma$ *of* $\mathscr{F}_v^*$ *with* $S(\Gamma) = 0$.

**Proof.** Choose $v \in (0, 1)$. (The cases $v = 0, 1$ are trivial.) To establish the first assertion we must show that $F^*$ is the lsc envelope of $F$ over

$$P_v := \{\text{parts } \Gamma: \text{vol}\,(\Gamma) = v\},$$

with topology induced by (2.5). By (3.1) and (3.4),

$$F^*(\Gamma) = F(\Gamma) - (\lambda - 1)\, S(\Gamma), \tag{3.5}$$

and, by $(2.7)_3$ and (3.1),

$$S(\Gamma) \geqq 0; \tag{3.6}$$

hence

$$F^*(\Gamma) \leqq F(\Gamma). \tag{3.7}$$

Choose $\Gamma \in P_v$ and let $\{\Gamma_n\}$ be a sequence in $P_v$ converging to $\Gamma$. Then, by (2.6), (3.4), and (3.7),

$$F^*(\Gamma) \leqq \liminf F^*(\Gamma_n) \leqq \liminf F(\Gamma_n).$$

On the other hand, if we choose $\{\Gamma_n\}$ to be the sequence (2.8),

$$F(\Gamma_n) = \operatorname{per}(\Gamma_n\ \Omega) = \operatorname{per}(\Gamma_n) \to \operatorname{per}(\Gamma) = F^*(\Gamma);$$

hence $F^*$ is the lsc envelope of $F$ over $P_v$.

To prove the last assertion of Theorem 3.2 we must show that

$$\Gamma \text{ solves } \mathscr{F}_v \Leftrightarrow \Gamma \text{ solves } \mathscr{F}_v^* \text{ and } S(\Gamma) = 0.$$

Suppose $\Gamma$ solves $\mathscr{F}_v$. Then by Proposition 2.2(i, iii), $F(\Gamma) = F^*(\Gamma)$ and (3.5) yields $S(\Gamma) = 0$. Conversely, if $\Gamma$ solves $\mathscr{F}_v^*$ and $S(\Gamma) = 0$, then $F(\Gamma) = F^*(\Gamma) = \inf F^* = \inf F$ and $\Gamma$ solves $\mathscr{F}_v$. □

---

[1] *Cf.* the experiments of HEADY & CAHN [1973], which indicate this phenomenon, as well as the discussion of CAHN [1977] for the corresponding one-dimensional problem. See also MODICA [1986].

[2] The first assertion is the content of a letter of mine to U. MASSARI & L. PEPE written in January, 1985. Concerning the second assertion, MASSARI & PEPE [1974b] show that solutions of $\mathscr{F}_v$ necessarily have $S(\Gamma) = 0$.

**Remark 3.1.** By Theorem 3.2, $\mathcal{F}_v$ will have a solution if and only if $\mathcal{F}_v^*$ has a solution $\Gamma$ with $S(\Gamma)=0$. For $v$ small enough, $\mathcal{F}_v^*$ will have (at least some) solutions which are balls $\Gamma$ with $\Gamma \subset\subset \Omega$. Such $\Gamma$ trivially have $S(\Gamma)=0$ and therefore solve $\mathcal{F}_v$. On the other hand, for $v$ close to $\operatorname{vol}(\Omega)=1$ one would *generally* expect solutions of $\mathcal{F}_v^*$ to have nontrivial contact with the container walls ($S(\Gamma) \neq 0$). But this is not *always* true: for $\Omega$ an open ball and $v \in (0, 1)$, $\mathcal{F}_v^*$ is solved by any ball in $\Omega$ with volume $v$. There are less trivial regions in which $\mathcal{F}_v$ has a solution for each $v \in (0, 1)$. For example[1], the region $\Omega \subset \mathbb{R}^2$ shown in Figure 5; $\Omega$ is the union of four discs of equal radii, arranged symmetrically.

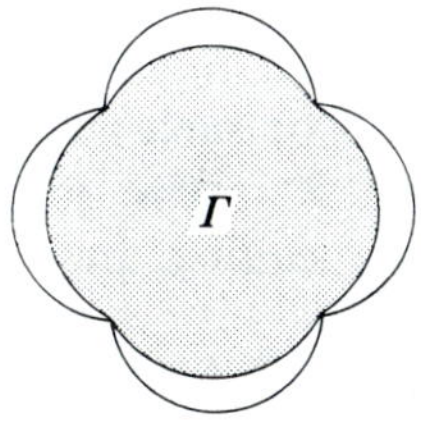


Fig. 5. An example of a region $\Omega$ in which the liquid-drop problem $\mathcal{F}_v$ with $\lambda > 1$ has a solution for all $v$. The shaded area illustrates a typical solution $\Gamma$ for $v$ large

**Open Problem.** Delineate the class $\mathcal{M}$ of nonspherical regions $\Omega$ with the following property; for each $v \in (0, 1)$ some solution $\Gamma$ of $\mathcal{F}_v^*$ has $S(\Gamma)=0$. (This is exactly the class of nonspherical regions in which the liquid drop problem $\mathcal{F}_v$ with $\lambda > 1$ has a solution for all $v \in (0, 1)$.)

**Conjecture.** $\mathcal{M}$ is properly contained in the family of nonconvex $\Omega$ with $\partial\Omega$ the union of spherical surface elements of equal radii.

## 4. The two-phase problem $\mathcal{P}_m$

### *4a. Preliminaries*

We continue to assume that (A1) and (A2) are satisfied, but now adjoin the additional hypotheses:

(A3) $A$ and $B$ are disjoint, open intervals with $A < B$.

(A4) $\psi \in C^2(A \cup B)$; $\psi$ is strictly convex on $A$ and on $B$.

(A5) $\psi(u)$ and $|\psi'(u)|$ tend to $+\infty$ as $u$ tends to the boundary[2] of $A$ or $B$.

(A6) there exist $\alpha_0 \in A$, $\beta_0 \in B$ such that

$$\psi(\alpha_0) = \psi(\beta_0) = 0, \quad \psi(u) > 0 \text{ otherwise}.$$

(A7) $m \in (\inf A, \sup B)$.

---

[1] This example was arrived at in discussions with W. HRUSA & V. MIZEL.

[2] If $A$ {respectively, $B$} is unbounded, then $-\infty$ {respectively, $+\infty$} is considered a boundary point.

Assumption (A7) is a requisite of the constraint $M(u) = m$ in conjunction with our agreement that $\text{vol}\,(\Omega) = 1$.

Assumptions (A3)–(A6) imply that $\psi$ have the form shown in Figure 2. In view of (A4) and (A5), for any $\mu \in \mathbb{R}$ the equation $\psi'(u) = \mu$ has exactly two solutions, one in $A$ and one in $B$; labelling these by $\alpha(\mu)$ and $\beta(\mu)$, respectively, (*cf.* Figure 2)

$$\alpha : \mathbb{R} \to A, \qquad \beta : \mathbb{R} \to B,$$

and

$$\psi'(\alpha(\mu)) = \psi'(\beta(\mu)) = \mu. \tag{4.1}$$

By an **admissible field** we mean a function[1] $u \in \text{BV}(\Omega)$ with $\psi(u) \in L^1(\Omega)$ and $u(x) \in A \cup B$ for all $x \in \Omega$. For such fields the sets $\Omega_i(u)$ $(i = A, B)$ are parts and the interfacial and contact areas, $I(u)$ and $S_i(u)$, introduced in (1.2) and (1.3), can be given a precise meaning. Indeed, using (3.1) as our guide, we define

$$\begin{aligned} I(u) &:= \text{per}\,(\Omega_A(u), \Omega) = \text{per}\,(\Omega_B(u), \Omega), \\ S_i(u) &:= \text{per}\,(\Omega_i(u)) - I(u), \qquad i = A, B. \end{aligned} \tag{4.2}$$

Then, defining the superficial energy by (1.5) and the (total) energy $E(u)$ by (1.6), we have the following statement of the **two-phase problem**:

$$\mathcal{P}_m \begin{cases} \text{minimize } E(u) \text{ over all admissible} \\ u \text{ that satisfy } M(u) = m. \end{cases}$$

The following terminology will be useful. Let $u$ be admissible. Then the pair $\Omega_A(u)$, $\Omega_B(u)$ constitutes the **phase configuration** of $u$, while

$$\text{vol}_A\,(u) := \text{vol}\,(\Omega_A(u)), \qquad \text{vol}_B\,(u) := \text{vol}\,(\Omega_B(u))$$

are the **phase volumes** of $u$. Further:

(1) $u$ is **single phase** if either $\text{vol}_A(u) = 0$ or $\text{vol}_B(u) = 0$, otherwise $u$ is **two phase.**

(2) $u$ is **phasewise constant** if there exist constants $a \in A$, $b \in B$ such that

$$u(x) = \begin{cases} a, & x \in \Omega_A(u) \\ b, & x \in \Omega_B(u), \end{cases}$$

in which case $a$ and $b$ are the **phase values** of $u$. (Trivially, $u$ is phasewise constant and single phase if and only if $u(x) = \text{constant}$, so that $u$ has only one phase value.)

(3) $u$ **minimizes superficial energy** if $\Omega_A(u)$ solves the liquid-drop problem $\mathcal{F}_v$ with $v = \text{vol}\,(\Omega_A(u))$.

---

[1] Roughly speaking, $\text{BV}(\Omega)$ is the set of functions $u \in L^1(\Omega)$ whose distributional gradients are measures. For that reason, the elements of $\text{BV}(\Omega)$ are not functions, but rather equivalence classes of functions that coincide almost everywhere. Since we make only elementary use of BV-theory, we shall not make this distinction; hence many of our statements are modulo a rearrangement on a set of zero volume.

Note that, by (1.10) and the sentence containing (1.8),

$$\Phi(u) = \sigma F(\Omega_A(u)) + C; \tag{4.3}$$

thus fields consistent with (3) in fact have least superficial energy when compared to other fields with phase volumes equal to those of $u$.

It is useful to consider the reduced problem:

$$\mathcal{R}_m \begin{cases} \text{minimize } E(u) \text{ over all phasewise} \\ \text{constant } u \text{ that satisfy } M(u) = m. \end{cases}$$

**Proposition 4.1.**

(i) *$u$ solves $\mathcal{P}_m$ if and only if $u$ solves $\mathcal{R}_m$.*

(ii) *Every solution of $\mathcal{P}_m$ minimizes superficial energy.*

**Proof.** The following two facts, valid whenever $u$ and $w$ are admissible, are obvious consequences of the previous denfinitions:

(a) $\Phi(u) = \Phi(w)$ if $u$ and $w$ have the same phase configuration;

(b) $M(u) = M(w)$ and $\Psi(u) = \Psi(w)$ if $u$ and $w$ are phasewise constant with the phase values and phase volumes of $u$ equal to those of $w$.

To prove (i) it suffices to show that

$$\left.\begin{array}{l} \text{given an admissible } w \text{ that is not phasewise} \\ \text{constant, there is a phasewise constant } u \\ \text{with } M(u) = M(w) \text{ and } E(u) < E(w). \end{array}\right\} \tag{4.4}$$

Let $w$, as in (4.4), be two phase, and let $u$ be the phasewise-constant field which has the same phase configuration as $w$ and phase values $a$ and $b$ equal to the average values of $w$ over $\Omega_A(w)$ and $\Omega_B(w)$, respectively. Then $M(u) = M(w)$ and, by (a), $\Phi(u) = \Phi(w)$; hence we have only to show that $\Psi(u) < \Psi(w)$. Let $\Omega_i = \Omega_i(u)$, $i = A, B$. Since $\Psi$ is convex on $A$,

$$\int_{\Omega_A} [\psi(w) - \psi(u)] = \int_{\Omega_A} [\psi(w) - \psi(a) - \psi'(a)(w - a)] \geqq 0$$

(which is Jensen's inequality). Similarly,

$$\int_{\Omega_B} [\psi(w) - \psi(u)] \geqq 0.$$

Moreover, since $w$ is not phasewise constant, one of the above inequalities is strict and $\Psi(u) < \Psi(w)$.

The proof of (i) for $w$ single phase is analogous.

Consider (ii). Suppose that $u$ solves $\mathcal{P}_m$ but does not minimize superficial energy. Then there is a part $\Gamma$ with $\operatorname{vol}(\Gamma) = \operatorname{vol}_A(u)$ and

$$F(\Gamma) < F(\Omega_A(u)).$$

By (i), $u$ is phasewise constant. Let $w$ be the phasewise-constant field whose phase values equal those of $u$ and which has $\Omega_A(w) = \Gamma$, $\Omega_B(w) = \Omega \setminus \Gamma$. Then, by (4.3), $\Phi(w) < \Phi(u)$, and, in view of (b), $M(w) = M(u)$, $E(w) < E(u)$, which contradicts the hypothesis that $u$ solve $\mathcal{P}_m$. □

### 4b. Solution of $\mathscr{P}_m$ for $|\sigma_A - \sigma_B| \leqq \sigma$

We assume throughout this subsection that

$$|\sigma_A - \sigma_B| \leqq \sigma. \tag{4.5}$$

By (3.2) and (4.3), if $u$ has minimal interface, then

$$\Phi(u) = \varphi\,(\mathrm{vol}_A\,(u))$$

with

$$\varphi(v) = \sigma f(v) + C. \tag{4.6}$$

Thus, in view of Proposition 4.1, Problem $\mathscr{P}_m$ reduces to minimizing

$$\Psi(u) + \varphi(\mathrm{vol}_A\,(u)) \tag{4.7}$$

over all

$$\begin{gathered}\text{phasewise constant fields } u \text{ that have minimal}\\ \text{superficial energy and satisfy } M(u) = m,\end{gathered} \tag{4.8}$$

and on this class (4.7) gives the total energy $E(u)$.

Assume, for convenience, that

$$m \in A;$$

the analysis for $m \notin A$ is no different.

Let $u$, consistent with (4.8), be single phase. Then, since $\mathrm{vol}\,(\Omega) = 1$,

$$u(x) \equiv m$$

and

$$E(u) = \psi(m) + \varphi(1). \tag{4.9}$$

If $u$ is two phase, with phase values $a$, $b$ and phase volumes $\nu_A$, $\nu_B$, then $M(u) = m$ and $\mathrm{vol}\,(\Omega) = 1$ yield

$$a\nu_A + b\nu_B = m,$$

$$\nu_A + \nu_B = 1,$$

and these equations imply that $a < m$ and

$$\begin{aligned}\nu_A &= \frac{b-m}{b-a} =: v_m(a, b),\\ \nu_B &= \frac{m-a}{b-a} = 1 - v_m(a, b);\end{aligned} \tag{4.10}$$

hence

$$E(u) = e_m(a, b) := \psi_m(a, b) + \varphi(v_m(a, b)) \tag{4.11}$$

with

$$\psi_m(a, b) := v_m(a, b)\,\psi(a) + [1 - v_m(a, b)]\,\psi(b). \tag{4.12}$$

The function $e_m$ is well defined on the set

$$T_m := \{(a, b) \in A \times B : a \leqq m\}.$$

For $a = m$ and $b \in B$ arbitrary, (4.11) reduces to (4.9). Thus $\mathscr{P}_m$ reduces to minimizing $e_m$ over $T_m$, provided we identify the entire boundary line $\{(m, b): b \in B\}$ with the single-phase field $u(x) \equiv m$.

We say that an admissible field $u$ has **constant chemical potential** $\mu$ if

$$\psi'(u(x)) = \mu \ (= \text{constant})$$

for all $x \in \Omega$. We now show that solutions of $\mathscr{P}_m$ have this property by showing that the minimum of $e_m$ over $T_m$ occurs on the curve

$$\mathscr{C} := \{(\alpha(\mu), \beta(\mu)) : \mu \in \mathbb{R}\} = \{(a, b) \in A \times B : \psi'(a) = \psi'(b)\}$$

(*cf.* (4.1)). In view of (4.1) and (A4)–(A6) (cf. Figure 2):

(1) $\alpha' > 0$, $\beta' > 0$;

(2) $(\alpha(\mu), \beta(\mu))$ tends to $(\inf A, \inf B)$ or $(\sup A, \sup B)$ according as $\mu$ tends to $-\infty$ or $+\infty$;

(3) $\psi'(b) > \psi'(a)$ for $(a, b)$ above $\mathscr{C}$, $\psi'(b) < \psi'(a)$ for $(a, b)$ below $\mathscr{C}$ ("above" and "below" in the obvious sense);

and we have the situation shown in Figure 6.

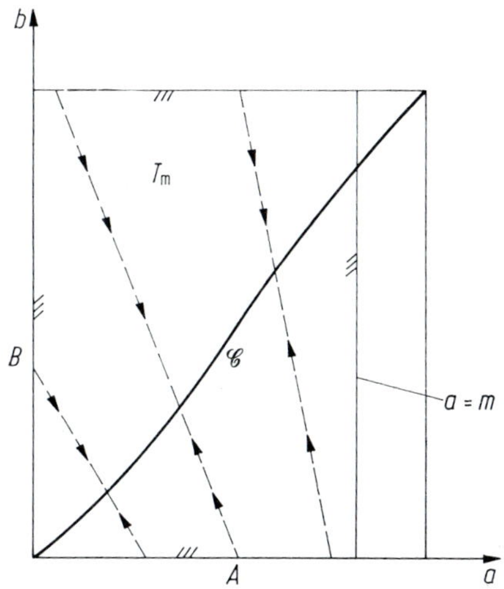


Fig. 6. The set $T_m$ and the curve $\mathscr{C}$ of constant chemical-potential; $v_m$ is constant on the dashed lines; the arrows give the direction of decreasing $\psi_m$

By (4.10), for each $c \in (0, 1)$ the set

$$\ell_c := \{(a, b) \in T_m : v_m(a, b) = c\} \tag{4.13}$$

is a straight line-segment of the form

$$b = b_c(a) = -ka + K, \qquad k = c/(1-c) > 0, \quad K = m/(1-c). \tag{4.14}$$

It is clear from the properties of $\mathscr{C}$ that: each such $\ell_c$ crosses $\mathscr{C}$ exactly once, but does not intersect the vertical line $b = m$; the family of lines $\ell_c$, $c \in (0, 1)$, completely fills the interior of $T_m$.

Next, fix $c \in (0, 1)$ and consider $b$ and $\psi_m$ on $\ell_c$ as functions $b = b_c(a)$ and $\psi_c(a) = \psi_m(a, b_c(a))$ of $a$. Then, by (4.12)–(4.14),

$$\frac{d\psi_c(a)}{da} = c[\psi'(a) - \psi'(b)],$$

and we conclude from (3) that on $\ell_c$, $\psi_m(a, b)$ has a strict minimum at $\ell_c \cap \mathscr{C}$. Thus, since $\varphi(v_m(a, b))$ is constant on $\ell_c$, $e_m(a, b)$, restricted to $\ell_c$, also has a strict minimum at $\ell_c \cap \mathscr{C}$. Moreover, $e_m(a, b)$ is constant on the boundary line $b = m$. Thus minimizing $e_m(a, b)$ over $T_m$ is equivalent to minimizing the function

$$\mathscr{E}_m(\mu) := e_m(\alpha(\mu), \beta(\alpha)) = \psi_m(\alpha(\mu), \beta(\mu)) + \varphi(v_m(\alpha(\mu), \beta(\mu))) \tag{4.15}$$

on the set $(-\infty, \mu_0]$, where $\mu_0$ is the value of $\mu$ at which $\alpha(\mu) = m$.

Write $\psi_m(\mu) = \psi_m(\alpha(\mu), \beta(\mu))$, $v_m(\mu) = v_m(\alpha(\mu), \beta(\mu))$; then

$$\psi_m(\mu) = v_m(\mu)\,\psi(\alpha(\mu)) + [1 - v_m(\mu)]\,\psi(\beta(\mu)).$$

By hypothesis (A5) and (2), $\psi(\alpha(\mu))$ and $\psi(\beta(\mu))$ tend to $+\infty$ as $\mu \to -\infty$. Further, in this limit $\beta(\mu) \to \inf B > \sup A > -\infty$, so that, by (4.10), $1 - v_m(\mu)$ is bounded away from zero as $\mu \to -\infty$, even for $A$ unbounded. Thus $\psi_m(\mu) \to -\infty$ as $\mu \to -\infty$. We therefore conclude from (4.15), (4.6) and the lower semicontinuity of $f$ (Theorem 3.1) that $\mathscr{E}_m(\mu)$ is lower semicontinuous on $(-\infty, \mu_0]$ with $\mathscr{E}_m(\mu) \to +\infty$ as $\mu \to -\infty$. Thus $\mathscr{E}_m(\mu)$ has a minimum on $(-\infty, \mu_0]$.

An almost identical analysis holds for the remaining case, $m \in A$, the only essential change being to replace $(-\infty, \mu_0]$ by

$$U_m := \{\mu \in \mathbb{R} : \alpha(\mu) \leqq m \leqq \beta(\mu)\}$$

(which equals $(-\infty, \mu_0]$ when $m \in A$).

The conclusions of this section are summarized in the next theorem, which is our main result.

**Theorem 4.1.** *Problem $\mathscr{P}_m$ has a solution, and each solution $u$ has the following properties:*

(i) *$u$ minimizes superficial energy;*

(ii) *$u$ has constant chemical potential $\mu$ with $\mu$ a minimizer of $\mathscr{E}_m$ over $U_m$;*

(iii) *$E(u) = \mathscr{E}_m(\mu)$;*

(iv) *$u$ is phasewise constant: if $u$ is single phase, then $u(x) \equiv m$; if $u$ is two phase, then $u$ has phase values $\alpha(\mu)$ and $\beta(\mu)$, phase volumes*

$$\frac{\beta(\mu) - m}{\beta(\mu) - \alpha(\mu)} \quad \text{and} \quad \frac{m - \alpha(\mu)}{\beta(\mu) - \alpha(\mu)}.$$

*Conversely, any minimizer of $\mathscr{E}_m$ over $U_m$ is the chemical potential of a solution of $\mathscr{P}_m$.*

*4c. The case $|\sigma_A - \sigma_B| > \sigma$*

Suppose without, loss in generality, that

$$\sigma_A > \sigma + \sigma_B;$$

Then a discussion identical to that given in the paragraph containing (3.3) leads to the conclusion that the "correct" problem[1] to study is not $\mathscr{P}_m$ but rather the problem $\mathscr{P}_m^*$ in which $\sigma_A$ is replaced by

$$\sigma_A^* = \sigma + \sigma_B; \tag{4.16}$$

that is, the problem

$$\mathscr{P}_m^* \begin{cases} \text{minimize } E^*(u) \text{ over all admissible} \\ u \text{ that satisfy } M(u) = m \end{cases}$$

with

$$\begin{aligned} E^*(u) &:= \Psi(u) + \Phi^*(u), \\ \Phi^*(u) &:= \sigma I(u) + \sigma_A^* S_A(u) + \sigma_B S_B(u) \end{aligned} \tag{4.17}$$

(*cf.* (4.2)). The next theorem makes this assertion precise; there the underlying topological space is

$$\mathscr{A}_m := \{\text{admissible } u \colon M(u) = m\}$$

with topology induced by $L^1(\Omega)$.

**Theorem 4.2.** *Assume that $\sigma_A > \sigma + \sigma_B$. Then Problem $\mathscr{P}_m^*$ is the natural regularization of Problem $\mathscr{P}_m$. Moreover, the solution set of $\mathscr{P}_m$ consists exactly of those solutions $u$ of $\mathscr{P}_m^*$ with $S_A(u) = 0$.*

**Proof.** By (4.2) and $(2.7)_2$,

$$S_A(u) + S_B(u) = \operatorname{per}(\Omega), \qquad S_A(u) + I(u) = \operatorname{per}(\Omega_A(u)), \tag{4.18}$$

and therefore (1.5), (1.6), (4.16), and (4.17) yield (cf. (3.6) and (3.7))

$$\begin{aligned} E^*(u) &= E(u) - (\sigma_A - \sigma - \sigma_B)\, S_A(u) \leqq E(u), \\ E^*(u) &= \Psi(u) + \sigma \operatorname{per}(\Omega_A(u)) + C, \qquad C = \sigma_B \operatorname{per}(\Omega). \end{aligned} \tag{4.19}$$

To show that $\mathscr{P}_m^*$ naturally regularizes $\mathscr{P}_m$ we must show that

$$E^* \text{ is the lsc envelope of } E \text{ over } \mathscr{A}_m. \tag{4.20}$$

Thus let $\{u_n\}$ be a sequence in $\mathscr{A}_m$ converging to $u \in \mathscr{A}_m$. Since $\psi \geqq 0$, we may use Fatou's Lemma to conclude that

$$\Psi(u) \leqq \liminf \Psi(u_n). \tag{4.21}$$

[1] For dimension $N = 1$ this conclusion was reached independently by T. Pence (private communication).

Next, let $\chi$ and $\chi_n$, respectively, denote the characteristic functions of $\Omega_A(u)$ and $\Omega_A(u_n)$. Since $u_n \to u$ in $L^1(\Omega)$, $\chi_n \to \chi$ almost everywhere in $\Omega$ (*cf.* (A3)), so that $\chi_n \to \chi$ in $L^1(\Omega)$, which is equivalent to

$$\Omega_A(u_n) \to \Omega_A(u) \tag{4.22}$$

(with respect to the metric (2.5)). Thus, by $(4.19)_2$, (4.21), and (2.6),

$$E^*(u) \leqq \liminf E^*(u_n)$$

and $E^*$ is lsc on $\mathscr{A}_m$. Further, since $E^* \leqq E$,

$$E^*(u) \leqq \liminf E^*(u_n) \leqq \liminf E(u_n). \tag{4.23}$$

Our next step in establishing (4.20) is to construct a sequence $\{u_n\}$ which leads to equality in (4.23). Thus choose $u \in \mathscr{A}_m$. We will construct a sequence $\{u_n\}$ in $\mathscr{A}_m$ with the following properties:

$$\begin{gathered} S_A(u_n) = 0, \qquad u_n \to u \quad \text{in } L^1(\Omega), \\ \Psi(u_n) \to \Psi(u), \qquad \operatorname{per}(\Omega_A(u_n)) \to \operatorname{per}(\Omega_A(u)). \end{gathered} \tag{4.24}$$

For such a sequence (4.19) yields

$$E(u_n) = E^*(u_n) \to E^*(u),$$

and this result, with (4.23), implies the desired conclusion (4.20).

To build the sequence (4.24), let $\Gamma = \Omega_A(u)$ and let $\{\Gamma_n\}$ denote the sequence (2.9). There is an $x \in \Omega$ such that the open ball $\mathscr{B}_\varrho$ with radius $\varrho$ centered at $x$ has the following property: for all sufficiently small $\varrho$, say $0 < \varrho < \varrho_0$,

$$\mathscr{B}_\varrho \subset\subset \Omega, \qquad \operatorname{vol}((\Omega \setminus \Gamma) \cap \mathscr{B}_\varrho) > 0. \tag{4.25}$$

Choose $a \in A$, $b \in B$. Then, by (4.25), for each sufficiently large $n$, say $n > n_0$, we can choose a member $\Theta_n$ from the family of sets of the form $(\Omega \setminus \Gamma) \cap \mathscr{B}_\varrho$, $\varrho \in (0, \varrho_0)$ such that

$$\int_{\Gamma \setminus \Gamma_n} (b - u) = \int_{\Theta_n} (u - a), \qquad \operatorname{vol}(\Theta_n) \to 0, \tag{4.26}$$

where we have used the facts: $\Gamma_n \to \Gamma$, $u(x) \in A$ on $\Gamma \setminus \Gamma_n$, $u(x) \in B$ on $\Theta_n$.

For convenience, remove the first $n_0$ terms of $\{\Gamma_n\}$ and relabel the $\Gamma_n$ and $\Theta$ so that (4.26) holds for all $n$. Choose

$$u_n(x) = \begin{cases} a, & x \in \Theta_n \\ b, & x \in \Gamma \setminus \Gamma_n \\ u(x), & \text{otherwise}. \end{cases} \tag{4.27}$$

Then each $u_n$ is admissible. Moreover, (4.26) yields $M(u_n) = M(u)$, so that $u_n \in \mathscr{A}_m$. Further, since $\operatorname{vol}(\Gamma \setminus \Gamma_n)$ and $\operatorname{vol}(\Theta_n)$ tend to zero, and since $\psi(u) \in L^1(\Omega)$ (because $u$ is admissible), we may use (4.27) to conclude that

$$u_n \to u \text{ in } L^1(\Omega), \qquad \Psi(u_n) \to \Psi(u).$$

Next,

$$\Omega_A(u_n) = \Theta_n \cup \Gamma_n, \tag{4.28}$$

and, by (2.9) and (4.25), $\Theta_n \cup \Gamma_n \subset\subset \Omega$; thus, in view of $(2.7)_4$ and (4.2), $S_A(u_n) = 0$. Further, using (2.9) and an argument of MASSARI & PEPE [1974b; (3.18)–(3.20)],

$$\text{per}\,(\Theta_n \cup \Gamma_n) \to \text{per}\,(\Gamma)$$

which, with (4.28), yields the last condition in (4.24). This completes the proof of (4.20).

The final assertion of the theorem follows from $(4.19)_1$ and an argument the same as that given in the last paragraph of the proof of Theorem 3.2. □

As noted in Remark 3.1, one would expect that, for each $\Omega$ in a large class of regions, and for $v$ large (close to one), each solution of $\mathcal{F}_v^*$ would nontrivially contact $\partial\Omega$. For each $\Omega$ in this class the liquid-drop problem $\mathcal{F}_v$ for $\lambda > 1$ has no solution for large $v$. The next theorem relates this loss of existence to the analogous issue for the two-phase problem.

**Theorem 4.3.** *Assume that* $\sigma_A > \sigma + \sigma_B$, *and that* $\Omega$ *is such that the liquid-drop problem* $\mathcal{F}_v$ *for* $\lambda > 1$ *has no solution for* $v$ *sufficiently close to one. Assume further that* $\inf A > -\infty$. *Then for all* $m \in A$ *sufficiently close to* $\inf A$, *Problem* $\mathcal{P}_m$ *has no solution.*

**Proof.** Let $v_0 \in (0, 1)$ be such that, for each $v \in (v_0, 1)$, $\mathcal{F}_v$ has no solution, and hence such that, for each $v \in (v_0, 1)$, every solution $\Gamma$ of $\mathcal{F}_v^*$ has $S(\Gamma) \neq 0$. Then, by Theorem 4.1(i) (applied to $\mathcal{P}_m^*$) in conjunction with Theorem 4.2,

$$\text{every solution } u \text{ of } \mathcal{P}_m \text{ solves } \mathcal{P}_m^* \text{ and has } \operatorname{vol}_A(u) \leqq v_0. \tag{4.29}$$

Let $m \in A$ and define

$$g_m(\mu) = \frac{\beta(\mu) - m}{\beta(\mu) - \alpha(\mu)},$$

so that, by Theorem 4.1(iv), for any solution $u$ of $\mathcal{P}_m^*$, $\operatorname{vol}_A(u)$ belongs to the set

$$V_m := \{g_m(\mu) : \mu \in \mathbb{R}, \alpha(\mu) \leqq m\}.$$

Thus, in view of (4.29), it suffices to show that for $m$ sufficiently close to $\inf A$,

$$V_m \subset (v_0, 1]. \tag{4.30}$$

To verify (4.30) we note first that

$$g_m' = (\beta - \alpha)^{-1}\,[(1 - g_m)\,\beta' + g_m \alpha'];$$

hence, appealing to properties (1) and (2) of $\mathcal{C}$ (*cf.* the paragraph preceeding (4.13)), $g_m' > 0$ and

$$V_m = \left(\frac{\inf B - m}{\inf B - \inf A}, 1\right)$$

(with all infima finite since $\inf B > \inf A > -\infty$). Therefore (4.30) is satisfied for $m$ sufficiently close to $\inf A$. □

**Remark 4.1.** Consider Theorem 4.3. If $\sigma_B > \sigma + \sigma_A$, then we replace the hypothesis $\inf A > -\infty$ by $\sup B < +\infty$ and conclude that $\mathscr{P}_m$ has no solution for $m \in B$ sufficiently close to $\sup B$.

*Acknowledgment.* This work was supported by the Army Research Office. I gratefully acknowledge valuable discussions with L. CAFFARELLI, W. HRUSA, V. MIZEL, and T. PENCE.

## Appendix: Proof of Theorem 3.1(ii)

By (3.1) and $(2.7)_{1,2}$.

$$S(\Gamma) + S(\Omega \setminus \Gamma) = \operatorname{per}(\Omega), \qquad I(\Gamma) = I(\Omega \setminus \Gamma), \tag{1}$$

hence, writing $f(v, \lambda)$ to denote the dependence of $f(v)$ on $\lambda$,

$$\begin{aligned} f(v, \lambda) &= \inf\{I(\Gamma) + \lambda S(\Gamma): \operatorname{vol}(\Gamma) = v\} \\ &= \inf\{I(\Omega \setminus \Gamma) + \lambda S(\Omega \setminus \Gamma): \operatorname{vol}(\Gamma) = 1 - v\} \\ &= \inf\{I(\Gamma) - \lambda S(\Gamma): \operatorname{vol}(\Gamma) = 1 - v\} + \lambda \operatorname{per}(\Omega) \\ &= f(1 - v, -\lambda) + \lambda \operatorname{per}(\Omega). \end{aligned}$$

Thus it suffices to consider the case

$$\lambda \leqq 0, \tag{2}$$

and since $\operatorname{per}(\Omega, \Omega) = 0$ and $S(\Omega) = \operatorname{per}(\Omega)$, (1) and (3.1) yield

$$F(\Gamma) - F(\Omega) = I(\Gamma) - \lambda S(\Omega \setminus \Gamma) \geqq 0. \tag{3}$$

For the remainder of the proof $\mathscr{B}$ and $\mathscr{B}_\varepsilon$ denote open balls. Trivially,

$$\operatorname{per}(\mathscr{B}) = \gamma \operatorname{vol}(\mathscr{B})^q,$$

where $\gamma = \gamma_N > 0$ is constant and $q = (N - 1)/N$.

Let $\Gamma$ solve $\mathscr{F}_v$ for some $v$ and let $\mathscr{B}_\varepsilon \subset\subset (\Omega \setminus \Gamma)^\circ$ with $\operatorname{vol}(\mathscr{B}_\varepsilon) = \varepsilon$. Then, for $p(\cdot)$ equal to $\operatorname{per}(\cdot)$ or $\operatorname{per}(\cdot, \Omega)$,

$$p(\Gamma \cup \mathscr{B}_\varepsilon) = p(\Gamma) + p(\mathscr{B}_\varepsilon) = p(\Gamma) + \gamma \varepsilon^q;$$

hence

$$F(\Gamma \cup \mathscr{B}_\varepsilon) = F(\Gamma) + \gamma \varepsilon^q. \tag{4}$$

Similarly, for $\mathscr{B}_\varepsilon \subset\subset \mathring{\Gamma}$ with $\operatorname{vol}(\mathscr{B}_\varepsilon) = \varepsilon$,

$$F(\Gamma \setminus \mathscr{B}_\varepsilon) = F(\Gamma) + \gamma \varepsilon^q. \tag{5}$$

Choose $v \in [0, 1)$ and let $\Gamma$ solve $\mathscr{F}_v$. Then for $\varepsilon > 0$ small, there is a ball $\mathscr{B}_\varepsilon \subset\subset (\Omega \setminus \Gamma)^\circ$ with $\operatorname{vol}(\mathscr{B}_\varepsilon) = \varepsilon$, and $Q_\varepsilon := \Gamma \cup \mathscr{B}_\varepsilon$ has $\operatorname{vol}(Q_\varepsilon) = v + \varepsilon$. Thus $f(v + \varepsilon) \leqq F(Q_\varepsilon)$ and, since $f(v) = F(\Gamma)$, (4) yields

$$f(v + \varepsilon) \leqq f(v) + \gamma \varepsilon^q. \tag{6}$$

Similarly, choosing $\mathscr{B}_\varepsilon \subset\subset \mathring{\Gamma}$ and using (5),

$$f(v-\varepsilon) \leqq f(v) + \gamma\varepsilon^q. \tag{7}$$

Next, each solution $\Gamma$ of $\mathscr{F}_v$ has $\operatorname{vol}(\Gamma) = v$ and $\partial\Gamma \cap \Omega$ a $C^\infty$ surface. Thus given any $v_0 \in (0, 1)$ there is a $\delta > 0$ such that, for $v_0 \leqq v \leqq 1$, each solution $\Gamma$ of $\mathscr{F}_v$ has $\mathscr{B}_\varepsilon \subset\subset \mathring{\Gamma}$ for some ball $\mathscr{B} = \mathscr{B}(\Gamma)$ with $\operatorname{vol}(\mathscr{B}) = \delta$. Thus, given $v \in (0, 1)$, for all sufficiently small $\varepsilon > 0$ some solution $\Gamma_{v+\varepsilon}$ of $\mathscr{F}_{v+\varepsilon}$ has $\mathscr{B}_\varepsilon \subset\subset \mathring{\Gamma}_{v+\varepsilon}$ for some ball $\mathscr{B}_\varepsilon$ with $\operatorname{vol}(\mathscr{B}_\varepsilon) = \varepsilon$. If $Q_\varepsilon := \Gamma_{v+\varepsilon} \setminus \mathscr{B}_\varepsilon$, it follows that $\operatorname{vol}(Q_\varepsilon) = v$; thus $f(v) \leqq F(Q_\varepsilon)$ and, since $f(v+\varepsilon) = F(\Gamma_{v+\varepsilon})$, (5) yields

$$f(v) \leqq f(v+\varepsilon) + \gamma\varepsilon^q. \tag{8}$$

A similar argument leads to

$$f(v) \leqq f(v-\varepsilon) + \gamma\varepsilon^q \tag{9}$$

for $v \in (0, 1)$ and $\varepsilon > 0$ small. The inequalities (6)–(9) clearly yield the continuity of $f$ on (0, 1).

We have only to show that $f(v)$ is continuous at $v = 0$. Let $\{v_n\}$ be a sequence with $v_n > 0$ and $v_n \to 0$. Since $f(0) = 0$, (6) at $v = 0$ yields

$$\limsup f(v_n) \leqq 0,$$

and the continuity of $f(v)$ at $v = 0$ will follow if we can show that

$$\liminf f(v_n) \geqq 0. \tag{10}$$

For each $n$ let $\Gamma_n$ solve $\mathscr{F}_{v_n}$. Then, since $\lambda \leqq 0$ and $|\lambda| \leqq 1$,

$$f(v_n) = I(\Gamma_n) + \lambda S(\Gamma_n) \geqq I(\Gamma_n) - S(\Gamma_n),$$

and letting $\Lambda_n = \Omega \setminus \Gamma_n$ we conclude, with the aid of (2), that

$$f(v_n) = \operatorname{per}(\Lambda_n) - \operatorname{per}(\Omega).$$

Further, since $\Lambda_n \subset \Omega$ and $\operatorname{vol}(\Lambda_n) \to 1$, $\Lambda_n \to \Omega$ with respect to the metric (2.5), and we conclude from (2.6) that (10) holds. □

## References


[1954] De Giorgi, E., Su una teoria generale della misura $(r-1)$-dimensionale in uno spazio a $r$ dimensioni. Ann. Mat. Pura Appl. (4) **36**, 191–213.

[1955] De Giorgi, E., Nuovi teoremi relativi alle misure $(r-1)$-dimensionali in uno spazio a $r$ dimensioni. Richerche Mat. **4**, 95–113.

[1969] Federer, H., *Geometric Measure Theory*. Springer-Verlag. Berlin Heidelberg New York.

[1973] Heady, R. B., & J. W. Cahn, Experimental test of classical nucleation theory in a liquid-liquid miscibility gap system. J. Chem. Phys. **58**, 896–910.

[1973] MIRANDA, M., Existence and regularity of hypersurfaces of $\mathbb{R}^n$ with prescribed mean curvature. Proc. Symp. Pure Math. **23**.

[1974] MASSARI, U., Esistenza e regolarità delle ipersuperfici di curvature media assegnata in $\mathbb{R}^n$. Arch. Rational Mech. Anal. **55**, 357–382.

[1974a] MASSARI, U., & L. PEPE, Sull'approssimazione degli aperti lipschitziani di $\mathbb{R}^n$ con varietà differenziabili. Boll. Unione Mat. Ital. **10**, 532–544.

[1974b] MASSARI, U., & L. PEPE, Su di una impostazione parametrica del problema dei capillari. Ann. Univ. Ferrara **20**, 21–31.

[1976] ALMGREN, F., Existence and regularity almost everywhere of solutions to elliptic variational problems with constraints. Mem. Amer. Math. Soc. No. 165.

[1977] CAHN, J. W., Critical point wetting. J. Chem. Phys. **66**, 3667–3672.

[1981] GIUSTI, E., The equilibrium configuration of liquid drops. J. reine angew. Math. **321**, 53–63.

[1983] GONZALEZ, E., U. MASSARI, & I. TAMANINI, On the regularity of boundaries of sets minimizing perimeter with a volume constraint. Indiana U. Math. J. **32**, 25–37.

[1984] GIUSTI, E., *Minimal Surfaces and Functions of Bounded Variation*. Birkhäuser-Verlag: Basel.

[1984] MASSARI, U., & M. MIRANDA, *Minimal Surfaces of Codimension One*. North Holland; Amsterdam.

[1985] GURTIN, M. E., On a theory of phase transitions with interfacial energy. Arch. Rational Mech. Anal. **87**, 187–212.

[1986] MODICA, L., Gradient theory of phase transitions with boundary contact energy. Forthcoming.

Department of Mathematics
Carnegie-Mellon University
Pittsburgh, Pennsylvania

*(Received March 3, 1986)*

# *Spectral Properties of the Laplacian in the Complement of a Deformed Cylinder*

WALTER LITTMAN

*Dedicated to James Serrin*

## Abstract

We prove the absence of eigenvalues in the continuous spectrum for the Dirichlet problem for the Laplacian in the complement of a deformed cylinder.

## Introduction

In the study of wave propagation by means of spectral decomposition it is of great interest to know the location of eigenvalues, in particular whether eigenvalues exist which are embedded in the continuous spectrum. This paper obtains some results which can be used to conclude that there are no positive eigenvalues $\lambda$ for which a square-integrable solution $u$ of the Dirichlet problem for the equation $\Delta u + \lambda u = 0$ exists in a domain which is the complement to a set which in the half space $x_n \geqq 0$ is a cylinder of finite cross section, with generator parallel to the $x_n$ axis, and below the plane $x_n = 0$ is just about anything, as long as the set is contained in a large "circular" cylinder, again with generator parallel to the $x_n$ axis. Indeed we prove a stronger theorem.

*Notation.* $R^{n-1}$ denotes the space of $(n-1)$-tuples $(x_1, \ldots, x_{n-1})$; $R^n$ denotes the space of $n$-tuples $(x_1, \ldots, x_{n-1}, z)$; $B_r$ denotes an open ball of radius $r$ in $R^{n-1}$, centered at the origin. A bar "—" will denote closure. A prime will denote "complement of". $H^m(\Omega)$ denotes the usual Sobolev space, while $H_0^m(\Omega)$ is, as usual, the closure of smooth functions with compact support in $\Omega$ in the norm of $H^m$. "$\partial$" means "boundary of". $\bar{S}'$ denotes the complement of the closure of $S$.

Let $\Omega$ be a bounded $C^3$ domain in $R^{n-1}$ with connected complement such that $\bar{\Omega} \subset B_R$ for a particular value of $R$. Let $C_U$ be the semi-infinite cylinder $\bar{\Omega} \times \{z \geqq 0\}$ and $C_L$ the "larger" cylinder

$$C_L = \bar{B}_R \times \{z \leqq 0\}.$$

Set $G = R^n \setminus \{C_L \cup C_U)$.

**Theorem.** *Let* $u = u(x_1, \ldots, x_{n-1}, z)$ *be a solution to the equation*

$$\Delta u + k^2 u = 0 \quad (k^2 > 0) \text{ in } G$$

*such that* $u \in L^2(G)$, *u is continuous in* $\bar{G}$, *and on* $\partial C_U$, $u \to 0$ *exponentially as* $z \to \infty$, *uniformly in x. Then* $u \equiv 0$ *in* $\bar{G}$.

*Note.* No condition need be imposed on the boundary values for $z < 0$. Also, note that without loss of generality we may take $k = 1$, and indeed we shall. We take $L = \Delta + 1$. We proceed with the proof in several steps. A subscript on a function will usually denote the section where it is introduced.

**1.** In this section we smooth out $u$ in the $z$ direction, to facilitate subsequent arguments. For $0 < \varepsilon < 1$, let

$$u_1(x, z) = \varepsilon^{-1} \int u(x, z') \gamma\left(\frac{z - z'}{\varepsilon}\right) dz'$$

where $\gamma$ is a $C^\infty$ function with support in $(-1, 1)$. $u_1$ is now defined in $G_1$ where $G_1$ is the result of translating $G$ one unit in the positive $u$ direction. $u_1$ satisfies the same equation in $G_1$, is continuous in $\bar{G}_1$, is $L^2(G_1)$ and approaches zero exponentially on $\partial\Omega \times \{z \geqq 1\}$ as $z \to \infty$, uniformly in $x$, together with all $z$ derivatives. We are going to show that $u_1 \equiv 0$ for every $\varepsilon$ $(0 < \varepsilon < 1)$, from which the stated theorem follows.

**2.** In this section we modify the definition of $u_1$ near the boundary of the upper cylinder so that the new function satisfies a zero Dirichlet condition there at the expense of satisfying a non-homogeneous equation, where the right hand side approaches zero exponentially as $z \to \infty$.

Let $R_1 < R_2$ be positive numbers such that $\bar{\Omega} \subset B_{R_1}$ and $R_2 < R$ ($R$ is defined before the theorem). For each *fixed* $z \geqq 1$, let $v(x, z)$ be the solutions to the following Dirichlet problem in $R^{n-1}$:

$$\Delta_x v(x, z) = 0 \text{ in } B_{R_2} \setminus \bar{\Omega}$$

$$v(x, z) = u_1(x, z) \text{ on } \partial\Omega$$

$$v(x, z) = 0 \text{ on } \partial B_{R_2}.$$

$v$ is continuous in $(\bar{B}_{R_2} \setminus \Omega) \times \{z \geqq 1\}$ and is $C^2$ in the interior of the same set. From the maximum principle and step 1 we see that $v \to 0$ exponentially as $z \to \infty$ (uniformly in $\bar{B}_{R_2} \setminus \Omega$), together with all $z$ derivatives. Let $\varphi(r)$ be an infinitely differentiable function of a single variable such that $\varphi(r) = 1$ near the set $\{r \leqq R_1\}$ and $\varphi(r) = 0$ near the set $\{r \geqq R_2\}$.

Now define $u_2$ as follows: For $z > 1$, $r = (x_1^2 + \ldots + x_{n-1}^2)^{\frac{1}{2}}$, $x \in \Omega'$, let

$$u_2(x, z) = u_1 - \varphi(r)\, v(x, z).$$

For $z \leqq 1$, $x \in B'_R$, let $u_2(x, z) = u_1(x, z)$ for $x \in B'_R$.

Let us now compute $Lu_2 \equiv (\Delta + 1)u$. For $z < 1$, $Lu_2 = 0$ for $x \in \bar{B}'_R$, while for $z > 1$, $x \in \bar{\Omega}'$,

$$Lu_2 = -2\nabla_x \varphi \, \Delta_x v - \nabla_x \varphi \cdot v - \varphi v - \varphi \frac{\partial^2 v}{\partial z^2} \equiv f_2$$

(we do not define $f_1$). From the properties of $v$ enumerated, it is clear that $f_2 \to 0$ exponentially as $z \to \infty$ uniformly in $\Omega'$. Furthermore $u_2 \in L^2(G_1)$.

**3.** In this section we modify $u_2$ in such a way that the new function $u_3$ satisfies a zero Dirichlet condition on $\partial\Omega$ for all $z$ as well as the nonhomogeneous equation $Lu_3 = f_3$, where $f_3 \in L^2(\Omega' \times R^1)$ and where for each $z$, $(\operatorname{supp} f_3(x, z)) \subset B_{R_4}$, with $f_3 \to 0$ exponentially as $z \to \infty$.

Pick $R_3$ and $R_4$ such that $R_4 > R_3 > R$. Let $\alpha(r)$ be a smooth function of the single variable $r$ such that $\alpha = 1$ for $r \leqq R_3$, $\alpha(r) = 0$ for $r \geqq R_4$ and $0 < \alpha(r) < 1$ for $R_3 < r < R_4$. Let $\beta(z)$ be a smooth function of $z$ such that $\beta(z) = 1$ for $z \leqq 2$, $0 < \beta(z) < 1$ for $2 < z < 3$, $\beta(z) = 0$ for $z \geqq 3$. Now define $u_3$ by

$$u_3(x, z) = u_2(x, z)\,(1 - \alpha(r)\,\beta(z)) \qquad \text{where } \alpha\beta \neq 1$$

$$u_3(x, z) = 0 \qquad \text{where } \alpha(r)\,\beta(r) = 1\,.$$

Here again $r = (r_1^2 + \ldots + r_{n-1}^2)^{\frac{1}{2}}$. It follows that, on $\partial\Omega$, $u_3(x, z) = 0$ for all $z$. Let $f_3 \equiv Lu_3$. From our construction, we see that for large positive values of $z$, $f_3$ and $f_2$ agree, hence $f_3 \to 0$ exponentially as $z \to \infty$. For $z \leqq 1$, we see that each derivative of $u_3$ with respect to $x$ or $z$ is square integrable in $\{z \leqq 1\}$. Hence $f_3 \in L^2(\Omega' \times R^1)$. From the construction it is further clear that the support of $f_3$ is contained in the infinite cylinder $0 \leqq r \leqq R_4$.

**4.** Let $A \equiv \bar{\Omega}' \times R^1$. $u_3$ is a classical solution to

$$Lu_3 = f_3 \qquad \text{in } A, \text{ with } u_3 \in L^2(A).$$

Furhermore, $u_3$ is continuous in $\bar{A}$ and $u_3 = 0$ on $\partial A$.

From the local boundary regularity theory it follows that $u_3$ belongs to $H_0^1$ locally near the boundary of $A$. Furthermore from the interior and boundary $L^2$ estimates it follows that in the interior the norm of the first derivatives (that is all we need) of the solution over a ball of radius $r$ can be estimated in terms of the $L^2$ norms of $f_3$ and $u_3$ in a ball of radius $2r$, multiplied by a constant depending on $r$. Near the boundary (in the present situation) a similar estimate exists for sufficiently small $r$ for approximate half-balls near the boundary. By utilizing a locally finite covering of $A$ by such balls and half balls, of fixed and sufficiently small radius $r$, we may conclude that $u_3 \in H_0^1(A)$ globally.

**5.** From the fact $u_3 \in H_0^1(A)$ it follows, in particular, that

$$u_3 \in L^2(R^1;\, H_0^1(\Omega'))\,,$$

*i.e.*, for almost every $z$, $u_3(\cdot, z) \in H_0^1(A)$, such that

$$\int \|u_3(\cdot, z)\|^2_{H^1(\Omega')}\, dz < \infty\,.$$

We now take Fourier transforms of $u_3$ and $f_3$ with respect to $z$ and call the results $U(x, \zeta)$ and $F(x, \zeta)$. We will have $U \in L^2(R^1; H_0^1(\Omega'))$, where $R^1$ is now the $\zeta$ axis and $R^{n-1}$ is as before. Also, $F \in L^2(A)$ (where $A$ is now $\bar{\Omega}' \times \{-\infty < \zeta < \infty\}$) and $(\operatorname{supp} F) \subset B_{R_4} \times R^1$.

**Lemma.** *In the present context, for almost every (real) $\zeta$, $U(x, \zeta)$ belongs to $H_0^1(\Omega')$, and satisfies the equation*

$$\Delta_x U + (1 - \zeta^2)\, U = F(x, \zeta)$$

*in a generalized sense.*

*Note.* We define the Fourier transform by $\hat{g}(\zeta) = \int e^{iz\zeta} g(z)\, dz$.

**Proof.** We have for all $\psi \in C_0^\infty(A)$,

$$\int_A\!\!\int \left(-\nabla_x u_3 \cdot \nabla \psi + u_3 \frac{\partial^2 \psi}{\partial^2 z} + u_3 \psi - f_3 \psi\right) dx\, dz = 0.$$

Let $\psi(x, z) = \alpha(x)\, \beta(z)$ ($\alpha$ and $\beta$ here are not the same as in previous sections), where $\alpha \in C_0^\infty(\Omega')$ and $\beta \in C_0^\infty(R^1)$ or by approximation, to the Schwartz space $S(R^1)$. Taking Fourier transforms, we find that

$$\int_A\!\!\int (-\nabla_x \alpha(x)\, \nabla_x U(x, \zeta) + \alpha(x)\,(1 - \zeta^2)\, U - \alpha(x)\, F(x, \zeta))\, \overline{\hat{\beta}(\zeta)}\, dx\, dz = 0$$

or $\int I(\zeta)\, \overline{\hat{\beta}(\zeta)}\, d\zeta = 0$ for a dense set of $\hat{\beta}(\zeta)$, where

$$I(\zeta) = \int_{\Omega'} (-\nabla_x \alpha \cdot \nabla_x U + \alpha(1 - \zeta^2)\, U - \alpha F)\, dx.$$

Hence $I(\zeta) = 0$ except for an exceptional null set. That this null set can be made independent of the test function $\alpha$ follows from the separability of the space $C_0^\infty(\Omega')$. The conclusion of the lemma follows.

**6.** In this section we show that $U(x, \zeta)$ vanishes almost everywhere in the set $\bar{B}'_{R_4} \times \{-1 < \zeta < 1\}$.

Let $\zeta$ be a one of the non-exceptional values in the previous section, and let $1 < \zeta < 1$. Then $U$ is a generalized solution in $H_0^1(A)$ of

$$\Delta_x U + (1 - \zeta^2)\, U = F \tag{6.1}$$

but is a strict solution to the corresponding homogeneous equation in $R^{n-1} \setminus B_{R_4}$. Furthermore, for each non-exceptional value of $\zeta$, $U(\cdot, \zeta)$ belongs to $L^2(R^{n-1})$. By Rellich's growth theorem for the reduced wave equation, $U(x, \zeta) \equiv 0$ in $R^{n-1} \setminus B_{R_4}$, for that value of $\zeta$. Thus $U(x, \zeta) = 0$ almost everywhere in the set $\bar{B}'_{R_4} \times \{1 < \zeta < 1\}$. Let us note here that since, for each non exceptional $\zeta$ in $1 < \zeta < 1$,

$U(x, \zeta)$ has compact support (in $x$) it satises fiboth of the Sommerfeld radiation conditions at infinity:

$$\frac{\partial U}{\partial r} \pm iU = 0\left(\frac{1}{r^{\frac{n-2}{2}}}\right) \quad \text{as } r \to \infty \ (n-1 \text{ is the dimension}). \tag{6.2}$$

Here "$-$" indicates an "outgoing", and "$+$" an "incoming" radiation condition.

**7.** In this section we review the principle of limiting absorption as it pertains to the present situation.

If we introduce the complex parameter $\lambda = 1 - \zeta^2$. For real $\zeta$ this leads us to consider the Dirichlet problem

$$\Delta_x W + \lambda W = G(x) \quad \text{in } \bar{\Omega}', \qquad W = 0 \quad \text{on } \partial\Omega. \tag{7.1}$$

For real $\lambda < 0$ and $G \in L^2(\bar{\Omega}')$, there is a unique solution $W_\lambda \equiv R_\lambda G$ which can be continued as a bounded operator

$$R_\lambda : L^2(\bar{\Omega}') \to L^2(\bar{\Omega}')$$

for all complex values of $\lambda$ off the closure of the positive real axis, varying analytically with $\lambda$. Furthermore, in a restricted sense, $R_\lambda$ will have limits $R_\lambda^\pm$ as $\lambda$ approaches compact subintervals of the positive $\lambda$ axis from above (respectively below). To be precise, for square integrable $G$ vanishing outside a large ball $B_a$, $R_\lambda$ will approach these limits in a new operator topology where $R_\lambda$ is now considered a mapping $\tilde{R}_\lambda$ from the space $L_0^2(B_a)$ of functions with compact support in $B_a$ to the space with $L^2$ convergence on bounded subsets of the region $\Omega'$. For $\lambda$ restricted to a bounded open set $D_+$ in the upper half $\lambda$ plane, bounded away from the origin, and having an interval on the positive $\lambda$ axis as part of its boundary, $R_\lambda$ gives rise to an operator $\tilde{R}_\lambda^+$ bounded and continuous in $\bar{D}_+$ in the new operator topology. (Similarly with $\tilde{R}_\lambda^-$ and $\bar{D}_-$.) For real positive $\lambda$, $W_\lambda^\pm \equiv \tilde{R}_\lambda^\pm(G)$ is the unique solution to the Sommerfeld radiation problem consisting in the Dirichlet problem (7.1) supplemented by one of two Sommerfeld radiation conditions (6.2).

For the basic facts outlined above we refer to [1] where however only the Neumann problem is considered. But with minor changes the Dirichlet problem can be treated similarly.

**8.** In this section we derive some properties of the $z$-Fourier transform $F(x, \zeta)$ of $f_3(x, z)$.

From the construction in Section 3 we saw that $f_3(x, z)$, in addition to having compact support in $x$, also was in $L^2(A)$ and tended exponentially to zero as $z \to \infty$ (uniformly in $x$). This implies that $f_3$ can be decomposed into two square integrable functions: $f_+$ with support in $\{z \geqq 0\} \to 0$ exponentially as $z \to \infty$, and $f_-$, with support in $\{z \leqq 0\}$. The Fourier transform of $f_+$ has an analytic extension to an open strip about the real $\zeta$ axis, while $f_-$ can be analytically extended to the lower half $\zeta$ plane. Thus the sum, $F = \hat{f}_3$ has an analytic extension to a strip, say $\eta_0 < \eta < 0$ (where $\zeta = \xi + i\eta$) in the $\zeta$ plane. For almost all (fixed) $x$, $F(x, \zeta)$ can be considered an analytical function in this strip with $L^2$ boundary values. From the Paley-Wiener theory we get the estimate for $\eta_0 < \eta < 0$,

$$\int\int_A |F(x, \xi + i\eta)|^2 \, dx \, d\xi \leq \text{Const} \left( |f_3|^2_{L^2(A)} + |f_3 \cdot e^{-\eta_0 z}|^2_{2L^2(A)} \right), \tag{8.1}$$

the integration extending over $A = \bar{\Omega}' \times R^1$. This inequality follows by integrating the appropriate inequality for a fixed $x$, with respect to $x$.

We claim that in addition to (8.1) the following relationship holds:

$$\int\int_A |F(x, \xi + i\eta) - F(x, \xi)|^2 \, dx \, d\xi \to 0 \quad \text{as } \eta \to 0. \tag{8.2}$$

For continuous $F$ this is obvious. For general square integrable $F$ replace $f_3$ (and the corresponding $F$) by a truncated approximation $f_3^h$ (belonging to $C_0^\infty(A)$), and its continuous $z$-Fourier transform $F^h$, respectively, and apply (8.1) to the error, in the usual fashion.

**9.** If we now set $\lambda = 1 - \zeta^2$ we see that the set of $\lambda$ where $R_\lambda$ fails to be analytic is the closure of the positive real axis. In the $\zeta$ plane this means that $R_{1-\zeta^2}$ is analytic away from the union of the imaginary $\zeta$ axis and the interval $[-1, 1]$ on the real $\zeta$ axis. Let us pick $\varphi(x) \in C_0^\infty(B'_{R_4})$ and consider the function

$$\Phi(\zeta) = \int \varphi(x) \, R_{1-\zeta^2} \, F(x, \zeta) \, dx.$$

This function of complex $\zeta$ will be analytic in the strip $\eta_0 < \eta < 0$ with the imaginary axis removed, dividing this strip into a right part $S_R$ and a left part $S_L$. As $\zeta$ approaches a non-exceptional real value $\zeta_*$, $0 < \zeta_* < 1$, in $S_R$, $\lambda = 1 - \zeta^2$ will approach the real value $1 - \zeta_*^2 = \lambda_* > 0$ from above. Hence it seems reasonable that $\Phi(\zeta)$ will approach

$$\Phi^*(\zeta_*) \equiv \int \varphi(x) \, U_{\lambda^*}^+(x) \, dx$$

in an appropriate sense, where $U_{\lambda^*}^+(x) \equiv \tilde{R}_{\lambda^*}^+ F(x, \zeta_*)$ is the unique solution to the outgoing Sommerfeld radiation problem with $G(x) = F(x, \zeta_*)$. In section 6 we saw that this solution vanishes outside $B_{R_4}$, and by choice of $\varphi$, $\Phi^+(\zeta_*) = 0$. Thus $\Phi(\xi_* + i\eta)$ will approach zero in a certain sense (to be made precise) as $\eta \downarrow 0$, for almost all $\xi_*$ in an interval on the positive real axis, making the vanishing of $\Phi^+$ in $S_R$ very likely. Similarly, as $\zeta$ approaches $\zeta_* \in (-1, 0)$ through $S_L$, $\lambda = 1 - \zeta^2$ will approach $1 - \zeta_*^2 = \lambda_*$ from below, yielding the conclusions that $\Phi = 0$ in the whole strip.

The only missing link in this argument is the determination of the exact manner in which $\Phi(\zeta)$ approaches its limiting values as $\zeta$ approaches the intervals (0, 1) and $(-1, 0)$ on the real axis from above.

**10.** In this section we shall show that $\Phi(\xi + i\eta) \to \Phi^+(\xi)$ in $L^2$ of a compact subinterval $J$ of $0 < \xi < 1$ as $\eta \downarrow 0$, an analogous relationship holding for $\Phi^-$ and $-1 < \xi < 0$.

To see this, let $D^+$ be the set $J\times\{0<\eta<\eta_0\}$ and let $\zeta=\xi+i\eta$ vary in $D^+$. We assume that the support of $\varphi$ is contained in $B_{R_5}$ and denote by $C_1$ a constant possibly depending on $\varphi$ and $R_5$. Then

$$\begin{aligned}\int_J |\Phi(\xi+i\eta)-\Phi^+(\xi)|^2\,d\xi &\leq C_1 \int_{B_{R_5}}\int_J |\tilde{R}^+_{1-\zeta^2}F(x,\zeta)-\tilde{R}^+_{1-\xi^2}F(x,\xi)|^2\,dx\,d\xi\\ &\leq C_1\iint |(\tilde{R}^+_{1-\zeta^2}(F(x,\zeta)-F(x,\zeta)))|^2\,dx\,d\xi\\ &+ C_1\iint |(\tilde{R}^+_{1-\zeta^2}-\tilde{R}^+_{1-\xi^2})\,F(x,\xi)|^2\,dx\,d\xi\\ &\leq C_1\cdot\text{const.}\iint |F(x,\zeta)-F(x,\xi)|^2\,dx\,d\xi\\ &+ C_1\,\delta(\eta)\iint |(F(x,\xi)|^2\,dx\,d\xi,\end{aligned}$$

where $\delta(\eta)\to 0$ as $\eta\downarrow 0$. Taking (8.2) into account, we see that as $\eta\downarrow 0$, $\xi\in J$, each term approaches zero, and hence the left hand side $\to 0$. From this, and the fact that $\Phi^+(\xi)=0$ a.e. on $J_1$ it follows that $\Phi(\zeta)=0$ in $S_R$, and similarly in $S_L$. A variation of the argument just given shows that $\Phi(\xi)=0$ a.e. on $|\xi|>1$. Since $\varphi$ is arbitrary, with support in $\bar{B}'_{R_4}$, we see that $U(x,\xi)$ vanishes a.e. in $B'_{R_4}\times R^1$. The same holds for $u_3(x,z)$ and hence for $u(x,z)$. By unique continuation $w\equiv 0$ in $\bar{G}$.

*Acknowledgment*. The research reported here was supported by NSF 84-13129.

## Bibliography

1. Wilcox, C. Scattering theory for the d'Alembert equation in exterior domains, Lecture Notes in Mathematics 442, Springer-Verlag, Berlin, 1975.

Department of Mathematics
University of Minnesota
Minneapolis

*(Received January 2, 1986)*

# *A New Proof of Moser's Parabolic Harnack Inequality Using the old Ideas of Nash*

E. B. FABES & D. W. STROOCK

*Dedicated to J. Serrin on the occasion of his 60<sup>th</sup> birthday*

## Introduction

In 1958 NASH published his fundamental work on the local Hölder continuity of solutions of second-order parabolic equations with coefficients which need not be smooth ([8]). The primary purpose of that work was to study the properties of the fundamental solution corresponding to the parabolic operator and from these properties to derive regularity for a general solution. Though the work is often cited in the literature about weak solutions of elliptic and parabolic equations, one feels that NASH's ideas were never fully understood (and maybe still are not) and that because of this the more understandable and seemingly more fruitful ideas of DE GIORGI ([4]) and MOSER ([6], [7]) were subsequently adopted.

In the present article, we return to NASH's ideas. In particular, by modifying and persuing his arguments, we establish directly what we feel is the logical goal of this line of reasoning, namely: the estimates for the fundamental solution first proved by D. G. ARONSON. (See [1] and [2].) From ARONSON's estimates the parabolic Harnack inequality of MOSER ([7]) and, consequently (as was shown by MOSER [7, p. 108]), NASH's local HÖLDER continuity of weak solutions to parabolic equations follow. That is, our approach reverses the chronological order in which these results were derived originally.

To make the above statements mathematically precise we introduce the basic notations and definitions to be used throughout this work. We shall study parabolic operators of the form

$$L = \sum_{i,j=1}^{n} D_{x_i}(a_{ij}(t, x)\, D_{x_j}) - D_t$$

where $t$ is a real number and $x = (x_1, \ldots, x_n) \in R^n$. Our basic assumptions on the matrix $a(t, x) \equiv (a_{ij}(t, x))$ are symmetry, (i.e. $a_{ij} = a_{ji}$), and the existence of a number $\lambda \in (0, 1]$ such that for all $(t, x) \in R^{n+1}$ and all $\xi \in R^n$

$$\lambda\, |\xi|^2 \leqq a(t, x)\, \xi \cdot \xi \equiv \sum_{i,j=1}^{n} a_{ij}(t, x)\, \xi_i \xi_j \leqq \lambda^{-1}\, |\xi|^2 .$$

We may and do make the qualitative assumption that the matrix $a(t, x)$ is smooth; however, we emphasize that all quantitative estimates are only allowed to depend on the dimension and the number $\lambda$. Besides, $x$, the letters $y$ and $\xi$ will be used to denote points in $R^n$ and the letters $t$, $s$, and $r$ will be reserved for real numbers.

We let $\Gamma(t, x; s, y) \equiv \Gamma_a(t, x; s, y)$ denote the fundamental solution of the parabolic operator $L$. As stated above, the purpose of this paper is to use the ideas of NASH to obtain the following estimates: for $s < t$

$$(*) \qquad \frac{\exp\left\{-C\dfrac{|x-y|^2}{t-s}\right\}}{C(t-s)^{n/2}} \leqq \Gamma(t, x; s, y) \leqq \frac{C\exp\left\{-\dfrac{|x-y|^2}{C(t-s)}\right\}}{(t-s)^{n/2}}$$

where $C$ depends only on $\lambda$ and $n$. The inequalities * were first obtained by ARONSON in [1]. (See also [2].) His proof, however, relied on MOSER's parabolic HARNACK inequality ([7]). Our point here is first to establish the estimates (*) and then to derive the Harnack inequality as an easy consequence. The outline of the paper is simple; the upper bound is obtained in Section 1, the lower bound in Section 2, and Harnack's inequality in Section 3.

## Section 1. The Upper Bound

Our proof of the upper bound for the fundamental solution is essentially due to NASH. In particular, NASH used the same proof to derive the right side of (*) without the exponential factor. There are various ways of passing from his result to the one including the exponential factor. The one which we have adopted is based on a method which was introduced in this context by E. B. DAVIES ([3]).

(We point out that ARONSON's original proof ([1]) of the upper bound in (*), like the one given here, does not depend on Harnack's inequality. Our reasons for presenting a proof here are completeness and unification of the arguments. At the same time, it should be emphasized that the upper bound itself is an important tool for our understanding and simplification of those ideas of NASH needed to obtain the lower bound in Section 2.

Fix an element $\vec{\alpha} \in R^n$ and set $\psi(x) = \vec{\alpha} \cdot x \equiv \sum_{i=1}^{n} \alpha_i x_i$. Let $A_t = \sum_{i,j=1}^{n} D_{x_i}(a_{ij}(t, x)\, D_{x_j})$ and $A_t^\psi = \exp(-\psi)\, A_t \exp\psi$. If $f \in S(R^n; (0, \infty))$ (*i.e.*, $f$ is a positive function from the Schwartz test function space)

$$\begin{aligned}(A_t^\psi f, f^{2p-1}) &\equiv \int A_t^\psi f(x) f^{2p-1}(x)\, dx \\ &= \int (\vec{\alpha} \cdot a(t, x)\, \vec{\alpha})\, f^{2p}(x)\, dx - 2(p-1) \int (\vec{\alpha} \cdot a(t, x)\, \nabla f(x))\, f^{2p-1}(x)\, dx \\ &\quad - (2p-1) \int (\nabla f(x) \cdot a(t, x)\, \nabla f(x))\, f^{2p-2}(x)\, dx.\end{aligned}$$

Hence, setting $\|f\|_p = \left( \int_{R^n} |f|^p \right)^{1/p}$, one gets

$$(A_t^\psi f, f^{2p-1}) \leqq -\frac{\lambda}{p} \int |\nabla f^p|^2 + \frac{\alpha^2}{\lambda} p \|f\|_{2p}^{2p}, \quad p \geqq 1,$$

where $\alpha = |\vec{\alpha}|$. (This observation is the key to DAVIES's analysis.) At the same time, $(2\pi)^n \|f\|_2^2 \leqq C[R^n \|f\|_1^2 + R^{-2} \|\nabla f\|_2^2]$ for all $R > 0$; and so

$$\|f\|_2^{2+\beta} \leqq C_n(\int |\nabla f|^2)(\int |f|)^\beta$$

with $\beta = 4/n$. (This inequality is the one from which NASH's upper bound comes.) Combining these, we arrive at

$$(1.1) \qquad (A_t^\psi f, f^{2p-1}) \leqq -\frac{\varepsilon}{2p} \frac{\|f\|_{2p}^{2p+\beta p}}{\|f\|_p^{\beta p}} + \frac{\alpha^2 p}{\lambda} \|f\|_{2p}^{2p}, \quad p \geqq 1,$$

for some $\varepsilon > 0$ which depends only on $n$ and $\lambda$. Finally, let $f \in S(R^n; (0, \infty))$ and define $f_t$ by

$$f_t(x) = \exp(-\psi(x)) \int f(y)\, \Gamma(t, x; 0, y) \exp(\psi(y))\, dy,$$

where $\Gamma = \Gamma_a$. Then $t \in [0, \infty) \to f_t \in S(R^n; (0, \infty))$ is smooth and, for $p \geqq 1$:

$$\frac{d}{dt} \|f_t\|_{2p}^{2p} = 2p(A_t^\psi f_t, f_t^{2p-1}).$$

Hence, by (1.1), for $p \in [1, \infty)$:

$$(1.2) \qquad \frac{d}{dt} \|f_t\|_{2p} \leqq -\frac{\varepsilon}{2p} \|f_t\|_{2p}^{1+\beta p} \|f_t\|_p^{-\beta p} + \frac{\alpha^2 p}{\lambda} \|f_t\|_{2p}, \quad t \geqq 0.$$

In particular,

$$(1.3) \qquad \|f_t\|_2 \leqq e^{\alpha^2 t/\lambda} \|f\|_2, \quad t \geqq 0.$$

**(1.4) Lemma.** *Let* $w: [0, \infty) \to [0, \infty)$ *be a continuous non-decreasing function and suppose that* $u \in C^1([0, \infty))$ *is a positive function which satisfies*

$$u'(t) \leqq -\frac{\varepsilon}{2p} \left( \frac{t^{(p-2)/\beta p}}{w(t)} \right)^{\beta p} u^{1+\beta p}(t) + \frac{\alpha^2 p}{\lambda} u(t), \quad t \geqq 0.$$

*where* $p \in [2, \infty)$. *Then, for each* $\delta > 0$ *there is a* $K = K(\varepsilon, \delta) < \infty$ *such that*

$$u(t) \leqq (Kp^2)^{1/\beta p} w(t)\, e^{\delta \alpha^2 t/\lambda p}\, t^{(1-p)/\beta p}, \quad t \geqq 0.$$

**Proof.** Set $v = e^{-\alpha^2 p t/\lambda} u$. Then

$$(v^{-\beta p})' \geqq \frac{\varepsilon\beta}{2} \exp(\beta\alpha^2 p^2 t/\lambda)\, t^{p-2}/w^{\beta p};$$

and so

$$\frac{\exp(\beta\alpha^2 p^2 t/\lambda)}{u(t)^{\beta p}} \geqq \frac{\varepsilon\beta}{2w(t)^{\beta p}} \int_0^t \exp(\beta\alpha^2 p^2 s/\lambda)\, s^{p-2}\, ds.$$

Note that

$$\int_0^t \exp(\beta\alpha^2 p^2 s/\lambda)\, s^{p-2}\, ds = \left(\frac{\lambda t}{\beta\alpha^2 p^2}\right)^{p-1} \int_0^{\beta\alpha^2 p^2/\lambda} e^{ts} s^{p-2}\, ds$$

$$\geqq \left(\frac{\lambda t}{\beta\alpha^2 p^2}\right)^{p-1} \exp\,[(\beta\alpha^2 p^2 - \delta\beta\alpha^2)\, t/\lambda] \int_{\beta\alpha^2 p^2(1-\delta/p^2)/\lambda}^{\beta\alpha^2 p^2/\lambda} s^{p-2}\, ds$$

$$= \frac{t^{p-1}}{p-1}\,[1-(1-\delta/p^2)^{p-1}]\,\exp\,[(\beta p^2\alpha^2 - \delta\beta\alpha^2)\, t/\lambda].$$

Combining these, we get

$$\frac{e^{\beta\alpha^2 p^2 t/\lambda}}{u(t)^{\beta p}} \geqq \frac{t^{p-1}}{Kp^2 w(t)^{\beta p}} \exp\,[(\beta p^2\alpha^2 - \delta\beta\alpha^2)\, t/\lambda],$$

where

$$\frac{1}{K} \equiv \frac{\varepsilon\beta}{2} \inf_{p\geqq 2} \{p[1-(1-\delta/p^2)^{p-1}]\} > 0. \quad \square$$

Now set $p_k = 2^k$, $u_k(t) = \|f_t\|_{p_k}$, and $w_k(t) = \max\{s^{(p_k-2)/\beta p_k} u_k(s): 0 \leqq s \leqq t\}$. By (1.3), if $\|f\|_2 = 1$, $u_1(t) \leqq \exp\{\alpha^2 t/\lambda\}$, and by (1.2) and the lemma,

$$\frac{w_{k+1}(t)}{w_k(t)} \leqq (4^k K)^{1/\beta 2^k} \exp\,(\delta\alpha^2 t/2^k\lambda).$$

Hence, there is a $C < \infty$, depending only on $n$, $\lambda$ and $\delta > 0$, such that

(1.5) $$\sup_k w_k(t) \leqq C \exp\,[(1+\delta)\,\alpha t/\lambda].$$

**(1.6) Theorem.** *There is a* $C < \infty$, *depending only on* $n$ *and* $\lambda$, *such that*

$$\Gamma_a(t, x; s, y) \leqq \frac{C}{t^{n/2}} \exp\,(-|y-x|^2/Ct)$$

*for all* $0 \leqq s < t < \infty$ *and* $x, y \in R^n$.

**Proof.** Since $\Gamma_a(t, x; s, y) = \Gamma_{a_s}(t-s, x; 0, y)$ where $a_s(r, \xi) = a(s+r, \xi)$, we may and will take $s = 0$. Now define

$$P_t^\psi f(x) = \exp(-\psi(x)) \int f(y)\, \Gamma_a(t, x; 0, y) \exp(-\psi(y))\, dy \qquad (\text{i.e. } = f_t).$$

Then, from the preceding (with $\delta = 1$),

$$\|P_t^\psi f\|_\infty \leqq \frac{C}{t^{n/4}} \exp\,(2\alpha^2 t/\lambda)\, \|f\|_2$$

for each $t > 0$. At the same time, it is clear that the adjoint $(P_t^\psi)^*$ of $P_t^\psi$ is given by

$$(P_t^\psi)^* f(y) = \exp(\psi(y)) \int \Gamma_{\tilde a}(t, y; 0, x) \exp(-\psi(x))\, f(x)\, dx,$$

where $\tilde{a}(r, \xi) = a(t - r, \xi)$. Since $(P_t^\psi)^*$ has the same form as $P_t^\psi$,

$$\|(P_t^\psi)^* f\|_\infty \leqq \frac{C}{t^{n/4}} \exp(2\alpha^2\, t/\lambda)\, \|f\|_2;$$

and so, by duality,

$$\|P_t^\psi f\|_2 \leqq \frac{C}{t^{n/4}} \exp(2\alpha^2\, t/\lambda)\, \|f\|_1.$$

Finally, note that $P_{2t} = Q_t^\psi \circ P_t^\psi$, where

$$Q_t^\psi f(x) = \exp(-\psi(x)) \int f(y)\, \Gamma_{a_t}(t, x; 0, y) \exp(\psi(y))\, dy,$$

and $a_t(\cdot, \cdot) = a(\cdot + t, \cdot)$. Hence,

$$\|P_{2t}^\psi f\|_\infty \leqq \frac{C^2}{t^{n/2}} \exp(4\alpha^2\, t/\lambda)\, \|f\|_1,$$

which is equivalent to

$$\Gamma_a(2t, x; 0, y) \leqq \frac{C^2}{t^{n/2}} \exp(4\alpha^2\, t/\lambda + \vec{\alpha} \cdot (x - y)).$$

We now get our estimate upon taking $\vec{\alpha} = \dfrac{\lambda}{8t}(y - x)$. $\square$

It is clear that we have not fully utilized the estimate (1.5) since we simply took $\delta = 1$. Had we carried $\delta$ through the proof of Theorem (1.6), we would have arrived at

$$\Gamma_a(t, x; s, y) \leqq \frac{C(\delta)}{t^{n/2}} \exp(-|y - x|^2/(1 + \delta)\, \Lambda_t t)$$

for each $\delta > 0$, where $\Lambda_t \equiv \max\{\eta \cdot a(r, \xi)\, \eta : (r, \xi) \in [0, t] \times R^n \text{ and } \eta \in S^{n-1}\}$. It is the power of his method to get such precise exponential estimates that justifies the word "explicit" in the title of DAVIES's article [3].

## Section 2. The Lower Bound

In this section we will establish the lower bound for the fundamental solution (*i.e.* the left-hand side of (*)). Our procedure is once again basically due to NASH. However the upper bound just established allows us to simplify his argument and to carry it to completion.

**Lemma (2.1).** (NASH's Lower Bound). *There is a constant $B < \infty$ depending only on $\lambda$ such that for all $|x| \leqq 1$*

$$\int e^{-\pi|y|^2} \log \Gamma_a(1, x; 0, y)\, dy \geqq -B.$$

**Proof.** Observe that

$$\Gamma_a(t, x; t-s, y) = \Gamma_{a_t}(s, y; 0, x)$$

where $a_t = a(t - \cdot, \cdot)$. In particular

$$\Gamma_a(1, x; 0, y) = \Gamma_{a_1}(1, y; 0, x).$$

Set $u(s, y) = \Gamma_{a_1}(s, y; 0, x)$ and $G(s) = \int e^{-\pi|y|^2} \log u(s, y)\, dy$.

Since $\int u(s, y)\, dy = 1$, $G(s) < 0$ and our goal is to estimate $G(1)$ from below. We will obtain this from a differential inequality satisfied by $G$. Namely,

$$\begin{aligned} G'(\tau) &= -\int \nabla_y \left(\frac{e^{-\pi|y|^2}}{u(\tau, y)}\right) \cdot a_1 \nabla_y u(\tau, y))\, dy \\ &= 2\pi \int e^{-\pi|y|^2} y \cdot a_1(\nabla_y \log u(\tau, y))\, dy + \int e^{-\pi|y|^2} \nabla_y \log u \cdot a_1(\nabla_y \log u)\, dy \\ &= -2\pi^2 \int e^{-\pi|y|^2} y \cdot a_1\, dy + \tfrac{1}{2} \int e^{-\pi|y|^2} (2\pi y + \nabla_y \log u) \cdot a_1(y + \nabla_y \log u)\, dv \\ &\quad + \int \tfrac{1}{2} e^{-\pi|y|^2} (\nabla_y \log u) \cdot a_1(\nabla_y \log u)\, dy. \end{aligned}$$

Hence

$$G'(s) \geqq -A + \frac{\lambda}{2} \int e^{-\pi|y|^2} |\nabla_y \log u|^2\, dy. \tag{2.2}$$

In particular $G(s) + As$ is nondecreasing on $[\frac{1}{2}, 1]$. Also since

$$\int e^{-\pi|y|^2} (\log u(s, y) - G(s))^2\, dy \leqq c \int e^{-\pi|y|^2} |\nabla_y \log u|^2\, dy$$

we have

$$G'(s) \geqq -A + B \int e^{-\pi|y|^2} (\log u(s, y) - G(s))^2\, dy \tag{2.3}$$

for constants $A$ and $B$ depending only on $\lambda$.

Next observe that $\dfrac{(\log u - G(s))^2}{u}$ is non-increasing as a function of $u$ in $[e^{2+G(s)}, \infty)$. Also from Theorem 1.9 $\sup\limits_{1/2 \leqq s \leqq 1} u(s, y) \leqq K$, an absolute constant. Combined with (2.3), this implies

$$G'(s) \geqq -A + B \frac{(\log K - G(A))^2}{K} \int\limits_{u(s,y) \geqq e^{2+G(s)}} e^{-\pi|y|^2} u(s, y)\, dy \tag{2.4}$$

for $s \in [\frac{1}{2}, 1]$. At the same time

$$\begin{aligned} \int\limits_{u(s,y) \geqq e^{2+G(s)}} e^{-\pi|y|^2} u(s, y)\, dy &\geqq \int e^{-\pi|y|^2} u(s, y)\, dy - e^{2+G(s)} \\ &\geqq e^{-\pi R^2} \int\limits_{|y|<R} u(s, y)\, dy - e^{2+G(s)} \\ &= e^{-\pi R^2} \left[1 - \int\limits_{|y| \geqq R} u(s, y)\, dy\right] - e^{2+G(s)}. \end{aligned}$$

Note that, by Theorem (1.4), there is an $R_\lambda$, depending only on $\lambda$, such that $\sup\limits_{1/2 \leqq s \leqq 1} \int\limits_{|y| > R_\lambda} u(s, y)\, dy \leqq \frac{1}{2}$.

Applying these last remarks to (2.3) and also remembering that $G(s) + As$ is nondecreasing on $[\frac{1}{2}, 1]$, we can conclude that there are a $\delta_\lambda$ and an $M_\lambda$, both depending only on $\lambda$, such that

$$G'(s) \geqq \delta_\lambda G(s)^2 \quad \text{for} \quad s \in [\tfrac{1}{2}, 1] \tag{2.5}$$

provided $G(1) \leqq -M_\lambda$. But if (2.5) holds then $G(1) \geqq -2/\delta_\lambda$. That is, we have proved that

$$G(1) \geqq -\max\left(\frac{2}{\delta_\lambda}, M_\lambda\right). \quad \square$$

**Lemma (2.6).** *There is a $C$, depending only on $\lambda$, such that*

$$\Gamma_a(t, x; s, y) \geqq \frac{1}{C(t-s)^{n/2}}$$

*for all $x$ and $y$ satisfying $|x - y| \leqq \sqrt{t - s}$.*

**Proof.** By rescaling, we may take $s = 0$ and $t = 2$. We write

$$\Gamma_a(2, x; 0, y) = \int \Gamma_a(1, \xi; 0, y)\, \Gamma_{\tilde{a}}(1, x; 0, \xi)\, d\xi$$

where $\tilde{a} = a(\cdot + 1, \cdot)$. Clearly, this leads to

$$\Gamma_a(2, x; 0, y) \geqq \int \Gamma_a(1, \xi; 0, y)\, \Gamma_{\tilde{a}}(1, x; 0, \xi)\, e^{-\pi|\xi|^2}\, d\xi$$

and by Jensen's inequality

$$\begin{aligned} \log\,[\Gamma_a(2, x; 0, y)] &\geqq [\int e^{-\pi|\xi|^2} \log \Gamma_{\tilde{a}}(1, x; 0, \xi)\, d\xi \\ &\qquad + \int e^{-\pi|\xi|^2} \log \Gamma_a(1, \xi; 0, y)\, d\xi] \\ &\geqq -c_\lambda \end{aligned}$$

by Lemma (2.1). (Remember $\Gamma_a(1, \xi; 0, y) = \Gamma_{a_1}(1, y; 0, \xi)$ where $a_1 = a(1 - \cdot, \cdot)$.) $\square$

We are now ready to prove the lower estimate for the fundamental solution.

**Theorem (2.7).** (Aronson). *There is a $C$, depending only on $\lambda$ and $n$, such that*

$$\Gamma_a(t, x; s, y) \geqq \frac{1}{C(t-s)^{n/2}} \exp\left(-C\,|x - y|^2/(t - s)\right).$$

**Proof.** Again we may assume $s = 0$, $t = 1$, and this time we also assume, as we may, that $y = 0$. That is, we wish to show that $\Gamma_a(1, x; 0, 0) \geqq \frac{1}{C} \exp(-C|x|^2)$. Because of Lemma (2.6), we may also assume $|x| > 1$.

Given $x \in R^n$ with $|x| > 1$, let $k$ be the smallest integer dominating $4|x|^2$ and set $S = \prod_{l=1}^{k-1} B\left(\frac{lx}{k}, \frac{1}{2\sqrt{k}}\right)$ $(B(y, r) \equiv \{\xi \in R^n : |\xi - y| < r\})$. Then, for $(\xi_1, \ldots, \xi_{k-1}) \in S : |\xi_1| < \frac{1}{\sqrt{k}}, \max_{1<l<k} |\xi_l - \xi_{l-1}| < \frac{1}{\sqrt{k}}$, and $|x - \xi_{k-1}| < \frac{1}{\sqrt{k}}$. Hence, by Lemma (2.6),

$$\Gamma(1, x; 0, 0) = \int \ldots \int \Gamma\left(1, x; \frac{k-1}{k}, \xi_{k-1}\right) \Gamma\left(\frac{k-1}{k}, \xi_{k-1}; \frac{k-2}{k}, \xi_{k-2}\right)$$

$$\ldots \Gamma\left(\frac{1}{k}, \xi_1; 0, 0\right) d\xi_1 \ldots d\xi_{k-1}$$

$$\geqq \int \ldots \int \Gamma\left(1, x; \frac{k-1}{k}, \xi_{k-1}\right) \Gamma\left(\frac{k-1}{k}, \xi_{k-1}; \frac{k-2}{k}, \xi_{k-2}\right)$$

$$\ldots \Gamma\left(\frac{1}{k}, \xi_1; 0, 0\right) d\xi_1 \ldots d\xi_{k-1}$$

$$\geqq \left(\frac{k^{n/2}}{C}\right)^k |S| = \left(\frac{k^{n/2}}{C}\right)^k (\Omega_n (2k)^{-n/2})^{k-1}$$

$$= \frac{(2k)^{n/2}}{\Omega_n} \left(\frac{\Omega_n}{2^{n/2} C}\right)^k.$$

Clearly the required estimate is immediate from this.

## Section 3

In this section we show how to derive from (*) both NASH's theorem on continuity and MOSER's Harnack principle. Actually, there are several ways to do this. Our choice has been dictated by our desire to show that (*), and nothing more, suffices. The proof here is modelled on the argument given by KRYLOV ([9]) (*cf.* [5] for a similar derivation of the Harnack principle for solutions to certain degenerate equations).

In what follows, $\Gamma_a^{(\xi,R)}(t, x; s, y)$ denotes the fundamental solution to $Lu = 0$ with zero boundary data on $\partial B(\xi, R)$. That is, if $(s, y) \in (0, \infty) \times B(\xi, r)$ and $u(t, x) = \Gamma_a^{(\xi,R)}(t, x; s, y)$, then $Lu = 0$ in $(s, \infty) \times B(\xi, R)$, $u = 0$ on $[s, \infty) \times \partial B(\xi, R)$, and $u(s, x) = \delta(x - y)$.

**(5.1) Lemma.** *For each* $\delta, \gamma \in (0, 1)$ *there is an* $\varepsilon = \varepsilon(n, \lambda, \delta, \gamma) > 0$ *such that*

$$\Gamma_a^{(\xi,R)}(t, x; s, y) \geqq \frac{\varepsilon}{|B(\xi, \delta R)|}$$

*for all,* $x\, y \in B(\xi, \delta R)$ *and* $s < t$ *satisfying* $\gamma R^2 \leqq t - s \leqq R^2$.

**Proof.** By rescaling and translation, we may and will assume that $\xi = 0$ and $R = 1$. For convenience, we use $\hat{\Gamma}$ to denote $\Gamma_a^{(0,1)}$. Clearly we need only treat the case when $s = 0$.

Note that

$$\hat{\Gamma}(t, x; 0, \gamma) = \Gamma(t, x; 0, y) - \int_{[0,t)\times\partial B(0,1)} \Gamma(t, x; r, \xi)\, \mu_{0,y}(d\tau \times d\xi)$$

where $\mu_{0,y}$ is a non-negative measure with total mass less than or equal to one. Hence, by (*)

$$\hat{\Gamma}(t, x; 0, y) \geqq \frac{1}{Ct^{n/2}} \exp\left[-C\,|y - x|^2/t\right] - C \sup_{0\leqq\tau\leqq t} \frac{1}{\tau^{n/2}} \exp\left[-(1 - \delta)^2/C\tau\right]$$

for $x \in B(0, \delta)$, $t > 0$, and $y \in B(0, 1)$. In particular, there is an $r \in (0, 1 - \delta)$ depending only on $C$ and $\delta$, such that

$$\hat{\Gamma}(t, x; 0, y) \geqq \frac{1}{2Ct^{n/2}} \exp\left[-C\,|y - x|^2/t\right]$$

for all $x \in B(0, \delta)$, $t \in (0, r^2]$, and $y$ with $|y - x| < r$.

Finally, we use the reproducing property of $\hat{\Gamma}$ to conclude from the above that

$$\hat{\Gamma}(t, x; 0, y) \geqq \frac{\alpha}{t^{n/2}} \exp\left[-K\,|y - x|^2/t\right]$$

for some $\alpha > 0$ and $K < \infty$, depending only on $r$ and $C$, for all $t \in (0, 1]$ and $x, y \in B(0, \delta)$ (*cf.* the argument used in Section 2 to pass from the lower bound of $\Gamma(t, x; s, y)$ for $|y - x|^2/(t - s)$ small to the general result.) Obviously, our estimate follows immediately. □

In the following we use the notation $\operatorname{Osc}(u; s, \xi, R)$ to denote $\sup\{|u(t, x) - u(t', x')| : s - R^2 \leqq t, t' \leqq s \text{ and } x, x' \in B(\xi, R)\}$.

**(5.2) Lemma.** *For each* $\delta \in (0, 1)$ *there is a* $\varrho = \varrho(n, \lambda, \delta) \in (0, 1)$ *such that for all* $(s, \xi) \in R^1 \times R^N$ *and* $R > 0$:

$$\operatorname{Osc}(u; s, \xi, \delta R) \leqq \varrho \operatorname{Osc}(u; s, \xi, R)$$

*whenever* $u \in C^\infty([s - R^2, s] \times \overline{B}(\xi, R))$ *satisfies* $Lu = 0$ *in* $(s - R^2, s) \times B(\xi, R)$.

**Proof.** Let $m(r)$ and $M(r)$ denote, respectively, the minimum and maximum values of $u$ on $[s - r^2, s] \times \overline{B(\xi, r)}$.

Set $S = \{x \in B(\xi, \delta R): u(s - R^2, x) \geqq (M(R) + m(R))/2\}$, and assume that $|S|/|B(\xi, \delta R)| \geqq \frac{1}{2}$. Then, for $(t, x) \in [s - \delta^2 R^2, s] \times B(\xi, \delta R)$:

$$u(t, x) - m(R) \geqq \int (u(s - R^2, y) - m(R))\, \Gamma_a^{(\xi,R)}(t, x; s - R^2, y)\, dy$$

$$\geqq \frac{M(R) - m(R)}{2} \int_S \Gamma_a^{(\xi,R)}(t, x; s - R^2, y)\, dy$$

$$\geqq \varepsilon(M(R) - m(R))/4;$$

and so $m(\delta R) \geqq \varepsilon M(R)/4 + (1 - \varepsilon/4)\, m(R))$. Hence

$$M(\delta R) - m(\delta R) \leqq M(R) - m(\delta R) \leqq (1 - \varepsilon/4)\,(M(R) - m(R)).$$

In other words, we can take $\varrho = 1 - \varepsilon/4$.

**(5.3) Theorem** (NASH). *For each $\delta \in (0, 1)$ there are a $C = C(n, \lambda\ \delta) < \infty$ and a $\beta = \beta(n, \lambda, \delta) \in (0, 1)$ such that for all $(s, \xi) \in R^1 \times R^n$ and $R > 0$:*

$$|u(t, x) - u(t', x')| \leqq C\, \|u\|_{C_b([s-R^2, R^2] \times \bar{B}(\xi, R))} \left(\frac{|t - t'|^{1/2} + |x - x'|}{R}\right)^\beta$$

*for $(t, x), (t', x') \in [s - (1 - \delta^2) R^2, s] \times \bar{B}(\xi, (1 - \delta) R)$ whenever $u \in C^\infty([s - R^2, s] \times \bar{B}(\xi, R))$ satisfies $Lu = 0$ in $(s - R^2, s) \times B(\xi, R)$.*

**Proof.** Let $(t, x), (t', x') \in [s - (1 - \delta^2) R^2, s] \times \bar{B}(\xi, (1 - \delta) R)$ with $t' \leqq t$ be given, and set $l = (t - t')^{\frac{1}{2}} + |x - x'|$. If $l \geqq \delta R$, then there is nothing to do. If $l < \delta R$, choose $k \in Z^+$ so that $\delta^{k+1} \leqq l/R \leqq \delta^k$. Then $[t - (l\, \delta^{-k+1})^2, t] \times B(x, \delta^{-k+1}\, l) \subseteq [s - R^2, s] \times \bar{B}(\xi, R)$ and $(t, x') \in [t - l, t] \times B(x, l)$. Hence:

$$|u(t, x) - u(t', x')| \leqq \operatorname{Osc}(u; t, x, l) \leqq \varrho^{k-1} \operatorname{Osc}(u; t, x, \delta^{-k+1}\, l)$$

$$\leqq 2\, \|u\|_{C_b([s-R^2, s] \times \bar{B}(\xi, R))}\, \varrho^{k-1}.$$

Finally, define $\beta$ by $\varrho = \delta^\beta$. Then

$$|u(x, t) - u(t', x')| \leqq 2\varrho^{-2}\, \|u\|_{C_b([s-R^2, s] \times \bar{B}(\xi, R))}\, (\delta^{k+1})^\beta$$

$$\leqq 2\varrho^{-2}\, \|u\|_{C_b([s-R^2, s] \times \bar{B}(\xi, R)}\, l^\beta. \quad \square$$

We can now prove the following statement of the Harnack principle for the operator $L$. Although our statement is not precisely the one given by MOSER, it can be used to easily prove his Theorem (2) in [7].

**(5.4) Theorem.** *Let $0 < \alpha < \beta < 1$ and $\gamma \in (0, 1)$ be given. Then there is an $M = M(n, \lambda, \alpha, \beta, \gamma) < \infty$ such that for all $(s, x) \in R^1 \times R^N$, all $R > 0$, and all non-negative $u \in C^\infty([s - R^2, s] \times \bar{B}(x, R))$ satisfying $Lu = 0$, one has*

$$u(t, y) \leqq M u(s, x)$$

*for all $(t, y) \in [s - \beta R^2, s - \alpha R^2] \times \bar{B}(x, \delta R)$.*

**Proof.** By translation and rescaling we may and will assume that $(s, x) = (0, 0)$ and $R = 1$. Also, we assume that $u(0, 0) = 1$.

From Lemma (5.1) we know that there is an $\varepsilon = \varepsilon(n, \lambda, \alpha) > 0$ such that for all $r \in [-1, \alpha]$ and $\lambda > 0$:

$$1 = u(0, 0) \geqq \int \Gamma_a^{(0,1)}(0, 0; r, \eta)\, u(r, \eta)\, d\eta$$

$$\geqq \varepsilon\lambda\, |S(r, \lambda)|$$

where $S(r, \lambda) \equiv \{\eta \in B(0, \tfrac{1}{2}(1 + \delta)) : u(r, \eta) \geqq \lambda\}$.

Next, let $\varrho = \varrho(n, \lambda, \tfrac{1}{2})$ be the constant in Lemma (5.2) and set $\sigma = \tfrac{1}{2}(1 - \varrho)$ and $K = \tfrac{1}{2}(1 + 1/\varrho)$. Also define $r(\lambda) = (2/\Omega_n \varepsilon\sigma\lambda)^{1/n}$ for $\lambda > 0$, where $\Omega_n = |B(0, 1)|$. Now suppose that $(t, y) \in (-1, -\alpha) \times B(0, \tfrac{1}{2}(1 + \delta))$ and $\lambda > 0$ are such that $u(t, y) \geqq \lambda$ and $[t - 4r(\lambda)^2, t] \times B(y, 2r(\lambda)) \subseteq [-1, \alpha] \times \bar{B}(0, \tfrac{1}{2}(1 + \delta))$. Since, for $r \in [-1, \alpha]$, $|S(r, \lambda\sigma)| \leqq 1/\varepsilon\sigma\lambda$ and $|B(y, r(\lambda))| = 2/\varepsilon\sigma\lambda$, there is an $\eta \in B(y, r(\lambda))$ such that $u(r, \eta) < \sigma\lambda$. Hence $\operatorname{Osc}(u; t, y, r(\lambda)) \geqq u(t, y) - u(r, \eta) > (1 - \sigma)\lambda$; and so, by Lemma (5.2), $\operatorname{Osc}(u; t, y, 2r(\lambda)) > \dfrac{1}{\varrho}(1 - \sigma)\lambda = K\lambda$. In particular, there is a $(t', y') \in [t - 4r(\lambda)^2, t] \times B(y, 2r(\lambda))$ such that $u(t', y') \geqq K\lambda$.

Finally, define $M$ by the relation

$$r(M) = \tfrac{1}{2}(1 - \beta)(1 - \delta)(1 - 1/K^{1/n});$$

and suppose that there were a $(t, y) \in [-\beta, -\alpha] \times B(0, \delta)$ such that $u(t, y) \geqq M$. Then, by the preceding paragraph, we could inductively find $(t_m, y_m)$, $m \geqq 0$, such that $(t_0, y_0) = (t, y)$, $(t_{m+1}, y_{m+1}) \in [t_m - 4r(K^m M), t_m] \times \bar{B}(y_m, r(K^m M)) \subseteq (-1, -\alpha) \times B(0, \tfrac{1}{2}(1 + \delta)$, and $u(t_m, y_m) \geqq K^m M$. But this would mean that $u$ is unbounded in $[-\beta, -\alpha] \times \bar{B}(0, \tfrac{1}{2}(1 + \delta))$, and so no such $(t, y)$ exists.

*Acknowledgements.* The research and preparation for this work were done while the authors were visiting members of the Institute for Mathematics and its Applications. E. B. Fabes was partially supported by NSF/DMS 8421377 and D. W. Stroock was partially supported by NSF/DMS 8415211 and ARO-DAAG29-84-K-0005.

## References

1. D. G. Aronson, Bounds for the fundamental solution of a parabolic equation, Bulletin of the American Mathematical Society **73** (1967), 890–896.
2. D. G. Aronson & J. Serrin, Local behavior of solutions of quasilinear parabolic equations, Arch. Rational Mech. Anal. **25** (1967), 81–122.
3. E. B. Davies, Explicit constants for Gaussian upper bounds on heat kernels, to appear.
4. E. De Giorgi, Sulle differentiabilità e l'analiticità degli integrali multipli regolari, Mem. Accad. Sci. Torino Cl. Sci. Fis. Mat. Natur. (III) (1957), 25–43.
5. S. Kusuoka & D. W. Stroock, Applications of the Malliavin Calculus, Part III, to appear.

6. J. MOSER, On Harnack's theorem for elliptic differential equations, Comm. Pure Appl. Math. **14** (1961), 47–79.
7. J. MOSER, A Harnack inequality for parabolic differential equations, Comm. Pure Appl. Math. **17** (1964), 101–134; Correction to "A Harnack inequality for parabolic differential equations", Comm. Pure Appl. Math. **20** (1960), 232–236.
8. J. NASH, Continuity of solutions of parabolic and elliptic equations, Amer. J. math. **80** (1958), 931–954.
9. M. N. SAFANOV, Harnack's inequality for elliptic equations and the Hölder property of their solutions, J. Soviet Mathematics **21** (1983), 851–863.

School of Mathematics
University of Minnesota
Minneapolis
and
Department of Mathematics
Massachusetts Institute of Technology
Cambridge 02139

(*Received February 19, 1986*)

# *Weakly Decaying Energy Separation and Uniqueness of Motions of an Elastic-Plastic Oscillator with Work-Hardening*

DAVID R. OWEN

*Dedicated to James Serrin on the occasion of his 60th birthday*

## Contents

1. Introduction . . . . . . . . 95
2. Elastic-plastic oscillators . . . . . . . . 97
3. Initial value problems and an energy identity . . . . . . . . 102
4. Structure of the proof of uniqueness of solutions . . . . . . . . 106
5. Proof of Lemma 4.1 . . . . . . . . 107

## 1. Introduction

A useful and natural first step in the analysis of motions of heavy objects supported by a variety of structural members is to ignore many of the geometrical characteristics of the supporting members. For example, the motion of a block suspended from a helical spring usually is studied first by regarding the spring as a one-dimensional structural element whose change in length determines the force that it exerts on the block. When the force varies linearly with change in length, the resulting spring-mass system is the familiar harmonic oscillator. If structural members undergo significant permanent deformation, it is standard practice to employ one-dimensional structural elements which, unlike the idealized spring, are capable of exhibiting hysteresis. The elastic-perfectly plastic element provides an example of this type in which there is an upper bound for the magnitude of the force that the element can provide. Below this upper "yield force", the element behaves elastically, as would a spring. When the upper bound would be surpassed through further elastic behavior, the element continues to deform but with no further increase in the force it provides. The resulting large permanent deformations might destabilize a structure consisting of elastic-plastic elements, but this tendency sometimes can be offset by the ability of the elements to dissipate mechanical energy as they undergo permanent deformations.

In the article [1979, 1], BUHITE & OWEN studied an ordinary differential equation that describes the motion of a mass supported by an elastic-perfectly plastic element and subjected to an external force general enough to require an analysis based on a generalized notion of solution for ordinary differential equations. This requirement arises from the switching rule that determines when the element behaves elastically and from the fact that switches between elastic and plastic (*i.e.* yielding) behavior can accumulate forward and backward in time. The study in the paper [1979, 1] emphasized the use of FILIPPOV's theory of ordinary differential equations with discontinuous right-hand sides [1960, 1] but pointed out also that the theory of monotone operators can be applied to obtain results on existence, uniqueness, and continuous dependence of solutions on initial data and forcing functions.

In this article, I study the case where the structural element has a yield force that can increase with the amount of permanent deformation. This element is called here an *elastic-plastic element with work-hardening*, following the terminology in the extensive literature on the theory of plasticity. (See, for example, the books [1950, 1], [1968, 1].) The forced motions of a mass supported by an element of this type are described by a system of three ordinary differential equations along with an inequality constraint. Techniques based on FILIPPOV's theory of differential equations can be adapted from [1979, 1] to obtain the existence of classical solutions on a preassigned interval of time with given initial values. I choose not to present these results here, because the problem of establishing uniqueness of solutions is more interesting, and resolution of the problem of uniqueness requires what I believe to be a new strategy. The principal difficulty in proving uniqueness of solutions lies in the failure of the operator determined by the differential equations to be monotone or locally Lipschitzian, thus precluding the use of standard arguments based on an energy separation of solutions that decreases in time or that satisfies a Gronwall inequality. The strategy that I formulate and carry out here requires only that the energy separation of solutions *decay weakly* in time in the sense that the initial energy separation never is exceeded by its subsequent values. Thus, the energy separation can decay weakly and still increase on certain intervals, as long as each increase is compensated by a prior decrease of the same or greater magnitude. (Here, the term "energy" for a single solution would denote the kinetic energy of the mass plus the potential energy of the structural element.)

It should be noted that the problem of establishing continuous dependence of solutions upon initial data and forcing functions can be resolved by using the uniqueness theorem in this paper, a theorem of FILIPPOV on passage to the limit in differential equations ([1960, 1], Theorem 3), and the fact that the concepts of classical and Filippov solutions are identical in the present problem (when the latter concept of solution is judiciously applied).

Another approach to modelling systems with hysteresis that includes some models of elastic-plastic behavior was introduced by KRASNOSELSKII and his co-workers in the last decade (see [1976, 1] and references cited there) and is developed extensively in the book on hysteresis by KRASNOSELSKII & POKROVSKII [1983, 1]. It is of interest to note that, on the one hand, I make crucial use here of approximation by continuous, piecewise linear functions in order to establish the property of weak decay, and, on the other hand, approximation by similar functions is a primary ingredient in the theory developed in the preceding reference. At the time of writing of this article, I do not know whether the specific model of elastic-plastic behavior that I study here or the method and result that I present are covered by the analysis in [1983, 1].

The reasons for the success of the strategy adopted here for resolving the question of uniqueness in the present problem are masked to some extent by the many details inherent in all problems in plasticity. Therefore, it is appropriate to close this introduction with a brief description of a simpler problem in which assumptions are given without full precision that permit a similar strategy to be carried out. We consider the initial value problem

$$\left.\begin{aligned}\dot{x}(t) &= f(x(t), y(t))\\ \dot{y}(t) &= g(x(t), y(t))\end{aligned}\right\} \quad \text{for } 0 \leqq t \leqq T, \tag{1.1}$$

$$x(0),\ y(0) \text{ given} \tag{1.2}$$

where $f$ and $g$ are continuous real-valued functions on an open subset of $\mathbb{R}^2$ such that

1. $g$ is non-negative;
2. if $(x, y)$ and $(x, \bar{y})$ are solutions of (1.1) and (1.2), then $y = \bar{y}$ (so that the problem of uniqueness centers on $(1.1)_1$); moreover, if solutions $(x, y)$ and $(\bar{x}, \bar{y})$ of (1.1) and $t_0 \in [0, T]$ are such that $y(t_0) = \bar{y}(t_0)$ and the graphs of $y$ and $\bar{y}$ "tangle" on at least one side of the line $t = t_0$, then $x(t_0) = \bar{x}(t_0)$;
3. there is a concave, non-decreasing function $G$ such that $(x - \bar{x})(f(x, y) - f(\bar{x}, \bar{y})) + (G(y) - G(\bar{y}))(g(x, y) - g(\bar{x}, \bar{y})) \leqq 0$ throughout the domain of $f$ and $g$.

Given these conditions on $f$ and $g$, if one puts

$$e(t) := \tfrac{1}{2}(x(t) - \bar{x}(t))^2 \quad \text{for } t \in [0, T]$$

with $e$ called the energy separation of $(x, y)$ and $(\bar{x}, \bar{y})$, then the first statement in condition 2 can be used to show that $(x, y) = (\bar{x}, \bar{y})$ whenever the energy separation of the solutions $(x, y)$ and $(\bar{x}, \bar{y})$ is identically zero. The second statement in condition 2 assures that $e(t_0) = 0$ at a time when the graphs of the second components of solutions $(x, y)$ and $(\bar{x}, \bar{y})$ have crossings that accumulate at $(t_0, y(t_0)) = (t_0, \bar{y}(t_0))$. Finally, conditions 1 and 3 permit one to show that $e(t_1) \geqq e(t_2)$ on an interval $[t_1, t_2]$ on which the graphs of $y$ and $\bar{y}$ bifurcate from their common point $(t_1, y(t_1)) = (t_1, \bar{y}(t_1))$. Putting these results together, one can show that

$$0 = e(0) \geqq e(t) \quad \text{for all } t \in [0, T]$$

whenever $(x, y)$ and $(\bar{x}, \bar{y})$ are solutions of (1.1) and (1.2), and this yields $x = \bar{x}$ and $y = \bar{y}$.

## 2. Elastic-plastic oscillators

If a mass $m$ is supported from above by a vertical element with force $\sigma(t)$ and is perturbed by a downward force $f(t)$, then the acceleration $\dot{v}(t)$ of the mass at time $t$ is determined by the relation

$$m\dot{v}(t) = f(t) - \sigma(t). \tag{2.1}$$

For a more concrete description of the vertical element, one may regard the element as a rod undergoing homogeneous motions with negligible change in cross section, in which case the velocity $v(t)$ of the mass is related to the (uniform) strain rate $\dot{\lambda}(t)$ for the rod in the following simple way:

$$v(t) = L\dot{\lambda}(t), \tag{2.2}$$

where $L$ is the length of the rod in a reference configuration.

The vertical element is characterized by specifying a relation between the force $\sigma$ in the element and the strain $\lambda$. For example, the relation

$$\sigma(t) = \mu\lambda(t), \tag{2.3}$$

with $\mu$ a positive constant, specifies a (*linearly*) *elastic element*, and relations (2.1)–(2.3) yield the system

$$\begin{aligned} m\dot{v}(t) &= f(t) - \sigma(t) \\ \dot{\sigma}(t) &= \frac{\mu}{L} v(t). \end{aligned} \tag{2.4}$$

For the case of an elastic element, the motion of the mass often is referred to as *forced motion of a harmonic oscillator*. By contrast, the vertical element is called an *elastic-perfectly plastic element* if the magnitude of the force $\sigma$ cannot exceed a fixed level $\sigma_y$ (its "yield value"),

$$|\sigma(t)| \leqq \sigma_y, \tag{2.5}$$

and if the rates $\dot{\sigma}(t)$ and $\dot{\lambda}(t)$ are related in the following manner:

$$\dot{\sigma}(t) = \begin{cases} 0 & \text{if } |\sigma(t)| = \sigma_y \quad \text{and } \sigma(t)\,\dot{\lambda}(t) \geqq 0, \\ \mu\dot{\lambda}(t) & \text{otherwise}. \end{cases} \tag{2.6}$$

Note that $(2.6)_2$ is compatible with the relation (2.3) for an elastic element, whereas $(2.6)_1$ is necessitated by the bound (2.5) on $|\sigma(t)|$. Switching rules of a form similar to the one in (2.6) are typical of models of elastic-plastic behavior. The relations (2.1), (2.2), (2.5), and (2.6) yield the system

$$\begin{aligned} m\dot{v}(t) &= f(t) - \sigma(t) \\ |\sigma(t)| &\leqq \sigma_y \\ \dot{\sigma}(t) &= \begin{cases} 0 & \text{if } |\sigma(t)| = \sigma_y \quad \text{and } \sigma(t)\,v(t) \geqq 0 \\ \dfrac{\mu}{L} v(t) & \text{otherwise} \end{cases} \end{aligned} \tag{2.7}$$

which is said to describe *forced motions of an elastic-perfectly plastic oscillator*.

The final example of a vertical element given in this section is the object of the research described in this paper. In this example, the yield value $\sigma_y$ in (2.5) is allowed to increase at times when $(2.6)_2$ is violated according to the amount of work done by the element during such times. Specifically, the element is called *elastic-plastic with* (*isotropic*) *work-hardening* if there is a positive-valued, twice continuously differentiable function $H$ such that $H' > 0$, $H'' \leqq 0$, and

$$|\sigma(t)| \leqq \sqrt{2H(w(t))}, \tag{2.8}$$

$$\dot{\sigma}(t) = \begin{cases} \dfrac{\mu H'(w(t))}{\mu + H'(w(t))}\dot{\lambda}(t) & \text{if } |\sigma(t)| = \sqrt{2H(w(t))}, \quad \sigma(t)\,\dot{\lambda}(t) \geqq 0, \\ \mu\dot{\lambda}(t) & \text{otherwise}, \end{cases} \tag{2.9}$$

$$\dot{w}(t) = \begin{cases} \dfrac{\mu\sigma(t)}{\mu + H'(w(t))}\dot{\lambda}(t) & \text{if } |\sigma(t)| = \sqrt{2H(w(t))}, \quad \sigma(t)\,\dot{\lambda}(t) \geqq 0, \\ 0 & \text{otherwise}. \end{cases} \tag{2.10}$$

When the strain-rate $\dot{\lambda}$ is given on a specified time interval, one may regard (2.8)–(2.10) as a system of equations for the unknowns $\sigma$ and $w$; this system contains a switching rule that distinguishes elastic behavior (the "otherwise" case) from plastic behavior of the element. From (2.10) we conclude that $\dot{w}(t) \geqq 0$ and that $\dot{w}(t) = 0$ during elastic behavior; the monotonicity of $H$ then tells us that the yield value $\sqrt{2H(w(t))}$ cannot decrease. The quantity $w(t_2) - w(t_1)$ is called the *plastic work* done by the element during the interval $[t_1, t_2]$, and the monotonicity of $t \mapsto \sqrt{2H(w(t))}$ is the source of the term "work-hardening".

The term plastic work is natural in view of the standard use of the term *plastic strain* for the quantity

$$\lambda_p = \lambda - (\sigma/\mu) \tag{2.11}$$

and in view of the relation

$$\dot{w}(t) = \sigma(t)\,\dot{\lambda}_p(t) \tag{2.12}$$

which follows from (2.9)–(2.11). The plastic strain represents the strain remaining in the element when the force $\sigma$ is reduced to zero. (The terms *residual strain* and *permanent strain* also are used.) It is important to note that $\dot{\lambda}_p(t)$ may be positive or negative, even though $\dot{w}(t)$ is non-negative, and the relationship between $\dot{\lambda}_p(t)$ and $\dot{w}(t)$ is akin to the relationship between velocity and speed.

A system analogous to (2.7) is obtained from (2.1), (2.2), (2.8), (2.9), and (2.10), and this system is said to describe *forced motions of an elastic-plastic oscillator with work-hardening*. Our study of initial value problems for this oscillator will be simplified through the following discussion of changes of variables in (2.9) and (2.10). On a time interval on which the force $\sigma$ is positive and on which there holds $|\sigma| = \sqrt{2H(w)}$, one can use the monotonicity of $H$ to express $w$ in terms of $\sigma$,

$$w = G(\sigma) \tag{2.13}$$

with $G$ continuously differentiable and $G' > 0$, so that if $\dot{\lambda}(t)$ also is not negative, $H'(w)$ can be replaced by $\sigma G'(\sigma)^{-1}$ when we compute $\dot{\sigma}(t)$ from $(2.9)_1$:

$$\dot{\sigma}(t) = \frac{\mu\sigma(t)}{\mu G'(\sigma(t)) + \sigma(t)}\dot{\lambda}(t). \tag{2.14}$$

From (2.11) and (2.14) we obtain

$$\dot{\sigma}(t) = H'(G(\sigma(t)))\,\dot{\lambda}_p(t), \tag{2.15}$$

and it is then easy to show that the force $\sigma$ can be expressed as a strictly increasing function of the plastic strain $\lambda_p$ when (2.15) holds on a time interval:

$$\sigma = S(\lambda_p). \tag{2.16a}$$

In fact, the function $S$ will be a solution of the differential equation

$$\frac{d\sigma}{d\lambda_p} = H'(G(\sigma)) \tag{2.16b}$$

subject to an initial condition of the form $S(\lambda_p(t_0)) = \sigma(t_0)$. Because $H$ is twice continuously differentiable, so are $G$ and $S$, and (2.15)–(2.16) yield

$$S'(\lambda_p) = \frac{d\sigma}{d\lambda_p} = H'(G(\sigma)) \tag{2.17}$$

$$S''(\lambda_p) = \frac{d^2\sigma}{d\lambda_p^2} = H''(G(\sigma))\,G'(\sigma)\frac{d\sigma}{d\lambda_p} = H''(G(\sigma))\,H'(G(\sigma))\,G'(\sigma). \tag{2.18}$$

It follows that $S''$ and $H''$ have the same sign; in particular, because $H$ is a concave function, so is $S$. Differentiating (2.11) and using (2.17), we can rewrite (2.15) as

$$\dot{\sigma}(t) = S'(\lambda_p(t))\left(\dot{\lambda}(t) - \frac{\dot{\sigma}(t)}{\mu}\right)$$

and (2.2) then implies that

$$\dot{\sigma}(t) = \frac{S'(\lambda_p(t))}{\mu + S'(\lambda_p(t))}\,\frac{\mu v(t)}{L}. \tag{2.19}$$

In conclusion, we may employ (2.19) in place of $(2.9)_1$ at times $t$ when $\dot{\lambda}(t) \geqq 0$ and $\sigma(t) = \sqrt{2H(w(t))}$. Under these conditions, equations (2.19), (2.15) and (2.17) tell us that the relation

$$\dot{\lambda}_p(t) = \frac{\dot{\sigma}(t)}{S'(\lambda_p(t))} = \frac{\mu}{\mu + S'(\lambda_p(t))}\,\frac{v(t)}{L} \tag{2.20}$$

can replace $(2.10)_1$. We note also that $(2.9)_2$ and (2.11) imply that $\dot{\lambda}_p(t) = 0$ for the "otherwise" case, *i.e.* during intervals of elastic behavior of the element.

It is important to recall the fact that (2.19) and (2.20) were derived from (2.9) and (2.10) by constructing a positive solution $\sigma = S(\lambda_p)$ of an initial value problem for (2.16b) with given initial condition. A particular solution $\sigma = S(\lambda_p)$ can no longer be used in (2.19) and (2.20) when $\sigma$ becomes negative and equal to $-\sqrt{2H(w)}$. Moreover, as often is the case, a particluar solution $\sigma = S(\lambda_p)$ may no longer be used after $\sigma$ becomes positive again, because increases in $w$ can occur when $\sigma$ is negative that increase the (positive) values of $\sigma$ at which the yield condition $\sigma(t) = \sqrt{2H(w(t))}$ can hold. Therefore, (2.19) and (2.20) will be used only to obtain information of a local nature about motions of an elastic-plastic oscillator with work-hardening.

When $\sigma = -\sqrt{2H(w)}$ on an interval of time, then (2.13) is replaced by

$$w = G(-\sigma), \tag{2.21}$$

with $G$ the same function as in (2.13), and $H'(w)$ can be replaced by $(-\sigma)\, G'(-\sigma)^{-1}$ when we compute $\dot{\sigma}(t)$ from $(2.9)_1$ and (2.11):

$$\begin{aligned}\dot{\sigma}(t) &= \frac{\mu\sigma(t)}{-\mu G'(--\sigma(t)) + \sigma(t)}\dot{\lambda}(t)\\ &= \frac{\mu(-\sigma(t))}{\mu G'(-\sigma(t)) + (-\sigma(t))}\dot{\lambda}(t)\\ &= \frac{-\sigma(t)}{G'(-\sigma(t))}\dot{\lambda}_p(t)\\ &= H'(G(-\sigma(t)))\,\dot{\lambda}_p(t).\end{aligned} \tag{2.22}$$

Thus, when $\sigma = -\sqrt{2H(w)}$, (2.16b) and (2.18) are replaced by

$$\frac{d\sigma}{d\lambda_p} = H'(G(-\sigma)) \tag{2.23}$$

$$\begin{aligned}\frac{d^2\sigma}{d\lambda_p^2} &= H''(G(-\sigma))\, G'(-\sigma)\left(-\frac{d\sigma}{d\lambda_p}\right)\\ &= -H''(G(-\sigma))\, H'(G(-\sigma))\, G'(-\sigma),\end{aligned} \tag{2.24}$$

so that $\sigma$ may be regarded as an increasing, convex function of $\lambda_p$ when $\sigma = -\sqrt{2H(w)}$.

We summarize the results on changes in variables in the following remark. (In the next section, a precise concept of solution is prescribed for the systems mentioned in this remark.)

**Remark 2.1.** *Each forced motion of an elastic-plastic oscillator with work-hardening in which the force $\sigma$ in the element is positive is not only a solution of the system* (2.1), (2.2), (2.8)–(2.10) *but also a solution of the system*

$$m\dot{v}(t) = f(t) - \sigma(t), \tag{2.25}$$

$$0 < \sigma(t) \leqq S_+(\lambda_p(t)), \tag{2.26}$$

$$\dot{\sigma}(t) = \begin{cases} \dfrac{S'_+(\lambda_p(t))}{\mu + S'_+(\lambda_p(t))}\dfrac{\mu v(t)}{L} & \text{if } \sigma(t) = S_+(\lambda_p(t)) \text{ and } v(t) \geqq 0,\\ \dfrac{\mu v(t)}{L} & \text{otherwise,}\end{cases} \tag{2.27}$$

$$\dot{\lambda}_p(t) = \begin{cases} \dfrac{\mu}{\mu + S'_+(\lambda_p(t))}\dfrac{v(t)}{L} & \text{if } \sigma(t) = S_+(\lambda_p(t)) \text{ and } v(t) \geqq 0,\\ 0 & \text{otherwise.}\end{cases} \tag{2.28}$$

*The function $S_+$ and its derivative $S'_+$ are positive-valued, and $S''_+$ has the same sign as $H''$, so that $S_+$ is a concave function. For motions in which $\sigma$ is negative, the following system is satisfied:*

$$m\dot{v}(t) = f(t) - \sigma(t), \tag{2.29}$$

$$S_-(\lambda_p(t)) \leqq \sigma(t) < 0, \tag{2.30}$$

$$\dot{\sigma}(t) = \begin{cases} \dfrac{S'_-(\lambda_p(t))}{\mu + S'_-(\lambda_p(t))} \dfrac{\mu v(t)}{L} & \text{if } \sigma(t) = S_-(\lambda_p(t)) \text{ and } v(t) \leqq 0 \\ \dfrac{\mu v(t)}{L} & \text{otherwise}, \end{cases} \tag{2.31}$$

$$\dot{\lambda}_p(t) = \begin{cases} \dfrac{\mu}{\mu + S'_-(\lambda_p(t))} \dfrac{v(t)}{L} & \text{if } \sigma(t) = S_-(\lambda_p(t)) \text{ and } v(t) \leqq 0 \\ 0 & \text{otherwise}. \end{cases} \tag{2.32}$$

*The function $S_-$ is negative-valued, its derivative $S'_-$ is positive-valued, and $S''_-$ has the opposite sign as $H''$, so that $S_-$ is a convex function. The function $S_+$ is uniquely determined through equation* (2.16b) *and the relation $S_+(\lambda_p(t_0)) = \sigma(t_0)$, where $t_0$ is any given time at which $\sigma(t_0) = \sqrt{2H(w(t_0))}$. $S_-$ is determined through* (2.22) *and the relation $S_-(\lambda_p(t_0)) = \sigma(t_0)$ where $t_0$ is chosen so that $\sigma(t_0) = -\sqrt{2H(w(t_0))}$. Moreover, the values of $\lambda_p$ can be computed to within a constant from the values of $\sigma$ and $v$ by means of the relation*

$$\dot{\lambda}_p = \frac{v}{L} - \frac{\dot{\sigma}}{\mu}.$$

## 3. Initial value problems and an energy identity

We begin by giving precise statements of initial value problems associated with forced motions of an elastic-plastic oscillator with work-hardening. We suppose given positive numbers $m$, $L$, $\mu$, and $T$, a twice continuously differentiable function $H: [0, \infty) \to \mathbb{R}$ with $H > 0$, $H' > 0$, and $H'' \leqq 0$, an integrable function $f: [0, T] \to \mathbb{R}$, and a triple $(v_0, \sigma_0, w_0) \in \mathbb{R}^3$ with $w_0 \geqq 0$ and $|\sigma_0| \leqq \sqrt{2H(w_0)}$; we define the *basic initial value problem* as that of finding absolutely continuous functions

$$v: [0, T] \to \mathbb{R}$$

$$\sigma: [0, T] \to \mathbb{R}$$

$$w: [0, T] \to [0, \infty)$$

that satisfy the initial conditions

$$v(0) = v_0, \quad \sigma(0) = \sigma_0 \quad \text{and} \quad w(0) = w_0, \tag{3.1}$$

the bound

$$|\sigma(t)| \leqq \sqrt{2H(w(t))} \quad \text{for all } t \in [0, T], \tag{3.2}$$

and, for almost every $t$ in $[0, T]$, the system of differential equations:

$$m\dot{v}(t) = f(t) - \sigma(t), \tag{3.3}$$

$$\dot{\sigma}(t) = \begin{cases} \dfrac{\mu H'(w(t))}{\mu + H'(w(t))} \dfrac{v(t)}{L} & \text{if } |\sigma(t)| = \sqrt{2H(w(t))} \text{ and } \sigma(t)\, v(t) \geqq 0, \\ \mu \dfrac{v(t)}{L} & \text{otherwise}, \end{cases} \tag{3.4}$$

$$\dot{w}(t) = \begin{cases} \dfrac{\mu\sigma(t)}{\mu + H'(w(t))} \dfrac{v(t)}{L} & \text{if } |\sigma(t)| = \sqrt{2H(w(t))} \text{ and } \sigma(t)\, v(t) \geqq 0, \\ 0 & \text{otherwise}. \end{cases} \tag{3.5}$$

We call such a triple $(v, \sigma, w)$ of functions a (*classical*) *solution of the basic initial value problem*.

As was mentioned in the introduction, the existence of solutions of the basic initial value problem can be established by means of straightforward modifications of the arguments used by BUHITE & OWEN [1979, 1] to establish existence of solutions for the elastic-perfectly plastic oscillator (there called the "dangling spider problem"). Our interest here is in the question of uniqueness of solutions of the basic initial value problem, and we observe first that when $|\sigma_0| < \sqrt{2H(w_0)}$ or when

$$|\sigma_0| = \sqrt{2H(w_0)} \quad \text{and} \quad \sigma_0 v_0 < 0$$

there exists $\tau \in (0, T]$ such that the basic initial value problem has exactly one solution on $[0, \tau]$. In fact, the above conditions on $\sigma_0$, $v_0$ and $w_0$ tell us that the "otherwise" case must hold for each solution for a short interval of time. Hence, two initially distinct solutions would provide distinct solutions of the initial value problem for forced motions of a harmonic oscillator, and it is an elementary fact that this situation cannot occur. We conclude that a necessary condition in order that two solutions of the basic initial value problem differ on a set of times that accumulate at zero is the pair of relations:

$$|\sigma_0| = \sqrt{2H(w_0)} \quad \text{and} \quad \sigma_0 v_0 \geqq 0. \tag{3.6}$$

It will be evident from our subsequent discussion and from the analysis leading to Remark 2.1 that it suffices to consider only the case

$$\sigma_0 = \sqrt{2H(w_0)} \quad \text{and} \quad v_0 \geqq 0. \tag{3.7}$$

We assume that $v_0$, $\sigma_0$, and $w_0$ are given such that (3.7) holds, we put

$$t_0 := 0, \quad \sigma(t_0) := \sigma_0, \quad \lambda_p(t_0) = 0: \tag{3.8}$$

in the auxiliary system (2.25)–(2.28), and we write $S$ for the function $S_+$ determined by (2.16b) and the initial condition $S_+(0) = \sigma_0$. This leads us to define the *auxiliary initial value problem* as that of finding $T_a > 0$ and absolutely continuous functions

$$v : [0, T_a] \to \mathbb{R}$$

$$\sigma : [0, T_a] \to \mathbb{R}$$

$$\lambda_p : [0, T_a] \to \mathbb{R}$$

that satisfy the initial conditions

$$v(0) = v_0, \quad \sigma(0) = \sigma_0, \quad \lambda_p(0) = 0, \tag{3.9}$$

the bounds

$$0 < \sigma(t) \leqq S(\lambda_p(t)) \quad \text{for all } t \in [0, T_a], \tag{3.10}$$

and, for almost every $t$ in $[0, T_a]$, the system of differential equations:

$$m\dot{v}(t) = f(t) - \sigma(t), \tag{3.11}$$

$$\dot{\sigma}(t) = \begin{cases} \dfrac{S'(\lambda_p(t))}{\mu + S'(\lambda_p(t))} \dfrac{\mu v(t)}{L} & \text{if } \sigma(t) = S(\lambda_p(t)) \text{ and } v(t) \geqq 0 \\ \dfrac{\mu v(t)}{L} & \text{otherwise} \end{cases} \tag{3.12}$$

$$\dot{\lambda}_p(t) = \begin{cases} \dfrac{\mu}{\mu + S'(\lambda_p(t))} \dfrac{v(t)}{L} & \text{if } \sigma(t) = S(\lambda_p(t)) \text{ and } v(t) \geqq 0 \\ 0 & \text{otherwise}. \end{cases} \tag{3.13}$$

A triple $(v, \sigma, \lambda_p)$ satisfying (3.9)–(3.13) is called a (*classical*) *solution of the auxiliary initial value problem*. We recall from Remark 2.1 that $S: [0, \infty) \to \mathbb{R}$ is twice continuously differentiable and satisfies

$$S > 0, \quad S' > 0, \quad \text{and} \quad S'' \leqq 0. \tag{3.14}$$

The arguments given in Section 2 leading to Remark 2.1 yield the next result.

**Remark 3.1.** *For each* $(v_0, \sigma_0, w_0)$ *satisfying* (3.7) *and each solution* $(v, \sigma, w)$ *of the basic initial value problem, there exists* $T_a \in (0, T]$ *such that the restrictions of the functions* $v$ *and* $\sigma$ *to* $[0, T_a]$ *together with the function* $\lambda_p: [0, T_a] \to \mathbb{R}$ *defined by*

$$\lambda_p(t) = \int_0^t \left( v - \frac{\dot{\sigma}}{\mu} \right) \quad \text{for } t \in [0, T_a], \tag{3.15}$$

*form a solution of the auxiliary initial value problem. Moreover,* (2.12) *and* (3.15) *tell us that distinct solutions of the basic initial value problem yield distinct solutions of the auxiliary problem. Consequently, uniqueness of solutions of the auxiliary initial value problem implies uniqueness of solutions of the basic initial value problem.*

Our goal now is to show that for a given $T_a > 0$ the auxiliary initial value problem (3.9)–(3.13) has at most one solution. In subsequent discussions, we use (3.7) in the form

$$\sigma(0) = S(\lambda_p(0)) \quad \text{and} \quad v(0) \geqq 0, \tag{3.16}$$

which suggests from examination of the switching rule in (3.12) and (3.13) that solutions of the auxiliary initial value problem are solutions of $(3.12)_1$ and $(3.13)_1$ on an initial interval of positive length, so that one might try to study separately uniqueness of solutions of $(3.12)_1$ and $(3.13)_1$. However, the fact that a corresponding strategy does not suffice in the case of an elastic-perfectly plastic oscillator leads us to seek a proof of uniqueness of solutions that allows for the possibility that switching times for the switching rule in (3.12) and (3.13) can accumulate backward to the initial time zero.

The key to our proof of uniqueness of solutions of the auxiliary problem is the following easily verified *energy identity* which holds for almost every $t \in [0, T_a]$ for triples $(v, \sigma, \lambda_p)$ and $(\bar{v}, \bar{\sigma}, \bar{\lambda}_p)$ satisfying (3.10)–(3.13) almost everywhere in $[0, T_a]$:

$$\left[\frac{m}{2L}(v-\bar{v})^2 + \frac{1}{2\mu}(\sigma-\bar{\sigma})^2\right]^{\cdot}(t) = -[(\sigma-\bar{\sigma})(\dot{\lambda}_p - \dot{\bar{\lambda}}_p)](t). \tag{3.17}$$

Verification of (3.17) consists of checking this relation in the four different cases that arise because of the switching rule in (3.12) and (3.13). We call the quantity

$$e(t) := \left[\frac{m}{2L}(v-\bar{v})^2 + \frac{1}{2\mu}(\sigma-\bar{\sigma})^2\right](t), \tag{3.18}$$

whose derivative appears as the left-hand member of (3.17), the *energy separation of* $(v, \sigma, \lambda_p)$ *and* $(\bar{v}, \bar{\sigma}, \bar{\lambda}_p)$ at time $t \in [0, T_a]$. For the elastic-perfectly plastic oscillator, the energy separation of solutions cannot increase, and this leads immediately to uniqueness of solutions of the initial value problem for the system (2.7). (See Theorems 4.1, 5.1 in [1979, 1].)

The difficulty in dealing with the case of work-hardening lies in the fact that the quantity

$$\omega_p(t) := (\sigma(t) - \bar{\sigma}(t))(\dot{\lambda}_p(t) - \dot{\bar{\lambda}}_p(t)) \tag{3.19}$$

in the right-hand side of (3.17) can be negative on sets of positive measure. For example, it can occur that on a non-trivial interval $I$ the solution $(v, \sigma, \lambda_p)$ behaves *elastically* (*i.e.*, the "otherwise" case applies in (3.12) and (3.13)) whereas the solution $(\bar{v}, \bar{\sigma}, \bar{\lambda}_p)$ behaves *plastically* (the opposite case in (3.12) and (3.13)). In this situation $\dot{\lambda}_p(t) = 0$, $\dot{\bar{\lambda}}_p(t) \geqq 0$ and $\omega_p(t) = (\sigma(t) - \bar{\sigma}(t))(-\dot{\bar{\lambda}}_p(t))$, for almost every $t$ in $I$, and the relations $\sigma(t) > \bar{\sigma}(t)$ and $\dot{\bar{\lambda}}_p(t) > 0$ would then imply that $\omega_p(t) < 0$ for almost every $t$ in $I$. (For the elastic-perfectly plastic oscillator, $\sigma(t) > \bar{\sigma}(t)$ cannot occur under the circumstances described above, so that $\omega_p(t) \geqq 0$ almost everywhere on $I$.)

Our task then is to use the fact that, in proving uniqueness of solutions of the auxiliary problem (3.9)–(3.13), we consider solutions $(v, \sigma, \lambda_p)$ and $(\bar{v}, \bar{\sigma}, \bar{\lambda}_p)$ of (3.10)–(3.13) with the same initial value $(v_0, \sigma_0, 0)$, so that, in particular, $e(0) = 0$.

This additional fact will permit us to show that the energy separation $e$ *decays weakly* in the sense that

$$e(0) \geqq e(t) \quad \text{for all } t \in [0, T_a]. \tag{3.20}$$

The condition of weak decay permits $e$ to increase on some intervals, as long as such increases are offset by prior decreases in such a way that (3.20) holds. Of course, $e(t)$ is non-negative for every $t$ in $[0, T_a]$, so that weak decay of $e$ on $[0, T_a]$ and the condition $e(0) = 0$ would tell us that $e$ is identically zero on $[0, T_a]$. From this fact, we could conclude that $\bar{v} = v$ and $\bar{\sigma} = \sigma$; the relations (3.15) and (3.9) would then yield equality of $\bar{\lambda}_p$ and $\lambda_p$ and, hence, the equality of the triples $(v, \sigma, \lambda_p)$ and $(\bar{v}, \bar{\sigma}, \bar{\lambda}_p)$.

## 4. Structure of the proof of uniqueness of solutions

Let $(v, \sigma, \lambda_p)$ and $(\bar{v}, \bar{\sigma}, \bar{\lambda}_p)$ be solutions of the auxiliary initial value problem (3.9)–(3.13). Our argument focuses on the set of points where $\lambda_p$ and $\bar{\lambda}_p$ agree:

$$\mathcal{T}_0 := \{\tau \in [0, T_a] \mid \lambda_p(\tau) - \bar{\lambda}_p(\tau) = 0\}. \tag{4.1}$$

$\mathcal{T}_0$ is a closed subset of $[0, T_a]$ that contains 0, and we single out two types of boundary points $\tau$ of $\mathcal{T}_0$:

*Type 1*: $\tau$ is isolated from $\mathcal{T}_0$ from the right, *i.e.*, there exists $\delta > 0$ such that $(\tau, \tau + \delta) \cap \mathcal{T}_0 = \emptyset$;

*Type 2*: The number $\tau$ satisfies at least one of the two conditions:

(1) $(\tau - \delta, \tau) \cap \mathcal{T}_0 \neq \emptyset$ and $(\tau - \delta, \tau) \cap ([0, T_a] \setminus \mathcal{T}_0) \neq \emptyset$ for every $\delta > 0$;

(2) $(\tau, \tau + \delta) \cap \mathcal{T}_0 \neq \emptyset$ and $(\tau, \tau + \delta) \cap ([0, T_a] \setminus \mathcal{T}_0) \neq \emptyset$ for every $\delta > 0$.

Of course, a boundary point $\tau$ will be of both types if it is isolated from $\mathcal{T}_0$ from the right but not from either $\mathcal{T}_0$ or its complement from the left. Each point of type 1 corresponds to a common point on the graphs of $\lambda_p$ and $\bar{\lambda}_p$ from which the graphs bifurcate (as time increases). Each point $\tau$ of type 2 corresponds to a common point on the graphs of $\lambda_p$ and $\bar{\lambda}_p$ at which the graphs are tangled on at least one side of the line $t = \tau$.

The following lemma is crucial to our argument; its proof is long and will be deferred to the next section.

**Lemma 4.1.** *Let* $\tau \in [0, T_a)$ *be given. If* $\tau$ *is a boundary point of* $\mathcal{T}_0$ *of type* 1, *then the energy separation decays weakly on the interval* $[\tau, \tau + d(\tau)]$, *with* $d(\tau) := \min\{|\tau - t| \mid t \in (\tau, T_a] \cap \mathcal{T}_0$ *or* $t = T_a\}$, i.e.,

$$e(\tau) \geqq e(t), \quad \textit{for all } t \in [\tau, \tau + d(\tau)]. \tag{4.2}$$

*If* $\tau$ *is a boundary point of* $\mathcal{T}_0$ *of type* 2, *then there holds* $(v(\tau), \sigma(\tau), \lambda_p(\tau)) = (\bar{v}(\tau), \bar{\sigma}(\tau), \bar{\lambda}_p(\tau))$, *so that* $e(\tau) = 0$.

In the terminology given before the statement of the lemma, the conclusions in Lemma 4.1 mean that bifurcation of the graphs of $\lambda_p$ and $\bar{\lambda}_p$ entails weak decay of the energy separation during the bifurcation and tangling of the graphs entails zero energy separation.

Let us now use Lemma 4.1 to prove that $e$ decays weakly on $[0, T_a]$. To do so, it suffices to show that for every $t$ in $(0, T_a]$, there exists $t_1$ in $[0, t)$ with $e(t_1) \geqq e(t)$. (In fact, if there were then to hold $e(0) < e(t^*)$ for some $t^*$ in $(0, T_a]$, the continuity of $e$ would imply that the set of points $\tilde{t}$ in $[0, t^*]$ such that $e(\tilde{t}) \geqq e(t^*)$ would have minimum 0, so that $e(0) \geqq e(t^*)$, a contradiction.) We let $t \in (0, T_a]$ be given and we suppose first that for every $t' \in [0, t)$, there holds $\mathscr{T}_0 \cap (t', t) \neq \emptyset$ and $(]0, T_a] \setminus \mathscr{T}_0) \cap (t', t) \neq \emptyset$. We may conclude immediately that $t$ is a boundary point of $\mathscr{T}_0$ of type 2, and Lemma 4.1 yields $e(t) = 0$. Because $e(t) = 0$, we may put $t_1$ equal to 0 and so obtain $e(t_1) \geqq e(t)$. Next, we suppose there is a number $t'$ in $[0, t)$ such that $(t', t) \subset \mathscr{T}_0$. In this case, $\lambda_p = \bar{\lambda}_p$ on $[t', t]$, and (3.17) and (3.18) tell us that $e$ is constant on $[t', t]$, so we may chose $t_1$ to be any such time $t'$. Finally if there is a $t'$ in $[0, t)$ such that $(t', t) \cap \mathscr{T}_0 = \emptyset$, then we put $t_1$ equal to the infimum of such times $t'$. Because 0 is in $\mathscr{T}_0$, it follows that $t_1$ is a boundary point of $\mathscr{T}_0$ of type 1 and, by Lemma 4.1, $e(t_1) \geqq e(t)$. This concludes the proof that $e$ decays weakly on $[0, T_a]$, and the remarks at the end of Section 3 tell us that the triples $(v, \sigma, \lambda_p)$ and $(\bar{v}, \bar{\sigma}, \bar{\lambda}_p)$ are identical. The remainder of the paper is devoted to proving the key result, Lemma 4.1.

## 5. Proof of Lemma 4.1

We first prove the second assertion of Lemma 4.1: *if $\tau$ is a boundary point of $\mathscr{T}_0$ of type 2, then there holds $(v(\tau), \sigma(\tau), \lambda_p(\tau)) = (\bar{v}(\tau), \bar{\sigma}(\tau), \bar{\lambda}_p(\tau))$, so that* $e(\tau) = 0$.

**Proof.** Let a boundary point $\tau$ of $\mathscr{T}_0$ of type 2 be given. Observe that this condition on $\tau$ excludes the possibility that both solutions behave elastically in an open interval containing $\tau$, for then $\lambda_p - \bar{\lambda}_p$ would vanish identically on that interval and $\tau$ would not be a boundary point of $\mathscr{T}_0$. We suppose next that there exists $\delta > 0$ such that both $(v, \sigma, \lambda_p)$ and $(\bar{v}, \bar{\sigma}, \bar{\lambda}_p)$ behave plastically on $(\tau - \delta, \tau + \delta) \cap [0, T_a]$, *i.e.*, both solutions satisfy $(3.12)_1$ and $(3.13)_1$ on that interval. We note in this case that $\sigma(\tau) = \bar{\sigma}(\tau) = S(\lambda_p(\tau)) = S(\bar{\lambda}_p(\tau))$ and, because the switching rule is not invoked on that interval and $S'$ is continuous, the right-hand sides of (3.12)–(3.13) as employed in this case are continuous in the velocity $v$, force $\sigma$ in the vertical element, and plastic strain $\lambda_p$. Therefore, $\lambda_p$ and $\bar{\lambda}_p$ are absolutely continuous functions whose derivatives are restrictions of continuous functions to a subset of $(\tau - \delta, \tau + \delta) \cap [0, T_a]$ of full measure. By the Fundamental Theorem of Calculus, $\lambda_p$ and $\bar{\lambda}_p$ are continuously differentiable on that interval; moreover, $\lambda_p(\tau) = \bar{\lambda}_p(\tau)$. Thus, by $(3.13)_1$,

$$\dot{\lambda}_p(\tau) - \dot{\bar{\lambda}}_p(\tau) = \frac{\mu}{L(\mu + S'(\lambda_p(\tau)))} (v(\tau) - \bar{v}(\tau)), \tag{5.1}$$

and the hypothesis that $\tau$ is of type 2 implies that $v(\tau) = \bar{v}(\tau)$. In fact, if $v(\tau) \neq \bar{v}(\tau)$, (5.1) and the continuity of $\dot{\lambda}_p$ and $\dot{\bar{\lambda}}_p$ would imply that the graphs of $\lambda_p$ and $\bar{\lambda}_p$ have a simple crossing at time $\tau$, so that $\tau$ would be an isolated point of $\mathscr{T}_0$, a contradiction. Thus, we have shown that $(v(\tau), \sigma(\tau). \lambda_p(\tau)) = (\bar{v}(\tau), \bar{\sigma}(\tau), \bar{\lambda}_p(\tau))$ in the case where both solutions behave plastically in a neighborhood of the time $\tau$ of type 2. Suppose now that one of the two solutions, say $(v, \sigma, \lambda_p)$, has intervals of elastic behavior *and* intervals of plastic in every neighborhood of $\tau$. This condition implies that the switching rule is invoked at time $\tau$ or that it is invoked on a sequence of distinct times approaching $\tau$, so that in either case $\sigma(\tau) = S(\lambda_p(\tau))$ and $v(\tau) = 0$. If the other solution $(\bar{v}, \bar{\sigma}, \bar{\lambda}_p)$ has the same property, then $\bar{\sigma}(\tau) = S(\bar{\lambda}_p(\tau)) = S(\lambda_p(\tau)) = \sigma(\tau)$ and $\bar{v}(\tau) = 0 = v(\tau)$, i.e., $(v(\tau), \sigma(\tau), \lambda_p(\tau)) = (\bar{v}(\tau), \bar{\sigma}(\tau), \bar{\lambda}_p(\tau))$. If $(\bar{v}, \bar{\sigma}, \bar{\lambda}_p)$ were to behave only elastically in a neighborhood of time $\tau$, then for almost every $t$ on that neighborhood

$$\dot{\lambda}_p(t) - \dot{\bar{\lambda}}_p(t) = \dot{\lambda}_p(t) - 0 \geqq 0,$$

so that $\lambda_p - \bar{\lambda}_p$ would be non-decreasing on the same neighborhood, contradicting the hypothesis that $\tau$ is of type 2. If $(\bar{v}, \bar{\sigma}, \bar{\lambda}_p)$ were to behave only plastically in a neighborhood $\mathscr{N}$ of time $\tau$, then $\bar{\sigma}(\tau) = S(\bar{\lambda}_p(\tau)) = S(\lambda_p(\tau)) = \sigma(\tau)$, and

$$L(\dot{\lambda}_p(t) - \dot{\bar{\lambda}}_p(t)) = \begin{cases} 0 - \dfrac{\mu \bar{v}(t)}{\mu + S'(\bar{\lambda}_p(t))} \\ \text{or} \\ \dfrac{\mu v(t)}{\mu + S'(\lambda_p(t))} - \dfrac{\mu \bar{v}(t)}{\mu + S'(\bar{\lambda}_p(t))} \end{cases} \tag{5.2}$$

for almost every $t$ in $\mathscr{N}$. If $\bar{v}(\tau) \neq 0$, then because $v(\tau) = 0$, the relation (5.2) and the continuity of $v$ and $\bar{v}$ would yield a neighborhood $\mathscr{N}'$ of $\tau$ included in $\mathscr{N}$ such that $\dot{\lambda}_p(t) - \dot{\bar{\lambda}}_p(t) < 0$ holds for almost every $t$ in $\mathscr{N}'$, and this fact also would contradict the hypothesis that $\tau$ is of type 2. Hence $\bar{v}(\tau) = 0$ in this case, so that again $(v(\tau), \sigma(\tau), \lambda_p(\tau)) = (\bar{v}(\tau), \bar{\sigma}(\tau), \bar{\lambda}_p(\tau))$, and the proof of the second assertion of Lemma 4.1 is complete.

We consider next the first assertion in Lemma 4.1:

*if $\tau$ is a boundary point of $\mathscr{T}_0$ of type* 1 *and if we put*

$$d(\tau) := \min \{t - \tau \mid t \in (\tau, T_a] \cap \mathscr{T}_0 \text{ or } t = T_a\}, \tag{5.3}$$

*then the energy separation of* $(v, \sigma. \lambda_p)$ *and* $(\bar{v}, \bar{\sigma}, \bar{\lambda}_p)$ *decays weakly on* $[\tau, \tau + d(\tau)]$, i.e., *for all* $t \in [\tau, \tau + d(\tau)]$,

$$e(\tau) \geqq e(t). \tag{5.4}$$

**Proof.** We observe from (3.17) and (3.18) that (5.4) is equivalent to the relation

$$\int_\tau^t (\sigma - \bar{\sigma})(\dot{\lambda}_p - \dot{\bar{\lambda}}_p) \geqq 0 \quad \text{for all } t \in [\tau, \tau + d(\tau)]. \tag{5.5}$$

Our first step in verifying (5.5) is to show that for every $t \in [\tau, T_a]$,

$$\int_\tau^t (\sigma - \bar{\sigma})(\dot{\lambda}_p - \dot{\bar{\lambda}}_p) \geqq \int_\tau^t (S(\lambda_p) - S(\bar{\lambda}_p))(\dot{\lambda}_p - \dot{\bar{\lambda}}_p). \tag{5.6}$$

We verify (5.6) by showing that the integrands satisfy the corresponding inequality

$$(\sigma - \bar{\sigma})(\dot{\lambda}_p - \dot{\bar{\lambda}}_p) \geqq (S(\lambda_p) - S(\bar{\lambda}_p))(\dot{\lambda}_p - \dot{\bar{\lambda}}_p) \tag{5.7}$$

almost everywhere on $[0, T_a]$. Let $t \in [0\ T_a]$ be given. If $(v, \sigma, \lambda_p)$ and $(\bar{v}, \bar{\sigma}, \bar{\lambda}_p)$ both behave elastically at time $t$, then $(3.13)_2$ yields (5.7) (with equality):

$$(\sigma(t) - \bar{\sigma}(t))(\dot{\lambda}_p(t) - \dot{\bar{\lambda}}_p(t)) = (S(\lambda_p(t) - S(\bar{\lambda}_p(t)))(\dot{\lambda}_p(t) - \dot{\bar{\lambda}}_p(t)) = 0.$$

If $(v, \sigma, \lambda_p)$ and $(\bar{v}, \bar{\sigma}, \bar{\lambda}_p)$ both behave plastically at time $t$, then $(3.13)_1$ yields (5.7) (with equality):

$$(\sigma(t) - \bar{\sigma}(t))(\dot{\lambda}_p(t) - \dot{\bar{\lambda}}_p(t)) = (S(\lambda_p(t)) - S_p(\bar{\lambda}_p(t)))(\dot{\lambda}_p(t) - \dot{\bar{\lambda}}_p(t)).$$

If $(v, \sigma, \lambda_p)$ behaves plastically and $(\bar{v}, \bar{\sigma}, \bar{\lambda}_p)$ behaves elastically at time $t$, then $(3.13)_1$ and $(3.13)_2$ yield (5.7):

$$\begin{aligned}(\sigma(t) - \bar{\sigma}(t))(\dot{\lambda}_p(t) - \dot{\bar{\lambda}}_p(t)) &= (S(\lambda_p(t)) - \bar{\sigma}(t))\,\dot{\lambda}_p(t) \\ &\geqq (S(\lambda_p(t)) - S(\bar{\lambda}_p(t))(\dot{\lambda}_p(t) - \dot{\bar{\lambda}}_p(t)),\end{aligned}$$

because $\sigma(t) = S(\lambda_p(t))$, $\bar{\sigma}(t) \leqq S(\bar{\lambda}_p(t))$, $\dot{\bar{\lambda}}_p(t) = 0$ and $\dot{\lambda}_p(t) \geqq 0$. The remaining case is verified in the same manner, so that (5.7) and (5.6) are valid. Thus we have reduced the verification of (5.5) to showing that for every $t \in [\tau, \tau + d(\tau)]$

$$\int_\tau^t (S(\lambda_p) - S(\bar{\lambda}_p))(\dot{\lambda}_p - \dot{\bar{\lambda}}_p) \geqq 0 \tag{5.8}$$

or, equivalently,

$$\int_\tau^t (S(\bar{\lambda}_p + u) - S(\bar{\lambda}_p))\dot{u} \geqq 0 \tag{5.9}$$

where

$$u := \lambda_p - \bar{\lambda}_p. \tag{5.10}$$

Because the time $\tau$ is of type 1 and $d(\tau)$ is given by (5.3), we may assume that

$$u(t) > 0 \quad \text{for all } t \in (\tau, \tau + d(\tau)); \tag{5.11}$$

of course, also

$$u(\tau) = 0 \quad \text{and} \quad u(\tau + d(\tau)) \geqq 0. \tag{5.12}$$

Thus, on $(\tau, \tau + d(\tau))$, the graph of $\lambda_p$ lies strictly above that of $\bar{\lambda}_p$; the graphs intersect at time $\tau$ and, if $\tau + d(\tau) \in \mathscr{T}_0$, also at time $\tau + d(\tau)$. If we ignore the details of how (5.9) was obtained, we are led to the following problem: *given $S$ of class $C^2$ on $[0, \infty)$ with $S > 0$, $S' > 0$ and $S'' \leqq 0$ and given a non-decreasing, absolutely continuous function $\bar{\lambda}_p : [\tau, \tau + d(\tau)] \to [0, \infty)$, show that* (5.9) *holds for every absolutely continuous function* $u\colon [\tau, \tau + d(\tau)] \to \mathbb{R}$ *that satisfies* (5.11) *and* (5.12). We can simplify this problem using the following observations: for each $t \in [\tau, \tau + d(\tau)]$, the mapping

$$u \mapsto \int_\tau^t (S(\bar{\lambda}_p + u) - S(\bar{\lambda}_p))\, \dot{u} \tag{5.13}$$

from absolutely continuous functions on $[\tau, \tau + d(\tau)]$ into $\mathbb{R}$ is continuous with respect to the standard norm

$$\|u\|_{AC} := |u(\tau)| + \int_\tau^{\tau + d(\tau)} |\dot{u}|, \tag{5.14}$$

and the continuous, piecewise linear functions $u$ that satisfy (5.11) and (5.12) are dense in the set of absolutely continuous functions $u$ that satisfy (5.11) and (5.12). The continuity of the mapping in (5.13) follows from the regularity assumptions on $S$ and $\bar{\lambda}_p$ and the fact that closeness of functions in the norm in (5.14) implies closeness in the uniform norm. The statement on denseness of piecewise linear functions can be verified by approximating an absolutely continuous function $u$ on a short interval $[\tau, \tau + \delta]$ by the linear function

$$\ell_\delta(t) = u(\tau + \delta)\,(t - \tau) \qquad \text{for } \tau \leqq t \leqq \tau + \delta, \tag{5.15}$$

and by approximating $\dot{u}$ on $[\tau + \delta, \tau + d(\tau)]$ by a step function $s_\delta$ chosen so that the function $t \mapsto \delta u(\tau + \delta) + \int_{\tau+\delta}^t s_\delta$ is close to $u$ on $[\tau + \delta, \tau + d(\tau)]$ with respect to the norm in (5.14) and also satisfies (5.11) on $[\tau + \delta, \tau + d(\tau)]$.

In summary, the problem stated in italics reduces to proving that

$$\int_\tau^t (S(\bar{\lambda}_p + \pi) - S(\bar{\lambda}_p))\, \dot{\pi} \geqq 0 \tag{5.16}$$

for every $t \in [\tau, \tau + \delta(\tau)]$ and for every continuous, piecewise linear function $\pi\colon [\tau, \tau + \delta(\tau)] \to \mathbb{R}$ satisfying

$$\pi(\tau) = 0 \tag{5.17}$$

and

$$\pi(t) > 0 \qquad \text{for all } t \in (\tau, \tau + d(\tau)). \tag{5.18}$$

(As above, $\bar{\lambda}_p$ is a given non-decreasing, absolutely continuous function and $S$ is a given $C^2$ mapping on $[0, \infty)$ such that $S > 0$, $S' > 0$ and $S'' < 0$.) Let $t \in (\tau, \tau + d(\tau))$ be given, and for each $m \in \mathbb{N}$ define $P(m)$ to be the assertion:

For every continuous, piecewise linear function $\pi\colon [\tau, t] \to \mathbb{R}$ satisfying

(a) $\pi(\tau) = 0$,

(b) $\pi(\xi) > 0$ for all $\xi \in (\tau, t]$,

(c) the number of strict local maxima of $\pi$ in $(\tau, t)$ *plus* the number of non-trivial, maximal closed intervals in $(\tau, t)$ on each of which $\pi$ is constant and at both of whose endpoints $\pi$ has a local maximum equals the number $m$, there holds

$$\int_{\tau} (S(\bar{\lambda}_p + \pi) - S(\bar{\lambda}_p))\, \dot{\pi} \geqq 0. \tag{5.19}$$

Condition (c) is the assertion that the number of "peaks" plus the number of "mesas" in the graph of $\pi$ is the number $m$.

*Proof that P*(0) *holds.* Let a continuous, piecewise linear function $\pi$ be given satisfying (a), (b) and (c) with $m = 0$. Because (a) and (b) hold, it follows from (c) that $\dot{\pi}(\xi) \geqq 0$ for all but a finite number of points $\xi$ in $[\tau, t]$. Moreover, (a), (b) and the monotonicity of $S$ imply that

$$S(\bar{\lambda}_p(\xi) + \pi(\xi)) - S(\bar{\lambda}_p(\xi)) \geqq 0$$

for every $\xi \in [\tau, t]$, and (5.19) follows immediately.

*Proof that for every* $m \in \mathbb{N}$, $P(m) \Rightarrow P(m+1)$. Let $m \in \mathbb{N}$ be given and let $\pi: [\tau, t] \to \mathbb{R}$ be given satisfying (a), (b), and (c), with $m$ replaced by $m + 1$. In particular, the graph of $\pi$ has at least one peak or mesa. We modify $\pi$ by choosing one peak (or one mesa, if there are no peaks), and cutting off that peak or mesa (and no other peaks or mesas) with a horizontal segment of minimum length such that, when we replace the chosen peak or mesa on the graph of $\pi$ by the horizontal cutting segment, the new graph so obtained is the graph of a continuous, piecewise linear function $\pi_c: [\tau, t] \to \mathbb{R}$ that satisfies (a), (b), and, for the given value of $m$, (c). In other words, while the graph of $\pi$ has $m + 1$ peaks and mesas, the graph of $\pi_c$ has $m$ peaks and mesas. Moreover, if $[\tau_1, \tau_2]$ denotes the interval lying below the horizontal cutting segment, we have the relation

$$\int_{\tau}^{t} (S(\bar{\lambda}_p + \pi) - S(\bar{\lambda}_p))\, \dot{\pi} =$$

$$\int_{\tau}^{t} (S(\bar{\lambda}_p + \pi_c) - S(\bar{\lambda}_p))\, \dot{\pi}_c + \int_{\tau_1}^{\tau_2} (S(\bar{\lambda}_p + \pi) - S(\bar{\lambda}_p))\, \dot{\pi}. \tag{5.20}$$

Indeed, (5.20) follows from the fact that $\pi_c$ is constant on $[\tau_1, \tau_2]$ and agrees with $\pi$ elsewhere on $[\tau, t]$. Because we may assume that $P(m)$ is true, we conclude that the first integral on the right-hand side of (5.20) is not negative, and it remains to show the same for the second integral. By the choice of the interval $[\tau_1, \tau_2]$, the restriction of $\pi$ to $[\tau_1, \tau_2]$ satisfies

(a)′ $\pi(\tau_1) = \pi(\tau_2)$,

(b)′ $\pi(\xi) > 0$ for all $\xi \in (\tau_1, \tau_2)$,

and

(c)′ the graph of $\pi$ on $[\tau_1, \tau_2]$ has either one peak or one mesa.

Thus, the induction step leads us to verify the inequality

$$\int_{\tau_1}^{\tau_2} (S(\bar{\lambda}_p + \pi) - S(\bar{\lambda}_p))\, \dot{\pi} \geqq 0 \tag{5.21}$$

whenever $\pi$ is a continuous, piecewise linear function on $[\tau_1, \tau_2]$ that satisfies (a)′, (b)′, and (c)′. However, the argument which provided us with this reduction suggests that we successively introduce additional horizontal cuts in the graph of $\pi$: each cut starts or ends at a corner on the graph that is not the peak or an endpoint of the mesa. This procedure allows us to write the left-hand side of (5.21) as a sum of integrals of the same form in each of which $\pi$ satisfies not only (a)′, (b)′, and (c)′, but also the condition

(d)′ the graph of $\pi$ on $[\tau_1, \tau_2]$ has exactly two maximal segments that are not horizontal and at most one maximal horizontal segment.

Because of (c)′, the graph of $\pi$ on $[\tau_1, \tau_2]$ has one maximal segment with positive slope over an interval of the form $[\tau_1, \tau_+]$, one maximal segment with negative slope over an interval of the form $[\tau_-, \tau_2]$, and

$$\tau_+ \leqq \tau_- \tag{5.22}$$

and

$$\pi(\xi) = \pi(\tau_+) \qquad \text{for all } \xi \in [\tau_+, \tau_-]. \tag{5.23}$$

If we write $s^+$ for the slope of the graph on $[\tau_1, \tau_+]$ and $s_-$ for the slope of the graph on $[\tau_-, \tau_2]$, then we have

$$\int_{\tau_1}^{\tau_2} (S(\bar{\lambda}_p + \pi) - S(\bar{\lambda}_p))\, \dot{\pi} = \int_{\tau_1}^{\tau^+} (S(\bar{\lambda}_p + \pi) - S(\bar{\lambda}_p))\, s^+ + \int_{\tau_-}^{\tau_2} (S(\bar{\lambda}_p + \pi) - S(\bar{\lambda}_p))\, s^- . \tag{5.24}$$

Moreover, on $[\tau_1, \tau_+]$ we have

$$\pi(\xi) = \pi(\tau_1) + (\xi - \tau_1)\, s^+ \tag{5.25}$$

and on $[\tau_-, \tau_2]$

$$\pi(\xi) = \pi(\tau_-) + (\xi - \tau_-)\, s^-, \tag{5.26}$$

so that

$$\xi = \tau_1 + \frac{\pi(\xi) - \pi(\tau_1)}{s^+} \tag{5.27}$$

or

$$\xi = \tau_- + \frac{\pi(\xi) - \pi(\tau_-)}{s^-} \tag{5.28}$$

on these two intervals. If on $[\tau_1, \tau_+]$ we use the change of variables

$$\xi = \tau_1 + \frac{z - \pi(\tau_1)}{s^+} =: t_+(z) \tag{5.29}$$

in the first integral on the right hand side of (5.24) and on $[\tau_-, \tau_2]$ the change of variables

$$\xi = \tau_- + \frac{z - \pi(\tau_-)}{s^-} =: t_-(z) \tag{5.30}$$

in the second integral, then we have

$$\int_{\tau_1}^{\tau_2} (S(\bar{\lambda}_p + \pi) - S(\bar{\lambda}_p))\, \dot{\pi}$$

$$\begin{aligned}
&= \int_{\pi(\tau_1)}^{\pi(\tau_+)} \{S(\bar{\lambda}_p(t_+(z)) + z) - S(\bar{\lambda}_p(t_+(z)))\}\, dz \\
&\quad + \int_{\pi(\tau_-)}^{\pi(\tau_2)} \{S(\bar{\lambda}_p(t_-(z)) + z)) - S(\bar{\lambda}_p(t_-(z)))\}\, dz \\
&= \int_{\pi(\tau_1)}^{\pi(\tau_+)} \{[S(\bar{\lambda}_p(t_+(z)) + z) - S(\bar{\lambda}_p(t_+(z)))] \\
&\quad - [S(\bar{\lambda}_p(t_-(z)) + z) - S(\bar{\lambda}_p(t_-(z)))]\}\, dz,
\end{aligned} \tag{5.31}$$

where we have used (5.23) and (a)′ in the last step of this calculation. It follows from (3.13) that $\bar{\lambda}_p$ is monotone and from the definitions of $t_+$ and $t_-$ in (5.29) and (5.30) that for all $z \in [\pi(\tau_1), \pi(\tau_+)]$

$$t_+(z) \leqq t_-(z), \tag{5.32}$$

vo that the inequality

$$\bar{\lambda}_p(t_+(z)) \leqq \bar{\lambda}_p(t_-(z)) \tag{5.33}$$

holds on $[\pi(\tau_1), \pi(\tau_+)]$. Thus, we may interpret each value of the integrand in the last integral in (5.31) as the difference between the increases in $S$ over two intervals, $[\bar{\lambda}_p(t_+(z)), \bar{\lambda}_p(t_+(z)) + z]$ and $[\bar{\lambda}_p(t_-(z)), \bar{\lambda}_p(t_-(z)) + z]$. These intervals both have length $z$, and by (5.33), the first has its left-hand endpoint on or to the left of the left-hand endpoint of the second. Therefore, we can invoke from (3.14) the concavity of $S$ to conclude that the integrand is not negative, and this completes the verification of the induction step and the proof of Lemma 4.1.

*Acknowledgment.* I wish to thank W. O. Williams for valuable discussions. This research was supported under grants from the National Science Foundation.

## References

1950 1. Hill, R., *The Mathematical Theory of Plasticity*, Oxford University Press.

1960 1. Filippov, A. F., Differential equations with discontinuous right-hand side (in Russian). Mat. Sbornik **51**; A.M.S. Translations (2) **42** (1964), 199–231.

1968 1. Mendelson, Alexander, *Plasticity: Theory and Application*, Macmillan, New York.

1976 1. Krasnoselskii, M. A., A mathematical description of the oscillations of a material point on an elastic-plastic element, A.M.S. Translation (2) **105**, 206–210.

1979 1. BUHITE, J. L., & D. R. OWEN, An ordinary differential equation from the theory of plasticity, Arch. Rational Mech. Anal. **71**, 357–383.
1983 1. KRASNOSELSKII, M. A., & A. V. POKROVSKII, *Systems with Hysteresis* (in Russian), Moskva, "Nauka".

Department of Mathematics
Carnegie-Mellon University
Pittsburgh

*(Received July 2, 1986)*

# *On Effects of Virtual Inertia during Diffusion of a Dispersed Medium in a Suspension*

G. CAPRIZ & P. GIOVINE

*Dedicated to James Serrin on his sixtieth birthday*

**Abstract**

Equations of motion of a binary mixture (where virtual inertia effects due to diffusion have relevance) are derived from an Eulerian variational principle. The results are consistent with general axioms of the theory of mixtures.

## 1. Introduction

The equations of motion for a mixture of two immiscible fluids (or a suspension of solid particles in a fluid) are still a matter for discussion. Indeed the fundamental choice of an appropriate expression for the inertia terms is itself not trivial, because virtual mass effects have an important role during diffusion and are not easily modelled within the scheme of a continuum.

It is usually claimed that the additional inertia force $f$ due to those effects is proportional to a relative acceleration; however, the volume fraction itself may vary in time and from place to place and these changes are not irrelevant for the correct evaluation of $f$. Some authors require that $f$ be objective (see, for instance, DREW, CHENG & LAHEY (1979)); others do not respect such a requirement. On these and other matters, ample references, detailed comparative analyses and interesting comments can be found in an extensive essay by BEDFORD & DRUMHELLER (1983). As in Appendix A of that essay, it seems easier, and perhaps less controversial, to start with the proposal of an expression for the virtual kinetic energy, following suggestions from classical analyses of the motion of a sphere (or of a disc) in a fluid, and then to formulate an appropriate variational principle. If one chooses a variational principle of material type, however, one must, perhaps implicitly, assume a mechanism for the transport of the virtual kinetic energy (see again the essay cited, App. A, and DRUMHELLER & BEDFORD (1980)); thus, in a sense, one begs at least part of the question. We prefer to adopt here a spatial variational principle along lines followed within classical contexts in CAPRIZ (1984).

Because we wish to concentrate attention on virtual mass effects due to diffusion, we examine only the most elementary type of two-phase flow that implies them: a suspension of incompressible particles, or bubbles, in an incompressible perfect fluid. After a brief justification of the choice of the expression for the total kinetic energy density, we introduce the variational principle. We give some details of the developments that lead to the balance equations of momentum. We check the consistency of those equations with TRUESDELL'S "metaphysical" principles for mixtures and we conclude with the study of a special problem of sedimentation.

The equations of motion we obtain seem consistent with those that have been derived within an extremely general context in the paper by BEDFORD & DRUMHELLER (1980) already cited. In view of the simplicity of the situation envisaged here, the details of the derivation can be followed explicitly; a constant effort is also made to assure physical insight into the analytical developments.

## 2. The variational principle

In this section we propose a continuum model of a two-phase system where the particles of the dispersed phase are large enough that virtual inertia effects cannot be disregarded. As is usual in theories of continua with microstructure, the idea is to obtain hints for an appropriate expression of the energy densities from results of analyses of simple motions in classical continua.

One can imagine an element of our continuum as a spherical drop of an incompressible perfect fluid containing a concentric spherical inclusion (the dispersed phase), the latter moving with an instantaneous purely translational relative speed $z$ with respect to the drop. If $v$ is the volume of the element, $\beta$ the ratio of the volume of the inclusion to $v$, $v_1$ the absolute speed of the mass centre of the fluid, $v_2 = v_1 + z$ the absolute speed of the inclusion, $\varrho_1$ the density of the fluid and $\varrho_2$ the density of the inclusion, then the total kinetic energy of the element turns out to be (see LAMB (1932) Sect. 93) $v$ times the expression

$$\tfrac{1}{2}\left(\varrho_1(1-\beta)\,v_1^2 + \varrho_2\beta v_2^2 + \varrho_1\psi(\beta)\,z^2\right), \tag{2.1}$$

$$\psi(\beta) = \tfrac{1}{2}\,\beta\,\frac{1+2\beta}{1-\beta}. \tag{2.2}$$

The expression (2.1) will be taken below as the density of kinetic energy per unit volume of the mixture; the special choice (2.2) for the function $\psi$, presumably adequate only when $\beta$ is sufficiently small, need not be made.

*Remark.* The model suggested above for an elementary drop of the mixture could be sharpened to include the effects of relative rotation of the inclusion. One could also, perhaps more appropriately in certain situations, attribute to the suspended particle an ellipsoidal shape, perhaps even the shape of a disc. But then the analysis would become much more complex, because the suspended phase would have to be modelled through a continuum with microstructure, the variable to represent which would be the unit vector of the axis of symmetry of the ellipsoid. Also, the effects of the interference between micromotions in neighboring elements could be brought to bear; then a non-local expression for the kinetic energy density would be required (a very interesting notion, in principle), or at least an expression involving the gradients of $\beta$ and $z$; for instance, formulae (16) of Sect. 98 and (7) of Sect. 99 of LAMB (1932) hint at expressions of the latter type. Finally the idea, (embodied in (2.2) and which conforms with the spirit of the paper) that the flow between the two spheres is that pertaining to a perfect fluid, could be forfeited in favor of a similar hypothesis but involving a viscous fluid; but then (as far as concerns formula (2.1)) only the dependence of $\psi$ on $\beta$ would change.

To reduce developments to essentials, we consider only the case when the mixture is contained in a fixed rigid vessel with impermeable walls; we call $B$ the region of space delimited by the vessel and $\partial B$ the boundary. Kinematic compatibility imposes the conditions

$$v_1 \cdot n = 0, \qquad v_2 \cdot n = 0 \qquad \text{on } \partial B, \tag{2.3}$$

if $n$ is the unit exterior vector normal to $\partial B$.

Again to simplify matters radically, both phases are assumed to be incompressible; $\varrho_1$ and $\varrho_2$ are then constant. Also mass exchanges between phases are excluded. Thus conservation of mass requires that

$$\left.\begin{aligned} &\frac{\partial((1-\beta)\,\varrho_1)}{\partial\tau} + \operatorname{div}((1-\beta)\,\varrho_1 v_1) = 0,\\ &\frac{\partial(\beta\varrho_2)}{\partial\tau} + \operatorname{div}(\beta\varrho_2 v_2) = 0, \end{aligned}\right\} \quad \text{in } B;$$

or

$$\left.\begin{aligned} &-\frac{\partial\beta}{\partial\tau} + \operatorname{div}((1-\beta)\,v_1) = 0,\\ &\frac{\partial\beta}{\partial\tau} + \operatorname{div}(\beta v_2) = 0. \end{aligned}\right\} \quad \text{in } B. \tag{2.4}$$

As for the external body forces, they are taken here to be conservative with a potential energy $\omega$ per unit mass. The potential energy of internal actions ($\sigma$, per unit volume) is taken to depend at most on $\beta$.

A mechanical process of duration $\bar{\tau}$ in the mixture is portrayed through the assignment at each instant $\tau$ in $[0, \bar{\tau}]$ of the fields $v_1$, $v_2$ and $\beta$ over $B$, subject to (2.3). Among all these *virtual* processes the *natural* process is distinguished because it satisfies a variational principle and the balance conditions (2.4). To state the principle the usual minor technical definitions are required.

Let $\{\hat{v}_1(x, \tau), \hat{v}_2(x, \tau), \hat{\beta}(x, \tau)\}$ be the natural process and $\{v_1(x, \tau, \varepsilon), v_2(x, t, \varepsilon),$

$\beta(x, \tau, \varepsilon)\}$ a family of virtual processes depending smoothly on a parameter $\varepsilon$, for $\varepsilon$ in a neighborhood $N_\varepsilon$ of the origin, and such that

$$\{v_1(x, \tau, 0), v_2(x, \tau, 0), \beta(x, \tau, 0)\} \equiv \{\hat{v}_1(x, \tau), \hat{v}_2(x, \tau), \hat{\beta}(x, \tau)\} \quad \forall\, x \in B, \quad \forall\, \tau \in [0, \bar{\tau}],$$

and

$$\{v_1(x, 0, \varepsilon), v_2(x, 0, \varepsilon), \beta(x, 0, \varepsilon)\} \equiv \{\hat{v}_1(x, 0), \hat{v}_2(x, 0), \hat{\beta}(x, 0)\},$$

$$\{v_1(x, \bar{\tau}, \varepsilon), v_2(x, \bar{\tau}, \varepsilon), \beta(x, \bar{\tau}, \varepsilon) \equiv \{\hat{v}_1(x, \bar{\tau}), \hat{v}_2(x, \bar{\tau}), \hat{\beta}(x, \bar{\tau})\} \quad \forall\, \varepsilon \in N_\varepsilon, \quad \forall\, x \in B.$$

The variation $\delta\Gamma$ of any quantity $\Gamma$ defined on a process class is given by

$$\delta\Gamma := \left.\frac{\partial \Gamma}{\partial \varepsilon}\right|_{\varepsilon=0}. \tag{2.5}$$

The conditions imposed upon $v_1(x, \tau, \varepsilon), v_2(x, \tau, \varepsilon), \beta(x, \tau, \varepsilon)$ assure us that

$$\delta v_1 = 0, \quad \delta v_2 = 0, \quad \delta\beta = 0 \quad \text{for } \tau = 0, \tau = \bar{\tau}, \tag{2.6}$$

whereas

$$\delta v_1 \cdot n = \delta v_2 \cdot n = 0 \quad \text{on } \partial B. \tag{2.7}$$

The variational principle asserts that, during the natural motion of the body, the equality

$$\begin{aligned} \delta \int_0^{\bar{\tau}} d\tau \int_B \Big( & \tfrac{1}{2}(\varrho_1(1-\beta)\, v_1^2 + \varrho_2\beta v_2^2 + \varrho_1\psi(\beta)\, z^2) - (\varrho_1(1-\beta) + \varrho_2\beta)\,\omega \\ & - \sigma + \phi\left(\frac{\partial \beta}{\partial \tau} + \operatorname{div}(\beta v_2)\right) + \gamma\left(-\frac{\partial \beta}{\partial \tau} + \operatorname{div}((1-\beta)\, v_1)\right)\Big)\, dB = 0 \end{aligned} \tag{2.8}$$

holds for all virtual processes; here $\phi$ and $\gamma$ are Lagrange multipliers of the constraints imposed upon the natural process by mass balance.

### 3. The balance equations for momentum

In view of the restrictive hypothesis made on $B$, reflected in the relations (2.3), the transport theorem leads, for any $\tau$, to the equalities

$$\delta \int_0^{\bar{\tau}} d\tau \int_B \Gamma\, dB = \int_0^{\bar{\tau}} d\tau \int_B \delta\Gamma\, dB,$$

$$\int_0^{\bar{\tau}} dt \int_B \frac{\partial \Gamma}{\partial \tau}\, dB = \left[\int_B \Gamma\, dB\right]_0^{\bar{\tau}},$$

and these, as well as repeated recourse to integration by parts in the usual manner, lead to the following consequences of (2.8)

$$\begin{aligned} &\tfrac{1}{2}(\varrho_2 v_2^2 - \varrho_1 v_1^2) + \tfrac{1}{2}\varrho_1 \frac{d\psi}{d\beta} z^2 + (\varrho_1 - \varrho_2)\,\omega - \frac{d\sigma}{d\beta} \\ &= \frac{\partial\phi}{\partial\tau} + v_2 \cdot \operatorname{grad}\phi - \frac{\partial\gamma}{\partial\tau} - v_1 \cdot \operatorname{grad}\gamma, \end{aligned} \tag{3.1}$$

$$\begin{aligned} \varrho_1 v_1 - \varrho_1 \frac{\psi z}{(1-\beta)} &= \operatorname{grad}\gamma, \\ \varrho_2 v_2 + \varrho_1 \frac{\psi z}{\beta} &= \operatorname{grad}\phi. \end{aligned} \tag{3.2}$$

The similarity of these consequences with those obtained from a variational principle in classic cases is obvious (see CAPRIZ (1984)): (3.1) has the form of a Bernoullian theorem, whereas (3.2) give $\gamma$ and $\phi$ the meaning of potentials. One can also proceed similarly to eliminate $\phi$ and $\gamma$ by cross-differentiation and obtain

$$\begin{aligned} &\operatorname{grad}\left((\varrho_1 - \varrho_2)\,\omega + \tfrac{1}{2}\varrho_1 \frac{d\psi}{d\beta} z^2 - \frac{d\sigma}{d\beta}\right) \\ &= \varrho_2 a_2 - \varrho_1 a_1 + \frac{\varrho_1}{\beta}\left(\frac{\partial(\psi z)}{\partial\tau} + (\operatorname{grad} v_2)^T \psi z\right) + \frac{\varrho_1}{1-\beta}\left(\frac{\partial(\psi z)}{\partial\tau} + (\operatorname{grad} v_1)^T \psi z\right) \\ &+ \left\{\varrho_1 \psi z \frac{\partial}{\partial\tau}\left(\frac{1}{\beta}\right) + \left(\operatorname{grad}\left(\frac{\varrho_1 \psi z}{\beta}\right)\right)^T v_2 - v_2 \times \operatorname{rot}\left(\frac{\varrho_1 \psi z}{\beta}\right)\right\} \\ &+ \left\{\varrho_1 \psi z \frac{\partial}{\partial\tau}\left(\frac{1}{1-\beta}\right) + \left(\operatorname{grad}\left(\frac{\varrho_1 \psi z}{1-\beta}\right)\right)^T v_1 - v_1 \times \operatorname{rot}\left(\frac{\varrho_1 \psi z}{1-\beta}\right)\right\}, \end{aligned} \tag{3.3}$$

where the peculiar accelerations $a_1$, $a_2$ have been introduced

$$a_i = \frac{\partial v_i}{\partial\tau} + \operatorname{grad}\frac{v_i^2}{2} - v_i \times \operatorname{rot} v_i, \qquad i = 1, 2$$

and obvious consequences of (3.2) have been exploited, *i.e.*,

$$\operatorname{rot}\varrho_1 v_1 = \operatorname{rot}\varrho_1 \frac{\psi z}{(1-\beta)}, \qquad \operatorname{rot}\varrho_2 v_2 = -\operatorname{rot}\varrho_1 \frac{\psi z}{\beta}. \tag{3.4}$$

Now one can use (2.4) to show that the two terms between curly brackets in (3.3) are equal respectively to

$$\frac{1}{\beta}\operatorname{div}(\varrho_1 \psi z \otimes v_2) \quad \text{and} \quad \frac{1}{1-\beta}\operatorname{div}(\varrho_1 \psi z \otimes v_1),$$

with the conclusion that

$$\beta\left(\varrho_1(1-\beta)\left(a_1+\operatorname{grad}\left(\omega+\tfrac{1}{2}\frac{d\psi}{d\beta}z^2\right)\right)\right.$$

$$\left.-\varrho_1\left(\frac{\partial\psi z}{\partial\tau}+(\operatorname{grad}v_1)^T\psi z\right)-\varrho_1\operatorname{div}(\psi z\otimes v_1)\right)$$

$$=(1-\beta)\left(\beta\left(\varrho_2 a_2+\varrho_2\operatorname{grad}\omega+\operatorname{grad}\frac{d\sigma}{d\beta}\right)\right.$$

$$\left.+\varrho_1\left(\frac{\partial\psi z}{\partial\tau}+(\operatorname{grad}v_1)^T\psi z\right)+\varrho_1\psi\operatorname{grad}\frac{z^2}{2}+\varrho_1\operatorname{div}(\psi z\otimes v_2)\right). \qquad (3.5)$$

If one calls $\beta(1-\beta)u$ the common value of the right-hand and left-hand sides of (3.5) (with $u$ an appropriate vector), one discovers, by way of (2.4) and (3.4), that the vector $u-\beta\operatorname{grad}\frac{d\sigma}{d\beta}$ is irrotational. Then, introducing a scalar $\pi$ (the physical meaning of which will be clear in the sequel), such that

$$-u+\beta\operatorname{grad}\frac{d\sigma}{d\beta}=\operatorname{grad}\left(\pi-\tfrac{1}{2}\varrho_1 z^2\left(\psi+(1-\beta)\frac{d\psi}{d\beta}\right)\right), \qquad (3.6)$$

one is led to the balance equations of momentum for the two constituents of the mixture

$$\varrho_1(1-\beta)a_1=-\varrho_1(1-\beta)\operatorname{grad}\omega-(1-\beta)\operatorname{grad}\pi-\operatorname{div}(\varrho_1\psi z\otimes z)+k,$$
$$\varrho_2\beta a_2=-\varrho_2\beta\operatorname{grad}\omega-\beta\operatorname{grad}\pi-k, \qquad (3.7)$$

where $\pi$ has the obvious meaning of a pressure, so that the terms which involve it express the Archimedean buoyancy forces, and $k$, which can be interpreted as the interaction force between the two phases, has the expression

$$k=(1-\beta)\operatorname{grad}\left(\tfrac{1}{2}\varrho_1 z^2\left(\psi-\beta\frac{d\psi}{d\beta}\right)\right)+\varrho_1\left(\frac{\partial(\psi z)}{\partial\tau}+(\operatorname{grad}v_1)^T(\psi z)\right)$$

$$+\operatorname{div}(\varrho_1\psi z\otimes v_2)+\beta(1-\beta)\operatorname{grad}\frac{d\sigma}{d\beta}. \qquad (3.8)$$

DRUMHELLER & BEDFORD (1980) have already remarked that the terms appearing in the right-hand sides of the two equations (3.7) and containing $\psi$ (*i.e.*, the terms which derive from the extra kinetic energy) do not add up to zero. Our interpretation, supported also by some developments below, is that $-\varrho_1\psi z\otimes z$ has the meaning of an additional Reynolds stress acting on the fluid.

Actually, there are many other forms into which equations (3.7), (3.8) can be put; one variant shows very clearly that they obey the "metaphysical" principles set by TRUESDELL (1969) as the basis of any theory of mixtures; the variant is as follows

$$\begin{aligned}\varrho_1(1-\beta)a_1&=-\varrho_1(1-\beta)\operatorname{grad}\omega+\operatorname{div}T_1+h,\\ \varrho_2\beta a_2&=-\varrho_2\beta\operatorname{grad}\omega+\operatorname{div}T_2-h,\end{aligned} \qquad (3.9)$$

with

$$T_1 = \varrho_1 \psi z \otimes v_1 - (1-\beta)\left(\pi - \tfrac{1}{2}\varrho_1 z^2 \left(\psi - \beta \frac{d\psi}{d\beta}\right)\right) 1,$$
$$T_2 = -\varrho_1 \psi z \otimes v_2 - \left(\beta\pi + \tfrac{1}{2}\varrho_1 (1-\beta) z^2 \left(\psi - \beta \frac{d\psi}{d\beta}\right)\right) 1, \tag{3.10}$$

and

$$h = k - \operatorname{div}\left(\varrho_1 \psi z \otimes v_2 + \tfrac{1}{2}(1-\beta)\varrho_1 z^2 \left(\psi - \beta \frac{d\psi}{d\beta}\right) 1\right) - \pi \operatorname{grad} \beta. \tag{3.11}$$

Thus each constituent satisfies the balance equation of momentum of an ordinary continuum; besides, if one sums term by term either (3.7) or (3.9), introduces mixture density $\varrho$, velocity $v$ and acceleration $a$

$$\varrho = \varrho_1(1-\beta) + \varrho_2\beta,$$
$$\varrho v = \varrho_1(1-\beta) v_1 + \varrho_2 \beta v_2, \tag{3.12}$$
$$a = \frac{\partial v}{\partial \tau} + (\operatorname{grad} v)\, v,$$

and remembers the kinematic relation (TRUESDELL (1969), formula (5.16))

$$\varrho a = \varrho_1(1-\beta) a_1 + \varrho_2 \beta a_2 - \operatorname{div}\left(\varrho_1(1-\beta) v_1 \otimes v_1 + \varrho_2 \beta v_2 \otimes v_2 - \varrho v \otimes v\right), \tag{3.13}$$

one is led to the balance equation for the mixture:

$$\varrho a = -\varrho \operatorname{grad} \omega + \operatorname{div} T, \tag{3.14}$$

where

$$T = -\varrho_1(1-\beta)(v_1 - v)\otimes(v_1 - v) - \varrho_2\beta(v_2 - v)\otimes(v_2 - v) - \varrho_1 \psi z \otimes z - \pi 1. \tag{3.15}$$

Again here TRUESDELL'S general requirement is satisfied, and, besides, $T$ differs from the expression (5.14) appearing in TRUESDELL (1969) only by the term $-\varrho_1 \psi z \otimes z$ already mentioned.

## 4. A problem of sedimentation

As we have repeatedly declared, our aim was to derive, from a single (and perhaps less disputable) principle, an expression for the interaction force due to virtual mass effects. For a discussion of practical problems, however, our balance equations are far too special. In particular the absence of terms accounting for viscosity is unrealistic; of course, such terms could easily be added, but the process of adapting (3.7) to a more concrete form will not be pursued.

Here, just for the sake of showing what can be expected from our analysis, we treat a particular and, as already admitted, rather artificial sedimentation problem. We consider a mixture occupying the half-space $\zeta \geqq 0$, subject to the effect of gravity ($\omega = -|g|\,\zeta$; $g$, acceleration due to gravity). We look for steady solutions of (3.7), (3.8) in which (i) all unknown functions depend only on $\zeta$; (ii) more specifically

$$v_1 = \alpha(\zeta)\,c, \qquad v_2 = \gamma(\zeta)\,c, \tag{4.1}$$

($c$, unit vector of the $\zeta$-axis); (iii) the mean velocity vanishes.

Conservation of mass (*i.e.*, equations (2.4)) requires

$$\alpha = (1-\beta_0)\,\alpha_0(1-\beta)^{-1}, \qquad \gamma = \beta_0\gamma_0\beta^{-1}, \tag{4.2}$$

if $\beta_0$, $\alpha_0$, $\gamma_0$ are values of $\beta$, $\alpha$, $\gamma$ at $\zeta = 0$, for instance. Hypothesis (iii) imposes the relation

$$\varepsilon(1-\beta_0)\,\alpha_0 = -\beta_0\gamma_0, \tag{4.3}$$

if $\varepsilon = \varrho_1/\varrho_2$.

By putting (4.2) into (3.7) we obtain a system of two ordinary differential equations of the first order in $\beta(\zeta)$ and $\pi(\zeta)$. Actually $\pi$ can be easily eliminated; in fact one could refer to (3.5) and obtain directly a single equation in $\beta$.

We spare the reader the details of the algebraic manipulations and record below results valid for small $\beta$ for the case when $\psi$ is taken to be zero:

$$\beta = \beta_0\left(1 + \frac{2(1-\varepsilon)\,|g|\,\zeta}{\gamma_0^2}\right)^{-\frac{1}{2}}$$

and for the case when $\psi$ is given by (2.2)

$$\beta = \beta_0\left(1 + \frac{4(1-\varepsilon)\,|g|\,\zeta}{(2+\varepsilon)\,\gamma_0^2}\right)^{-\frac{1}{2}}.$$

*Acknowledgement.* This work is part of a research program under the auspices of the Italian Ministry of Education. G. C. acknowledges also the support of the Institute of Mathematics and its Applications of the University of Minnesota.

## References

1. BEDFORD, A., & DRUMHELLER, D. S. (1983), Theories of immiscible and structured mixtures. Int. J. Engng. Sci. **21**, 863–960.
2. CAPRIZ, G. (1984), Spatial variational principles in continuum mechanics. Arch. Rational Mech. Analysis, **85**, 99–109.
3. DREW, D., CHENG, L. & LAHEY, R. T. (1979), The analysis of virtual mass effects in two-phase flow. Int. J. Multiphase Flow, **5**, 233–242.
4. DRUMHELLER, D. S., & BEDFORD, A. (1980), A thermomechanical theory for reacting immiscible mixtures. Arch. Rational Mech. Analysis, **73**, 257–284.
5. LAMB, H. (1932), Hydrodynamics. Cambridge University Press.
6. TRUESDELL, C. (1969), Rational Thermodynamics, McGraw-Hill, New York.

Dipartimento di Matematica
Università di Pisa

*(Received July 1, 1986)*

# *Existence of Positive Solutions for Semilinear Elliptic Equations in General Domains*

JOEL SMOLLER & ARTHUR WASSERMAN

*To James Serrin for his sixtieth birthday*

## § 1. Introduction

The main purpose of this paper is to prove some new existence theorems for positive solutions to the Dirichlet problem

$$\Delta u(x) + f(u(x)) = 0, \quad x \in \Omega, \tag{1.1}$$

$$u(x) = 0, \quad x \in \partial\Omega. \tag{1.2}$$

Here $\Omega$ is a bounded domain in $\mathbb{R}^n$, with smooth boundary, and $f$ is a continuous function on $\mathbb{R}$.

We introduce in § 2 the notion of "eccentricity", $e(\Omega)$, of a domain $\Omega \subset \mathbb{R}^n$, which has the property that $1 \leqq e(\Omega) < \infty$ and $e(\Omega) = 1$ if and only if $\Omega$ is a ball. We then define for any function $f$ on $\mathbb{R}$ the "nonlinearity" of $f$, $N(f)$. If $f$ is a linear function, $N(f) = 1$; if $f(u) = u^k$, $k < 1$, $N(f) = \infty$; if $f(u) = u^k$, $k > 1$, $N(f) < 1$. The main theorem of the paper, Theorem 2.2, states that if $N(f) > e(\Omega)$ then there exist positive solutions to (1.1), (1.2) on all domains $\lambda\Omega$ if $\lambda$ is sufficiently large. Equivalently, if $N(f) > e(\Omega)$, then positive solutions to the Dirichlet problem for $\Delta u + \mu f(u) = 0$, exist on $\Omega$, for some range of $\mu$. Our method of proof is a variation of the technique of "upper and lower solutions", *cf.* [14, Ch. 10], whereby we use techniques developed in [10, 13] (where $\Omega$ is an $n$-ball) in order to construct these "solutions". We note that our methods apply to any elliptic operator; we consider (1.1) for definiteness. We then give in § 3, several applications of this theorem. For example, if $f(u)/u \to 0$ as $u \to \infty$, and[1] $F(u_0) > 0$ for some $u_0 > 0$ (where $F' = f$, $F(0) = 0$) then positive solutions exist on $\lambda\Omega$ if $\lambda$ is sufficiently large. Again, if $f(u_0) = 0$ for some $u_0 > 0$, and $F(u_0) > 0$, then the same conclusion holds. In both cases, we require no conditions on the behavior of $f$ near zero.

In § 4 we take $\Omega$ to be an $n$-ball, and we prove existence of positive solutions

---

[1] This is always a necessary condition as follows from POHOZAEV's identity; see [2] and [9].

of (1.1), (1.2) on some ball, provided only that $f$ satisfies an inequality of the form

$$c_1 \leqq f(u)/u^k \leqq c_2 \quad \text{as } u \to \infty. \tag{1.3}$$

Here $c_1$ and $c_2$ are positive constants, and $0 < k < n/(n-2)$. This result is a consequence of some general existence theorems whereby $f$ is required to satisfy an inequality of the form $cu^\alpha \leqq f(u) \leqq du^\beta$, for large $u$, with $c$, $d$, $\alpha$ and $\beta$ being positive constants.

In § 5 we study the Neumann problem for (1.1) on $n$-balls, *i.e.* solutions of (1.1) which satisfy homogeneous Neumann boundary conditions

$$du(x)/dn = 0, \; x \in \partial\Omega, \tag{1.4}$$

where $d/dn$ denotes differentiation in the radial direction on $\partial\Omega$. Thus, for example, if $f$ satisfies (1.3), where $0 < k < n/(n-2)$, and, in addition, there exists an $\eta$ such that $f(u) < 0$ for $u \leqq \eta < 0$, and $F(u) \to \infty$ as $u \to -\infty$, we show that the Neumann problem has a monotone radial solution on some $n$-ball.

We note that the existence of positive solutions to the Dirichlet problem has been obtained if $f(u)$ is required to be positive, or if $f(0) = 0$ and $f'(0) > 0$; see [1, 2, 3, 5, 7, 8]. We allow $f$ to be an arbitrary function near zero. Such freedom is necessary if one is interested in questions of "symmetry-breaking" (see [11, 12]). Indeed, as we have shown in [11], the symmetry cannot break on a positive radial solution of (1.1), (1.2) unless $f(0) < 0$.

We remark that if $f(u) = u^k$, then it is well known [8] that there are no positive solutions to the Dirichlet problem on $n$-balls, if $k \geqq \dfrac{n+2}{n-2} = \dfrac{n}{n-2} + \dfrac{2}{n-2}$, but that there are such solutions if $k < \dfrac{n+2}{n-2}$; see [7, 9]. Thus our result given above is probably not the best possible. It would be interesting to know if there are positive solutions to the Dirichlet problem on $n$-balls if $f(u) = u^k - \sigma$, where $\sigma$ is a positive constant, and $\dfrac{n}{n-2} < k < \dfrac{n+2}{n-2}$.

Throughout this paper, $F$ will denote the primitive of $f$ satisfying $F(0) = 0$, and $D_R^n$ will be the $n$-ball of radius $R$ centered at $0 \in \mathbb{R}^n$. Finally prime, ($'$), will denote differentiation with respect to $r$.

## § 2. The Dirichlet Problem on General Domains

In this section we shall show that the existence of positive solutions on $n$-balls can be used to prove the existence of positive solutions on general domains. We introduce the notion of the eccentricity of a domain, and we use it to prove our general, (somewhat abstract), existence theorem. In the next section we shall how it applies to extend known results.

Thus let $\Omega$ be a smooth bounded domain in $\mathbb{R}^n$, and let $B_r(p)$ denote an $n$-ball of radius $r$ centered at $p$. We begin with a purely geometrical result.

**Lemma 2.1.** *There exists a* $\varrho > 0$ *such that if* $x \in \Omega$, *then* $x \in \overline{B_\varrho(y)} \subset \Omega$ *for some* $y \in \Omega$.

**Proof.** We start by mimicking the standard proof of the tubular neighborhood theorem; *cf.* [6]. Thus define a function $\phi: \partial\Omega \times \mathbb{R} \to \mathbb{R}^n$ by

$$\phi(p, t) = p + t\bar{n},$$

where $\bar{n}$ is the outward normal at $p$. Since $d\phi(p, 0)$ is invertible, $\phi$ is a local diffeomorphism. But $\phi$ is a diffeomorphism on the compact manifold $\partial\Omega \times \{0\}$; thus $\phi$ is, in fact, a diffeomorphism on some closed neighborhood of $\partial\Omega$ in $\partial\Omega \times \mathbb{R}$ (see [6]). Let $\partial\Omega \times [-\varrho, \varrho]$ be such a neighborhood, $\varrho > 0$. We claim that this $\varrho$ "works". Thus, let $x \in \Omega$. If $x \notin \phi(\partial\Omega \times (-\varrho, 0])$, then $\overline{B_\varrho(x)} \subset \Omega$; otherwise we could find $z \in B_\varrho(x) \cap \partial\Omega$ with $x = z + t\bar{n}$ for some $t$, $|t| < \varrho$, (namely $z$ is the point on $\partial\Omega$ closest to $x$), and so $x \in \phi(\partial\Omega \times (-\varrho, 0])$. On the other hand, if $x \in \phi(\partial\Omega \times (-\varrho, 0])$, then $x = p + t\bar{n}$ for some $p \in \partial\Omega$, and $-\varrho < t \leqq 0$. It follows that $x \in \overline{B_\varrho(p - \varrho\bar{n})}$, and the proof is complete. □

Note that lemma 2.1 requires $\Omega$ to have a smooth boundary; if $\partial\Omega$ has a cusp, then $\varrho = 0$.

In view of this lemma, we may define the number

$$\varrho(\Omega) = \sup\{\varrho > 0 : x \in \Omega \Rightarrow \exists\, y \in \Omega \text{ with } \overline{B_\varrho(y)} \subset \Omega\}. \tag{2.1}$$

Next, we define the *eccentricity*, $e(\Omega)$, of the region $\Omega$ by

$$e(\Omega) = \inf\left\{\frac{R}{\varrho(\Omega)} : \Omega \subset \overline{B_R(p)} \text{ for some } p \in \mathbb{R}^n, \text{ and some } R > 0\right\}. \tag{2.2}$$

Observe that $e(\Omega) \geqq 1$, and $e(\Omega) = 1$ if and only if $\Omega$ is an $n$-ball. To get a feel for $e(\Omega)$, note that if $\Omega \subset \boldsymbol{R}^2$ is the ellipse $a^2x^2 + b^2y^2 \leqq a^2b^2$, where $b > a$, then $\varrho(\Omega) = a^2/b$, and since $\Omega \subset \overline{B_b(0, 0)}$, we see that $e(\Omega) = b^2/a^2$.

Consider 4-tuples $(u_1, R_1, u_2, R_2)$ where $R_1, R_2 > 0$ and $u_i: D^n_{R_i} \to \mathbb{R}$. We say that a 4-tuple is admissible if there exist functions $f_i(u)$ with $f_1(u) \leqq f(u) \leqq f_2(u)$ such that

i) $\Delta u_i(x) + f_i(u_i(x)) = 0 \quad x \in D^n_{R_i}$

ii) $u_1(x) = 0$ and $u_1'(x) < 0 \quad$ if $|x| = R_1$

iii) $u_2(x) \geqq u_1(y) \geqq 0 \quad$ for all $x \in D^n_{R_2}$, $y \in D^n_{R_1}$.

Note that $u_1$ is a lower solution of 1.1 and $u_2$ is an upper solution of 1.1. See Figure 1. ($D^n_R$ is the $n$-ball of radius $R$ centered at the origin.)

We define $N(f)$, the "nonlinearity" of $f$, by $N(f) = \sup \dfrac{R_2}{R_1}$ where the sup is taken over all admissible 4-tuples; if there is no admissible 4-tuple we set $N(f) = 0$.

*Remark*. There exists an admissible 4-tuple for $f$ iff $F(p) > 0$ for some $p > 0$. We show (Lemma 3.1) that $F(p) > 0$ for some $p > 0$ is a sufficient condition for the existence of a pair $(u_1, R_1)$ satisfying i). and ii). We can complete this to an admissible 4-tuple by setting $f_2(u) = \sup\{f(u) : 0 \leqq u \leqq 2u_1(0)\}$ and letting $u_2(r)$ be the radial solution of $\Delta u + f_2(u) = 0$ satisfying $u_2(0) = 2u_1(0)$. If we define $R_2$ by $u_2(R_2) = u_1(0)$ we have that $(u_1, R_1, u_2, R_2)$ is an admissible 4-tuple and $N(f) > 0$.

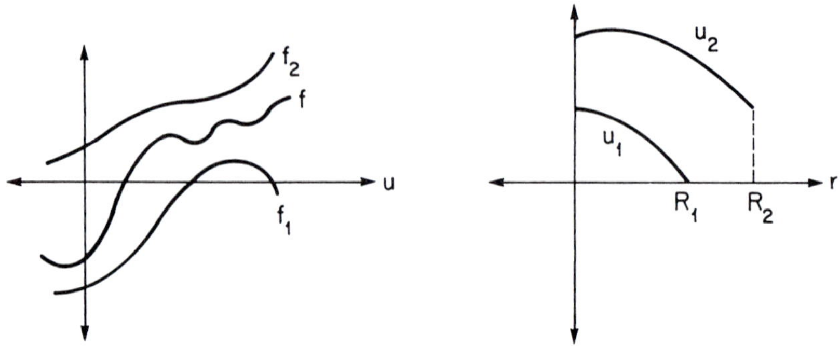


Fig. 1

On the other hand the condition $F(p) > 0$ for some $p > 0$ is a necessary condition for the existence of positive solutions of (1.1), (1.2); see [2]. Thus if $u(x_1) = \max\{u(x) : x \in \Omega\}$, then $\Delta u(x_1) < 0$ so $f(u(x_1)) > 0$. The well-known Pohozaev identity (see [9]),

$$2n \int_\Omega F(u)\, dx - (n-2) \int_\Omega uf(u)\, dx = \int_{\partial\Omega} \left(\frac{du}{dn}\right)^2 (r \cdot n)\, ds,$$

implies that $F$ must be positive on some ball centered at $x_1$. Thus, if $(u_1, R_1, U_2, R_2)$ is an admissible 4-tuple, $F_1(p) > 0$ for some $p > 0$ and $F(p) \geqq F_1(p) > 0$. Hence, there exists an admissible 4-tuple if and only if $F(p) > 0$ for some $p > 0$ if and only if $N(f) > 0$.

To gain some feeling for $N(f)$ we sketch an argument to show that $N(f) = 1$ if $f(u) = \gamma u$, $\gamma > 0$.

If $u(r)$ is a radial solution to (1.1), (1.2) on $\Omega = D_R^n$ set $f_1 = f_2 = f$, $u_1 = u$, $u_2 = cu$, $R_1 = R$ and define $R_2$ by $u_2(R_2) = u_1(0)$ i.e. $cu(R_2) = u(0)$. Note that $R_2 < R = R_1$ but as $c \to \infty$, $R_2 \to R$, and hence $R_2/R_1 \to 1$ and $N(f) \geqq 1$.

Next we show $N(f) \leqq 1$. Suppose $(u_1, R_1, u_2, R_2)$ is an admissible 4-tuple with $R_2 > R_1$. Then $\Delta u_1 + f_1(u_1) = 0$, $\Delta u_2 + f_2(u_2) = 0$, and hence by multiplying the first equation by $u_2$, the second by $u_1$, subtracting and integrating over $D_{R_1}^n = \Omega$, we get $\int_\Omega (u_2 \Delta u_1 - u_1 \Delta u_2) + \int_\Omega u_2 f_1(u_1) - u_1 f_2(u_2) = 0$. Applying Green's theorem twice we get $\int_{\partial\Omega} u_2 \dfrac{du_1}{dn} + \int_\Omega u_2 f_1(u_1) - u_1 f_2(u_2) = 0$; but $f_1(u_1) \leqq \gamma u_1$ and $f_2(u_2) \geqq \gamma u_2$ hence $u_2 f_1(u_1) - u_1 f_2(u_2) \leqq 0$. Also $u_2 > 0$ on $\partial\Omega$ and $\dfrac{du_1}{dn} = u_1' < 0$ by ii). Thus $0 = \int_\partial u_2 u_1' + \int_\Omega u_2 f_1(u_1) - u_1 f_2(u_2) < 0$. Hence $R_1 \geqq R_2$ and $N(f) = 1$.

With a little more effort one can show that the function

$$f(u)=\begin{cases}\mu_1 u & u \leqq 1\\ \mu_2(u-1)+\mu_1 & u \geqq 1\end{cases} \quad \text{where } \mu_1>\mu_2>0 \text{ has } N(f)=\sqrt{\frac{\mu_1}{\mu_2}}.$$

Similarly the function $f(u)=u^k$ has $N(f)<1$ if $k>1$ and $N(f)=\infty$ if $k<1$ (see Theorem 3.3).

**Theorem 2.2.** *If*

$$N(f)>e(\Omega), \tag{2.3}$$

*then there exists* $\lambda>0$ *and a positive solution to*

$$\Delta u(x)+f(u(x))=0, \quad x\in\lambda\Omega,$$

$$u(x)=0, \quad x\in\partial(\lambda\Omega).$$

*Equivalently if* (2.3) *holds, then there exists* $\mu>0$ *such that the problem*

$$\Delta u(x)+\mu f(u(x))=0, \quad x\in\Omega$$

$$u(x)=0, \quad x\in\partial\Omega,$$

*has a positive solution.*

**Proof.** We shall give the proof of statement i) since the two statements are easily seen to be equivalent through a simple scaling argument.

We first need the following lemma.

**Lemma 2.3.** *Let* $f\in C^1$, *and let* $u_1$ *be a non-negative solution of the problem*

$$\begin{aligned}&u''+\frac{n-1}{r}u'+f(u)=0, \quad 0<r<R,\\ &u(0)=p>0, \quad u'(0)=0, \quad u(R_1)=0.\end{aligned} \tag{2.3}$$

*Suppose that* $u'(R_1)=q<0$. *Then there exists an* $\tilde{f}\in C^1$ *such that the following hold*:

(a) $\tilde{f}(u)=f(u)$ if $u\geqq 0$,
(b) *If* $\tilde{u}$ *solves* (2.3) *with* $\tilde{f}$ *replaced by* $f$, *then* $\tilde{u}'(r)<0$ *for all* $r>0$ (so $\tilde{u}(r)<0$ if $r>R_1$).

**Proof.** Choose $B>0$ such that $f(0)>-B$. Then let $a>0$ satisfy

$$\tfrac{3}{4}q^2>\frac{4(n-1)}{|q|\,R_1}(q^2+2aB)\,a+2aB; \tag{2.4}$$

note that this can be achieved for small $a > 0$. Let $\tilde{f} \in C^1$ satisfy (*cf.* Figure 2):

i) $\tilde{f}(u) = f(u), \quad u \geqq 0$
ii) $|\tilde{f}(u)| < B, \quad u \leqq 0$
iii) $\tilde{f}(u) \equiv 0$ if $u \leqq -a$.

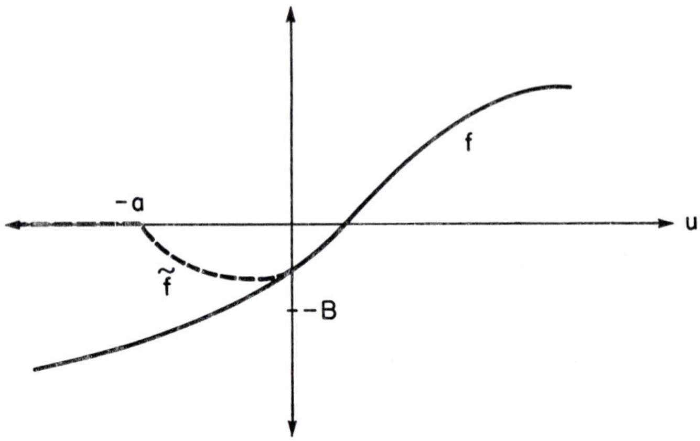


Fig. 2

We now verify that $\tilde{f}$ meets our requirement; i.e. $\tilde{u}'(r) < 0$ if $r > R_1$. As a first step, we will prove that

$$\tilde{u}'(r) < q/2 \quad \text{if } R_1 < r \leqq R_1 + \frac{2a}{|q|}. \tag{2.5}$$

Thus, set

$$\tilde{F}(u) = \int_0^u \tilde{f}(s)\, ds;$$

then

$$|\tilde{F}(u)| \leqq aB \quad \text{if } -a \leqq u \leqq 0. \tag{2.6}$$

We write, as usual, $H(u, v) = v^2/2 + \tilde{F}(u)$. Then since $H' = -\frac{n-1}{r} v^2 \leqq 0$, we see that for $r \geqq R_1$, $\tilde{v}^2 + 2\tilde{F}(\tilde{u}) \leqq q^2$ (where $\tilde{v} = \tilde{u}'$). Thus

$$\tilde{v}(r)^2 \leqq q^2 + 2aB, \quad r \geqq R_1. \tag{2.7}$$

But since

$$\tilde{v}^2 + 2\tilde{F}(\tilde{u}) = q^2 - 2(n-1) \int_{R_1}^{r} \frac{\tilde{v}^2(t)}{t}\, dt,$$

we have that

$$\begin{aligned}\tilde{v}^2 &\geqq q^2 - 2\tilde{F}(u) - 2(n-1)\int_{R_1}^{r} \frac{q^2 + 2aB}{R_1}\,dt \\ &= q^2 - 2\tilde{F}(u) - 2(n-1)(r - R_1)\frac{(q^2 + 2aB)}{R_1} \\ &\geqq q^2 - 2aB - 2(n-1)\frac{2a}{|q|}\frac{(q^2 + 2aB)}{R_1} \\ &\geqq q^2/4,\end{aligned}$$

in view of (2.6) and (2.4). This proves (2.5) Now if $\theta = R_1 + \frac{2a}{|q|}$, (2.5) gives

$$\begin{aligned}\tilde{u}(\theta) &= \tilde{u}(R_1) + \int_{R_1}^{\theta} \tilde{u}'(t)\,dt \\ &\leqq \frac{2a}{|q|}\cdot\frac{q}{2} = -a.\end{aligned}$$

But in the region $u \leqq -a$, $\tilde{f}(u) \equiv 0$ so $(r^{n-1}\tilde{u}') = 0$ Thus for $r > \theta$, $r^{n-1}\tilde{u}'(r) = \theta^{n-1}\tilde{u}'(\theta) < 0$, so $\tilde{u}'(r) < 0$ if $r > \theta$ It follows that $\tilde{u}'(r) < 0$ if $r \geqq R_1$, and the proof is complete. □

We can now complete the proof of Theorem 2.2. Let $\varrho(\Omega)$ be as defined in (2.1), and assume $\Omega \subset B_R(p)$ for some $p \in \mathbb{R}^n$, where $R$ is minimal[2] By assumption, we have pairs $(u_i, R_i)$, $i = 1, 2$, such that $u_1$ and $u_2$ are solutions of (1.1), $u_1(R_1) \leqq 0$, $u_2(R_2) > u_1(0)$, $u_1'(R_1) < 0$, where $0 < R_1 \leqq \varrho(\Omega)$, and $R_2 > R_1$.

Define $\lambda = R_2/R$; then $(\lambda\Omega) \subset B_{\lambda R}(p) = B_{R_2}(p)$. Also $R_2/R_1 \geqq e(\Omega) = R/\varrho(\Omega)$, so $\lambda\varrho(\Omega) = R_2\varrho(\Omega)/R = R_2/e(\Omega) \geqq R_1$, and hence

$$R_1 \leqq \varrho(\lambda\Omega). \tag{2.8}$$

Now let $x \in (\lambda\Omega)$; then in view of (2.8), there is some $q \in (\lambda\Omega)$ such that $x \in \overline{B_{R_1}(q)} \subset (\lambda\Omega)$. It follows that

$$x \in \overline{B_{R_1}(q)} \subset (\lambda\Omega) \subset B_{R_2}(p) \subset B_{2R_2}(q), \tag{2.9}$$

since $\operatorname{dist}(p, q) < R_2$; see Figure 3.

We shall construct our solution by the method of upper and lower solutions; see [14]. To this end we define

$$u_0(x) = \tilde{u}_1(2R_2), \qquad x \in (\lambda\Omega),$$

where $\tilde{u}_1$ is obtained from $u_1$ according to Lemma 2.3. Since $2R_2 > R_1$, $\tilde{u}_1(2R_0) < 0$, (so $u_0$ is not a solution); in fact, $\tilde{u}_1 < 0$ on the complement of

[2] Such an $R$ exists since the function $\psi\colon \bar{\Omega} \to \mathbb{R}$ defined by $\psi(q) = \min\{R\colon B_R(q) \supset \Omega\}$ is continuous and $\bar{\Omega}$ is compact.

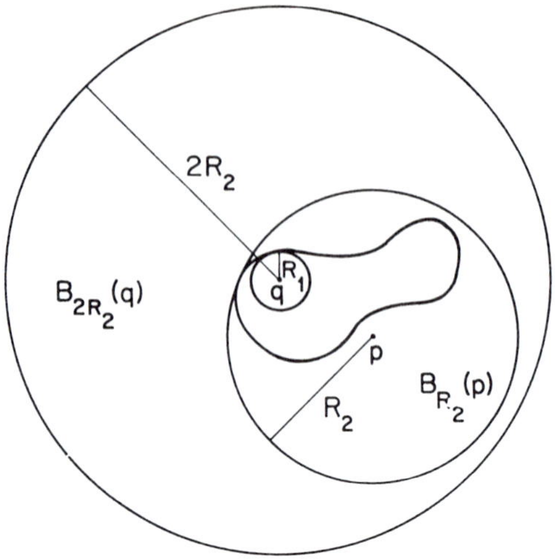


Fig. 3

$B_{R_1}(q)$. It follows that $u_0$ is a lower solution to

$$\Delta u(x) + \tilde{f}_1(u(x)) = 0, \quad x \in (\lambda\Omega) \tag{2.9}$$

$$u(x) = 0, \quad x \in \partial(\lambda\Omega), \tag{2.10}$$

(where $\tilde{f}$ is obtained from $f$ according to Lemma 2.3).

To construct an upper solution to (2.9), (2.10), we define

$$U(x) = u_2(|x - p|), \quad x \in (\lambda\Omega) \subset B_{R_2}(p).$$

Then $U$ satisfies (2.9), and for $x \in \partial(\lambda\Omega)$, $U(x) = u_2(|x - p|) > u_1(0) > 0$. We thus see that $U$ is an upper solution (and not a solution), to the problem (2.9), (2.10). Hence by a well-known theorem ([14, Ch. 10]), the problem (2.9), (2.10) has solutions $u$ satisfying

$$u_0(x) \leqq u(x) \leqq U(x), \quad x \in (\lambda\Omega). \tag{2.11}$$

Let $\bar{u}$ be a maximal solution with respect to the ordering in (2.11).

To complete the proof, it suffices to show that

$$\bar{u}(x) > 0 \quad \text{if } x \in (\lambda\Omega). \tag{2.12}$$

For, if this were true, then since $\tilde{f}(u) = f(u)$ for $u \geqq 0$, $\bar{u}$ would be a positive solution of (1.1).

To prove (2.12), let $x_0 \in (\lambda\Omega)$. Then as above, there is some $q \in (\lambda\Omega)$ with

$$x_0 \in B_{R_1}(q) \subset (\lambda\Omega) \subset B_{R_2}(p) \subset B_{2R_2}(q).$$

Define

$$w(x) = \tilde{u}_1(|x - q|), \quad x \in (\lambda\Omega).$$

We shall show that $w$ satisfies the following three properties:

(1) $u_0(x) \leqq w(x) \leqq U(x), \quad x \in (\lambda\Omega)$;
(2) $w$ is a lower solution of (2.9), (2.10);
(3) $w(x_0) > 0$.

If these hold, then from (1) and (2), there is a solution $\bar{\bar{u}}$ of (2.9), (2.10) with $u_0 \leqq w \leqq \bar{\bar{u}} \leqq U$ on $\lambda\Omega$. Thus by the maximality of $\bar{u}$, $\bar{u} \geqq \bar{\bar{u}}$ on $\lambda\Omega$, so that using (3), $\bar{u}(x_0) \geqq \bar{\bar{u}}(x_0) \geqq w(x_0) > 0$.

We now show that properties (1)–(3) hold. First, recall that $w(x) = \tilde{u}_1(|x-q|)$ with $|x-q| \leqq 2R_2$. But since $w$ is a decreasing function, (Lemma 2.3), it follows that $w(x) \geqq \tilde{u}_1(2R_2) = u_0(x)$. Next, since $U(x) = u_2(|x-p|)$, for $|x| < R_2$, we have $U(x) \geqq u_2(R_2) > u_1(0) = \tilde{u}_1(0) \geqq \tilde{u}_1(|x-q|) = w(x)$ if $x \in (\lambda\Omega)$; this proves (1). Now $w$ satisfies (2.9), and since $\tilde{u}_1(r) < 0$ for $r \geqq R_1$, (Lemma 2.3), we see that for $x \in \partial(\lambda\Omega)$, $|x-q| \geqq R_1$, so $w(x) = \tilde{u}_1(|x-q|) < 0$; thus $w$ is a lower solution. Finally, $x_0 \in B_{R_1}(q)$ implies $|x_0 - q| < R_1$ so $w(x_0) = \tilde{u}_1(|x_0 - q|) > 0$. Thus (3) holds and the proof is complete □

We remark that this theorem can be easily extended to equations of the form $\Delta u(x) + \lambda f(u,x) = 0$, with e.g, $f(u,x) = g(x)\,h(u)$, where $g(x) > 0$ and $H(u_0) > 0$ for some $u_0 > 0$. This is done by comparison with the equations $\Delta u = \lambda M h(u) + 0$ and $\Delta u + \lambda m h(u) = 0$, (where $M$ and $m$ are, respectively, the sup and inf of $g$ in $\bar{\Omega}$), and yields existence theorems for some range of $\lambda$; see [3] for related results.

Finally, the reader will notice that our technique is actually to construct an upper solution on a ball contaning $\Omega$, and a lower solution on a ball in $\Omega$. Such a strategy is certainly applicable to more general elliptic operators.

## § 3. Applications

In this section we shall illustrate the use of Theorem 2.2. Before doing this, however, it is useful to have the following lemma.

**Lemma 3.1.** *Let $F$ be the primitive of $f$ satisfying $F(0) = 0$. Suppose $F(\bar{p}) > 0$ for some $\bar{p} > 0$. Then there exists a function $f_1$ with $f_1(u) \leqq f(u)$ and an $R > 0$ and a non-negative solution $u$ of $\Delta u + f_1(u) = 0$ on $D_R^n$ with $u(R) \leqq 0$ and $u'(R) < 0$. (That is, $u$ is a "lower solution" of the Dirichlet problem (see* [13]), *satisfying $u'(R) < 0$.)*

**Proof.** Since $F(\bar{p}) > 0$, we can find a point $p_0 > 0$ with $F(p) < F(p_0)$ for all $p$, $0 \leqq p < p_0$; e.g. $p_0 = \inf F^{-1}(F(\bar{p}))$. If $f(p_0) = 0$, then the conclusion follows from our result in [13]. If $f(p_0) > 0$, we let $f_1(u)$ be a $C^1$-function satisfying $f_1(u) = f(u)$ for $0 \leqq u \leqq p_0$, $f_1(p_1) = 0$ for some $p_1 > p_0$, and $f_1(u) > 0$ on $p_0 < u < p_1$; *cf.* figure 4. Again using the result in [13], we can find positive solutions $u_1$ to $\Delta u + f_1(u) = 0$ on $D_R^n$ for some $R > 0$, with $u_1(R), u_1'(R) < 0$. But then $u_1$ restricted to the interval $0 \leqq u < p_0$ is the desired lower solution. □

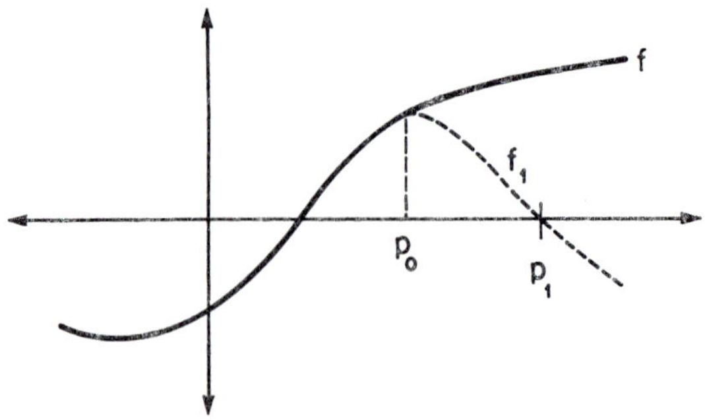


Fig. 4

We remark that the condition $F(p) > 0$ for some $p > 0$ is a necessary condition for the existence of positive solutions of (1.1), (1.2); see [2]. Thus if $u(x_1) = \max\{u(x): x \in \bar{\Omega}\}$, then $\Delta u(x_1) < 0$ so $f(u(x_1)) > 0$. The well-known Pohozaev identity (see [8]),

$$2n \int_\Omega F(u)\,dx - (n-2) \int_\Omega uf(u)\,dx = \int_{\partial\Omega} \left(\frac{du}{d\vec{n}}\right)^2 (r \cdot \vec{n})\,ds,$$

implies that $F(u(x_1)) \geqq \frac{n-2}{2n} u(x_1) f(u(x_1)) > 0$.

As a first application of Theorem 2.2, we have the following theorem, which extends the main results in [5] and [13].

**Theorem 3.2.** *Let $\Omega \subset \mathbb{R}^n$ be a bounded open set with smooth boundary, and suppose that there is a $p_0 > 0$ for which $f(p_0) = 0$ and $F(p_0) > 0$. Then the conclusion of Theorem 2.2 is valid.*

**Proof.** We shall show that (2.2) holds. Let $\bar{p} = \inf F^{-1}(F(p_0))$; then $F(\bar{p}) > F(p)$ if $0 \leqq p < \bar{p}$. If $p_1 = \inf\{p \geqq \bar{p} : f(p) = 0\}$, then $f(p_1) = 0$, and $F(p_1) > F(p)$ if $0 \leqq p < p_1$. From the theorem in [13], there is an $R_1 > 0$ for which there exists a positive radial solution $u_1$ to the problem (1.1), (1.2) on the domain $D^n_{R_1}$, with $u_1(0) < p_0$ and $u_1'(R_1) < 0$. Choose any $R_2 > R_1$. Setting $u_2(x) \equiv p_0$ on $D^n_{R_2}$ and $f_2 = f$, we see that $N(f) \geqq R_2/R_1$, for every such $R_2$. It follows that $N(f) = +\infty$ and this gives (2.2). □

Our next theorem extends to arbitrary domains $\Omega$ a result in [10, Theorem 18]; see also [3].

**Theorem 3.3.** *Let $\Omega$ be a bounded smooth domain. If $f(u)/u \to 0$ as $u \to \infty$, and $F(p_0) > 0$ for some $p_0 > 0$, then the conclusion of Theorem 2.2 is valid.*

Thus for example, $f(u)$ can be as depicted in the figure below, where $k < 1$.

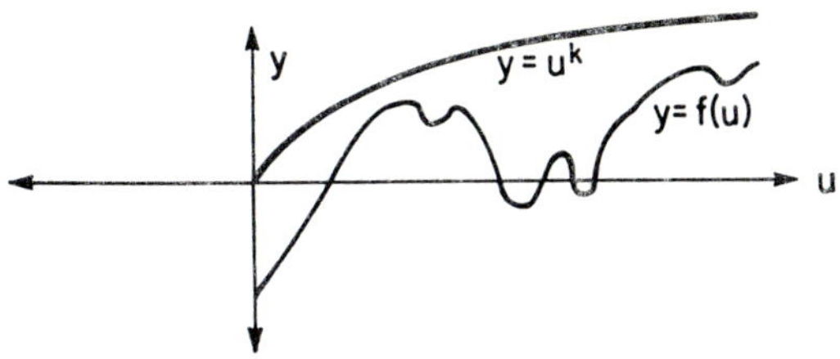


Fig. 5

**Proof.** If $f(p_0)=0$, then the result follows from the last theorem. If $f(p_0)<0$, then we can find $\tilde{p}<p_0$ with $F(\tilde{p})>0$ and $f(\tilde{p})=0$, so this case is reduced to the previous one. Suppose $f(p_0)>0$. Here there are two cases: (i) $f(u)>0$ for all $u>p_0$ or (ii) $f(p)=0$ for some $p>p_0$. Since case (ii) is easily reduced to the previous ones, we may assume that we are in case (i).

Since $F(p_0)>0$, we may apply Lemma 3.1 to conclude the existence of a non-negative solution $u_1$ of (1.1), (1.2) on some ball $D^n_{R_1}$, with $u_1'(R_1)<0$. Let $u_1(0)=p_1>0$. If $\varepsilon>0$ is given, we can find a point $p_\varepsilon>0$ such that $f(u)<\varepsilon^2 u$ if $u \geqq p_\varepsilon$. We may assume that $p_\varepsilon>p_1$. Now choose $p>2p_\varepsilon$, and let $p_\varepsilon<u<p$. Then on this range $F(p)-F(u)=\frac{\varepsilon^2}{2}(p^2-u^2)$, so with $v=u'$, we have $v^2 \leqq 2F(p)-2F(u)=\varepsilon^2(p^2-u^2)$. If $t(p)$ is defined by $u(t(p),p)=p_\varepsilon$ (such a function is easily shown to exist by the methods in [10]), then

$$t(p)=\int_{u_\varepsilon}^{p}\frac{du}{-v}\geqq\int_{u_\varepsilon}^{p}\frac{du}{\varepsilon\sqrt{p^2-u^2}}=\frac{1}{\varepsilon}\sin^{-1}\left(\frac{u}{p}\right)\Big|_{p_\varepsilon}^{p}$$

$$=\frac{1}{\varepsilon}\left[\pi/2-\sin^{-1}\left(\frac{p_\varepsilon}{p}\right)\right]\geqq\frac{\text{const}}{\varepsilon}.$$

This shows that $t(p)\to\infty$ as $\varepsilon\to 0$. Now as above, we can find solutions $u_2(r,p)$ of (1.1) on $0\leqq r<R_2(p)$ with $u_2(R_2(p),p)>p_1$ where $R_2(p)>t(p)$. It follows that $N(f)\geqq t(p)/R_1$ so that $N(f)=\infty$ and (2.2) holds. □

As a final application, we have the following result (compare with [7, Theorem 2.1]).

**Theorem 3.4.** *Let $f(0)>0$ and assume that $F(p)>0$ for some $p>0$. If $\Omega$ is any bounded domain, the conclusions of Theorem 2.2 are valid.*

**Proof.** From Lemma 3.1, there exists an $R>0$ and a non-negative solution $u_1$ of (1.1), (1.2) on $D^n_R$ with $u_1'(R)<0$. Since $f(0)>0$, we can find $p_k\to 0$ and solutions $u_k$ with $u_k(0)=p_k$, on $D^n_{R_k}$, $R_k\to 0$ (see [10, Th. 18]). Thus $N(f)\geqq R/R_k$ so $N(f)=+\infty$ and the result follows. □

## § 4. The Dirichlet Problem on n-Balls

We recall the celebrated theorem of GIDAS, NI & NIRENBERG [4], which assert. that a positive solution of (1.1), (1.2) on an $n$-ball must be radially symmetrics It thus satisfies the ordinary differential equation

$$u''(r) + \frac{n-1}{r} u'(r) + f(u(r)) = 0, \quad (r = |x|), \tag{4.1}$$

and the initial conditions

$$u(0) = p, \quad u'(0) = 0, \tag{4.2}$$

for some $p > 0$. The solution of (4.1), (4.2) will be denoted by $u(r, p)$. We define (*cf.* [9])

$$T_D(p) = \min\{r > 0: u(r, p) = 0\},$$

whenever the set on the right-hand side is not empty. Thus $T_D(p)$ is the smallest zero of $u(\cdot, p)$, and if $T_D(p) = R$, then $u(\cdot, p)$ is a solution of (1.1), (1.2). In this context, it is convenient to allow the radius $R$ of the $n$-ball $D_R^n$ to vary with $p$, and we shall thus consider the quantity $p$ as a parameter; see [10–13].

Throughout this section, we shall always assume that $f$ satisfies the following growth condition:

$$cu^\alpha \leqq f(u) \leqq du^\beta \quad \text{for } u \geqq b, \tag{4.3}$$

where $\alpha, \beta, c, d$ and $b$ are positive constants. If we take $p \geqq 2b$, then $f(u)$ is bounded away from zero on the intervals $[p/2, p]$, and $[b, p]$, so there are numbers $\tau = \tau(p)$, and $T = T(p)$ that satisfy[3]

$$u(\tau(p), p) = p/2, \quad u(T(p), p) = b. \tag{4.4}$$

Whenever (4.4) holds, we may define the quantity $q = q(p)$ by

$$q(p) = u'(T(p), p). \tag{4.5}$$

Now as we have shown in [10, Theorem 8], in order to prove the existence of positive solutions to the Dirichlet problem, it suffices to show that

$$\lim_{p \to \infty} q(p)\, T(p) = -\infty. \tag{4.6}$$

(In the appendix to this section we give a simpler proof of this result.) If (4.6) holds, then in fact we can find a point $p_0 > 0$ such that if $p \geqq p_0$, the function $u(r, p)$ solves (1.1), (1.2) with $R = T_D(p)$.

In order to prove (4.6), we write

$$-qT = -qT^{n-1}/T^{n-2},$$

and we estimate separately the numerator and denominator in terms of powers of $p$. We then collect these powers, and if the resulting exponent is positive, (4.6) holds and so positive solutions exist. We need a few lemmas before giving our main results (Theorems 4.6, 4.8 and 4.9, below).

---

[3] In [10] it is shown that such numbers exist.

**Lemma 4.1.** $-qT^{n-1} \geqq \text{const.}\, p^{\alpha}\tau^{n}$.

**Proof.** From (4.1), we may write $-(r^{n-1}u')' = r^{n-1}f(u)$; thus integrating from $r=0$ to $r=T(p)$ gives

$$\begin{aligned}
-qT^{n-1} &= \int_0^T r^{n-1} f(u(r))\, dr \\
&\geqq \int_0^T r^{n-1} cu^{\alpha}\, dr \\
&\geqq c\int_0^{\tau} r^{n-1} u^{\alpha}\, dr \\
&\geqq c\int_0^{\tau} r^{n-1}(p/2)^{\alpha}\, dr = \frac{c}{n}\frac{\tau^n}{2^{\alpha}}p^{\alpha}. \qquad \square
\end{aligned}$$

Next we estimate $\tau(p)$.

**Lemma 4.2.** *There are positive constants $k_1$ and $k_2$ such that*

$$k_1 p^{1-\beta} \leqq \tau(p)^2 \leqq k_2 p^{1-\alpha}.$$

**Proof.** For $u \geqq b$, we have

$$cr^{n-1}u^{\alpha} \leqq -(r^{n-1}u')' = r^{n-1}f(u) \leqq dr^{n-1}u^{\beta}.$$

But $p/2 \leqq u \leqq p$ when $0 \leqq r \leqq \tau$; thus on this range

$$cr^{n-1}(p/2)^{\alpha} \leqq -(r^{n-1}u')' \leqq dr^{n-1}p^{\beta}.$$

If we integrate this from 0 to $r \leqq \tau$, we get

$$\frac{c}{2^{\alpha}n}p^{\alpha}r \leqq -u'(r) \leqq \frac{d}{n}p^{\alpha}r.$$

Now integrate this from $r=0$ to $r=\tau$; this gives

$$\frac{c}{2^{\alpha}n}p^{\alpha}\frac{\tau^2}{2} \leqq \frac{p}{2} \leqq \frac{d}{n}p^{\beta}\frac{\tau^2}{2},$$

or

$$\frac{n}{d}p^{1-\beta} \leqq \tau^2 \leqq \frac{2^{\alpha}n}{c}p^{1-\alpha},$$

as desired. $\square$

Combining these lemmas gives

$$\begin{aligned}
-qT &= \frac{-qT^{n-1}}{T^{n-2}} \geqq \text{const.}\frac{p^{\alpha}\tau^n}{T^{n-2}} \\
&\geqq \frac{\text{const.}}{T^{n-2}}p^{\alpha}(p^{(1-\beta)/2})^n,
\end{aligned}$$

or

$$-qT \geqq \frac{\text{const.}}{T^{n-2}} p^{(2\alpha+n-n\beta)/2}. \tag{4.7}$$

We turn our attention now to the problem of estimating $T = T(p)$ from above. This will follow as a consequence of estimates obtained for linear equations, together with a simple monotonicity theorem.

**Lemma 4.3.** *Let $u$ be a positive solution of the (linear) equation*

$$u''(r) + \frac{n-1}{r} u'(r) + \lambda u(r) = 0, \quad 0 < r < S(\lambda), \quad \lambda > 0,$$

*together with the boundary conditions* $u(0) = p$, $u'(0) = u(S(\lambda)) = 0$. *Then* $S(\lambda) = \lambda^{-\frac{1}{2}} S(1)$.

**Proof.** This statement follows from a simple scaling argument. Thus, defining $w(r) = u(cr)$, where $c$ is a constant, we have $w'(r) = cu'(r)$, $w''(r) = c^2 u''(r)$ so if $c^2 = \lambda^{-1}$,

$$\begin{aligned} w''(r) + \frac{n-1}{r} w'(r) + w(r) &= c^2 u''(cr) + \frac{n-1}{cr} c^2 u'(cr) + u(cr) \\ &= c^2 \left[ u''(cr) + \frac{n-1}{cr} u'(cr) + \lambda u(cr) \right] \\ &= 0. \end{aligned}$$

Also, $w'(0) = 0$, $w(0) = p$, so $u(S(1)) = w(S(1))$, by uniqueness. Thus $u(cS(1)) = w(S(1)) = u(S(1)) = 0$ so $cS(1) = S(\lambda)$ and $S(\lambda) = \lambda^{-1/2} S(1)$. □

We next have the following monotonicity theorem; it's really a variant of the classical Sturm theorem.

**Lemma 4.4.** *Consider the two equations* $(-r^{n-1} u_i')' = r^{n-1} f_i(u)$, *together with the conditions* $u_i'(0) = 0$, *and* $u_i(0) \in (0, p]$, $i = 1, 2$. *Suppose that* $f_2(u)/u > f_1(w)/w$ *whenever* $0 < u$, $w < p$. *If $R_i$ is defined to be the first zero of $u_i$,* $i = 1, 2$, *then* $R_1 > R_2$.

**Proof.** Let $w = u_1 v_2 - u_2 v_1$, where $v_i = u_i'$, $i = 1, 2$. Then

$$(r^n w)' = r^{n-1} u_1 u_2 \left\{ \frac{f_1(u_1)}{u_1} - \frac{f_2(u_2)}{u_2} \right\} < 0, \tag{4.8}$$

if $0 < r < \min(R_1, R_2)$. Now if $R_2 \geqq R_1$, we can integrate (4.8) from $r = 0$ to $r = R_1$ to get

$$0 > R_1^{n-1}(u_1(R_1) v_2(R_1) - u_2(R_1) v_1(R_1)) = -R_1^{n-1} u_2(R_1) v_1(R_1).$$

On the other hand $R_2 \geqq R_1$ also implies that $u_2(R_1) \geqq 0$, and since $v_1(R_1) \leqq 0$, the above inequality is violated. Thus $R_1 > R_2$ as desired. □

We are now able to estimate $T(p)$ from above. We distinguish two cases: $\alpha \leqq 1$ and $\alpha > 1$.

**Lemma 4.5.** *Suppose that* $f(u) \geqq cu^\alpha$ *for* $u \geqq b$, *where* $\alpha \leqq 1$, *and* $c > 0$. *Then* $T(p) \leqq \text{const.}\, p^{(1-\alpha)/2}$.

**Proof.** Fix $p > b$ and define $f_1(u) = cp^{\alpha-1}u$. Let $f_2(u)$ be a $C^1$-function for which $f_2(u) = f(u)$ if $u \geqq b$, and which also satisfies the inequality $f_2(u)/u > cp^{\alpha-1}$ on $0 \leqq u \leqq b$. Then if $b \leqq u < p$, and $0 \leqq w < p$, we have

$$\frac{f_2(u)}{u} = \frac{f(u)}{u} \geqq cu^{\alpha-1} > cp^{\alpha-1} = \frac{f_1(w)}{w},$$

while if $0 \leqq u \leqq b$, and $0 \leqq w < p$, we have

$$\frac{f_2(u)}{u} > cp^{\alpha-1} = \frac{f_1(w)}{w}.$$

Thus the hypotheses of the last lemma are satisfied so (in the obvious notation) $R_1 > R_2$. Now from Lemma 4.3,

$$R_1 = \frac{S(1)}{\sqrt{\lambda}} = S(1)\, c^{-\frac{1}{2}}\, p^{\frac{1-\alpha}{2}},$$

and since $u_2(R_2, p) = 0$, and $u_2(T, p) = u(T, p) = b$, we have $R_2 > T$. It follows that $T(p) < R_2 < R_1 = \text{const.}\, p^{\frac{1-\alpha}{2}}$. □

We may now combine Lemmas 4.1 and 4.5 to obtain the following result.

**Theorem 4.6.** *Assume that* $\alpha \leqq 1$, *and* $\beta - \alpha < 2/n$. *Then for all sufficiently large* $p$, *there are positive solutions* $u(\cdot, p)$ *of* (1.1) *on* $D^n_{T_D(p)}$. *In particular, if* $\alpha = \beta \leqq 1$, *there are solutions on* $D^n_{T_D(p)}$ *for all sufficiently large* $p$.

**Proof.** Using (4.7) and the last lemma, we have

$$\begin{aligned} -qT &\geqq \text{const.}\, p^{(2\alpha+n-n\beta)/2}\, T^{2-n} \\ &\geqq \text{const.}\, p^{(2\alpha+n-n\beta)/2}\, p^{(n-2)(\alpha-1)/2} \\ &= \text{const.}\, p^{[2-n(\beta-\alpha)]/2}, \end{aligned}$$

so that $-qT \to \infty$ as $p \to \infty$ if $\beta - \alpha < 2/n$. □

We now consider the case where $\alpha > 1$. Again we shall crudely estimate $T$ from above and use (4.7).

**Lemma 4.7.** *If* $f(u) \geqq cu^\alpha$ *for* $u \geqq b$, *where* $\alpha > 1$, *and* $c > 0$, *then* $T(p)$ *is bounded.*

**Proof.** If $u \geqq b$, $f(u)/u \geqq cu^{\alpha-1} \geqq cb^{\alpha-1}$. Let $f > b$; then if we define $f_1(u) = cb^{\alpha-1} u$, and let $f_2$ be a $C^1$-function such that $f_2(u) = f(u)$ on $u \geqq b$, and which satisfies the inequality $f_2(u)/u \geqq cp^{\alpha-1}$ on $0 \leqq u \leqq b$, then Lemma 4.4, implies that $R_1 > R_2$. Now as in the proof of Lemma 4.5, $R_2 > T(p)$ and since $f_1$ is linear, $R_1$ is a constant, independent of $p$. It follows that $T(p) \leqq$ const. □

We use this estimate to prove

**Theorem 4.8.** *Let $\alpha > 1$, and suppose that $2\alpha > n(\beta - 1)$. Then for all sufficiently large p, there are positive solutions $u(\cdot, p)$ of (1.1) on $D^n_{T_D(p)}$. In particular, if $1 < \alpha = \beta < n/(n-2)$ there are solutions on $D^n_{T_D(p)}$ for all sufficiently large p.*

**Proof.** Using (2.7) and the last lemma, we have

$$-qT \geqq \text{const.}\, p^{(n+2\alpha-n\beta)/2},$$

where the exponent is positive; thus $-qT \to \infty$ as $p \to \infty$. □

We may combine Theorems 4.6 and 4.8 to obtain the following theorem.

**Theorem 4.9.** *Suppose that $f(u) = O(u^k)$ as $u \to \infty$, where $0 < k < n/(n-2)$. Then there exist positive solutions to the Dirichlet problem on $D^n_{T_D(p)}$ for all sufficiently large p.*

Since $1 < n/(n-2)$ for all $n \geqq 2$, we obtain at once the following corollary, (which corrects an error in [10; Corollary 15]).

**Corollary 4.10.** *If $f(u) = O(u)$ as $u \to \infty$, then for every positive integer n, there exists a point $p_n > 0$ such that if $p \geqq p_n$, there exist positive solutions to the Dirichlet problem on $D^n_{T_D(p)}$.*

We close this section with the following observation. Namely, if one were able to strengthen Lemma 4.7 to the statement that if $\alpha > 1$, $T(p) \leqq \text{const.}\, p^{\frac{1}{2-n}}$, that is, if we could estimate the rate of decay of $T(p)$, then existence of positive solutions to the Dirichlet problem would hold for all sufficiently large $p$, whenever $n + 2 > n\beta - 2\alpha$. Thus in particular if $\alpha = \beta = k$, there would be such solutions whenever $0 < k < (n+2)/(n-2)$. This would be the optimal result in this generality.

## Appendix

In this section we shall give a short proof of our basic result in [10]; in particular we shall avoid the use of comparison functions and certain tedious estimates.

**Theorem.** *Suppose that $f(u) \geqq m$ for $u \geqq A$. Then the following statements hold:*

(i) *For any* $p > A$, *there is a* $T(p)$ *with* $u(T(p), p) = A$.

(ii) *If* $q(p) = u'(T(p), p)$, *then* $-q(p)/T(p) \geqq m/n$.

(iii) *If* $q(p)\,T(p) \to -\infty$ *as* $p \to \infty$, *then there is a* $\bar{p} > 0$ *such that if* $u(0) \geqq \bar{p}$, *the problem* (4.1), (4.2) *has a positive solution, with* $u' > 0$.

**Proof.** We only prove (iii); parts (i), and (ii) are easily proved as in [9].

Choose $B > 0$ such that $-B \leqq f(u) \leqq B$, $0 \leqq u \leqq A$; then if $F' = f$, $F(0) = 0$, we have

$$(A_1) \qquad |F(u)| \leqq BA, \quad 0 \leqq u \leqq A.$$

Set $H(r) = v(r)^2/2 + F(u(r))$; then $H$ decreases on orbits so $v^2 + 2F(u) \leqq q^2 + 2F(A)$, *if* $0 \leqq u \leqq A$. Using ($A_1$), we find

$$(A_2) \qquad v^2(r) \leqq q^2 + 4BA, \quad 0 \leqq u \leqq A.$$

We shall use this inequality to show

$$(A_3) \qquad v(r, p) < q/2 \quad \text{for } T(p) \leqq r \leqq T(p) + \frac{2A}{|q|},$$

if $-q(p)\,T(p) \gg 1$; *i.e.* if $p \gg 1$.

Thus, we have

$$\begin{aligned} v^2/2 + F(u) - (q^2/2 + F(A)) &= \int_T^r H'(t)\,dt \\ &= -(n-1)\int_T^r \frac{v^2}{t}\,dt \\ &\geqq -(n-1)\int_T^r \frac{q^2 + 4BA}{t}\,dt. \end{aligned}$$

It follows that

$$\begin{aligned} v^2 &\geqq q^2 + 2(F(A) - F(u)) - (n-1)\int_T^r \frac{q^2 + 4BA}{t}\,dt \\ &\geqq q^2 - 4BA - (n-1)\int_T^r \frac{q^2 + 4BA}{T}\,dt \\ &= q^2 - 4BA - \frac{(n-1)}{T}(q^2 + 4BA)(r - T) \\ &\geqq q^2 - 4BA - \frac{(n-1)}{T}(q^2 + 4BA)\frac{2A}{|q|}. \end{aligned}$$

Thus

$$v^2/q^2 \geqq 1 - 4BA/q^2 - \frac{2A(n-1)}{|q|\,T}(1 + 4BA/q^2).$$

But $qT \to -\infty$ and $-q/T \geqq m/n$ imply that $q^2 \to \infty$ as $p \to \infty$. Thus for large $p$, we see that the last inequality shows that we can make $v^2 \geqq q^2/4$ on the requisite interval; this implies $(A_3)$.

From this the proof of (iii) follows rather easily; namely, if $\theta = T(p) + 2A/|q|$, (where $p$ is chosen so large as to make $(A_3)$ valid), we have

$$u(\theta) = A + \int_{T(p)}^{\theta} v(t)\,dt$$

$$< A + \frac{q}{2}\frac{2A}{|q|} = 0.$$

Hence $u$ becomes negative on this interval and since $v < q/2 < 0$, on this interval, we see that a solution exists.

## § 5. The Neumann Problem

We shall show here how the methods of the precedingection can be extended to yield monotone solutions of the Neumann (and in fact, any other linear) boundary-value problem. We continue to assume that inequalities (4.3) hold.

**Theorem 5.1.** *Suppose $\Omega$ is an $n$-ball, and that $f$ satisfies* (4.3), *where either* i) $\alpha \leqq 1$ *and* $\beta - \alpha < 2/n$, *or* ii) $\alpha > 1$ *and* $2\alpha > n(\beta - 1)$. *Assume in addition there is an* $\eta < 0$ *such that*

$$f(u) < 0 \quad \text{if } u \leqq \eta, \tag{5.1}$$

*and that*

$$\lim_{u \to \infty} F(u) = \infty. \tag{5.2}$$

*Then there is a* $p_0 > 0$ *such that if* $u(0) \geqq p_0$, *there is a monotone radial solution of the Neumann problem*

$$\begin{aligned} \Delta u(x) + f(u(x)) &= 0, \quad x \in D^n_{T_N(p)} \\ du(x)/dn &= 0, \quad x \in \partial D^n_{T_N(p)}. \end{aligned} \tag{5.3}$$

Recall that $u(r, p)$ is the solution of (4.1) and (4.2); here $T_N(p)$ is defined by

$$T_N(p) = \min\{r > 0\colon du(r, p)/dn = 0\},$$

whenever the set on the right is not empty.

**Corollary 5.2.** *If* (5.1) *and* (5.2) *hold, and in addition* $f(u) = O(u^k)$ *as* $u \to \infty$, $0 < k < n/(n-2)$, *then the conclusion of Theorem* 5.1 *is valid.*

We remark that if $f(u) > 0$ for all $u$, then there cannot be any radial solutions to the Neumann problem (5.3) since the $u$-axis would be a repellor for the system (equivalent to (4.1)),

$$u' = v, \qquad v' = -\frac{n-1}{r} v - f(u). \tag{5.4}$$

Thus some conditions like (5.1) and (5.2) are necessary.

**Proof.** There are two steps in the proof; namely

(i) There exists a $p_1 > 0$ such that if $p \geqq p_1$, there is a $\tilde{T}(p)$ satisfying $u(\tilde{T}(p), p) = \eta$.

(ii) For $p \geqq p_1$, $u(r, p)$ is a monotone solution of the Neumann problem.

We begin by showing that (ii) holds. As usual, let $H$, the "total energy", be defined by

$$H(r) \equiv H(v(r), u(r)) = F(u(r)) + v(r)^2/2, \qquad v = u'. \tag{5.5}$$

Then $H' = -(n-1) r^{-1} v^2$, so that $H$ decreases on orbits $(u(r, p), v(r, p))$ of (5.4), where, as usual, $p = u(0, p)$.

Now let $p > p_1$, and suppose $u(\bar{r}, p) > 0$, and $H(\bar{r}) = F(u(\bar{r}, p)) + v(\bar{r}, p)^2/2$. Using (5.2), we can find $\bar{\eta} < \eta$ for which $F(u) \geqq H(\bar{r}) + 1$, if $u \leqq \bar{\eta}$. Thus, since $H$ is a decreasing function, and $F(u) \leqq H(u, v)$, we see that $u(r, p) > \bar{\eta}$ for all $r \geqq 0$.

Suppose that the function $u(\cdot, p)$ does not yield a solution of the Neumann problem. Then both $u'(r, p) < 0$, and $\bar{\eta} < u(r, p) \leqq p$, for all $r \geqq 0$. Thus we see that $u(\cdot, p)$ is a bounded, monotone decreasing function and thus has a limit; say

$$\lim_{r \to \infty} u(r, p) = u_1. \tag{5.6}$$

Note that $u(\tilde{T}, p) = \eta$ implies that $u_1 < \eta$, (since $u' < 0$), so that (5.1) gives $f(u_1) < 0$. Thus there is a neighborhood of $u_1$ on which $f$ is negative; say $|u - u_2| < \delta$ implies $-B \leqq f(u) \leqq -B/2$, for some $B > 0$. Using (5.6), we can find an $R_1 > 0$ such that $u_1 < u(r, p) < u_1 + \delta$, if $r \geqq R_1$. It follows that $-B \leqq f(u(r, p)) \leqq -B/2$ if $r \geqq R_1$. Using these facts, we have, for $r \geqq R_1$,

$$-(r^{n-1} v)' = r^{n-1} f(u) \leqq -r^{n-1} B/2,$$

and integrating from $R_1$ to $r$ gives

$$-r^{n-1} v(r, p) + q_1 R_1^{n-1} \leqq -\frac{r^n B}{2n} + \frac{R_1^n B}{2n},$$

where $q_1 = v(R_1, p)$. Thus

$$-v(r, p) \leqq \frac{-q_1 R_1^{n-1}}{r^{n-1}} - \frac{rB}{2n} + \frac{BR_1}{2n} \left(\frac{R_1}{r}\right)^{n-1},$$

and this shows that $-v(r, p) \to -\infty$ as $r \to \infty$. Thus $v(r, p) > 0$ for large $r$, so that $v(r, p)$ must vanish somewhere. This is the desired contradiction, and so the proof of statement (ii) is complete.

To prove (i), we set $\bar{u} = u - \eta$ and then write the equation in terms of $\bar{u}$:

$$-(r^{n-1}\,\bar{u}')' = r^{n-1}\,\bar{f}(\bar{u}), \tag{5.7}$$

where $\bar{f}(\bar{u}) = f(\bar{u} + \eta)$. Now if $u \geqq b$, *cf.* (4.3), we have the bounds

$$c(\bar{u} + \eta)^{\alpha} \leqq \bar{f}(\bar{u}) \leqq d(\bar{u} + \eta)^{\beta}, \quad \bar{u} \geqq b - \eta. \tag{5.8}$$

Since $\eta > 0$, $(\bar{u} + \eta) \leqq \bar{u}$, so (5.8) gives

$$\bar{f}(\bar{u}) \leqq d\bar{u}^{\beta} \quad \text{if } \bar{u} \geqq b - \eta. \tag{5.9}$$

On the other hand, $b - \eta > -\eta$ so we can find a constant $c_1$, $0 < c_1 < 1$ for which $(1 - c_1)(b - \eta) \geqq -\eta$. Thus, if $\bar{u} \geqq b - \eta$, $(1 - c_1)\,\bar{u} \geqq (1 - c_1)(b - \eta) \geqq -\eta$, and hence $\bar{u} + \eta \geqq c_1\bar{u}$. Thus from (5.8) we get $cc_1^{\alpha}\bar{u}^{\alpha} \leqq \bar{f}(\bar{u})$, for $\bar{u} \geqq b - \eta$. Combining this with (5.9), we see that $\bar{f}(\bar{u})$ satisfies the same inequalities as does $f(u)$, for large $\bar{u}$. We have seen in § 4 that under the conditions which we have imposed on $\alpha$ and $\beta$, there are radial solutions of the Dirichlet problem. That is, there is a $p_1 > 0$ such that if $p \geqq p_1$, there is a $\bar{T}(p)$ for which $\bar{u}(\bar{T}(p), p) = 0$; thus $u(\bar{T}(p), p) = \eta$. This proves (i) and completes the proof of the theorem. □

We close this section with the observation that any monotone decreasing radial solution to the Neumann problem on an $n$-ball guarantees the existence of a monotone decreasing solution satisfying any linear boundary conditions. Thus the two classical boundary conditions, Dirichlet and Neumann, are actually the most general ones as far as questions of existence are concerned.

*Acknowledgement.* We thank XABIER GARAIZER for pointing out an error in our original proof of Corollary 4.10 in [10]. We also thank JOHAN DUPONT for showing us the simple proof of Lemma 2.1. This research was supported in part by NSF Contract # MCS-800-2337. The second author thanks the Mathematics Institute of Aarhus University for supporting this research.

Note added in proof: P. CLÉMENT & G. SWEERS (C. R. Acad. Sc. Paris, **302**, 681–683 (1986)), using similar techniques, have obtained results related to ours in § 3.

## References


1. AMANN, H., Fixed point equations and nonlinear eigenvalue problems in Banach spaces, SIAM Rev., **18**, 620–709 (1976).
2. BERESTYCKI, H., & P. L. LIONS, Existence of stationary states in nonlinear scalar field equations, in Bifurcation Phenomena in Mathematical Physics and Related Topics, ed. by C. BARDOS & D. BESSIS, D. Reidel Publ. Co.: Doredrecht, Boston, London, (1980).
3. BREZIS, H., & L. OSWALD, Remarks on sublinear elliptic equations, (preprint).
4. GIDAS, B., NI, W. M., & L. NIRENBERG, Symmetry of positive solutions of nonlinear elliptic equations in $\mathbb{R}^n$, Comm. Math. Phys., **68**, 209–243 (1975).

5. HESS, P., On multiple positive solutions of nonlinear elliptic eigenvalue problems, Comm. P.D.E., **6**, 951–961 (1981).
6. HIRSCH, M. W., Differential Topology, Springer-Verlag: Berlin, Heidelberg, New York, (1976).
7. LIONS, P. L., On the existence of positive solutions of semilinear elliptic equations, SIAM, Rev. **24**, 441–467 (1982).
8. NUSSBAUM, R., Positive solutions of nonlinear elliptic boundary-value problems, J. Math. Anal. Appl. **51**, 461–482 (1975).
9. POHOZAEV, S. I., Eigenfunctions of the equation $\Delta u + \lambda f(u) = 0$, Sov. Math. Dok. **5**, 1408–1411 (1965).
10. SMOLLER, J., & A. WASSERMAN, Existence, uniqueness, and nondegeneracy of positive solutions of semilinear elliptic equations, Comm. Math. Phys. **95**, 129–159, (1984).
11. SMOLLER, J., & A. WASSERMAN, Symmetry-breaking for positive solutions of semilinear elliptic equations, Arch. Rational Mech. Anal. **95**, 217–225 (1986).
12. SMOLLER, J., & A. WASSERMAN, Symmetry-breaking for semilinear elliptic equations with general boundary conditions, Comm. Math. Phys. **105**, 415–441 (1986).
13. SMOLLER, J., & A. WASSERMAN, An existence theorem for positive solutions of semilinear elliptic equations, Arch. Rational Mech. Anal. **95**, 211–216 (1986).
14. SMOLLER, J., Shock Waves and Reaction-Diffusion Equations, Springer-Verlag: Berlin, Heidelberg, New York, (1983).

The University of Michigan
Ann Arbor

*(Received June 22, 1986)*

# *Assembling a Rearrangement*

GIORGIO TALENTI

*A James Serrin, con stima ed amicizia*

**1.** Rearrangements of functions have proved to be a fairly interesting tool in analysis. Systematically introduced by HARDY & LITTLEWOOD, they have been used by a number of authors in real and harmonic analysis, in investigations about singular integrals, function spaces and interpolation theory. See [2, 3, 5–8] for instance. PÓLYA & SZEGÖ and their followers have demonstrated a good many isoperimetric theorems and inequalities by means of rearrangements—[9] is a source book for this matter. More recent investigations have shown that rearrangements of functions also fit well into the theory of elliptic second-order partial differential equations. See [4, 11], for example, and the references cited therein.

In this paper I point out a property of rearrangements. Loosely speaking, I show that rearranging a real-valued function of one real variable in decreasing order amounts to solving a free-boundary problem for a (*sui generis*) first-order quasilinear partial differential equation.

Let $u$ be a real-valued measurable function defined in an interval $[0, l]$. The *decreasing rearrangement* of $u$, usually denoted by $u^*$, is characterized by the following properties:

(i) $u^*$ maps $[0, l]$ into $[0, \infty[$;

(ii) $u^*$ is decreasing and right-continuous;

(iii) $u^*$ is equimeasurable (or equidistributed) with $|u|$, *i.e.*

$$\text{measure of } \{s \in [0, l] : u^*(s) > y\} = \text{measure of } \{x \in [0, l] : |u(x)| > y\}$$

for all $y \geqq 0$.

Suppose $u$ is not too wild—piecewise monotonic and Lipschitz continuous, say. Conceivably, $u^*$ can be formed by cutting the graph of $u$ into several suitable pieces, remodelling some of these pieces, and then sewing up again. A detailed, though somewhat informal, description of this procedure is our concern here and reads as follows.

Generate $v$, a smooth (*i.e.* Lipschitz-continuous) map from $[0, l]\times]-\infty,\infty[$ into $[0,\infty[$, subject to the following equation:

(1)
$$\frac{\partial v}{\partial t}(t, s)+\frac{\partial v}{\partial s}(t, s)\times\begin{cases}1 & \text{if } v(t, s)<|u(t)| \\ 0 & \text{if } v(t, s)>|u(t)|\end{cases}=0$$

and the following auxiliary conditions:

(2a)
$$v(t, s) \geqq|u(t)| \quad \text{and} \quad \frac{\partial v}{\partial t}(t, s) \geqq 0 \quad \text{if } s<0,$$

(2b)
$$v(t, s) \leqq|u(t)| \quad \text{and} \quad \frac{\partial v}{\partial t}(t, s)+\frac{\partial v}{\partial s}(t, s) \leqq 0 \quad \text{if } s>t.$$

Look at $v(l, \cdot)$, the restriction of $v$ to the line where $t=l$. The decreasing rearrangement of $u$ is then found, that is:

(3)
$$v(l, s)=u^*(s) \quad \text{for } 0 \leqq s \leqq l.$$

Proofs are sketched in section 6. Sections 3 and 4 are devoted to considerations about rearrangements, which may motivate the above statements. Section 2 contains preparatory comments and an example.

**2.** Equation (1) and condition (2a) result in

$$v(t, s) \geqq|u(t)|,$$

$$v_t(t, s) \geqq 0,$$

$$v_t(t, s)=0 \text{ if } t \text{ is such that } v(t, s)>|u(t)|$$

when $s<0$. Hence

$$v(t, s)=\max \{v(0, s), \max \{|u(x)|: 0 \leqq x \leqq t\}\}$$

for $s \leqq 0$, as an easy argument shows. Analogously, equation (1) and condition (2b) give

$$v(t, s)=\min \{v(0, s-t), \min \{|u(x)|: 0 \leqq x \leqq t\}\}$$

for $s \geqq t$. Conditions (2) impiy

(4)
$$v(0,0)=|u(0)|.$$

Therefore, equation (1) and conditions (2) *uniquely determine both* $[0, l] \ni t \rightarrow v(t, 0)$, the restriction of $v$ to the $t$-axis, *and* $[0, l] \ni s \rightarrow v(s, s)$, the restriction of $v$ to the diagonal. The former is the smallest increasing function above $|u|$, the latter is the largest decreasing function below $|u|$:

(5a)
$$v(t, 0)=\max \{|u(x)|: 0 \leqq x \leqq t\},$$

(5b)
$$v(s, s)=\min \{|u(x)|: 0 \leqq x \leqq s\}.$$

The core of our matter bears upon the restriction of $v$ to the following triangle:

(6) $$\{(t, s): 0 \leqq t \leqq l, 0 \leqq s \leqq t\}.$$

Later we shall show that equations (1) and (5) uniquely determine such a restriction. Equation (1) will tell us how $v$ develops; equations (5) will play the role of (redundant) *initial-boundary conditions*.

Note that equation (1) reads

$$\frac{\partial v}{\partial t} + \chi \frac{\partial v}{\partial s} = 0,$$

a first-order quasilinear partial differential equation with a bizarre coefficient $\chi$: the *characteristic function* of

(7) $$\{(t, s): v(t, s) < |u(t)|\}.$$

Thus, $v$ is constant along broken lines—the relevant *characteristic curves*—whose pieces are parallel either to the $t$-axis or to the diagonal. The vertices of these broken lines span (together with segments of the $t$-axis and/or the diagonal)

(8) $$\text{boundary of } \{(t, s): v(t, s) < |u(t)|\},$$

the *free boundary* of our problem.

An example may help to grasp the situation. Let $l = 3$ and

$$u(x) = 2x^3 - 9x^2 + 12x \quad \text{if } 0 \leqq x \leqq 3,$$

as plotted in Figure 1a. Figure 1b shows the main features of the problem stated by (1) and (2) in the present special case. Note that $u^*$, whose graph is displayed in Figure 1c, obeys

$$u^*(s) = \begin{cases} u(3 - s) & \text{if } \ 0 \leqq s \leqq 1/2 \quad \text{or } 5/2 \leqq s \leqq 3 \\ \frac{1}{4}s^3 - \frac{9}{8}s^2 + \frac{15}{16}s + \frac{153}{32} & \text{if } \ 1/2 < s < 5/2, \end{cases}$$

in agreement with formula (3).

**3.** The decreasing rearrangement is a *non-expansive* mapping (*i.e.* a *contraction*) in Lebesgue spaces $L^p(0, l)$, or, more generally, in Orlicz spaces; see [1]. Hence an *algorithm* for computing—within some tolerance—the decreasing rearrangement of a given function $u$ may consist in selecting an appropriate approximation of $u$, then computing the—authentic—decreasing rearrangement of such an approximation. This is an algorithm indeed, provided the decreasing rearrangement of the approximating function can actually be computed in a finite number of steps. Approximations by *step functions* are quite convenient in the present context. In fact, the very definitions tell us that the decreasing rearrangement of a step function is still a step function: just the step function which is obtained by assembling the oringinal blocks in order of decreasing height. Thus, rearranging a step function, whose steps all have *equal length*, is exactly the same as arranging, in decreasing order, the list of (the absolute values of) its values.

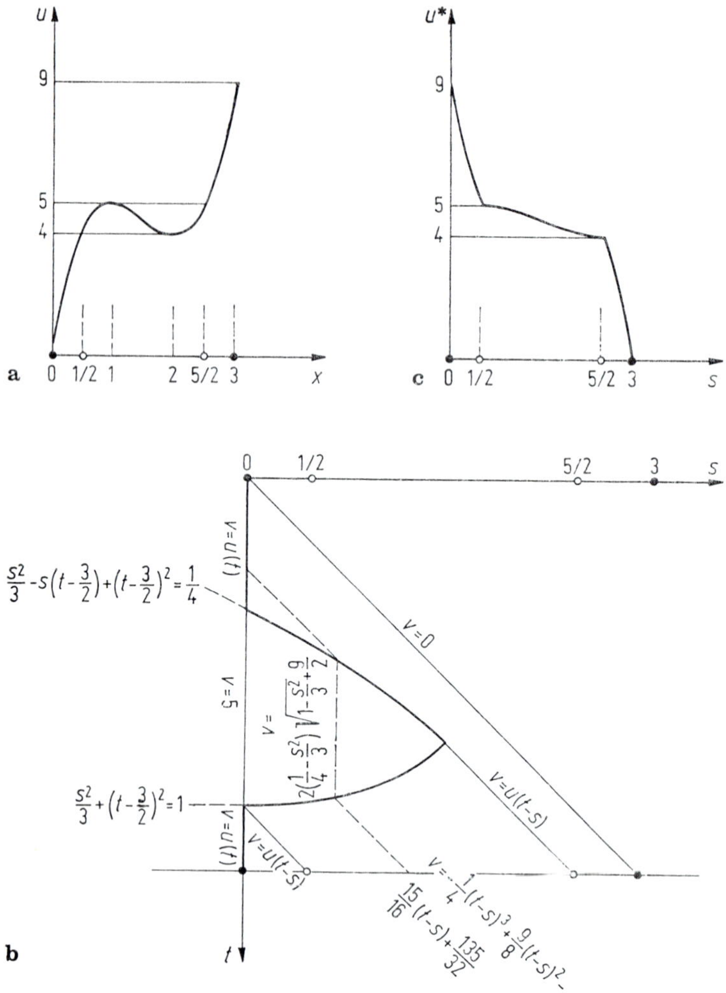


Fig. 1a–c

To be more specific, let us suppose $u$ is absolutely continuous in $[0, l]$ and $u'$ belongs to $L^q(0, l)$ for some $q \geqq 1$. Let us split $[0, l]$ into $n$ subintervals by *equidistant points*

$$x_i = i(l/n) \quad (i = 0, 1, 2, \ldots, n), \tag{10a}$$

then sample $u$ at the *mid-point* of each subinterval, and then name stpfn the step function whose steps are our subintervals and whose values are the sample values of $u$. In formula

$$\text{stpfn} = \sum_{=1}^{n} u\left(\frac{x_{i-1} + x_i}{2}\right) \mathbb{1}_{[x_{i-1}, x_i[}\,, \tag{10b}$$

where $\mathbb{1}$ stands for characteristic function.

As is easy to see,

$$\|u - \text{stpfn}\|_{L^p(0,l)} \leqq C(h/2)^{1+1/p-1/q} \|u'\|_{L^q(0,l)},$$

where $p$ is any exponent $\geqq q$,

$$h = l/n \ (= \text{mesh size}),$$

$$C = \left(1 + \frac{q}{p(q-1)}\right)^{1/q} \left(1 + p\left(1 - \frac{1}{q}\right)\right)^{-1/p} \frac{p}{B(1/p, 1 - 1/q)},$$

$B$ denotes the Euler beta function.

Therefore

$$\|u^* - (\text{stpfn})^*\|_{L^p(0,l)} \leqq C(\tfrac{1}{2}h)^{1+1/p-1/q} \|u'\|_{L^q(0,l)}. \tag{12}$$

Thus (stpfn)* comes arbitrarily close to $u^*$ in terms of the metric of $L^p(0, l)$ as the number $n$ of the mesh points is made large enough. The quality of such an approximation depends on $n$ and the smoothness of $u$ only.

A convenient algorithm for computing $u^*$ may amount to rearranging the step function stpfn defined above. An implementation of this algorithm is available.

It is worth mentioning that basically similar (though differently organized) arguments were introduced by C. Somigliana in a paper of 1899 [10] and enabled him to demonstrate the following. Suppose a liquid sits in a cylindrical vessel and its density, is, at some instant, constant along any cross section, *i.e.* a function $u$ of height $x$ only (see Fig. 2). Then eventually, when the liquid has reached a stable equilibrium, the density must be $u^*$, the decreasing rearrangement of the initial density $u$.

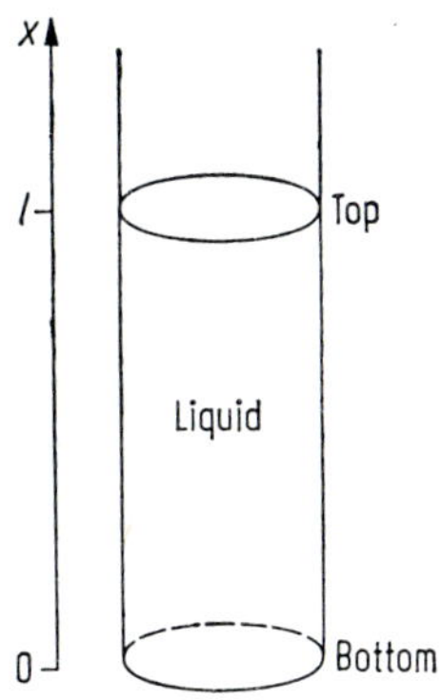


Fig. 2

**4.** In this section we shall examine in detail a procedure for rearranging a *list of n numbers* in *decreasing order*. Let

$$u(1), u(2), \ldots, u(n) \tag{13}$$

be the input. The process we have in mind results in

$$(v(i, j))_{i,j=0,1,\dots,n} \tag{14}$$

an $(n+1)\times(n+1)$ matrix, and

$$s(1), s(2), \dots, s(n), \tag{15}$$

a sequence of indices from 1 to $n$. Specifications follow.

Define

$$v(0, 0),$$

the top left entry of our matrix, and

$$v(0, 1), v(0, 2), \dots, v(0, n),$$

the remaining entries from row no. 0, to be

$$\text{ub} = \text{any } \textit{upper} \text{ bound of } u(1), u(2), \dots, u(n) \tag{16a}$$

and

$$\text{lb} = \text{any } \textit{lower} \text{ bound of } u(1), u(2), \dots, u(n) \tag{16b}$$

respectively.

For $i = 1, 2, \dots, n$ do the following. Suppose

$$v(i-1, 0), v(i-1, 1), \dots, v(i-1, n),$$

row no. $(i-1)$, is already set forth (and is an arrangement in decreasing order). Pick $u(i)$. Start scanning the above list backwards and test whether thc entry being used exceeds $u(i)$. Go on scanning until the test is positive, *i.e.* an entry is met which exceeds $u(i)$. Then stop scanning and produce

$$v(i, 0), v(i, 1), \dots, v(i, n),$$

the entries from row no. $i$, by:

$$v(i, j) = \begin{cases} v(i-1, j-1) & \text{if } j \text{ is such that } v(i-1, j-1) < u(i) \\ v(i-1, j) & \text{if } j \text{ is such that } v(i-1, j) \geqq u(i) \\ u(i) & \text{otherwise}. \end{cases} \tag{20}$$

In particular,

$$v(i, 0) = \text{ub}, \quad v(i, j) = \text{lb} \quad \text{if } j > i. \tag{21}$$

Define also

$$s(i) = \min\{j = 0, 1, \dots, n : v(i-1, j) < u(i)\}. \tag{22}$$

the site from row no. $i$ where we set up $u(i)$.

End.

Clearly,

$$v(n, 1), v(n, 2), \ldots, v(n, n)$$

is

$$u^*(1), u^*(2), \ldots, u^*(n),$$

the decreasing rearrangement of $u(1), u(2), \ldots, u(n)$. The sequence

$$s(1), s(2), \ldots, s(n)$$

gives evidence of how the process above evolves. See Figure 3.

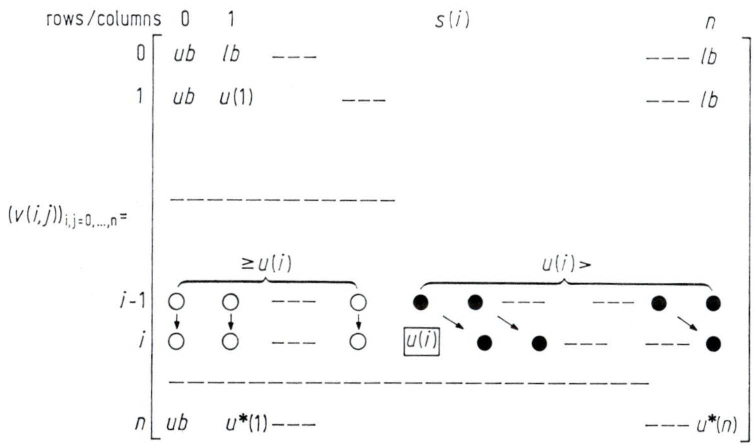


Fig. 3

**5.** Now we are in a position to motivate the statements from section 1.

Suppose one starts as in section 3, rearranges the values of $|\text{stpfn}|$ as in section 4, then lets the mesh size $h = l/n$ tend to zero.

Heuristically speaking, one might guess that matrix $(v(i,j))_{i,j=0,\ldots,n}$—*i.e.* a function whose values at points of a plane grid are the entries of such a matrix—approaches a solution $v$ to the problem stated by (1) and (2). Simultaneously, the set of points whose coordinates are

$$t = ih \qquad s = s(i)\, h,$$

should approach the free boundary.

In fact, discretizing equation (1) by a standard finite difference scheme leads precisely to recurrence equations (20). On the other hand, auxiliary conditions (2) match with equations (21), as equation (5) show.

We omit the relevant convergence proofs, which might follow from the contents of section 3 and the following section 6, and limit our effort to writing a code for approximating the free boundary.

**6.** To prove equation (3) we may demonstrate that any solution $v$ of the problem stated by (1) and (2) enjoys the following properties:

(i) For every $t$ from $[0, l]$,

$$[0, t] \in s \to v(t, s) \tag{25a}$$

is a decreasing positive function.
(ii) For every $t$ from $[0, l]$ and any test function $\varphi$ (*i.e.* any infinitely differentiable, compactly supported map $\varphi$ from $]-\infty, \infty[$ into $]-\infty, \infty[$),

$$\int_0^t \varphi(v(t, s))\, ds = \int_0^t \varphi(|u(x)|)\, dx. \tag{26}$$

Property (ii) guarantees that function (25a) is equidistributed with

$$[0, t] \ni x \to u(x), \tag{25b}$$

the restriction of $u$ to the interval $[0, t]$. Therefore properties (i) and (ii) yield

$$\text{function (25a)} = \text{the decreasing rearrangement of function (25b)} \tag{27}$$

for every $t$ in $[0, l]$.

Equation (27) is a statement of uniqueness. Clearly, equation (27) implies (3).

Proof of (i). The following set

$$\{t \in [0, l] : v(t, s_1) \geqq v(t, s_2) \text{ if } 0 \leqq s_1 \leqq s_2 \leqq t\} \tag{28])}$$

is closed and not empty. We claim that (28) is open in the relative topology of $[0, l]$.

Firstly, we show that (28) contains $[0, h]$, a relative neighborhood of $t = 0$, if $h$ is sufficiently small. In fact, let $h$ be such that the restriction of $|u|$ to $[0, h]$ is monotonic, an assumption consistent with our hypotheses on the function $u$. Equations (1) and (5) tell us that either

$$v(t, s) = |u(t - s)|$$

or

$$v(t, s) = |u(s)|$$

as $0 \leqq t \leqq h$ and $0 \leqq s \leqq t$, according to whether $|u|$ increases or decreases. Hence every $t$ in $[0, h]$ belongs to (28).

Secondly, we show that, if $t$ is not 0 and belongs to (28), then a relative neighborhood of $t$ is included in (28). A proof rests on similar arguments as above, and the following remark. If $v$ satisfies

$$\frac{\partial v}{\partial t} + \chi \frac{\partial v}{\partial s} = 0,$$

then

$$w = \frac{\partial v}{\partial s}$$

is a (weak) solution of

$$\frac{\partial w}{\partial t} + \frac{\partial}{\partial s}(\chi w) = 0,$$

a conservation law which *preserves positivity*.

The claim follows. Hence, (i) follows too. □

Proof of (ii). Let $\psi$ be any test function whose support lies in the interior of $[0, l]$. Green's formulas and equation (1) yield

$$-\int_0^l dt\, \psi'(t) \int_0^t \varphi(v(t,s))\, ds = \int_0^l \psi(s)\, \varphi(v(s,s))\, ds - \oint \psi(t)\, \varphi(v(t,s))\, dt$$

where the last integral, which is an integral of a differential form, runs (counterclockwise) over the boundary of

$$\{(t,s) : 0 \leqq t \leqq l,\, 0 \leqq s \leqq t,\, v(t,s) < |u(t)|\,.$$

Thanks to property (i) and equations (5), this set intersects every line where $t =$ constant exactly along a segment; the terminals of such a segment are on the diagonal where $s = t$ and the free boundary where $v(t,s) = |u(t)|$. Hence

$$-\int_0^l dt\, \psi'(t) \int_0^t \varphi(v(t,s))\, ds = \int_0^l \psi(t)\, \varphi(|u(t)|)\, dt\,.$$

In other words,

$$\frac{d}{dt} \int_0^t \varphi(v(t,s))\, ds = \varphi(|u(t)|)\,,$$

so that equation (26) follows. □

## References


1. G. Chiti, Rearrangements of functions and convergence in Orlicz spaces (Appl. Anal. **9**, 1979).
2. Hardy, Littlewood, Pólya, Inequalities (Cambridge Univ. Press, 1964).
3. C. Herz, The Hardy-Littlewood maximal theorem (Symposium on Harmonic Analysis, Univ. of Warwick, 1968).
4. B. Kawohl, Rearrangements and convexity of level sets in PDE (Lecture Notes in Math. **1150**, Springer-Verlag, 1985).
5. R. O'Neil, Adjoint operators and interpolation of linear operators (ISNM vol. **10**, 1969).
6. R. O'Neil, Convolution operators and $L(p,q)$ spaces (Duke Math. J. **30**, 1963).
7. R. O'Neil, Integral transforms and tensor products on Orlicz and $L(p,q)$ spaces (J. Analyse Math. **21**, 1968).
8. R. O'Neil & G. Weiss, The Hilbert transform and rearrangement of function (Studia Math. **23**, 1963).
9. G. Pólya & G. Szegö, Isoperimetric inequalities in mathematical physics (Princeton Univ. Press, 1951).
10. C. Somigliana, Sulle funzioni reali d'una variabile (Rendiconti R. Accademia dei Lincei, vol. **8**, 1899).
11. G. Talenti, Linear elliptic p.d.e.'s: level sets, rearrangements and a priori estimates of solutions (Boll. U.M.I., 4–B, 1985).


Istituto Matematico
Università di Firenze

*(Received January 4, 1986)*

# *Asymptotic Behaviour of Solutions of Semi-Linear Elliptic Equations in* $\mathbb{R}^n$

KEVIN McLEOD

*Dedicated to James Serrin on his 60th birthday*

## 1. Introduction

In this paper we discuss the asymptotic behavior, as $|x| \to \infty$, of solutions of the problem

$$\begin{gathered} \Delta u + f(u) = 0 \quad \text{in } \Omega \\ u \to 0 \quad \text{as } |x| \to \infty \\ u \in C^2(\Omega), \end{gathered} \tag{1}$$

where $\Omega$ is the complement of some closed ball in $\mathbb{R}^n$. We will always assume that $f$ is continuous and $f(0) = 0$. Since the behavior of $u$ is known if $f$ has a negative derivative at the origin (see [2]), we will concentrate on the case where $f(u) = o(|u|)$ as $u \to 0$. Our results will apply in particular to classical solutions of

$$\Delta u + f(u) = 0 \quad \text{in } \mathbb{R}^n \tag{2}$$

$$u \to 0 \quad \text{as } |x| \to \infty.$$

In section 2 we prove our main tool, a comparison lemma, and obtain general bounds on the asymptotic behavior.

In section 3 we consider the case when $f$ is *quasi-decreasing* at 0. By this, we mean that for some $\gamma > 0$,

$$f(u) \leqq 0 \quad \text{when } u \in (0, \gamma), \tag{3}$$

$$f(u) \geqq 0 \quad \text{when } u \in (-\gamma, 0). \tag{4}$$

(We say that $f$ is *quasi-increasing* at 0 if these inequalities are reversed). The case $f(u) \equiv -|u|^{q-1} u$ $(q > 1)$ was studied by VERON in [3]. His results applied not to classical solutions, but to solutions of the class described in [1]; for $n \geqq 3$, this class is

$$S \equiv \{u \in M^{\frac{n}{n-2}}(\mathbb{R}^n) \mid \Delta u \in L^1(\mathbb{R}^n)\}.$$

(The definition of $M^P(\mathbb{R}^n)$ can be found in the Appendix.) It will follow from our results that in many cases solutions of (1) can be modified on a compact set so that they belong to the class $S$. We may then apply VERON's results to obtain more accurate information in the case $f(u) \equiv -|u|^{q-1}u$.

For simplicity, we will always assume that $n \geqq 3$, although it will be clear that many of our results (including the comparison lemma) remain valid when $n = 2$.

## 2. Two Comparison Lemmas

In this section we will prove our basic comparison lemma (Lemma 2). We will not use Lemma 1 in the later sections, but we include it for completeness.

**Lemma 1.** *Let $\Omega$ be an open set in $\mathbb{R}^n$, and suppose $u, v \in C^2(\Omega)$ satisfy*

$$\Delta u + f(u) = 0,$$

$$\Delta v + g(v) = 0$$

*in $\Omega$. We assume that $f$ and $g$ are both defined on $E =$ range $(u) \cup$ range $(v)$ and, on $E$, one of the following conditions is satisfied:*

(i) *$g > f$, and either $f$ or $g$ is decreasing, or*
(ii) *$g \geqq f$, and either $f$ or $g$ is strictly decreasing.*
*Then $v - u$ cannot have a negative minimum in $\Omega$.*

**Proof.** Set $w = v - u$. If $w$ has a negative minimum at $x_0 \in \Omega$, then $u > v$ at $x_0$, and so in case (i) we obtain $f(u) \leqq f(v) < g(v)$, or $f(u) < g(u) \leqq g(v)$ (depending on whether $f$ or $g$ is decreasing). In case (ii), we have $f(u) < f(v) \leqq g(v)$, or $f(u) \leqq g(u) < g(v)$ and so, under any of our hypotheses, $f(u) < g(v)$ at $x_0$. Thus

$$\Delta w = f(u) - g(v) < 0$$

at $x_0$, which contradicts the fact that $w(x_0)$ is a minimum.

**Lemma 2.** *Let $\Omega$ be a domain in $\mathbb{R}^n$, and suppose $u, v \in C^2(\Omega) \cap C(\bar{\Omega})$ satisfy*

$$\Delta u + f(u) = 0,$$

$$\Delta v + g(v) = 0$$

*in $\Omega$, where $g \geqq f$ and either $f$ or $g$ is decreasing on $E =$ range $(u) \cup$ range $(v)$. If $v \geqq u$ on $\partial\Omega$ and*

$$\liminf_{\substack{|x| \to \infty \\ x \in \Omega}} (v - u) \geqq 0, \tag{5}$$

*then $v \geqq u$ in $\Omega$.*

**Proof.** Put $w = v - u$, and define $\Omega_- = \{x \in \Omega \mid w(x) < 0\}$. We assume for contradiction that $\Omega_-$ is non-empty. Since $w \geqq 0$ on $\partial\Omega$, while $\underline{\lim}\, w(x) \geqq 0$ as $|x| \to \infty$, $x \in \Omega$, it follows that $w$ has a negative interior minimum, say at $x_0 \in \Omega_-$. Note that if $x \in \Omega_-$, then $v(x) < u(x)$ and so either $g(v) \geqq f(v) \geqq f(u)$, or $g(v) \geqq g(u) \geqq f(u)$ (depending on whether $f$ or $g$ is decreasing). In either case, we have

$$\Delta w = f(u) - g(v) \leqq 0$$

in $\Omega_-$, and so $w$ is superharmonic in $\Omega_-$. Let $\Omega_1$ be the component of $\Omega_-$ which contains $x_0$. Then $w$ is superharmonic in $\Omega_1$, and attains an interior minimum at $x_0 \in \Omega_1$. It follows that $w$ is constant in $\Omega_1$:

$$w(x) = w(x_0) < 0, \qquad x \in \Omega_1. \tag{6}$$

Since $w \geqq 0$ on $\partial\Omega$, it follows from (6) that $\partial\Omega_1 \cap \partial\Omega = \emptyset$. If $x_1 \in \partial\Omega_1 \cap \Omega$ then $w(x_1) = 0$ (by continuity of $w$) and again (6) gives a contradiction. Thus $\partial\Omega_1 \cap \Omega = \emptyset$. Since $\partial\Omega_1 = (\partial\Omega_1 \cap \partial\Omega) \cap (\partial\Omega_1 \cap \Omega)$, we see that $\partial\Omega_1 = \emptyset$; in particular $\Omega_1$ is unbounded. We may thus pick a sequence $\{x_n\} \subset \Omega_1$ such that $|x_n| \to \infty$ as $n \to \infty$. By (6), $w(x_n) = w(x_0) < 0$ for all $n$, which contradicts (5).

Our assumption that $\Omega_- \neq \emptyset$ has led to a contradiction, and so we conclude that $\Omega_- = \emptyset$. Thus $w \geqq 0$, and $v \geqq u$ in $\Omega$.

As an example of the type of results that can be obtained using Lemma 2, we will find a general bound on solutions of (1). This result will be refined in section 3 when we consider more specific behavior of the nonlinearity $f$.

**Theorem 1** (quasi-decreasing $f$). *Let* $\Omega = \{x \in \mathbb{R}^n \mid |x| > R\}$, *and let $u$ be a solution of* (1). *Suppose that $f$ satisfies* (3) *and* (4) *for some* $\gamma > 0$. *Then there is a constant* $c_1 > 0$ *such that*

$$|u(x)| \leqq \frac{c_1}{|x|^{n-2}} \qquad \text{as } |x| \to \infty.$$

**Proof.** Since $u \to 0$ as $|x| \to \infty$, there is some $R_1 > R$ such that $|u(x)| < \gamma$ for $|x| > R_1$. Define $\Omega_1 = \{x \in \mathbb{R}^n \mid |x| > R_1\}$ and put

$$v(x) = \frac{c_1}{|x|^{n-2}}$$

where $c_1 > 0$ is chosen so that $v(x) \geqq |u(x)|$ on $\partial\Omega_1$.

Since $|u| < \gamma$ on $\Omega_1$, we may redefine $f$ arbitrarily for $|u| \geqq \gamma$ without affecting the equation $\Delta u + f(u) = 0$ on $\Omega_1$. We define $f$ so that $f$ remains continuous, $f(u) \leqq 0$ whenever $u > 0$ and $f(u) \geqq 0$ whenever $u < 0$. Now define $g(v)$ by

$$g(v) = \begin{cases} \sup\limits_{t \in [v,0]} f(t), & v < 0 \\ 0, & v \geqq 0 \end{cases}.$$

(Note that $g$ is continuous, since $f$ is.) Then $g \geqq f$ on range $(u|_{\bar{\Omega}_1}) \cup$ range $(v|_{\bar{\Omega}_1})$, $g$ is decreasing and $v$ satisfies $\Delta v + g(v) = 0$ on $\Omega_1$. Applying Lemma 2 on $\Omega_1$, we see that $v \geqq u$ on $\Omega_1$.

Since $-u$ satisfies the equation

$$\Delta\psi + \tilde{f}(\psi) = 0 \quad \text{in } \Omega_1,$$

where $\tilde{f}(\psi) \equiv -f(-\psi)$ satisfies the same conditions as $f$, we may apply Lemma 2 again to conclude that $v \geqq -u$ on $\Omega_1$.

In future applications of Lemma 2 we will not always extend the functions $f$ and $g$ so that they are defined and satisfy the conditions of the lemma over the entire set $E$. It will be clear by comparison with the previous proof how this can be done in any particular case.

In case $f$ is quasi-increasing at 0, there is a lower bound for positive solutions of (1). The proof is similar to that of Theorem 1 and is omitted.

**Theorem 2** (quasi-increasing $f$). *Let* $\Omega = \{x \in \mathbb{R}^n \mid |x| > R\}$ *and let $u$ be a positive solution of* (1). *Assume that* $f(u) \geqq 0$ *when* $u \in (0, \gamma)$ *for some* $\gamma > 0$. *Then there is a constant* $c_2 > 0$ *such that*

$$u(x) \geqq \frac{c_2}{|x|^{n-2}} \quad \text{as } |x| \to \infty.$$

*Remark*. The results of this section depend only on the maximum principle and the behavior of the fundamental solution of the Laplace operator. They can thus be easily generalized to solutions of the problem

$$Lu + f(u) = 0 \quad \text{in } \Omega$$

$$u \to 0 \quad \text{as } |x| \to \infty$$

$$u \in c^2(\Omega),$$

where $L \equiv a^{ij}(x)\, D^2_{ij} + b^i(x)\, D_i$ is a uniformly elliptic operator with sufficiently smooth coefficients.

### 3. f Quasi-Decreasing at 0

We will assume throughout this section that $f$ satisfies

$$f(u) \leqq 0 \qquad u \in (0, \gamma) \tag{3}$$

$$f(u) \geqq 0, \qquad u \in (-\gamma, 0) \tag{4}$$

for some $\gamma > 0$. We note, however, that in any result concerning positive solutions we need only assume (3), and extend $f$ to negative $u$ as an odd function. We will also impose conditions on $f$ of the form $|f(u)| \leqq k\,|u|^q$, for some $q > 1$.

After a change of scale in $\mathbb{R}^n$ we may assume that $k = 1$, and we will make this assumption in the statements of our theorems.

Following VERON [3, 4] we distinguish three cases.

*a. Weak nonlinearity*. We assume that $f$ satisfies

$$|f(u)| \leqq |u|^q \quad \text{in } (-\gamma, \gamma), \text{ for some } q > \frac{n}{q-2}. \tag{7}$$

**Theorem 3.** *Assume $f$ satisfies* (3) *and* (7), *and let $u$ be a positive solution of* (1). *Then there are constants $c_1, c_2 > 0$ such that*

$$\frac{c_1}{|x|^{n-2}} \geqq u(x) \geqq \frac{c_2}{|x|^{n-2}}.$$

**Proof.** We already know from Theorem 1 that $u(x) \leqq \frac{c_1}{|x|^{n-2}}$. In order to obtain the second inequality, we write the equation satisfied by $u$ as

$$\Delta u + K(x)\, u = 0,$$

where $K(x) = f(u(x))/u(x)$, and so $|K(x)| \leqq u(x)^{q-1} \leqq c_1^{q-1}\, |x|^{-(n-2)(q-1)}$ as $|x| \to \infty$.

Since $q > \frac{n}{n-2}$, we have

$$(n-2)\,(q-1) > (n-2)\left(\frac{n}{n-2} - 1\right) = 2$$

and so the desired lower bound on $u(x)$ follows from the Hopf Lemma ($\mathrm{H}_1'$) in [2].

**Corollary 1.** *Suppose that $f$ satisfies* (3) *and* (7) *with $q > \frac{n+1}{n-2}$, and that $f$ can be written in the form $f = f_1 + f_2$ with $f_1 \in C^1$ and $f_2$ continuous and non-decreasing. Then any positive $C^2$ solution of* (2) *is radially symmetric about some point in $\mathbb{R}^n$.*

**Proof.** We will apply Theorem 1 of [2]. Since $u(x) \leqq \frac{c_1}{|x|^{n-2}}$, we can take $m = n-2$ in that theorem. We must show that $q > \max\left(\frac{n+1}{m}, \frac{2}{m} + 1\right) = \max\left(\frac{n+1}{n-2}, \frac{n}{n-2}\right) = \frac{n+1}{n-2}$, which is precisely our hypothesis on $q$.

We remark that it was proved in Theorem 2 of [2] that if $f'(0) < 0$ then any positive, $C^2$ solution of (2) is radial. It is an open question whether this result is true for any function $f$ which is decreasing near the origin.

In case $f$ is decreasing, we may apply a result of VERON to obtain further information.

**Corollary 2.** *Suppose $f$ satisfies* (7) *and $f$ is decreasing on some interval $(-\gamma, \gamma)$. Then $|x|^{n-2} u(x)$ converges to some real number as $|x| \to \infty$. This limit is positive if and only if $u$ is ultimately positive.*

**Proof.** By Theorem 1.1 of **[4]**, $|x|^{n-2} u(x)$ converges to some real number $l$. If $u(x_n) \leqq 0$ for some sequence $x_n$ with $|x_n| \to \infty$, then $l \leqq 0$. If $u > 0$ in the complement of some ball then, by Theorem 3, $l \geqq c_2 > 0$.

*b. Strong non-linearity.* We now assume that $f$ satisfies (3), (4) and

$$|f(u)| \geqq |u|^q \quad \text{in } (-\gamma, \gamma), \quad 1 < q < \frac{n}{n-2}. \tag{8}$$

Our results will depend strongly on the fact that for $q > 1$ the function $v(x)$ defined by

$$v(x) \equiv \lambda |x|^{\overline{-q-1}^{\,2}} \quad (\lambda > 0)$$

is a solution of

$$\Delta v - \lambda^{-(q-1)} \left(\frac{2}{q-1}\right) \left(\frac{2q}{q-1} - n\right) v^q = 0$$

in $\mathbb{R}^n \setminus \{0\}$. Denoting the coefficient of $v^q$ in this last equation by $-C_\lambda$, we see that for fixed $q \in \left(1, \frac{n}{n-2}\right)$, $C_\lambda$ is a positive, decreasing function of $\lambda$ and $C_\lambda \to 0^+(+\infty)$ as $\lambda \to +\infty(0^+)$.

Since the behavior of $u(x)$ at $\infty$ depends on the behavior of $f(u)$ near $u = 0$ (as can be seen from the explicit solution $v(x)$ exhibited above), we will have to bound $f(u)$ from below in order to obtain lower bounds on positive solutions of (1).

**Theorem 4.** *Suppose that $f$ satisfies* (3), (4) *and* (8), *and let $u$ be a solution of* (1). *Then there is a constant $c_1 > 0$ such that*

$$|u(x)| \leq \frac{c_1}{|x|^{2/(q-1)}} \quad \text{as } |x| \to \infty.$$

*If $u > 0$ and $f$ satisfies* (3) *and*

$$f(u) \geqq -u^q, \quad 1 < q < \frac{n}{n-2} \tag{9}$$

*on $(0, \gamma)$, then there is a constant $c_2 > 0$ such that*

$$u(x) \geqq \frac{c_2}{|x|^{2/(q-1)}} \quad \text{as } |x| \to \infty.$$

**Proof.** Put $v(x) = \lambda |x|^{-2/(q-1)}$, and let $g(v) = -C_\lambda v^q$ for $v \geqq 0$, where

$$C_\lambda = \lambda^{-(q-1)} \left(\frac{2}{q-1}\right) \left(\frac{2q}{q-1} - n\right).$$

Choose $R_1 > R$ (where $\Omega = \{x \mid |x| > R\}$) so large that $|u(x)| < \gamma$ in $\Omega_1 = \{x \in \mathbb{R}^n \mid |x| > R_1\}$, and choose $\lambda > 0$ so large that $v(x) \geqq |u(x)|$ on $\partial\Omega_1$ and $C_\lambda < 1$. Then by (8), $g \geqq f$ on $(0, \gamma)$ and we extend $g$ and $f$ (as in the proof of Theorem 1) so that this inequality remains true on range $(u|_{\bar{\Omega}_1}) \cup$ range $(v|_{\bar{\Omega}_1})$, with $g$ decreasing. By Lemma 2, $u \leqq v$ on $\Omega_1$. The estimate $-u \leqq v$ in $\Omega_1$ follows from the observation that $-u$ satisfies an equation of the same type as $u$.

In case $u > 0$ and $f$ satisfies (3) and (9), we may choose $\lambda$ so small that $v \leqq u$ on $\partial\Omega_1$, while $g \leqq f$ on $(0, \lambda)$. Lemma 2 (with $u$ and $v$ interchanged) then shows that $v \leqq u$ on $\Omega_1$.

If $f(u) \equiv -|u|^{q-1}u$ near $u = 0$, we can apply VERON's results in [3] to obtain a more accurate description of the asymptotic behavior. Before doing so, however, we must verify that our classical solution $u$ lies in he space of functions to which the results of [3] apply; that is $u \in M^{\frac{n}{n-2}}(\mathbb{R}^n)$, $\Delta u \in L^1(\mathbb{R}^n)$. (At least after $u$ has been modified on a compact set.)

Let $u$ be a solution of (1), and choose a function $\theta \in C^\infty(\mathbb{R})$ such that $\theta(r) = 0$ if $r < \frac{1}{2} R_1$ and $\theta(r) = 1$ if $r > R_1$, where $R_1 > 2R$ is chosen so large that $|u(x)| < \gamma$ if $|x| > R_1$. We put $\tilde{u}(x) = \theta(|x|)\,u(x)$, and define $\psi(x)$ by the equation

$$\Delta\tilde{u} - |\tilde{u}|^{q-1}\tilde{u} = \psi(x). \tag{10}$$

If $|x| > R_1$, then $\tilde{u}(x) = u(x)$, $|u(x)| < \gamma$ and $f(u) = |u|^{q-1}u$. It follows that $\Delta\tilde{u} - |\tilde{u}|^{q-1}\tilde{u} = 0$ if $|x| > R_1$ and so $\operatorname{supp}(\psi) \subset \overline{B(0, R_1)}$. Since $\tilde{u} \in C^2(\mathbb{R}^n)$, $\psi \in C_0(\mathbb{R}^n)$ and so equation (10) is of the type considered in [1]. We will now show that $\tilde{u}$ is the solution of (10) as defined in that paper.

It follows from Theorem 1 that for some constant $c > 0$,

$$|\tilde{u}(x)| \leqq \frac{c}{|x|^{n-2}},$$

and so by the Lemma in the Appendix, $\tilde{u} \in M^{\frac{n}{n-2}}(\mathbb{R}^n)$. (We may note that this result applies to any classical solution of (1) whenever $f$ satisfies (3) and (4)). Next from (10),

$$|\Delta\tilde{u}| \leqq |\psi(x)| + |\tilde{u}|^q.$$

From Theorem 4 we obtain

$$|\tilde{u}|^q \leqq \frac{c}{|x|^{2q/(q-1)}} \quad \text{as } |x| \to \infty.$$

Since $\tilde{u}$ is bounded and $2q/(q-1) > n$ if $1 < q < n/(n-2)$, this shows that $\tilde{u}^q \in L^1(\mathbb{R}^n)$. Since also $\psi \in L^1(\mathbb{R}^n)$, we see that $\Delta u \in L^1(\mathbb{R}^n)$. We may thus apply Theorem 2.1 of [3] and obtain

**Corollary 3.** *Suppose that* $f(u) \equiv -|u|^{q-1}u$ *in some interval* $(-\gamma, \gamma)$, *and that* $u$ *is a solution of* (1). *Then if*

$$\frac{n+1}{n-1} \leqq q < \frac{n}{n-2},$$

$|x|^{\frac{2}{q-1}} u(x)$ *converges, as* $|x| \to \infty$, *to an isotropic limit which can only take the values* 0 *or* $\pm l_{q,n}$, *where*

$$l_{q,n} = \left(\left(\frac{2}{q-1}\right)\left(\frac{2q}{q-1} - n\right)\right)^{\frac{1}{q-1}}.$$

*If* $u > 0$ *and* $1 < q < \frac{n}{n-2}$, *then* $|x|^{\frac{2}{q-1}} u(x) \to l_{q,n}$ *as* $|x| \to \infty$.

(Veron does not explicitly state that the limit must be $l_{q,n}$ in the case $u > 0$, but it follows easily from his arguments, or from our Theorem 4.)

*c. Critical non-linearity.* We still assume that $f$ satisfies (3) and (4), and we impose one of the conditions

$$|f(u)| \geqq |u|^{\frac{n}{n-2}} \quad \text{on } (-\gamma, \gamma) \tag{11}$$

or

$$|f(u)| \leqq |u|^{\frac{n}{n-2}} \quad \text{on } (0, \gamma), \tag{12}$$

for some $\gamma > 0$.

**Theorem 5.** *Suppose* $f$ *satisfies* (3), (4) *and* (11), *and let* $u$ *be a solution of* (1). *Then there is a constant* $c_1 > 0$ *such that*

$$|u(x)| \leqq \frac{c_1}{|x|^{n-2} (\log |x|)^{\frac{n-2}{2}}} \quad \text{as } |x| \to \infty.$$

*If* $u > 0$, *and* $f$ *satisfies* (3) *and* (12), *then there is a constant* $c_2 > 0$ *such that*

$$u(x) \geqq \frac{c_2}{|x|^{n-2} (\log |x|)^{\frac{n-2}{2}}} \quad \text{as } |x| \to \infty.$$

**Proof.** Choose $R_1 > \max(1, R)$ and so large that $|u(x)| < \lambda$ for $|x| > R_1$ and

$$\tfrac{1}{2}(n-2)^2 \leqq \tfrac{1}{2}(n-2)^2 + \frac{n(n-2)}{4 \log r} \leqq (n-2)^2 \quad \text{when } r > R_1. \tag{13}$$

If $\lambda > 0$, put

$$v_\lambda(r) = \lambda r^{2-n} (\log r)^{\frac{2-n}{2}}, \quad r \geqq R_1$$

and define $r_\lambda(v)$ by

$$r_\lambda(v) = \begin{cases} \text{inverse function to } v_\lambda(r), & v \in (0, v_\lambda(R_1)) \\ R_1, & v \geqq v_\lambda(R_1). \end{cases}$$

Note that for any $\lambda > 0$, $v_\lambda(r)$ is decreasing on $[R_1, \infty)$ and so $r_\lambda$ is decreasing on $(0, \infty)$. Also, for any $\lambda > 0$, $r_\lambda \geqq R_1$ on $(0, \infty)$.

In $\Omega_1 = \{x \mid |x| > R_1\}$, $v_\lambda$ satisfies the equation $\Delta v + g_\lambda(v) = 0$, where

$$g_\lambda(v) = -\lambda^{-\frac{2}{n-2}} \left[\tfrac{1}{2}(n-2)^2 + \frac{n(n-2)}{4 \log r_\lambda(v)}\right], \quad v > 0.$$

Since $r_\lambda$ is a decreasing function of $v$ on $(0, \infty)$, so is $g_\lambda$. Since $r_\lambda(v) \to \infty$ as $v \downarrow 0$, we see that $g_\lambda(v) \to 0$ as $v \downarrow 0$, and so $g$ may be extended to negative values of $v$ (as in the proof of Theorem 1), and will in fact be continuous for such $v$. Finally, since $r_\lambda \geqq R_1$ for all $\lambda > 0$, we see from (13) that

$$-(n-2)^2 \lambda^{-\frac{2}{n-2}} v^{\frac{n}{n-2}} \leqq g_\lambda(v) \leqq -\tfrac{1}{2}(n-2)^2 \lambda^{-\frac{2}{n-2}} v^{\frac{n}{n-2}}$$

on $(0, \infty)$. Since the power of $\lambda$ in these inequalities is negative, the rest of the proof may proceed as in Theorem 4: if $f$ satisfies (11), choose $\lambda$ large to obtain the bounds in $|u(x)|$, while if $f$ satisfies (12), and $u > 0$, choose $\lambda$ small and positive to bound $u(x)$ from below.

Once again, we can obtain a more accurate result in case $f(u) \equiv -|u|^{\frac{2}{n-2}} u$ near $u = 0$ by applying VERON's results. We first modify $u$ on a compact set, as in the remarks leading up to Corollary 3, to a function $\tilde{u} \in C^2(\mathbb{R}^n)$ which satisfies

$$\Delta \tilde{u} - |\tilde{u}|^{\frac{2}{n-2}} \tilde{u} = \psi(x) \tag{14}$$

where $\psi \in C_0(\mathbb{R}^n)$. By Theorem 1, $\tilde{u} \in M^{\frac{n}{n-2}}(\mathbb{R}^n)$. By Theorem 5,

$$|\tilde{u}(x)|^{\frac{n}{n-2}} \leqq \frac{c_1}{|x|^n (\log |x|)^{\frac{n}{2}}} \quad \text{as } |x| \to \infty,$$

and since $n \geqq 3$,

$$\int_{|x|>R} |x|^{-n} (\log |x|)^{-\frac{n}{2}} dx = c_n \int_R^\infty \frac{1}{r} (\log r)^{-\frac{n}{2}} dr < \infty.$$

It now follows easily from (14) that $\Delta \tilde{u} \in L^1(\mathbb{R}^n)$, and so we may apply Theorem 3.1 of [3] to obtain

**Corollary 4.** *Suppose that* $f(u) \equiv -|u|^{\frac{2}{n-2}} u$ *on some interval* $(-\gamma, \gamma)$, *and that* $u$ *is a solution of* (1). *Then* $|x|^{n-2} (\log |x|)^{\frac{n-2}{n}} u(x)$ *converges, as* $|x| \to \infty$, *to an isotropic limit which can only take the values* 0 *or* $\pm((n-2)/\sqrt{2})^{n-2}$. *If* $u > 0$, *then*

$$|x|^{n-2}(\log|x|)^{\frac{n-2}{2}}u(x)\to\left(\frac{n-2}{\sqrt{2}}\right)^{n-2}\qquad\text{as } |x|\to\infty.$$

*Remarks.* Our results indicate that when $f$ is quasi-decreasing, the behaviour of a positive solution of (1) as $|x|\to\infty$ is determined solely by the behaviour of $f(u)$ near $u=0$. This is not true for arbitrary $f$. There is an example of a function $f$ such that

(i) $f\in C^1([0,+\infty))$

(ii) $f(u)\equiv u^q$, $q>(n+2)/(n-2)$, on some interval $(0,\gamma)$, and $f$ is strictly monotonic increasing on $(0,\infty)$.

(iii) The problem (2) has two positive, radial solutions $u_1$ and $u_2$, with the asymptotic behaviors

$$c_1|x|^{2-n}\leqq u_1(x)\leqq c_2|x|^{2-n},$$

$$u_2(x)\equiv c|x|^{-2/(q-1)},$$

as $|x|\to\infty$.

## Appendix

We collect here those results about the spaces $M^p(\mathbb{R}^n)$ which we use in section 2. These facts and much other information may be found in the appendix to [1].

**Definition.** Let $u$ be a measurable function on $\mathbb{R}^n$, $1<p<\infty$ and $1/p'+1/p=1$. Then

$$\|u\|_{M_p}=\inf\left\{C\in[0,\infty]\;\middle|\;\int_K|u(x)|\,dx\leqq C\,(\text{meas }K)^{1/p'}\right.$$
$$\left.\text{for all measurable } K\subset\mathbb{R}^n\right\}.$$

$M^p(\mathbb{R}^n)$ is the set of measurable functions $u$ on $\mathbb{R}^n$ for which $\|u\|_{M_p}<\infty$.

$M^p(\mathbb{R}^n)$ is a Banach space with the norm $\|\cdot\|_{M_p}$, and (Hölder's inequality) $L^p(\mathbb{R}^n)\subset M^p(\mathbb{R}^n)$. That this inclusion is strict is shown by the following lemma.

**Lemma** (Lemma A.3 of [1]). *For $n>\alpha>0$, the function $|x|^{-\alpha}$ lies in $M^{n/\alpha}(\mathbb{R}^n)$. In particular, if $u\in L^\infty(\mathbb{R}^n)$ satisfies*

$$|u(x)|\leqq\frac{C}{|x|^{n-2}},\qquad|x|\to\infty$$

*then $u\in M^{\frac{n}{n-2}}(\mathbb{R}^n)$.*

*Acknowledgment*. This paper formed a part of my thesis, completed under the direction of JAMES SERRIN. It is a pleasure to be able to thank him for his support and friendship, both during my time as his student and after.

## Bibliography

1. BÉNILAN, PH., H. BRÉZIS, & M. C. CRANDALL, A semi-linear elliptic equation in $L^1(\mathbb{R}^n)$, *Ann. Scuola Norm. Sup. Pisa* **2**, 523–555 (1975).
2. GIDAS, B., W.-M. NI, & L. NIRENBERG, Symmetry of positive solutions of nonlinear elliptic equations in $\mathbb{R}^n$, *Math. Anal. Appl.* Part A. **7A**, 369–402 (1981).
3. VERON, L., Comportement asymptotique des solutions d'équations elliptiques semilinéaires dans $\mathbb{R}^n$, *Ann. Mat. Pura Appl.* **127**, 23–50 (1981).
4. VERON, L., Global behaviour and symmetry properties of singular solutions of nonlinear elliptic equations, preprint.

Department of Mathematics
Indiana University
Bloomington

*(Received March 21, 1986)*

# *Existence Theorems Concerning Simple Integrals of the Calculus of Variations for Discontinuous Solutions*

L. CESARI, P. BRANDI & A. SALVADORI

*Dedicated to James Serrin on his 60th birthday*

## 1. Introduction

In this paper we apply the direct method of the calculus of variations, based on lower semicontinuity and lower closure (see [5]), to prove the existence of optimal solutions $x(t) = (x^1, \ldots, x^n)$, $t_1 \leqq t \leqq t_2$, for which $\alpha$ components $y(t) = (x^1, \ldots, x^\alpha)$ are AC and $n - \alpha$ components $z(t) = (x^{\alpha+1}, \ldots, x^n)$ are BV and not necessarily AC. If $\alpha = 0$ all components of $x$ are BV, and in this situation no growth assumption is made on the integrand function. The cost functional $\mathscr{I}$ is of SERRIN type ([14]), *i.e.* it is obtained from the usual integral expression $I$ by means of a limit process, based on a topology $\tau$, of this integral $I$ over curves $x_k$ whose components are all AC. The topology $\tau$ that we use here is the topology of uniform convergence on the $y_k$ components and pointwise convergence almost everywhere on the $z_k$ components. This pointwise convergence almost everywhere has been used by CESARI in the study of area of discontinuous surfaces ([4], 1936) and in existence theorems concerning simple integrals for AC solutions (see [5], Chapt. 15 and the papers cited there).

In Section 2 we first prove a closure theorem (Theorem 1) for problems in which mere pointwise convergence almost everywhere is adopted. The closure theorem is used, in Section 3, for proving a lower semicontinuity theorem (Theorem 1′) based on the topology $\tau$. The same lower semicontinuity theorem allows us to prove that $\mathscr{I}$ is a true extension of $I$, in the sense that $\mathscr{I} = I$ whenever all components are AC.

In Section 3 we prove also an existence theorem for the absolute minimum of extended problems of the calculus of variations with constraints on the direction of the tangent. In Section 4 we derive, as a corollary, an existence theorem for the absolute minimum of problems of optimal control. For a different viewpoint connecting Serrin-type integrals, usual integrals, and Burkill-Cesari integrals we mention the paper [3] by CANDELORO & PUCCI, where also lower semicontinuity theorems are given for solutions which are only continuous and of bounded variation.

Elsewhere ([6b]) the present work on discontinuous solutions will be extended to multiple integrals of the calculus of variations and functions of $\nu > 1$ independent variables. Therefore, the BV concept in [6b] will be the one introduced by CESARI in 1936 ([4]) and shown by KRICKEBERG ([12]) to be equivalent to the one in terms of distributions. Later the functions of bounded variation in the sense of CESARI were briefly denoted as BVC by CONWAY & SMOLLER ([8]), DAFERMOS ([9]) and DIPERNA ([10]). The functions of bounded variation defined in the equivalent terms of distributions were briefly denoted as BV by VOLPERT ([16]) and others. In order that the present work, which concerns functions of one variable, be in harmony with [6b], we use the notations from [6b].

## 2. A closure theorem with components converging only pointwise

Let $A$ be a subset of the $(t, x)$-space $R^{n+1}$ whose projection on the $t$-axis contains the fixed interval $[t_1, t_2]$. Let $Q(t, x)$, $(t, x) \in A$, $Q(t, x) \subset R^n$, or $Q: A \to R^n$, be a given set valued function.

Following CESARI [5] we shall say that the set function $Q$ has *property* $(Q)$ at the point $(\bar{t}, \bar{x})$, with respect to $(t, x)$, if

$$Q(\bar{t}, \bar{x}) = \bigcap_{\delta > 0} \operatorname{cl} \operatorname{co} \bigcup_{(t,x) \in B(\bar{t}, \bar{x}; \delta)} Q(t, x)$$

where $B(\bar{t}, \bar{x}; \delta) = \{(t, x) \in A : |(t, x) - (\bar{t}, \bar{x})| \leqq \delta\}$. Let $Q(\bar{t}, \bar{x}; \delta) = \bigcup Q(t, x)$ for $(t, x) \in B(\bar{t}, \bar{x}; \delta)$.

Analogously, $Q$ is said to have *property* $(Q)$ at the point $(\bar{t}, \bar{x})$, with respect to $x$ only, if

$$Q(\bar{t}, \bar{x}) = \bigcap_{\delta > 0} \operatorname{cl} \operatorname{co} \bigcup_{x \in B'(\bar{t}, \bar{x}; \delta)} Q(\bar{t}, x)$$

where $B'(\bar{t}, \bar{x}; \delta) = \{(t, x) \in A : |x - \bar{x}| \leqq \delta\}$. The corresponding Kuratowski properties (K) are obtained by writing only cl, instead of cl co, in the relations above.

We mention here that a summable function $x(t)$ from $[t_1, t_2]$ into $R^n$, or $x: [t_1, t_2] \to R^n$, is said to be of *bounded variation in the sense of Cesari*, briefly BVC, if it is equivalent to a BV function $\tilde{x}: [t_1, t_2] \to R^n$. It may well occur that $x$ is equivalent to infinitely many BV functions $\tilde{x}$. In this case, at every point $t_0 \in (t_1, t_2)$ of (first kind) discontinuity for $\tilde{x}$, we may take $\tilde{x}(t_0)$ so that $\tilde{x}(t_0 - 0) \leqq \tilde{x}(t_0) \leqq \tilde{x}(t_0 + 0)$ or the same relations with the sign $\geqq$. Also, we may take $\tilde{x}(t_1) = \tilde{x}(t_1 + 0)$, $\tilde{x}(t_2) = \tilde{x}(t_2 - 0)$. With this choice for $\tilde{x}$ the variation $V(\tilde{x})$ is uniquely determined and it has the minimum value for all $\tilde{x}$ equivalent to $x$. We take, by definition of *generalized variation* $V^*(x)$ the number $V^*(x) = V(\tilde{x})$, for $\tilde{x}$ chosen as stated. Moreover we take, by definition $x' = \tilde{x}'$ (a.e. in $[t_1, t_2]$).

Analogously, $x$ is said to be *absolutely continuous in the generalized sense*, briefly ACg if $x$ is equivalent to an AC function $\tilde{x}$. In this case $\tilde{x}$ is uniquely defined, and for the generalized variation we take $V^*(x) = V(\tilde{x})$.

For further properties of such functions see [4], [6a], [1], [2], [13].

We shall consider the orientor field equation

$$(t, x(t)) \in A, \quad x'(t) \in Q(t, x(t)), \quad \text{a.e. in } [t_1 \ t_2]. \tag{1}$$

that is, the problem of determining a BVC function $x$ satisfying these relations.

We state and prove now a closure theorem which replaces, in the present situation, the closure theorem 15.2.i of [5].

**Theorem 1** (A closure theorem). *Let us assume that* (i) *A is closed*; (ii) *the set valued function Q has closed and convex values*; (iii) *the set valued function Q has property* (Q), *with respect to* $(t, x)$, *at every point* $(\bar{t}, \bar{x}) \in A$, *with the exception perhaps of a set of points whose t-coordinate lies in a set H of measure zero in* $[t_1, t_2]$.

*Let* $x_k : [t_1, t_2] \to R^n$, $k \in N$, *be a sequence of* ACg *solutions of the orientor field* (3), *and assume that* $V^*(x_k) \leqq V_0$, $k \in N$, *and that* $x_k \to x$ *pointwise a.e. in* $[t_1, t_2]$, *with* $x \in$ BVC. *Then the function x is a solution of the orientor field relation* (3).

**Proof.** (a) By the hypotheses it follows that

$$(t, x(t)) \in A, \quad \text{a.e. in } [t_1 \ t_2], \tag{2}$$

so we have only to prove that

$$x'(t) \in Q(t, x(t)) \cdot \quad \text{a.e. in } [t_1, t_2]. \tag{3}$$

In order to see that, without loss of generality we can suppose that $x$ is BV and $x_k$ is AC, $k \in N$.

Let $T_0 \subset [t_2, t_2]$ be a set of measure zero such that in $[t_1, t_2] - T_0$ we have

$$\lim_{k \to \infty} x_k(t) = x(t) \quad \text{and} \quad x'(t) = x_a'(t)$$

where $x = x_a + x_s$ denotes the Jordan decomposition of $x$.

(b) Now for every $m \in N$, we divide $[t_1, t_2]$ into $m$ equal parts $I_r^{(m)}$, $r = 1, \ldots, m$, each of length $(t_2 - t_1) m^{-1} = Tm^{-1}$; and denote by $T_1 \subset [t_1, t_2]$ the set of all points of subdivision, so that $T_1$ is denumerable and, therefore, has measure zero.

Let $m \in N$ and $\varepsilon > 0$ be fixed. For every $k \in N$, we consider those intervals $I_r^{(m)}$ if any, such that $\omega(x_k, I_r^{(m)}) \geqq \varepsilon$, where $\omega(x_k, I)$ denotes the oscillation of $x_k$ over $I$.

Let $S_k^{(m)}$ be the system of such intervals, or

$$S_k^{(m)} = \{I_r^{(m)}, r = 1, \ldots, m : \omega(x_k, I_r^{(m)}) \geqq \varepsilon\}.$$

We now proceed to the determination of a suitable set $\Sigma^{(m)}$ and to the extraction of a suitable subsequence of $(x_k)_{k \in N}$. First, if $I_1^{(m)} \in S_k^{(m)}$ for all $k$ sufficiently large, we put $I_1^{(m)}$ in $\Sigma^{(m)}$; if not then there are infinitely many $k \in N$ such that $I_1^{(m)} \notin S_k^{(m)}$ and we denote by $(k_{1s})_{s \in N}$ such sequence, i.e. $I_1^{(m)} \notin S_{k_{1s}}^{(m)}$, $s \in N$. If $I_2^{(m)} \in S_{k_{1s}}^{(m)}$ for all $s$ sufficiently large, we put $I_2^{(m)}$ in $\Sigma^{(m)}$; if not then there are infinitely many $s \in N$ such that $I_2^{(m)} \notin S_{k_{1s}}^{(m)}$. We denote such a sequence by $(k_{2s})_{s \in N}$; then $(k_{2s}) \subset (k_{1s})$ and $I_2^{(m)} \notin S_{k_{2s}}^{(m)}$, $s \in N$.

We proceed as indicated for $I_3^{(m)}, \ldots, I_m^{(m)}$.

At the end we have a set $\Sigma^{(m)}$ made up of all points of certain intervals $I_r^{(m)}$, $r = 1, \ldots, m$, say, for simplicity $\Sigma^{(m)} = \{I_i^{(m)}, i = 1, \ldots, \nu\}$, and a final sequence $(k_{ms})_{s \in N}$ with $(k_{ms})_{s \in N} \subset (k_{m-1,s})_{s \in N} \subset \cdots \subset (k_{1s})_{s \in N}$.

Note that for all $k_{ms}$ sufficiently large we have

$$\omega(x_{k_{ms}}, I_i^{(m)}) \geqq \varepsilon, \quad i = 1, \ldots, \nu,$$

and hence

$$\nu\varepsilon \leqq \sum_{i=1}^{\nu} \omega(x_{k_{ms}}, I_i^{(m)}) \leqq \sum_{r=1}^{m} \omega(x_{k_{ms}}, I_r^{(m)}) \leqq V(x_{k_{ms}}) \leqq V_0,$$

i.e. $\nu \leqq V_0 \varepsilon^{-1}$.

This implies that

$$\text{meas}\,(\Sigma^{(m)}) = \sum_{i=1}^{\nu} \text{meas}\,(I_i^{(m)}) = \nu T m^{-1} \leqq V_0 T/\varepsilon m.$$

Hence, for every $\varepsilon > 0$ we can choose an integer $m_\varepsilon$ sufficiently large that

$$\text{meas}\,(\Sigma^{(m_\varepsilon)}) \leqq V_0 T/\varepsilon m_\varepsilon < \varepsilon.$$

Now we take $\varepsilon$ ranging in succession over the values $(1/2^\lambda)_{\lambda \in N}$. Thus, for $\lambda = 1$ then $\varepsilon = 1/2$ and, starting from the original sequence $(k)_{k \in N}$, we obtain from the above an integer $m_\varepsilon$, which we denote by $m_1$, a set $\Sigma^{(m_\varepsilon)}$, which we denote by $\Sigma^{(1)}$, and a subsequence $(k_{ms})_{s \in N}$ that we denote by $(k_s^1)_{s \in N}$.

For $\lambda = 2$ then $\varepsilon = 1/2^2$ and, starting from the sequence $(k_s^1)_{s \in N}$ we obtain, as before, an integer $m_2$, a set $\Sigma^{(2)}$ and a sequence $(k_s^2)_{s \in N}$.

Proceeding as indicated for the generic $\lambda \in N$, we see that $\varepsilon = 1/2^\lambda$ and, starting from the sequence $(k_s^{\lambda-1})_{s \in N}$, we obtain an integer $m_\lambda$, a set $\Sigma^{(\lambda)}$ and a subsequence $(k_s^\lambda)_{s \in N}$ as before.

It is not restrictive to assume that $(m_\lambda)_{\lambda \in N}$ is an increasing sequence. We consider now the sets

$$\Sigma_n = \bigcup_{\lambda=n}^{\infty} \Sigma^{(\lambda)}, \quad n \in N \quad \text{and} \quad \Sigma_0 = \bigcap_{n \in N} \Sigma_n.$$

We have,

$$\text{meas}\,\Sigma_n \leqq \sum_{\lambda=n}^{\infty} \text{meas}\,(\Sigma^{(\lambda)}) \leqq \sum_{\lambda=n}^{\infty} 1/2^\lambda = 1/2^{n-1}1, \qquad \text{meas}\,\Sigma_0 = 0.$$

(c) Let us now take any point $t_0 \in [t_1\ t_2] - (\Sigma_0 \cup T_0 \cup T_1 \cup H)$. Then there is a real $\sigma > 0$ and an integer $n_0$ such that $t_1 < t_0 - \sigma < t_0 < t_0 + \sigma < t_2$ and $t_0 \notin \Sigma^{(\lambda)}$ for every $\lambda \geqq n_0$.

For every given $\varepsilon > 0$ we take $\lambda \in N$ sufficiently large that $1/2^\lambda < \varepsilon/2$ and $t_0 \notin \Sigma^{(\lambda)}$. Consequently $t_0 \in (I^{(\lambda)})^0$ with $I^{(\lambda)} \notin S_{k_s^\lambda}^{(\lambda)}$, $s \in N$; hence

$$\omega(x_{k_s^\lambda}, I^{(\lambda)}) < 1/2^\lambda < \varepsilon/2 \quad \text{for every } s \in N. \tag{4}$$

Since $\lambda$ is fixed now, for simplicity we shall write $(k_s^\lambda)_{s\in N} = (k_s)_{s\in N}$. For every $0 < h < \sigma$ we consider the averages

$$m_h = h^{-1} \int_0^h x'(t_0 + \tau)\, d\tau = h^{-1}[x_a(t_0 + h) - x_a(t_0)]$$

$$m_{k_s h} = h^{-1} \int_0^h x'_{k_s}(t_0 + \tau)\, d\tau = h^{-1}[x_{k_s}(t_0 + h) - x_{k_s}(t_0)].$$

Now, for an arbitrary fixed $\eta > 0$ and for all $0 < h < \sigma$ sufficiently small we have

$$|m_h - x'_a(t_0)| < \eta/2 \quad \text{and} \quad |x_s(t_0 + h) - x_s(t_0)| < h\eta/4. \tag{5}$$

Thus we fix $0 < h < \min(\varepsilon, \sigma)$ in such a way that relation (5) holds and moreover $t_0 + h \notin T_0 \cup T_1$ and $[t_0, t_0 + h] \subset I^{(\lambda)}$. From (4), for every $t_0 \leqq t \leqq t_0 + h$, we have

$$|x_{k_s}(t) - x_{k_s}(t_0)| \leqq \omega(x_{k_s}, I^{(\lambda)}) < \varepsilon/2, \qquad s \in N. \tag{6}$$

Since $t_0, t_0 + h \notin T_0$, we can find an integer $s$ such that we have

$$|x_{k_s}(t_0) - x(t_0)| \leqq \min\{\varepsilon/2, \eta h/8\} \tag{7}$$

and

$$|x_{k_s}(t_0 + h) - x(t_0 + h)| < \eta h/8.$$

Therefore from (6) and (7), for every $t_0 \leqq t \leqq t_0 + h$, we have

$$|x_{k_s}(t) - x(t_0)| \leqq |x_{k_s}(t) - x_{k_s}(t_0)| + |x_{k_s}(t_0) - x(t_0)| < \varepsilon/2 + \varepsilon/2 = \varepsilon. \tag{8}$$

By hypothesis we have

$$x'_{k_s}(t) \in Q(t, x_{k_s}(t)) \qquad \text{for a.a. } t_0 \leqq t \leqq t_0 + h;$$

hence, from (8) and because $h < \varepsilon$, we have

$$x'_{k_s}(t) \in \operatorname{cl}\operatorname{co} \bigcup_{\substack{t \in [t_0, t_0 + \varepsilon] \\ x \in B(x(t_0), \varepsilon)}} Q(t, x) = \operatorname{cl}\operatorname{co} Q(t_0, x(t_0), \varepsilon) \tag{9}$$

for a.a. $t_0 \leqq t \leqq t_0 + h$.

Finally we observe that the average $m_{k_s h}$ is also a point of the same closed and convex set ([5], p. 288), i.e.

$$m_{k_s h} \in \operatorname{cl}\operatorname{co} Q(t_0, x(t_0), \varepsilon). \tag{10}$$

Now by virtue of (7) and (5) we derive that

$$\begin{aligned} |m_{k_s h} - m_h| &= h^{-1} |(x_{k_s}(t_0 + h) - x_{k_s}(t_0)) - (x_a(t_0 + h) - x_a(t_0))| \\ &\leqq h^{-1} |x_{k_s}(t_0 + h) - x(t_0 + h)| + h^{-1} |x_{k_s}(t_0) - x(t_0)| \\ &\quad + h^{-1} |x_s(t_0 + h) - x_s(t_0)| \\ &\leqq h^{-1} \eta h/8 + h^{-1} \eta h/8 + h^{-1} \eta h/4 = \eta/2. \end{aligned} \tag{11}$$

Thus, by (5) and (11), it follows that,

$$|x_a'(t_0) - m_{k_sh}| \leqq |x_a'(t_0) - m_h| + |m_h - m_{k_sh}| \leqq \eta/2 + \eta/2 = \eta; \tag{12}$$

and, from (10) and (12), that

$$x_a'(t_0) = x'(t_0) \in [\text{cl co}\, Q(t_0, x(t_0), \varepsilon)]_\eta .$$

Because $\eta$ is arbitrary, it follows that for every $\varepsilon > 0$

$$x'(t_0) \in \text{cl co}\, Q(t_0, x(t_0), \varepsilon) . \tag{13}$$

Now the function $Q$ satisfies property (Q) at $(t_0, x(t_0))$; hence from (13) we derive

$$x'(t_0) \in \bigcap_{\varepsilon > 0} \text{cl co}\, Q(t_0, x(t_0), \varepsilon) = Q(t_0, x(t_0)) .$$

This completes the proof of (2).

*Remark 1.* In Theorem 1 condition (Q) cannot be replaced by the weaker condition (K) as the following example, from Section 8.8 of [5], shows. We report the example here with some simplifications for the convenience of the reader. Let $n = 1$ and $A = [0, 1] \times R$, let $C$ be a closed Cantor subset of [0, 1] whose measure $|C|$ is positive, and let $C' = [0, 1] - C$. Then $C'$ is the countable union of disjoint subintervals of [0, 1], $C' = \bigcup_{n \in N} I_n$. Let $\sigma(t): C' \to R^+$ be a continuous and integrable function, which tends to $+\infty$ whenever $t$ tends to an end point of any interval $I_n$. Moreover let us suppose that $\lim_{n \to +\infty} \min \sigma/I_n = +\infty$.

Let $Q(t) = \{-1\}$ if $t \in C$, and $Q(t) = \{z \in R : z \geqq \sigma(t)\}$ if $t \in C'$. Let us extend the function $\sigma$ by taking $\sigma(t) = 0$, for $t \in C$, and consider the decomposition of [0, 1] into $k$ intervals of equal length: $J_k^s = [t_{k,s-1}, t_{k,s}]$, $s = 1, \ldots, k$, $t_{k,s} = s/k$. Define $\xi_k$ by taking $\xi_k(t) = \sigma(t) + \nu_k(t)$, where $\nu_k(t) = -1$ if $t \in C$, and $\nu_k(t) = |C \cap J_k^s|/|C' \cap J_k^s|$ if $t \in C' \cap J_k^s$. Then $\xi_k$ is integrable in [0, 1], and $\xi_k(t) \in Q(t)$ for every $t \in [0, 1]$ and $k \in N$.

Let $x_k(t) = \int_0^t \xi_k(\tau)\, d\tau$, $0 \leqq t \leqq 1$, or $x_k(t) = x(t) + y_k(t) = \int_0^t \sigma(\tau)\, d\tau + \int_0^t \nu_k(\tau)\, d\tau$. Here $\int_{J_k^s} \nu_k(t)\, dt = 0$; hence $y_k(t_{k,s}) = 0$ for all $s$ and $k$, and $|y_k(t)| \leqq 2/k$. Hence $x_k \to x$ uniformly on [0, 1], as $k \to +\infty$; moreover all $x_k$ and $x$ are AC with $x_k'(t) \in Q(t)$, $t \in [0, 1]$. Now $x'(t) = 0$ a.e. in $C$, while $Q(t) = \{-1\}$ for $t \in C$. Thus $x'(t) \notin Q(t)$ on a subset $C$ of positive measure in [0, 1]. Note that $\int_{J_k^s} |\nu_k(t)|\, dt = 2\, |J_k^s \cap C|$; hence $V(y_k) \leqq 2$ and $V(x_k) \leqq \int_0^1 \sigma(t)\, dt + 2 = V_0$, a constant, for all $k$. Here the sets $Q(t, x)$ have property (K) on [0, 1]; moreover they have property (Q) both on $C$ and on $C'$ but not on [0, 1].

## 3. An existence theorem of the calculus of variations

### 3a. The integral $\mathscr{I}$

Let $\alpha, n$ be integers such that $0 \leqq \alpha \leqq n$, $n \geqq 1$, and for every $x \in R^n$ we write $x = (y, z)$ with $y \in R^\alpha$, $z \in R^{n-\alpha}$. Let $A \subset R^{n+1}$ and $Q: A \to R^n$ be defined as before. Let $M \subset R^{2n+1}$ denote the set $M = \{(t, x, \zeta): (t, x) \in A, \zeta \in Q(t, x)\}$, and let $F_0(t, x, \zeta)$, or $F_0: M \to R$ be a given function.

Let $\Omega$ be a class of admissible functions, *i.e.* functions $x(t) = (y(t), z(t))$, or $x: [t_1, t_2] \to R^n$, such that (i) $y$ is ACg and $z$ is BVC; (ii) $(t, x(t)) \in A$, $x'(t) \in Q(t, x(t))$ a.e. in $[t_1, t_2]$; (iii) $F_0(\cdot, x(\cdot), x'(\cdot)) \in L_1$.

We consider the functional $\mathscr{I}: \Omega \to \tilde{R}$ defined by

$$\begin{aligned}\mathscr{I}(x) = \mathscr{I}(y, z) &= \inf_{\Gamma(x)} \varliminf_{k\to\infty} \int_{t_1}^{t_2} F_0(t, y_k(t), z_k(t), y_k'(t), z_k'(t))\, dt \\ &= \inf_{\Gamma(x)} \varliminf_{k\to\infty} I(y_k, z_k),\end{aligned}$$

where $\Gamma(x)$ denotes the class of all sequences $(x_k)_{k\in N}$ such that (a) $x_k = (y_k, z_k) \in$ ACg $\cap\, \Omega$, $k \in N$; (b) $y_k \to y$ uniformly and $z_k \to z$ pointwise a.e. in $[t_1, t_2]$, and where, as stated in (i), $y$ is ACg and $z$ is BVC. If $\Gamma(x) = \emptyset$ we put $\mathscr{I}(x) = +\infty$. We may think of $F_0$ as extended to all of $R^{2n+1}$ by taking $F_0 = +\infty$ on $R^{2n+1} - M$. We denote by $\tilde{R}$ the reals with the addition of $+\infty$. Note that if $x \in$ ACg $\cap\, \Omega$, then $\Gamma(x) \neq \emptyset$. Moreover, if $A$ is convex, $Q(t, x) = R^n$, $(t, x) \in A$, and the integral means $(x_h)_{h>0}$ belong to $\Omega$, then $\Gamma(x) \neq \emptyset$.

The class $\Omega$ is said to be *closed* if $\Omega$ has the following property (c): if $x_k(t) = (y_k, z_k)$ is any sequence of (admissible) pairs in $\Omega$ satisfying (a) and (b) above, and if $x = (y, z)$ is admissible, then $x$ belongs to $\Omega$.

The functional $\mathscr{I}$ is modeled on Lebesgue area theory for nonparametric discontinuous surfaces (see Cesari [4]) and it is also close to the concept of integral in the sense of Serrin [14] when the present mixed convergence is used, uniform on $y$ and pointwise a.e. on $z$.

Note that the generalized weighted variation and length for a BVC curve are particular cases of the functional $\mathscr{I}$. In fact, let $F_0(t, v): [t_1, t_2] \times R^n \to R_0^+$ be defined by $F_0(t, v) = |\varphi(t) \cdot v|$, where $\varphi(t): [t_1, t_2] \to R^n$ is continuous. Then (see [1], [13])

$$\mathscr{I}(z) = \inf_{\Gamma(x)} \varliminf_{k\to\infty} \int_{t_1}^{t_2} |z_k'(t) \cdot \varphi(t)|\, dt = \inf_{\Gamma(x)} \varliminf_{k\to\infty} V_\varphi^*(z_k) \geqq V_\varphi^*(z),$$

and moreover

$$\mathscr{I}(z) \leqq \lim_{h\to 0} \int_{t_1}^{t_2} |z_h'(t) \cdot \varphi(t)|\, dt = V_\varphi^*(z),$$

where $z_h(t) = \frac{1}{h} \int_0^h z(t+\tau)\, d\tau$ is the integral mean of $z$. Analogous considerations hold for the length.

*Remark 2.* In order to deal with the minimization of the functionals under consideration when initial and terminal values for $x$ are involved, we carry over the definition of the functional $\mathscr{I}$ to the following setting.

Let $\mathscr{N}$ denote a family of subsets $N \subset [t_1, t_2]$, with $|N| = 0$, which is closed under countable unions. We shall write briefly $\mathscr{N}$-a.e. when we refer to a neglected null set $N \in \mathscr{N}$. Moreover we shall denote by $\mathscr{N}$-AC and $\mathscr{N}$-BVC the family of all the functions which are $\mathscr{N}$-a.e. equal to an AC function or a BV function, respectively.

We consider now the class $\Omega_{\mathscr{N}}$ of all the functions $x(t) = (y(t), z(t))$, such that

i) $y \in \mathscr{N}$-AC and $z \in \mathscr{N}$-BVC,
ii) $(t, x(t)) \in A$, $\mathscr{N}$-a.e. and $x'(t) \in Q(t, x(t))$, a.e. in $[t_1, t_2]$,
iii) $F_0(\cdot, x(\cdot), x'(\cdot)) \in L_1$.

For every $x \in \Omega_{\mathscr{N}}$, we shall denote by $\Gamma_{\mathscr{N}}(x)$ the class of all the sequences $(x_k)_{k \in N}$ such that

(a) $x_k = (y_k, z_k) \in \mathscr{N}$-AC, $k \in N$,
(b) $y_k \to y$ uniformly and $z_k \to z$ pointwise $\mathscr{N}$-a.e.

Thus we shall take $\mathscr{I}_{\mathscr{N}} : \Omega_{\mathscr{N}} \to \tilde{R}$ with

$$\mathscr{I}_{\mathscr{N}} = \inf_{\Gamma_{\mathscr{N}}(x)} \underline{\lim}_{k \to \infty} I(x_k) \quad \text{if } \Gamma_{\mathscr{N}}(x) \neq \emptyset; \quad \mathscr{I}_{\mathscr{N}} = +\infty \text{ otherwise}.$$

Note that, if $\mathscr{N}$ is the family of *all* the null sets in $[t_1, t_2]$, then $\mathscr{I}_{\mathscr{N}}$ is exactly the functional $\mathscr{I}$. Furthermore, if $\mathscr{N}_1 \subset \mathscr{N}_2$, then $\Omega_{\mathscr{N}_1} \subset \Omega_{\mathscr{N}_2}$, $\Gamma_{\mathscr{N}_1}(x) \subset \Gamma_{\mathscr{N}_2}(x)$ and $\mathscr{I}_{\mathscr{N}_1}(x) \geqq \mathscr{I}_{\mathscr{N}_2}(x)$, for every $x \in \Omega_{\mathscr{N}_1}$. Note that $>$ sign may hold, as the Example 2 in Section 3 shows. In this way we get a "spectrum" of integral functionals whose lower and upper lines are $\mathscr{I}$ and $\mathscr{I}_{\{\emptyset\}}$, respectively. Observe that, if $F_0(v) = |v|$, then $\{\mathscr{I}_{\mathscr{N}}\}_{\mathscr{N}}$ is the "spectrum of variations" whose lower and upper lines are the generalized variation and the classic one, respectively.

In the following, for simplicity, we shall deal with the functional $\mathscr{I}$, but all our results hold for any other functional $\mathscr{I}_{\mathscr{N}}$, as well. In fact, we shall make systematic use of Helly's theorem which guarantees convergence at *all* points $t \in [t_1, t_2]$.

Note that, in this way, we treat also minimization which involves given initial and terminal data for $x$, say $x(t_1) \in B_1$, $x(t_2) \in B_2$, with $B_1$, $B_2$ closed sets in $R^n$. This is the case when the family $\mathscr{N}$ is such that $\bigcup_{N \in \mathscr{N}} N = (t_1, t_2)$. Thus, in the computation of the variation $V^*(x)$ of $x$ in $[t_1, t_2]$ we always take note of the values of $x$ at $t_1$ and $t_2$. Concerning the convergence of the trace operator, see the known results mentioned in [6a].

*Remark 3.* We shall see now that, for every $x = (y, z) \in \Omega$, with $\Gamma(x) \neq \emptyset$, there is a sequence $(x_k)_{k \in N} \in \Gamma(x)$ such that

$$\lim_{k \to \infty} I(x_k) = \mathscr{I}(x).$$

If $\mathscr{I}(x) \in R$, the proof is analogous to that for the case in which $\mathscr{I}(x) = +\infty$. Observe that, by definition, for every $m \in N$ there is a sequence $(y_n^m, z_n^m)_{n \in N} \in$

$\Gamma(x)$ such that

$$\left|\lim_{n\to\infty} I(y_n^m, z_n^m) - \mathscr{I}(x)\right| < 1/m;$$

moreover, by virtue of Severini-Egoroff theorem, we know that $z_n^m \xrightarrow[n\to\infty]{} z$ almost uniformly, $m \in N$. Thus we can find a set $T_m \subset [t_1, t_2]$ and an integer $n_m$ such that $\operatorname{meas}(T_m) > (t_2 - t_1) - 1/2^m$ and, for every $n \geqq n_m$, we have $|I(y_n^m, z_n^m) - \mathscr{I}(x)| < 2/m$, $|y_n^m(t) - y(t)| < 1/m$, $t \in [t_1, t_2]$ and $|z_n^m(t) - z(t)| < 1/m$, $t \in T_m$.

Let us denote by $(x_k)_{k\in N}$ the sequence $x_k = (y_k, z_k) = (y_{n_k}^k, z_{n_k}^k)$, $k \in N$. We shall see that $(x_k)_{k\in N}$ is the sequence we were looking for. To do so, we put $T^k = \bigcap_{m\geqq k} T_m$ and $T = \bigcup_{k\in N} T^k$, then $\operatorname{meas}(T^k) > (t_2 - t_1) - 1/2^{k-1}$ and $\operatorname{meas}(T) = t_2 - t_1$. Thus, for every fixed $\varepsilon > 0$ and $t \in T$, there is an integer $\bar{k} > 1/\varepsilon$ such that $t \in T_k$ for every $k \geqq \bar{k}$ and therefore $|z_k(t) - z(t)| < 1/k < \varepsilon$. In other words $z_k \to z$ pointwise on $T$. Obviously $y_k \to y$ uniformly in $[t_1, t_2]$ and hence $(x_k)_{k\in N} \in \Gamma(x)$. Finally, having fixed $\varepsilon > 0$ and having taken $k_\varepsilon \in N$ such that $k_\varepsilon > 2/\varepsilon$, then for every $k \geqq k_\varepsilon$, we have $|I(x_k) - \mathscr{I}(x)| < 2/k_\varepsilon < \varepsilon$. The proof is complete.

### 3b. *A lower semicontinuity property of $I$ and $\mathscr{I}$*

As is well known ([5]), closure theorems can be reworded into lower closure theorems and into lower semicontinuity theorems. From the closure Theorem 1 of Section 2 we derive here a lower semicontinuity theorem for the integral $I$ and the relevant inequality $I(x) \leqq \mathscr{I}(x)$, under the assumption $V^*(x_k) \leqq V_0$, $k \in N$, and the topology under consideration, namely uniform convergence on the components $y^i$ and pointwise convergence almost everywhere on the components $z^j$.

For the lower semicontinuity theorem we shall need the auxiliary sets

$$\tilde{Q}(t, x) = \{(z^0, \xi) \in R^{n+1} : z^0 \geqq F_0(t, x, \xi), \xi \in Q(t, x)\}, \quad (t, x) \in A, \tag{1}$$

or "augmented" set-valued function $\tilde{Q}: A \to R^{n+1}$.

**Theorem 1'** (*A lower semicontinuity theorem*). *Let* $1 \leqq \alpha \leqq n - 1$, *and assume that*: (i) *$A$ is closed*; (ii) *the sets $\tilde{Q}(t, x)$ are closed, convex and have property* (Q) *with respect to $(t, x)$ at every point $(\bar{t}, \bar{x}) \in A$, with the exception perhaps of a set of points whose $t$-coordinate lies in a set $H$ of measure zero*; (iii) *$F_0(t, y, z, u, v)$ is lower semicontinuous in $M$ and $\lambda \in L_1$ exists such that $F_0(t, y, z, u, v) \geqq \lambda(t)$ for all $(t, y, z, u, v) \in M$*; (iv) *a sequence of vector functions is given*

$$x(t) = (y, z),\ x_k(t) = (y_k, z_k),\ t \in [t_1, t_2],\ y, y_k, z_k \in \mathrm{ACg},\ z \in \mathrm{BVC},$$

*$y_k \to y$ uniformly, $z_k \to z$ pointwise a.e. in $[t_1, t_2]$, such that $(t, x_k(t)) \in A$, $x_k'(t) \in Q(t, x_k(t))$ a.e. in $[t_1, t_2]$*; (v) *$V^*(x_k) \leqq V_0$ for all $k$ and some constant $V_0$. Then $(t, x(t)) \in A$, $x'(t) \in Q(t, x(t))$, a.e. in $[t_1, t_2]$ and $\varliminf_{k\to+\infty} I(x_k) \geqq I(x)$. Thus,*

*if* $(x_k)_{k\in N}$ *is a sequence as in the definition of* $\mathscr{I}(x)$, i.e. $x_k \in \mathrm{ACg} \cap \Omega$, *and* $V^*(x_k) \leq V_0$, $k \in N$, *then*

$$\varliminf_{k\to\infty} I(x_k) \geqq \mathscr{I}(x) \geqq I(x).$$

*For* $\alpha = 0$ *this statement concerns sequences* $z_k(t) = (z_k^1, \ldots, z_k^n)$, $z(t) = (z^1, \ldots, z^n)$, $t \in [t_1, t_2]$, $z_k \in \mathrm{ACg}$, $z \in \mathrm{BVC}$, *and the conclusions are still valid.*
*For* $\alpha = n$, *this statement concerns sequences* $y_k(t) = (y_k^1, \ldots, y_k^n)$, $y(t) = (y^1, \ldots, y^n)$, $t \in [t_1, t_2]$, $y, y_k \in \mathrm{ACg}$, *and the problem reduces to those discussed in Theorem* 10.8.i *of* [5].

*Remark 4.* As for Theorem 1, condition (Q) cannot be replaced by the weaker condition (K), as the following example shows. This example is only a modification of the one in Remark 1. Indeed, we take $n = 1$, $F_0 = 0$ and $\tilde{Q}(t) = \{(z^0, z) : z^0 \geqq 0, z \in Q(t)\}$, where the sets $Q(t)$ are defined in Remark 1. Then, for $x$, $x_k$ as in Remark 1, we have $x_k'(t) \in Q(t)$, but $x'(t) \notin Q(t)$, for $t \in C$, a set of positive measure.

**Proof** of Theorem 1′. Without loss of generality we can suppose that $i = \varliminf_{k\to+\infty} I(x_k) = \lim_{k\to+\infty} I(x_k) < +\infty$, where $i$ is finite because of (iii). Take $F_k(t) = F_0(t, y_k(t), z_k(t), y_k'(t), z_k'(t))$, $t \in [t_1, t_2]$, and note that for $F_k(t) = F_k^+(t) - F_k^-(t)$, $|F_k(t)| = F_k^+ + F_k^-$, $F_k^+, F_k^- \geqq 0$, we have $0 \leqq F_k^-(t) \leqq |\lambda(t)|$; hence $0 \leqq \int_{t_1}^{t_2} F_k^-(t)\, dt \leqq \int_{t_1}^{t_2} |\lambda(t)|\, dt$. Since $i$ is finite, we have that $I(x_k) \leqq W_0$ for all $k$ and some constant $W_0$. Finally,

$$\int_{t_1}^{t_2} |F_k(t)|\, dt = \int_{t_1}^{t_2} (F_k^+ + F_k^-)(t)\, dt = \int_{t_1}^{t_2} F_k(t)\, dt + 2\int_{t_1}^{t_2} F_k^-(t)\, dt$$

$$\leqq W_0 + 2\int_{t_1}^{t_2} |\lambda(t)|\, dt < +\infty.$$

Let us consider the AC functions $z_k^0 : [t_1, t_2] \to R$ defined by $z_k^0(t) = \int_{t_1}^{t} F_0(\tau, x_k(\tau), x_k'(\tau))\, d\tau$, $k \in N$. Then $z_k^0(t_1) = 0$ and

$$V(z_k^0) = \int_{t_1}^{t_2} |F_k(t)|\, dt \leqq W_0 + 2\int_{t_1}^{t_2} |\lambda(t)|\, dt = V,$$

$V$ a constant. By Helly's theorem there is a subsequence, say still $(k)$, such that $z_k^0(t) \to z^0(t)$ pointwise everywhere in $[t_1, t_2]$, with $z^0 \in \mathrm{BV}$, $z^0$ not necessarily continuous, and $z^0(t_1) = 0$. Note that the functions $(z_k^0, x_k)_{k\in N}$ are ACg solutions of the orientor field

$$(t, x_k(t)) \in A,\ (z_k^{0\prime}(t), x_k'(t)) \in \tilde{Q}(t, x_k(t)), \text{ a.e. in } [t_1, t_2], \tag{2}$$

where $\tilde{Q} : A \to R^{n+1}$ is the set-valued function defined by (1). As an application of Theorem 1 we now prove that the limit function $(z^0, y, z) = (z^0, x)$:

$[t_1, t_2] \to R^{n+1}$ is again a solution of the orientor field (2), *i.e.*

$$(t, x(t)) \in A, \quad (z^{0\prime}(t), x'(t)) \in \tilde{Q}(t, x(t)) \text{ a.e. in } [t_1, t_2],$$

or

$$(t, x(t)) \in A, \quad x'(t) \in Q(t, x(t)), \quad z^{0\prime}(t) \geqq F_0(t, x(t), x'(t)) \text{ a.e. in } [t_1, t_2].$$

Note that $z^{0\prime}(t) \geqq F_0(t) \geqq \lambda(t)$; hence $F_0(t, x(t), x'(t))$ is summable in $[t_1, t_2]$ because it lies between summable functions, and $x: [t_1, t_2] \to R^n$ is admissible. Finally, if we take

$$Z^0(t) = z^0(t) - \int_{t_1}^{t} \lambda(\tau)\, d\tau, \qquad t \in [t_1, t_2],$$

we see that $Z^0(t)$ is monotone non-decreasing in $[t_1, t_2]$; hence

$$\begin{aligned} Z^0(t_2) &= Z^0(t_2) - Z^0(t_1) = z^0(t_2) - \int_{t_1}^{t_2} \lambda(t)\, dt \\ &= V(Z^0) \geqq \int_{t_1}^{t_2} |z^{0\prime}(t) - \lambda(t)|\, dt = \int_{t_1}^{t_2} (z^{0\prime}(t) - \lambda(t))\, dt \\ &\geqq \int_{t_1}^{t_2} (F_0(t) - \lambda(t))\, dt = \int_{t_1}^{t_2} F_0(t)\, dt - \int_{t_1}^{t_2} \lambda(t)\, dt. \end{aligned}$$

Hence $z^0(t_2) \geqq \int_{t_1}^{t_2} F_0(t)\, dt$, and finally

$$I(x) = \int_{t_1}^{t_2} F_0(t)\, dt \leqq z^0(t_2) = \lim_{k \to +\infty} z_k^0(t_2) = \lim_{k \to +\infty} I(x_k).$$

Theorem 1′ is thereby proved.

Theorem 1′ has an important consequence concerning the concept of integral $\mathscr{I}(x)$ defined at the beginning of Section 3. Indeed, as long as we define $\mathscr{I}(x)$ by means of sequences $x_k = (y_k, z_k) \in \text{ACg}$, $k \in N$, with equibounded variation, then we can well say that $\mathscr{I}$ is an extension to BVC of the functional $I$; in other words, if $x = (y, z)$, $y, z$ both ACg, then $\mathscr{I}(x) = I(x)$. In fact, from Theorem 1′ we have[1] $I(x) \leqq \mathscr{I}(x)$, but $\Gamma(x)$ contains now the sequence of repetitions $x_k = (y, z)$, $k \in N$; hence $\mathscr{I}(x) \leqq I(x)$, and finally $I(x) = \mathscr{I}(x)$.

### *3c. The existence theorem*

We now state and prove an existence theorem of the calculus of variations for the integral $\mathscr{I}$. In other words we have to prove that $\mathscr{I}$ has an absolute minimum in $\Omega$. That is we have to prove, under the assumptions that

(a) the infimum $i$ of $I(x)$ in $\text{ACg} \cap \Omega$ is finite; hence there are minimizing sequences $x_k = (y_k, z_k)$, $k \in N$, in $\Omega$, both $y_k, z_k \in \text{ACg}$, such that $I(x_k) \to i$;

(b) for some subsequence, say $(k)$ again, and elements $x = (y, z) \in \Omega$, $y \in \text{ACg}$, $z \in \text{BVC}$, we have $y_k \to y$ uniformly, $z_k \to z$ pointwise a.e. in $[t_1, t_2]$.

---

[1] This is not true, in general, if assumption (v) of Theorem 1′ is dropped, as Example 3 below shows.

Since we shall assume that there are such sequences $x_k = (y_k, z_k)$ with equibounded variations $V(x_k)$, then by Theorem 1′ we know that $I(x) \leqq \mathcal{I}(x) = i$.[2]

To state and prove our existence theorem, we denote by $(\gamma_1)$, $(\gamma_2)$, $(\gamma_3)$ the following alternative assumptions on the function $F_0$.

$(\gamma_1)$ There is a scalar function $\phi(\zeta)$, $0 \leqq \zeta < +\infty$, or $\phi: R_0^+ \to R$, bounded below, with $\phi(\zeta)/\zeta \to +\infty$ as $\zeta \to +\infty$, such that $F_0(t, y, z, u, v) \geqq \phi(|u|)$ for all $(t, y, z, u, v) \in M$.

$(\gamma_2)$ For any $\varepsilon > 0$ there is an integrable scalar function $\psi_\varepsilon(t) \geqq 0$, or $\psi_\varepsilon: [t_1, t_2] \to R_0^+$, such that $|u| \leqq \psi_\varepsilon(t) + \varepsilon F_0(t, y, z, u, v)$ for all $(t, y, z, u, v) \in M$.

$(\gamma_3)$ For every $\alpha$-vector $p \in R^\alpha$ there is an integrable scalar function $\phi_p(t) \geqq 0$, or $\phi_p: [t_1, t_2] \to R_0^+$, such that $F_0(t, y, z, u, v) \geqq \langle p, u \rangle - \phi_p(t)$ for all $(t, y, z, u, v) \in M$.

Note that under condition $(\gamma_1)$ certainly $\phi(\zeta) \geqq \lambda$ for some real constant $\lambda$, and then $F_0(t, y, z, u, v) \geqq \phi(|u|) \geqq \lambda$ for all $(t, y, z, u, v)$. Under condition $(\gamma_2)$ and $\varepsilon = 1$ we have $|u| \leqq \psi_1(t) + F_0(t, y, z, u, v)$; hence $F_0(t, y, z, u, v) \geqq -\psi_1(t)$, a summable function in $[t_1, t_2]$. Under condition $(\gamma_3)$ and $p = 0$, we have $F_0(t, y, z, u, v) \geqq -\phi_0(t)$, a summable function in $[t_1, t_2]$.

**Theorem 2** (An existence theorem). *Let* $1 \leqq \alpha \leqq n-1$, *and assume that* (i) *A is compact and M is closed*; (ii) *the sets* $\tilde{Q}(t, x)$ *are closed and convex and have property* (Q) *with respect to* $(t, x)$ *at every point* $(t, x)$ *of A* (*with the exception perhaps of a set of points whose t-coordinate lies on a set of measure zero on the t-axis*); (iii) $F_0(t, y, z, u, v)$ *is lower semicontinuous in M*; (iv) $F_0$ *satisfies one of the growth conditions* $(\gamma_1)$, *or* $(\gamma_2)$, *or* $(\gamma_3)$. *Also we assume that the class* $\Omega$ *is nonempty and closed, and* (v) *there exists a constant* $W_0$ *such that for every element* $x = (y, z) \in$ ACg $\cap\ \Omega$, *then* $V^*(z) \leqq W_0$.

*Then the functional* $\mathcal{I}$ *has an absolute minimum* $x = (y, z)$ *in* $\Omega$, $y \in$ ACg, $z \in$ BVC, and $I(x) \leqq \mathcal{I}(x) = i$.

*For* $\alpha = 0$, *then* $x = z$, *requirements* $(\gamma_1)$, *or alternatively* $(\gamma_2)$, $(\gamma_3)$ *do not apply, yet the conclusion is still valid if we know that* (iv)′ *there is a summable scalar function* $\lambda: [t_1, t_2] \to R$ *such that* $F_0(t, z, \zeta) \geqq \lambda(t)$ *for all* $(t, z, \zeta) \in M$.

*For* $\alpha = n$, *then* $x = y$, $\Omega$ *is a nonempty and closed class of* ACg *functions* $y(t) = (y^1, \ldots, y^n)$, $t \in [t_1, t_2]$, *condition* (v) *does not apply, and the problem reduces essentially to those discussed in* Theorems 11.1.i *and* ii *of* [5].

Note that the condition in (ii) concerning property (Q) for the sets $\tilde{Q}(t, x)$, not only implies that the same sets $\tilde{Q}(t, x)$ are closed and convex, but also that their projections, the sets $Q(t, x)$, also are convex, and that $F_0(t, y, z, u, v)$ is convex in $(u, v)$.

---

[2] See Note 1.

Note that for $0 \leqq \alpha \leqq n-1$, if (v') there are scalar functions $\psi_i \in L_1([t_1, t_2])$, $i = \alpha+1, \ldots, n$, such that $(t, y, z, u, v) \in M$ implies $v^i \geqq \psi_i(t)$ a.e. in $[t_1, t_2]$, then (v) certainly holds. (*Cf.* part (g) of proof of Theorem 2 below).

Note that for $1 \leqq \alpha \leqq n-1$, the sets $\tilde{Q}(t, x)$ closed and convex, and $(\gamma_1)$ holds, if (v'') there are constants $L_i$ such that, a.e. in $[t_1, t_2]$, $(t, y, z, u, v) \in M$ implies $v^i \geqq L_i$, $i = \alpha+1, \ldots, n$, then both (ii) and (v) hold. (*Cf.* part (h) of proof of Theorem 2 below).

*Remark 5.* If $\alpha = n$, $x = y$, note that Theorems 11.1,i, ii of [5] are proved under weaker assumptions on the function $F_0$ and definitively less information on the sets $\tilde{Q}(t, x)$. In particular in Theorem 11.1.ii, under none of the assumptions $(\gamma_1)$, $(\gamma_2)$, $(\gamma_3)$ is it needed to verify that the sets $\tilde{Q}(t, x)$ have property (Q). Indeed, a different topology is used on the functions $y_k$, namely $y_k \to y$ uniformly and $y'_k \to y'$ weakly in $L_1$. Then, in terms of the equivalence theorem ([5], Theorem 10.3.i; see also CESARI & PUCCI [7]), these sets $\tilde{Q}(t, y)$ are shown to have augmented sets $\tilde{\tilde{Q}}(t, y)$ which have property (Q) with respect to $y$, a.e. in $t$ (see [5], proof of Theorem 10.7.i.).

**Proof** of Theorem 2. Without loss of generality we can suppose that there is an element $x \in \Omega$ such that $\Gamma(x) \neq \emptyset$. Let $i = \inf_{\Omega} \mathscr{I}(x)$; then $-\infty \leqq i < +\infty$. Let $(\bar{x}_k)_{k \in N}$ be a minimizing sequence, *i.e.* $\mathscr{I}(\bar{x}_k) \to i$ as $k \to \infty$. We divide the proof into parts.

(a) First note that we can find a sequence $(x_k)_{k \in N}$ in ACg $\cap\, \Omega$ such that $I(x_k) \to i$ as $k \to \infty$. In order to see this, note that for every $k \in N$, there is an integer $n_k$ such that $|\mathscr{I}(\bar{x}_{n_k}) - i| < 1/k$, and moreover (see Remark 2) there is a sequence $(x_n^k)_{n \in N}$ in ACg $\cap\, \Omega$ such that $I(x_n^k) \to \mathscr{I}(\bar{x}_{n_k})$ as $n \to \infty$. Thus, for every $k \in N$, there is an integer $\bar{n} = \bar{n}(k)$ such that $|I(x_{\bar{n}}^k) - i| < 2/k$ (or $I(x_{\bar{n}}^k) < k$ if $i = -\infty$). Let $x_k = x_{\bar{n}}^k$, $k \in N$. Then $x_k \in$ ACg $\cap\, \Omega$, and $|I(x_k) - i| < 2/k$, $k \in N$. Without loss of generality we can suppose that $x_k$ is AC, $k \in N$, and, since $A$ is compact, the sequence $(x_k)_{k \in N}$, $x_k = (y_k, z_k)$, is equibounded.

(b) By virtue of hypothesis $(\gamma_1)$, we prove, as in Theorem 10.4.i of [5] that the sequence $(y_k)_{k \in N}$ is equi-absolutely continuous. Thus, $V(y_k) \leqq V_1$ for all $k \in N$ and some constant $V_1$. Moreover, since $(y_k)_{k \in N}$ is equibounded, there is a subsequence, say still $(y_k)_{k \in N}$, such that $y_k \to y$ uniformly in $[t_1, t_2]$, with $y \in$ AC.

Moreover, the assumption (v) shows that $V(z_k) \leqq W_0$ for all $k \in N$. Hence, by Helly's theorem (see [5], Theorem 15.1.i) there is a subsequence, say still $(z_k)_{k \in N}$, such that $z_k(t) \to z(t)$ pointwise everywhere in $[t_1, t_2]$ with $z \in$ BV, not necessarily continuous.

(c) Now the function $\phi$ in $(\gamma_1)$ is bounded below, say $\phi(\zeta) \geqq \lambda$, $\lambda \in R$; hence $F_k(t) = F(t, y_k(t), z_k(t), y'_k(t), z'_k(t)) \geqq \phi(|y'_k(t)|) \geqq \lambda$ for $k \in N$ and $t \in [t_1, t_2]$. Consequently, for $F_k(t) = F_k^+(t) - F_k^-(t)$, $F_k^- \geqq 0$, $F_k^+ \geqq 0$, $|F_k(t)| = F_k^+(t) +$

$F_k^-(t)$, we have $F_k(t) \geqq \lambda$, $F_k^-(t) \leqq |\lambda|$. Since $\int_{t_1}^{t_2} F_k(t)\,dt \to i$ as $k \to \infty$, we derive $i \geqq \lambda(t_2 - t_1)$, and thus $i$ is finite. Without loss of generality we can assume that $\int_{t_1}^{t_2} F_k(t)\,dt \leqq i + 1$ for all $k \in N$, and then

$$\int_{t_1}^{t_2} F_k^+(t)\,dt = \int_{t_1}^{t_2} F_k(t)\,dt - \int_{t_1}^{t_2} F_k^-(t)\,dt \leqq i + 1 + |\lambda|\,(t_2 - t_1),$$

$$\int_{t_1}^{t_2} |F_k(t)|\,dt \leqq i + 1 + 2\,|\lambda|\,(t_2 - t_1).$$

Let us consider now the AC functions $z_k^0 : [t_1, t_2] \to R$ defined by $z_k^0(t) = \int_{t_1}^{t} F_0(\tau, x_k(\tau), x_k'(\tau))\,d\tau$, $k \in N$. Then $V(z_k^0) = \int_{t_1}^{t_2} |F_k(t)|\,dt \leqq i + 1 + 2\,|\lambda|\,(t_2 - t_1) = V$, $k \in N$. Again, by Helly's theorem there is a subsequence, say still $(z_k^0)_{k\in N}$, such that $z_k^0(t) \to z^0(t)$ pointwise everywhere in $[t_1, t_2]$, with $z^0 \in \mathrm{BV}$ (not necessarily continuous).

(d) Note that the functions $(z_k^0, x_k)_{k\in N}$ are AC solutions of the orientor field $(t, x_k(t)) \in A$, $(z_k^{0\prime}(t), x_k'(t)) \in \tilde{Q}(t, x_k(t))$, a.e. in $[t_1, t_2]$, where $\tilde{Q} : A \to R^{n+1}$ is the set-valued function defined by $\tilde{Q}(t, x) = \{(r, \zeta) : \zeta \in Q(t, x),\ r \geqq F_0(t, x, \zeta)\}$ $= \operatorname{epi} F_0(t, x, \cdot)_{/Q(t,x)}$. By (ii) the sets $\tilde{Q}(t, x)$ have property $(Q)$ with respect to $(t, x)$ in $A$, for a.e. $t$.

(e) As an application of Theorem 1, we see that the limit function $(z^0, y, z) = (z^0, x) : [t_1, t_2] \to R^{n+1}$ is again a solution of the orientor field $(t, x(t)) \in A$, $(z^{0\prime}(t), x'(t)) \in \tilde{Q}(t, x(t))$, a.e. in $[t_1, t_2]$, i.e. $(t, x(t)) \in A$, $x'(t) \in Q(t, x(t))$, $z^{0\prime}(t) \geqq F(t, x(t), x'(t))$, a.e. in $[t_1, t_2]$. Note that $z^{0\prime}(t) \geqq F_0 \geqq \lambda$; hence $F_0(t, x(t), x'(t))$ is summable in $[t_1, t_2]$ and the function $x : [t_1, t_2] \to R^n$ is admissible. Since the class $\Omega$ is closed, we conclude that $x \in \Omega$.

(f) Finally, by definition, $(x_k)_{k\in N} \in \Gamma(x)$ and $i \leqq \mathcal{I}(x) \leqq \lim_{k\to\infty} I(x_k) = i$, or $\mathcal{I}(x) = i$ and the proof is complete.

(g) Let us prove now that, for $0 \leqq \alpha \leqq n - 1$, $A$ compact and property (v′), then $V(z_k) \leqq V_2$, $k \in N$, for some constant $V_2$. Indeed, since $z_k$ is AC, we have (writing $z_k$ for any of its components $z_k^i$, $i = \alpha + 1, \ldots, n$),

$$z_k(t_2) - z_k(t_1) = \int_{t_1}^{t_2} z_k'^+(t)\,dt - \int_{t_1}^{t_2} z_k'^-(t)\,dt$$

and

$$V(z_k) = \int_{t_1}^{t_2} z_k'^+(t)\,dt + \int_{t_1}^{t_2} z_k'^-(t)\,dt.$$

Thus, by condition (v′), we have

$$V(z_k) = 2\int_{t_1}^{t_2} z_k'^{-}(t)\,dt + z_k(t_2) - z_k(t_1) \leqq 2\int_{t_1}^{t_2} \psi^{-}(t)\,dt + \operatorname{diam} A.$$

(h) Let us prove that, if $1 \leqq \alpha \leqq n-1$, $A$ compact, the sets $\tilde{Q}(t, x)$ closed and convex, and $(\gamma_1)$ and (v″), then not only (v) but also (ii) holds, that is, the sets $\tilde{Q}(t, x)$ have property (Q) with respect to $(t, x)$. Indeed, we have $F_0(t, x, u, v) \geqq \phi(|u|)$, $v^i \geqq L_i$, $i = \alpha+1, \ldots, n$, for all $(t, x, u, v) \in M$, and by virtue of Theorem 10.5.ii (second version) of [5] (with the variable $x$ replaced by $(t, x)$), the thesis follows.

*Remark 6*. The hypothesis (v) can be replaced by the weaker assumption (v‴) the level sets $L_K = \{x = (y, z) \in \mathrm{ACg} \cap \Omega : I(x) \leqq K\}$ are equibounded in variation. In fact, as can be seen by the proof, we use condition (v) only to guarantee that every minimizing sequence of ACg curves $x = (y, z)$ is equibounded in variation.

We shall illustrate now three situations in which even condition (v‴) can be dropped.

a) Let $F_0 \colon R \to R_0^+$ be a convex function then $F_0$ is the least upper bound of its support straight lines, i.e.

$$F_0(v) = \sup\{\varphi(v) : \varphi(w) = aw + b \leqq F_0(w),\ w \in R\}.$$

Thus, either $F_0(v) = \text{const}$, and then Theorem 2 is trivial, or $F_0(v) \geqq av + b$, $v \in R$, $a \neq 0$. But in this last case, it is easy to see that every minimizing sequence of AC curves is equibounded in variation.

b) Let consider now an integrand $F_0$ which does not depend on the variable $z$ and does satisfy the condition

$$|v_1| \leqq |v_2| \text{ implies } F_0(t, v_1) \leqq F_0(t, v_2).$$

Then condition (v‴) in Theorem 2 can be omitted, provided we suppose that the sets $Q(t, x)$ are such that if $v \in Q(t, x)$ and $|w| \leqq |v|$, then $w \in Q(t, x)$. In fact, in the present case, we can find a minimizing sequence of AC curves with equibounded variation. In order to see this, given any sequence of AC curves such that $I(z_k) \to i$ as $k \to +\infty$, it is sufficient to alter the sequence $(z_k)_{k \in N}$ in the following way. For simplicity we write $z_k$ for any of its components $z_k^i$, $i = 1, \ldots, n$. Let us suppose first that $z_k(t_1) < z_k(t_2)$, take

$$\bar{t} = \max\{t \in [t_1, t_2] : z_k(t) = z_k(t_1)\},\ \bar{\bar{t}} = \min\{t \in [\bar{t}, t_2] : z_k(t) = z_k(t_2)\}$$

and

$$\bar{z}_k(t) = \begin{cases} z_k(t_1), & t \in [t_1, \bar{t}], \\ \max\{z_k(\tau), \quad \tau \in [\bar{t}, t]\}, & t \in [\bar{t}, \bar{\bar{t}}], \\ z_k(t_2), & t \in [\bar{\bar{t}}, t_2]. \end{cases} \tag{3}$$

If $z_k(t_1) > z_k(t_2)$, we define $\bar{z}_k$ analogously by substituting min for max in (3). Finally, if $z_k(t_1) = z_k(t_2)$, we take $\bar{z}_k(t) = z_k(t_1)$, $t \in [t_1, t_2]$. Observe that, in any case, $\bar{z}_k$ is again AC and moreover it is monotone and $|\bar{z}'_k(t)| \leqq |z'_k(t)|$, a.e. in $[t_1, t_2]$. Therefore $V(\bar{z}_k) \leqq \operatorname{diam} A$, $k \in N$, and, by virtue of the assumption on the integrand $F$, we have $I(\bar{z}_k) \leqq I(z_k)$, $k \in N$. This proves that $(\bar{z}_k)_{k \in N}$ is still a minimizing sequence.

c) Let $F_0(t, z, v): M \to R^+$ subjected to the growth condition $F_0(t, z, v) \geqq a|v| + b(t)$, with $a > 0$ and $b \in L_1$. In this case condition (v‴) is trivially satisfied.

*Remark 6′*. Note that we may drop the requirement that $A$ be bounded if we know that there is a minimizing sequence $x_k = (y_k, z_k)$, $y_k, z_k \in \mathrm{AC}$, with $I(x_k) \to i$, which is equibounded. Thus the assumptions "$A$ compact and (v)" can be replaced by the weaker conditions: $A$ closed and

(v′) the level sets $L_K = \{x = (y, z) \in \mathrm{ACg} \cap \Omega : I(x) \leqq K\}$ are bounded in the norm $\|x\| = |x_e(t_1)| + V^*(x)$, where $x_e(t_1) = \lim\operatorname{ess}_{t \to t_1^+} x(t)$.

*Remark 7*. Note that we consider the infimum $i$ of $I(x)$ in the class $\mathrm{ACg} \cap \Omega$ and we prove in Theorem 2 under the hypotheses that there is some element $x = (y, z)$ in $\Omega$, $y \in \mathrm{ACg}$, $z \in \mathrm{BVC}$, and some sequence $x_k = (y_k, z_k)$, $y_k$, $z_k \in \mathrm{ACg}$, $k \in N$, in $\Omega$ with $I(x_k) \to i$, and $I(x) \leqq \mathcal{I}(x) = i$.

In other words, under the assumptions of Theorem 2, the infimum $i$ is attained by $\mathcal{I}$, or $\mathcal{I}(x) = i$, while $I(x)$ may have a value equal to or less than $i$. In Examples 4 and 5 below $I(x) = \mathcal{I}(x) = i$. However, it may well happen that $I(x) < \mathcal{I}(x) = i$ as Example 1 below shows. Note that if we denote by $i_0$ the infimum of $I(x)$ in the class $\Omega$, then $\mathrm{ACg} \cap \Omega \subset \Omega$; hence $i_0 \leqq i$. We shall see in Example 2 below that possibly $i_0 < i$, and that both can be attained, say $I(x) = i_0$ and $I(\bar{x}) \leqq \mathcal{I}(\bar{x}) = i$, possibly by different $x, \bar{x} \in \Omega$. Also note that for $\bar{x}$ optimal for $\mathcal{I}$ under the assumptions of the present paper, we certainly have $i_0 \leqq I(\bar{x}) \leqq \mathcal{I}(\bar{x}) = i$.

*Example 1*. Let us show that, if $i$ is the infimum of $I(x)$ in $\mathrm{ACg} \cap \Omega$ (and therefore the infimum of $\mathcal{I}(x)$ in $\Omega$), and $x \in \Omega$ is a minimizing element, then it may happen that $I(x) < \mathcal{I}(x) = i$.

Let us consider the problem of minimizing the length of the plane curves $z^1 = z^1(t)$, $z^2 = z^2(t)$, $0 \leqq t \leqq 1$, joining two given points, say (0, 0) and (1, 1), or

$$I(x) = \int_0^1 [(z^{1'}(t))^2 + (z^{2'}(t))^2]^{1/2}\, dt,$$

$$z^1(0) = 0, \quad z^2(0) = 0, \quad z^1(1) = 1, \quad z^2(1) = 1.$$

Here, for $z^1, z^2 \in \mathrm{AC}$, the infimum of $I$ is $i = \sqrt{2}$, and this infimum is attained not only by the obvious solution $z^1(t) = z^2(t) = t$, $0 \leqq t \leqq 1$, but also by the infinitely many solutions $z^1(t) = z^2(t) = \xi(t)$, $0 \leqq t \leqq 1$, $\xi \in \mathrm{AC}$, monotone nondecreasing with $\xi(0) = 0$, $\xi(1) = 1$; hence $z^1(0) = 0$, $z^2(0) = 0$, $z^1(1) = 1$,

$z^2(1) = 1$, and

$$I(z) = \sqrt{2} \int_0^1 |\xi'| \, dt = \sqrt{2} \int_0^1 \xi' \, dt = \sqrt{2} = i.$$

On the other hand, let us consider the usual ternary Cantor function $\varphi(t)$, $0 \leqq t \leqq 1$, $\varphi(0) = 0$, $\varphi(1) = 1$, continuous, monotone non decreasing, with derivative zero a.e. in $[0, 1]$, $\varphi$ BV and not AC. Let $\xi_k(t)$, $0 \leqq t \leqq 1$, $k \in N$, be a sequence of monotone nondecreasing AC approximations of $\varphi$ with $\xi_k(0) = 0$, $\xi_k(1) = 1$, and $\xi_k \to \varphi$ uniformly in $[0, 1]$. Now we take the sequence of AC functions $z_k^1(t) = z_k^2(t) = \xi_k(t)$, $0 \leqq t \leqq 1$, $k \in N$. For $z_k = (z_k^1, z_k^2) = (\xi_k, \xi_k)$, $z = (z^1, z^2) = (\varphi, \varphi)$, we have

$$I(z_k) = \sqrt{2} \int_0^1 |\xi_k'(t)| \, dt = \sqrt{2} \int_0^1 \xi_k'(t) \, dt = \sqrt{2};$$

$z_k \to z$, i.e. $z_k^1 \to z^1$, $z_k^2 \to z^2$ uniformly; hence pointwise in $[0, 1]$

$$I(z_k) \to i = \sqrt{2}.$$

Thus, in the terms of the beginning of Section 3, $\mathscr{I}(\varphi) = i = \sqrt{2}$, $I(\varphi) = 0$; hence $I(\varphi) < \mathscr{I}(\varphi) = i$.

*Example 2.* Suppose

$$A = [0, 1]^2 \cap \{(t, x) \in R^2 : t - \tfrac{1}{3} \leqq z \leqq t + \tfrac{1}{3}\}, Q(t, x) = R, n = 1,$$

and let $F_0(v): R \to R$ be defined by $F_0(v) = |v|$. In this case (see Remark 2) the functional $\mathscr{I}$ is the generalized variation. Let $\varphi(t): [0, 1] \to [0, 1]$ be the usual ternary Cantor function. Then $\varphi$ is continuous, BV and not ACg, with $\varphi'(t) = 0$, a.e. in $[0, 1]$, and graph $\varphi \subset A$. Thus $I(\varphi) = 0$, *i.e.*; if $i_0$ denotes the infimum of $I(x)$ in $\Omega$, then $I(\varphi) = i_0 = 0$. On the other hand, if $i$ denotes the infimum of $I(x)$ in $\mathrm{ACg} \cap \Omega$ then by Theorem 2, $\mathscr{I}$ also attains its infimum at some minimizing element $\bar{z} \in \Omega$, and $\mathscr{I}(\bar{z}) = i$. Certainly $i > 0$ (hence $i > i_0$), since $i = 0$ would imply $\bar{z}'(t) = 0$ a.e., $\bar{z}(t) =$ const. a.e., and this is not possible given the shape of the set $A$. It is easy to see that $i = 2/3$ and that a minimizing element is $\bar{z}(t) = 0$ for $0 \leqq t \leqq 1/3$, $\bar{z}(t) = t - 1/3$ for $1/3 < t \leqq 1$, with $I(\bar{z}) = \mathscr{I}(\bar{z}) = 2/3$. Note that if we consider the same problem with boundary data $z(0) = 0$, $= z(1) = 1$, then $\varphi$ satisfies the same data;h ence $I(\varphi) = i_0 = 0$ as before. On the other hand, it is easy to see that the new infimum is now $i = 1$, and that a minimizing element is $\bar{z}(t) = 0$ for $0 \leqq t \leqq 1/3$, $\bar{z}(t) = t - \frac{1}{3}$ for $1/3\ t < 1$, $\bar{z}(1 - 0) = 2/3$, $\bar{z}(1) = 1$ with a jump of $1/3$ at $t = 1$, and $\mathscr{I}(\bar{z}) = 1 = i$, $I(\bar{z}) = 2/3$ and again $I(\bar{z}) < \mathscr{I}(\bar{z}) = 1$ as in Example 1. Thus $0 = i_0 = I(\varphi) < I(\bar{z}) < \mathscr{I}(\bar{z}) = i = 1$.

*Example 3.* We show here an example in which occurs the "non natural" situation $i = \mathscr{I}(x) < I(x)$. In this example $A$ is compact, but the sets $\tilde{Q}$ do not have the property $(Q)$, and for a minimizing sequence of AC functions $x_k = (y_k, z_k)$, $k \in N$, the total variations $V^*(z_k)$ are not equibounded, and the sequence $z_k'$, $k \in N$, is not equibounded below.

Let $A = [0, 2\pi]\times[-1, 1]^2$, $Q(t, x) = R^2$, $(t, x)\in A$, and let $F_0(z, v)$: $[-1, 1]^2\times R^2 \to R^+$ be defined by

$$F_0(z_1, z_2, v_1, v_2) = \exp(z_1 v_2 - z_2 v_1).$$

We consider now the sequence $(z_k)_{k\in N}$ given by

$$z_k^1(t) = r_k \sin kt, \quad z_k^2(t) = r_k \cos kt, \quad 0 \leqq t \leqq 2\pi, k \in N,$$

where $r_k = k^{-1/3}$. Note that $z_k^{1'}(t) = r_k k \cos kt$, $z_k^{2'}(t) = -r_k k \sin kt$, and $z_k^1 z_k^{2'} - z_k^2 z_k^{1'} = -r_k^2 k = -k^{1/3} \to -\infty$. Therefore we have

$$0 \leqq \mathscr{I}(z_k) \leqq I(z_k) = \int_0^{2\pi} F_0(z_k(t), z_k'(t))\, dt = 2\pi \exp(-k^{1/3}) \to 0,$$

as $K\to\infty$. Thus $i = 0$. Note that $V(z_k^1) = V(z_k^2) = 4r_k k = 4k^{2/3} \to +\infty$ as $k\to\infty$. Since $r_k \to 0$ as $k\to\infty$, if we take $z = (z^1, z^2)$ with $z^1(t) = z^2(t) = 0$, $t\in[t_1, t_2]$, then $z$ is AC in $[0, 2\pi]$ and $z_k \to z$ uniformly, hence $\mathscr{I}(z) = 0$. But $I(z) = \int_0^{2\pi} \exp(0)\, dt = 2\pi > 0$. Moreover, for every $z\in$ BVC we certainly have

$$I(z) = \int_0^{2\pi} \exp(z_1(t)\, z_2'(t) - z_2(t)\, z_1'(t))\, dt > 0.$$

We shall give now two examples which illustrate Theorem 2 and show that, in general, the minimum of $\mathscr{I}$ is attained by a BVC function, not necessarily ACg.

*Example 4.* Let $A = [0, 2]\times[0, 1]$, $n = 1$, $\alpha = 0$, $\psi(t) = 0$. $Q(t, x) = [0, +\infty)$ for $(t, x)\in A$, $M = A\times[0, +\infty)$, with boundary conditions $x(0) = 0$, $x(2) = 1$. Let $F_0(t, v)$ be defined by $F_0(t, v) = |1 - t|\,|v|$ for $(t, v)\in M$. Thus the functional $\mathscr{I}(x)$ is nonnegative. Note that, for the sequence $z_k: [0, 2]\to R$, $k\in N$, defined by $z_k(t) = 0$ for $t\in[0, 1 - 1/k]$; $z_k(t) = 1$ for $t\in[1, 2]$, $z_k(t) = 1 - k + kt$ for $t\in(1 - 1/k, 1)$, we have

$$0 \leqq \mathscr{I}(z_k) = I(z_k) = \int_{1-1/k}^{1} (1 - t)\, k\, dt = 1/2k \to 0 \text{ as } k\to\infty.$$

Thus the infimum $i$ of $\mathscr{I}$ is zero, and $z_k$ is a minimizing sequence. The minimum is attained by the discontinuous function $z: [0, 2]\to R$ defined by $z(t) = 0$ for $t\in[0, 1)$, $z(t) = 1$ for $t\in[1, 2]$. In other words $I(z) = \mathscr{I}(z) = 0$.

*Example 5.* Let $A = [-1, 1]\times[0, 1]$, $n = 1$, $\alpha = 0$, $\psi(t) = 0$, $Q(t, x) = [-1, +\infty)$, $M = A\times[-1, +\infty)$, $F_0(t, v) = |t|\, v^2$, $F_0 \geqq 0$. with boundary conditions $x(-1) = 0$, $x(1) = 1$. The functional $\mathscr{I}(z)$ is nonnegative. Note that for the sequence $z_k: [-1, 1]\to R$, $k\in N$, defined by $z_k(t) = 0$ for $t\in[-1, 1/k]$; $z_k(t) = (\log k)^{-1}\log t + 1$, for $t\in(1/k, 1]$, we have

$$0 \leqq \mathscr{I}(z_k) = I(z_k) = \int_{1/k}^{1} (\log k)^{-2}/t\, dt = (\log k)^{-1} \to 0 \quad \text{as } k\to\infty.$$

Thus the infimum $i$ of $\mathscr{I}$ is zero and $(z_k)_{k\in N}$ is a minimizing sequence. The minimum is attained by the discontinuous function $z:[-1,1]\to R$, defined by $z(t)=0$ for $t\in[-1,0]$, $z(t)=1$ for $t\in(0,1]$. In other words, $I(z)=\mathscr{I}(z)=i$. For this example *cf.* [5], Section 1.1, no. 4.

## 4. An existence theorem for problems of optimal control

As above, let $\alpha, n$, $0\leqq\alpha\leqq n$, be given integers and, for every $x\in R^n$, let $x=(y,z)$ with $y\in R^\alpha$ and $z\in R^{n-\alpha}$. Let $A$ be a compact subset of the $(t,x)$-space such that its projection onto the $t$-axis contains the fixed interval $[t_1,t_2]$. Let $U(t,x)$, $(t,x)\in A$, $U(t,x)\subset R^m$, or $U:A\to R^m$ be a given set-valued function and let $M_0$ denote the set

$$M_0=\{(t,x,w):(t,x)\in A,\ w\in U(t,x)\}\subset R^{1+n+m}.$$

Let $f_0(t,x,w), f(t,x,w)=(f_1,\dots,f_n)$ be given functions defined on $M_0\subset R^{1+n+m}$. Let $\Omega_0$ be a class of admissible systems $(y(t),z(t),w(t))$, $t\in[t_1,t_2]$, i.e. functions $x(t)=(y(t),z(t))$, or $x:[t_1,t_2]\to R^n$, $w:[t_1,t_2]\to R^m$, such that (i) $y\in \mathrm{ACg}$, $z\in \mathrm{BVC}$, $w$ is measurable; (ii) $(t,y(t),z(t))\in A$, $w(t)\in U(t,y(t),z(t))$, a.e. in $[t_1,t_2]$; (iii) $x'(t)=f(t,x(t),w(t))$, a.e. in $[t_1,t_2]$, $f_0(\cdot,x(\cdot),w(\cdot))\in L_1([t_1,t_2])$.

We consider the functional $\mathscr{I}_0:\Omega_0\to R$ defined by

$$\mathscr{I}_0(x)=\mathscr{I}_0(y,z)=\inf_{\Gamma_0(x)}\ \underline{\lim}_{k\to\infty}\int_{t_1}^{t_2} f_0(t,y_k(t),z_k(t),w_k(t))\,dt$$

$$=\inf_{\Gamma_0(x)}\ \underline{\lim}_{k\to\infty} I_0(y_k,z_k,w_k),$$

where $\Gamma_0(x)$ denotes the class of all sequences $(x_k,w_k)_{k\in N}$ in $\Omega_0$ such that (a) $x_k=(y_k,z_k)\in \mathrm{ACg}$, $k\in N$; (b) $y_k\to y$ uniformly and $z_k\to z$ pointwise a.e. in $[t_1,t_2]$. If $\Gamma_0(x)=\emptyset$ we take $\mathscr{I}_0(x)=+\infty$. The class $\Omega_0$ is said to be *closed* if it has the following property (c): If $(y_k,z_k,w_k)_{k\in N}$ is a sequence of admissible systems, all in $\Omega_0$, satisfying (a) and (b), and if there exists a measurable $w$ such that $(y,z,w)$ is an admissible system, then $(y,z,w)$ belongs to $\Omega_0$.

Note that, if the problem of minimizing the functional above involves given initial or terminal values for $x$, say $x(t_1)\in B_1$, $x(t_2)\in B_2$, then we will proceed as illustrated in Remark 2 of Section 3.

It is well known (see [5], Section 1.13) that the problem of optimal control described above can be deparametrized, and essentially reduced to a problem of calculus of variations as discussed in Section 3. For every $(t,x)\in A$ let $Q(t,x)$ denote the set

$$Q(t,x)=\{\zeta\in R^n:\zeta=f(t,x,w),\ w\in U(t,x)\},$$

and take

$$M=\{(t,x,\zeta)\in R^{2n+1}:(t,x)\in A,\ \zeta\in Q(t,x)\}.$$

Let $F_0(t, x, \zeta)$ denote the scalar function defined on $M$ by taking

$$F_0(t, x, \zeta) = \inf\{z^0 \in R;\ z^0 \geqq f_0(t, x, w),\ \zeta = f(t, x, w),\ w \in U(t, x)\}. \tag{1}$$

If for some $\zeta$ the set in brackets is empty, we take $F_0 = +\infty$. If in (1) inf is actually a minimum for all $(t, x, \zeta) \in M$, then we may replace the problem of optimal control with the problem of the calculus of variations studied in Section 3, concerning the integral functional $\mathscr{I}$ relative to the integrand $F_0$, with constraints $(t, x(t)) \in A$, $x'(t) \in Q(t, x(t))$, a.e. in $[t_1, t_2]$, and where $x = (y, z)$, $y \in \mathrm{ACg}$, $z \in \mathrm{BVC}$. We will apply Theorem 2 of Section 3 to the present problem of the calculus of variations. Of course, we shall assume that the sets $Q(t, x)$ are nonempty and convex and that the scalar function $F_0(t, x, \zeta)$ is lower semicontinuous in $(t, x, \zeta)$ and convex in $\zeta$. Moreover, once we have a solution $x = (y, z)$ of the deparametrized problem, or problem of the calculus of variations, we shall need to know that there exists some measurable function $w(t)$, or $w: [t_1, t_2] \to R^m$ such that

$$w(t) \in U(t, x(t)),\ f_0(t, x(t), w(t)) = F_0(t, x(t), x'(t)), f(t, x(t), w(t)) = x'(t), \quad \text{a.e. in } [t_1, t_2]. \tag{2}$$

This is a consequence of the implicit function theorems. For instance, if $f_0$ and $f$ are continuous on the closed set $M_0$, then the existence of a measurable $w(t)$ satisfying (2) follows from the McShane-Warfield implicit function theorem ([5], Theorem 8.2.iii). In [5], Sections 8.2., 8.3, a great many situations are depicted for which some implicit function theorem applies. Concerning the $n$-vector function $f(t, x, w) = (f_1, \ldots, f_n)$, we write $\tilde{f}_1 = (f_1, \ldots, f_\alpha)$ and $\tilde{f}_2 = (f_{\alpha+1}, \ldots, f_n)$. We shall need the following alternative assumptions:

($g_1$) There is a scalar function $\phi(\zeta)$, $0 \leqq \zeta < +\infty$, or $\phi: R_0^+ \to R$ bounded below, such that $\phi(\zeta)/\zeta \to +\infty$, as $\zeta \to +\infty$, and $f_0(t, x, w) \geqq \phi(|\tilde{f}_1(t, x, w)|)$ for all $(t, x, w) \in M_0$.

($g_2$) For every $\varepsilon > 0$ there is a summable scalar function $\psi_\varepsilon(t) \geqq 0$ such that $|\tilde{f}_1(t, x, w)| \leqq \psi_\varepsilon(t) + \varepsilon f_0(t, x, w)$ for all $(t, x, w) \in M_0$.

($g_3$) For any $\alpha$-vector $p \in R^\alpha$ there is a summable scalar function $\phi_p(t) \geqq 0$, such that $f_0(t, x, w) \geqq \langle p, \tilde{f}_1(t, x, w)\rangle - \phi_p(t)$, for all $(t, x, w) \in M_0$.

Note that, under condition ($g_1$), certainly $\phi(\zeta) \geqq \lambda$ for some real constant $\lambda$, and then $f_0(t, x, w) \geqq \phi(|\tilde{f}_1(t, x, w)|) \geqq \lambda$ for all $(t, x, w) \in M_0$. Under condition ($g_2$) and $\varepsilon = 1$, we have $|\tilde{f}_1| \leqq \psi_1(t) + f_0(t, x, w)$; hence $f_0(t, x, w) \geqq -\psi_1(t)$, a summable function. Under condition ($g_3$) and $p = 0$, we have $f_0(t, x, w) \geqq -\phi_0(t)$, a summable function.

**Theorem 3** (An existence theorem for problems of Optimal Control). *Let* $1 \leqq \alpha \leqq n - 1$, *and assume that* (i) *$A$ is compact and $M_0$ is closed*; (ii) *the sets* $\tilde{Q}(t, x)$ *are closed, convex and satisfy property* (Q) *with respect to* $(t, x)$ *at every point* $(t,x)$ *of $A$ (with the exception perhaps of a set of points whose t-coordinate lies on a set of measure zero on the t-axis)*; (iii) *the functions $f$ and $f_0$ are continuous and satisfy one of the growth conditions* ($g_1$), ($g_2$), ($g_3$). *Also we assume that* *the class* $\Omega_0$ *is nonempty and closed, and* (iv) *there is a constant* $W_0$ *such that for every element* $x = (y, z) \in \Omega_0 \cap \mathrm{ACg}$, *then* $V^*(z) \leqq W_0$. *Then the functional* $\mathcal{I}_0$ *has an absolute minimum* $x = (y, z)$ *in* $\Omega_0$.

*For* $\alpha = 0$, *then* $x = z$, *the requirements* ($g_1$) *or* ($g_2$) *or* ($g_3$) *do not apply, yet the conclusion is still valid if we know that* (iii′) *there is a summable scalar function* $\lambda: [t_1, t_2] \to R$ *such that* $f_0(t, x, w) \geqq \lambda(t)$, *for all* $(t, x, w) \in M_0$.

*For* $\alpha = n$, *then* $x = y$, $\Omega_0$ *is a nonempty and closed class of* ACg *functions* $y(t) = (y^1, \ldots, y^n)$, $t \in [t_1, t_2]$. *condition* (iv) *does not apply, and the problem reduces essentially to those discussed in Theorems* 11.4.i. *and* ii *of* [5].

Statement 3 is a corollary of Theorem 2.

Note that for $0 \leqq \alpha \leqq n - 1$, if (iv′) there are scalar functions $\psi_i \in L_1([t_1, t_2]$, $i = \alpha + 1, \ldots, n$, such that $(t, y, z, u, v) \in M_0$ implies $v^i \geqq \psi_i(t)$ a.e. in $[t_1, t_2]$, then (iv) certainly holds.

Note that, for $1 \leqq \alpha \leqq n - 1$, the sets $\tilde{Q}(t, x)$ are closed and convex, ($g_1$) holds, and if (iv″) there exist constants $L_i$ such that a.e. in $[t_1, t_2]$, $(t, y, z, u, v) \in M_0$ implies $v^i \geqq L_i$, $i = \alpha + 1, \ldots, n$, then both (ii) and (iv) hold.

See also Remarks 6 and 6′.

*Acknowledgment*. This research was partially supported by the G.N.A.F.A.–C.N.R.

## References

1. M. Boni, *Variazione generalizzata con peso e quasi additività*, Atti Sem. Mat. Fis. Univ. Modena **25** (1976), 195–210.
2. M. Boni & P. Brandi, *Variazione classica e generalizzata con peso*, Atti Sem. Mat. Fis. Univ. Modena **23** (1974), 286–307.
3. D. Candeloro & P. Pucci, *L'integrale di Burkill-Cesari come integrale del Calcolo delle Variazioni*, Boll. Un. Mat. Ital. (5) **18-B** (1981), 1–24.
4. L. Cesari, *Sulle funzioni a variazione limitata*, Ann. Scuola Norm. Sup. Pisa **5** (1936), 299–313.
5. L. Cesari, *Optimization-Theory and Applications*, Springer Verlag (1983).
6. L. Cesari, P. Brandi & A. Salvadori, (a) *Discontinuos solutions in problems of optimization*, Ann. Scuola Norm. Sup. Pisa, to appear. – (b) *Existence theorems for multiple integrals of the calculus of variations for discontinuous solutions*, Annali Mat. pura appl., to appear.
7. L. Cesari & P. Pucci, *An elementary proof of an equivalence theorem relevant in the theory of optimization*, J. Optimization Theor. Appl. **46** (1985), 215–226.
8. E. Conway & J. Smoller, *Global solutions of the Cauchy problem for quasi-linear first-order equations in several space variables*, Comm. Pure Appl. Math. **19** (1966), 95–105.
9. C. M. Dafermos, *Generalized characteristics and the structure of solutions of hyperbolic conservation laws*, Indiana Univ. Math. J. **26** (1977), 1097–1119.
10. R. J. di Perna, *Singularities of solutions of nonlinear hyperbolic systems of conservation laws*, Arch. Rational Mech. Anal. **60** (1975), 75–100.
11. C. Goffman, *Lower semicontinuity and area functionals. I. The non parametric case*, Rend. Circ. Mat. Palermo **2** (2) (1953), 203–235.
12. K. Krickeberg, *Distributionen, Funktionen beschränkter Variation und Lebesguescher Inhalt nichtparametrischer Flächen*, Ann. Mat. Pura Appl. **44** (1957), 105–133.

13. A. SALVADORI, *Sulla convergenza in lunghezza con peso per una successione di curve parametriche*, Rend. Circ. Mat. Palermo **36** (2) (1977), 195–228.
14. J. SERRIN, *On the definition and properties of certain variational integrals*, Trans. Amer. Math. Soc. **101** (1961), 139–167.
15. C. VINTI, *L'integrale di Weierstrass-Burkill*, Sem. Mat. Fis. Univ. Modena **18** (196), 295–316.
16. A. L. VOLPERT, *The spaces BV and quasilinear equations*, Mat. Sb. **73** (1967), 225–267.

University of Michigan
Ann Arbor
and
Università degli Studi
Perugia

*(Received April 25, 1986)*

# *On the Uniqueness of Flow of a Navier-Stokes Fluid Due to a Stretching Boundary*

J. B. McLEOD & K. R. RAJAGOPAL

*To James Serrin, on the occasion of his sixtieth birthday, and in recognition of his fundamental contributions both to analysis and to applied mathematics in its finest sense*

## 1. Introduction

When a sheet of polymer is extruded continuously from a die, it entrains the ambient fluid and a boundary layer develops. Such a boundary layer is markedly different from that in the Blasius flow past a flat plate in that the boundary layer grows in the direction of the motion of the sheet, starting at the die. SAKIADIS [1]–[3] was the first to study such a boundary layer flow due to a rigid flat continuous surface moving in its own plane. Later, ERICKSON, FAN & FOX [4] studied the boundary layer due to the motion of a porous flat plate when the transverse velocity at the surface is non-zero.

There are situations when the extruded polymer sheet is being stretched as it is being extruded (*cf.* MCCORMACK & CRANE [5]) and this is bound to alter significantly the boundary layer characteristics of the flow considered by SAKIADIS [1]. A detailed analysis of the boundary layer flow of a Navier-Stokes fluid due to a stretching sheet has been carried out recently by DANBERG & FANSLER [6]. They assume that the free-stream velocity of the fluid is a constant, while the sheet is being stretched by a velocity proportional to $x$, where $x$ is the coordinate in the direction of stretch. The problem of the sheet velocity being a constant, while the free-stream velocity is proportional to $x$, had been studied earlier by ROTT [7].

Let us consider the two-dimensional flow of a Navier-Stokes fluid due to the stretching of a flat sheet coinciding with the plane $y = 0$, the fluid being confined to $y > 0$. A steady uniform stress leading to equal and opposite forces is applied along the $x$-axis so that the sheet is stretched keeping the origin fixed. Thus, the appropriate boundary conditions are

$$u = cx, \quad v = 0, \quad \text{at} \quad y = 0 \quad (c > 0) \tag{1.1}$$

and

$$u \to U_\infty \quad \text{as} \quad y \to \infty. \tag{1.2}$$

Here $u$ and $v$ denote the velocity components in the $x$ and $y$ directions, respectively, and $U_\infty$ is the free-stream velocity.

We introduce the dimensionless quantities

$$\bar{x} = \frac{cx}{U_\infty}, \quad \bar{y} = \frac{cy}{U_\infty}\sqrt{R}, \quad \bar{u} = \frac{u}{U_\infty}, \quad \bar{v} = \frac{v}{U_\infty}\sqrt{R}, \tag{1.3}$$

where

$$R = \frac{U_\infty^2}{c\nu} \tag{1.4}$$

is the Reynolds number. We shall seek a solution of the form

$$\bar{u} = \bar{x}f'(\bar{y}) + g'(\bar{y}) \quad \text{and} \quad \bar{v} = -f(\bar{y}). \tag{1.5}$$

Notice that the above velocity field is isochoric or volume-preserving. Substituting (1.5) into the Navier-Stokes equations (with the pressure independent of $x$) leads to the following system of coupled ordinary differential equations:

$$f''' + ff'' - f'^2 = 0, \tag{1.6}$$

$$g''' + fg'' - f'g' = 0. \tag{1.7}$$

The appropriate boundary conditions are

$$f(0) = 0, \quad f'(0) = 1, \quad f'(\bar{y}) \to 0 \quad \text{as} \quad \bar{y} \to \infty, \tag{1.8}$$

and

$$g(0) = 0, \quad g'(0) = 0, \quad g'(\bar{y}) \to 1 \quad \text{as } \bar{y} \to \infty. \tag{1.9}$$

It is interesting to note that for the assumed form of the velocity field (1.5), the boundary layer equations and the Navier-Stokes equations are the same. Thus, for the velocity field under consideration, the terms which are dropped out of the Navier-Stokes equations in deriving the boundary layer approximations happen to be identically zero.

Equation (1.6), subject to (1.8), has the following simple solution:

$$f(\bar{y}) = 1 - e^{-\bar{y}}. \tag{1.10}$$

Using this solution for $f$ in (1.7), Danberg & Fansler were able to exhibit an explicit solution for $g$, by using the method of variation of parameters.

Since equation (1.6) is nonlinear, it is possible that (1.6), subject to (1.8), possesses other solutions in addition to (1.10). This, then, would imply the possible existence of a boundary layer with a structure different from that exhibited by (1.10).In this paper we prove that (1.10) is the unique solution to (1.6) and (1.8). An alternative treatment of this is given in [8].

Recently, Rajagopal, Na & Gupta [9], [10], have studied the boundary layer due to a stretching sheet, in the case of an incompressible homogeneous fluid of second grade. Using the velocity field (1.5), they show that the equations of motion are given by

$$f''' + ff'' - f'^2 + k\{2f'''f' - f''^2 - f^{iv}\} = 0, \qquad (1.11$$

and

$$g''' + fg'' - f'g' + k\{g'''f' + f'''g' - f''g'' - fg^{iv}\} = 0, \qquad (1.12)$$

where $k$ is a non-Newtonian parameter. They study the problem by means of a perturbation in $k$, their zero$^{\text{th}}$ order equation corresponding to the equations (1.6) and (1.7). Thus the uniqueness established here bears on their study also.

The boundary value problem under consideration also illustrates a pitfall which one encounters while solving nonlinear equations in unbounded domains numerically, namely, the specification of infinity. Fixing infinity at what seems a sufficiently large distance and solving the boundary value problem using a gears method suggests the existence of two distinct solutions, one of which is (1.10) (*cf.* SZERI [11]). Shifting the location of infinity moderately produces no change, even in the seventh decimal place in the numerical solution, and lends credence to the existence of a second solution. However, in the light of the uniqueness theorem proved in this paper, the problem was re-examined. It was found (*cf.* SZERI [11]) that, as the location of infinity is moved significantly further, the apparent second solution in fact diverges.

The underlying reason for uniqueness is that equation (1.6) is invariant under two simple transformations, first a translation of the independent variable, and secondly the basic transformation

$$f \to \alpha f, \qquad x \to \alpha^{-1} x.$$

This implies that we can in effect reduce the order of the equation by two, so that we deal with a first-order equation. It is then a matter of looking at the phase plane for this first-order equation, from which uniqueness of the solution can be deduced quite easily.

The plan of the paper is as follows. In section 2 we prove some easy preliminary lemmas on the asymptotic signs of possible solutions. In section 3 we give the transformation which reduces the equation to a form in which it is amenable to phase plane treatment, and in section 4 we discuss the phase plane. The uniqueness proof is then completed in section 5.

## 2. Some Preliminary Lemmas

**Lemma 1.** *The solution* (1.10) *is the only solution of* (1.6) *and* (1.8) *for which* $f' \geqq 0$.

**Proof.** First observe that a solution of (1.6) and (1.8) which satisfies $f' \geqq 0$ also satisfies $f'' \leqq 0$. For (1.6) implies that

$$\left\{f'' \exp\left(\int_0^x f(t)\, dt\right)\right\}' = f'^2 \exp\left(\int_0^x f(t)\, dt\right),$$

so that

$$f'' \exp\left(\int_0^x f(t)\,dt\right)$$

cannot deerease, and so $f''$ vanishes at most once and, if it does vanish, is positive after the zero. But if also $f' \geqq 0$, then because $f''$ is ultimately positive, $f'$ tends to a strictly positive limit at infinity, which contradicts $f'(\infty) = 0$. Hence $f'' \leqq 0$.

Now suppose for contradiction that there are two solutions $f_1$, $f_2$, where $\phi = f_2 - f_1$. Then $\phi$ satisfies

$$\phi''' + f_2\phi'' - (f_1 + f_2)'\,\phi' + f_1''\phi = 0, \tag{2.1}$$

with

$$\phi(0) = 0, \quad \phi'(0) = 0, \quad \phi'(\infty) = 0.$$

Suppose without loss of generality that $\phi''(0) > 0$. Then initially $\phi > 0$, $\phi' > 0$, $\phi'' > 0$, and so long as these inequalities are maintained,

$$\phi'' \exp\left(\int_0^x f(t)\,dt\right)$$

is nondecreasing since $(f_1 + f_2)' \geqq 0$, $f_1'' \leqq 0$. Hence $\phi$, $\phi'$, $\phi''$ never vanish, which contradicts $\phi'(\infty) = 0$.

**Lemma 2.** *For any second solution of* (1.6) *and* (1.8) $f'$, $f''$, $f'''$ *all vanish precisely once, at* $x_1, x_2, x_3$, *say, with* $x_1 < x_2 < x_3$. *Also, ultimately,* $f > 0$, $f' < 0$, $f'' > 0$, $f''' < 0$.

**Proof.** Certainly, from Lemma 1, $f'$ must have at least one zero, and so $f''$ must have at least one zero, since $f'(\infty) = 0$.

Hence, from the proof of Lemma 1, $f''$ has precisely one zero, at $x_2$, say, and $f'$ is decreasing before $x_2$ and increasing after. Since $f'(0) = 1$ and $f'(\infty) = 0$, it is clear that $f'$ also has precisely one zero, at $x_1$, say, with $x_1 < x_2$. (Note that a double zero of $f'$ is impossible, for $f' = f'' = 0$ implies $f''' = 0$ and so $f =$ constant; similarly, $f''$ cannot have a double zero.)

Finally, $f(x) > 0$ for $x \leqq x_1$, and so, from (1.6), $f'''(x) > 0$ for $x \leqq x_1$. Since

$$f^{iv} + f'''f = f'f'',$$

it follows that in $(x_1, x_2)$, where $f'f'' > 0$, we must have

$$f''' \exp\left(\int_0^x f(t)\,dt\right) \tag{2.2}$$

nondecreasing, and so $f'''(x_1) > 0$ implies $f'''(x_2) > 0$. For $x > x_2$, (2.2) is decreasing, and so $f'''$ has at most one zero, and indeed exactly one since $f''(x_2) = 0$ and $f''(\infty) = 0$. So $f'''(x_3) = 0$ with $x_3 > x_2$.

The above argument shows that ultimately $f' < 0$, $f'' > 0$, $f''' < 0$. It is also true that $f(x) > 0$ for all $x > 0$. For $f$ is first increasing (from zero) and then decreasing. But

$$f'' \exp\left(\int_0^x f(t)\,dt\right)$$

never decreases. Hence ultimately, when $f'''$ is positive decreasing, wemust have $\exp\left(\int_0^x f(t)\,dt\right)$ increasing, so that $f > 0$. Hence $f > 0$ ultimately, and so always.

## 3. The Transformation of the Equation

As long as $f > 0$, $f' < 0$, we can make the change of variables

$$z = -\log f, \qquad u = f'/f^2, \qquad \delta u = \frac{du}{dz},$$

and (1.6) becomes

$$\delta^2 u + \left\{\frac{(\delta u)^2}{u} - 7\delta u - \frac{\delta u}{u}\right\} + 6\,(u + \tfrac{1}{6}) = 0. \tag{3.1}$$

The proof is routine. For

$$\delta u = -\frac{du}{d(\log f)} = -f\frac{du}{dx}\frac{dx}{df} = -\frac{f}{f'}\left(\frac{ff'' - 2f'^2}{f^3}\right), \tag{3.2}$$

$$f'' = -ff'(\delta u - 2u) = -e^{-3z}\,u(\delta u - 2u),$$

$$\begin{aligned} f''' &= -\frac{d}{dz}\{e^{-3z}\,u(\delta u - 2u)\}\frac{dz}{dx} \\ &= \frac{f'}{f}\,e^{-3z}\{-3(u\,\delta u - 2u^2) + u\,\delta^2 u + (\delta u)^2 - 4u\,\delta u\}, \end{aligned}$$

and substitution in (1.6) gives the answer.

Note that Equation (3.1) is second-order autonomous, so that it could be reduced to first order, but it is easier to work with it as it stands. Since it is autonomous, it is amenable to phase plane methods, and we study the phase plane in the next section. The fact that equation (1.6) can be reduced to a first-order equation can be observed at once from the fact that it is invariant under both the transformations

$$f \to f, \qquad x \to x + c;$$

$$f \to \alpha f, \qquad x \to \alpha^{-1} x.$$

## 4. The Phase Plane

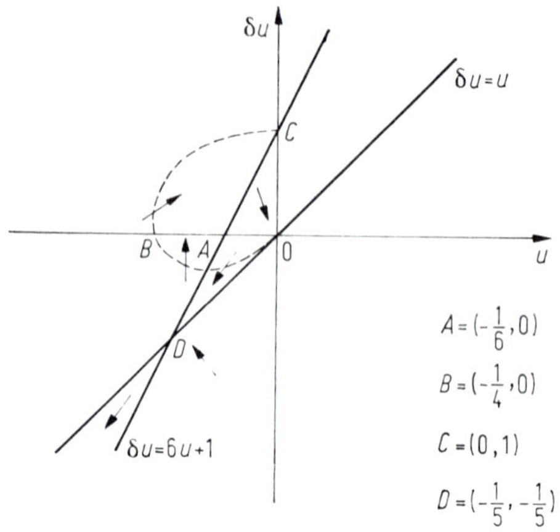


Fig. 1

The arrows denote the direction of a trajectory lying in the appropriate area of the phase plane. Thus

$$\nearrow \text{ implies both } \delta u,\ \delta^2 u > 0,$$

$$\searrow \text{ implies } \delta u > 0,\ \delta^2 u < 0$$

and so on. The appropriate areas are determined by observing that $\delta^2 u = 0$ when

$$(\delta u)^2 - 7u\,\delta u - \delta u + 6u^2 + u = 0,$$

*i.e.* when

$$(\delta u - u)(\delta u - 6u - 1) = 0.$$

The only rest-point ($\delta u = \delta^2 u = 0$) is the point $A(-\frac{1}{6}, 0)$ although the origin is also some sort of singular point. (The equation is singular at $u = 0$.)

We established the validity of the transformation leading to the phase plane only when $u < 0$, so that arrows are not entered in the right-hand half of the diagram.

The dotted line denotes one special trajectory. It is trivial to verify that

$$f = a + be^{-ax} \tag{4.1}$$

is a solution of (1.6) for every $a$, $b$. (Our interest will be in $a > 0$, $b > 0$). In fact, because of the invariance of the equation under the transformations previously mentioned, all solutions of the form (4.1) give the same trajectory in the phase plane, and it is given by the dotted line. Note the following points:

(i) as $x \to \infty$,

$$f \sim a, \quad f' \sim -abe^{-ax}, \quad f'' \sim a^2be^{-ax},$$

so that

$$u \to 0, \quad \delta u \to 1;$$

(ii) as $x \to -\infty$,

$$f \sim be^{-ax}, \quad f' \sim -abe^{-ax}, \quad f'' \sim a^2be^{-ax},$$

so that, since (3.2) leads to

$$\delta u - 2u = -\frac{f''}{ff'} \sim \frac{a}{b}\, e^{ax} \sim -u,$$

we have $\delta u \sim u$;

(iii) when $\delta = 0$, we have

$$f'' = 2f'^2,$$

which gives

$$(a + be^{-ax})\, a^2be^{-ax} = 2a^2b^2e^{-2ax}$$

so that

$$a = be^{-ax}, \qquad u = -\tfrac{1}{4}.$$

## 5. The Proof of Nonexistence of a Second Solution

Let $C$ denote the closed curve consisting of the dotted curve and the portion of the $\delta u$-axis. Suppose for contradiction that we do have a second solution to (1.6) and (1.8).

Suppose first that for some value of $x$ the trajectory of this second solution lies inside $C$. Then, as we decrease $x$ (or $z$), we cannot get outside $C$. For we cannot cross the dotted portion of $C$, since that itself is a trajectory and trajectories do not cross without being identical. (The form (4.1) corresponding to the dotted trajectory does not give us a second solution since it cannot satisfy the inequalities of Lemma 2, which would imply $a > 0$, $b > 0$, and also $f(0) = 0$.) Also, the direction of the arrows shows that we cannot cross $\delta u = 0$. As we decrease $x$, therefore, the trajectory always lies inside $C$, so that $u < 0$, $f' < 0$, $f$ is increasing (for $x$ decreasing), and so $f(0) = 0$ is impossible.

The only other possibility is that the trajectory lies outside $C$. But then the arrows show that, as we increase $x$, we eventually increase $\delta u$, and so eventually we must have $\delta u > 0$, $u = 0$. But Lemma 1 says that for any second solution we must have $u < 0$ if $x$ is sufficiently large, and this again gives a contradiction. (Note that the line $\delta u = u$ corresponds to $f''' = 0$, so that, from Lemma 1, we have to be above $\delta u = u$ when $x$ is sufficiently large.)

This completes the proof, but there are a number of remarks that can be made.

*Remark 1.* We have used the fact that a trajectory, in $C$ and going backwards in $x$, does not cross $C$. It can also be shown that a trajectory, in $C$ and going forwards in $x$, does not cross $C$, although the argument is a little more delicate. But the final conclusion is that no trajectory can cross $C$, and it is interesting to ask what happens to those trajectories which lie always inside $C$.

It is just a matter of linearization to prove that a trajectory starting sufficiently close to the rest-point $A$ spirals out, and although we have not carried out the calculations completely, it seems that a trajectory starting sufficiently closc to $C$ spirals in. Hence there must be a limit-cycle inside $C$, which corresponds to a periodic solution of the equation (3.1), periodic, that is, in $z$, not in $x$.

*Remark 2.* It is possible to make a similar investigation of the equation

$$f''' + ff'' = \lambda f'^2,$$

for varying values of $\lambda\,(> 0)$. We have not carried out all the details, but in (3.1) the only change is in the constant term, and the main consequence as far as the phase plane is concerned is that, at $\lambda = 8/7$, the rest-point changes from unstable to stable and merges with the limit-cycle. For $\lambda > 8/7$, therefore, the limit-cycle disappears and we just have, inside $C$, spirals spiralling in to the stable rest-point. Of course, we no longer have the explicit form (1.10) for the solutions that correspond to the trajectory $C$, but we can establish the existence of such solutions by shooting arguments.

*Remark 3.* It is interesting that, with $\delta u = y$, $u = x$, the equation (3.1) becomes of the form

$$\frac{dy}{dx} = -\frac{6x^2 - 7xy + y^2 + x - y}{xy},$$

which is the type of equation that is discussed in Hilbert's sixteenth problem, with reference to the maximum number of limit-cycles that such an equation may possess.

*Acknowledgement.* This research is supported in part by the Solid Mechanics Program of the National Science Foundation.

## References

1. Sakiadis, B. C., Boundary-layer behavior on continuous solid surfaces I: Boundary-layer equations for two-dimensional and axisymmetric flow, Amer. Inst. Ch. Eng. J. **7** (1961), 26–28.
2. Sakiadis, B. C., Boundary-layer behavior on continuous solid surfaces II: The boundary layer on a continuous flat surface, Amer. Inst. Ch. Eng. J. **7** (1961), 221–225.
3. Sakiadis, B. C., Boundary-layer behavior on continuous solid surfaces III: The boundary layer on a continuous cylindrical surface, Amer. Inst. Ch. Eng. J. **7** (1961), 467–472.
4. Erickson, L. E., L. T. Fan, & V. G. Fox, Heat and mass transfer on a moving continuous flat plate with suction or injection, Ind. and Eng. Ch. Fund. **5** (1966), 19–25.
5. McCormack, P. D., & L. Crane, Physical Fluid Dynamics, Academic Press, New York (1973).
6. Danberg, J. E., & K. S. Fansler, A nonsimilar moving-wall boundary-layer problem, Quart. Appl. Math. **34** (1976), 305–309.

7. Rott, N.. Unsteady viscous flow in the vicinity of a stagnation point, Quart. Appl. Math. **13** (1956), 444–451.
8. Troy, W. C., E. A. Overman, G. B. Ermentrout & J. P. Keener, Uniqueness of flow of a second order fluid past a stretching sheet, Quart. Appl. Math., to appear.
9. Rajagopal, K. R., T. Y. Na, & A. S. Gupta, Flow of a viscoelastic fluid over a stretching sheet, Rheologica Acta **23** (1984), 213–215.
10. Rajagopal, K. R., T. Y. Na, & A. S. Gupta, A nonsimilar boundary layer on a stretching sheet in a non-Newtonian fluid with uniform free stream, submitted.
11. A. Z. Szeri, Private communication (1985).

Wadham College
Oxford
and
Department of Mechanical Engineering
University of Pittsburgh

*(Received February 28, 1986)*

# *On the Behavior at Infinity of Solutions of Elliptic Systems with a Finite Energy Integral*

V. A. KONDRATIEV & O. A. OLEINIK

*Dedicated to J. Serrin on his* 60<sup>th</sup> *birthday*

Asymptotic behavior in a neighborhood of infinity is studied in this paper for solutions of an elliptic system of order $2m$ with constant complex coefficients. It is supposed that the weighted Dirichlet integral is bounded. Our considerations include solutions with finite energy for the system of linear elasticity (see Theorem 3). A class of solutions periodic in some independent variables is also studied in this paper (the E. SANCHEZ-PALENCIA problem).

Problems connected with asymptotics at infinity of solutions of elliptic equations and systems form the subject of numerous papers. In the case of second-order elliptic equations the behavior of solutions at infinity was studied by J. SERRIN & H. WEINBERGER in [1]. Similar problems under other assumptions are treated in [2], [3]. An asymptotic expansion at infinity for a class of solutions of elliptic systems with constant coefficients is given in [4]. This problem is considered also for some elliptic systems in [5], for polyharmonic equations in [6].

Consider a system of equations

$$\sum_{j=1}^{N} l_{kj}(\mathscr{D}_z)\, u_j(z) = f_k(z), \qquad k = 1, \ldots, N, \tag{1}$$

where $l_{kj}(\mathscr{D}_z)$ are linear homogeneous differential operators of order $2m$ with constant complex coefficients, $z = (x, y)$, $x = (x_1, \ldots, x_n)$, $y = (y_1, \ldots, y_q)$, $z = (z_1, \ldots, z_p)$, $n + q = p$, $n \geqq 2$, $q \geqq 0$, $\mathscr{D}_z = \left(\frac{\partial}{\partial z_1}, \ldots, \frac{\partial}{\partial z_p}\right)$, $\mathscr{D}_x = \left(\frac{\partial}{\partial x_1}, \ldots, \frac{\partial}{\partial x_n}\right)$, $\mathscr{D}_y = \left(\frac{\partial}{\partial y_1}, \ldots, \frac{\partial}{\partial y_q}\right)$.

Let us rewrite system (1) in vector form:

$$\mathscr{L}(\mathscr{D}_x, \mathscr{D}_y)\, u = f, \qquad u = (u_1, \ldots, u_N), \tag{2}$$

$$f = (f_1, \ldots, f_N).$$

Suppose that $f \in C^\infty(R^p)$ and $f \equiv 0$ for $|x| > A$, $A = \text{const}$, $u \in C^\infty(R^p)$, that $u(z)$ is a solution of system (2); and that $u(z)$ and $f(z)$ are vector-valued functions, $2\pi$-periodic in $y_1, \ldots, y_q$.

System (1) is assumed to be elliptic and therefore, for any $\xi \in R^p$, $\xi \neq 0$,

$$\det \| l_{kj}(\xi) \| \neq 0. \tag{3}$$

Set $Q = \{z : x \in R^n, 0 < y_j < 2\pi, j = 1, \ldots, q\}$. We shall study the behavior of solutions $u(z)$ of system (1) as $|x| \to \infty$ assuming that $u(z)$ is $2\pi$-periodic in $y$ and that

$$J(u) = \sum_{j=1}^{N} \sum_{|\alpha|=m} \int_Q (1 + |x|)^{2\sigma} |\mathcal{D}_z^\alpha u_j|^2 \, dz < \infty, \tag{4}$$

where $\alpha = (\alpha_1, \ldots, \alpha_p)$, $|\alpha| = \alpha_1 + \ldots + \alpha_p$, $\mathcal{D}_{z_j} \equiv \dfrac{\partial}{\partial z_j}$, $\mathcal{D}_z^\alpha \equiv \mathcal{D}_{z_1}^{\alpha_1} \ldots \mathcal{D}_{z_p}^{\alpha_p}$, $|x|^2 = x_1^2 + \ldots + x_n^2$, $\sigma = \text{const} \leq 0$.

This problem for $q > 0$ arises in homogenization theory. The case $N = 1$, $m = 1$, $\sigma = 0$ is considered in [2] where it is supposed that the elliptic equation is of divergence form and its coefficients are bounded measurable functions, periodic in $y$. Similar problems for second order parabolic equations are studied in [3]. The case $n = 1$, $q > 0$ for a second-order elliptic equation is considered in [7], [8] and for the stationary system of linear elasticity with variable coefficients in [9].

The main results of the present paper are formulated in Theorems 1–3. Theorem 2 gives a complete asymptotic expansion at infinity for a solution of system (1) under condition (4) as well as an estimate for the remainder term. Theorem 1 is a special case of Theorem 2 for $q = 0$ and it is used in the proof of Theorem 2.

We also consider in this paper a special class of elliptic systems

$$\sum_{j=1}^{N} m_{kj}(\mathcal{D}_z) \, u_j(z) = f_k(z), \qquad k = 1, \ldots, N, \tag{5}$$

$p = N$, $q \geqq 0$, $z = (z_1, \ldots, z_N)$, $u = (u_1, \ldots, u_N)$, where $m_{kj}(\mathcal{D}_z) = m_{kj}(\mathcal{D}_x, \mathcal{D}_y)$ are linear second-order homogeneous differential operators with constant coefficients. The system of linear elasticity for a homogeneous medium is a particular case of system (5). The integral of the form

$$\int_Q \sum_{k,j=1}^{N} \left( \frac{\partial u_k}{\partial z_j} + \frac{\partial u_j}{\partial z_k} \right)^2 dz \tag{6}$$

corresponds to the energy integral in the theory of elasticity. In Theorem 3 we obtain an asymptotic expansion in a neighborhood of infinity for a solution $u(z)$ of system (5) and in particular of the system of linear elasticity, assuming that $u(z)$ is $2\pi$-periodic in $y$ and the integral (6) is finite. A generalization of the Korn inequality is used (see formula (53)).

To prove Theorems 1–3 we shall need a number of auxiliary results concerning elliptic systems with constant coefficients.

**1.** In this section we consider in $R_x^n$ systems of the form

$$L(\mathscr{D}_x)\,u = F, \qquad F \in C_0^\infty(R^n), \qquad x = (x_1, \ldots, x_n), \tag{7}$$

where $u(x) = (u_1(x), \ldots, u_N(x))$, $F = (F_1(x), \ldots, F_N(x))$, $L(\mathscr{D}_x) \equiv \|L_{kj}(\mathscr{D}_x)\|$, $k, j = 1, \ldots, N$, $L_{kj}$ are linear differential operators of order $2m$ with constant coefficients. We assume that for any $\xi \in R^n$

$$K_1(1 + |\xi|)^{2mN} \leqq |\det \|L_{kj}(i\xi)\|| \leqq K_2(1 + |\xi|)^{2mN}, \tag{8}$$

$K_1, K_2 = \text{const} > 0$. Condition (8) obviously implies the ellipticity of systems (7).

Let us define $W_{s,\gamma}$ as a space of vector-valued functions $(u_1, \ldots, u_N)$ such that $u_j(x)$, $j = 1, \ldots, N$, has in $R^n$ weak derivatives up to order $s$ and

$$\|u\|^2_{W_{s,\gamma}} \equiv \sum_{j=1}^{N} \sum_{|\alpha| \leqq s} \int_{R^n} \exp\{2\gamma |x|\} |\mathscr{D}_x^\alpha u_j|^2 \, dx < \infty, \qquad \gamma = \text{const}.$$

The following lemma describes some properties of solutions of system (7).

**Lemma 1.** *Suppose that the operator $L$ of system* (7) *satisfies conditon* (8). *Then there is a positive constant $\gamma_0$ such that for all $\gamma$ with $|\gamma| \leqq \gamma_0$ the operator $L$ effects an injective mapping of $W_{s+2m,\gamma}$ into $W_{s,\gamma}$ and $\gamma_0$, $\|L\|$, $\|L^{-1}\|$ depend only on $K_1$, $K_2$, $n$, $s$ and the maximum modulus of the coefficients of $L$.*

**Proof.** Applying the Fourier transformation in $x$ to system (7), we conclude in a usual way that a solution of system (7) exists and

$$\|u\|_{W_{s+2m,0}} \leqq C_1 \|F\|_{W_{s,0}}, \qquad C_1 = \text{const}. \tag{9}$$

Thus the operator $L$ is a bounded invertible operator from $W_{s+2m,0}$ to $W_{s,0}$.

Let $|\gamma| < 1$ and let $\theta(|x|) > 0$ be a function in $C^\infty(R^n)$ such that $\theta(|x|) = |x|$ for $|x| \geqq 1$. We set $u = v \exp\{-\gamma\theta(|x|)\}$ in (7). As a result we get a new system

$$L(\mathscr{D}_x)\,v + \gamma L_1(\gamma, x, \mathscr{D}_x)\,v = \exp\{\gamma\theta(|x|)\}\,F \equiv F^1(x), \tag{10}$$

where the elements of the matrix $L_1$ are differential operators of order not larger than $2m - 1$ whose coefficients together with all their derivatives are bounded in $R^n$. According to the well known existence theorem, there is a constant $\gamma_0$ such that $L + \gamma L_1$ for $|\gamma| < \gamma_0$ is a bounded invertible operator from $W_{2m+s,0}$ to $W_{s,0}$. Let $F \in W_{s,\gamma}$. Then $Fe^{\gamma\theta} \in W_{s,0}$ and if $|\gamma| \leqq \gamma_0$ there is a solution $v(x)$ of system (10) such that $v \in W_{2m+s,0}$, $\|v\|_{W_{2m+s,0}} \leqq C_2 \|e^{\gamma\theta} F\|_{W_{s,0}} \leqq C_3 \|F\|_{W_{s,\gamma}}$, $C_2, C_3 = \text{const}$. Therefore $u = v\exp\{-\gamma\theta\} \in W_{2m+s,\gamma}$ and

$$\|u\|_{W_{2m+s,\gamma}} \leqq C_4 \|F\|_{W_{s,\gamma}}, \qquad C_4 = \text{const}.$$

The unique solvability of (7) in $W_{2m+s,\gamma}$ for $F \in W_{s,\gamma}$ follows from the unique solvability of (10) in $W_{2m+s,0}$ for $F^1 \in W_{s,0}$. The lemma is proved.

**Corollary of Lemma 1.** *For any integer $s \geqq 0$ there are constants $\gamma_0 > 0$, $C > 0$ such that if $u(x)$ is a solution of system* (7) *and $u \in W_{0,\gamma}$, $|\gamma| \leqq \gamma_0$, $\gamma < 0$,*

*$F \in C_0^\infty(R^n)$, then $u \in W_{2m+s,\tau}$ for any $s \geqq 0$, $\tau \in [0, \gamma_0]$ and*

$$\|u\|_{W_{2m+s,\tau}} \leqq C\,\|F\|_{W_{s,\tau}} \tag{11}$$

*where the constants $C$, $\gamma_0$ do not depend on $F$, $u$ and $\tau$.*

To prove this proposition, it is sufficient to note that due to the interior estimates for solutions of elliptic systems we have $u \in W_{2m+s,\gamma}$ for any $s \geqq 0$, if $u \in W_{0,\gamma}$. (We apply interior estimates [10] in every cube of $R^n$ with vertices having integer coordinates, then we multiply each of these estimates by $\exp\{\gamma\theta(|\hat{x}|)\}$ where $\hat{x}$ is the center of the cube, and take the sum over all disjoint cubes. Thus we obtain an estimate for the norm of $u$ in $W_{2m+s,\gamma}$). According to Lemma 1 system (7) has a unique solution in $W_{2m+s,\gamma}$ with $|\gamma| \leqq \gamma_0$ and satisfies (11).

**2.** In this section we consider an elliptic system of the form

$$B(\mathscr{D}_x)\,u = F, \tag{12}$$

where

$$B = \|B_{kj}(\mathscr{D}_x)\|, \qquad k, j = 1, \ldots, N,\ B_{kj}(\mathscr{D}_x)$$

are homogeneous differential operators of order $2m$ with constant complex coefficients and

$$K_3\,|\xi|^{2mN} \leqq |\det \|B_{kj}(i\xi)\|\,| \leqq K_4\,|\xi|^{2mN}, \tag{13}$$

where $K_3, K_4 = \text{const} > 0$. Condition (13) follows from the ellipticity of system (12).

We study the asymptotic behavior at infinity of solutions of system (12) with a finite integral of the form

$$J_1(u) \equiv \int_{R^n} \sum_{j=1}^{N} \sum_{|\alpha|=m} (1 + |x|)^{2\sigma}\,|\mathscr{D}_x^\alpha u_j|^2\,dx$$

and $F \in C_0^\infty(R^n)$.

**Lemma 2.** *Suppose that $F \in C_0^\infty(R^n)$, $|u(x)| \leqq C(1 + |x|)^k$, $k \geqq 0$, $C = \text{const} > 0$, $u(x)$ is a solution of system (12) in $R^n$. Then for $|x| > 1$*

$$u(x) = P_k(x) + \sum_{h_1 \leqq |\alpha| \leqq h} \mathscr{D}^\alpha \Gamma(x)\,C_\alpha + u^h(x), \tag{14}$$

*where $C_\alpha = (C_1^\alpha, \ldots, C_N^\alpha)$, $C_j^\alpha = \text{const.}$, $h_1 = 2m - n - k$, $\Gamma(x)$ is a fundamental matrix of solutions of system (12) such that*

$$\Gamma(x) = p(x) \ln |x| + \Phi_1(x)|x|^{2m-n}, \tag{15}$$

*if $2m - n \geqq 0$ and $n$ is even, and*

$$\Gamma(x) = \Phi_2(x)\,|x|^{2m-n}, \tag{16}$$

*if $2m - n < 0$ or $n$ is odd. Here $p(x)$ is a matrix, its elements are polynomials of order $2m - n$, the elements of matrix $\Phi_1(x)$ and $\Phi_2(x)$ are homogeneous* *functions of order zero,*

$$|\mathscr{D}_x^\beta u^h(x)| \leqq C_{\beta h} |x|^{2m-n-h-1-|\beta|}, \qquad C_{\beta h} = \text{const}, \tag{17}$$

*for* $|x| \geqq 1$ *and any multi-index* $\beta$; $h \geqq 0$, *if* $n$ *is odd or* $2m - n < 0$; $h \geqq 2m - n$, *if* $2m - n \geqq 0$ *and* $n$ *is even*; $P_k(x) = (P_1^k(x), \ldots, P_N^k(x))$, $P_j^k(x)$ *is a polynomial of order not greater than* $k$.

**Proof.** The fundamental solution of the form (15), (16) for an equation of order $2m$ with constant coefficients is constructed in [11], [12], p. 213. Using such a fundamental solution of an equation, one can construct a fundamental matrix of solutions of the form (15), (16) for system (12). In fact, it is easy to see that if $\Gamma_0(x)$ is a fundamental solution of the form (15), (16) for the equation

$$\det \| B_{kj}(\mathscr{D}_x) \| \, u = 0,$$

then $\Gamma(x) = M^*(\mathscr{D}_x)\,\Gamma_0$ is the fundamental matrix of solutions of the form (16), (15) for system (12), where $M(\mathscr{D}_x)$ is a matrix, its elements are cofactors for matrix $B(\mathscr{D}_x)$; $M^*(\mathscr{D}_x)$ is a conjugate matrix for $M(\mathscr{D}_x)$.

Consider the Taylor series for $\Gamma(x - \eta)$ in a neighborhood of $\eta = 0$ for $|\eta| \leqq a$, $|x| \geqq 4a$, $a = \text{const} > 0$. We have

$$\Gamma(x - \eta) = \sum_{|\alpha| \leqq h} \frac{1}{\alpha!} \mathscr{D}_x^\alpha \Gamma(x) (-\eta)^\alpha + \gamma_h(x, \eta), \tag{18}$$

where $\eta^\alpha = \eta_1^{\alpha_1}, \ldots \eta_{in}^{\alpha_n}$, $\alpha = (\alpha_1, \ldots, \alpha_n)$, $\eta \in R^n$,

$$|\mathscr{D}_x^\beta \gamma_h^{lj}(x, \eta)| \leqq C_{\beta h}^1 |x|^{2m-n-h-1-|\beta|}, \qquad |x| > 1,$$

$C_{\beta h}^1 = \text{const}$, $\gamma_h^{lj}$ are elements of matrix $\gamma_h$, $h \geqq 2m - n$ if $2m - n \geqq 0$ and $n$ is even, and $h \geqq 0$ if $2m - n < 0$ or $n$ is odd.

Suppose that $\operatorname{supp} F \subset \{x : |x| < a\}$. We set

$$v(x) = \int_{R^n} \Gamma(x - \eta) F(\eta)\, d\eta. \tag{19}$$

Using (18), we get for $|x| > 4a$

$$v(x) = \sum_{|\alpha| \leqq h} \mathscr{D}_x^\alpha \Gamma(x) C_\alpha + v^h(x),$$

where $C_\alpha = (C_1^\alpha, \ldots, C_N^\alpha)$, $C_j^\alpha = \text{const}$, $j = 1, \ldots, N$,

$$|\mathscr{D}_x^\beta v^h(x)| \leqq M_{\beta h} |x|^{2m-n-h-1-|\beta|}, \qquad M_{\beta h} = \text{const}. \tag{20}$$

The vector function $V(x) = u(x) - v(x)$ is a solution of the system $B(\mathscr{D}_x) V = 0$ in $R^n$ and for $|x| \geqq 4a$

$$|V(x)| \leqq C' |x|^{k+|2m+n|+1}, \qquad C' = \text{const}.$$

Therefore, the components of $V(x)$ are polynomials (for examples see [13]). Since $u(x) = V(x) + v(x)$, $|u(x)| \leqq C(1 + |x|)^k$, and since estimate (20) holds, we have (14). The lemma is proved.

**Theorem 1.** *Suppose that* $F \in C_0^\infty(R^n)$, $u(x)$ *is a solution of system* (12) *in* $R^n$ *and*

$$J_1(u) = \int_{R^n} \sum_{j=1}^{N} \sum_{|\alpha|=m} (1+|x|)^{2\sigma} |\mathscr{D}_x^\alpha u_j|^2 \, dx < \infty$$

*with* $\sigma = \text{const} \leqq 0$. *Then for* $|x| > 1$

$$u(x) = \mathscr{P}(x) + \sum_{\omega < |\alpha| \leqq h} \mathscr{D}_x^\alpha \Gamma(x) C_\alpha + u^h(x), \tag{21}$$

*where*

$$\mathscr{P}(x) = (\mathscr{P}_1(x), \ldots, \mathscr{P}_N(x)), \mathscr{P}_j(x)$$

*is a polynomial of order less than* $m_0 = \max(m, m - n/2 - \sigma)$, $j = 1, \ldots, N$; $C_\alpha = (C_1^\alpha, \ldots, C_N^\alpha)$, $C_j^\alpha = \text{const}$, $\Gamma(x)$ *is the fundamental matrix of solutions of system* (12), *described in Lemma* 1, $\omega = m - n/2 + \sigma$, $h$ *any integer, if* $n$ *is odd or* $2m - n < 0$, *and* $h \geqq 2m - n$, *if* $2m - n \geqq 0$, $n$ is even,

$$|\mathscr{D}^\beta u^h(x)| \leqq C_{\beta h} |x|^{2m-n-h-1-|\beta|}, \quad C_{\beta h} = \text{const}, \tag{22}$$

*for any multi-index* $\beta$ *and* $|x| > 1$.

**Proof.** First we suppose that $2\sigma$ is not an integer. Consider $v(x) = \theta(x) u(x)$, where $\theta \in C^\infty(R^n)$, $\theta(x) = 1$ for $|x| > 2$ and $\theta(x) = 0$ for $|x| \leqq 1$. It is evident that $J_1(v) < \infty$. Applying successively the Hardy inequality (see [2]), we conclude that there are polynomials $q_j(x)$ of order $m-1$ $(j = 1, \ldots, N)$ such that

$$\int_{|x|>1} |x|^{2\sigma-2m} |v - q(x)|^2 \, dx < \infty, \tag{23}$$

where $q(x) = (q_1(x), \ldots, q_N(x))$. It is easy to see that

$$w(x) = v(x) - q(x)$$

satisfies system (12) with $F \in C_0^\infty(R^n)$.

From *a priori* estimates for solutions of elliptic systems [10] and embedding theorems it follows that

$$|w(x)|^2 \leqq C_1 \lambda^{-n} \int_{\lambda/2 < |\xi| < 2\lambda} |w(\xi)|^2 \, d\xi, \tag{24}$$

where $C_1 = \text{const}$, $\lambda < |x| < 3\lambda/2$, $F \equiv 0$ for $|x| > \lambda/2$. Taking into account (24), we show for $\lambda < |x| < 3\lambda/2$ that

$$|w(x)|^2 \leqq C_2 |x|^{-n-2\sigma+2m} \int_{\lambda/2 < |\xi| < 2\lambda} |\xi|^{2\sigma-2m} |w(\xi)|^2 \, d\xi.$$

Since (23) is valid we obtain from the last inequality

$$|w(x)|^2 \leqq \varepsilon(|x|) \, |x|^{-n-2\sigma+2m} \tag{25}$$

for $|x| > 1$, where $\varepsilon(|x|) \to 0$ as $|x| \to \infty$.

According to Lemma 2 and (25) we have

$$w(x) = \mathscr{P}(x) + \sum_{h_1 < |\alpha| \leqq h} \mathscr{D}^\alpha \Gamma(x)\, C_\alpha + w^h(x), \tag{26}$$

where $\mathscr{P}(x)$ is a vector function, $\mathscr{P}(x) = (\mathscr{P}_1(x), \dots, \mathscr{P}_N(x))$, $\mathscr{P}_j(x)$ is a polynomial of order less than $m - \sigma - n/2$; $h \geqq 0$ and $h \geqq 2m - n$, if $2m - n \geqq 0$ and $n$ is even, $h_1 = m - n/2 + \sigma$,

$$|\mathscr{D}_x^\beta w^h(x)| \leqq C_{h\beta} |x|^{2m-n-h-1-|\beta|}$$

for $|x| > 1$. Since $u(x) = v(x) = w(x) + q(x)$ for $|x| > 2$, the theorem follows from (26) if $2\sigma$ is not an integer.

Now, let $2\sigma$ be an integer. The finiteness of the integral

$$J_1(u) = \int_{R^n} \sum_{j=1}^N \sum_{|\alpha|=m} (1 + |x|)^{2\sigma} |\mathscr{D}_x^\alpha u_j|^2 \, dx$$

implies that

$$J_1'(u) = \int_{R^n} \sum_{j=1}^N \sum_{|\alpha|=m} (1 + |x|)^{2\sigma'} |\mathscr{D}_x^\alpha u_j|^2 \, dx < \infty$$

for any $\sigma'$ such that $\sigma' < \sigma$. Therefore, (21) is valid for $\sigma'$ instead of $\sigma$, if $2\sigma'$ is not an integer and $\sigma' < \sigma$.

It is easy to see that both $m - \sigma - n/2$ and $m + \sigma - n/2$ are integers or both are not. If they are not integers, we choose $\sigma'$ so close to $\sigma$ that $[m - \sigma' - n/2] = [m - \sigma - n/2]$ and $[m + \sigma' - n/2] = [m + \sigma - n/2]$, where $[\mu]$ is the integral part of $\mu$. Then one can take $\omega = m + \sigma - n/2$, $m_0 = \max\{m, m - \sigma - n/2\}$ in (21) and in this case Theorem 1 is proved.

If $m - \sigma - n/2$ and $m + \sigma - n/2$ are integers, we consider for $|x| > 1$ the vector function

$$V(x) = \sum_{|\alpha| = m+\sigma-n/2} \mathscr{D}_x^\alpha \Gamma(x)\, C_\alpha + \mathscr{P}'(x),$$

where the components of $\mathscr{P}'(x)$ are those terms of the corresponding components of $\mathscr{P}(x)$ in (21) with $\sigma'$ instead of $\sigma$ whose degree is $m - \sigma - n/2$.

Consider the case $\Gamma(x) = |x|^{2m-n} \Phi_2(x)$. Then $V(x)$ is a homogeneous vector function of order $m - \sigma - n/2$. Since $J_1(u) < \infty$ and $J_1(u - V) < \infty$, we have

$$\int_{|x|>1} |x|^{2\sigma} \sum_{j=1}^N \sum_{|\alpha|=m} |\mathscr{D}_x^\alpha V_j|^2 \, dx < \infty. \tag{27}$$

If the function

$$W(x) = \sum_{j=1}^N \sum_{|\alpha|=m} |x|^{2\sigma} |\mathscr{D}_x^\alpha V_j|^2, \qquad V = (V_1, \dots, V_N),$$

is not equal to zero at least at one point, then, being a smooth homogeneous function of order $-n$, this function fails to vanish in a cone $G$ and for any integer $s \geqq 1$

$$\int_{G \cap \{x : s < |x| < 2s\}} W(x)\, dx = \int_{G \cap \{x' : 1 < |x'| < 2\}} W(x')\, dx', \qquad \int_{G \cap \{x : |x| > 1\}} W(x)\, dx = \infty.$$

The last equality contradicts to (27). Therefore, $W \equiv 0$ in $R^n$. This means that $\mathcal{D}_x^\alpha V_j \equiv 0$ for $|\alpha| = m$ and $V_j$, $j = 1, \ldots, N$, is a polynomial of order $m-1$. Thus $V(x)$ can be included in $\mathcal{P}(x)$ and in (21) we can set $\omega = m + \sigma - n/2$, $m_0 = \max(m, m - \sigma - n/2)$.

If $\Gamma(x)$ has the form (15), then $W(x) = W_1(x) + \ln|x|\, W_2(x) + (\ln|x|)^2 W_3(x)$, where $W_1$, $W_2$, $W_3$ are homogeneous functions of order $-n$. Let $G$ be a cone with the vertex at the origin. For $s \geqq 1$

$$\begin{aligned} \int\limits_{G \cap \{x: s < |x| < 2s\}} W(x)\, dx &= \int\limits_{G \cap \{x': 1 < |x'| < 2\}} (W_1(x') + \ln|x'|\, W_2(x') \\ &+ (\ln|x'|)^2 W_3(x'))\, dx' + \ln s \int\limits_{G \cap \{x': 1 < |x'| < 2\}} (W_2(x') \\ &+ 2 \ln|x'|\, W_3(x'))\, dx' + \int\limits_{G \cap \{x': 1 < |x'| < 2\}} (\ln s)^2\, W_3(x')\, dx' < \infty, \end{aligned} \tag{28}$$

because (27) is valid. It follows from (28) that

$$\int\limits_{G \cap \{x': 1 < |x'| < 2\}} W_3(x')\, dx' = 0$$

and $W_3(x) \equiv 0$, since $G$ is an arbitrary cone with a vertex at the origin and $W_3$ is a homogeneous function. Similarly we obtain that $W_2(x) \equiv 0$ and then that $W_1(x) \equiv 0$. The theorem is proved.

**3.** Using Theorem 1, we shall prove a more general theorem 2 which gives an asymptotic expansion in a neighborhood of infinity for solutions of system (1), satisfying condition (4) (the E. Sanchez-Palencia problem).

**Theorem 2.** *If $f \in C^\infty(R^p)$, $f(z) \equiv 0$ for $|x| \geqq M$, $M = \text{const} > 0$, and $u(z)$ is a solution of system* (2) *satisfying condition* (4), *and $u(z)$, $f(z)$ are $2\pi$-periodic in $y$, then*

$$u(z) = \mathcal{P}(x) + \sum_{\omega < |\alpha| \leqq h} \mathcal{D}^\alpha \Gamma(x)\, C_\alpha + u^h(z), \qquad |x| > M, \tag{29}$$

*where the components of the vector function $\mathcal{P}(x)$ are polynomials in $x$ of order less than $m_0 = \max(m, m - \sigma - n/2)$, $C_\alpha = (C_1^\alpha, \ldots, C_N^\alpha)$, $C_j^\alpha = \text{const}$, $\omega = m + \sigma - n/2$, $h \geqq 0$ and $h \geqq 2m - n$, if $2m - n \geqq 0$ and $n$ is even,*

$$|\mathcal{D}_z^\beta u^h(z)| \leqq C_{\beta h} |x|^{2m-n-h-1-|\beta|}, \qquad C_{\beta h} = \text{const},$$

*$\Gamma(x)$ is the fundamental matrix of solutions of the system*

$$\mathcal{L}(\mathcal{D}_x, 0)\, u = 0, \tag{30}$$

*obtained from system* (2) *and $\Gamma(x)$ has the form* (15), (16).

**Proof.** Let us write the Fourier series in $y$ for $u(z)$ and $f(z)$. We get

$$u(z) = \sum_{\nu \in \mathbb{Z}} u^\nu(x)\, e^{i(\nu, y)}, \qquad f(z) = \sum_{\nu \in \mathbb{Z}} f^\nu(x)\, e^{i(\nu, y)}, \tag{31}$$

$$\nu = (\nu_1, \ldots, \nu_q), \qquad (\nu, y) = \nu_1 y_1 + \ldots + \nu_q y_q,$$

$$u^\nu(x) = (2\pi)^{-q} \int\limits_g u(x, y)\, e^{-i(\nu, y)}\, dy, \qquad f^\nu(x) = (2\pi)^{-q} \int\limits_g f(x, y)\, e^{-i(\nu, y)}\, dy,$$

where $\mathbb{Z}$ is the set of vectors with integer components, $g = \{y: 0 < y_j < 2\pi, j = 1, \ldots, q\}$. On every compact set in $R_x^n$ the integrals in (31) converge uniformly with respect to $x$ and $y$, and one can differentiate them term by term any number of times in $x$ and $y$. If we substitute (31) into system (2), we obtain

$$\mathscr{L}(\mathscr{D}_x, i\nu)\, u^\nu(x) = f^\nu(x), \quad \nu \in \mathbb{Z}. \tag{32}$$

Condition (4) yields

$$J(u) = (2\pi)^q \int_{R^n} (1 + |x|)^{2\sigma} \sum_{\substack{\nu \in \mathbb{Z} \\ |\alpha| + |\beta| = m}} |\nu^\beta|^2 \, |\mathscr{D}_x^\alpha u^\nu(x)|^2 \, dx < \infty, \tag{33}$$

where $\nu^\beta = \nu_1^{\beta_1} \ldots \nu_q^{\beta_q}$, $\beta = (\beta_1, \ldots, \beta_q)$, $\nu^\beta = 1$ for $\nu = (0, .., 0)$ and for $\beta = (0, \ldots, 0)$.

Consider system (32) for $\nu \neq 0$. Set $x' = |\nu|\, x$, where $|\nu| = |\nu_1| + \ldots \ldots + |\nu_q|$. Then system (32) takes the form

$$\mathscr{L}\left(\mathscr{D}_{x'}, i\frac{\nu}{|\nu|}\right) u^\nu = |\nu|^{-2m} f^\nu, \quad f^\nu \in C_0^\infty(R^n). \tag{34}$$

It follows from (33) that $u^\nu \in W_{0,\gamma}$ for any $\gamma < 0$. System (34) satisfies the conditions of Lemma 1 and by virtue of inequality (11) we have for $|\gamma| \leqq \gamma_0$

$$\int_{R^n} \exp\{2\gamma\, |x'|\}\, |u^\nu|^2 \, dx' \leqq C' \int_{R^n} |\nu|^{-4m} \exp\{2\gamma\, |x'|\}\, |f^\nu|^2 dx'. \tag{35}$$

Set $\gamma = \gamma_0/|\nu|$, $x' = |\nu|\, x$. From inequality (35) we obtain

$$\int_{R^n} \exp\{2\gamma_0\, |x|\}\, |u^\nu|^2 \, dx \leqq C' \int_{R^n} |\nu|^{-4m} \exp\{2\gamma_0\, |x|\}\, |f^\nu|^2 \, dx, \quad \nu \neq 0. \tag{36}$$

Summing (36) on all $\nu \neq 0$, we find

$$\int_Q \exp\{2\gamma_0\, |x|\}\, |w(z)|^2 \, dz \leqq C' \int_Q \exp\{2\gamma_0\, |x|\}\, |f|^2 \, dz, \tag{37}$$

where $C' = \text{const}$, $w(z) = \sum_{\nu \in \mathbb{Z}, \nu \neq 0} u^\nu(x)\, e^{i(\nu, y)}$. The vector function $w(z)$ is a solution of the system

$$\mathscr{L}(\mathscr{D}_x, \mathscr{D}_y)\, w = \sum_{\nu \in \mathbb{Z}, \nu \neq 0} f^\nu e^{i(\nu, y)} = f^*. \tag{38}$$

Let us estimate $\mathscr{D}_z^\alpha w$. Consider in $R_z^p$ a domain $\Omega(\lambda/2, 2\lambda) = \{x, y: \lambda/2 < |x| < 2\lambda, -2\lambda < y_i < 2\lambda, j = 1, \ldots, q\}$ and assume that $\lambda$ is so large that $f \equiv 0$ for $\lambda/2 > |x|$. The change of variables $z = z'\lambda$ in system (38) in $\Omega(\lambda/2, 2\lambda)$ leads to the system

$$\mathscr{L}(\mathscr{D}_{x'}, \mathscr{D}_{y'})\, w' = 0, \quad z' = (x', y'), \quad z' \in \Omega(1/2, 2), \quad w'(z') = w(\lambda z').$$

It follows from the interior estimates for solutions of elliptic systems [10] that

$$|\mathscr{D}_{z'}^\alpha w'(z')|^2 \leqq C^\alpha \int_{\Omega(1/2, 2)} |w'(z')|^2 \, dz', \quad C^\alpha = \text{const}, \quad |z'| = 1. \tag{39}$$

We rewrite (39) in terms of the variable $z = \lambda z'$. Taking into account the periodicity of $w$ in $y$, we get

$$\begin{aligned} |\mathscr{D}_z^\alpha w(z)|^2 &\leqq C^\alpha \lambda^{-\mu} \int_{\Omega(\lambda/2,2\lambda)} \lambda^{\mu-n-2|\alpha|} |w(z)|^2 \, dz \\ &\leqq C_*^\alpha \lambda^{-\mu} \int_{Q \cap \Omega(\lambda/2,\lambda)} \lambda^{\mu-n-2|\alpha|+q} |w(z)|^2 \, dz, \end{aligned} \tag{40}$$

$|x| = \lambda$, $\mu = \text{const} > 0$. Inequalities (40) and (37) yield

$$|\mathscr{D}_z^\alpha w(z)|^2 \leqq C_{\alpha\mu} |x|^{-\mu}, \qquad C_{\alpha\mu} = \text{const}, \tag{41}$$

for any $\alpha$ and $\mu > 0$.

The function $u^0$ in expansion (31) is a solution of the system

$$\mathscr{L}(\mathscr{D}_x, 0)\, u^0 = f^0 \tag{42}$$

in $R^n$. System (42) satisfies condition (13). It follows from (33) that $J_1(u^0) < \infty$. Therefore according to Theorem 1 $u^0(x)$ can be written in the form (21). Since $u(z) = u^0(x) + w(z)$, the theorem follows from (21) and (41).

**4.** Let us consider a special class of systems (1), which includes the system of linear elasticity. We shall study the behavior at infinity of solutions of system (5) having a finite energy integral (6).

**Lemma 3.** *Let* $u = (u_1, \ldots, u_N)$, $u_j = u_j(x_1, \ldots, x_N)$, $u_j \in L^2_{\text{loc}}(R^n)$, $j = 1, \ldots, N$, $\dfrac{\partial u_j}{\partial x_k} \in L^2_{\text{loc}}(R^n)$,

$$\int_{|x|<\lambda} \sum_{k,j=1}^{N} \left(\frac{\partial u_j}{\partial x_k} + \frac{\partial u_k}{\partial x_j}\right)^2 dx \leqq C_1(1+\lambda)^{m_1}, \qquad C_1, m_1 = \text{const}, \tag{43}$$

*for any* $\lambda > 0$. *Then there are constants* $C_2, m_2$ *such that for any* $\lambda > 0$

$$\int_{|x|<\lambda} \sum_{j=1}^{N} |\text{grad } u_j|^2 \, dx \leqq C_2(1+\lambda)^{m_2}, \tag{44}$$

*where* $C_2$ *does not depend on* $\lambda$; $m_2$ *depends on* $m_1$ *and* $N$ *only.*

**Proof.** Set $S_{a,b} = \{x : a < |x| < b\}$, $S_b = S_{0,b}$. From the classical Korn inequality we get

$$\int_{S_1} \sum_{j=1}^{N} |\text{grad } u_j|^2 \, dx \leqq C_3 \left( \int_{S_1} E(u)\, dx + \int_{S_1} |u|^2 \, dx \right), \qquad C_3 = \text{const}, \tag{45}$$

where $C_3$ does not depend on $u$ and

$$E(u) = \sum_{k,j=1}^{N} \left(\frac{\partial u_j}{\partial x_k} + \frac{\partial u_k}{\partial x_j}\right)^2.$$

We first prove that for any $\varepsilon = \text{const} > 0$ there are constants $\delta(\varepsilon) \in (0, 1)$ and $C(N) > 0$ such that

$$\int_{S_1} |u|^2\, dx \leqq \varepsilon \int_{S_1} \sum_{j=1}^{N} |\operatorname{grad} u_j|^2\, dx + C(N) \int_{S_{1-\delta(\varepsilon)}} |u|^2\, dx. \tag{46}$$

Let $\delta$ be an arbitrary positive number, $\delta \leqq 1/4$, $u \in C^\infty$ and let $(r, \omega) = (r, \omega_1, \ldots, \omega_{N-1})$ be polar coordinates in $R^N$, $0 \leqq \eta \leqq 1/2$. Then for $1 - \eta \leqq r \leqq 1$ we get

$$|u_j(r, \omega)|^2 \leqq 2\,|u_j(1 - \eta, \omega)|^2 + 2\eta \int_{1/2}^{1} \left|\frac{\partial u_j}{\partial r}\right|^2 \partial r. \tag{47}$$

According to the mean value theorem there is an $\eta^*$ such that

$$\int_{|x|=1-\eta^*} |u_j|^2\, d\sigma = \frac{1}{\delta} \int_{S_{1-2\delta,1-\delta}} |u_j|^2\, dx, \qquad \delta \leqq \eta^* \leqq 2\delta. \tag{48}$$

Setting $\eta = \eta^*$ in (47) and integrating the obtained inequality over the sphere $\sigma_1$ with radius 1 and with center at the origin, we find

$$\int_{\sigma_1} |u_j(r, \omega)|^2\, d\sigma_1 \leqq 2 \int_{\sigma_1} |u_j(1 - \eta^*, \omega)|^2\, d\sigma_1 + 4\delta \cdot 2^{N-1} \int_{S_{1/2,1}} |\operatorname{grad} u_j|^2\, dx. \tag{49}$$

Using (48) and integrating (49) with respect to $r$ from $1 - \eta^*$ to 1, we get

$$\int_{S_{1-\eta^*,1}} |u_j|^2\, dx \leqq 2\delta^{-1}(1 - \eta^*)^{1-N}\eta^* \int_{S_{1-2\delta,1-\delta}} |u_j|^2\, dx + 4\delta\, 2^{N-1}\eta^* \int_{S_1} |\operatorname{grad} u_j|^2\, dx. \tag{50}$$

Since $1 - \eta^* > 1/2$, it follows from (50) that

$$\int_{S_{1-\delta,1}} |u_j|^2\, dx \leqq 2^{N+1} \int_{S_{1-\delta}} |u_j|^2\, dx + \delta^2\, 2^{N+2} \int_{S_1} |\operatorname{grad} u_j|^2\, dx. \tag{51}$$

If we take $2^{N+2}\,\delta^2 = \varepsilon$, we get from (51)

$$\int_{S_1} |u_j|^2\, dx \leqq (1 + 2^{N+1}) \int_{S_{1-\delta}} |u_j|^2\, dx + \varepsilon \int_{S_1} |\operatorname{grad} u_j|^2\, dx, \qquad j = 1, \ldots, N. \tag{52}$$

Summing up these inequalities with respect to $j$ from 1 to $N$, we obtain (45). From (45) and (46), by taking $\varepsilon = 1/2$ we find for some $\delta \in (0, 1)$ that

$$\int_{S_1} \sum_{j=1}^{N} |\operatorname{grad} u_j|^2\, dx \leqq C_4 \left( \int_{S_1} E(u)\, dx + \int_{S_{1-\delta}} |u|^2\, dx \right). \tag{53}$$

Let $\bar{u} = (\bar{u}_1, \ldots, \bar{u}_N)$, in which $\bar{u}_j$ is the mean value of $u_j$ over $S_{1-\delta}$. Then according to (53) we have

$$\int_{S_1} \sum_{j=1}^{N} |\operatorname{grad} u_j|^2\, dx \leqq C_4 \left( \int_{S_1} E(u)\, dx + \int_{S_{1-\delta}} |u - \bar{u}|^2\, dx \right).$$

Applying the Poincaré inequality to estimate the last integral, we get

$$\int_{S_1} \sum_{j=1}^{N} |\operatorname{grad} u_j|^2\, dx \leqq C_5 \left( \int_{S_1} E(u)\, dx + \int_{S_{1-\delta}} \sum_{j=1}^{N} |\operatorname{grad} u_j|^2\, dx \right), \quad C_4, C_5 = \text{const}.$$

Using the transformation of variables $x' = \lambda x$, $\lambda = \text{const} > 1$, one can get from the last inequality that

$$\int_{S_\lambda} \sum_{j=1}^{N} |\operatorname{grad} u_j|^2\, dx \leqq C_5 \left( \int_{S_\lambda} E(u)\, dx + \int_{S_{\lambda(1-\delta)}} \sum_{j=1}^{N} |\operatorname{grad} u_j|^2\, dx \right), \quad C_5 > 1.$$

Applying this inequality successively $k - 1$ times to estimate the last integral of its right-hand side, we obtain

$$\int_{S_\lambda} \sum_{j=1}^{N} |\operatorname{grad} u_j|^2\, dx \leqq C_5^k\, k \int_{S_\lambda} E(u)\, dx + C_5^k \int_{S_{\lambda(1-\delta)^k}} \sum_{j=1}^{N} |\operatorname{grad} u_j|^2\, dx. \tag{54}$$

Let us choose $k$ so large that $\lambda(1-\delta)^k < 1$. Set $k = [a \ln \lambda] + 1$, where $a = \text{const} > 0$, $[\mu]$ is the integral part of $\mu$ and $1 + a \ln(1-\delta) < 0$. Then $\lambda(1-\delta)^k < 1$ for $\lambda > 1$ and it follows from (54) that

$$\int_{S_\lambda} \sum_{j=1}^{N} |\operatorname{grad} u_j|^2\, dx \leqq C_6(1+\lambda)^{m_2},$$

where the constants $C_6$, $m_2$ do not depend on $\lambda$. The lemma is proved.

**Theorem 3.** *Suppose that* $u(z) = (u_1(z), \dots, u_N(z))$ *is a solution of system* (5), $f_k \equiv 0$ *for* $|x| > l$, $l = \text{const.}$, $k = 1, \dots, N$, $f_k \in C^\infty(R^n)$,

$$\int_Q E(u)\, dz \equiv \int_Q \sum_{k,j=1}^{N} \left( \frac{\partial u_k}{\partial z_j} + \frac{\partial u_j}{\partial z_k} \right)^2 dz < \infty, \tag{55}$$

$u_k(z)$, $f_k(z)$ $(k = 1, \dots, N)$ *are* $2\pi$*-periodic functions in* $y$. *Then*

$$u_j(z) = \sum_{k=1}^{n} c_{jk} x_k + c_j + \sum_{\omega \leqq |\alpha| \leqq h} \mathscr{D}^\alpha \Gamma(x)\, C_\alpha + u_{jh}(z), \tag{56}$$

*where* $\|c_{jk}\|$ *is a skew-symmetric matrix,* $c_{jk}$, $c_j$ *are constants,* $k, j = 1, \dots, N$; $c_{jk} = 0$ *for* $j > n$ *and* $\omega = 0$ *for* $n > 2$, $\omega = 1$ *for* $n = 2$,

$$|\mathscr{D}_z^\beta u_{jh}(z)| \leqq M_{\beta h} |x|^{2-n-|\beta|-h-1}, \quad |x| > 1,$$

*for any* $\beta$ *and* $h$; $M_{\beta h} = \text{const.}$

**Proof.** It follows from (55) that

$$\int_{|z|<\lambda} E(u)\, dz \leqq M_1(1+\lambda)^q, \quad M_1 = \text{const}, \quad \lambda > 1, \tag{57}$$

since $u(z)$ is $2\pi$-periodic vector function in $y$. According to Lemma 3 we obtain from (57) that

$$\int_{|z|<\lambda} \sum_{j=1}^{N} |\operatorname{grad} u_j|^2\, dz \leqq M_2(1+\lambda)^{m_3}, \quad M_2, m_3 = \text{const}, \lambda > 1,$$

$M_2$, $m_3$ do not depend on $\lambda$. Because $u(z)$ is $2\pi$-periodic in $y$, the last inequality yields

$$\int_{Q\cap\{z:|x|<\lambda\}} \sum_{j=1}^{N} |\operatorname{grad} u_j|^2 \, dz \leqq M_3(1+\lambda)^{m_4}, \qquad M_3, m_4 = \text{const} > 0.$$

Therefore

$$\int_{Q\cap\{z:\lambda<|x|<2\lambda\}} (1+|x|)^{-m_4-2} \sum_{j=1}^{N} |\operatorname{grad} u_j|^2 \, dz \leqq M_4(1+\lambda)^{-2}, \qquad M_4 = \text{const}. \tag{58}$$

Summing inequalities (58) over all integers $\lambda \geqq 1$, we find that

$$\int_Q (1+|x|)^{-m_4-2} \sum_{j=1}^{N} |\operatorname{grad} u_j|^2 \, dz < \infty.$$

Thus condition (4) of Theorem 2 is satisfied with $2\sigma = -m_4 - 2$. According to Theorem 2 we have

$$u_j(z) = \mathscr{P}_j(x) + \sum_{l=1}^{N} \sum_{0\leqq|\alpha|\leqq h} \mathscr{D}_x^\alpha \Gamma_{lj}(x)\, C_\alpha^l + u_{jh}(z), \tag{59}$$

for $|x| > 1$, where $C_\alpha^l = \text{const}$, $\Gamma_{lj}(x)$ is a fundamental matrix of solutions of the system

$$\sum_{j=1}^{N} m_{kj}(\mathscr{D}_x, 0)\, u_j(x) = 0, \qquad k = 1, \dots, N,$$

obtained from system (5); $\mathscr{P}_j(x)$ is a polynomial, $j = 1, \dots, N$;

$$|\mathscr{D}_z^\beta u_{jh}(z)| \leqq M_{\beta h} |x|^{2-n-1-h-|\beta|}, \; M_{\beta h} = \text{const} > 0,$$

for $|x| > 1$ and any $\beta$.

Consider the case $n > 2$. Then $|\mathscr{D}^\alpha \Gamma_{lj}(x)| \leqq M_\alpha |x|^{2-n-|\alpha|}$. Set $V_j = u_{jh} + \sum_{l=1}^{N} \sum_{0\leqq|\alpha|\leqq h} \mathscr{D}_x^\alpha \Gamma_{lj} C_\alpha^l$, $V = (V_1, \dots, V_N)$. Let us prove that $\mathscr{P}_j(x)$ is a polynomial of degree 1. Set $\mathscr{P}(x) = (\mathscr{P}_1(x), \dots, \mathscr{P}_N(x))$. Since $\int_Q E(u)\, dz < \infty$, $\int_{Q\cap\{z:|x|>1\}} E(V)\, dz < \infty$, it follows from (59) that $\int_{Q\cap\{z:|x|>1\}} E(\mathscr{P})\, dz < \infty$ and therefore $\int_{R^n\cap\{x:|x|>1\}} E(\mathscr{P})\, dx < \infty$. It is easy to see that $E(\mathscr{P}) \equiv 0$. In fact,

$$\int_{\{x:s<|x|<2s\}} E(\mathscr{P})\, dx$$

is a polynomial with respect to $s$; using the change of variables $x = sx'$, we get $\int_{\{x:1<|x|<2\}} E(\mathscr{P})\, dx = 0$ and therefore $E(\mathscr{P}) \equiv 0$, since $E(\mathscr{P}) \geqq 0$. Thus

$$\frac{\partial \mathscr{P}_j}{\partial z_l} + \frac{\partial \mathscr{P}_l}{\partial z_j} = 0, \qquad j, l = 1, \dots, N.$$

Moreover, $\mathscr{P}_j(x) = \text{const}$ for $j > n$, since $\frac{\partial \mathscr{P}_j}{\partial x_l} = 0$ for $j > n$ and any $l \leqq n$. It is evident that

$$\frac{\partial^2 \mathscr{P}_j}{\partial x_k\, \partial x_l} = -\frac{\partial^2 \mathscr{P}_l}{\partial x_k\, \partial x_j} = \frac{\partial^2 \mathscr{P}_k}{\partial x_j\, \partial x_l} = -\frac{\partial^2 \mathscr{P}_j}{\partial x_k\, \partial x_l}.$$

Thus $\frac{\partial^2 \mathscr{P}_j}{\partial x_k\, \partial x_l} = 0$ for any $k, l, j = 1, \ldots, n$. Therefore, $\mathscr{P}_j(x)$ is a polynomial of degree not larger than 1, $\mathscr{P}_j(x) = \sum_{k=1}^{n} c_{jk} x_k + c_j$; $c_{jk}, c_j = \text{const}$. The matrix $\| c_{jk} \|$ is a skew-symmetric, because of the condition $\frac{\partial \mathscr{P}_j}{\partial x_k} = -\frac{\partial \mathscr{P}_k}{\partial x_j}$, $\mathscr{P}_j = \text{const}$ for $j > n$.

Consider the case $n = 2$. Then $\Gamma_{lj}(x) = C_{lj} \ln |x| + \Phi_{lj}(x)$, $C_{lj} = \text{const}$, $\Phi_{lj}(x)$ are homogeneous functions of degree 0. From (59) it follows that

$$\int_{R^n \cap \{x : |x| > 1\}} E(\mathscr{P} + \Gamma C_0)\, dx < \infty,$$

$C_0 = (C_0', \ldots, C_0^N)$. Since $\left| \frac{\partial \Gamma_{lj}}{\partial x_k} \right| \leqq M_0 |x|^{-1}$, $M_0 = \text{const} > 0$, considering $\int_{\{x : s < |x| < 2s\}} E(\mathscr{P} + \Gamma C_0)\, dx$ and making the change of variables $x = x's$, we conclude as before that $E(\mathscr{P}) \equiv 0$, $\mathscr{P} = (\mathscr{P}_1, \ldots, \mathscr{P}_N)$, $\mathscr{P}_j = \sum_{k=1}^{n} c_{jk} x_k + c_j$, $c_{jk}$ is a skew-symmetric $n \times n$-matrix, $l, j = 1, \ldots, n$; $c_j = \text{const}$, $c_{jk} = 0$ for $j > n$.

Since $\int_{Q \cap \{z : |x| > 1\}} E(\mathscr{P})\, dz < \infty$ and $\int_{Q \cap \{z : |x| > 1\}} E(u - u^h - \sum_{1 \leqq |\alpha| \leqq h} \mathscr{D}^\alpha \Gamma C_\alpha)\, dz < \infty$, then $\int_{Q \cap \{z : |x| > 1\}} E(\Gamma C_0)\, dz < \infty$.

Let $\Gamma C_0 = (A_1, A_2)$. We have

$$\int_{R^n \cap \{x : |x| > 1\}} \left( \left( \frac{\partial A_1}{\partial x_1} \right)^2 + \left( \frac{\partial A_2}{\partial x_2} \right)^2 + \left( \frac{\partial A_1}{\partial x_2} + \frac{\partial A_2}{\partial x_1} \right)^2 \right) dx < \infty. \tag{60}$$

Using polar coordinates and the explicit form of $\Gamma(x)$, one can easily find from (60) that $E(\Gamma C_0) \equiv 0$. Thus $\Gamma C_0$ is a vector function of the same form as $\mathscr{P}(x)$. Therefore, (56) is valid. The theorem is proved.

The case $n = 1$, $q > 0$ in Theorems 2 and 3 can be treated in a similar way.

## References


1. J. SERRIN & H. WEINBERGER, Isolated singularities of linear elliptic equations. *Amer. Math. Journ.*, 1966, v. 88, Nr. 1, p. 258–272.
2. V. A. KONDRATIEV & O. A. OLEINIK, Sur un probléme de E. Sanchez-Palencia. C. R. Acad. Sci. Paris, 1984, v. 299, ser. 1, Nr. 15, p. 745–748.
3. V. A. KONDRATIEV & O. A. OLEINIK, On periodic in the time solutions of a second order parabolic equation in exterior domains. Vestnik of the Moscow University, Math. Mech., ser. 1, 1985, Nr. 4, p. 38–47.

4. B. R. WAINBERG, On solutions of elliptic equations with constant coefficients and a right hand side growing at the infinity. Vestnik of the Moscow University. Math. Mech., ser. 1, 1968, Nr. 1, p. 41–48.
5. JA. B. LOPATINSKY, The behaviour at infinity of solutions of systems of differential equations of the elliptic type. Dokl. AN Ukr. SSR, 1959, Nr. 9, p. 931–935.
6. S. L. SOBOLEV, Introduction to the theory of cubature formulas. Moscow, Nauka, 1974.
7. E. M. LANDIS & G. P. PANASENKO, On a variant of the Phragmen-Lindelöf type theorem for elliptic equations with coefficients periodic in all variables except one. Trudi seminara imeni I. G. Petrovsky, 1979, v. **5**, p. 105–136.
8. O. A. OLEINIK & G. A. YOSIFIAN, On the behavior at infinity of solutions of second order elliptic equations in domains with noncompact boundary. Matem. Sbórnik, 1980, v. **112**, Nr. 4, p. 588–610.
9. O. A. OLEINIK & G. A. YOSIFIAN, On the asymptotic behavior at infinity of solutions in linear elasticity. Archive Rational Mech. and Analysis, 1982, v. **78**, Nr. 1, p. 29–53.
10. A. DOUGLIS & L. NIRENBERG, Interior estimates for elliptic systems of partial differential equations. Comm. Pure Appl. Math., 1955, v. **8**, p. 503–538.
11. JA. B. LOPATINSKY, Fundamental solutions of a system of differential equations of elliptic type. Ukr. mat. journal, 1951, v. **3**, Nr. 1, p. 3–38.
12. L. BERS, F. JOHN, & M. SCHECHTER, Partial differential equations, Interscience publishers, New York, 1964.
13. I. M. GELFAND & G. E. SHILOV, Generalized functions, v. 3, Moscow, 1958.


Moscow University

*(Received April 22, 1986)*

# *A Class of Quasilinear Differential Inequalities whose Solutions are Ultimately Constant*

RAY REDHEFFER

*To James Serrin, in celebration of his 60th birthday*

**Abstract**

Let $\Omega$ be a domain in $R^n$ with compact complement and let $T$ be a quasilinear elliptic or degenerate elliptic operator associated with functions $u \in C^2(\Omega)$. This paper is a study of solutions of $(\operatorname{sgn} u)\, Tu \geqq f(|u|, |\operatorname{grad} u|)$ where $f$ belongs to a class of functions here termed bifurcation functions. The main condition on $f$ is that uniqueness fails for the ordinary differential equation $y'' = f(y, y')$ with the initial condition $y(0) = y'(0) = 0$. The conclusion is that $u$ is constant for large $|x|$ and hence, under mild supplementary hypotheses, $u$ has compact support. Examples show that the results fail if the assumptions on $f$ are only slightly weakened, so that the class of $f$ is, essentially, the largest class for which the results can be stated truly.

## 1. Introduction

Let us begin by placing this paper in a proper historical and philosophical context. The general area of concern is the study of nonlinear problems whose solutions, even in unbounded regions, have compact support. A comprehensive and up-to-date bibliography of over 300 references in this field can be found in the excellent monograph [9], and a representative sample, chosen to illustrate the diversity of the problems, has been included here. Only a few of these references are truly germane to our results, however, because of the sharp distinction between priorities in the theory of differential equations and differential inequalities. An example of this distinction is seen when we compare the Nagumo monotonicity theorems for the general parabolic operator

$$Tu = u_t - f(x, t, u, u_x, u_{xx})$$

on the one hand, and classical existence theory for the equation $u_t = u_{xx}$ on the other. The former does not include the latter because it says nothing about existence and the latter does not include the former, even if the proof is so managed as to include monotonicity, because the operator $Tu = u_t - u_{xx}$ is far too special. Another difference is that in the equation $u_t = u_{xx}$ one is preoccupied with the question whether initial and boundary values can be prescribed arbitrarily, while in the Nagumo theory this does not matter at all.

Similar remarks apply to a broader class of inequalities and equations, including those which form the subject of the present paper; we used the Nagumo theory as a framework for the views expressed above because it provides perhaps the earliest and best-known example. The distinction between inequalities and equations must be kept in mind when our results are compared with those in the literature.

Like a number of other investigations, this one and its predecessor [21] were inspired by an early paper of BRÉZIS [8]. The latter considers set inclusions of the form $Tu \in \beta(u)$ where $T$ is a linear operator analogous to the nonlinear operator $T$ introduced below and where $\beta(u)$ is a maximal monotone graph with suitable properties. The underlying idea is to obtain a superfunction $v(x) = \mu(|x|)$ for $u(x)$, where $\mu(t)$ has compact support and satisfies an ordinary differential inequality

$$\mu''(t) \leqq f(t, \mu(t), |\mu'(t)|).$$

In [21] it is found that a certain growth condition given in [8] can be replaced by a sharper condition involving divergence of an integral. (That the latter is sharp follows from the fact that, when it fails, there is a radial counterexample.) The integral condition is also given in [2], except that the conditions for $n = 1$ and $n = 2$ have been inadvertently interchanged. (The same error appears in [9].) In any case, not much importance can be attached to this refinement insofar as it pertains to compactness of support, since it requires only minor modification of the argument in the original paper of BRÉZIS.

More significant is the fact that the growth conditions on the coefficients of the differential operator are much weaker in [21] than in [8]. In particular, it is found that the relevant parameter is not the dimension $n$ of the space but the effective dimension $N$ in the sense of MEYERS & SERRIN [19]. Similar considerations apply to the problems considered here.

The most important improvement is that the comparison equation is used only at points where $|u'|$ is arbitrarily small. As a consequence, the growth conditions and even the condition of ellipticity $a(x, u, u') \geqq 0$ are needed only subject to this restriction. Thus the equation is allowed to *change type*. The comparison functions that naturally present themselves do not have a small gradient, and a major part of the difficulty is to construct other, less natural, comparison functions that do. This shift of emphasis is clearly explained in [21]. In the present paper the difficulty is exaggerated. Not only do solutions of the comparison equation fail to have a small gradient, they may even fail to exist except over an interval of minutely small length. These problems are dealt with in Lemmas 2, 3, 4, 5 below.

In many results obtained hitherto the comparison equation is linear in $\mu, \mu', \mu''$ and a dominant role is taken by the inhomogeneous part $f(t, 0, 0)$. In the following results a dominant role is taken by the nature of the nonlinearity in $f$. In the autonomous case, which is considered here, the comparison equation has the form

$$A\mu''(t) + B\,|\mu'(t)| \leqq f(\mu(t), |\mu'(t)|).$$

It will be seen that the theory acquires a distinctive flavor from the fact that the above equation is invariant under reflection and translation. For instance, our principal result does not assert that the solution has compact support, but that it must be constant for large $|x|$. Compactness of support then follows from weak additional assumptions, such as

$$\liminf_{x\to\infty} |u(x)| = 0.$$

Whether or not such conditions are imposed, the main theorem reduces the global existence problem to a series of local existence problems, one in $|x| < R$ for each $R$. Since such reduction is a principal goal of the BRÉZIS theory, we conclude this introduction by a brief discussion.

Taking a naive view, one would expect to be able to use any ready-to-hand existence theorem in the part of the domain where $|x| < R$, and the compactness of support would immediately give a global extension. However, because of the nonlinearities associated with the compactness part of the problem, the local existence theorem requires a very careful formulation in terms of set inclusions. This approach was pioneered by BRÉZIS and has been carried further by him and others, notably by BÉNILAN, BRÉZIS & CRANDALL in [2]. The subtlety and delicacy of the arguments in [2] point to the difficulty of getting a corresponding formulation in the context of our Theorem 1. The latter allows a singular behavior in grad $u$ as well as in $u$, and the set-inclusion formulation would have to take account of this.

Another difficulty is connected with the fact that, as also in [21], we allow nonlinear boundary conditions of mixed type. The difficulty of developing a comprehensive existence theory in this setting is underlined by SERRIN's work on HILBERT's 20$^{\text{th}}$ problem [23], where it is seen that conditions on the boundary curvature play a role even for Dirichlet boundary conditions of the simplest possible form.

The most serious obstacle to existence theory (in a generality appropriate to the rest of this paper) is the fact noted above, that we allow the differential equation to change type. In the terminology of fluid flow, any proposed existence theorems would have to allow a change from a subsonic to a supersonic regime, where the locus of the two regimes depends on the unknown function $u$. To my knowledge no existence theorems in this degree of generality are currently available.

## 2. Notation

Except in Section 9, where a more general case is considered, $\Omega$ denotes an exterior region in $E^n$; namely, an open connected set whose complement is compact. For sufficiently large $R$

$$\Omega(R) = \{x : x \in \Omega, |x| < R\}$$

and $u$ denotes a function which is of class $C$ in $\bar{\Omega}$ or $\overline{\Omega(R)}$, as the case may be, and of class $C^2$ in the interior. The inner normal derivative of $u$ is denoted by $u_\nu$ and the boundary $\delta\Omega$ by $\Gamma$.

We denote the gradient by $u'$ and the Hessian by $u''$. Although this notation is by no means universal, it has been used at least since 1967 and it is used in the paper [21] of which the present paper is a continuation. For comparison with other notations,

$$u' = Du = u_x = \text{Vector}\left(\frac{\partial u}{\partial x_i}\right), \qquad u'' = D^2u = u_{xx} = \text{Matrix}\left(\frac{\partial^2 u}{\partial x_i\, \partial x_j}\right).$$

We agree further that $a$ denotes a real symmetric $n$ by $n$ matrix, $b$ is a real $n$ vector, $au''$ and $bu'$ denote the obvious contraction, and an expression such as $xax$ denotes a quadratic form. For example the operator $T$ defined by

$$Tu = a(x, u, u')\, u'' - b(x, u, u')\, u'$$

can be expressed more explicitly by

$$Tu = \sum_{i,j=1}^{n} a_{ij}(x, u, Du) \frac{\partial^2 u}{\partial x_i\, \partial x_j} - \sum_{i=1}^{n} b_i(x, u, Du) \frac{\partial u}{\partial x_i}\,.$$

The matrix inequalities $a > 0$ or $a \geqq 0$ are interpreted by quadratic forms and the norms $|a|$, $|b|$, $|x|$ are Euclidean. The word *increasing* always means *weakly increasing*, in contrast to *strictly increasing* which needs no explanation. Our use of the Landau $O$ symbol pertains to the behavior as $|x| \to \infty$; for example, $B(x) \leqq O(|x|)$ means there are constants $R$ and $K$ such that $B(x) \leqq K|x|$ for $|x| \geqq R$. The maximum modulus is denoted by

$$M(R) = \sup\{|u(x)| : |x| = R\}.$$

Depending on context, letters such as $u$, $u_\nu$, $u'$, $u''$, $a$, $b$ denote either functions or functional values. In this latter use the category is

$$u \in R, \quad v_\nu \in R, \quad u' \in R^n, \quad u'' \in S^n, \quad a \in S^n, \quad b \in R^n$$

where $S^n$ is the class of real symmetric $n$ by $n$ matrices. Similar remarks apply to $c$ and $v$. We use $p \in R^n$ as an independent variable though for many purposes $p$ need not be distinguished from $u'(x)$. The independent variable corresponding to $u(x)$ is denoted by $u$.

If $c(x) = \frac{1}{2}|x|^2$ then according to the above conventions $c' = x$ and $c'' = I$, the $n$ by $n$ identity matrix. Hence we can, and do, consider that

$$x' = I, \qquad ax' = \text{trace}\, a\,.$$

This seemingly capricious notation is an aid to computation, as seen in the course of the proofs.

## 3. Two definitions

As was pointed out by WOLFGANG WALTER over twenty years ago, introduction of the inner normal derivative $u_\nu$ does not presuppose any smoothness of the boundary $\Gamma$. However, we shall impose the following regularity condition, which is discussed more fully in Section 11:

**Definition 1.** *The boundary $\Gamma$ is said to be regular if there exists a real-valued function $c$ of class $C^2$ in a neighborhood of $\Gamma$ such that $c_\nu > 0$ on $\Gamma$.*

Another definition which we shall find useful is:

**Definition 2.** *The function $f\colon R \times R \to R$ is said to be a bifurcation function if $f$ is continuous, $f(r, s)$ is increasing in $r$ for each $s$, and the problem*

$$\begin{cases} \sigma'' \leqq f(\sigma, |\sigma'|), & \sigma'' \text{ increasing} \\ \sigma(0) = \sigma'(0) = 0, & \sigma(t_0) > 0 \end{cases}$$

*has a solution on some interval $0 \leqq t \leqq t_0$.*

## 4. Formulation of the main theorem

From now on we agree that $f$ is a bifurcation function in the sense of Definition 2 and $\Gamma$ is regular in the sense of Definition 1. A quasilinear operator $T$ is defined by

$$Tu = a(x, u, u')\, u'' - b(x, u, u')\, u'$$

where $a$ and $b$ are continuous functions of the form

$$a: \bar{\Omega} \times R \times R^n \to S^n, \qquad b: \bar{\Omega} \times R \times R^n \to R^n.$$

Associated with this operator $T$ are two problems, global and local. The *global problem* is to find a function $u \in C(\bar{\Omega}) \cap C^2(\Omega)$ such that

(a) $(\operatorname{sgn} u)\, Tu \geqq 0, \qquad x \in \Omega, \quad u \neq 0$

(b) $(\operatorname{sgn} u)\, Tu \geqq f(|u|, |u'|), \qquad x \in \Omega, \quad u \neq 0, \quad |x| > R_0$

(c) $a(x, u, 0) > 0, \quad a(x, u, p) \geqq 0, \qquad x \in \Omega, \quad |p| < \varepsilon.$

Here $R_0$ and $\varepsilon$ are positive constants with $R_0$ large and $\varepsilon$ small. Conditions (ab) are self-explanatory and (c) asserts that the operator $T$ is, in a limited region, either strongly elliptic or degenerate elliptic. When the side conditions $p = 0$ or $|p| < \varepsilon$ fail, no hypothesis of ellipticity is imposed.

In addition to this interior hypothesis (abc) we require a boundary condition on $\Gamma$ and at $\infty$, namely,

(d) $|u| > M_0 \Rightarrow u u_\nu \geqq 0, \quad x \in \Gamma; \quad \liminf_{R \to \infty} M(R) \leqq M_1.$

Here $M_0$ and $M_1$ are nonnegative constants. The second condition needs no explanation and the first is discussed in Section 11.

The *local problem* is to find a function $u = u(x; R)$ of class $C(\overline{\Omega(R)}) \cap C^2(\Omega(R))$ which satisfies (abc) with $\Omega$ replaced by $\Omega(R)$ and (d) with the second condition replaced by $M(R) \leqq M_1$. The local problem is *solvable* if its solution can be found for arbitrarily large $R$. A simplified version of our main theorem is then as follows:

**Theorem 0.** *Suppose $a(x, u, p)$ and $b(x, u, p)$ are bounded whenever $|u| \leqq \max(M_0, M_1)$ and $|p| < \varepsilon$. Then the problem* (abcd) *is globally solvable if, and only if, it is locally solvable, and in that case every solution of the global problem is constant for large $|x|$.*

The concluding statement means that there exist constants $K$ and $R_1$ such that $u(x) = K$ for $|x| \geqq R_1$. Since (d) gives $|K| \leqq M_1$ the passage from global to local solution is immediate once it is known that $K$ exists. Although $K$ has different values for different solutions, in general, it will be seen that one and the same $R_1$ will do for all solutions.

The condition of boundedness of $a$ and $b$ in Theorem 0 is far too restrictive and has been introduced only because of its simplicity. A sharper theorem, which is the main result of this paper, is given next:

**Theorem 1.** *Suppose the growth conditions*

$$xa(x, u, p)\, x \leqq O(|x|^2), \qquad |xb(x, u, p) - x'a(x, u, p)| \leqq O(|x|)$$

*hold whenever $|u| \leqq \max(M_0, M_1)$ and $|p| < \varepsilon$. Then the conclusion of Theorem 0 holds.*

## 5. An a priori bound

We shall establish the following:

**Lemma 1.** *Let $u$ be a local or global solution of the problem corresponding to* (acd). *Then $|u| \leqq \max(M_0, M_1)$.*

For proof, let $h$ be a function of class $C^2$ such that $h = 1$ in a sufficiently small neighborhood of $\Gamma$ and $h = 0$ outside another sufficiently small neighborhood. Consideration of $hc$ in Definition 2 then gives a function of class $C^2(R^n)$ which satisfies $(hc_\nu) = c_\nu > 0$ on $\Gamma$, and hence we may as well consider that $c$ itself has this property. For any given $R$ the singularities of $c$ (the points where $c' = 0$) can be moved, by a retract, to a location outside of $|x| = R$. Since $c_\nu > 0$ implies that there are no singularities near $\Gamma$, we can do this without changing $c$ in a neighborhood of $\Gamma$. The net result is that we get a function, call it $c$ again, such that

$$c_\nu > 0 \quad \text{on } \Gamma, \qquad c' \neq 0, \qquad c \in C^2$$

in $\overline{\Omega(R)}$. This is the first step.

The rest of the argument follows a standard pattern and will be presented briefly. Suppose $u = u(x, R)$ is a local solution in $\Omega(R)$, where $R$ is so large that $\Gamma \subset \Omega(R)$. We assume that $\max u > M = \max(M_0, M_1)$ and obtain a contradiction. To this end let us consider

$$w = u + \alpha e^{\beta c}$$

where $\alpha$ and $\beta$ are positive constants. We first choose $\beta$ large, as described below, and we then choose $\alpha$ sufficiently small. In particular, $\alpha$ is so small that the condition $u > M$ occurs at the maximum of $w$. In that case the maximum cannot be on $\Gamma$, since the condition $u > M_0$ at the maximum ensures

$$w_\nu = u_\nu + \alpha\beta e^{\beta c} c_\nu \geqq \alpha\beta e^{\beta c} c_\nu > 0.$$

If $\alpha$ is small the maximum also does not occur on $|x| = R$, as is easily seen, and hence it is an interior maximum.

Since $a(x, u, p)$ is continuous the hypothesis $a(x, u, 0) > 0$ shows that there exist positive constants $\lambda$, $\delta$ such that

$$a(x, u, u') \geqq \lambda I \quad \text{whenever } |u'| < \delta, \quad x \in \Omega(R).$$

At the maximum of $w$ the condition $w' = 0$ gives

$$u' = -\alpha\beta e^{\beta c} c'$$

and hence, in fact, $|u'| < \delta$ if $\alpha$ is sufficiently small. This is now imposed as a further condition on $\alpha$.

If $n = 1$ the inequality $w'' \leqq 0$ at the maximum would give

$$u'' \leqq -(\alpha e^{\beta c})'' = -\alpha\beta^2 e^{\beta c} c' c' - \alpha\beta e^{\beta c} c''$$

and the condition $a \geqq 0$ would then give

$$a u'' \leqq -\alpha\beta^2 e^{\beta c} c' a c' - \alpha\beta e^{\beta c} a c''.$$

Our notation has been so chosen that, in view of the lemma of SCHUR & FEJÉR, this is correct for arbitrary $n$. Hence, at the maximum of $w$,

$$Tu \leqq -\alpha\beta e^{\beta c}(\beta c' a c' + a c'' - b c').$$

Since $a = a(x, u, p) \geqq \lambda I$, we get a contradiction if

$$\beta\lambda |c'|^2 > b c' - a c''.$$

But $\inf |c'| > 0$ in $\Omega(R)$, since the singularities of $c$ have been moved outside $\overline{\Omega(R)}$, and hence $\beta$ can be chosen suitably. Once this is done we choose $\alpha$ so small that the other conditions also hold. This gives a contradiction, showing that the assumption $u > M$ was not tenable.

Since $-u$ satisfies almost the same hypothesis as $u$, we deduce $-u \leqq M$ as well; or, if preferred, this case can be dealt with by considering the minimum of the function

$$u - \alpha e^{\beta c}.$$

The result is $|u| \leqq M$, as desired.

If $u$ is a global solution then by a suitable choice of $R$ we can ensure

$$|u| \leqq M_1 + \eta, \qquad |x| = R$$

where $\eta$ is any prescribed positive number. Thus we get a local solution corresponding to $M_0$ and $M_1 + \eta$, and the foregoing argument gives $|u| \leqq M + \eta$. Since $R$ can be arbitrarily large and $\eta$ can be arbitrarily small, it follows that the global solution $u$ actually satisfies $|u| \leqq M$ in every disk $|x| \leqq R_0$. This gives $|u| \leqq M$ and completes the proof of Lemma 1.

## 6. A family of comparison functions

The following elementary remark facilitates the proof of Lemma 3 below:

**Lemma 2.** *The function $\sigma$ in Definition 2 can be chosen so as to satisfy the additional conditions $\sigma > 0$, $\sigma' > 0$, $\sigma'' > 0$ for $0 < t \leqq t_0$.*

For proof let $\sigma$ satisfy the condition in Definition 2. Our first objective is to show that $f(0, 0) \geqq 0$ is necessary, and that $\sigma$ can be chosen so as to satisfy the additional condition $0 \leqq \sigma''(0) \leqq f(0, 0)$. If $\sigma''(0) \geqq 0$ there is no problem, since the differential inequality gives $\sigma''(0) \leqq f(0, 0)$ automatically. Suppose then that $\sigma''(0) < 0$. In this case $\sigma(t)$ must have a negative minimum at some value $t_1$, $0 < t_1 < t_0$. Thus,

$$\sigma(t_1) = m < 0, \qquad \sigma'(t_1) = 0, \qquad \sigma''(t_1) \geqq 0.$$

The inequality $\sigma'' \leqq f(\sigma, |\sigma'|)$ at $t_1$ together with the monotonicity of $f$ gives $f(0, 0) \geqq 0$ and the function $\sigma_1(t) = \sigma(t + t_1) - m$ satisfies $\sigma_1''(0) \geqq 0$ together with the other condition associated with Definition 2. The interval, however, is reduced to $[0, t_0 - t_1]$.

Returning to the notation of Definition 2, we can suppose, then, that $\sigma''(0) \geqq 0$. Since $\sigma''$ is increasing it follows that $\sigma$, $\sigma'$ and $\sigma''$ are all $\geqq 0$ on the interval $(0, t_0]$. We want to construct a solution for which these inequalities are strict. To this end let $t_2$ be the first point at the left of $t_0$ at which $\sigma(t) = 0$. Thus, $0 \leqq t_2 < t_0$ and $\sigma(t_2) = 0$, $\sigma(t) > 0$ for $t_2 < t \leqq t_0$. If $t_2 > 0$ then $\sigma'(t_2) = 0$; otherwise $\sigma'(t_2) > 0$ and we would have $\sigma(t) < 0$ at some $t$ near $t_2-$, which is not the case. The condition $\sigma'(t_2) = 0$ also holds, by hypothesis, if $t_2 = 0$.

It must be true that $\sigma''(t_i) > 0$ at a sequence of points $t_i \to t_2+$. Otherwise $\sigma''(t) = 0$ at all points near $t_2+$, hence $\sigma(t) = \sigma'(t) = 0$ near $t_2+$, and this contradicts the fact that $t_2$ was the first zero at the left of $t_0$. Since $\sigma''(t) \geqq \sigma''(t_i)$ for $t \geqq t_i$, we conclude that in fact $\sigma''(t) > 0$ for $t > t_2$. Hence $\sigma'(t) > 0$ and $\sigma(t) > 0$ for $t > t_2$ also. The function $\sigma_2(t) = \sigma(t + t_2)$ satisfies the desired supplementary conditions of Lemma 2, as well as the other conditions associated with Definition 2. However, $t_0$ is replaced by $t_0 - t_2$, giving a second reduction.

**Lemma 3.** *Let $f$ be a bifurcation function and let $A$, $B$, $\varepsilon$ be positive constants. Then there exist a function $\mu$ and a positive constant $T$ such that*

$$\mu(0) = \mu'(0) = 0, \qquad 0 \leqq \mu''(0) \leqq f(0, 0)$$

*and such that, for* $0 < t \leqq T$,

$$A\mu'' + B\mu' < f(\mu, \mu'), \qquad \mu > 0, \qquad \mu' > 0, \qquad \mu'' > 0, \qquad \mu' < \varepsilon.$$

For proof let $\sigma$ satisfy the conditions of Lemma 2 and let $\mu(t) = k\sigma(t/k)$ where $k$ is constant, $k \geqq 1$. By the monotonicity of $f$ it is readily checked that

$$k\mu'' \leqq f(\mu, |\mu'|), \qquad 0 \leqq t \leqq t_0 k,$$

and we take $T = \min(t_0 k, 1)$. Since $\sigma''$ is increasing the same is true of $\mu''$ for $t \geqq 0$ and the mean-value theorem shows that $|\mu'| \leqq \mu''$ on the relevant interval. Thus, if $k > A + B$ we have

$$A\mu'' + B\mu' \leqq (A + B)\,\mu'' < k\mu'' \leqq f(\mu, \mu'), \qquad 0 < t \leqq T.$$

It should be noticed that the last step entails no reduction in the length of the interval, but $\mu(T)$ may be much smaller than $\sigma(t_0)$ if $k$ is large. To get $\mu' < \varepsilon$ all we have to do is to diminish $T$ so that $\mu'(T) < \varepsilon$. This leads to an even smaller value for $\mu(T)$, in general, but we still have $\mu(T) > 0$.

**Lemma 4.** *Let the function $\mu$ of Lemma 3 be extended by the definition*

$$\mu(t) = \mu(T) + k(t - T), \qquad t > T$$

*where $k$ is constant, and extended further by the definition $\mu(-t) = \mu(t)$. Then the constant $k$ can be so chosen that the extended function satisfies*

$$|\mu'| < \varepsilon, \qquad A\mu'' + B\,|\mu'| < f(\mu, |\mu'|) \qquad (t \neq 0, \qquad t \neq -T, \qquad t \neq T)$$

*and, for the left and right derivatives,*

$$D_L\mu(t) > D_R\mu(t), \qquad |t| = T.$$

If we take $k = \mu'(T)$ the differential inequality of Lemma 3 at $t = T$ would give

$$Bk < A\mu'' + Bk < f(\mu(T), k)$$

and by continuity this holds for some $k$ satisfying $0 < k < \mu'(T)$. Such a $k$ is chosen now. Then the extended function $\mu$ satisfies

$$A\mu'' + B\,|\mu'| = Bk < f(\mu(T), k) \leqq f(\mu(t), |\mu'(t)|), \qquad t > T,$$

where in the last step the monotonicity of $f$ was used. The remaining statements in Lemma 4 are easily verified. The graph of $\mu$ is shown in Figure 1.

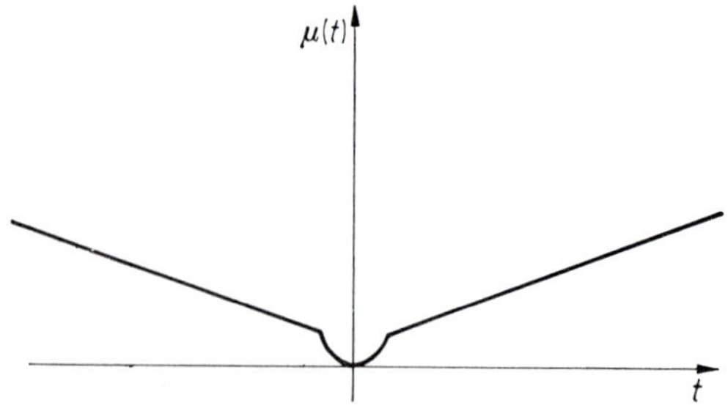


Fig. 1. Graph of $\mu$.

## 7. Condition for a function to be constant

We shall establish the following:

**Lemma 5.** *Let $J$ be an open interval of the real axis, let $U$ be a continuous function $J \to \mathbb{R}$, and let $V$ be a continuous function $\mathbb{R} \to \mathbb{R}$ satisfying the additional condition $V'(0) = 0$. Whenever $R \in J$ and $U(R) > 0$, suppose the maximum of the function*

$$F(r) = U(r) - V(r - R), \qquad r \in J$$

*is attained at the value $r = R$. Then $U \leqq 0$ on $J$, or $U$ is constant on $J$.*

For proof, let $R \in J$ and $u(R) > 0$. Since $F(r)$ is maximum at the interior point $r = R$ the lower left and upper Dini derivates satisfy, respectively,

$$D_{-}F(r) \geqq 0, \qquad D^{+}F(r) \leqq 0, \qquad r = R.$$

In view of the fact that $V'(0) = 0$ we have $D_{-}U(r) \geqq 0$, $D^{+}U(r) \leqq 0$ at $r = R$. Since $J$ is open and $U$ is continuous the assumption $U(R) > 0$ gives $U(R^*) > 0$ in some open interval

$$I = \{R^* : |R^* - R| < \delta\}, \qquad I \subset J.$$

The two inequalities

$$D_{-}U(r) \geqq 0, \qquad D^{+}U(r) \leqq 0$$

hold whenever $r = R^*$, by the above argument, and hence they hold in $I$. The inequalities show that $U(r)$ is both increasing and decreasing on $I$ and hence is constant. Thus, the set $\{r : U(r) = U(R)\}$ is open. Since this set is also closed relatively to $J$, it must be empty or all of $J$. This completes the proof.

## 8. Proof of Theorem 1

Referring to the growth condition in Theorem 1, choose positive constants $A$, $B$, $R_0$ so that the inequalities

$$xa(x,u,p)\,x \leqq A\,|x|^2, \qquad \left| xb(x,u,p) - x'a(x,u,p) + \frac{xa(x,u,p)\,x}{|x|^2} \right| \leqq B\,|x|$$

hold for $|x| \geqq R_0$, $|p| < \varepsilon$, and $|u| \leqq M$, where $M = \max(M_0, M_1)$. We assume that $u$ is a global solution and define a family of comparison functions by

$$v(x) = \mu(|x| - R)$$

where $R$ is a sufficiently large positive constant and where $\mu$ is the function in Lemma 4, associated with the above values of $A$, $B$, $\varepsilon$. The constant $k$ in Lemma 4 satisfies $0 < k < \varepsilon$ and $\mu$ satisfies

$$\mu(t) = \mu(T) + k(|t| - T), \qquad |t| > T.$$

Let us choose values $R_i$ such that

$$R_0 + T + \frac{M}{k} = R_1 < R_2 < R_3 = R_2 + T + \frac{M}{k}$$

and let us introduce the one-sided maximum modulus,

$$U(r) = \sup\{u(x) : |x| = r\}.$$

If $U(R) > 0$ at some value $R$, $R_1 < R < R_2$, it will be shown that the function

$$w(x) = u(x) - v(x)$$

attains its maximum in the ring $R_0 \leqq |x| \leqq R_3$ at a value $x$ satisfying $|x| = R$. Once this is done, Lemma 5 enables us to complete the proof. See Figure 2.

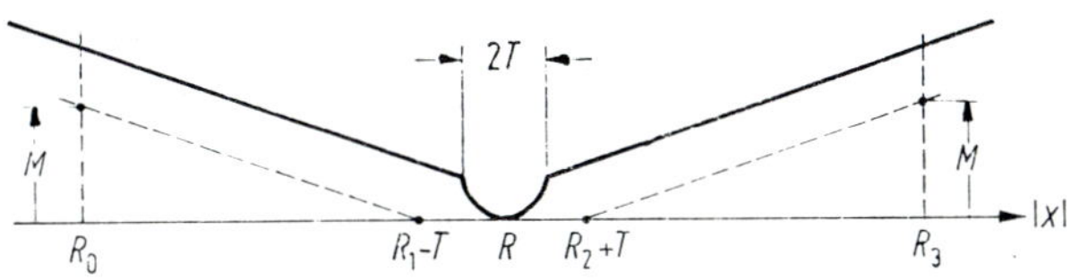


Fig. 2. Profile of the function $v(x)$.

Suppose, then, that $U(R) > 0$. The function $w = u - v$ satisfies $w = u$ when $|x| = R$, since $\mu(0) = 0$, hence $w(x) = U(R)$ for some $x$ on $|x| = R$, and this shows that $w$ attains a positive maximum at some point of the closed ring $R_0 \leqq |x| \leqq R_3$. On $|x| = R_0$ we have

$$v(x) = \mu(R_0 - R) = \mu(T) + k(R - R_0 - T) > k(R_1 - R_0 - T) = M$$

and similarly on $|x| = R_3$. This shows that $w < 0$ on the outer boundary of the ring, and hence the positive maximum is not attained there.

Let us show that the maximum cannot occur at a point where $|x| - R = T$ or $-T$. This is tolerably clear from Figure 2 and is proved by considering the radial variation. Namely, if the maximum of $u - v$ occurs at $x_0$, consider the function

$$F(t) = u(tx_0) - v(tx_0) = u(tx_0) - \mu(t\,|x_0| - R) = G(t) - H(t)$$

where $G$ and $H$ are defined as suggested by the previous equation. Since $F$ has a maximum at $t = 1$, the left and right hand derivatives satisfy, respectively,

$$D_L F(t) \geqq 0, \qquad D_R F(t) \leqq 0, \qquad t = 1.$$

If $|x_0| - R = T$ this gives

$$G'(1) \geqq (D_L \mu)(T), \qquad G'(1) \leqq (D_R \mu)(T)$$

which is incompatible with the conditions on $\mu$ at $T$. Similarly, a contradiction is obtained if $|x_0| - R = -T$ at the maximum.

Let us see next whether the maximum can occur at a point where $|x_0| \neq R$, so that the strict inequality for $\mu$ is available. If we set $s = |x| - R$, for short, then at the maximum

$$u'(x) = v'(x) = \mu'(s)\frac{x}{|x|}$$

and hence $|u'(x)| < \varepsilon$. If $n = 1$ we would have

$$u''(x) \leqq v''(x) = \mu'(s)\left(\frac{x'}{|x|} - \frac{x\,x}{|x|^3}\right) + \mu''(s)\frac{x\,x}{|x|^2}$$

where $x' = 1$, and this suggests that

$$au'' - bu' \leqq \mu''(s)\frac{xax}{|x|^2} + \mu'(s)\left(\frac{x'a - xb}{|x|} - \frac{xax}{|x|^3}\right) \tag{1}$$

where $a = a(x, u, u')$, $b = b(x, u, u')$. This actually follows from our notation, in view of the lemma of SCHUR & FEJÈR; recall that $x'a = \text{trace}\, a$. In any case the above calculation is valid and gives

$$Tu \leqq A\mu''(s) + B\,|\mu'(s)| < f(\mu(s), |\mu'(s)|) \leqq f(u, |u'|).$$

In the last step we used the monotonicity of $f$ together with the fact that $u > v$ at a positive maximum of $u - v$. This contradicts the basic inequality for $u$.

The only possible location for the maximum, then, is on $|x| = R$. The result can be interpreted as follows: If $U(R) > 0$ for some $R$ satisfying $R_1 < R < R_2$ then the maximum of the function

$$\phi(r) = U(r) - \mu(r - R), \qquad R_0 \leqq r \leqq R_3,$$

is taken on only when $r = R$. Naturally, the maximum of $\phi(r)$ for $R_1 < r < R_2$ is also taken on only when $r = R$. Lemma 5 shows, then, that $U(r)$ is constant on this interval. Since $R_2$ can be as large as desired, we find that $U(r)$ is constant for all large $r$. That $u(x)$ is constant for all large $|x|$ now follows from the strong maximum principle, which holds even under the weak hypothesis of Theorem 1.

Similar analysis applies if the minimum modulus

$$\inf\{u(x): |x| = r\}$$

is negative at some value $r = R$, or this case can be reduced to the previous one by consideration of $-u$. This completes the proof that a global solution in Theorem 1 must be constant for large $|x|$.

Similar methods apply to a local solution $u(x; R_3)$, say, provided $R_3$ is as large as the $R_3$ in the foregoing discussion. The discussion does not give any conclusion for large $|x|$, since $u(x; R_3)$ is defined only for $|x| \leqq R_3$, but it does give a region of the form $R_1 \leqq |x| \leqq R_2$ on which $u(x; R_3)$ is constant. If we now define $u(x)$ to agree with $u(x; R_3)$ for $|x| \leqq R_2$, and to be constant for $|x| \geqq R_2$, we get a global solution from the local solution $u(x; R_3)$. Thus the problem is globally solvable if it is locally solvable, as stated in Theorem 1.

## 9. Compactness of support

When the conclusion of Theorem 1 holds it has already been mentioned that the condition

$$\liminf_{|x|\to\infty} |u(x)| = 0 \tag{2}$$

ensures that $u$ has compact support. Condition (2) means that $\lim u(x_i) = 0$ for a sequence of points with $\lim |x_i| = \infty$. One way to ensure such a condition is to allow the boundary $\Gamma$ to be unbounded, and to require the Dirichlet boundary condition $u(\Gamma) = 0$ on the part of $\Gamma$ exterior to some sphere. It is not, of course, necessary that $\Gamma$ should be connected—a sequence of points $x_i \in \Gamma$, $|x_i| \to \infty$ is enough. This modification of Theorem 1 entails no significant change in the proof.

A more subtle condition leading to compactness of support is

$$f(s, 0) \neq 0, \qquad s > 0. \tag{3}$$

To see why, let $u(x) = C$ for $|x| > R_0$, as stated in the conclusion of Theorem 1, and suppose $C \neq 0$. The differential inequality (b) gives $f(|C|, 0) \leqq 0$, that together with (3) gives $f(|C|, 0) < 0$, and the monotonicity of $f$ now gives $f(0, 0) < 0$. But this is incompatible with the hypothesis that $f$ is a bifurcation function, as was seen in the proof of Lemma 2. Hence the only possible value for $C$ is $C = 0$ and $u$ must have compact support.

The most natural way to construct solutions of compact support is to take $M_1 = 0$ in a family of local solutions, so that the latter satisfy $u(x) = 0$ on $|x| = R$. When applied to a global solution the condition $M_1 = 0$ means

$$\liminf_{R\to\infty} M(R) = 0 \tag{4}$$

which is, of course, stronger than (2). It will be seen that this stronger hypothesis allows a significant weakening of the hypothesis in Theorem 1:

**Supplement to Theorem 1.** *If* $M_1 = 0$ *the bound for* $|x'a - xb|$ *in Theorem 1 can be replaced by a one-sided bound*

$$xb(x, u, p) - x'a(x, u, p) \leqq O(|x|) \qquad (|u| \leqq M, \quad |p| < \varepsilon)$$

*without invalidating the conclusion.*

It is to be shown that local solutions satisfying $M(R) = 0$ lead to a global solution having compact support, and that every global solution satisfying (4) does, in fact, have compact support. The second statement is somewhat harder to establish than the first and is discussed next.

Within the framework of the proof of Theorem 1 let $M(R) = 2\delta > 0$ and choose $R_3 > R_2$ so that $M(R_3) < \delta$. Choose constants $B$ and $C$ so that

$$xb(x, u, p) - x'a(x, u, p) \leqq B|x|, \qquad x'a(x, u, p) - xb(x, u, p) \leqq C|x|$$

for $|p| < \varepsilon$, $|u| \leqq M$, $x \in \Omega(R_3)$. The distinction between $B$ and $C$ is that $B$ can be chosen independently of $R$ while $C$ may tend to infinity as $R_3$ increases without bound. We use the same construction of $\mu$ as formerly except that it is based on the bound $B$ at the left of $R$ and on $C$ at the right. Only the one-sided bounded is needed at the left, since $\mu' \leqq 0$ there; see (1).

Next we add a constant to $\mu$ so that $\mu(0) = \delta$ instead of $\mu(0) = 0$. This is possible because $f(s, t)$ is monotone with respect to $s$. Because $C$ may be uncontrollably large, the slope of the linear part of the graph of $\mu$ at the right of $R$ may be arbitrarily small. It is positive however, and since $\mu(0) = \delta$ the comparison function satisfies $v(x) > u(x)$ on $|x| = R_3$. Once this is known, the rest of the proof is similar to that in Theorem 1. See Figure 3.

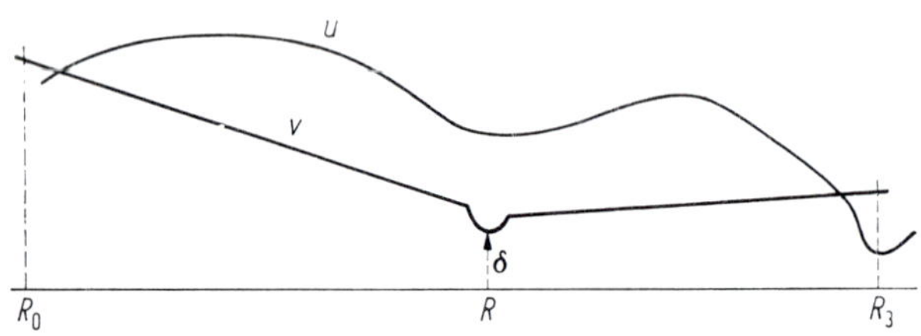


Fig. 3. Profile of $u$ and $v$.

In the case of a local solution the corresponding condition is $M(R_3) = 0$ instead of $M(R_3) < \delta$. Here it is not necessary to have $\mu(0) = \delta$ and the proof simplifies.

Under the hypothesis of the Supplement one can sometimes relax the monotonicity condition on $\sigma''$ without requiring an artificially strong restriction on the coefficient $b$. Suppose, for example, that

$$xax \leqq |x|^2, \quad xb \leqq x'a - \frac{xax}{|x|^2}. \tag{5}$$

Then the conditions for $\sigma$ can be replaced by

$$\sigma'' < f(\sigma, |\sigma'|), \quad \sigma'' \geqq 0, \quad \sigma(t_0) > 0.$$

The comparison based on (1) with $\mu = \sigma$ then gives

$$Tu \leqq \mu'' < f(\mu, |\mu'|) \leqq f(u, |u'|) \tag{7}$$

as before. If the monotonicity of $f$ is strict, $\sigma'' \leqq f(\sigma, \sigma')$ suffices in (6). The analysis bypasses Lemma 4, which is no longer available.

As an illustration let $f(s, t) = f(s)\, g(t)$ so that

$$(\operatorname{sgn} u)\, Tu \geqq f(|u|)\, g(|u'|).$$

It is supposed that $f$ and $g$ are continuous on $[0, \infty)$, positive on $(0, \infty)$, and that $f(s)$ is strictly increasing as a function of $s$. We set

$$F(s) = \int_0^s f(t)\, dt, \quad G(s) = \int_0^s \frac{t\, dt}{g(t)}$$

where the latter integral is assumed convergent, and we impose the condition

$$\int_0^1 \frac{ds}{G^{-1}[F(s)]} < \infty.$$

Then under the hypothesis (5) and $M_1 = 0$, it follows that $u$ has compact support. This is Theorem 4 of [21], where the condition that $f$ be monotone (which is needed) was inadvertently omitted. Since the trace $x'a$ is the sum of the eigenvalues of $a$, while the quotient on the right of (5) is dominated by the largest eigenvalue, the condition $xb \leqq 0$ suffices. Note that the eigenvalues are nonnegative, since $a \geqq 0$.

If we make the further assumption that $\sigma''$ is increasing, (5) can be replaced by the much weaker condition in the Supplement. This fact was not established in [21] but depends on Lemmas 2–4 above. Finally, if the two-sided bound of Theorem 1 is given, together with monotonicity of $\sigma''$ in Definition 1, we can drop the assumption that $M_1 = 0$. The conclusion, then, is that $u(x)$ is constant for large $|x|$.

## 10. Examples and counterexamples

One of the most interesting features of Theorem 1 is that $f$ need not be positive; that is why (a) was imposed as a separate condition. For example let $f(s, t) = s^{\frac{1}{3}} - t^{\frac{1}{2}}$ so that the main hypothesis is

$$(\operatorname{sgn} u)\, Tu \geqq |u|^{\frac{1}{3}} - |u'|^{\frac{1}{2}}.$$

The function $\sigma(t) = t^4$ shows that $f$ is a bifurcation function and hence Theorem 1 applies. If condition (a) is dropped, the conclusion would be that any bounded solution of the global problem has $M(R)$ constant for large $R$. However, the maximum principle is needed for the *a priori* bound, and it is also needed to go from constancy of $M(R)$ to constancy of $u(x)$.

For subsequent examples it is convenient to impose the conditions

$$xa(x)\,x = |x|^2, \qquad x'a(x) - xb(x) = 1$$

taking $a$ and $b$ to be functions of $x$ alone. These ensure $Tu = v''$ when $u = v(r)$, $|x| = r$.

If $f(|C|, 0) = 0$ for some constant $C \neq 0$ one can not conclude that $u$ has compact support, since $u = C$ satisfies the hypothesis. Thus, condition (3) is sharp. We give an example to show that $u$ may be constant only for large $|x|$, as allowed in the conclusion of Theorem 1. Let

$$v(r) = (2-r)^3 + 1, \quad 1 \leqq r \leqq 2; \quad v(r) = 1, \quad 2 \leqq r < \infty.$$

Then, with $a$ and $b$ as above, and $r = |x|$, $u = v(|x|)$,

$$v(r)^{\frac{1}{3}}\, Tu = 2\sqrt{3}\, |u|^{\frac{1}{3}}\, |u'|^{\frac{1}{2}}, \qquad |x| > 1.$$

We have introduced the factor $v(r)^{\frac{1}{3}}$ on the left to get an equation instead of an inequality; note that $1 \leqq v \leqq 2$, so that the modification of $a$ and $b$ introduced by this factor is harmless. That $f(s,t) = s^{\frac{1}{3}} t^{\frac{1}{2}}$ is a bifurcation function follows from the choice $\sigma(t) = t^{10}$, and hence every solution must be constant for large $|x|$. The above solution satisfies $u(\Gamma) = 2$, $u(\infty) = 1$.

Since Definition 1 involves $f(s, t)$ only when $s$ and $t$ are small, while the value of $u$ may be large, it is tolerably clear that the monotonicity hypothesis on $f$ cannot be dropped. If a simple example is desired, let $f(s)$ be a smooth, nonnegative function satisfying

$$f(s) = s^{\frac{1}{3}}, \quad |s| \text{ small}; \quad f(s) = 2(s-1)^3, \quad s \geqq 1.$$

Then with $a$ and $b$ as above, $Tu = f(u)$ has a solution $u(x) = 1 + 1/|x|$ even though $f$ is obviously a bifurcation function. If the additional condition $f(s) \neq 0$ for $s \neq 0$ is imposed the requirements for $u$ get entangled with those for $\sigma$ and the construction of a counterexample seems to be difficult. Whether the monotonicity of $f$ can be weakened or dropped under this hypothesis is left as an open problem.

In our next example let us show that the one-sided condition of the Supplement does not suffice if $M_1 > 0$. Instead of working with $a$ and $b$ as above we take $a = I$, $n = 3$, so that $au'' = \Delta u$, the 3-dimensional Laplacian. The equation

$$\Delta u - (|x|^2 - |x|^3)^{\frac{1}{3}}\, xu' = u^{\frac{1}{3}}, \qquad |x| > 1$$

has a solution $u(x) = 1 - 1/r$, $r = |x|$. Nevertheless $f(t) = t^{\frac{1}{3}}$ is a bifurcation function and $xb$ admits a one-sided bound

$$xb(x) = |x|^2\,(|x|^2 - |x|^3)^{\frac{1}{3}} < 0.$$

Since the trace $x'a = 3$ this gives (5) with a good deal to spare. The trouble is that $-xb(x)$ grows like $|x|^3$, which is too large.

Let us show next that the hypothesis "$\sigma''$ increasing" in Definition 1 is necessary for construction of $\mu$ in the subsequent lemmas. To see this consider the problem

$$\mu'' + B|\mu'| = f(|\mu'|). \tag{8}$$

The function $f(s)$ is assumed to be continuous for $s \geqq 0$, positive for $s > 0$, and it oscillates between the curves $y = s^{\frac{1}{2}}$ and $y = s^2$ in such a way that

$$\int_0^1 \frac{ds}{f(s)} < \infty, \qquad \liminf_{s \to 0-} \frac{f(s)}{s^2} = 1. \tag{9}$$

This can be accomplished by having the sojourn of $f(s)$ near $y = s^2$ sufficiently brief, as suggested by Figure 4.

If we had $f(s) = s^{\frac{1}{2}}$ for all $s > 0$ then $\sigma(t) = t^3/12$ would satisfy

$$\sigma'' = f(|\sigma'|), \quad \sigma(0) = \sigma'(0) = \sigma''(0) = 0, \quad \sigma'' \geqq 0, \quad \sigma(1) > 0. \tag{10}$$

and $\sigma''$ is obviously increasing. Because of the convergence of the integral in (9) one can find a solution of (10) even with the more complicated function $f$ shown in Figure 4. However, $\sigma''$ would not be increasing, so that the proof of Lemma 3 no longer goes through.

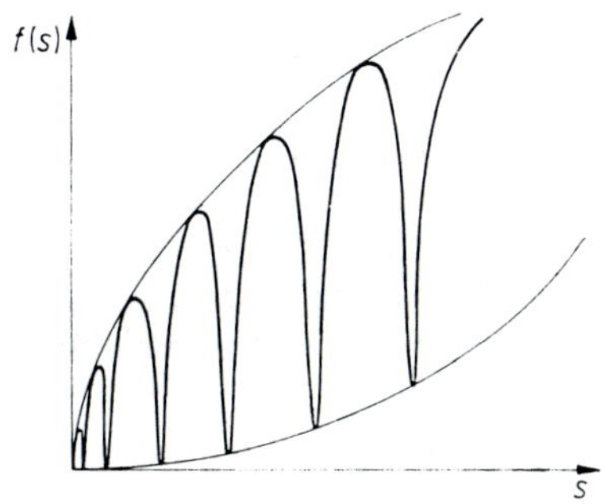


Fig. 4. Function behaving like $\sqrt{s}$ when integrated and like $s^2$ on a sequence.

We show that Lemma 3 is in fact false for this function even if $A = 1$ and the positive constant $B$ is arbitrarily small. This is true because the function

$$h(s) = f(|s|) - B|s|$$

associated with (8) has infinitely many zeros in every neighborhood of 0, as seen by Figure 4. OSGOOD's uniqueness theorem applied to $p = \mu'$ shows that $\mu'(0) = 0$ implies $\mu = 0$, and the same conclusion follows from [24]. Hence, a suitable $\mu$ does not exist.

As our final example we shall show that Theorem 1 fails when applied to the equation

$$(\operatorname{sgn} u)\, u'' \geqq h(|u'|), \qquad B > 0, \qquad n = 1.$$

In other words, there is a bounded solution $u$ which is not constant for large $x$. The restriction $n = 1$ has no importance, since a solution for that case produces a radial solution for arbitrary $n$ as seen above.

Since a constant can be added to $u$ we assume $u > 0$ so that $p = u'$ satisfies $p' \geqq h(|p|)$. Besides boundedness, the only property of $h$ we shall use is that $h$ has infinitely many zeros near $0+$ and has opposite signs on adjacent intervals separated by these zeros.

Referring to Figure 5, we let the intervals of negativity be $(a_n, b_n)$ so that

$$h(|s|) < 0, \quad a_n < s < b_n; \quad h(|s|) > 0, \quad b_n < s < a_{n+1}$$

We take $y = p(x)$ negative and increasing. The graph intersects the lines $p = a_n$ and $p = b_n$ at $x = c_n$ and $x = d_n > c_n$, respectively. We want to construct the curve in such a way that

$$\sum (d_n - c_n) = \infty, \quad \sum a_n(d_n - c_n) < \infty, \quad \sum b_n(c_{n+1} - d_n) < \infty.$$

The first condition ensures that the graph extends to $x = \infty$ without ever reaching the $x$ axis and the second and third conditions ensure that the (unsigned) area is finite. Since $u$ is obtained by integrating $p$, we shall have $u$ bounded on ([0, ∞). Also $u$ is not constant for large $x$, as is evident.

On the intervals $b_n < p < a_{n+1}$ corresponding to $d_n < x < c_{n+1}$ the curve must be steep enough to give $p' \geqq h(p)$, the latter being positive. This allows us to make $c_{n+1} - d_n$ as small as we please, though not as large as we please. We make $c_{n+1} - d_n < 1/n^2$, for instance, so that the third of the above series converges.

On the intervals $a_n < p < b_n$ corresponding to $c_n < x < d_n$ we have $h(p) < 0$ and the sole constraint on the curve is that $p' \geqq 0$. Hence, we can make $d_n - c_n$ be anything we like. There is no difficulty in joining the curves in a smooth fashion to get $u \in C^2$. Since $\lim a_n = 0$, we can satisfy the first two of the above conditions by judicious choice of $d_n - c_n$, and this completes the construction.

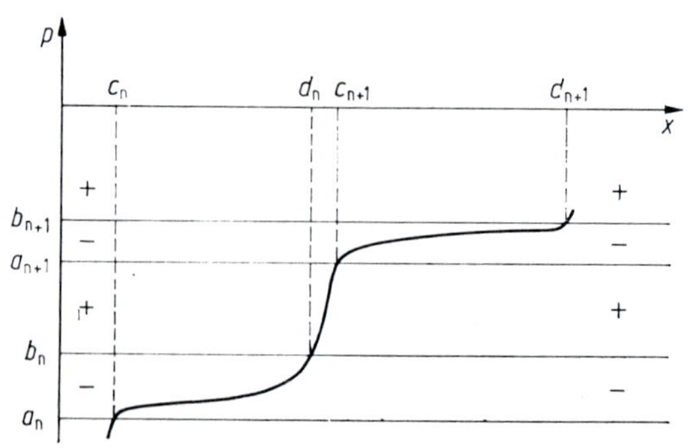


Fig. 5. Solution of $dp/dx \geqq h(|p|)$ as dependent upon the intervals of positivity of $h$.

## 11. Remarks on the boundary conditions

Existence of the function $c(x)$ in Definition 1 is a mild condition which allows corners; it does not presuppose that the boundary has bounded curvature. For example, $c(x) = -|x|^2$ satisfies $c_\nu > 0$ when the region is a cube centered at the origin. Often the function $|x|^2$ can be applied locally in the neighborhood of any corner and the results pieced together to get $c$. As another illustration, if $\Gamma$ has finitely many components $\Gamma_i$, each of which is bounded and admits a function $c_i$ satisfying $(c_i)_\nu > 0$ on $\Gamma_i$, then the functions can be used to construct a single function $c$ for $\Gamma$.

The boundary condition "$|u| > M_0 \Rightarrow uu_\nu \geqq 0$" is satisfied if $|u| \leqq M_0$ and if $u_\nu = 0$; these correspond to the Dirichlet and Neumann conditions, respectively. More generally it holds if

$$\alpha u = \beta u_\nu$$

where $\alpha$ and $\beta$ are any functions of $(x, u, u_\nu)$ which, for $|u| > M_0$, satisfy

$$\alpha \geqq 0, \quad \beta \geqq 0, \quad \alpha\beta > 0.$$

If both $\alpha$ and $\beta$ are nonvanishing this is a nonlinear version of Robin's condition. Mixed boundary conditions are allowed when $\alpha = 0$ or $\beta = 0$ on part of the boundary. Closer to the full generality allowed by our boundary condition is the hypothesis

$$K(x, u) = M(x, u, u_\nu), \qquad x \in \Gamma,$$

where for $|s| > M_0$ and for all real $t$

$$sK(x, s) \geqq 0, \quad tM(x, s, t) \geqq 0 \quad (t \neq 0)$$

at least one of these inequalities being strict at each relevant $(x, s, t)$.

## 12. Refinements of Theorem 1

Since the hypothesis $(\operatorname{sgn} u)\, Tu \geqq 0$ in (a) is used only in the proof of Lemma 1, it can be replaced by the weaker hypothesis

(a′) $$(\operatorname{sgn} u)\, Tu \geqq -k(x)\,|u'|$$

where $k(x)$ is any function of class $C(\bar{\Omega})$. Such a function is bounded in $\Omega(R)$ for each $R$ and hence the perturbation introduced by $k(x)\,|u'|$ can be absorbed into the perturbation produced by the term $bu'$.

Theorem 1 has two parts. The first pertains to the construction of a global solution from a family of local solutions, while the second is an assertion about the qualitative behavior of any given global solution. The refinement discussed next is not appropriate to the first and pertains only to the second. What we shall finally have is a theorem of Phragmén-Lindelöf type which, instead of giving the familiar conclusion that $u$ is bounded, asserts the stronger conclusion that $u$ is ultimately constant.

Suppose, then, that $u$ is a global solution of (a) or (a') and (b) and (c) which satisfies, instead of (d), a growth condition of Phragmén-Lindelöf type:

$$\text{(d')} \qquad \liminf_{R\to\infty} \frac{M(R)}{R} = 0.$$

Suppose further that the condition $|u| \leqq M$ in the hypothesis of Theorem 1 is dropped, so that the hypotheses on $xax$ and on $x'a - bx$ are required to hold without restriction on $u$. (The restriction $|p| < \varepsilon$ remains in force.) Then it still follows that $u(x)$ is constant for large $|x|$. To see this, note that the continuous function $u$ is bounded on some sphere $|x| = R_0$ and the bound so obtained can replace the bound formerly associated with $\Gamma$. To be sure, the bound depends on $u$, but this does no harm, since $u$ is fixed. The growth condition (d') suffices for the rest of the argument, since $\mu(t)$ is a linear function, of small but positive slope, for $t$ large.

Condition (d') is clearly sharp, since when it fails the choice $u(x) = \alpha|x|$ with $\alpha > 0$ is permitted. This gives a counterexample whenever $f(s, \alpha) = 0$ for all $s \geqq 0$, and the latter condition is in no conflict with the hypothesis that $f$ is a bifurcation function.

In our third refinement we continue to consider a specific global solution, but we return to the original hypothesis (d). By reducing $T$ one can ensure that the comparison function used in the proof of Theorem 1 satisfies

$$|v'(x)| \leqq \frac{2M}{R_3 - R_0 - 1}.$$

This is so because we merely require that the linear part exceed $M$ when $x = R_0$ and $x = R_3$, and we can locate $R$ near the center of the interval. Since $u'(x) = v'(x)$ at the maximum we see that $|u'|$ has the order $O(1/R_3)$ at the relevant point, as $R_3 \to \infty$, and *a fortiori* $|u'| = O(1/r)$. Hence all the side conditions of the type $|p| < \varepsilon$ can be replaced by $|p| \leqq O(1/r)$ in this part of the argument.

As an illustration let

$$Tu = (1 + A(x)|u'|^2)\Delta u - B(x)|u'|^2 xu', \qquad A(x) \geqq 0.$$

The growth hypothesis in Theorem 1 requires

$$A(x) \leqq O(1), \qquad |B(x)| \leqq O\left(\frac{1}{|x|}\right).$$

But since the bounds for $a$ and $b$ are needed only when $|u'| = O(1/|x|)$ it suffices to have

$$A(x) \leqq O(|x|^2), \qquad |B(x)| \leqq O(|x|).$$

This is a significant improvement.

Another refinement is connected with the hypothesis of continuity of $a$ and $b$, which was imposed for $x \in \bar{\Omega}$ rather than for $x \in \Omega$ as might seem more natural. This was needed to show that $a$ and $b$ are bounded in $\Omega(R)$ and also to get a condition of form

$$ya(x, u, p)\, y \geqq \lambda |y|^2, \qquad x \in \Omega(R), \qquad |p| < \varepsilon \tag{11}$$

from $a(x, u, 0) > 0$ in (c). We want to mention that, instead of continuity, one could assume the local boundedness in $\Omega(R)$ together with a condition of Bernstein-Serrin type,

$$pa(x, u, p)\, p \geqq \lambda\, |p|^2, \qquad x \in \Omega(R), \qquad |p| < \varepsilon. \tag{12}$$

Here $u$ is fixed, $\lambda$ and $\varepsilon > 0$ can depend on $R$, and $\lambda > 0$ if $\varepsilon$ is sufficiently small. This replaces the former condition $a(x, u, 0) > 0$. Hypotheses of that kind were introduced in early work of S. BERNSTEIN and were revived in SERRIN's study of HILBERT's 20<sup>th</sup> problem [23]. When $a$ actually involves $p$, Condition (12) is much less restrictive than (11).

Another refinement goes through most smoothly when $T$ is linear, so that

$$a(x, u, p) = a(x), \qquad b(x, u, p) = b(x)$$

independently of $(u, p)$. Let the basic inequality (b) be replaced by

$$(\operatorname{sgn} u)\, Tu \geqq \frac{xa(x)\, x}{|x|^2} f(|u|, |u'|) \tag{13}$$

and suppose $xa(x)\, x > 0$ for $|x|$ large. In an obvious fashion (13) reduces to the inequality considered in Theorem 1, with a different $a$ and $b$, and we conclude that the hypothesis of the theorem or of the supplement holds, respectively, if

$$|D(x)| = O(|x|) \qquad \text{or} \qquad -D(x) \leqq O(|x|)$$

where

$$D(x) = \frac{x'a(x) - xb(x)}{xa(x)\, x} |x|^2.$$

When finite the quantity $N = \inf D(x)$ is the *effective dimension* in the sense of MEYERS & SERRIN [19]. The interest of this formulation is that the extra factor $xa(x)\, x$ on the right of the inequality compensates for the growth of $a(x)$, no matter how large it may be, so that only a single growth condition is needed. A somewhat more elaborate formulation applies to the nonlinear case but will not be discussed here.

Next we mention that the inequality (b) could be replaced by

$$\text{(b}') \qquad (\operatorname{sgn} u)\, Tu \geqq f(u, |u'|)$$

where $f(s, t)$ coincides with one bifurcation function for $s > 0$ and with another for $s < 0$; thus,

$$f(s, t) = f_1(s, t), \qquad f(-s, t) = f_2(s, t), \qquad s > 0$$

where $f_1$ and $f_2$ are bifurcation functions. This ensures that $-u$ satisfies the same sort of hypothesis as $u$, which is all we need. The nature of the improvement is most easily discerned when the differential inequality is replaced by an equation, so that the factor $\operatorname{sgn} u$ can be transferred to the right. The result is

$$Tu = (\operatorname{sgn} u)\, f(u, |u'|) = g(u, |u'|)$$

where $g$ is defined by the equation. In contrast to $f$, which is symmetrically monotone, $g(s, t)$ is monotone for each $t$ and for $-\infty < s < \infty$. Our initial formulation of the basic inequality (b) makes $g$ odd, which is an irrelevant restriction.

In conclusion, we mention a reformulation of the basic relations which is of special interest when the emphasis shifts from qualitative behavior (as in this paper) to the related theories of existence and uniqueness. Ever since the pioneering work of BRÉZIS and CRANDALL it has been known that the truly appropriate hypothesis, for certain nonlinear problems, is best expressed neither by differential inequalities nor by equations, but by set inclusions. For present purposes

$$Tu \in C(x, u, u'), \qquad u_\nu \in K(x, u)$$

and the conditions expressing the main hypothesis (b) would be

$$c \in C(x, u, p), \quad |x| > R_0, \quad 0 < |u| \leqq M, \quad |p| < \varepsilon$$
$$\Rightarrow c \operatorname{sgn} u \geqq g(u, |p|).$$

When $|x| > R_0$ but the constraints on $(u, p)$ fail, and $C(x, u, p) = (-\infty, \infty)$, so that $c$ and hence $Tu$ are unrestricted. The monotonicity of $g$, discussed above, is consistent with the often-made assumption that $C(x, u, p)$ is a complete monotone graph as a function of $u$.

*Acknowledgement.* Theorem I was obtained in 1977–1978 under auspices of NSF Grant MCS 77-0350. The results were presented at the University of Karlsruhe in Spring, 1985 in a series of lectures sponsored by the Deutsche Forschungsgemeinschaft.

## References

1. ANTONCEV, S. N., On the localization of solutions of nonlinear degenerate elliptic and parabolic equations, *Soviet. Math. Dokl.* **260**, 1981, 6; translation *ibid.* **24**, 1981, 2, 420–424.
2. BÉNILAN, PHILLIPPE, H. BRÉZIS & M. CRANDALL, A semilinear equation in $L^1(R^n)$, *Ann. Scuola Norm. Sup. Pisa Cl. Sci.* (4) **2**, 1975, 523–555.
3. BENSOUSSAN, ALAIN, HAIM BRÉZIS & AVNER FRIEDMAN, Estimates on the free boundary for quasivariational inequalities, *Comm. PDE* **2**, 1977, 297–321.
4. BENSOUSSAN, ALAIN, & AVNER FRIEDMAN, On the support of the solution of a system of quasi-variational inequalities, *J. Math. Anal. App.* **65**, 1978, 660–674.
5. BERKOWITZ, L. D., & HARRY POLLARD, A nonclassical variational problem arising from an optimal filter problem, *Arch. Rational Mech. Anal.* **26**, 1967, 281–304.
6. BERKOWITZ, L. D., & HARRY POLLARD, A nonclassical variational problem arising from an optimal filter problem, II, *ibid.* **38**, 1970, 161–172.
7. BERNIS, F., Compactness of support for some nonlinear elliptic problems of arbitrary order in dimension $n$, *Comm. PDE* **9**, 1984, 271–312.
8. BRÉZIS, HAIM, Solutions with compact support of variational inequalities, *Uspehi Mat. Nauk* **29** (1974), 103–108 = *Russian Math. Surveys* **29** (1974) 103–108.
9. DÍAZ, J. I., *Nonlinear partial differential equations and free boundaries*, Vol. I, *Elliptic equations*, Pitman Advanced Publishing Program, Boston, London Melbourne, Research Note 106, 1985.
10. DÍAZ, GREGORIS, & ILDEFONSO DÍAZ, Finite extinction time for a class of nonlinear parabolic equations, *Comm. PDE 4*, **1979**, 1213–1231.

11. Díaz, J. Ildefonso, & M. A. Herrero, Estimates on the support of the solutions of some nonlinear elliptic and parabolic problems, *Proc. Roy. Soc. Edinburgh* **89 A**, 1981, 249–258.
12. Díaz, J. Ildefonso, & L. Veron, Compacité du support des solutions d'équations quasilinéares elliptiques ou paraboliques, *C. R. Acad. Sci. Paris* **297**, 1983, 149–152.
13. Evans, Lawrence C., & Barry F. Knerr, Instantaneous shrinking of the support of non-negative solutions to certain nonlinear parabolic equations and variational inequalities, *Ill. J. Math.* **23**, 1979, 153–166.
14. Herero, M. A., & J. L. Vazquez, Asymptotic behavior of solutions of a strongly nonlinear parabolic problem, *Ann. Fac. Sci. Toulouse Math.* **3**, 1981, 113–127.
15. Hestenes, Magnus, & Ray Redheffer, On the minimization of certain quadratic functionals, I, *Arch. Rational Mech. and Anal.* **56**, 1974, 1–14.
16. Hestenes, W. Magnus, & Ray Redheffer, On the minimization of certain quadratic functionals, II, *ibid.*, 15–33.
17. Johanssen, D. E., Solution of a linear mean square problem when process statistics are undefined, *IEEE Trans. Aut. Control* **11**, 1966, 20–30.
18. Kalashnikov, A. S., The propagation of disturbances in problems of nonlinear heat conduction and absorption, *USSR Comm. Math. and Math. Phys.* **14**, 1974, 70–85.
19. Meyers, Norman, & James Serrin, The exterior Dirichlet problem for second-order elliptic partial differential equations, *J. Math. Mech.* **9**, 1960, 513–538.
20. Redheffer, Ray, On a nonlinear functional of Berkowitz and Pollard, *Arch. Rational Mech. and Anal.* **50**, 1973, 1–9.
21. Redheffer, Ray, Nonlinear differential inequalities and functions of compact support, *Trans. Amer. Math. Soc.* **220**, 1976, 133–157.
22. Redheffer, Ray, & Reinhard Redlinger, Quenching in time-delay systems: a summary and a counterexample, *Siam J. Math. Anal* **15**, 1984, 1114–1124.
23. Serrin, J., The problem of Dirichlet for quasilinear elliptic differential equations with many independent variables, *Phil. Trans. Royal Soc.* (London) **264**, 1969, 413–496.
24. Wallach, Sylvan, The differential equation $y' = f(y)$, *Am. J. Math.* **70**, 1948, 345–350.

Department of Mathematics
University of California
Los Angeles

*(Received October 6, 1986)*

# *Singular Solutions for some Semilinear Elliptic Equations*

H. BREZIS & L. OSWALD

*Dedicated to James Serrin on his sixtieth birthday*

## 1. Introduction

Let $B_R = \{x \in \mathbb{R}^N; |x| < R\}$ with $N \geqq 2$. Consider a function $u$ which satisfies

$$\begin{gathered} u \in C^2(B_R \setminus \{0\}), \quad u \geqq 0 \quad \text{on } B_R \setminus \{0\}, \\ -\Delta u + u^p = 0 \quad \text{on } B_R \setminus \{0\}. \end{gathered} \tag{1}$$

We are concerned with the behavior of $u$ near $x = 0$. There are two distinct cases:
1) When $p \geqq N/(N-2)$ and $(N \geqq 3)$ it has been shown by BREZIS & VÉRON [9] that $u$ must be smooth at 0 (See also BARAS & PIERRE [1] for a different proof). In other words, isolated singularities are *removable*.
2) When $1 < p < N/(N-2)$ there are solutions of (1) with a singularity at $x = 0$. Moreover all singular solutions have been classified by VÉRON [22]. We recall his result:

**Theorem 1.** *Assume that* $1 < p < N/(N-2)$ *and that* $u$ *satisfies* (1). *Then one of the following holds:*

(i) *either* $u$ *is smooth at* 0,

(ii) or $\lim_{x \to 0} u(x)/E(x) = c$ *where* $c$ *is a constant which can take any value in the interval* $(0, \infty)$,

(iii) or $\lim_{x \to 0} \left| u(x) - l(p, N) |x|^{-2/(p-1)} \right| = 0$.

Here $E(x)$ denotes the fundamental solution of $-\Delta$ and $l = l(p, N)$ is the (unique) positive constant $C$ such that $C|x|^{-2/(p-1)}$ satisfies (1)—more precisely

$$l = l(p, N) = \left[\frac{2}{(p-1)}\left(\frac{2p}{p-1} - N\right)\right]^{1/(p-1)}.$$

We shall first present a proof of Theorem 1 which is simpler than the original proof of Véron. In particular, it does not make use of FOWLER's results [10] for the Emden differential equation. Instead, it relies on some simple *scaling argument* (see the proof of Lemma 5) which is similar to the one used by KAMIN & PELETIER [12] for parabolic equations.

Next, we emphasize that a *singular behavior* such as (ii) or (iii) *can be prescribed* together with a boundary condition, and these determine uniquely the solution.

More precisely, let $\Omega$ be a smooth bounded domain in $\mathbb{R}^N$ with $0 \in \Omega$ and let $\varphi \geqq 0$ be a smooth function defined on $\partial\Omega$. We consider the problem

$$\begin{gathered} u \in C^2(\bar{\Omega} \setminus \{0\}), \quad u \geqq 0 \quad \text{on } \Omega \setminus \{0\}, \\ -\Delta u + u^p = 0 \quad \text{on } \Omega \\ u = \varphi \quad \text{on } \partial\Omega. \end{gathered} \tag{2}$$

**Theorem 2.** *Assume* $1 < p < N/(N-2)$. *Then*

(i) *There is a unique solution* $u_0$ *of* (2) *which belongs to* $C^2(\bar{\Omega})$.

(ii) *Given any constant* $c \in (0, +\infty)$ *there is a unique solution* $u_c$ of (2) *which satisfies*

$$\lim_{x \to 0} u(x)/E(x) = c.$$

(iii) *There is a unique solution* $u_\infty$ *of* (2) *which satisfies*

$$\lim_{x \to 0} |x|^{2/(p-1)} u(x) = l(p, N)$$

*In addition,* $\lim\limits_{c \downarrow 0} u_c = u_0$ *and* $\lim\limits_{c \uparrow \infty} u_c = u_\infty$.

Singular solutions of (1) occur in the THOMAS-FERMI theory with $N = 3$ and $p = 3/2$ (see e.g. [13] for a detailed exposition). Other results dealing with singular solutions of nonlinear elliptic equations have been obtained by a number of authors: J. SERRIN [20], [21], VERON and VAZQUEZ (See the exposition in [23]), P. L. LIONS [14], W. M. NI & J. SERRIN [16]. Semilinear parabolic equations with isolated singularities have been considered by BREZIS & FRIEDMAN [5], BREZIS & PELETIER & TERMAN [8], KAMIN & PELETIER [12], OSWALD [18].

## 2. Some preliminary facts

We recall some known results dealing with functions $u$ satisfying (1). Set $\alpha = 2/(p-1)$ (for $1 < p < \infty$).

**Lemma 1.** *Assume* $u \in C^2(B_R)$ *satisfies* (1).
*Then*

$$u(0) \leqq C(p, N)/R^\alpha$$

*where* $C(p, N)$ *is defined by* $C(p, N) = \text{Max}\,\{2\alpha N, 4\alpha(\alpha + 1)\}^{1/(p-1)}$.

The proof of Lemma 1 uses a comparison function $U$ of the same type as in OSSERMAN [17] (or LOEWNER & NIRENBERG [15]), namely set

$$U(x) = \frac{C(p, n)\, R^{\alpha}}{(R^2 - |x|^2)^{\alpha}} \text{ on } B_R\,.$$

A direct computation shows that

$$-\Delta U + U^p \geqq 0 \quad \text{on } B_R.$$

By the maximum principle we see that

$$u \leqq U \quad \text{on } B_R$$

and in particular $u(0) \leqq U(0)$.

**Lemma 2.** *Assume $u$ satisfies* (1) *with* $1 < p < N/(N-2)$. *Then, for* $0 < |x| < R/2$,

$$u(x) \leqq \frac{l(p, N)}{|x|^{\alpha}} \left(1 + \frac{C(p, N)}{l(p, N)} \left(\frac{|x|}{R}\right)^{\beta}\right)$$

*where* $\beta = 2\alpha + 2 - N > \alpha$.

Lemma 2 is established in BREZIS & LIEB [6] (Proposition A.4) for the special case where $N = 3$ and $p = 3/2$. The proof in the general case is just the same.

**Lemma 3.** *Assume* $1 < p < N/(N-2)$ *and let* $c > 0$ *be a constant. Then there is a unique function $u$ satisfying*

$$\begin{aligned} &u \in L^p(\mathbb{R}^N) \cap C^2(\mathbb{R}^N \setminus \{0\}),\\ &u \geqq 0 \quad \text{on } R^N \setminus \{0\},\\ &-\Delta u + u^p = c\delta \quad \text{on } \mathbb{R}^N \end{aligned} \tag{3}$$

*We set* $u = W_c$.

Lemma 3, as well as Lemma 4 below, are due to BENILAN & BREZIS (unpublished); the ingredients for the proofs may be found in [2], [3], [4] (and also [1] and [11]).

Finally, we assume that $\Omega$ is a smooth bounded domain in $\mathbb{R}^N$ with $0 \in \Omega$ and that $\varphi \geqq 0$ is a smooth function defined as $\partial\Omega$.

**Lemma 4.** *Assume* $1 < p < N/(N-2)$ *and let* $c > 0$ *be a constant. Then, there is a unique function $u$ satisfying*

$$\begin{aligned} &u \in L^p(\Omega) \cap C^2(\overline{\Omega} \setminus \{0\})\\ &u \geqq 0 \quad \text{on } \Omega \setminus \{0\}\\ &-\Delta u + u^p = c\delta \quad \text{on } \Omega\\ &u = \varphi \quad \text{on } \partial\Omega. \end{aligned} \tag{4}$$

## 3. A Scaling Argument

An important step in the proof of Theorem 1 is the following

**Lemma 5.** *Assume* $1 < p < N/(N-2)$. *Then*

$$\lim_{c\uparrow\infty} W_c(x) = l\,|x|^{-\alpha} = W_\infty(x).$$

**Proof.** It is clear (by comparison) that $W_c(x)$ is a nondecreasing function of $c$. Moreover we have

$$W_c(x) \leqq l\,|x|^{-\alpha}$$

(by letting $R \to \infty$ in Lemma 2). Therefore $\lim\limits_{c\uparrow\infty} W_c(x) = W_\infty(x)$ exists pointwise (for $x \neq 0$) and $W_\infty(x) \leqq l\,|x|^{-\alpha}$. The uniqueness of the solution of (3) implies that $W_c(x)$ is radial and so is $W_\infty(x)$. Next, we observe that the function

$$u(x) = k^\alpha W_c(kx) \qquad (k > 0)$$

satisfies

$$-\Delta u(x) + u^p(x) = k^{\alpha p} c\delta(kx) = k^{\alpha p - N} c\delta(x).$$

It follows, again by uniqueness, that

$$k^\alpha W_c(kx) = W_{ck^{\alpha p - N}}(x).$$

As $c \uparrow \infty$ we see that

$$k^\alpha W_\infty(kx) = W_\infty(x).$$

Choosing $k = 1/|x|$ we obtain

$$W_\infty(x) = W_\infty\left(\frac{x}{|x|}\right)|x|^{-\alpha} = C\,|x|^{-\alpha}$$

where $C > 0$ is some constant. Finally we note that since

$$-\Delta W_c + W_c^p = 0 \quad \text{in } \mathscr{D}'(\mathbb{R}^N \setminus \{0\})$$

and

$$W_c \to W_\infty \quad \text{in } L^p_{\text{loc}}(\mathbb{R}^N \setminus \{0\}),$$

it follows that

$$-\Delta W_\infty + W_\infty^p = 0 \quad \text{in } \mathscr{D}'(\mathbb{R}^N \setminus \{0\}).$$

This determines the value of the constant $C$ to be $C = l$.

There is a similar result in balls: Set $u = V_c$ to be the unique solution of problem (4) with $\Omega = B_R$.

**Lemma 6.** *Assume* $1 < p < N/(N-2)$. *Then* $V_\infty(x) \equiv \lim_{c \uparrow \infty} V_c(x)$ *exists pointwise on* $B_R \setminus \{0\}$ *and moreover*

$$W_\infty(x) - lR^{-\alpha} \leqq V_\infty(x) \leqq W_\infty(x) \quad \text{on } B_R.$$

**Proof.** It is again clear (by comparison) that $V_c(x)$ is a nondecreasing function of $c$. Also we have

$$0 \leqq V_c(x) \leqq W_c(x). \tag{5}$$

It follows from (4) and (5) that

$$-\Delta(W_c - V_c) \leqq 0 \quad \text{on } B_R,$$

and consequently $\operatorname{Sup}_{B_R} (W_c - V_c) \leqq \operatorname{Sup}_{\partial B_R} (W_c - V_c) \leqq \operatorname{Sup}_{\partial B_R} W_\infty = lR^{-\alpha}$. The conclusion follows by letting $c \to \infty$.

## 4. Proof of Theorem 1

Throughout this section we suppose $1 < p < N/(N-2)$. Assume $u$ satisfies (1) and set

$$c = \limsup_{x \to 0} u(x)/E(x).$$

We distinguish three cases:

Case (i) $c = 0$

Case (ii) $0 < c < \infty$

Case (iii) $c = \infty$.

*Cases* (i) *and* (ii). Here the main ingredient is the following:

**Lemma 7.** *In cases* (i) *and* (ii) *the function u belongs to* $L^p_{loc}(B_R)$ *and satisfies*

$$-\Delta u + u^p = c_0 \delta \quad \text{in } \mathcal{D}'(B_R)$$

*for some constant* $c_0$.

**Proof.** It is clear that $u \in L^p_{loc}(B_R)$ since $E \in L^p_{loc}(B_R)$ and $c < \infty$. We now use the same argument as in [7]: set

$$T = -\Delta u + u^p \in \mathcal{D}'(B_R).$$

Since the support of $T$ is contained in $\{0\}$, it follows from a classical result about distributions (see [19]) that

$$T = \sum_{0 \leqq |\alpha| \leqq m} c_\alpha D^\alpha(\delta). \tag{6}$$

We claim $c_\alpha = 0$ when $|\alpha| \geqq 1$. Indeed let $\zeta \in \mathscr{D}(B_R)$ be any fixed function such that $(-1)^{|\alpha|} D^\alpha \zeta(0) = c_\alpha$ for every $\alpha$ with $|\alpha| \leqq m$. Multiplying (6) through by $\zeta_\varepsilon(x) = \zeta(x/\varepsilon)$ we obtain

$$-\int u \Delta\zeta_\varepsilon + \int u^p \zeta_\varepsilon = \sum_{0 \leqq |\alpha| \leqq m} c_\alpha^2 \varepsilon^{-|\alpha|}.$$

An easy computation—using the estimate $u \leqq CE$—shows that

$$\left|\int u \Delta\zeta_\varepsilon\right| \leqq C \quad \text{when } N \geqq 3$$

$$\left|\int u \Delta\zeta_\varepsilon\right| \leqq C |\log \varepsilon| + C \quad \text{when } N = 2.$$

Since $\int u^p \zeta_\varepsilon \to 0$ as $\varepsilon \to 0$, we conclude that $c_\alpha = 0$ for $|\alpha| \geqq 1$. Therefore we obtain

$$-\Delta u + u^p = c_0 \delta \quad \text{in } \mathscr{D}'(B_R).$$

We conclude the proof of Theorem 1 in cases (i) and (ii) with the help of the following:

**Lemma 8.** *Assume* $u \in C^2(B_R \setminus \{0\}) \cap L^p_{\text{loc}}(B_R)$ *satisfies*

$$\begin{cases} u \geqq 0 & \text{on } B_R, \\ -\Delta u + u^p = c_0 \delta & \text{in } \mathscr{D}'(B_R) \end{cases}$$

*for some constant* $c_0$. *Then*

(i) if $c_0 = 0$, *then* $u$ *is smooth on* $B_R$,

(ii) if $c_0 \neq 0$, *then* $\lim_{x \to 0} u(x)/E(x) = c_0$.

**Proof.**
(i) Assume $c_0 = 0$. Since $u$ is subharmonic it follows that $u \in L^\infty_{\text{loc}}(B_R)$ and thus $\Delta u \in L^\infty_{\text{loc}}(B_R)$. We deduce that $u \in C^1(B_R)$ and then $u \in C^2(B_R)$. In fact $u \in C^\infty(B_R)$ since, by the strong maximum principle, we have either $u \equiv 0$ or $u > 0$ or $B_R$.

(ii) Assume $c_0 \neq 0$. By the maximum principle we have

$$u \leqq c_0 E + C \quad \text{on } B_{R/2}$$

and therefore

$$\begin{aligned} -\Delta u &\geqq c_0 \delta - (c_0 E + C)^p \\ &\geqq c_0 \delta - C(E^p + 1) \quad \text{on } B_{R/2} \end{aligned}$$

An elementary computation leads to

$$u(x) \geqq c_0 E - o(E) \quad \text{as } x \to 0,$$

and we conclude that $\lim_{x \to 0} u(x)/E(x) = c_0$.

*Remark 1.* Assume $c_0 \neq 0$. The argument above provides in fact an estimate for $|u - c_0 E|$ as $x \to 0$. More precisely we have

a) If $N = 2$ and $1 < p < \infty$ or $N = 3$ and $1 < p < 2$, then

$$|u - c_0 E| \leqq C \quad \text{on } B_{R/2}$$

b) If $N = 3$ and $p = 2$, then

$$|u(x) - c_0 E(x)| \leqq C(|\log |x|| + 1) \quad \text{on } B_{R/2}$$

c) If $N = 3$ and $2 < p < 3$ or $N \geqq 4$ and $1 < p < N/(N-2)$ then

$$|u(x) - c_0 E(x)| \leqq C\,|x|^{2-(N-2)p} \quad \text{on } B_{R/2}$$

and consequently

$$\left|\frac{u(x)}{E(x)} - c_0\right| \leqq C\,|x|^{\nu} \quad \text{on } B_{R/2}$$

with $\nu = N - (N-2)\,p > 0$.

**Proof of Theorem 1 in Case (iii).** We first recall a result of VÉRON [22] (Lemma 1.5):

**Lemma 9.** *Assume u satisfies* (1). *Then there is a constant V (depending only on p and N) such that*

$$\operatorname*{Sup}_{|x|=r} u(x) \leqq C \operatorname*{Inf}_{|x|=r} u(x) \quad \textit{for } 0 < r < R/2.$$

The conclusion of Lemma 9 is a simple consequence of Harnack's inequality and the estimate of Lemma 1, see [22] for the details.

We may now complete the proof of Theorem 1 with the help of the following:

**Lemma 10.** *Assume u satisfies* (1) and $\limsup_{x \to 0} u(x)/E(x) = \infty$. *Then*

$$|u(x) - l\,|x|^{-\alpha}| \leqq C\,|x|^{\gamma} \quad \text{on } B_{R/2}$$

*for some constants* $C = C(p, N, R)$ *and* $\gamma = \gamma(p, N) > 0$.

**Proof.** By Lemma 2 we already have the estimate

$$u(x) \leqq l\,|x|^{-\alpha} + C\,|x|^{\gamma} \quad \text{on } B_{R/2}$$

with

$$\gamma = \beta - \alpha = \alpha + 2 - N > 0.$$

We now establish an estimate from below. Let $x_n \to 0$ be such that $\lim u(x_n)/E(x_n) = \infty$. Set $r_n = |x_n|$, so that we obtain from Lemma 9

$$\operatorname*{Inf}_{|x|=r_n} u(x)/E(x) \underset{n\to\infty}{\to} \infty. \tag{7}$$

We recall that $V_c$ is the unique solution of (4) when $\Omega = B_R$, so that $V_c \leqq cE$ on $B_R$.

Given any constant $c > 0$, we see (by (7)) that

$$u(x) \geqq cE(x) \quad \text{for } |x| = r_n \text{ and } n \text{ large enough.}$$

Therefore

$$u(x) \geqq V_c(x) \quad \text{for } |x| = r_n \text{ and } n \text{ large enough.}$$

Applying the maximum principle in the domain $\{x \in \mathbb{R}^N; r_n < |x| < R\}$ we find that

$$u(x) \geqq V_c(x) \quad \text{for } r_n < |x| < R \text{ and } n \text{ large enough.}$$

As $n \to \infty$ we conclude that

$$u(x) \geqq V_c(x) \quad \text{on } B_R \setminus \{0\}$$

and as $c \to \infty$ we see that

$$u(x) \geqq V_\infty(x) \quad \text{on } B_R \setminus \{0\}.$$

In Lemma 6 we had the estimate

$$V_\infty(x) \geqq l(|x|^{-\alpha} - R^{-\alpha}).$$

However it is not good enough to deduce conclusion (iii) of Theorem 1. We need a better estimate from below for $V_\infty(x)$; we claim that

$$V_\infty(x) \geqq l\,|x|^{-\alpha}\left(1 - \left(\frac{|x|}{R}\right)^\beta\right) \quad \text{on } B_R, \tag{8}$$

where $\beta$ is defined in Lemma 2.

Clearly, it suffices to establish (8) for $R = 1$. The function $V_\infty$ is radial and so we write $V_\infty(r)$. We define the function $v$ on (0, 1) by the relation

$$v(r^\beta) = l^{-1} r^\alpha V_\infty(r)$$

so that $0 \leqq v \leqq 1$ on (0, 1), $v(1) = 0$ and $v(0) = 1$. Using the relation $-\Delta V_\infty + V_\infty^p = 0$ it is easy to deduce (as in the proof of Proposition A.4 [6]) that

$$-\beta^2 t^2 v''(t) + l^{p-1} v(t)\,(v^{p-1}(t) - 1) = 0 \quad \text{for } t \in (0, 1).$$

Consequently $v$ is concave and thus we have

$$v(t) \geqq 1 - t \qquad t \in (0, 1),$$

which is (8).

*Remark 2.* VÉRON [22] obtains in case (iii) an estimate of the form $|u(x) - l|x|^{-\alpha}| \leqq C|x|^\delta$ with an exponent $\delta$ which is better than $\gamma = \beta - \alpha$.

## 5. Proof of Theorem 2

*Case* (*i*) is classical.

*Case* (*ii*). The existence of a solution follows from Lemma 4 and 8. Suppose now $u$ satisfies (2) and $\lim_{x\to 0} u(x)/E(x) = c$. We deduce from Lemma 7 and 8 that $-\Delta u + u^p = c\delta$; uniqueness follows from Lemma 4.

*Case* (*iii*). We denote by $u_c$ the unique solution of (4) given by Lemma 4. We claim that $u_\infty = \lim_{c\uparrow\infty} u_c$ has all the required properties.

Indeed $u_c(x)$ is a nondecreasing function of $c$. Fix $R > 0$ such that $2R <$ dist $(0, \partial\Omega)$. By Lemma 1 we have

$$u_c(x) \leqq C(p, N)\, R^{-\alpha} \quad \text{for } |x| = R.$$

The maximum principle applied in the region

$$\Omega_R = \{x \in \Omega;\ |x| > R\}$$

shows that, in $\Omega_R$,

$$u_c(x) \leqq \operatorname{Max}\{\sup_{\partial\Omega} \varphi,\ C(p, N)\, R^{-\alpha}\}.$$

Therefore $u_\infty(x) = \lim_{c\uparrow\infty} u_c(x)$ exists and $u_\infty$ satisfies (2). By comparison on $B_R$ we have

$$V_c \leqq u_c \quad \text{on } B_R$$

and as $c \to \infty$ we obtain $V_\infty \leqq u_\infty$ on $B_R$. It follows that $\lim_{x\to 0} |u_\infty(x) - l|x|^{-\alpha}| = 0$ (by Lemma 6 and Theorem 1).

We turn now to the question of uniqueness. Suppose $u_1$ and $u_2$ satisfy (2) and $\lim_{x\to 0} |x|^\alpha u_i(x) = l$ for $i = 1, 2$. Lemma 10 implies that

$$|u_1(x) - u_2(x)| \leqq C\,|x|^\gamma \quad \text{on } B_R.$$

On the other hand

$$-\Delta(u_1 - u_2) + u_1^p - u_2^p = 0 \quad \text{on } \Omega \setminus \{0\}.$$

Applying the maximum principle in $\Omega_R$ we obtain

$$\underset{\Omega_R}{\operatorname{Max}} |u_1 - u_2| \leqq \underset{\partial B_R}{\operatorname{Max}} |u_1 - u_2| \leqq CR^{\gamma}$$

and then we let $R \to 0$ to conclude that $u_1 = u_2$.

The work of Brezis was partially supported by the United States Army under Contract No. DAAG29-80-C-0041.

## References

1. P. Baras & M. Pierre, Singularités éliminables pour des équations semi-linéaires, Ann. Inst. Fourier **34** (1984) pp. 185–206.
2. Ph. Benilan & H. Brezis & M. Crandall, A semilinear elliptic equation in$L^1(\mathbb{R}^N)$, Ann. Sc. Norm. Sup. Pisa **2** (1975) pp. 523–555.
3. H. Brezis, Some variational problems of the Thomas-Fermi type, in *Variational inequalities*, Cottle, Gianessi, Lions ed., Wiley (1980) pp. 53–73.
4. H. Brezis, Semilinear equations in $\mathbb{R}^N$ without condition at infinity, Applied Math. and Opt. **12** (1984) pp. 271–282.
5. H. Brezis & A. Friedman, Nonlinear parabolic equations involving measures as initial conditions, J. Math. Pures et Appl. **62** (1983) pp. 73–97.
6. H. Brezis & E. Lieb, Long range potentials in Thomas-Fermi theory, Comm. Math. Phys. **65** (1979) pp. 231–246.
7. H. Brezis & P. L. Lions, A note on isolated singularities for linear elliptic equations, in *Mathematical Analysis and Applications*, Part A, a volume, dedicated to L. Schwartz, L. Nachbin ed., Acad. Press (1981) pp. 263–266.
8. H. Brezis & L. Peletier & D. Terman, A very singular solution of the heat equation with absorption, Archive Rational Mech. Anal. **95** (1986) pp. 185–209.
9. H. Brezis & L. Véron, Removable singularities of some nonlinear elliptic equations, Archive Rational Mech. Anal. **75** (1980) pp. 1–6.
10. R. H. Fowler, Further studies on Emden's and similar differential equations, Quarterly J. Math. **2** (1931) pp. 259–288.
11. Th. Gallouet & J. M. Morel, Resolution of a semilinear equation in $L^1$, Proc. Roy. Soc. Edinburgh, **96A** (1984) pp. 275–288.
12. S. Kamin & L. Peletier, Singular solutions of the heat equation with absorption (to appear).
13. E. Lieb, Thomas-Fermi and related theories of atoms and molecules, Reviews of Modern Phys. **53** (1981) pp. 603–641.
14. P. L. Lions, Isolated singularities in semilinear problems, J. Diff. Eq. **38** (1980) pp. 441–450.
15. C. Loewner & L. Nirenberg, Partial differential equations invariant under conformal or projective transformations in *Contributions to Analysis*, Acad. Press. (1974) pp. 245–272.
16. W. M. Ni & J. Serrin, Non-existence theorems for singular solutions of quasilinear partial differential equations, Comm. Pure Appl. Math. **34** (1986) pp. 379–399.
17. R. Osserman, On the inequality $\Delta u \geqq f(u)$, Pacific J. Math. **7** (1957) pp. 1641–1647.
18. L. Oswald, Isolated singularities for a nonlinear heat equation, Houston J. Math. (to appear).
19. L. Schwartz, *Théorie des distributions*, Hermann (1966).

20. J. SERRIN, Local behavior of solutions of quasilinear equations, Acta Math. **111** (1964), pp. 247–302.
21. J. SERRIN, Isolated singularities of solutions of quasilinear equations, Acta Math. **113** (1965), pp. 219–240.
22. L. VÉRON, Singular solutions of some nonlinear elliptic equations, Nonlinear Anal. **5** (1981), pp. 225–242.
23. L. VÉRON, Weak and strong singularities of nonlinear elliptic equations, in *Nonlinear Functional Analysis and its Applications*, F. BROWDER ed. Proc. Symp. Pure Math. Vol. 45, Amer. Math. Soc. (1986).

(This paper was completed and appeared first as an MRC report in August 1985, but the copy sent for publication to this *Archive* in September 1985 was lost in the mail.)

Département de Mathématiques
Université Pierre et Marie Curie
Paris

*(Received February 6, 1987)*

# *Eventual $C^\infty$-Regularity and Concavity for Flows in One-Dimensional Porous Media*

D. G. ARONSON & J. L. VAZQUEZ

*For James B. Serrin on his sixtieth birthday.*

**Abstract**

We study the regularity and the asymptotic behavior of the solutions of the initial value problem for the porous medium equation

$$u_t = (u^m)_{xx} \quad \text{in } Q = \mathbb{R}\times(0,\infty),$$

$$u(x, 0) = u_0(x) \quad \text{for } x \in \mathbb{R},$$

with $m > 1$ and, $u_0$ a continuous, nonnegative function.

It is well known that, across a moving interface $x = \zeta(t)$ of the solution $u(x, t)$, the derivatives $v_t$ and $v_x$ of the pressure $v = (m/(m-1))\, u^{m-1}$ have jump discontinuities. We prove that each moving part of the interface is a $C^\infty$-curve and that $v$ is $C^\infty$ on each side of the moving interface (and up to it). We also prove that for solutions with compact support the pressure becomes a concave function of $x$ after a finite time. This fact implies sharp convergence rates for the solution and the interfaces as $t \to \infty$.

## Introduction

Let $u = u(x, t)$ represent the density of an ideal gas flowing isentropically in an infinite one-dimensional porous medium. If $u$ is appropriately scaled its evolution is governed by the equation

$$u_t = (u^m)_{xx} \quad \text{in } Q_T = \mathbb{R}\times(0, T) \tag{0.1}$$

together with the initial condition

$$u(\cdot, 0) = u_0 \quad \text{in } \mathbb{R}. \tag{0.2}$$

Here $m > 1$ and $T > 0$ are constants, while the initial distribution $u_0$ is a given nonnegative function. Because equation (0.1) is of parabolic type at the points where $u > 0$ but degenerates if $u = 0$, problem (0.1), (0.2) does not possess a classical solution if $u_0$ vanishes in some interval of $\mathbb{R}$. However the problem does possess a unique weak solution which is continuous everywhere in $Q_T$ under quite mild hypothesis on the initial data. Two of the main properties of the solutions of problem (0.1), (0.2), to which an extensive literature has been devoted, are the regularity and the asymptotic behavior, see [A3] for a survey of results. It is the aim of this paper to address the basic open problems for both properties.

One of the most important consequences of the degeneracy of equation (0.1) is the *finite speed* of propagation of disturbances from zero. For example, this implies that if $u_0$ is compactly supported then so is $u(\cdot, t)$ for every $t > 0$, and an *interface* or free boundary appears which separates the sets $[u > 0]$ and $[u = 0]$. Across a moving part $\Gamma$ of this interface $u$ is not smooth, since both $(u^{m-1})_x$ and $(u^{m-1})_t$ are bounded functions with jump discontinuities on $\Gamma$. In the first part of this paper we prove that $\Gamma$ is a $C^\infty$-curve, and also that $u^{m-1}$ can be continued up to $\Gamma$ from each side as a $C^\infty$-function. A full description of the results with reference to related works in given in Section 1, while the main estimates are proved in Sections 2 and 3.

The last section of the paper is devoted to the asymptotic behavior of the solutions whose initial data $u_0$ have compact support. It is known that any such solution approaches, as $t \to \infty$, a concave profile given by an explicit solution $\bar{u}$ (the BARENBLATT solution) which satisfies

$$(\bar{u}^{m-1})_{xx} = -\frac{m-1}{m(m+1)\,t}.$$

Recently BÉNILAN & VAZQUEZ [BV] have proved that whenever $u^{m-1}$ has compact support and is concave in its support, then as $t \to \infty$

$$(u^{m-1})_{xx} = -\frac{m-1}{m(m+1)\,t}\left(1 + O\left(\frac{1}{t}\right)\right) \tag{0.3}$$

inside the support of $u$. From (0.3) strong convergence theorems follow for the solution and its interface.

In Section 4 we prove that any solution $u$ whose initial data has compact support is *spatially concave* in its support after a finite time so that (0.3) and its consequences hold without the restriction of initial concavity.

## 1. Regularity. Statement of results

Let us recall some of the main facts about problem (0.1), (0.2), and introduce convenient definitions and notations. The existence, uniqueness and well-posedness of weak solutions for an optimal class of (nonnegative) initial data has been discussed in the papers [AC], [BCP], [DK], among others, even for the $N$-dimensional analogue $u_t = \Delta(u^m)$ with $N \geqq 1$. Without loss of generality we may assume that $u_0$ is a continuous nonnegative real function which grows as $|x| \to \infty$ at most like $O(|x|^\alpha)$ with $\alpha = 2/(m-1)$. In that case a unique continuous nonnegative function $u$ exists in a strip $Q_T$ with $0 < T \leqq \infty$ which satisfies (0.1) in the sense of distributions in $Q_T$ and (0.2) in the sense of trace on $\mathbb{R} \times \{0\}$. The solution is global, *i.e.* $T = \infty$, precisely if $u_0(x) = o(|x|^\alpha)$ [BCP]. To avoid unnecessary complications we assume in the sequel that $T = \infty$. See, *e.g.*, [V2] for details about the blow up if $T < \infty$. We set $Q = Q_\infty$.

Since (0.1) is strictly parabolic in the *positivity* set of a solution $u$, defined as

$$\mathscr{P}[u] = \{(x, t) \in Q : u(x, t) > 0\}, \tag{1.1}$$

it follows that $u \in C^\infty(\mathscr{P})$, [OKC]. In case $u_0$ is, say, a continuous positive function, then $\mathscr{P} = Q$ and $u$ is a classical solution of (0.1). However, if $u_0$ vanishes on some open interval, then, because of the finite propagation property, the set $\mathscr{P}[u]$ is a proper subset of $Q$ and an interface appears which separates $\mathscr{P}[u]$ from the region where $u = 0$. We do *not* require that $u_0$ have compact support.

The basic properties of the interface are by now well understood. Without essential loss of generality we may consider the case where $u_0$ vanishes on $\mathbb{R}^-$ and is a continuous positive function, at least, on an interval $(0, a)$ with $a > 0$. Then $\mathscr{P}[u]$ is bounded to the left in the $(x, t)$-plane by the *left interface* curve $x = \zeta(t)$, where

$$\zeta(t) = \inf\{x \in \mathbb{R} : u(x, t) > 0\}. \tag{1.2}$$

It is known that $\zeta(t)$ is a Lipschitz continuous nonincreasing function on $[0, \infty)$ with $\zeta(0) = 0$. Moreover, there is a time $t^* \in [0, \infty]$, called the *waiting time*, such that $\zeta(t) = 0$ for $0 \leqq t \leqq t^*$ and $\zeta(t) < 0$ for $t > t^*$. $t^*$ can be zero or positive depending on the local behavior of $u_0$ near 0, *cf.* [K], [ACK], [V2] for related results. CAFFARELLI & FRIEDMAN [CF] have shown that $\zeta \in C^1(t^*, T)$ and $\zeta'(t) < 0$ for every $t > t^*$, *i.e.*, a moving interface never stops. In deriving these results they also established *lateral* $C^1$ regularity for the *pressure* variable

$$v = \frac{m}{m-1} u^{m-1}, \tag{1.3}$$

which satisfies the equation

$$v_t = (m-1)\, v v_{xx} + v_x^2. \tag{1.4}$$

In fact, the functions $v_x(x, t)$ and $v_t(x, t)$ admit limits as $(x, t) \in \mathscr{P}$ tends to a point $(\zeta(t_0), t_0)$ of the moving part of the interface. Moreover, the extended functions thus obtained are continuous in the set

$$D = \{(x, t) : t > t^*, \zeta(t) \leqq x \leqq 0\}. \tag{1.5}$$

The relation between these results is given by the following equation which holds on the interface ([A2], [K], [CF])

$$\zeta'(t) = -v_x(\zeta(t), t). \tag{1.6}$$

Finally it is proved in [CF] that $v_t = v_x^2$ on the moving interface. Observe that (1.6) means that $v_x$ has a jump across the moving interface, since $\lim v_x$ is zero as $(x, t) \to (\zeta(t_0), t_0)$ with $x < \zeta(t)$.

After these results, the main regularity question for the solutions of problems (0.1), (0.2) can be formulated as follows. How smooth is the interface and how smooth is the solution near the interface? Our answer to this question is the following

**Theorem 1.** *$v$ is a $C^\infty$ function in $D$ and $\zeta(t)$ is a $C^\infty$ function in $(t^*, \infty)$.*

In case $t^* = 0$ the interface is then $C^\infty(\mathbb{R}^+)$. If $t^* > 0$ we have $\zeta(t) = 0$ for $0 < t < t^*$ so a lack of regularity of $\zeta$ can only occur at $t = t^*$. In fact it has been proved in [ACV] that for some initial data, *e.g.*, if $u_0(x) = o(|x|^{2/m-1})$ as $x \downarrow 0$, we have $t^* > 0$ and $D^+\zeta(t^*) < 0$ so that $\zeta$ has a corner at $t = t^*$. Lipschitz continuity is therefore the optimal global regularity for interfaces with a positive waiting time. However, there are solutions with a positive waiting time and a $C^1$ interface, *cf.* [ACK] and [LOT]. In this last work self-similar solutions are constructed with $t^* > 0$ and $\zeta(t) = -(t - t^*)^\alpha$ for $t > t^*$, where $\alpha$ is larger than 1. Therefore $\zeta \in C^1(\mathbb{R}^+)$ in those cases.

The main step in the proof of Theorem 1 consists in improving the known regularity for $v$ by establishing a bound for $v_{xx}$ near the points of the moving interface.

**Lemma 1.1.** *$v_{xx}$ is locally bounded in $D$.*

Part of this estimate is well known. In fact, the lower bound

$$v_{xx} \geqq -\frac{1}{(m+1)\,t} \quad \text{in } \mathscr{D}'(Q_T), \tag{1.7}$$

established in [AB], is one of principal tools used in the theory of the porous medium equation. (The analogous estimate also holds in several space dimensions). Since $v$ is a $C^\infty$ functions in $\mathscr{P}[u]$, it suffices to derive local upper bounds near points of the interface. At those points we construct (see Section 2) an explicit *local barrier* for $v_{xx}$ by taking advantage of the fact that the possible blow-up of $v_{xx}$ at the boundary is controlled by

$$v v_{xx} = \frac{1}{(m-1)}(v_t - v_x^2) \to 0 \tag{1.8}$$

if $(x, t) \in D$ and $(x, t) \to (\zeta(t_0), t_0)$ with $t_0 > t^*$. Again this bound cannot be extended to $t = t^*$. In fact, if $t^* > 0$ and $\gamma = D^+\zeta(t^*) < 0$ then

$$\limsup_{(x,t)\to(0,t^*)} v_{xx}(x, t) = +\infty. \tag{1.9}$$

Moreover $v_{xx}$ blows up as $(x, t) \to (0, t^*)$ along the line $x = \gamma(t - t^*)$ faster than $O(1/x)$, *cf.* [ACV].

Bounds for the higher order derivatives of $v$ with respect to $x$, i.e. $v^{(k)} \equiv (\partial/\partial x)^k v$ for $k \geqq 3$, are obtained by a similar process in Section 3. We first derive the differential equation for $v^{(k)}$ and obtain a preliminary estimate near the boundary by a rescaling argument together with interior estimates for non-degenerate equations. Finally we apply an iterative barrier argument to get a bound in $L^\infty_{\mathrm{loc}}(D)$. The barrier argument is different from, but related to the one used for $k = 2$. Since bounds for the mixed derivatives $(\partial/\partial t)^l (\partial/\partial x)^k v$ with $l \geqq 1$ and $k \geqq 0$ can be obtained from bounds for $v^{(k)}$ with $k \geqq 0$ by applying the differential operator $\left(\frac{\partial}{\partial t}\right)^{l-1} \left(\frac{\partial}{\partial x}\right)^k$ to the equation (1.4), it follows that $v$ is $C^\infty$ smooth in $D$ (*cf.* [A1]). The $C^\infty$ smoothness of $\zeta$ is then easily obtained by repeatedly differentiating equation (1.6). Thus to prove Theorem 1 it suffices to derive bounds for $v^{(k)}$.

The consideration of moving parts of a right hand interface, if such exists, or of inner interfaces if the initial support contains holes, offers no difficulties due to the local character of our results. Observe in particular, that any moving inner interface must end after a finite time at a point where it meets another inner interface, and $v_x$ cannot be continuous at the point of intersection; *cf.* [V1].

After we proved Theorem 1 we learned that the result had been obtained previously by Höllig & Kreiss [HK]. Their proof requires the additional restriction that the initial data have compact support and it is based on a technique of iterative weighted norms. A partial $C^\infty$ result had been previously obtained by Höllig & Pilant [HP]. More recently Angenent [An] has announced a proof of the analytic regularity of the free boundaries for a solution with compact support that exhibits a left hand and a right hand interface. Analyticity results for $t$ larger than the maximum of both waiting times. His proof uses our Lemma 1.1 and is based on nonlinear semigroup theory.

## 2. The Upper Bound for $v_{xx}$

Let $v = v(x, t)$ be the pressure corresponding to a solution $u = u(x, t)$ of the porous medium equation (0.1). Then $v = \frac{m}{m-1} u^{m-1}$ and it satisfies the equation (1.4):

$$v_t = (m-1)\, v v_{xx} + v_x^2$$

in the positivity set $\mathscr{P}[u]$. Let $q = (x_0, t_0)$ be a point on the left interface, so that $x_0 = \zeta(t_0)$, $v(x, t_0) = 0$ for all $x \leqq \zeta(t_0)$, and $v(x, t_0) > 0$ for all sufficiently small $x > \zeta(t_0)$. We assume that the interface is moving at $q$. Thus $t_0 > t^*$, where $t^*$ is the waiting time and

$$\zeta'(t_0) = -v_x(x, t_0) \equiv -a < 0. \tag{2.1}$$

We shall use the notation

$$R_{\delta,\eta} = R_{\delta,\eta}(t_0) \equiv \{(x, t) \in \mathbb{R}^2 \colon \zeta(t) < x \leqq \zeta(t) + \delta,\ t_0 - \eta \leqq t \leqq t_0 + \eta\}$$

(see Figure 1).

We prove the following precise version of Lemma 1.1.

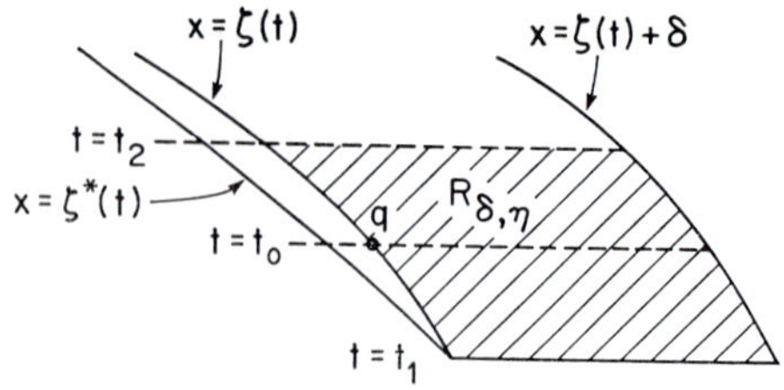


Fig. 1

**Lemma 2.1.** *Let $q$ be a point on the left interface and assume that* (2.1) *holds. There exist positive constants $C$, $\delta$ and $\eta$ depending only on $m$, $q$, and $u$ such that*

$$v_{xx} \leqq C \quad \text{in } R_{\delta,\eta/2}. \tag{2.2}$$

**Proof.** As we said above, [CF] proved that $v_t$, $v_x$ and $vv_{xx}$ are continuous in a closed neighborhood in $\overline{\mathscr{P}}[u]$ of any point on a moving interface, and that

$$(vv_{xx})(x, t) \to 0 \quad \text{as} \quad \mathscr{P}[u] \ni (x, t) \to (\zeta(\tau), \tau) \tag{2.3}$$

for any $\tau > t^*$. Choose now an $\varepsilon > 0$ such that

$$(a - 4m\varepsilon)(a - \varepsilon) \geqq 4(m + 1)\varepsilon > 0. \tag{2.4}$$

Then there exists a $\delta = \delta(\varepsilon) > 0$ and $\eta = \eta_1(\varepsilon) \in (0, t_0 - t^*)$ such that $R_{\delta,\eta} \subset \mathscr{P}[u]$,

$$a - \varepsilon < v_x < a + \varepsilon, \tag{2.5}$$

and

$$vv_{xx} \leqq \varepsilon \tag{2.6}$$

in $R_{\delta,\eta}$. In view of (2.5) we have

$$(a - \varepsilon)(x - \zeta(t)) < v(x, t) < (a + \varepsilon)(x - \zeta(t)) \quad \text{in } R_{\delta,\eta} \tag{2.7}$$

and

$$-(a + \varepsilon) < \zeta'(t) < -(a - \varepsilon) \quad \text{in } [t_1, t_2] \tag{2.8}$$

where $t_1 = t_0 - \eta$ and $t_2 = t_0 + \eta$. We set

$$\zeta^*(t) \equiv \zeta_1 - b(t - t_1), \tag{2.9}$$

where $\zeta_1 = \zeta(t_1)$ and $b = a + 2\varepsilon$. Clearly $\zeta(t) > \zeta^*(t)$ in $(t_1, t_2]$.

On $\mathscr{P}[u]$, $p \equiv v_{xx}$ satisfies

$$\mathscr{L}(p) \equiv p_t - (m - 1)vp_{xx} - 2mv_xp_x - (m + 1)p^2 = 0. \tag{2.10}$$

We shall construct a *barrier* for $p$ in $R_{\delta,\eta}$ of the form

$$\phi(x, t) \equiv \frac{\alpha}{x - \zeta(t)} + \frac{\beta}{x - \zeta^*(t)}, \tag{2.11}$$

where $\alpha$ and $\beta$ are positive constants. To accomplish this we must show that $\alpha$ and $\beta$ can be chosen such that $\mathscr{L}(\phi) \geqq 0$ in $R_{\delta,\eta}$ and $\phi \geqq p$ on the parabolic boundary of $R_{\delta,\eta}$. Indeed, we shall show that this can be done for arbitrarily small $\alpha$ so that, in the limit as $\alpha \downarrow 0$, $\phi$ gives a finite bound for $p$ in $R_{\delta,\eta/2}$.

It is easy to verify that

$$\mathscr{L}(\phi) = \frac{\alpha}{(x-\zeta)^2}\left\{\zeta' - 2(m-1)\frac{v}{x-\zeta} + 2mv_x\right\}$$

$$+\frac{\beta}{(x-\zeta^*)^2}\left\{\zeta^{*\prime} - 2(m-1)\frac{v}{x-\zeta^*} + 2mv_x\right\} - (m+1)\,\phi^2$$

$$\geqq \frac{\alpha}{(x-\zeta)^2}\left\{\zeta' - 2(m-1)\frac{v}{x-\zeta} + 2mv_x - 2(m+1)\,\alpha\right\}$$

$$+\frac{\beta}{x-\zeta^*}\left\{\zeta^{*\prime} - 2(m-1)\frac{v}{x-\zeta^*} + 2mv_x - 2(m+1)\,\beta\right\}.$$

From the estimates (2.5), (2.7), (2.8) and the definition (2.9) of $\zeta^*$ we conclude that

$$\mathscr{L}(\phi) \geqq \frac{\alpha}{(x-\zeta)^2}\{a - (4m-1)\,\varepsilon - 2(m+1)\,\alpha\}$$

$$-\frac{\beta}{(x-\zeta^*)^2}\{a - 4m\varepsilon - 2(m+1)\,\beta\}.$$

Set

$$\beta = (a - 4m\varepsilon)/2(m+1) \tag{2.12}$$

and note that (2.4) implies that $\beta > 0$. Then $\mathscr{L}(\phi) \geqq 0$ in $R_{\delta,\eta}$ for all $\alpha \in (0, \alpha_0]$, where $\alpha_0 = \{a - (4m-1)\,\varepsilon\}/2(m+1)$.

Let us now compare $p$ and $\phi$ on the parabolic boundary of $R_{\eta,\delta}$. In view of (2.6) and (2.7) we have

$$v_{xx} < \frac{\varepsilon}{(a-\varepsilon)(x-\zeta)} \quad \text{in } R_{\delta,\eta}$$

so that, in particular,

$$v_{xx}(\zeta(t)+\delta, t) \leqq \frac{\varepsilon}{(a-\varepsilon)\,\delta} \quad \text{in } [t_1, t_2].$$

By the mean value theorem and (2.8) it follows that for some $\tau \in (t_1, t_2)$

$$\zeta(t) + \delta - \zeta^*(t) = \delta + (a+2\varepsilon)(t-t_1) + \zeta'(\tau)(t-t_1)$$

$$\leqq \delta + 3\varepsilon(t-t_1) \leqq \delta + 6\varepsilon\eta.$$

Now set

$$\eta \equiv \min\{\eta_1(\varepsilon), \delta(\varepsilon)/6\varepsilon\}.$$

Since $\varepsilon$ satisfies (2.4) and $\beta$ is given by (2.12) it follows that

$$\phi(\zeta+\delta, t) \geqq \frac{\beta}{2\delta} \geqq \frac{\varepsilon}{(a-\varepsilon)\,\delta} \geqq v_{xx}(\zeta+\delta, t) \quad \text{on } [t_1, t_2].$$

Moreover,

$$\phi(x, t_1) \geqq \frac{\beta}{x-\zeta_1} > \frac{\varepsilon}{(a-\varepsilon)\,(x-\zeta_1)} > v_{xx}(x, t_1) \quad \text{on } (\zeta_1, \zeta_1+\delta].$$

Let $\Gamma = \{(x,t) \in \mathbb{R}^2 : x = \zeta(t),\ t_1 \leqq t \leqq t_2\}$. Clearly $\Gamma$ is a compact subset of $\mathbb{R}^2$. Fix $\alpha \in (0, \alpha_0)$. For each point $s \in \Gamma$ there is an open ball $B_s$ centered at $s$ such that

$$(vv_{xx})\,(x, t) \leqq \alpha(a-\varepsilon) \quad \text{in } B_s \cap \mathscr{P}[u].$$

In view of (2.7) we have

$$\phi(x, t) \geqq \frac{\alpha}{x-\zeta} \geqq v_{xx}(x, t) \quad \text{in } B_s \cap \mathscr{P}[u].$$

Since a finite number of these balls suffice to cover $\Gamma$ it follows that there is a $\gamma = \gamma(\alpha) \in (0, \delta)$ such that

$$\phi(x, t) \geqq p(x, t) \quad \text{in } R_{\gamma,\eta}.$$

Thus for every $\alpha \in (0, \alpha_0)$, $\phi$ is a barrier for $p$ in $R_{\delta,\eta}$.

By the comparison principle for parabolic equations [LSU] we conclude that

$$v_{xx}(x, t) \leqq \frac{\alpha}{x-\zeta(t)} + \frac{\beta}{x-\zeta^*(t)} \quad \text{in } R_{\delta,\eta},$$

where $\beta$ is given by (2.12) and $\alpha \in (0, \alpha_0)$ is arbitrary. Now let $\alpha \downarrow 0$ to obtain

$$v_{xx}(x, t) \leqq \frac{\beta}{x-\zeta^*(t)} \leqq \frac{2\beta}{\varepsilon\eta} \quad \text{in } \mathbb{R}. \qquad \square$$

*Remark.* The above argument shows that $C$, $\delta$ and $\eta$ may be chosen such as to vary continuously with the point $q = (x_0, t_0)$, where $x_0 = \zeta(t_0)$ and $t_0 > t^*$.

## 3. Bounds for $\left(\frac{\partial}{\partial x}\right)^j v$

We now turn to the problem of estimating the derivatives of the form

$$v^{(j)} \equiv \left(\frac{\partial}{\partial x}\right)^j v$$

for $j \geqq 3$. Theorem 1 easily follows from these estimates as we explained in the Introduction. Here we shall use a barrier similar to the one employed in § 2 in the estimation of $v^{(2)}$, but there are some crucial differences. In the estimate for

$v^{(2)}$ we were able to let $\alpha \downarrow 0$ because we knew *a priori* that

$$v^{(2)}(x, t) \leqq o(1/d)$$

as $x \downarrow \zeta(t)$, where $d = d(x, t) \equiv x - \zeta(t)$. For $j \geqq 3$ the only *a priori* information we can get is the weaker estimate

$$|v^{(j)}(x, t)| \leqq O(1/d)$$

(Lemma 3.2). However to compensate for this, when $j \geqq 3$ the equation for $v^{(j)}$ is linear, and so there will be no restriction on the size of $\beta$. We shall exploit this fact in an iterative barrier argument which enables us to derive a finite bound starting with an estimate which blows up on the interface (Lemma 3.4).

For $j \geqq 3$, the $v^{(j)}$ satisfy the equation

$$\begin{aligned} \mathscr{L}_j v^{(j)} \equiv v_t^{(j)} - (m-1)\, v v_{xx}^{(j)} - (2 + j(m-1))\, v_x v_x^{(j)} - c_{mj} v_{xx} v^{(j)} \\ - \sum_{l=3}^{j^*} d_{mj}^l\, v^{(l)} v^{(j+2-l)}. \end{aligned} \tag{3.1}$$

in $\mathscr{P}[u]$, where $j^* = [j/2] + 1$, and the $c_{mj}$ and $d_{mj}^l$ are positive constants which depend only on their indices, but whose precise values are irrelevant. Observe that the sum in $\mathscr{L}_j$ involves only derivatives of order $< j$ and it is omitted for $j = 3$ since in that case $j^* = 2 < 3$. Our result is

**Proposition 3.1.** Let $q = (x_0, t_0)$ *be a point on the interface for which* (2.1) *holds. For each integer* $j \geqq 2$ *there exist constants* $C_j$, $\delta$ *and* $\eta$ *depending only on* $m$, $j$, $q$, *and* $u$ *such that*

$$\left| \left( \frac{\partial}{\partial x} \right)^j v \right| \leqq C_j \quad \text{in } R_{\delta, \eta/2}. \tag{3.2}$$

The proof proceeds by induction on $j$. Suppose that $q = (x_0, t_0)$ is a point on the left interface for which (2.1) holds. Fix $\varepsilon \in (0, a)$ and take $\delta_0 = \delta_0(\varepsilon) > 0$ and $\eta_0 = \eta_0(\varepsilon) \in (0, t_0 - t^*)$ such that $R_0 \equiv R_{\delta_0, \eta_0}(t_0) \subset \mathscr{P}[u]$ and (2.5) holds. Thus we also have (2.7) and (2.8) in $R_0$. Assume that there are constants $C_k \in \mathbb{R}^+$ for $k = 2, 3, \ldots, j-1$ such that

$$|v^{(k)}| \leqq C_k \quad \text{on } R_0 \quad \text{for } k = 2, \ldots, j-1. \tag{3.3}$$

Observe that, by Lemma 2.1 and the estimate (1.7), (3.3) holds for $k = 2$.

Rescaling and interior estimates allow us to obtain a first estimate near $\Gamma$.

**Lemma 3.2.** *There are constants* $K \in \mathbb{R}^+$, $\delta \in (0, \delta_0)$ *and* $\eta \in (0, \eta_0)$, *depending only on* $q$, $m$ *and the* $C_k$ *for* $k \in [2, j-1]$ *with* $j \geqq 3$, *such that*

$$|v^{(j)}(x, t)| \leqq K/(x - \zeta(t)) \quad \textit{in } R_{\delta, \eta}.$$

**Proof.** Set

$$\delta = \min\{2\delta_0/3, 2s\eta_0\},$$

$$\eta = \eta_0 - \frac{\delta}{4s},$$

and define

$$R(\bar{x}, \bar{t}) \equiv \left\{(x, t) \in \mathbb{R}^2 : |x - \bar{x}| < \frac{\lambda}{2}, \bar{t} - \frac{\lambda}{4s} < t \leqq \bar{t}\right\}$$

for $(\bar{x}, \bar{t}) \in R_{\delta,\eta}$, where $s = a - \varepsilon$ and $\lambda = \bar{x} - \zeta(\bar{t})$. Then $(\bar{x}, \bar{t}) \in R_{\delta,\eta}$ implies that $R(\bar{x}, \bar{t}) \subset R_0$. Also observe that for each $(\bar{x}, \bar{t}) \in R_{\delta,\eta}$, $R(\bar{x}, \bar{t})$ lies to the right of the line $x = \zeta(\bar{t}) + s(\bar{t} - t)$.

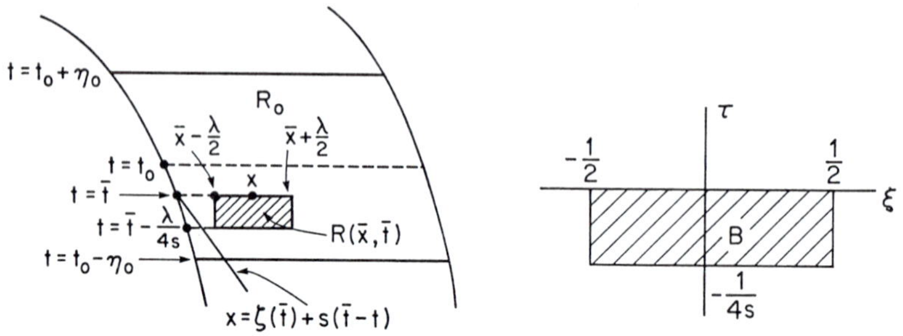


Fig. 2

Set $x = \lambda\xi + \bar{x}$ and $t = \lambda\tau + \bar{t}$. The function

$$V^{(j-1)}(\xi, \tau) \equiv v^{(j-1)}(\lambda\xi + \bar{x}, \lambda\tau + \bar{t}) = v^{(j-1)}(x, t)$$

satisfies the equation

$$\begin{aligned} V_\tau^{(j-1)} = &\left\{(m-1)\frac{v}{\lambda} V_\xi^{(j-1)} + (2 + (j-1)(m-1))\, v_x V^{(j-1)}\right\}_\xi \\ &- (m-1)\, v_x V_\xi^{(j-1)} + \lambda(c_{m,j-1} - 2 - (j-1)(m-1))\, v_{xx} V^{(j-1)} \\ &+ \lambda \sum_{l=3}^{(j-1)^*} d_{m,j-1}^l v^{(l)} v^{(j+1-l)} \end{aligned} \tag{3.4}$$

in the region

$$B \equiv \left\{(\zeta, \tau) \in \mathbb{R}^2 : |\xi| \leqq \tfrac{1}{2}, -\frac{1}{4s} < \tau \leqq 0\right\},$$

and $|V^{(j-1)}| \leqq C_{j-1}$ in $B$ (see Fig. 2). In view of (2.7) and (2.8)

$$(a - \varepsilon)\frac{x - \zeta(t)}{\lambda} \leqq \frac{v(x, t)}{\lambda} \leqq (a + \varepsilon)\frac{x - \zeta(t)}{\lambda}$$

and

$$\zeta(\bar{t}) \leqq \zeta(t) \leqq \zeta(\bar{t}) + s(\bar{t} - t) \leqq \zeta(\bar{t}) + \frac{\lambda}{4}.$$

Therefore

$$\frac{\lambda}{4} = \bar{x} - \frac{\lambda}{2} - \zeta(\bar{t}) - \frac{\lambda}{4} \leqq x - \zeta(t) \leqq \bar{x} + \frac{\lambda}{2} - \zeta(\bar{t}) = \frac{3\lambda}{2}$$

which implies

$$\frac{a-\varepsilon}{4} \leqq \frac{v}{\lambda} \leqq \frac{3(a+\varepsilon)}{2},$$

that is, equation (3.4) is uniformly parabolic in $B$. Moreover, it follows from (2.5) and (3.3) that $V^{(j-1)}$ satisfies all of the hypotheses of Theorem 5.3.1 of [LSU]. Thus we conclude that there is a constant $K = K(a, m, C_1, \ldots, C_{j-1}) > 0$ such that

$$\left| \frac{\partial}{\partial \xi} V^{(j-1)}(0, 0) \right| \leqq K;$$

that is,

$$|v^{(j)}(\bar{x}, \bar{t})| \leqq K/\lambda.$$

Since $(\bar{x}, \bar{t}) \in R_{\delta,\eta}$ is arbitrary this proves the lemma. □

We now turn to the barrier construction. If $\gamma \in (0, \delta)$ we will use the notation

$$R^{\gamma}_{\delta,\eta} = R^{\gamma}_{\delta,\eta}(t_0) \equiv \{(x, t) \in \mathbb{R}^2 : \zeta(t) + \gamma \leqq x \leqq \zeta(t) + \delta,\ t_0 - \eta \leqq t \leqq t_0 + \eta\}.$$

**Lemma 3.3.** *Let $R_{\delta_1,\eta_1}$ be the region constructed in the proof of Lemma 2.1. For $j \geqq 3$ and $(x, t) \in R^{\gamma}_{\delta_1,\eta_1}$ let*

$$\phi_\gamma(x, t) \equiv \frac{\alpha}{x - \zeta(t) - \gamma/3} + \frac{\beta}{x - \zeta^*(t)} \tag{3.5}$$

*where $\zeta^*$ is given by (2.9), and $\alpha$ and $\beta$ are positive constants. There exist $\delta \in (0, \delta_1)$ and $\eta \in (0, \eta_1)$ depending only on $a, m, C_1, \ldots, C_{j-1}$ such that*

$$\mathscr{L}_j(\phi_j) \geqq 0 \quad \text{in } R^{\gamma}_{\delta,\eta}$$

*for all $\gamma \in (0, \delta)$*

**Proof.** Choose $\varepsilon$ such that

$$0 < \varepsilon < \frac{a}{4 + 2j(m-1)}. \tag{3.6}$$

There exist $\delta_2 \in (0, \delta_1)$ and $\eta \in (0, \eta_1)$ such that (2.5), (2.7) and (2.8) hold in $R_{\delta_2,\eta}$. Fix $\gamma \in (0, \delta_2)$. For $(x, t) \in R^{\gamma}_{\delta_2,\eta}$

$$\begin{aligned}
\mathscr{L}_j(\phi_j) = {} & \frac{\alpha}{(x-\zeta-\gamma/3)^2}\Big\{\zeta' - \frac{2(m-1)\,v}{x-\zeta-\gamma/3} + (2 + j(m-1))\,v_x \\
& - c_{mj}(x-\zeta-\gamma/3)\,v_{xx} - (x-\zeta-\gamma/3)^2 \sum_{l=3}^{*} d^l_{mj}\, v^{(l)} v^{(j+2-l)}\Big\} \\
& + \frac{\beta}{(x-\zeta^*)^2}\Big\{\zeta^{*\prime} - \frac{2(m-1)\,v}{x-\zeta^*} + (2 + j(m-1))\,v_x \\
& - c_{mj}(x-\zeta^*)\,v_{xx} - (x-\zeta^*)^2 \sum_{l=3}^{j*} d^l_{mj}\, v^{(l)} v^{(j+2-l)}\Big\}.
\end{aligned}$$

From (2.7) and (3.6), together with the fact that $x - \zeta^* \geqq x - \zeta - \gamma/3$ we conclude that

$$\frac{v}{x-\zeta^*} \leqq \frac{v}{x-\zeta-\gamma/3} \leqq (a+\varepsilon)\frac{x-\zeta}{x-\zeta-\gamma/3} \leqq (a+\varepsilon)\frac{\gamma}{\gamma-\gamma/3} = \tfrac{3}{2}(a+\varepsilon).$$

Thus it follows from (2.5), (2.8) and (3.3) that

$$\mathcal{L}_j(\phi_j) \geqq \frac{\alpha}{(x-\zeta-\gamma/3)^2}\Big\{a - (3+2j(m-1))\,\varepsilon - \delta_2\Big(c_{mj}C_2 + \delta_2 \sum_{l=3}^{j^*} d_{mj}^l\, C_l C_{j+2-l}\Big)\Big\} + \frac{\beta}{(x-\zeta^*)^2} \times$$
$$\times \Big\{a - (4+2j(m-1))\,\varepsilon - \delta_2\Big(c_{mj}C_2 + \delta_2 \sum_{l=3}^{j^*} d_{mj}^l\, C_l C_{j+2-l}\Big)\Big\}.$$

Since $\varepsilon$ satisfies (3.6) we can choose $\delta = \delta_2(\varepsilon, m, C_2, \ldots, C_{j-1}) > 0$ so small that $\mathcal{L}_j(\phi_j) \geqq 0$ in $R^\gamma_{\delta,\eta}$. □

**Lemma 3.4.** (*Barrier Transformation*). *Let $\delta$ and $\eta$ be as in Lemma 3.3 with the additional restriction that*

$$\eta < \delta/6\varepsilon, \tag{3.7}$$

*where $\varepsilon$ satisfies* (3.6). *Suppose that for some nonnegative constants $\alpha$ and $\beta$*

$$v^{(j)}(x,t) \leqq \frac{\alpha}{x-\zeta(t)} + \frac{\beta}{x-\zeta^*(t)} \quad \text{in } R_{\delta,\eta}. \tag{3.8}$$

*Then $v^{(j)}$ also satisfies*

$$v^{(j)}(x,t) \leqq \frac{2\alpha/3}{x-\zeta(t)} + \frac{\beta+2\alpha/3}{x-\zeta^*(t)} \quad \text{in } R_{\delta,\eta}. \tag{3.9}$$

**Proof.** By Lemma 3.3, for any $\gamma \in (0, \delta)$ the function

$$\phi_j(x,t) = \frac{2\alpha/3}{x-\zeta-\gamma/3} + \frac{\beta+2\alpha/3}{x-\zeta^*}$$

satisfies $\mathcal{L}_j(\phi_j) \geqq 0$ in $R^\gamma_{\delta,\eta}$. On the other hand, on the parabolic boundary of $R^\gamma_{\delta,\eta}$ we have $\phi_j \geqq v^{(j)}$. In fact, for $t = t_1$ and $\zeta_1 + \gamma \leqq x \leqq \zeta_1 + \delta$, with $\zeta_1 = \zeta(t_1)$, we have

$$\phi_j(x,t_1) = \frac{2\alpha/3}{x-\zeta_1-\gamma/3} + \frac{\beta+2\alpha/3}{x-\zeta_1} > \frac{4\alpha/3}{x-\zeta_1} + \frac{\beta}{x-\zeta_1} > v^{(j)}(x,t_1),$$

while for $x = \zeta + \delta$ and $t_1 \leqq t \leqq t_2$ we get, in view of (3.7),

$$\phi_j(\zeta+\delta,t) \geqq \frac{2\alpha/3}{\delta-\gamma/3} + \frac{\beta}{\zeta+\delta-\zeta^*} + \frac{2\alpha/3}{\delta+6\varepsilon\eta}$$
$$\geqq \frac{2\alpha/3}{\delta} + \frac{\beta}{\zeta+\delta-\zeta^*} + \frac{\alpha/3}{\delta} \geqq v^{(j)}(\zeta+\delta,t).$$

Finally, for $x = \zeta + \gamma$, $t_1 \leqq t \leqq t_2$ we have

$$\phi_j(\zeta + \gamma, t) = \frac{2\alpha/3}{\gamma - \gamma/3} + \frac{\beta + 2\alpha/3}{\zeta + \gamma - \zeta^*} \geqq \frac{\alpha}{\gamma} + \frac{\beta}{\zeta + \gamma - \zeta^*} \geqq v^{(j)}(\zeta + \gamma, t).$$

By the comparison principle we get

$$\phi_j \geqq v^{(j)} \quad \text{in } R^{\gamma}_{\delta,\eta}$$

for any $\gamma \in (0, \delta)$, and (3.9) follows by letting $\gamma \downarrow 0$.

**Completion of proof of Proposition 3.1.** By Lemma 3.2. we have an estimate for $v^{(j)}$ of the form (3.8) with $\alpha = K$ and $\beta = 0$. Iterating this estimate by the Barrier Transformation Lemma we obtain the sequence of estimates

$$v^{(j)}(x, t) \leqq \frac{\alpha_n}{x - \zeta(t)} + \frac{\beta_n}{x - \zeta^*(t)}$$

with $\alpha_n = (2/3)^n K$ and $\beta_n = \{(\frac{2}{3}) + \dots + (\frac{2}{3})^n\} K$. Thus if we let $n \to \infty$ we obtain an upper bound for $v^{(j)}$ of the form

$$v^{(j)}(x, t) \leqq \frac{2K}{x - \zeta^*(t)} \quad \text{in } R_{\delta,\eta}. \tag{3.10}$$

As in the proof of Lemma 2.1, this implies that $v^{(j)}$ is bounded above in $R_{\delta,\eta/2}$.

Since the equation (3.1) for $v^{(j)}$ is linear, a similar lower bound can be obtained in the same way and the induction step is complete. □

## 4. Eventual concavity and asymptotic behavior

In this section we assume that $u_0$ has compact support, say the interval $I = [a, b]$. Without loss of generality we may assume that $u_0$ is continuous and $u_0(x) > 0$ for $x \in (a, b)$. According to the results of [V1] there are in that case two expanding interfaces, $\zeta_-(t)$ and $\zeta_+(t)$, such that

$$\mathscr{P}[u] = \{(x, t) \in Q : \zeta_-(t) < x < \zeta_+(t)\},$$

and as $t \to \infty$ we have the expansion

$$\zeta_\pm(t) = \pm r(t) + x_0 + o(1), \tag{4.1}$$

where $r(t) = c_m(M^{m-1}t)^{\frac{1}{m+1}}$, and $c_m$ is constant that depends only on $m$,

$$M = \int u_0(x)\, dx, \tag{4.2}$$

and

$$x_0 = M^{-1} \int x u_0(x)\, dx. \tag{4.3}$$

The *total mass* $M$ and *center of mass* $x_0$ are two invariants of the evolution. Moreover, $x = \pm r(t) + x_0$ is precisely the interface of a particular solution of

(0.1), called the Barenblatt solution, whose pressure is given by

$$\bar{v}(x, t) = \frac{(r^2(t) - |x - x_0|^2)_+}{2(m+1)\, t}. \tag{4.4}$$

We remark that the Barenblatt solution takes initial values $u_0(x) = M\,\delta(x - x_0)$ with $M$ given by (4.2) and $\delta$ being the Dirac measure concentrated at $x_0$. The asymptotic closeness of $v$ and $\bar{v}$ is not restricted to formula (4.1). In fact we also have as $t \to \infty$,

$$\zeta'_\pm(t) = \pm r'(t) + o(1/t) \quad \text{and } |\zeta'_\pm(t)| \leqq r'(t), \tag{4.5}$$

$$v(x, t) - \bar{v}(x, t) = o(t^{-\frac{m}{m+1}}) \quad \text{uniformly in } x, \tag{4.6}$$

and

$$v_x(x, t) + \frac{x - x_0}{(m+1)\, t} = o\left(\frac{1}{t}\right) \quad \text{uniformly in } (\zeta_-(t), \zeta_+(t)). \tag{4.7}$$

The convergence rates in the above formulas are not optimal. Optimal rates are obtained in [V1] under the assumption of spatial symmetry. In [BV] the case of a solution whose initial pressure $v_0$ is concave in its support (i.e., $v'' \leqq -C$ in $\mathscr{D}'(a, b)$) is considered, and it is shown that $v(\cdot, t)$ is a concave function in $(\zeta_-(t), \zeta_+(t))$ for every $t > 0$ and also that

$$v_{xx} = -\frac{1}{(m+1)\, t} + O\left(\frac{1}{t^2}\right) \tag{4.8}$$

uniformly in $(\zeta^-(t), \zeta^+(t))$. From (4.8) sharp estimates for $v$ and $v_x$ can be derived by integration and, using formula (1.6), for $\zeta_\pm$, $\zeta'_\pm$ and $\zeta''_\pm$ (see [BV]). In particular,

$$\zeta''_\pm = \pm r''(t)\left(1 + O\left(\frac{1}{t}\right)\right). \tag{4.9}$$

Here we show that the assumption of initial concavity can be removed in the above results. We prove

**Theorem 2.** *For every solution $u$ of* (0.1), (0.2) *which is compactly supported in space there exists a finite time $T$ after which the pressure $v(x, t)$ is a concave function with respect to the space variable in $\mathscr{P}[u]$, the functions $|\zeta_\pm(t)|$ are concave in $(T, \infty)$, and formulas* (4.8), (4.9) *hold as $t \to \infty$.*

We divide the proof of Theorem 2 into a series of Lemmas. In the first two Lemmas we show that (4.8) holds locally in $\mathscr{P}[u]$ as $t \uparrow \infty$. The remaining Lemmas show that this estimate can be made global. The key step is to show that the estimates in Sections 2 and 3 hold uniformly as $t \uparrow \infty$ (Lemma 4.4) and for this we need a uniform rate of convergence of $vv_{xx}$ to zero at the interface (Lemma 4.3).

Without loss of generality we set $M = 1$ and $x_0 = 0$. We also introduce the sets

$$\mathscr{P}_\alpha = \{(x, t) \in Q : |x| \leqq \alpha r(t)\}$$

for $\alpha \in (0, 1)$. Since $|\zeta_\pm(t)| \geqq r(t)$ (*cf.* [V1: Theorem A]) we have $\mathscr{P}_\alpha \subset \mathscr{P}[u]$.

**Lemma 4.1.** *For every* $\alpha \in (0, 1)$ *there are positive constants* $C_\alpha$ *and* $t_\alpha$ *such that*

$$|v_{xxx}(x, t)| \leqq C_\alpha t^{-\frac{m+2}{m+1}} \tag{4.10}$$

*if* $(x, t) \in \mathscr{P}_\alpha$ *with* $t \geqq t_\alpha$.

**Proof.** For every $\lambda > 0$ the function

$$u^\lambda(x, t) \equiv \lambda u(\lambda x, \lambda^{m+1} t)$$

is again a solution of (0.1) with $M = 1$ and $x_0 = 0$. If $\zeta_\pm(t)$ and $v$ denote respectively the interfaces and pressure for $u$, then the corresponding quantities for $u^\lambda$ are

$$\zeta_\pm^\lambda(t) = \frac{1}{\lambda} \zeta_\pm(\lambda^{m+1} t) \tag{4.11}$$

and

$$v^\lambda(x, t) = \lambda^{m-1} v(\lambda x, \lambda^{m+1} t),$$

where $v^\lambda$ satisfies

$$v_t^\lambda = (m - 1)\, v^\lambda v_{xx}^\lambda + (v_x^\lambda)^2 \quad \text{in } \mathscr{P}[u^\lambda].$$

From (4.6) and (4.7) we have

$$v^\lambda(x, t) = \frac{\{r^2(t) - x^2\}_+}{2(m + 1)\, t} + o\left(\frac{1}{\lambda}\right) \quad \text{in } \mathbb{R}_\times [\tfrac{1}{2}, 2] \tag{4.12}$$

and

$$v_x^\lambda(x, t) = -\frac{x}{(m + 1)\, t} + o\left(\frac{1}{\lambda}\right) \quad \text{in } \mathscr{P}[u^\lambda] \cap \{t \in [\tfrac{1}{2}, 2]\} \tag{4.13}$$

uniformly as $\lambda \uparrow \infty$. Set $\beta = (1 + \alpha)/2$. Then in view of (4.1), (4.5) and (4.11)

$$|\zeta_\pm^\lambda(t)| \geqq \frac{1}{\lambda} r(\lambda^{m+1} t) = r(t) > \beta r(t) > \alpha \gamma(t) \quad \text{on } [\tfrac{1}{2}, 2].$$

It follows from (4.12) and (4.13) that there is a constant $\lambda_\alpha \in \mathbb{R}^+$ such that for $\lambda \geqq \lambda_\alpha$ we have $v^\lambda$ strictly positive and bounded, and $|v_x^\lambda|$ bounded in $\mathscr{P}_\beta$ for $t \in [\frac{1}{2}, 2]$. We now apply the interior regularity argument used in the proof of Lemma 1 of [A1] to conclude that for $\lambda \geqq \lambda_\alpha$ all of the derivatives of $v^\lambda$ are bounded independently of $\lambda$ in $\mathscr{P}_\alpha$ for $t \in [\frac{3}{4}, 2]$. In particular, there is a constant $C_\alpha \in \mathbb{R}^+$ depending only on $\alpha$ (and $m$) such that

$$|v_{xxx}^\lambda(y, 1)| \leqq C_\alpha \quad \text{if } |y| \leqq \alpha r(1) \text{ and } \lambda \geqq \lambda_\alpha.$$

Since $v_{xxx}^\lambda(y, 1) = \lambda^{m+2} v_{xxx}(\lambda y, \lambda^{m+1})$ and $t^{\frac{1}{m+1}} r(1) = r(t)$, (4.10) follows. □

We can now estimate $v_{xx}$ in $\mathscr{P}_\alpha$.

**Lemma 4.2.** *As* $t \to \infty$

$$v_{xx}(x,t) = -\frac{1}{(m+1)\,t} + o(t^{-\gamma}), \quad \left(\gamma = 1 + \frac{1}{2(m+1)}\right), \tag{4.14}$$

*uniformly in* $x$ *for* $(x,t) \in \mathscr{P}_\alpha$, *with* $\alpha \in (0,1)$

**Proof.** The function

$$h(x,t) \equiv v_{xx}(x,t) + \frac{1}{(m+1)\,t}$$

is of class $C^\infty$ in $\mathscr{P}[u]$ for $t > 0$ and, by the estimate (1.7), is nonnegative. It therefore suffices to estimate $h$ from above in $\mathscr{P}_\alpha$.

For $t$ large we have

$$\int\limits_{\zeta_-(t)}^{\zeta_+(t)} h(x,t)\,dx = v_x(\zeta_+(t),t) - v_x(\zeta_-(t),t) + \frac{\zeta_+(t) - \zeta_-(t)}{(m+1)\,t}$$

$$= -\zeta_+'(t) + \zeta_-'(t) + o\left(\frac{1}{t}\right) + \frac{\zeta_+(t) - \zeta_-(t)}{(m+1)\,t} = o\left(\frac{1}{t}\right), \tag{4.15}$$

by virtue of (4.7). Therefore $h(\cdot,t) \in L^1(\zeta_-(t), \zeta_+(t))$ with $\|h(\cdot,t)\|_1 = o(1/t)$ as $t \to \infty$. On the other hand, by Lemma 4.1, we have $h_x = v_{xxx} = O(t^{-\mu})$ as $t \to \infty$, where $\mu = (m+2)/(m+1)$. We claim that $h(\cdot,t) \in L^\infty_{\mathrm{loc}}(\zeta_-(t), \zeta_+(t))$ and $\sup_x h(x,t) = o(t^{-\gamma})$ in sets of form $|x| \leqq \alpha r(t)$. In fact, assume that there is a constant $c_1 > 0$ and a sequence $(x_n, t_n) \in \mathscr{P}_\alpha$ with $t_n \to \infty$, such that $h(x_n, t_n) \geqq c_1 t^{-\gamma}$. Then, by the estimate on $h_x$, there exists $c_2$ such that

$$h(x, t_n) \geqq c_1 t_n^{-\gamma} - c_2 t^{-\mu}(x - x_n).$$

This means that $h(x,t_n) \geqq c_1 t^{-\gamma/2}$ in the interval $I_n$ centered at $x_n$ with length

$$l_n = \frac{c_1}{c_2} t^{\mu-\gamma} = \frac{c_1}{c_2} t_n^{\frac{1}{2(m+1)}} - o(r(t_n)).$$

Hence

$$\int\limits_{I_n} h(x,t_n)\,dx \geqq \left(\frac{c_1}{2} t_n^{-\gamma}\right)\left(\frac{c_1}{c_2} t_n^{\frac{1}{2(m+1)}}\right) = \frac{c_1^2}{2c_2} t_n^{-1},$$

which contradicts (4.15). □

The next step in the proof of Theorem 2 is to rederive the convergence of $vv_{xx}$ to zero at the interface [CF] so as to obtain a rate of convergence which is uniformly valid for all large $t$. For convenience of calculation we state our results instead for $v^\lambda(x,t)$ at $t = 1$ for large $\lambda$ and only consider the right-hand interface $\zeta^\lambda(t) \equiv \zeta_+^\lambda(t)$.

**Lemma 4.3.** *There exists a constant* $C > 0$ *independent of* $\lambda$ *such that*

$$v_t^\lambda(x,1) - (v_x^\lambda(x,1))^2 \leqq C(\zeta^\lambda(1) - x)^{1/2} + O(\lambda^{-1/2}) \tag{4.16}$$

*for all* $x < \zeta^\lambda(1)$ *such that* $\zeta^\lambda(1) - x$ *is small, independently of* $\lambda$

**Proof.** To simplify the notation we temporarily drop the superscript $\lambda$ from $v^\lambda$ and $\zeta^\lambda$. Set

$$k \equiv c_m/(m+1).$$

From (4.1) and (4.5) we have

$$\zeta(1) = c_m + o(1),$$

$$\zeta'(1) = k = O(1/\lambda) \quad \text{and} \quad |\zeta'(t)| \leqq kt^{-\frac{m}{m+1}} \quad \text{in } \mathbb{R}^+, \tag{4.17}$$

so that (*cf.* [V3: Theorem 1])

$$|v_x(x, t)| \leqq kt^{-\frac{m}{m+1}} \quad \text{in } Q. \tag{4.18}$$

We shall follow the argument of [CF: Lemma 4.4]. Let $\alpha$ and $\varepsilon$ be small positive constants and take $\lambda$ so large that $\zeta(1) > c_m/2$. Set $x_0 = \zeta(1) - \varepsilon$ and $t_0 = 1 + \alpha\varepsilon$. Then, by the Taylor's theorem,

$$v(x_0, t_0) = v(x_0, 1) + v_t(x_0, 1)\,\alpha\varepsilon + \tfrac{1}{2}\alpha^2\varepsilon^2 v_{tt}(x_0, t_0 + \theta\alpha\varepsilon)$$

for some $\theta \in (0, 1)$. By [CF: Lemma 4.2]

$$v_{tt}(x, t) \leqq \frac{C_1}{d(x, t)} \quad \text{in } N_\delta,$$

where $N_\delta$ is a neighborhood of $(\zeta(1), 1)$ in $\mathscr{P}^\lambda = \{(x, t) \in Q : u^\lambda(x, t) > 0\}$, $d(x, t)$ is the distance from $(x, t) \in N_\delta$ to the interface $x = \zeta(t)$ and $C_1$ is a positive constant. It can easily be verified that $C_1$ and $\delta$ do not depend on $\lambda$. Therefore, if $\varepsilon$ is small we get

$$v(x_0, t_0) \geqq v(x_0, 1) + \alpha\varepsilon v_t(x_0, 1) - \tfrac{1}{2}\alpha^2\varepsilon C_2, \tag{4.19}$$

where $C_2$ does not depend on $\lambda$. Moreover, since

$$-v_x(\zeta(1), 1) = \zeta'(1) = k + o(1/\lambda)$$

and $v_{xx}(x, 1) \geqq -1/(m+1)$ we have

$$v(x_0, 1) \geqq (k + o(1/\lambda))\,\varepsilon - \frac{\varepsilon^2}{2(m+1)}.$$

Substituting this into (4.19) we get

$$v(x_0, t_0) \geqq (k + o(1/\lambda))\,\varepsilon - \frac{\varepsilon^2}{2(m+1)} - \tfrac{1}{2}\alpha^2\varepsilon C_2 + \alpha\varepsilon v_t(x_0, 1). \tag{4.20}$$

On the other hand (4.18) implies that

$$v(x_0, t_0) \leqq kt_0^{-\frac{m}{m+1}}(\zeta(t_0) - x_0),$$

and from (4.17) it follows that $\zeta(t_0) - \zeta(1) \leqq k(t_0 - 1) = \alpha\varepsilon k$; hence

$$v(x_0, t_0) \leqq (\alpha\varepsilon k^2 + \varepsilon k)\, t_0^{-\frac{m}{m+1}}. \tag{4.21}$$

Combining (4.20) and (4.21) gives an estimate for $v_t$:

$$v_t(x_0, 1) \leqq \left(k^2 + \frac{k}{\alpha}\right)(1 + \alpha\,\varepsilon)^{-\frac{m}{m+1}} - \frac{1}{\alpha}\left(k + o\left(\frac{1}{\lambda}\right)\right) + \frac{\varepsilon}{2(m+1)\,\alpha} + \tfrac{1}{2}\alpha C_2$$

$$\leqq k^2 + \frac{1}{\alpha}\left\{o\left(\frac{1}{\lambda}\right) + \frac{\varepsilon}{2(m+1)}\right\} + \tfrac{1}{2}\alpha C_2.$$

Now let $\alpha = o(\lambda^{-1/2}) + \varepsilon^{1/2}$ to obtain

$$v_t(x_0, 1) \leqq k^2 + o(\lambda^{-1/2}) + C_3 \varepsilon^{1/2} \tag{4.22}$$

where $C_3$ does not depend on $\lambda$. On the other hand, it follows from (4.7) that for $t \geqq 1$

$$v_x(x, t) + \frac{x}{(m+1)\,t} = o\left(\frac{1}{\lambda}\right), \tag{4.23}$$

hence

$$v_x(x_0, 1) = k - \frac{\varepsilon}{m+1} + o\left(\frac{1}{\lambda}\right). \tag{4.24}$$

Since $\varepsilon = \zeta(1) - x_0$, (4.16) follows from (4.22) and (4.24). □

*Remark.* A lower bound for $v_t - v_x^2$ is easier to obtain. In fact, by (1.7),

$$v_t - v_x^2 = (m-1)\, v v_{xx} \geqq -\frac{(m-1)\,v}{(m+1)\,t}.$$

Therefore as $x \to \zeta^\lambda(1)$, $v_t^\lambda(x, 1) - (v^\lambda(x, 1))^2 \geqq O(\zeta^\lambda(1) - x)$, uniformly in $\lambda$.

**Lemma 4.4.** *There are bounds $C_k$ for the derivatives $(\partial/\partial k)^k\, v^\lambda(x, t)$ in a neighborhood of $(\zeta^\lambda(1), 1)$ in $\mathscr{P}[u^\lambda]$ which are independent of $\lambda$.*

**Proof.** We have only to reexamine the proofs in Sections 2 and 3, and observe that all of the estimates can be made uniform in $\lambda$, for $\lambda$ sufficiently large. Thus to obtain the bound (2.2) for $v_{xx}^\lambda$ we must show that given $\varepsilon \in (0, a/4m)$ satisfying (2.4) we can choose $\delta$ and $\eta$ independent of $\lambda$ for large $\lambda$ so that (2.5) and (2.6) hold in $R_{\delta,\eta}$. The fact that $\delta$ and $\eta$ can be chosen in this manner is a consequence of the estimates derived in the proof of Lemma 4.3, namely, (4.18) and (4.23) for $v_x^\lambda$ and (4.16) for

$$(m-1)\, v^\lambda v_{xx}^\lambda = v_t^\lambda - (v_x^\lambda)^2.$$

Similarly, the estimates for higher derivatives in Section 3 are also independent of $\lambda$. □

We may now extend the validity of the estimates of Lemmas 4.1 and 4.2 uniformly up to the boundary.

**Lemma 4.5.** *For every $t_0 > \max(t_-^*, t_+^*)$ there is a constant $C_0 \in \mathbb{R}^+$ depending only on $t_0$ such that*

$$|v_{xxx}(x, t)| \leqq C_0 t^{-\frac{m+2}{m+1}} \quad \text{in } \mathscr{P}[u] \cap \{t \geqq t_0\}.$$

**Proof.** It is clear from Lemmas 4.1 and 4.4 that there is a $\lambda_0 \in \mathbb{R}^+$ such that $\{v^\lambda_{xxx}(x, 1)\}$ is bounded in $(\zeta^\lambda_-(1), \zeta^\lambda_+(1))$ independently of $\lambda$ for $\lambda \geqq \lambda_0$. The assertion is an easy consequence of this observation.

With this result it is easy to see that the argument in the proof of Lemma 4.2 can now be made global. Thus we obtain

**Lemma 4.6.** *As* $t \to \infty$, (4.14) *holds uniformly for* $x \in \mathscr{P}[u]$. *In particular, there exists a time* $T > 0$ *such that* $v(\cdot, T)$ *is a strictly concave function in its support.*

More precisely for every $\varepsilon > 0$ we may obtain from (4.14) a time $T_\varepsilon$ such that

$$v_{xx}(x, t) \leqq -\frac{1-\varepsilon}{(m+1)\,t}$$

if $(x, t) \in \mathscr{P}[u]$ and $t \geqq T_\varepsilon$. The sharp asymptotic formulas (4.8) and (4.9) are now a consequence of Theorems 1 and 2 of [BV]. This completes the proof of Theorem 2.

*Acknowledgment*. The work of both authors was partially supported by USA-Spain Cooperation Agreement under Joint Research Grant CCB-8402023. ARONSON's work was partially supported also by the National Science Foundation Grant No. DMS 83-01247. VAZQUEZ was a long-term visitor at the Institute for Mathematics and its Applications, Minnesota in 1985.

## References

[An] S. ANGENENT, Analyticity of the interface of the porous media equation after waiting time. Proc. Am. Math. Soc., to appear.

[A1] D. G. ARONSON, Regularity properties of flows through porous media: a counterexample, SIAM J. Appl. Math., **19** (1970), 299–307.

[A2] D. G. ARONSON, Regularity properties of flows through porous media: the interface, Arch. Rational Mech. Anal. **37** (1970), 1–10.

[A3] D. G. ARONSON, The porous medium equation, in *Nonlinear Diffusion Problems*. Lecture Notes in Math. 1224. C.I.M.E. Series, Springer-Verlag, Berlin, 1986.

[AB] D. G. ARONSON & PH. BÉNILAN, Régularité des solutions de l'équation des milieux poreux dans $\mathbb{R}^N$, C.R. Acad. Sci. Paris, A-B **288** (1979), 103–105.

[AC] D. G. ARONSON & L. A. CAFFARELLI, The initial trace of a solution of the porous medium equation, Trans. Amer. Math. Soc. **280** (1983), 351–366.

[ACK] D. G. ARONSON, L. A. CAFFARELLI & S. KAMIN, How an initially stationary interface begins to move in porous medium flow, SIAM J. Math. Anal. **14** (1983), 639–658.

[ACV] D. G. ARONSON, L. A. CAFFARELLI & J. L. VAZQUEZ, Interfaces with a corner point in one-dimensional porous medium flow, Comm. Pure Appl. Math. **38** (1985), 375–404.

[BCP] PH. BÉNILAN, M. G. CRANDALL & M. PIERRE, Solutions of the porous medium equation under optimal conditions on initial values, Indiana Univ. Math. J. **33** (1984), 51–87.

[BV] PH. BÉNILAN & J. L. VAZQUEZ, Concavity of solutions of the porous medium equation, Trans. Amer. Math. Soc. **299** (1987), 81–93.

[CV] L. A. CAFFARELLI & A. FRIEDMAN, Regularity of the free boundary for the one-dimensional flow of gas in a porous medium, Amer. J. Math. **101** (1979), 1193–1218.

[DK] B. E. J. DAHLBERG & C. E. KENIG, Non-negative solutions of the porous medium equation, Comm. P.D.E. **9** (1984), 409–437.

[HK] K. HÖLLIG & H. O. KREISS, $C^\infty$-Regularity for the porous medium equation, Math. Z., **192** (1986), 217–224.

[HP] K. HÖLLIG & M. PILANT, Regularity of the free boundary for the porous medium equation, Indiana Univ. Math. J., **34** (1985), 723–732.

[K] B. F. KNERR, The porous medium equation in one dimension, Trans. Amer. Math. Soc. **234** (1977), 381–415.

[LOT] A. A. LACEY, J. R. OCKENDON & A. B. TAYLER, 'Waiting time' solutions of a nonlinear diffusion equation, SIAM J. Appl. Math. **42** (1982), 1252–1264.

[LSU] O. A. LADYZHENSKAYA, V. A. SOLONNIKOV & N. N. URAL'CEVA, *Linear and Quasilinear Equations of Parabolic Type*, Translations of Math. Monographs, Amer. Math. Soc. 1968.

[OKC] O. A. OLEINIK, A. S. KALASHNIKOV & CZHOU Y. L., The Cauchy problem and boundary problems for equations of the type of nonstationary filtration. Izv. Akad. Nauk SSSR Ser. Mat. **22** (1958), 667–704 (in Russian).

[V1] J. L. VAZQUEZ, Asymptotic behavior and propagation properties of the one-dimensional flow of a gas in a porous medium, Trans. Amer. Math. Soc. **277** (1983), 507–527.

[V2] J. L. VAZQUEZ, The interface of one-dimensional flows in porous media, Trans. Amer. Math. Soc. **285** (1984), 717–737.

[V3] J. L. VAZQUEZ, Behavior of the velocity of one-dimensional flows in porous media, Trans. Amer. Math. Soc. **286** (1984), 787–802.



School of Mathematics
University of Minnesota
Minneapolis

División de Matemáticas
Universidad Autónoma de Madrid
Spain


*(Received January 5, 1987)*

# *Fine Phase Mixtures as Minimizers of Energy*

*Dedicated to James Serrin*

J. M. BALL & R. D. JAMES

**Table of Contents**

1. Introduction . . . . . . . . . . . . . . . . . . . . . . . . . . . . 13
2. Internally Twinned Martensite . . . . . . . . . . . . . . . . . . . . 15
3. The Free Energy Functional and Minima . . . . . . . . . . . . . . . . 17
4. Compatibility, Almost Compatibility and Minimizing Sequences . . . . . . . 21
5. Materials Which Can Form Internally Twinned Martensite . . . . . . . . . 31
6. Surface Energy and Scaling . . . . . . . . . . . . . . . . . . . . . . 42
7. Other Similar Phenomena . . . . . . . . . . . . . . . . . . . . . . . 44

a. Fine twins in a minimization problem with no absolute minimum . . . . . 44
b. Strongly elliptic energy with minimizers having fine boundary wrinkles . . . 46
c. Minimizers of energy having a finer and finer mixture of phases as an interface is approached from one side . . . . . . . . . . . . . . . . . 48

## 1. Introduction

Solid-solid phase transformations often lead to certain characteristic microstructural features involving fine mixtures of the phases. In martensitic transformations one such feature is a plane interface which separates one homogeneous phase, austenite, from a very fine mixture of twins of the other phase, martensite. In quartz crystals held in a temperature gradient near the $\alpha$-$\beta$ transformation temperature, the $\alpha$-phase breaks up into triangular domains called Dauphiné twins which become finer and finer in the direction of increasing temperature. In this paper we explore a theoretical approach to these fine phase mixtures based on the minimization of free energy.

In simplified terms the idea is the following. Suppose that for energetic reasons, a body prefers to be deformed, say, in three states specified by three constant deformation gradients $\mathbf{1}$, $\boldsymbol{F}^-$ and $\boldsymbol{F}^+$. Assume that conditions of geometric compatibility are satisfied across an interface separating regions deformed with gradients $F^+$ and $F^-$, *i.e.* that there are vectors $\boldsymbol{a}$ and $\boldsymbol{n}$ such that

$$F^+ - F^- = \boldsymbol{a} \otimes \boldsymbol{n},$$

but that compatibility cannot be maintained across an interface separating $\mathbf{1}$ and $F^+$ or $\mathbf{1}$ and $F^-$, *i.e.*

$$F^{\pm} - \mathbf{1} \neq \text{a rank-one matrix}.$$

Thus, while it is possible to construct a continuous piecewise affine deformation consisting of layers having deformation gradients $F^+/F^-/F^+/F^-, \ldots$, it is not possible to construct a continuous piecewise affine deformation using all three matrices $\mathbf{1}$, $F^-$ and $F^+$. However, we show that it is possible (for certain choices of $F^+$ and $F^-$) to arrange a very fine mixture of the layers $F^+/F^-/F^+/F^-, \ldots$, on one side of an appropriately oriented interface so that the "average" deformation gradient of these layers does approximately satisfy conditions of compatibility with $\mathbf{1}$. The approximation gets better as the distribution of layers gets finer and finer. We argue that this is the essential reason for fineness in some martensitic transformations.

The energetic interpretation of these configurations is in terms of minimizing sequences rather than minimizers. In fact, each of the minimizing sequences we study converges weakly to a deformation which is not itself a minimizer of the total free energy. Thus, the total free energy functional is not lower semicontinuous with respect to weak convergence in the Sobolev space $W^{1,p}(\Omega, \mathbb{R}^3)$, $p \geqq 1$. Our calculations in Section 5 show that this failure of lower semicontinuity is a typical property of free energy functionals for solids which change phase and results from a failure of ellipticity of these functionals. From the point of view of comparison of theory with experiment, the detailed structure of minimizing sequences appears to be as important in these problems as the minimizers.

In fact, it is well known from the pioneering work of L. C. YOUNG [52] that, in the absence of ellipticity conditions, integrals of the calculus of variations do not attain a minimum among ordinary functions, but can be thought of as attaining a minimum in a space of "generalized curves". Such generalized curves are the limits of minimizing sequences that necessarily oscillate more and more finely. The finely twinned configurations of martensite described above can be viewed as approximations of generalized curves. Another example from elasticity is the "infinitesimal wrinkling" of membranes studied by PIPKIN [38].

Our calculations are related to those involved in what is known as the *crystallographic theory of martensite* in the metallurgical literature and to emerging methods of *homogenization theory* in the mathematical literature. Treatments of the crystallographic theory of martensitic transformations are found in the books by CHRISTIAN [15], NISHIYAMA [35], and WAYMAN [49]. The theory was first put forth by BOWLES & MACKENZIE [11] and WECHSLER, LIEBERMAN & READ [50]. Our calculations of Section 5 are similar to those of the crystallographic theory. However, by developing the theory on the basis of a free energy minimization, we achieve several advantages. First, by looking at minimizers and minimizing sequences, we predict both the twinned martensite interface, with the observed twin planes, and the austenite/finely twinned martensite interface. Along the way, we clarify the role of fineness in energetic terms. Also, since our free energy accounts for general three-dimensional changes of shape, it can be used in conjunction with various loading potentials to study the effect of multiaxial loads on transformation temperatures.

There is some similarity between the observed geometrical configurations of martensite and the arrangement of constituent materials used to achieve optimal bounds and designs in homogenization theory*. For a striking example of this similarity, compare Figure 1 of a recent paper by MILTON [32] with the photographs of a "mishandled" crystal of InTl shown by BASINSKI & CHRISTIAN [10, plate III]. Our problem is different in that the material itself makes the "composite". Another major difference is that compatibility does not play an essential role in homogenization theory, while the fineness in our configurations is a consequence of the material striving to achieve compatibility.

We conclude the paper with some different examples of fineness in energy minimizers. In Section 7a we give an example which, while not applying directly to martensitic transformations because it does not satisfy the appropriate invariance requirements, suggests strongly that fine twinning can be initiated by a temperature or concentration gradient, this leading to a minimization problem which when interfacial energy is neglected only has a minimizer in the sense of generalized curves. In Section 7b we give an example of a strongly elliptic material which has potential wells and which supports configurations with very fine boundary wrinkles. In the Section 7c we return to the observations of Dauphiné twinning in quartz. We give an example of a configuration involving five deformation gradients in which compatibility is achieved by a self-similar mixing of smaller and smaller triangles on one side of an interface. We believe that this configuration is related to fine triangular domains in quartz observed by VAN TENDELOO, VAN LANDUYT & AMELINCKX [48], but our example is based on a simplified free energy function and therefore we are not able to make a quantitative comparison.

## 2. Internally Twinned Martensite

Our approach in Section 2 through Section 5 is suggested by observations of internally twinned martensite. One of the most studied of the alloys which form internally twinned martensite, because of its accessibility to low power optical microscopy and its simple crystal structure, is Indium–Thallium. The alloy consists of a substitutional solid solution of Tl in a crystalline matrix of In which at high temperature is a face-centered cubic.

If a single crystal of InTl is cooled to its transformation temperature (105 °C for In-18.5% Tl, 25 °C for In-23% Tl), it undergoes a diffusionless reversible change from a face-centered cubic to a face-centered tetragonal structure. The transformation is made evident by the movement of one or more interfaces across the

---

* See, for example, KLOSOWICZ & LURIE [27], KOHN & STRANG [28], LAVROV, LURIE & CHERKAEV [29], LURIE, CHERKAEV & FEDEROV [30], MILTON [32], MURAT & TARTAR [33], RAITUM [40] and TARTAR [45, 46].

specimen. A typical observation at the transformation temperature, redrawn from the photomicrograph of BASINSKI & CHRISTIAN [10, Figure 5], is shown in Figure 1. The cubic austenitic phase (to the right in Figure 1) is stable above the transformation temperature while the tetragonal martensite is stable below the transformation temperature. The twin spacing in the martensite is on the order of 20 $\mu$m.

Sometimes the transformation produces a more complicated arrangement of the phases; a fairly common observation is the $X$-shaped interface in Figure 2a (BASINSKI & CHRISTIAN [10, Figures 11–15], BURKART & READ [13, Figure 3]).

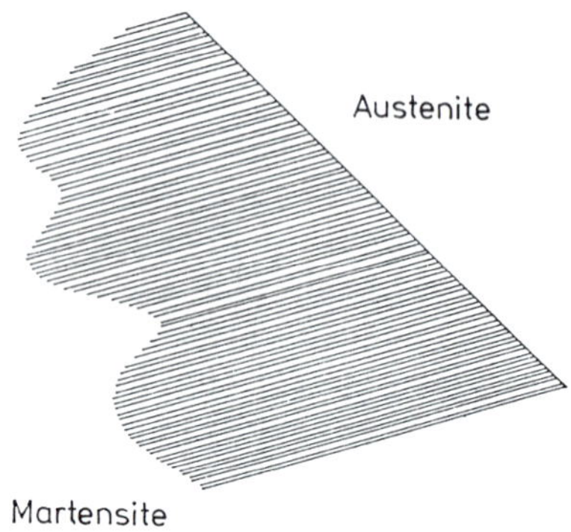


Fig. 1. Single interface transformation in InTl

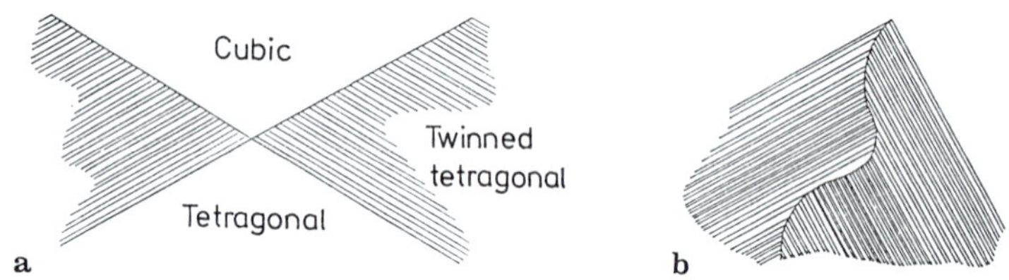


Fig. 2a. Transformation by an $X$-interface; b Curved martensite/martensite interface

At the bottom of Figure 2a is a single crystal of martensite. The single and $X$-interfaces are planar, but intriguing curved interfaces which separate twinned martensite from twinned martensite also are seen (BURKART & READ [13, Figure 3 and Figure 2b]). If the temperature is lowered further the phase boundaries move so as to eliminate all the austenite, often leaving a twinned crystal of martensite. Austenite and martensite co-exist in a crystal over a temperature range of about $3\frac{1}{2}$ °C so actually there is not a single transformation temperature.*

In all cases the twin planes arise from the $\{1\ 1\ 0\}$ family of planes in the austenite. That is, if we adopt a reference configuration interpreted as the undistorted

* We return to this observation in Section 6.

austenite just above the transformation temperature, the six planes in the reference configuration with normals (1 1 0), (1 0 1), (0 1 1), (1 −1 0), (−1 0 1), (0 1 −1) relative to a basis parallel to the cubic axes are deformed into the twin planes by the transformation. Within each twin band the material is a tetragonal single crystal, but neighboring bands are oriented differently. Also, all the interfaces separating austenite from finely twinned martensite are observed to be very nearly {1 1 0} planes. Where the austenite/martensite interface meets the boundary of the body, this boundary bends sharply through a small angle.

Below the transformation temperature, the body is a single or partly twinned crystal of martensite. A partly twinned crystal is extremely flexible in that small applied loads easily change the spacing of the twinned layers. Below the transformation temperature, the general tendency of uniform loads on the faces of a crystal is to drive the twins out and leave a single crystal of martensite.

If an unloaded crystal of martensite, either twinned or not, is heated to the transformation temperature, the austenite/martensite interfaces reappear. Further heating to above the transformation temperature causes the crystal to return to a single untwinned crystal of austenite.

## 3. The Free Energy Functional and Minima

We now propose a free energy for materials which undergo reversible martensitic transformations and work out the details for InTl.

The change in crystal structure associated with the transformation in InTl is from face-centered cubic to face-centered tetragonal. Consider a regular reference configuration $\Omega \subset \mathbb{R}^3$ which is interpreted as the undistorted austenite at the transformation temperature $\theta_0 = \text{const.}$ For later use we assume $\boldsymbol{o}$ belongs to the interior of $\Omega$. The change of shape in going from fcc to fct can be described by a deformation $\boldsymbol{y} = \boldsymbol{U}_0\boldsymbol{x}$, $\boldsymbol{x} \in \Omega$, $\boldsymbol{U}_0$ being the constant positive-definite symmetric matrix given by

$$\boldsymbol{U}_0 = \eta_1 \mathbf{1} + (\eta_2 - \eta_1)\, \boldsymbol{e} \otimes \boldsymbol{e} \tag{3.1}$$

$|\boldsymbol{e}| = 1$, for some positive constants $\eta_1 \neq \eta_2$ (Burkart & Read [13]). For InTl $\eta_1 \doteq 1 - \varepsilon$, $\eta_2 \doteq 1 + 2\varepsilon$ with $\varepsilon \doteq .013$ for concentrations near 20% Tl. We refer to $\boldsymbol{U}_0$ as the *transformation strain*.

Since the transformation leads to a change of shape at a certain temperature, we are led to assume the existence of a free energy $\phi$ which depends on the change of shape, measured by the deformation gradient $\boldsymbol{F}$, and the temperature $\theta$. Thus at a certain temperature $\theta$ let a deformation $\boldsymbol{y}\colon \Omega \to \mathbb{R}^3$ with gradient $\boldsymbol{F} = D\boldsymbol{y}(\boldsymbol{x})$ have a free energy per unit volume in $\Omega$ given by

$$\phi(\boldsymbol{F}, \theta). \tag{3.2}$$

Assume $\phi$ is defined and continuous for all $\boldsymbol{F} \in \mathscr{D} = \{\boldsymbol{F} \in M^{3\times 3} \mid \det \boldsymbol{F} > 0\}$ and for all $\theta$ in a neighborhood of the transformation temperature $\theta_0$. Here $M^{m\times n}$ denotes the set of real $m \times n$ matrices.

We assume that the free energy is Galilean invariant: for all $F \in \mathscr{D}$, all $\theta$ near $\theta_0$, and each orthogonal $R$ with $\det R = 1$ (we call such $R$ *rotations*),

$$\phi(RF, \theta) = \phi(F, \theta). \tag{3.3}$$

The restriction (3.3) implies that

$$\phi(F, \theta) = \phi(U, \theta)\big|_{U=(F^TF)^{\frac{1}{2}}}. \tag{3.4}$$

Let $\mathscr{D}^s$ be the subset of $\mathscr{D}$ consisting of positive definite symmetric matrices. Since we have interpreted $\Omega$ as undistorted austenite at the temperature $\theta_0$, we assume that there is a finite group of rotations $P^\nu$ of order $\nu$, representing the symmetry of austenite, such that

$$\phi(RUR^T, \theta) = \phi(U, \theta) \tag{3.5}$$

holds for all $U \in \mathscr{D}^s$, all $R \in P^\nu$ and all $\theta$ near $\theta_0$. For the InTl alloy $\nu = 24$ and $P^{24}$ consists of the 24 rotations which map a cube into itself. Also, one of the 4-fold rotations in $P^{24}$ has the axis $e$ of (3.1).

In an unloaded body the austenite is observed above and the martensite below the transformation temperature. We shall therefore presume an exchange of stability, in the sense that for all $U$ in $\mathscr{D}^s$ which are unequal to $\mathbf{1}$, $U_0$ or any matrix of the form $RU_0R^T$, $R \in P^\nu$,

$$\phi(U, \theta_0) > \phi(\mathbf{1}, \theta_0) = \phi(U_0, \theta_0). \tag{3.6}$$

For $\theta > \theta_0$, we assume that some symmetric matrix $U_a(\theta)$ near $\mathbf{1}$ minimizes $\phi(\cdot, \theta)$, whereas for $\theta < \theta_0$, we assume that some symmetric matrix $U_m(\theta)$ near $U_0$ minimizes $\phi(\cdot, \theta)$. Here, nearness means that $|U_a - \mathbf{1}| \ll |U_0 - \mathbf{1}|$ and $|U_m - U_0| \ll |U_0 - \mathbf{1}|$. We use the notation $|A| = (\operatorname{tr} AA^T)^{\frac{1}{2}}$ for any $A \in M^{m\times n}$. See JAMES [23] for a fuller discussion of these kinds of energy functions. Note that this assumption means that the symmetry group of $\phi$ is smaller than that considered by ERICKSEN [16] and FONSECA [20], and, in particular, does not contain certain nontrivial shears.

In general, we have assumed that $\phi(\cdot, \theta_0)$ has (up to) $\nu + 1$ potential wells with minima at the matrices $\mathbf{1}, R_1U_0R_1^T, \ldots, R_\nu U_0R_\nu^T$, in which $R_1, \ldots, R_\nu$ is an enumeration of the point group $P^\nu$. Each distinct potential well with minimum of the form $R_iU_0R_i^T$ is associated with a *variant* of the martensite. For $U_0$ of the form (3.1) with $\eta_1 \neq \eta_2$, there are only three variants because some of the matrices of the form $R_iU_0R_i^T$ coincide.

The condition $U_0 = R_iU_0R_i^T$ in InTl is

$$\eta_1 \mathbf{1} + (\eta_2 - \eta_1) R_ie \otimes R_ie = \eta_1 \mathbf{1} + (\eta_2 - \eta_1)\, e \otimes e \tag{3.7}$$

or simply

$$R_ie = \pm e \tag{3.8}$$

which is satisfied by eight members of $P^{24}$ (three rotations with axis $e$, four 180° rotations with axes perpendicular to $e$ and the identity). Since $P^{24}$ is a group, there are precisely $24/8 = 3$ distinct matrices of the form $R_iU_0R_i^T$, and it is easily seen that these are $\eta_1 \mathbf{1} + (\eta_2 - \eta_1)\, \hat{e} \otimes \hat{e}$, $\hat{e}$ being a 4-fold axis of $P^{24}$.

To make a connection with calculations of the effect of stress on transformation temperatures in [25], we observe that $\mathbf{1}$ can be written as the convex combination

$$\tfrac{1}{3}\,\boldsymbol{U}_0 + \tfrac{1}{3}\,\boldsymbol{R}\boldsymbol{U}_0\boldsymbol{R}^T + \tfrac{1}{3}\,\boldsymbol{R}^2\boldsymbol{U}_0\boldsymbol{R}^{2T}, \tag{3.9}$$

$\boldsymbol{R}$ being any 3-fold rotation in $P^{24}$. Thus, in the notation of [25], $\mathbf{1} \in \delta\mathscr{H}_0$.

We define the total energy functional by

$$\mathscr{I}[y] = \int_\Omega \phi(Dy(\boldsymbol{x}), \theta_0)\, d\boldsymbol{x}. \tag{3.10}$$

(We shall only be concerned with stable configurations at the transformation temperature.) A deformation will be termed *stable* if it minimizes the total free energy. In precise terms the deformation

$$\tilde{y} \in \mathscr{A} = \{y \in W^{1,1}(\Omega, \mathbb{R}^3) \mid Dy \in \mathscr{D} \text{ a.e.}\}$$

is stable if

$$\mathscr{I}[\tilde{y}] \leqq \mathscr{I}[y] \quad \forall y \in \mathscr{A}. \tag{3.11}$$

This stability criterion is appropriate for an unloaded body at the transformation temperature. Here and below $W^{1,p}(\Omega, \mathbb{R}^m)$ denotes the Sobolev space of mappings $y : \Omega \to \mathbb{R}^m$ such that $\|y\|_{1,p} < \infty$, where

$$\|y\|_{1,p} = \begin{cases} \left(\int_\Omega (|y|^p + |Dy|^p)\, d\boldsymbol{x}\right)^{1/p}, & 1 \leqq p < \infty, \\ \operatorname*{ess\,sup}_{\boldsymbol{x} \in \Omega} (|y(\boldsymbol{x})| + |Dy(\boldsymbol{x})|), & p = \infty. \end{cases}$$

See Adams [1] for information on these spaces. We write $W^{1,p}(\Omega)$ for $W^{1,p}(\Omega, \mathbb{R}^1)$.

We now describe all stable deformations. A necessary and sufficient condition that $\tilde{y}$ be stable is that $D\tilde{y}(\boldsymbol{x})$ minimize the integrand $\phi(\cdot, \theta_0)$ for almost every $\boldsymbol{x}$. Because of the property (3.6), $D\tilde{y}(\boldsymbol{x})$ minimizes the integrand if and only if for almost every $\boldsymbol{x}$ in $\Omega$, the function defined by

$$\tilde{U}(\boldsymbol{x}) = (D\tilde{y}(\boldsymbol{x})^T D\tilde{y}(\boldsymbol{x}))^{\frac{1}{2}} \tag{3.12}$$

takes on one of the values

$$\mathbf{1},\ \boldsymbol{R}_1\boldsymbol{U}_0\boldsymbol{R}_1^T, \ldots, \boldsymbol{R}_\nu\boldsymbol{U}_0\boldsymbol{R}_\nu^T. \tag{3.13}$$

A geometric characterization of all $\tilde{y} \in \mathscr{A}$ satisfying the preceding condition does not appear to be available in the literature. Thus, we focus first on smooth interfaces separating austenite from itself, martensite from itself or martensite from austenite.

Let $\tilde{y}$ be a stable, continuous and piecewise differentiable deformation and suppose a smooth interface separates two regions on which $D\tilde{y}$ has constant values $\boldsymbol{F}^+$ and $\boldsymbol{F}^-$. It is well known that if $\boldsymbol{F}^+ \neq \boldsymbol{F}^-$ then the interface is a plane (with say reference normal $\boldsymbol{n}$, $|\boldsymbol{n}| = 1$) and that for some nonzero vector $\boldsymbol{a}$

$$\boldsymbol{F}^+ - \boldsymbol{F}^- = \boldsymbol{a} \otimes \boldsymbol{n}. \tag{3.14}$$

Let $F^+ = R^+U^+$ and $F^- = R^-U^-$ be the polar decompositions of $F^+$ and $F^-$. Since $\tilde{y}$ is stable, $U^+$ and $U^-$ must each take on one of the values given in (3.13). We work out the possible interfaces below for $U_0$ given by (3.1).

(i) *Austenite/Austenite Interfaces.* These are governed by the condition

$$R^+ - R^- = a \otimes n, \quad a \neq 0, \tag{3.15}$$

which implies that

$$R = 1 + a' \otimes n, \tag{3.16}$$

where $R = R^{-T}R^+$ and $a' = R^{-T}a$. Equation (3.16) shows that $R$ has two linearly independent axes ($\perp$ to $n$) which in turn implies that $R = 1$ and $R^+ = R^-$. Thus, there are no austenite/austenite interfaces. According to a theorem of RESHETNYAK [41, Corollary of Lemma 3], if $y \in W^{1,\infty}(\Omega, \mathbb{R}^3)$ is such that $Dy(x)$ is a rotation for almost all $x \in \Omega$ then $Dy$ is necessarily a constant rotation. This implies the stronger result that the body cannot be inhomogeneously deformed in the austenite phase. Obviously, this result is independent of the choice of point group or transformation strain.

(ii) *Martensite/Martensite Interfaces.* These are governed by the equation

$$\hat{R}^+U_0R_i - \hat{R}^-U_0R_j = \hat{a} \otimes \hat{n}, \tag{3.17}$$

with $\hat{a} \neq 0$ and $|\hat{n}| = 1$.

Premultiply (3.17) by $\hat{R}^{-T}$ and postmultiply by $R_j^T$. Then, (3.17) becomes

$$RU_0\bar{R} - U_0 = a \otimes n, \tag{3.18}$$

where $\bar{R} = R_iR_j^T \in P^{24}$, $R = \hat{R}^{-T}\hat{R}^+$, $a = \hat{R}^{-T}\hat{a} \neq 0$, and $n = R_j\hat{n}$. Let $\{e, e_1, e_2\}$ be an orthonormal basis with $e_1$ and $e_2$ also four-fold axes of rotation in $P^{24}$. $\bar{R}e$ equals one of the vectors $\pm e_1, \pm e_2$ or $\pm e$ because of the structure of $P^{24}$. The case $\bar{R}e = \pm e$ yields no solutions of (3.18) by a quick calculation using the fact that $\bar{R}^{-T}U_0\bar{R} = U_0$. (This would correspond to an interface between one variant of martensite and itself.) There are various strategies for completing the calculation. The methods of ERICKSEN [18] and GURTIN [22] can be applied to the remaining cases $\bar{R}e = \pm e_1$ or $\pm e_2$. Alternately, our Proposition 4 of Section 5 can be used with $C = U_0^{-1}\bar{R}^TU_0^2\bar{R}U_0^{-1}$. The results are:

*Twins*

$$\begin{aligned} n &= \frac{1}{\sqrt{2}}(\bar{e} + e) \\ a &= \frac{\sqrt{2}(\eta_2^2 - \eta_1^2)}{\eta_2^2 + \eta_1^2} q_1 \\ \bar{R} &= -1 + 2n \otimes n \\ R &= -1 + \frac{2}{|q_2|^2} q_2 \otimes q_2 \end{aligned} \tag{3.19}$$

in which

$$q_1 = \eta_1 \bar{e} - \eta_2 e, \tag{3.20}$$

$$q_2 = \eta_2 \bar{e} + \eta_1 e,$$

and

$$\bar{e} \in \{e_1, -e_1, e_2, -e_2\}. \tag{3.21}$$

The equations (3.19) through (3.21) give all solutions of the equation (3.18) with $U_0$ given, $|n| = 1$ and $\bar{R} \in P^{24}$ in the sense that $a$ and $n$ are given by $(3.19)_{1,2}$ up to the replacement $a \to -a$ and $n \to -n$. The expressions for $R$ and $\bar{R}$ are not unique. The various values of $R$ and $\bar{R}$, which give rise to the $a$ and $n$ of (3.19) are associated with Type I and Type II twins (normal and parallel twins in the terminology of GURTIN [22]). The twins given by (3.19) are so-called compound twins, which means that they can be represented as both type I and type II twins. The expressions of $R$ and $\bar{R}$ in (3.19) correspond to the type I description. All solutions of the original equation (3.17) are obtained by reversing the argument which leads from (3.17) to (3.18). The twin planes are all of the $\{1\ 1\ 0\}$ family and agree exactly with the observed twin planes described in Section 2. Solutions* of (3.18) for general $U_0$ andf or a more general family of groups than point groups are given by ERICKSEN [17].

(iii) *Austenite/Martensite Interfaces.* These are governed by the equation

$$R U_0 R_k - \mathbf{1} = a \otimes n, \tag{3.22}$$

with $a \neq 0$, $|n| = 1$, which implies that

$$R_k^T U_0^2 R_k = (\mathbf{1} + n \otimes a)(\mathbf{1} + a \otimes n), \tag{3.23}$$

which in turn implies that a vector perpendicular to both $a$ and $n$ is an eigenvector of $R_k^T U_0^2 R_k$ with eigenvalue equal to 1. But this is impossible unless one of the eigenvalues $\eta_1$ or $\eta_2$ of $U_0$ equals 1. In particular, there are no austenite/martensite interfaces.

The latter conclusion does not agree with observations like those shown in Figure 1, which clearly show *some kind* of austenite/martensite interface.

## 4. Compatibility, Almost Compatibility, and Minimizing Sequences

In this section we explore the idea that the austenite/finely twinned martensite interface is modelled by certain minimizing sequences for the total free energy (3.10). The corresponding deformations are essentially piecewise affine, but a small correction is necessary close to the interface so as to render the deformation gradients compatible.

---

* It would be necessary to consider the more general groups to describe martensitic transformations involving slip, which does not occur under small loads in the internally twinned martensites.

We begin by giving a version of the Hadamard jump condition for deformations whose gradients take only two values, there being no assumption on the structure of the sets where these values are taken.

**Proposition 1.** *Let* $\Omega \in \mathbb{R}^n$ *be open and connected. Let* $\boldsymbol{y} \in W^{1,\infty}(\Omega, \mathbb{R}^m)$ *satisfy*

$$\begin{aligned} Dy(\boldsymbol{x}) &= \boldsymbol{A}, \qquad \text{a.e. } \boldsymbol{x} \in \Omega_A, \\ Dy(\boldsymbol{x}) &= \boldsymbol{B}, \qquad \text{a.e. } \boldsymbol{x} \in \Omega_B, \end{aligned} \tag{4.1}$$

*where* $\boldsymbol{A}, \boldsymbol{B} \in M^{m \times n}$ *and* $\Omega_A, \Omega_B$ *are disjoint measurable sets with* $\Omega = \Omega_A \cup \Omega_B$, meas $\Omega_A > 0$, meas $\Omega_B > 0$. *Then*

$$\boldsymbol{A} - \boldsymbol{B} = \boldsymbol{c} \otimes \boldsymbol{n} \tag{4.2}$$

*for some* $\boldsymbol{c} \in \mathbb{R}^m$, $\boldsymbol{n} \in \mathbb{R}^n$, $|\boldsymbol{n}| = 1$, *and*

$$\boldsymbol{y}(\boldsymbol{x}) = \boldsymbol{y}_0 + \boldsymbol{B}\boldsymbol{x} + \theta(\boldsymbol{x})\,\boldsymbol{c}, \qquad \boldsymbol{x} \in \Omega, \tag{4.3}$$

*where* $\boldsymbol{y}_0 \in \mathbb{R}^m$, $\boldsymbol{y}_0 \cdot \boldsymbol{c} = 0$, $\theta \in W^{1,\infty}(\Omega)$ *satisfies* $D\theta(\boldsymbol{x}) = \chi_A(\boldsymbol{x})\,\boldsymbol{n}$ a.e., *and* $\chi_A$ *denotes the characteristic function of* $\Omega_A$.

**Proof.** Let $\boldsymbol{z}(\boldsymbol{x}) = \boldsymbol{y}(\boldsymbol{x}) - \boldsymbol{B}\boldsymbol{x}$, $\boldsymbol{C} = \boldsymbol{A} - \boldsymbol{B}$, so that $D\boldsymbol{z} = \chi_A \boldsymbol{C}$. Since $\chi_A$ is not constant, there exists $\varrho \in C_0^\infty(\Omega)$ such that

$$\boldsymbol{n} \stackrel{\text{def}}{=} \int_{\Omega_A} D\varrho\, dx = \int_{\Omega} \chi_A\, D\varrho\, dx$$

is nonzero, and clearly we may suppose that $|\boldsymbol{n}| = 1$. But

$$\begin{aligned} 0 &= \int_{\Omega} (z^i_{,\alpha}\varrho_{,\beta} - z^i_{,\beta}\varrho_{,\alpha})\, dx \\ &= C^i_\alpha n_\beta - C^i_\beta n_\alpha, \end{aligned}$$

and hence (4.2) holds with $\boldsymbol{c} = \boldsymbol{C}\boldsymbol{n}$. To obtain (4.3) we note that if $\boldsymbol{b} \cdot \boldsymbol{c} = 0$ then $D(\boldsymbol{z}(\boldsymbol{x}) \cdot \boldsymbol{b}) = \boldsymbol{0}$ a.e., so that $\boldsymbol{z}(\boldsymbol{x}) - \boldsymbol{z}(\boldsymbol{x}_0)$ is parallel to $\boldsymbol{c}$ for a.e. $\boldsymbol{x} \in \Omega$, where $\boldsymbol{x}_0 \in \Omega$ is fixed. Assuming without loss of generality that $\boldsymbol{c} \neq \boldsymbol{0}$, it follows from (4.3) that $D\theta(\boldsymbol{x}) = \chi_A(\boldsymbol{x})\,\boldsymbol{n}$ a.e., completing the proof. □

From (4.3) we see that, on any convex subset $E$ of $\Omega$, $\boldsymbol{y}$ has the form

$$\boldsymbol{y}(\boldsymbol{x}) = \boldsymbol{y}_0 + \boldsymbol{B}\boldsymbol{x} + f_E(\boldsymbol{x} \cdot \boldsymbol{n})\,\boldsymbol{c}, \tag{4.4}$$

where $f_E$ is Lipschitz with derivative 0 or 1 a.e.. Thus $\Omega_A \cap E$ and $\Omega_B \cap E$ consist of parallel layers normal to $\boldsymbol{n}$. However, if $\Omega$ is not convex, there may be no representation (4.4) with $f_E$ independent of $E$; for example, $\Omega_A$ could have the form of the shaded set in Figure 3.

We next consider deformations whose gradients to a good approximation take only two values.

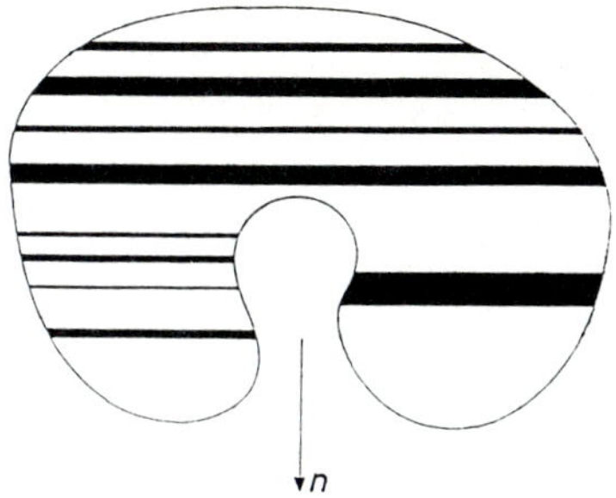


Fig. 3. Distribution of the sets $\Omega_A$ and $\Omega_B$ consistent with the hypotheses of Proposition 1

**Proposition 2.** *Let* $\Omega \subset \mathbb{R}^n$ *be bounded, open and connected. Let* $p > 2$, *and let* $\boldsymbol{A}, \boldsymbol{B} \in M^{m\times n}$ *be distinct. Let* $\boldsymbol{y}^{(j)} \rightharpoonup \boldsymbol{y}$ *in* $W^{1,p}(\Omega, \mathbb{R}^m)$ *and suppose that for every* $\varepsilon > 0$

$$\lim_{j\to\infty} \operatorname{meas}\{\boldsymbol{x} \in \Omega : |D\boldsymbol{y}^{(j)}(\boldsymbol{x}) - \boldsymbol{A}| > \varepsilon \quad \text{and} \quad |D\boldsymbol{y}^{(j)}(\boldsymbol{x}) - \boldsymbol{B}| > \varepsilon\} = 0. \tag{4.5}$$

*Then*

$$D\boldsymbol{y}(\boldsymbol{x}) = \lambda(\boldsymbol{x})\,\boldsymbol{A} + (1 - \lambda(\boldsymbol{x}))\,\boldsymbol{B}, \qquad \text{a.e. } \boldsymbol{x} \in \Omega, \tag{4.6}$$

*for some measurable function* $\lambda$ *satisfying* $0 \leqq \lambda(\boldsymbol{x}) \leqq 1$ a.e., *and one of the following possibilities holds*:

(i) $\lambda(\boldsymbol{x}) = 1$ a.e. *and* $D\boldsymbol{y}^{(j)}(\boldsymbol{x}) \to \boldsymbol{A}$ *in measure,*

(ii) $\lambda(\boldsymbol{x}) = 0$ a.e. *and* $D\boldsymbol{y}^{(j)}(\boldsymbol{x}) \to \boldsymbol{B}$ *in measure,*

(iii) $\lambda$ *equals neither* 0 a.e. *nor* 1 a.e. *and*

$$\boldsymbol{A} - \boldsymbol{B} = \boldsymbol{c} \otimes \boldsymbol{n} \tag{4.7}$$

*for some* $\boldsymbol{c} \in \mathbb{R}^m$, $\boldsymbol{n} \in \mathbb{R}^n$, $|\boldsymbol{n}| = 1$.

**Proof.** Let $\Omega_A^{j,\varepsilon} = \{\boldsymbol{x} \in \Omega : |D\boldsymbol{y}^{(j)}(\boldsymbol{x}) - \boldsymbol{A}| \leqq \varepsilon\}$, $\Omega_B^{j,\varepsilon} = \{\boldsymbol{x} \in \Omega : |D\boldsymbol{y}^{(j)}(\boldsymbol{x}) - \boldsymbol{B}| \leqq \varepsilon\}$, and let $\chi_A^{j,\varepsilon}, \chi_B^{j,\varepsilon}$ denote the characteristic functions of $\Omega_A^{j,\varepsilon}, \Omega_B^{j,\varepsilon}$ respectively. Then for $\varepsilon$ sufficiently small that $\Omega_A^{j,\varepsilon}$, $\Omega_B^{j,\varepsilon}$ are disjoint,

$$\begin{aligned} D\boldsymbol{y}^{(j)}(\boldsymbol{x}) &= \chi_A^{j,\varepsilon}(\boldsymbol{x})\,(\boldsymbol{A} + \boldsymbol{\theta}^{j,\varepsilon}(\boldsymbol{x})) + \chi_B^{j,\varepsilon}(\boldsymbol{x})\,(\boldsymbol{B} + \boldsymbol{\psi}^{j,\varepsilon}(\boldsymbol{x})) \\ &\quad + (1 - \chi_A^{j,\varepsilon}(\boldsymbol{x}) - \chi_B^{j,\varepsilon}(\boldsymbol{x}))\,D\boldsymbol{y}^{(j)}(\boldsymbol{x}), \qquad \text{a.e. } \boldsymbol{x} \in \Omega, \end{aligned} \tag{4.8}$$

where $\boldsymbol{\theta}^{j,\varepsilon}(\boldsymbol{x}) \stackrel{\text{def}}{=} D\boldsymbol{y}^{(j)}(\boldsymbol{x}) - \boldsymbol{A}$, $\boldsymbol{\psi}^{j,\varepsilon}(\boldsymbol{x}) \stackrel{\text{def}}{=} D\boldsymbol{y}^{(j)}(\boldsymbol{x}) - \boldsymbol{B}$. Since $\chi_A^{j,\varepsilon}$, $\chi_B^{j,\varepsilon}$ are uniformly bounded there exists a subsequence, again denoted $\boldsymbol{y}^{(j)}$, such that $\chi_A^{j,\varepsilon} \stackrel{*}{\rightharpoonup} \lambda_A^\varepsilon$, $\chi_B^{j,\varepsilon} \stackrel{*}{\rightharpoonup} \lambda_B^\varepsilon$ in $L^\infty(\Omega)$. Since, by (4.5), $\lim_{j\to\infty} \int_\Omega (1 - \chi_A^{j,\varepsilon} - \chi_B^{j,\varepsilon})\,d\boldsymbol{x} = 0$, it follows that

$$0 \leqq \lambda_A^\varepsilon(\boldsymbol{x}) = 1 - \lambda_B^\varepsilon(\boldsymbol{x}) \leqq 1, \qquad \text{a.e. } \boldsymbol{x} \in \Omega. \tag{4.9}$$

Also

$$\int_\Omega (1 - \chi_A^{j,\varepsilon} - \chi_B^{j,\varepsilon}) \, |Dy^{(j)}(x)| \, dx$$

$$\leqq \left( \int_\Omega (1 - \chi_A^{j,\varepsilon} - \chi_B^{j,\varepsilon}) \, dx \right)^{1/p'} \left( \int_\Omega |Dy^{(j)}(x)|^p \, dx \right)^{1/p},$$

where $1/p + 1/p' = 1$, so that the last term in (4.8) tends to zero as $j \to \infty$ strongly in $L^1$. Passing to the limit $j \to \infty$ in (4.8) we obtain

$$Dy(x) = \lambda_A^\varepsilon(x) \, A + (1 - \lambda_A^\varepsilon(x)) \, B + H^\varepsilon(x), \quad \text{a.e. } x \in \Omega, \tag{4.10}$$

where $|H^\varepsilon(x)| \leqq \varepsilon$ a.e.. From (4.9) there exists a subsequence $\varepsilon_k \to 0$ such that $\lambda_A^{\varepsilon_k} \overset{*}{\rightharpoonup} \lambda(\cdot)$ in $L^\infty(\Omega)$, where $0 \leqq \lambda(x) \leqq 1$ a.e.. Passing to the limit in (4.10) we obtain (4.6).

Suppose that $\lambda(x) = 1$ a.e.. We claim that $Dy^{(j)} \to A$ in measure. If not, there would exist $\varepsilon > 0$ and a subsequence $y^{(\mu)}$ such that $\text{meas } \Omega_B^{\mu,\varepsilon} = \int_\Omega \chi_B^{\mu,\varepsilon} \, dx \geqq \delta > 0$. Applying the preceding argument to $y^{(\mu)}$ gives

$$Dy(x) = \bar{\lambda}(x) \, A + (1 - \bar{\lambda}(x)) \, B, \quad \text{a.e. } x \in \Omega,$$

where $\int_\Omega (1 - \bar{\lambda}(x)) \, dx \geqq \delta$, contradicting (4.6). A similar argument shows that if $\lambda(x) = 0$ a.e., then $Dy^{(j)} \to B$ in measure.

Suppose that $\lambda$ equals neither 0 a.e. nor 1 a.e.. If $\lambda$ takes only the values 0 and 1 a.e., then (4.7) follows from Proposition 1 applied to $y$. Hence we need only consider the case when $0 < \lambda(x) < 1$ on a set $\bar{\Omega}$ of positive measure. Since (4.7) says nothing if $m = 1$ or $n = 1$, we suppose also that $m \geqq 2$, $n \geqq 2$. If $M \in M^{m \times n}$ we denote by $J(M)$ some $2 \times 2$ minor of $M$. Since $p > 2$, we have that

$$J(Dy^{(j)} - A) \rightharpoonup J(Dy - A) \quad \text{in } L^{p/2}(\Omega) \tag{4.11}$$

(see RESHETNYAK [41], BALL [4], BALL, CURRIE & OLVER [5]). But, since $\chi_A^{j,\varepsilon}$ and $\chi_B^{j,\varepsilon}$ are characteristic functions of disjoint sets,

$$\begin{aligned} J(Dy^{(j)}(x) - A) &= \chi_A^{j,\varepsilon}(x) \, J(\theta^{j,\varepsilon}(x)) + \chi_B^{j,\varepsilon}(x) \, J(B - A + \psi^{j,\varepsilon}(x)) \\ &\quad + (1 - \chi_A^{j,\varepsilon}(x) - \chi_B^{j,\varepsilon}(x)) \, J(Dy^{(j)}(x) - A), \quad \text{a.e. } x \in \Omega. \end{aligned} \tag{4.12}$$

Using a similar argument as for (4.8), we deduce from (4.11) and (4.12) that

$$J(Dy - A) = (1 - \lambda(x)) \, J(B - A), \quad \text{a.e. } x \in \Omega,$$

and hence from (4.6) that

$$\lambda(x) \, (1 - \lambda(x)) \, J(B - A) = 0 \quad \text{a.e. } x \in \Omega.$$

Since $\bar{\Omega}$ has positive measure it follows that $J(B - A) = 0$. Since a matrix all of whose $2 \times 2$ minors vanish is of rank one, this completes the proof. □

*Remarks:*

1. The argument using the minors $J$ is only needed to handle the case when $\lambda(\boldsymbol{x}) \in (0, 1)$ is constant. If $\lambda(\boldsymbol{x})$ is not constant a.e. then (4.7) follows from (4.6) using an argument similar to that in Proposition 1.
2. In case (i) (respectively (ii)) it is easily shown that $\boldsymbol{y}^{(j)} \to \boldsymbol{y}_0 + \boldsymbol{A}\boldsymbol{x}$ (respectively $\boldsymbol{y}^{(j)} \to \boldsymbol{y}_0 + \boldsymbol{B}\boldsymbol{x}$) strongly in $W^{1,q}(\Omega, \mathbb{R}^m)$ for $1 \leqq q < p$, where $\boldsymbol{y}_0 \in \mathbb{R}^m$.
3. We do not know if Proposition 2 holds for $1 \leqq p \leqq 2$.
4. Suppose for simplicity that $\boldsymbol{y}^{(j)} \overset{*}{\rightharpoonup} \boldsymbol{y}$ in $W^{1,\infty}(\Omega; \mathbb{R}^m)$. Then the proof of Proposition 2 can be adapted to show that $\boldsymbol{y}^{(j)}$ converges in the sense of generalized curves (*cf.* TARTAR [46, Section 4]), so that for any continuous function $f$: $M^{m \times n} \to \mathbb{R}$
$$f(D\boldsymbol{y}^{(j)}) \overset{*}{\rightharpoonup} \langle \nu_{\boldsymbol{x}}, f \rangle \quad \text{in } L^\infty(\Omega),$$
and that the Young measure $\nu_{\boldsymbol{x}}$ is given by
$$\nu_{\boldsymbol{x}} = \lambda(\boldsymbol{x})\, \delta_{\boldsymbol{A}} + (1 - \lambda(\boldsymbol{x}))\, \delta_{\boldsymbol{B}} \quad \text{a.e. } \boldsymbol{x} \in \Omega,$$
where $\delta_{\boldsymbol{A}}$, $\delta_{\boldsymbol{B}}$ denote Dirac masses at $\boldsymbol{A}$, $\boldsymbol{B}$ respectively. See CHIPOT & KINDERLEHRER [14] and KINDERLEHRER [26] for further remarks in this direction.
5. By applying Proposition 2 we can strengthen the statement made in Section 3 concerning the nonexistence of austenite/martensite interfaces to the assertion that there is no sequence of deformations $\boldsymbol{y}^{(j)} \rightharpoonup \boldsymbol{y}$ in $W^{1,p}(\Omega; \mathbb{R}^3)$, $p > 2$, which, in the sense of Proposition 2, have gradients taking to a good approximation only the two values $\boldsymbol{R}$ and $\bar{\boldsymbol{R}} \boldsymbol{U}_0 \boldsymbol{R}_k$, where $\boldsymbol{R}$, $\bar{\boldsymbol{R}}$ are rotations and $\boldsymbol{R}_k \in P^\nu$, unless $D\boldsymbol{y}^{(j)} \to \boldsymbol{R}$ in measure or $D\boldsymbol{y}^{(j)} \to \bar{\boldsymbol{R}} \boldsymbol{U}_0 \boldsymbol{R}_k$ in measure.

We can now address the case of deformations whose gradients to a good approximation take only two values on one side of an interface $\mathscr{S}$, and a third value on the opposite side. Here one goal is to understand why the austenite/finely twinned martensite interface is flat.

**Theorem 3.** *Let* $m \geqq 2$, $n \geqq 2$, $p > 2$. *Let* $\Omega \subset \mathbb{R}^n$ *be bounded, open and connected, and suppose that* $\Omega$ *can be written in the form* $\Omega = \Omega_{A,B} \cup \Omega_C \cup \mathscr{S}$, *where* $\Omega_{A,B}$ *and* $\Omega_C$ *are disjoint, open and connected, and where* $\mathscr{S} = \partial\Omega_{A,B} \cap \Omega = \partial\Omega_C \cap \Omega$. *Assume either that* meas $\mathscr{S} = 0$ *or that* $p > n$. *Let* $\boldsymbol{A}, \boldsymbol{B}, \boldsymbol{C} \in M^{m \times n}$ *be distinct, and suppose that neither* $\boldsymbol{C} - \boldsymbol{A}$ *nor* $\boldsymbol{C} - \boldsymbol{B}$ *is of rank one. Let* $\boldsymbol{y}^{(j)} \rightharpoonup \boldsymbol{y}$ *in* $W^{1,p}(\Omega, \mathbb{R}^m)$ *satisfy for every* $\varepsilon > 0$

$$\lim_{j \to \infty} \text{meas}\,\{\boldsymbol{x} \in \Omega_{A,B} \colon |D\boldsymbol{y}^{(j)}(\boldsymbol{x}) - \boldsymbol{A}| > \varepsilon \ \text{ and } \ |D\boldsymbol{y}^{(j)}(\boldsymbol{x}) - \boldsymbol{B}| > \varepsilon\} = 0, \tag{4.13}$$

*and*

$$\lim_{j \to \infty} \text{meas}\,\{\boldsymbol{x} \in \Omega_C \colon |D\boldsymbol{y}^{(j)}(\boldsymbol{x}) - \boldsymbol{C}| > \varepsilon\} = 0. \tag{4.14}$$

*Then the interface* $\mathscr{S}$ *is necessarily planar*, i.e., *there exists a unit vector* $\boldsymbol{m} \in \mathbb{R}^n$ *and* $k \in \mathbb{R}$ *with*

$$\mathscr{S} \subset \{\boldsymbol{x} \in \mathbb{R}^n \colon \boldsymbol{x} \cdot \boldsymbol{m} = k\}, \tag{4.15}$$

*and*

$$A - B = c \otimes n, \tag{4.16}$$

$$C - B = -b \otimes m + \lambda c \otimes n, \tag{4.17}$$

*where* $n \in \mathbb{R}^n$ *is a unit vector that is not parallel to* $m$, *where* $b, c \in \mathbb{R}^m$ *are not parallel, and where* $0 < \lambda < 1$. *Furthermore, each point* $x_0 \in \mathscr{S}$ *has an open neighborhood* $N(x_0)$ *such that*

$$Dy(x) = B + \lambda c \otimes n, \quad x \in \Omega_{A,B} \cap N(x_0), \tag{4.18}$$

$$Dy(x) = C, \quad x \in \Omega_C \cap N(x_0). \tag{4.19}$$

*Conversely, suppose that* $\mathscr{S}$, $A$, $B$ *and* $C$ *have the forms* (4.15) *through* (4.17). *Then there exist sequences* $y^{(j)}$ *converging weak * in* $W^{1,\infty}(\Omega, \mathbb{R}^m)$ *to some* $y$ *and satisfying* (4.13) *and* (4.14). *If* $m = n$, $\det A > 0$, $\det B > 0$ *and* $\det C > 0$, *then* $y^{(j)}$ *can be chosen so that* $\det Dy^{(j)}(x) \geqq \delta > 0$ a.e. $x \in \Omega$, *for some constant* $\delta > 0$.

**Proof.** Applying Proposition 2 to $\Omega_{A,B}$, we see that one of the following three possibilities holds a.e. in $\Omega_{A,B}$:

(i) $y(x) = y_0 + Ax$ for some $y_0 \in \mathbb{R}^m$,

(ii) $y(x) = y_0 + Bx$ for some $y_0 \in \mathbb{R}^m$,

(iii) $A - B = c \otimes n$ for some $c \in \mathbb{R}^m$, $n \in \mathbb{R}^n$, $|n| = 1$,

and

$$Dy(x) = B + \lambda(x)\, c \otimes n, \tag{4.20}$$

where $0 \leqq \lambda(x) \leqq 1$ and $\lambda(x)$ equals neither 1 a.e. nor 0 a.e.; in this case, we have, using the argument at the end of the proof of Proposition 1, that

$$y(x) = y_0 + Bx + \theta(x)\, c, \tag{4.21}$$

where $y_0 \in \mathbb{R}^m$ and $\theta \in W^{1,\infty}(\Omega_{A,B})$, $D\theta(x) = \lambda(x)\, n$.

Similarly, we have

$$y(x) = y_1 + Cx, \quad \text{a.e. } x \in \Omega_C, \tag{4.22}$$

for some $y_1 \in \mathbb{R}^m$.

If $p > n$ then an appropriate choice of representative $y$ is continuous on $\Omega$, and the same holds if meas $\mathscr{S} = 0$ since then $y \in W^{1,\infty}(\Omega, \mathbb{R}^m)$. Pick $x_0$ and suppose for contradiction that $(x_0 + r_i) \in \mathscr{S}$ for $n$ linearly independent vectors $r_i$. We suppose (4.21) holds in $\Omega_{A,B}$; the cases (i) and (ii) are handled similarly. Since $c \neq o$, by (4.21) $\theta$ has a continuous extension, again denoted $\theta$, to $\partial\Omega_{A,B} \cap \Omega$. By the continuity of $y$ we therefore have

$$(C - B)\, x_0 = y_0 - y_1 + \theta(x_0)\, c,$$

$$(C - B)(x_0 + r_i) = y_0 - y_1 + \theta(x_0 + r_i)\, c, \quad i = 1, \ldots, n.$$

Subtracting, we find that $(C - B)\, r_i$ is parallel to $c$ for $i = 1, \ldots, n$, contradicting our assumption that $C - B$ is not of rank one. Hence (4.15) holds for some

$\boldsymbol{m}$ and $k$. In the cases (i) (ii), (4.15) and Proposition 1 imply that either $\boldsymbol{C}-\boldsymbol{A}$ or $\boldsymbol{C}-\boldsymbol{B}$ is of rank one; thus these cases are impossible and (4.21) holds in $\Omega_{A,B}$.

Given $\boldsymbol{x}_0 \in \mathscr{S}$, choose $r>0$ sufficiently small that the open ball $B(\boldsymbol{x}_0, r)$ with center $\boldsymbol{x}_0$ and radius $r$ is contained in $\Omega$. Let

$$B^+(\boldsymbol{x}_0, r) = \{\boldsymbol{x} \in B(\boldsymbol{x}_0, r) : \boldsymbol{x} \cdot \boldsymbol{m} > k\},$$

$$B^-(\boldsymbol{x}_0, r) = \{\boldsymbol{x} \in B(\boldsymbol{x}_0, r) : \boldsymbol{x} \cdot \boldsymbol{m} < k\}.$$

Then $\Omega_{A,B} \cap B(\boldsymbol{x}_0, r) = B^+(\boldsymbol{x}_0, r)$, $\Omega_C \cap B(\boldsymbol{x}_0, r) = B^-(\boldsymbol{x}_0, r)$ or *vice versa*, and $\mathscr{S} \cap B(\boldsymbol{x}_0, r) = \{\boldsymbol{x} \in B(\boldsymbol{x}_0, r) : \boldsymbol{x} \cdot \boldsymbol{m} = k\}$. Since $D\theta(\boldsymbol{x}) = \lambda(\boldsymbol{x})\,\boldsymbol{n}$ in $\Omega_{A,B}$ it follows that

$$\theta(\boldsymbol{x}) = f(\boldsymbol{x} \cdot \boldsymbol{n}) \quad \text{for all } \boldsymbol{x} \in \overline{\Omega}_{A,B} \cap B(\boldsymbol{x}_0, r), \tag{4.23}$$

for some $f \in W^{1,\infty}(\mathbb{R})$. For $\boldsymbol{x} \in B(\boldsymbol{x}_0, r)$ denote by $\boldsymbol{x}'$ the orthogonal projection of $\boldsymbol{x}$ onto $\mathscr{S}$. Then $\boldsymbol{x}' = \boldsymbol{x} + (k - (\boldsymbol{x} \cdot \boldsymbol{m}))\,\boldsymbol{m}$ and

$$(\boldsymbol{C}-\boldsymbol{B})\,\boldsymbol{x}' = \boldsymbol{y}_0 - \boldsymbol{y}_1 + f(\boldsymbol{x}' \cdot \boldsymbol{n})\,\boldsymbol{c},$$

so that

$$(\boldsymbol{C}-\boldsymbol{B})\,\boldsymbol{x} = \boldsymbol{y}_0 - \boldsymbol{y}_1 + f(\boldsymbol{x}' \cdot \boldsymbol{n})\,\boldsymbol{c} + ((\boldsymbol{x} \cdot \boldsymbol{m}) - k)\,\boldsymbol{d}, \tag{4.24}$$

where $\boldsymbol{d} = (\boldsymbol{C}-\boldsymbol{B})\,\boldsymbol{m}$. Taking the derivative of (4.23) with respect to $\boldsymbol{x}$, we find, using the chain rule for Lipschitz maps (*cf.* MARCUS & MIZEL [31, Lemma 2.1]) that

$$\boldsymbol{C}-\boldsymbol{B} = \boldsymbol{d} \otimes \boldsymbol{m} + f'(\boldsymbol{x}' \cdot \boldsymbol{n})\,\boldsymbol{c} \otimes [\boldsymbol{n} - (\boldsymbol{m} \cdot \boldsymbol{n})\,\boldsymbol{m}], \tag{4.25}$$

a.e. in $B(\boldsymbol{x}_0, r)$. Since $\boldsymbol{C}-\boldsymbol{B}$ is not of rank one, $\boldsymbol{m}$ is not parallel to $\boldsymbol{n}$, $(\boldsymbol{m} \cdot \boldsymbol{n})^2 \neq 1$, and $\boldsymbol{c} \neq \boldsymbol{o}$. Taking the inner product of (4.25) with $\boldsymbol{n}$, we thus deduce that $f'(\boldsymbol{x}' \cdot \boldsymbol{n}) = \lambda =$ constant a.e. in $B(\boldsymbol{x}_0, r)$. Since $\boldsymbol{m}$ is not parallel to $\boldsymbol{n}$, it follows that $f'(t) = \lambda$ for $t$ in a neighborhood of $\boldsymbol{x}_0 \cdot \boldsymbol{n}$ and hence from (4.20), (4.21) and (4.23) that $\lambda(\boldsymbol{x}) = \lambda$, $0 < \lambda < 1$, for $\boldsymbol{x} \in \Omega_{A,B} \cap N(\boldsymbol{x}_0)$, where $N(\boldsymbol{x}_0)$ is some open neighborhood of $\boldsymbol{x}_0$, possibly smaller than $B(\boldsymbol{x}_0, r)$. The relation (4.17) follows from (4.25), $\boldsymbol{b}$ not being parallel to $\boldsymbol{c}$ since $\boldsymbol{C}-\boldsymbol{B}$ is not of rank one.

Conversely, suppose that $\mathscr{S}$, $\boldsymbol{A}$, $\boldsymbol{B}$ and $\boldsymbol{C}$ have the forms (4.15) through (4.17). We suppose without loss of generality that $k=0$. Let $\theta : \mathbb{R} \to \mathbb{R}$ be the 1-periodic function satisfying

$$\theta(t) = \begin{cases} (1-\lambda)(t+\lambda) & \text{for } -\lambda \leqq t < 0, \\ -\lambda(t-1+\lambda) & \text{for } 0 \leqq t < 1-\lambda. \end{cases} \tag{4.26}$$

Let $\mu > 0$ and

$$\boldsymbol{D}^{\pm} = \boldsymbol{C} + \left(\boldsymbol{b} \pm \frac{\lambda(1-\lambda)}{\mu}\,\boldsymbol{c}\right) \otimes \boldsymbol{m}. \tag{4.27}$$

For $x \in \mathbb{R}^n$ define

$$z(x) = \begin{cases} (\lambda A + (1-\lambda) B)\, x + \theta(x \cdot n)\, c & \text{for } |x \cdot m| > \dfrac{\mu}{\lambda(1-\lambda)}\,\theta(x \cdot n), \\ D^+ x & \text{for } 0 \leqq x \cdot m \leqq \dfrac{\mu}{\lambda(1-\lambda)}\,\theta(x \cdot n), \\ D^- x & \text{for } -\dfrac{\mu}{\lambda(1-\lambda)}\,\theta(x \cdot n) \leqq x \cdot m \leqq 0. \end{cases} \tag{4.28}$$

It is easily verified that $z \in W^{1,\infty}_{\text{loc}}(\mathbb{R}^n, \mathbb{R}^m)$, that $Dz$ takes almost everywhere only the four values $A$, $B$, $D^+$, $D^-$ and that $z(x) = Cx$ for $x \cdot m = 0$. For $j = 1, 2, \ldots,$ define for $x \in \Omega$

$$y^{(j)}(x) = \begin{cases} j^{-1} z(jx) & \text{for } x \in \Omega_{A,B}, \\ Cx & \text{otherwise}, \end{cases} \tag{4.29}$$

(see Figure 4b). Since $y^{(j)}$ is continuous, $y^{(j)}(\mathbf{0}) = \mathbf{0}$ and $Dy^{(j)}$ takes almost everywhere only the five values $A$, $B$, $C$ and $D^{\pm}$, it follows that $y^{(j)}$ is a bounded sequence in $W^{1,\infty}(\Omega, \mathbb{R}^m)$. Since also

$$Dy^{(j)}(x) = \lambda A + (1-\lambda) B + \theta'(jx \cdot n)(A - B)$$

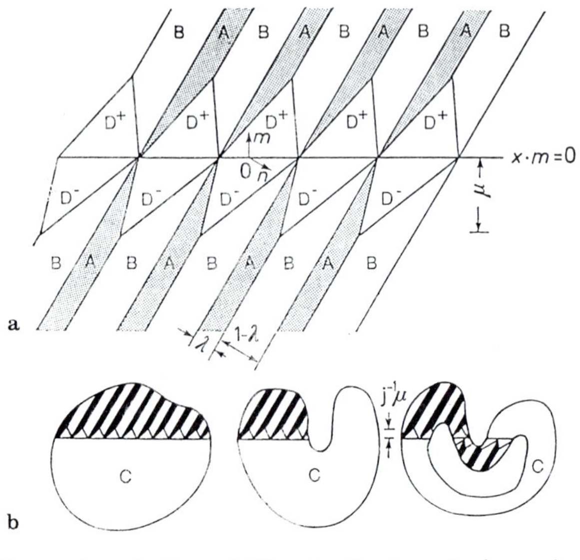


Fig. 4a. The orthogonal projection of $\mathbb{R}^n$ onto the $(m, n)$ plane, showing the values taken in various regions by $Dz$ when $z$ is given by (4.28); b Possible two-dimensional cross sections of $\Omega$ parallel to the $(m, n)$ plane, showing the division of $\Omega$ into regions where $Dy^{(j)}$ takes different values when $y^{(j)}$ is given by (4.29)

for a.e. $x \in \Omega_{A,B}$ satisfying $|x \cdot m| > j^{-1}\mu$, it is easily shown that $y^{(j)} \overset{*}{\rightharpoonup} y$ in $W^{1,\infty}(\Omega, \mathbb{R}^m)$, where

$$y(x) = \begin{cases} (\lambda A + (1-\lambda) B)\, x & \text{for } x \in \Omega_{A,B}, \\ Cx & \text{otherwise}, \end{cases} \tag{4.30}$$

and that (4.13) and (4.14) hold. Finally, if $m = n$ and $\det A > 0$, $\det B = 0$, $\det C > 0$, then by (4.16)

$$\det(\lambda A + (1-\lambda) B) = \lambda \det A + (1-\lambda) \det B.$$

Since

$$\det D^{\pm} = \det\left(\lambda A + (1-\lambda) B \pm \frac{\lambda(1-\lambda)}{\mu} c \otimes m\right),$$

it follows that $\det D^{\pm} > 0$ if $\mu$ is chosen sufficiently large, and in this case $\det Dy^{(j)}(x) \geqq \min\{\det A, \det B, \det C, \det D\} > 0$ a.e. for all $j$. □

*Remarks:*

1. Let $m = n$, $\det A > 0$, $\det B > 0$, $\det C > 0$ and let $\mu$ be chosen sufficiently large that $\det D^{\pm} > 0$. Then for domains $\Omega$ such as those in Figure 4b (ii) and (iii), the deformation $y$ given by (4.30) need not be invertible, and hence $y^{(j)}$ given by (4.29) need not be invertible. Sufficient conditions for $y^{(j)}$ to be invertible are that $\Omega_{A,B} \subset \{x \in \mathbb{R}^n : x \cdot m > k\}$ and $\Omega_C \subset \{x \in \mathbb{R}^n : x \cdot m < k\}$.
2. It follows from the proof of Proposition 2 that

$$\lim_{\varepsilon \to 0} \lim_{j \to \infty} \frac{\operatorname{meas}\{x \in N(x_0) : |Dy^{(j)}(x) - A| < \varepsilon\}}{\operatorname{meas}(N(x_0) \cap \Omega_{A,B})} = \lambda,$$

$$\lim_{\varepsilon \to 0} \lim_{j \to \infty} \frac{\operatorname{meas}\{x \in N(x_0) : |Dy^{(j)}(x) - B| < \varepsilon\}}{\operatorname{meas}(N(x_0) \cap \Omega_{A,B})} = 1 - \lambda,$$

so that $\lambda$, $1 - \lambda$ denote the asymptotic proportions of $A$ and $B$ respectively in $N(x_0) \cap \Omega_{A,B}$.

We now return to the minimization problem (3.11) with the free energy function described in Section 3, not necessarily specialized to InTl. Assume that $F^+$ and $F^-$ satisfy for some rotations $R^+$ and $R^-$

$$F^{\pm} = R^{\pm} U_0. \tag{4.31}$$

Recall that $\phi$ is a Galilean invariant, that $U_0$ is at the minimum of a potential well and that $\phi(\mathbf{1}, \theta_0) = \phi(U_0, \theta_0) = 0$. We apply Theorem 3 with $m = n = 3$, $A = F^+$, $B = F^-$ and $C = \mathbf{1}$. Let $y^{(j)}$ be a minimizing sequence for $\mathscr{I}$ in $\mathscr{A}$, so that

$$\lim_{j \to \infty} \mathscr{I}(y^{(j)}) = 0, \tag{4.32}$$

and suppose that the hypotheses of the necessity part of Theorem 3 holds. (The condition that a minimizing sequence is weakly convergent in $W^{1,p}(\Omega; \mathbb{R}^3)$ is satisfied for some subsequence of that sequence provided $\phi$ satisfies an estimate of the form

$$\phi(F) \geqq K_0 |F|^p + K_1 \quad \text{for all } F \in \mathscr{D} \tag{4.33}$$

for some constants $K_0 > 0$, $K_1$.) From Theorem 3 we deduce that the interface $\mathscr{S}$ is *flat* and, by (4.16) and (4.17), that

$$\begin{aligned} F^+ &= \mathbf{1} + (1 - \lambda)\, c \otimes n + b \otimes m, \\ F^- &= \mathbf{1} - \lambda c \otimes n + b \otimes m, \end{aligned} \tag{4.34}$$

for nonparallel vectors $b, c \in \mathbb{R}^3$, nonparallel unit vectors $m, n \in \mathbb{R}^3$ and some $\lambda \in (0, 1)$. Conversely, if the conditions (4.34) hold and $\mathscr{S}$ has the form (4.15) then the $y^{(j)}$ constructed in the theorem and satisfying (4.13) and (4.14) will be a minimizing sequence for $\mathscr{J}$ in $\mathscr{A}$, since $\phi$ is continuous, $Dy^{(j)}$ bounded in $L^\infty$ and $\det Dy^{(j)}(x) \geqq \delta > 0$ a.e. in $\Omega$. In fact, since $F^+$ and $F^-$ are at the minima of potential wells,

$$\begin{aligned} \int_\Omega \phi(Dy^{(j)}, \theta_0)\, dx &= \int_{\{x \mid Dy^{(j)}(x) = D^+\}} \phi(D^+, \theta_0)\, dx + \int_{\{x \mid Dy^{(j)}(x) = D^-\}} \phi(D^-, \theta_0)\, dx \\ &\to 0 \quad \text{as } j \to \infty. \end{aligned} \tag{4.35}$$

That is, the only contribution to the total energy of $y^{(j)}$ is from the layer of triangular prisms pictured in Figure 4b whose total volume tends to zero as $j \to \infty$. The existence of rotations $R^\pm$ such that $F^\pm$ given by (4.31) satisfy (4.34) will be established for InTl in Section 5.

Simple examples show that if we drop the hypothesis in Theorem 3 that the open set $\Omega_{A,B}$ is connected, then $\mathscr{S}$ need not be contained in a plane. The $X$-interface in Figure 2a provides an example where the finely twinned region is a disconnected open set and the austenite/martensite interface $\mathscr{S}$ is not contained in a plane. Note, however, that the deformation gradient in this configuration takes to a good approximation four values. The $X$-interface is easily understood by patching together two deformations of the type given by Theorem 3. The curved martensite/martensite interface shown in Figure 2b is obviously not covered by Theorem 3. In principle, it should be possible to relate the orientation of this interface to the local twin concentrations $\lambda(x)$ and $\lambda'(x)$ on each side of the interface.

Of course the weak limit $y$ of a minimizing sequence $y^{(j)}$ is not in general a minimizer; for example, with $y^{(j)} \rightharpoonup y$ as above, (4.18) becomes

$$Dy(x) = \lambda F^+ + (1 - \lambda) F^-, \tag{4.36}$$

and $\lambda F^+ + (1 - \lambda) F^- = \mathbf{1} + b \otimes m$ does not in general, or in particular for InTl, yield one of the stretch matrices $\mathbf{1}, R_i U_0 R_i^T$, $i = 1, \ldots, \nu$. Hence $\mathscr{J}$ is generally not sequentially weakly lower semicontinuous (swlsc) in $W^{1,q}(\Omega, \mathbb{R}^3)$ for any $q \geqq 1$. This typical feature of multidimensional phase change problems results from the failure of strong ellipticity of $\phi$ (*cf.* ERICKSEN [16], BALL [4]) and contrasts with certain models of rubber-like materials for which $\mathscr{J}$ *is* swlsc and for which the direct method of the calculus of variations can consequently be applied to establish the existence of minimizers (BALL [3]). When $\phi$ is not $W^{1,1}$-quasiconvex, a condition closely related to strong ellipticity, an argument of BALL & MURAT [8, Theorem 5.1] in fact shows that for appropriate boundary conditions and body forces the total free energy does *not* attain a minimum (see also the examples in Section 7).

## 5. Materials Which Can Form Internally Twinned Martensite

Given a transformation strain and a family of energy minimizing twins, we now consider the algebraic problem of whether the equations $(4.34)_{1,2}$ can be satisfied. If so, we can construct minimizing sequences by the methods given in Section 4. The critical physical question is whether the austenite/martensite planes work out correctly. These calculations are closely related to those of the crystallographic theory of martensite, although necessary and sufficient conditions for the existence of solutions of $(4.34)_{1,2}$ in the case of a general transformation strain and point group appear to be absent from the literature. We restrict attention to the case $m = n = 3$ throughout this section.

According to the analysis of Section 3, all classical interfaces between minimizing deformation gradients are martensite/martensite twins. The deformation gradients associated with these twins, $\boldsymbol{F}^+$ and $\boldsymbol{F}^-$, have the forms

$$\begin{aligned} \boldsymbol{F}^+ &= \hat{\boldsymbol{R}}^+ \boldsymbol{U}_0 \boldsymbol{R}_i, \\ \boldsymbol{F}^- &= \hat{\boldsymbol{R}}^- \boldsymbol{U}_0 \boldsymbol{R}_j, \end{aligned} \tag{5.1}$$

with $\boldsymbol{R}_i$ and $\boldsymbol{R}_j$ in $P^\nu$. It is sufficient* for our purposes to take $\boldsymbol{R}_j = \boldsymbol{1}$ and to rewrite (5.1) in the form

$$\begin{aligned} \boldsymbol{F}^+ &= \hat{\boldsymbol{R}} \boldsymbol{R} \boldsymbol{U}_0 \bar{\boldsymbol{R}}, \\ \boldsymbol{F}^- &= \hat{\boldsymbol{R}} \boldsymbol{U}_0, \end{aligned} \tag{5.2}$$

where

$$\boldsymbol{F}^+ - \boldsymbol{F}^- = \boldsymbol{c} \otimes \boldsymbol{n}, \quad \boldsymbol{c} = \hat{\boldsymbol{R}} \boldsymbol{a}. \tag{5.3}$$

For InTl, the values of $\boldsymbol{R}$, $\bar{\boldsymbol{R}}$, $\boldsymbol{a}$ and $\boldsymbol{n}$ satisfying the equations (5.2) and (5.3) are obtained from (3.19) with $\eta_1 = 1 - \varepsilon$ and $\eta_2 = 1 + 2\varepsilon$, $\varepsilon \doteq .013$. In this section we allow $\boldsymbol{U}_0$ to be an arbitrary positive-definite symmetric matrix and the twins in (5.2) to be general in that they are subject only to (3.18), that is

$$\boldsymbol{R} \boldsymbol{U}_0 \bar{\boldsymbol{R}} = \boldsymbol{U}_0 + \boldsymbol{a} \otimes \boldsymbol{n}, \tag{5.4}$$

where $\boldsymbol{R}$, $\bar{\boldsymbol{R}}$ are rotations, $\boldsymbol{a} \neq \boldsymbol{0}$ and $|\boldsymbol{n}| = 1$. We note the relations

$$\boldsymbol{U}_0^{-1} \boldsymbol{a} \cdot \boldsymbol{n} = 0, \tag{5.5}$$

$$2 \boldsymbol{U}_0 \boldsymbol{a} \cdot \boldsymbol{n} + |\boldsymbol{a}|^2 = 0, \tag{5.6}$$

$$2 \boldsymbol{U}_0^{-2} \boldsymbol{a} \cdot \boldsymbol{U}_0^{-1} \boldsymbol{n} - |\boldsymbol{U}_0^{-1} \boldsymbol{a}|^2 \, |\boldsymbol{U}_0^{-1} \boldsymbol{n}|^2 = 0, \tag{5.7}$$

---

* The twins represented by (5.1) can be obtained from those represented by (5.2) and (5.3) by replacing $\boldsymbol{F}^+$ and $\boldsymbol{F}^-$ in (5.1) by $\boldsymbol{R}_j^T \boldsymbol{F}^+ \boldsymbol{R}_j$ and $\boldsymbol{R}_j^T \boldsymbol{F}^- \boldsymbol{R}_j$. Note that these replacements do not alter the forms of $(4.34)_{1,2}$.

which follow from (5.4) by taking determinants and by calculating tr $(RU_0^2R^T)$ and tr $(RU_0^{-2}R^T)$.

To decide whether the twins (5.2) can participate in an austenite/finely twinned martensite interface, we consider the algebraic problem of whether $F^+$ and $F^-$ given by (5.2) can assume the forms $(4.34)_{1,2}$ for some choice of $\hat{R}$. We view $U_0$, $a$ and $n$ as given, consistent with (3.18). Since $(4.34)_1$ follows immediately from $(4.34)_2$ and (5.3), we only need to consider $(4.34)_2$ which becomes

$$F^- = \hat{R}U_0 = 1 - \lambda\hat{R}a \otimes n + b \otimes m, \tag{5.8}$$

or equivalently,

$$U_0 + \lambda a \otimes n = \hat{R}^T(1 + b \otimes m), \tag{5.9}$$

which is to be solved for $\lambda \in (0, 1)$, $b$, the unit vector $m$ and the rotation $\hat{R}$.

Let $C_0(\lambda)$ be defined by

$$C_0(\lambda) \stackrel{\text{def}}{=} (U_0 + \lambda n \otimes a)(U_0 + \lambda a \otimes n). \tag{5.10}$$

According to the polar decomposition theorem, the basic equation (5.9) with $\det(U_0 + \lambda a \otimes n) > 0$ is equivalent to

$$C_0(\lambda) = (1 + m \otimes b)(1 + b \otimes m), \tag{5.11}$$

together with the restriction that $\det(1 + b \otimes m) = 1 + b \cdot m > 0$. Hence, we begin with a characterization of $b$ and $m$ satisfying (5.11).

**Proposition 4.** *Necessary and sufficient conditions for a symmetric $3 \times 3$ matrix $C$ with eigenvalues $\lambda_1 \leqq \lambda_2 \leqq \lambda_3$ to be expressible in the form*

$$C = (1 + m \otimes b)(1 + b \otimes m) \tag{5.12}$$

*for nonzero $b$ and $m$ are that $\lambda_1 \geqq 0$ (i.e., $C \geqq 0$) and $\lambda_2 = 1$.*

*The solutions are given by*:

a) $C \neq 1$,

$$b = \varrho\left(\sqrt{\frac{\lambda_3(1-\lambda_1)}{\lambda_3-\lambda_1}}\, e_1 + \varkappa\sqrt{\frac{\lambda_1(\lambda_3-1)}{\lambda_3-\lambda_1}}\, e_3\right), \tag{5.13}$$

$$m = \varrho^{-1}\left(\frac{\bar{\varkappa}\sqrt{\lambda_3} - \sqrt{\lambda_1}}{\sqrt{\lambda_3-\lambda_1}}\right)\left(-\bar{\varkappa}\sqrt{1-\lambda_1}\, e_1 + \varkappa\sqrt{\lambda_3-1}\, e_3\right),$$

*where $\varrho \neq 0$ is a constant, and $e_1$, $e_3$ are normalized eigenvectors of $C$ corresponding to $\lambda_1$, $\lambda_3$ respectively, and where each of $\varkappa$, $\bar{\varkappa}$ can take the values $\pm 1$. For these solutions $1 + b \cdot m = \bar{\varkappa}\sqrt{\lambda_1\lambda_3}$.*

b) $C = 1$,

$$b = \varrho e, \qquad m = -2\varrho^{-1} e, \tag{5.14}$$

*where $\varrho \neq 0$ is a constant and $|e| = 1$. For these solutions $1 + b \cdot m = -1$.*

**Proof.** Necessity. Let $\boldsymbol{p}$ be perpendicular to $\boldsymbol{b}$ and $\boldsymbol{m}$. Then $\boldsymbol{C}\boldsymbol{p} = \boldsymbol{p}$ so that one eigenvalue of $\boldsymbol{C}$ equals 1. If $\boldsymbol{b}$ and $\boldsymbol{m}$ are linearly dependent, then the eigenvalue 1 has a multiplicity of at least two, so $\lambda_2 = 1$. Suppose that $\boldsymbol{b}$ and $\boldsymbol{m}$ are linearly independent. Consider

$$\begin{aligned} \boldsymbol{x} \cdot \boldsymbol{C}\boldsymbol{x} - |\boldsymbol{x}|^2 &= 2(\boldsymbol{x} \cdot \boldsymbol{b})(\boldsymbol{x} \cdot \boldsymbol{m}) + (\boldsymbol{x} \cdot \boldsymbol{m})^2 |\boldsymbol{b}|^2 \\ &= (\boldsymbol{x} \cdot \boldsymbol{m})[2(\boldsymbol{x} \cdot \boldsymbol{b}) + (\boldsymbol{x} \cdot \boldsymbol{m})|\boldsymbol{b}|^2]. \end{aligned} \tag{5.15}$$

If we choose $\boldsymbol{x} \cdot \boldsymbol{b} > 0$ and $\boldsymbol{x} \cdot \boldsymbol{m} > 0$, the expression (5.15) is positive. If we choose $\boldsymbol{x} \cdot \boldsymbol{b} = -1$ and $\boldsymbol{x} \cdot \boldsymbol{m}$ small and positive, the expression (5.15) is negative. Hence $\lambda_1 < 1 < \lambda_3$. Finally, $\lambda_1\lambda_3 = \det \boldsymbol{C} = (1 + \boldsymbol{b} \cdot \boldsymbol{m})^2 \geqq 0$, so that $\lambda_1 \geqq 0$.

Sufficiency. First suppose that $\boldsymbol{C} \neq \mathbf{1}$, so that $\boldsymbol{b} \neq \mathbf{0}$, and suppose that $\lambda_1 > 0$. If $\boldsymbol{b}$ and $\boldsymbol{m}$ satisfy (5.12), we have

$$\begin{aligned} \boldsymbol{C}\boldsymbol{b} &= (\mathbf{1} + \boldsymbol{m} \otimes \boldsymbol{b})\,\boldsymbol{b}(1 + \boldsymbol{b} \cdot \boldsymbol{m}), \\ &= \pm(\det \boldsymbol{C})^{1/2}(\boldsymbol{b} + |\boldsymbol{b}|^2 \boldsymbol{m}). \end{aligned} \tag{5.16}$$

Hence,

$$\boldsymbol{m} = [\pm(\det \boldsymbol{C})^{-1/2}\boldsymbol{C} - \mathbf{1}]\left(\frac{\boldsymbol{b}}{|\boldsymbol{b}|^2}\right). \tag{5.17}$$

In view of (5.17) a necessary and sufficient condition that nonzero vectors $\boldsymbol{b}$ and $\boldsymbol{m}$ satisfy (5.12) is that $\boldsymbol{b} \neq \mathbf{0}$ satisfies

$$\boldsymbol{C} = \mathbf{1} - \frac{\boldsymbol{b} \otimes \boldsymbol{b}}{|\boldsymbol{b}|^2} + \frac{\boldsymbol{C}\boldsymbol{b} \otimes \boldsymbol{C}\boldsymbol{b}}{|\boldsymbol{b}|^2 \det \boldsymbol{C}}. \tag{5.18}$$

In the orthonormal basis of eigenvectors $\{\boldsymbol{e}_1, \boldsymbol{e}_2, \boldsymbol{e}_3\}$, we have say $\boldsymbol{b} = (b_1, b_2, b_3)$ and $\boldsymbol{C}\boldsymbol{b} = (\lambda_1 b_1, b_2, \lambda_3 b_3)$. Then (5.18) is equivalent to

$$\begin{aligned} &\begin{bmatrix} \lambda_1 - 1 & 0 & 0 \\ 0 & 0 & 0 \\ 0 & 0 & \lambda_3 - 1 \end{bmatrix} \\ &= \frac{1}{|\boldsymbol{b}|^2} \begin{bmatrix} b_1^2\left(-1 + \frac{\lambda_1}{\lambda_3}\right) & b_1 b_2\left(-1 + \frac{1}{\lambda_3}\right) & 0 \\ b_1 b_2\left(-1 + \frac{1}{\lambda_3}\right) & b_2^2\left(-1 + \frac{1}{\lambda_1\lambda_3}\right) & b_2 b_3\left(-1 + \frac{1}{\lambda_1}\right) \\ 0 & b_2 b_3\left(-1 + \frac{1}{\lambda_1}\right) & b_3^2\left(-1 + \frac{\lambda_3}{\lambda_1}\right) \end{bmatrix}. \end{aligned} \tag{5.19}$$

Since by assumption $\lambda_1$ and $\lambda_3$ are not both 1, (5.19) holds if and only if $b_2 = 0$ and

$$\frac{b_1^2}{|\boldsymbol{b}|^2} = \frac{\lambda_3(1 - \lambda_1)}{\lambda_3 - \lambda_1}, \quad \frac{b_3^2}{|\boldsymbol{b}|^2} = \frac{\lambda_1(\lambda_3 - 1)}{\lambda_3 - \lambda_1}. \tag{5.20}$$

The equations (5.20) are consistent and so (5.18) holds if and only if $\boldsymbol{b}$ has the form $(5.13)_1$ with $\varrho \neq 0$. Then we get $(5.13)_2$ from (5.17) as required.

If $\boldsymbol{C} = \mathbf{1}$ then (5.12) is satisfied by nonzero vectors $\boldsymbol{b}$ and $\boldsymbol{m}$ if and only if

$$\boldsymbol{b} = \tau \boldsymbol{m} \qquad \text{with } 2\tau + \tau^2 \, |\boldsymbol{m}|^2 = 0, \tag{5.21}$$

so that the nonzero solutions of (5.12) are given by (5.14).

If $\lambda_1 = 0$, then from (5.16), which holds without the restriction $\lambda_1 > 0$, we have $\boldsymbol{C}\boldsymbol{b} = \mathbf{0}$, so $\boldsymbol{b} = \varrho \boldsymbol{e}_1$. If we write $\boldsymbol{m} = m_1 \boldsymbol{e}_1 + m_2 \boldsymbol{e}_2 + m_3 \boldsymbol{e}_3$, we can write (5.12) in the form

$$\begin{pmatrix} -1 & 0 & 0 \\ 0 & 0 & 0 \\ 0 & 0 & \lambda_3 - 1 \end{pmatrix} = \varrho \begin{pmatrix} 2m_1 & m_2 & m_3 \\ m_2 & 0 & 0 \\ m_3 & 0 & 0 \end{pmatrix} + \varrho^2 \begin{pmatrix} m_1^2 & m_1 m_2 & m_1 m_3 \\ m_2 m_1 & m_2^2 & m_2 m_3 \\ m_3 m_1 & m_3 m_2 & m_3^2 \end{pmatrix}, \tag{5.22}$$

that is,

$$\begin{aligned} (1 + \varrho m_1)^2 &= 0, \\ \varrho m_2 (1 + \varrho m_1) &= 0, \\ \varrho m_3 (1 + \varrho m_1) &= 0, \\ \varrho^2 m_2^2 &= 0, \\ \varrho^2 m_2 m_3 &= 0, \\ \varrho^2 m_3^2 &= \lambda_3 - 1, \end{aligned} \tag{5.23}$$

with solutions

$$m_1 = -\frac{1}{\varrho}, \qquad m_2 = 0, \qquad m_3 = \frac{\pm \sqrt{\lambda_3 - 1}}{\varrho}, \tag{5.24}$$

which are already covered by (5.13). □

*Remarks.*

1. Consider solutions of (5.12) with $1 + \boldsymbol{b} \cdot \boldsymbol{m} \geqq 0$. If $\lambda_1 < \lambda_2 = 1 < \lambda_3$, then there are two essentially distinct such solutions $\boldsymbol{b}^- \otimes \boldsymbol{m}^-$ and $\boldsymbol{b}^+ \otimes \boldsymbol{m}^+$, consistent with the analysis of [24, Appendix 1]. These are related by a rotation $\tilde{\boldsymbol{R}}$ in the sense that $\mathbf{1} + \boldsymbol{b}^+ \otimes \boldsymbol{m}^+ = \tilde{\boldsymbol{R}}(\mathbf{1} + \boldsymbol{b}^- \otimes \boldsymbol{m}^-)$; this follows from the polar decomposition theorem in the case $\lambda_1 > 0$ and by an explicit calculation if $\lambda_1 = 0$. If $\lambda_1$ or $\lambda_3$ equals 1, there is only one solution. If $\lambda_1$ and $\lambda_3$ both equal 1, there is no solution.
2. Note that if $\boldsymbol{m}^\perp$ is perpendicular to $\boldsymbol{m}$ then

$$\boldsymbol{m}^\perp \cdot \boldsymbol{C} \boldsymbol{m}^\perp = |\boldsymbol{m}^\perp|^2, \tag{5.25}$$

so that $\boldsymbol{m}$ is normal to an "undistorted" plane for $\boldsymbol{C}$. Similarly, if $\boldsymbol{b}^\perp$ is perpendicular to $\boldsymbol{b}$, then

$$\boldsymbol{b}^\perp \cdot \boldsymbol{C}^{-1} \boldsymbol{b}^\perp = |\boldsymbol{b}^\perp|^2 \tag{5.26}$$

so that $\boldsymbol{b}$ is normal to an "undistorted" plane for $\boldsymbol{C}^{-1}$.

3. The formula for $\boldsymbol{b}$ can be written

$$\boldsymbol{b} = \frac{\varrho}{\sqrt{\lambda_1^{-1} - \lambda_3^{-1}}} \left(\sqrt{\lambda_1^{-1} - 1}\, \boldsymbol{e}_1 + \varkappa \sqrt{1 - \lambda_3^{-1}}\, \boldsymbol{e}_3\right) \tag{5.27}$$

which has the same form as the formula for $\boldsymbol{m}$.

Proposition 4 makes it clear in particular that in order to solve the basic equation (5.11) we must show that $\boldsymbol{C}_0(\lambda)$ has an eigenvalue equal to 1 for some $\lambda$, so we consider the following proposition:

**Proposition 5.** *Let the* $3 \times 3$ *nonsingular matrix* $\boldsymbol{U}_0 = \boldsymbol{U}_0^T$, *the vector* $\boldsymbol{a}$, *the unit vector* $\boldsymbol{n}$ *and rotations* $\boldsymbol{R}$ *and* $\bar{\boldsymbol{R}}$ *be given subject to the twinning relations* (5.4). *Let* $\boldsymbol{C}_0(\lambda) = (\boldsymbol{U}_0 + \lambda \boldsymbol{n} \otimes \boldsymbol{a})(\boldsymbol{U}_0 + \lambda \boldsymbol{a} \otimes \boldsymbol{n})$ *and let*

$$g(\lambda) = \det(\boldsymbol{C}_0(\lambda) - \boldsymbol{1}). \tag{5.28}$$

*Then* $g(\lambda)$ *is a quadratic function of* $\lambda$ *which satisfies* $g(\lambda) = g(1 - \lambda)$.

**Proof.** To show that $g(\lambda)$, which appears to be a sixth order polynomial, is in fact only quadratic, note that by (5.5)

$$\det(\boldsymbol{U}_0 + \lambda \boldsymbol{a} \otimes \boldsymbol{n}) = \det \boldsymbol{U}_0 \neq 0. \tag{5.29}$$

Hence

$$\begin{aligned} g(\lambda) &= \det[(\boldsymbol{U}_0 + \lambda \boldsymbol{n} \otimes \boldsymbol{a})(\boldsymbol{U}_0 + \lambda \boldsymbol{a} \otimes \boldsymbol{n}) - \boldsymbol{1}] \\ &= \det \boldsymbol{U}_0 \det[(\boldsymbol{U}_0 + \lambda \boldsymbol{a} \otimes \boldsymbol{n}) - (\boldsymbol{U}_0 + \lambda \boldsymbol{n} \otimes \boldsymbol{a})^{-1}] \\ &= \det \boldsymbol{U}_0 \det[(\boldsymbol{U}_0 - \boldsymbol{U}_0^{-1}) + \lambda(\boldsymbol{a} \otimes \boldsymbol{n} + \boldsymbol{U}_0^{-1}\boldsymbol{n} \otimes \boldsymbol{U}_0^{-1}\boldsymbol{a})]. \end{aligned} \tag{5.30}$$

Since the matrix multiplying $\lambda$ is singular, the right-hand side of $(5.30)_3$ is at most quadratic in $\lambda$.

Since

$$g(1) = \det(\bar{\boldsymbol{R}}^T(\boldsymbol{U}_0^2 - 1)\bar{\boldsymbol{R}}) = \det(\boldsymbol{U}_0^2 - 1) = g(0),$$

if follows that $g(\lambda) = g(1 - \lambda)$. □

From Proposition 5 we can see how to make one eigenvalue of $\boldsymbol{C}_0(\lambda)$ equal to 1 at some $\lambda$ so as to satisfy part of the conditions in Proposition 4. For the remaining part we need to show that the other two eigenvalues of $\boldsymbol{C}_0(\lambda)$ bound 1 above and below, using the following proposition:

**Proposition 6.** *Suppose that for some* $\lambda$, $\boldsymbol{C}_0(\lambda)$ *has the unordered triple of eigenvalues* $1, \lambda_1, \lambda_3$. *Then*

$$(1 - \lambda_1)(\lambda_3 - 1) = \operatorname{tr} \boldsymbol{U}_0^2 - \det \boldsymbol{U}_0^2 - 2 + (\lambda^2 - \lambda)|\boldsymbol{a}|^2. \tag{5.31}$$

**Proof.** Using (5.6) we obtain

$$\begin{aligned} 1 + \lambda_1 + \lambda_3 &= \operatorname{tr} C_0(\lambda) \\ &= \operatorname{tr} U_0^2 + 2\lambda(U_0 a \cdot n) + \lambda^2 |a|^2 \\ &= \operatorname{tr} U_0^2 + (\lambda^2 - \lambda)|a|^2. \end{aligned} \tag{5.32}$$

Since also

$$\lambda_1\lambda_3 = \det C_0(\lambda) = \det U_0^2, \tag{5.33}$$

the result follows. □

We now combine Propositions 4 through 6 to get an existence theorem for the original equation (5.9). For the purpose of Theorem 7 a triple $(\hat{R}, \lambda, b \otimes m)$ consisting of a rotation $\hat{R}$, a scalar $\lambda \in (0, 1)$ and a rank-one matrix $b \otimes m$ such that

$$U_0 + \lambda a \otimes n = \hat{R}^T(1 + b \otimes m) \tag{5.34}$$

will be termed a *solution* of (5.34).

**Theorem 7.** *Let the positive-definite symmetric matrix $U_0$ satisfy the twinning relation*

$$R U_0 \bar{R} = U_0 + a \otimes n \tag{5.35}$$

*for some pair of rotations $R$ and $\bar{R}$ and for vectors $a \neq 0$ and $n$, $|n| = 1$.*

I. *Assume $U_0$ does not have an eigenvalue equal to* 1. *Necessary and sufficient conditions that* (5.34) *has a solution are that*

$$1 + \tfrac{1}{2}\delta^* \leqq 0 \tag{5.36}$$

*and that*

$$\operatorname{tr} U_0^2 - \det U_0^2 - 2 + \frac{1}{2\delta^*}|a|^2 \geqq 0, \tag{5.37}$$

*where*

$$\delta^* = a \cdot U_0(U_0^2 - 1)^{-1} n. \tag{5.38}$$

*If further*

$$1 + \tfrac{1}{2}\delta^* < 0, \tag{5.39}$$

*then strict inequality holds also in* (5.37) *and there are exactly four distinct solutions of* (5.34), *these having the form*

$$\begin{aligned} &(\hat{R}_1, \lambda^*, b_1^+ \otimes m_1^+), \\ &(\hat{R}_2, \lambda^*, b_1^- \otimes m_1^-), \\ &(\hat{R}_3, 1 - \lambda^*, b_2^+ \otimes m_2^+), \\ &(\hat{R}_4, 1 - \lambda^*, b_2^- \otimes m_2^-), \end{aligned} \tag{5.40}$$

*where*

$$\lambda^* = \tfrac{1}{2}\left(1 - \sqrt{1 + \frac{2}{\delta^*}}\right), \tag{5.41}$$

*so that* $0 < \lambda^* < 1/2$. *If*

$$1 + \tfrac{1}{2}\delta^* = 0, \tag{5.42}$$

*then all solutions have* $\lambda = 1/2$; *if strict inequality holds in* (5.37) *then there are exactly two distinct solutions, while if equality holds in* (5.37) *there is just one solution.*

II. *Assume* $\boldsymbol{U}_0$ *has an eigenvalue equal to* 1. *A necessary and sufficient condition that* (5.34) *has a solution is that*

$$\mu^* \stackrel{\text{def}}{=} \operatorname{tr} \boldsymbol{U}_0^2 - \det \boldsymbol{U}_0^2 - 2 > 0. \tag{5.43}$$

*All solutions are given as follows*:

*If* $\mu^* > \dfrac{|\boldsymbol{a}|^2}{4}$ *then for each* $\lambda \in (0, 1)$ *there are exactly two distinct solutions*

$$\begin{aligned} &(\hat{\boldsymbol{R}}_\lambda^+, \lambda, \boldsymbol{b}_\lambda^+ \otimes \boldsymbol{m}_\lambda^+), \\ &(\hat{\boldsymbol{R}}_\lambda^-, \lambda, \boldsymbol{b}_\lambda^- \otimes \boldsymbol{m}_\lambda^-). \end{aligned} \tag{5.44}$$

*If* $0 < \mu^* \leqq \dfrac{|\boldsymbol{a}|^2}{4}$ *and* $\bar{\lambda} \stackrel{\text{def}}{=} \tfrac{1}{2}\left(1 - \sqrt{1 - \dfrac{4\mu^*}{|\boldsymbol{a}|^2}}\right)$, *so that* $0 < \bar{\lambda} \leqq \tfrac{1}{2}$, *then for each* $\lambda \in (0, \bar{\lambda}) \cup (1 - \bar{\lambda}, 1)$ *there are exactly two distinct solutions of the form* (5.44), *while if* $\lambda = \bar{\lambda}$ *or* $1 - \bar{\lambda}$ *with either* $\bar{\lambda} \neq \tfrac{1}{2}$ $\left(\text{i.e., } \mu^* < \dfrac{|\boldsymbol{a}|^2}{4}\right)$ *or* $\bar{\lambda} = \tfrac{1}{2}$ *and* $\det \boldsymbol{U}_0 \neq 1$, *then there is one solution* $(\hat{\boldsymbol{R}}_\lambda, \lambda, \boldsymbol{b}_\lambda \otimes \boldsymbol{m}_\lambda)$.

*In all the cases above, formulas for* $\boldsymbol{b} \otimes \boldsymbol{m}$ *associated with a solution* $(\hat{\boldsymbol{R}}, \lambda, \boldsymbol{b} \otimes \boldsymbol{m})$ *are given by* (5.13) *evaluated at the ordered eigenvalues of* $\boldsymbol{C}_0(\lambda)$.

**Proof.** By Proposition 4 and the polar decomposition theorem, necessary and sufficient conditions that the basic equation (5.34) has a solution with $\lambda = \lambda^*$ are that the eigenvalues $0 < \lambda_1 \leqq \lambda_2 \leqq \lambda_3$ of $\boldsymbol{C}_0(\lambda^*)$ satisfy $\lambda_2 = 1$ and $(\lambda_1 - 1)^2 + (\lambda_3 - 1)^2 \neq 0$. (The condition $\lambda_1 > 0$ is automatically satisfied because $\det(\boldsymbol{U}_0 + \lambda \boldsymbol{a} \otimes \boldsymbol{n}) = \det \boldsymbol{U}_0 > 0$ for all $\lambda$.)

*Part I.* By Proposition 5, $g$ can be written in the form

$$g(\lambda) = a(\lambda - \tfrac{1}{2})^2 + b, \tag{5.45}$$

for some constants $a$ and $b$ with $b = g(\tfrac{1}{2})$ and

$$g(0) = \tfrac{1}{4}a + b, \quad g'(0) = -a. \tag{5.46}$$

The condition that $\boldsymbol{C}_0(\lambda^*)$ has an eigenvalue equal to 1 for some $\lambda^* \in (0, 1)$ is that $g(\lambda^*) = \det(\boldsymbol{C}_0(\lambda^*) - \boldsymbol{1}) = 0$ for some $\lambda^* \in (0, 1)$, which holds if and only if $g(0)\, g(\tfrac{1}{2}) \leqq 0$ with strict inequality if $\lambda^* \neq \tfrac{1}{2}$.

By direct calculation,

$$\begin{aligned} g(0) &= \det (U_0^2 - \mathbf{1}), \\ g'(0) &= 2\mathbf{a} \cdot U_0 \operatorname{adj} (U_0^2 - \mathbf{1})\, \mathbf{n}, \end{aligned} \tag{5.47}$$

so that by (5.46) the inequality $g(0)\, g(\tfrac{1}{2}) \leqq 0$ is equivalent to (5.36). If $g(\lambda^*) = 0$, $\lambda^* \in (0, 1)$, then by (5.45) through (5.47)

$$\lambda^{*2} - \lambda^* = \frac{1}{2\delta^*}, \tag{5.48}$$

with $\delta^*$ defined by (5.38). Hence, by Proposition 6, the eigenvalues of $C_0(\lambda^*)$ satisfy $\lambda_1 \leqq \lambda_2 = 1 \leqq \lambda_3$ if and only if (5.36) and (5.37) hold. The eigenvalues of $C_0(\lambda^*)$ are not all 1, since $C_0(\lambda^*) = \mathbf{1}$ implies that $C_0(\lambda^*)\, \mathbf{e} = U_0^2 \mathbf{e} = \mathbf{e}$ for $\mathbf{e}$ perpendicular to $\mathbf{n}$ and $U_0 \mathbf{a}$, and $U_0^2$ does not have an eigenvalue 1.

If (5.39) holds then $g(0)\, g(\tfrac{1}{2}) < 0$, so that $g$ has the two roots $\lambda^*$, $1 - \lambda^*$ with $\lambda^*$ given by (5.41). By Proposition 6 equality holds in (5.37) if and only if 1 is a double eigenvalue of $C_0(\lambda^*)$. But this is impossible, since there would then exist a corresponding eigenvector $\mathbf{e}$ with $\mathbf{e} \cdot \mathbf{n} = 0$, so that

$$U_0^2 \mathbf{e} + \lambda^* (U_0 \mathbf{a} \cdot \mathbf{e})\, \mathbf{n} = \mathbf{e}, \tag{5.49}$$

which with (5.47) implies that

$$g(0)\, (U_0 \mathbf{a} \cdot \mathbf{e}) = -\tfrac{1}{2} \lambda^* g'(0)\, (U_0 \mathbf{a} \cdot \mathbf{e}). \tag{5.50}$$

The quantity $U_0 \mathbf{a} \cdot \mathbf{e}$ does not vanish because of (5.49) and the assumption that $U_0$ does not have an eigenvalue equal to 1. Hence (5.50) implies that

$$g(0) = -\tfrac{1}{2} \lambda^* g'(0), \tag{5.51}$$

which immediately gives $\lambda^* = \tfrac{1}{2}$, a contradiction. Hence, by Proposition 4 there are four distinct solutions of (5.34) as claimed.

If (5.42) holds then by (5.48) $\lambda^* = \tfrac{1}{2}$ and by Proposition 4 there are two solutions if strict inequality holds in (5.38) and only one otherwise.

*Part II.* If $U_0$ has an eigenvalue equal to 1, then $g(0) = g(1) = \det (U_0^2 - \mathbf{1}) = 0$. Hence, if there is a solution of (5.34) with $\lambda = \lambda^*$ (which means that $\lambda^* \in (0, 1)$), then $g(\lambda) = 0$ for all $\lambda$. Hence, one eigenvalue of $C_0(\lambda)$ equals 1 for each $\lambda \in (0, 1)$ and it remains to examine the other two eigenvalues. Let

$$\theta(\lambda) = \operatorname{tr} U_0^2 - \det U_0^2 - 2 + (\lambda^2 - \lambda)\, |\mathbf{a}|^2. \tag{5.52}$$

A necessary and sufficient condition that $\theta(\lambda) \geqq 0$ for some $\lambda \in (0, 1)$ is that $\mu^* > 0$. If $\theta(\lambda) > 0$, $\lambda \in (0, 1)$, by Proposition 6 the eigenvalues of $C_0(\lambda)$ satisfy $\lambda_1 < \lambda_2 = 1 < \lambda_3$ and hence there are two distinct solutions of the form (5.44); this case occurs if $\mu^* > |\mathbf{a}|^2/4$ or if $0 < \mu^* \leqq |\mathbf{a}|^2/4$ and $\lambda \in (0, \bar{\lambda}) \cup (1 - \bar{\lambda}, 1)$. If $\theta(\lambda) = 0$, that is if $\lambda = \bar{\lambda}$ or $1 - \bar{\lambda}$, there is a single solution $(\hat{\mathbf{R}}_\lambda, \lambda, \mathbf{b}_\lambda \otimes \mathbf{m}_\lambda)$ if and only if $C_0(\lambda) \neq \mathbf{1}$. But $C_0(\lambda) = \mathbf{1}$ implies by (5.5), (5.6) that

$$0 = U_0^{-1} \mathbf{a} \cdot C_0(\lambda)\, \mathbf{n} = (\lambda - \tfrac{1}{2})\, |\mathbf{a}|^2 \tag{5.53}$$

and hence that $\lambda = \frac{1}{2}$. Hence there is a single solution if $\bar{\lambda} \neq \frac{1}{2}$. If $\bar{\lambda} = \frac{1}{2}$ then $\boldsymbol{C}_0(\bar{\lambda}) = \mathbf{1}$ implies $\det \boldsymbol{U}_0 = (\det \boldsymbol{C}_0(\bar{\lambda}))^{1/2} = 1$, while $\det \boldsymbol{U}_0 = 1$ implies by (5.32) and $\theta(\bar{\lambda}) = 0$ that $\operatorname{tr} \boldsymbol{C}_0(\bar{\lambda}) = 3$, which together with the fact that $\boldsymbol{C}_0(\bar{\lambda})$ has two eigenvalues equal to 1 gives $\boldsymbol{C}_0(\bar{\lambda}) = \mathbf{1}$. Hence there is a solution if and only if $\det \boldsymbol{U}_0$ is not equal to one. □

*Remarks:*

1. An alternative method to that in Proposition 5 for evaluating $g(\lambda)$ is to write

$$\begin{aligned} g(\lambda) &= \det (\boldsymbol{C}_0(\lambda) - \mathbf{1}) \\ &= \det \boldsymbol{C}_0(\lambda) - \operatorname{tr} \operatorname{adj} \boldsymbol{C}_0(\lambda) + \operatorname{tr} \boldsymbol{C}_0(\lambda) - 1 \end{aligned} \tag{5.54}$$

and note that

$$\begin{aligned} \operatorname{tr} \operatorname{adj} \boldsymbol{C}_0(\lambda) &= \det \boldsymbol{C}_0(\lambda) \operatorname{tr} \boldsymbol{C}_0^{-1}(\lambda) \\ &= \det \boldsymbol{U}_0^2 \operatorname{tr} (\mathbf{1} - \lambda \boldsymbol{U}_0^{-1}\boldsymbol{a} \otimes \boldsymbol{n}) \boldsymbol{U}_0^{-2}(\mathbf{1} - \lambda \boldsymbol{n} \otimes \boldsymbol{U}_0^{-1}\boldsymbol{a}) \\ &= \det \boldsymbol{U}_0^2(\operatorname{tr} \boldsymbol{U}_0^{-2} + (\lambda^2 - \lambda) \, |\boldsymbol{U}_0^{-1}\boldsymbol{a}|^2 \, |\boldsymbol{U}_0^{-1}\boldsymbol{n}|^2), \end{aligned} \tag{5.55}$$

where we have used (5.7). Thus, from (5.32),

$$g(\lambda) = \det (\boldsymbol{U}_0^2 - \mathbf{1}) + (\lambda^2 - \lambda) (|\boldsymbol{a}|^2 - (\det \boldsymbol{U}_0^2) \, |\boldsymbol{U}_0^{-1}\boldsymbol{a}|^2 \, |\boldsymbol{U}_0^{-1}\boldsymbol{n}|^2), \tag{5.56}$$

and hence

$$\delta^* = \frac{\det \boldsymbol{U}_0^2 \, |\boldsymbol{U}_0^{-1}\boldsymbol{a}|^2 \, |\boldsymbol{U}_0^{-1}\boldsymbol{n}|^2 - |\boldsymbol{a}|^2}{2 \det (\boldsymbol{U}_0^2 - 1)}. \tag{5.57}$$

2. A different way of writing the necessary and sufficient conditions in Theorem 7 that there be a solution of (5.34) with $\lambda \neq 1/2$ can be obtained by noting that if $\boldsymbol{C}_0(\lambda)$ has the unordered triple of eigenvalues $1, \lambda_1, \lambda_3$, and if $\boldsymbol{e}$ is any eigenvector of $\boldsymbol{C}_0(\lambda)$ corresponding to the eigenvalue 1, then

$$\begin{aligned} &(1 - \lambda_1)(\lambda_3 - 1)(\boldsymbol{e} \cdot \boldsymbol{n})^2 \\ &\qquad = \operatorname{tr} \boldsymbol{C}_0(\lambda) - \boldsymbol{n} \cdot \boldsymbol{C}_0(\lambda)\boldsymbol{n} - \boldsymbol{n} \cdot \operatorname{adj} \boldsymbol{C}_0(\lambda)\boldsymbol{n} - 1. \end{aligned} \tag{5.58}$$

Since $\operatorname{tr} \boldsymbol{C}_0(\lambda) - \boldsymbol{n} \cdot \boldsymbol{C}_0(\lambda)\boldsymbol{n} = \operatorname{tr} \boldsymbol{U}_0^2 - \boldsymbol{n} \cdot \boldsymbol{U}_0^2\boldsymbol{n}$ and $\boldsymbol{n} \cdot \operatorname{adj} \boldsymbol{C}_0(\lambda)\boldsymbol{n} = \boldsymbol{n} \cdot \operatorname{adj} \boldsymbol{U}_0^2\boldsymbol{n}$, we have that

$$(1 - \lambda_1)(\lambda_3 - 1)(\boldsymbol{e} \cdot \boldsymbol{n})^2 = \operatorname{tr} \boldsymbol{U}_0^2 - \boldsymbol{n} \cdot \boldsymbol{U}_0^2\boldsymbol{n} - \boldsymbol{n} \cdot \operatorname{adj} \boldsymbol{U}_0^2\boldsymbol{n} - 1. \tag{5.59}$$

Since, as is shown in the proof of Theorem 7, $\boldsymbol{e} \cdot \boldsymbol{n}$ can vanish only if $\lambda = 1/2$, it follows that necessary and sufficient conditions that (5.34) has a solution with $\lambda \neq 1/2$ are that (5.39) holds and

$$\operatorname{tr} \boldsymbol{U}_0^2 - \boldsymbol{n} \cdot \boldsymbol{U}_0^2\boldsymbol{n} - \boldsymbol{n} \cdot \operatorname{adj} \boldsymbol{U}_0^2\boldsymbol{n} - 1 > 0. \tag{5.60}$$

From the physical point of view, it is rare to have a material whose measured transformation strain has an eigenvalue equal to 1 within experimental error. We have not been able to find any such examples* in the literature after an extensive search, although even in some relatively common alloys the transforma-

* See Notes added in proof.

tion strain has not been measured. Nevertheless, the transformation strain is generally a continuous function of composition and while the composition of an alloy does not change during the martensitic transformation, it can be adjusted when the crystal is originally grown. Thus, it seems possible that for very special compositions, alloys could be made that have an eigenvalue of $\boldsymbol{U}_0$ equal to 1 in addition to the property (5.43). Such alloys would be interesting because of the great variety of austenite/martensite interfaces possible, *i.e.*, those given by (5.44). However, austenite/martensite interfaces in a cubic-to-tetragonal transformation are only possible when $\boldsymbol{U}_0$ does not have an eigenvalue equal to 1, as is evident from (5.43) applied to (3.1) with either $\eta_1 = 1$ or $\eta_2 = 1$. The possibility mentioned above can occur in cubic-to-orthorhombic transformations, for example.

We now specialize the calculations to the cubic-to-tetragonal transformation and then to the specific case of InTl. Let $\boldsymbol{U}_0$ have the form (3.1) with $\eta_2 \neq 1$ and $\eta_1 \neq 1$ and let $\boldsymbol{a}$ and $\boldsymbol{n}$ be given by the twinning formulas $(3.19)_{1,2}$. Since $\boldsymbol{U}_0$ in this case does not have an eigenvalue equal to 1, we turn to Part I of Theorem 7. There is a solution of the basic equation (5.34) with $\lambda^* \neq 1/2$ if and only if (5.39) and (5.37) are satisfied with strict inequality. This pair of inequalities is equivalent to the two conditions

$$\begin{aligned} &\eta_1 < 1 < \eta_2 \quad \text{and} \quad \frac{1}{\eta_1^2} + \frac{1}{\eta_2^2} < 2, \quad \text{or} \\ &\eta_2 < 1 < \eta_1 \quad \text{and} \quad \eta_1^2 + \eta_2^2 < 2. \end{aligned} \tag{5.61}$$

The inequalities (5.61) delineate the hatched region of Figure 5. There are also solutions for which $\lambda = 1/2$ that occur when the conditions (5.37) and (5.42)

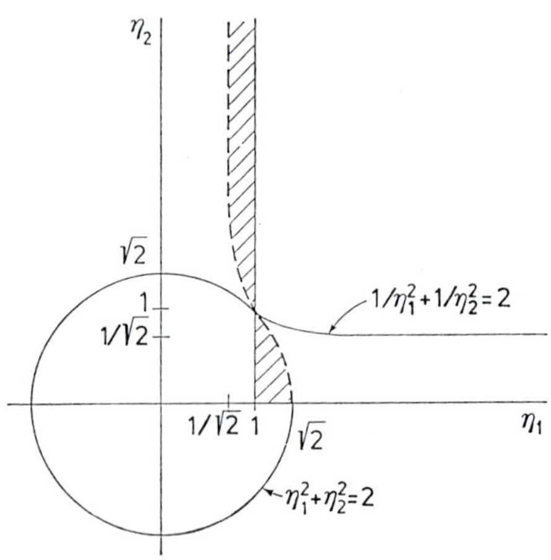


Fig. 5. Values of $\eta_1 > 0$ and $\eta_2 > 0$ for which a solution of (5.34) exists with $\lambda \neq 1/2$ are contained in the hatched region, not including its boundary. Solutions of (5.34) with $\lambda = 1/2$ exist for $\eta_1$ and $\eta_2$ on the dashed curve, not including the point (1, 1).

are satisfied. These conditions are

$$\begin{aligned} &\eta_1 < 1 \quad \text{and} \quad \frac{1}{\eta_1^2} + \frac{1}{\eta_2^2} = 2, \quad \text{or} \\ &\eta_1 > 1 \quad \text{and} \quad \eta_1^2 + \eta_2^2 = 2, \end{aligned} \tag{5.62}$$

which are satisfied on the dashed curves of Figure 5. No solutions are possible when either $\eta_1 = 1$ or $\eta_2 = 1$. We now give formulas for the solutions under the conditions (5.61). The value of $\lambda^*$ is given by (5.41), which becomes

$$\lambda^* = \tfrac{1}{2}\,[1 - (2(\eta_2^2 - 1)\,(\eta_1^2 - 1)\,(\eta_1^2 + \eta_2^2)\,(\eta_2^2 - \eta_1^2)^{-2} + 1)^{1/2}]. \tag{5.63}$$

To complete the calculation we need to find the eigenvalues and eigenvectors of $\boldsymbol{C}_0(\lambda^*)$, according to the last statement of Theorem 7. Once these are found, the vectors $\boldsymbol{b}_{\mathrm{I}}^{\pm}$ and $\boldsymbol{m}_{\mathrm{I}}^{\pm}$ of (5.40) are determined by (5.13). To find the eigenvalues $\lambda_1 \leqq \lambda_2 = 1 \leqq \lambda_3$ of $\boldsymbol{C}_0(\lambda^*)$, it is helpful to note that by (5.32) and (5.33)

$$\begin{aligned} 1 + \lambda_1 + \lambda_3 &= 1 + \eta_1^2 + \eta_1^2\eta_2^2, \\ \lambda_1\lambda_3 &= \eta_1^4\eta_2^2. \end{aligned} \tag{5.64}$$

Hence $\{\lambda_1, \lambda_3\} = \{\eta_1^2, \eta_1^2\eta_2^2\}$. Let $\{\boldsymbol{e}, \bar{\boldsymbol{e}}, \bar{\bar{\boldsymbol{e}}}\}$ be the orthonormal basis introduced in (3.1) and (3.19) to (3.21). To find the eigenvectors of $\boldsymbol{C}_0(\lambda^*)$, notice that by a direct calculation $\bar{\bar{\boldsymbol{e}}}$ is an eigenvector of $\boldsymbol{C}_0(\lambda^*)$ corresponding to the eigenvalue $\eta_1^2$. The remaining two eigenvectors are found by a brute force computation which is simplified by the identity

$$(2\eta_1^2\eta_2^2 - \eta_1^2 - \eta_2^2)\,(\eta_1^2 + \eta_2^2 - 2) = 2(\eta_1^2 - 1)\,(\eta_2^2 - 1)\,(\eta_1^2 + \eta_2^2) + (\eta_2^2 - \eta_1^2)^2. \tag{5.65}$$

It is also helpful to change to the variables $\delta$ and $\tau$ defined by

$$\begin{aligned} \delta &= [(\eta_2^2 + \eta_1^2 - 2)\,(1 - \eta_1^2)^{-1}]^{1/2}, \\ \tau &= [(2\eta_1^2\eta_2^2 - \eta_1^2 - \eta_2^2)\,(1 - \eta_1^2)^{-1}]^{1/2}. \end{aligned} \tag{5.66}$$

Then, in the basis $\{\boldsymbol{e}, \bar{\boldsymbol{e}}, \bar{\bar{\boldsymbol{e}}}\}$, the other two eigenvectors are

$$\frac{1}{(2(\tau^2 + \delta^2))^{1/2}}\,(\tau \pm \delta, \tau \mp \delta, 0). \tag{5.67}$$

Therefore, by Proposition 4 with $\varrho$ chosen to make the third component of $\boldsymbol{m}_{\mathrm{I}}^{\pm}$ equal to 1 for simplicity, we get the formulas

$$\begin{aligned} \boldsymbol{b}_{\mathrm{I}}^{\pm} &= (\mp\tfrac{1}{2}\,\zeta(\delta + \tau), \pm\tfrac{1}{2}\,\zeta(\delta - \tau), \beta), \\ \boldsymbol{m}_{\mathrm{I}}^{\pm} &= (\mp\tfrac{1}{2}\,(\delta + \tau), \pm\tfrac{1}{2}\,(\delta - \tau), 1) \end{aligned} \tag{5.68}$$

where $\delta$ and $\tau$ are given by (5.66) and

$$\begin{aligned} \zeta &= (1 - \eta_1^2)\,(1 + \eta_2)^{-1}, \\ \beta &= \eta_2(\eta_1^2 - 1)\,(1 - \eta_2)^{-1}. \end{aligned} \tag{5.69}$$

The signs are taken in parallel (either all upper or all lower). The expressions for $b_2^{\pm}$ and $m_2^{\pm}$ associated with $(1 - \lambda^*)$ are obtained by changing the signs just before $\tau$ on the right-hand sides of $(5.69)_{1,2}$.

The expressions (5.68) are equivalent to the formulas given by WECHSLER, LIEBERMAN & READ [50] and summarized elsewhere as the crystallographic theory of martensite, which is not based on energy considerations as here. The formulas (5.63) and (5.68) simplify considerably in the case $\operatorname{tr} U_0^2 = 3$ and have been given by ERICKSEN [19]; this case arises naturally from his theory of constrained crystals. With $\eta_1 = 1 - \varepsilon$ and $\eta_2 = 1 + 2\varepsilon$, all quantities become functions of $\varepsilon$ and as $\varepsilon \to 0$, $\lambda^* \to 1/3$ and $m_1^{\pm} \to (\mp 1, 0, 1)$. For InTl, with $\varepsilon = .013$, we get

$$\begin{aligned} \lambda^* &\doteq .338, \\ m_1^{\pm} &\doteq (\mp .993, \pm .0265, 1). \end{aligned} \tag{5.70}$$

This expression for each $m_1^{\pm}$ differs by about $1°$ from a member of the $\{1\ 1\ 0\}$ family of planes. Note that the acute angle between $n$ and $m_1^{\pm}$ is very nearly $60°$, which can be seen from the pictures of BASINSKI & CHRISTIAN [10]. Finally, if we go back and use the full set of symmetry related twins represented by (3.17), we get $24(= 6 \times 4)$ distinct austenite/finely twinned martensite interfaces.

## 6. Surface Energy and Scaling

The analysis of the minimizing sequences yields the observed austenite/martensite interfaces but otherwise has an obvious flaw. That is, the sequences suggest infinite fineness whereas the spacing of the observed twins is small but nonzero. In this section we explore the idea that interfacial energies, which have been excluded from our total free energy, can account for limited fineness without nullifying the overall conclusions we have reached.

Twin boundaries contribute a small free energy, not accounted for by our $\phi$, which is most simply introduced as an energy per unit area assigned on twin boundaries (GIBBS [21, p. 314–328]). For simplicity, we assume that the interfacial free energy per unit area is a constant $\sigma$. This assumption is open to question near places where the twin boundaries meet the boundary of the body or the austenite/martensite interface. Whether or not an interfacial energy should be assigned to the austenite/martensite interface is debatable. (Recall that our calculation of Section 4 does assign a bulk energy to this interface of $O(j^{-1})$, *cf.* equation (4.35).) An electron micrograph of an austenite/martensite interface by NAKANISHI [34] does not suggest a surface of discontinuity of the deformation gradient, so we omit the interfacial energy there.

Since the twin boundaries are essentially plane and parallel, we do not expect that the introduction of interfacial energy will change our expressions for the bulk deformation gradients, stresses or energies. Hence, we assume $F^+$ and $F^-$ are given by the equations $(4.34)_{1,2}$, and we repeat the construction in Theorem 3, except that now we assume that the total free energy is

$$\bar{\mathscr{I}}[y^{(j)}] = \int_\Omega \phi(Dy^{(j)}, \theta_0)\, dx + \sigma A(j), \tag{6.1}$$

$A(j)$ being the total area of twin boundaries in $\boldsymbol{y}^{(j)}(\Omega)$. Here we allow $j$ to take any nonnegative real value, defining $\boldsymbol{y}^{(j)}$ by (4.29) for $j > 0$ and $\boldsymbol{y}^{(0)} = \boldsymbol{y}$ given by (4.30). $A(j)$ implicity depends on $\Omega$ and the deformation $\boldsymbol{y}^{(j)}$. In any case,

$$A(j) \begin{cases} = \text{const.} & \text{at } j = 0, \\ \to \infty & \text{as } j \to \infty. \end{cases} \tag{6.2}$$

With a suitable choice of $\Omega$, $A(j)$ is continuous. The first term in (6.1) tends to zero as $j \to \infty$ and is a continuous function of $j$. Thus $\bar{\mathscr{I}}[\boldsymbol{y}^{(j)}]$ attains its minmum at some $j_0 < \infty$. A quantitative calculation of $j_0$ based on linear elasticity is given by BURKART & READ [13].

In this calculation we have assumed a certain interpolation of the austenite/martensite interface involving the deformation gradients $D^+$ and $D^-$ (see Figure 4). This choice was made so as to simplify the task of showing that each $\boldsymbol{y}^{(j)}(\boldsymbol{x})$ is invertible. It appears that this particular interpolation does not have any special physical significance. Other interpolations might lead to a uniformly lower value of the bulk energy and therefore a higher value of $j_0$. It appears that with any fixed interpolation and realistic free energy functions we would always get limited fineness by this calculation.

It is interesting to speculate on the reasons why some materials that are very similar to InTl do not form internally twinned martensite. The high temperature $A$-15 superconductors (see ERICKSEN [16]) undergo a reversible cubic to tetragonal transformation,* the martensite is found twinned on the $\{1\ 1\ 0\}$ planes, but the austenite and martensite do *not* co-exist at equilibrium as in InTl. In this connection we note that the deformations found in this section are at best metastable with regard to the total energy given by (6.1) since any of the linear deformations $\boldsymbol{y} = \boldsymbol{F}^+\boldsymbol{x}$, $\boldsymbol{y} = \boldsymbol{F}^-\boldsymbol{x}$, $\boldsymbol{y} = \boldsymbol{x}$ have less energy than twinned deformations. Whether the twinned deformations can really be some kind of relative minima of a total energy, which includes interfacial energy in a general way, appears to be a delicate matter.

Another possibly significant fact is that any single crystal of an alloy in solid solution, which is grown from the melt, inevitably contains a slight concentration gradient due to segregation during growth. Also, it is impossible to eliminate completely temperature gradients in a heat bath. Both concentration and temperature gradients can be modelled by explicit dependence of the free energy on $\boldsymbol{x}$. We explore the consequences of this in Section 7(a), where we show that it is a possible mechanism for the initiation of fine twinning.

The reasoning summarized by the total energy (6.1) suggests that places in $\Omega$ where the cross-section

$$\{\boldsymbol{x} \in \Omega_{F^+,F} : \boldsymbol{x} \cdot \boldsymbol{n} = \text{const.}\} \tag{6.3}$$

has a small area should contain a larger twin density than in regions where this area is large. To explore this suggestion via a heuristic calculation, we consider a specimen with uniform length and width but with a variable thickness and suppose an austenite/martensite interface divides the specimen perpendicular to the thickness direction. Assume $\boldsymbol{m} \wedge \boldsymbol{n}$ is parallel to the width. We use the

* However, it is not clear that this transformation is really of 1st order ($\boldsymbol{U}_0 \neq \boldsymbol{1}$).

calculation of bulk energy given in Section 4 directly and therefore ignore the fact that with limited fineness the boundary of the body will not quite be free of traction. The bulk energy per unit length is given approximately by (const. $j^{-1}$) while the surface energy per unit length is given approximately by (const. $jA$), $A$ being the area of the set given in (6.3) and the constants being positive. The total energy per unit length is therefore

$$\frac{\text{const.}}{j} + \text{const. } Aj \tag{6.4}$$

which is minimized among all $j > 0$ at

$$j = \text{const. } A^{-1/2}. \tag{6.5}$$

Since $j^{-1}$ is proportional to the twin spacing (see Figure 4), equation (6.5) gives an inverse square root relation between the fineness and the cross-sectional area. Apparently, experiments to test this relation are not available.

OTSUKA & SHIMIZU [37] observe that "the reason for the absence of internal twins in 'small' martensites is not known presently". In the spirit of the calculation just above, consider a cube of side $L$ divided by an austenite/martensite interface which is parallel to a pair of faces. Again by use of expressions from Section 4 and the total energy (6.1), the bulk energy in this situation is approximately (const. $L^2 j^{-1}$) while the surface energy is approximately (const. $L^3 j$), the constants being positive and independent of $L$. The total energy is minimized as a function of $j > 0$ when

$$j = \text{const. } L^{-1/2}. \tag{6.6}$$

The twin spacing is therefore proportional to $L^{1/2}$. However, the cube is of side $L$. Hence, if $L$ is sufficiently small the twin spacing will be larger than a side of the cube, suggesting that small crystals containing both austenite and martensite will not be stable, as is observed.

## 7. Other Similar Phenomena

### *a. Fine twins in a problem with no absolute minimizer*

The minimization problem we have studied so far has linear absolute minima in addition to the minimizing sequences described in Section 4. By allowing the free energy $\phi$ to depend explicitly on $\boldsymbol{x}$, we now construct a similar example in which the total free energy does not have an absolute minimizer in $W^{1,1}(\Omega, \mathbb{R}^3)$.

One way to think of doing this is to put the body in a temperature gradient so as to introduce dependence on $\boldsymbol{x}$ through the composition $\phi(\boldsymbol{F}, \theta(\boldsymbol{x}))$. Within the context of a thermodynamic theory based on the Planck inequality and a Fourier Law of heat conduction, BALL & KNOWLES [9] justify the criterion of stability

$$\min_{y} \int_{\Omega} \phi(D\boldsymbol{y}(\boldsymbol{x}), \theta(\boldsymbol{x}))\, d\boldsymbol{x} \tag{7.1}$$

as appropriate for an unloaded body having a steady temperature distribution $\theta(\boldsymbol{x})$.

To avoid technicalities, we ignore the Galilean invariance of the free energy and consider a linear temperature distribution. Specifically, let a smooth function $\tilde{\phi}: \mathscr{D} \times \mathbb{R} \to \mathbb{R}$ satisfy for each $\theta < 0$

$$\tilde{\phi}(\boldsymbol{F}, \theta) > \tilde{\phi}(\boldsymbol{F}^+, \theta) = \tilde{\phi}(\boldsymbol{F}^-, \theta) \quad \text{for } \boldsymbol{F} \neq \boldsymbol{F}^{\pm}, \tag{7.2}$$

while for $\theta > 0$, let

$$\tilde{\phi}(\boldsymbol{F}, \theta) > \tilde{\phi}(\boldsymbol{1}, \theta) \quad \text{for } \boldsymbol{F} \neq \boldsymbol{1}. \tag{7.3}$$

Assume $\boldsymbol{F}^{\pm}$ are given by (4.34) for appropriate choices of the vectors. As done there, assume $\boldsymbol{F}^{\pm} - \boldsymbol{1}$ is not a rank-one matrix. Suppose $\boldsymbol{o} \in \Omega$. Let

$$\tilde{\phi}(\boldsymbol{F}, \boldsymbol{x}) = \tilde{\phi}(\boldsymbol{F}, \boldsymbol{x} \cdot \boldsymbol{m}) \tag{7.4}$$

and consider the total free energy

$$\mathscr{I}[\boldsymbol{y}] = \int_{\Omega} \tilde{\phi}(D\boldsymbol{y}(\boldsymbol{x}), \boldsymbol{x})\, d\boldsymbol{x}. \tag{7.5}$$

As before let $\mathscr{A} = \{\boldsymbol{y} \in W^{1,1}(\Omega, \mathbb{R}^3): D\boldsymbol{y} \in \mathscr{D} \text{ a.e.}\}$. Then

$$\inf_{\boldsymbol{y} \in \mathscr{A}} \mathscr{I}[\boldsymbol{y}] = \int_{\Omega \cap \{\boldsymbol{x} \cdot \boldsymbol{m} < 0\}} \tilde{\phi}(\boldsymbol{F}^{\pm}, \boldsymbol{x} \cdot \boldsymbol{m})\, dx + \int_{\Omega \cap \{\boldsymbol{x} \cdot \boldsymbol{m} > 0\}} \tilde{\phi}(\boldsymbol{1}, \boldsymbol{x} \cdot \boldsymbol{m})\, d\boldsymbol{x},$$

and examples of minimizing sequences are given by the family of functions $\boldsymbol{y}^{(j)}(\boldsymbol{x})$ constructed in Theorem 3.

We claim that the absolute minimum of $\mathscr{I}$ in $\mathscr{A}$ is not attained. This in fact follows from Theorem 3. That is, any absolute minimizer $\boldsymbol{y} \in \mathscr{A}$ must have the property

$$D\boldsymbol{y} = \boldsymbol{F}^+ \quad \text{or} \quad \boldsymbol{F}^- \text{ a.e. on } \Omega \cap \{\boldsymbol{x} \cdot \boldsymbol{m} < 0\} \tag{7.6}$$

and the property

$$D\boldsymbol{y} = \boldsymbol{1} \text{ a.e. on } \Omega \cap \{\boldsymbol{x} \cdot \boldsymbol{m} \geqq 0\}. \tag{7.7}$$

Any such $\boldsymbol{y}$ belongs to $W^{1,\infty}(\Omega, \mathbb{R}^3)$ and is therefore continuous on $\Omega$. Let $B$ be an open ball with center $\boldsymbol{o}$ and contained in $\Omega$. On $\Omega \cap \{\boldsymbol{x} \cdot \boldsymbol{m} \geqq 0\}$ any such $\boldsymbol{y}$ satisfies

$$\boldsymbol{y}(\boldsymbol{x}) = \boldsymbol{x} + \boldsymbol{c}_1, \tag{7.8}$$

for some $\boldsymbol{c}_1 = \text{const.}$, while on $E = \Omega \cap \{\boldsymbol{x} \cdot \boldsymbol{m} \leqq 0\}$ $\boldsymbol{y}$ is given by the expression

$$\boldsymbol{y}(\boldsymbol{x}) = \boldsymbol{c}_0 + \boldsymbol{F}^- \boldsymbol{x} + f_E(\boldsymbol{x} \cdot \boldsymbol{n})\, \boldsymbol{c} \tag{7.9}$$

according to (4.4), where $f_E$ is Lipschitz with derivative 0 or 1 a.e. and $\boldsymbol{c}_0 = \text{const.}$ However, it is easily seen using (4.34) that no continuous $\boldsymbol{y}$ satisfies both (7.9) and (7.8) on $\Omega \cap \{\boldsymbol{x} \cdot \boldsymbol{m} = 0\}$. Hence there are no absolute minimizers of $\mathscr{I}$ in $\mathscr{A}$.

Similar arguments show that for appropriate boundary conditions of place, say $\boldsymbol{y}(\boldsymbol{x}) = \boldsymbol{x} + \boldsymbol{b}(\boldsymbol{x} \cdot \boldsymbol{m})$ for $\boldsymbol{x} \in \partial\Omega$, $\boldsymbol{x} \cdot \boldsymbol{m} < 0$, a minimizer of the original problem (3.11) fails to exist.

*b. Strongly elliptic energies with minimizers having fine boundary wrinkles*

We now consider free energy functions for isotropic, $n$-dimensional elastic materials of the type analyzed by BALL [6, Section 6.4] and BALL & MARSDEN [7]. Let $1 < \alpha < n$, $0 < \lambda < \mu < \infty$. Let $\phi_1 : (0, \infty) \to (0, \infty)$ be a smooth function satisfying

$$\phi_1' > 0, \qquad \phi_1'' > 0, \tag{7.10}$$

and

$$\phi_1(\gamma) = \gamma^\alpha, \qquad \lambda \leqq \gamma \leqq \mu. \tag{7.11}$$

Now choose a smooth function $\phi_2 : (0, \infty) \to \mathbb{R}$ with the properties

$$\begin{aligned} &\phi_2'' > 0, \\ &\phi_2(\tau) = -n\tau^{\alpha/n}, \qquad && \tau \in [\lambda^n, \mu^n], \\ &\phi_2(\tau) > -n\phi_1(\tau^{1/n}), \qquad && \tau \notin [\lambda^n, \mu^n]. \end{aligned} \tag{7.12}$$

There are functions $\phi_2$ satisfying (7.12) because $(\tau^{\alpha/n})'' < 0$ for $\tau \in [\lambda^n, \mu^n]$. Define

$$\phi_3(\gamma_1, \dots, \gamma_n) = \sum_{i=1}^n \phi_1(\gamma_i) + \phi_2\left(\prod_{i=1}^n \gamma_i\right) \tag{7.13}$$

and

$$\phi(F) = \phi_3(\gamma_1, \dots, \gamma_n), \tag{7.14}$$

where $\gamma_i = \gamma_i(F)$, $i = 1, \dots, n$, are the eigenvalues of $(F^TF)^{1/}$ .

From (7.10) it follows that $\phi_1$ is strictly convex in $\log \gamma$, and hence that the minimum of $\sum_{i=1}^n \phi_1(\gamma_i)$ subject to $\gamma_i > 0$, $\prod_{i=1}^n \gamma_i = \tau$, is attained exactly when $\gamma_1 = \gamma_2 = \dots = \gamma_n = \tau^{1/n}$. Since by construction the nonnegative function $n\phi_1(\tau^{1/n}) + \phi_2(\tau)$ is zero only when $\tau \in [\lambda^n, \mu^n]$, we have shown that the absolute minima of $\phi$ subject to $\det F > 0$ are given precisely by $F$ with

$$\gamma_1 = \gamma_2 = \dots = \gamma_n = \gamma \in [\lambda, \mu]. \tag{7.15}$$

We have thus constructed a free energy function $\phi$ for an isotropic nonlinear elastic material which has a continuous line of absolute minimizers at dilatations. Surprisingly, $\phi$ is also strictly polyconvex (see BALL [6]) and strongly elliptic, unlike our energies for crystals. Furthermore, by suitably choosing $\phi_1$ and $\phi_2$, $\phi(F)$ can be chosen to grow as fast as desired as $|F| \to \infty$.

Consider the problem appropriate to an unloaded body,

$$\min_{y \in \mathscr{A}} \int_\Omega \phi(Dy(x))\, dx, \tag{7.16}$$

where $\Omega \subset \mathbb{R}^n$ is bounded and open and $\mathscr{A} = \{y \in W^{1,1}(\Omega, \mathbb{R}^n) : \det Dy(x) > 0$ a.e. $x \in \Omega\}$. The absolute minimizers for this problem are those $y \in \mathscr{A}$ having the property that

$$(Dy(x))^T Dy(x) = \gamma(x)^2\, \mathbf{1} \qquad \text{a.e. } x \in \Omega, \tag{7.17}$$

where $\gamma(\boldsymbol{x}) \in [\lambda, \mu]$ a.e.. Equivalently,

$$Dy(\boldsymbol{x}) = \gamma(\boldsymbol{x})\, \boldsymbol{R}(\boldsymbol{x}) \quad \text{a.e. } \boldsymbol{x} \in \Omega, \tag{7.18}$$

for some rotation-valued measurable function $\boldsymbol{R}(\boldsymbol{x})$. The condition (7.18) says that $\boldsymbol{y}$ is *conformal.*

We first consider the case $n = 2$. The conformal mappings are representable by analytic functions $w = f(z)$ with the correspondence $\gamma(x_1, x_2) = |f'(x_1 + ix_2)|$. Thus any function analytic in $\Omega$ and such that $\lambda \leqq |f'(z)| \leqq \mu$ generates an absolute minimizer of the problem (7.16). We first take the example

$$f(z) = \gamma z + \varepsilon \varrho e^{iz/\varepsilon}, \tag{7.19}$$

with $\Omega = (0, 1)^2$ and $\gamma \in (\lambda, \mu)$. We pick $\varrho > 0$ sufficiently small that $\gamma \in [\lambda + \varrho, \mu - \varrho]$, and take $\varepsilon > 0$. Since $|e^{iz/\varepsilon}| = |e^{-x_2/\varepsilon} e^{ix_1/\varepsilon}| \leqq 1$ for $x_2 \geqq 0$ we have that $\lambda \leqq |f'(z)| \leqq \mu$ in $\Omega$. Note that $f$ is invertible for $\operatorname{Re} z \geqq 0$, since $|e^{iz} - e^{iw}| \leqq |z - w|$ for $\operatorname{Re} z$, $\operatorname{Re} w \geqq 0$ and $\gamma > \varrho$. The mapping $\boldsymbol{y}^\varepsilon$ corresponding to (7.19) is given by

$$\begin{aligned} y_1^\varepsilon &= \gamma x_1 + \varepsilon \varrho e^{-x_2/\varepsilon} \cos(x_1/\varepsilon), \\ y_2^\varepsilon &= \gamma x_2 + \varepsilon \varrho e^{-x_2/\varepsilon} \sin(x_1/\varepsilon). \end{aligned} \tag{7.20}$$

As $\varepsilon \to 0$, $\boldsymbol{y}^\varepsilon$ has finer and finer oscillations near $x_2 = 0$. Note that $\boldsymbol{y}^\varepsilon \to \gamma \boldsymbol{x}$ as $\varepsilon \to 0$ strongly in $W^{1,p}(\Omega, \mathbb{R}^2)$ for $1 \leqq p < \infty$, but that $\boldsymbol{y}^\varepsilon|_{x_2=0}$ only converges weakly to $\gamma \boldsymbol{x}$ in $W^{1,p}((0, 1), \mathbb{R}^2)$ (weak * if $p = \infty$). As a second example, we let

$$f(z) = z^{1+i}, \tag{7.21}$$

taking the principal value, with $\Omega$ the unit disc centered at $z = i$. Thus with suitable choices of $\lambda, \mu$ we obtain a minimizer $\boldsymbol{y} \in W^{1,\infty}(\Omega, \mathbb{R}^2)$ which is smooth except at $\boldsymbol{o} \in \Omega$ where there is a spiral singularity with $Dy$ discontinuous. It is easily shown that $\boldsymbol{y}$ is invertible on $\bar{\Omega}$. It is probably significant that the equilibrium equations, when linearized about the deformation $\boldsymbol{y} = \gamma \boldsymbol{x}$, $\gamma \in [\lambda, \mu]$, fail to satisfy the complementing condition of AGMON, DOUGLIS & NIRENBERG [2] with respect to boundary conditions of null traction. SIMPSON & SPECTOR [43] discuss in detail exactly this linearized problem within the context of elasticity theory. See also SIMPSON & SPECTOR [44] for a discussion of the complementing condition in the context of nonlinear elasticity.

We turn to the case $n \geqq 3$. The conformal transformations are now characterized by Liouville's theorem as products of inversions. Under our regularity assumptions (*i.e.*, $\boldsymbol{y} \in W^{1,\infty}(\Omega, \mathbb{R}^3)$ by (7.18)) an appropriate version of Liouville's theorem has been proved by RESHETNYAK [42]. For $n$ odd an example is given by

$$\boldsymbol{y}(\boldsymbol{x}) = -\frac{\boldsymbol{x}}{|\boldsymbol{x}|^2}. \tag{7.22}$$

If $\boldsymbol{o} \in \bar{\Omega}$, then $\boldsymbol{y}$ satisfies (7.18) with $\gamma(\boldsymbol{x}) = |\boldsymbol{x}|^{-2}$. Note that when $\Omega$ is convex this furnishes an example of a nontrivial deformation which is an absolute minimizer of the total energy for a strictly polyconvex isotropic unloaded material, and thus bears on a conjecture of NOLL [36] (see also TRUESDELL [47]) to the effect that for rubber-like materials, the absolute minimum is homogeneous and unique up to rigid-body translation and rotation.

### c. *Minimizers of energy having a finer and finer mixture of phases as an interface is approached from one side*

VAN TENDELOO, VAN LANDUYT & AMELINCKX [48, Figure 8] show arrays of triangular Dauphiné twins in quartz which get finer and finer in the direction of increasing temperature. This suggests another phenomenon whereby compatibility at an interface is achieved by mixing different deformations in triangles which themselves get finer and finer as the interface is approached from one side.

We choose a simplified free energy function which is not intended to model the behavior of quartz. Also, we ignore the temperature gradient. A free energy function which accounts for the $\alpha$-$\beta$ transformation and Dauphiné twinning has been given by JAMES [23].

Let $\{e_1, e_2, e_3\}$ be an orthonormal basis and consider a Galilean invariant free energy function $\phi(F)$ whose point group contains a 180° rotation about $e_1$. That is, for all $F$ with $\det F > 0$, assume

$$\phi(F) = \phi(U)\big|_{U=(F^TF)^{1/2}} = \phi(RUR^T) \tag{7.23}$$

with $R = -1 + 2e_1 \otimes e_1$. Let $\varepsilon \in (0, 1)$ and $\gamma \neq 0$ be given constants. Assume that $\phi(U)$ has absolute minima at the three positive symmetric matrices $U_1$, $U_2$, $U_3$ where

$$\begin{aligned} U_1^2 &= (\tfrac{1}{4}\varepsilon^2)\, e_1 \otimes e_1 + (\tfrac{1}{2}\varepsilon)\,(e_1 \otimes e_2 + e_2 \otimes e_1) + 1, \\ U_2^2 &= (\tfrac{1}{4}\varepsilon^2)\, e_1 \otimes e_1 + (\tfrac{1}{2}\varepsilon)\,(1-\varepsilon)\,(e_1 \otimes e_2 + e_2 \otimes e_1) + \varepsilon(\varepsilon - 2)\, e_2 \otimes e_2 + 1, \\ U_3^2 &= \gamma(e_1 \otimes e_2 + e_2 \otimes e_1) + \gamma^2 e_2 \otimes e_2 + 1. \end{aligned} \tag{7.24}$$

Referring to Figure 6(a), consider a deformation $\tilde{y}(x)$ defined on the indicated reference configuration $\Omega$ by

$$\tilde{y}(x) = \begin{cases} x & \text{at circled nodes}, \\ x + 2^{-(i+1)}\varepsilon e_2 & \text{at uncircled nodes } x \text{ satisfying} \\ & \quad x \cdot e_2 = (1 - 2^{-i}),\ i = 0, 1, 2, \ldots, \\ x + \gamma e_1[(x \cdot e_2) - 1] & \text{for } x \cdot e_2 > 1, \end{cases} \tag{7.25}$$

and which is linear in each triangle. Assume also $D\tilde{y}e_3 = e_3$. Constructed in this way $D\tilde{y}$ has a.e. only the five distinct values

$$\begin{aligned} F_1 &= 1 + \gamma e_1 \otimes e_2, \\ F_2 &= 1 + \tfrac{1}{2}\varepsilon e_2 \otimes e_1, \\ F_3 &= 1 - \tfrac{1}{2}\varepsilon e_2 \otimes e_1, \\ F_4 &= 1 + \tfrac{1}{2}\varepsilon e_2 \otimes e_1 - \varepsilon e_2 \otimes e_2, \\ F_5 &= 1 - \tfrac{1}{2}\varepsilon e_2 \otimes e_1 - \varepsilon e_2 \otimes e_2. \end{aligned} \tag{7.26}$$

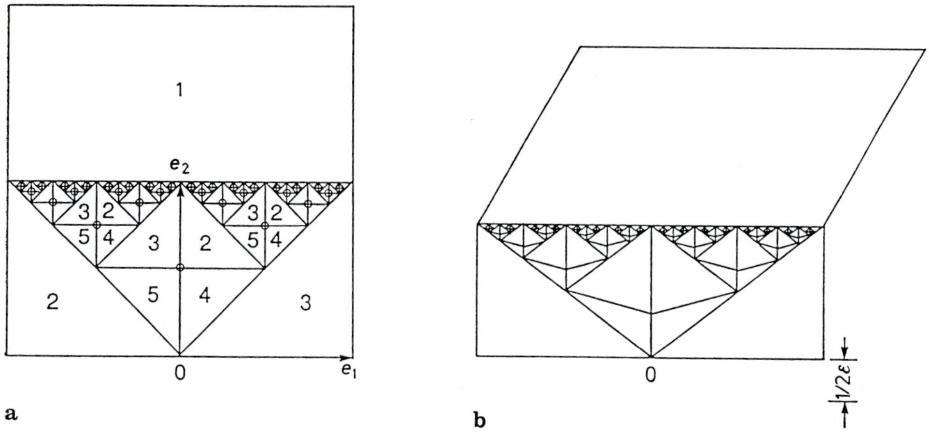


Fig. 6a and b. Compatibility at an interface achieved by a fine mixture of five deformation gradients. a Reference configuration; b Deformed configuration

The subscripts on the deformation gradients in (7.26) correspond to the numbered regions in Figure 6a. Clearly $\tilde{\boldsymbol{y}}(\boldsymbol{x})$ is in $W^{1,\infty}(\Omega, \mathbb{R}^3)$ and is globally invertible.

The function $\tilde{\boldsymbol{y}}$ we have constructed is an absolute minimizer of the total free energy

$$\int_{\Omega} \phi(D\boldsymbol{y})\, d\boldsymbol{x} \tag{7.27}$$

in the class $W^{1,1}(\Omega, \mathbb{R}^3)$ because

$$\begin{aligned} \boldsymbol{F}_1^T\boldsymbol{F}_1 &= \boldsymbol{U}_3^2, \\ \boldsymbol{F}_2^T\boldsymbol{F}_2 &= \boldsymbol{R}^T\boldsymbol{F}_3^T\boldsymbol{F}_3\boldsymbol{R} = \boldsymbol{U}_1^2, \\ \boldsymbol{F}_4^T\boldsymbol{F}_4 &= \boldsymbol{R}^T\boldsymbol{F}_5^T\boldsymbol{F}_5\boldsymbol{R} = \boldsymbol{U}_2^2. \end{aligned} \tag{7.28}$$

However, it fails to satisfy the classical conditions of compatibility because

$$\boldsymbol{F}_1 - \boldsymbol{F}_i \neq \text{a rank-one matrix}, \quad i = 2, \ldots, 5. \tag{7.29}$$

As in the example of finely twinned martensite, here compatibility is achieved by mixing a fine distribution of "phases" near an interface. However, here the minimum is achieved in $W^{1,\infty}(\Omega, \mathbb{R}^3)$ by the fine phase mixture rather than merely approached by a minimizing sequence.

The example raises the question of what are the conditions of compatibility at an "interface". In this regard we note that both the example presented in this section and the one involving fine twins have the property that

$$D\tilde{\boldsymbol{y}}(\boldsymbol{x}_2) - D\tilde{\boldsymbol{y}}(\boldsymbol{x}_1) = \text{a rank-two matrix} \tag{7.30}$$

for almost every $\boldsymbol{x}_1$ and $\boldsymbol{x}_2$ on opposite sides of what one would think of as the interface. We also note that in the example of this section

$$\left(\frac{1}{4}\sum_{i=2}^{5} \boldsymbol{F}_i\right) - \boldsymbol{F}_1 = \text{a rank-one matrix}, \tag{7.31}$$

an analog of the equation

$$(\lambda F^{+} + (1 - \lambda) F^{-}) - \mathbf{1} = \boldsymbol{b} \otimes \boldsymbol{m} \tag{7.32}$$

found in Section 4.

The example also raises the question of what are the fewest number of deformation gradients of a function in $W^{1,\infty}(\Omega, \mathbb{R}^3)$ such that at least one of them does not differ from any of the others by a rank-one matrix. We conjecture that the answer is four and we are pursuing this and related questions in a further study of fine phase mixtures.

*Acknowledgements*. We thank J. W. CAHN for telling us about reference [48].

The research of J.M.B. was supported by an SERC Senior Fellowship and by the Institute of Mathematics and its Applications, University of Minnesota. The research of R.D.J. was supported by an SERC Visiting Fellowship and by the National Science Foundation under Grant No. MSM-8612420.

*Note 1 added in proof.* Professor J. W. CHRISTIAN has shown us an example of a cubic to orthorhombic transformation in titanium-tantalum alloys which exhibits an exact interface between cubic and orthorhombic phases (K. A. BYWATER & J. W. CHRISTIAN, Martensitic transformations in titanium-tantulum alloys, Phil. Mag. **25** (1972), p. 1249-1272). To get the exact interfaces, these authors adjusted the concentration of tantulum during preparation of the alloy in order to make one eigenvalue of the transformation strain equal to 1, while maintaining the condition that the other two eigenvalues are greater and less than 1.

*Note 2 added in proof.* By Proposition 4 the conditions in Part II of Theorem 1 that $U_0$ have an eigenvalue equal to 1 and that $\mu^* > 0$ are also sufficient that there be an exact austenite/martensite interface.

## References

1. R. A. ADAMS, *Sobolev Spaces*. New York: Academic Press, 1975.
2. S. AGMON, A. DOUGLIS & L. NIRENBERG, Estimates near the boundary for solutions of elliptic partial differential equations satisfying general boundary conditions, II. Comm. Pure Appl. Math. **17** (1964), 35–92.
3. J. M. BALL, Convexity conditions and existence theorems in nonlinear elasticity. Arch. Rational Mech. Anal. **63** (1977), 337–403.
4. J. M. BALL, Strict convexity, strong ellipticity, and regularity in the calculus of variations. Math. Proc. Camb. Phil. Soc. **87** (1980), 501–513.
5. J. M. BALL, J. C. CURRIE & P. J. OLVER, Null Lagrangians, weak continuity and variational problems of arbitrary order. J. Functional Anal. **41** (1981), 135–174.
6. J. M. BALL, Discontinuous equilibrium solutions and cavitation in nonlinear elasticity. Phil. Trans. Roy. Soc. London A **306** (1982), 557–611.
7. J. M. BALL & J. E. MARSDEN, Quasiconvexity, positivity of the second variation, and elastic stability. Arch. Rational Mech. Anal. **86** (1984), 251–277.
8. J. M. BALL & F. MURAT, $W^{1,p}$-quasiconvexity and variational problems for multiple integrals. J. Funct. Anal. **58** (1984), 225–253.
9. J. M. BALL & G. KNOWLES, Liapunov functions for thermomechanics with spatially varying boundary temperatures. Arch. Rational Mech. Anal. **92** (1986), 193–204.

10. Z. S. Basinski & J. W. Christian, Crystallography of deformation by twin boundary movements in indium-thallium alloys. Acta Met. **2** (1954), 101–116; also, Experiments on the martensitic transformation in single crystals of indium-thallium alloys. Acta Met. **2** (1954), 148–166.
11. J. S. Bowles & J. K. MacKenzie, The crystallography of martensitic transformations I and II. Acta Met. **2** (1954), 129–137, 138–147.
12. D. A. G. Bruggeman, Berechnung verschiedener physikalischer Konstanten, von Heterogenen Substanzen. Ann. Phys. **5** (1935), 636–664.
13. M. W. Burkart & T. A. Read, Diffusionless phase change in the indium-thallium system. Trans. AIME J. Metals **197** (1953), 1516–1524.
14. M. Chipot & D. Kinderlehrer, Equilibrium configurations of crystals, to appear.
15. J. W. Christian, *The Theory of Transformations in Metals and Alloys.* Pergamon Press, 1975.
16. J. L. Ericksen, Some phase transitions in crystals. Arch. Rational Mech. Anal. **73** (1980), 99–124.
17. J. L. Ericksen, Constitutive theory for some constrained elastic crystals. IMA Preprint # 123, Institute for Mathematics and its Applications, University of Minnesota.
18. J. L. Ericksen, Some surface defects in unstressed thermoelastic solids. Arch. Rational Mech. Anal. **88** (1985), 337–345.
19. J. L. Ericksen, Some Constrained Elastic Crystals, in *Material Instabilities in Continuum Mechanics*, (ed. J. M. Ball). Oxford University Press, to appear.
20. I. Fonseca, Variational methods for elastic crystals. Arch. Rational Mech. Anal., **97** (1987), 189–220.
21. J. W. Gibbs, On the equilibrium of heterogeneous substances, in *The Scientific Papers of J. Willard Gibbs*, Vol. 1. Dover Publications, New York, 1961.
22. M. E. Gurtin, Two-phase deformations of elastic solids. Arch. Rational Mech. Anal. **84** (1983), 1–29.
23. R. D. James, Finite deformation by mechanical twinning. Arch. Rational Mech. Anal. **77** (1981), 143–176.
24. R. D. James, The stability and metastability of quartz, in *Metastability and Incompletely Posed Problems*, IMA, Vol. 3 (ed. S. Antman, J. L. Ericksen, D. Kinderlehrer & I. Müller) Springer-Verlag, 1987, 147-176.
25. R. D. James, Displacive phase transformations in solids. J. Mech. Phys. Solids **34** (1986), 359–394.
26. D. Kinderlehrer, Remarks about equilibrium configurations of crystals, in *Material Instabilities in Continuum Mechanics*, (ed. J. M. Ball). Oxford University Press, to appear.
27. B. Klosowicz & K. A. Lurie, On the optimal nonhomogeneity of a torsional elastic bar. Arch. of Mech. **24** (1971), 239–249.
28. R. V. Kohn & G. Strang, Explicit relaxation of a variational problem in optimal design. Bull. Amer. Math. Soc. **9** (1983), 211–214.
29. N. A. Lavrov, K. A. Lurie & A. V. Cherkaev, Nonuniform rod of extremal torsional stiffness. Mech. of Sol. **15** (1980), 74–80.
30. K. A. Lurie, A. V. Cherkaev & A. V. Fedorov, Regularization of optimal design problems for bars and plates I, II. J. Opt. Th. Appl. **37** (1982), 499–522 and 523–543; also, On the existence of solutions to some problems of optimal design for bars and plates. J. Opt. Th. Appl. **42** (1984), 247–281.
31. M. Marcus & V. J. Mizel, Absolute continuity on tracks and mappings of Sobolev spaces. Arch. Rational Mech. Anal. **45** (1972), 294–320.
32. G. W. Milton, Modelling the properties of composites by laminates, in *Proc.*

*Workshop on Homogenization and Effective Moduli of Materials and Media*, 1984, to appear.

33. F. MURAT & L. TARTAR, Calcul des variations et homogénéisation, in *Ecole d'été d'homogénéisation*, Bureau sans Nappe, Eyrolles, Paris, July 1983, to appear.
34. N. NAKANISHI, Characteristics of stress-strain behaviour associated with thermoelastic martensitic transformation. Arch. Mech. **35** (1983), 37–62.
35. Z. NISHIYAMA, *Martensitic Transformation*. Academic Press, 1978.
36. W. NOLL, A general framework for problems in the statics of finite elasticity, in *Contemporary Developments in Continuum Mechanics and Partial Differential Equations* (ed. G. M. DE LA PENHA & L. A. MEDEIROS), North-Holland, 1978.
37. K. OTSUKA & K. SHIMIZU, Morphology and crystallography of thermoelastic $\gamma'$ Cu–Al–Ni martensite. Japanese J. Appl. Phys. **8** (1969), 1196–1204.
38. A. C. PIPKIN, Some examples of crinkles, in *Homogenization and Effective Moduli of Materials and Media*. IMA, Vol. 1 (ed. J. L. ERICKSEN, D. KINDERLEHRER, R. KOHN & J. L. LIONS), Springer-Verlag, 1986.
39. U. E. RAITUM, The extension of extremal problems connected with a linear elliptic equation. Soviet Math. Dokl. **19** (1978), 1342–1345.
40. U. E. RAITUM, On optimal control problems for linear elliptic equations. Soviet Math. Dokl. **20** (1979), 129–132.
41. YU. G. RESHETNYAK, On the stability of conformal mappings in multidimensional spaces. Siberian Math. J. **8** (1967), 69–85.
42. YU. G. RESHETNYAK, Liouville's theorem on conformal mappings under minimal regularity assumptions. Siberian Math. J. **8** (1967), 631–653.
43. H. C. SIMPSON & S. J. SPECTOR, On failure of the complementing condition and non-uniqueness in linear elastostatics. J. Elasticity **15** (1985), 229–231.
44. H. C. SIMPSON & S. J. SPECTOR, On the sign of the second variation in finite elasticity, Arch. Rational Mech. Anal. **98** (1987), 1–30.
45. L. TARTAR, The compensated compactness method applied to systems of conservation laws, in *Systems of nonlinear partial differential equations* (ed. J. M. BALL). Reidel, 1983.
46. L. TARTAR, Étude des oscillations dans les équations aux dérivées partielles non-linéaires, in *Lecture Notes in Physics 195*. Springer-Verlag, 1984, 384–412.
47. C. TRUESDELL, Some challenges offered to analysis by rational thermomechanics in *Contemporary Developments in Continuum Mechanics and Partial Differential Equations* (ed. G. M. DE LA PENHA & L. A. J. MEDEIROS), North-Holland, 1978.
48. G. VAN TENDELOO, J. VAN LANDUYT & S. AMELINCKX, The $\alpha$-$\beta$ phase transition in quartz and $AlPO_4$ as studied by electron microscopy and diffraction. Phys. Stat. Sol. a**33** (1976), 723–735.
49. C. M. WAYMAN, *Introduction to the Crystallography of Martensitic Transformations*. MacMillan, 1964.
50. M. S. WECHSLER, D. S. LIEBERMAN & T. A. READ, On the theory of the formation of martensite. Trans. AIME J. Metals, **197** (1953), 1503–1515.
51. H. WEYL, *The Classical Groups*. Princeton, 1946.
52. L. C. YOUNG, *Lectures on the Calculus of Variations and Optimal Control Theory*. Chelsea, 1980.

Heriot-Watt University
Edinburgh

University of Minnesota
Minneapolis

*(Received April 20, 1987)*

# *Fit Regions and Functions of Bounded Variation*

WALTER NOLL & EPIFANIO G. VIRGA

*Dedicated to* JAMES B. SERRIN *on the occasion of his sixtieth birthday*

## Contents

Introduction . . . . . 1
1. Functions of bounded variation . . . . . 4
2. Mollifications . . . . . 6
3. Two basic theorems about functions of bounded variation . . . . . 9
4. Regularly open sets and sets of finite perimeter . . . . . 12
5. Fit regions . . . . . 14
6. Reduced boundaries . . . . . 16
7. Counterexamples . . . . . 18
References . . . . . 21

## Introduction

Ever since one of us (W. N.) began the attempt to analyze some of the basic concepts of continuum physics in terms of precise mathematical structures about thirty years ago (see [N1]), it was clear that a concept of "fit region" in a Euclidean space was needed. Only such "fit regions" should be sets fit to be occupied by continuous bodies and their subbodies.

At first, it seemed that one could define a "fit region" as a set with piecewise smooth boundary. But then the union or intersection of two fit regions would not necessarily be a fit region. For example, if we define the subsets $\mathscr{C}$ and $\mathscr{D}$ of $\mathbb{R}^2$ by (see Figure 1):

$$\mathscr{C} := \{(s, t) \in \mathbb{R}^2 \mid t \in ]0, 1[, s \in ]0, 1[\}$$

and

$$\mathscr{D} := \left\{(s, t) \in \mathbb{R}^2 \,\middle|\, t \in ]0, 1[, s \in ]-1, e^{-\frac{1}{t^2}} \sin\left(\frac{2}{t}\right)\right\}$$

then $\mathscr{C}$ and $\mathscr{D}$ have piecewise-$C^\infty$ boundaries but $\mathscr{C} \cap \mathscr{D}$ consists of infinitely many pieces and can certainly not be regarded as a set with a piecewise smooth boundary.

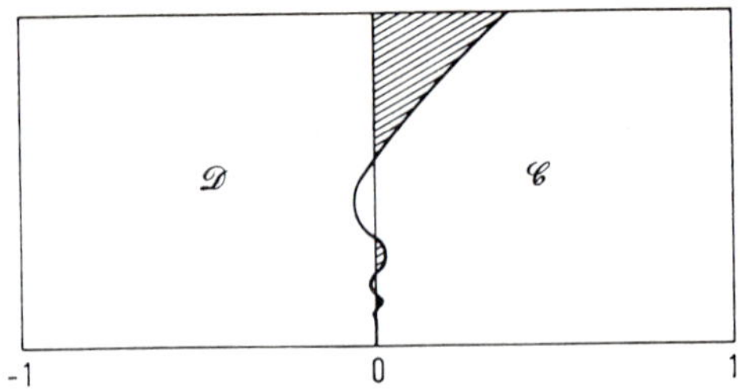


Fig. 1

As was pointed in [N4] the class of fit regions should meet certain requirements which ensure that what is usually done in continuum physics can be justified in precise mathematical terms. First, the set of all fit subregions of a given fit region should satisfy the axioms of a "material universe" as described in [N3], Appendix, or in [T], Chapter I, Sects. 2 and 3, or in [N2] and [N4]. Second, the class of fit regions should be invariant under *transplacements*, which should include adjustments to fit regions of smooth diffeomorphisms from one Euclidean space to another. Third, each fit region should have a *surface-like boundary* for which a form of the Integral-Gradient Theorem (also called Gauss-Green Theorem) should be valid. It is also desirable that the class of fit regions include all that can possibly be imagined by an engineer but exclude those that can be dreamt up *only* by an ingenious mathematician.

The first of the requirements is *not* met by the class of all regions with a piecewise smooth boundary because this class fails to be stable under intersection as shown by the example above. One cannot remove this difficulty by considering only a subclass of the class of regions with piecewise smooth boundary. Indeed, if rectangles belong to this subclass (as they surely should) a set such as $\mathcal{D}$ above must also belong because $\mathcal{D}$ can be obtained from a rectangle by a $C^\infty$-transplacement. Therefore it is impossible to exclude sets as pathological as the set $\mathcal{C} \cap \mathcal{D}$ described above, and the desideratum stated at the end of the preceding paragraph cannot be met.

One is left with the problem of delimiting a class of regions that satisfies the three requirements above, includes at least balls and boxes, and is otherwise as small as possible. It has been clear for several years to most people who have thought about the problem that such a delimitation would have to involve the concept of a "set with finite perimeter". It seems that BANFI & FABRIZIO [BF] were the first to propose, in the context of continuum physics, a class of sets involving this concept. However, their class does not meet the first of the requirements mentioned. Also, it is unnecessarily large. The class of sets used recently by GURTIN, WILLIAMS, & ZIEMER [GWZ] does satisfy the three requirements, but it is also unnecessarily large. For example, the set described in Example (3) of Sect. 7 belongs to their class but not to ours. The study of their class requires the use of the concepts of "measure-theoretic" interior, closure, and boundary of a set. We do not need these concepts here.

It is the purpose of this paper to define a class Fr of fit regions (Def. 7 of Sect. 5) which satisfies the three requirements mentioned and cannot easily be made any smaller. We believe that the class Fr should be used as a basis for mathematical theories of continuum physics.

A set of finite perimeter is most easily defined as a set whose characteristic function has bounded variation (see Def. 4 of Sect. 4). In Sects. 1–3 we give a nearly self-contained treatment of those aspects of the theory of functions of bounded variation that are needed in Sects. 4 and 5 to prove that the class Fr defined in Def. 7 of Sect. 5 indeed satisfies the first and second requirement mentioned above (see Theorems 4 and 5). It seems that Theorem 2 of Sect. 3 is actually new, at least in the form given here.

To show that our class Fr also satisfies the third requirement, one needs deeper aspects of the theory of sets with finite perimeter, most of which have their origin in the work of De Giorgi in the 1950s. We state the appropriate definitions and results precisely, but must refer the reader to the textbooks for justifications. (See, *e.g.*, [VH], Chapter 5 or [DCP], or [F].) It is interesting that the "surface-like" boundary of a fit region need not be the (topological) boundary, but it is the *reduced boundary* as defined by Def. 8. It is this reduced boundary that can be expected to play a crucial role in the theory of contact interactions as sketched in [N4].

In Sect. 7 we list some examples (concocted by W. N.) intended to illuminate the conditions of Def. 7, which define our class Fr.

**Notation and Terminology.** Generally we use the notation and terminology of [FDS]. Specifically:

$\mathbb{N}$ denotes the set of all natural numbers including zero. If $n \in \mathbb{N}$, then $n^{[} := \{k \in \mathbb{N} \mid 0 \leqq k \leqq n-1\}$.

$\mathbb{P}$ denotes the set of all positive reals including zero. A superscript cross, as in $\mathbb{P}^{\times}$, indicates that zero has been taken out. The *extended* set of positive reals is $\overline{\mathbb{P}} := \mathbb{P} \cup \{\infty\}$.

Let $f$ be a mapping. Then $\operatorname{Rng} f := \{f(x) \mid x \in \operatorname{Dom} f\}$ denotes the *range* (set of all values) of $f$. If $U$ is a subset of the domain $\operatorname{Dom} f$ of $f$ and $V$ a subset of the codomain $\operatorname{Cod} f$ of $f$, then $f_{>}(U) := \{f(x) \mid x \in U\}$ denotes the *image* of $U$ under $f$ and $f^{<}(V) := \{x \in \operatorname{Dom} f \mid f(x) \in V\}$ denotes the *pre-image* of $V$ under $f$. If $f$ is invertible we denote its inverse by $f^{\leftarrow}$ (except when $f$ is linear, in which case $f^{-1}$ is used). If the codomain of $f$ contains a zero-element 0, then $\operatorname{Supt}(f) := \{x \in \operatorname{Dom} f \mid f(x) \neq 0\}$ is called the *support* of $f$. The identity mapping of a set $E$ is denoted by $1_E$. If $\mathscr{D}$ is a subset of a given set $\mathscr{E}$, then the characteristic function $\operatorname{ch}_{\mathscr{D}} : \mathscr{E} \to \mathbb{R}$ of $\mathscr{D}$ is defined by $\operatorname{ch}_{\mathscr{D}}(x) = 1$ if $x \in \mathscr{D}$ and $\operatorname{ch}_{\mathscr{D}}(x) = 0$ if $x \in \mathscr{E} \setminus \mathscr{D}$.

When we use the terms "inner-product space" and "Euclidean space", it is understood that they are *genuine* in the sense of the definitions in Chapt. 4 of [FDS]. The *unit ball* of an inner-product space $\mathscr{W}$ is denoted by $\operatorname{Ubl} \mathscr{W} := \{\boldsymbol{v} \in \mathscr{W} \mid |\boldsymbol{v}| < 1\}$ and the unit sphere by $\operatorname{Usph} \mathscr{W} := \{\boldsymbol{v} \in \mathscr{W} \mid |\boldsymbol{v}| = 1\}$. If $x$ is a point in a Euclidean space $\mathscr{E}$ with translation space $\mathscr{V}$, then $\operatorname{Ball}_{x,\varrho}(\mathscr{E}) := x + \varrho \operatorname{Ubl} \mathscr{V}$ is the ball in $\mathscr{E}$ of radius $\varrho \in \mathbb{P}^{\times}$ centered at $x$.

Lebesgue-volume measures are denoted by vol, and Hausdorff-area measures (of dimension one less than the dimension of the space in question) are denoted by ar. To distinguish between integrals with respect to volume or area we write dv or da, respectively, after the integrand if confusion might otherwise arise.

## 1. Functions of bounded variation

We deal here with measurable functions whose domains are Euclidean spaces and whose codomains are inner-product spaces. The following symbols will be used to denote certain classes of such functions:

L: (Lebesgue-) integrable functions.
$\mathrm{L}_{\mathrm{loc}}$: functions whose restrictions to bounded measurable subsets of the domain are integrable.
$\mathrm{C}^r$, $r \in \mathbb{N} \cup \{\infty\}$: functions that have continuous gradients of order $r$ if $r \in \mathbb{N}$ or of all orders if $r = \infty$.
bR: functions with bounded range.
bS: functions with bounded support.
U: unit-bounded functions, *i.e.* functions whose range is included in the closed unit ball of the codomain.

We use juxtaposition to denote intersection of the corresponding classes. For example, $\mathrm{bSC}^0$ denotes the class of all functions that are continuous and have bounded support. Note the following inclusions:

$$\mathrm{bSC}^0 \subset \mathrm{bSbR} \subset \mathrm{L}. \tag{1.1}$$

If $\mathscr{E}$ is a Euclidean space and $\mathscr{W}$ an inner-product space and if J is a juxtaposition of symbols from the list above, then $\mathrm{J}(\mathscr{E}, \mathscr{W})$ denotes the set of all functions from $\mathscr{E}$ to $\mathscr{W}$ that belong to the class J. For example:

$$\mathrm{bRC}^r\mathrm{L}(\mathscr{E}, \mathscr{W}) := \{\boldsymbol{h} \in \mathrm{Map}\,(\mathscr{E}, \mathscr{W}) \mid \boldsymbol{h} \text{ is measurable,}$$
$$\mathrm{Rng}\,\boldsymbol{h} \text{ is bounded, } \boldsymbol{h} \text{ is of class } \mathrm{C}^r, \textstyle\int |\boldsymbol{h}| < \infty\}.$$

If $\mathscr{W} := \mathbb{R}$, we omit writing it. For example we abbreviate $\mathrm{bRC}^r(\mathscr{E}) := \mathrm{bRC}^r(\mathscr{E}, \mathbb{R})$.

**Definition 1.** *Let a Euclidean space $\mathscr{E}$ with translation space $\mathscr{V} := \mathscr{E} - \mathscr{E}$ be given. Then*

$$\mathrm{var}: L_{\mathrm{loc}}(\mathscr{E}) \to \overline{\mathbb{P}} \tag{1.2}$$

*is defined by*

$$\mathrm{var}\,(f) := \sup \{\textstyle\int f \,\mathrm{div}\, \boldsymbol{h} \mid \boldsymbol{h} \in \mathrm{UbSC}^1(\mathscr{E}, \mathscr{V})\}. \tag{1.3}$$

*We call* var $(f)$ *the* **variation** *of* $f \in \mathrm{L}_{\mathrm{loc}}(\mathscr{E})$ *and say that* $f$ *is of* **bounded variation** *if* $\mathrm{var}\,(f) < \infty$.

We add the following symbol to the list above:

bV: functions of bounded variation (with codomain $\mathbb{R}$).

We now assume that a Euclidean space $\mathscr{E}$ is given and put $\mathscr{V} := \mathscr{E} - \mathscr{E}$. The following result is an immediate consequence of Def. 1.

**Proposition 1.** *We have*

$$\operatorname{var}(\alpha f) = |\alpha| \operatorname{var}(f) \tag{1.4}$$

*and*

$$\operatorname{var}(f+g) \leqq \operatorname{var}(f) + \operatorname{var}(g) \tag{1.5}$$

*for all* $f, g \in \mathrm{L_{loc}}(\mathscr{E})$, $\alpha \in \mathbb{R}$.

It follows from Prop. 1 that $\mathrm{bV}(\mathscr{E})$ is a sub*space* of the function space $\mathrm{L_{loc}}(\mathscr{E})$

**Lemma 1.** *We have*

$$\left(\int f \operatorname{div} \boldsymbol{h}\right) - \operatorname{var}(g) \leqq \max\left(\operatorname{Rng}|\operatorname{div} \boldsymbol{h}|\right) \int |f-g| \tag{1.6}$$

*for all* $f \in \mathrm{L_{loc}}(\mathscr{E})$, $g \in \mathrm{bV}(\mathscr{E})$, $\boldsymbol{h} \in \mathrm{UbSC}^1(\mathscr{E}, \mathscr{V})$.

**Proof.** It follows from (1.3) that

$$\begin{aligned}\left(\int f \operatorname{div} \boldsymbol{h}\right) - \operatorname{var}(g) &\leqq \int f \operatorname{div} \boldsymbol{h} - \int g \operatorname{div} \boldsymbol{h} = \int (f-g) \operatorname{div} \boldsymbol{h} \\ &\leqq \int |f-g|\,|\operatorname{div} \boldsymbol{h}| \leqq \max\left(\operatorname{Rng}|\operatorname{div} \boldsymbol{h}|\right) \int |f-g|\end{aligned}$$

for all $f$, $g$, and $\boldsymbol{h}$ as indicated. □

**Proposition 2.** *Let* $f \in \mathrm{L_{loc}}(\mathscr{E})$ *and a family* $(k_\sigma \mid \sigma \in \mathbb{P}^\times)$ *in* $\mathrm{bV}(\mathscr{E})$ *be given such that*

$$\lim_{\sigma \to 0} \int |k_\sigma - f| = 0. \tag{1.7}$$

*Then*

$$\operatorname{var}(f) \leqq \liminf_{\sigma \to 0} \operatorname{var}(k_\sigma). \tag{1.8}$$

**Proof.** *Case 1.* $\operatorname{var}(f) = \infty$.

Let $\gamma \in \mathbb{P}^\times$ be given. By (1.3) we may choose $\boldsymbol{h} \in \mathrm{UbSC}^1(\mathscr{E}, \mathscr{V})$ such that

$$\gamma - \int f \operatorname{div} \boldsymbol{h} \leqq -1. \tag{1.9}$$

Note that $|\operatorname{div} \boldsymbol{h}| \in \mathrm{bSC}^0(\mathscr{E})$ and hence that $|\operatorname{div} \boldsymbol{h}|$ attains a maximum, which cannot be zero by (1.9). We put

$$\delta := \frac{1}{\max \operatorname{Rng}|\operatorname{div} \boldsymbol{h}|}. \tag{1.10}$$

By (1.7) we may choose $\tau \in \mathbb{P}^\times$ such that

$$\int |f - k_\sigma| \leqq \delta \quad \text{for all } \sigma \in ]0, \tau]. \tag{1.11}$$

Hence, by Lemma 1, (1.10), and (1.11), we have

$$(\int f \operatorname{div} \boldsymbol{h}) - \operatorname{var}(k_\sigma) \leqq \frac{1}{\delta} \int |f - k_\sigma| \leqq 1 \quad \text{for all } \sigma \in ]0, \tau].$$

Adding this inequality to (1.9), we obtain

$$\gamma - \operatorname{var}(k_\sigma) \leqq 0 \quad \text{for all } \sigma \in ]0, \tau].$$

Since $\gamma \in \mathbb{P}^\times$ was arbitrary, we conclude that $\lim_{\sigma \to 0} \operatorname{var}(k_\sigma) = \infty$ which is equivalent to (1.8) when $\operatorname{var}(f) = \infty$.

*Case 2.* $\operatorname{var}(f) < \infty$.

Let $\varepsilon \in \mathbb{P}^\times$ be given. By (1.3) we may choose $\boldsymbol{h} \in \operatorname{UbSC}^1(\mathscr{E}, \mathscr{V})$ such that

$$\operatorname{var}(f) - \int f \operatorname{div} \boldsymbol{h} \leqq \frac{\varepsilon}{2}. \tag{1.12}$$

If $\operatorname{div} \boldsymbol{h} = 0$ we put $\delta := 1$ and if $\operatorname{div} \boldsymbol{h} \neq 0$ we put

$$\delta := \frac{\varepsilon}{2 \max \operatorname{Rng} |\operatorname{div} \boldsymbol{h}|}. \tag{1.13}$$

By (1.7) we may choose $\tau \in \mathbb{P}^\times$ such that

$$\int |f - k_\sigma| \leqq \delta \quad \text{for all } \sigma \in ]0, \tau]. \tag{1.14}$$

Hence, by Lemma 1, (1.13) and (1.14), we have

$$(\int f \operatorname{div} \boldsymbol{h}) - \operatorname{var}(k_\sigma) \leqq \frac{\varepsilon}{2\delta} \int |f - k_\sigma| \leqq \frac{\varepsilon}{2} \quad \text{for all } \sigma \in ]0, \tau].$$

Adding this inequality to (1.12), we obtain

$$\operatorname{var}(f) - \operatorname{var}(k_\sigma) \leqq \varepsilon \quad \text{for all } \sigma \in ]0, \tau].$$

Since $\varepsilon \in \mathbb{P}^\times$ was arbitrary, (1.8) follows. □

## 2. Mollifications

We assume again that a Euclidean space $\mathscr{E}$ with translation space $\mathscr{V} := \mathscr{E} - \mathscr{E}$ is given. We now choose a **mollifier**, by which we mean a function $\mu \in \operatorname{bSC}^\infty(\mathscr{V})$ such that

(i) $\operatorname{Supt} \mu \subset \operatorname{Ubl} \mathscr{V}$,
(ii) $\int \mu = 1$,
(iii) $\mu \circ (-1_\mathscr{V}) = \mu$.

For example, we may assume that $\mu$ is defined by

$$\mu(\boldsymbol{v}) := \begin{Bmatrix} \varkappa e^{-\frac{1}{1-|\boldsymbol{v}|^2}} & \text{if } \boldsymbol{v} \in \operatorname{Ubl} \mathscr{V}, \\ 0 & \text{if } \boldsymbol{v} \in \mathscr{V} \setminus \operatorname{Ubl} \mathscr{V} \end{Bmatrix},$$

where $\varkappa \in \mathbb{P}^\times$ is determined such that (ii) holds.

**Definition 2.** *For each inner-product space $\mathscr{W}$, we define the family of* **mollifications** $(M_\sigma^{\mathscr{W}} \mid \sigma \in \mathbb{P}^\times)$,

$$M_\sigma^{\mathscr{W}} : \mathrm{L}_{\mathrm{loc}}(\mathscr{E}, \mathscr{W}) \to C^\infty(\mathscr{E}, \mathscr{W}) \quad \text{for all } \sigma \in \mathbb{P}^\times \tag{2.1}$$

*by*

$$(M_\sigma^{\mathscr{W}} \boldsymbol{h})(x) := \int_{\mathscr{V}} \mu(\boldsymbol{v}) \, \boldsymbol{h}(x + \sigma \boldsymbol{v}) \, \mathrm{d}\boldsymbol{v}$$

$$= \frac{1}{\sigma^{\dim \mathscr{E}}} \int_{\mathscr{E}} \mu\left(\frac{x - z}{\sigma}\right) \boldsymbol{h}(z) \, \mathrm{d}z \tag{2.2}$$

*for all* $\sigma \in \mathbb{P}^\times$, $\boldsymbol{h} \in \mathrm{L}_{\mathrm{loc}}(\mathscr{E}, \mathscr{W})$, *and* $x \in \mathscr{E}$.

The following properties of mollification are fairly easy consequences of Def. 2 and will be listed without proof. The properties are valid for each inner-product space $\mathscr{W}$ and every $\sigma \in \mathbb{P}^\times$.

(a) $M_\sigma^{\mathscr{W}}$ is linear.
(b) $(M_\sigma^{\mathscr{W}})_>(\mathrm{UL}_{\mathrm{loc}}(\mathscr{E}, \mathscr{W})) \subset \mathrm{UC}^\infty(\mathscr{E}, \mathscr{W})$.
(c) $\operatorname{Clo}\operatorname{Supt}(M_\sigma^{\mathscr{W}} \boldsymbol{h}) \subset (\mathrm{d}_{\operatorname{Supt}\boldsymbol{h}})^<(]0, \sigma])$ for all $\boldsymbol{h} \in \mathrm{L}_{\mathrm{loc}}(\mathscr{E}, \mathscr{W})$, where $\mathrm{d}_{\mathscr{A}} : \mathscr{E} \to \mathbb{P}$, for each $\mathscr{A} \subset \mathscr{E}$, is defined by

$$\mathrm{d}_{\mathscr{A}}(x) := \inf\{|x - z| \mid z \in \mathscr{A}\}. \tag{2.3}$$

It follows that

$$(M_\sigma^{\mathscr{W}})_>(\mathrm{bSL}_{\mathrm{loc}}(\mathscr{E}, \mathscr{W})) \subset \mathrm{bSC}^\infty(\mathscr{E}, \mathscr{W}).$$

(d) $M_\sigma^{\operatorname{Lin}(\mathscr{V},\mathscr{W})}(\nabla \boldsymbol{h}) = \nabla(M_\sigma^{\mathscr{W}} \boldsymbol{h})$ for all $\boldsymbol{h} \in C^1(\mathscr{E}, \mathscr{W})$.
(e) $M_\sigma^{\mathbb{R}}(\operatorname{div} \boldsymbol{h}) = \operatorname{div}(M_\sigma^{\mathscr{V}} \boldsymbol{h})$ for all $\boldsymbol{h} \in C^1(\mathscr{E}, \mathscr{V})$.
(f) $\int f(\mathrm{M}_\sigma^{\mathbb{R}} g) = \int g(\mathrm{M}_\sigma^{\mathbb{R}} f)$ for all $f, g \in \mathrm{L}_{\mathrm{loc}}(\mathscr{E})$.
(g) $\int |\mathrm{M}_\sigma^{\mathscr{W}} \boldsymbol{h}| \leq \int |\boldsymbol{h}|$ for all $\boldsymbol{h} \in \mathrm{L}_{\mathrm{loc}}(\mathscr{E}, \mathscr{W})$
and hence $(\mathrm{M}_\sigma^{\mathscr{W}})_>(\mathrm{L}(\mathscr{E}, \mathscr{W})) \subset \mathrm{LC}^\infty(\mathscr{E}, \mathscr{W})$.

From now on, to avoid clutter, we omit the superscript and write $\mathrm{M}_\sigma$ instead of $\mathrm{M}_\sigma^{\mathscr{W}}$. Such an omission is not likely to lead to confusion.

**Proposition 3.** *Given an inner-product space $\mathscr{W}$ we have*

$$\lim_{\sigma \to 0} \int |(\mathrm{M}_\sigma \boldsymbol{h}) - \boldsymbol{h}| = 0 \quad \text{for all } \boldsymbol{h} \in \mathrm{L}(\mathscr{E}, \mathscr{W}). \tag{2.4}$$

The proof, which will not be given here, is based on the non-trivial fact that $\mathrm{bSC}^0(\mathscr{E}, \mathscr{W})$ is dense in $\mathrm{L}(\mathscr{E}, \mathscr{W})$ with respect to the $\mathrm{L}^1$-topology of $\mathrm{L}(\mathscr{E}, \mathscr{W})$. (See *e.g.* [A], pp. 28–30.)

**Proposition 4.** *For every* $f \in \mathrm{bSC}^1(\mathscr{E})$ *we have*

$$\operatorname{var}(f) = \int |\nabla f|. \tag{2.5}$$

**Proof.** Let $f \in \mathrm{bSC}^1(\mathscr{E})$ be given. Using the Divergence Theorem and the Product Rule (in the form (67.3) of [FDS]) we find that

$$0 = \int \operatorname{div}(f\boldsymbol{h}) = \int f \operatorname{div} \boldsymbol{h} + \int \nabla f \cdot \boldsymbol{h} \tag{2.6}$$

and hence, by the Inner-product Inequality, that

$$\int f \operatorname{div} \boldsymbol{h} = -\int \nabla f \cdot \boldsymbol{h} \leqq \int |\nabla f|\,|\boldsymbol{h}| \leqq \int |\nabla f|$$

holds for all $\boldsymbol{h} \in \mathrm{UbSC}^1(\mathscr{E}, \mathscr{V})$. In view of Def. 1, we conclude that $\operatorname{var}(f) \leqq \int |\nabla f|$.

We now define $\boldsymbol{g} : \mathscr{E} \to \mathscr{V}$ by

$$\boldsymbol{g}(x) := \left\{ \begin{array}{ll} -\dfrac{\nabla_x f}{|\nabla_x f|} & \text{if } x \in \operatorname{Supt} \nabla f \\[2ex] 0 & \text{if } x \in \mathscr{E} \setminus \operatorname{Supt} \nabla f \end{array} \right\}. \tag{2.7}$$

Since $\nabla f$ is continuous. Supt $\nabla f$ is open and hence $\boldsymbol{g}$ is measurable. Therefore we have

$$\boldsymbol{g} \in \mathrm{UbS}(\mathscr{E}, \mathscr{V}) \subset \mathrm{L}(\mathscr{E}, \mathscr{V}). \tag{2.8}$$

Let $\sigma \in \mathbb{P}^{\times}$ be given. The properties (b) and (c) of mollifications ensure that $M_\sigma \boldsymbol{g} \in \mathrm{UbSC}^\infty(\mathscr{E}, \mathscr{V})$. Hence, using (2.6) with $\boldsymbol{h} := \mathrm{M}_\sigma \boldsymbol{g}$ and Def. 1 we get

$$-\int \nabla f \cdot (\mathrm{M}_\sigma \boldsymbol{g}) = \int f \operatorname{div}(\mathrm{M}_\sigma \boldsymbol{g}) \leqq \operatorname{var}(f). \tag{2.9}$$

Noting that $|\nabla f| \in \mathrm{bSC}^0(\mathscr{E})$ attains a maximum, we see that

$$\left|\int \nabla f \cdot (\mathrm{M}_\sigma \boldsymbol{g}) - \int \nabla f \cdot \boldsymbol{g}\right| \leqq \max \operatorname{Rng} |\nabla f| \int |(\mathrm{M}_\sigma \boldsymbol{g}) - \boldsymbol{g}|. \tag{2.10}$$

Since (2.10) is valid for all $\sigma \in \mathbb{P}^{\times}$ and since (2.8) holds, we can apply Prop. 3 and conclude that

$$\lim_{\sigma \to 0} \int \nabla f \cdot (\mathrm{M}_\sigma \boldsymbol{g}) = \int (\nabla f) \cdot \boldsymbol{g},$$

and hence, by (2.9), that $-\int (\nabla f) \cdot \boldsymbol{g} \leqq \operatorname{var}(f)$. By definition (2.7) of $\boldsymbol{g}$, it follows that $\int |\nabla f| \leqq \operatorname{var}(f)$. $\square$

**Proposition 5.** *Given* $f \in \mathrm{L}_{\mathrm{loc}}(\mathscr{E})$ *we have*

$$\operatorname{var}(f) \geqq \operatorname{var}(\mathrm{M}_\sigma f) \quad \text{for all } \sigma \in \mathbb{P}^{\times}. \tag{2.11}$$

*Moreover, if* $f \in \mathrm{L}(\mathscr{E})$, *we have*

$$\lim_{\sigma \to 0} \operatorname{var}(\mathrm{M}_\sigma f) = \operatorname{var}(f). \tag{2.12}$$

**Proof.** Let $\sigma \in \mathbb{P}^\times$ and $\boldsymbol{h} \in \mathrm{UbSC}^1(\mathscr{E}, \mathscr{V})$ be given. Then $\operatorname{div} \boldsymbol{h} \in \mathrm{bSC}^0(\mathscr{E}) \subset \mathrm{L}(\mathscr{E})$. Using the properties (f) and (e) of mollifications we obtain

$$\int (\mathrm{M}_\sigma f) \operatorname{div} \boldsymbol{h} = \int f(\mathrm{M}_\sigma(\operatorname{div} \boldsymbol{h})) = \int f \operatorname{div} (\mathrm{M}_\sigma \boldsymbol{h}). \tag{2.13}$$

Since $\mathrm{M}_\sigma \boldsymbol{h} \in \mathrm{UbSC}^1(\mathscr{E}, \mathscr{V})$ by (b) and (c) we conclude from (2.13) and Def. 1 that

$$\int (\mathrm{M}_\sigma f) \operatorname{div} \boldsymbol{h} \leqq \operatorname{var}(f). \tag{2.14}$$

Since $\boldsymbol{h} \in \mathrm{UbSC}^1(\mathscr{E}, \mathscr{V})$ was arbitrary, it follows from (2.14) and Def. 1 that (2.11) holds.

If $f \in \mathrm{L}(\mathscr{E})$ we can apply Props. 3 and 2 to conclude that

$$\operatorname{var}(f) \leqq \liminf_{\sigma \to 0} \operatorname{var}(\mathrm{M}_\sigma f).$$

This result, together with (2.11), shows that (2.12) holds. □

## 3. Two basic theorems about functions with bounded variation

We assume again that a Euclidean space $\mathscr{E}$ with translation space $\mathscr{V} := \mathscr{E} - \mathscr{E}$ is given.

**Theorem 1.** *If $f$ and $g$ belong to* $\mathrm{bVbSU}(\mathscr{E})$, *so does the value-wise product $fg$ and*

$$\operatorname{var}(fg) \leqq \operatorname{var}(f) + \operatorname{var}(g). \tag{3.1}$$

**Proof.** Let $f, g \in \mathrm{bVbSU}(\mathscr{E})$ be given. It is clear that $fg$ belongs to $\mathrm{bSU}(\mathscr{E})$. Let $\sigma \in \mathbb{P}^\times$ and $\boldsymbol{h} \in \mathrm{bSUC}^1(\mathscr{E}, \mathscr{V})$ be given. By the properties (b) and (c) of mollifications, we have $\mathrm{M}_\sigma g \in \mathrm{bSUC}^\infty(\mathscr{E})$ and hence $(\mathrm{M}_\sigma g)\, \boldsymbol{h} \in \mathrm{bSUC}^1(\mathscr{E}, \mathscr{V})$. Using the Product Rule and Def. 1 we obtain

$$\begin{aligned} \int (f(\mathrm{M}_\sigma g)) \operatorname{div} \boldsymbol{h} &= \int f \operatorname{div} ((\mathrm{M}_\sigma g)\, \boldsymbol{h}) - \int f \boldsymbol{h} \cdot \nabla(\mathrm{M}_\sigma g) \\ &\leqq \operatorname{var}(f) + \int |f|\, |\boldsymbol{h}|\, |\nabla(\mathrm{M}_\sigma g)| \leqq \operatorname{var}(f) + \int |\nabla(\mathrm{M}_\sigma g)|. \end{aligned} \tag{3.2}$$

Since $\boldsymbol{h} \in \mathrm{bSUC}^1(\mathscr{E}, \mathscr{V})$ was arbitrary, it follows from (3.2) and Def. 1 that

$$\operatorname{var}(f(\mathrm{M}_\sigma g)) \leqq \operatorname{var}(f) + \int |\nabla(\mathrm{M}_\sigma g)|. \tag{3.3}$$

Using Props. 4 and 5, we find that

$$\int |\nabla(\mathrm{M}_\sigma g)| = \operatorname{var}(\mathrm{M}_\sigma g) \leqq \operatorname{var}(g)$$

and hence, by (3.3), that

$$\operatorname{var}(f(\mathrm{M}_\sigma g)) \leqq \operatorname{var}(f) + \operatorname{var}(g). \tag{3.4}$$

On the other hand, since $|f| \leqq 1$, we have

$$\int |f(\mathrm{M}_\sigma g) - fg| \leqq \int |f|\, |(\mathrm{M}_\sigma g) - g| \leqq \int |(\mathrm{M}_\sigma g) - g|. \tag{3.5}$$

Since $\sigma \in \mathrm{P}^\times$ was arbitrary and since $g \in \mathrm{L}(\mathscr{E})$, we can apply Prop. 3 and conclude from (3.5) that

$$\lim_{\sigma \to 0} \int |f(\mathrm{M}_\sigma g) - fg| = 0. \tag{3.6}$$

Using Prop. 2 for the case when $k_\sigma := f(\mathrm{M}_\sigma g)$ for all $\sigma \in \mathrm{P}^\times$ we conclude from (3.6) that

$$\operatorname{var}(fg) \leqq \liminf_{\sigma \to 0} \operatorname{var}(f(\mathrm{M}_\sigma g)). \tag{3.7}$$

Since (3.4) holds for all $\sigma \in \mathrm{P}^\times$ the desired result (3.1) follows from (3.7). □

**Corollary.** *If $f$ and $g$ belong to* bVbSbR($\mathscr{E}$), *so does the value-wise product $fg$ and*

$$\operatorname{var}(fg) \leqq (\sup \operatorname{Rng} |f|) \operatorname{var}(g) + (\sup \operatorname{Rng} |g|) \operatorname{var}(f). \tag{3.8}$$

**Proof.** It suffices to apply Theorem 1 to the case when $f$ and $g$ are replaced by $f/\sup \operatorname{Rng} |f|$ and $g/\sup \operatorname{Rng} |g|$, respectively. □

**Theorem 2.** *Let $\varphi : \mathscr{E} \to \mathscr{E}'$ be a $\mathrm{C}^1$-diffeomorphism from the given Euclidean space $\mathscr{E}$ to a Euclidean space $\mathscr{E}'$. For every $f \in$ bVbSL($\mathscr{E}$) we then have $f \circ \varphi^{\leftarrow} \in$ bVbSL($\mathscr{E}'$) and*

$$\operatorname{var}(f \circ \varphi^{\leftarrow}) \leqq \varkappa(f, g) \operatorname{var}(f), \tag{3.9}$$

*where*

$$\varkappa(f, \varphi) := \sup \{|\det \nabla_x \varphi| \, \|(\nabla_x \varphi)^{-1}\| \mid x \in \operatorname{Gsp}(f)\} \tag{3.10}$$

*in which the* **gradient-support** Gsp ($f$) *of $f$ is defined by*

$$\operatorname{Gsp}(f) := \bigcap_{\sigma \in \mathbf{P}^\times} \operatorname{Supt}(\nabla(M_\sigma f)). \tag{3.11}$$

(In (3.10) $\| \ \|$ denotes the operator norm defined according to (52.19) of [FDS].)

**Proof.** Let $f \in$ bVbSL($\mathscr{E}$) and $\sigma \in \mathrm{P}^\times$ be given and put

$$k := \mathrm{M}_\sigma f \in \mathrm{bSC}^\infty(\mathscr{E}) \tag{3.12}$$

(see property (c) of mollifications). The Theorem on Transformation of Integrals (see vol. II of [FDS]) implies that

$$\int_{\mathscr{E}'} g = \int_{\mathscr{E}} (g \circ \varphi) \, |\det \circ \nabla \varphi| \tag{3.13}$$

holds for all $g \in$ bSL($\mathscr{E}'$).

By the Chain Rule (see (63.18) of [FDS]) we have

$$\nabla k = \nabla((k \circ \varphi^{\leftarrow}) \circ \varphi) = (\nabla \varphi)^T (\nabla(k \circ \varphi^{\leftarrow}) \circ \varphi)$$

and hence

$$|\nabla(k \circ \varphi^{\leftarrow})| \circ \varphi = |((\nabla \varphi)^{\mathrm{T}})^{-1} (\nabla k)|. \tag{3.14}$$

It is clear from (3.12) that

$$|\nabla(k \circ \varphi^{\leftarrow})| \in \mathrm{bSC}^0(\mathscr{E}') \subset \mathrm{bSL}(\mathscr{E}').$$

Hence we can apply (3.13) to $g := |\nabla(k \circ \varphi^{\leftarrow})|$ and obtain, by (3.14),

$$\int_{\mathscr{E}'} |\nabla(k \circ \varphi^{\leftarrow})| = \int_{\mathscr{E}} |((\nabla\varphi)^{-1})^{\mathrm{T}} (\nabla k)| \, |\det \circ \nabla\varphi|. \tag{3.15}$$

Since $|((\nabla_x\varphi)^{-1})^{\mathrm{T}} \nabla_x k| \leqq \|(\nabla_x\varphi)^{-1}\| \, |\nabla_x k|$ for all $x \in \mathscr{E}$ by (52.19) and (52.17) of [FDS] it follows from (3.15) that

$$\int_{\mathscr{E}'} |\nabla(k \circ \varphi^{\leftarrow})| \leqq \sup\{|\det \nabla_x\varphi| \, \|(\nabla_x\varphi)^{-1}\| \mid x \in \mathrm{Supt}\, \nabla k\} \int_{\mathscr{E}} |\nabla k|.$$

Hence, by Prop. 4, (3.12) and (2.11), we conclude that

$$\mathrm{var}\,((\mathrm{M}_\sigma f) \circ \varphi^{\leftarrow}) \leqq \sup\{|\det \nabla_x\varphi| \, \|(\nabla_x\varphi)^{-1}\| \mid x \in \mathrm{Supt}\, \nabla(\mathrm{M}_\sigma f)\} \,\mathrm{var}\,(f). \tag{3.16}$$

By property (c) of mollifications we have

$$\mathrm{Supt}\,(\mathrm{M}_\sigma f) \subset (\mathrm{d}_{\mathrm{Supt}(f)})^{<}(]0, \sigma]) \subset (\mathrm{d}_{\mathrm{Supt}(f)})^{<}(]0, 1]) \tag{3.17}$$

for all $\sigma \in ]0, 1]$, where the definition (2.3) applies. Hence, if we put

$$\gamma := \sup\{|\det \nabla_x\varphi| \mid x \in (\mathrm{d}_{\mathrm{Supt}(f)})^{<}(]0, 1])\}$$

and apply (3.13) to the case when

$$g := |(M_\sigma f) - f| \circ \varphi^{\leftarrow} = |(\mathrm{M}_\sigma f) \circ \varphi^{\leftarrow} - f \circ \varphi^{\leftarrow}| \in \mathrm{bSL}(\mathscr{E}'),$$

we find that

$$\int_{\mathscr{E}'} |(\mathrm{M}_\sigma f) \circ \varphi^{\leftarrow} - f \circ \varphi^{\leftarrow}| = \int_{\mathscr{E}} |(\mathrm{M}_\sigma f) - f| \, |\det \circ \nabla\varphi| \leqq \gamma \int_{\mathscr{E}} |(\mathrm{M}_\sigma f) - f|$$

holds for all $\sigma \in ]0, 1]$. Since $f \in \mathrm{L}(\mathscr{E})$ it follows from Prop. 3 that

$$\lim_{\sigma \to 0} \int_{\mathscr{E}'} |(\mathrm{M}_\sigma f) \circ \varphi^{\leftarrow} - f \circ \varphi^{\leftarrow}| = 0.$$

By applying Prop. 2 to the case when $k_\sigma := (\mathrm{M}_\sigma f) \circ \varphi^{\leftarrow}$, which belongs to $\mathrm{bV}(\mathscr{E}')$ by (3.16) for all $\sigma \in \mathbb{P}^\times$, we conclude that

$$\mathrm{var}\,(f \circ \varphi^{\leftarrow}) \leqq \liminf_{\sigma \to 0} \mathrm{var}\,((\mathrm{M}_\sigma f) \circ \varphi^{\leftarrow}).$$

This result and (3.16) yield the desired result (3.10). □

**Corollary.** *We have*

$$\mathrm{Gsp}(f) \subset \mathrm{Clo}\,\mathrm{Supt}\,(f) \quad \text{for all } f \in \mathrm{L}_{\mathrm{loc}}(\mathscr{E}) \tag{3.18}$$

*and hence the conclusion of Theorem 2 remains valid if, in the definition* (3.10) *of* $\varkappa(f, \varphi)$, *the gradient support* Gsp $(f)$ *of* $f$ *is replaced by* Clo Supt $(f)$.

**Proof.** Let $f \in \mathrm{L}_{\mathrm{loc}}(\mathscr{E})$ be given. Since $\mathrm{Supt}\,(\nabla h) \subset \mathrm{Clo}\,\mathrm{Supt}\,(h)$ for all $h \in C^1(\mathscr{E})$, it follows from (3.11) that

$$\mathrm{Gsp}\,(f) \subset \bigcap_{\sigma \in \mathbf{P}^\times} \mathrm{Clo}\,\mathrm{Supt}\,(\mathrm{M}_\sigma f). \tag{3.19}$$

Using the property (c) of mollifications and Prop. 6 of Sect. 56 of [FDS], we conclude from (3.19) that

$$\operatorname{Gsp}(f) \subset \bigcap_{\sigma \in \mathbf{P}^\times} (\mathrm{d}_{\operatorname{Supt}(f)})^{<}(]0, \sigma]) = (\mathrm{d}_{\operatorname{Supt}(f)})^{<}(\{0\}) = \operatorname{Clo} \operatorname{Supt}(f). \quad \square$$

## 4. Regularly open sets and sets of finite perimeter

We assume again that a Euclidean space $\mathscr{E}$ with translation space $\mathscr{V}$ is given.

**Definition 3.** *We say that a subset* $\mathscr{D}$ *of* $\mathscr{E}$ *is* **regularly open** *if* $\operatorname{Int}\operatorname{Clo}\mathscr{D} = \mathscr{D}$.

The following facts are easily established:

(A) If $\mathscr{C}$ is any subset of $\mathscr{E}$, then $\operatorname{Int}\operatorname{Clo}\mathscr{C}$ is regularly open.
(B) A subset $\mathscr{D}$ of $\mathscr{E}$ is regularly open if and only if it is open and $\operatorname{Bdy}\mathscr{D} = \operatorname{Bdy}\operatorname{Clo}\mathscr{D}$.
(C) The intersection of two regularly open sets is again regularly open.
(D) If $\mathscr{C}$ and $\mathscr{D}$ are regularly open subsets of $\mathscr{E}$ then

$$\mathscr{C} \vee \mathscr{D} := \operatorname{Int}\operatorname{Clo}(\mathscr{C} \cup \mathscr{D}) \tag{4.1}$$

is the smallest regularly open set that includes both $\mathscr{C}$ and $\mathscr{D}$. We call $\mathscr{C} \vee \mathscr{D}$ the **join** of $\mathscr{C}$ and $\mathscr{D}$.
(E) If $\mathscr{C}$ and $\mathscr{D}$ are regularly open subsets of $\mathscr{E}$, then

$$\mathscr{C} \mathbin{\triangle} \mathscr{D} := \operatorname{Int}(\mathscr{C} \setminus \mathscr{D}) \tag{4.2}$$

is the largest regularly open set included in $\mathscr{C}$ and disjoint from $\mathscr{D}$. We call $\mathscr{C} \mathbin{\triangle} \mathscr{D}$ the **difference-region** of $\mathscr{C}$ and $\mathscr{D}$.

**Remark 1.** Def. 3 remains meaningful and the statements (A)–(E) remain facts if $\mathscr{E}$ is an arbitrary topological space. $\square$

**Definition 4.** *The* **perimeter** *of a measurable subset* $\mathscr{D}$ *of* $\mathscr{E}$ *is defined by*

$$\operatorname{per}(\mathscr{D}) := \operatorname{var}(\mathrm{ch}_{\mathscr{D}}). \tag{4.3}$$

*We say that* $\mathscr{D}$ *has* **finite perimeter** *if* $\operatorname{per}(\mathscr{D}) < \infty$, i.e. *if* $\mathrm{ch}_{\mathscr{D}}$ *has bounded variation.*

**Definition 5.** *We say that two subsets* $\mathscr{D}$ and $\mathscr{D}'$ *of* $\mathscr{E}$ *are* **volume-equivalent,** *and we write* $\mathscr{D} \sim_{\mathrm{vol}} \mathscr{D}'$, *if the symmetric difference*

$$\mathscr{D} \ominus \mathscr{D}' := (\mathscr{D} \setminus \mathscr{D}') \cup (\mathscr{D}' \setminus \mathscr{D}) \tag{4.4}$$

*has volume-measure zero.*

It is clear that $\mathscr{D} \sim_{\mathrm{vol}} \mathscr{D}'$ if and only if $\mathrm{ch}_{\mathscr{D}}$ differs from $\mathrm{ch}_{\mathscr{D}'}$ only on a set of volume-measure zero. By Def. 1, the variation of a function remains the same if that function is altered only on a set of volume-measure zero. Hence it follows from Defs. 4 and 5 that

$$\mathscr{D} \sim_{\mathrm{vol}} \mathscr{D}' \quad \Rightarrow \quad \mathrm{per}\,(\mathscr{D}) = \mathrm{per}\,(\mathscr{D}'). \tag{4.5}$$

**Proposition 6.** *If the subsets* $\mathscr{D}$ and $\mathscr{D}'$ *of* $\mathscr{E}$ *are regularly open and if* $\mathscr{D} \sim_{\mathrm{vol}} \mathscr{D}'$, *then* $\mathscr{D} = \mathscr{D}'$.

**Proof.** Since a set cannot have volume-measure zero if its interior is non empty, $\mathscr{D} \sim_{\mathrm{vol}} \mathscr{D}'$ implies $\mathrm{Int}\,(\mathscr{D} \ominus \mathscr{D}') = \emptyset$. It is an easy exercise to prove that $\mathrm{Int}\,(\mathscr{D} \ominus \mathscr{D}') = \emptyset$ implies $\mathscr{D} = \mathscr{D}'$ if both $\mathscr{D}$ and $\mathscr{D}'$ are regularly open. □

**Proposition 7.** *If the subsets* $\mathscr{C}$ *and* $\mathscr{D}$ of $\mathscr{E}$ *are bounded and have finite perimeter, so do* $\mathscr{C} \cap \mathscr{D}$, $\mathscr{C} \setminus \mathscr{D}$, *and* $\mathscr{C} \cup \mathscr{D}$, *and we have*

$$\mathrm{per}\,(\mathscr{C} \cap \mathscr{D}) \leqq \mathrm{per}\,(\mathscr{C}) + \mathrm{per}\,(\mathscr{D}), \tag{4.6}$$

$$\mathrm{per}\,(\mathscr{C} \setminus \mathscr{D}) \leqq 2\,\mathrm{per}\,(\mathscr{C}) + \mathrm{per}\,(\mathscr{D}), \tag{4.7}$$

$$\mathrm{per}\,(\mathscr{C} \cup \mathscr{D}) \leqq 2\,\mathrm{per}\,(\mathscr{C}) + 2\,\mathrm{per}\,(\mathscr{D}). \tag{4.8}$$

**Proof.** It is clear that

$$\begin{gathered} \mathrm{ch}_{\mathscr{C}\cap\mathscr{D}} = \mathrm{ch}_{\mathscr{C}}\,\mathrm{ch}_{\mathscr{D}}, \qquad \mathrm{ch}_{\mathscr{C}\setminus\mathscr{D}} = \mathrm{ch}_{\mathscr{C}} - \mathrm{ch}_{\mathscr{C}}\,\mathrm{ch}_{\mathscr{D}}, \\ \mathrm{ch}_{\mathscr{C}\cup\mathscr{D}} = \mathrm{ch}_{\mathscr{C}} + \mathrm{ch}_{\mathscr{D}} - \mathrm{ch}_{\mathscr{C}}\,\mathrm{ch}_{\mathscr{D}}. \end{gathered} \tag{4.9}$$

If $\mathscr{C}$ and $\mathscr{D}$ are bounded and have finite perimeter, then $\mathrm{ch}_{\mathscr{C}}, \mathrm{ch}_{\mathscr{D}} \in \mathrm{bVbSU}(\mathscr{E})$ and hence Theorem 1 of Sect. 3 and Def. 4 can be applied to show that $\mathrm{var}\,(\mathrm{ch}_{\mathscr{C}}\,\mathrm{ch}_{\mathscr{D}}) \leqq \mathrm{per}\,(\mathscr{C}) + \mathrm{per}\,(\mathscr{D})$. Using also Prop. 1 of Sect. 1. it follows from (4.9) that (4.6)–(4.8) hold. □

**Remark 2.** The inequalities (4.7) and (4.8) remain valid after the factor 2 has been omitted. In fact, the sharper inequality

$$\mathrm{per}\,(\mathscr{C} \cap \mathscr{D}) + \mathrm{per}\,(\mathscr{C} \cup \mathscr{D}) \leqq \mathrm{per}\,(\mathscr{C}) + \mathrm{per}\,(\mathscr{D}) \tag{4.10}$$

is valid (see [DCP], Theorem 3.5), but the results of Sect. 3 are insufficient for the proof of (4.10). The proof of (4.10) in [DCP] is based on results about approximation of sets of finite perimeter by polyhedra. □

**Theorem 3.** *Let* $\varphi : \mathscr{E} \to \mathscr{E}'$ *be a* $\mathrm{C}^1$*-diffeomorphism from the given Euclidean space* $\mathscr{E}$ *onto a Euclidean space* $\mathscr{E}'$ *and let* $\mathscr{D}$ *be a bounded subset of* $\mathscr{E}$ *that has finite perimeter. Then the subset* $\varphi_>(\mathscr{D})$ *of* $\mathscr{E}'$ *is again bounded and has finite perimeter and*

$$\mathrm{per}\,(\varphi_>(\mathscr{D})) \leqq \gamma(\mathscr{D}, \varphi)\,\mathrm{per}\,(\mathscr{D}), \tag{4.11}$$

*where*

$$\gamma(\mathscr{D}, \varphi) := \sup\{|\det \nabla_x\varphi|\,\|(\nabla_x\varphi)^{-1}\| \mid x \in \mathrm{Bdy}\,\mathscr{D}\}. \tag{4.12}$$

**Proof.** It is clear that $\mathrm{ch}_{\mathscr{D}}$ belongs to $\mathrm{bVbSU}(\mathscr{E})$ and hence Theorem 2 of Sect. 3 can be applied to $f := \mathrm{ch}_{\mathscr{D}}$. Since $\mathrm{ch}_{\mathscr{D}} \circ \varphi^{\leftarrow} = \mathrm{ch}_{\varphi_>(\mathscr{D})}$ (see (03.17) of [FDS]), it follows from Theorem 2 that $\varphi_>(\mathscr{D})$ is bounded and has finite perimeter. The formula (3.9) becomes

$$\operatorname{per}(\varphi_>(\mathscr{D})) \leqq \varkappa(\mathrm{ch}_{\mathscr{D}}, \varphi) \operatorname{per}(\mathscr{D}). \tag{4.13}$$

Now, since the restriction of $\mathrm{ch}_{\mathscr{D}}$ to $\operatorname{Int}\mathscr{D}$ is the constant 1, it follows from (2.2) that, for every $\sigma \in \mathbb{P}^\times$, the restriction of $\mathrm{M}_\sigma \mathrm{ch}_{\mathscr{D}}$ to $\operatorname{Int}\mathscr{D} \cap (d_{\mathrm{Bdy}\mathscr{D}})^<(]\sigma, \infty[)$ is the constant 1 and hence that

$$\operatorname{Supt} \nabla(\mathrm{M}_\sigma \mathrm{ch}_{\mathscr{D}}) \subset (d_{\mathrm{Bdy}\mathscr{D}})^<([0, \sigma]). \tag{4.14}$$

Therefore, the definition (3.11) of the gradient-support gives

$$\operatorname{Gsp}(\mathrm{ch}_{\mathscr{D}}) \subset \bigcap_{\sigma \in \mathbb{P}^\times} (d_{\mathrm{Bdy}\mathscr{D}})^<([0, \sigma]) = (d_{\mathrm{Bdy}\mathscr{D}})^<(\{0\}) = \operatorname{Bdy}\mathscr{D}. \tag{4.15}$$

Comparing (4.12) with (3.10), we conclude from (4.15) that $\varkappa(\mathrm{ch}_{\mathscr{D}}, \varphi) \leqq \gamma(\mathscr{D}, \varphi)$ and hence that (4.11) follows from (4.13). □

**Remark 3.** Let $\varphi : \mathscr{E} \to \mathscr{E}'$ be as in the statement of Theorem 3 and let $\mathscr{S}$ be an orientable $C^1$-surface in $\mathscr{E}$ with unit normal field $\boldsymbol{n}_{\mathscr{S}} : \mathscr{S} \to \operatorname{Usph}\mathscr{V}$ and finite area. Recall the version of the Transformation Theorem for Surface Integrals (see vol. II of [FDS]) which states that

$$\int_{\varphi_>(\mathscr{S})} g \, \mathrm{da} = \int_{\mathscr{S}} (g \circ \varphi) \left|(\det \circ \nabla\varphi)\,((\nabla\varphi)^{-1})^{\mathrm{T}} \boldsymbol{n}_{\mathscr{S}}\right| \mathrm{da} \tag{4.16}$$

holds for all continuous functions $g : \mathscr{S} \to \mathbb{R}$. Noting that

$$\left|(\det \nabla_x\varphi)\,((\nabla_x\varphi)^{-1})^{\mathrm{T}} \boldsymbol{n}_{\mathscr{S}}(x)\right| \leqq |\det \nabla_x\varphi| \, \|(\nabla_x\varphi)^{-1}\|$$

for all $x \in \mathscr{S}$, it follows from (4.16) with the choice $g := 1$ that

$$\operatorname{ar}(\varphi_>(\mathscr{S})) \leqq \sup\{|\det \nabla_x\varphi| \, \|(\nabla_x\varphi)^{-1}\| \mid x \in \mathscr{S}\} \operatorname{ar}(\mathscr{S}). \tag{4.17}$$

When $\mathscr{D}$ is an open bounded subset of $\mathscr{E}$ whose boundary is a $C^1$-surface $\mathscr{S} := \operatorname{Bdy}\mathscr{D}$, then $\operatorname{per}(\mathscr{D}) = \operatorname{ar}(\mathscr{S})$ (see Prop. 8 below) and (4.11) and (4.12) reduce to (4.17). □

## 5. Fit Regions

We assume again that a Euclidean space $\mathscr{E}$ with translation space $\mathscr{V}$ is given. The following definition is equivalent to one given in vol. II of [FDS]:

**Definition 6.** *We say that a subset $\mathscr{S}$ of $\mathscr{E}$ is* **negligible** *if for every $\varepsilon \in \mathbb{P}^\times$ one can find a finite collection of balls which cover $\mathscr{S}$ and the sum of whose volumes does not exceed $\varepsilon$.*

The following facts follow easily from Def. 6:

(α) Negligible sets are bounded.

(β) Every subset of a negligible set is negligible.

($\gamma$) The union of a finite collection of negligible sets is again negligible.
($\delta$) The closure of a negligible set is again negligible.

**Definition 7.** *We say that a subset* $\mathscr{D}$ *of* $\mathscr{E}$ *if a* **fit region** *in* $\mathscr{E}$ *if it*

(i) *is bounded,*
(ii) *is regularly open,*
(iii) *has finite perimeter,*
(iv) *has a negligible boundary.*

*The set of all fit regions in* $\mathscr{E}$ *will be denoted by* $\mathrm{Fr}(\mathscr{E})$ *and the class of all fit regions (in all Euclidean spaces) by* Fr.

**Remark 4.** Every negligible set has volume-measure zero. If the suset $\mathscr{S}$ of $\mathscr{E}$ is compact and has volume-measure zero, then $\mathscr{S}$ is negligible. Therefore, since the boundary of a bounded set is compact, condition (iv) of Def. 7 can be replaced by "has a boundary of volume-measure zero" without change of substance. □

**Theorem 4.** *If* $\mathscr{C}$ *and* $\mathscr{D}$ *are fit regions in* $\mathscr{E}$, *so are the intersection* $\mathscr{C} \cap \mathscr{D}$, *the join* $\mathscr{C} \vee \mathscr{D}$ *as defined by* (4.1), *and the difference-region* $\mathscr{C} \diagdown \mathscr{D}$ *as defined by* (4.2).

**Proof.** Let $\mathscr{C}, \mathscr{D} \in \mathrm{Fr}(\mathscr{E})$ be given. It is easily seen that $\mathscr{C} \cap \mathscr{D}$, $\mathscr{C} \vee \mathscr{D}$, and $\mathscr{C} \diagdown \mathscr{D}$ are bounded and it follows from the facts (C), (D), and (E) of Sect. 4 that they are regularly open. It is easily seen, also, that $\mathrm{Bdy}(\mathscr{C} \cap \mathscr{D})$, $\mathrm{Bdy}(\mathscr{C} \cup \mathscr{D})$ and $\mathrm{Bdy}(\mathscr{C} \setminus \mathscr{D})$ are included in $\mathrm{Bdy}\,\mathscr{C} \cup \mathrm{Bdy}\,\mathscr{D}$ and hence, by the facts ($\beta$) and ($\gamma$) stated above, negligible. Since the boundary of a set includes the boundaries of its interior and closure, it follows from ($\beta$) and (4.1) and (4.2) that $\mathscr{C} \cap \mathscr{D}$, $\mathscr{C} \vee \mathscr{D}$, and $\mathscr{C} \diagdown \mathscr{D}$ all have negligible boundary.

It follows from Prop. 7 of Sect. 3 that $\mathscr{C} \cap \mathscr{D}$, $\mathscr{C} \cup \mathscr{D}$, and $\mathscr{C} \setminus \mathscr{D}$ all have finite perimeter. The symmetric difference of $\mathscr{C} \vee \mathscr{D}$ and $\mathscr{C} \cup \mathscr{D}$ is included in the boundary of $\mathscr{C} \cup \mathscr{D}$ and hence negligible, so that $\mathscr{C} \vee \mathscr{D} \sim_{\mathrm{vol}} \mathscr{C} \cup \mathscr{D}$. By (4.5), we see that $\mathscr{C} \vee \mathscr{D}$ has the same finite perimeter as $\mathscr{C} \cup \mathscr{D}$. In a similar way, one sees that $\mathscr{C} \diagdown \mathscr{D}$ has the same finite perimeter as $\mathscr{C} \setminus \mathscr{D}$. □

It follows from the facts (C), (D), and (E) stated in Sect. 3 and from Theorem 4 that the set $\mathrm{Fr}(\mathscr{D})$ of all fit subregions of a given fit region $\mathscr{D}$ in $\mathscr{E}$ satisfies the axioms of a "material universe" as described in the Appendix of [N3] when the inclusion is taken to be the "is a part of" relation.

**Proposition 8.** *Let* $\mathscr{C}$ *be a subset of* $\mathscr{E}$ *that satisfies the conditions* (i), (iii), *and* (iv) *of Def. 7. Then there is exactly one fit region* $\mathscr{D}$ *in* $\mathscr{E}$ *that is volume-equivalent to* $\mathscr{C}$ *in the sense of Def. 5. In fact,* $\mathscr{D}$ *is given by* $\mathscr{D} := \mathrm{Int}\,\mathrm{Clo}\,\mathscr{C}$.

**Proof.** Since the boundary of $\mathscr{C}$ includes the boundary of $\mathrm{Int}\,\mathrm{Clo}\,\mathscr{C}$, it follows that $\mathrm{Int}\,\mathrm{Clo}\,\mathscr{C}$ has negligible boundary and hence is volume-equivalent to $\mathscr{C}$. In view of (4.5), $\mathscr{C}$ and $\mathrm{Int}\,\mathrm{Clo}\,\mathscr{C}$ have the same finite perimeter.

Since $\mathrm{Int}\,\mathrm{Clo}\,\mathscr{C}$ is regularly open by (A) of Sect. 3, it belongs to $\mathrm{Fr}(\mathscr{E})$. The uniqueness assertion follows from Prop. 6. □

Prop. 8 shows that the condition (ii) of Def. 7 becomes redundant if one does not distinguish between volume-equivalent sets. We will show in Sect. 6 that none of the conditions (i), (iii), and (iv) is redundant by giving examples of regularly open sets that satisfy two but not the third of these three conditions.

**Theorem 5.** *Let $\varphi : \mathscr{E} \to \mathscr{E}'$ be a $C^1$-diffeomorphism from the given Euclidean space $\mathscr{E}$ onto a Euclidean space $\mathscr{E}'$. Then the image under $\varphi$ of every fit region in $\mathscr{E}$ is a fit region in $\mathscr{E}'$.*

**Proof.** Let $\mathscr{D} \in \mathrm{Fr}(\mathscr{E})$ be given. Since $\varphi$ is a homeomorphism, the image $\varphi_>(\mathscr{D})$ of the regularly open set $\mathscr{D}$ under $\varphi$ is again regularly open. Also, we have $\varphi_>(\mathrm{Bdy}\,\mathscr{D}) = \mathrm{Bdy}\,\varphi_>(\mathscr{D})$. It is a (non-trivial) result of the theory of integration (see vol. II of [FDS]) that the image under $\varphi$ of a negligible set is negligible. Hence $\varphi_>(\mathscr{D})$ has a negligible boundary.

Since $\mathscr{D}$ is bounded and has finite perimeter, it follows from Theorem 3 that $\varphi_>(\mathscr{D})$ also is bounded and has finite perimeter. □

**Remark 5.** Sometimes it is useful in continuum physics to consider bodies that occupy regions that are unbounded and hence violate condition (i) of Def. 7. We believe that such bodies should be considered as "improper bodies" and should be treated in terms of proper bodies in a way which parallels the treatment of improper integrals in terms of proper integrals in elementary analysis. □

**Remark 6.** One may consider the subset $\mathrm{Fr}'(\mathscr{E})$ of $\mathrm{Fr}(\mathscr{E})$ that is obtained by replacing the condition (iv) of Def. 7 by the following weaker one:

(iv′) $\mathscr{D}$ has a boundary of finite area-measure.

The subclass Fr′ of Fr defined in this way is strictly smaller than Fr and still meets the three requirements mentioned in the introduction. However, given a subset of a Euclidean space, it is usually easier to decide whether the boundary is negligible than whether the boundary has finite area-measure. Also, the definition of the class Fr and the basic Theorems 4 and 5 can be understood in terms of elementary mathematics without concepts from measure-theory (although the *proofs* of the theorems *do* use a lot of measure theory). It is for these reasons that we believe that the class Fr is more fitting. □

## 6. Reduced boundaries

We assume that a fit region $\mathscr{D}$ in a Euclidean space $\mathscr{E}$ with translation space $\mathscr{V}$ is given and put $n := \dim \mathscr{V} = \dim \mathscr{E}$.

**Definition 8.** *We say that $\boldsymbol{n} \in \mathrm{Usph}\,\mathscr{V}$ is an* **outer unit normal** *to $\mathscr{D}$ at a given point $x \in \mathscr{E}$ if*

$$\lim_{\varrho \to 0} \frac{1}{\varrho^n} \mathrm{vol}\,(\mathscr{D} \cap \mathrm{Ball}_{x,\varrho}(\mathscr{E}) \cap (x + \mathrm{Hs}(\boldsymbol{n}))) = 0 \tag{6.1}$$

*and*

$$\lim_{\varrho \to 0} \frac{1}{\varrho^n} \operatorname{vol}\left((\mathscr{E} \setminus \mathscr{D}) \cap \operatorname{Ball}_{x,\varrho}(\mathscr{E}) \cap (x - \operatorname{Hs}(\boldsymbol{n}))\right) = 0, \tag{6.2}$$

*where the half-space* $\operatorname{Hs}(\boldsymbol{n})$ *is defined by*

$$\operatorname{Hs}(\boldsymbol{n}) := \{\boldsymbol{v} \in \mathscr{V} \mid \boldsymbol{n} \cdot \boldsymbol{v} \geqq 0\}. \tag{6.3}$$

*The set*

$$\operatorname{Rby} \mathscr{D} := \{x \in \mathscr{E} \mid \text{there exists an outer normal to } \mathscr{D} \text{ at } x\} \tag{6.4}$$

*is called the* **reduced boundary** *of* $\mathscr{D}$.

It is easily seen that the reduced boundary must be a subset of the boundary of $\mathscr{D}$:

$$\operatorname{Rby} \mathscr{D} \subset \operatorname{Bdy} \mathscr{D} \tag{6.5}$$

and that for each $x \in \operatorname{Rby} \mathscr{D}$, there is *exactly one* outer unit normal to $\mathscr{D}$ at $x$. Hence we may consider the mapping

$$\boldsymbol{n}_{\mathscr{D}}: \operatorname{Rby} \mathscr{D} \to \operatorname{Usph} \mathscr{V} \tag{6.6}$$

which assigns to each $x \in \operatorname{Rby} \mathscr{D}$ the (unique) outer unit normal $\boldsymbol{n}_{\mathscr{D}}(x)$ to $\mathscr{D}$ at $x$. We call $\boldsymbol{n}_{\mathscr{D}}$ the **outer-normal field** of $\mathscr{D}$.

The difference between the reduced boundary and the ordinary boundary may be significant because one can give examples of fit regions $\mathscr{D}$ for which $\operatorname{ar}(\operatorname{Bdy} \mathscr{D} \setminus \operatorname{Rby} \mathscr{D}) > 0$ (see Sect. 6). However, one can prove that $\operatorname{Clo} \operatorname{Rby} \mathscr{D} = \operatorname{Bdy} \mathscr{D}$ for every fit region $\mathscr{D}$. (This fact was pointed out to us by Prof. De Giorgi.)

We now mention, without proof, a few facts concerning reduced boundaries that are of great importance in the applications to continuum physics.

**Proposition 9.** *The reduced boundary* $\operatorname{Rby} \mathscr{D}$ *has finite area-measure and*

$$\operatorname{ar}(\operatorname{Rby} \mathscr{D}) = \operatorname{per}(\mathscr{D}). \tag{6.7}$$

*Moreover, the outer-normal field* $\boldsymbol{n}_{\mathscr{D}}$ *is area-integrable on* $\operatorname{Rby} \mathscr{D}$.

**Proposition 10.** $\operatorname{Rby} \mathscr{D}$ *differs from the union of a countable collection of compact subsets of* $C^1$*-surfaces only by a set of area-measure zero.*

**Theorem 6 (Integral-Gradient Theorem).** *For every continuous function* $f: \operatorname{Clo} \mathscr{D} \to \mathbb{R}$ *for which* $f|_{\mathscr{D}}$ *is differentiable and* $\nabla(f|_{\mathscr{D}})$ *integrable on* $\mathscr{D}$, *we have*

$$\int_{\mathscr{D}} \nabla(f|_{\mathscr{D}})\,\mathrm{dv} = \int_{\operatorname{Rby}\mathscr{D}} f \boldsymbol{n}_{\mathscr{D}}\,\mathrm{da}. \tag{6.8}$$

If $\mathscr{D}$ is a region with piecewise $C^1$-boundary, then this theorem reduces to the classical Integral-Gradient Theorem (see vol. II of [FDS]).

**Remark 7.** In order to develop a theory of contact interactions, one of us (W. N.) introduced the notion of *contact* of separate subbodies (see [N4]). Adapted to the present context this notion of contact corresponds to the intersection of the boundaries of two disjoint fit regions. It has now become clear that it may be more useful to define the contact of two disjoint fit regions to be the intersection of their *reduced* boundaries. In view of Prop. 9 one can then be assured that contacts are area-measurable. □

## 7. Counterexamples

The examples are all subsets of the Euclidean space $\mathcal{E} := \mathbb{R}^2$. Then $\mathrm{Ball}_{(q,\varrho)}(\mathcal{E})$ denotes the open disc of radius $\varrho$ centered at $q \in \mathcal{E}$. The measures ar and vol on $\mathcal{E}$ have the meanings of arclength and area, respectively, in common parlance.

(1) *A set* $\mathcal{D} \in \mathrm{Fr}(\mathcal{E})$ *such that* $\mathrm{ar}(\mathrm{Bdy}\,\mathcal{D} \setminus \mathrm{Rby}\,\mathcal{D}) > 0$:

Let $d \in \mathbb{P}^\times$ be given. For each $m \in \mathbb{N}^\times$ and each odd integer $k$ between 0 and $2^m$ we consider the disc of radius $\frac{d}{4^m}$ centered at the point $\left(\frac{k}{2^m}, \frac{d}{2^m}\right) \in \mathbb{R}^2$. Any two discs obtained in this way are disjoint (see Fig. 2). We denote the union of all these discs by $\mathcal{D}_d$, *i.e.*

$$\mathcal{D}_d := \bigcup \{\mathrm{Ball}_{\left(\frac{1}{2^m}(k,d), \frac{1}{4^m}d\right)}(\mathcal{E}) \mid m \in \mathbb{N}^\times, k \in (2^m)^{[} \cap (2\mathbb{N}+1)\}. \tag{7.1}$$

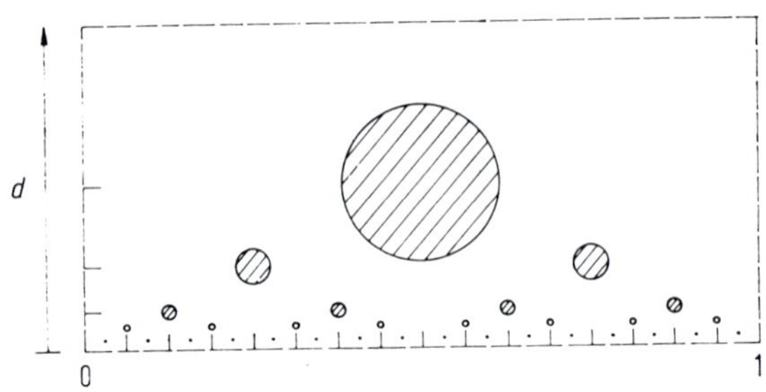


Fig. 2

It is easily seen that $\mathcal{D}_d$ is a fit region. The reduced boundary of $\mathcal{D}_d$ is the union of all the boundary-circles of the discs described above. To each $m \in \mathbb{N}^\times$ correspond $2^{m-1}$ such circles, all of length $2\pi \frac{d}{4^m}$. Hence the sum of their lengths is $2^{m-1}\left(2\pi \frac{d}{4^m}\right) = \frac{\pi d}{2^m}$. Therefore, in view of Prop. 9 of Sect. 6, the perimeter of $\mathcal{D}_d$ is given by

$$\mathrm{per}(\mathcal{D}_d) = \mathrm{ar}(\mathrm{Rby}\,\mathcal{D}_d) = \sum_{m \in \mathbb{N}^\times} \frac{\pi d}{2^m} = \pi d. \tag{7.2}$$

The boundary of $\mathscr{D}_d$ is the union of Rby $\mathscr{D}_d$ and the line segment $\{0\}\times[0, 1]$, which has length 1. Hence

$$\operatorname{ar}(\operatorname{Bdy}\mathscr{D}_d \setminus \operatorname{Rby}\mathscr{D}_d) = 1 > 0, \tag{7.3}$$

and $\mathscr{D} := \mathscr{D}_d$ has the desired properties.

(2) *A subset $\mathscr{C}$ of $\mathscr{E}$ that fails to be bounded but satisfies* (ii), (iii), *and* (iv) *of Def. 7.*

We may take $\mathscr{C} := \mathscr{E}$. Since $\mathscr{E} = \operatorname{Clo}\mathscr{E} = \operatorname{Int}\operatorname{Clo}\mathscr{E}$ and $\operatorname{Bdy}\mathscr{E} = \emptyset$, (ii), (iii) and (iv) are trivially valid. We may also take $\mathscr{C}$ to be the complement in $\mathscr{E}$ of any set belonging to Fr $(\mathscr{E})$.

(3) *A subset $\mathscr{G}$ of $\mathscr{E}$ whose boundary fails to be negligible but which satisfies* (i), (ii), *and* (iii) *of Def. 7.*

We consider an open subset $D$ of $[0, 1]$ that is dense in $[0, 1]$ but has a length-measure strictly less than 1. (For example, given $\alpha \in ]0, 1[$, we may take

$$D := \bigcup_{k\in\mathbb{N}^\times} \left]r_k - \frac{\alpha}{2^k}, r_k + \frac{\alpha}{2^k}\right[ \cap ]0, 1[, \tag{7.4}$$

where $(r_k \mid k \in \mathbb{N}^\times)$ is an enumeration of the rational numbers in $]0, 1[$. Then $0 < \operatorname{length}(D) \leqq \alpha < 1$.) Like every open subset of $\mathbb{R}$, $D$ is the union of a disjoint countable collection Oip $(D)$ of open intervals. Let $I \in \operatorname{Oip}(D)$ be given. We then have $I = ]s_I, s_I + d_I[$ for suitable $s_I \in ]0, 1[$, $d_I \in \mathbb{P}^\times$. Consider the subset of $]0, 1[\times I$ given by $(0, s_I) + \mathscr{D}_{d_I}$, where $\mathscr{D}_{d_I}$ is constructed according to Example (1) (see Fig. 3).

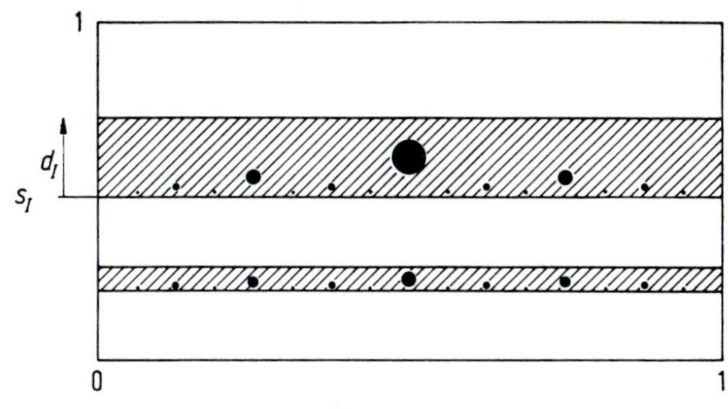


Fig. 3

Consider the union

$$\mathscr{G} := \bigcup \{(0, s_I) + \mathscr{D}_{d_I} \mid I \in \operatorname{Oip}(D)\} \tag{7.5}$$

of all sets obtained in this manner. It is easily seen that $\mathscr{G}$ is bounded, regularly open, and of finite perimeter. Its reduced boundary is the union of all the reduced boundaries of the sets $(0, s_I) + \mathscr{D}_{d_I}$, $I \in \operatorname{Oip}(D)$, and hence is a union of a countable collectlon of circles. It follows from (7.2) that

$$\operatorname{per}(\mathscr{G}) = \operatorname{ar}(\operatorname{Rby}\mathscr{G}) = \operatorname{per}\left(\bigcup_{I\in\operatorname{Oip}(D)} ((0, s_I) + \mathscr{D}_{d_I})\right)$$

$$= \sum_{I\in\operatorname{Oip}(D)} \operatorname{per}(\mathscr{D}_I) = \pi \sum_{I\in\operatorname{Oip}(D)} d_I = \pi \operatorname{length}(D). \tag{7.6}$$

The boundary of $\mathscr{G}$ is the union of Rby $\mathscr{G}$ and the union of all segments of the form $\{s\}\times[0, 1]$ with $s\in[0, 1[\setminus D$, *i.e.*,

$$\operatorname{Bdy}\mathscr{G}\setminus\operatorname{Rby}\mathscr{G} = ([0, 1[\setminus D)\times[0, 1]. \tag{7.7}$$

We have $\operatorname{vol}(\operatorname{Rby}\mathscr{G}) = 0$ but

$$\operatorname{vol}(\operatorname{Bdy}\mathscr{G}) = \operatorname{vol}(\operatorname{Bdy}\mathscr{G}\setminus\operatorname{Rby}\mathscr{G})$$

$$= \operatorname{length}([0, 1[\setminus D) = 1 - \operatorname{length}(D) > 0, \tag{7.8}$$

and hence Bdy $\mathscr{G}$ fails to be negligible.

**Remark 8.** One can show that $\operatorname{Clo}\mathscr{G} = \mathscr{G}\cup\operatorname{Bdy}\mathscr{G}$ does *not* have finite perimeter. □

(4) *A subset $\mathscr{H}$ of $\mathscr{E}$ that does not have finite perimeter but satisfies* (i), (ii), *and* (iv) *of Def. 7.*
It is not hard to see that

$$\mathscr{H} := \left\{(s, t)\in\mathbb{R}^2 \mid s\in\,]0, 1[,\ t\in\left]\sin\left(\frac{1}{s}\right), 2\right[\right\}$$

meets the requirements (see Fig. 4).

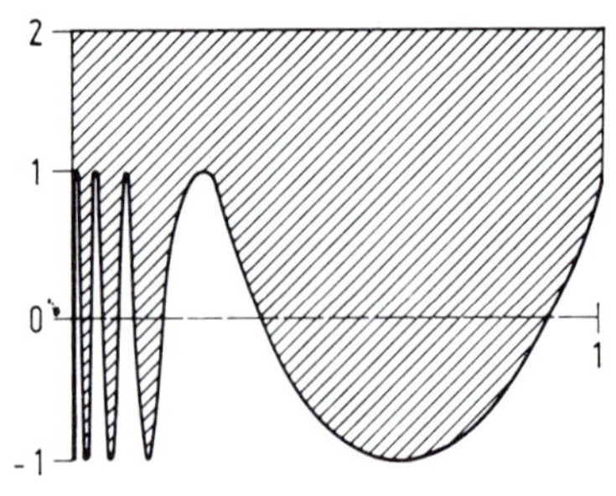


Fig. 4

*Acknowledgement.* The research leading to this paper was partially supported by a grant from the Italian Ministry of Education for the Research Project entitled: "Termomeccanica dei Continui".

## References


[A] ADAMS, R. A., "Sobolev Spaces", Academic Press, 1975.

[BF] BANFI, C. & M. FABRIZIO, "Sul Concetto di Sottocorpo nella Meccanica dei Continui", Rendiconti della Classe di Scienze fisiche, matematiche e naturali, Accademia Nazionale dei Lincei, Vol. LXVI, 136–142, 1979.

[DCP] DE GIORGI, E., F. COLOMBINI, & L. C. PICCININI, "Frontiere Orientate di Misura Minima e Questioni Collegate", Scuola Normale Superiore, Pisa, 1972.

[F] FEDERER, H., "Geometric Measure Theory", Springer Verlag, 1969.

[FDS] NOLL, W., "Finite-Dimensional Spaces: Algebra, Geometry, and Analysis, Vol. I", Martinus Nijhoff Publishers, 1987. Vol. II, to be published.

[GWZ] GURTIN, M. E., W. O. WILLIAMS & W. P. ZIEMER, "Geometric Measure Theory and the Axioms of Continuum Thermodynamics", Arch. Rational Mech. Anal. **92**, 1–22, 1986.

[N1] NOLL, W., "The foundations of classical mechanics in the light of recent advances in continuum mechanics", *Proceedings of the Berkeley Symposium on the Axiomatic Method*, 226–281, Amsterdam, 1959.

[N2] NOLL, W., "The foundations of mechanics", *Non-Linear Continuum Theories, C.I.M.E. Lectures*, 1965, pp. 159–200, Rome, 1966.

[N3] NOLL, W., "Lectures on the foundations of continuum mechanics and thermodynamics", Arch. Rational Mech. and Anal. **52**, 62–92, 1973.

[N4] NOLL, W, "Continuum Mechanics and Geometric Integration Theory", *Categories in Continuum Physics, Lecture Notes in Mathematics 1174*, 17–29, Springer Verlag, 1986.

[T] TRUESDELL, C, "A First Course in Rational Continuum Mechanics, Vol. I", Academic Press, 1977.

[VH] VOL'PERT, A. I., & S. I. HUDJAEV, "Analysis in Classes of Discontinuous Functions and Equations of Mathematical Physics", Martinus Nijhoff Publishers, 1985.



Department of Mathematics
Carnegie-Mellon University
Pittsburgh

and

Dipartimento di Matematica
Università di Pavia


*(Received August 10, 1987)*

# *The Lavrentiev Phenomenon for Invariant Variational Problems*

A. C. HEINRICHER & V. J. MIZEL

*Dedicated to* JAMES B. SERRIN

## 1. Introduction

In this paper we consider the problem of minimizing

$$\bar{I}[y] := \int_a^b \bar{f}(x, y(x), y'(x))\, dx \tag{1}$$

in the set $\bar{\mathscr{A}}$ of absolutely continuous functions $y(\cdot): [a, b] \to \mathbb{R}$ satisfying the end conditions

$$y(a) = \alpha, \quad y(b) = 0, \tag{2}$$

where $\alpha$ is a prescribed constant. In (1), $[a, b]$ is a bounded interval, "$'$" denotes differentiation with respect to $x$, and the integrand $\bar{f} = \bar{f}(x, y, p)$ is assumed to be smooth ($C^3$) and nonnegative. In addition, $\bar{f}$ is assumed to satisfy:

$$\bar{f}_{pp} \geqq 0. \tag{3}$$

The significance of this convexity is that, as is well known, it ensures the existence of at least one absolute minimizer for $\bar{I}$ in $\bar{\mathscr{A}}$, provided that $\bar{f}$ also satisfies an appropriate growth condition with respect to $p$.

We consider here problems for which the exact specification of the admissible class is crucial to the problem. If we restrict attention to trajectories in $\bar{\mathscr{A}}$ whose first derivative is an element of $L^{p_0}(a, b)$ for a certain $p_0 > 1$, call this smaller admissible class $\bar{\mathscr{A}}_{p_0}$, then the minimum cost is *strictly larger than the value attained by the absolute minimizer in* $\bar{\mathscr{A}}$. In fact, for each $\alpha \neq 0$, the strict inequality

$$\min \{\bar{I}[y] : y \in \bar{\mathscr{A}}\} < \min \{\bar{I}[y] : y \in \bar{\mathscr{A}}_{p_0}\}, \tag{4}$$

holds despite the fact that $\bar{\mathscr{A}}_{p_0}$ is dense in $\bar{\mathscr{A}}$.

The first example of a gap such as (4) between the *infimum* of the values of a variational integral on a dense subset $S$ of $\bar{\mathscr{A}}$ and its minimum value on $\bar{\mathscr{A}}$, was introduced by M. A. LAVRENTIEV [13] in 1926. In LAVRENTIEV'S example, $S$ was the set of continuously differentiable trajectories. MANIÀ [17] provided a simpler example where the dense subset $S$ was the class of Lipschitz ($W^{1,\infty}$) trajectories. (MANIÀ'S example is discussed in [6, Chapter 18] and in the recent paper of LOEWEN [14].) The next results in this direction were obtained by BALL & MIZEL [4, 5] during an investigation into the validity of necessary conditions for minimizers in nonlinear elasticity. Their work yielded the first construction of strictly regular ("elliptic") integrands in one dimension for which the Lavrentiev phenomenon occurs. (Recently DAVIE [8] has produced a profound improvement of the original examples of BALL & MIZEL [5].) HEINRICHER & MIZEL [9, 10] provided examples of stochastic control problems which exhibit the Lavrentiev phenomenon.

ANGELL [1] (see also CESARI [6]) has given conditions which are sufficient to preclude the Lavrentiev phenomenon. LOEWEN [14] has also considered this problem and extended ANGELL'S results. CLARKE & VINTER [7] show, in particular, that the Lavrentiev phenomenon cannot occur when the variational integrand $\bar{f} = \bar{f}(x, y, p)$ is independent of $x$.

The Lavrentiev phenomenon provides a serious obstacle for numerical schemes of minimization: the cost of any sequence in the smaller admissible class is bounded away from the true minimum value. Also, when a minimizer in the smaller admissible class exists, as in the examples we consider, the approximation scheme typically converges to this wrong solution. BALL & KNOWLES [3] have succeeded in the development of numerical approximation schemes which detect the lower energy "singular minimizers".

The present work aims to develop further understanding of the properties of a variational integrand in one dimension which lead to the Lavrentiev phenomenon. It will be shown that integrands endowed with a certain homogeneity are in a sense the borderline between integrands exhibiting the Lavrentiev phenomenon and those for which this phenomenon is absent.

Our examples are different from the "usual" examples of the Lavrentiev phenomenon in that it is possible to demonstrate the *existence* of minimizers in the smaller admissible class $S$, hereafter referred to as *pseudominimizers* (so that we really can write "min" on both sides of the inequality (4)). The usual arguments, as presented for example in [6, Chapter 18], demonstrate that Lipschitz trajectories must cross a certain region and pay a penalty for this crossing, while the absolute minimizer need not cross this region. BALL & MIZEL [5, Theorem 5.7] verify the Lavrentiev gap through this type of argument, but they also show that the minimum in the dense subset $S$ is actually attained. In [11] we considered problems where the Lavrentiev phenomenon could be demonstrated by solving the Hamilton-Jacobi equation and obtaining two distinct value functions, one for each class of admissible trajectories. (The absolute minimizer and the pseudominimizer could both be computed explicitly.) The methods employed here supplement those utilized in [5] by making use of field theory, in conjunction with NOETHER'S theorem for invariants of Lie groups, to replace the delicate phase plane analysis of the Euler-Lagrange equation sometimes used there.

The Lavrentiev phenomenon in the calculus of variations is a surprising phenomenon which is contrary to intuition. We take the time now to construct a (degenerate) example exhibiting the Lavrentiev phenomenon. This discussion will help motivate the properties required for our variational integrands, in particular their homogeneity. It also indicates why the problems described in Sections 2 and 3 of this paper are in a sense the borderline cases for the occurrence of the phenomenon.

Begin with a very simple fixed-endpoint problem: Minimize[1]

$$I[y] := \int_s^b |y'(x)|^\alpha \, dx,$$

where we take $s < b$ and $\alpha > 1$. It follows from Jensen's inequality that the minimizer in the admissible class

$$\mathscr{A}(s, z) := \{y(\cdot) \in W^{1,1}(s, b) : y(s) = z, y(b) = 0\}$$

is the straight line joining the prescribed endpoints. If we think of this as a family of problems indexed by the initial position $(s, z)$, then the *value function*

$$V(s, z) := \inf\{I[y] : y(\cdot) \in \mathscr{A}(s, z)\}$$

is constant along the semiparabolas

$$z = \delta(b - s)^{(1-1/\alpha)}, \quad s < b, \quad \delta \in \mathbb{R}.$$

In fact, the minimum cost everywhere on $z = \delta(b - s)^{(1-1/\alpha)}$ is just $|\delta|^\alpha$.

Define $\gamma := 1 - 1/\alpha$, fix $0 < \gamma_0 \leqq \gamma$ and $\delta_0 > 0$, and consider the curve

$$\Gamma_0 := \{(s, z) : z = \delta_0(b - s)^{\gamma_0}, \quad s < b\}.$$

Since $\gamma_0 \leqq \gamma$, the value function $V(\cdot, \cdot)$ is a *nondecreasing* function along this semiparabola. (If $\gamma_0 = \gamma$, $V(\cdot, \cdot)$ is constant on $\Gamma_0$, while $\gamma_0 < \gamma$ implies that $V(\cdot, \cdot)$ is strictly increasing along $\Gamma_0$, becoming infinite as $s \to 0^-$.) Now introduce a weight function $a = a(x, y)$ into the variational integrand which vanishes on the curve $\Gamma_0$ and is positive everywhere else, and consider the problem of minimizing

$$J[s, z; y] = J[y] := \int_s^b a(x, y(x)) \, |y'(x)|^\alpha \, dx$$

over the admissible class $\mathscr{A}(s, z)$ defined above. If we take

$$a(x, y) := \begin{cases} 0, & (x, y) \in \Gamma_0, \\ 1, & \text{otherwise}, \end{cases} \tag{5}$$

it is easy to see that this problem will exhibit the Lavrentiev phenomenon. For initial data $(s, z)$ such that $0 < z < \delta_0(b - s)^{\gamma_0}$, there is a trajectory in $\mathscr{A}(s, z)$ which attains zero cost:[2]

$$y_0(x) := z \wedge \delta_0(b - x)^{\gamma_0}, \quad x \in [s, b].$$

---

[1] We use the symbol ":=" to denote an equality in which the left-hand side is defined by the right-hand side.

[2] We use $a \wedge b$ to denote the minimum of $a$ and $b$, and $a \vee b$ to denote the maximum.

This trajectory is an element of $W^{1,p}(s,b)$ for $1 \leqq p < p_0$, where $p_0 := (1-\gamma_0)^{-1}$. Now if we restrict attention to the smaller class

$$\mathscr{A}(s,z) \cap W^{1,p_0} := \{y(\cdot) \in W^{1,p_0}(s,b) : y(s) = z,\ y(b) = 0\},$$

then the absolute minimizer $y_0(\cdot)$ is no longer admissible. In fact, the minimizer in the new admissible class is the straight line:

$$y_1(x) = \left(\frac{z}{b-s}\right)(b-x), \qquad x \in [s,b].$$

We have the Lavrentiev phenomenon:

$$\inf_{y(\cdot)\in\mathscr{A}(s,z)} J[s,z;y] = J[y_0] = 0,$$

while

$$\inf_{x\in\mathscr{A}(s,z)\cap W^{1,p_0}} J[s,z;y] = J[y_1] = z^{\alpha}(b-s)^{1-\alpha} > 0.$$

This simple example gives a hint of a more serious form of the Lavrentiev gap. Not only is there a lower bound on the cost of any approximating sequence, but it may happen that the cost along any sequence approximating the absolute minimizer is unbounded. For example, assume that $\gamma_0 < \gamma$ and consider the following approximating sequence:

$$w_n(x) := \begin{cases} y_0(x), & s \leqq x \leqq b - 1/n, \\ \delta_0(n)^{1-\gamma_0}(b-x), & b - 1/n < x \leqq b. \end{cases}$$

(That is, follow the absolute minimizer up to $x = b - 1/n$ and then switch to follow the straight line to $(b, 0)$.) This is an approximating sequence of Lipschitz functions converging uniformly to $y_0(\cdot)$ on $[s, b]$, and (for $n$ sufficiently large)

$$\begin{aligned} J[s,z;w_n] &= \int_{b-1/n}^{b} |\delta_0(n)^{1-\gamma_0}|^{\alpha}\,dx \\ &= \delta_0^{\alpha}\cdot(n)^{\alpha(1-\gamma_0)-1} \\ &= \delta_0^{\alpha}\cdot(n)^{\frac{\gamma-\gamma_0}{1-\gamma}}. \end{aligned}$$

Hence, since $\gamma_0 < \gamma$, we have

$$\lim_{n\to+\infty} J[s,z;w_n] = +\infty.$$

In fact, the cost will become infinite along any approximating sequence in $W^{1,p_0}(s,b)$ which converges to $y_0(\cdot)$ pointwise on $(s,b)$. We demonstrate this strong form of the Lavrentiev phenomenon for a class of problems in Section 4 (*cf.* Theorem C.2). This class includes MANIÀ'S example as well as the fully regular problem described by BALL & MIZEL [5].

The example described above is not interesting in itself; it *should* be possible to construct strange examples if one allows discontinuous integrands. The example does provide insight into the mechanism behind the Lavrentiev phenomenon.

The problems we discuss in this paper retain the following key properties of the example:

1. There are zero-cost curves, curves which are in $W^{1,p}$ only for $1 \leqq p < p_0$. The absolute minimizers follow these zero-cost curves on some left-neighborhood of the final time and they are not admissible when the trajectories are restricted to $W^{1,p_0}$.
2. A value function is constant along curves of the form

$$z = \delta(b - s)^\gamma, \quad s < b, \quad \delta \in \mathbb{R},$$

and this value function will be *nondecreasing* along the zero-cost curves.

We will obtain the second of these properties for problems with smooth integrands by requiring the following *homogeneity*:

$$(\overline{\mathbf{HP}}) \quad \bar{f}(\lambda s + b, \lambda^\gamma y, \lambda^{\gamma-1} p) = \lambda^{-1} \bar{f}(s + b, y, p), \quad \lambda > 0, s < 0, y, p \in \mathbb{R}, \tag{6}$$

for some $\gamma \in (0, 1)$. We show that $(\overline{\mathbf{HP}})$ implies that the value function is constant along curves of the form $y = \delta(b - x)^\gamma$. (See Lemma 2.1 and the remarks following.)

The first property follows when we require that $\bar{f}(x, y, p)$ vanish on some (finitely many) curves of the form $y = \delta_j(b - x)^\gamma$, $\delta_j \in \mathbb{R}$. Notice in particular that in terms of the discontinuous example described above, we are in the borderline case $\gamma_0 = \gamma$, where the value function is *constant* along the zero-cost curves.

In addition to the Lagrange problem (1), (2) we will also demonstrate the occurrence of the Lavrentiev gap (4) for certain Bolza problems of the form

$$\bar{I}[y] := \int_a^b \bar{f}(x, y(x), y'(x))\, dx + \varphi(y(b)), \tag{7}$$

$$y(a) = \alpha, \tag{8}$$

where $\varphi(\cdot) : \mathbb{R} \to \mathbb{R}$ is lower semicontinuous and constant away from $y = 0$. The Lagrange problem is the limiting case of this latter problem with

$$\varphi(y) = \begin{cases} 0, & y = 0, \\ +\infty, & \text{otherwise}. \end{cases}$$

The paper is organized as follows. Section 2 is devoted to the analysis of the Lagrange problem. The main results of this section are Lemma 2.4, where we construct the Mayer field, and Theorem A, where we prove the existence of minimizers and pseudominimizers.

Section 3 is devoted to the Bolza problem. For terminal costs of special form, the results of Theorem A extend almost immediately to the Bolza problem.

In section 4 we study perturbations of the original Lagrange problem. For some of these perturbations, the Lavrentiev phenomenon persists, while for some perturbations the phenomenon disappears. By considering additive perturbations, we obtain a large class of *fully regular* problems exhibiting the Lavrentiev gap. By considering multiplicative perturbations, we obtain a class of problems exhibiting the strong form of the Lavrentiev phenomenon described above, where the cost for any sequence approximating the absolute minimizer (pointwise) must become unbounded. (BALL & MIZEL [5] demonstrate this strong form of the Lavrentiev phenomenon for their examples. LOEWEN [14] demonstrates the strong form for MANIÀ's example.) By considering another class of multiplicative perturbations, we eliminate the Lavrentiev gap. This indicates that the invariant integrals discussed here are, in a sense, the borderline cases for the Lavrentiev phenomenon.

## 2. The Lagrange Problem

### 2.1. Introduction

In this section we provide a complete analysis of the (fixed-endpoint) Lagrange problem. We prove the existence of distinct optimal trajectories in two different admissible classes (*i.e.*, absolute minimizers and pseudominimizers) and verify the existence of the Lavrentiev phenomenon.

It will be convenient to change the origin by the substitution $s = x - b$ and to denote

$$z(s) = y(s + b), \quad f(s, z, p) = \bar{f}(s + b, z, p)$$

for $s \leqq 0$, $z, p \in \mathbb{R}$. If then $s_0 := a - b$, the original problem (1), (2) becomes one of minimizing

$$I[z] := \int_{s_0}^{0} f(s, z(s), z'(s))\, ds \tag{9}$$

over the class $\mathscr{A} = \mathscr{A}(s_0, z_0)$ of absolutely continuous functions $z(\cdot) : [s_0, 0] \to \mathbb{R}$ satisfying

$$z(s_0) = z_0, \quad z(0) = 0. \tag{10}$$

We consider this problem for initial data $(s_0, z_0)$ in the set

$$Q := \{(s, z) : s < 0, z \in \mathbb{R}\},$$

and denote by $Q_+$ and $Q_-$ the intersection of $Q$ with $z > 0$ and $z < 0$, respectively. If we wish to emphasize the dependence on the initial data we write

$$I[z] = I[s_0, z_0; z].$$

### 2.2. Basic Technical Assumptions and their Consequences

The hypotheses made on $\bar{f}$ imply that $f = f(s, z, p)$ is a nonnegative, $C^3$ function of its arguments with

$$f_{pp}(s, z, p) \geqq 0, \quad (s, z) \in Q, p \in \mathbb{R}. \tag{11}$$

Homogeneity $(\overline{\mathrm{HP}})$ becomes

**(HP)** $\quad f(\lambda s, \lambda^{\gamma} z, \lambda^{\gamma-1} p) = \lambda^{-1} f(s, z, p), \qquad \forall\, \lambda > 0, \quad (s, z) \in Q,\, p \in \mathbb{R}.$

Notice that this implies

$$f(s, z, 0) = f_p(s, z, 0) = 0 \quad \text{for all } (s, z) \in Q. \tag{12}$$

Also, if we take $\lambda = (-s)^{-1}$ and define the new (scaled) variables

$$Z := (-s)^{-\gamma} z, \qquad P := (-s)^{1-\gamma} p,$$

it follows that

$$f(s, z, p) = (-s)^{-1} f(-1, Z, P) =: (-s)^{-1} e(Z, P).$$

The properties of $f$ imply immediately that $e: \mathbb{R}^2 \to \mathbb{R}$ is nonnegative, $C^3$ in its arguments, and satisfies

$$e_{PP}(Z, P) \geqq 0, \qquad Z, P \in \mathbb{R}.$$

Property (12) provides

$$e(Z, 0) = e_P(Z, 0) = 0, \qquad Z \in \mathbb{R}. \tag{13}$$

Convexity and (13) imply the growth condition

$$e(Z, P) \leqq P e_P(Z, P), \qquad Z, P \in \mathbb{R}.$$

We impose a strengthened form of this condition which, in particular, implies at most algebraic growth for the function $e(Z, \cdot)$:

$$K' e(Z, P) \leqq P e_P(Z, P) \leqq K e(Z, P), \qquad Z, P \in \mathbb{R}, \tag{14}$$

for some constants $K \geqq K' > 1$.

As mentioned in the introduction, the curves $Z = \delta$ (i.e., $z(s) = \delta(-s)^{\gamma}$, $s < 0$), for $\delta$ some constant, play a crucial role in our analysis. Notice that along such curves $P = -\gamma\,\delta$ and so

$$I[z] = \int_{s_0}^{0} (-s)^{-1}\, e(\delta, -\gamma\,\delta)\, ds = +\infty$$

*unless* $e(\delta, -\gamma\,\delta) = 0$. Our demonstration of the Lavrentiev phenomenon requires the existence of $\delta \neq 0$ such that $e(\delta, -\gamma\,\delta) = 0$. In particular, *we require that* $e(\cdot, \cdot)$ *possess a nonempty finite set of zeros along the line* $P = -\gamma Z$ (in addition to the zero $(0, 0)$):

$$\begin{aligned} \mathscr{Z} &:= \{\delta \in \mathbb{R} : e(\delta, -\gamma\,\delta) = 0\} \\ &= \{\delta_j : j = -m', \ldots, -1, +1, \ldots, m\}, \end{aligned} \tag{15}$$

where $m, m' \geqq 1$, $\delta_{-m'} < \ldots < \delta_{-1} < 0 < \delta_1 < \ldots < \delta_m$. The loci $Z = \delta_j$, $\delta_j \in \mathscr{Z}$ correspond to curves which incur zero cost. For this reason, we refer to $\mathscr{Z}$ as the *free zone*.

We assume that a strict form of (11) holds away from $\mathscr{Z}$,

$$e_{PP}(Z, P) > 0, \qquad Z \notin \mathscr{Z}, P \neq 0,$$

and that for each $d > 0$,

$$c_d(P) := \inf\{e(Z, P) : \operatorname{dist}(Z, \mathscr{Z}) \geqq d\} > 0, \qquad P \neq 0. \tag{16}$$

Finally, the following ratio is important in our construction of the Mayer field (*cf.* Lemma 4.2, equation (34)):

$$\frac{e(Z, P)}{Pe_P(Z, P)}.$$

This ratio is not defined for $Z \in \mathscr{Z}$, but we assume that it may be extended to the free zone to obtain a continuous function. That is, we assume that

$$(Z, P) \mapsto \frac{e(Z, P)}{Pe_P(Z, P)} \tag{17}$$

is the restriction of a Lipschitz-continuous function on all of $\mathbb{R}^2$.

These properties can be motivated by considering some special types of variational integrands and some examples. In the case of product integrands

$$f(s, z, p) = a(s, z)\, b(s, p),$$

the function $e(\cdot, \cdot)$ takes the form

$$e(Z, P) = d(Z)\, g(P),$$

where

$$d(Z) := a(-1, Z), \qquad g(P) := b(-1, P).$$

The properties of $e(\cdot, \cdot)$ described above reduce to

$$0 \leqq d(Z), 0 \leqq g(P), \quad g''(P) > 0 \quad \text{for } P \neq 0,$$
$$K'g(P) \leqq Pg'(P) \leqq Kg(P),$$
$$\mathscr{Z} = \{\delta \neq 0 : d(\delta) = 0\} \neq \emptyset \quad \text{is finite}.$$

Notice that the ratio mentioned in (17) is independent of $Z$:

$$\frac{e(Z, P)}{Pe_P(Z, P)} = \frac{g(P)}{Pg'(P)}.$$

HEINRICHER & MIZEL [9, 10, 11] have considered product integrands of the form

$$f(s, z, p) = a(s, z)\, |p|^{\alpha},$$

where $\alpha > 2$ and **(HP)** holds with $\gamma = 1/2$. Notice that

$$\frac{e(Z, P)}{Pe_P(Z, P)} \equiv \frac{1}{\alpha},$$

and so (17) is trivially satisfied, and (14) is satisfied with $K' = K = \alpha$. A specific example is given by

$$f(s, z, p) := (z^2 + s)^2 p^6, \quad (s, z) \in Q, p \in \mathbb{R}, \tag{18}$$

so that

$$e(Z, P) = (Z^2 - 1)^2 P^6.$$

The free zone is simply

$$\mathscr{Z} = \{-1, +1\},$$

and the two semiparabolas $z(s) = \pm\sqrt{-s}$, $s < 0$, are curves incurring zero cost.

MANIÀ'S example, as described by CESARI [6, Section 18.4] but with time reversed so that $s = 0$ is the final time, takes the following form:

$$f(s, z, p) = (z^3 + s)^2 |p|^\alpha, \quad (s, z) \in Q, p \in \mathbb{R},$$

where $\alpha \geqq 6$. This integrand satisfies the conditions specified with the smaller exponent $\alpha = 9/2$ (and $\gamma = 1/3$). We show in Section 4 that the Lavrentiev phenomenon exists for all $\alpha \geqq 9/2$, with the strong form of the phenomenon exhibited for $\alpha > 9/2$.

BALL & MIZEL [5, section 5] consider the integrand

$$f(s, z, p) = (z^6 - s^4)^2 |p|^\beta + \varepsilon |p|^2, \quad (s, z) \in Q, p \in \mathbb{R}, \tag{19}$$

for $\beta \geqq 27$ and $\varepsilon > 0$. When $\beta = 27$ and $\varepsilon = 0$, this integrand satisfies **(HP)** with $\gamma = 2/3$. The free zone is once again

$$\mathscr{Z} = \{-1, +1\},$$

and the two semiparabolas $z(s) = \pm(-s)^{2/3}$, $s < 0$, are curves incurring zero cost.

We state two lemmas which are consequences of homogeneity **(HP)** and the growth condition (14).

**Lemma 2.1.** *Given* $(s_0, z_0) \in Q$, *and* $z(\cdot) \in \mathscr{A}(s_0, z_0)$, *let* $Z_0 := (-s_0)^{-\gamma} z_0$. *Then*

$$I[s_0, z_0; z] = I[-1, Z_0; \hat{z}], \tag{20}$$

*where*

$$\hat{z}(\bar{s}) := (-s_0)^{-\gamma} z((-s_0)\bar{s}), \quad -1 \leqq \bar{s} \leqq 0. \tag{21}$$

**Proof.** Defining $s := \lambda\bar{s}$, with $\lambda = (-s_0)$, we have

$$\hat{z}'(\bar{s}) = \lambda^{1-\gamma} z'(s).$$

Homogeneity **(HP)** provides

$$f(\bar{s}, \hat{z}(\bar{s}), \hat{z}'(\bar{s}))\frac{d\bar{s}}{ds} = f(\lambda^{-1} s, \lambda^{-\gamma} z(s), \lambda^{1-\gamma} z'(s))\, \lambda^{-1}$$

$$= f(s, z(s), z'(s)).$$

The result follows by integrating. □

*Remark.* This lemma can be used to show that the value function, $V = V(s, z)$, is a function only of the ratio $Z = (-s)^{-\gamma} z$: given any admissible trajectory starting at $(s_0, z_0)$, there is another admissible trajectory, obtaining the same cost, which starts at $(-1, Z_0)$.

The next lemma provides a lower estimate for the cost incurred when an admissible trajectory crosses a strip $[Z_1, Z_1']$, where $0 < Z_1 < Z_1'$. In particular, since any admissible trajectory which happens to be an element of $W^{1,p_0}$, with $p_0 := (1-\gamma)^{-1}$, must cross each such strip, we obtain a (positive) lower bound for the cost of such trajectories.

**Lemma 2.2.** *Given* $Z_0 = z_0(-s_0)^{-\gamma} > 0$, *set*

$$\delta = \delta(Z_0) := \begin{cases} \max\{(0, Z_0) \cap \mathscr{Z}\}, & \text{if } (0, Z_0) \cap \mathscr{Z} \neq \emptyset, \\ Z_0, & \text{if } (0, Z_0) \cap \mathscr{Z} = \emptyset, \end{cases}$$

*and select* $Z_1, Z_1' \in (0, \delta)$ *satisfying* $Z_1 < Z_1'$, $[Z_1, Z_1'] \cap \mathscr{Z} = \emptyset$. *Then for all* $z(\cdot) \in \mathscr{A}(s_0, z_0) \cap W^{1,p_0}$,

$$I[z] \geqq k_1 > 0,$$

*where* $k_1$ *is explicitly computable in terms of* $Z_1, Z_1'$ *and* $e(\cdot, \cdot)$.

**Proof.** We first show that $e(Z, P)$ is bounded from below by a nonnegative convex function of $P$ (vanishing only at $P = 0$) as long as $Z$ is away from the free zone.

Consider the growth condition (14). For $P_0 = \pm 1$, one concludes by integrating from $P_0$ to $P$, with $P \cdot P_0 > 0$, that

$$\begin{aligned} e(Z, P_0)\,|P|^{K'} &\leqq e(Z, P) \leqq e(Z, P_0)\,|P|^{K}, \quad P/P_0 \geqq 1, \\ e(Z, P_0)\,|P|^{K'} &\geqq e(Z, P) \geqq e(Z, P_0)\,|P|^{K}, \quad 0 < P/P_0 < 1. \end{aligned} \tag{22}$$

Define $c_d^* := \min(c_d(-1), c_d(+1))$, which is strictly positive by (16). It follows from (22) that the set of convex functions $\{e(Z, \cdot) : \operatorname{dist}(Z, \mathscr{Z}) \geqq d\}$ satisfies

$$c_d^*\,(|P|^{K'} \wedge |P|^{K}) \leqq e(Z, P), \quad P \in \mathbb{R}.$$

Thus if we let $\psi_d(\cdot)$ denote the (strictly) monotone convex envelope of the functions

$$r \mapsto c_d^*\,|r|^{K'}, \quad r \mapsto c_d^*\,|r|^{K}, \quad r \in \mathbb{R},$$

we have

$$\psi_d(P) \leqq e(Z, P), \quad P \in \mathbb{R}, \quad \operatorname{dist}(Z, \mathscr{Z}) \geqq d. \tag{23}$$

Let $z(\cdot) \in \mathscr{A}(s_0, z_0) \cap W^{1,p_0}$ be given. It follows from Hölder's inequality that this trajectory crosses the strip $[Z_1, Z_1']$, and consequently we may choose $s_1, s_1'$, with $s_0 < s_1' < s_1 < 0$, satisfying

$$Z(s_1') = z(s_1')\,(-s_1')^{-\gamma} = Z_1', \quad Z(s_1) = z(s_1)\,(-s_1)^{-\gamma} = Z_1,$$

and

$$Z(s) = z(s)\,(-s)^{-\gamma} \in [Z_1, Z_1'] \quad \text{for all } s \in [s_1', s_1].$$

Now set $d := \operatorname{dist}([Z_1, Z_1'], \mathscr{Z})$. The lower bound (23) holds for all $Z \in [Z_1, Z_1']$ and we have

$$I[z] \geqq \int_{-s_1'}^{s_1} (-s)^{-1}\, e(z(s)\,(-s)^{-\gamma}, z'(s)\,(-s)^{1-\gamma})\, ds$$

$$\geqq \int_{-s_1'}^{s_1} (-s)^{-1}\, \psi_d(z'(s)\,(-s)^{1-\gamma})\, ds.$$

Setting $-s =: e^{-\sigma}$ and using Jensen's inequality we obtain

$$\int_{-s_1'}^{s_1} (-s)^{-1}\, \psi_d(z'(s)\,(-s)^{1-\gamma})\, ds = \int_{-\sigma_1'}^{\sigma_1} \psi_d(z'(s(\sigma))\, \mathrm{e}^{-(1-\gamma)\sigma})\, d\sigma$$

$$\geqq (\sigma_1 - \sigma_1')\, \psi_d\left(\frac{1}{(\sigma_1 - \sigma_1')} \int_{-\sigma_1'}^{\sigma_1} \frac{d}{d\sigma}(z(s(\sigma)))\, \mathrm{e}^{\gamma\sigma}\, d\sigma\right)$$

$$= (\sigma_1 - \sigma_1')\, \psi_d\left(\frac{1}{(\sigma_1 - \sigma_1')} \int_{-\sigma_1'}^{\sigma_1} \left[\frac{d}{d\sigma}(Z(s(\sigma))\, \mathrm{e}^{\gamma\sigma}) - \gamma Z(s(\sigma))\right] d\sigma\right)$$

$$= (\sigma_1 - \sigma_1')\, \psi_d\left(-\frac{Z_1' - Z_1}{(\sigma_1 - \sigma_1')} - \gamma \bar{Z}\right),$$

where $\bar{Z} \in [Z_1, Z_1']$ is the mean of the function $z(s(\cdot))\,(-s(\cdot))^{-\gamma}$. Because $\psi_d(\cdot)$ is monotonic, we can replace $\bar{Z}$ by $Z_1$ and we have

$$I[z] \geqq (Z_1' - Z_1) \inf_{r>0} \frac{\psi_d(-\gamma Z_1 - r)}{r}.$$

But, by the properties of $\psi_d(\cdot)$,

$$\frac{\psi_d(-\gamma Z_1 - r)}{r} = \frac{\psi_d(\gamma Z_1 + r)}{r}$$

$$= \frac{\psi_d(\gamma Z_1 + r)}{\gamma Z_1 + r}\left(1 + \frac{\gamma Z_1}{r}\right)$$

$$\geqq \frac{\psi_d(\gamma Z_1)}{\gamma Z_1}.$$

Defining

$$k_1 = k_1(\gamma, Z_1, Z_1') := (Z_1' - Z_1)\frac{\psi_d(\gamma Z_1)}{\gamma Z_1}, \tag{24}$$

we have the desired lower bound. □

*Remark*. Lemma 2.2 is the basis for the standard proof of the Lavrentiev phenomenon: any $W^{1,p_0}$ trajectory must pay a positive penalty $k_1$, while $W^{1,1}$ trajectories can avoid this penalty.

### 2.3. The Mayer Field

We make use of field theory arguments (see, for example [6], [12]) in our analysis of the Lagrange problem (9), (10). Homogeneity **(HP)** allows us to construct a Mayer field. The key tool is the theorem of E. NOETHER concerning invariant variational problems [19], (see also [20], [21], or [15]).

Homogeneity was used in Lemma 2.1 to show that the cost starting at $(s_0, z_0)$ could be computed by shifting the initial position to $(-1, Z_0)$, where $Z_0 := (-s_0)^{-\gamma} z_0$. We now restate this result as an invariance under a one-parameter (Lie) group of transformations $\mathscr{G} = \{G_\varepsilon : \mathbb{R}^2 \to \mathbb{R}^2, \varepsilon \in \mathbb{R}\}$, where $G_\varepsilon$ is defined by

$$G_\varepsilon(s, z) := (\phi(s, z; \varepsilon), \psi(s, z; \varepsilon)),$$

with

$$\phi(s, z; \varepsilon) := e^{\varepsilon} s, \tag{25}$$

$$\psi(s, z; \varepsilon) := e^{\gamma\varepsilon} z, \qquad (s, z) \in \mathbb{R}^2, \varepsilon \in \mathbb{R}.$$

Thus $G_0 : \mathbb{R}^2 \to \mathbb{R}^2$ is the identity mapping and the group generator of $\mathscr{G}$ is given by

$$\tau(s, z) = \phi_\varepsilon(s, z; 0) = s, \qquad \xi(s, z) = \psi_\varepsilon(s, z; 0) = \gamma z, \qquad (s, z) \in \mathbb{R}^2. \tag{26}$$

The group $\mathscr{G}$ has as *first-order prolongation* (see [21] for this and related notions) the group $\hat{\mathscr{G}} = \{\hat{G}_\varepsilon : \mathbb{R}^3 \to \mathbb{R}^3, \varepsilon \in \mathbb{R}\}$ which is given by

$$\hat{G}_\varepsilon(s, z, p) = (\phi(s, z; \varepsilon), \psi(s, z; \varepsilon), \chi(s, z, p; \varepsilon)), \qquad (s, z, p) \in \mathbb{R}^3,$$

with $\phi$, $\psi$ as above and $\chi$ determined as follows. Given any function $z(\cdot) \in C^1([c_1, d_1])$, $[c_1, d_1] \subset (\infty, 0]$, let $\hat{z}(\cdot)$ denote the function whose graph is the image under $G_\varepsilon$ of the graph of $z(\cdot)$. That is,

$$\hat{z}(\bar{s}) = e^{\gamma\varepsilon} z(e^{-\varepsilon} \bar{s}), \qquad \bar{s} \in [e^{\varepsilon} c_1, e^{\varepsilon} d_1] =: [\bar{c}_1, \bar{d}_1],$$

so that

$$\hat{p}(\bar{s}) := \frac{d}{d\bar{s}} \hat{z}(\bar{s}) = e^{(\gamma-1)\varepsilon} z'(s) = e^{(\gamma-1)\varepsilon} p(s), \qquad s = e^{-\varepsilon}\bar{s}, \bar{s} \in [\bar{c}_1, \bar{d}_1].$$

Then $\chi$ is defined by

$$\chi(s, z, p; \varepsilon) := e^{(\gamma-1)\varepsilon} p,$$

and the group generator of $\hat{\mathscr{G}}$ is given by

$$\tau(s, z) = s, \qquad \xi(s, z) = \gamma z, \qquad \chi_\varepsilon(s, z, p; 0) = (\gamma - 1) p, \qquad (s, z, p) \in \mathbb{R}^3.$$

Lemma 2.1 can be interpreted as the statement that the variational integral is *absolutely invariant*[3] under the prolonged group $\hat{\mathscr{G}}$.

---

[3] We actually have something stronger than absolute invariance, which only requires that the change in the cost under $\hat{G}_\varepsilon$ be $o(\varepsilon)$.

The absolute invariance of the variational integral ensures by a theorem of E. NOETHER ([19], [21, section 4.4]) that the first-order expression

$$\varkappa(s, z, p) := \tau(s, z)\,[f - pf_p]\,(s, z, p) + \xi(s, z)\,f_p(s, z, p) \tag{27}$$

is a first integral of the Euler-Lagrange equation. That is, on any subinterval of $[s_0, 0)$ where a solution of the integrated form of the Euler-Lagrange equation

$$f_p(s, z(s), z'(s)) = \int_{s_0}^{s} f_z(t, z(t), z'(t))\,dt + \text{const.}$$

is $C^2$, it is a solution of the first-order equation

$$\varkappa(s, z(s), z'(s)) = \varkappa_0, \tag{28}$$

for some value of the constant $\varkappa_0$.

We proceed to show that any minimizers of (9), (10) which are smooth on $[s_0, 0)$ satisfy (28) with $\varkappa_0 = 0$.

**Lemma 2.3.** *If* $z(\cdot) \in \mathscr{A}(s_0, z_0)$ *is a minimizer for* (9), (10) *and if for some interval* $[-c_1, 0) \subset [-s_0, 0)$, $z(\cdot) \in C^1([-c_1, 0))$, *then* $z(\cdot)$ *satisfies the differential equation* (28) *with* $\varkappa_0 = 0$.

**Proof.** Notice that if at some point $s \in [s_0, 0)$, $p(s) = z'(s) = 0$, then $\varkappa_0 = 0$ (since (12) implies that $\varkappa(s, z, 0) = 0$). Thus we can and will assume that $p(s) \neq 0$ throughout $[s_0, 0)$. Next notice that (12) ensures that each minimizer of (9), (10) is monotone since any nonmonotone portion of a trajectory could be replaced by a horizontal line segment, attaining a strictly lower cost. Hence, we can assume that

$$z(s) \cdot z'(s) < 0, \qquad s \in [s_0, 0). \tag{29}$$

Finally, observe that $I[z] < +\infty$ requires $F(\cdot) \in L^1(s_0, 0)$, where

$$F(s) := f(s, z(s), z'(s)) = (-s)^{-1}\, e(Z(s), P(s)).$$

Rewriting (28) in terms of $Z$ and $P$ (suppressing their dependence on $s$), we have

$$\varkappa_0 - (Pe_P(Z, P) - e(Z, P)) = \gamma Z e_P(Z, P). \tag{30}$$

If we multiply both sides by $P \cdot Z$, then

$$PZ[\varkappa_0 - (Pe_P(Z, P) - e(Z, P))] = \gamma Z^2 P e_P(Z, P),$$

and by (14), the right hand side is nonnegative. By (29), $P \cdot Z < 0$ and it follows that

$$\varkappa_0 - (Pe_P(Z, P) - e(Z, P)) \leqq 0.$$

But then the growth condition (14) provides

$$\begin{aligned} \varkappa_0 &\leqq (K - 1)\, e(Z, P) \\ &= (K - 1)\,(-s)\,F(s), \end{aligned}$$

and since $F(\cdot) \in L^1(s_0, 0)$, the final term can be made arbitrarily small with an appropriate choice of $s$ (near 0). It follows that $\varkappa_0 \leqq 0$.

Assume now that $\varkappa_0 < 0$. Multiplying both sides of (30) by $z'(s)$, we have

$$\begin{aligned}[\varkappa_0 - (Pe_P(Z, P) - e(Z, P))]\, z'(s) &= (\gamma Z e_P(Z, P))\, z'(s) \\ &= (\gamma(-s)^{-\gamma}\, e(Z, P))\, z(s) \cdot z'(s) \\ &= (\gamma(-s)^{-1}\, Pe_P(Z, P))\, z(s)\end{aligned}$$

and so,

$$z'(s) = \frac{\gamma(-s)^{-1}\, Pe_P\,(Z, P)}{\varkappa_0 - (Pe_P(Z, P) - e(Z, P))}\, z(s).$$

That is, $z(\cdot)$ satisfies a linear ordinary differential equation,

$$z'(s) = -g(s)\, z(s) \qquad z(0-) = 0, \tag{31}$$

with $g(\cdot)$ defined in the obvious way. Now $\varkappa_0 < 0$ implies that $g(\cdot) \in L^1(s_0, 0)$ since

$$\begin{aligned}0 \leqq g(s) &= \frac{\gamma(-s)^{-1}\, Pe_P(Z, P)}{-\varkappa_0 + (Pe_P(Z, P) - e(Z, P))} \\ &\leqq \frac{\gamma}{-\varkappa_0}(-s)^{-1}\, Pe_P(Z, P) \\ &\leqq \frac{\gamma K}{-\varkappa_0} F(s).\end{aligned}$$

But then (31) has a unique solution $z(s) \equiv 0$, and this contradicts the assumption that $p(s) = z'(s) \neq 0$ for all $s \in [s_0, 0)$. Hence $\varkappa_0 = 0$. $\square$

We next show that equation (28) (with $\varkappa_0 = 0$) has a unique $C^1$ solution for each $(s_0, z_0) \in Q$. These solutions define a Mayer field for the Lagrange problem and will be used in Theorem A to construct the pseudominimizers.

**Lemma 2.4.** *For each $(s_0, z_0) \in Q$, there is a $C^1$ solution for the differential equation*

$$\varkappa(s, z(s), z'(s)) = 0 \tag{32}$$

*which is an element of $\mathcal{A}(s_0, z_0) \cap W^{1,p_0}$ $(p_0 = (1-\gamma)^{-1})$. Moreover, these solutions have the following properties:*

(i) *For some $d, d' > 0$ (depending on $s_0$ and $z_0$)*

$$d'(-s)^{\gamma\left(\frac{K'}{K'-1}\right)} \leqq |z(s)| \leqq d(-s)^{\gamma\left(\frac{K}{K-1}\right)}, \qquad s \in [s_0, 0).$$

(ii) *In terms of the scaled variables, these solutions can all be obtained by a change of time scale:*

$$Z(s; \lambda) = Z(s) = \begin{cases} Z_+(\lambda s), & Z \geqq 0, \\ Z_-(\lambda s), & Z \leqq 0, \end{cases}$$

*for some $\lambda > 0$, where $Z_+(\cdot)$ and $Z_-(\cdot)$ are fixed positive and negative solutions, respectively.*

**Proof.** We focus on the case $(s_0, z_0) \in Q_+$ for simplicity. The arguments for $(s_0, z_0) \in Q_-$ are analogous. It will also simplify the discussion to work primarily with the scaled variables $(Z, P)$ where $Z := z/(-s)^\gamma$, $P = p(-s)^{\gamma-1}$.

We begin by showing that the differential equation (32) has a unique, local solution for initial data $(s_0, z_0) \in Q_+$. Rewriting (32) in terms of $e(\cdot, \cdot)$ and the scaled variables, we have

$$P(s)\, e_P(Z(s), P(s)) - e(Z(s), P(s)) = -\gamma Z(s)\, e_P(Z(s), P(s)),$$

or,

$$-\gamma Z(s) = \frac{P(s)\, e_P(Z(s), P(s)) - e(Z(s), P(s))}{e_P(Z(s), P(s))}. \tag{33}$$

Since

$$e_{PP}(Z, P) > 0, \qquad P \neq 0, \qquad Z \notin \mathscr{Z},$$

it follows that for each $Z$ such that $Z \notin \mathscr{Z}$, the function

$$P \mapsto \frac{Pe_P(Z, P) - e(Z, P)}{e_P(Z, P)} \tag{34}$$

is strictly monotone (take its value at $P = 0$ to be 0). By (14) and the Lipschitz-continuity condition (17), this is still true for $Z \in \mathscr{Z}$. Denoting the inverse of the function (34) by

$$W \mapsto m(Z, W),$$

we conclude that (33) is equivalent to

$$P(s) = m(Z(s), -\gamma Z(s)), \tag{35}$$

with $m(\cdot, \cdot)$ a continuously differentiable function where it is defined.

Since $Z'(s) = (-s)^{-1}[P(s) + \gamma Z(s)]$, we can rewrite (35) as a differential equation in standard form for $Z(\cdot)$:

$$Z'(s) = (-s)^{-1}[m(Z(s), -\gamma Z(s)) + \gamma Z(s)], \qquad Z(s_0) = z_0(-s_0)^{-\gamma} \tag{36}$$

for all $s \in [s_0, 0)$.

The existence of a local, smooth solution for (36) is a standard result from the theory of ordinary differential equations. The estimates in condition (i) show that this solution is defined for all $s \in [s_0, 0)$.

To verify condition (i), use the growth condition (14) to obtain

$$(1 - 1/K)\, P \leqq \frac{Pe_P(Z, P) - e(Z, P)}{e_P(Z, P)} \leqq (1 - 1/K')\, P, \qquad P < 0.$$

It follows that the inverse function $m(Z, \cdot)$ satisfies

$$\frac{K'}{K' - 1} W \leqq m(Z, W) \leqq \frac{K}{K - 1} W, \qquad W < 0,$$

and (36) yields the following differential inequalities:

$$-\frac{\gamma K'}{K' - 1}(-s)^{-1} Z(s) \leqq Z'(s) \leqq -\frac{\gamma K}{K - 1}(-s)^{-1} Z(s), \qquad Z \geqq 0.$$

From these inequalities, we obtain

$$d'(-s)^{\gamma/(K'-1)} \leqq Z(s) \leqq d(-s)^{\gamma/(K-1)}, \qquad Z \geqq 0, \qquad s \in [s_0, 0),$$

for some constants $d', d > 0$. Condition (i) follows.

Condition (ii) follows from the form of the differential equation (36): if $Z_+(\cdot)$ solves (36), then so does $Z(s; \lambda) := Z_+(\lambda s)$, for any $\lambda > 0$. For example, let $Z_+(\cdot)$ be the solution satisfying $Z(-1) = +1$. Given $s_0$ and $Z_0$, there is a unique $s_+ < 0$ such that $Z_+(s_+) = Z_0$. If $\lambda := s_+/s_0$, $Z(\cdot; \lambda)$ is the solution for (36) satisfying the initial condition $Z(s_0) = Z_0$. $\square$

Hereafter, we take $Z_+(\cdot)$ to be the unique solution to (36) satisfying the initial condition $Z_+(-1) = +1$, while $Z_-(\cdot)$ will denote the unique solution satisfying $Z_-(-1) = -1$. Given $(s_0, z_0) \in Q_+$, there is a unique $\lambda_0 = \lambda_0(s_0, z_0) > 0$ such that $Z(s) = Z(s; \lambda_0) := Z_+(\lambda_0 s)$ satisfies the initial condition $Z(s_0) = Z_0$ (as well as (36)). The family

$$\{Z(\cdot; \lambda) : \lambda > 0\}, \tag{37}$$

defines the Mayer field in $Q_+$. In terms of the original variables:

$$z(s; \lambda) := Z(s; \lambda)\,(-s)^{\gamma}, \qquad s < 0. \tag{38}$$

The analogous construction (working with $Z_-(\cdot)$) provides the field in $Q_-$.

*Remark*. It may be illuminating to think of an alternate construction for the field trajectories provided in this lemma. Begin by constructing the solution *away* from the free zone $\mathscr{Z}$. The properties of the function in (34) imply that there is a unique, smooth solution to (36) defined, forward and backward, until the trajectory intersects the free zone. (Condition (i) guarantees that these intersections occur.) We construct such solutions on each interval $\delta_{j-1} < Z < \delta_j$, and then use condition (ii) to patch these segments together. This can obviously be done to obtain a continuous solution (with the hitting time from below on $Z = \delta_j$ equal to the hitting time from above). The joint continuity condition (17) is introduced to ensure that the resulting trajectory is actually smooth across each free zone curve $Z = \delta_j$.

In addition to the possible smooth minimizers of (9), (10), which as we have seen satisfy (32) and (36), consider the curves of the free zone associated with the zeros of $e(\cdot, \cdot)$ given in (15). Each such zero $(\delta_j, -\gamma\,\delta_j)$ corresponds to a curve

$$Z(s) = \delta_j, \qquad s \in [s_0, 0) \qquad \text{for } \delta_j \in \mathscr{Z},$$

which, by Lemma 2.4 (*i*), enters the origin more steeply than curves of the Mayer field (37). Moreover, along such curves, the integrand in (9) is identically zero. Note that by (13), (14)

$$e(\delta_j, P) = 0, \qquad e_P(\delta_j, P) = 0 \qquad \text{for all } P. \tag{39}$$

Therefore, it is clear that each curve of the free zone is a solution of (32), though not of (36) since (39) ensures that $m(\delta_j, \cdot)$ is not defined.

We can define infinitely many "solutions" to (32) if we are willing to admit *piecewise* $C^1$ trajectories which follow the free zone on some subinterval of $[s_0, 0)$. For each $(s_0, z_0) \in Q \setminus \mathscr{Z}$, there is a unique solution for (32): $z(\cdot; \lambda_0)$, with $\lambda_0 = \lambda(s_0, z_0)$. If $Z_0 = (-s_0)^{-\gamma} z_0$ satisfies $|Z_0| > |\delta_j|$ for some $\delta_j \in \mathscr{Z}$, then this trajectory intersects $Z = \delta_j$ at some $\sigma_0 < 0$. We can define a new solution to (32) by switching at this point of intersection to follow the free zone up to the final time:

$$\hat{z}(s) := \begin{cases} z(s; \lambda_0), & s_0 \leqq s \leqq \sigma_0, \\ \delta_j(-s)^{-\gamma}, & \sigma_0 \leqq s \leqq 0. \end{cases} \tag{40}$$

We can construct still more trajectories satisfying (32) (but not (36)) if we allow multiple switches between members of the family (38) and segments of the free zone. For example, we could choose any $\sigma_1$ greater than $\sigma_0$ and define the trajectory

$$\hat{z}(s) = \hat{z}(s; s_0, z_0, \sigma_1) := \begin{cases} z(s; \lambda_0), & s_0 \leqq s \leqq \sigma_0, \\ \delta_j(-s)^{-\gamma}, & \sigma_0 \leqq s \leqq \sigma_1, \\ z(s; \lambda_1), & \sigma_1 \leqq s \leqq 0, \end{cases} \tag{41}$$

where $\lambda_1$ is chosen so that $z(\sigma_1; \lambda_1) = \delta_j(\sigma_1)^{-\gamma}$.

For a concrete example, consider the polynomial integrand defined in (18). It is easily verified that (32) and (36) have the following forms:

$$(Z^2(s) - 1)^2 \, [5P^6(s) + 3Z(s)\, P^5(s)] = 0,$$
$$Z'(s) = -\tfrac{1}{10}(-s)^{-1} Z(s), \qquad Z(0-) = 0.$$

The nonzero solutions of the latter equation, which automatically satisfy (32), are given by

$$Z(s; \lambda) = \begin{cases} Z_+(\lambda s) = (-\lambda s)^{1/10}, & Z > 0, \\ Z_-(\lambda s) = -(-\lambda s)^{1/10}, & Z < 0, \lambda > 0. \end{cases}$$

Equivalently, they form the family

$$z(s; \lambda) = (-s)^{1/2} Z(s; \lambda) = \begin{cases} (-s)^{1/2}\, Z_+(\lambda s) = \lambda^{1/10}(-s)^{3/5}, & z > 0, \\ (-s)^{1/2}\, Z_-(\lambda s) = -\lambda^{1/10}(-s)^{3/5}, & z < 0, \lambda > 0. \end{cases}$$

On the other hand, the free zone curves

$$Z(s) = \pm 1 \qquad (z(s) = \pm(-s)^{1/2}),$$

are obviously solutions of (32) but do not satisfy (36). Note further that the greater steepness of the free zone at $s = 0$ is reflected in the fact that

$$z(\cdot\,; \lambda) \in W^{1,p_0} \quad \text{for all } \lambda > 0, \quad z(\cdot) \notin W^{1,p_0}, \quad p_0 := \frac{1}{1-\gamma} = 2.$$

## 2.4. Theorem A

It follows from the preceding analysis that, in general, possible minimizers for (9), (10) can be of at least two types: those which are solutions of Noether's equation in the form (36) and hence intersect each free zone $Z(s) = \delta_j$, $\delta_j \in \mathscr{Z}$, in at most one point, and those which lie along one of the free zones on some subinterval of $[s_0, 0)$.

We consider the problem of minimizing the cost function

$$I[z] = \int_{s_0}^{0} f(s, z(s), z'(s))\, ds.$$

If the class of admissible trajectories is

$$\mathscr{A}(s_0, z_0) = \{w \in W^{1,1}(s_0, 0) : w(s_0) = z_0,\ w(0) = 0\},$$

we refer to this as the problem **(P1)**. When we consider the smaller admissible class

$$\mathscr{A}(s_0, z_0) \cap W^{1,p_0} = \{w \in W^{1,p_0}(s_0, 0) : w(s_0) = z_0,\ w(0) = 0\},$$

where $p_0 := (1-\gamma)^{-1}$, we refer to the problem **(P2)**.

The observations in the previous sections make the following result plausible.

**Theorem A.** *Suppose that $f = f(s, z, p)$ is a smooth, nonnegative function satisfying the growth condition*

$$K' f(s, z, p) \leqq p f_p(s, z, p) \leqq K f(s, z, p), \qquad (s, z) \in Q,\ p \in \mathbb{R}, \tag{42}$$

*for some $K \geqq K' > 1$. Assume that for some $\gamma \in (0, 1)$:*

**(HP)** $\quad f(\lambda s, \lambda^{\gamma} z, \lambda^{\gamma-1} p) = \lambda^{-1} f(s, z, p), \qquad \lambda > 0, \qquad (s, z) \in Q,\ p \in \mathbb{R},$

*and that the "free zone"*

$$\mathscr{Z} := \{\delta \neq 0 : f(-1, \delta, -\gamma\,\delta) = 0\}$$

*is finite, with*

$$\mathscr{Z} \cap (0, +\infty) \neq \emptyset \quad \text{and} \quad \mathscr{Z} \cap (-\infty, 0) \neq \emptyset. \tag{43}$$

*Assume also that $f$ satisfies*

$$f_{pp} \geqq 0, \quad \text{with} \quad f_{pp}(s, z, p) > 0 \quad \text{provided} \quad (-s)^{-\gamma} z \notin \mathscr{Z},\ p \neq 0, \tag{44}$$

*and admits the lower bound*

$$c_d(P) := \inf\{f(-1, Z, P) : \operatorname{dist}(Z, \mathscr{Z}) \geqq d\} > 0, \qquad P \neq 0,\ d > 0. \tag{45}$$

*Finally, assume that the function*

$$(Z, P) \mapsto \frac{f(-1, Z, P)}{P f_p(-1, Z, P)} \tag{46}$$

*is the restriction of a Lipschitz continuous function on* $\mathbb{R}^2$. *Under these assumptions, the problems* **(P1)** *and* **(P2)** *both possess one or more minimizers. Moreover, when* $z_0 \neq 0$, *the following Lavrentiev phenomenon holds:*

$$m_1 := \min\{I[z] : z \in \mathscr{A}(s_0, z_0)\} < m_{p_0} := \min\{I[z] : z \in \mathscr{A}(s_0, z_0) \cap W^{1,p_0}\}.$$

*Remarks.*

1. Note that the growth condition (42) implies the *existence* of a minimizer for **(P1)** by results of TONELLI [22] for convex integrands obeying superlinear growth conditions outside a "slender" set ([16], [6]). The proof below utilizes field theory instead. By way of contrast, we are not aware of any way to prove existence of a minimizer for **(P2)** *except* by the field theory techniques that are given below.

2. It will be seen from the proof that if (43) is modified to permit $\mathscr{Z} \cap (0, \infty) = \emptyset$, then the conclusion remains valid for those $(s_0, z_0)$ with $z_0 < 0$. Corresponding results apply when $\mathscr{Z} \cap (-\infty, 0) = \emptyset$.

3. The solutions obtained are in the form of a "regular synthesis" of feedback controls in the sense of control theory. Similarly, the minimizers are almost explicit.

4. One could replace the lower bound (45) with

$$\liminf_{Z \to \infty} f(-1, Z, P) = \hat{c}(P) > 0, \qquad P \neq 0.$$

**Proof.** Observe that solutions of Noether's equation (32) given in Lemma 2.4 provide a Mayer field on the half-plane $Q = \{(s, z) \in \mathbb{R}^2 : s < 0\}$. For simplicity, we focus on the case of data in the upper quadrant

$$Q_+ := \{(s, z) : s < 0, z > 0\}.$$

The same approach applies to the lower quadrant $Q_-$.

In the case of $Q_+$, the curves (38) yield a family of solutions of (32) which cover $Q_+$ simply, by the strict monotonicity of $Z_+(\cdot)$. Thus, associated to each point $(s_0, z_0) \in Q_+$, is a slope

$$\hat{p}(s_0, z_0) := z'(s_0; \lambda_0), \tag{47}$$

where $\lambda_0 = \lambda(s_0, z_0)$ is the unique value of $\lambda$ for which the curve $s \mapsto z(s; \lambda)$ passes through $(s_0, z_0)$. Now the smoothness of the mapping $(s, z) \mapsto \lambda(s, z)$ is a consequence of the inverse function theorem, while $Z_+(\cdot)$ is a smooth solution of (36). Threfore, the smoothness of the slope field $\hat{p}(\cdot, \cdot)$ follows from (38).

It now follows that the Hilbert integrand (*cf.* [6], [12])

$$L^*(s, z, z') := f(s, z, \hat{p}(s, z)) + (z' - \hat{p}(s, z))\, f_p(s, z, \hat{p}(s, z))$$

yields an exact differential ("null-Lagrangian" in the sense of [2]) on the simply connected domain $Q^+$ [6, page 75]. In fact, the function $W: Q_+ \to \mathbb{R}$ given by

$$\begin{aligned} W(s_0, z_0) &= \int_{s_0}^{0} L^*(s, z(s;\lambda_0), z'(s;\lambda_0))\, ds \\ &= \int_{s_0}^{0} f(s, z(s;\lambda_0), z'(s;\lambda_0))\, ds \\ &= \int_{s_0}^{0} (-s)^{-1}\, e(Z_+(\lambda_0 s), m(Z_+(\lambda_0 s), -\gamma Z_+(\lambda_0 s)))\, ds, \end{aligned} \tag{48}$$

is one function whose differential is $L^*(s, z, z')$.

Given $(s_0, z_0) \in Q_+$, consider $z(\cdot) \in \mathscr{A}(s_0, z_0)$ satisfying:

(a) $I[z] < \infty$,
(b) for all $d > 0$ sufficiently small, the graph of $z(\cdot)$ intersects the curve $Z(s) = d$ at some point in $Q_+$.

Notice that, in particular, (b) is valid for all $z \in \mathscr{A}(s_0, z_0) \cap W^{1,p_0}$, $p_0 := (1-\gamma)^{-1}$, since by Hölder's inequality all such functions satisfy

$$z(s) = o((-s)^\gamma) \quad \text{as } s \to 0-.$$

In preparation for the comparison of $I[z]$ with $I[z(\cdot;\lambda_0)]$ recall that the convexity (44) ensures that the Weierstrass excess function be nonnegative:

$$\mathscr{E}(s, z, p, q) = f(s, z, q) - f(s, z, p) - (q-p) f_p(s, z, p) \geqq 0,$$

where $(s, z) \in Q_+$, and $p, q \in \mathbb{R}$. Now put $d_0 := Z_0$ and choose a monotone sequence $\{d_n\}_{n \geqq 1}$, $d_n \searrow 0$, with $d_1 < d_0$. Let $\sigma_n$ denote the first time that the graph of $z(\cdot)$ meets that of the graph $Z = d_n$:

$$\sigma_n := \inf\{s \geqq s_0 : z(s) = d_n(-s)^\gamma\}, \quad n \geqq 1.$$

Note that $\sigma_n < 0$ by condition (b). Finally, use the exactness of $L^*$ to carry out the following decomposition for each $n \geqq 1$:

$$\begin{aligned} I[z] - I[z(\cdot;\lambda_0)] &= \int_{s_0}^{\sigma_n} [f - L^*]\,(s, z(s), z'(s))\, ds \\ &\quad + \int_{\sigma_n}^{0} f(s, z(s), z'(s))\, ds - W(\sigma_n, z(\sigma_n)) \\ &= \int_{s_0}^{\sigma_n} \mathscr{E}(s, z(s), \hat{p}(s, z(s)), z'(s))\, ds \\ &\quad + \int_{\sigma_n}^{0} f(s, z(s), z'(s))\, ds - W(\sigma_n, z(\sigma_n)). \end{aligned}$$

As $n \to +\infty$, the second term on the right approaches zero by (a). The third term on the right may be rewritten

$$W(\sigma_n, z(\sigma_n)) = \int_{\sigma_n}^{0} (-s)^{-1} e(Z_+(\lambda_n s), m(Z_+(\lambda_n s), -\gamma Z_+(\lambda_n s)) ds$$

$$= \int_{\lambda_n \sigma_n}^{0} (-r)^{-1} e(Z_+(r), m(Z_+(r), -\gamma Z_+(r)) dr$$

with $\lambda_n = \lambda(\sigma_n, z(\sigma_n))$. It follows from the relation

$$Z_+(\lambda_n \sigma_n) = (-\sigma_n)^{-\gamma} z(\sigma_n; \lambda_n) = (-\sigma_n)^{-\gamma} z(\sigma_n) = d_n,$$

that $\lambda_n \sigma_n \to 0$ (as $d_n \to 0$). This yields

$$I[z] - I[z(\cdot; \lambda_0)] = \int_{s_0}^{0} \mathscr{E}(s, z(s), \hat{p}(s, z(s)), z'(s)) ds \geqq 0. \tag{49}$$

Since conditions (a) and (b) hold for each $z(\cdot) \in \mathscr{A}(s_0, z_0) \cap W^{1,p_0}$, while, by Lemma 2.4, $z(\cdot; \lambda) \in \mathscr{A}(s_0, z_0) \cap W^{1,p_0}$, (49) yields, for $(s_0, z_0) \in Q_+$,

$$m_{p_0} = \inf \{I[z] : z \in \mathscr{A}(s_0, z_0) \cap W^{1,p_0}\} = I[z(\cdot; \lambda_0)] = W(s_0, z_0). \tag{50}$$

Thus problem **(P1)** possesses at least one minimizer, as claimed.

Notice that the analysis above implies the following homogeneity:

$$W(\lambda s, \lambda^\gamma z) = W(s, z), \quad \lambda > 0,$$

so that, taking $\lambda := (-s)^{-1}$, one obtains

$$W(s, z) = W(-1, (-s)^{-\gamma} z) = W(-1, Z). \tag{51}$$

(Recall also the remark after Lemma 2.1.)

Next, the analysis of **(P1)** will hinge on the fact that the free zones

$$Z(s) = \delta_j, \quad \delta_j \in \mathscr{Z},$$

have the property that for any $s$-interval where an admissible curve coincides with the free zone, the contribution to $I[z]$ is zero. Likewise, since the variational integrand vanishes when $z = \text{constant}$, such segments also contribute nothing to the cost. These facts imply that for each $(s_0, z_0) \in Q_+$ satisfying

$$(-s_0)^{-\gamma} z_0 < \delta_j \quad \text{for some } \delta_j \in \mathscr{Z} \cap (0, \infty),$$

one has

$$\inf \{I[z] : z \in \mathscr{A}(s_0, z_0)\} = 0.$$

Namely, the following recipe yields an element $\bar{z}(\cdot) \in \mathscr{A}(s_0, z_0)$ with $I[\bar{z}] = 0$:

$$\bar{z}(s) := z_0 \wedge \delta_j(-s)^\gamma, \quad s \in [s_0, 0). \tag{52}$$

Clearly $\bar{z}(\cdot)$ is therefore a minimizer for **(P1)**.

Consider now $(s_0, z_0)$ satisfying $(-s_0)^{-\gamma} z_0 > \delta_{max}$ and put $\lambda_0 := \lambda(s_0, z_0)$. Let $\bar{z}(\cdot)$ denote the curve which follows $z(\cdot; \lambda_0)$ until it first intersects the free zone $Z = \delta_{max}$, thereafter following the free zone to the origin. That is,

$$\bar{z}(s) = z(s; \lambda_0) \vee (\delta_{max}(-s)^\gamma) = \begin{cases} z(s; \lambda_0), & s \in [s_0, \bar{\sigma}) \\ \delta_{max}(-s)^\gamma, & s \in [\bar{\sigma}, 0), \end{cases} \tag{53}$$

where $\bar{\sigma} > 0$ is given by $z(\bar{\sigma}; \lambda_0) = \delta_{max}(-\bar{\sigma})^\gamma$. Since $\bar{z}(\cdot)$ makes zero contribution to $I[\bar{z}]$ on the interval $[\bar{\sigma}, 0)$, we can express $I[\bar{z}]$ in terms of $W$ as follows:

$$I[\bar{z}] = W(s_0, z_0) - W(\bar{\sigma}, \delta_{max}(-\bar{\sigma})^\gamma). \tag{54}$$

We proceed to the proof that $\bar{z}(\cdot)$ as defined in (53) is a minimizer for $I[\cdot]$ in $\mathscr{A}(s_0, z_0)$. Given $z(\cdot) \in \mathscr{A}(s_0, z_0)$, put

$$\sigma := \inf\{s \geqq s_0 : z(s) = \delta_{max}(-s)^\gamma\} \leqq 0.$$

If $\sigma < 0$, then one can suppose without loss of generality that $z(\cdot)$ lies along the free zone $Z = \delta_{max}$ throughout $[\sigma, 0]$ since any other behavior would incur positive cost on that interval. Then, by (54) and (51),

$$\begin{aligned} I[z] - I[\bar{z}] &= \int_{s_0}^{\sigma} f(s, z(s), z'(s))\, ds - (W(s_0, z_0) - W(\bar{\sigma}, \delta_{max}(-\bar{\sigma})^\gamma)) \\ &= \int_{s_0}^{\sigma} f(s, z(s), z'(s))\, ds - (W(s_0, z_0) - W(\sigma, \delta_{max}(-\sigma))). \end{aligned}$$

Now by the smoothness of $W$ and the absolute contuinity of $z(\cdot)$, the function $s \mapsto W(s, z(s))$ is locally absolutely continuous with derivative given by the chain rule ([23], [18]):

$$\begin{aligned} \frac{d}{ds} W(s, z(s)) &= f(s, z(s), \hat{p}(s, z(s)) - \hat{p}(s, z(s)) f_p(s, z(s), \hat{p}(s, z(s)) \\ &\quad - z'(s) f_p(s, z(s), \hat{p}(s, z(s)) \\ &= -L^*(s, z(s), z'(s)), \qquad \text{a.e. } s \in [s_0, 0). \end{aligned}$$

Hence

$$\begin{aligned} I[z] - I[\bar{z}] &= \int_{s_0}^{\sigma} [f(s, z(s), z'(s)) - L^*(s, z(s), z'(s))]\, ds \\ &= \int_{s_0}^{\sigma} \mathscr{E}(s, z(s), \hat{p}(s, z(s)), z'(s))\, ds \geqq 0. \end{aligned}$$

On the other hand, if $\sigma = 0$, we appeal to the following lemma whose proof will be provided below.

**Lemma 2.5.** *Let* $(s_0, z_0) \in Q_+$ *be such that* $Z_0 := z_0(-s)^{-\gamma} > \delta_{max}$. *Let* $z(\cdot) \in \mathscr{A}(s_0, z_0)$ *be an admissible trajectory and define* $Z(s) := (-s)^{-\gamma} z(s)$, $s_0 \leqq s < 0$. *If* $I[z] < +\infty$, *then*

$$\limsup_{s \to 0-} Z(s) \leqq \delta_{max}.$$

It follows from this lemma that each $z(\cdot) \in \mathscr{A}(s_0, z_0)$ satisfying

(a) $I[z] < \infty$

also satisfies

(b′) for all $d > 0$ sufficiently small, the graph of $z(\cdot)$ intersects the curve $Z = \delta_{\max} + d$ at some point $(\sigma_d, z(\sigma_d)) \in Q_+$.

Thus putting $d_0 := (-\sigma_0)^{-\gamma} z_0 - \delta_{\max}$, take a decreasing sequence $\{d_n\}_{n \geqq 1}$, $d_n \searrow 0$, with $d_1 < d_0$. Setting

$$\sigma_n := \inf\{s \geqq s_0 : (-s)^{-\gamma} z(s) = \delta_{\max} + d_n\} < 0,$$

we can imitate the decomposition of the preceding paragraph:

$$\begin{aligned} I[z] - I[\bar{z}] &= \int_{s_0}^{\sigma_n} f(s, z(s), z'(s))\, ds - (W(s_0, z_0) - W(\sigma_n, (\delta_{\max} + d_n)(-\sigma_n)^\gamma)) \\ &\quad + \int_{\sigma_n}^{0} f(s, z(s), z'(s))\, ds - W(\sigma_n, (\delta_{\max} + d_n)(-\sigma_n)^\gamma) \\ &= \int_{s_0}^{\sigma_n} [f((s, z(s), z(s)) - L^*(s, z(s), z'(s))]\, ds \\ &\quad + \int_{\sigma_n}^{0} f(s, z(s), z'(s))\, ds - W(\sigma_n, (\delta_{\max} + d_n)(-\sigma_n)^\gamma). \end{aligned}$$

Again, by (a), (b′), and (51), it follows that the last two terms approach zero as $n \to \infty$, whence

$$I[z] - I[\bar{z}] = \int_{s_0}^{0} \mathscr{E}(s, z(s), \hat{p}(s, z(s)), z'(s))\, ds \geqq 0.$$

To summarize, we have shown that for each $(s_0, z_0) \in Q_+$ minimizers $\bar{z}(\cdot)$ exist for problem **(P1)**. The analysis for initial data in the lower quadrant $Q_-$ is identical to the analysis for $Q_+$, and so the optimal cost (value function) for arbitrary $(s_0, z_0)$ is given by

$$m_1 = I[s_0, z_0; \bar{z}] = \begin{cases} 0, & (-s_0)^{-\gamma} z_0 \in [\delta_{\min}, \delta_{\max}]; \\ W(s_0, z_0) - W(\bar{\sigma}, \delta_{\max}(-\bar{\sigma})^\gamma), & (-s_0)^{-\gamma} z_0 > \delta_{\max}; \\ W(s_0, z_0) - W(\underline{\sigma}, \delta_{\min}(-\underline{\sigma})^\gamma), & (-s_0)^{-\gamma} z_0 < \delta_{\min}; \end{cases} \tag{55}$$

where

$$\bar{\sigma} := \inf\{s \geqq s_0 : (-s)^{-\gamma} z(s; \lambda_0) > \delta_{\max}\},$$

and

$$\underline{\sigma} := \inf\{s \geqq s_0 : (-s)^{-\gamma} z(s; \lambda_0) < \delta_{\min}\}.$$

Since the optimal cost for problem **(P2)** (recall (50)) is given by

$$m_{p_0} = I[z(\cdot; \lambda_0)] = W(s_0, z_0),$$

we have verified the Lavrentiev phenomenon for $z_0 \neq 0$. □

We conclude by providing the proof of Lemma 2.5.

**Proof.** (Lemma 2.5). Let $(s_0, z_0)$ and $z(\cdot) \in \mathscr{A}(s_0, z_0)$ be given as specified in the lemma and assume that

$$\limsup_{s \to 0-} Z(s) \geqq \delta_{\max} + 2d,$$

for some $d > 0$. We may choose then a monotone increasing sequence $\{s_n\}_{n \geq 1}$ converging to 0 such that

$$Z_n := (-s_n)^{-\gamma} z(s_n) \geqq \delta_{\max} + d, \qquad n = 1, 2, 3, \ldots$$

Then

$$\begin{aligned} I[s_0, z_0; z] &= \int_{s_0}^{s_n} f(s, z(s), z'(s))\, ds + I[s_n, z(s_n); z] \\ &= \int_{s_0}^{s_n} f(s, z(s), z'(s))\, ds + I[-1, Z_n; \hat{z}] \\ &\geqq \int_{s_0}^{s_n} f(s, z(s), z'(s))\, ds + k_1(\gamma, d) \end{aligned}$$

where the (positive) constant $k_1(\gamma, d)$ is obtained by taking $s_1 = -1$ and $Z_0 = \delta_{\max} + d$ in Lemma 2.2, and the second equality follows from Lemma 2.1 (where $\hat{z}(\cdot)$ is defined). Now letting $n \to \infty$ we obtain

$$I[z] \geqq I[z] + k_1(\gamma, d).$$

But this is not possible unless $I[z] = +\infty$, and the proof is complete. □

*Remark*. It is not obvious *a priori* that the function $W(\cdot, \cdot)$ defined in equation (48) is finite. A proof of this fact can be obtained by an analysis of the Hamilton-Jacobi partial differential equation associated with the minimization problem.

*Summary*. The minimizers and pseudominimizers constructed in the proof of Theorem A are not the unique minimizers for our problem, but all minimizers are constructed entirely from segments of the free zone, horizontal segments, and the field trajectories defined in (38).

The absolute minimizers for $(s_0, z_0) \in Q_+$ are given in equations (52) and (53). When the initial position is outside all or all but one of the free zone curves, equations (52) and (53) define the *unique* absolute minimizer. (If the free zone consists in a single curve $(\mathscr{Z} \cap (0, +\infty) = \{\delta_1\})$, then the absolute minimizer is unique.) Otherwise, there are infinitely many absolute minimizers.

The pseudominimizer is unique only when the initial position is strictly below all free zone curves. Otherwise, there is an infinite family of pseudominimizers constructed from solutions to the field equations and portions of the free zone (*cf.* (41)).

## 3. The Bolza Problem

The Lagrange problem described in the previous section can be viewed as the limiting case of a family of Bolza problems. In these Bolza problems, there is a running cost (the variational integral considered in (9)) *plus* a terminal cost. The Lagrange problem may be viewed as the limiting case that results as the terminal cost for missing the final state $z = 0$ becomes infinite. The problem considered is one of minimizing the cost function

$$\tilde{I}[z] = \tilde{I}[s_0, z_0; z] = \int_{s_0}^{0} f(s, z(s), z'(s))\, ds + \varphi(z(0)), \tag{56}$$

over the admissible class

$$\tilde{\mathscr{A}}(s_0, z_0) = \{w \in W^{1,1}(s_0, 0) : w(s_0) = z_0\}. \tag{57}$$

The integrand $f$ is assumed to have the properties specified in Theorem A and the *terminal cost* function $\varphi : \mathbb{R} \to \mathbb{R}$ is given by

$$\varphi(z) := \begin{cases} N_1, & z > 0, \\ 0, & z = 0, \\ N_2, & z < 0, \quad N_1, N_2 \geqq 0. \end{cases} \tag{58}$$

The choice $\varphi(0) = 0$ is a convenient normalization since in the limiting case, equation (58) becomes $\varphi(0) = 0$, $\varphi(0\pm) = +\infty$.

Let $W(\cdot, \cdot)$ be the function defined in (48), so that for each $(s_0, z_0) \in Q$, the scalar $W(s_0, z_0) = W(-1, z_0(-s_0)^{-\gamma})$ denotes the minimal cost associated with problem **(P2)**. As noted in section 2,

$$\begin{aligned} dW(s) &= [\hat{p}(s, z) f_p(s, z, \hat{p}(s, z)) - f(s, z, \hat{p}(s, z))]\, ds - f_p(s, z, \hat{p}(s, z))\, dz \\ &= (-s)^{-1} [\hat{P}(s, z)\, e_P(Z, \hat{P}(s, z)) - e(Z, \hat{P}(s, z))]\, ds \\ &\quad - (-s)^{-\gamma} e_P(Z, \hat{P}(s, z))\, dz, \end{aligned}$$

with $\hat{p}$ as in (47). Since $\hat{P}$ satisfies

$$\hat{P}(s_0, z_0) = (-s)^{\gamma-1} z'(s_0; z_0) \begin{cases} < 0, & (s_0, z_0) \in Q_+, \\ > 0, & (s_0, z_0) \in Q_-, \end{cases}$$

it follows that

$$W_s(s, z) = (-s)^{-1} (\hat{P} e_P(Z, \hat{P}) - e(Z, \hat{P})) > 0 \quad (s, z) \in Q_+ \cup Q_-.$$

Hence

$$\begin{gathered} s \mapsto W(s, z) \text{ is strictly increasing, } (s, z) \in Q_+ \cup Q_-, \\ W(-\infty, z) = W(-1, 0) = 0, \qquad W(0-, z) = W(-1, \infty) = \infty. \end{gathered} \tag{59}$$

For brevity, we confine our discussion to points $(s_0, z_0) \in Q_+$. Exactly similar arguments apply when $(s_0, z_0) \in Q_-$.

Now define subregions $\mathscr{D}_1, \mathscr{D}_2 \subset Q_+$ as follows. Let $Z_1, Z_2$ respectively, be the solutions of

$$W(-1, Z) - W(-1, \delta_{\max}) = N_1, \qquad W(-1, Z) = N_1, \tag{60}$$

and let $\mathscr{C}_1, \mathscr{C}_2$ denote the curves

$$\mathscr{C}_1 : \{(s, z) : z(-s)^{-\gamma} = Z_1\}, \qquad \mathscr{C}_2 : \{(s, z) : z(-s)^{-\gamma} = Z_2\}.$$

Then $\mathscr{D}_1, \mathscr{D}_2$ denote the regions between these curves and the $z$-axis. Thus by (55), (60), $(s, z) \in \mathscr{D}_1$ if and only if the minimum cost for problem **(P1)** of reaching the origin from $(s, z)$ exceeds $N_1$. Similarly, $(s, z) \in \mathscr{D}_2$ if and only if the minimum cost for problem **(P2)** of reaching the origin from $(s, z)$ exceeds $N_1$. Note that by (59), (60), it follows that $\mathscr{D}_1$ is strictly contained in $\mathscr{D}_2$.

We can now formulate the results available for the Bolza problem (56), (57).

**Theorem B.** *Suppose that* $f = f(s, z, p)$ *is a smooth function satisfying the conditions imposed in Theorem A. In addition, suppose that* $\varphi(\cdot) = \varphi_{N_1,N_2}(\cdot)$ *is the function defined in* (58). *The problems of minimizing*

$$\tilde{I}[z] = \int_{s_0}^{0} f(s, z(s), z'(s))\, ds + \varphi(z(0)),$$

*over the admissible classes*

$$(\tilde{\mathbf{P}}\mathbf{1}) \qquad z \in \tilde{\mathscr{A}}(s_0, z_0) := \{w \in W^{1,1}(s_0, 0) : w(s_0) = z_0\},$$

*or*

$$(\tilde{\mathbf{P}}\mathbf{2}) \qquad z \in \tilde{\mathscr{A}}(s_0, z_0) \cap W^{1,p_0} = \{w \in W^{1,p_0}(s_0, 0) : w(s_0) = z_0\}, \qquad p_0 := (1 - \gamma)^{-1},$$

*both possess minimizers. Moreover, for* $(s_0, z_0) \in Q^+ \setminus \mathscr{D}_1$, *the Lavrentiev phenomenon holds:*

$$\tilde{m}_1 := \inf\{\tilde{I}[z] : z(\cdot) \in \tilde{\mathscr{A}}(s_0, z_0)\} < \tilde{m}_{p_0} := \inf\{\tilde{I}[z] : z(\cdot) \in \tilde{\mathscr{A}}(s_0, z_0) \cap W^{1,p_0}\},$$

*while for* $(s_0, z_0) \in \mathscr{D}_1$, $(\tilde{\mathbf{P}}\mathbf{1})$ and $(\tilde{\mathbf{P}}\mathbf{2})$ *both have the horizontal segment* $z = z_0$, $s \in [s_0, 0]$ *as minimizer, attaining the value* $\varphi(z_0)$.

*Remarks.*

1. It will be seen from the proof that if (43) is modified to permit $\mathscr{Z} \cap (0, \infty) = \emptyset$, then the conclusions of the theorem remain valid for points $(s_0, z_0)$ with $z_0 < 0$. Corresponding results remain valid for when $\mathscr{Z} \cap (-\infty, 0) = \emptyset$.

2. The minimizers are described almost explicitly since the proof provides a "regular synthesis" for feedback controls in the sense of control theory (Mayer field).

**Proof.** Examine $(\tilde{\mathbf{P}}\mathbf{1})$ first. As shown (by field theory arguments) in the proof of Theorem A, among those $W^{1,1}$ trajectories $z(\cdot)$ starting at $(s_0, z_0)$ for which $z(0-) = 0$, the minimal $\tilde{I}$ value is attained by the trajectory $z(\cdot)$ specified in

(52) and (53). On the other hand, the present problem also possesses finite cost trajectories $z(\cdot)\,s$ satisfying $z(0-)>0$. Since for all such trajectories $\varphi(z(0-)))=N_1$, it is clear that any trajectory of this class for which the integral term in $\tilde{I}[z]$ happens to be zero will have minimal cost, namely $N_1$. Now it follows from (12) that $\tilde{z}(s)=z_0,\ s\in[s_0,0]$ is a trajectory of this sort.

The above reasoning shows on examination of (54) that for points $(s_0,z_0)\in Q_+$ satisfying $W(s_0,z_0)-W(-1,\delta_{\max})<N_1$, the trajectory $\bar{z}(\cdot)$ defined in (52) and (53) minimizes $\tilde{I}$ over $\tilde{\mathscr{A}}(s_0,z_0)$, while for those $(s_0,z_0)\in Q_+$ at which $W(s_0,z_0)-W(-1,\delta_{\max})>N_1$, the horizontal trajectory $\bar{z}(\cdot)$ described above has minimal cost. By (60) the dichotomy considered here is just that based on whether $(s_0,z_0)\notin\mathscr{D}_1$ or $(s_0,z_0)\in\mathscr{D}_1$.

A completely analogous analysis applies to **($\tilde{\mathbf{P}}$2)**, with the crucial condition now being

$$W(s_0,z_0)<N_1,\qquad \text{or } W(s_0,z_0)>N_1\,.$$

Thus for points $(s_0,z_0)\notin\mathscr{D}_2$, the minimal value for $\tilde{I}$ over $\tilde{\mathscr{A}}(s_0,z_0)\cap W^{1,p_0}$ is given by the pseudominimizer $z(\cdot;\lambda_0)$ through $(s_0,z_0)$ described in (38), for which $I[z(\cdot;\lambda_0)]=W(s_0,z_0)$, $\varphi(z(\cdot;\lambda_0))=0$. On the other hand for points $(s_0,z_0)\in\mathscr{D}_2$, the minimizer in $\tilde{\mathscr{A}}(s_0,z_0)\cap W^{1,p_0}$ is given by the horizontal trajectory $\tilde{z}(\cdot)$ described earlier, for which $\tilde{I}[\tilde{z}]=N_1$.

Combining these observations with the results of Theorem A completes the argument. □

## 4. Perturbations of the Integrand

In this section we provide an analysis, for the (fixed-endpoint) Lagrange problem, of how the presence of the Lavrentiev phenomenon is affected by various perturbations of the integrand $f$. In particular, it will be shown that the Lavrentiev phenomenon persists even in cases where the modified integrand $\tilde{f}$ satisfies the strict form of Tonelli's regularity and growth conditions:

$$\tilde{f}_{pp}(s,z,p)>0\quad\text{and}\quad \tilde{f}(s,z,p)>\phi(p),\quad\text{where}\quad \liminf_{|p|\to+\infty}\frac{\phi(p)}{|p|}=+\infty. \tag{61}$$

Furthermore, it will be shown that integrands to which Theorem A applies are in a sense the boundary of the class of integrands for which the Lavrentiev phenomenon occurs.

### 4.1. *Additive Perturbations*

Consider first variational integrands of the form

$$\tilde{f}(s,z,p)=f(s,z,p)+g(s,z,p),\qquad g\geqq 0 \text{ and smooth}, \tag{62}$$

where $f$ satisfies the hypotheses of Theorem A. Setting

$$\tilde{I}[z] := \int_{s_0}^{0} \tilde{f}(s, z(s), z'(s))\, ds,$$

we see by (62) that

$$\inf_{z \in \mathscr{A}(s_0, z_0) \cap W^{1,p_0}} \tilde{I}[z] \geqq \inf_{z \in \mathscr{A}(s_0, z_0) \cap W^{1,p_0}} I[z] = \mathrm{m}_{p_0} > \mathrm{m} = I[\bar{z}],$$

where $\bar{z}(\cdot) \in \mathscr{A}(s_0, z_0)$ is the absolute minimizer for problem **(P1)** specified in the proof of Theorem A (*cf.* (52), (53)). It follows that the Lavrentiev phenomenon will continue to occur when $f$ is replaced by $\tilde{f}$, provided that

$$\tilde{I}[\bar{z}] - I[\bar{z}] = \int_{s_0}^{0} g(s, \bar{z}(s), \bar{z}'(s))\, ds < \mathrm{m}_{p_0} - \mathrm{m}_1, \tag{63}$$

since this estimate implies that

$$\inf_{z \in \mathscr{A}(s_0, z_0)} \tilde{I}[z] \leqq \tilde{I}[\bar{z}] < \mathrm{m}_{p_0} \leqq \inf_{z \in \mathscr{A}(s_0, z_0) \cap W^{1,p_0}} \tilde{I}[z]. \tag{64}$$

For the special case $g(s, z, p) = \varepsilon g_0(p)$, we have the following theorem.

**Theorem C.1.** *Suppose that* $f = f(s, z, p)$ *is a smooth function satisfying the conditions of Theorem A. Let*

$$\tilde{f}(s, z, p) := f(s, z, p) + \varepsilon g_0(p),$$

*where* $\varepsilon > 0$ *and* $g_0(\cdot)$ *is a smooth, nonnegative function satisfying*

$$g_0''(p) > 0, \quad p \in \mathbb{R} \quad \text{and} \quad \liminf_{|p| \to +\infty} \frac{g_0(p)}{|p|} = +\infty.$$

*If*

$$\limsup_{|p| \to +\infty} \frac{g_0(p)}{|p|^q} < \infty \quad \text{for some } q \in (1, p_0), \tag{65}$$

*then, for sufficiently small* $\varepsilon$, *the problem of minimizing*

$$\tilde{I}[z] := \int_{s_0}^{0} \tilde{f}(s, z(s), z'(s))\, ds$$

*exhibits the Lavrentiev phenomenon* (64).

**Proof.** Recall from the definition of $\bar{z}(\cdot)$ in equations (52), (53) that $\bar{z}(s) = \delta(-s)^{\gamma}$ for some $\delta \in \mathscr{Z}$ and $s$ in some left-neighborhood of zero. The result follows immediately from the remarks above since (65) implies that

$$\int_{s_0}^{0} g_0(\bar{z}'(s))\, ds < +\infty,$$

and so we may choose $\varepsilon > 0$ such as to make (63) hold. □

It should be noted that the permissible size of $\varepsilon$ varies with the value of $Z_0 = z_0(-s_0)^{-\gamma}$. In fact, if $g_0(0) = 0$, then for each $d > 0$ there is a value $\bar{\varepsilon} = \bar{\varepsilon}(d)$ such that the Lavrentiev phenomenon holds for $\tilde{I}$ whenever the initial data $(s_0, z_0)$ satisfy $Z_0 = z_0(-s)^{-\gamma} \geqq d$. On the other hand, because of the strict regularity condition $(g_0'' > 0)$, it follows from field theory that such a result is the best that can be achieved ([5, page 376], [6], [12]). That is, for fixed $\varepsilon > 0$, the Lavrentiev phenomenon does not occur for $Z_0$ sufficiently small.

*Examples.* Recall the example discussed in Section 2:

$$f(s, z, p) = (z^2 + s)^2 \, |p|^6.$$

This integrand satisfies **(HP)** with $\gamma = 1/2$, so $p_0 = 2$. The perturbed problem

$$\tilde{I}[z] = \int_{s_0}^{0} [f(s, z(s), z'(s)) + \varepsilon \, |z'(s)|^q] \, ds$$

will exhibit the Lavrentiev phenomenon for $1 \leqq q < 2$ and $\varepsilon$ sufficiently small. A perturbation satisfying the strong regularity condition is given by

$$g_0(p) := \frac{p^2}{(1 + p^2)^\nu} \qquad \text{for } 0 < \nu < 8/17.$$

For integrands $f$ satisfying **(HP)** with $\gamma \in (1/2, 1)$, we can take $g_0(p) = |p|^2$ so that

$$\tilde{I}[z] = \int_{s_0}^{0} [f(s, z(s), z'(s)) + \varepsilon \, |p|^2] \, ds.$$

In this way, we obtain a fully regular problem exhibiting the Lavrentiev phenomenon. For example,

$$f(s, z, p) = (z^6 - s^4)^2 \, |p|^{27}$$

satisfies the conditions of Theorem A with $\gamma = 2/3$ in **(HP)**. This is a special case of the example discussed in [5, section 5].

### *4.2. Multiplicative Perturbations*

We next study the effect of the following type of perturbation:

$$\tilde{f}(s, z, p) = f(s, z, p) \, g_0(p), \qquad s \leqq 0, z, p \in \mathbb{R}, \tag{66}$$

where $f$ obeys the conditions of Theorem A. Our first results are concerned with a class of functions $g_0(\cdot)$ for which the perturbed variational problem retains the Lavrentiev phenomenon; if $g_0(p)$ is nondecreasing as $|p|$ increases, then the Lavrentiev phenomenon persists. We also show that if $\lim_{|p| \to +\infty} g_0(p) = +\infty$, then a stronger form of the Lavrentiev phenomenon is present. In this case, not only is it impossible to approximate the absolute minimizer and its cost simultaneously, but the cost of any ($W^{1,p_0}$) approximating sequence must become infinite. For instance, we are interested in perturbations of the form

$$g_0(p) = |p|^q \quad \text{for some } q > 0.$$

Note that with this class of perturbations, we must abandon the assumption that the variational integrand is $C^3$ in its arguments. But Tonelli's existence theorem requires only *continuity* of the integrand in conjunction with convexity and growth conditions, and so the smoothness is not necessary.

Our second result concerns functions $g_0(\cdot)$ for which the Lavrentiev phenomenon disappears. We show that if $g_0(p) \to 0$ as $|p| \to +\infty$ in such a way as to make the existence theorem of Tonelli applicable, then the infimum over absolutely continuous trajectories is the same as the infimum over the class of Lipschitz-continuous trajectories. Thus, the invariant problems considered in Section 2 are "borderline" cases for the occurrence of the Lavrentiev phenomenon.

**Theorem C.2.** *Suppose that* $f = f(s, z, p)$ *is a smooth function satisfying the conditions of Theorem A. Let*

$$\tilde{f}(s, z, p) := f(s, z, p)\, g_0(p), \qquad s \leqq 0,\ z, p \in \mathbb{R},$$

*where* $g_0(\cdot)$ *is continuous, nonnegative (not identically zero), symmetric and nondecreasing on* $[0, +\infty)$. *Then the Lavrentiev phenomenon persists:*

$$\tilde{m}_1 = \min\{\tilde{I}[z] : z \in \mathscr{A}(s_0, z_0)\} < \tilde{m}_{p_0} = \inf\{\tilde{I}[z] : z \in \mathscr{A}(s_0, z_0) \cap W^{1,p_0}\}. \tag{67}$$

*If in addition,* $g_0(\cdot)$ *satisfies*

$$g_0(cp) \geqq \alpha(c)\, g_1(p), \qquad p \in \mathbb{R},\ c \geqq 1, \tag{68}$$

*where* $g_1(\cdot)$ *satisfies the original hypotheses on* $g_0(\cdot)$ *and*

$$\lim_{c \to +\infty} \alpha(c) = +\infty, \tag{69}$$

*then the Lavrentiev phenomenon holds in the following strong form: if* $\bar{\bar{z}}(\cdot)$ *is the absolute minimizer, then for each sequence* $\{w_n(\cdot)\} \subset \mathscr{A}(s_0, z_0) \cap W^{1,p_0}$ *such that* $w_n(s) \to \bar{\bar{z}}(s)$ *for each* $s$ *in some set* $S \subset (s_0, 0)$ *containing points arbitrarily close to zero, one necessarily has*

$$\lim_{n \to +\infty} \tilde{I}[w_n] = +\infty.$$

**Proof.** We recall that it follows by a theorem of Tonelli [22, Theorem VI] (see Remark 1 after the statement of Theorem A) that for each $(s_0, z_0) \in Q$ there exists an absolute minimizer $\bar{\bar{z}}(\cdot) \in \mathscr{A}(s_0, z_0)$ for $\tilde{I}$. Now $\bar{\bar{z}}(\cdot)$ can be explicitly identified whenever $(s_0, z_0)$ is such that $Z_0 = z_0(-s_0)^{-\gamma}$ lies between the endpoints $\delta_{\min}$ and $\delta_{\max}$ of $\mathscr{Z}$. For instance, when $z_0 > 0$ and $\delta \in \mathscr{Z}$ is such that $Z_0 \in (0, \delta]$, the following trajectory

$$\bar{\bar{z}}(s) := z_0 \wedge \delta(-s)^{\gamma}, \qquad s \in [s_0, 0],$$

obviously satisfies $\tilde{I}[\bar{\bar{z}}] = 0$. For values of $Z_0$ outside the interval $[\delta_{\min}, \delta_{\max}]$, say $Z_0 > \delta_{\max}$, we simply observe that $\bar{\bar{z}}(\cdot)$ does not cross below the curve $Z = \delta_{\max}$ since after reaching $Z = \delta_{\max}$ it is possible to reach the origin with zero additional cost by following the free zone.

We first prove that the Lavrentiev phenomenon (67) persists. Let $(s_0, z_0) \in Q_+$ and $z(\cdot) \in \mathscr{A}(s_0, z_0) \cap W^{1,p_0}$ be given. By Lemma 2.1, we may assume without loss that $s_0 \geqq -1$. If $Z_0 \leqq \delta_{\max}$, choose $Z_1$ and $Z_1'$ as in Lemma 2.2, and it follows that, taking $s_1$, $s_1'$, and $\psi_d(\cdot)$ as in Lemma 2.2,

$$\begin{aligned} \tilde{I}[z] - \tilde{I}[\bar{\bar{z}}] &\geqq \int_{s_1'}^{s_1} (-s)^{-1}\, e(Z(s), P(s))\, g_0(z'(s))\, ds \\ &\geqq \int_{s_1'}^{s_1} (-s)^{-1}\, \psi_d(P(s))\, g_0((-s)^{\gamma-1}\, P(s))\, ds \qquad (70) \\ &\geqq \int_{s_1'}^{s_1} (-s)^{-1}\, \psi_d(P(s))\, g_0(P(s))\, ds > 0. \end{aligned}$$

For $Z_0 > \delta_{\max}$, we define for each $z(\cdot) \in \mathscr{A}(s_0, z_0) \cap W^{1,p_0}$ a new trajectory $\hat{z}(\cdot) \in \mathscr{A}(s_0, z_0)$ by

$$\hat{z}(s) := z(s) \vee \delta_{\max}(-s)^{\gamma}, \qquad s \in [s_0, 0].$$

Then, just as above, we have

$$\tilde{I}[z] - \tilde{I}[\hat{z}] \geqq \int_{s_1'}^{s_1} (-s)^{-1}\, \psi_d(P(s))\, g_0(P(s))\, ds. \qquad (71)$$

Now the properties of $g_0(\cdot)$ guarantee that the product $\psi_d(\cdot) \cdot g_0(\cdot)$ inherits the properties of $\psi_d(\cdot)$ used in the proof of Lemma 2.2. In particular, it is symmetric, nonnegative and convex. Hence the integral term on the right side in (70) and (71) is bounded below by a positive constant $\tilde{k}_1$ depending only on $Z_1, Z_1', e(\cdot,\cdot)$, and $g_0(\cdot)$, namely

$$\int_{s_1'}^{s_1} (-s)^{-1}\, [\psi_d \cdot g]\, (P(s))\, ds \geqq (Z_1' - Z_1) \inf_{r>0} \frac{[\psi_d \cdot g_0](-\gamma Z_1 - r)}{r} =: \tilde{k}_1 > 0.$$

Thus we obtain from the preceding calculations the estimate

$$\tilde{I}[z] \geqq \tilde{k}_1 + \tilde{I}[\bar{\bar{z}}] \quad \text{for all } z \in \mathscr{A}(s_0, z_0) \cap W^{1,p_0},$$

and the Lavrentiev phenomenon (67) is verified.

To verify the strong form of the Lavrentiev phenomenon, assume that $g_0(\cdot)$ satisfies the growth condition (68), (69). Returning to the estimates (70) and (71), we have

$$\begin{aligned} \tilde{I}[z] - \tilde{I}[\bar{\bar{z}}] &\geqq \int_{s_1'}^{s_1} (-s)^{-1}\, \psi_d((-s)^{\gamma-1}\, P(s))\, g_0(z'(s))\, ds \qquad (72) \\ &\geqq \alpha((-s_1')^{\gamma-1}) \int_{s_1'}^{s_1} (-s)^{-1}\, \psi_d(P(s))\, g_1(P(s))\, ds. \end{aligned}$$

To complete the proof, it remains only to point out that for $\{w_n\}_{n\geq 1} \subset \mathscr{A}(s_0, z_0) \cap W^{1,p_0}$ the estimate (72) applies with $s_1' \in S$, provided $n$ is sufficiently large (depending on $s_0$). Hence for each $s \in S$

$$\tilde{I}[w_n] \geqq \alpha((-s_1')^{\gamma-1})\, \tilde{k}_1 \geqq \alpha((-s)^{\gamma-1})\, \tilde{k}_1 \quad \text{for } n \text{ sufficiently large},$$

where $\tilde{k}_1 = \tilde{k}_1(Z_1, Z_1', e(\cdot, \cdot), g_0(\cdot))$ is defined as above. □

*Remark*. It is evident that $\tilde{f}$ can be modified as in Theorem C.1 to provide an integrand obeying the Tonelli conditions (61) yet still satisfying the strengthened Lavrentiev phenomenon.

*Examples*. Recall MANIÀ'S example, as described in section 2. The variational integrand takes the form

$$f(s, z, p) = (z^3 + s)^2\, |p|^{\alpha}, \qquad (s, z) \in Q,\ p \in \mathbb{R}.$$

As noted earlier, this satisfies the conditions of Theorem A when $\alpha = 9/2$. Applying Theorem C.2 with $g_0(p) := |p|^{3/2}$, we obtain MANIÀ'S original example (as presented in CESARI [6]). We have also deduced the strong form of the Lavrentiev phenomenon, which was demonstrated by LOEWEN [14].

Theorems C.1 and C.2 also contain the example presented in BALL & MIZEL [5, section 5]. They consider the integrand

$$f(s, z, p) = (z^6 - s^4)^2\, |p|^{\beta} + \varepsilon\, |p|^2.$$

for $\beta \geqq 27$. When $\beta = 27$ and $\varepsilon = 0$, this integrand satisfies the homogeneity property with $\gamma = 2/3$. Theorem C.2 implies the strong form of the Lavrentiev phenomenon for $\beta > 27$, by taking $g_0(p) := |p|^{\beta-27}$. The strong form of the Lavrentiev phenomenon is retained, as shown by the argument given in Theorem C.1, with the addition of the perturbation $\varepsilon\, |p|^2$.

Finally, we examine the effect of the modification

$$\tilde{f}(s, z, p) = f(s, z, p)\, g_0(p), \qquad (s, z) \in Q,\ p \in \mathbb{R},$$

when $g_0(\cdot)$ is a continuous, symmetric, nonnegative function (slowly) approaching zero at $\pm\infty$, and decreasing on $(0, +\infty)$. Obviously, if this convergence is too rapid, then the superlinearity condition ensuring the existence of an absolute minimizer will be violated. We need to retain the following properties:

$$\tilde{f}_{pp}(s, z, p) > 0, \qquad \lim_{|p| \to +\infty} \frac{\tilde{f}(s, z, p)}{|p|} = +\infty \text{ whenever } Z \notin \mathscr{Z},$$

uniformly on $\{Z : \operatorname{dist}(Z, \mathscr{Z}) \geqq d\}$ for each $d > 0$. (Note that the superlinearity follows from (42) when

$$\lim_{|p| \to +\infty} g_0(p)\, |p|^{K'-1} = +\infty.$$

For instance,

$$g_0(p) = [\ln(1 + p^2)]^{-1}$$

is such a function.) We have the following result.

**Theorem C.3.** *Suppose that* $f = f(s, z, p)$ *is a smooth function satisfying the conditions of Theorem A, as well as*

$$zf_z(s, z, p) \geqq 0 \quad \textit{whenever} \quad z \cdot (-s)^{-\gamma} \notin [\delta_{\min}, \delta_{\max}].$$

*Let*

$$\tilde{f}(s, z, p) = f(s, z, p)\, g_0(p), \qquad (s, z) \in Q,\ p \in \mathbb{R},$$

*where* $g_0(\cdot)$ *is a continuous, symmetric, nonnegative function decreasing on* $(0, +\infty)$ *such that*

$$\lim_{|p| \to +\infty} g_0(p) = 0,$$

*and*

$$\tilde{f}_{pp}(s, z, p) > 0, \qquad \lim_{|p| \to +\infty} \frac{\tilde{f}(s, z, p)}{|p|} = +\infty \ \textit{whenever}\ Z \notin \mathscr{Z}, \tag{73}$$

*these conditions being uniform on* $\{Z : \operatorname{dist}(Z, \mathscr{Z}) \geqq d\}$ *for each* $d > 0$.

*Then for each* $(s_0, z_0)$ *the variational integral* $\tilde{I}$ *possesses an absolute minimizer* $\bar{\bar{z}}(\cdot) \in \mathscr{A}(s_0, z_0)$ *and the Lavrentiev phenomenon is absent:*

$$\tilde{m}_1 = \inf\{\tilde{I}[z] : z \in \mathscr{A}(s_0, z_0)\} = \tilde{m}_\infty = \inf\{\tilde{I}[z] : z \in \mathscr{A}(s_0, z_0) \cap W^{1,\infty}\}.$$

**Proof.** By the theorem of Tonelli cited earlier, condition (73) ensures the existence for each $(s_0, z_0) \in Q$ of an absolute minimizer $\bar{\bar{z}}(\cdot) \in \mathscr{A}(s_0, z_0)$ for $\tilde{I}$. Furthermore, when $Z_0 = z_0(-s_0)^{-\gamma}$ lies in the interval $[\delta_{\min}, \delta_{\max}]$ spanned by $\mathscr{Z}$, $\bar{\bar{z}}(\cdot)$ can be explicitly identified. For instance, if $(s_0, z_0) \in Q_+$ and $\delta_j \geqq T_0$ for some $\delta \in \mathscr{Z}$, then

$$\bar{\bar{z}}(s) := z_0 \wedge \delta_j(-s)^\gamma, \qquad s \in [s_0, 0], \tag{74}$$

is a trajectory such that $\tilde{I}[\bar{\bar{z}}] = 0$, and hence is an absolute minimizer. Therefore, all absolute minimizers have zero cost and are of the form (74) for some $\delta_j \in \mathscr{Z}$. We first provide a sequence $\{z_m\} \in \mathscr{A}(s_0, z_0) \cap W^{1,\infty}$ satisfying

$$\tilde{I}[z_m] \to \tilde{I}[\bar{\bar{z}}] = 0,$$

for such points $(s_0, z_0)$. Let $\bar{\bar{s}}_j$ be defined by

$$z_0 = \delta_j(-\bar{\bar{s}}_j)^\gamma.$$

Given any sequence $\{s_m\} \subset [\bar{\bar{s}}_j, 0)$ satisfying $s_m \uparrow 0$, put

$$z_m(s) := \begin{cases} \bar{\bar{z}}(s), & s \in [s_0, s_m], \\ \bar{\bar{z}}(s_m)\,(s/s_m), & s \in [s_m, 0]. \end{cases} \tag{75}$$

Then

$$\tilde{I}[z_m] = \tilde{I}[z_m] - \tilde{I}[\bar{\bar{z}}] \tag{76}$$

$$= \int_{s_m}^{0} (-s)^{-1} e(\delta_j(s/s_m)^{1-\gamma}, -\delta_j(s/s_m)^{1-\gamma}) g_0(-\delta_j(-s_m)^{\gamma-1}) \, ds.$$

Now it follows from the growth condition (14) that

$$\check{e}_j(P) := \sup_{Z \in [0,\delta_j)} e(Z, P)$$

is convex and satisfies

$$\check{e}(P) \leqq C_j(|P|^K \vee |P|^{K'}), \qquad P \in \mathbb{R}, \tag{77}$$

where $C_j := \sup\{e(Z, 1), e(Z, -1) : 0 \leqq Z \leqq \delta_j\}$. Hence we can appraise (76) as follows:

$$\tilde{I}[z_m] - \tilde{I}[\bar{\bar{z}}] \leqq g_0(-\delta_j(-s_m)^{\gamma-1}) \int_{s_m}^{0} (-s)^{-1} \check{e}_j(-\delta_j(s/s_m)^{1-\gamma}) \, ds$$

$$\leqq C_j \cdot g_0(-\delta_j(-s_m)^{\gamma-1}) \int_{s_m}^{0} (-s)^{-1} [(\delta_j(s/s_m)^{1-\gamma})^K \vee (\delta_j(s/s_m)^{1-\gamma})^{K'}] \, ds$$

$$\leqq (\text{const.}) \, g_0(-\delta_j(-s_m)^{\gamma-1}), \tag{78}$$

since the integral term is equal to

$$\int_0^1 u^{-1}((\delta_j u^{1-\gamma})^K \vee (\delta_j u^{1-\gamma})^{K'}) \, du.$$

It follows that $\tilde{I}[z_m] - \tilde{I}[\bar{\bar{z}}] \to 0$ as claimed.

We now consider the case in which $Z_0 = z_0(-s_0)^{-\gamma}$ is outside the interval $[\delta_{\min}, \delta_{\max}]$, say $Z_0 > \delta_{\max}$. It is clear that for such initial points $\bar{\bar{z}}(\cdot)$ never crosses below the curve $Z = \delta_{\max}$. Moreover, as we shall show below, any minimizing trajectory must approach the curve $Z = \delta_{\max}$:

**Lemma 4.1.** *If* $Z_0 = z_0(-s_0)^{-\gamma} > \delta_{\max}$ *and* $\bar{\bar{z}}(\cdot) \in \mathscr{A}(s_0, z_0)$ *is a minimizer, then*

$$\liminf_{s \to 0-} \bar{\bar{z}}(s)\,(-s)^{-\gamma} = \delta_{\max}.$$

Now select $s_m \uparrow 0$ such that

$$\bar{\bar{z}}(s_m)\,(-s_m)^{-\gamma} \leqq \delta_{\max} + 1,$$

and define $z_m(\cdot)$ as in (75). Notice that

$$\check{e}(P) := \check{e}_{(\delta_{\max}+1)}(P) := \sup\{e(Z, P) : 0 \leqq Z \leqq \delta_{\max} + 1\}$$

satisfies (77) with $C_j = C := \sup\{e(Z, 1), e(Z, -1) : 0 \leqq Z \leqq \delta_{\max} + 1\}$. Then, just as in (78),

$$\begin{aligned}
\tilde{I}[z_m] - \tilde{I}[\bar{\bar{z}}] &= \int_{s_m}^{0} (-s)^{-1}\, e(\bar{\bar{z}}(s_m)(s/s_m)^{1-\gamma}, -\bar{\bar{z}}(s_m)(s/s_m)^{1-\gamma})\, g_0(-\bar{\bar{z}}(s_m)(-s_m)^{-1})\, ds \\
&\quad - \int_{s_m}^{0} \tilde{f}(s, \bar{\bar{z}}(s), \bar{\bar{z}}'(s))\, ds \\
&\leqq g_0(-\bar{\bar{z}}(s_m)\,(-s_m)^{-1}) \int_{s_m}^{0} (-s)^{-1}\, \check{e}_{(\delta_{\max}+1)}(-\bar{\bar{z}}(s_m)\,(s/s_m)^{1-\gamma})\, ds \\
&\quad - \int_{s_m}^{0} \tilde{f}(s, \bar{\bar{z}}(s), \bar{\bar{z}}'(s))\, ds \\
&\leqq C \cdot g_0(-\delta_{\max}(-s_m)^{\gamma-1}) \\
&\quad \times \int_{s_m}^{0} (-s)^{-1}\, [|(\delta_{\max}+1)\,(s/s_m)^{1-\gamma}|^{K} \vee |(\delta_{\max}+1)\,(s/s_m)^{1-\gamma}|^{K'}]\, ds \\
&\quad - \int_{s_m}^{0} \tilde{f}(s, \bar{\bar{z}}(s), \bar{\bar{z}}'(s))\, ds.
\end{aligned}$$

The first integral term is bounded by a constant independently of $m$, and $g_0(-\delta_{\max}(-s_m)^{\gamma-1}) \to 0$, while the second integral term approaches 0 as $m \to +\infty$ since $\tilde{I}[\bar{\bar{z}}] < \infty$. □

It remains only to prove Lemma 4.1.

**Proof.** (Lemma 4.1). The proof is by contradiction. Thus, suppose not only that $\bar{\bar{Z}}(s) > \delta_{\max}$ for each $s \in [s_0, 0)$, but that $\liminf_{s \to 0-} \bar{\bar{Z}}(s) > \delta_{\max}$. Then

$$\inf_{s \in [s_0, 0)} \bar{\bar{Z}}(s) = \delta_{\max} + d \qquad \text{for some } d > 0.$$

Putting

$$A := \{(s, z) : z(-s)^{-\gamma} \geqq \delta_{\max} + d\},$$

we deduce that

$$\tilde{f}_{pp}(s, z, p) > 0, \quad \tilde{f}_z(s, z, p) \geqq 0, \quad (s, z) \in A, \tag{79}$$

and

$$\liminf_{p \to -\infty} \tilde{f}_p(s, z, p) = -\infty \qquad \text{uniformly for } (s, z) \in A. \tag{80}$$

Now (79) ensures that Tonelli's partial regularity theorem ([22], [5, Theorem 2.7]) holds for $\bar{\bar{z}}(\cdot)$. That is, $\bar{\bar{z}}(\cdot)$ is smooth on each interval $J$ belonging to a certain open subset of $(s_0, 0)$ whose complement has measure zero, and the Euler-Lagrange equation holds on each such interval:

$$\frac{d}{ds}\tilde{f}_p(s, \bar{\bar{z}}(s), \bar{\bar{z}}'(s)) = \tilde{f}_z(s, \bar{\bar{z}}(s), \bar{\bar{z}}'(s)) \geqq 0 \quad \text{for all } s \in J. \tag{81}$$

Moreover, at the right endpoint of each such interval one must have $\bar{\bar{z}}'(\cdot) = -\infty$ and hence $\tilde{f}_p = -\infty$ (by (80)). However, the latter possibility is contradicted by (81).

*Acknowledgements.* Part of this research was carried out during a visit by MIZEL to the Department of Mathematics, Heriot-Watt University during July of 1986. His research was partially supported by the U.S. National Science Foundation under grant DMS 86-02954 and DMS 87-04530.

## References


1. T. S. ANGELL, "A note on approximation of optimal solutions of free problems of the calculus of variations", *Rend. Circ. Mat. Palermo* **2** (1979), 258–272.
2. J. M. BALL, J. C. CURRIE, & P. J. OLVER, "Null Lagrangians, weak continuity, and variational problems of arbitrary order", *J. Funct. Anal.* **41** (1981), 135–174.
3. J. M. BALL & G. KNOWLES, "A numerical method for detecting singular minimizers", *Numer. Math.* **51** (1987), 181–197.
4. J. BALL & V. J. MIZEL, "Singular minimizers for regular one-dimensional problems in the calculus of variations", *Bull. Am. Math. Soc.* **11** (1984), 143–146.
5. J. BALL & V. J. MIZEL, "One-dimensional variational problems whose minimizers do not satisfy the Euler-Lagrange equation", *Arch. Rational Mech. Anal.* **90** (1985), 325–388.
6. L. CESARI, *Optimization—Theory and Applications*, Berlin Heidelberg New York Tokyo: Springer-Verlag, 1983.
7. F. H. CLARKE & R. B. VINTER, "Regularity properties of solutions to the basic problem in the calculus of variations", *Trans. Am. Math. Soc.* **291** (1985), 73–98.
8. A. M. DAVIE, "Singular minimizers in the calculus of variations in one dimension", *Arch. Rational Mech. Anal.* (1987) (to appear).
9. A. C. HEINRICHER, "A singular stochastic control problem arising from a deterministic problem with non-Lipschitzian minimizers", Ph. D. Dissertation, Department of Mathematics, Carnegie-Mellon University (1986).
10. A. C. HEINRICHER & V. J. MIZEL, "A stochastic control problem with different value functions for singular and absolutely continuous control", Proc. 1986 IEEE Conf. Decision and Control, Athens, Greece.
11. A. C. HEINRICHER & V. J. MIZEL, "A new example of the Lavrentiev phenomenon", SIAM Journal for Control and Optimization, to appear.
12. M. R. HESTENES, *Calculus of Variations and Optimal Control Theory*, Krieger Publishing, 1980.
13. M. LAVRENTIEV, "Sur quelques problèmes du calcul des variations", *Ann. Mat. Pura Appl.* **4** (1926), 107–124.
14. P. D. LOEWEN, "On the Lavrentiev Phenomenon", *Can. Math. Bull.* **30** (1987), 102–108.
15. J. D. LOGAN, *Invariant Variational Principles*, New York London: Academic Press, 1977.

16. E. J. McShane, "Existence theorems for problems in the calculus of variations", *Duke Math. J.* **4** (1938), 132–156.
17. B. Manià, *Sorpra un esempio di Lavrentieff*, Bull. Un. Mat. Ital. **13** (1934), 146–153.
18. M. Marcus & V. Mizel, "Absolute continuity on tracks and mappings of Sobolev spaces", *Arch. Rational Mech. Anal.* **45** (1972), 294–320.
19. E. Noether, "Invariante Variationsprobleme", *Nachr. Akad. Wiss. Göttingen, Math.-Phys. Kl. II* (1918), 235–257.
20. E. Noether, "Invariant Variational Problems", *Transport Theory of Statis. Physics* **1**, 186–207 (1971) (Translation by M. A. Tavel of the original article).
21. P. J. Olver, *Applications of Lie Groups to Differential Equations*, Berlin Heidelberg New York Tokyo: Springer-Verlag, 1986.
22. L. Tonelli, "Sugli integrali del calcolo delle variazioni in forma ordinaria", *Ann. R. Scuola Norm. Sup. Pisa* **3** (1934), 401–450 (in L. Tonelli *Opere Scelte* vol. III # 105, Edizioni Cremonese, Roma, 1961).
23. Ch. De La Vallée Poussin, "Sur l'intégrale de Lebesgue", *Trans. Am. Math. Soc.* **16** (1915), 435–501.

Department of Mathematics
University of Kentucky
Lexington

and

Department of Mathematics
Carnegie Mellon University
Pittsburgh

*(Received September 2, 1987)*

# *Steady, Structured Shock Waves. Part 1: Thermoelastic Materials*

J. E. DUNN & R. L. FOSDICK

*To James Serrin,
with affection and admiration*

**Contents**

1. Introduction . . . . . . . . . . 295
2. Preliminary Notions . . . . . . . . . . 304
3. Thermoelastic Materials and the Speeds of Sound . . . . . . . . . . 309
4. Departures, Arrivals, and the Structure of the Hugoniot . . . . . . . . . . 325
5. A Priori Inequalities . . . . . . . . . . 334
6. Blocked Rayleigh Curves, Bifurcated Hugoniots, and the Curve of Thermal Extremes . . . . . . . . . . 340
References . . . . . . . . . . 364

## 1. Introduction

One dimensional plane shock waves have been widely used to study the mechanical response of solids to high velocity impact*. For many materials, under a certain range of impact pressures, a two-wave structure arises, the first wave of which, the so-called elastic precursor, travels at a velocity near to the speed of sound, while the second wave, often called a plastic shock wave, travels at a slower speed which increases with impact pressure. While the full two-wave structure is thus not steady (does not propagate without change of form), each component wave may be treated as steady after having propagated sufficiently far.† As indicated by the experimental data presented in Figure 1, in some materials the elastic precursor is generally much steeper than the plastic shock wave. Indeed, for some materials, the elastic precursor is infinitely steep to within current techniques of data resolution. Accordingly, it is usually modelled as a singular surface. The plastic shock, however, typically exhibits a smooth variation, albeit still highly localized in time and space. Thus, these shocks are *structured*, and a

---

* Convenient references are [1, 2].

† See, for example, the study of JOHNSON & BARKER [3].

central question for constitutive modelling concerns those aspects of material response that can serve as structuring agencies in the shock‡.

In 1968 BARKER [4], using a new technique of velocity interferometry, obtained the wave profile for the particle velocity in shock loaded aluminium. BARKER noticed that if $p_D$ is the impact or driving pressure needed to create a steady, structured plastic shock wave and if $\dot{\varepsilon}_M$ is the peak strain rate (more precisely, the maximum of the absolute value of the strain rate) in that wave, then a log-log plot of $p_D$ *vs.* $\dot{\varepsilon}_M$ for his data was fit by a straight line. Thus, in essence, BARKER established that the relation between $p_D$ and $\dot{\varepsilon}_M$ for plastic shock waves in aluminum was a simple power law. In fact, his data yields the explicit *shock thickness relation*

$$p_D = k\,|\dot{\varepsilon}_M|^{1/4}, \tag{1.1}$$

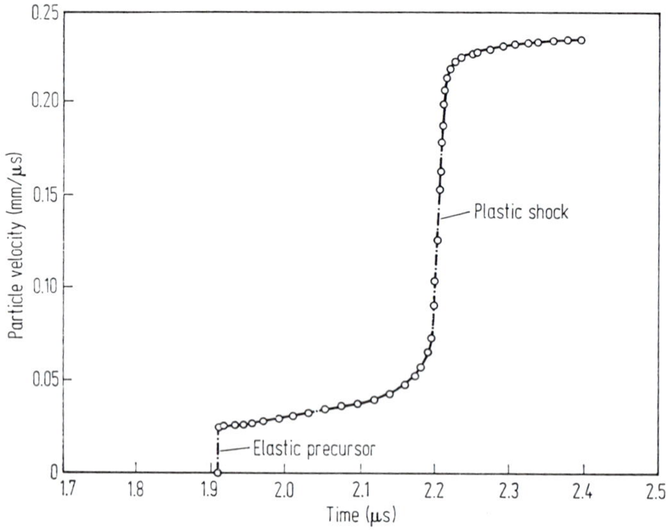


Fig. 1. Shot No. 927 from JOHNSON & BARKER [3] showing a two wave shock structure in 6061-T6 aluminum. The impact stress, the target thickness, the elastic shock speed, and the plastic shock speed were, respectively, 3.7 GPa, 12.2 mm, 6.39 KM/s, and 5.66 KM/s.

where $k$ is a material constant. Subsequently, GRADY [5] studied the constitutive implications of BARKER's observation within the class of materials given by

$$\sigma = f(\varepsilon) + g(\dot{\varepsilon}), \tag{1.2}$$

‡ While Figure 1 shows an extremely sharp elastic shock for 6061-T6 aluminium, other materials—indeed, even other alloys of aluminum—evince more complex response: the elastic precursor may itself show shock structure, even evolving shock structure. Thus, while we have in mind here mainly the plastic shock, the considerations of shock structure in our work could also be brought to bear on the elastic precursor in many materials.

where $\sigma$ is the axial stress and where $f(\varepsilon)$ and $g(\dot{\varepsilon})$ are certain functions of, respectively, the uniaxial strain $\varepsilon$ and its material time rate $\dot{\varepsilon}$. GRADY found that (1.1), along with a standard assumed form of $f(\cdot)$, uniquely determined the viscous stress function $g(\cdot)$, and indeed, to within a first approximation,

$$g(\dot{\varepsilon}) = \mu_0 \operatorname{sgn}(\dot{\varepsilon}) |\dot{\varepsilon}|^{1/2}, \tag{1.3}$$

where $\mu_0$ is a material constant. Later, SWEGLE & GRADY [6] noted that the empirical relation (1.1) in fact seemed to hold not only for the plastic waves in aluminum but also for the steady, plastic shock waves in uranium, beryllium, magnesium oxide, iron, fused silica, copper, and bismuth. They further examined the full wave profiles in these materials using the constitutive form (1.3) of GRADY within a linear theory of plasticity. DUNN & GRADY [7] then studied the constitutive implications of (1.1) when (1.2) is replaced by the more general constitutive class

$$\sigma = f(\varepsilon) + g(\varepsilon, \dot{\varepsilon}),$$

where $g(\varepsilon, 0) = 0$. They showed that, contrary to the uniqueness found by GRADY, the shock thickness relation (1.1) was then compatible with an infinity of viscous stress forms $g(\varepsilon, \dot{\varepsilon})$. Indeed, it turned out that even within the class of power laws, $g(\varepsilon, \dot{\varepsilon}) = \operatorname{sgn}(\dot{\varepsilon})\, h(\varepsilon)\, |\dot{\varepsilon}|^m$, an infinity of constitutive forms are compatible with (1.1): for every strain rate sensitivity $m > 2/9$, there is a strain dependent viscosity $h_m(\varepsilon)$ such that (1.1) is satisfied. The power law model that gave the best fit to the full wave profiles in aluminum occurred for $m \approx .69$, and, interestingly, the corresponding viscosity function, $h_{.69}(\cdot)$, induced by (1.1) suggested a high degree of strain softening. More recently, DUNN [8] studied the relation (1.1) for the shock thickness within a nonlinear viscoplasticity model that (i) introduces an internal state variable for the "plastic strain", (ii) postulates a certain decomposition of total strain into "elastic" and "plastic" parts, and (iii) lays down a companion rate law for the evolution of the plastic strain. For shear driven plastic flow and a relatively standard constitutive equation for the stress, it was found there that BARKER's relation (1.1) completely determined the structure of the rate law. Moreover, it did this in a way that demonstrated far more rapid evolution of plastic strain than is commonly found in, say, dislocation models of plastic deformation.

All of the above analyses examine BARKER's discovery (1.1) for the shock thickness in the context of pure mechanics. A main goal of the present work, of which this is the first of three parts, is to explore (1.1) in the setting of a full thermomechanical model that takes proper account of thermal as well as mechanical constitutive response. In particular, we will study the constitutive implications of (1.1) within the general theory of thermoviscous response, being especially interested in both the capacity of heat conduction and/or viscosity to structure shock waves and the detailed thermal dependencies of the response functions necessitated by (1.1). Our work is thus related to—but is constitutively more general than—the work, for example, of GILBARG & PAOLUCCI [9], GRAD [10], and BAND [11], all of whom where interested in the structure of shock waves but, unfortunately, lacked quantitative information like (1.1). Addionally, our work has required us to develop a rather detailed treatment of steady, structured shock waves in general. As we shall see, this has led to the discovery of several aspects of shock waves, structured or not, that seemingly have been heretofore overlooked or insufficiently examined.

In Parts 1 and 2 of our work we consider only heat conduction as a structuring agency in the shock. Thus, we initially consider only nonlinear thermoelasticity as a constitutive setting and put off until Part 3 consideration of the additional structuring mechanism of viscous dissipation. For reasons we will discuss shortly, thermoelasticity is not usually thought of as an appropriate framework to discuss structured shock waves of arbitrary intensity. A major aim of this the first part of our work is to demonstrate that this prejudice is insufficiently refined, being based on a (as we shall see) far too special analysis due to RAYLEIGH. Indeed, structured shocks in thermoelastic materials would appear to be a much richer area of study than has heretofore been thought. Additionally, there are certain concepts and features of structured shock waves that are more or less independent of the structuring agency of the shock and thus turn out to be most simply addressed in the restricted setting of thermoelasticity.

Thus, in Section 2 we begin by setting notation and deriving some of the basic principles of steady, structured shock waves that hold across broad classes of material models. These principles include the so-called *Rayleigh line construction* for stress-deformation pairs in the shock, the increase in entropy due to the passage of the shock, the introduction of the *Hugoniot set* of those entropy-deformation pairs that are allowable back states in a shock with a given front state, and the introduction of the relationship between the *shock velocity* and the (*back state*) *particle velocity*, the, perhaps, main product of experimental shock wave work in condensed matter.

In Section 3 we introduce thermoelastic materials. For them the Rayleigh line construction of Section 2 gives rise to the *Rayleigh curve*. This is a curve, here given by a simple algebraic relation, on which must fall all of the entropy-deformation pairs experienced by a particle as a shock passes through it. The Rayleigh curve for a given shock, of course, intersects the Hugoniot set at both its front and back states but, at intermediate states in the shock, the two are quite distinct. Although the entropy at the back of a shock must be at least as large as the entropy at the front, a fundamental aspect of the Rayleigh curve for thermoelastic materials (and, indeed, for thermoviscous materials also) is that, typically, there are places in the wave where a particle experiences entropies considerably larger than that of its back state, *i.e.*, the entropy increase through a structured shock wave is, in general, *not monotone and has an interior absolute maximum**. This phenomenon of *entropy overshoot* in steady, structured shock waves turns out to be intimately connected to (indeed, for many materials, tantamount to) the *supersonic-subsonic* property of the shock velocity, according to which the

* The above observations are for materials that heat up during the passage of a shock wave. For materials that are *cooled* by the passage of a shock wave there are, instead, places in the center of the wave where a particle experiences entropies much *smaller* than that at its *front* state. Thus, the entropy increase in a structured shock in these materials is again not monotone with now an interior absolute *minimum*. We make no effort here to find the earliest statement of this non-monotone growth. See, for example, the interesting discussions by LANDAU & LIFSCHITZ [28, § 87], by SERRIN & WHANG [30], and by ZEL'DOVICH & RAIZER [29, Chapter I, § 23].

speed of a shock must exceed the adiabatic sound speed of its front state and be exceeded by the adiabatic sound speed of its back state.

Ever since publication of the classic works of BETHE [12] and WEYL [13] it has become traditional to prove the supersonic-subsonic property of the shock velocity by appealing to rather strong *a priori* inequalities that, typically, are required to hold over the entire constitutive domain of the material†. Here we offer two different approaches to the supersonic-subsonic condition. The first, presented in Section 3‡, frees this condition from entanglement with any global *a priori* inequalities at all. We show in Theorem 1 that the underlying principles of momentum balance, energy balance, and entropy imbalance are themselves sufficient to ensure that in any steady, structured shock wave the adiabatic sound speed at the back state will be at least as large as the actual shock speed if the specific heat there is positive. Indeed, our methods give a sharper bound: at the back state the shock speed is trapped between the *isothermal* sound speed below and the adiabatic sound speed above. Similarly, our Theorem 3 demonstrates that momentum balance, energy balance, and entropy imbalance are alone sufficient to ensure that any steady, structured shock wave having a front state with a positive specific heat must necessarily have a shock speed that is either at least as large as the adiabatic sound speed of the front state or no larger than the isothermal sound speed of the front state. The first of these two alternative bounds on the shock speed is easily combined with our Theorem 1 to give, in essence, the supersonic-subsonic condition. The second alternative, in which a shock can move at a speed less than even the isothermal sound speed of its front state, is rather strange. We call such shocks *ultra-slow*, and, in Section 5 and in the final part of Section 6, we give strong reasons why ultra-slow shocks would be rather rare if, indeed, they are possible at all.

Theorems 2 and 4 treat the borderline cases when the shock speed equals the adiabatic sound speed of either the back state or the front state. While these cases apparently cannot be universally ruled out, useful information on the structure of such shocks still can be extracted from the dissipation of energy through them.

In Section 4 we take up questions of the local structure of the Hugoniot set and any given Rayleigh curve. At points where they are both regular (and so the Hugoniot set is locally curve-like), we study in some detail the types of intersections these two sets can have. Our initial analysis is purely geometric but our main aim is Theorem 5 and its Corollary in which we show that the thermodynamics of steady, structured shock waves in general forbids that portion of a Rayleigh curve swept out in a shock process from crossing the Hugoniot set at any points other than its front and back states. This result, discussed also by HAGAN & SERRIN [17], not only makes a statement about the structure of a shock wave, it also has the character of a uniqueness theorem, forbidding, as it does, the existence of more than one compressive shock and more than one expansive shock corresponding to a given front state and shock speed. At points where a Rayleigh curve just grazes ths Hugoniot the tools of our anaysis fail. However,

† See, for example, the exposition of COURANT & FRIEDRICHS [14] or of SERRIN [15].

‡ And also discussed by DUNN & FOSDICK in [16].

as we shall show, thermodynamics still has a bit to say about the "flow" of the shock along the Rayleigh curve near such states.

In Section 5 we present the second of our approaches to the supersonic-subsonic condition. Our ideas in this section are close in spirit to some of the methods of BETHE [12], and, unlike our first approach in Section 3 which exploited energy dissipation across the structure of the wave, the results here are also applicable to the theory of shock waves as singular surfaces. Essentially, we have merely tried to lighten as much as possible the usual three *a priori* constitutive inequalities of shock wave analysis. In particular, the much weaker point condition $(5.4)_{1,2}$ with which we replace the usual global condition on the third derivative of the energy, *i.e.*, $(3.7)_3$, serves not only to provide an interesting interpretation of that inequality but also emphasizes its extremely special and limiting character. The *a priori* (indeed, almost *ad hoc*) nature of the usual three constitutive inequalities was pointed out long ago by BETHE [12] who felt himself compelled to verify them directly for no less than six special constitutive models, stating at the outset that no known physical principle secured them. While we do not regard the usual *a priori* inequalities (or our substitutes for them) as intrinsic to the shock process, they do provide an interesting complement to it. Thus, besides using our inequalities to obtain hypotheses which, though weaker than usual, still ensure the supersonic-subsonic property of the shock speed, we also illustrate their effectiveness in forbidding points of grazing or osculatory contact of the Hugoniot and a Rayleigh curve.

Our final section, Section 6, takes up matters that bear on not only the qualitative behavior of steady, structured shocks in thermoelastic materials but also on the existence of such shocks. All three of the issues we study there bring out the significance of the *Grüneisen coefficient* for the analyses of shock waves. This coefficient, defined below by (6.1), measures the sensitivity of the stress to energetic influences and, for reasons of thermostatic stability, has the same sign as the coefficient of thermal expansion wherever this latter exists. For real materials, then the Grüneisen coefficient is typically positive, but it can be a complicated function of, say, the entropy and the deformation and can, indeed, even change its sign over different ranges of material behavior. Unfortunately, while the detailed functional behavior of the Grüneisen coefficient can have a profound effect on both the existence and the qualitative character of steady, structured shock waves, it turns out to be extremely difficult for experimenters to measure it accurately over the full range of shock response.

The first existence issue we treat in Section 6 is rather special and probably would be of little significance were it not for the ease with which it can arise in what is called the *Mie-Grüneisen* class of thermoelastic materials. It turns out that if, for some fixed deformation, the Grüneisen coefficient vanishes for all values of the entropy (equivalently, if the stress at some fixed deformation is a constant, independent of the entropy), then in the entropy-deformation plane this line of fixed deformation in general forms an impenetrable barrier to steady, structured shocks: no steady, strutured shock wave whose front and back states are on opposite sides of this line can exist.

All earlier shock wave studies of which we are aware have treated the Hugoniot set as if it were *everywhere* curve-like. The second existence issue we take up in

Section 6 calls into question this easy habit. We find rather natural and mild conditions on the Grüneisen coefficient and the adiabatic sound speed along the Hugoniot that allow the Hugoniot to *bifurcate* at select points. A fairly complete local analysis of the bifurcated branches of the Hugoniot turns out to be possible, and, for example, we find that, in a deleted neighborhood of one of its bifurcation points, the requirement that the shock speed be subsonic with respect to the sound speed of the back state can be met only on a *kinked* (*i.e.*, non-smooth) portion of the Hugoniot[§]. The existence of shocks with these back states then turns out to hinge on the orientation of this kinked portion of the Hugoniot with respect to the front state. We close this subsection with some remarks on Mie-Grüneisen materials, since materials of this class happen to be especially susceptible to bifurcated Hugoniots.

The final subsection of Section 6 treats, firstly, some interesting details of shock structure induced by the sign and/or sign changes experienced by the Grüneisen coefficient through the shock structure. Specifically, we delineate the occurrence of compressive shock heating as well as the less familiar phenomenon of compressive shock *cooling*. Moreover, we find conditions on the Grüneisen coefficient that, in contrast with the usual situation, allow the growth of the specific volume through the shock to be *non-monotone*; indeed, we even find conditions that can ensure that the specific volume through the shock, like the entropy, undershoots or overshoots its values at the end states.

More significant than the above issues of shock structure, however, is an argument due originally to RAYLEIGH [18] which is frequently used to suggest that only very weak steady, structured shock waves can be supported in thermoelastic materials. In the final subsection of Section 6 we confront, generalize, and deepen RAYLEIGH'S analysis. Contrary to the apparent beliefs of its proponents, we find RAYLEIGH'S argument wanting and largely ineffective in proving that steady, structured shocks of arbitrary intensity in general thermoelastic materials are impossible. Indeed, we shall see that even for his own highly specialized class of materials (*i.e.*, for ideal gases), RAYLEIGH'S result is overstated. We remark briefly below and at length in the conclusion of Section 6 about the mathematical detail of RAYLEIGH'S original argument. The physics that underlies it, although not discussed by RAYLEIGH, is interesting: In a thermoelastic material one seeks to use heat conduction and the energy equation in which it appears as a structuring agency in the shock wave. This means that a central role is given to the temperature and its gradient through the shock. In particular, it becomes essential that the temperature gradient, the driving force of the heat conduction, not be permitted to vanish except at the end states of the shock. However, the states of temperature which a particle experiences as a steady, structured shock sweeps past it are foreordained to be those seen on the Rayleigh curve, a curve which is determined by the shock speed, the constitutive equations, and momentum balance alone. Thus arises the potential for conflict between the principles of energy balance and momentum balance. If there is a non-monotone variation of temperature along the stretch of the Rayleigh curve joining the front state to a putative back state of a structured shock and if the specific heat is of fixed sign, then it turns out that

[§] See Figure 8.

the energy equation cannot be satisfied by single-valued functions for the entropy and the specific volume. Since its then multi-valued "solutions" cannot correspond to the situation in a real material, we see that in such a case no steady, structured shock can exist at all.

To address formally the above potential conflict between the energy equation and the momentum equation, we introduce in Section 6 the *curve of thermal extremes*. This is the set of all points in the entropy-deformation plane where the temperature on the Rayleigh curve through that point has a local extremum. The problem RAYLEIGH noticed can be avoided anytime the curve of thermal extremes and the segments of Rayleigh curves swept out in shock processes do not intersect; *a fortiori*, it can be avoided whenever the curve of thermal extremes and the segments of Rayleigh curves swept out in shock processes keep to opposite sides of the Hugoniot. Additionally, when the specific heat is positive, it turns out that the curve of thermal extremes and a given one of these Rayleigh curve segments fail to intersect if and only if the associated shock speed exceeds the isothermal sound speed *throughout the structure of the shock*. As we have seen, this is in sharp contrast to the situation for the adiabatic sound speed which typically must, at the back state, *exceed* the shock speed and, at the front state, *be exceeded* by the shock speed.

The placement of the curve of thermal extremes relative to the Hugoniot and the Rayleigh curves is interestingly complex. After giving a few examples of the rich variety of this placement, we formulate conditions which ensure that most thermoelastic materials will admit mild enough steady, structured shocks. We thus generalize the result RAYLEIGH obtained for a certain subclass of gases. We close by showing that with a proper choice of constitutive constants even the ideal gases of RAYLEIGH will be such that the curve of thermal extremes is never crossed by any Rayleigh curve in its transit from the front state to its back state: *these materials now admit of arbitrarily intense steady, structured shock waves.*

## *Historical Miscellanea*
## *Earnshaw, Rankine, Jouguet, Zemplén, Rayleigh*

The first study of steady, structured shock waves (or, as they were called then, *waves of permanent regime* or *type*) that allowed for thermal effects and took account of dissipative forces was RANKINE'S [19] paper of 1869. There, RANKINE obtained the fundamental first integral of the equations of compatibility and momentum balance that was to become known subsequently, for reasons not clear to us, as the *Rayleigh line construction**. Properly accounting for energy balance

* See our (2.11). Actually, (2.11) was first obtained by EARNSHAW [20] in 1859 and was so credited to him and to RANKINE by RAYLEIGH [18] who wrote much later. Unfortunately, EARNSHAW studied wave propagation in essentially a purely mechanical context and thus took (2.11) to be a *constitutive equation* requiring the stress to vary inversely with the density. Since such a variance conflicted with Boyle's law, EARNSHAW dismissed the possibility of steady, structured shock waves, concluding that waves of permanent type "have no real existence where Boyle's law is strictly true". RANKINE, on the other hand, drew explicit attention to the fact that (2.11) requires that in thermo-

through the wave, RANKINE also, in essence, obtained in his paper what has now become known as the *Rankine-Hugoniot equation* †. Moreover, RANKINE addressed explicitly the mechanism of heat conduction as a structuring agency in shock wave profiles, and, as RAYLEIGH remarks "although there are one or two serious, deficiencies, ..., the memoir marks a very definite advance".

The "deficiencies" to which RAYLEIGH refers went undiscovered for more than thirty years. Indeed, the remedy for the first "deficiency" drew on a physical principle unrecognized when RANKINE wrote. It was stated for the first time by JOUGUET [21] in 1901 in the context of propagating first order discontinuities in fluids ‡ and is the requirement, imposed by the "principle of Clausius", that the passage of a shock wave must leave in its wake material whose entropy is no smaller than the entropy of the material before the arrival of the shock. Had RANKINE had access to this principle he would have been able to argue that only steady waves of condensation are possible in an ideal gas and so anticipate the analyses of JOUGUET [21, 22], ZEMPLÉN [23], and RAYLEIGH [18], all of whom showed that the increase of entropy through the shock precluded the possibility of steady rarefaction waves in such a material.

The second "deficiency" RAYLEIGH found with RANKINE'S work involves matters of some subtlety. Briefly, RAYLEIGH examined the differential equation whose solution gives the form of the pressure through a structured shock wave in RANKINE'S materials and noted that, for this profile to be single-valued, the ratio of the pressures at the back state and the front state must be less than or equal to

$$\frac{1+\mathfrak{R}}{3-\mathfrak{R}} (\approx 1.61 \textit{ for Rayleigh}), \tag{*}$$

where $\mathfrak{R}$ is the ratio of specific heats. Thus, RAYLEIGH concluded, and it has since become widely accepted, that only relatively mild structured shocks can be supported by heat conduction alone.

Unfortunately, RAYLEIGH'S analysis is itself a bit deficient: While one cannot doubt the truth of the bound (*), RAYLEIGH gives little insight into its origins. One is left wondering to what extent RAYLEIGH'S conclusion depends on his special choice of materials, and such speculations are only reinforced when one notices that the bound (*) becomes empty if $\mathfrak{R} > 3$. As our Section 6 reveals, the issues here are more complex than the equations of gas dynamics might suggest.

---

elastic materials, in addition to the deformation, "there must be both a change of temperature and a conduction of heat" in order that "permanency of type may be possible in a wave of longitudinal disturbance". RANKINE, unlike EARNSHAW, thus made the key move of interpreting (2.11) as a statement about a *process*, rather than a constitutive equation.

† Although not rendered in the now standard form of our (2.16).

‡ And stated by him again in 1904, [22], this time in the context of propagating "quasi-waves" in a perfect gas with viscosity. The subsequent note [23] by ZEMPLÉN a year later seems to us inferior when it is not derivative, and in this light the comment [24] by HADAMARD, which takes no notice of JOUGUET but praises the contribution of ZEMPLÉN, seems peculiar. In this regard, it is also noteworthy that all four of these papers appeared in the same journal over a period of only four years.

## 2. Preliminary Notions

Let $\mathscr{B}$ be a one-dimensional body whose particles in a fixed reference configuration are labelled $X$, $X \in (-\infty, \infty)$. If $x = \chi(X, t)$ is a *motion* of $\mathscr{B}$, then let

$$\mathrm{v} = \mathrm{v}(X, t) \equiv \dot{\chi}(X, t),$$
$$F = F(X, t) \equiv \chi_{,X}(X, t),$$

denote, respectively, the *velocity* of $X$ and the *deformation gradient* at $X$ at time $t$. To ensure the invertibility of $\chi(\cdot, t)$, we assume that $F(X, t)$ is positive. Further, for smooth enough motions, we have the condition of *kinematic compatibility*, *i.e.*,

$$\mathrm{v}_{,X} = \dot{F}. \tag{2.1}$$

We also recall that the balance laws of linear momentum and energy, as well as the imbalance law of entropy, can be written for $\mathscr{B}$ in the respective forms

$$\sigma_{,X} = \varrho_R \dot{\mathrm{v}}, \tag{2.2}$$

$$\varrho_R \dot{e} = \sigma \mathrm{v}_{,X} - q_{,X}, \tag{2.3}$$

$$\varrho_R \dot{\eta} \geqq -\left(\frac{q}{\theta}\right)_{,X}. \tag{2.4}$$

Here, at the particle $X$ at time $t$, $\sigma = \sigma(X, t)$ is the *axial stress*, $e = e(X, t)$ is the *specific internal energy* per unit mass, $q = q(X, t)$ is the *heat flux*, $\eta = \eta(X, t)$ is the *specific entropy* per unit mass, $\theta = \theta(X, t)$ is the (positive) *absolute temperature*, and $\varrho_R = \varrho_R(X)$ is the *density* at $X$ in the given reference configuration (and is here taken to be independent of $X$). We remark (i) that (2.2)–(2.4) are written under the assumption that $\mathscr{B}$ is experiencing no external body force or radiant heating, and (ii) that our sign convention is such that in compression, $\sigma < 0$. We remark also that, within the framework here, mass conservation is just the assertion that $\varrho = \varrho(X, t) \equiv \varrho_R / F(X, t)$ is the current density at $X$ at time $t$ during the motion $\chi$, or, equivalently, that $v = v(X, t) \equiv v_R F(X, t)$, $v_R \equiv 1/\varrho_R$, is the current *specific volume*.

A *steady, structured shock wave* for $\mathscr{B}$ is any smooth 7-tuple $(F, \mathrm{v}, \sigma, e, q, \eta, \theta)$ of functions satisfying (2.1)–(2.4) and such that

$$\begin{aligned}
F &= \hat{F}(X - \mathscr{V} t),\\
\mathrm{v} &= \hat{\mathrm{v}}(X - \mathscr{V} t),\\
\sigma &= \hat{\sigma}(X - \mathscr{V} t),\\
e &= \hat{e}(X - \mathscr{V} t),\\
q &= \hat{q}(X - \mathscr{V} t),\\
\eta &= \hat{\eta}(X - \mathscr{V} t),\\
\theta &= \hat{\theta}(X - \mathscr{V} t).
\end{aligned} \tag{2.5}$$

Here the number $\mathscr{V}$ is called the *material* or *referential wave speed*, or, more briefly, the *shock speed* or *shock velocity*; it evidently gives the span of material, $\Delta X$, in the reference configuration which is traversed in unit time by any disturbance intensity in the wave (*e.g.*, by $\sigma^* \equiv \hat{\sigma}(\xi^*)$, $\xi^* \in (-\infty, \infty)$ arbitrary). We assume $\mathscr{V} > 0$, so the material at $+\infty$ is *ahead* of the wave, while the material at $-\infty$ is *in back* of the wave. The wave is said to be *compressive* if $F^- < F^+$, *expansive* if $F^- > F^+$, where $F^- \equiv \hat{F}(-\infty)$ and $F^+ \equiv \hat{F}(+\infty)$.

We now insert $(2.5)_{1,2}$ into (2.1), $(2.5)_{2,3}$ into (2.2), $(2.5)_{2-5}$ into (2.3), and $(2.5)_{5-7}$ into (2.4). In this way we obtain

$$\hat{\mathrm{v}}' = -\mathscr{V}\hat{F}',$$

$$\hat{\sigma}' = -\varrho_R\mathscr{V}\hat{\mathrm{v}}',$$

$$\varrho_R\mathscr{V}\hat{e}' = -\hat{\sigma}\hat{\mathrm{v}}' + \hat{q}',$$

$$\varrho_R\mathscr{V}\hat{\eta}' \leqq \left(\frac{\hat{q}}{\hat{\theta}}\right)',$$

where $\hat{\mathrm{v}}' = \dfrac{d}{d\xi}\hat{\mathrm{v}}(\xi)$, *etc.* Equivalently,

$$\hat{\mathrm{v}}' = -\mathscr{V}\hat{F}', \tag{2.6}$$

$$\hat{\sigma}' = \varrho_R\mathscr{V}^2\hat{F}', \tag{2.7}$$

$$\varrho_R\mathscr{V}\hat{e}' = \mathscr{V}\hat{\sigma}\hat{F}' + \hat{q}', \tag{2.8}$$

$$\varrho_R\mathscr{V}\hat{\eta}' \leqq \left(\frac{\hat{q}}{\hat{\theta}}\right)'. \tag{2.9}$$

If we integrate the first and second of these between $+\infty$ and $\xi$, we see that

$$\hat{\mathrm{v}}(\xi) - \mathrm{v}^+ = -\mathscr{V}(\hat{F}(\xi) - F^+), \tag{2.10}$$

$$\hat{\sigma}(\xi) - \sigma^+ = \varrho_R\mathscr{V}^2(\hat{F}(\xi) - F^+), \tag{2.11}$$

where $\mathrm{v}^+ \equiv \hat{\mathrm{v}}(+\infty)$, *etc.* From (2.11) it follows that the combination $\hat{\sigma}\hat{F}'$ occurring in the energy equation (2.8) is explicitly integrable. As a consequence the energy equation may be put in the form

$$\varrho_R\mathscr{V}(\hat{e}(\xi) - e^+) = \mathscr{V}\{\tfrac{1}{2}\varrho_R\mathscr{V}^2(\hat{F}(\xi) - F^+)^2 + \sigma^+(\hat{F}(\xi) - F^+)\} + \hat{q}(\xi),$$

where we have assumed that $q^+ \equiv \hat{q}(+\infty)$ vanishes. But, by (2.11)

$$\begin{aligned}\tfrac{1}{2}\varrho_R\mathscr{V}^2(\hat{F} - F^+)^2 + \sigma^+(\hat{F} - F^+) &= \{\tfrac{1}{2}\varrho_R\mathscr{V}^2(\hat{F} - F^+) + \sigma^+\}(\hat{F} - F^+),\\ &= \{\tfrac{1}{2}\hat{\sigma} + \tfrac{1}{2}\sigma^+\}(\hat{F} - F^+),\end{aligned}$$

and thus we have

$$\varrho_R\mathscr{V}(\hat{e}(\xi) - e^+) = \tfrac{1}{2}\mathscr{V}(\hat{\sigma}(\xi) + \sigma^+)(\hat{F}(\xi) - F^+) + \hat{q}(\xi). \tag{2.12}$$

Lastly, integration of the entropy inequality (2.9) from $+\infty$ to $\xi$ yields

$$\varrho_R \mathscr{V}(\hat{\eta}(\xi) - \eta^+) \geqq \frac{\hat{q}(\xi)}{\hat{\theta}(\xi)}, \tag{2.13}$$

where again we have taken $q^+ = 0$.

Several remarks are in order concerning the system (2.10)–(2.13). First, observe that (2.10) asserts that the velocity-deformation gradient pairs experienced throughout $\mathscr{B}$ in a steady, structured shock wave fall on a straight line in the v–$F$ plane of slope $(-\mathscr{V})$. Similarly, (2.11) is the assertion, frequently called the *Rayleigh line construction*, that in a steady, structured shock wave the stress-deformation gradient pairs experienced throughout $\mathscr{B}$ fall on a straight line in the $\sigma$–$F$ plane of slope $\varrho_R \mathscr{V}^2$.

Second, let us set $\xi = -\infty$ in (2.10)–(2.13). We thereby find the *front state-back state relations*:

$$\mathrm{v}^- - \mathrm{v}^+ = -\mathscr{V}(F^- - F^+), \tag{2.14}$$

$$\sigma^- - \sigma^+ = \varrho_R \mathscr{V}^2(F^- - F^+), \tag{2.15}$$

$$\varrho_R(e^- - e^+) = \tfrac{1}{2}(\sigma^- + \sigma^+)(F^- - F^+), \tag{2.16}$$

$$\eta^- \geqq \eta^+, \tag{2.17}$$

where, in deriving (2.16) and (2.17), we have assumed that $\hat{q}(-\infty) = 0$, and where $\mathrm{v}^- \equiv \hat{\mathrm{v}}(-\infty)$, *etc.* These relations are, of course, familiar from the singular surface theory of shock propagation; (2.14)–(2.16) are usually referred to as the *Rankine-Hugoniot conditions*, (2.16) being the celebrated *Rankine-Hugoniot equation*. The condition (2.17), stated first by JOUGUET and then by ZEMPLÉN early in this century, is the requirement that after the passage of a shock wave the entropy of a particle will not have decreased.

Now in a shock experiment the front state is typically known. Thus, the 6-tuple $(F^+, \mathrm{v}^+, \sigma^+, e^+, \eta^+, \theta^+)$ is known, and (2.14)–(2.17) provide restrictions on the allowable back states and the shock speed $\mathscr{V}$, *i.e.*, restrict the 7-tuple $(F^-, \mathrm{v}^-, \sigma^-, e^-, \eta^-, \theta^-, \mathscr{V})$. Complementing the system (2.14)–(2.17) are *equilibrium** *constitutive equations*, and for many materials these may be taken to be of the form

$$\begin{aligned} e &= e^{=}(\eta, F), \\ \sigma &= \sigma^{=}(\eta, F), \\ \theta &= \theta^{=}(\eta, F). \end{aligned} \tag{2.18}$$

Here, $e^{=}$, $\sigma^{=}$, and $\theta^{=}$ are material dependent functions, defined and smooth over an open set in $R^2$, giving, respectively, the equilibrium values of the energy, stress, and temperature in terms of the equilibrium values of entropy and deformation.

* At the present level of generality, the notion of "equilibrium" is rather vague, and, indeed, there is no obvious reason why the material at the front and back states of a shock should be "equilibrated". Here we follow the usual practice of assuming this to be so, and, of course, to some extent we had this in mind when we took $q^{\pm} = 0$ and when we assumed the existence of the limits $F^{\pm}$, $\mathrm{v}^{\pm}$, $\sigma^{\pm}$, *etc.*

Through the use of (2.18), we see that $\eta^+$ and $F^+$ determine $e^+$, $\sigma^+$, and $\theta^+$. Further, given $\eta^+$, $F^+$ and $\mathrm{v}^+$, the augmented system (2.14)–(2.18) may generally be solved for any six of the seven quantities $F^-$, $\mathrm{v}^-$, $\sigma^-$, $e^-$, $\eta^-$, $\theta^-$, and $\mathscr{V}$, provided the remaining one is specified (so specifying an *intensity* of the shock wave).

A standard method for carrying out the above procedure is to use $(2.18)_{1,2}$ to define the *Hugoniot function*

$$\mathscr{H}^\#(\eta, F) \equiv \varrho_R(e^\#(\eta, F) - e^+) - \tfrac{1}{2}\,(\sigma^\#(\eta, F) + \sigma^+)\,(F - F^+), \tag{2.19}$$

and to observe that the Rankine-Hugoniot equation (2.16) requires

$$\mathscr{H}^\#(\eta^-, F^-) = 0, \qquad (2.20)_1$$

or, equivalently

$$(\eta^-, F^-) \in {}^0\mathscr{H}^\#, \qquad (2.20)_2$$

where ${}^0\mathscr{H}^\#$ is the zero level set of $\mathscr{H}^\#$. The set ${}^0\mathscr{H}^\#$ is commonly called the *Hugoniot set* (more briefly, the *Hugoniot*) based at $(\eta^+, F^+)$. Evidently it is the set, frequently a curve, of all those states that are potentially attainable as back states in a shock process which has $(\eta^+, F^+)$ as a front state. At those points of ${}^0\mathscr{H}^\#$ which are regular, *i.e.*, at those points of ${}^0\mathscr{H}^\#$ where not both of the derivatives $\mathscr{H}^\#_\eta$ and $\mathscr{H}^\#_F$ vanish, ${}^0\mathscr{H}^\#$ is indeed curve-like, and we may solve $(2.20)_1$ for either $\eta^- = \eta^H(F^-)$ or $F^- = F^H(\eta^-)$. When, say, the first of these solutions and $(2.18)_2$ are entered into (2.15), we find a relation that determines $F^-$ when $\mathscr{V}$ is given or $\mathscr{V}$ when $F^-$ is given. Then (2.14) determines $\mathrm{v}^-$, the function $\eta^H(\cdot)$ allows us to find $\eta^-$, and $(2.18)_{1,2,3}$ gives $e^-$, $\sigma^-$, and $\theta^-$. The condition (2.17) of JOUGUET and ZEMPLÉN restricts our attention to those portions of ${}^0\mathscr{H}^\#$ in the half-plane $\{(\eta, F) \mid \eta \geqq \eta^+\}$.

As we said, the above is a standard method for determining the back state data *if the constitutive functions $e^\#$, $\sigma^\#$, and $\theta^\#$ are already known*. Frequently, however, shock experiments are used to elicit information about the unknown forms of $e^\#$, $\sigma^\#$, and $\theta^\#$. Most often this is done by using the Rankine-Hugoniot equations in a slightly different way that hinges on exploiting an experimentally determined relation between the *shock velocity* and the (*back state*) *particle velocity* of the form $\mathscr{V} = \bar{\mathscr{V}}(\mathrm{v}^-)$†. Indeed, with (2.14), this gives us

$$\mathscr{V} = \bar{\mathscr{V}}(\mathrm{v}^+ - \mathscr{V}\,(F^- - F^+)).$$

Upon solving this for $\mathscr{V}$ we reach $\mathscr{V} = \mathscr{V}(F^-)$, a relation giving the shock velocity $\mathscr{V}$ in terms of the deformation $F^-$ at the back state. This determined, (2.15) then gives us the stress on the Hugoniot

$$\sigma^- = \sigma^\#(\eta^H(F^-), F^-) = \sigma^H(F^-) \equiv \sigma^+ + \varrho_R \mathscr{V}(F^-)^2\,(F^- - F^+), \tag{2.21}$$

which in turn can be entered into (2.16) to compute the energy on the Hugoniot

$$e^- = e^\#(\eta^H(F^-), F^-) = e^H(F^-) \equiv e^+ + \frac{\sigma^+}{\varrho_R}(F^- - F^+) + \tfrac{1}{2}\mathscr{V}\,(F^-)^2\,(F^- - F^+)^2. \tag{2.22}$$

† See, for example, [1] or [2].

In Part 2 of our paper we will consider explicit functions $\bar{\mathscr{V}}(\cdot)$ connecting the shock velocity $\mathscr{V}$ and the particle velocity $v^-$ and study in some detail the constitutive implications tacit in (2.21) and (2.22).

As (2.10) demonstrates, the compatibility condition (2.1) implies that in a steady, structured shock wave the functions $\hat{F}(\cdot)$ and $\hat{v}(\cdot)$ of $(2.5)_{1,2}$ are not independent but, instead, must satisfy

$$\hat{v}(\xi) + \mathscr{V}\hat{F}(\xi) = \nu, \tag{A}$$

for some fixed number $\nu$. It follows that the motion $\chi(X, t)$ associated with any steady, structured shock wave satisfies

$$\dot{\chi} + \mathscr{V}\chi_{,X} = \nu; \tag{B}$$

and this is easily integrated to reveal that $\chi(X, t)$ must be of the form

$$x = \chi(X, t) = \nu t + \hat{\phi}(X - \mathscr{V}t),$$

for some, necessarily invertible, function $\hat{\phi}(\cdot)$. We can thus write

$$X - \mathscr{V}t = \hat{\phi}^{-1}(x - \nu t),$$

which, when entered into $(2.5)_{1-7}$, shows that *all of the underlying thermomechanical variables in a steady, structured shock wave also may be considered functions of* $x - \nu t$, i.e., $F = \check{F}(x - \nu t)$, $v = \check{v}(x - \nu t)$, $\sigma = \check{\sigma}(x - \nu t)$, etc.

Conversely, suppose that $(F, v, \sigma, e, q, \eta, \theta) = (\check{F}(x - \nu t), \check{v}(x - \nu t), \check{\sigma}(x - \nu t), \dots)$ is a smooth 7-tuple of functions which satisfies not only the condition of kinematic compatibility (2.1) in its spatial form

$$v_{,x}F = F_{,t} + F_{,x}v,$$

$\left(\text{here } F_{,t} \equiv \dfrac{\partial F(x, t)}{\partial t} \text{ is the spatial time derivative of } F\right)$ but also (2.2)–(2.4) in their respective spatial forms. In this case it is straightforward to show that kinematic compatibility does not allow $\check{F}(\cdot)$ and $\check{v}(\cdot)$ to be independent; instead, they must be such that

$$\frac{\hat{v}(\zeta) - \nu}{\check{F}(\zeta)} = -\mathscr{V}, \tag{C}$$

for some fixed number $\mathscr{V}$. Thus, here too the motion $\chi(X, t)$ is seen to satisfy the differential equation (B). As a consequence, we again have that

$$x = \chi(X, t) = \nu t + \hat{\phi}(X - \mathscr{V}t),$$

where $\hat{\phi}(\cdot)$ is invertible. It follows that $F = \check{F}(x - \nu t) = \check{F}(\hat{\phi}(X - \mathscr{V}t)) = \hat{F}(X - \mathscr{V}t)$, for some function $\hat{F}(\cdot)$, and similarly for $v$, $\sigma$, $q$, $\eta$, and $\theta$: *we have recovered the structure* (2.5) *of steady, structured shock waves.*

The above simple remarks perform a dual service. First, they connect our treatment, framed in terms of material variables, with alternate analyses of structured shocks couched in terms of spatial variables. Additionally, they provide us with an interpretation of the constant $\nu$ appearing in (A): $\nu$ gives the distance in space, $\Delta x$, that any fixed disturbance intensity travels in unit time. Accordingly, we call $\nu$ the *spatial wave speed* of our structured shock. As is clear from (A), it will generally differ in value from the material wave speed $\mathscr{V}$. In fact, $\nu = \mathscr{V}$ if and only if there is a place $\xi^\dagger$ in the wave where the velocity $\mathrm{v}^\dagger$ and the *strain* $\varepsilon^\dagger\,(\equiv F^\dagger - 1)$ meet $\mathrm{v}^\dagger + \mathscr{V}\varepsilon^\dagger = 0$. *A fortiori*, $\nu = \mathscr{V}$ for those special, structured shock waves that enter unstrained material at rest.

Finally, it is worth noting that in spatial treatments of shocks the relation (C), for us a consequence of kinematic compatibility, usually makes its appearance in the form it takes upon multiplication by $\varrho_R$, *i.e.*, in the form

$$\varrho(\mathrm{v} - \nu) = -\varrho_R \mathscr{V} \equiv m,$$

defining the *mass flux m* through the shock. In this guise (C) is typically identified with conservation of mass.

## 3. Thermoelastic Materials and the Speeds of Sound

Except for the introduction of the equilibrium response functions in (2.18), the discussion of Section 2 did not explicitly involve the constitutive structure of the material composing $\mathscr{B}$. We now suppose $\mathscr{B}$ to be composed of simple, *thermoelastic* material. In this case, relations like (2.18) hold not only in equilibrium but throughout every process $\mathscr{B}$ undergoes, and, moreover, the energy is a potential for the stress and temperature, *i.e.*,

$$\begin{aligned} e &= \bar{e}(\eta, F), \\ \sigma &= \bar{\sigma}(\eta, F) = \varrho_R \bar{e}_F(\eta, F), \qquad (3.1)_{1,2,3} \\ \theta &= \bar{\theta}(\eta, F) = \bar{e}_\eta(\eta, F). \end{aligned}$$

Additionally, the heat flux is given by a constitutive relation of the form

$$q = \bar{q}(\eta, F, G), \qquad (3.1)_4$$

where $G \equiv \theta_{,X}$ is the (material) temperature gradient. Further, the balance laws (2.2)–(2.4), along with the constitutive equations $(3.1)_{1,2,3}$, now imply the *heat conduction inequality*

$$q\theta_{,X} \leqq 0, \qquad (3.1)_5$$

which, together with the continuity of $\bar{q}(\eta, F, \cdot)$, has the easy implication that

$$\bar{q}(\eta, F, 0) = 0, \qquad (3.1)_6$$

*i.e.*, entropy and deformation alone cannot induce the flow of heat—a temperature gradient is required.

For a thermoelastic material, the Rayleigh line construction (2.11) becomes the requirement that *all* of the entropy-deformation gradient pairs $(\hat{\eta}, \hat{F})\,(\equiv (\hat{\eta}(\xi), \hat{F}(\xi)))$ experienced in a steady, structured shock wave satisfy the *purely algebraic* condition

$$\bar{\sigma}(\hat{\eta}, \hat{F}) - \sigma^+ = \varrho_R \mathscr{V}^2(\hat{F} - F^+), \qquad (3.2)_1$$

*i.e.*, they must lie in the set $\mathscr{R}(\mathscr{V})$ in the $\eta$–$F$ plane given by

$$\mathscr{R}(\mathscr{V}) = \{(\eta, F) \mid \bar{\sigma}(\eta, F) - \sigma^+ = \varrho_R \mathscr{V}^2(F - F^+)\}. \tag{3.2$_2$}$$

Since $\bar{\sigma}_\eta|_{\mathscr{R}(\mathscr{V})}$ and $(\bar{\sigma}_F - \varrho_R \mathscr{V}^2)|_{\mathscr{R}(\mathscr{V})}$ do not usually vanish simultaneously, the points of $\mathscr{R}(\mathscr{V})$ are typically regular. Thus, $\mathscr{R}(\mathscr{V})$ is usually curve-like, and we call it the *Rayleigh curve* based at $(\eta^+, F^+)$ determined by $\mathscr{V}$. Observe that (i) two different Rayleigh curves, each based at $(\eta^+, F^+)$, can intersect at most along the vertical ray $(-\infty, \infty) \times \{F^+\}$, and (ii) such intersections, other than the trivial one at $(\eta^+, F^+)$, are ruled out if, say, $\bar{\sigma}(\cdot, F^+)$ is monotone. The *Rayleigh trajectory* through $(\eta^+, F^+)$ determined by $\mathscr{V}$ is the ordered pair of functions $(\hat{\eta}, \hat{F})(\cdot)$ of $(2.5)_{1,6}$ which, along with $\hat{v}$, $\hat{\sigma}$, $\hat{e}$, $\hat{q}$, and $\hat{\theta}$, actually gives the *structure* of our steady shock wave. The Rayleigh trajectory, of course, typically sweeps out only a portion of $\mathscr{R}(\mathscr{V})$; nevertheless, we will slightly abuse language and notation and call $(\hat{\eta}, \hat{F})(-\infty, \infty)$ a Rayleigh curve and denote it too by $\mathscr{R}(\mathscr{V})$.

The rate at which the Rayleigh curve is traversed in a steady, structured shock wave—that is, in essence, the Rayleigh trajectory—is determined by the energy equation (2.12) which is now the differential equation‡

$$\bar{q}(\hat{\eta}, \hat{F}, \hat{\theta}') = \mathscr{V} \mathscr{H}(\hat{\eta}, \hat{F}), \tag{3.3}$$

where the Hugoniot function is given by

$$\mathscr{H}(\eta, F) \equiv \varrho_R(\bar{e}(\eta, F) - e^+) - \tfrac{1}{2}(\bar{\sigma}(\eta, F) + \sigma^+)(F - F^+). \tag{3.4}$$

Typical Rayleigh curves for compressive shocks entering an already pre-compressed material modelling aluminum are shown in Figure 2, along with an appropriate portion of the Hugoniot ${}^0\mathscr{H}$. A remarkable feature of the Rayleigh curves exhibited in Figure 2 is that, even though the front state-back state condition, $\eta_i^- \geqq \eta^+$, is satisfied, *the entropy does* not *vary monotonically along* $\mathscr{R}(\mathscr{V}_i)$, $i = 1, 2, 3$, or 4. For each of the $\mathscr{R}(\mathscr{V}_i)$, there are places in the center of the corresponding wave that see considerably higher entropies than the entropy $\eta_i^-$ of the back state. This phenomenon of *entropy overshoot* suggests interesting ideas about entropy-activated processes. Further, when we overlay Figure 2 with the phase boundaries of aluminum (see Figure 3), we see that, not only can certain shocks terminate in the unstable region of the $\eta$–$F$ plane, but also that this lack of monotonicity of $\mathscr{R}(\mathscr{V})$ means that for certain intensity shocks the front and back states can both correspond to the solid phase but the center portion of the shock can pass through the unstable region and even into the liquid phase region of the material. These possibilities appear to have interesting consequences for shock stability, for the evolution of shock-induced microstructure, and for shock-induced phase transformations.

‡ Since (3.3) is autonomous, the existence of one trajectory ensures, by translation of the independent variable, the existence of an infinite number of trajectories. When convenient to do so, we consistently ignore this (trivial) multiplicity of solutions and, instead, fix our attention on the single one of them satisfying some convenient, if tacit, normalization.

The material model on which Figure 2 is based is for an aluminum alloy and has relatively standard thermomechanical properties near its front state. In some of its more compressed states, however, the model has a Grüneisen coefficient (see Section 6) which is somewhat higher than is normally assumed for aluminum. While thermal effects through the shock process are thus enhanced, the qualitative features of the figure are in fact little affected. Indeed, as we shall see in Part 3, the qualitative features of Figure 2 are not even much affected by our bringing in viscosity as an additional dissipative mechanism (although the simple algebraic definition (3.2) of Rayleigh curves will then need to be replaced by a differential equation).

Let $(\eta, F)$ be a regular point of a Rayleigh curve $\mathscr{R}(\mathscr{V})$. If $\bar{\sigma}_\eta \neq 0$ at $(\eta, F)$ then locally we may solve $(3.2)_1$ for $\eta = \eta^R(F)$, *i.e.*, locally $\mathscr{R}(\mathscr{V})$ is the graph of $\eta^R(\cdot)$, and

$$\frac{d\eta^R}{dF} = \frac{\varrho_R \mathscr{V}^2 - \bar{\sigma}_F}{\bar{\sigma}_\eta}. \tag{3.5$_1$}$$

Similarly, if $\bar{\sigma}_F \neq \varrho_R \mathscr{V}^2$ at $(\eta, F)$, then locally $\mathscr{R}(\mathscr{V})$ is the graph of a smooth function $F^R(\cdot)$, $F = F^R(\eta)$, with

$$\frac{dF^R}{d\eta} = \frac{\bar{\sigma}_\eta}{\varrho_R \mathscr{V}^2 - \bar{\sigma}_F}. \tag{3.5$_2$}$$

Now given, as is depicted in Figure 2, that *the trajectory* $(\hat{\eta}, \hat{F})(\cdot)$ *leaves* $(\eta^-, F^-)$ *to the right and comes into* $(\eta^+, F^+)$ *from the left* (and we will shortly find conditions which, for compressive shocks, ensure this), the phenomenon of

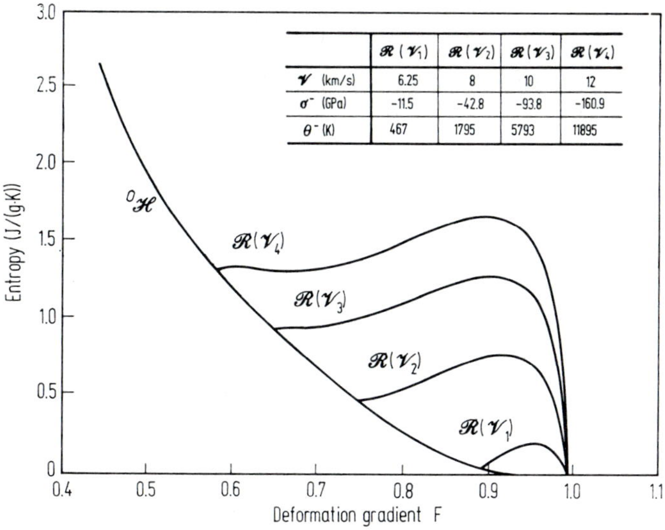


Fig. 2. A portion of the Hugoniot ${}^0\mathscr{H}$ of back states for shocks with front state $(\eta^+, F^+) = (0, .9952)$. The front state stress $\sigma^+$ is $-.526$ GPa, and the front state temperature $\theta^+$ is 298° K. Also shown are various Rayleigh curves $\mathscr{R}(\mathscr{V}_i)$, $i = 1, 2, 3$, and 4, based at $(\eta^+, F^+)$. The inset gives their associated shock speeds and their associated back state stresses and temperatures.

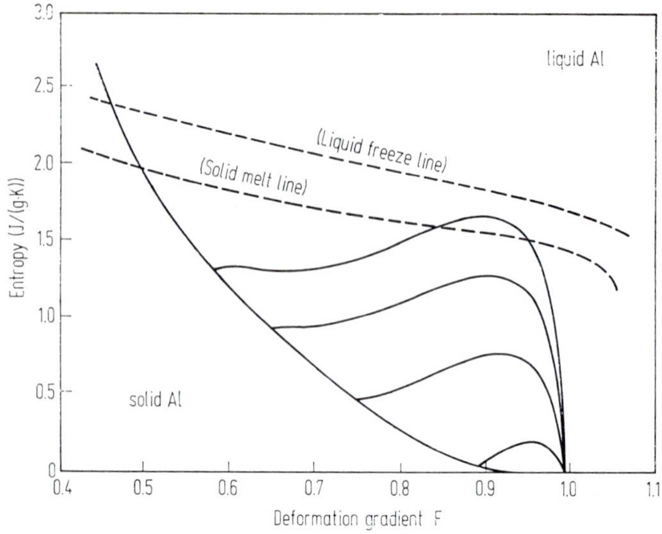


Fig. 3. Phase boundaries of aluminum and the way they are crossed by certain Rayleigh curves. Phase boundaries courtesy of G. KERLEY, Sandia National Labs.

(immediate) entropy overshoot of $\eta^-$ as well as the companion phenomenon of "(initial) entropy increase away from $(\eta^+, F^+)$" are, respectively, the geometric assertions that

*the slope of* $\mathscr{R}(\mathscr{V})$ *is positive at* $(\eta^-, F^-)$,

and

*the slope of* $\mathscr{R}(\mathscr{V})$ *is negative at* $(\eta^+, F^+)$.

From $(3.5)_1$ these are easily seen to be equivalent to

$$\frac{\varrho_R \mathscr{V}^2 - \sigma_F^-}{\sigma_\eta^-} > 0,$$

and

$$\frac{\varrho_R \mathscr{V}^2 - \sigma_F^+}{\sigma_\eta^+} < 0,$$

respectively, if $\sigma_\eta^\pm \neq 0$. If $\bar{\sigma}_\eta < 0$ at the front and back states,* these last two inequalities reduce to

$$A^+ < \varrho_R \mathscr{V}^2 < A^-, \tag{3.6}$$

where $A = A(\eta, F) \equiv \bar{\sigma}_F(\eta, F)$ is the *adiabatic modulus*. But, when $A$ is non-negative (see below), $\sqrt{A/\varrho_R}$ is the *adiabatic material sound speed* in $\mathscr{B}$ at the state $(\eta, F)$. Thus (3.6) is just the well-known statement that the shock speed $\mathscr{V}$ is sub-

* The assumption that $\bar{\sigma}_\eta < 0$, even globally, is a common one in shock physics. We shall soon see in some detail what happens if this requirement is dropped.

sonic with respect to the material in back of the wave and is supersonic with respect to the material in front of the wave[†].

The above discussion has thus revealed conditions which, along with the subsonic condition $\varrho_R \mathscr{V}^2 < A^-$, suffice to ensure not just entropy overshoot but also that this overshoot happens arbitrarily near $(\eta^-, F^-)$. However, when we turn to the standard proofs[‡] of the supersonic-subsonic condition (3.6), we find that they rely on the hypothesis that various *a priori* inequalities on the second and third derivatives of $\bar{e}(\cdot, \cdot)$ hold globally. Commonly,[§] these may be written as

$$\bar{\sigma}_\eta < 0, \quad \bar{\sigma}_F > 0, \quad \text{and} \quad \bar{\sigma}_{FF} < 0. \tag{3.7$_{1,2,3}$}$$

Whatever the merit of these inequalities locally (and it is hard to resist the suspicion that $\bar{\sigma}_\eta < 0$ and $\bar{\sigma}_{FF} < 0$ are not too special), there seems little reason to believe that they hold over the entire $\eta$–$F$ plane. Indeed, $\bar{\sigma}_\eta < 0$ fails to hold for even so common a substance as water near its freezing point, and $\bar{\sigma}_F > 0$ itself breaks down in the unstable region. We now give a rather different proof of relations like (3.6). It rests, in essence, on just the thermodynamic principles of Section 2 and thereby depends only on the shock process itself. Our results show that shock properties like the supersonic-subsonic condition (3.6) are in fact consequences of the dissipation principle embodied in the law of entropy imbalance.

As they will be helpful in interpreting and assessing our results, we begin by recalling some of the basic facts of Gibbsian thermostatics. First, according to GIBBS, a necessary condition for a point $(\eta, F)$ to be *thermostatically stable* is that it be a *point of convexity* for $\bar{e}(\cdot, \cdot)$, i.e.,

$$\bar{e}(\eta, F) + (\tilde{\eta} - \eta)\,\bar{e}_\eta(\eta, F) + (\tilde{F} - F)\,\bar{e}_F(\eta, F) \leqq \bar{e}(\tilde{\eta}, \tilde{F}), \tag{3.8}$$

for all $(\tilde{\eta}, \tilde{F})$ in the domain of $\bar{e}(\cdot, \cdot)$. This in turn requires the matrix

$$\begin{pmatrix} \bar{e}_{\eta\eta} & \bar{e}_{\eta F} \\ \bar{e}_{F\eta} & \bar{e}_{FF} \end{pmatrix}$$

to be positive semi-definite at $(\eta, F)$, so that $\bar{e}_{\eta\eta} \geqq 0$, $\bar{e}_{FF} \geqq 0$, and $\bar{e}_{\eta\eta}\bar{e}_{FF} - \bar{e}^2_{\eta F} \geqq 0$. By $(3.1)_{2,3}$, it thus follows that at a thermostatically stable point we have

$$\bar{\theta}_\eta \geqq 0, \quad \bar{\sigma}_F \geqq 0, \quad \text{and} \quad \bar{\theta}_\eta \bar{\sigma}_F - \bar{\theta}_F \bar{\sigma}_\eta \geqq 0. \tag{3.9$_{1,2,3}$}$$

---

[†] In the special case of ideal gases, this property of the shock velocity was shown as long ago as 1904 by JOUGUET [22].

[‡] See, for example, the original papers by BETHE [12] and WEYL [13] or the later expositions by COURANT & FRIEDRICHS [14] and SERRIN [15].

[§] The inequalities (3.7) are exactly those adopted by WEYL [13], COURANT & FRIEDRICHS [14], and SERRIN [15]. BETHE [12], on the other hand, while adopting $(3.7)_3$ and (tacitly) $(3.7)_2$, did not demand anything so strict as $(3.7)_1$. Instead, he asked only that $\bar{\sigma}_\eta$ always be less than $2\varrho_R\theta/F$, *i.e.*, (see our Section 6) the *Grüneisen coefficient* always must be larger than $-2$. As BETHE notes, unlike $(3.7)_1$, his weaker requirement is met by liquid water near its freezing point.

We say that a point $(\eta, F)$ is *infinitesimally superstable* if the inequalities in (3.9) are strict at $(\eta, F)$. Thus, by $(3.9)_2$, at a thermostatically stable point the adiabatic material sound speed $\sqrt{A/\varrho_R}$ is always well defined; it is positive at any infinitesimally superstable point.

Next, consider any point in the $\eta$–$F$ plane at which $\bar{\theta}_\eta \neq 0$. In a neighborhood of such a point, we can invert the relation $\theta = \bar{\theta}(\eta, F)$ to obtain $\eta = \tilde{\eta}(\theta, F)$ and thus determine the *specific heat at fixed deformation* $\mathrm{c_d} \equiv \theta \tilde{\eta}_\theta(\theta, F) = \bar{\theta}/\bar{\theta}_\eta$. Additionally, paralleling our introduction of the adiabatic modulus $A = \bar{\sigma}_F$, we can now set $\sigma = \tilde{\sigma}(\theta, F) \equiv \bar{\sigma}(\tilde{\eta}(\theta, F), F)$ and define the *isothermal modulus* $I$ by

$$I = I(\eta, F) = \tilde{I}(\theta, F)|_{\theta=\bar{\theta}(\eta,F)} \quad \text{where} \quad \tilde{I}(\theta, F) \equiv \tilde{\sigma}_F(\theta, F).$$

When $I$ is non-negative, the quantity $\sqrt{I/\varrho_R}$ gives the *isothermal material sound speed* in $\mathscr{B}$ at the state $(\eta, F)$. In addition, a simple calculation yields

$$I = A - \bar{\sigma}_\eta \left(\frac{\bar{\theta}_F}{\bar{\theta}_\eta}\right) = A - \varrho_R \left(\frac{\bar{\theta}_F^2}{\bar{\theta}_\eta}\right), \tag{3.10}$$

so that $I \leqq A$ $(I \geqq A)$ whenever $\bar{\theta}_\eta > 0$ $(\bar{\theta}_\eta < 0)$, with equality if and only if $\bar{\theta}_F \left(= \frac{1}{\varrho_R} \bar{\sigma}_\eta\right)$ vanishes. By $(3.9)_{1,3}$, we note that $\mathrm{c_d}$ and $\sqrt{I/\varrho_R}$ are well defined and positive at any infinitesimally superstable point.

Similarly, at any point in the $\eta$–$F$ plane at which $\bar{\theta}_\eta \bar{\sigma}_F - \bar{\theta}_F \bar{\sigma}_\eta \neq 0$, we may locally invert the system $\theta = \bar{\theta}(\eta, F)$ and $\sigma = \bar{\sigma}(\eta, F)$ to obtain $\eta = \bar{\bar{\eta}}(\theta, \sigma)$ and $F = \bar{\bar{F}}(\theta, \sigma)$. In this way we can define both the *specific heat at fixed stress*, $\mathrm{c_s} \equiv \theta \bar{\bar{\eta}}_\theta(\theta, \sigma) = \bar{\theta} \bar{\sigma}_F / (\bar{\theta}_\eta \bar{\sigma}_F - \bar{\theta}_F \bar{\sigma}_\eta)$, and the *coefficient of thermal expansion*, $\varphi \equiv \bar{\bar{F}}_\theta(\theta, \sigma)/F = -\bar{\sigma}_\eta / F(\bar{\theta}_\eta \bar{\sigma}_F - \bar{\theta}_F \bar{\sigma}_\eta)$. Clearly, at any infinitesimally superstable point both $\mathrm{c_s}$ and $\varphi$ are well defined with $\mathrm{c_s} \geqq \mathrm{c_d} > 0$. More generally, at any point where neither $\bar{\theta}_\eta$ nor $\bar{\theta}_\eta \bar{\sigma}_F - \bar{\theta}_F \bar{\sigma}_\eta$ vanishes, we have the useful equation

$$\frac{\mathrm{c_s}}{\mathrm{c_d}} = \frac{A}{I}. \tag{3.11}$$

When $A$ is positive, the adiabatic material sound speed $\mathscr{C}_A \equiv \sqrt{A/\varrho_R}$ gives the span of material, $\Delta X$, in the reference configuration which an infinitesimal mechanical disturbance, imposed on a motionless body in a state $(\eta, F)$, traverses in unit time. Paralleling the relation between the material wave speed $\mathscr{V}$ and the spatial wave speed $v$, we call the number

$$c_A \equiv F \mathscr{C}_A = \sqrt{\frac{-\partial \check{\sigma}(\eta, \varrho)}{\partial \varrho}},$$

the *adiabatic spatial sound speed*. It gives the rate of advance of the infinitesimal, mechanical disturbance across distance in space. Analogous remarks apply to the isothermal material sound speed $\mathscr{C}_I \equiv \sqrt{I/\varrho_R}$ and its spatial cognate

$$c_I \equiv F \mathscr{C}_I = \sqrt{\frac{-\partial \check{\tilde{\sigma}}(\theta, \varrho)}{\partial \varrho}},$$

the *isothermal spatial sound speed.* Note that, in writing the right hand sides of the last two equations, we have replaced the deformation measure $F$ in each respective constitutive equation for $\sigma$ with the density $\varrho = \varrho_R/F$.

Now, along the trajectory $(\hat{\eta}, \hat{F})(\cdot)$ of a steady, structured shock wave we have from $(3.2)_1$ that

$$\bar{\sigma}_\eta \hat{\eta}' + \bar{\sigma}_F \hat{F}' = \varrho_R \mathscr{V}^2 \hat{F}'.$$

Thus, at any point in its traverse where $A(\eta, F) - \varrho_R \mathscr{V}^2 \neq 0$, we may write

$$\begin{aligned} \hat{\theta}' &= \bar{\theta}_\eta \hat{\eta}' + \bar{\theta}_F \hat{F}', \\ &= \left(\bar{\theta}_\eta - \bar{\theta}_F \left(\frac{\bar{\sigma}_\eta}{\bar{\sigma}_F - \varrho_R \mathscr{V}^2}\right)\right) \hat{\eta}', \qquad (3.12)_1 \\ &= \bar{\theta}_\eta \left(\frac{I - \varrho_R \mathscr{V}^2}{A - \varrho_R \mathscr{V}^2}\right) \hat{\eta}'. \end{aligned}$$

Similarly, at any point along $(\hat{\eta}, \hat{F})(\cdot)$ where $\bar{\sigma}_\eta(\eta, F) \neq 0$, we have

$$\begin{aligned} \hat{\theta}' &= \bar{\theta}_\eta \hat{\eta}' + \bar{\theta}_F \hat{F}', \\ &= \left(\bar{\theta}_\eta \left(\frac{\varrho_R \mathscr{V}^2 - \bar{\sigma}_F}{\bar{\sigma}_\eta}\right) + \bar{\theta}_F\right) \hat{F}', \qquad (3.12)_2 \\ &= \frac{-\bar{\theta}_\eta}{\bar{\sigma}_\eta} (I - \varrho_R \mathscr{V}^2)\, \hat{F}'. \end{aligned}$$

In writing $(3.12)_{1,2}$ we have used the fact that, unlike $I$ itself, the product $\bar{\theta}_\eta (I - \varrho_R \mathscr{V}^2)$ $(= \bar{\theta}_\eta (A - \varrho_R \mathscr{V}^2) - \bar{\theta}_F \bar{\sigma}_\eta)$ is well defined even when $\bar{\theta}_\eta$ vanishes.

We need one more ingredient to derive our analogs of the supersonic-subsonic conditions (3.6). To acquire it we observe that the energy equation (2.8) and the thermoelastic relations (3.1) allow us to write

$$\begin{aligned} \hat{q}' + \mathscr{V} \hat{\sigma} \hat{F}' &= \varrho_R \mathscr{V} (\bar{e}_\eta \hat{\eta}' + \bar{e}_F \hat{F}'), \\ &= \varrho_R \mathscr{V} \hat{\theta} \overline{(\hat{\eta} - \eta^-)}' + \mathscr{V} \hat{\sigma} \hat{F}', \\ &= \varrho_R \mathscr{V} \overline{\hat{\theta}(\hat{\eta} - \eta^-)}' - \varrho_R \mathscr{V} \hat{\theta}' (\hat{\eta} - \eta^-) + \mathscr{V} \hat{\sigma} \hat{F}'. \end{aligned}$$

Thus

$$\overline{\hat{q} - \varrho_R \mathscr{V} \hat{\theta}(\hat{\eta} - \eta^-)}' + \varrho_R \mathscr{V} \hat{\theta}'(\hat{\eta} - \eta^-) = 0,$$

and so, since $q^- = 0$,

$$\hat{q}(\xi) - \varrho_R \mathscr{V} \hat{\theta}(\xi)\, (\hat{\eta}(\xi) - \eta^-) + \varrho_R \mathscr{V} \int_{-\infty}^{\xi} \hat{\theta}'(\lambda)\, (\hat{\eta}(\lambda) - \eta^-)\, d\lambda = 0$$

for any $\xi \in (-\infty, \infty)$. But the entropy inequality (2.9), along with $q^- = 0$, gives

$$\hat{q}(\xi) - \varrho_R \mathscr{V} \hat{\theta}(\xi)\,(\hat{\eta}(\xi) - \eta^-) \geqq 0,$$

and therefore we have demonstrated that

$$\varrho_R \mathscr{V} \int_{-\infty}^{\xi} \hat{\theta}'(\lambda)\,(\hat{\eta}(\lambda) - \eta^-)\,d\lambda \leqq 0 \tag{3.13}$$

for all $\xi$.

We can now prove

**Theorem 1.** *Let the back state of a steady, structured shock wave in a thermoelastic material satisfy*

$$A^- \neq \varrho_R \mathscr{V}^2.$$

*Then,*

(i) *if* $\theta_\eta^- > 0$, $\sigma_\eta^-$ *cannot vanish and the shock speed* $\mathscr{V}$ *satisfies*

$$\varrho_R \mathscr{V}^2 \in [I^-, A^-),$$

(ii) *if* $\theta_\eta^- < 0$, *the shock speed* $\mathscr{V}$ *satisfies*

$$\varrho_R \mathscr{V}^2 \notin [A^-, I^-).$$

At back states with $\theta_\eta^- > 0$ the result (i) of Theorem 1 is a good deal stronger* than the usual subsonic condition $(3.6)_2$: not only must $\varrho_R \mathscr{V}^2$ be less than $A^-$, it also must be at least as large as $I^-$. Observe also that (i) does not require $I^-$ to be nonnegative. *A fortiori*, the back states of (i) need not be infinitesimally superstable (much less, thermostatically stable), although those that are not must be rather special: they fail the three criteria for infinitesimal superstability—(3.9) with inequalities strict—only by failing to satisfy $(3.9)_3$.

**Proof.** Since $A^- \neq \varrho_R \mathscr{V}^2$, the point $(\eta^-, F^-)$ is a regular point of $\mathscr{R}(\mathscr{V})$. There thus is an interval $\mathscr{I}_\delta = (\eta^- - \delta, \eta^- + \delta)$, $\delta > 0$, and a smooth function $F^R(\cdot): \mathscr{I}_\delta \to R$ such that locally $\mathscr{R}(\mathscr{V})$ is the graph of $F^R(\cdot)$. Additionally, since $A^- \neq \varrho_R \mathscr{V}^2$, there is a neighborhood $\mathscr{N}$ of $(\eta^-, F^-)$ in which $\bar{\sigma}_F - \varrho_R \mathscr{V}^2 \neq 0$.

Let $\delta$ be so small that $F^R(\cdot)$ is bounded on $\mathscr{I}_\delta$, set $F_m = \inf_{\mathscr{I}_\delta} F^R$, $F_M = \sup_{\mathscr{I}_\delta} F^R$, and $\mathscr{N}^- = \mathscr{N} \cap (\mathscr{I}_\delta \times (F_m, F_M))$. Since we assume $(\eta^-, F^-) \neq (\eta^+, F^+)$, we can suppose $\mathscr{N}$ to be so small that $(\eta^+, F^+) \notin \mathscr{N}$. Consequently, $(\eta^+, F^+) \notin \mathscr{N}^-$, and the number

$$\xi^\dagger = \sup \{\xi \mid (\hat{\eta}, \hat{F})\,((-\infty, \xi)) \subseteqq \mathscr{N}^-\}$$

is well defined and finite. Clearly, $(\hat{\eta}, \hat{F})\,(\xi^\dagger)$ is a boundary point of $\mathscr{N}^-$, and, since $(\hat{\eta}, \hat{F})\,(\cdot)$ satisfies the Rayleigh line construction, we have that $\hat{F}(\xi) = F^R(\hat{\eta}(\xi))$ for all $\xi < \xi^\dagger$.

* Except we have not yet shown that $A^- = \varrho_R \mathscr{V}^2$ is impossible.

It now follows from $(3.12)_1$ and the above that, for $\xi < \xi^*$, we may write (3.13) in the form

$$\varrho_R \mathscr{V} \int_{-\infty}^{\xi} \bar{\theta}_{,\eta} \left( \frac{I - \varrho_R \mathscr{V}^2}{A - \varrho_R \mathscr{V}^2} \right) \Bigg|_{\substack{F = F^R(\hat{\eta}(\lambda)) \\ \eta = \hat{\eta}(\lambda)}} (\hat{\eta}(\lambda) - \eta^-) \, \hat{\eta}'(\lambda) \, d\lambda \leqq 0,$$

or, by a change of variables,

$$\varrho_R \mathscr{V} \int_{\eta^-}^{\hat{\eta}(\xi)} \bar{\theta}_{,\eta} \left( \frac{I - \varrho_R \mathscr{V}^2}{A - \varrho_R \mathscr{V}^2} \right) \Bigg|_{F = F^R(\eta)} (\eta - \eta^-) \, d\eta \leqq 0. \tag{3.14}$$

Now we need to know that (3.14) is not trivial, *i.e.*, that there are values of $\xi$ for which it holds with $\hat{\eta}(\xi) \neq \eta^-$. But this is clear: if $\hat{\eta}(\cdot)$ is constant on $(-\infty, \xi^*)$, its constant value can only be $\eta^-$. Further, since $\bar{\sigma}_{,\eta} \hat{\eta}' = (\varrho_R \mathscr{V}^2 - \bar{\sigma}_F) \hat{F}'$ with $\varrho_R \mathscr{V}^2 - \bar{\sigma}_F \neq 0$ in $\mathscr{N}^-$, the constancy of $\hat{\eta}(\cdot)$ would imply the constancy of $\hat{F}(\cdot)$ on $(-\infty, \xi^*)$, i.e., $\hat{F}(\cdot) = F^-$ on $(-\infty, \xi^*)$. It would now follow that $(\hat{\eta}, \hat{F})(\xi^*) = (\eta^-, F^-)$, in contradiction to the fact that $(\hat{\eta}, \hat{F})(\xi^*)$ is a boundary point of $\mathscr{N}^-$. It now follows, by the intermediate value theorem for continuous functions, that $\hat{\eta}(\xi)$ in (3.14) can be replaced with any $\eta$ in an interval of one of the forms $(\eta^- - \delta, \eta^-)$, $(\eta^-, \eta^- + \delta)$, or $(\eta^- - \delta, \eta^- + \delta)$, $\delta > 0$.

We have thus shown that there is a non-empty interval, which has $\eta^-$ either as an endpoint or in its interior, on which

$$J(\check{\eta}) \equiv \varrho_R \mathscr{V} \int_{\eta^-}^{\check{\eta}} \bar{\theta}_{,\eta} \left( \frac{I - \varrho_R \mathscr{V}^2}{A - \varrho_R \mathscr{V}^2} \right) \Bigg|_{F = F^R(\eta)} (\eta - \eta^-) \, d\eta$$

satisfies

$$J(\check{\eta}) \leqq 0.$$

Since $J(\eta^-) = J'(\eta^-) = 0$, we see that it is thus necessary for $J''(\eta^-)$ to be non-positive, *i.e.*, for

$$\varrho_R \mathscr{V} \theta_{,\eta}^- \left( \frac{I^- - \varrho_R \mathscr{V}^2}{A^- - \varrho_R \mathscr{V}^2} \right) \leqq 0. \tag{3.15}$$

Now suppose first that $\theta_{,\eta}^- > 0$. Then, since $\mathscr{V}$ is positive and $A^- - \varrho_R \mathscr{V}^2$ is assumed not to vanish, the above inequality requires either

$$A^- - \varrho_R \mathscr{V}^2 < 0 \quad \text{and} \quad I^- - \varrho_R \mathscr{V}^2 \geqq 0,$$

or

$$A^- - \varrho_R \mathscr{V}^2 > 0 \quad \text{and} \quad I^- - \varrho_R \mathscr{V}^2 \leqq 0.$$

But the first of these is impossible since it would imply that $A^- < I^-$ which, as we have seen, can never occur at a point where $\bar{\theta}_{,\eta}$ is positive. Thus, only the second assertion can hold, *i.e.*, we have shown that

$$A^- \neq \varrho_R \mathscr{V}^2 \quad \text{and} \quad \theta_{,\eta}^- > 0 \Rightarrow I^- \leqq \varrho_R \mathscr{V}^2 < A^-.$$

By (3.10), it now also follows that $\sigma_{,\eta}^-$ cannot vanish. This completes the proof of the first half of the theorem. The second half, for $\theta_{,\eta}^- < 0$, is proven by a similar analysis of the inequality (3.15). □

When $\theta_\eta^- > 0$, *i.e.*, at back states with a positive specific heat $c_d$, Theorem 1 tells us conditions under which the Rayleigh curve can be written as a smooth function of both $\eta$ (i.e., $F = F^R(\eta)$) and $F$ (i.e., $\eta = \eta^R(F)$) near $(\eta^-, F^-)$. Moreover, since

$$\frac{d\eta^R(F^-)}{dF} = \frac{-(A^- - \varrho_R \mathscr{V}^2)}{\sigma_\eta^-},$$

with $A^- - \varrho_R \mathscr{V}^2 > 0$ and $\sigma_\eta^- \neq 0$, we see that near $(\eta^-, F^-)$ the Rayleigh curve looks like (a) or (b) of Figure 4, according as $\sigma_\eta^-$ is negative or positive.

Now, with $c_d^- > 0$, consider the direction from which the Rayleigh trajectory $(\hat{\eta}, \hat{F})(\cdot)$ approaches $(\eta^-, F^-)$. Suppose first that the shock is *compressive*, so that not only is $\eta^- \geqq \eta^+$ by the Jouguet-Zemplén condition (2.17), but also $F^- < F^+$. We are thus considering back states $(\eta^-, F^-)$ that, in the $\eta$–$F$ plane, are above and to the left of $(\eta^+, F^+)$. Now suppose that in this case $(\hat{\eta}, \hat{F})(\cdot)$ does not approach $(\eta^-, F^-)$ from the *left*. This is not only very reasonable (its violation would mean that there are particles in the center of the wave that experience even more intense compressions than $F^-$, i.e., *strain overshoot* would occur), it can also be generally and rigorously ruled out by noting one fact and making one assumption. As fact (see the Corollary to Theorem 5 in Section 4), that, if $\theta_\eta^-$ and $I^- - \varrho_R \mathscr{V}^2$ do not vanish, then $A^- \neq \varrho_R \mathscr{V}^2$ implies that the the Rayleigh trajectory $(\hat{\eta}, \hat{F})(\cdot)$ cannot cross and recross $(\eta^-, F^-)$, *i.e.*, $(\hat{\eta}, \hat{F})(\xi)$ can be at $(\eta^-, F^-)$ only when $\xi = -\infty$. Thus, since $(\hat{\eta}, \hat{F})(\xi)$ is to the right of $(\eta^-, F^-)$ at $\xi = +\infty$, if it is ever to the left of $(\eta^-, F^-)$ it must get there by crossing the ray $(-\infty, \infty) \times \{F^-\}$ at some place other than $\eta^-$, *i.e.*, the Rayleigh curve $\mathscr{R}(\mathscr{V})$ must intersect the ray $(-\infty, \infty) \times \{F^-\}$ at least twice. But this possibility is easily struck down by making the common assumption that $\bar{\sigma}(\cdot, F^-)$ is *monotone*.

In any case, given that $(\hat{\eta}, \hat{F})(\cdot)$ does not approach $(\eta^-, F^-)$ from the left, we see from (a) of Figure 4 that entropy overshoot is ensured if (as is common) $\sigma_\eta^- < 0$. On the other hand, if $\sigma_\eta^- > 0$, then (b) of Figure 4 shows that, even though entropy overshoot is not forbidden, no manifestation of it is possible near $(\eta^-, F^-)$.

Conversely, the above line of reasoning applied to *expansive* shocks reveals that, if we again rule out strain overshoot, then the Rayleigh trajectory $(\hat{\eta}, \hat{F})(\cdot)$ must approach $(\eta^-, F^-)$ from the left. As a glance at (b) of Figure 4 shows, entropy overshoot is then ensured if $\sigma_\eta^-$ is positive. On the other hand, (a) of Figure 4 indicates that no entropy overshoot near $(\eta^-, F^-)$ is possible when $\sigma_\eta^-$ is negative.

Next, consider the case omitted in Theorem 1, *i.e.*, suppose the back state is such that $A^- = \varrho_R \mathscr{V}^2$. Also suppose, however, that $\sigma_\eta^-$ does not vanish. In this case, near $(\eta^-, F^-)$, the Rayleigh curve can be written as a smooth function of $F$. A simple calculation then yields (recall $(3.5)_1$)

$$\frac{d\eta^R(F^-)}{dF} = 0, \qquad \frac{d^2\eta^R(F^-)}{dF^2} = \frac{-\sigma_{FF}^-}{\sigma_\eta^-}.$$

Setting aside the situation when $\sigma_{FF}$ vanishes, we thus see that near $(\eta^-, F^-)$ the Rayleigh curve will look like (c) or (d) of Figure 4 according as the ratio $\sigma_{FF}^-/\sigma_\eta^-$ is positive or negative.

We now have the following complement to Theorem 1. As we shall see, it gives a partial answer to the question of existence of structured shocks with $A^- = \varrho_R \mathscr{V}^2$.

**Theorem 2.** *Let the back state of a steady, structured shock wave in a thermoelastic material satisfy*

$$A^- = \varrho_R \mathscr{V}^2, \quad \sigma_\eta^- \neq 0, \quad \text{and} \quad \theta_\eta^- \neq 0.$$

*Then, as* $\xi \to -\infty$,

$$\text{(i)}\ \sigma_{FF}^- < 0 \Rightarrow \hat{F}(\xi) \nearrow F \quad \text{and} \quad \hat{F}(\xi) \nearrow F \Rightarrow \sigma_{FF}^- \leqq 0,$$

$$\text{(ii)}\ \sigma_{FF}^- > 0 \Rightarrow \hat{F}(\xi) \searrow F \quad \text{and} \quad \hat{F}(\xi) \searrow F \Rightarrow \sigma_{FF}^- \geqq 0,$$

$$\text{(iii)}\ \begin{cases} \sigma_{FF}^- = 0 \Rightarrow \sigma_{FFF}^- \geqq 0, \\ \hat{F}(\xi) \to F^- \text{ from above and below} \Rightarrow \sigma_{FF}^- = 0. \end{cases}$$

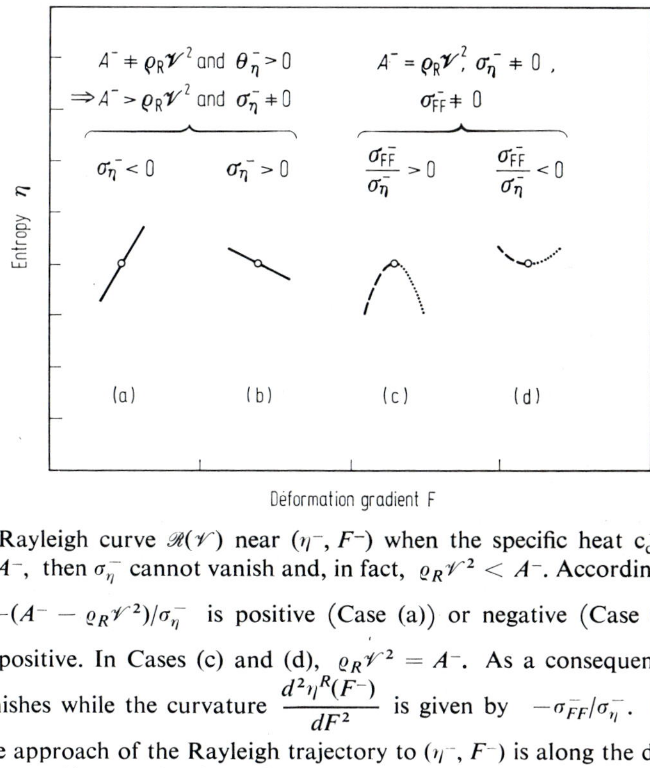


Fig. 4. The Rayleigh curve $\mathscr{R}(\mathscr{V})$ near $(\eta^-, F^-)$ when the specific heat $c_d^-$ is positive. If $\varrho_R \mathscr{V}^2 \neq A^-$, then $\sigma_\eta^-$ cannot vanish and, in fact, $\varrho_R \mathscr{V}^2 < A^-$. Accordingly, the slope $\dfrac{d\eta^R(F^-)}{dF} = -(A^- - \varrho_R \mathscr{V}^2)/\sigma_\eta^-$ is positive (Case (a)) or negative (Case (b)) as $\sigma_\eta^-$ is negative or positive. In Cases (c) and (d), $\varrho_R \mathscr{V}^2 = A^-$. As a consequence, the slope $\dfrac{d\eta^R(F^-)}{dF}$ vanishes while the curvature $\dfrac{d^2\eta^R(F^-)}{dF^2}$ is given by $-\sigma_{FF}^-/\sigma_\eta^-$. In these last two cases the approach of the Rayleigh trajectory to $(\eta^-, F^-)$ is along the dashed branch if $\sigma_{FF}^- < 0$, along the dotted branch if $\sigma_{FF}^- > 0$.

Unlike Theorem 1, Theorem 2 thus gives conditions which determine the direction of approach of the trajectory $(\hat{\eta}, \hat{F})(\cdot)$ to $(\eta^-, F^-)$. For example, if $\sigma_{FF}^- < 0$, then the trajectory $(\hat{\eta}, \hat{F})(\cdot)$ can go to $(\eta^-, F^-)$ only along the left hand branches of Cases (c) and (d) of Figure 4. As we have seen, however, this is impossible for compressive shocks unless there is strain overshoot in the wave. Barring this (see below), Theorem 2 tells us that thermodynamics does not allow compressive shocks which satisfy

$$A^- = \varrho_R \mathscr{V}^2, \qquad \sigma_{FF}^- < 0, \qquad \sigma_\eta^- \neq 0, \qquad \text{and} \qquad \theta_\eta^- \neq 0.$$

Such shocks, if they exist at all, must be expansive. Similarly, there cannot be a expansive shock whose back state is such that

$$A^- = \varrho_R \mathscr{V}^2, \qquad \sigma_{FF}^- > 0, \qquad \sigma_\eta^- \neq 0, \qquad \text{and} \qquad \theta_\eta^- \neq 0.$$

If such a shock existed, thermodynamics would require it to be compressive. We remark, however, that the rigorous barring of strain overshoot required for these results is more subtle than was the case for the shocks of Theorem 1. Specifically, while an assumption of monotonicity for $\bar{\sigma}(\cdot, F^-)$ is sufficient to bar $\mathscr{R}(\mathscr{V})$ from crossing the ray $(-\infty, \infty) \times \{F^-\}$ at any place other than $\eta^-$, we can no longer use Theorem 5 and its Corollary to rule out the possibility of multiple crossings of $(\eta^-, F^-)$ by the Rayleigh trajectory $(\hat{\eta}, \hat{F})(\cdot)$. Instead, since now $A^- = \varrho_R \mathscr{V}^2$, we need the final result of Section 4, *i.e.*, (4.14), which tells us that near $(\eta^-, F^-)$ the function $\hat{F}(\cdot)$ cannot increase if $\sigma_{FF}^- < 0$, nor can it decrease if $\sigma_{FF}^- > 0$.

Since it is much like that of Theorem 1, we only sketch the proof of Theorem 2.

**Proof.** Since $\sigma_\eta^- \neq 0$, we can express the Rayleigh curve near $(\eta^-, F^-)$ in terms of a smooth function of $F$, *i.e.*, $\eta = \eta^R(F)$. Then, by $(3.12)_2$, we can express (3.13) in the form

$$-\varrho_R \mathscr{V} \int_{-\infty}^{\xi} \frac{\bar{\theta}_\eta}{\bar{\bar{\sigma}}_\eta} (I - \varrho_R \mathscr{V}^2)(\eta - \eta^-) \Bigg|_{\substack{\eta = \eta^R(\hat{F}(\lambda)) \\ F = \hat{F}(\lambda)}} \hat{F}'(\lambda)\, d\lambda \leqq 0,$$

and so, by a change of variables, arrive at

$$\varrho_R \mathscr{V} \int_{F^-}^{\hat{F}(\xi)} \frac{\bar{\theta}_\eta}{\bar{\bar{\sigma}}_\eta} (I - \varrho_R \mathscr{V}^2)(\eta - \eta^-) \Bigg|_{\eta = \eta^R(F)} dF \geqq 0.$$

As in Theorem 1, we can now argue that this last implies that

$$K(\check{F}) = \varrho_R \mathscr{V} \int_{F^-}^{\check{F}} \frac{\bar{\theta}_\eta}{\bar{\bar{\sigma}}_\eta} (I - \varrho_R \mathscr{V}^2)(\eta - \eta^-) \Bigg|_{\eta = \eta^R(F)} dF,$$

satisfies

$$K(\check{F}) \geqq 0$$

for all $\check{F}$ in an interval of the form $(F^- - \delta, F^-)$, $(F^-, F^- + \delta)$, or $(F^- - \delta, F^- + \delta)$, $\delta > 0$, depending, respectively, on whether $\hat{F}(\cdot)$ approaches $F^-$ from below, from above, or from below and above. But, an easy calculation shows that

$K(F^-) = K'(F^-) = K''(F^-) = 0$. Thus,

$$K(F) = \tfrac{1}{6}K'''(F^-)(F - F^-)^3 + o(|F - F^-|^3),$$

$$= -\tfrac{1}{6}\varrho_R\mathscr{V}\frac{\theta_\eta^-(I^- - \varrho_R\mathscr{V}^2)}{(\sigma_\eta^-)^2}\sigma_{FF}^-(F - F^-)^3 + o(|F - F^-|^3).$$

But $\sigma_\eta^- \neq 0$, $A^- = \varrho_R\mathscr{V}^2$, and $\theta_\eta^- \neq 0$. As a consequence, the product $\theta_\eta^-(I^- - \varrho_R\mathscr{V}^2)$ must be negative. Thus, if $\sigma_{FF}^- \neq 0$, we see that near $F^-$ the sign of $K(\cdot)$ is the sign of $\sigma_{FF}^-((\cdot) - F^-)^3$. If therefore $\sigma_{FF}^- < 0$, then the condition $K(\check{F}) \geqq 0$ for $\check{F}$ near $F^-$ can only hold for $\check{F} \in (F^- - \delta, F^-)$, *i.e.*, $\hat{F}(\cdot)$ can approach $F^-$ only from below. We have proven the first half of our assertion (i); the remaining portions of Theorem 2 are proved similarly. □

Paralleling our treatment in Theorems 1 and 2 of the back state $(\eta^-, F^-)$, we can study the implications of dissipation through a steady, structured shock wave for the properties at the front state $(\eta^+, F^+)$. These are summarized in Theorems 3 and 4 whose proofs we omit.

**Theorem 3.** *Let the front state of a steady, structured shock wave in a thermoelastic material satisfy*

$$A^+ \neq \varrho_R\mathscr{V}^2.$$

*Then,*

(i) *if* $\theta_\eta^+ > 0$, *the shock speed* $\mathscr{V}$ *satisfies*

$$\varrho_R\mathscr{V}^2 \notin (I^+, A^+],$$

(ii) *if* $\theta_\eta^+ < 0$, $\sigma_\eta^+$ *cannot vanish and the shock speed* $\mathscr{V}$ *satisfies*

$$\varrho_R\mathscr{V}^2 \in (A^+, I^+].$$

Theorem 3 is thus rather different from its analog, Theorem 1. Indeed, in the more usual case, $\theta_\eta^+ > 0$, we do *not* find that $\sigma_\eta^+$ cannot vanish, and, moreover, the shock strength $\varrho_R\mathscr{V}^2$, instead of being trapped between $I$ and $A$ as it is at $(\eta^-, F^-)$, is here forbidden to lie between the values of these two local moduli at $(\eta^+, F^+)$: shocks of strength $\varrho_R\mathscr{V}^2$ larger than $A^+$ (the usual type of shock) and shocks of strength $\varrho_R\mathscr{V}^2$ no larger than $I^+$ are both consistent with local energy dissipation in a structured shock wave near a front state with $\theta_\eta^+ > 0$. Since the term "weak" was assigned long ago to shocks satisfying $\varrho_R\mathscr{V}^2 \searrow A^+$, we will call *ultra slow* those shocks satisfying

$$\varrho_R\mathscr{V}^2 \leqq I^+ \leqq A^+, \quad \textit{equality holding at most once}.$$

The existence and character of ultra slow shocks is discussed a bit in Sections 5 and 6.

**Theorem 4.** *Let the front state of a steady, structured shock wave in a thermoelastic material satisfy*

$$A^+ = \varrho_R\mathscr{V}^2, \quad \sigma_\eta^+ \neq 0, \quad \textit{and} \quad \theta_\eta^+ \neq 0.$$

*Then, as* $\xi \to \infty$,

$$\text{(i) } \sigma^+_{FF} < 0 \Rightarrow \hat{F}(\xi) \searrow F^+ \quad \textit{and} \quad \hat{F}(\xi) \searrow F^+ \Rightarrow \sigma^+_{FF} \leqq 0,$$

$$\text{(ii) } \sigma^+_{FF} > 0 \Rightarrow \hat{F}(\xi) \nearrow F^+ \quad \textit{and} \quad \hat{F}(\xi) \nearrow F^+ \Rightarrow \sigma^+_{FF} \geqq 0,$$

$$\text{(iii) } \begin{cases} \sigma^+_{FF} = 0 \Rightarrow \sigma^+_{FFF} \leqq 0, \\ \hat{F}(\xi) \to F^+ \textit{ from above and below} \Rightarrow \sigma^+_{FF} = 0. \end{cases}$$

When $A^+ \neq \varrho_R \mathscr{V}^2$ and $\sigma^+_\eta \neq 0$, the Rayleigh curve near $(\eta^+, F^+)$ looks like (a) or (b) of Figure 5. Indeed, since then

$$\frac{d\eta^R(F^+)}{dF} = \frac{-(A^+ - \varrho_R \mathscr{V}^2)}{\sigma^+_\eta},$$

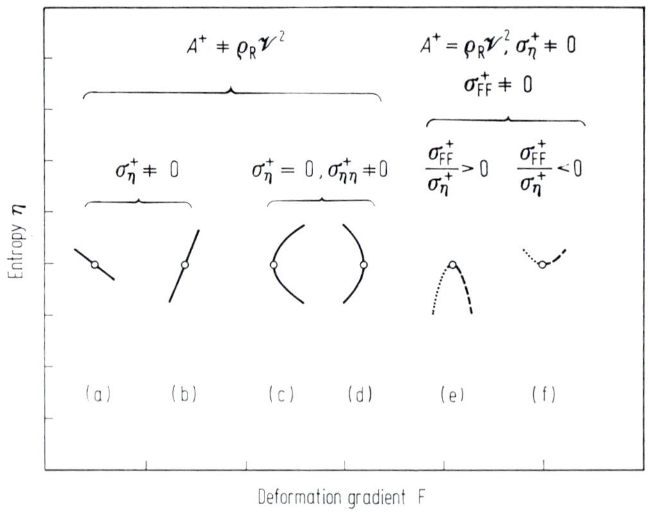


Fig. 5. The Rayleigh curve $\mathscr{R}(\mathscr{V})$ near $(\eta^+, F^+)$ when the specific heat $c^+_d$ is positive. In Cases (a), (b), (e), and (f), the slope $\dfrac{d\eta^R(F^+)}{dF}$ is well defined and given by $-(A^+ - \varrho_R \mathscr{V}^2)/\sigma^+_\eta$. In Cases (e) and (f) it is zero, and the curvature $\dfrac{d^2\eta^R(F^+)}{dF^2} = -\sigma^+_{FF}/\sigma^+_\eta$. The Rayleigh trajectory approaches $(\eta^+, F^+)$ along the dashed branch of (e) and (f) if $\sigma^+_{FF} < 0$, along the dotted branch if $\sigma^+_{FF} > 0$. In Cases (c) and (d), $\varrho_R \mathscr{V}^2 \neq A^+$ but $\sigma^+_\eta$ vanishes. Accordingly, the slope $\dfrac{dF^R(\eta^+)}{d\eta}$ is zero, while the curvature $\dfrac{d^2F^R(\eta^+)}{d\eta^2}$ is just $-\sigma^+_{\eta\eta}/(A^+ - \varrho_R \mathscr{V}^2)$.

we see that Case (a) corresponds to $\sigma^+_\eta$ and $A^+ - \varrho_R \mathscr{V}^2$ being of the same sign, while Case (b) corresponds to $\sigma^+_\eta$ and $A^+ - \varrho_R \mathscr{V}^2$ being oppositely signed. If therefore, as we did in our discussion of Theorems 1 and 2, we rule out the possibility of strain overshoot (say by assuming $\bar{\sigma}(\cdot, F^+)$ to be monotone and by again appealing to Thorem 5 and its Corollary to now rule out the trajectory's crossing and recrossing of $(\eta^+, F^+)$) we see from (a) of Figure 5 that for compressive shocks

$$\left\{\begin{matrix} \sigma_\eta^+ < 0 \quad \textit{and} \quad \varrho_R \mathscr{V}^2 > A^+, \\ \textit{or} \\ \sigma_\eta^+ > 0 \quad \textit{and} \quad \varrho_R \mathscr{V}^2 < A^+ \end{matrix}\right\} \Rightarrow \left\{\begin{matrix}\textit{initial entropy increase in the} \\ \textit{shock away from } (\eta^+, F^+).\end{matrix}\right\}$$

Perhaps more interesting (if less common) is the situation presented in (b) of Figure 5. It illustrates conditions under which the particles in a steady, structured shock wave will, for a time, experience *smaller* entropies than the entropy $\eta^+$ of the front state. Precisely, for a compressive shock,

$$\left\{\begin{matrix} \sigma_\eta^+ > 0 \quad \textit{and} \quad \varrho_R \mathscr{V}^2 > A^+, \\ \textit{or} \\ \sigma_\eta^+ < 0 \quad \textit{and} \quad \varrho_R \mathscr{V}^2 < A^+ \end{matrix}\right\} \Rightarrow \left\{\begin{matrix}\textit{initial entropy decrease in the shock} \\ \textit{away from } (\eta^+, F^+)\text{, i.e., } \textit{there is} \\ \textbf{entropy undershoot } \textit{in the wave.}\end{matrix}\right\}$$

Conversely, for expansive shocks we see from (a) of Figure 5 that

$$\left\{\begin{matrix} \sigma_\eta^+ < 0 \quad \textit{and} \quad \varrho_R \mathscr{V}^2 > A^+, \\ \textit{or} \\ \sigma_\eta^+ > 0 \quad \textit{and} \quad \varrho_R \mathscr{V}^2 < A^+ \end{matrix}\right\} \Rightarrow \left\{\begin{matrix}\textit{initial entropy decrease in the shock} \\ \textit{away from } (\eta^+, F^+)\text{, i.e., } \textit{there is} \\ \textbf{entropy undershoot } \textit{in the wave,}\end{matrix}\right\}$$

while from (b) of Figure 5

$$\left\{\begin{matrix} \sigma_\eta^+ < 0 \quad \textit{and} \quad \varrho_R \mathscr{V}^2 < A^+, \\ \textit{or} \\ \sigma_\eta^+ > 0 \quad \textit{and} \quad \varrho_R \mathscr{V}^2 > A^+ \end{matrix}\right\} \Rightarrow \left\{\begin{matrix}\textit{initial entropy increase in the} \\ \textit{shock away from } (\eta^+, F^+).\end{matrix}\right\}$$

When $\sigma_\eta^+ = 0$ and $A^+ \neq \varrho_R \mathscr{V}^2$, the Rayleigh curve near $(\eta^+, F^+)$ looks like (c) or (d) of Figure 5; Case (c) if $\sigma_{\eta\eta}^+$ and $A^+ - \varrho_R \mathscr{V}^2$ are signed oppositely, Case (d) if they are of the same sign. Again setting aside the possibility of strain overshoot, we see that Case (c) would necessarily correspond to an expansive shock, Case (d) a compressive shock. Observe also that, while certainly not common, these two cases present us with a rather striking possibility for non-uniqueness of steady, structured shocks with front state $(\eta^+, F^+)$: Both the upper and the lower branches of $\mathscr{R}(\mathscr{V})$ illustrated in Cases (c) and (d) could intersect the Hugoniot ${}^0\mathscr{H}$ based at $(\eta^+, F^+)$ at back states $(\eta_u^-, F_u^-)$ and $(\eta_l^-, F_l^-)$, respectively, and corresponding to each of these branches we might expect a trajectory that solves (3.3).

Cases (e) and (f) of Figure 5 represent the appearance of the Rayleigh curve near $(\eta^+, F^+)$ under the hypothesis of Theorem 4. By that theorem, for example, we see that the condition $\sigma_{FF}^+ < 0$ allows the trajectory $(\hat{\eta}, \hat{F})(\cdot)$ to approach $(\eta^+, F^+)$ only along the right hand branch of (e) or of (f). Once again setting aside the possibility of strain overshoot, Theorem 4 is thus seen to forbid the existence of compressive (expansive) shock waves satisfying

$$A^+ = \varrho_R \mathscr{V}^2, \quad \sigma_{FF}^+ < (>)\, 0, \quad \sigma_\eta^+ \neq 0, \quad \textit{and} \quad \theta_\eta^+ \neq 0,$$

*i.e.*, steady, structured shocks that satisfy these four conditions are necessarily expansive (compressive).

It is worth noting that there is a special process and/or material degeneracy we wish to avoid in our analysis. Specifically, suppose that for a certain structured shock the condition (2.17) of JOUGUET and ZEMPLÉN was satisfied as an *equality*, *i.e.*, $\eta^- = \eta^+$, so that there was no net entropy increase due to the passage of this particular shock. Since the energy equation for a thermoelastic material can easily be put in the form $\varrho_R \bar{e}_{,\eta} \dot{\eta} = -q_{,X}$, we see that for steady, structured shocks

$$\varrho_R \mathscr{V} \theta \hat{\eta}' = \hat{q}'.$$

Thus, $\varrho_R \mathscr{V} \hat{\eta}' = (\hat{q}/\hat{\theta})' + \hat{q}\hat{\theta}'/\hat{\theta}^2$, and, by integrating this over $(-\infty, \infty)$, we find that

$$\int_{-\infty}^{\infty} \frac{\hat{q}\hat{\theta}'}{\hat{\theta}^2}\, d\xi = 0,$$

since $q^\pm = 0$ and $\eta^- = \eta^+$. By virtue of the heat conduction inequality $(3.1)_5$, this implies that

$$\hat{q}(\xi)\, \hat{\theta}'(\xi) = \bar{q}(\hat{\eta}(\xi), \hat{F}(\xi), \hat{\theta}'(\xi))\, \hat{\theta}'(\xi) \equiv 0 \qquad \forall\, \xi \in (-\infty, \infty),$$

*i.e.*, either the heat flux or the temperature gradient or both must vanish at each point in the wave. But, by $(3.1)_6$, this means that $\hat{q}(\xi)$ must vanish everywhere, and then, by the energy equation, so too must $\hat{\eta}'(\xi)$. Hence, a steady, structured shock in a thermoelastic material causes no *net* entropy increase if and only if it causes no entropy variation *anywhere* in the wave. Equivalently, the Rayleigh trajectory for such a wave corresponds to at most a flow along the isentrope $\eta = \eta^+$. If the accompanying deformation gradient field is also so trivial, *i.e.*, if $\hat{F}(\cdot) = F^+$, then we really have no dynamic process at all. This case corresponds to the case of $\eta^- = \eta^+$ and $F^- = F^+$ in the singular surface description of shocks and is essentially empty. We ignore it.

Only slightly more interesting is the case when such a uniform entropy wave has associated with it a non-constant deformation wave $\hat{F}(\cdot)$. In this case $\hat{F}((-\infty, \infty))$ is an interval containing $F^+$ with a non-empty interior and the definition (3.2) of the Rayleigh curve amounts to delimiting the maximal interval $[F_1, F_2]$ containing $F^+$ over which $\bar{\sigma}(\eta^+, \cdot)$ is *linear*†, *i.e.*,

$$\bar{\sigma}(\eta^+, F) = \bar{\sigma}(\eta^+, F^+) + \varrho_R \mathscr{V}^2 (F - F^+) \qquad \forall\, F \in [F_1, F_2].$$

The graph of the Rayleigh trajectory, $\{\eta^+\} \times \hat{F}((-\infty, \infty))$, then must lie in the segment $\{\eta^+\} \times [F_1, F_2]$ of the isentrope $\eta = \eta^+$. As a consequence, the adiabatic sound speed $\sqrt{A/\varrho_R}$, $A = \bar{\sigma}_F(\eta^+, F)$, will be constant through the wave and the shock speed $\mathscr{V}$ will be exactly $\sqrt{A^+/\varrho_R}$. During the shock process the material is thus effectively *linearly elastic*. Accordingly, steady wave shapes $\hat{F}(\cdot)$ of arbitrary structure, untailored by the material, can be supported but they all must move at the same speed $\sqrt{A^+/\varrho_R}$, the speed of sound waves. The linearity of $\bar{\sigma}(\eta^+, \cdot)$ over $[F_1, F_2]$ also makes it easy to show that $\mathscr{H}(\eta^+, \cdot)$

† While $\mathscr{R}(\mathscr{V})$ may contain points not in $\{\eta^+\} \times [F_1, F_2]$, they will be of no relevance to our uniform entropy wave.

vanishes over the same interval. Thus, for this material the Hugoniot set ${}^0\mathscr{H}$ based at the given front state will include exactly that section of the isentrope $\eta = \eta^+$ that contains the graph of the Rayleigh trajectory. This renders satisfaction of the energy equation in the form (3.3) trivial. We ignore this case also‡. Thus, we may always assume that $\eta^- > \eta^+$: the passage of a steady, structured shock wave always strictly increases the entropy.

## 4. Departures, Arrivals, and the Structure of the Hugoniot

The results of Section 3 on the ordering between the shock strength $\varrho_R \mathscr{V}^2$ and the moduli $A$ and $I$ at the front and back states in a steady, structured shock wave rested, in essence, on the principle of energy dissipation along the trajectory of the wave. In particular, we did not require any explicit information on the local or global structure of the Hugoniot, and we found no need to appeal to the usually postulated inequalities (3.7)*. Indeed, we did not even employ JOUGUET'S and ZEMPLÉN'S condition (2.17),

$$\eta^- \geqq \eta^+,$$

as a necessary condition on allowable back states. In this and the following section, we consider certain refinements and generalizations of the results of Section 3 which follow if some of these additional considerations are brought into play.

In the present section, we discuss the geometric structure possessed by the Hugoniot ${}^0\mathscr{H}$ and a Rayleigh curve $\mathscr{R}(\mathscr{V})$ in a neighborhood of any one of their intersection points where they are both regular. While the front and back states, $(\eta^+, F^+)$ and $(\eta^-, F^-)$, are included in our discussion, we also wish to consider the possibility of intermediate intersection points of ${}^0\mathscr{H}$ and $\mathscr{R}(\mathscr{V})$. The tools for our study are quite simple, being really no more than the definitions of ${}^0\mathscr{H}$ and $\mathscr{R}(\mathscr{V})$, the differential equation (3.3), and the heat conduction inequality $(3.1)_5$. In Section 5, we finally explore the implications for shock waves of certain *a priori* inequalities. Instead of presenting the most general treatment our methods would allow†, we focus chiefly on materials in which compressive shocks are the norm and on weakening the rather stringent classical inequality $(3.7)_3$.

We begin by noting a few simple geometric properties of the Hugoniot ${}^0\mathscr{H}$. From (3.1) and (3.4) we see that

$$\begin{aligned} \mathscr{H}_F &= \tfrac{1}{2}\{\sigma - \sigma^+ - \bar{\sigma}_F(F - F^+)\}, \\ \mathscr{H}_\eta &= \varrho_R \theta - \tfrac{1}{2}\bar{\sigma}_\eta (F - F^+). \end{aligned} \qquad (4.1)_{1,2}$$

‡ Say, by assuming that $\bar{\sigma}(\eta^+, \cdot)$ is *not* linear near enough to $F^+$ or by requiring the heat flux function to be *strictly conductive* and the *Grüneisen coefficient* to be nontrivial, as is discussed in the last subsection of Section 6.

* Note, however, that the terms involved in the inequalities of (3.7), $\bar{\sigma}_\eta, \bar{\sigma}_F (= A)$ and $\bar{\sigma}_{FF}$, have appeared repeatedly in our analysis: By (i) of Theorem 1, $\sigma_\eta^-$ cannot vanish and $\varrho_R \mathscr{V}^2 < A^-$; by (i) of Theorem 3, either $A^+ < \varrho_R \mathscr{V}^2$ or $\varrho_R \mathscr{V}^2 \leqq I^+ \leqq A^+$; by (i) of Theorem 2, $\sigma_{FF}^- < 0$ forces $\hat{F}(\cdot)$ to approach $F^-$ from below.

† Straightforward duals of many of our results follow at once by simply reversing the inequalities of our hypotheses.

It follows from $(4.1)_2$ that on the ray $(-\infty, \infty)\times\{F^+\}$

$$\mathscr{H}_\eta = \mathscr{H}_\eta(\eta, F^+) = \varrho_R\bar{\theta}(\eta, F^+) > 0,$$

*i.e.*, $\mathscr{H}(\cdot, F^+)$ is strictly monotone increasing. As a consequence, *the Hugoniot set* ${}^0\mathscr{H}$ *intersects the ray* $(-\infty, \infty)\times\{F^+\}$ *only at* $(\eta^+, F^+)$. Equivalently, if we set aside the degenerate case‡ $(\eta^-, F^-) = (\eta^+, F^+)$, there are no shocks whose back state $(\eta^-, F^-)$ meets $F^- = F^+$. All shocks are either compressive or expansive.

Now, let $(\eta, F)$ be a regular point of ${}^0\mathscr{H}$. In a neighborhood of $(\eta, F)$ we can thus solve $\mathscr{H}(\eta, F) = 0$ for either $F = F^H(\eta)$ or $\eta = \eta^H(F)$, depending on which of $\mathscr{H}_F$ and $\mathscr{H}_\eta$ does not vanish at $(\eta, F)$. In the first case we have

$$\frac{dF^H}{d\eta} = \frac{-\{\varrho_R\theta - \frac{1}{2}\bar{\sigma}_\eta(F - F^+)\}}{\frac{1}{2}\{\sigma - \sigma^+ - \bar{\sigma}_F(F - F^+)\}}, \tag{4.2$_1$}$$

while in the second

$$\frac{d\eta^H}{dF} = \frac{-\frac{1}{2}\{\sigma - \sigma^+ - \bar{\sigma}_F(F - F^+)\}}{\varrho_R\theta - \frac{1}{2}\bar{\sigma}_\eta(F - F^+)}. \tag{4.2$_2$}$$

Observe that (4.1) has been used in deriving (4.2).

Equation $(4.2)_2$ makes at once apparent the well known result that $\dfrac{d\eta^H(F^+)}{dF} = 0$, *i.e.*, at $(\eta^+, F^+)$ the Hugoniot is tangent to the isentrope $\eta = \eta^+$. Moreover, upon differentiating $(4.2)_2$ with respect to $F$, we find the also well known result that $\dfrac{d^2\eta^H(F^+)}{dF^2} = 0$. Thus, the tangency is at least "second order", and, as Figure 2 suggests, ${}^0\mathscr{H}$ is rather flat near $(\eta^+, F^+)$.

Next, let us consider the possible geometric structure at a point where the Hugoniot intersects a Rayleigh curve. Thus, suppose that ${}^0\mathscr{H}$ intersects a Rayleigh curve $\mathscr{R}(\mathscr{V})$ at a point $(\eta^*, F^*)$, and suppose also that $(\eta^*, F^*)$ is a regular point for both ${}^0\mathscr{H}$ and $\mathscr{R}(\mathscr{V})$. Then use of (3.2) reveals that $(4.2)_1$ and $(4.2)_2$ may be put in the respective forms

$$\frac{dF^H}{d\eta} = \frac{\bar{\sigma}_\eta - \dfrac{2\varrho_R\theta}{(F - F^+)}}{\varrho_R\mathscr{V}^2 - \bar{\sigma}_F} \quad \text{at } (\eta^*, F^*), \tag{4.3$_1$}$$

and

$$\frac{d\eta^H}{dF} = \frac{\varrho_R\mathscr{V}^2 - \bar{\sigma}_F}{\bar{\sigma}_\eta - \dfrac{2\varrho_R\theta}{(F - F^+)}} \quad \text{at } (\eta^*, F^*), \tag{4.3$_2$}$$

as long as $(\eta^*, F^*) \neq (\eta^+, F^+)$ since, as we have seen, $F^* = F^+$ for $(\eta^*, F^*) \in {}^0\mathscr{H}$ is then impossible. Further, since the intersection point $(\eta^*, F^*)$ is also a regular point of $\mathscr{R}(\mathscr{V})$, it follows that locally $\mathscr{R}(\mathscr{V})$ is the graph of a smooth function

‡ Recall the discussion at the end of Section 3.

$\eta^R(\cdot)$ or a smooth function $F^R(\cdot)$. By $(3.5)_{1,2}$,

$$\frac{dF^R}{d\eta} = \frac{\bar{\sigma}_\eta}{\varrho_R \mathscr{V}^2 - \bar{\sigma}_F} \quad \text{at } (\eta^*, F^*), \tag{4.4$_1$}$$

and

$$\frac{d\eta^R}{dF} = \frac{\varrho_R \mathscr{V}^2 - \bar{\sigma}_F}{\bar{\sigma}_\eta} \quad \text{at } (\eta^*, F^*), \tag{4.4$_2$}$$

depending respectively on whether $\varrho_R \mathscr{V}^2 - \bar{\sigma}_F$ or $\bar{\sigma}_\eta$ does not vanish at $(\eta^*, F^*)$.

Assume first that $\varrho_R \mathscr{V}^2 \neq A^*$. If $(\eta^*, F^*) \neq (\eta^+, F^+)$, then an examination of $(4.3)_1$ and $(4.4)_1$ reveals that, since $\theta^* \neq 0$, at $(\eta^*, F^*)$ *the slopes* $\frac{dF^H}{d\eta}$ *and* $\frac{dF^R}{d\eta}$ *of* ${}^0\mathscr{H}$ *and* $\mathscr{R}(\mathscr{V})$, *respectively, are both finite and distinct.* If $(\eta^*, F^*) = (\eta^+, F^+)$, then $(4.2)_2$ and $(4.4)_1$ show, respectively, that *the slope* $\frac{d\eta^H(F^+)}{dF}$ *vanishes* and *the slope* $\frac{dF^R(\eta^+)}{d\eta} = \frac{\sigma_\eta^+}{\varrho_R \mathscr{V}^2 - A^+}$ *is finite*. In either case, it is clear that $\varrho_R \mathscr{V}^2 \neq A^*$ is a sufficient condition for ${}^0\mathscr{H}$ and $\mathscr{R}(\mathscr{V})$ to, in fact, *cross* at the intersection point $(\eta^*, F^*)$. Indeed, they even cross at a nonzero angle.

Assume now that $\varrho_R \mathscr{V}^2 = A^*$. Then, by the regularity of $(\eta^*, F^*)$, neither $\bar{\sigma}_\eta$ nor $\varrho_R \theta - \frac{1}{2} \bar{\sigma}_\eta (F - F^+)$ can vanish at $(\eta^*, F^*)$. Moreover, *the respective slopes,* $\frac{d\eta^H}{dF}$ *and* $\frac{d\eta^R}{dF}$, *of* ${}^0\mathscr{H}$ *and* $\mathscr{R}(\mathscr{V})$ *at* $(\eta^*, F^*)$ *are both zero,* while (by an easy calculation)

$$\frac{d^2\eta^H}{dF^2} = \frac{-\bar{\sigma}_{FF}}{\bar{\sigma}_\eta - \dfrac{2\varrho_R \theta}{(F - F^+)}}, \tag{4.5$_1$}$$

and

$$\frac{d^2\eta^R}{dF^2} = \frac{-\bar{\sigma}_{FF}}{\bar{\sigma}_\eta}, \tag{4.5$_2$}$$

at $(\eta^*, F^*) \neq (\eta^+, F^+)$. *Clearly these curvatures are finite and distinct unless* $\sigma_{FF}^*$ *vanishes.* Here then $\sigma_{FF}^* \neq 0$ is a condition sufficient for ${}^0\mathscr{H}$ and $\mathscr{R}(\mathscr{V})$ to *not* cross at the intersection point $(\eta^*, F^*)$. Instead, ${}^0\mathscr{H}$ and $\mathscr{R}(\mathscr{V})$ (more precisely, $\eta^H(\cdot)$ and $\eta^R(\cdot)$) each have a strict local extremum at $(\eta^*, F^*)$.

Suppose that, in addition to $\varrho_R \mathscr{V}^2 = \sigma_F^*$, we also had $\frac{\partial^i \bar{\sigma}}{\partial F^i} = 0$ at $(\eta^*, F^*) \neq (\eta^+, F^+)$, for $i = 2, 3, \ldots, n-1$. Then it is easy to show that $\frac{d^i\eta^H}{dF^i} = \frac{d^i\eta^R}{dF^i} = 0$ at $F^*$ for $i = 1, 2, \ldots, n-1$, while

$$\frac{d^n\eta^H}{dF^n} = \frac{-\dfrac{\partial^n \bar{\sigma}}{\partial F^n}}{\bar{\sigma}_\eta - \dfrac{2\varrho_R \theta}{(F - F^+)}},$$

and

$$\frac{d^n\eta^R}{dF^n} = \frac{-\dfrac{\partial^n\bar{\sigma}}{\partial F^n}}{\bar{\sigma}_\eta},$$

at $F^*$. Thus, given sufficient differentiability, the crossing or osculatory character of ${}^0\mathscr{H}$ and $\mathscr{R}(\mathscr{V})$ is not hard to assess at any intersection point where they are both regular.

In Figure 6, Cases (a) and (b) depict the "kissing" type of contact entailed by $\varrho_R\mathscr{V}^2 = A^*$ and $\sigma^*_{FF}$ not zero. Note that a kissing like that shown in Case (c) of Figure 6 is impossible—it would require the absolute temperature to vanish at the contact point.

The following lemma deals with one more purely geometric aspect of the intersection of the Hugoniot ${}^0\mathscr{H}$ with a Rayleigh curve $\mathscr{R}(\mathscr{V})$. We omit its proof, which is essentially straightforward.

**Lemma 1.** *Let $(\eta^*, F^*)$ be an intersection point of the Hugoniot ${}^0\mathscr{H}$ and a Rayleigh curve $\mathscr{R}(\mathscr{V})$, and suppose that $\varrho_R\mathscr{V}^2 \neq A^*$. Then $(\eta^*, F^*)$ is a regular point for both ${}^0\mathscr{H}$ and $\mathscr{R}(\mathscr{V})$, and it has a neighborhood $\mathscr{N}$ that contains no other intersection point of ${}^0\mathscr{H}$ and $\mathscr{R}(\mathscr{V})$. Additionally, we have*

$$\mathscr{N} = \mathscr{H}^- \cup ({}^0\mathscr{H} \cap \mathscr{N}) \cup \mathscr{H}^+,$$

*where $\mathscr{H}^- \equiv \{(\eta, F) \in \mathscr{N} \mid \mathscr{H}(\eta, F) < 0\}$, $\mathscr{H}^+ \equiv \{(\eta, F) \in \mathscr{N} \mid \mathscr{H}(\eta, F) > 0\}$, and where neither $\mathscr{H}^-$ nor $\mathscr{H}^+$ is empty. In particular,*

$$\mathscr{R}(\mathscr{V}) \cap \mathscr{N} = \mathscr{R}^- \cup \{(\eta^*, F^*)\} \cup \mathscr{R}^+,$$

*where $\mathscr{R}^\pm \equiv \mathscr{R}(\mathscr{V}) \cap \mathscr{H}^\pm$, and where neither $\mathscr{R}^-$ nor $\mathscr{R}^+$ is empty. Moreover,*

$$\eta^* > \eta \;\forall\, (\eta, F) \in \mathscr{R}^-, \quad \text{while } \eta^* < \eta \;\forall\, (\eta, F) \in \mathscr{R}^{+\dagger}.$$

We now wish to consider those consequences (as it happens, restrictions) imposed by thermo*dynamics* on the contact of a Rayleigh trajectory $(\hat{\eta}, \hat{F})(\cdot)$ with the Hugoniot. We will see that "most" contact points must either be "arrival points" or "departure points" for the trajectory $(\hat{\eta}, \hat{F})(\cdot)$. It will follow that in general the Rayleigh trajectory is forbidden, *by thermodynamic principles alone*, to touch the Hugoniot at intermediate points in its transit from $(\eta^-, F^-)$ to $(\eta^+, F^+)$. To establish this, we will need one more lemma. The result we have in mind is almost geometrically obvious and makes precise the notions of an *arrival point* and *departure point* on the graph of a trajectory.

**Lemma on Arriving and Departing.** *Let $\mathscr{I}$ be an interval, and let $\boldsymbol{t}(\cdot)$ be a continuous trajectory in $R^n$ defined over $\mathscr{I}$. Let $\boldsymbol{x}^*$ be a point in the graph of $\boldsymbol{t}(\cdot)$ and suppose $\boldsymbol{t}(\cdot)$* **arrives** at $\boldsymbol{x}^*$, i.e., *there exist $\bar{s}$ and $s^* > \bar{s}$ such that $\boldsymbol{t}(\bar{s}) \neq \boldsymbol{x}^*$ and $\boldsymbol{t}(s^*) = \boldsymbol{x}^*$. Then there is an interval $[s_1, s_2] \subseteq [\bar{s}, s^*]$ on which $\boldsymbol{t}(\cdot)$ reaches $\boldsymbol{x}^*$*

† The reverse inequalities would hold if we were to allow $\theta^* < 0$.

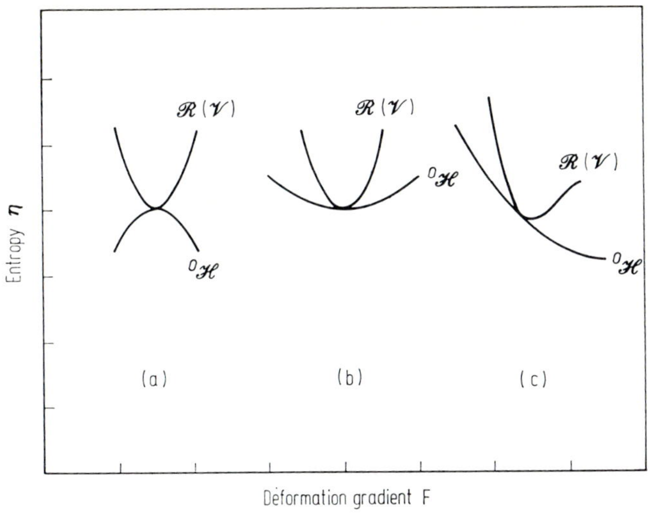


Fig. 6. Cases (a) and (b) illustrate allowed osculatory contact of $^0\mathscr{H}$ and $\mathscr{R}(\mathscr{V})$ at a point $(\eta^*, F^*) \neq (\eta^+, F^+)$ when $\varrho_R \mathscr{V}^2 = A^*$ but $\sigma^*_{FF} \neq 0$. Osculatory contact like that of Case (c) is impossible unless we permit $\theta^* = 0$.

*only at* $s_2$, i.e.,

$$\boldsymbol{x}^* \notin \boldsymbol{t}([s_1, s_2)) \quad \textit{and} \quad \boldsymbol{x}^* = \boldsymbol{t}(s_2).$$

*Similarly, let* $\boldsymbol{x}^*$ *be a point in the graph of* $\boldsymbol{t}(\cdot)$ *and suppose* $\boldsymbol{t}(\cdot)$ **departs** *from* $\boldsymbol{x}^*$, i.e., *there exist* $\bar{s}$ *and* $s^* < \bar{s}$ *such that* $\boldsymbol{t}(\bar{s}) \neq \boldsymbol{x}^*$ *and* $\boldsymbol{t}(s^*) = \boldsymbol{x}^*$. *Then there is an interval* $[s_3, s_4] \subseteq [s^*, \bar{s}]$ *on which* $\boldsymbol{t}(\cdot)$ *was at* $\boldsymbol{x}^*$ *only at* $s_3$, i.e.,

$$\boldsymbol{x}^* = \boldsymbol{t}(s_3) \quad \textit{and} \quad \boldsymbol{x}^* \notin \boldsymbol{t}((s_3, s_4]).$$

Assuming they exist, it is clear that at its endpoints $\boldsymbol{t}(\cdot)$ can only *depart from* $\boldsymbol{t}$ (inf $\mathscr{I}$) and can only *arrive at* $\boldsymbol{t}$ (sup $\mathscr{I}$).

**Proof.** We prove the lemma for arriving; the case for departing is completely analogous. Since $\boldsymbol{t}(\bar{s}) \neq \boldsymbol{x}^*$, there is a neighborhood $\mathscr{N}$ of $\boldsymbol{x}^*$ such that $\boldsymbol{t}(\bar{s}) \notin \mathscr{N}$. Let $\mathrm{A} = \{s \mid s < s^* \textit{ and } \boldsymbol{t}([s, s^*]) \subseteq \mathscr{N}\}$. Since $\boldsymbol{t}(\cdot)$ is continuous, A is not empty. Moreover, A is bounded below by $\bar{s}$. Thus, $s_1 \equiv \inf \mathrm{A}$ exists, and evidently $\boldsymbol{t}(s_1) \in \partial\mathscr{N}$ with $\boldsymbol{t}((s_1, s^*]) \subseteq \mathscr{N}$.

Let $\mathrm{B} = \{s \mid s \in [s_1, s^*] \textit{ and } \boldsymbol{t}(s) = \boldsymbol{x}^*\}$. It is clear that B is bounded below, and since $s^* \in \mathrm{B}$, it is also clear that B is non-empty. Consequently, $s_2 \equiv \inf \mathrm{B}$ exists. Since $\boldsymbol{t}(s_2) = \boldsymbol{x}^*$ and $s_2 \in [s_1, s^*]$, it is clear that $s_1 < s_2$. Thus, $[s_1, s_2]$ is a sub-interval of $[\bar{s}, s^*]$, and there can be no $s \in [s_1, s_2)$ with $\boldsymbol{t}(s) = \boldsymbol{x}^*$, since this would contradict the fact that $s_2$ is the infimum of B. □

Observe now that for a steady, structured shock wave the heat conduction inequality $(3.1)_5$ implies

$$\hat{q}(\xi)\hat{\theta}'(\xi) \leq 0,$$

throughout the wave. By (3.3) this means that

$$\mathscr{V}\mathscr{H}(\hat{\eta}(\xi), \hat{F}(\xi))\,\hat{\theta}'(\xi) \leqq 0, \tag{4.6}$$

for all $\xi$. By use of $(3.12)_1$, we thus see that

$$\mathscr{V}\bar{\theta}_\eta \left(\frac{I - \varrho_R \mathscr{V}^2}{A - \varrho_R \mathscr{V}^2}\right) \mathscr{H}\hat{\eta}' \leqq 0, \tag{4.7$_1$}$$

at any point $(\hat{\eta}, \hat{F})(\xi)$ where $\varrho_R \mathscr{V}^2 - A$ is not zero, while with $(3.12)_2$ we get

$$-\mathscr{V}\frac{\bar{\theta}_\eta}{\bar{\sigma}_\eta}(I - \varrho_R \mathscr{V}^2)\,\mathscr{H}\hat{F}' \leqq 0, \tag{4.7$_2$}$$

at any point $(\hat{\eta}, \hat{F})(\xi)$ where $\bar{\sigma}_\eta$ is not zero.

**Theorem 5.** *Let the Hugoniot* ${}^0\mathscr{H}$ *intersect a Rayleigh curve* $\mathscr{R}(\mathscr{V})$ *at a point* $(\eta^*, F^*)$, *suppose that*

$$A^* \neq \varrho_R \mathscr{V}^2, \qquad \theta_\eta^* \neq 0, \qquad \textit{and} \qquad I^* \neq \varrho_R \mathscr{V}^2,$$

*and set*

$$\omega^* \equiv \mathscr{V}\theta_\eta^* \left(\frac{I^* - \varrho_R \mathscr{V}^2}{A^* - \varrho_R \mathscr{V}^2}\right).$$

*Let* $(\hat{\eta}, \hat{F})(\cdot)$ *be a Rayleigh trajectory on* $\mathscr{R}(\mathscr{V})$ *corresponding to a steady, structured shock wave in a thermoelastic material, and suppose that* $(\hat{\eta}, \hat{F})(\xi^*) = (\eta^*, F^*)$. *Then, if* $\omega^* > 0$, $(\hat{\eta}, \hat{F})(\cdot)$ *can never depart from* $(\eta^*, F^*)$; *if* $\omega^* < 0$, *it can never arrive at* $(\eta^*, F^*)$.

If we limit our attention to intersection points of the type described in Theorem 5, we have the following

**Corollary.** *The Rayleigh trajectory of a steady, structured shock wave touchcs the Hugoniot only at* $(\eta^-, F^-)$ *and* $(\eta^+, F^+)$. *Moreover, it can only depart from* $(\eta^-, F^-)$ *and can only arrive at* $(\eta^+, F^+)$. *In particular, it cannot cross and recross its front state or its back state.*

Observe that, since $(\eta^-, F^-)$ must be a departure point for $(\hat{\eta}, \hat{F})(\cdot)$, Theorem 5 with $\mathscr{V} > 0$ requires $\theta_\eta^- \left(\frac{I^- - \varrho_R \mathscr{V}^2}{A^- - \varrho_R \mathscr{V}^2}\right)$ to be negative. Thus, for $\theta_\eta^- > 0$ we get $\varrho_R \mathscr{V}^2 \in (I^-, A^-)$; while, for $\theta_\eta^- < 0$, we get $\varrho_R \mathscr{V}^2 \notin [A^-, I^-]$, *i.e.*, Theorem 5 essentially implies Theorem 1 (and also Theorem 3).

Theorem 5 also provides an effective tool to answer certain *uniqueness* questions for steady, structured shock waves; indeed, it even provides useful necessary conditions for the *existence* of such waves: Suppose we are given a putative front state $(\eta^+, F^+)$ and a putative shock speed $\mathscr{V} > 0$ for which the number $\omega^+ \equiv \mathscr{V}\theta_\eta^+ \left(\frac{I^+ - \varrho_R \mathscr{V}^2}{A^+ - \varrho_R \mathscr{V}^2}\right)$ is well defined. For a corresponding steady, structured shock wave to exist at all, Theorem 5 tells us that $\omega^+$ must be nonnegative. Let us assume it to be positive and venture out along $\mathscr{R}(\mathscr{V})$ in one or the other direction from $(\eta^+, F^+)$. If it does not intersect the Hugoniot ${}^0\mathscr{H}$ on that particular half of $\mathscr{R}(\mathscr{V})$, then that branch of $\mathscr{R}(\mathscr{V})$ is unavailable as any part of a steady, structured shock wave into $(\eta^+, F^+)$ at speed $\mathscr{V}$. Suppose, on the other hand, that there are multiple intersections of the selected half branch of $\mathscr{R}(\mathscr{V})$ with the Hugoniot ${}^0\mathscr{H}$, and let us look at the "first" of them, *i.e.*, the one which is closest to $(\eta^+, F^+)$ along the selected half of $\mathscr{R}(\mathscr{V})$. Call it $(\eta^*, F^*)$, and let us suppose that it satisfies the hypotheses of Theorem 5 and has $\omega^* > 0$. Since no steady, structured shock wave of speed $\mathscr{V}$ can ever depart from $(\eta^*, F^*)$, it is not available as a back state: the corresponding steady, structured shock into $(\eta^+, F^+)$ at speed $\mathscr{V}$ cannot exist nor can any other shock exist that has a part of its trajectory on the segment of $\mathscr{R}(\mathscr{V})$ between $(\eta^+, F^+)$ and $(\eta^*, F^*)$. If $\omega^*$ is negative, however, then a steady, structured shock into $(\eta^+, F^+)$ at speed $\mathscr{V}$ with back state $(\eta^*, F^*)$ is indeed possible‡. Now, no later intersection point of the given half branch of $\mathscr{R}(\mathscr{V})$ with ${}^0\mathscr{H}$ is ever available as a back state in a steady, structured shock that contains the segment of $\mathscr{R}(\mathscr{V})$ between $(\eta^+, F^+)$ and $(\eta^*, F^*)$§.

Crucial to the above discussion of uniqueness and existence was the hypothesis that the desired shock was moving into material in the state $(\eta^+, F^+)$. Indeed, what our discussion suggests, and what is in fact true, is that a steady, structured shock wave of speed $\mathscr{V}$ with its trajectory on $\mathscr{R}(\mathscr{V})$ can be found which connects *any* two points in ${}^0\mathscr{H} \cap \mathscr{R}(\mathscr{V})$ as long as (i) it is possible to proceed along $\mathscr{R}(\mathscr{V})$ from one to the other of them without encountering any other points of ${}^0\mathscr{H} \cap \mathscr{R}(\mathscr{V})$, *i.e.*, in loose terms, the two points should be adjacent points of ${}^0\mathscr{H} \cap \mathscr{R}(\mathscr{V})$, (ii) one of the points, call it $(\eta_1, F_1)$, is an arrival point so that

$$\omega = \mathscr{V}\bar{\theta}_\eta\left(\frac{I - \varrho_R\mathscr{V}^2}{A - \varrho_R\mathscr{V}^2}\right)$$

is positive there, while the other point, $(\eta_2, F_2)$, is a departure point so that at it $\omega$ is negative, and (iii) the stretch of $\mathscr{R}(\mathscr{V})$ between the two points does not cross the curve of thermal extremes (see Section 6). It is, of course, necessary for this construction that $(\eta_2, F_2)$ lie in the Hugoniot set ${}^0\mathscr{H}_1$ based at $(\eta_1, F_1)$. This, however, follows easily from the fact that $(\eta_1, F_1)$ and $(\eta_2, F_2)$ both lie on the Rayleigh curve $\mathscr{R}(\mathscr{V})$ based at $(\eta^+, F^+)$ and both lie in the Hugoniot set ${}^0\mathscr{H}$ based at $(\eta^+, F^+)$. An easy way of producing such pairs of points will become clear in Section 6 with our discussion of bifurcated Hugoniots. Indeed, points like $(\eta_1, F_1)$ and $(\eta_2, F_2)$ will occur naturally in a small enough neighborhood of a bifurcation point of ${}^0\mathscr{H}$.

We now turn to the proof of Theorem 5.

‡ Although not assured; see the final part of Section 6 on the *curve of thermal extremes*.

§ For thermo*viscous* materials that are also constitutively sensitive to strain gradients, a result like this was announced but not formally proven in [17]. For HAGAN & SERRIN, multiple intersections of the Hugoniot and certain Rayleigh curves occurred as a natural consequence of the material's being able to experience a phase transition.

**Proof.** By hypothesis, the function

$$\omega = \omega(\eta, F) \equiv \mathscr{V}\bar{\theta}_\eta \left( \frac{I - \varrho_R \mathscr{V}^2}{A - \varrho_R \mathscr{V}^2} \right)$$

is well defined in a neighborhood of $(\eta^*, F^*)$ and does not vanish at $(\eta^*, F^*)$. Let $\mathscr{N}$ be a neighborhood of $(\eta^*, F^*)$ small enough that (i) $\omega$ vanishes nowhere in $\mathscr{N}$, and (ii) $\mathscr{N}$ satisfies the conditions affirmed by Lemma 1. For definiteness, let us suppose that $\omega$ is positive on $\mathscr{N}$. In this case, $(4.7)_1$ allows us to write

$$\mathscr{H}(\hat{\eta}(\xi), \hat{F}(\xi))\, \hat{\eta}'(\xi) \leqq 0 \qquad \forall\, \xi \;\; \textit{such that} \;\; (\hat{\eta}, \hat{F})(\xi) \in \mathscr{N}. \tag{4.8}$$

To reach a contradiction, we now suppose that the trajectory $(\hat{\eta}, \hat{F})(\cdot)$, which is at $(\eta^*, F^*)$ when $\xi = \xi^*$, both arrives at and departs from $(\eta^*, F^*)$. That is, we suppose the existence of $\bar{\xi}$ and $\bar{\bar{\xi}}$, $\bar{\xi} < \xi^* < \bar{\bar{\xi}}$, such that

$$(\hat{\eta}, \hat{F})(\bar{\xi}) \neq (\eta^*, F^*) \neq (\hat{\eta}, \hat{F})(\bar{\bar{\xi}}).$$

We next appeal to our Lemma on Arriving and Departing. It tells us that there is an interval $[\xi_1, \xi_2] \subseteqq [\bar{\xi}, \xi^*]$ and an interval $[\xi_3, \xi_4] \subseteqq [\xi^*, \bar{\bar{\xi}}]$ such that, on the "arriving interval" $[\xi_1, \xi_2]$,

$$(\eta^*, F^*) \notin (\hat{\eta}, \hat{F})([\xi_1, \xi_2)) \quad \text{and} \quad (\eta^*, F^*) = (\hat{\eta}, \hat{F})(\xi_2),$$

and, on the "departing interval" $[\xi_3, \xi_4]$,

$$(\eta^*, F^*) = (\hat{\eta}, \hat{F})(\xi_3) \quad \text{and} \quad (\eta^*, F^*) \notin (\hat{\eta}, \hat{F})((\xi_3, \xi_4]).$$

Moreover, it is clear that we may suppose $\xi_1$ big enough and $\xi_4$ small enough that $(\hat{\eta}, \hat{F})([\xi_1, \xi_2])$ and $(\hat{\eta}, \hat{F})([\xi_3, \xi_4])$ are both in $\mathscr{N}$. By Lemma 1 and the continuity of $(\hat{\eta}, \hat{F})(\cdot)$, we now see that

$$(\hat{\eta}, \hat{F})((\xi_1, \xi_2)) \subseteqq \mathscr{R}^- \quad \text{or} \quad (\hat{\eta}, \hat{F})((\xi_1, \xi_2)) \subseteqq \mathscr{R}^+$$

and

$$(\hat{\eta}, \hat{F})((\xi_3, \xi_4)) \subseteqq \mathscr{R}^- \quad \text{or} \quad (\hat{\eta}, \hat{F})((\xi_3, \xi_4)) \subseteqq \mathscr{R}^+.$$

Now observe that (4.8) allows us to write

$$(\hat{\eta}, \hat{F})(\xi) \in \mathscr{R}^- \Rightarrow \hat{\eta}' \geqq 0, \text{ while } (\hat{\eta}, \hat{F})(\xi) \in \mathscr{R}^+ \Rightarrow \hat{\eta}' \leqq 0. \tag{4.9}$$

Thus, if we now integrate $\hat{\eta}'$ from $\xi \in (\xi_1, \xi_2)$ to $\xi_2$, the above considerations make clear that for the arriving interval $[\xi_1, \xi_2]$

$$\begin{aligned} &(\hat{\eta}, \hat{F})((\xi_1, \xi_2)) \subseteqq \mathscr{R}^- \Rightarrow \eta^* \geqq \hat{\eta}(\xi)\; \forall\, \xi \in (\xi_1, \xi_2), \\ &(\hat{\eta}, \hat{F})((\xi_1, \xi_2)) \subseteqq \mathscr{R}^+ \Rightarrow \eta^* \leqq \hat{\eta}(\xi)\; \forall\, \xi \in (\xi_1, \xi_2). \end{aligned} \tag{4.10}$$

In the same way we may integrate $\hat{\eta}'$ from $\xi_3$ to $\xi \in (\xi_3, \xi_4)$ to find that for the departing interval $[\xi_3, \xi_4]$

$$\begin{aligned} &(\hat{\eta}, \hat{F})((\xi_3, \xi_4)) \subseteqq \mathscr{R}^- \Rightarrow \eta^* \leqq \hat{\eta}(\xi)\; \forall\, \xi \in (\xi_3, \xi_4), \\ &(\hat{\eta}, \hat{F})((\xi_3, \xi_4)) \subseteqq \mathscr{R}^+ \Rightarrow \eta^* \geqq \hat{\eta}(\xi)\; \forall\, \xi \in (\xi_3, \xi_4). \end{aligned} \tag{4.11}$$

But, *while both of the assertions of* (4.10) *are compatible with the final assertion of Lemma 1, neither assertion of* (4.11) *is*. We thus conclude that, of (4.10) and (4.11), only (4.10) is possible, *i.e.*, when $\omega^*$ is positive, the trajectory $(\hat{\eta}, \hat{F})(\cdot)$ can never depart from $(\eta^*, F^*)$. When $\omega^*$ is negative, the inequalities of (4.8)–(4.11) all reverse, and we conclude that $(\hat{\eta}, \hat{F})(\cdot)$ can never arrive at $(\eta^*, F^*)$. □

Now, suppose that the Hugoniot has an osculatory or kissing contact with a Rayleigh curve $\mathscr{R}(\mathscr{V})$. Two such cases are depicted in (a) and (b) of Figure 6 and, as we have seen, occur if $\varrho_R \mathscr{V}^2 = A^*$ but $\sigma^*_{FF} \neq 0$. In such cases, even though ${}^0\mathscr{H}$ still locally divides the plane into "+" and "−" components,† $\mathscr{H}^-$ and $\mathscr{H}^-$, the Rayleigh curve $\mathscr{R}(\mathscr{V})$, in contrast with the situation in Theorem 5, will no longer typically be itself segmented by ${}^0\mathscr{H}$ into two components, one in $\mathscr{H}^+$, the other in $\mathscr{H}^-$. Additionally, note that $(\eta^*, F^*)$ need no longer be an isolated intersection point of ${}^0\mathscr{H}$ and $\mathscr{R}(\mathscr{V})$; indeed, ${}^0\mathscr{H} \cap \mathscr{R}(\mathscr{V})$ could contain a line segment, $\{\eta^*\} \times [F_1, F_2]$, $F^* \in [F_1, F_2]$. These facts appear to us to make an extension of Theorem 5 to the case of kissing contact exceedingly difficult unless one is willing to impose rather stringent (and, from the point of view of material modeling, artificial) analytical conditions on the functions $\bar{q}(\cdot, \cdot, \cdot)$ and $\mathscr{H}(\cdot, \cdot)$ appearing in the differential equation (3.3). Lacking such additional hypotheses, we thus expect that a point of kissing contact between ${}^0\mathscr{H}$ and $\mathscr{R}(\mathscr{V})$ can be both an arrival point and a departure point for the Rayleigh trajectory $(\hat{\eta}, \hat{F})(\cdot)$.

The above remarks notwithstanding, at the present level of generality a little can be asserted with regard to kissing contact between the Hugoniot set and a given Rayleigh curve. For concreteness, we assume that $\sigma^*_{FF} \neq 0$; nothing of importance in our discussion would change if $\dfrac{\partial^i \sigma^*}{\partial F^i} = 0$, $i = 2, 3, \ldots, 2n-1$, and $\dfrac{\partial^{2n} \sigma^*}{\partial F^{2n}} \neq 0$, $n \geqq 2$. First, since $\varrho_R \mathscr{V}^2 = A^*$ and $\sigma^*_\eta \neq 0$, it follows directly from (3.10) that

$$\frac{-\theta^*_\eta}{\sigma^*_\eta}(I^* - \varrho_R \mathscr{V}^2) = \frac{\sigma^*_\eta}{\varrho_R}, \tag{4.12}$$

when $\theta^*_\eta \neq 0$. Next, it is clear from the curvature conditions $(4.5)_{1,2}$ that in the present special circumstance $(\eta^*, F^*)$ is an isolated intersection point of ${}^0\mathscr{H}$ and $\mathscr{R}(\mathscr{V})$. Moreover, since $\dfrac{d\eta^R(F^*)}{dF} = 0$ and $A^* = \varrho_R \mathscr{V}^2$, we see that, along with $\mathscr{H}(\eta^*, F^*)$, the derivative $\dfrac{d}{dF}\mathscr{H}(\eta^R(F), F)|_{F=F^*}$ also vanishes. As a consequence,

---

† Of course, we are still supposing that the intersection point $(\eta^*, F^*)$ is a point at which both ${}^0\mathscr{H}$ and $\mathscr{R}(\mathscr{V})$ are regular. Then, since $\varrho_R \mathscr{V}^2 - A^*$ vanishes, $\mathscr{H}^*_\eta = \varrho_R \theta^* - \frac{1}{2}\sigma^*_\eta(F^* - F^+)$ cannot vanish. Thus, locally

$\mathscr{H}^+$ *is above* (*below*) ${}^0\mathscr{H}$, *and* $\mathscr{H}^-$ *is below* (*above*) ${}^0\mathscr{H}$,

according as $\mathscr{H}^*_\eta$ is positive (negative).

an easy calculation gives

$$\begin{aligned}\mathscr{H}(\eta^R(F), F) &= \tfrac{1}{2} \frac{d^2}{dF^2} \mathscr{H}(\eta^R(F), F)\big|_{F=F^*} (F - F^*)^2 + o(|F - F^*|^2), \\ &= \tfrac{1}{2} \varrho_R \theta^* \left(\frac{-\sigma^*_{FF}}{\sigma^*_\eta}\right) (F - F^*)^2 + o(|F - F^*|^2),\end{aligned} \tag{4.13}$$

for $F$ close enough to $F^*$. Near enough to $(\eta^*, F^*)$ then, the Rayleigh curve lies in $\mathscr{H}^+$ or $\mathscr{H}^-$ according as $\sigma^*_{FF}/\sigma^*_\eta$ is negative or positive, respectively. Further, (4.12) and (4.13) together are now seen to yield

$$-\mathscr{V} \frac{\bar{\theta}_\eta}{\bar{\sigma}_\eta} (I - \varrho_R \mathscr{V}^2)\, \mathscr{H}\big|_{\eta=\eta^R(F)} = -\tfrac{1}{2} \mathscr{V} \theta^* \sigma^*_{FF} (F - F^*)^2 + o(|F - F^*|^2).$$

Thus, in a small enough neighborhood of the contact point $(\eta^*, F^*)$, *the sign of the quantity* $-\mathscr{V}\bar{\theta}_\eta(I - \varrho_R \mathscr{V}^2)\, \mathscr{H}/\bar{\sigma}_\eta$ *on* $\mathscr{R}(\mathscr{V})$ *is completely determined by the sign of* $\sigma^*_{FF}$, *being positive if* $\sigma^*_{FF}$ *is negative, negative if* $\sigma^*_{FF}$ *is positive*. It follows at once from this observation and $(4.7)_2$ that the trajectory $(\hat{\eta}, \hat{F})(\cdot)$ satisfies

$$\begin{aligned}\hat{F}'(\cdot) &\leqq 0 \quad \text{if} \quad \sigma^*_{FF} < 0, \\ \hat{F}'(\cdot) &\geqq 0 \quad \text{if} \quad \sigma^*_{FF} > 0,\end{aligned} \tag{4.14}$$

whenever it is near $(\eta^*, F^*)$. Thus, while we cannot rule out the possibility that $(\hat{\eta}, \hat{F})(\cdot)$ both arrive at and depart from $(\eta^*, F^*)$, we are able to assert that the flow of the trajectory $(\hat{\eta}, \hat{F})(\cdot)$ near $(\eta^*, F^*)$ is to the left if $\sigma^*_{FF} < 0$, to the right if $\sigma^*_{FF} > 0$. This evidently complements and extends the results of Theorems 2 and 4. In the next section we will deal with points of kissing contact between ${}^0\mathscr{H}$ and $\mathscr{R}(\mathscr{V})$ in another way: we will find structural properties of the energy surface which prevent their existence by ensuring that $\varrho_R \mathscr{V}^2 - A^*$ never vanishes.

## 5. A Priori Inequalities

We now turn to some of those special features of steady, structured shock waves that hold under the imposition of certain *a priori* constitutive inequalities. Rather than attempt a complete, general treatment, we select only a few items that bear most directly on the shock relations and Rayleigh trajectory behavior discussed earlier. As we will see, even though the inequalities we explore are relatively weak, their consequences for shock waves are beguilingly strong. It is therefore important to keep in mind the very much different character of the results of the present section and those of Sections 3 and 4: In Sections 3 and 4, we were able to show that much about the shock process followed from the underlying balance and dissipation principles alone. Thus, for example, these underlying physical principles alone are sufficient to ensure that, if the specific heat there is positive, the only back states $(\eta^-, F^-)$ that will ever be observed (*i.e.*, that will ever occur in a physically possible steady, structured shock wave) are those that satisfy $\varrho_R \mathscr{V}^2 \in [I^-, A^-]$, with $\sigma^-_\eta \neq 0$ for those with $\varrho_R \mathscr{V}^2 \in [I^-, A^-)$. Such a result manifestly has a character much different from our finding below that *if* the energy function has a certain structure or growth, *then* $\varrho_R \mathscr{V}^2 < A^-$ follows automatically. While the two results complement one another, the first, growing as it does out of basic physical principles, is clearly more fundamental to the shock process. This point is of all the more moment when it is kept in mind that, unlike the *a priori* inequalities (3.8) and $(3.9)_{1,2,3}$ of Gibbsian thermostatics, no known physical principle secures for us, even locally, either of the inequalities $\bar{\sigma}_\eta < 0$ or $\bar{\sigma}_{FF} < 0$. Even more unsecured, then, are the structural properties of the energy function induced by the global versions of such inequalities‡.

To motivate the weaker hypotheses with which we will replace it, we begin by discussing the classical condition $(3.7)_3$. It asserts that

$$\bar{\sigma}_{FF}(\eta, F) < 0, \tag{3.7$_3$}$$

and, typically, is intended to hold for all $(\eta, F)$ pairs in some (convex) domain of interest. As is well known, it implies that $\bar{\sigma}(\eta, \cdot)$ is a *concave* function and so lies everywhere *below* its tangent lines. Thus, for each $\eta$ and for all $F$ and $\tilde{F}$, $(3.7)_3$ implies that

$$\bar{\sigma}(\eta, F) + (\tilde{F} - F)\,\bar{\sigma}_F(\eta, F) \geqq \bar{\sigma}(\eta, \tilde{F}), \tag{5.1}$$

with equality holding only if $\tilde{F} = F$.

Now recall $(3.1)_2$, *i.e.*, $\sigma = \varrho_R \bar{e}_F$. With this, integrate (5.1) with respect to $\tilde{F}$, first over an interval $[\tilde{F}_1, F]$, and then over an interval $[F, \tilde{F}_2]$. We thus find that $(3.7)_3$ also implies that, for each $\eta$ and for all $\tilde{F}_1$, $F$ and $\tilde{F}_2$, $\tilde{F}_1 \leqq F \leqq \tilde{F}_2$,

$$\begin{aligned} \bar{e}(\eta, \tilde{F}_1) &\geqq \bar{e}(\eta, F) + (\tilde{F}_1 - F)\,\bar{e}_F(\eta, F) + \tfrac{1}{2}(\tilde{F}_1 - F)^2\,\bar{e}_{FF}(\eta, F), \\ \bar{e}(\eta, F) + (\tilde{F}_2 - F)\,\bar{e}_F(\eta, F) &+ \tfrac{1}{2}(\tilde{F}_2 - F)^2\,\bar{e}_{FF}(\eta, F) \geqq \bar{e}(\eta, \tilde{F}_2), \end{aligned} \tag{5.2}$$

with equality in $(5.2)_{1,2}$ only if $\tilde{F}_1 = F$ and $\tilde{F}_2 = F$, respectively.

While $(5.2)_{1,2}$ is not standard, there is no difficulty in interpreting it: Just as (5.1) is the requirement that, along the ray $\{\eta\} \times (0, \infty)$, $\bar{\sigma}(\eta, \cdot)$ lies everywhere below its best fitting straight line at $F$, so $(5.2)_{1,2}$ is the requirement that $\bar{e}(\eta, \cdot)$ lies everywhere *above* its best fitting *parabola* at $F$ on the half-ray $\{\eta\} \times (0, F]$, everywhere *below* it on the half-ray $\{\eta\} \times [F, \infty)$. Thus, along the ray $\{\eta\} \times (0, \infty)$, $(3.7)_3$ forces the energy to grow super-quadratically to the left of each point $(\eta, F)$, sub-quadratically to the right of $(\eta, F)$. To see that this is rather more stringent than we ought to be content with, suppose that a given point $(\eta, F)$ is thermo-

‡ Although seemingly now forgotten, this discomfort with the *ad hoc* character of inequalities like those of (3.7) goes back at least as far as BETHE [12]. Indeed, with regard to the three inequalities used in his work, BETHE remarked that "Analysis shows that none of the three conditions is required by any general thermodynamic or statistical argument because it can be shown that for each one of the three conditions there exist some substances for which the condition is violated at certain temperatures and densities. Therefore we can only ascertain the range of validity of the conditions by investigating a sufficient number of different physical states." BETHE then went on to examine in detail some six specific material models in order to verify the fitness of his three conditions.

statically stable. By (3.8), with $\tilde{\eta}$ taken equal to $\eta$, this implies that

$$\bar{e}(\eta, F) + (\tilde{F} - F)\,\bar{e}_F(\eta, F) \leqq \bar{e}(\eta, \tilde{F})$$

for all $\tilde{F}$, *i.e.*, along the ray $\{\eta\}\times(0,\infty)$, $\bar{e}(\eta,\cdot)$ lies everywhere *above* the straight line that best fits it at $F$. Hence, if it is adopted, $(3.7)_3$ limits us to energy surfaces that, from the perspective of thermostatic stability, must grow faster than necessary on half-rays $\{\eta\}\times(0,F]$ and slower than might be interesting on half-rays $\{\eta\}\times[F,\infty)$.

The above remarks are intended to suggest that the inequality $(3.7)_3$ is not innocuous: many interesting energy surfaces are excluded by its imposition. Here we loosen the squeeze of this restriction a little by considering judiciously applied point analogs of its consequences (5.1) and (5.2). We say that $F$ *is a point of concavity for* $\bar{\sigma}(\eta,\cdot)$ if

$$\bar{\sigma}(\eta, F) + (\tilde{F} - F)\,\bar{\sigma}_F(\eta, F) \geqq \bar{\sigma}(\eta, \tilde{F})\ \forall\ \tilde{F}; \tag{5.3}$$

$F$ is a point of *strict* concavity for $\bar{\sigma}(\eta,\cdot)$, if this inequality is strict for $\tilde{F} \neq F$. We also say that $\bar{e}(\eta,\cdot)$ *grows super-quadratically to the left (right) of F* if, for all $\tilde{F} \leqq (\geqq)\, F$,

$$\bar{e}(\eta, F) + (\tilde{F} - F)\,\bar{e}_F(\eta, F) + \tfrac{1}{2}(\tilde{F} - F)^2\,\bar{e}_{FF}(\eta, F) \leqq \bar{e}(\eta, \tilde{F}), \qquad (5.4)_1$$

and $\bar{e}(\eta,\cdot)$ *grows sub-quadratically to the right (left) of F*, if for all $\tilde{F} \geqq (\leqq)\, F$,

$$\bar{e}(\eta, \tilde{F}) \leqq \bar{e}(\eta, F) + (\tilde{F} - F)\,\bar{e}_F(\eta, F) + \tfrac{1}{2}(\tilde{F} - F)^2\,\bar{e}_{FF}(\eta, F). \qquad (5.4)_2$$

We say $\bar{e}(\eta,\cdot)$ grows *strictly* super-quadratically or sub-quadratically according as $(5.4)_1$ or $(5.4)_2$ is strict unless $\tilde{F} = F$. As we have seen, we have the implication

$$\left\{\begin{array}{l}\textit{F a point of}\\ \textit{concavity for } \bar{\sigma}(\eta,\cdot)\end{array}\right\} \Rightarrow \left\{\begin{array}{l}\bar{e}(\eta,\cdot)\ \textit{grows super-quadratically to the left}\\ \textit{of F, sub-quadratically to the right,}\end{array}\right\}$$

and, indeed, it is not hard to establish the sharper result

$$\left\{\begin{array}{l}\textit{F a point of concavity for}\\ \bar{\sigma}(\eta,\cdot)\ \textit{with (5.3) strict in a}\\ \textit{deleted neighborhood}\\ (F-\delta, F)\cup(F, F+\delta),\ \delta>0\end{array}\right\} \Rightarrow \left\{\begin{array}{l}\bar{e}(\eta,\cdot)\ \textit{grows strictly super-}\\ \textit{quadratically to the left of F,}\\ \textit{strictly sub-quadratically to the}\\ \textit{right of F.}\end{array}\right\}$$

The additional implications

$$\left\{\begin{array}{l}\textit{F a point of concavity}\\ \textit{for } \bar{\sigma}(\eta,\cdot)\end{array}\right\} \Rightarrow \bar{\sigma}_{FF} \leqq 0\ \textit{at}\ (\eta, F),$$

and

$$\left\{\begin{array}{l}\bar{e}(\eta,\cdot)\ \textit{grows either super-}\\ \textit{quadratically to the left of}\\ \textit{F or sub-quadratically to the}\\ \textit{right of F}\end{array}\right\} \Rightarrow \bar{\sigma}_{FF} \leqq 0\ \textit{at}\ (\eta, F),$$

are also not difficult to demonstrate.

Now, instead of $(3.7)_3$ globally, suppose that merely one point $(\eta, F)$ is such that $F$ is a point of concavity for $\bar{\sigma}(\eta, \cdot)$. Since the adiabatic modulus $A$ is just $\bar{\sigma}_F$, this means by (5.3) that

$$\bar{\sigma}(\eta, F) + (\tilde{F} - F)\, A(\eta, F) \geqq \bar{\sigma}(\eta, \tilde{F})$$

for all $\tilde{F}$. It follows that for those deformation gradients $\tilde{F}$ for which $\bar{\sigma}(\cdot, \tilde{F})$ is monotone decreasing* we may write

$$\bar{\sigma}(\eta, F) + (\tilde{F} - F)\, A(\eta, F) \geqq \bar{\sigma}(\tilde{\eta}, \tilde{F}),$$

for any $\tilde{\eta} \geqq \eta$. Moreover, the inequality is strict if either $F$ is a point of strict concavity for $\bar{\sigma}(\eta, \cdot)$ and $\tilde{F} \neq F$, or if $\bar{\sigma}(\cdot, \tilde{F})$ is strictly monotone decreasing and $\tilde{\eta} > \eta$. Now suppose also that $(\eta, F)$ and $(\tilde{\eta}, \tilde{F})$ belong to the same Rayleigh curve, $\mathscr{R}(\mathscr{V})$, based at a point $(\eta^+, F^+)$. Then, since $\bar{\sigma}(\tilde{\eta}, \tilde{F}) = \bar{\sigma}(\eta, F) + \varrho_R \mathscr{V}^2(\tilde{F} - F)$, we see at once that the above inequality yields

$$(A(\eta, F) - \varrho_R \mathscr{V}^2)\,(\tilde{F} - F) \geqq 0,$$

where the inequality is strict if either $F$ is a point of strict concavity for $\bar{\sigma}(\eta, \cdot)$ and $\tilde{F} \neq F$, or if $\bar{\sigma}(\cdot, \tilde{F})$ is strictly monotone decreasing and $\tilde{\eta} > \eta$.

The above observations form the heart the next three propositions. Proposition 1 gives conditions that restrict the relative positions of certain points on a Rayleigh curve according as the speed of sound of a lower entropy point exceeds or is exceeded by the shock speed. Proposition 2 bears on the character of contact points between $\mathscr{R}(\mathscr{V})$ and ${}^0\mathscr{H}$. The result (i) of Proposition 3 is, of course, classcial; the hypotheses we require for it here seem about as light as possible.

**Proposition 1.** *Let* $(\eta_1, F_1)$ *and* $(\eta_2, F_2)$ *be any two points on a Rayleigh curve* $\mathscr{R}(\mathscr{V})$ *that satisfy*

(i) $\eta_2 > \eta_1$ *and* $F_2 \neq F_1$,

(ii) $F_1$ *is a point of (strict) concavity for* $\bar{\sigma}(\eta_1, \cdot)$,

(iii) $\bar{\sigma}(\cdot, F_2)$ *is (strictly) monotone decreasing on* $[\eta_1, \eta_2]$,

*the condition of strictness holding in* (ii) *and/or* (iii). *Then,*

$$A_1 > \varrho_R \mathscr{V}^2 \Leftrightarrow F_2 > F_1, \qquad A_1 < \varrho_R \mathscr{V}^2 \Leftrightarrow F_2 < F_1.$$

**Proof.** Follows at once from the immediately preceding discussion. □

For points $(\eta_1, F_1)$ and $(\eta_2, F_2)$ that answer to its hypotheses, we see by Proposition 1 that, for example, when the adiabatic sound speed at $(\eta_1, F_1)$ is larger (smaller) than the shock speed, those points $(\eta_2, F_2)$ on $\mathscr{R}(\mathscr{V})$ whose entropy exceeds $\eta_1$ are necessarily to the right (left) of $(\eta_1, F_1)$ in the $\eta$-$F$ plane.

With a slight change of emphasis, Proposition 1 gives us

* The reader will note that this is similar to but rather weaker than the global inequality $(3.7)_1$, $\bar{\sigma}_\eta < 0$.

**Proposition 2.** *Let* $(\eta, F)$ *be any point on a Rayleigh curve* $\mathscr{R}(\mathscr{V})$ *such that*

(i) $F$ *is a point of* (*strict*) *concavity for* $\bar{\sigma}(\eta, \cdot)$,

(ii) *there is a point* $(\tilde{\eta}, \tilde{F})$ *on* $\mathscr{R}(\mathscr{V})$ *with* $\tilde{\eta} > \eta$, $\tilde{F} \neq F$, *and such that* $\bar{\sigma}(\cdot, \tilde{F})$ *is* (*strictly*) *monotone decreasing on* $[\tilde{\eta}, \eta]$.

*Further, let the condition of strictness hold in* (i) *and/or* (ii). *Then, it is impossible for* $\varrho_R \mathscr{V}^2 - A(\eta, F)$ *to vanish, and,* a fortiori, *if* $\mathscr{R}(\mathscr{V})$ *touches* ${}^0\mathscr{H}$ *at* $(\eta, F)$, *the two cut at a nonzero angle.*

Our discussion at the end of Section 3 concerning the degenerate behavior necessitated if $\eta^-$ ever equals $\eta^+$, along with the Jouguet-Zemplén condition (2.17), means that Proposition 1 also has the straightforward specialization

**Proposition 3.** *Let* $(\eta^+, F^+)$ *and* $(\eta^-, F^-)$ *be, respectively, the front and back states of a steady, structured shock wave in a thermoelastic material. If* $F^+$ *is a point of concavity for* $\bar{\sigma}(\eta^+, \cdot)$, *and if* $\bar{\sigma}(\cdot, F^-)$ *is monotone decreasing on* $[\eta^+, \eta^-]$, *then*

(i) *shocks with* $\varrho_R \mathscr{V}^2 > A^+$ *are compressive,*

(ii) *shocks with* $\varrho_R \mathscr{V}^2 < A^+$ *are expansive.*

*Moreover, if either* $F^+$ *is a point of strict concavity for* $\bar{\sigma}(\eta^+, \cdot)$ *or* $\bar{\sigma}(\cdot, F^-)$ *is strictly monotone decreasing, then, unless the material response is essentially linearly elastic and the wave is one of uniform entropy,*

(iii) *shocks with* $\varrho_R \mathscr{V}^2 = A^+$ *are impossible.*

**Proof.** Our discussion above has made clear that to establish (i) and (ii) we need only show that $F^- = F^+$ is impossible. If $\eta^- > \eta^+$ this is clear: as we saw in Section 4, the only intersection of ${}^0\mathscr{H}$ and $\mathscr{R}(\mathscr{V})$ on the ray $(-\infty, \infty) \times \{F^+\}$ is at $(\eta^+, F^+)$. On the other hand, as we saw earlier, the condition $\eta^- = \eta^+$ means that there are no steady, structured shocks for which $\varrho_R \mathscr{V}^2$ differs from $A^+$.

To establish (iii), we note that the initial hypotheses on $\bar{\sigma}(\eta^+, \cdot)$ and $\bar{\sigma}(\cdot, F^-)$, the Jouguet-Zemplén condition (2.17), the assumption that $\varrho_R \mathscr{V}^2 = A^+$, and the fact that $(\eta^\pm, F^\pm) \in \mathscr{R}(\mathscr{V})$ allow us to write

$$\begin{aligned} \sigma^+ + \varrho_R \mathscr{V}^2 (F^- - F^+) &\geqq \sigma(\eta^+, F^-), \\ &\geqq \sigma(\eta^-, F^-) = \sigma^+ + \varrho_R \mathscr{V}^2 (F^- - F^+). \end{aligned}$$

Thus, the intermediate inequalities can only be equalities. The first of these would force $F^- = F^+$ if $\bar{\sigma}(\eta^+, \cdot)$ in fact had $F^+$ as a point of strict concavity; it would thus force $\eta^- = \eta^+$ and so lead to the linearly elastic behavior and wave of uniform entropy discussed in Section 3. The second of our inequalities, compelled to be an equality, would again force $\eta^- = \eta^+$ if $\bar{\sigma}(\cdot, F^-)$ were strictly monotone decreasing. □

The light of Proposition 3 not only reveals the classical result (i), it also illuminates two issues that have arisen earlier: The *a priori* inequalities of the hypotheses of Proposition 3 render shocks of the critical strength $\varrho_R \mathscr{V}^2 = A^+$ impossible unless the material is fairly degenerate. Additionally, while they do not rule out ultra slow shocks, the *a priori* inequalities of Proposition 3 do imply that such shocks are necessarily *expansive*. Since $\eta^-$ must be greater than $\eta^+$, ultra slow shocks are seen to now require the rather non-standard (but probably not impossible) presence of a portion of the Hugoniot ${}^0\mathscr{H}$ in the upper right quadrant $(\eta^+, \infty) \times (F^+, \infty)^\dagger$.

For compressive shocks there is a partial converse to (i) of Proposition 3 that sheds additional light on the possibility of these shocks' having the critical strength $\varrho_R \mathscr{V}^2 = A^+$ or $\varrho_R \mathscr{V}^2 = A^-$, whenever the material is assumed to satisfy certain *a priori* constitutive inequalities. Moreover, the key move used in establishing the results we have in mind yields yet one more set of conditions which ensure the impossibility of

$$\varrho_R \mathscr{V}^2 = A^*$$

at select contact points $(\eta^*, F^*)$ of ${}^0\mathscr{H}$ and the Rayleigh curve $\mathscr{R}(\mathscr{V})$. At these contact points our results thus make moot the possibility of a kissing contact between ${}^0\mathscr{H}$ and a Rayleigh curve: such osculatory contact is impossible.

Let $(\eta, F)$ then be a point on the Hugoniot such that $\bar{e}(\eta, \cdot)$ grows sub-quadratically to the right of $F$. If $F^+ \geqq F$, this allows us to write

$$\bar{e}(\eta, F^+) \leqq \bar{e}(\eta, F) + (F^+ - F)\, \bar{e}_F(\eta, F) + \tfrac{1}{2}(F^+ - F)^2\, \bar{e}_{FF}(\eta, F).$$

In addition, if $\eta^+ \leqq \eta$, we may combine the above inequality with $\bar{e}(\eta^+, F^+) \leqq \bar{e}(\eta, F^+)$ (recall that $\theta = \bar{e}_\eta$ is positive) to find

$$e^+ \leqq \bar{e}(\eta, F) + (F^+ - F)\, \bar{e}_F(\eta, F) + \tfrac{1}{2}(F^+ - F)^2\, \bar{e}_{FF}(\eta, F).$$

Equivalently,

$$\varrho_R(e - e^+) + (F^+ - F)\,\sigma + \tfrac{1}{2}(F^+ - F)^2 A \geqq 0,$$

for $(\eta, F)$ on that portion of ${}^0\mathscr{H}$ which is in the quadrant $[\eta^+, \infty) \times (0, F^+]$. Moreover, the inequality clearly is strict if either $\eta > \eta^+$ or $\bar{e}(\eta, \cdot)$ grows *strictly* sub-quadratically to the right of $F$ and $F \neq F^+$.

Now, since $(\eta, F) \in {}^0\mathscr{H}$ means that

$$\varrho_R(e - e^+) = \tfrac{1}{2}(\sigma + \sigma^+)(F - F^+),$$

we see that our last inequality and $F < F^+$ implies the capital inequality

$$\bar{\sigma}(\eta, F) - \sigma^+ + (F^+ - F)\, A(\eta, F) \geqq 0, \tag{5.5}$$

for any point $(\eta, F) \in {}^0\mathscr{H} \cap ([\eta^+, \infty) \times (0, F^+))$ such that $\bar{e}(\eta, \cdot)$ grows sub-quadratically to the right of $F$. Further, the inequality (5.5) is strict if either the growth is strict or $\eta > \eta^+$.

---

$^\dagger$ As we have seen, the hypothesis of Proposition 3 that $F^+$ is a point of concavity for $\bar{\sigma}(\eta^+, \cdot)$ implies that $\sigma^+_{FF} \leqq 0$. In a small enough neighborhood of $(\eta^+, F^+)$, it is well known that the stronger (and, in practice, fairly common) condition $\sigma^+_{FF} < 0$ requires the Hugoniot ${}^0\mathscr{H}$ to lie in the upper left quadrant $(\eta^+, \infty) \times (0, F^+)$ and in the lower right quadrant $(-\infty, \eta^+) \times (F^+, \infty)$.

The left hand side of (5.5) is precisely $2\mathscr{H}_F(\eta, F)$. Whenever it is positive we know that near $(\eta, F)$ we may express the Hugoniot set ${}^0\mathscr{H}$ as the graph of a smooth function $F^R(\cdot)$ defined on $(\eta - \delta, \eta + \delta)$, $\delta > 0$. Moreover, if a Rayleigh curve $\mathscr{R}(\mathscr{V})$ touches ${}^0\mathscr{H}$ at $(\eta, F)$, then $\mathscr{H}_F(\eta, F) = \frac{1}{2}(\varrho_R \mathscr{V}^2 - A)(F - F^+)$, and we see that strict inequality in (5.5) precludes the possibility of $\varrho_R \mathscr{V}^2 = A(\eta, F)$. As a consequence, $\mathscr{R}(\mathscr{V})$ will actually *cut* ${}^0\mathscr{H}$ at $(\eta, F)$.

Specialized to the back state, the line of reasoning that led to (5.5) gives

**Proposition 4.** *Let $(\eta^+, F^+)$ and $(\eta^-, F^-)$ be, respectively, the front and back states of a steady, structured shock wave of shock velocity $\mathscr{V}$ in a thermoelastic material, and assume the shock to be compressive. Then,*

(i) *if $\bar{e}(\eta^+, \cdot)$ grows super-quadratically to the left of $F^+$,*

$$A^+ \leqq \varrho_R \mathscr{V}^2;$$

(ii) *if $\bar{e}(\eta^-, \cdot)$ grows sub-quadratically to the right of $F^-$,*

$$\varrho_R \mathscr{V}^2 \leqq A^-.$$

*Moreover, equality is impossible in both of* (i) *and* (ii) *if either the growth is strict or $\eta^- > \eta^+$.*

**Proof.** To prove (ii) one only need replace $(\eta, F)$ in (5.5) with $(\eta^-, F^-)$ and then note that $\sigma^- - \sigma^+ = \varrho_R \mathscr{V}^2 (F^- - F^+)$. The proof of (i) is similar to that of (ii); it requires the analog to (5.5)

$$\bar{\sigma}(\eta, F) - \sigma^+ + (F^+ - F)\, A^+ \leqq 0,$$

and this is straightforward. □

## 6. Blocked Rayleigh Curves, Bifurcated Hugoniots, and the Curve of Thermal Extremes

All of our results up to now have constituted, in effect, necessary conditions on steady, structured shock waves in thermoelastic materials. Thus, in Sections 3 and 4, we dealt with the consequences necessitated by both thermal dissipation and the thermal structuring of the wave as embodied in the heat conduction inequality $(3.1)_5$ and the energy equation (3.3), respectively. Similarly, Section 5 was essentially a study of the conditions necessitated by certain *a priori* inequalities laid down on the structure of the energy surface. In the present section, our interests are different. Here, we take up three features of thermoelastic materials that can preclude certain intensity structured shocks from existing at all.

In presenting our results, it is convenient to introduce the *Grüneisen coefficient*. This is the dimensionless function $\gamma = \bar{\gamma}(\eta, F)$ given by

$$\gamma \equiv -v\check{\sigma}_e = -\frac{v}{\theta}\bar{\sigma}_\eta = -\frac{F}{\theta}\bar{\theta}_F. \tag{6.1}$$

The equivalence of these three expressions for $\gamma$ is straightforward. We need only remark that the positivity of the temperature was utilized to invert $\bar{e}(\cdot, F)$, which allowed us to write $\eta = \check{\eta}(e, F), \sigma = \check{\sigma}(e, F) \equiv \bar{\sigma}(\check{\eta}(e, F), F)$. We see that $\gamma$ may thus be viewed as measuring either the sensitivity of stress to thermal influences or, alternatively, the sensitivity of temperature to mechanical influences. In terms of $\gamma$, the identity (3.11) may be augmented: at any point where neither $\bar{\theta}_\eta$ nor $\bar{\theta}_\eta \bar{\sigma}_F - \bar{\theta}_F \bar{\sigma}_\eta$ vanishes,

$$\frac{A}{I} = \frac{c_s}{c_d} = 1 + \varphi\gamma\theta, \tag{6.2}$$

where $\varphi$ is the coefficient of thermal expansion introduced in Section 3. It is also not hard to establish the additional identities

$$\varphi = \frac{\varrho_R \gamma c_d}{F^2 I} = \frac{\varrho_R \gamma c_s}{F^2 A},$$

which are sometimes useful.

Concerning the Grüneisen coefficient, it is worth noting that, by virtue of any of the last three identities, $\gamma$ and $\varphi$ are of the same sign at any point where either of the products $c_d\, I$ or $c_s\, A$ is positive. Typically then one expects $\gamma$ to be positive, and for most metals under standard conditions $\gamma$ ranges between about 1 and 3, generally being either slightly less than or slightly more than 2. However, materials in which $\varphi$ (and so $\gamma$) changes sign are not uncommon, perhaps the most familiar being water near its freezing point. Other materials which have a negative coefficient of thermal expansion are compound semiconductors at low temperatures and certain iron-nickel alloys. Moreover, the recent work reported in [26] and [27] indicates that the Grüneisen coefficient for nitrogen changes sign when subjected to intense enough shocks.

A much studied class of materials in shock physics is that of MIE and GRÜNEISEN‡. A Mie-Grüneisen material is a thermoelastic material for which $\bar{\gamma}_\eta \equiv 0$, *i.e.*, $\gamma$ depends only on the deformation gradient $F$. Since

$$c_d^{-1} = (\ln \bar{\theta})_\eta \quad \text{and} \quad \gamma = -F(\ln \bar{\theta})_F,$$

we see that

$$\bar{\gamma}_\eta = -F(\bar{c}_d^{-1})_F \tag{6.3}$$

anywhere $\bar{\theta}_\eta$ is nonzero (so that $c_d = \bar{c}_d(\eta, F)$ is defined). Thus, it follows that a Mie-Grüneisen material may be equivalently defined as one whose specific heat $c_d$ is independent of deformation. *A fortiori*, any thermoelastic material with constant specific heat $c_d$ is of Mie-Grüneisen type. More generally, with $(3.1)_{1,2,3}$ we observe that the condition $\bar{\gamma}_\eta \equiv 0$ is easily integrated to show that a material is of Mie-Grüneisen type if and only if

$$e = \bar{e}(\eta, F) = \alpha(F)\, N(\eta) + \beta(F), \tag{6.4}$$

‡ See, for example, [1, 2].

for arbitrary* functions $\alpha$ and $\beta$ of $F$ and for an arbitrary* function $N$ of $\eta$. In Parts 2 and 3 of our work, where we take up specific material models, Mie-Grüneisen materials will figure prominently; here they serve merely as an elementary class of thermoelastic materials that can exhibit each of three constitutive features whose presence can interfere with the existence of steady, structured shock waves.

For the first two matters we wish to discuss, it is also useful to recall a few facts from the theory of singular points of plane curves. Thus, let $f: R \times R \to R$ be smooth and consider its zero level set

$$^0 f \equiv \{(x, y) \mid f(x, y) = 0\}.$$

A point $(x, y) \in {}^0 f$ is *regular* if $f_x^2 + f_y^2 \neq 0$ there; otherwise it is said to be *singular*. For a singular point $(\tilde{x}, \tilde{y})$ of $^0 f$, let the discriminant

$$f_{xy}^2(\tilde{x}, \tilde{y}) - f_{xx}(\tilde{x}, \tilde{y}) f_{yy}(\tilde{x}, \tilde{y})$$

be denoted by $\tilde{D}$. Then

$$\tilde{D} < 0 \Rightarrow (\tilde{x}, \tilde{y}) \textit{ is an isolated point of } {}^0 f,$$

while

$$\tilde{D} > 0 \Rightarrow \begin{cases} \textit{near enough to } (\tilde{x}, \tilde{y}), \ {}^0 f \textit{ is the} \\ \textit{union of two smooth arcs that} \\ \textit{intersect at a nonzero angle.} \end{cases}$$

Thus, $\tilde{D}$ positive is a sufficient condition for $^0 f$ to *bifurcate* at the singular point $(\tilde{x}, \tilde{y})$. Moreover, in this case, if $f_{yy}(\tilde{x}, \tilde{y}) \neq 0$, then the slopes $\frac{dy}{dx}$ of the two arcs of $^0 f$ through $(\tilde{x}, \tilde{y})$ are given by

$$\frac{dy}{dx} = \frac{-f_{xy}(\tilde{x}, \tilde{y}) \pm \sqrt{\tilde{D}}}{f_{yy}(\tilde{x}, \tilde{y})}.$$

Finally, if $\tilde{D}$ vanishes the structure of $^0 f$ near $(\tilde{x}, \tilde{y})$ can become rather complex; we do not state general results in this case.

## *Blocked Rayleigh Curves*

Suppose that the Grüneisen coefficient, $\gamma = \bar{\gamma}(\eta, F)$, is such that $\bar{\gamma}(\cdot, \tilde{F})$ vanishes on the entire ray $(-\infty, \infty) \times \{\tilde{F}\}$, for some $\tilde{F} \neq F^+$.** Since $\bar{\sigma}_\eta(\eta, \tilde{F})$ then vanishes for every $\eta$, we see that

$$\bar{\sigma}(\cdot, \tilde{F}) \textit{ is a constant, say } \tilde{\sigma}, \textit{ on the ray } (-\infty, \infty) \times \{\tilde{F}\}.$$

---

* Since the temperature $\theta = \bar{e}_\eta$ must be positive, the product $\alpha(F) \frac{dN(\eta)}{d\eta}$ must evidently be positive also. Other than this, the functions $\alpha$, $\beta$, and $N$ may be selected freely.

** Note that this is precisely what happens in any Mie-Grüneisen material whenever the function $\alpha(\cdot)$ of (6.4) is such that $\frac{d\alpha(\tilde{F})}{dF} = 0$ for some $\tilde{F}$.

The number $(\tilde{\sigma} - \sigma^+)/(\tilde{F} - F^+)$ is thus well defined. If it is negative, it is clear that the ray $(-\infty, \infty)\times\{\tilde{F}\}$ forms a barrier which no Rayleigh curve can touch. Consequently, those front states $(\eta^+, F^+)$ and those back states $(\eta^-, F^-)$ on the Hugoniot† baed at $(\eta^+, F^+)$ that are on opposite sides of the ray $(-\infty, \infty)\times\{\tilde{F}\}$ cannot be connected by a steady, structured shock wave. Put another way, steady, structured shocks with front states and back states so separated cannot exist in thermoelastic materials.

If the number $(\tilde{\sigma} - \sigma^+)/(\tilde{F} - F^+)$ is positive, then we may define a wave speed $\tilde{\mathscr{V}}$ through

$$\varrho_R \tilde{\mathscr{V}}^2 = \frac{\sigma - \sigma^+}{\tilde{F} - F^+}$$

and be assured that $\mathscr{R}(\tilde{\mathscr{V}})$ is not empty: it contains at least $(-\infty, \infty)\times\{\tilde{F}\}$. But, although $\mathscr{R}(\tilde{\mathscr{V}})$ itself may contain a branch that pierces $(-\infty, \infty)\times\{\tilde{F}\}$ (see below), it is clear that no other Rayleigh curve $\mathscr{R}(\mathscr{V})$, $\mathscr{V} \neq \tilde{\mathscr{V}}$, can pass through $(-\infty, \infty)\times\{\tilde{F}\}$. This ray again blocks back states on one side of it and front states on the other side of it from ever being connected by a steady, structured shock wave. The only possible exception to this is a structured shock moving at the exact speed $\tilde{\mathscr{V}}$, and this is possible only if $\mathscr{R}(\tilde{\mathscr{V}})$ contains a branch, other than $(-\infty, \infty)\times\{\tilde{F}\}$, that passes through $(-\infty, \infty)\times\{\tilde{F}\}$.

The existence of a second branch of $\mathscr{R}(\tilde{\mathscr{V}})$ which could cut $(-\infty, \infty)\times\{\tilde{F}\}$ requires the existence of singular points of $\mathscr{R}(\tilde{\mathscr{V}})$ on $(-\infty, \infty)\times\{\tilde{F}\}$ where such cutting could take place. Since $\bar{\sigma}_\eta$ already vanishes all along $(-\infty, \infty)\times\{\tilde{F}\}$, our attention is thus directed to solutions $\tilde{\eta}$ of

$$\bar{\sigma}_F(\tilde{\eta}, \tilde{F}) = \varrho_R \tilde{\mathscr{V}}^2.$$

This gives all of the singular points $(\tilde{\eta}, \tilde{F})$ of $\mathscr{R}(\tilde{\mathscr{V}})$ that lie on the ray $(-\infty, \infty)\times\{\tilde{F}\}$. Equivalently, it gives the only possible points where other branches of $\mathscr{R}(\tilde{\mathscr{V}})$ could cut this ray. Moreover, since the discriminant $\bar{\sigma}^2_{\eta F} - \bar{\sigma}_{\eta\eta}\bar{\sigma}_{FF}$ is just $\bar{\sigma}^2_{\eta F}$ on $(-\infty, \infty)\times\{\tilde{F}\}$, any such point $(\tilde{\eta}, \tilde{F})$ will actually be a bifurcation point of $\mathscr{R}(\tilde{\mathscr{V}})$ if merely $\bar{\sigma}_{\eta F}(\tilde{\eta}, \tilde{F}) \neq 0$.

### *Bifurcated Hugoniots*

So far in our analysis we have made the (seemingly universal) tacit assumption that the Hugoniot set ${}^0\mathscr{H}$ is curve-like, *i.e.*, that each of its points is regular. We consider now the situation where some point $(\eta^*, F^*)$ of ${}^0\mathscr{H}$ is singular, so that both $\mathscr{H}^*_F$ and $\mathscr{H}^*_\eta$ vanish. As we shall see, this may happen more easily than might be suspected. What are its implications?

---

† In general, the Hugoniot set ${}^0\mathscr{H}$ is not greatly affected by the vanishing of $\gamma$ on $(-\infty, \infty)\times\{\tilde{F}\}$, *i.e.*, by the constancy of stress on this ray. In particular, unlike Rayleigh curves, ${}^0\mathscr{H}$ can easily pierce the ray $(-\infty, \infty)\times\{\tilde{F}\}$.

Consider first the vanishing of $\mathscr{H}^*_\eta$. By $(4.1)_2$, we see that this cannot occur along the ray $(-\infty, \infty)\times\{F^+\}$. We may therefore write

$$\mathscr{H}^*_\eta = \varrho_R \theta^* - \tfrac{1}{2}\sigma^*_\eta (F^* - F^+),$$
$$= \frac{\varrho_R \theta^*(F^* - F^+)}{2F^*}\left\{\gamma^* - \frac{2F^*}{F^+ - F^*}\right\},$$

where we have used the definition (6.1) of the Grüneisen coefficient. It follows that

$$\mathscr{H}_{,\eta}(\eta^*, F^*) = 0 \quad \text{if and only if} \quad \bar{\gamma}(\eta^*, F^*) = \frac{2F^*}{F^+ - F^*};$$

*i.e.*, $\mathscr{H}^*_\eta$ vanishes if and only if $\bar{\gamma}(\cdot,\cdot)$ intersects the cylinder given by

$$\gamma_{\text{crit}} = \bar{\gamma}_{\text{crit}}(\eta, F) \equiv \frac{2F}{F^+ - F}$$

at $(\eta^*, F^*)$. The projection of $\gamma_{\text{crit}}$ onto the $\gamma$–$F$ plane is illustrated in Figure 7 so that putative Grüneisen functions $\bar{\gamma}(\cdot,\cdot)$ might be easily appraised for such intersections. We note, in particular, that the choice $\gamma$ equal a positive constant, as well as the choice $\gamma = \left(\dfrac{\gamma^+}{F^+}\right) F$ with $\gamma^+ > 2$, always yields an intersection with $\gamma_{\text{crit}}$ on $(0, F^+)$.

When they occur, intersections of $\gamma$ and $\gamma_{\text{crit}}$ have rather severe implications for the structure of ${}^0\mathscr{H}$: If $\mathscr{H}^*_\eta = 0$ but $\mathscr{H}^*_F \neq 0$, then $(4.2)_2$ reveals that the slope $\left|\dfrac{d\eta^H}{dF}\right|$ is $\infty$ at $F^*$, *i.e.*, the Hugoniot has a vertical tangent at $(\eta^*, F^*)$. If, in addition, $\gamma$ is not just grazing $\gamma_{\text{crit}}$ at $(\eta^*, F^*)$, but is, instead, actually piercing

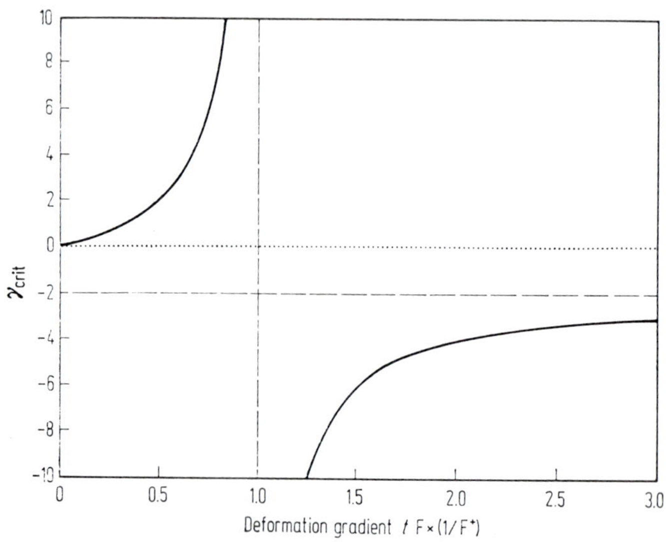


Fig. 7. Projection of the cylinder $\bar{\gamma}_{\text{crit}}(\eta, F) = 2F/(F^+ - F)$ onto the $\gamma$-$F$ plane. Note the vertical asymptote at $F = F^+$ and the horizontal asymptote at $\gamma = -2$.

or cutting $\gamma_{\text{crit}}$ at $(\eta^*, F^*)$ in such a way that $\mathscr{H}_\eta$ changes sign on ${}^0\mathscr{H}$ at $(\eta^*, F^*)$‡, then the derivative $\dfrac{d\eta^H}{dF}$ will itself change sign at $F^*$. Equivalently, *the Hugoniot* ${}^0\mathscr{H}$*, as well as the stress-deformation curve* $\sigma^H(\cdot) \equiv \bar{\sigma}(\eta^H(\cdot), \cdot)$ *along the Hugoniot, bends back on itself*§.

Now, if $\mathscr{H}^*_\eta = 0$ and $\mathscr{H}^*_F = 0$, then not only does $\gamma$ intersect $\gamma_{\text{crit}}$ at $(\eta^*, F^*)$, but the point $(\eta^*, F^*)$ is additionally a singular point of ${}^0\mathscr{H}$. By easy calculation, at such points $(\eta^*, F^*)$

$$\begin{aligned}\mathscr{H}^*_{\eta\eta} &= \frac{\varrho_R\theta^*_\eta(F^* - F^+)}{2F^*}\left\{\gamma^* - \frac{2F^*}{F^+ - F^*}\right\} + \frac{\varrho_R\theta^*(F^* - F^+)}{2F^*}\gamma^*_\eta, \\ &= \frac{\varrho_R\theta^*(F^* - F^+)}{2F^*}\gamma^*_\eta, \\ \mathscr{H}^*_{F\eta} &= -\tfrac{1}{2}\varrho_R\left\{\frac{\theta^*\gamma^*}{F^*} + \theta^*_{FF}\,(F^* - F^+)\right\},\end{aligned}$$

and

$$\mathscr{H}^*_{FF} = -\tfrac{1}{2}\sigma^*_{FF}(F^* - F^+).$$

Thus, at $(\eta^*, F^*)$ the discriminant $D^* = \mathscr{H}^{*2}_{\eta F} - \mathscr{H}^*_{\eta\eta}\mathscr{H}^*_{FF}$ is given by

$$D^* = \tfrac{1}{4}\varrho_R^2\left\{\frac{\theta^*\gamma^*}{F^*} + \theta^*_{FF}(F^* - F^+)\right\}^2 + \tfrac{1}{4}\varrho_R\left\{\frac{\theta^*(F^* - F^+)^2}{F^*}\right\}\sigma^*_{FF}\gamma^*_\eta.$$

We see at once that *either of the conditions*

$$\gamma^*_\eta\sigma^*_{FF} \geqq 0 \quad \textit{and} \quad \mathscr{H}^*_{F\eta} \neq 0,$$

*or*

$$\gamma^*_\eta\sigma^*_{FF} > 0 \quad \textit{and} \quad \mathscr{H}^*_{F\eta} = 0,$$

*suffices for the Hugoniot set* ${}^0\mathscr{H}$ *to bifurcate at singular points* $(\eta^*, F^*)$. A fortiori, *in any Mie-Grüneisen material bifurcation will occur at every singular point of* ${}^0\mathscr{H}$ *where* $\mathscr{H}^*_{F\eta} \neq 0$. Indeed, as we shall shortly see, for Mie-Grüneisen materials much more can be said about the structure of ${}^0\mathscr{H}$ and its possible branches.

In the above case, when bifurcation on ${}^0\mathscr{H}$ occurs because $D^*$ is positive, one further general observation is interesting. Assuming that $\sigma^*_{FF}$ is not zero, we see that the slopes $\dfrac{dF^H}{d\eta}$ of the two smooth branches of the Hugoniot at $(\eta^*, F^*)$

‡ It is not hard to show that a sufficient condition for such a sign change of $\mathscr{H}_\eta$ is $\gamma^*_\eta \neq 0$. This condition, of course, is *not* satisfied by Mie-Grüneisen materials, and, indeed, for them one can show that Hugoniots that bend back on themselves are impossible unless the energy fails to be a monotone function of the entropy, *i.e.*, unless there exists states of positive *and* states of negative temperature.

§ That Hugoniots need not be globally expressible as $\eta = \eta^H(F)$ was noted long ago by Bethe [12]. The existence of Hugoniot curves for real materials, *viz.*, porous metals, was established by Kormer, Funtikov, Urlin, & Kolesnikova [25].

are given by

$$\frac{dF^H}{d\eta} = \frac{-\mathscr{H}^*_{F\eta} \pm \sqrt{D^*}}{\mathscr{H}^*_{FF}}.$$

As a consequence, we easily compute that on each of these branches

$$\frac{d\mathscr{H}_F}{d\eta} = \frac{d\mathscr{H}_F(\eta, F^H(\eta))}{d\eta} = \pm\sqrt{D^*} \quad \text{at} \quad (\eta^*, F^*).$$

Thus, on one of the branches $\mathscr{H}_F$ is locally strictly increasing, on the other locally strictly decreasing. To see the import of this, recall that on ${}^0\mathscr{H}$ a typical point $(\eta^-, F^-)$ satisfies

$$\begin{aligned}\mathscr{H}_F^- &= \tfrac{1}{2}(\sigma^- - \sigma^+ - \sigma_F^-(F^- - F^+)),\\ &= \tfrac{1}{2}(\varrho_R \mathscr{V}^2 - A^-)(F^- - F^+),\end{aligned}$$

where in writing the second line we have introduced the wave speed $\mathscr{V}$ of the Rayleigh curve that intersects ${}^0\mathscr{H}$ at $(\eta^-, F^-)$. Thus, the necessary conditions of Section 3 on the back states of a steady, structured shock wave may be framed in terms of $\mathscr{H}_F^-$: For back states with a positive specific heat $c_d$, compressive (expansive) shocks must have $\mathscr{H}_F^-$ non-negative (non-positive). Equivalently, putative back states on ${}^0\mathscr{H}$ with positive $c_d$ which are compressed (expanded) relative to $(\eta^+, F^+)$ are not consistent with energy dissipation if $\mathscr{H}_F^-$ is negative (positive)—thermodynamically such states are *not accessible* by means of steady, structured shock waves; while they may lie on Rayleigh curves that connect them with $(\eta^+, F^+)$, they will never lie on Rayleigh trajectories.

The strict local monotonicity of $\mathscr{H}_F$ on each of the smooth branches of ${}^0\mathscr{H}$ through the singular point $(\eta^*, F^*)$ now has a clear interpretation: Since $\mathscr{H}_F^* = 0$, one half of each branch locally consists of back states that satisfy the subsonic condition $\varrho_R\mathscr{V}^2 \leqq A^-$ (indeed, satisfy $\varrho_R\mathscr{V}^2 < A^-$ away from $(\eta^*, F^*)$), on the other half of each branch this fails and the back states there are not accessible by means of steady, structured shocks. Additionally, since on one branch $\mathscr{H}_F$ is increasing with $\eta$, while on the other $\mathscr{H}_F$ is decreasing with $\eta$, the two "good" portions of each branch, *i.e.*, the two portions of each branch where the subsonic condition is satisfied, form a continuous, if kinked, portion of ${}^0\mathscr{H}$ along which the entropy varies monotonically. (See Figure 8.) The particular Rayleigh curve, call it $\mathscr{R}(\mathscr{V}^*)$, that reaches the kink point $(\eta^*, F^*)$ will find itself satisfying $\varrho_R\mathscr{V}^{*2} = A^*$; the Rayleigh curves $\mathscr{R}(\mathscr{V})$ to either side of it will be such that $\varrho_R\mathscr{V}^2 < A^-$ where they contact ${}^0\mathscr{H}$ on either of its locally "good" half branches and will be such that $\varrho_R\mathscr{V}^2 > A^-$ where they contact ${}^0\mathscr{H}$ on either of its locally "bad" half branches. Since $I^* \neq \varrho_R\mathscr{V}^{*2}$ (note that $\gamma^* = \gamma^*_{\text{crit}} \Rightarrow \sigma^*_\eta \neq 0$), it is easy to see that near $(\eta^*, F^*)$ all of these contact points, good or bad, will satisfy $\varrho_R\mathscr{V}^2 \neq I^-$ for the Rayleigh curve $\mathscr{R}(\mathscr{V})$ that cuts through them. Thus, if $\theta^*_\eta \neq 0$, Theorem 5 allows us to assert that if the Rayleigh curves near to $\mathscr{R}(\mathscr{V}^*)$ are to correspond to actual steady, structured shock waves with front state $(\eta^+, F^+)$—if they are to correspond to Rayleigh trajectories with $(\hat{\eta}, \hat{F})(\infty) = (\eta^+, F^+)$—then the Hugoniot ${}^0\mathscr{H}$ must be positioned so that its good branches are not screened from $(\eta^+, F^+)$ by its bad branches. Put another way, it must be possible to parameterize each $\mathscr{R}(\mathscr{V})$ smoothly so as to pass continuously from $(\eta^+, F^+)$ to the "good" back state $(\eta^-, F^-)$ without crossing the "bad" portion of ${}^0\mathscr{H}$.

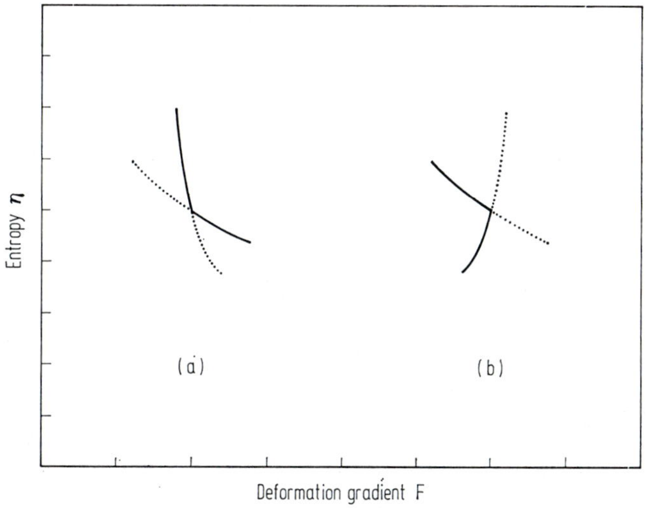


Fig. 8. Smooth branches of the Hugoniot near one of its bifurcation points. The back state condition $\varrho_R \mathscr{V}^2 < A^-$ fails on the dotted section of each branch; it is satisfied by all of the back states on the kinked union formed by the solid portions of each branch.

As we shall see in Part 2, the above observations have some importance in constitutive modeling. Indeed, it is quite possible to combine the Rankine-Hugoniot conditions (2.14)–(2.16) with certain experimental data to produce the partial Hugoniot data $e^H = e^H(F) (\equiv \bar{e}(\eta^H(F), F))$ and $\sigma^H = \sigma^H(F) (\equiv \bar{\sigma}(\eta^H(F), F))$. Then, by an injudicious choice of a response function (*e.g.*, $\bar{\gamma}(\cdot, \cdot)$), one may easily (and unknowingly) obtain a material model whose Hugoniot ${}^0\mathscr{H}$ is bifurcated and has whole segments of the branch corresponding to the data $e^H(\cdot)$ and $\sigma^H(\cdot)$ on which the subsonic condition $\varrho_R \mathscr{V}^2 \leqq A^-$ is violated. Of course, whether real materials with bifurcated Hugoniots actually exist is an open, experimental question.

Inasmuch as

$$\mathscr{H}_\eta(\eta, F) = \frac{\varrho_R \bar{\theta}(\eta, F)\,(F - F^+)}{2F} \left\{ \bar{\gamma}(\eta, F) - \frac{2F}{F^+ - F} \right\}$$

for $F \neq F^+$, we see straightway that for Mie-Grüneisen materials $\mathscr{H}_\eta(\cdot, \bar{F})$ vanishes for any $\bar{F}$ where $\gamma = \bar{\gamma}(\eta, \bar{F}) = \bar{\gamma}(\bar{F}) = 2\bar{F}/(F^+ - \bar{F})$, *i.e.*, for any $\bar{F}$ where the Grüneisen function contacts $\gamma_{\text{crit}}$. For such $\bar{F}$, the Hugoniot function is thus constant along the entire ray $(-\infty, \infty) \times \{\bar{F}\}$. Consequently, *every point on this ray is in* ${}^0\mathscr{H}$ *if one point is*. Such rays are thus prime candidates for branches of the Hugoniot set ${}^0\mathscr{H}$. However, while the Hugoniot for certain Mie-Grüneisen materials will have such vertical, straight branches, for others more subtle behavior is possible. For example, a more detailed analysis of the Hugoniot structure implied by (6.4) reveals that, with $\bar{\gamma} = \gamma_{\text{crit}}$ at $\bar{F}$, there are Mie-Grüneisen materials and front states for which (a) $(-\infty, \infty)\times\{\bar{F}\} \nsubseteq {}^0\mathscr{H}$, (b) the set $(-\infty, \infty)\times\{\bar{F}\} \subseteq {}^0\mathscr{H}$ with another bifurcating branch at some $(\bar{\eta}, \bar{F})$, (c) the set $(-\infty, \infty)\times\{\bar{F}\} \subseteq {}^0\mathscr{H}$ but isolated from the rest of ${}^0\mathscr{H}$, and (d) $(-\infty, \infty)\times\{\bar{F}\} \subseteq {}^0\mathscr{H}$ but with "bifurcation" at $\infty$, *i.e.*, $(-\infty, \infty)\times\{\bar{F}\}$ is a vertical asymptote for the remaining part of ${}^0\mathscr{H}$. Note that the presence of such vertical rays in ${}^0\mathscr{H}$ will, in general, form an uncrossable barrier to Rayleigh trajectories: back states on ${}^0\mathscr{H}$, but on the other side of $(-\infty, \infty)\times\{\bar{F}\}$ than $(\eta^+, F^+)$, cannot be reached by means of steady, structured shocks.

### *The Curve of Thermal Extremes*

The final feature of steady, structured shocks in thermoelastic materials which we discuss here is, in an incomplete way, the best known. Indeed, its discovery by RAYLEIGH [18] in the too special constitutive context of gas dynamics is probably responsible for the widespread (and, as we shall see, false) impression that structured shocks in thermoelastic materials are rather rare and peculiar. The matter is, in fact, much more complicated than the rather simple constitutive equations of gas dynamics would lead one to believe.

To expose the possible difficulty, let us begin by directing our attention to the differential equation (3.3) governing the structure of a shock,

$$\bar{q}(\hat{\eta}, \hat{F}, \hat{\theta}') = \mathscr{V}\mathscr{H}(\hat{\eta}, \hat{F}), \tag{3.3}$$

and to the constitutive equation $\bar{q}(\cdot, \cdot, \cdot)$ governing the flow of heat. Besides the heat conduction inequality $(3.1)_5$, little so far has been demanded of $\bar{q}(\cdot, \cdot, \cdot)$. Clearly this won't do: If, for example, $\bar{q} \equiv 0$ then, while $(3.1)_5$ is satisfied, the resulting form of (3.3) generally will *not* be compatible with the balance of momentum, expressed by the algebraic condition (3.2) defining Rayleigh curves. More generally, even if $\bar{q}(\cdot, \cdot, \cdot)$ vanished only over, say, a curve $\mathscr{Q}$ in the $\eta$–$F$ plane (for example, $\bar{q}(\eta, F, \cdot) = 0$ for $(\eta, F) \in \mathscr{Q}$) the placement of $\mathscr{Q}$ would have to be carefully arranged so as to prevent it from being crossed by Rayleigh curves in their passage from $(\eta^+, F^+)$ to their respective back states $(\eta^-, F^-) \in {}^0\mathscr{H}$. Indeed, no Rayleigh trajectory could cross $\mathscr{Q}$ before it reached ${}^0\mathscr{H}$ and also be consistent with (3.3).

To set all such problems as just described aside, we henceforth will assume that, in addition to the heat conduction inequality $(3.1)_5$, the heat flux function $\bar{q}(\cdot, \cdot, \cdot)$ is *strictly conductive* in the sense that

$$\bar{q}(\eta, F, G) = 0 \quad \Leftrightarrow \quad G = 0, \tag{6.5}$$

which is rather more stringent than the necessary condition $(3.1)_6$[†]. However, a key difficulty still remains and derives from a delicate interplay between the needs

† For materials whose heat flux function is strictly conductive a bit more can be said concerning the waves of uniform entropy that are necessitated whenever $\eta^- = \eta^+$. As we saw at the conclusion of Section 3, such waves required $\hat{q}(\xi)$ to vanish everywhere. It follows that, for $\bar{q}(\cdot, \cdot, \cdot)$ strictly conductive, the temperature gradient $\hat{\theta}'(\xi)$ must also vanish everywhere. Thus, the temperature, like the entropy, is necessarily constant

of the differential equation (3.3) and the demands of the algebraic condition (3.2) specifying the Rayleigh curve: Suppose that somewhere along a Rayleigh trajectory the temperature gradient $\hat{\theta}'$ vanished. By (6.5) we see that this would imply that the heat flux $\hat{q} = \bar{q}(\hat{\eta}, \hat{F}, \hat{\theta}')$ at such a point would vanish, and then, by (3.3), $\mathscr{H}(\hat{\eta}, \hat{F})$ would have to vanish also. Thus, the heat conduction inequality and the differential equation (3.3) allow the temperature gradient along a Rayleigh trajectory to vanish only when the trajectory touches $^0\mathscr{H}$. Off the Hugoniot $\hat{\theta}'$ can never vanish. But the Rayleigh trajectory merely traces out the Rayleigh curve, and this last is given by the algebraic relation (3.2) with no reference to either $\bar{q}(\cdot, \cdot, \cdot)$ or $\mathscr{H}(\cdot, \cdot)$. Thus, along the stretch of a Rayleigh curve $\mathscr{R}(\mathscr{V})$ between $(\eta^+, F^+)$ and $(\eta^-, F^-)$, it is quite possible for the temperature $\theta|_{\mathscr{R}(\mathscr{V})}$ to vary in a non-monotone manner. Suppose this occurs, so that a particle experiencing the $\eta$–$F$ pairs on $\mathscr{R}(\mathscr{V})$ would at some places be warming up, at others cooling down, *i.e.*, suppose there is local "overheating" or "underheating" along $\mathscr{R}(\mathscr{V})$. We set aside the possibility that any of the points on the stretch of $\mathscr{R}(\mathscr{V})$ connecting $(\eta^+, F^+)$ to $(\eta^-, F^-)$ are singular. Thus, at every point of $\mathscr{R}(\mathscr{V})$ one or the other of the derivatives $\dfrac{d\bar{\theta}(\eta, F^R(\eta))}{d\eta}$ or $\dfrac{d\bar{\theta}(\eta^R(F), F)}{dF}$ exists. Now, paralleling $(3.12)_{1,2}$, we have

$$\frac{d\bar{\theta}(\eta, F^R(\eta))}{d\eta} = \bar{\theta}_\eta \left(\frac{I - \varrho_R \mathscr{V}^2}{A - \varrho_R \mathscr{V}^2}\right) \quad \textit{wherever } A - \varrho_R \mathscr{V}^2 \textit{ does not vanish},$$

while

$$\frac{d\bar{\theta}(\eta^R(F), F)}{dF} = \frac{-\bar{\theta}_\eta}{\bar{\sigma}_\eta}(I - \varrho_R \mathscr{V}^2) \quad \textit{wherever } \bar{\sigma}_\eta \textit{ does not vanish}.$$

Thus, at any strict local extremum of temperature on $\mathscr{R}(\mathscr{V})$ the quantity

$$\bar{\theta}_\eta(I - \varrho_R \mathscr{V}^2)$$

must vanish and change sign.

Now, by (3.10) we have

$$\bar{\theta}_\eta(I - \varrho_R \mathscr{V}^2) = \bar{\theta}_\eta(A - \varrho_R \mathscr{V}^2) - \frac{1}{\varrho_R}\bar{\sigma}_\eta^2.$$

Thus, there are only two ways for $\bar{\theta}_\eta(I - \varrho_R \mathscr{V}^2)$ to vanish on $\mathscr{R}(\mathscr{V})$: either $\bar{\theta}_\eta$ and $\bar{\sigma}_\eta$ vanish *simultaneously*, or $I - \varrho_R \mathscr{V}^2$ vanishes. We examine only the second of these alternatives here. Indeed, for simplicity, we restrict our attention to Rayleigh curves on which $\bar{\theta}_\eta$ does not vanish at all. Thus, $I$ is always well defined

---

throughout the wave: $\bar{\theta}(\eta^+, F) = \bar{\theta}(\eta^+, F^+)$ for all $F$ in the (nontrivial) interval $\hat{F}((-\infty, \infty))$. Besides our earlier finding of the linearity of $\bar{\sigma}(\eta^+, \cdot)$ on $\hat{F}((-\infty, \infty))$, we thus reach the additional necessary condition that $\bar{\theta}_F(\eta^+, \cdot)$ (equivalently, the Grüneisen coefficient $\bar{\gamma}(\eta^+, \cdot)$) must now vanish over $\hat{F}((-\infty, \infty))$. *A fortiori*, such shocks can be ruled out by assuming the contrary behavior for the Grüneisen coefficient.

and along such $\mathscr{R}(\mathscr{V})$ we are interested in strict local extrema of temperature which occur due to the vanishing and change of sign of $I - \varrho_R \mathscr{V}^2$. The existence of such extrema has rather strong implications for the corresponding Rayleigh trajectory $(\hat{\eta}, \hat{F})(\cdot)$ we might hope to obtain by solving (3.3). Indeed, we have already seen that $\hat{\theta}'$ cannot vanish off ${}^0\mathscr{H}$. From $(3.12)_{1,2}$ it follows now that *at points on* $\mathscr{R}(\mathscr{V}) - {}^0\mathscr{H}$ *where* $I - \varrho_R \mathscr{V}^2$ *vanishes, the differential equation* (3.3) *forces* $\hat{\eta}'$ *and* $\hat{F}'$ *to be infinite*. To see this, note that if $I - \varrho_R \mathscr{V}^2$ vanishes then $A - \varrho_R \mathscr{V}^2 = \bar{\sigma}_\eta^2/(\varrho_R \bar{\theta}_\eta)$. Since we are not allowing singular points of $\mathscr{R}(\mathscr{V})$, this means that *neither* $A - \varrho_R \mathscr{V}^2$ *nor* $\bar{\sigma}_\eta$ vanishes at the point in question. Consequently, by $(3.12)_{1,2}$, the solution $(\hat{\eta}, \hat{F})(\cdot)$ of (3.3) must be such that, if $\hat{I} - \varrho_R \mathscr{V}^2 = I(\hat{\eta}, \hat{F}) - \varrho_R \mathscr{V}^2$ goes to zero as $\xi \to \xi^*$, then $\hat{F}'$ and $\hat{\eta}'$ must tend to infinity at exactly the correct rate to ensure that the limit

$$\hat{\theta}'(\xi^*) = \lim_{\xi \to \xi^*} \hat{\theta}_\eta \left( \frac{\hat{I} - \varrho_R \mathscr{V}^2}{\hat{A} - \varrho_R \mathscr{V}^2} \right) \hat{\eta}' = \lim_{\xi \to \xi^*} \frac{-\hat{\theta}_\eta}{\hat{\sigma}_\eta} (\hat{I} - \varrho_R \mathscr{V}^2) \hat{F}'$$

is finite and satisfies $\bar{q}(\hat{\eta}(\xi^*), \hat{F}(\xi^*), \hat{\theta}'(\xi^*)) = \mathscr{V} \mathscr{K}(\hat{\eta}(\xi^*), \hat{F}(\xi^*))$. Much worse than this infinite steepness of $\hat{\eta}(\cdot)$ and $\hat{F}(\cdot)$ is the fact that, *if* $\hat{I} - \varrho_R \mathscr{V}^2$ *also changes sign at* $\xi^*$, *then so too must* $\hat{\eta}'$ *and* $\hat{F}'$: otherwise, by $(3.12)_{1,2}$, $\hat{\theta}'$ would change sign and this is impossible off of ${}^0\mathscr{H}$. This means that the "solution" of (3.3) now delivers multi-valued $\hat{\eta}(\cdot)$ and $\hat{F}(\cdot)$ (see Figure 9), and, moreover, *unless there is another, compensating change of sign of* $\hat{\eta}'$ *and* $\hat{F}'$, i.e. *another compensating extremum of* $\bar{\theta}|_{\mathscr{R}(\mathscr{V})}$, no "solution" $(\hat{\eta}, \hat{F})(\cdot)$ of (3.3) will exist and be defined on the *entire* interval $(-\infty, \infty)$. There will be no steady, structured shock wave corresponding to the wave speed $\mathscr{V}$. Of course, even if the extrema of $\bar{\theta}|_{\mathscr{R}(\mathscr{V})}$ come in pairs, so that a "solution" of (3.3) over $(-\infty, \infty)$ exists, it is not clear what physical significance should be attached to such multi-valued relations $(\hat{\eta}, \hat{F})(\cdot)$. In any case, there will be no *single-valued* steady, structured shock wave corresponding to the wave speed.

The above considerations make clear that, for a steady, structured shock of speed $\mathscr{V}$ in a thermoelastic material, the sign of $I - \varrho_R \mathscr{V}^2$ cannot be permitted to change on that part of $\mathscr{R}(\mathscr{V})$ connecting $(\eta^+, F^+)$ to $(\eta^-, F^-)$. If $I$ is positive along $\mathscr{R}(\mathscr{V})$—as it will be if the points of $\mathscr{R}(\mathscr{V})$ are infinitesimally superstable—then the isothermal material sound speed $\mathscr{C}_I \equiv \sqrt{I/\varrho_R}$ is well defined on $\mathscr{R}(\mathscr{V})$. Thus, our observations on the sign of $I - \varrho_R \mathscr{V}^2$ imply that, if the wave speed $\mathscr{V}$ in a steady, structured shock ever exceeds $\mathscr{C}_I$ at some point in the wave, then it must meet $\mathscr{V} \geqq \mathscr{C}_I$ everywhere else in the wave. Conversely, if $\mathscr{V}$ is ever below $\mathscr{C}_I$ at some point in the wave, then $\mathscr{V} \leqq \mathscr{C}_I$ at all other places in the wave†. We will soon see that the case of equality in these last two assertions is rather rare. Thus, if the constitutive structure of the body is such that one or the other of the inequalities $\varrho_R \mathscr{V}^2 \leqq I$ or $\varrho_R \mathscr{V}^2 \geqq I$ holds along $\mathscr{R}(\mathscr{V})$ between $(\eta^+, F^+)$ and $(\eta^-, F^-)$, then typically that inequality holds in its strict form.

† If $I$ is negative somewhere on the portion of $\mathscr{R}(\mathscr{V})$ connecting $(\eta^+, F^+)$ to $(\eta^-, F^-)$, then $\mathscr{C}_I$ is not defined there. Clearly, in this case $I \leqq \varrho_R \mathscr{V}^2$ everywhere in the wave.

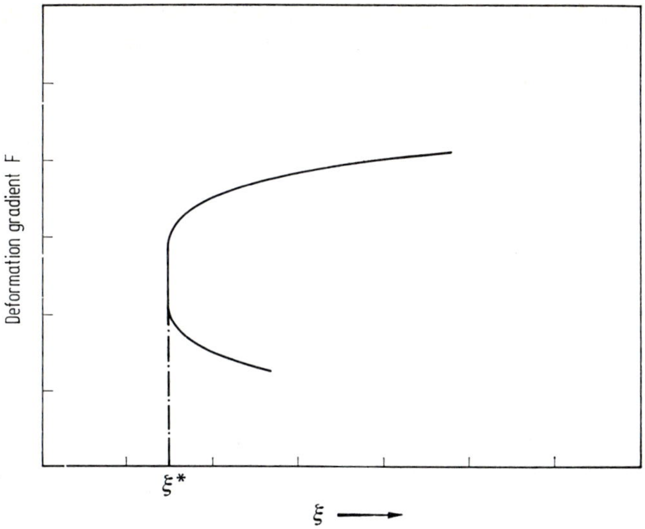


Fig. 9. A portion of the graph of the deformation gradient $\hat{F}(\cdot)$ as determined by the solution of the differential equation (3.3) when $I - \varrho_R \mathscr{V}^2$ changes sign along $\mathscr{R}(\mathscr{V})$.

These simple observations have interesting implications when connected with our results in Section 3. For example, from Theorem 3 we see that, if the front state has a positive specific heat $c_d^+$ and if $A^+ \neq \varrho_R \mathscr{V}^2$, then $\varrho_R \mathscr{V}^2 \notin (I^+, A^+]$. For $\sigma_\eta^+ \neq 0$, thus either $\varrho_R \mathscr{V}^2 \leqq I^+$ and the shock is ultra slow, or $\varrho_R \mathscr{V}^2 > A^+ (> I^+)$ and the shock speed is supersonic with respect to the material in front of it. Our discussion on the sign of $I - \varrho_R \mathscr{V}^2$ now makes clear that ultra slow shocks with $\varrho_R \mathscr{V}^2 < I^+$ must satisfy $\varrho_R \mathscr{V}^2 \leqq I$ throughout their structure. By Theorem 1 this means that, the singular case $\varrho_R \mathscr{V}^2 = I^-$ aside (see below), *ultra slow shocks with $\varrho_R \mathscr{V}^2 < I^+$ and possessing a positive specific heat throughout their structure cannot exist in a thermoelastic material.* On the other hand, for regular shocks the condition $\varrho_R \mathscr{V}^2 > A^+ (> I^+)$ fits nicely with condition (i) of Theorem 1 that, if $\varrho_R \mathscr{V}^2 \neq A^-$, then $I^- \leqq \varrho_R \mathscr{V}^2 < A^-$ whenever $c_d^- > 0$.

In a moment we will geometrize the condition that $\hat{I} - \varrho_R \mathscr{V}^2$ cannot vanish and change sign across the structure of a steady, structured shock wave in a thermoelastic material. However, before we do this it is worth noting that the ideas employed to establish this result have a bit more to tell us. Indeed, since $\hat{\theta}'$ cannot vanish and $\hat{I} - \varrho_R \mathscr{V}^2$ cannot change sign on $\mathscr{R}(\mathscr{V}) - {}^0\mathscr{H}$, we see from $(3.12)_2$ that $\frac{\hat{\theta}_\eta}{\hat{\sigma}_\eta} \hat{F}'$ cannot change sign either. Thus, any changes of sign in the ratio $\frac{\hat{\theta}_\eta}{\hat{\sigma}_\eta}$ along $\mathscr{R}(\mathscr{V})$ can occur only at the price of compensating changes of sign in $\hat{F}'$. On the other hand, if $\frac{\hat{\theta}_\eta}{\hat{\sigma}_\eta}$ does not change sign, then neither can $\hat{F}'$. To illustrate the implications of these simple facts, suppose that a shock is entering material which has a positive specific heat. In this case the supersonic condition $\varrho_R \mathscr{V}^2 > A^+$

implies that $\varrho_R \mathscr{V}^2 > I^+$, *i.e.*, the sign of $I - \varrho_R \mathscr{V}^2$ at $(\eta^+, F^+)$ is negative. This and $(3.12)_2$ with the substitutions $\bar{\theta}_\eta = \theta/c_d$ and $\bar{\sigma}_\eta = -(\varrho_R \gamma \theta)/F$ makes it clear that:

$\left\{\right.$ *Let a steady, structured shock wave satisfy $\varrho_R \mathscr{V}^2 > A^+$. Also, let $\gamma$ and $c_d$ be* **positive** *everywhere along $\mathscr{R}(\mathscr{V})$.* $\left.\right\}$ *Then,* $\left\{\right.$ *the deformation gradient $\hat{F}(\cdot)$ is monotone, and the passage of such a compressive (expansive) shock always heats (cools) the material behind it.* $\left.\right\}$

The above contains as a special case the familiar phenomenon of compressive shock heating. Less common, but by no means unknown‡, is the phenomenon of compressive shock *cooling*. Delineating its occurrence we have:

$\left\{\right.$ *Let a steady, structured shock wave satisfy $\varrho_R \mathscr{V}^2 > A^+$. Also, let $c_d$ be positive and let $\gamma$ be* **negative** *everywhere on $\mathscr{R}(\mathscr{V})$.* $\left.\right\}$ *Then,* $\left\{\right.$ *the deformation gradient $\hat{F}(\cdot)$ is monotone, and the passage of such a compressive (expansive) shock always cools (heats) the material behind it.* $\left.\right\}$

Now we have not considered (or allowed) the situation when $\bar{\theta}_\eta$ vanishes and/or changes sign along $\mathscr{R}(\mathscr{V})$. Suppose, however, that the Grüneisen coefficient $\gamma$ changes sign on $\mathscr{R}(\mathscr{V})$. It is now no longer straightforward to infer from $(3.12)_2$ whether the body warms or cools as a result of the passage of the shock. Instead, the impossibility that $\hat{\theta}'$ and $\hat{I} - \varrho_R \mathscr{V}^2$ change sign and the consequent impossibility of a change of sign in $\dfrac{\hat{\theta}_\eta}{\hat{\sigma}_\eta} \hat{F}'$ now gives us the interesting result that:

$\left\{\right.$ *Let a steady, structured shock wave have a specific heat $c_d$ of fixed sign throughout its structure. Assume, however, that $\gamma$ changes sign on $\mathscr{R}(\mathscr{V})$.* $\left.\right\}$ *Then,* $\left\{\right.$ *the deformation gradient $\hat{F}(\cdot)$ is* **not** *monotone through the shock and is locally* **flat**, i.e., $\hat{F}' = 0$, *at each place where $\gamma$ vanishes.* $\left.\right\}$

The only thing that possibly is not clear here is that $\hat{F}'(\cdot)$ and $\hat{\gamma}(\cdot) = \bar{\gamma}(\hat{\eta}(\cdot), \hat{F}(\cdot))$ must vanish together. Since $\gamma$ vanishes if and only if $\bar{\sigma}_\eta$ vanishes, we see from the regularity of $\mathscr{R}(\mathscr{V})$ that the vanishing of $\gamma$ means that $A - \varrho_R \mathscr{V}^2 \neq 0$. Additionally, by (3.10), the vanishing of $\gamma$ can be seen to imply that $I - \varrho_R \mathscr{V}^2 = A - \varrho_R \mathscr{V}^2$. Therefore, it is evident from $(3.12)_2$ that, as $\gamma$ approaches zero along $\mathscr{R}(\mathscr{V})$, the temperature gradient $\hat{\theta}'$ can stay finite and nonzero only if $\hat{F}'$ also approaches zero. Two points should be noted about this local flatness of $\hat{F}(\cdot)$. First, corresponding to the vanishing of $\gamma$ and $\hat{F}'$, the slope at the associated point on the Rayleigh curve, $\dfrac{d\eta^R}{dF}$, goes to infinity. Second, this local flatness of $\hat{F}(\cdot)$ generally is associated with change in the sign of $\hat{F}'$. Thus, the wave forms for $\hat{F}(\cdot)$ and, by

‡ See, for example, the experimental study [26, 27] on fluid nitrogen.

(2.11), $\hat{\sigma}(\cdot)$ do *not* have the characteristic stepped shape typically associated with multiple wave structures.[§]

A final observation on the consequences of $\gamma$ changing sign along $\mathscr{R}(\mathscr{V})$ is also interesting:

$$\left\{\begin{array}{l}\textit{Additionally, if } I^{\pm} - \varrho_R \mathscr{V}^2 \textit{ does}\\ \textit{not vanish and if } \gamma^- \gamma^+ < 0,\end{array}\right\} \quad \textit{then,} \quad \left\{\begin{array}{l}\textit{besides failing to be monotone,}\\ \hat{F}(\cdot) \textit{ is not everywhere}\\ \textit{between } F^+ \textit{ and } F^- \text{ i.e.,}\\ \textit{there is strain } \textbf{undershoot}\\ \textit{or strain } \textbf{overshoot} \textit{ at either}\\ \textit{the front or the back of the wave.}\end{array}\right\}$$

Now, since $\varrho_R \mathscr{V}^2 = (\bar{\sigma} - \sigma^+)/(F - F^+)$ on $\mathscr{R}(\mathscr{V})$, the requirement that $I - \varrho_R \mathscr{V}^2$ cannot be permitted to vanish and change sign on that part of $\mathscr{R}(\mathscr{V})$ connecting $(\eta^+, F^+)$ to $(\eta^-, F^-)$ is equivalent to the requirement that the quantity $I - (\bar{\sigma} - \sigma^+)/(F - F^+)$ cannot vanish and change sign on the portion of $\mathscr{R}(\mathscr{V})$ connecting $(\eta^+, F^+)$ to $(\eta^-, F^-)$. We are thus led to study the zero level set ${}^0\mathscr{T}$ of the function

$$\mathscr{T} = \mathscr{T}(\eta, F) \equiv \bar{\sigma}(\eta, F) - \sigma^+ - I(\eta, F)\,(F - F^+).$$

For $F \neq F^+$, the set ${}^0\mathscr{T}$ is exactly the locus of points where the quantity $I - (\bar{\sigma} - \sigma^+)/(F - F^+)$ vanishes. Further, if $\theta^+_\eta \neq 0$, it is clear that the point $(\eta^+, F^+)$ is in ${}^0\mathscr{T}$. Moreover, since

$$\mathscr{T}^+_\eta = \sigma^+_\eta \quad \textit{and} \quad \mathscr{T}^+_F = \frac{\sigma^{+2}_\eta}{\varrho_R \theta^+_\eta}, \tag{6.6}$$

the point $(\eta^+, F^+)$ is also a regular point of ${}^0\mathscr{T}$ whenever $\sigma^+_\eta \neq 0$. For simplicity, we shall assume that all the points of ${}^0\mathscr{T}$ are regular. Thus, ${}^0\mathscr{T}$ is everywhere curve-like, and $I - (\bar{\sigma} - \sigma^+)/(F - F^+)$ actually changes sign as ${}^0\mathscr{T}$ is crossed. As we have seen, whenever a Rayleigh curve $\mathscr{R}(\mathscr{V})$ intersects ${}^0\mathscr{T}$, the function $\bar{\theta}|_{\mathscr{R}(\mathscr{V})}$ has a strict, local minimum or maximum. Accordingly, we call ${}^0\mathscr{T}$ the *curve of thermal extremes* based at $(\eta^+, F^+)$, and we seek relatively mild and physically natural conditions on the constitutive functions $\bar{\sigma}(\cdot, \cdot)$ and $I(\cdot, \cdot)$, *i.e.*, on the energy function $\bar{e}(\cdot, \cdot)$, which would allow us to assess and control whether or not ${}^0\mathscr{T}$ is crossed by Rayleigh curves $\mathscr{R}(\mathscr{V})$ in their transit from $(\eta^+, F^+)$ to $(\eta^-, F^-) \in {}^0\mathscr{H}$[†], *i.e.*, to assess and control whether a steady, structured shock of a given speed $\mathscr{V}$ and entering material in the state $(\eta^+, F^+)$ is or is not possible in a given thermoelastic material.

Unfortunately, such conditions seem rather hard to formulate. Even within the class of Mie-Grüneisen materials, the placement and structure of ${}^0\mathscr{T}$ relative to

[§] The exception to this would be the situation where $\gamma$ vanishes somewhere along $\mathscr{R}(\mathscr{V})$ but does not change sign. In this case the shapes of $\hat{F}(\cdot)$ and $\hat{\sigma}(\cdot)$ would indeed have the appearance of a two (or more)-wave structure. However, the components of this structure, all moving at the speed $\mathscr{V}$, will not separate.

[†] Note that in terms of ${}^0\mathscr{T}$ the condition $\varrho_R \mathscr{V}^2 = I^-$ means that ${}^0\mathscr{T}$ and ${}^0\mathscr{H}$ must intersect at $(\eta^-, F^-)$. While this can certainly happen at select points of ${}^0\mathscr{H}$, it will never happen over an entire stretch of back states unless ${}^0\mathscr{T}$ and ${}^0\mathscr{H}$ are coincident on a finite sub-segment.

the Hugoniot ${}^0\mathscr{H}$ and the set of Rayleigh curves is rather complex. In Figures 10–12 we present the Hugoniot and curve of thermal extremes, all based at the same front state $(\eta^+, F^+)$, for three different Mie-Grüneisen materials. Note that, while the entropy has been normalized so that $\eta^+ = 0$, $\theta^+ = 298\,^\circ\mathrm{K}$ and $F^+ = .9952$. Thus, the material at the front state is only a little warmer than room temperature and is slightly compressed relative to the reference configuration where $F = 1$. The Hugoniot ${}^0\mathscr{H}$ is represented by a solid line in our figures, the curve of thermal extremes ${}^0\mathscr{T}$ by a dashed line. As can be seen, the Hugoniots in all three models are nearly the same, and, moreover, the Rayleigh curves for each of the three models have the same character as those depicted in Figures 2 and 3, *viz.*, from $(\eta^+, F^+)$ they sweep upwards and to the left to reach their associated back state $(\eta^-, F^-)$ on ${}^0\mathscr{H}$. Nevertheless, the curves of thermal extremes for the three models are vastly different.

Figure 10 presents us with a curve of thermal extremes ${}^0\mathscr{T}$ that blocks off or screens ${}^0\mathscr{H}$ from almost all of the Rayleigh curves based at $(\eta^+, F^+)$. Indeed, as the inset reveals, only Rayleigh curves with $F^-$ between about .968 and .9952 $(= F^+)$ can reach their associated back state on ${}^0\mathscr{H}$ from $(\eta^+, F^+)$ without first crossing ${}^0\mathscr{T}$. Only extremely mild steady, structured shock waves are possible in this material. The material of Figure 11 is more interesting. For it, the inset again reveals that Rayleigh curves with $F^- \in [.968, .9952)$ can reach ${}^0\mathscr{H}$ from $(\eta^+, F^+)$ without crossing ${}^0\mathscr{T}$. Each of them may again be associated with a (very mild) steady, structured shock wave travelling into $(\eta^+, F^+)$. Then, after an interval of compressions $F^-$ from .968 down to about .745 in which ${}^0\mathscr{T}$ screens ${}^0\mathscr{H}$ from Rayleigh curves and makes the associated shock waves impossible, Figure 11 shows that the ability of the material to support steady, structured shocks returns,

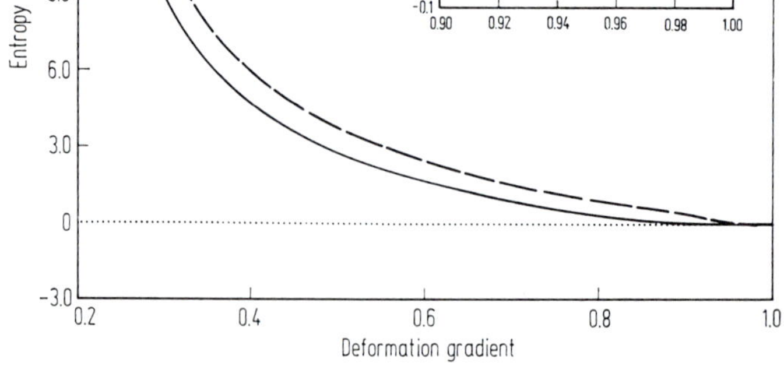


Fig. 10. A material and front state for which only a very small portion, $F^- \in [.968, .9952)$, of the Hugoniot (solid line) is not blocked by the curve of thermal extremes (dashed line) from Rayleigh curves based at $(\eta^+, F^+)$.

*i.e.*, structured shocks with $F^-$ in the range .5233 to .745 exist for this material. Finally, we note that Rayleigh curves with associated back state compressions $F^-$ below .5233 are again blocked from ${}^0\mathscr{H}$ by ${}^0\mathscr{T}$—steady, structured shocks with so compressed back states do not exist.

In Figure 12 we see that ${}^0\mathscr{T}$ can be composed of more than one branch. Further, we see from the inset that now ${}^0\mathscr{T}$, unlike the situation of Figures 10 and 11, fails to rise above ${}^0\mathscr{T}$ anywhere near $F^+$. As a consequence, steady, structured shocks with back state compressions $F^-$ anywhere in the interval (.58, .9952) can be supported by this Mie-Grüneisen material. Structured shocks with $F^- \leqq .58$, where the second branch of ${}^0\mathscr{T}$ cuts ${}^0\mathscr{H}$, cannot be supported.

Figure 13 shows the Hugoniot and the curve of thermal extremes based at $(\eta^+, F^+)$ for one more Mie-Grüneisen material. However, there is a crucial difference between the material of this example and those of Figures 10–12. Here we have selected our constitutive equation for the energy so that $\gamma$ has a constant value of 2.47. As a result, as is clear from Figure 7, $\gamma$ will now intersect and cross the cylinder $\bar{\gamma}_{\text{crit}}(\eta, F)$. Indeed, this happens for $(\eta, F)$ pairs such that

$$\frac{2F}{F^+ - F} = 2.47,$$

*i.e.*, for $F = .55$ and for $\eta$ arbitrary. As we saw in our discussion of bifurcated Hugoniots, it now follows from the Mie-Grüneisen character of the material that $\mathscr{H}(\cdot, \cdot)$ is constant along the entire ray $(-\infty, \infty) \times \{.55\}$. Since this ray is also cut by that part of ${}^0\mathscr{H}$ emanating from $(\eta^+, F^+)$ and curving upwards and to the left, it is itself part of ${}^0\mathscr{H}$. The Hugoniot for this material thus contains a bifur-

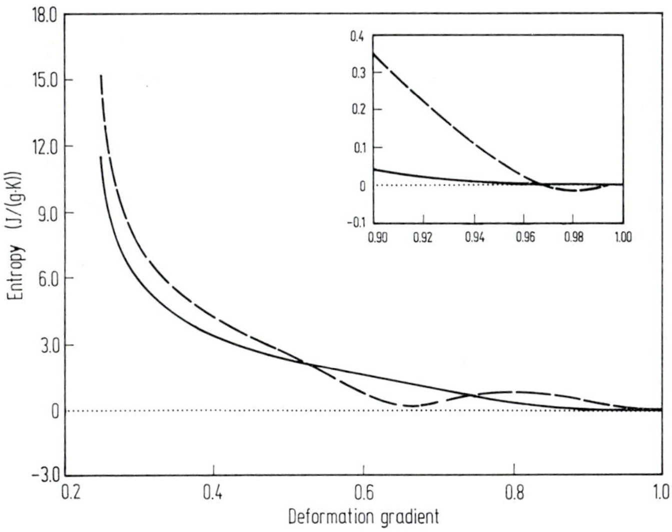


Fig. 11. A material and front state for which there are disjoint spans, $F^- \in [.5233, .745]$ and $F^- \in [.968, .9952)$, of back states $(\eta^-, F^-)$ on the Hugoniot (solid line) which are not screened by the curve of thermal extremes ${}^0\mathscr{T}$ (dashed line) from their associated Rayleigh curves based at $(\eta^+, F^+)$. Consequently, the corresponding steady, structured shock waves will all exist.

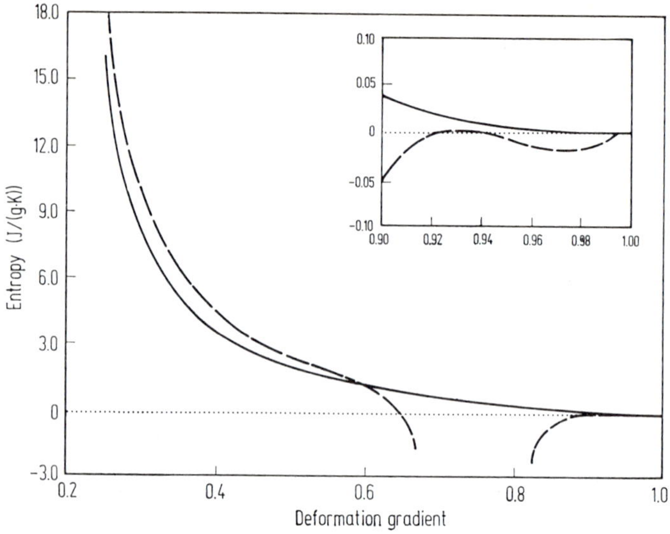


Fig. 12. A material and front state for which the curve of thermal extremes (dashed line) has two branches. The right hand branch never rises high enough to cut the Hugoniot (solid line). Consequently, steady, structured shocks with $F^-$ anywhere in the interval $[.58, .9952)$ can be supported by this material.

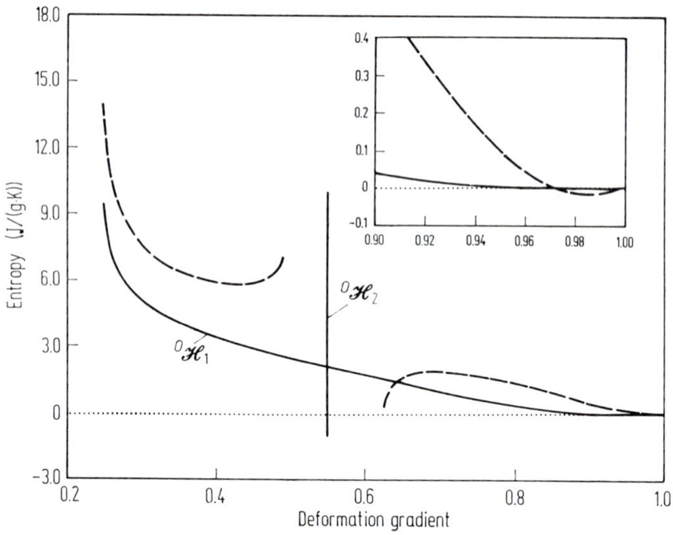


Fig. 13. A material and front state for which the curve of thermal extremes $^0\mathscr{T}$ (dashed line) consists of two branches. Additionally, the Hugoniot $^0\mathscr{H}$ (solid line) contains a bifurcation point at $F^- = .55$. One branch of $^0\mathscr{H}$, call it $^0\mathscr{H}_1$, is much like the Hugoniots of Figures 10–12; the other branch $^0\mathscr{H}_2$ is the vertical ray $(-\infty, \infty)\times\{.55\}$. As discussed earlier, the condition $\varrho_R \mathscr{V}^2 < A^-$ is met only by back states on the right hand portion of $^0\mathscr{H}_1$ and on the upper portion of $^0\mathscr{H}_2$. All of these back states, with the exception of those with $F^- \in [.62, .973]$, may be reached by their associated Rayleigh curves based at $(\eta^+, F^+)$ without being blocked by $^0\mathscr{T}$.

cation point (*cf.* Figure 8), and ${}^0\mathscr{H}$ is here composed of two smooth branches. One branch is very much like the Hugoniots of Figures 10–12; the other is the vertical ray through $F = .55$. The curve of thermal extremes for this material also possesses two branches. However, only the right hand branch affects the existence of steady, structured shock waves: any Rayleigh trajectory, sweeping upwards and to the left from $(\eta^+, F^+)$, will terminate when (if) it comes in contact with the vertical branch of ${}^0\mathscr{H}$ and never have the opportunity to cross the left hand branch of ${}^0\mathscr{T}$. For this material steady, structured shocks for which $F^-$ is in [.973, .9952) or in [.55, .62]‡ all exist. Observe, however, that all those beyond a certain level of shock intensity (measured, say, by $\mathscr{V}$) will have the *same* back state compression $F^- = .55$.

Figures 10–13 make clear that whether or not ${}^0\mathscr{T}$ is crossed by Rayleigh curves in their transit from $(\eta^+, F^+)$ to $(\eta^-, F^-) \in {}^0\mathscr{H}$ is a somewhat subtle affair, even for those special thermoelastic materials that are of Mie-Grüneisen type. While we do not give a complete treatment of this key problem here, one general result is suggested by the insets in Figures 10–13 and is worth mentioning. Thus, suppose that neither $\sigma_\eta^+$ nor $\theta_\eta^+$ vanishes. Then, as we have seen, the front state is a regular point of ${}^0\mathscr{T}$, and we may locally express ${}^0\mathscr{T}$ as the graph of a smooth function $\eta = \eta^T(F)$. In fact, by (6.6),

$$\frac{d\eta^T(F^+)}{dF} = \frac{-\mathscr{T}_F^+}{\mathscr{T}_\eta^+} = \frac{-\sigma_\eta^+}{\varrho_R \theta_\eta^+}.$$

But now $(\eta^+, F^+)$ also is a regular point of every Rayleigh curve based at it, and, by $(3.5)_1$,

$$\frac{d\eta^R(F^+)}{dF} = \frac{\varrho_R \mathscr{V}^2 - A^+}{\sigma_\eta^+}.$$

We see in particular that the slopes $\dfrac{d\eta^T(F^+)}{dF}$ and $\dfrac{d\eta^R(F^+)}{dF}$ have opposite signs, *i.e.*,

$$\frac{d\eta^T(F^+)}{dF}\,\frac{d\eta^R(F^+)}{dF} = \frac{A^+ - \varrho_R \mathscr{V}^2}{\varrho_R \theta_\eta^+} < 0,$$

whenever $\theta_\eta^+ > 0$ and $\varrho_R \mathscr{V}^2 > A^+$. Since the slope $\dfrac{d\eta^H}{dF}$ of the Hugoniot is zero at $(\eta^+, F^+)$, we thus see that the insets in Figures 10–13 are not accidental: *for front states with* $\sigma_\eta^+ \neq 0$ *and* $\theta_\eta^+ > 0$, *the curve of thermal extremes* ${}^0\mathscr{T}$ *and every supersonic Rayleigh curve are initially on opposite sides of* ${}^0\mathscr{H}$. Specifically, if $\theta_\eta^+ > 0$ and $\sigma_\eta^+ < (>)\, 0$, then to the left of $(\eta^+, F^+)$ the curve of thermal extremes ${}^0\mathscr{T}$ is locally below (above) and Rayleigh curves $\mathscr{R}(\mathscr{V})$, $\mathscr{V} > \sqrt{A^+/\varrho_R}$, are locally above (below) the Hugoniot ${}^0\mathscr{H}$; conversely, to the right of $(\eta^+, F^+)$ the curve ${}^0\mathscr{T}$ is locally above (below) and Rayleigh curves $\mathscr{R}(\mathscr{V})$, $\mathscr{V} > \sqrt{A^+/\varrho_R}$, are locally below (above) ${}^0\mathscr{H}$. Consequently, those supersonic Rayleigh curves

‡ While it is not apparent from Figure 13, not all of the Rayleigh curves with $F^-$ in [.55, .64] will avoid crossing ${}^0\mathscr{T}$.

based at $(\eta^+, F^+)$ which stay within a sufficiently small neighborhood of $(\eta^+, F^+)$ during their transit to their associated back state on ${}^0\mathscr{H}$ will complete that transit without crossing ${}^0\mathscr{T}$—mild enough steady, structured shocks which satisfy the supersonic condition $\varrho_R \mathscr{V}^2 > A^+$ and which have a front state with $\sigma_\eta^+ \neq 0$ and $\theta_\eta^+ > 0$ can be supported in any thermoelastic material. That even the simple constitutive equations of gas dynamics admit of materials that can support *arbitrarily intense* structured shocks is the content of our closing remarks.

As we observed in our Introduction, nearly eighty years ago RAYLEIGH presented an argument, since repeated in many places[§], that in certain, simple thermoelastic materials, intended to model gases, heat conduction alone can support steady, structured shock waves only if the ratio $\frac{\sigma^-}{\sigma^+}$ does not exceed $\frac{1+\Re}{3-\Re}$, where $\Re$ is the ratio of specific heats. Since typically $\Re$ is on the order of 1.5 for gases, RAYLEIGH thus showed that only relatively mild shocks, *i.e.*, $\frac{\sigma^-}{\sigma^+} \leqq 1.66$, could be supported by heat conduction alone.

Although he did not present it in these terms, RAYLEIGH's criterion in fact delimits for the materials he studied exactly those Rayleigh curves that are unblocked by ${}^0\mathscr{T}$ in their transit from $(\eta^+, F^+)$ to $(\eta^-, F^-)$. In spite of the extremely narrow constitutive structure within which he worked, RAYLEIGH's result seems to be responsible for a widespread belief that steady, structured shocks in thermoelastic materials are rare and peculiar. As we have seen in our general discussion, however, whether and how much of ${}^0\mathscr{H}$ is blocked by ${}^0\mathscr{T}$ is a rather complex matter. Moreover, as we now show, RAYLEIGH's result is in a major way incomplete, even within his own special constitutive setting.

RAYLEIGH's first constitutive assumption was that his materials obeyed the laws of BOYLE and CHARLES, *i.e.*,

$$\sigma = \tilde{\sigma}(\theta, F) = \frac{-R\theta}{F}, \tag{A}$$

where $R$ is a material constant. The coefficient of thermal expansion is thus given by

$$\varphi = \frac{1}{F} \overset{=}{F}_\theta(\theta, \sigma) = \frac{1}{\theta},$$

and consequently (6.2) tells us that for such materials the ratio of specific heats $\Re$ and the Grüneisen coefficient $\gamma$ are related by

$$\Re = 1 + \gamma. \tag{B}$$

In particular, we may re-write the criterion of RAYLEIGH in the form

$$\frac{\sigma^-}{\sigma^+} \leqq \frac{2+\gamma}{2-\gamma}. \tag{C}$$

By $(3.1)_{2,3}$ we can express the constitutive assumption (A) as $F\bar{e}_F + \frac{R}{\varrho_R}\bar{e}_\eta = 0$. This partial differential equation is easily integrated to reveal that the energy function $\bar{e}(\cdot, \cdot)$ must be of the form

$$e = \bar{e}(\eta, F) = E(\mathrm{e}^{a\eta} F^{-1}),$$

[§] Convenient references are LANDAU & LIFSCHITZ [28] and ZEL'DOVICH & RAIZER [29].

where $a \equiv \dfrac{\varrho_R}{R}$ and the function $E(\cdot)$ is arbitrary. We thus calculate that

$$\theta = \bar{e}_\eta = \frac{\varrho_R}{R} E'(z)\, z,$$

and

$$\gamma = -F \frac{\bar{\theta}_F}{\theta} = \frac{E'(z) + E''(z)z}{E'(z)},$$

where $z = e^{a\eta} F^{-1}$. It follows that the materials analyzed by RAYLEIGH are of Mie-Grüneisen type if and only if $\gamma$ (and so, $\mathfrak{R}$) is a constant. Moreover, in this case it then follows that, to within an arbitrary additive constant, the function $E(\cdot)$ must be of the form $E(z) = E_0 z^\gamma$, for a fixed number $E_0$. We thus find the familiar forms

$$e = \bar{e}(\eta, F) = E_0 F^{-\gamma} e^{a\gamma\eta} = E_0 F^{-\gamma} e^{\eta/c_d}, \tag{D$_1$}$$

$$\theta = \bar{\theta}(\eta, F) = \frac{1}{c_d} E_0 F^{-\gamma} e^{\eta/c_d}, \tag{D$_2$}$$

where $c_d \equiv \dfrac{R}{\gamma \varrho_R} = \dfrac{1}{a\gamma}$ is the specific heat a fixed deformation. The familiar relation

$$e = \tilde{e}(\theta, F) = c_d\, \theta$$

of course follows easily. Additionally, by $(3.1)_2$ and $(D)_1$, we find that the adiabatic modulus may be put in the form

$$A = \bar{\sigma}_F = -(1 + \gamma) \frac{\sigma}{F}, \tag{E$_1$}$$

while, by (A), we see that the isothermal modulus is given by

$$I = \tilde{\sigma}_F = \frac{R\theta}{F^2} = \frac{-\sigma}{F}. \tag{E$_2$}$$

Now RAYLEIGH was interested in modelling gases. Accordingly, for him $\mathfrak{R}$ values near 1.5 (*i.e.*, $\gamma$ values near .5) probably seemed completely general. For solids, however, $\gamma$ values greater than 2 are not uncommon, and, indeed, even negative values of $\gamma$ are possible. Since RAYLEIGH's criterion (C) manifestly is empty for such values of $\gamma$, it is clear, first, that some side condition has been left out in asserting (C) and, second, that as it stands (C) provides no insight whatsoever about the possibility of steady, structured shocks in even the special materials of RAYLEIGH when $\gamma \notin (0, 2)$, *i.e.*, when $\mathfrak{R} \notin (1, 3)$.

Perhaps the entire matter is exposed most clearly by re-establishing RAYLEIGH's result but using the ideas developed in our general discussion. Thus, for the materials here the Hugoniot function may easily be expressed in terms of $\theta$ and $F$, and we have

$$\begin{aligned}\mathscr{H} &= \varrho_R(\tilde{e} - e^+) - \tfrac{1}{2}(\tilde{\sigma} + \sigma^+)(F - F^+),\\ &= \varrho_R c_d (\theta - \theta^+) + \tfrac{1}{2} R \left\{\frac{\theta}{F} + \frac{\theta^+}{F^+}\right\}(F - F^+).\end{aligned}$$

On the Hugoniot therefore

$$\theta = \theta^H(F) = \theta^+ \frac{s(2 + \gamma - \gamma s)}{(2 + \gamma)s - \gamma}, \tag{F}$$

where $s \equiv F/F^+ (> 0)$ is the *compression ratio*. By (A) we then see that

$$\sigma = \sigma^H(F) = \sigma^+ \frac{(2+\gamma-\gamma s)}{(2+\gamma)s-\gamma}, \tag{G}$$

and thus any Rayleigh curve $\mathscr{R}(\mathscr{V})$ which intersects ${}^0\mathscr{H}$ at $F^- \neq F^+$ must be such that its shock speed $\mathscr{V}$ and its back state deformation gradient $F^-$ (back state compression ratio $s^-$) satisfy

$$\varrho_R \mathscr{V}^2 = \frac{\sigma^H(F^-)-\sigma^+}{F^- - F^+} = \frac{-2\sigma^+}{F^+} \frac{(1+\gamma)}{(2+\gamma)s^- - \gamma}. \tag{H}$$

By use of (E) and (G) this last may be put into the form(s)

$$\frac{2}{(2+\gamma)s^- - \gamma} A^+ = \varrho_R \mathscr{V}^2 = \frac{2s^-}{2+\gamma-\gamma s^-} A^-, \tag{I}$$

which relates in a useful way the back state compression ratio $s^-$, the adiabatic moduli $A^+$ and $A^-$, and the shock speed $\mathscr{V}$. Finally, (F) and $(D)_2$ imply that $\eta = \eta^H(F)$ satisfies

$$e^{\frac{(\eta^H - \eta^+)}{c_d}} = h(s) \equiv \frac{s^{\gamma+1}(2+\gamma-\gamma s)}{(2+\gamma)s-\gamma}. \tag{J}$$

Next, by $(E)_2$, the function $\mathscr{T}$ used to define the curve of thermal extremes is even more straightforward to find. We have

$$\begin{aligned} \mathscr{T} &= \tilde{\sigma} - \sigma^+ - I(F - F^+), \\ &= \frac{-R\theta}{F} + \frac{R\theta^+}{F^+} - \frac{R\theta}{F^2}(F - F^+), \\ &= \frac{F^+R}{F^2}[\theta^+ s^2 - \theta(2s-1)]. \end{aligned}$$

Consequently, along the curve of thermal extremes

$$\theta = \theta^T(F) = \theta^+ \frac{s^2}{2s-1}.$$

With $(D)_2$ this last shows that $\eta = \eta^T(F)$ satisfies

$$e^{\frac{(\eta^T - \eta^+)}{c_d}} = \frac{s^{\gamma+2}}{2s-1}. \tag{K}$$

Now the Jouguet-Zemplén condition (2.17) requires us to consider only back states for which $\eta^- = \eta^H(F^-)$ is at least as large as $\eta^+$, or, equivalently, for which

$$e^{\frac{(\eta^H - \eta^+)}{c_d}} \geqq 1.$$

Assume first that $\gamma$ is positive. In this case the function $h(\cdot)$ appearing in (J) is positive only on the the interval $\left(\frac{\gamma}{2+\gamma}, \frac{2+\gamma}{\gamma}\right)$. Thus, $\eta^H(F)$ (and so ${}^0\mathscr{H}$) is defined only for $F \in \left(\frac{\gamma}{2+\gamma} F^+, \frac{2+\gamma}{\gamma} F^+\right)$. Additionally, the function $h(\cdot)$ may now be shown to be monotone decreasing at any point in $(0, \infty)$ where it is defined. Since $h(1) = 1$, we see easily that the only points of ${}^0\mathscr{H}$ which are available as back states of steady, structured shocks are those with $s \in \left(\frac{\gamma}{2+\gamma}, 1\right]$, *i.e.*, those with $F \in \left(\frac{\gamma}{2+\gamma} F^+, F^+\right]$. All possible shocks in the material are thus compressive, and, by (I), those with $s^- \in \left(\frac{\gamma}{2+\gamma}, 1\right)$ satisfy the supersonic-subsonic condition $A^+ < \varrho_R \mathscr{V}^2 < A^-$. Additionally, as $F^- \searrow \frac{\gamma}{2+\gamma} F^+$, the back state entropy, $\eta^- = \eta^H(F^-)$, tends monotonically upwards to infinity.

Next, it is evident from (K) that the curve of thermal extremes ${}^0\mathscr{T}$ is defined only for $s > \frac{1}{2}$, *i.e.*, only for $F > \frac{1}{2} F^+$, with $\eta^T(F)$ tending monotonically to infinity as $F \searrow \frac{1}{2} F^+$. We also know from our general analysis that, since $\gamma$ is positive, near enough and to the left of $F^+$ the curve of thermal extremes ${}^0\mathscr{T}$ is locally *under* ${}^0\mathscr{H}$ while Rayleigh curves $\mathscr{R}(\mathscr{V})$, $\mathscr{V} > \sqrt{A^+/\varrho_R}$, are locally *above* ${}^0\mathscr{H}$. Steady, structured shocks can be supported by the material as long as this persists, *i.e.*, as long as $\eta^T(F) \leqq \eta^H(F)$. From (J) and (K) we find readily that ${}^0\mathscr{T}$ and ${}^0\mathscr{H}$ intersect if and only if

$$s \in (\tfrac{1}{2}, \infty) \cap \left(\frac{\gamma}{2+\gamma}, \frac{2+\gamma}{\gamma}\right) \quad \text{exists such that} \quad \frac{s}{2s-1} = \frac{2+\gamma-\gamma s}{(2+\gamma)\, s-\gamma}.$$

Since $\dfrac{s}{2s-1} = \dfrac{2+\gamma-\gamma s}{(2+\gamma)\, s-\gamma}$ if and only if $s = 1$ or $s = \dfrac{2+\gamma}{2+3\gamma}$, we see that, other than the front state where they are both based, ${}^0\mathscr{T}$ and ${}^0\mathscr{H}$ intersect if and only if

$$\frac{2+\gamma}{2+3\gamma} \in (\tfrac{1}{2}, \infty) \cap \left(\frac{\gamma}{2+\gamma}, \frac{2+\gamma}{\gamma}\right).$$

For $\gamma \geqq 2$ the number $\dfrac{2+\gamma}{2+3\gamma}$ is no larger than $\frac{1}{2}$ while the number $\dfrac{\gamma}{2+\gamma}$ is no smaller than $\frac{1}{2}$. Thus, when $\gamma$ is 2 or larger, the curves ${}^0\mathscr{T}$ and ${}^0\mathscr{H}$ intersect only at their base point $(\eta^+, F^+)$—the curve of thermal extremes ${}^0\mathscr{T}$ is everywhere *below* ${}^0\mathscr{H}$ on the interval $\left(\frac{\gamma}{2+\gamma} F^+, F^+\right)$ and never blocks any Rayleigh curve from reaching its associated back state. Nonsensical on its face for $\gamma \geqq 2$, RAYLEIGH's criterion (C) thus fails even to have an analog here: any of RAYLEIGH's materials with $\gamma \geqq 2$ can support steady, structured shock waves with arbitrarily high back state (compressive) stresses or arbitrarily high shock speeds $\mathscr{V}$.

For $\gamma$ positive but less than 2, the number $\dfrac{2+\gamma}{2+3\gamma} \in (\frac{1}{2}, 1)$ while the number $\dfrac{\gamma}{2+\gamma} \in (0, \frac{1}{2})$. Correspondingly, over the interval $\left(\frac{\gamma}{2+\gamma} F^+, F^+\right)$ the curves ${}^0\mathscr{T}$ and ${}^0\mathscr{H}$ now intersect (and cross) at not only their base point but also at the point $(\eta^*, F^*)$ given by $F^* = \dfrac{2+\gamma}{2+3\gamma} F^+$, $\eta^* = \eta^H(F^*) = \eta^T(F^*)$. Only those Rayleigh curves whose back state $(\eta^-, F^-)$ is such that

$$F^- \in \left[\frac{2+\gamma}{2+3\gamma} F^+, F^+\right),$$

are unblocked by ${}^0\mathscr{T}$ in their transit from $(\eta^+, F^+)$ to $(\eta^-, F^-)$, and only the steady, structured shocks associated with them can be supported by heat conduction alone. This is

RAYLEIGH's result, and, since the right hand side of (G) is monotone increasing¶ in $s$, we have

$$\frac{2+\gamma}{2+3\gamma} \leqq s^- < 1,$$

if and only if

$$\frac{2+\gamma}{2-\gamma}\sigma^+ \leqq \sigma^- < \sigma^+,$$

which bounds the back state stress attainable by steady, structured shock waves and also reproduces exactly RAYLEIGH's criterion (C). Further, since the right hand side of (F) is monotone decreasing in $s$ for $s \in \left(\frac{\gamma}{2+\gamma}, 1\right)$, we have also that $\frac{2+\gamma}{2+3\gamma} \leqq s^- < 1$ if and only if

$$\theta^+ < \theta^- \leqq \theta^+ \frac{(2+\gamma)^2}{(2-\gamma)(2+3\gamma)},$$

which bounds the temperature rise induced by the passage of the shock. For the shock speed $(\mathrm{I})_1$ gives

$$\sqrt{\frac{A^+}{\varrho_R}} < \mathscr{V} \leqq \sqrt{\frac{2+3\gamma}{(2-\gamma)(1+\gamma)}}\sqrt{\frac{A^+}{\varrho_R}}.$$

Consider now the situation when $\gamma$ is either negative or zero. Since $R = \gamma\varrho_R \mathrm{c_d}$, we see that this requires us to allow $R$ to be, respectively, either negative or zero in order for the specific heat $\mathrm{c_d}$ to be positive. However, it readily follows that, when $R$ is negative, the determinant $\bar{e}_{\eta\eta}\bar{e}_{FF} - \bar{e}^2_{F\eta}$ is everywhere negative—no point $(\eta, F)$ of the material is thermodynamically stable. Thus, when $\gamma$ is taken negative, the constitutive form studied by RAYLEIGH can hardly be expected to model the behavior of real materials over appreciable ranges of $\eta$ and $F$. Nevertheless, over certain restricted ranges—for example, inside the spinodal region of a real material—RAYLEIGH's form with negative $\gamma$ may offer an interesting prototype. For this reason and because they illustrate some additional subtleties of Hugoniot and Rayleigh curves, we end with a few remarks on them.

First, if $\gamma = 0$, then (A) shows that $\sigma \equiv 0$. It is therefore easy to see that (*i*) the Hugoniot based at any front state $(\eta^+, F^+)$ is just the isentrope $\{\eta^+\}\times(0, \infty)$, and (*ii*) every Rayleigh curve $\mathscr{R}(\mathscr{V})$, $\mathscr{V} \neq 0$, based at $(\eta^+, F^+)$ is the vertical ray $(-\infty, \infty) \times\{F^+\}$. Accordingly, we leave this degenerate case aside and consider only $\gamma < 0$. Equation (A) now shows that all states of stress in such a material must be *tensions*. Thus $(\mathrm{E})_1$ may now be seen to imply that the adiabatic modulus $A$ is positive at one point if and only if it is positive at every point, which holds if and only if $\gamma < -1$. We thus also leave aside the degenerate cases $\gamma \in [-1, 0)$ and examine only the situation when $\gamma$ lies in the interval $(-\infty, -1)$.

If $\gamma \in (-2, -1)$ the function $h(\cdot)$ appearing in (J) is easily shown to be positive and monotone decreasing for all $s \in (0, \infty)$. In this case then ${}^0\mathscr{H}$ is defined over the entire interval $0 < F < \infty$. However, only for $F \leqq F^+$, *i.e.*, only for $s \leqq 1$, is the criterion $\eta^H(F) \geqq \eta^+$ of JOUGUET and ZEMPLÉN satisfied. Thus, as in the case $\gamma > 0$, all possible shocks are compressive, and, by the use of (I), those with $s^- \in (0, 1)$ easily are seen to satisfy the supersonic-subsonic condition $A^+ < \varrho_R\mathscr{V}^2 < A^-$. For any $\gamma < 0$ the curve of thermal extremes ${}^0\mathscr{T}$ is monotone decreasing on its domain of existence $(\frac{1}{2}F^+, \infty)$. Moreover, for $\gamma \in (-2, -1)$, the putative root where ${}^0\mathscr{T}$ and ${}^0\mathscr{H}$ can inter-

¶ Note that by (A) all stresses are negative in this material.

sect, $s = \dfrac{2+\gamma}{2+3\gamma} = \frac{1}{3}\left(\dfrac{2+\gamma}{2/3+\gamma}\right)$, is negative. On the interval $(\frac{1}{2}F^+, F^+)$ the curve ${}^0\mathcal{T}$ this lies everywhere *above* ${}^0\mathcal{H}$ since, from our general analysis, it lies above ${}^0\mathcal{H}$ near enough and to the left of $F^+$. On the other hand, the Rayleigh curves $\mathcal{R}(\mathcal{V})$, $\mathcal{V} > \sqrt{A^+/\varrho_R}$, lie everywhere *below* ${}^0\mathcal{H}$ in their transit from $(\eta^+, F^+)$ to $(\eta^-, F^-)$. Thus, this material will support steady, structured, compressive shocks of arbitrary intensity $F^- \in (0, F^+)$. Note, however, that the shocks attainable in the material can never move more swiftly than

$$\sqrt{\left(\frac{-2}{\gamma}\right)\frac{A^+}{\varrho_R}},$$

the value of $\mathcal{V}$ given by $(\mathrm{I})_1$ when $s$ is set equal to zero. Additionally, we find easily from (G) that the back state stress must be such that

$$\left(\frac{2+\gamma}{-\gamma}\right)\sigma^+ < \sigma^- < \sigma^+,$$

which, although a bound on $\sigma^-$, is rather different from RAYLEIGH's (C).

If $\gamma \in (-\infty, -2)$ the function $h(\cdot)$ is positive only over the interval $\left(\dfrac{2+\gamma}{\gamma}, \dfrac{\gamma}{2+\gamma}\right)$. It vanishes at $s = \dfrac{2+\gamma}{\gamma}$, tends to infinity at $s = \dfrac{\gamma}{2+\gamma}$, and is monotone increasing everywhere in between. Thus ${}^0\mathcal{H}$ is now defined only over $\left(\dfrac{2+\gamma}{\gamma}F^+, \dfrac{\gamma}{2+\gamma}F^+\right)$, and JOUGUET's and ZEMPLÉN's criterion is satisfied only on $\left[F^+, \dfrac{\gamma}{2+\gamma}F^+\right)$. Thus here the only possible shocks are *expansive*, and again, by (I), they can all be seen to satisfy automatically the supersonic-subsonic condition $A^+ < \varrho_R\mathcal{V}^2 < A^-$ for $F^- \in \left(F^+, \dfrac{\gamma}{2+\gamma}F^+\right)$. Since $\eta^T(\cdot)$ is monotone decreasing for $\gamma < 0$ with $\eta^T(F^+) = \eta^+$, we see that ${}^0\mathcal{T}$ now lies everywhere *below* ${}^0\mathcal{H}$ on the interval $\left(F^+, \dfrac{\gamma}{2+\gamma}F^+\right)$. The Rayleigh curves $\mathcal{R}(\mathcal{V})$, $\mathcal{V} > \sqrt{A^+/\varrho_R}$, however lie everywhere *above* ${}^0\mathcal{H}$ in their transit from $(\eta^+, F^+)$, to $(\eta^-, F^-)$ and, so, are never blocked by ${}^0\mathcal{T}$. These materials will thus support steady, structured, *expansive* shock waves. Lastly we have the case $\gamma = -2$, which is intermediate between the expansive shocks of $\gamma \in (-\infty, -2)$ and the compressive shocks of $\gamma \in (-2, -1)$. As might be expected this case is rather degenerate. Indeed, the Hugoniot ${}^0\mathcal{H}$ is now exactly the isentrope $\{\eta^+\}\times(0, \infty)$ while the only Rayleigh curve $\mathcal{R}(\mathcal{V})$ that intersects ${}^0\mathcal{H}$ at a point other than the front state is that given by $\varrho_R\mathcal{V}^2 = \dfrac{\sigma^+}{F^+}$ $(= A^+)$. This Rayleigh curve is, moreover, congruent with ${}^0\mathcal{H}$!

*Acknowledgment*. Both of us thank our colleague, DENNIS GRADY, for many helpful discussions on shock structure. R. L. F. also gratefully acknowledges support of the National Science Foundation.

## References


1. RICE, M. H., R. G. MCQUEEN, & J. M. WALSH, Compression of Solids by Strong Shock Waves, in: *Solid State Physics*, **6**, 1–63, ed. F. SEITZ & D. TURNBULL, Academic Press, New York and London, 1958.
2. MCQUEEN, R. G., S. P. MARSH, J. W. TAYLOR, J. N. FRITZ, & W. J. CARTER, The Equation of State of Solids from Shock Wave Studies, in: *High-Velocity Impact Phenomena*, 294–417, ed. R. KINSLOW, Academic Press, New York and London, 1970.
3. JOHNSON, J. N., & L. M. BARKER, Dislocation dynamics and steady plastic wave profiles in 6061-T6 aluminum. *Jour. Appl. Phys.* **40**, 4321–4334 (1969).
4. BARKER, L. M., Fine Structure of Compression and Release Wave Shapes in Aluminum Measured by the Velocity Interferometer Technique, in: *Behavior of Dense Media under High Dynamic Pressure*, 483–504, Gordon and Breach, New York, 1968.
5. GRADY, D. E., Strain rate dependence of the effective viscosity under steady wave shock compression. *Appl. Phys. Lett.* **38**, 825–826 (1981).
6. SWEGLE, J. W., & D. E. GRADY, Shock viscosity and the prediction of shock wave risetimes. *J. Appl. Phys.* **58**, 692–701 (1985).
7. DUNN, J. E., & D. E. GRADY, Strain Rate Dependence in Steady Plastic Shock Waves, in: *Shock Waves in Condensed Matter*, 359–364, ed. Y. M. GUPTA, Plenum Press, New York, 1986.
8. DUNN, J. E., The Evolution of Plastic Strain in Steady, Plastic Shock Waves, in: *Shock Waves in Condensed Matter*—1987, 211–214, ed. S. C. SCHMIDT & N. C. HOLMES, North-Holland, Amsterdam, 1987.
9. GILBARG, D., & D. PAOLUCCI, The structure of shock waves in the continuum theory of fluids. *J. Rational Mech. Anal.* **2**, 617–642 (1953).
10. GRAD, H., The profile of a steady plane shock wave. *Comm. Pure Appl. Math.* **5**, 257–300 (1952).
11. BAND, W., Studies in the theory of shock propagation in solids. *J. Geophys. Res.* **65**, 695–719 (1960).
12. BETHE, H. A., The Theory of Shock Waves for an Arbitrary Equation of State. Office of Scientific Research and Development, Report No. 545 (1942).
13. WEYL, H., Shock waves in arbitrary fluids. *Comm. Pure Appl. Math.* **2**, 103–122 (1949).
14. COURANT, R., & K. O. FRIEDRICHS, Supersonic Flow and Shock Waves, Interscience, New York and London, 1948.
15. SERRIN, J., Mathematical Principles of Classical Fluid Dynamics. Flügge's *Handbuch der Physik*, VIII/1, Springer, Berlin-Heidelberg-New York, 1957.
16. DUNN, J. E., & R. FOSDICK, A Dissipation Principle and Its Consequences for Structured Shock Waves in Thermoelastic Materials, in: *Shock Waves in Condensed Matter*—1987, 215–218, ed. S. C. SCHMIDT & N. C. HOLMES, North-Holland, Amsterdam, 1987.
17. HAGAN, R., & J. SERRIN, One-Dimensional Shock Layers in Korteweg Fluids, in: *Phase Transformations and Material Instabilities in Solids*, 113–127, ed. M. E. GURTIN, Academic Press, New York and London, 1984.
18. RAYLEIGH, J., Aerial plane waves of finite amplitude. *Proc. Royal Soc. London* Series A, **84**, 247–284 (1910).
19. RANKINE, W. J. M., On the thermodynamic theory of waves of finite longitudinal disturbance. *Trans. Royal Soc. London* **160**, 277–288 (1870).
20. EARNSHAW, the REV. S., On the mathematical theory of sound. *Trans. Royal Soc. London* **150**, 133–148 (1860).

21. JOUGUET, E., Sur la propagation des discontinuitiés dans les fluides. *C. R. Acad. Sci.* Paris **132**, 673–676 (1901).
22. JOUGUET, E., Remarques sur la propagation des percussions dans les gaz. *C. R. Acad. Sci.* Paris **138**, 1685–1688 (1904).
23. ZEMPLÉN, G., Sur l'impossibilité des ondes choc négative dans les gaz. *C. R. Acad. Sci.* Paris **141**, 710–712 (1905).
24. HADAMARD, J., Remarque sur la note de M. Gyözö Zemplén. *C. R. Acad. Sci.* Paris **141**, 713 (1905).
25. KORMER, S. B., A. I. FUNTIKOV, V. D. URLIN, & A. N. KOLESNIKOVA, Dynamic compression of porous metals and the equation of state with variable specific heat at high temperatures. *Soviet Phys. JETP* **15**, 477–488 (1962).
26. NELLIS, W. J., N. C. HOLMES, A. C. MITCHELL, & M. VAN THIEL, Phase transition in fluid nitrogen at high densities and temperatures. *Phys. Rev. Lett.* **53**, 1661–1664 (1984).
27. RADOUSKY, H. B., W. J. NELLIS, M. ROSS, D. C. HAMILTON, & A. C. MITCHELL, Molecular dissociation and shock-induced cooling in fluid nitrogen at high densities and temperatures. *Phys. Rev. Lett.* **57**, 2419–2422 (1986).
28. LANDAU, L. D., & E. M. LIFSCHITZ, Fluid Mechanics. Pergamon Press, Oxford and New York, 1959.
29. ZEL'DOVICH, Y. B., & Y. P. RAIZER, Physics of Shock Waves and High-Temperature Hydrodynamic Phenomena. Vol. 1 & 2. Academic Press, New York and London, 1967.
30. SERRIN, J., & Y. C. WHANG, On the entropy change through a shock layer. *J. Aerospace Sci.* **28**, No. 12 (1961).

Solid Dynamics Department
Sandia National Laboratories
Albuquerque New Mexico

Department of Aerospace Engineering
and Mechanics
University of Minnesota
Minneapolis

*(Received September 30, 1987; revised, February 18, 1988)*

# *Published Works of James Serrin*

Bibliographic information at the end of an entry is enclosed in brackets when it refers to the full volume cited by the entry.


1. Free boundaries and jets in the theory of cavitation (with DAVID GILBARG). *Journal of Mathematics and Physics* **29** (1950), pp. 1–12.
2. Uniqueness theorems for two free boundary problems. *American Journal of Mathematics* **74** (1952), pp. 492–506.
3. Existence theorems for some hydrodynamical free boundary problems. *Journal of Rational Mechanics and Analysis* **1** (1952), pp. 1–48.
4. Two hydrodynamical comparison theorems. *Journal of Rational Mechanics and Analysis* **1** (1952), pp. 563–572.
5. On plane and axially symmetric free boundary problems. *Journal of Rational Mechanics and Analysis* **2** (1953), pp. 563–575.
6. A note on the wave equation. *Proceedings of the American Mathematical Society* **5** (1954), pp. 307–308.
7. Comparison theorems for subsonic flows. *Journal of Mathematics and Physics* **33** (1954), pp. 27–45.
8. On the Phragmén-Lindelöf theorem for elliptic partial differential equations. *Journal of Rational Mechanics and Analysis* **3** (1954), pp. 395–413.
9. A uniqueness theorem for the parabolic equation $u_t = a(x)\, u_{xx} + b(x)\, u_x + c(x)\, u$. *Bulletin of the American Mathematical Society* **60** (1954), p. 344. Abstract, only published version.
10. Uniqueness of axially symmetric subsonic flow past a finite body (with DAVID GILBARG). *Journal of Rational Mechanics and Analysis* **4** (1955), pp. 169–175.
11. A characterization of regular boundary points for second order linear differential equations. *Bulletin of the American Mathematical Society* **61** (1955), p. 224. [Abstract, only published version.]
12. On the Harnack inequality for linear elliptic equations. *Journal d'Analyse Mathématique* **4** (1956), pp. 292–308.
13. On isolated singularities of solutions of second order linear elliptic equations (with DAVID GILBARG). *Journal d'Analyse Mathématique* **4** (1956), pp. 309–340.

14. A note on harmonic functions defined in a half-plane. *Duke Mathematical Journal* **24** (1956), pp. 523–526.
15. On the Hölder continuity of quasi-conformal and elliptic mappings (with ROBERT FINN). *Transactions of the American Mathematical Society* **89** (1958), pp. 1–15.
16. Mathematical principles of classical fluid mechanics. *Handbuch der Physik*, vol. VIII/1 (1959), pp. 125–263.
16A. Russian Translation: Foreign Literature Publishing House, Moscow 1963 (256 pages).
17. On the stability of viscous fluid motions. *Archive for Rational Mechanics and Analysis* **3** (1959), pp. 1–13.
18. A note on the existence of periodic solutions of the Navier-Stokes equations. *Archive for Rational Mechanics and Analysis* **3** (1959), pp. 120–122.
19. On the uniqueness of compressible fluid motions. *Archive for Rational Mechanics and Analysis* **3** (1959), pp. 271–288.
20. On the derivation of stress-deformation relations for a Stokesian fluid. *Journal of Mathematics and Mechanics* **8** (1959), pp. 459–470.
21. Poiseuille and Couette flow of non-Newtonian fluids. *Zeitschrift für Angewandte Mathematik und Mechanik* **39** (1959), pp. 295–299.
22. On a fundamental theorem of the calculus of variations. *Acta Mathematica* **102** (1959), pp. 1–22.
23. A new definition of the integral for non-parametric problems in the calculus of variations. *Acta Mathematica* **102** (1959), pp. 23–32.
24. The exterior Dirichlet problem for second order elliptic equations (with NORMAN MEYERS). *Journal of Mathematics and Mechanics* **9** (1960), pp. 513–538.
25. On the area of curved surfaces. *American Mathematical Monthly* **68** (1961), pp. 435–440.
26. On the differentiability of functions of several variables. *Archive for Rational Mechanics and Analysis* **7** (1961), pp. 359–372.
27. On the definition and properties of certain variational integrals. *Transactions of the American Mathematical Society* **101** (1961), pp. 139–167.
28. On the entropy change through a shock layer (with Y. C. WHANG). *Journal of the Aerospace Sciences* **28** (1961), pp. 990–991.
29. Dirichlet's principle in the calculus of variations. *Proceedings of Symposia in Pure Mathematics (American Mathematical Society)* **IV**, pp. 17–22.
30. Interior estimates for solutions of the Navier-Stokes equations. *Partial Differential Equations and Continuum Mechanics*, edited by R. LANGER, pp. 376–378. University of Wisconsin Press, Madison 1961. [Conference proceedings.]
31. On the interior regularity of weak solutions of the Navier-Stokes equations. *Archive for Rational Mechanics and Analysis* **9** (1962), pp. 187–195.
32. Strong convergence in a product space. *Proceedings of the American Mathematical Society* **13** (1962), pp. 651–655.
33. The initial value problem for the Navier-Stokes equations. *Nonlinear problems*, edited by R. E. LANGER, pp. 69–98. University of Wisconsin Press, Madison 1963. [Conference proceedings.]
34. Variational problems of minimal surface type, I (with HOWARD JENKINS). *Archive for Rational Mechanics and Analysis* **12** (1963), pp. 185–212. See also entries 47 and 57.
35. Comparison and averaging methods in mathematical physics. *Proprietà di Media e Teoremi di Confronto in Fisica Matematica*, pp. 1–87. Centro Internazionale Matematico Estivo, Rome 1963.
36. A priori estimates for solutions of the minimal surface equation. *Archive for Rational Mechanics and Analysis* **14** (1963), pp. 376–383. See also entry 54.

37. A Harnack inequality for nonlinear equations. *Bulletin of the American Mathematical Society* **69** (1963), pp. 481–486. See also entries 38 and 44.
38. Local behavior of solutions of quasi-linear equations. *Acta Mathematica* **111** (1964), pp. 247–302.
39. Sublinear functions of measures and variational integrals (with CASPAR GOFFMAN). *Duke Mathematical Journal* **31** (1964), pp. 159–178.
40. H = W (with NORMAN MEYERS). *Proceedings of the National Academy of Sciences* (USA) **51** (1964), pp. 1055–1056.
41. Removable singularities of solutions of elliptic equations. *Archive for Rational Mechanics and Analysis* **17** (1964), pp. 67–78. See also entry 45.
42. Pathological solutions of elliptic differential equations. *Annali della Scuola Normale Superiore di Pisa, Classe di Scienze* (Serie II) **18** (1964), pp. 385–387.
43. Singularities of solutions of nonlinear equations. *Proceedings of Symposia in Applied Mathematics (American Mathematical Society)* **XVII** (1965), pp. 68–88.
44. Isolated singularities of solutions of quasi-linear equations. *Acta Mathematica* **113** (1965), pp. 219–240.
45. Removable singularities of solutions of elliptic differential equations, II. *Archive for Rational Mechanics and Analysis* **20** (1965), pp. 163–169.
46. The Dirichlet problem for the minimal surface equation, with infinite data (with HOWARD JENKINS). *Bulletin of the American Mathematical Society* **72** (1966), pp. 102–106. See also entry 57.
47. Variational problems of minimal surface type, II: Boundary value problems for the minimal surface equation (with HOWARD JENKINS). *Archive for Rational Mechanics and Analysis* **21** (1966), pp. 321–342.
48. Isolated singularities of solutions of linear elliptic equations (with H. F. WEINBERGER). *American Journal of Mathematics* **88** (1966), pp. 258–272.
49. Removable singularities of solutions of elliptic equations. *Notices of the American Mathematical Society* **13** (1966), pp. 123. Abstract, only published version.
50. Local behavior of solutions of quasilinear parabolic equations (with D. G. ARONSON). *Archive for Rational Mechanics and Analysis* **25** (1967), pp. 81–122.
51. A maximum principle for nonlinear parabolic equations (with D. G. ARONSON). *Annali della Scuola Normale Superiore di Pisa, Classe di Scienze (Serie II)* **21** (1967), pp. 291–305.
52. The Dirichlet problem for quasilinear elliptic equations with many independent variables. *Proceedings of the National Academy of Sciences (USA)* **58** (1967), pp. 1829–1835. See also entry 62.
53. On the asymptotic behavior of velocity profiles in the Prandtl boundary layer theory. *Proceedings of the Royal Society of London*, **A 299** (1967), pp. 491–507.
54. Addendum to "A priori estimates for solutions of the minimal surface equation". *Archive for Rational Mechanics and Analysis* **28** (1968), pp. 149–154. See also entry 36.
55. The Dirichlet problem for the minimal surface equation in higher dimensions. *Journal für die reine und angewandte Mathematik* **223** (1968), pp. 170–187.
56. On the mathematical basis of Prandtl's boundary layer theory: an example. *Archive for Rational Mechanics and Analysis* **28** (1968), pp. 217–225.
57. Variational problems of minimal surface type, III: The Dirichlet problem with infinite data (with HOWARD JENKINS). *Archive for Rational Mechanics and Analysis* **29** (1968), pp. 304–322.
58. The behavior of similar solutions in a compressible boundary layer (with J. B. MCLEOD). *Journal of Fluid Mechanics* **34** (1968), pp. 337–342.
59. A new proof in differentiation theory. *Notices of the American Mathematical Society* **15** (1968), p. 1036. Abstract; see also RUDIN, W.: *Real and Complex Analysis, Second Edition* (McGraw-Hill, 1974), pp. xii and 435, and pp. 162–167.

60. A general chain rule for derivatives (with D. E. VARBERG). *American Mathematical Monthly* **76** (1969), pp. 514–520.
61. The existence of similar solutions for some laminar boundary layer problems (with J. B. MCLEOD). *Archive for Rational Mechanics and Analysis* **31** (1969), pp. 288–303.
62. The problem of Dirichlet for quasilinear elliptic differential equations with many independent variables. *Philosophical Transactions of the Royal Society of London*, **A 264** (1969), pp. 413–496.
62A. The Dirichlet problem for nonuniformly elliptic partial differential equations. *Notices of the American Mathematical Society* **14**, (1967) p. 712.
62B. On the nonexistence of solutions of Dirichlet's problem. *Notices of the American Mathematical Society* **14** (1967), p. 841.
63. On surfaces of constant mean curvature which span a given space curve. *Mathematische Zeitschrift* **88** (1969), pp. 77–88.
64. Existence theorems for more compressible boundary layer problems. *Studies in Applied Mathematics (SIAM)* **5** (1970), pp. 35–42. [Conference proceedings.]
65. The Dirichlet problem for surfaces of constant mean curvature. *Proceedings of the London Mathematical Society* (3) **21** (1970), pp. 361–384.
66. On the strong maximum principle for nonlinear second order differential inequalities. *Journal of Functional Analysis* **5** (1970), pp. 184–193.
67. Recent developments in the mathematical aspects of boundary layer theory. *International Journal of Engineering Sciences* **9** (1971), pp. 233–240.
68. Curvature inequalities for surfaces over a disk (with H. F. WEINBERGER). *Some Problems of Mathematics and Mechanics*, pp. 242–250. "Nauka" Publishing House, Leningrad, 1970. [M. A. LAVRENTIEFF anniversary volume, Academy of Sciences USSR, in Russian.]
68A. English version: *American Mathematical Society Translations* **104** (1976), pp. 223–231.
69. Uniqueness and comparison theorems for nonlinear elliptic equations in divergence form (with JIM DOUGLAS, Jr., & TODD DUPONT). *Archive for Rational Mechanics and Analysis* **42** (1971), pp. 157–168.
70. Boundary curvatures and the solvability of Dirichlet's problem. *Proceedings of the International Congress of Mathematicians* (Nice **1970**), Volume 2, pp. 867–875. Gauthier-Villars, Paris 1971.
71. A symmetry problem in potential theory. *Archive for Rational Mechanics and Analysis* **43** (1971), pp. 304–318.
72. Gradient estimates for solutions of nonlinear elliptic and parabolic equations. *Contributions to Nonlinear Functional Analysis*, edited by E. ZARANTONELLO, pp. 565–601. Academic Press, New York, London 1971. [Conference proceedings.]
73. The swirling vortex. *Philosophical Transactions of the Royal Society of London*, **A 271** (1972), pp. 325–360.
74. Entire solutions of nonlinear Poisson equations. *Proceedings of the London Mathematical Society* (3) **24** (1972), pp. 348–366.
75. A note on the preceding paper of Amann. *Archive for Rational Mechanics and Analysis* **44** (1972), pp. 182–186.
76. Rectilinear steady flow of simple fluids (with R. L. FOSDICK). *Proceedings of the Royal Society of London*, **A 332** (1973), pp. 311–333.
77. *On the axioms of classical mechanics*. University of Minnesota, Minneapolis 1974 (74 pages, multiplied typescript).
78. Liouville theorems for nonlinear Poisson equations. *Accademia Nazionale dei Lincei, Problemi Attuali di Scienza e di Cultura* **217** (1975), pp. 207–215. [Conference proceedings.]

79. Is "Doomsday" on target? *Science* **189** (1975), pp. 86–87.
80. The solvability of boundary value problems: Hilbert's twentieth problem. *Proceedings of Symposia in Pure Mathematics (American Mathematical Society)* **XXVIII** (1976), pp. 507–524. [Anniversary volume for the Hilbert problems, edited by F. BROWDER.]
81. Global properties of continuum thermodynamic processes (with R. L. FOSDICK). *Archive for Rational Mechanics and Analysis* **59** (1975), pp. 97–109.
82. Liouville theorems and gradient bounds for quasilinear elliptic systems. *Archive for Rational Mechanics and Analysis* **66** (1977), pp. 295–310.
83. Optional shapes for brittle beams under torsion (with H. F. WEINBERGER). *Complex Analysis and its Applications*, pp. 88–91. "Nauka" Publishing House, Moscow 1978. [Anniversary volume in honor of I. Y. VEKUA, Steklov Mathematical Institute, Academy of Sciences USSR.]
84. Gradient bounds and Liouville theorems for quasilinear elliptic equations (with L. A. PELETIER). *Annali della Scuola Normale Superiore di Pisa, Classe di Scienza (Serie* III) **5** (1978), pp. 65–104.
85. The concepts of thermodynamics. *Contemporary Developments in Continuum Mechanics and Partial Differential Equations* (North-Holland Mathematical Studies 30), edited by G. M. DE LA PENHA & L. A. J. MEDEIROS, pp. 411–451. North-Holland, Amsterdam, 1978. [Conference proceedings.]
86. On the impossibility of linear Cauchy and Piola-Kirchhoff constitutive theories for stress in solids (with R. L. FOSDICK). *Journal of Elasticity* **9** (1979), pp. 83–89.
87. Conceptual analysis of the classical second laws of thermodynamics. *Archive for Rational Mechanics and Analysis* **70** (1979), 254–272.
88. Phase transitions and interfacial layers for van der Waals fluids. *Recent Methods in Nonlinear Analysis and Applications*, edited by A. CANFORA, S. RIONERO, C. SBORDONE & G. TROMBETTI, pp. 169–175. Liguori, Naples 1981. [Conference proceedings.]
89. The second law of thermodynamics for systems with approximate cycles (with BERNARD D. COLEMAN & DAVID R. OWEN). *Archive for Rational Mechanics and Analysis* **77** (1981), pp. 103–142.
90. Uniqueness of solutions of semilinear Poisson equations (with KEVIN MCLEOD). *Proceedings of the National Academy of Sciences (USA)* **78** (1981), pp. 6592–6598. See also entry 110.
91. Uniqueness of positive solutions of semilinear equations in $\mathbb{R}^n$ (with L. A. PELETIER). *Archive for Rational Mechanics and Analysis* **81** (1983), pp. 181–197.
92. The form of interfacial surfaces in Korteweg's theory of phase equilibria. *Quarterly of Applied Mathematics* **41** (1983), pp. 357–364.
93. The mechanical theory of fluid interfaces and Maxwell's rule (with ELIAS C. AIFANTIS). *Journal of Colloid and Interface Science* **96** (1983), pp. 519–547.
94. The structure and laws of thermodynamics. *Proceedings of the International Congress of Mathematicians (Warsaw 1983)*, Volume 2, pp. 1717–1728. North-Holland, Amsterdam 1984.
95. One-dimensional shock layers in Korteweg fluids (with R. HAGAN). *Phase Transformations and Material Instabilities in Solids*, edited by M. E. GURTIN, pp. 113–128. Academic Press, Orlando 1984. [Conference proceedings.]
96. Applied mathematics and scientific thought. *Nonlinear Analysis and Optimization (Springer Lecture Notes in Mathematics 1107)*, edited by C. VINTI, pp. 19–27. Springer-Verlag, Berlin, Heidelberg, New York 1984. [Proceedings of anniversary conference in honor of LAMBERTO CESARI.]
97. Extensions of the mountain pass theorem (with PATRIZIA PUCCI). *Journal of Functional Analysis* **59** (1984), pp. 185–210.

98. A mountain pass theorem (with PATRIZIA PUCCI). *Journal of Differential Equations* **60** (1985), pp. 142–149.
99. On the thermomechanics of interstitial working (with J. E. DUNN). *Archive for Rational Mechanics and Analysis* **88** (1985), pp. 95–133.
100. A continuum model for chemical mixture dynamics. *Developments in Mechanics (Proceedings of Midwestern Mechanics Conferences)* **13** (1985), pp. 59–64.
101. Non-existence theorems for quasilinear partial differential equations (with WEI-MING NI). *Supplemento ai Rendiconti del Circolo Matematico di Palermo* (Serie II) **8** (1985), pp. 171–185. [Proceedings of conference celebrating the centenary of the Circolo Matematico di Palermo.]
102. An outline of thermodynamical structure. *New Perspectives in Thermodynamics*, pp. 3–32. Springer-Verlag, Berlin, Heidelberg, New York 1986.
103. Dynamic changes of phase in a van der Waals fluids (with R. HAGAN). *New Perspectives in Thermodynamics*, pp. 241–260. Springer-Verlag, Berlin, Heidelberg, New York 1986.
104. Uniqueness of non-negative solutions of semilinear equations in $\mathbb{R}^n$ (with L. A. PELETIER). *Journal of Differential Equations* **61** (1986), pp. 380–397.
105. Existence and nonexistence theorems for ground states of quasilinear partial differential equations. The anomalous case (with WEI-MING NI). *Accademia Nazionale dei Lincei, Atti dei Convegni Lincei* **77** (1986), pp. 231–257. [Proceedings of conference celebrating the centenary of MAURO PICONE and LEONIDA TONELLI.]
106. Non-existence of singular solutions of quasilinear partial differential equations (with WEI-MING NI). *Communications on Pure and Applied Mathematics* **39** (1986), pp. 379–399.
107. A general variational identity (with PATRIZIA PUCCI). *Indiana University Mathematics Journal* **35** (1986), pp. 681–703.
108. Esistenza e unicità degli stati fondamentali per equazioni ellittiche quasilineari (with BRUNO FRANCHI & ERMANNO LANCONELLI). *Atti della Accademia Nazionale dei Lincei (Serie 8), Rendiconti: Classe di Scienze Fisiche, Matematiche e Naturali* **79** (1986), pp. 121–126. See also entry 116.
109. The structure of the critical set in the mountain pass theorem (with PATRIZIA PUCCI). *Transactions of the American Mathematical Society* **299** (1987), pp. 115–132.
110. Uniqueness of positive radial solutions of $\Delta u + f(u) = 0$ in $\mathbb{R}^n$ (with KEVIN MCLEOD). *Archive for Rational Mechanics and Analysis* **99** (1987), pp. 115–145.
111. Ground states for a prescribed mean curvature equation (with L. A. PELETIER). *Proceedings of the American Mathematical Society* **100** (1987), pp. 694–700.
112. The laws of thermodynamics. *Res Mechanica* **21** (1987), pp. 361–379.
113. Some strange cycles in thermodynamics. *Science and Pseudo-Science*, edited by A. LIBERMAN & R. ARIS, pp. 3–27. University of Minnesota, 1987.
114. Positive solutions of a prescribed mean curvature equation. *Calculus of Variations and Partial Differential Equations* (Springer Lecture Notes in Mathematics 1340), edited by S. HILDEBRANDT, D. KINDERLEHRER & M. MIRANDA, pp. 245–255. Springer-Verlag 1988. [Proceedings of conference in honor of HANS LEWY.]
115. Ground states for the prescribed mean curvature equation: The supercritical case (with F. V. ATKINSON & L. A. PELETIER). *Nonlinear Diffusion Equations and their Equilibrium States*, edited by J. SERRIN, W.-M. NI & L. A. PELETIER, pp. 51–74. MSRI Publications, vol. 12, Springer-Verlag 1988.
116. Existence and uniqueness of ground states for quasilinear elliptic equations (with B. FRANCHI & E. LANCONELLI). *Nonlinear Diffusion Equations and their Equilibrium States*, edited by J. SERRIN, W.-M. NI & L. A. PELETIER, pp. 293–300. MSRI Publications, vol. *12*, Springer-Verlag 1988.

117. Remarks on the first eigenspace for the polyharmonic operator in a ball (with PATRIZIA PUCCI). Atti del Seminario Matematico dell'University di Modena **36** (1988), pp. 106–117.

## In Press

118. Critical exponents and critical dimensions for the polyharmonic operator (with PATRIZIA PUCCI). Journal de Mathématiques Pures et Appliquées.
119. Asymptotic properties for solutions of strongly nonlinear ordinary differential equations (with PATRIZIA PUCCI). *Proceedings of Conference on Differential Equations and Geometry, Torino.*

## Lecture Notes

L1. *Mathematical Aspects of Boundary Layer Theory*, notes taken by H. K. WILSON. Department of Mathematics, University of Minnesota, Minneapolis 1963 (131 pages, multiplied typescript).
L2. *Theory of Differentiation*, notes taken by T. HATCHER. Department of Mathematics, University of Minnesota, Minneapolis 1965 (135 pages, multiplied typescript).
L3. *Nonlinear Elliptic Equations of Second Order*. Lectures at Symposium on Partial Differential Equations, Berkeley 1971 (57 pages, multiplied typescript).
L4. *Foundations of Classical Thermodynamics*. Department of Mathematics, University of Chicago 1975 (223 pages, multiplied typescript).
L5. *Foundations of Thermodynamics*. University of Naples 1979 (138 pages, multiplied typescript).

## Edited Volumes

*New Perspectives in Thermodynamics*. Springer-Verlag, Berlin, Heidelberg, New York 1986 (xvi + 260 pages).
*Nonlinear Diffusion Equations and their Equilibrium States*, I, II. MSRI Publications 12, 13. Springer-Verlag, New York, Berlin, Heidelberg, 1988 (xviii + 359 + 365 pages). [Co-edited with W.-M. NI & L. A. PELETIER.]

# *Places of First Publication*

of the articles reprinted above, all in
*Archive for Rational Mechanics and Analysis*


AMICK, C. J., & FRAENKEL, L. E., The Uniqueness of Hill's Spherical Vortex **92**, 91–119 (1986) . . . . . . . . . . . . 43

ARONSON, D. G., & VAZQUEZ, J. L., Eventual $C^\infty$-Regularity and Concavity for Flows in One-Dimensional Porous Media **99**, 329–348 (1987) . . . . . . . . . . . . 627

ATKINSON, F. V., & PELETIER, L. A., Ground States and Dirichlet Problems for $-\Delta u = f(u)$ in $R^2$ **96**, 147–165 (1986) . . . . . . . . . . . . 409

BAKELMAN, I. J., The Boundary Value Problems for Non-Linear Elliptic Equations and the Maximum Principle for Euler-Lagrange Equations **93**, 271–300 (1986) . . . . . . . . . . . . 181

BALL, J. M., & JAMES, R. D., Fine Phase Mixtures as Minimizers of Energy **100**, 13–52 (1987) . . . . . . . . . . . . 647

BALL, J. M., & KNOWLES, G., Lyapunov Functions for Thermomechanics with Spatially Varying Boundary Temperatures **92**, 193–204 (1986) . . . . . . . . . . . . 73

BREZIS, H., & OSWALD, L., Singular Solutions for Some Semilinear Elliptic Equations **99**, 249–259 (1987) . . . . . . . . . . . . 615

CAFFARELLI, L. A., & FRIEDMAN, A., Differentiability of the Blow-up Curve for One Dimensional Nonlinear Wave Equations **91**, 83–98 (1985) . . . . . . . . . . . . 1

CAPRIZ, G., & GIOVINE, P., On Effects of Virtual Inertia During Diffusion of a Dispersed Medium in a Suspension **98**, 115–122 (1987) . . . . . . . . . . . . 491

CESARI, L., BRANDI, P., & SALVADORI, A., Existence Theorems Concerning Simple Integrals of the Calculus of Variations for Discontinuous Solutions **98**, 307–328 (1987) . . . . . . . . . . . . 543

CHIPOT, M., KINDERLEHRER, D., & VERGARA CAFFARELLI, G., Smoothness of Linear Laminates **96**, 81–96 (1986). . . . . . . . . . . . . 383

DAFERMOS, C. M., Quasilinear Hyperbolic Systems with Involutions **94**, 373–389 (1986) .......... 291

DING, W.-Y., & NI, W.-M., On the Existence of Positive Entire Solutions of a Semilinear Elliptic Equation **91**, 283–308 (1986) .......... 17

DUNN, J. E., & FOSDICK, R. L., Steady, Structured Shock Waves. Part 1: Thermoelastic Materials **104**, 295–365 (1988) .......... 747

EDMUNDS, D. E., & EDMUNDS, R. M., Embeddings of Anisotropic Sobolev Spaces **94**, 245–252 (1986) .......... 269

ERICKSEN, J. L., Stable Equilibrium Configurations of Elastic Crystals **94**, 1–14 (1986) .......... 239

FABES, E. B., & STROOCK, D. W., A New Proof of Moser's Parabolic Harnack Inequality Using the Old Ideas of Nash **96**, 327–338 (1986) .......... 459

FIFE, P., HSIAO, L., & ZHANG, T., A Multiparameter Study of a Boundary Value Problem from Chemical Reactor Theory **92**, 307–335 (1986) .......... 107

GALDI, G. P., & MAREMONTI, P., Monotonic Decreasing and Asymptotic Behavior of the Kinetic Energy for Weak Solutions of the Navier-Stokes Equations in Exterior Domains **94**, 253–266 (1986) .......... 277

GIUSTI, E., On the Behavior of the Derivatives of Minimizers near Singular Points **96**, 137–146 (1986) .......... 399

GURTIN, M. E., On Phase Transitions with Bulk, Interfacial, and Boundary Energy **96**, 243–264 (1986) .......... 429

HEINRICHER, A. C., & MIZEL, V. J., The Lavrentiev Phenomenon for Invariant Variational Problems **102**, 57–93 (1988) .......... 709

JONES, M., & TOLAND, J., Symmetry and the Bifurcation of Capillary-Gravity Waves **96**, 29–53 (1986) .......... 357

KONDRATIEV, V. A., & OLEINIK, O. A., On the Behavior at Infinity of Solutions of Elliptic Systems with a Finite Energy Integral **99**, 75–89 (1987) .......... 575

LEVINE, H. A., & WEINBERGER, H. F., Inequalities Between Dirichlet and Neumann Eigenvalues **94**, 193–208 (1986) .......... 253

LITTMAN, W., Spectral Properties of the Laplacian in the Complement of a Deformed Cylinder **96**, 319–325 (1986) .................... 451

MCLEOD, K., Asymptotic Behaviour of Solutions of Semi-Linear Elliptic Equations in $\mathbb{R}^n$ **98**, 295–305 (1987) .................... 531

MCLEOD, J.B., & RAJAGOPAL, K.R., On the Uniqueness of Flow of a Navier-Stokes Fluid Due to a Stretching Boundary **98**, 385–393 (1987) .................... 565

NOLL, W., & VIRGA, E.G., Fit Regions and Functions of Bounded Variation **102**, 1–21 (1988) .................... 687

OWEN, D.R., Weakly Decaying Energy Separation and Uniqueness of Motions of an Elastic-Plastic Oscillator with Work-Hardening **98**, 95–114 (1987) .................... 471

PUCCI, P., A Semilinear Elliptic Problem Which Is Not Selfadjoint **92**, 265–285 (1986) .................... 85

RABINOWITZ, P.H., A Note on a Theorem of R. Duffin **93**, 91–102 (1986) .................... 169

REDHEFFER, R., A Class of Quasilinear Differential Inequalities Whose Solutions Are Ultimately Constant **99**, 165–187 (1987) .................... 591

ROYTBURD, V., & SLEMROD, M., Positively Invariant Regions for a Problem in Phase Transitions **93**, 61–79 (1986) .................... 149

SHIH, K.-G., & ANTMAN, S.S., Qualitative Properties of Large Buckled States of Spherical Shells **93**, 357–384 (1986) .................... 211

SMOLLER, J., & WASSERMAN, A., Existence of Positive Solutions for Semilinear Elliptic Equations in General Domains **98**, 229–249 (1987) .................... 499

TALENTI, G., Assembling a Rearrangement **98**, 285–293 (1987) .................... 521

TEMAM, R., A Generalized Norton-Hoff Model and the Prandtl-Reuss Law of Plasticity **95**, 137–183 (1986) .................... 309

ZIEMER, W.P., Boundary Regularity for Quasiminima **92**, 371–382 (1986) .................... 137